Lecture Notes in Computer Science   7484

Commenced Publication in 1973
Founding and Former Series Editors:
Gerhard Goos, Juris Hartmanis, and Jan van Leeuwen

Advanced Research in Computing and Software Science
Subline of Lectures Notes in Computer Science

Christos Kaklamanis
Theodore Papatheodorou   Paul G. Spirakis (Eds.)

Euro-Par 2012
Parallel Processing

18th International Conference, Euro-Par 2012
Rhodes Island, Greece, August 27-31, 2012
Proceedings

 Springer

Volume Editors

Christos Kaklamanis
University of Patras
Computer Technology Institute and Press "Diophantus"
N. Kazantzaki
26504 Rio, Greece
E-mail: kakl@ceid.upatras.gr

Theodore Papatheodorou
University of Patras
University Building B
26504 Rio, Greece
E-mail: tsp@hpclab.ceid.upatras.gr

Paul G. Spirakis
University of Patras
Computer Technology Institute and Press "Diophantus"
N. Kazantzaki
26504 Rio, Greece
E-mail: spirakis@cti.gr

ISSN 0302-9743   e-ISSN 1611-3349
ISBN 978-3-642-32819-0   e-ISBN 978-3-642-32820-6
DOI 10.1007/978-3-642-32820-6
Springer Heidelberg Dordrecht London New York

Library of Congress Control Number: 2012944429

CR Subject Classification (1998): D.1.3, D.3.3-4, C.1.4, D.4, C.4, C.2, G.1.0,
C.3, H.3, I.6, I.2.6, F.1.2, H.2.8

LNCS Sublibrary: SL 1 – Theoretical Computer Science and General Issues

Typesetting: Camera-ready by author, data conversion by Scientific Publishing Services, Chennai, India

Printed on acid-free paper

Springer is part of Springer Science+Business Media (www.springer.com)

Preface

Euro-Par is an annual series of international conferences dedicated to the promotion and advancement of all aspects of parallel and distributed computing.

Euro-Par covers a wide spectrum of topics from algorithms and theory to software technology and hardware-related issues, with application areas ranging from scientific to mobile and cloud computing.

Euro-Par provides a forum for the introduction, presentation, and discussion of the latest scientific and technical advances, extending the frontier of both the state of the art and the state of the practice.

The main audience of Euro-Par are researchers in academic institutions, government laboratories, and industrial organizations. Euro-Par's objective is to be the primary choice of such professionals for the presentation of new results in their specific areas. As a wide-spectrum conference, Euro-Par fosters the synergy of different topics in parallel and distributed computing. Of special interest are applications that demonstrate the effectiveness of the main Euro-Par topics.

In addition, Euro-Par conferences provide a platform for a number of accompanying, technical workshops. Thus, smaller and emerging communities can meet and develop more focussed topics or as-yet less established topics.

Euro-Par 2012 was the 18th conference in the Euro-Par series, and was organized by CTI (Computer Technology Institute and Press "Diophantus"). Previous Euro-Par conferences took place in Stockholm, Lyon, Passau, Southampton, Toulouse, Munich, Manchester, Padderborn, Klagenfurt, Pisa, Lisbon, Dresden, Rennes, Las Palmas, Delft, Ischia, and Bordeaux. Next year the conference will take place in Aachen, Germany. More information on the Euro-Par conference series and organization is available on the website http://www.europar.org

The conference was organized in 16 topics. The paper review process for each topic was managed and supervised by a committee of at least four persons: a Global Chair, a Local Chair, and two members. Some specific topics with a high number of submissions were managed by a larger committee with more members. The final decisions on the acceptance or rejection of the submitted papers were made in a meeting of the Conference Co-chairs and Local Chairs of the topics.

The call for papers attracted a total of 228 submissions, representing 44 countries (based on the corresponding authors' countries). A total of 873 review reports were collected, which makes an average of 3.83 review reports per paper. In total 75 papers were selected as regular papers to be presented at the conference and included in the conference proceedings, representing 29 countries from all continents, and yielding an acceptance rate of 32.9%. Three papers were selected as distinguished papers. These papers, which were presented in a separate session, are:

1. Ricardo J. Dias, Tiago M. Vale, and João M. S. Lourenço "Efficient Support for In-Place Metadata in Transactional Memory"

2. Wesley Bland, Peng Du, Aurelien Bouteiller, Thomas Herault, George Bosilca, and Jack Dongarra "A Checkpoint-on-Failure Protocol for Algorithm-Based Recovery in Standard MPI"
3. Konstantinos Christodoulopoulos, Marco Ruffini, Donal O'Mahony, and Kostas Katrinis "Topology Configuration in Hybrid EPS/OCS Interconnects"

Euro-Par 2012 was very happy to present three invited speakers of high international reputation, who discussed important developments in very interesting areas of parallel and distributed computing:

1. Ewa Deelman (Information Sciences Institute, University of Southern California, USA)
2. Burkhard Monien (University of Paderborn, Germany)
3. Thomas Schulthess (CSCS, ETH Zurich, Switzerland)

In this edition, 11 workshops were held in conjunction with the main track of the conference. These workshops were:

1. 10th International Workshop on Algorithms, Models and Tools for Parallel Computing on Heterogeneous Platforms (Heteropar)
2. 7th Workshop on Virtualization in High-Performance Cloud Computing (VHPC)
3. 5th Workshop on Unconventional High-Performance Computing (UCHPC)
4. 5th Workshop on Productivity and Performance (PROPER)
5. Third Workshop on High-Performance Bioinformatics and Biomedicine (HiBB)
6. Workshop on Resiliency in High-Performance Computing (Resilience)
7. CoreGRID/ERCIM Workshop on Grids, Clouds, and P2P Computing (CGWS)
8. First Workshop on Big Data Management in Clouds (BDMC)
9. Workshop on Architecture and Systems Software for Data Intensive Supercomputing
10. First Workshop on On-chip Memory Hierarchies and Interconnects: Organization, Management and Implementation (OMHI)
11. Paraphrase Workshop

The 18th Euro-Par conference in Rhodes was made possible thanks to the support of many individuals and organizations. Special thanks are due to the authors of all the submitted papers, the members of the Topic Committees, and all the reviewers in all topics, for their contributions to the success of the conference. We also thank the members of the Organizing Committee. We are grateful to the members of the Euro-Par Steering Committee for their support. We acknowledge the help we had from Emmanuel Jeannot of the organization of Euro-Par 2011. It was our pleasure and honor to organize and host Euro-Par 2012 in Rhodes. We hope all the participants enjoyed the technical program and the social events organized during the conference.

August 2012

Christos Kaklamanis
Theodore Papatheodorou
Paul Spirakis

Organization

Euro-Par Steering Committee

Chair

Chris Lengauer · University of Passau, Germany

Vice-Chair

Luc Bougé · ENS Cachan, France

European Representatives

José Cunha	New University of Lisbon, Portugal
Marco Danelutto	University of Pisa, Italy
Emmanuel Jeannot	LaBRI-INRIA, France
Christos Kaklamanis	Computer Technology Institute and Press "Diophantus", Greece
Paul Kelly	Imperial College, UK
Thomas Ludwig	University of Hamburg, Germany
Emilio Luque	Autonomous University of Barcelona, Spain
Tomàs Margalef	Autonomous University of Barcelona, Spain
Wolfgang Nagel	Dresden University of Technology, Germany
Rizos Sakellariou	University of Manchester, UK
Henk Sips	Delft University of Technology, The Netherlands
Domenico Talia	University of Calabria, Italy

Honorary Members

Ron Perrott	Queen's University Belfast, UK
Karl Dieter Reinartz	University of Erlangen-Nuremberg, Germany

Observer

Felix Wolf · RWTH Aachen, Germany

Euro-Par 2012 Organization

Conference Co-chairs

Christos Kaklamanis	CTI and University of Patras, Greece
Theodore Papatheodorou	University of Patras, Greece
Paul Spirakis	CTI and University of Patras, Greece

Workshop Co-chairs

Luc Bougé   ENS Cachan, France
Ioannis Caragiannis   CTI and University of Patras, Greece

Local Organizing Committee

Katerina Antonopoulou   CTI, Greece
Stavros Athanassopoulos   CTI and University of Patras, Greece
Rozina Efstathiadou   CTI, Greece
Lena Gourdoupi   CTI, Greece
Panagiotis Kanellopoulos   CTI and University of Patras, Greece
Evi Papaioannou   CTI and University of Patras, Greece

Euro-Par 2012 Program Committee

Topic 1: Support Tools and Environments

Global Chair

Omer Rana   Cardiff University, UK

Local Chair

Marios Dikaiakos   University of Cyprus, Cyprus

Members

Daniel Katz   University of Chicago, USA
Christine Morin   INRIA, France

Topic 2: Performance Prediction and Evaluation

Global Chair

Allen Malony   University of Oregon, USA

Local Chair

Helen Karatza   Aristotle University of Thessaloniki, Greece

Members

William Knottenbelt   Imperial College London, UK
Sally McKee   Chalmers University of Technology, Sweden

Topic 3: Scheduling and Load Balancing

Global Chair

Denis Trystram   Grenoble Institute of Technology, France

Local Chair

Ioannis Milis   Athens University of Economics and Business,
                                           Greece

Members

Zhihui Du   Tsinghua University, China
Uwe Schwiegelshohn   TU Dortmund, Germany

Topic 4: High-Performance Architecture and Compilers

Global Chair

Alex Veidenbaum   University of California, USA

Local Chair

Nektarios Koziris   National Technical University of Athens, Greece

Members

Avi Mendelson   Microsoft, Israel
Toshinori Sato   Kyushu University, Japan

Topic 5: Parallel and Distributed Data Management

Global Chair

Domenico Talia   University of Calabria, Italy

Local Chair

Alex Delis   University of Athens, Greece

Members

Haimonti Dutta   Columbia University, USA
Arkady Zaslavsky   Lulea University of Technology, Sweden
                                           and CSIRO, Australia

Topic 6: Grid, Cluster and Cloud Computing

Global Chair

Erik Elmroth   Umea University, Sweden

Local Chair

Paraskevi Fragopoulou   FORTH, Greece

Members

Artur Andrzejak	Heidelberg University, Germany
Ivona Brandic	Vienna University of Technology, Austria
Karim Djemame	University of Leeds, UK
Paolo Romano	INESC-ID, Portugal

Topic 7: Peer-to-Peer Computing

Global Chair

Alberto Montresor	University of Trento, Italy

Local Chair

Evaggelia Pitoura	University of Ioannina, Greece

Members

Anwitaman Datta	Nanyang Technological University, Singapore
Spyros Voulgaris	Vrije Universiteit Amsterdam, The Netherlands

Topic 8: Distributed Systems and Algorithms

Global Chair

Andrzej Goscinski	Deakin University, Australia

Local Chair

Marios Mavronicolas	University of Cyprus, Cyprus

Members

Weisong Shi	Wayne State University, USA
Teo Yong Meng	National University of Singapore, Singapore

Topic 9: Parallel and Distributed Programming

Global Chair

Sergei Gorlatch	University of Münster, Germany

Local Chair

Rizos Sakellariou	University of Manchester, UK

Members

Marco Danelutto	University of Pisa, Italy
Thilo Kielmann	Vrije Universiteit Amsterdam, The Netherlands

Topic 10: Parallel Numerical Algorithms

Global Chair

Iain Duff   Rutherford Appleton Laboratory, UK

Local Chair

Efstratios Gallopoulos   University of Patras, Greece

Members

Daniela di Serafino   Second University of Naples, Italy
Bora Ucar   ENS, France

Topic 11: Multicore and Manycore Programming

Global Chair

Eduard Ayguade   Technical University of Catalonia, Spain

Local Chair

Dionisios Pnevmatikatos   Technical University of Crete and FORTH,
                                   Greece

Members

Rudolf Eigenmann   Purdue University, USA
Mikel Luján   University of Manchester, UK
Sabri Pllana   University of Vienna, Austria

Topic 12: Theory and Algorithms for Parallel Computation

Global Chair

Geppino Pucci   University of Padova, Italy

Local Chair

Christos Zaroliagis   CTI and University of Patras, Greece

Members

Kieran Herley   University College Cork, Ireland
Henning Meyerhenke   Karlsruhe Institute of Technology, Germany

Topic 13: High-Performance Network and Communication

Global Chair

Chris Develder   Ghent University, Belgium

Local Chair

Emmanouel Varvarigos   CTI and University of Patras, Greece

Members

Admela Jukan   Technical University of Braunschweig, Germany
Dimitra Simeonidou   University of Essex, UK

Topic 14: Mobile and Ubiquitous Computing

Global Chair

Paolo Santi   IIT-CNR, Italy

Local Chair

Sotiris Nikoletseas   CTI and University of Patras, Greece

Members

Cecilia Mascolo   University of Cambridge, UK
Thiemo Voigt   SICS, Sweden

Topic 15: High-Performance and Scientific Applications

Global Chair

Thomas Ludwig   University of Hamburg, Germany

Local Chair

Costas Bekas   IBM Zurich, Switzerland

Members

Alice Koniges   Lawrence Berkeley National Laboratory, USA
Kengo Nakajima   University of Tokyo, Japan

Topic 16: GPU and Accelerators Computing

Global Chair

Alex Ramirez   Technical University of Catalonia, Spain

Local Chair

Dimitris Nikolopoulos   University of Crete and FORTH, Greece

Members

David Kaeli   Northeastern University, USA
Satoshi Matsuoka   Tokyo Institute of Technology, Japan

Euro-Par 2012 Referees

Luca Abeni
Shoaib Akram
Jay Alameda
Susanne Albers
Marco Aldinucci
Ahmed Ali-ElDin
Srinivas Aluru
Ganesh Ananthanarayanan
Nikos Anastopoulos
Eric Angel
Constantinos Marios Angelopoulos
Ashiq Anjum
Mohammad Ansari
Alexandros Antoniadis
Christos Antonopoulos
Filipe Araujo
Django Armstrong
Cedric Augonnet
Win Than Aung
Aurangeb Aurangzeb
Scott Baden
Rosa M. Badia
Hansang Bae
Enes Bajrovic
Henri Bal
Harish Balasubramanian
Michael Bane
Leonardo Bautista Gomez
Ewnetu Bayuh Lakew
Tom Beach
Vicenç Beltran
Julien Bernard
Carlo Bertolli
Rob Bisseling
Luiz Bittencourt
Filip Blagojevic
François Bodin
Erik Boman
Sara Bouchenak

Steven Brandt
Ivan Breskovic
Patrick Bridges
Christopher Bun
Daniele Buono
Kevin Burrage
Alfredo Buttari
Javier Cabezas
Rosario Cammarota
Sonia Campa
Edouard Canot
Paul Carpenter
Daniel Cederman
Eugenio Cesario
Nicholas Chaimov
Kang Chen
Zhongliang Chen
Mosharaf Chowdhury
Chunbo Chu
Murray Cole
Carmela Comito
Guojing Cong
Fernando Costa
Maria Couceiro
Adrian Cristal
Ruben Cuevas Rumin
Yong Cui
Pasqua D'Ambra
Gabriele D'Angelo
Patrizio Dazzi
Usman Dastgeer
Ewa Deelman
Karen Devine
Diego Didona
Ngoc Dinh
Julio Dondo Gazzano
Nam Duong
Alejandro Duran
Ismail El Helw

Kaoutar El Maghraoui
Robert Elsaesser
Vincent Emeakaroha
Toshio Endo
Daniel Espling
Montse Farreras
Renato Figueiredo
Salvatore Filippone
Wan Fokkink
Alexander Fölling
Agostino Forestiero
Edgar Gabriel
Giulia Galbiati
Saurabh Garg
Michael Garland
Rong Ge
Bugra Gedik
Isaac Gelado
Michael Gerndt
Aristotelis Giannakos
Lee Gillam
Luc Giraud
Sarunas Girdzijauskas
Harald Gjermundrod
Alfredo Goldman
Zeus Gómez Marmolejo
Georgios Goumas
Anastasios Gounaris
Vincent Gramoli
Clemens Grelck
Christian Grimme
Alessio Guerrieri
Ajay Gulati
Panagiotis Hadjidoukas
Eyad Hailat
Tim Harris
Piyush Harsh
Masae Hayashi
Jiahua He
Yuxiong He
Bruce Hendrickson
Pieter Hijma
Torsten Hoefler
Matthias Hofmann
Christian Hoge

Jonathan Hogg
Theus Hossmann
Nathanael Hübbe
Kevin Huck
Sascha Hunold
Felix Hupfeld
Nikolas Ioannou
Thomas Jahns
Klaus Jansen
Aubin Jarry
Bahman Javadi
Yvon Jegou
Ming Jiang
Hideyuki Jitsumoto
Fahed Jubair
Vana Kalogeraki
Yoshikazu Kamoshida
Takahiro Katagiri
Randy Katz
Richard Kavanagh
Stamatis Kavvadias
Kamer Kaya
Gabor Kecskemeti
Safia Kedad-Sidhoum
Ian Kelley
Paul Kelly
Markus Kemmerling
Rajkumar Kettimuthu
Le Duy Khanh
Peter Kilpatrick
Taesu Kim
Mariam Kiran
Nicolaj Kirchhof
Thomas Kirkham
Luc Knockaert
Takeshi Kodaka
Panagiotis Kokkinos
Charalampos Konstantopoulos
Ulrich Körner
Christos Kotselidis
Nektarios Kranitis
Michael Kuhn
Manaschai Kunaseth
Julian Martin Kunkel
Krzysztof Kurowski

Okwan Kwon
Felix Langner
Francis Lau
Adrien Lèbre
Chee Wai Lee
Pierre Lemarinier
Hermann Lenhart
Ilias Leontiadis
Dimitrios Letsios
Dong Li
Wubin Li
Youhuizi Li
John Linford
Luong Ba Linh
Nicholas Loulloudes
João Lourenço
Hatem Ltaief
Dajun Lu
Drazen Lucanin
Giorgio Lucarelli
Ewing Lusk
Spyros Lyberis
Maciej Machowiak
Sandya Mannarswamy
Osni Marques
Maxime Martinasso
Xavier Martorell
Naoya Maruyama
Lukasz Masko
Toni Mastelic
Carlo Mastroianni
Michael Maurer
Gabriele Mencagli
Massimiliano Meneghin
Mohand Mezmaz
George Michael
Milan Mihajlovic
Timo Minartz
Perhaad Mistry
Bernd Mohr
Matteo Mordacchini
Benjamin Moseley
Dheya Mustafa
Hironori Nakajo
Franco Maria Nardini

Rammohan Narendula
Sarfraz Nawaz
Zsolt Nemeth
Tung Nguyen
Bogdan Nicolae
Vincenzo Nicosia
Konstantinos Nikas
Andy Nisbet
Akihiro Nomura
Akira Nukada
Richard O'Keefe
Satoshi Ohshima
Stephen Olivier
Salvatore Orlando
Per-Olov Ostberg
Linda Pagli
George Pallis
Roberto Palmieri
Costas Panagiotakis
Harris Papadakis
Andreas Papadopoulos
Manish Parashar
Nikos Parlavantzas
Jean-Louis Pazat
Sebastiano Peluso
Raffaele Perego
Miquel Pericas
Dennis Pfisterer
Bernard Philippe
Vinicius Pinheiro
Timothy Pinkston
Polyvios Pratikakis
Jan Prins
Bart Puype
Nikola Puzovic
Thanh Quach
Jean-Noel Quintin
Yann Radenac
M. Mustafa Rafique
Kees Reeuwijk
Laurent Réveillère
Olivier Richard
Thomas Ropars
Mathis Rosenhauer
Barry Rountree

Mema Roussopoulos
Krzysztof Rzadca
Amit Sabne
P. Sadayappan
Putt Sakdhnagool
Friman Sánchez
Carlos Alberto Alonso Sanches
Martin Sandrieser
Vijay Saraswat
Hitoshi Sato
Kento Sato
Thomas Sauerwald
Philip Schill
Elad Schiller
Scott Schneider
Mina Sedaghat
Kenshu Seto
Aamir Shafi
Jawwad Shamsi
Rajesh Sharma
Sameer Shende
Jinsong Shi
Jun Shirako
Yogesh Simmhan
Fabrizio Sivestri
Martin Skutella
Georgios Smaragdakis
Ismael Solis Moreno
Michael Spear
Jochen Speck
Ivor Spence
Cliff Stein
Mark Stillwell
John Stone
John Stratton
Petter Svärd
David Swanson
Guangming Tan
Yoshio Tanaka
Osamu Tatebe
Marc Tchiboukdjian
Samuel Thibault
Alex Tiskin
Rubén Titos
Hiroyuki Tomiyama

Massimo Torquati
Raul Torres
Pedro Trancoso
Paolo Trunfio
Hong-Linh Truong
Konstantinos Tsakalozos
Tomoaki Tsumura
Bogdan Marius Tudor
Rafael Ubal
Yash Ukidave
Osman Unsal
Philipp Unterbrunner
Jacopo Urbani
Marian Vajtersic
Rob van Nieuwpoort
Ben van Werkhoven
Hans Vandierendonck
Xavier Vasseur
Luís Veiga
Ioannis Venetis
Salvatore Venticinque
Vassilis Verroios
Kees Verstoep
Lluís Vilanova
Frederic Vivien
David Walker
Edward Walker
John Walters
Luís Wanderley Góes
Jun Wang
Xinqi Wang
Ian Watson
Marc Wiedemann
Tong Wieqin
Adam Wierzbicki
Martin Wimmer
Justin Wozniak
Di Wu
Yong Xia
Wei Xing
Lei Xu
Gagarine Yaikhom
Ayse Yilmazer
Yitong Yin
Ossama Younis

Matei Zaharia
Mohamed Zahran
Jidong Zhai
Guoxing Zhan
Haibo Zhang
Sen Zhang
Yunquan Zhang

Zhao Zhang
Aqun Zhao
Dali Zhao
Gengbin Zheng
Wei Zheng
Eugenio Zimeo
Michaela Zimmer

Table of Contents

Topic 3: Scheduling and Load Balancing

Topic 4: High-Performance Architecture and Compilers

Topic 5: Parallel and Distributed Data Management

Topic 6: Grid, Cluster and Cloud Computing

Topic 7: Peer to Peer Computing

Topic 8: Distributed Systems and Algorithms

Topic 9: Parallel and Distributed Programming

Topic 10: Parallel Numerical Algorithms

Topic 11: Multicore and Manycore Programming

Topic 12: Theory and Algorithms for Parallel Computation

Topic 13: High Performance Network and Communication

Topic 14: Mobile and Ubiquitous Computing

Topic 15: High Performance and Scientific Applications

Topic 16: GPU and Accelerators Computing

Selfish Distributed Optimization

Burkhard Monien and Christian Scheideler

Faculty of Computer Science, Electrical Engineering and Mathematics,
University of Paderborn, Fürstenallee 11, 33102 Paderborn, Germany
{bm,scheideler}@uni-paderborn.de

Abstract. In this talk, we present a selection of important concepts
and results in *algorithmic game theory* in recent years, some of which received the 2012 Gödel Prize, along with some applications in distributed
settings.

A famous solution concept for non-cooperative games is the *Nash equilibrium*. In a Nash equilibrium, no selfish player can unilaterally deviate
from his current strategy and improve his profit. Nash dynamics is a
method to compute a Nash equilibrium. Here, in each round, a single
player is allowed to perform a selfish step, i. e. unilaterally change his
strategy and improve his cost. The Nash dynamics terminates if it does
not run into a cycle. This is always the case if the game has a potential
function. In this case, computing a Nash equilibrium is a $\mathcal{PLS}$ problem
(Polynomial Local Search) and belongs to the large class of well-studied
local optimization problems.

Inspired by real-world networks, *network congestion games* have been
under severe scrutiny for the last years. Network congestion games model
selfish routing of unsplittable units. These units may be weighted or unweighted. Weighted congestion games do not necessarily have a pure
Nash equilibrium. Conversely, an unweighted congestion game has a potential function. Computing a pure Nash equilibrium for an unweighted
congestion game is $\mathcal{PLS}$-complete.

The absence of a central coordinating authority can result in a loss
of performance due to the selfishness of the participants. This situation
is formalized in the notion *"Price of Anarchy"*. The Price of Anarchy is
defined to be the worst case ratio between the maximal social cost in
a Nash equilibrium and the optimal social cost. We present the recent
results for congestion games and for the special case of load balancing.

Classical game theory assumes that each player acts rationally and
wants to improve his profit. This is not realistic in a distributed setting
since it requires that each player has the complete information about the
state of the system. We introduce the concept of *selfish distributed load
balancing* and describe recent results.

We will also consider distributed algorithms for *network creation
games*. In the past, network creation games have mostly been studied
under the assumption that the players have a global view on the network, or more precisely, that the players are able to compute the average
distance or the maximum distance to the nodes they want to interact
with in the given network, depending on the objective function. A player
may then decide to add one or more edges for some extra cost or to drop

C. Kaklamanis et al. (Eds.): Euro-Par 2012, LNCS 7484, pp. 1–2, 2012.

an edge. We will look at network creation games from a different angle. In our case, the players have fixed distances to each other that are based on some underlying metric (determined by, for example, the geographic positions of the players), and the goal is to study the networks formed if players selfishly add and remove edges based on that metric. We show that for certain metrics like the line metric, tree metric, and the Euclidean metric, certain selfish behavior, that only requires a local view of the players on the network, will lead to stable networks that give a good approximation of the underlying metric.

Topic 1: Support Tools and Environments

Omer Rana, Marios Dikaiakos, Daniel S. Katz, and Christine Morin

Topic Committee

Despite an impressive body of research, parallel and distributed computing remains a complex task prone to subtle software issues that can affect both the correctness and the performance of the computation. It is interesting to note that this topic has always been listed as Topic 1 in the EuroPar conference series for some time now – emphasising its importance and focus in the parallel and distributed systems community. The increasing demand to distribute computing over large-scale parallel and distributed platforms, such as grids and large clusters, often combined with the use of hardware accelerators, overlaps with an increasing pressure to make computing more dependable. To address these challenges, the parallel and distributed computing community continuously requires better tools and environments to design, program, debug, test, tune, and monitor programs that must execute over parallel and distributed systems. This topic aims to bring together tool designers, developers and users to share their concerns, ideas, solutions, and products covering a wide range of platforms, including homogeneous and heterogeneous multi-core architectures. Contributions with solid theoretical foundations and experimental validations on production-level parallel and distributed systems were particularly valued. This year we encouraged submissions proposing intelligent monitoring and diagnosis tools and environments which can exploit behavioral knowledge to detect programming bugs or performance bottlenecks and help ensure correct and efficient parallel program execution.

Each paper was reviewed by at least three reviewers and we selected 4 papers for the conference. It was interesting to see papers focusing on emerging themes such as multi-core and GPUs, pattern-oriented parallel computing, deployment over Android platform, Cloud interoperability and the use of autonomic computing techniques along with papers that covered more established themes such as program profiling, performance analysis, debugging, workflow management and application tuning. The four selected papers cover program visualisation to support semi-automated parallelisation, a programming model and run time environment to support application development/deployment over multiple Cloud environments, detection of hand-crafted collective operations in MPI programs (rather than the use of functions already provided in the MPI standard) and a language extension (based on the use of a type system) for supporting programming over accelerator architectures. The four selected papers cover a combination of theoretical underpinnings and practical development and deployment.

We would like to thank the authors who submitted a contribution, as well as the Euro-Par Organizing Committee, and the referees who provided useful and timely comments.

C. Kaklamanis et al. (Eds.): Euro-Par 2012, LNCS 7484, p. 3, 2012.
© Springer-Verlag Berlin Heidelberg 2012

Tulipse: A Visualization Framework for User-Guided Parallelization

Yi Wen Wong[1], Tomasz Dubrownik[2], Wai Teng Tang[3], Wen Jun Tan[3],
Rubing Duan[4], Rick Siow Mong Goh[4], Shyh-hao Kuo[4],
Stephen John Turner[3], and Weng-Fai Wong[1]

[1] National University of Singapore, Singapore
[2] University of Warsaw, Poland
[3] Nanyang Technological University, Singapore
[4] Institute of High Performance Computing, A*Star, Singapore

Abstract. Parallelization of existing code for modern multicore processors is tedious as the person performing these tasks must understand the algorithms, data structures and data dependencies in order to do a good job. Current options available to the programmer include either automatic parallelization or a complete rewrite in a parallel programming language. However, there are limitations with these options. In this paper, we propose a framework that enables the programmer to visualize information critical for semi-automated parallelization. The framework, called Tulipse, offers a program structure view that is augmented with key performance information, and a loop-nest dependency view that can be used to visualize data dependencies gathered from static or dynamic analyses. Our paper will demonstrate how these two new perspectives aid in the parallelization of code.

1 Introduction

As multicore and multi-node architectures become more prevalent and widely available, programs have to be written using multiple threads to take full advantage of all the cores available to them. Unfortunately, the task of multithreaded programming remains a hard one. Programmers are required to take more factors into account to write code that is both correct and that has good performance at the same time. This places a great burden on application developers, not all of whom may be as proficient in parallel programming as would be required.

Furthermore, it is increasingly difficult for existing legacy programs to take advantage of the multicore capabilities of these chips without resorting to a partial or complete rewrite of the source code. However, the cost of rewriting software is prohibitive. This problem is exacerbated by the fact that many application domain experts who maintain the legacy codes are not parallel programmers. Without the domain knowledge that is required in certain applications, it may also be difficult for programmers outside the domain to convert them from sequential programs into multithreaded ones. This is because apart from understanding the algorithm, the programmer performing the code modification

C. Kaklamanis et al. (Eds.): Euro-Par 2012, LNCS 7484, pp. 4–15, 2012.

has to understand the data structures, control flow and dependencies within the existing application to be able to do a good job in refactoring the code.

Visualization tools such as SUIF Explorer [8] and ParaGraph [1] have been developed to help facilitate the task of converting sequential code to parallel code. Such tools can reduce the effort needed by the programmer to understand the program and to make the necessary changes to parallelize it. For example, the ParaGraph tool displays a control flow graph of a code fragment to the user, and augments it with data dependency edges in order to help the user understand the data relations between different program statements. From this visual information, the programmer is then able to decide if the code is parallelizable or if synchronization is needed for certain data structures. Many of these tools, such as the ParaScope Editor [6] or ParaGraph, are also front-ends for their corresponding parallelizing compilers. They typically provide a limited form of visualization, usually as text output organized using tables, or in some cases, a two-dimensional graph representation of the data of interest [1]. Furthermore, they do not support the typical workflow a programmer goes through during the code parallelization process. For example, the ParaScope Editor and ParaGraph attempt to parallelize all the loops found within a program regardless of their suitability.

In this paper, we will describe an integrated visualization framework for parallelization that we have developed called Tulipse. The guiding principle behind the design is to simplify the workflow for parallelizing a program through an integrated visualization environment, and to provide visually useful information for the parallelizing process. This can be accomplished by the following capabilities. First, it consists of a graphical view that allows the user to visualize the global structure of an application by displaying procedures and loops hierarchically. Profiler measurements which indicate code sections that take up a large amount of the execution time can be overlaid in the graphical view. Second, a three dimensional view is provided to help the programmer visualize the data dependencies within a code section. It also allows the user to navigate through the view interactively. Through these capabilities, the programmer would then make an informed decision to effect the necessary code changes that will enable the parallelization of the application.

The framework is designed to involve the programmer in the parallelization workflow. This is because the programmer's knowledge about the code will be useful during the parallelization process. In addition, he is ultimately responsible for maintaining the application, and therefore may want to have finer control over code changes. Tulipse provides the following features to support semi-automatic code parallelization: (1) it is a visualization framework written in Java for the Eclipse IDE. In doing so, not only are all the facilities of the Eclipse IDE available to the programmers, the framework can also be extended with new visualization plug-ins; (2) it integrates a profiling and measurement tool that can be used to instrument the application under study. It can, for example, be used to find out which loops dominate execution time or have many cache misses.

The programmer can then focus his attention on these hotspots, where even small improvements can have a significant impact on the overall execution time; (3) it provides a way to visualize data dependencies in the application through the use of a three-dimensional visualizer. The programmer can also animate, and walk-through the index space to obtain a better understanding of the data dependences. This allows the programmer to experiment with different ways of parallelizing the code; (4) it inserts OpenMP directives [9] into code which is found to be parallelizable using static and dynamic analyses, allowing the user to focus his attention on code sections which cannot be automatically parallelized.

The rest of the paper is organized as follows: Sect. 2 gives an overview of Tulipse and its capabilities. Section 3 presents examples on the usage of the visualization capabilities of Tulipse. Section 4 discusses prior work that is related and Sect. 5 concludes the paper.

2   Overview of Tulipse

The guiding principle behind the design of Tulipse is to offer to the programmer as much help as possible in the parallelization process. It does not attempt to anticipate the programmer's intent, but rather, allows the programmer to make the important decisions with respect to the code changes. It does this by providing sufficient visual feedback for the programmer to identify the best way to proceed with the parallelization. In this way, it gives the programmer control over the way the code is modified. In order to achieve these goals, two visualization capabilities are supported in Tulipse: the *Loop-Procedure View* and the *Data Dependency View*. Figure 1 shows these two views embedded in an instance of the Eclipse editor. The top panel displays the Loop-Procedure View, while the bottom-left panel shows a code editor and the bottom-right panel shows the Data Dependency View.

The workflow supported by Tulipse is shown in Fig. 2. An application is imported and loaded into the Eclipse IDE. Through the Loop-Procedure View, the user gets a hierarchical view of the whole application represented by procedures and loops that are defined in all of the project's source files. Next, the project is compiled and built with instrumentation automatically inserted into the application binary so that run-time statistics can be collected using hardware performance counters. The run-time statistics gathered from the profiling run is then overlaid onto this view. The user can choose to see, for example, which procedures or loops took up a large proportion of the overall execution time, or experienced a significant number of cache misses. The user can then focus his attention on these parts of the code using the Data Dependency View. This is a three-dimensional perspective of the data dependences within a code section. Through interactive visualization, it allows the user to decide whether the code can be effectively parallelized or tuned. The process can be repeated until the user is satisfied with the changes.

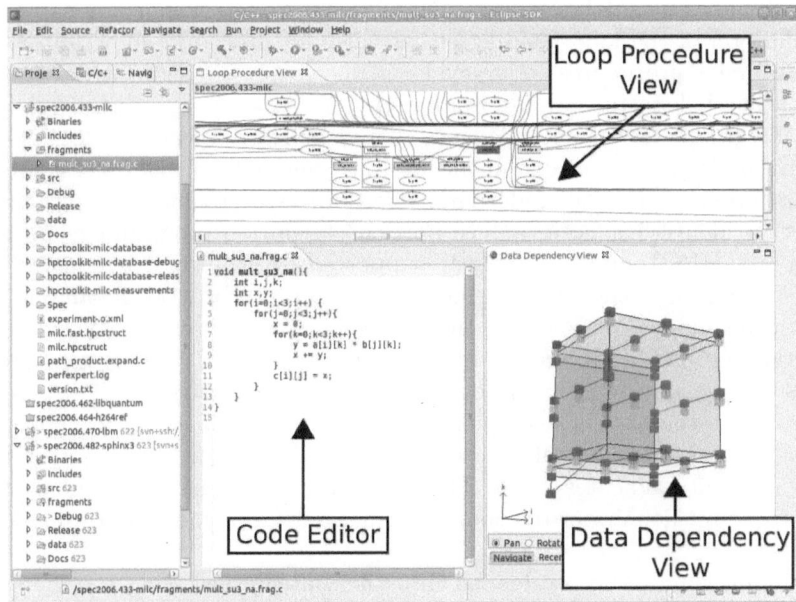

Fig. 1. A screenshot of the Tulipse plug-in in the Eclipse development platform

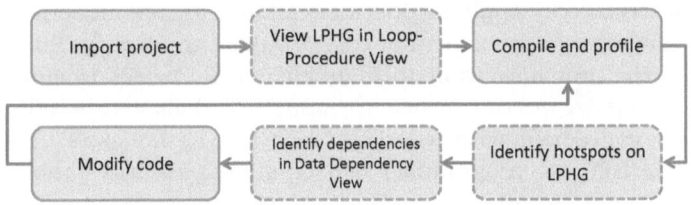

Fig. 2. Workflow of parallelization process

2.1   Loop-Procedure View

This view presents the user with visual information on the application's overall structure. In this view, a Loop-Procedure Hierarchy Graph (LPHG) is constructed for the entire application project. This is essentially a call-graph embedded with loop nest relations obtained from the application source files. For example, Fig. 3 shows the LPHG for the SPEC2006 470.lbm benchmark application [12]. The inset in the figure shows a zoomed-in image of the hotspots. The square-shaped nodes denote procedure definitions, and the ellipse-shaped nodes denote loops. Edges into a procedure node represent calling instances to that procedure. An edge from a procedure to a loop indicates that the loop is defined within the procedure. An edge from a loop to another loop indicates that the latter is nested in the former. There may be multiple incoming edges for a procedure, indicating multiple calling instances of the same procedure. However, there will only be exactly one incoming edge for a loop since there can only be one definition of the loop residing either within a procedure or within another loop. Recursive procedures create cycles in the graph. Nodes which are grouped within a box belong to the same source file.

We integrated HPCToolkit [4] into our visualization framework to simplify run-time performance measurement for the user. HPCToolkit uses hardware counters provided by the underlying microprocessor to measure performance metrics for identifying performance bottlenecks during program execution. Although only HPCToolkit is currently supported by Tulipse, it is relatively easy to add support for other measurement tools such as Tau [13]. Profile measurements of the application taken by the performance measurement component can be overlaid on the LPHG in the Loop-Procedure View. Different measurement metrics can be selected for the overlay. Customized metrics can also be constructed using the base measurement metrics. The metric values are normalized and mapped to a default color gradient from red to white, where red indicates 'hot' while white indicates 'cold'. The respective nodes on the LPHG, including both procedure definition and loop information, are colored according to this mapping.

Profile measurements are loaded into the view by accessing the view menu in the Loop-Procedure View. Different metrics can be overlaid on the LPHG through the `Load Overlay` menu, including base metrics from the profile measurements as well as user-specified derived metrics. A derived metric is essentially a formula constructed by applying operators on metrics and numerical constants. It is also possible to specify a custom color gradient to identify different ranges. For example, the user may want to highlight metric values ranging from 50% to 100% as hotspots, instead of just the top 10%. This can be adjusted using a different color gradient with a larger range for the hotspots. By inspecting the overlaid LPHG, the programmer will be able to identify problematic code regions quickly, and focus his attention on them. By accessing a context menu, we provide the user with the ability to target a code fragment of a problem node through the code editor, or to visualize it through the Data Dependency View.

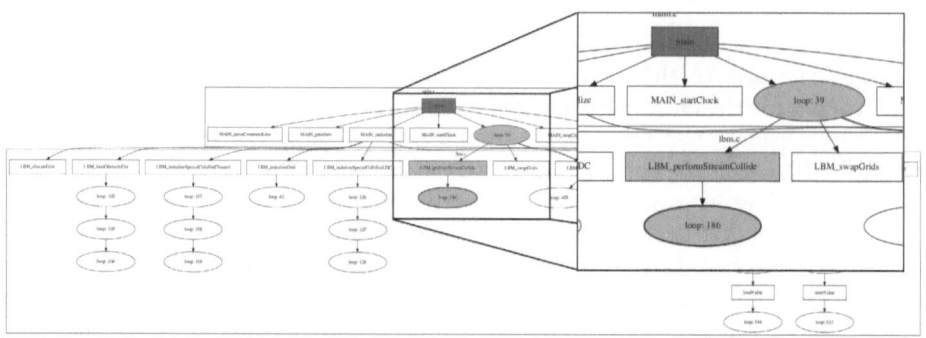

Fig. 3. Loop-Procedure View overlaid with performance measurements

2.2 Three-Dimensional Data Dependency View

The 3D Data Dependency View provides a way for programmers to interactively explore the code's data layout as well as the parallelization options for the loop nests. This view can be launched directly from the Loop-Procedure View by selecting a hotspot. In this view, the iteration spaces of loops within a procedure are visualized. Statements in the loops are mapped to a higher dimensional space based on the polyhedral model [3]. Each statement has a domain determined from its enclosing control statement. Its upper and lower bounds are extracted from the enclosing loops, and intersected with the domains of conditional statements to obtain a system of linear constraints that defines a polyhedron. The polyhedron in this space is projected into the 3D world space for visualization.

Loop nests do not necessarily have to be tightly nested and dependencies can cross loop boundaries. There is no limit on the level of nesting. By selecting a loop in the code panel, the visualizer will highlight the associated domains of the all enclosed statements within the loop. This reduces clutter in the visualization. The data dependencies are currently obtained by static analysis and dynamic analysis built into the framework. The analyses yield *flow dependence* (read after write), *output dependence* (write after write), and *anti-dependences* (write after read). Each dependency is represented by an arrow drawn between the projected coordinates of the source and target statement instance in the 3D world space. Static analysis of dependencies is conservative in order to guarantee correctness, thus dependencies may be over-reported. On the other hand, dependencies obtained by dynamic analysis are dependent on the execution instance and may

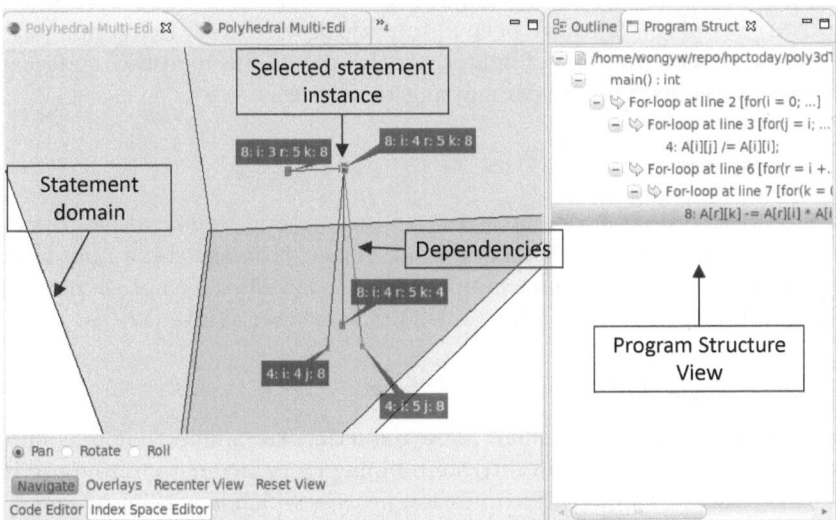

Fig. 4. Selection of a statement instance (line 8, iteration $(i = 4, r = 5, k = 8)$) in the Data Dependency View. Blue arrows indicate the dependences of this instance. Pink arrows show which statements are dependent on the selected instance.

be input-dependent. However, it can obtain dependencies in cases where static analysis is difficult or impossible, such as code involving pointer arithmetic.

Initially, the convex hulls corresponding to each statement domain are shown to the user, with each statement using a different color. As the user zooms in, statement instances and data dependencies (represented by cubes and arrows) become visible to the user. This constrains the amount of information that is shown to the user at any one time. Furthermore, users can use the mouse to obtain more information about each point of interest, e.g. the dependencies of a particular statement instance in the iteration space, or which specific statement instances are responsible for a particular dependency (see Fig. 4), by clicking on a node or an arrow in the view. This view also provides additional features that allow users to further understand the code, and reason about the parallelization options. For example, users can step through the iteration space either manually or via animation to visualize the execution order of the statement instances. This allows users to see the dependencies as they are generated during execution, and to visually inspect if parallelization of certain loops are safe.

As a convenience to users, hints that depict parallel loops are provided by the visualization engine by drawing a set of hyperplanes on top of the view. In general, the equations for each hyperplane are specified as a set of constraints such that statement instances that fall within the same plane can be executed by the same thread. These planes are projected down to 3D space, and drawn as an overlay on top of the polyhedral visualization. The framework also includes an OpenMP parallelization component that assists the user in parallelizing code. This component helps to insert OpenMP directives into the original source code by presenting a list of parallelizable loops for the user to choose from. After selecting the loop to parallelize, it then allows the user to modify a list of shared and private variables that have been automatically detected. Reduction patterns can also be detected and highlighted to the user. The modified code can be previewed by the user before committing the changes.

3   Examples

In this section, we shall demonstrate the use of our visualization framework on two examples. The first example is an image processing example found in many applications. The second example is from *482.sphinx3*, a speech recognition application taken from the SPEC CPU2006 benchmark suite [12].

3.1   Anisotropic Diffusion

Anisotropic diffusion is an image noise reduction technique that is commonly used in applications such as in ultrasound imaging or magnetic resonance imaging [10]. It simulates an iterative diffusion process which is non-linear and space-variant, and is aimed at removing image noise while at the same time preserving important image details, especially the edges in an image. The algorithm takes a noisy image and calculates for each pixel a set of eight values based on predefined kernels, and then accumulates the sum of their weighted differences.

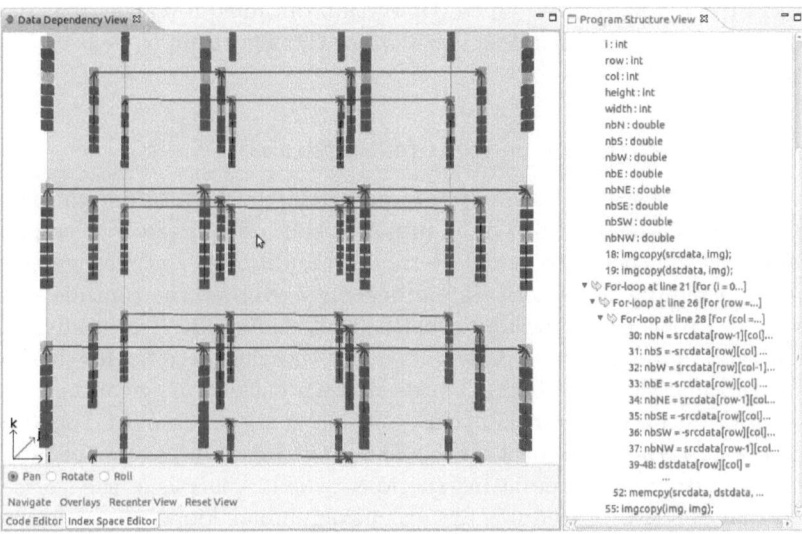

Fig. 5. Visualization of the code section in the `diffuse` procedure. The user can select statements using the Program Structure View on the right to highlight the enclosing domains. No arrows cross the planes normal to j and k axes, indicating that the j and k loops are parallelizable.

From the Loop-Procedure View, we determined that the code section in the `diffuse` procedure took up about 98% of the total execution time while the rest is mainly due to I/O operations. Figure 5 shows the corresponding polyhedral model of the code. Since there are three `for`-loops present, the 3D iteration space resembles a rectangular cuboid. Each colored cube represents a statement instance within the iteration space, and each arrow represents a producer-consumer relationship between statement instances. The i, j and k axes denote the direction of their respective loop iterations. As the implementation contains pointer arithmetic, the dependencies were obtained using dynamic analysis. The user can highlight a statement instance in the Data Dependency View by selecting the corresponding statement in the code panel on the right. By inspecting the polyhedral model in Fig. 5, the user is able to determine that only planes normal to the j and k axes do not have any arrows crossing them. Therefore, the corresponding j and k loops do not have any loop-carried dependencies and are fully parallelizable. The programmer may then select the outer loop and invoke the OpenMP parallelization component to insert OpenMP directives to the loop.

On a Intel Core 2 Extreme Q6850 processor running at 3.00GHz and with a 512 by 512-pixel image, the OpenMP version of the code yielded close to a 4 times speedup compared to the original single-threaded version. This is a significant improvement because the sequential version of the code is not suitable for many practical applications that demand real-time processing of images acquired from sensors. The sequential form of the code could only manage about 4.2 frames per second (fps), whereas the OpenMP version obtained a respectable 16.6 fps. This improvement is significant as it will allow the algorithm to be used in many real-time applications.

3.2   Speech Recognition System (482.sphinx3)

Sphinx-3 is a speech recognition system based on the Viterbi search algorithm using beam search heuristics. The inputs are read initially, and the application then processes the speech to calculate the probabilities at each recognition step in order to prune the set of active hypothesis. We overlaid the runtime statistics on the Loop-Procedure View and identified two hotspots. Essentially, the application is dominated by two procedures that together account for over half of the total runtime. The two hotspot procedures identified are mgau_eval, which accounted for 30.1% of the total cycles, and vector_gautbl_eval_logs3, which accounted for 24.5% of the total cycles. These two procedures also accounted for 92.9% of the total floating point instructions issued. Analysis of the source code revealed that the two procedures are executing similar loops. As such, we shall only present one of the procedures, mgau_eval, in this paper. The procedure consists of a two-dimensional loop with the inner loop accumulating a score for the search probabilities.

The polyhedral model generated by the Data Dependency View is shown in the Fig. 6. The polyhedra are two-dimensional planes, which correspond to the two-dimensional loop. The statement represented by the gray nodes has a flow dependence to the statement represented by the blue nodes. Within the j iteration, the red and blues nodes have a flow dependence, as well as a loop-carried data dependence. By inspecting Fig. 6, the user can see that it is possible

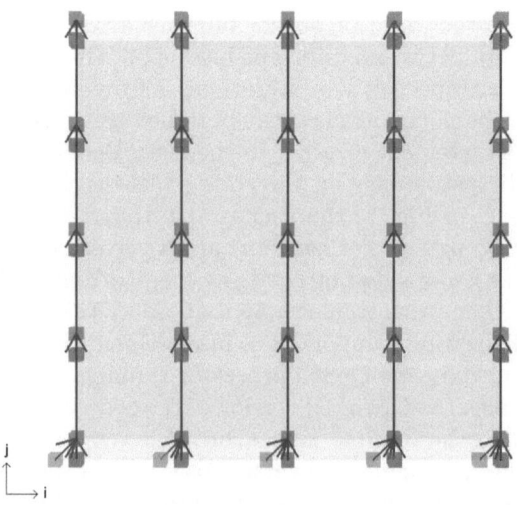

Fig. 6. Visualization of the mgau_eval procedure. Loop iterations along the i axis can be partitioned since there are no dependencies in the horizontal direction.

to partition the polyhedra in the i domain by cutting across the i axis. In other words, one can parallelize the i loop without violating the correctness of the code, since there are no arrows in the horizontal direction, and therefore, no dependencies in the i direction.

4   Related Work

A number of visualization tools that target the Fortran language have been developed over the past two decades. The ParaScope Editor [6], developed at Rice University, allows the user to step through each loop in the program and displays the relevant information in a two-panel window. The bottom panel displays a list of the detected data-dependences along with the dependence vectors, as well as other relevant details in a tabular format. The top panel displays the source code of the current loop under consideration. It then allows the user to select a transformation that is deemed safe to be applied to the program. NaraView introduced an interactive 3D visualization system for aiding the parallelization of sequential programs [11]. Of notable interest is the 3D Data Dependence View, which displays data accesses as colored cubes and dependences as poles connecting the cubes. However, NaraView does not perform program profiling to determine the important loops, and does not include the ability to animate a walk-through of the iteration space. Instead, it uses the z-axis to represent the iteration access time and the x-y plane to denote the data location. Therefore, visualization is limited to loops with a depth of at most two.

SUIF Explorer [8] is another interactive parallelization tool which targets both the Fortran and C languages. It includes a Loop Profile Analyzer that identifies the important loops that dominate the execution time. A useful feature of the tool is that it applies inter-procedural slicing to the program to display only the relevant lines of code to the programmer so that he can make the appropriate decisions. Annotations are added to the code, which are then checked for their validity by the built-in checkers. The annotations help to enable the parallelizing compiler to parallelize the loop. However, it does not make use of OpenMP directives. Other tools such as Merlin [7] provide a textual representation of the program analysis to the user. It compiles the code using the Polaris parallelizer, gathers static and dynamic execution data, then performs analysis using "performance maps", and finally presents diagnostics about the program as well as suggestions on how to improve the code to the user in a textual format. Currently, the only tool that supports 3D visualization for Fortran code is the Iteration Space Visualizer [15].

In comparison, there are fewer visualization tools that target the C language. The SUIF Explorer is able to perform inter-procedural analysis on C programs by building upon the SUIF parallelizing compiler, and presents the results to the user using program slicing. Another tool that supports C code parallelization is ParaGraph [1]. It makes use of the Cetus compiler [2] to automatically parallelize loops using OpenMP directives [9]. Alternatively, it also allows the user to specify the directives, which it validates before attempting parallelization.

To our knowledge, ParaGraph is currently the only visualization tool for C that runs as a Eclipse plug-in. However, apart from the source code outline and a properties tab, the visualization support provided to the user consists of only a control flow graph augmented with the dependency information in the form of directed and dashed arrows between control blocks. Another related work is VisualPolylib [14], which draws the polyhedral model supplied by the user. However, the user has to manually extract the model from the code and supply it to the tool. Apart from ParaGraph, most of the tools are stand-alone front-ends, and are not integrated with any IDEs. On the other hand, Tulipse is wholly integrated in the Eclipse software development environment, which makes it highly extensible and allows it to leverage many existing software engineering tools and plug-ins available in the Eclipse developer framework. In addition, in using our framework, parallelization opportunities can be identified even if the code cannot be analyzed using static approaches, such as code involving pointer arithmetic.

The Intel Parallel Advisor [5] is closest to what our framework offers in terms of capabilities and workflow. However there are a few major differences. First, Intel Parallel Advisor is mostly textual. Unlike our framework, it does not provide visualization capabilities. Secondly, the workflow is different. The Advisor requires the programmers to take a trial and error approach by guessing and annotating parts of the code which they believe can be parallelized. The Advisor then performs a trial run to detect data races. If races are detected, the user is notified and may again attempt to identify other parallelizable sections of the code. On the other hand, in our workflow, the data dependencies are first obtained and then projected to 3D space to allow the user to directly identify the parallelization opportunities.

5   Conclusions and Future Work

In this paper, we introduced Tulipse, a visualization framework for parallelization built on top of Eclipse, with the goal of enhancing program understanding and reducing the cognitive load of the developers in parallelizing applications.

In Tulipse, the programmer starts with a loop-procedural hierarchy graph of the entire application using the Loop-Procedure View. This gives the user a bird's eye view of the application. Color coding allows for the display of selectable performance data in this view, allowing the programmer to quickly zoom in to the hotspots or the problem areas. Zooming in to the loop level exposes the polyhedral view of the loop. In the Data Dependency View, the programmer can easily correlate data dependencies found in the code by static or dynamic analysis. This visualization also allows for the programmer to estimate the effort involved in attempting to parallelize the loop along a particular dimension of the iteration space. The user may invoke the OpenMP parallelization component to aid in adding OpenMP directives to the selected loop.

Performance is but one aspect of software development. By being integrated in a rich software development environment such as Eclipse, it gives developers access to a wide range of tools that support their various workflows. As future work, we shall be extending Tulipse with other parallelization and performance tuning visualizers that will grow its functionality, for example cache usage visualization that will help in identifying optimal data layout. We will also be investigating how the user can carry out code transformations such as loop skewing and tiling with the help of interactive visualization.

Acknowledgments. This work was supported by the Agency for Science, Technology and Research PSF Grant No. 102-101-0028. We are also grateful to the anonymous reviewers for their suggestions.

References

1. Bluemke, I., Fugas, J.: A tool supporting C code parallelization. Innovations Comput. Sci. Soft. Eng., 259–264 (2010)
2. Dave, C., Bae, H., Min, S.J., Lee, S., Eigenmann, R., Midkiff, S.: Cetus: A source-to-source compiler infrastructure for multicores. Computer 42, 36–42 (2009)
3. Girbal, S., Vasilache, N., Bastoul, C., Cohen, A., Parello, D., Sigler, M., Temam, O.: Semi-automatic composition of loop transformations for deep parallelism and memory hierarchies. Int. J. Parallel Program. 34, 261–317 (2006)
4. HPCToolkit, `http://www.hpctoolkit.org`
5. Intel Parallel Advisor, `http://software.intel.com/en-us/`
6. Kennedy, K., McKinley, K.S., Tseng, C.W.: Interactive parallel programming using the ParaScope editor. IEEE Trans. Parallel Distrib. Syst. 2, 329–341 (1991)
7. Kim, S.W., Park, I., Eigenmann, R.: A Performance Advisor Tool for Shared-Memory Parallel Programming. In: Midkiff, S.P., Moreira, J.E., Gupta, M., Chatterjee, S., Ferrante, J., Prins, J.F., Pugh, B., Tseng, C.-W. (eds.) LCPC 2000. LNCS, vol. 2017, pp. 274–288. Springer, Heidelberg (2001)
8. Liao, S.W., Diwan, A., Bosch Jr., R.P., Ghuloum, A., Lam, M.S.: SUIF Explorer: an interactive and interprocedural parallelizer. In: PPoPP, pp. 37–48 (1999)
9. OpenMP, `http://www.openmp.org`
10. Perona, P., Malik, J.: Scale-space and edge detection using anisotropic diffusion. IEEE Trans. Pattern Anal. Mach. Intell. 12, 629–639 (1990)
11. Sasakura, M., Joe, K., Kunieda, Y., Araki, K.: Naraview: An interactive 3D visualization system for parallelization of programs. Int. J. Parallel Program. 27, 111–129 (1999)
12. SPEC CPU 2006 v1.1, `http://www.spec.org`
13. Tau: `http://www.cs.uoregon.edu/research/tau/`
14. VisualPolylib, `http://icps.u-strasbg.fr/polylib/`
15. Yu, Y., D'Hollander, E.H.: Loop parallelization using the 3D iteration space visualizer. J. Visual Lang. Comput. 12, 163–181 (2001)

Enabling Cloud Interoperability with COMPSs

Fabrizio Marozzo[3], Francesc Lordan[1], Roger Rafanell[1], Daniele Lezzi[1],
Domenico Talia[3,4], and Rosa M. Badia[1,2]

[1] Barcelona Supercomputing Center - Centro Nacional de Supercomputación
(BSC-CNS)
{daniele.lezzi,francesc.lordan,roger.rafanell,rosa.m.badia}@bsc.es
[2] Artificial Intelligence Research Institute (IIIA),
Spanish Council for Scientific Research (CSIC)
[3] DEIS, University of Calabria, Rende (CS), Italy
{fmarozzo,talia}@deis.unical.it
[4] ICAR-CNR, Rende (CS), Italy

Abstract. The advent of Cloud computing has given to researchers the
ability to access resources that satisfy their growing needs, which could
not be satisfied by traditional computing resources such as PCs and
locally managed clusters. On the other side, such ability, has opened
new challenges for the execution of their computational work and the
managing of massive amounts of data into resources provided by different
private and public infrastructures.

COMP Superscalar (COMPSs) is a programming framework that provides a programming model and a runtime that ease the development of
applications for distributed environments and their execution on a wide
range of computational infrastructures. COMPSs has been recently extended in order to be interoperable with several cloud technologies like
Amazon, OpenNebula, Emotive and other OCCI compliant offerings.

This paper presents the extensions of this interoperability layer to
support the execution of COMPSs applications into the Windows Azure
Platform. The framework has been evaluated through the porting of
a data mining workflow to COMPSs and the execution on an hybrid
testbed.

Keywords: Parallel programming models, Cloud computing, Data mining, PaaS.

1   Introduction

The growth of cloud services and technologies has brought many advantages and
opportunities to scientific communities offering users efficient and cost-effective
solutions to their problems of lack of computational resources. Even though
the cloud paradigm does not address all the issues related to the porting and
execution of scientific applications on distributed infrastructures, it is widely
recognized that, through clouds, researchers can provision compute resources on
a pay-per-use basis, thus avoiding to enter in a procurement process that implies
investment costs for buying hardware or access procedures to supercomputers.

C. Kaklamanis et al. (Eds.): Euro-Par 2012, LNCS 7484, pp. 16–27, 2012.

Recently, several grid initiatives and distributed computing infrastructures [1] [2] [3] have started to develop cloud services in order to provide existing services through virtualized technologies for the dispatch of scientific applications. These technologies allow the deployment of hybrid computing environments where the provision of private clouds is backed up by public offerings such as Azure[4] or Amazon[5]. The VENUS-C[6] project in particular aims to support research and industry user communities to leverage cloud computing for their applications through the provision of a hybrid platform that provides commercial (Azure) and open source cloud services.

In such a hybrid landscape, there are technical challenges such as interoperability that need to be addressed from different points of view. The interoperability concept can refer to different things at many levels. It could mean the ability to keep the behaviour of an application when it runs on different environments such as a cluster, a grid or an IaaS provided infrastructure like Amazon instances. At lower level, it might refer to a single application running in many clouds being able to share information, which might require having a common set of interfaces and the ability of users to use the same management tools, server images and other software with a variety of Cloud computing providers and platforms. From a programming framework perspective these issues have to be solved also at different levels, developing the appropriate interfaces to interact with several cloud providers, ensuring that the applications are executed on different infrastructure without having to adapt them and handling data movements seamlessly amongst different cloud storages.

The COMP Superscalar[7] programming framework allows the programming of scientific applications and their execution on a wide number of distributed infrastructures. In cloud environments, COMPSs provides scaling and elasticity features allowing to adapt the number of available resources to the actual need of the execution. The availability of connectors for several providers makes possible the execution of scientific applications on hybrid clouds taking into account the above mentioned issues related to the porting of applications to a target cloud and their transparent execution with regards to the underlying infrastructure. This paper describes the developments for making COMPSs interoperable with Windows Azure Platform through the design of a specific adaptor.

The rest of the document is organized as follows. Section 2 describes the COMPSs framework. Section 3 details the developed Azure GAT Adaptor. Section 4 illustrates the porting of a data mining application to COMPSs. Section 5 evaluates the performance of the ported application. Section 6 discusses the related work. Section 7 presents the conclusions and the future work.

2   The COMPSs Framework

COMPSs is a programming framework, composed of a programming model and an execution runtime which supports it, whose main objective is to ease the development of applications for distributed environments.

On the one hand, the programming model aims to keep the programmers unaware of the execution environment and parallelization details. They are only required to create a sequential application and specify which methods of the application code will be executed remotely. This selection is done by providing an annotated interface where these methods are declared with some metadata about them and their parameters.

On the other hand, the runtime is in charge of optimizing the performance of the application by exploiting its inherent concurrency. The runtime intercepts any call to a selected method creating a representative task and finding the data dependencies with all the previous ones that must be considered along the application run. The task is added to a task dependency graph as a new node and such dependencies are represented by edges of the graph. Tasks with no dependencies enter the scheduling step and are assigned to available resources. This decision is made according to a scheduling algorithm that takes into account data locality, task constraints and the workload of each node. According to this decision the input data for the scheduled task are transferred to the selected host and the task is remotely submitted. Once a task finishes, the task dependency graph is updated, possibly resulting in new dependency-free tasks that can be scheduled.

In a previous work[8], some of the authors described how COMPSs could also benefit from Infrastructure-as-a-Service (IaaS) offerings. Through the monitoring of the workload of the application, the runtime determines the excess/lack of resources and turns to cloud providers enforcing a dynamic management of the resource pool. In order to make COMPSs interoperable with different providers, a common interface is used, which implements the specific cloud provider API. Currently, there exist connectors for Amazon EC2 and for providers that implement the Open Cloud Computing Interface (OCCI)[9] and the Open Virtualization Format (OVF)[10] specifications for resource management.

The contribution presented in this paper is an extension of the COMPSs runtime to make it interoperable with the Microsoft Azure Platform-as-a-Service (PaaS) offering. COMPSs virtualizes all the resources provided by Azure as a single logical machine where multiple tasks can be executed at the same time. These extensions do not affect the programming model, keeping existing COMPSs applications unchanged and the user unaware of the technical details. Users only have to deal with the deployment of some components, as described later, on their Azure account.

3   The Azure JavaGAT Adaptor

In order to solve the interoperability issues related to the execution of tasks using an heterogeneous pool of resources in distributed environments, COMPSs adopts JavaGAT[11] as the uniform interface to underlying Grid and Cloud middlewares implemented in several adaptors. Whenever a task has to be executed on a specific resource, COMPSs manages all the data transfers and submits the task using the proper adaptor. The Azure JavaGAT Adaptor here described enriches COMPSs with data management and execution capabilities that make it interoperable with Azure and implemented using two subcomponents.

Data management is supported by a subcomponent called Azure File Adaptor. It allows to read and write data on the Azure Blob Storage (*Blobs*), to deploy the *libraries* needed to execute on Azure and to store the input and output data (*taskdata*) for the tasks. The Azure Resource Broker Adaptor, on the other side, is responsible for the task submission. Following the Azure Work Queue pattern, this subcomponent adds into a *Task Queue* the tasks that must be executed on an Azure resource by a *Worker*. The implementation of these COMPSs workers as Worker Role instances is based on a previous work on a Data Mining Cloud App framework[12]. In order to keep the runtime informed about each task execution, the status of the tasks is updated in a *Task Status Table*. The whole architecture of the Azure JavaGAT Adaptor is depicted in Figure 1.

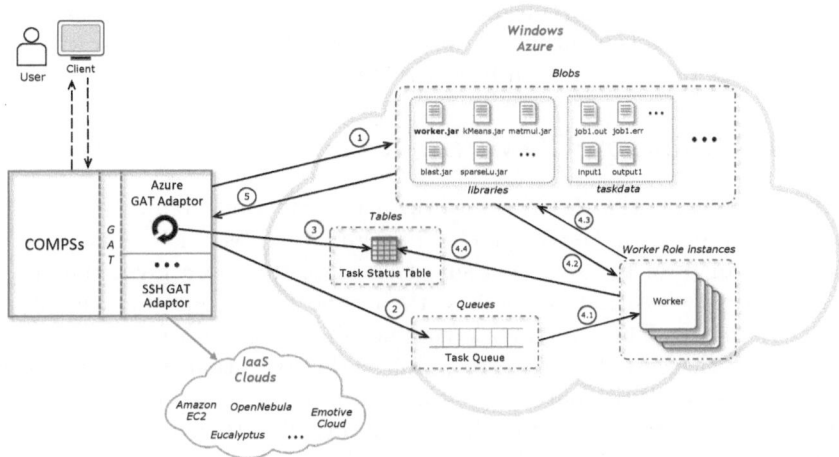

Fig. 1. The Azure GAT adaptor architecture

The numbered components in Figure 1 correspond to each item in the list below, which describes the different stages of a remote task execution on Azure. The whole process starts when the COMPSs runtime decides to execute a dependency-free task t in the platform following the next steps:

1. The Azure GAT adaptor, through the Azure File adaptor, prepares the execution environment uploading the input application files and libraries into the Blob containers, *taskdata* and *libraries*.
2. The adaptor, via the Azure Resource Broker, inserts a task t description into the *Task Queue*.
3. The adaptor sets the status of the task t to *Submitted* in the *Task Status Table* and polls periodically in order to monitor its status until it becomes *Done* or *Failed*.

4. An idle worker W takes the task t description from the queue and, after parsing all the parameters, it runs the task. This step can be divided in the following sub-steps:

   4.1. The worker W takes the task t from the *Task Queue* starting its execution on a virtual resource. The worker sets the status of t to *Running*.

   4.2. The worker gets the needed input data and the needed libraries according to the description of t. To this end, a file transfer is performed from the Blob, where the input data is located, to the local storage of the resource, and the task is executed.

   4.3. After a task completion, the worker W moves the resulting files in the *taskdata* Blob container.

   4.4. The worker updates the status of the task in the *Task Status Table* setting it to a final status that could be *Done* or *Failed*.

5. When the adaptor detects that the task t execution has finalized, it notifies the execution end to the runtime which looks for new dependency-freed tasks to be executed. If the output files are not going to be used by any other task, the runtime downloads them from the Azure Blob.

4 Data Mining on COMPSs: A Classifier-Based Workflow

In order to validate the described work, a data mining application has been adapted to run in a cloud environment through COMPSs. Such application runs multiple instances of the same classification algorithm on a given dataset, obtaining multiple classification models, then chooses the one that classify in a more accurate way. Thus, the aim is twofold: first, validate the implementation checking that the system is able to manage the execution on different Cloud deployments; second, compare the performance of the proposed solution on an hybrid cloud scenario. The rest of the section describes the data mining application as a workflow (Section 4.1), its Java implementation (Section 4.2) and the porting to COMPSs (Section 4.3).

4.1 The Application Workflow

Figure 2 depicts the four general steps of the classifier-based workflow:

1. **Dataset Partition:** the initial dataset is split into two parts: a training set, which trains the classifiers, and a test set to check the effectiveness of the achieved models.

2. **Classification:** during this step, the training dataset is analyzed in parallel using multiple instances of the same classifier algorithm with different parameters.

3. **Evaluation:** the quality of each classification model is measured using different performance metrics (e.g., number of misclassified items, precision and recall measure, F-measure).

4. **Model Selection:** finally, the best model is selected optimizing the chosen performance metrics.

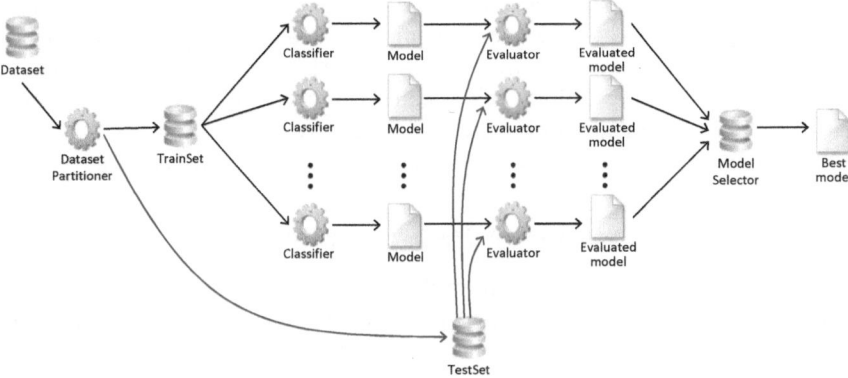

Fig. 2. The data mining application workflow

4.2 The Application Implementation

Following the described workflow, the initial dataset is divided in two parts: $2/3^{rd}$ are left as a training set and the remaining $1/3^{rd}$ is used as test set. The classification algorithm is the *J48*, provided in Weka[13] data mining toolkit, based on C4.5[14] algorithm. This algorithm builds a decision tree using the concept of information entropy to classify the different items in the training set. The different models are obtained varying the *confidence value* parameter of J48 in a range of values (i.e., from 0.05 to 0.50). Such range is divided in a certain number of intervals specified by the user as an application parameter. Each model is evaluated using, as a performance metric, the number of misclassified items. Listing 1.1 presents the main code of the application:

```
public static void main(String args[]) throws Exception {
    ...
    //Run remote method
    for (int i = 0; i < n_itvls; i++){
        c_val = c_min_val+i*(c_max_val-c_min_val)/(num_itvls-1);
        //***************** Remote methods ******************//
        models[i]=WorkflowImpl.learning(trainSet, c_val);
        reports[i]=WorkflowImpl.evaluate(models[i], testSet);
    }
    //Selection of the best model in binary tree way
    int n = 1;
    while (n < n_itvls){
        for (int i = 0; i < n_itvls; i+= 2 * n){
            if (i + n < n_itvls) {
                //***************** Remote method ******************//
                WorkflowImpl.getBestIndex(reports[i], reports[i+n]);
            }
        }
        n *= 2;
    }
    //Read best model
    J48 bestModel = models[reports[0].getIndex()];
}
```

Listing 1.1. Main application code

As described in the application workflow section, the methods in lines *7* and *8* correspond to the classification and evaluation steps of the workflow. The *c_min_val* and *c_max_val* are the limits of the confidence value range, and *num_itvls* is the number of intervals specified by user. The model selection step (lines 11 − 22) is performed in binary tree way in order to exploit the possibility to be parallelized by COMPSs as detailed in along the next section.

4.3   Parallelization with COMPSs: The Interface

The main step of the porting of an application to COMPSs includes the preparation of a Java annotated interface provided by the programmer in order to select which methods will be executed remotely. For each annotated method, the interface specifies some information like the name of the class that implements it, and the type (e.g., primitive, object, file) and direction (e.g., in, out or in/out) of its parameters. The user can add some additional metadata to define the resource features required to execute each method. The Listing 1.2 shows the annotated interface for the presented application.

```
1   public interface WorkflowItf {
2     @Constraints(processorCPUCount = 1, storageElemSize= 1.0)
3     @Method(declaringClass = "workflow.WorkflowImpl")
4     J48 learning(
5       @Parameter(type = Type.OBJECT, direction = Direction.IN)
6       Instances trainSet,
7       @Parameter(type = Type.FLOAT, direction = Direction.IN)
8       float confFactor);
9
10    @Method(declaringClass = "workflow.WorkflowImpl")
11    Report evaluate(
12      @Parameter(type = Type.INT, direction = Direction.IN)
13      int i,
14      @Parameter(type = Type.OBJECT, direction = Direction.IN)
15      J48 model,
16      @Parameter(type = Type.OBJECT, direction = Direction.IN)
17      Instances testSet);
18
19     @Method(declaringClass = "workflow.WorkflowImpl")
20     void getBestIndex(
21      @Parameter(type = Type.OBJECT, direction = Direction.INOUT)
22      Report rep0,
23      @Parameter(type = Type.OBJECT, direction = Direction.IN)
24      Report rep1);
25  }
```

Listing 1.2. Application Java interface

The COMPSs runtime intercepts the invocations in the main code to any method contained in this interface by generating a task-dependency graph. Figure 3 shows an example of the resulting dependency graph of the data mining application. The red circles corresponds to *learning* tasks which forwards their results to the *evaluate* tasks represented in yellow, creating dependencies between them. All these evaluations end in a reduction process implemented using *getBestIndex* tasks, colored as blue, which find the model that minimizes the number of classification errors.

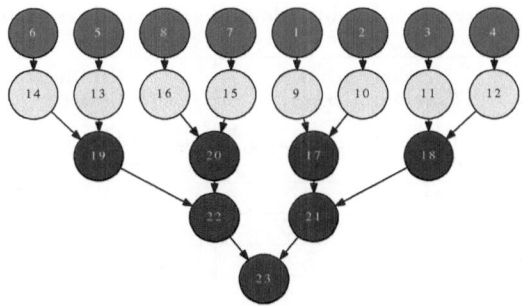

Fig. 3. Task dependency graph automatically generated by COMPSs

5 Performance Evaluation

In order to evaluate the performance of the workflow application, a set of experiments has been conducted using three different configurations: *i*) a private cloud environment managed by Emotive Cloud[15] middleware; *ii*) a public cloud testbed made of Azure instances; *iii*) an hybrid configuration using both private and public clouds.

The private cloud included a total of 96 cores available in the following way: 4 nodes with 12 Intel Xeon X5650 Six Core at 2.6GHz processors, 24GB of memory and 2TB of storage each, and 3 nodes with 16 AMD Opteron 6140 Eight Core at 2.6GHz processors, 32GB of memory and 2TB of storage each. The nodes were interconnected by a Gigabit Ethernet network and the storage was offered through a GlusterFS[16] distributed file system running in a replica configuration mode providing a total of 8TB of usable space. On this testbed 8 quad-core virtual instances with 8GB of memory and 2GB of disk space have been created running a fresh Debian Squeeze Linux distribution.

The public testbed based on Windows Azure was composed of up to 20 small virtual instances with 1.6GHz single core processor, 1.75GB of memory and 225GB of disk space each. In order to reduce the impact of data transfer on the overall execution time, the Azure's Affinity Group feature has been exploited allowing the storage and servers to be located in the same data center for performance reasons.

The *covertype*[1] dataset has been used as data source. This dataset contains information about forest cover type of a large number of sites in the United States. Each instance, corresponds to a site observation and contains 54 attributes that describe the main features of a site (e.g., elevation, aspect, slope, etc.). A subset with 290.000 instances has been taken from this dataset creating a new 36MB large one.

Table 1 presents the execution times and the speedup of an application run with 100 different models and up to 20 and 32 processors available in the public and private cloud deployments respectively. Table 2 presents the results of the

---

[1] http://kdd.ics.uci.edu/databases/covertype/covertype.html

Table 1. Private and public cloud deployment execution times

N. of cores	Private cloud (Emotive Cloud)		Public cloud (Microsoft Azure)	
	Execution time	Speedup	Execution time	Speedup
1	7:34:41	1	8:19:05	1
2	3:50:25	1.97	4:18:04	1.93
4	2:07:35	3.56	2:07:30	3.91
8	1:08:51	6.6	1:08:15	7.31
16	0:37:13	12.22	0:36:22	13.72
20	0:28:11	16.13	0:29:55	16.68
32	0:18:24	24.71	N/A	N/A

Table 2. Hybrid cloud deployment execution times

N. of cores	Private cloud + Azure	
	Execution time	Speedup
32 + 2	0:17:29	26.01
32 + 4	0:17:07	26.56
32 + 8	0:16:38	27.34
32 + 12	0:14:14	31.94
32 + 16	0:14:06	32.25
32 + 20	0:13:17	34.23

same experiment running on the hybrid cloud scenario; in this case, cloud out-sourcing is used to expand the computing pool out of the private cloud domain.

As depicted in Figure 4, execution times are similar in both cases where a single cloud provider is used: Emotive Cloud and Azure. The speedup keeps a quasi-linear gain along the execution up to the point where the outsourcing starts. The trend changes observed in the speedup curve are not originated by the usage of outsourced resources but by a workload unbalance due to the impossibility to adjust the total number of tasks (constrained by the specific use case) to the amount of available resources. When the number of resources allows a good load balancing, the speedup curve recovers some of the lost performance as depicted in the 32+12 case where the gain is increased over the ideal line. Generally, when the workload does not depend on the application input, the COMPSs runtime scheduler is able to adapt the number of tasks to the number of available resources.

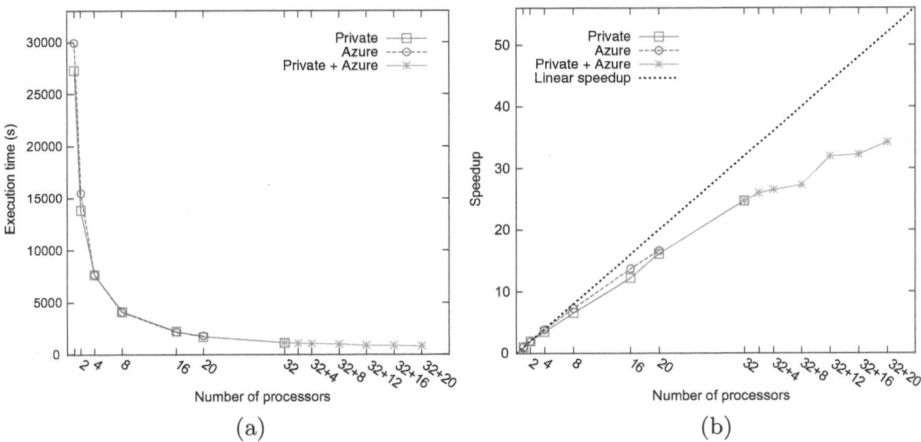

Fig. 4. Execution time and speedup values depending on the number of processors

6   Related Work

There already exist several frameworks that enable the programming and execution of applications in the Cloud and several research products are being developed to enhance the execution of applications in the Azure Platform. MapReduce [17], a widely-offered programming model, permits the processing of vast amounts of data by dividing it into many small blocks that are processed in parallel (i.e, map phase) and their results merged (i.e., reduce phase).

Hadoop[18] is an open source software platform which implements MapReduce using the Hadoop Distributed File System (HDFS). HDFS creates multiple replicas of data blocks for reliability and places them on compute nodes so they can be processed locally. Hadoop on Azure[19] is a new Apache Hadoop based distribution for Windows Server. Microsoft Daytona[20] presents an iterative MapReduce runtime for Windows Azure designed to support a wide class of data analytics and machine learning algorithms. Also Google supports MapReduce executions in its Google App Engine[21], which provides a set of libraries to invoke external services and queue units of work (tasks) for execution. Twister[22] is an enhanced MapReduce runtime with an extended programming model that supports iterative MapReduce computations efficiently. These public cloud platforms have a high level of user's intervention in the porting of the applications requiring the use of specific APIs and the deployment and execution of the applications on their own infrastructure thus avoiding to port the code to another platform. COMPSs, on the contrary, can execute the applications on any supported cloud provider without the need to adapt the original code to the specific target platform nor writing the map and reduce functions as in MapReduce frameworks.

Manjrasoft Aneka[23] platform provides a framework for the development of application supporting not only the MapReduce programming model but also a Task Programming and Thread Programming ones. The applications can be deployed on private or public clouds such as Windows Azure, Amazon EC2, and GoGrid Cloud Service. The user has to use a specific .NET SDK for the porting of the code also to enact legacy code. Microsoft Generic Worker[24] has been extended in the context of the VENUS-C project to ease the porting of legacy code in the Azure platform. Even if the user does not have to change the core of the code, the creation of workflows is not automated, as is in COMPSs, in any of them; but has to be explicitly enacted through separated executions. Moreover, an application executed through the Generic Worker, can not be ported to other platforms.

7   Conclusions and Future Work

This paper presents the extensions of COMPSs programming framework to make it able to execute e-Science applications also on the Azure Platform. The contribution include the development of a JavaGAT adaptor that allows the scheduling of COMPSs tasks on Azure instances taking care of the related data transfers, and the implementation of a set of components deployed on Azure to manage the execution of the tasks internally to the instances. The proposed approach has been validated through the execution of a data mining workflow ported to COMPSs and executed on an hybrid testbed composed of a private cloud managed by Emotive Cloud and Azure machines. The results demonstrates that the runtime is able to manage and schedule the tasks on different infrastructures in a transparent way, keeping the overall performance of the application.

Future work includes the creation of a new connector in COMPSs to support the dynamic resource provisioning in Azure and enhancements to the Azure JavaGAT adaptor to optimize data transfers among different clouds, and the possibility to specify input files already available on the Azure storage. The scheduling of the COMPSs runtime will be also optimized to better balance the execution of tasks taking also into account the required time to transfer data.

Acknowledgements. This work has been supported by the Spanish Ministry of Science and Innovation (contracts TIN2007-60625, CSD2007-00050 and CAC2007-00052), by Generalitat de Catalunya (contract 2009-SGR-980), and the European Commission (VENUS-C project, Grant Agreement Number: 261565). This work was also made possible using the computing use grant provided by Microsoft in the VENUS-C project.

References

1. European Grid Infrastructure, http://www.egi.eu
2. StratusLab, http://www.stratuslab.eu
3. European Middleware Initiative, http://www.eu-emi.eu
4. Microsoft Azure, http://www.microsoft.com/azure

5. Amazon Elastic Compute Cloud, `http://aws.amazon.com/es/ec2`
6. Virtual multidisciplinary ENvironments USing Cloud infrastructures, `http://www.venus-c.eu`
7. Tejedor, E., Badia, R.M.: COMP Superscalar: Bringing GRID superscalar and GCM Together. In: IEEE Int. Symposium on Cluster Computing and the Grid, Lyon, France (2008)
8. Lezzi, D., Rafanell, R., Carrion, A., Blanquer, I., Badia, R.M., Hernandez, V.: Enabling e-Science applications on the Cloud with COMPSs. Cloud Computing: Project and Initiatives (2011)
9. Open Cloud Computing Interface Working Group, `http://www.occi-wg.org`
10. Distributed Management Task Force Inc., Open Virtualization Format Specification v1.1. DMT Standar DSP0243 (2010)
11. Allen, G., Davis, K., Goodale, T., Hutanu, A., Kaiser, H., Kielmann, T., Merzky, A., van Nieuwpoort, R., Reinefeld, A., Schintke, F., Schütt, T.T., Seidel, E., Ullmer, B.: The Grid Application Toolkit: Towards Generic and Easy Application Programming Interfaces for the Grid. Proceedings of the IEEE 93(3) (March 2005)
12. Marozzo, F., Talia, D., Trunfio, P.: A Cloud Framework for Parameter Sweeping Data Mining Applications. In: 3rd IEEE Int. Conference on Cloud Computing Technology and Science (CloudCom 2011), Athens, Greece (2011)
13. Witten, H., Frank, E.: Data Mining: Practical machine learning tools with Java implementations. Morgan Kaufmann Publishers (2000)
14. Quinlan, J.R.: C4.5: Programs for Machine Learning. Morgan Kaufmann Publishers (1993)
15. Goiri, I., Guitart, J., Torres, J.: Elastic Management of Tasks in Virtualized Environments. In: XX Jornadas de Paralelismo (JP 2009), Coruña, Spain (2009)
16. GlusterFS Distributed Network File System, `http://www.gluster.org`
17. Dean, J., Ghemawat, S.: MapReduce: simplified data processing on large clusters. Commun. ACM 51, 107–113 (2008)
18. Apache Hadoop, `http://hadoop.apache.org`
19. Hadoop on Azure, `https://www.hadooponazure.com`
20. Project Daytona, `http://research.microsoft.com/en-us/projects/daytona`
21. Google App Engine, `http://code.google.com/intl/de/appengine`
22. Ekanayake, J., Li, H., Zhang, B., Gunarathne, T., Bae, S., Qiu, J., Fox, G.: Twister: A Runtime for Iterative MapReduce. In: 1st Int. Workshop on MapReduce and its Applications (MAPREDUCE 2010), Chicago, USA (2010)
23. Wei, Y., Sukumar, K., Vecchiola, C., Karunamoorthy, D., Buyya, R.: Aneka Cloud Application Platform and Its Integration with Windows Azure. CoRR, abs/1103.2590 (2011)
24. Simmhan, Y., Ingen, C., Subramanian, G., Li, J.: Bridging the Gap between Desktop and the Cloud for eScience Applications. In: 3rd IEEE Int. Conference on Cloud Computing (CLOUD 2010), Washington, USA (2010)

Pattern-Independent Detection
of Manual Collectives in MPI Programs

Alexandru Calotoiu[1,2], Christian Siebert[1,2], and Felix Wolf[1,2,3]

[1] German Research School for Simulation Sciences, 52062 Aachen, Germany
[2] RWTH Aachen University, Computer Science Department, 52056 Aachen, Germany
[3] Forschungszentrum Jülich, Jülich Supercomputing Centre, 52425 Jülich, Germany

Abstract. In parallel applications, a significant amount of communication occurs in a collective fashion to perform, for example, broadcasts, reductions, or complete exchanges. Although the MPI standard defines many convenience functions for this purpose, which not only improve code readability and maintenance but are usually also highly efficient, many application programmers still create their own, manual implementations using point-to-point communication. We show how instances of such hand-crafted collectives can be automatically detected. Matching pre- and post-conditions of hashed message exchanges recorded in event traces, our method is independent of the specific communication pattern employed. We demonstrate that replacing detected broadcasts in the HPL benchmark can yield significant performance improvements.

Keywords: MPI, collective operations, performance optimization, HPL.

1 Introduction

The most scalable parallel application codes today use message passing as their primary parallel programing model, which offers explicit communication primitives for the exchange of messages. While pair-wise communication is most common, the majority of applications require communication among larger groups of processes [5]. The latter is needed, for example, to distribute data, gather results, make collective decisions, or broadcast their outcomes. Although all those communication objectives can be mapped onto point-to-point messages between two processes, their efficient realization is often challenging.

For this reason, the Message Passing Interface (MPI) [10], the de-facto standard for message passing, defines 17 so-called *collective operations* to support the most common group exchange patterns. For example, sending data from one process to all other processes is encapsulated in the functionality of MPI_Bcast(). Although equivalent semantics could be achieved by sending the same piece of data iteratively to all processes, one at a time, using MPI_Bcast() is simpler, the resulting code looks cleaner and is easier to maintain. In addition, sophisticated implementations of MPI_Bcast() are likely to be much more efficient.

C. Kaklamanis et al. (Eds.): Euro-Par 2012, LNCS 7484, pp. 28–39, 2012.

In general, MPI collectives offer advantages in terms of simplicity, expressiveness, programmability, performance, and predictability [5]. In particular, they allow users to profit from both efficient algorithms [2,13,14] and platform-specific optimizations [8,9,7]. Such improvements have often been reported to make collective implementations several times faster. Some of them exploit hardware features such as multicast or utilize special networks for collectives not even accessible via MPI point-to-point communication. These advantages cannot be overemphasized as the efficient implementation of collective operations is a complex task, which requires detailed knowledge not only of parallel algorithms and programming but also of the specific physical properties of the target platform.

However, in spite of such benefits, not all applications today make consequent use of predefined collectives and still deploy hand-crafted ensembles of point-to-point messages instead. One way of encouraging their adoption, is to recognize manually-implemented collectives in existing codes and to suggest their replacement. Existing recognition methods available for this purpose rely on the specifics of the underlying message exchange pattern [11,3]. But given the multitude of ways collectives can be implemented, any such attempt is too restrictive.

In this paper, we show how to overcome these disadvantages using a novel approach based on a semantic detection. We propose a method for identifying patterns of point-to-point messages in compact communication traces of MPI applications that are semantically equivalent to predefined collective operations. Relying exclusively on pre- and postconditions derived from the specification of the collective operation, we do not make any assumptions regarding the specific characteristics of the pattern. Our method detects broadcasts and operations composed of broadcasts fully automatically. It detects more sophisticated collective operations with a certain degree of prior user instrumentation. Applying our method to the HPL benchmark [1] pinpoints all contained collective communication operations. Replacing those manual collectives with the corresponding MPI collectives improves the HPL performance by up to 44%.

The remainder of the paper is organized as follows: In Section 2 we formalize the semantics of collective operations and show how pre- and postconditions can be derived that can be verified based on trace data. Proving such conditions requires analyzing both the contents of messages and the paths along which they travel. How we store all the necessary information in trace files is explained in Section 3. There, we place special emphasis on the hash functions we apply to avoid excessive memory requirements and their structure-preserving properties. The actual identification of manual collectives is outlined in Section 4. A major part of it is devoted to the parallel message replay we need to track communication pathways and the additional challenges posed by more complex collective operations such as scatter or reduce. Experimental results demonstrating the benefits of our method are presented in Section 5. Finally, we compare our approach to related work in Section 6, before we conclude the paper with an outlook on future work in Section 7.

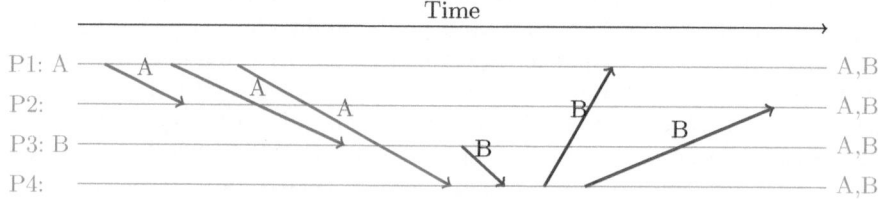

Fig. 1. Timeline diagram of two broadcast implementations

2 Semantics of Collective Operations

The MPI standard specifies only the semantics of collective operations, but does not dictate how they must be implemented. Therefore, any method capable of detecting a wide variety of manual collectives can not rely on any specific implementation. We will start our discussion with the simplest collective, the broadcast, and later move on to more challenging collectives such as scatter and reduce. For broadcast, the standard provides the following definition:

> MPI_BCAST broadcasts a message from the process with rank root to all processes of the group [...].

This definition does, however, not imply that a correct implementation needs to send a message from the root to all other processes directly. On the contrary, efficient implementations typically involve other processes to forward messages. By reversing this semantic definition, it is possible to infer pre- and postconditions that can tell whether or not a broadcast occurred. If at some point in time only one process owns a certain message and at a later time all processes within a group own the same message, then there must have been a collective communication that is semantically equivalent to a broadcast.

Figure 1 illustrates the behavior of two different broadcasts in the form of a timeline diagram. The diagram shows a timeline for every process and arrows between them to depict point-to-point communication. In addition, the letters A and B represent message contents. At the beginning, processes 1 and 3 own contents A and B, respectively. All processes receive further contents via messages as the time progresses. The communications with message A are semantically equivalent to a broadcast carried out using a simple centralized algorithm. The communications with message B are also semantically equivalent to a broadcast but in a hierarchical fashion. Identifying those different communication patterns as the same collective operation needs some deterministic rules. A precondition that needs to be true before the broadcast happens is that one of the processes, which is called the *root*, owns some data X. During the broadcast, X travels to the other processes in the group to which the broadcast applies. In other words, exactly one process in the group must **not** receive X before sending it. Although more receives are valid, a postcondition that needs to be fulfilled after the broadcast happened is that all processes in the group except the root must

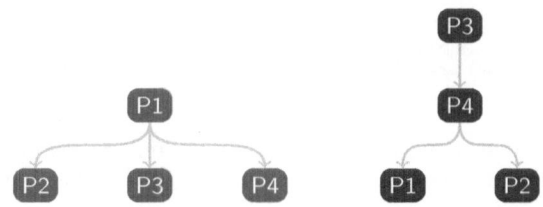

Fig. 2. Communication graphs for the two broadcast variants

have received X. Only then it is guaranteed that, at the end of the analyzed period of time, all processes share X. This fulfills the semantics of a broadcast.

Figure 2 maps the two broadcast variants from Figure 1 onto simpler communication graphs, ignoring temporal relationships. If any of the two communication graphs was not connected, then it would describe independent communications, which is in contradiction to the collective character of the operation. This implicit connectivity requirement among the processes of the group is checked during the analysis to prove the presence of a collective operation (see Section 4).

Defining pre- and postconditions for scatter and gather is more intricate, as messages can be split or concatenated. In the case of scatter, a message at the root is split into pieces to be distributed to all processes. Again, the precondition requires the original message to be located at the root and the postcondition requires non-overlapping parts of the message to be located at specific ranks. We use hashes with homomorphic properties to handle message splitting and concatenation. Even more challenging are collectives computation operations (a.k.a. reductions), where the messages are combined using, for example, arithmetic or logical operations. Nevertheless, even this can be formalized. While all such conditions can theoretically be verified under the assumption of unlimited access to the memory and message buffers used by the application, difficulties arise in practice when knowledge is restricted to manageable amounts of trace data. Two specific challenges need to be addressed:

1. For reasons of space efficiency, we only store message hashes in our trace files. Section 3 explains how we can still track many of the above-mentioned transformational relationships even with hashes.
2. In manual collectives, data destined to remain at the root may never appear in a message buffer and is thus not recorded in the traces. In Section 4, we suggest a method to make those visible again.

A further advantage of our technique, which does not require any knowledge of a collective operations' implementation, is that it can also be used to search for collective exchanges that do not have an existing primitive yet. This could motivate the standardization of new collective operations such as neighborhood or sparse collectives (under consideration for MPI-3.0). In this sense, it is not only an instrument for application optimizers but also for MPI developers.

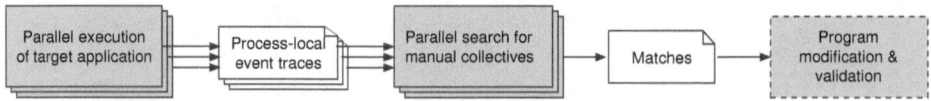

Fig. 3. Workflow of detecting manual collectives within parallel applications. Stacked boxes denote parallel programs. The user is in charge of the last step.

3   Analysis Workflow and Trace Generation

Figure 3 illustrates the workflow of detecting manual collectives. First, the target application is prepared for tracing by linking it to a library of PMPI interposition wrappers. Not to duplicate development efforts, we leverage the tracing infrastructure of the Scalasca toolset [4]. Only if more sophisticated collectives should be detected, the application needs further preparation as described in Section 4. The prepared application is then executed to generate a trace of happened communication events. As a next step, these traces are searched for manual collectives, which is done with the help of communication replays. This search is carried out by our analyzer, which is a parallel program in its own right. The analysis creates a list of matches, which the user then can decide to replace with predefined collectives. Since the matches characterize only a single run, the user not only needs to validate whether performance objectives are met but also has to ensure that a replacement does not violate the program's correctness.

In addition to the default information Scalasca stores with communication events, we record a hash of the message payload, the starting address of the message buffer, and the MPI data type. The first item is needed to track the path along which a particular message is forwarded, and the last two to support concatenation and split of messages, as explained in Section 4. Hashing message payloads avoids storing full messages, which would consume a prohibitive amount of storage space. A hash provides a fixed-length value regardless of the message size. If two messages have the same hash value, they are identical with high probability. If their hashes are different, the messages are different for sure. Although testing for equality is a fundamental application of hashes and sufficient to detect for example broadcasts, it is not enough to identify collectives such as scatter, gather, or reduce, which split, concatenate, or combine messages. For those, we exploit homomorphic properties of certain hash functions h that ensure the following conditions for operations $\oplus$ on messages m_1 and m_2:

$$h(m_1 \oplus m_2) = h(m_1) \oplus h(m_2)$$

As a default, we use the established *CRC-32* checksum from *zlib* because it is fast, needs only 32 bits per message, gives acceptably-low collision probability, and even supports split and concatenation. We also identified further hash functions to support arithmetic or logical reductions of certain data types. Many hash algorithms, however, entail difficult compromises. For example, cryptographic hashes have a lower collision probability but are much slower, need more memory and can neither be concatenated nor combined. In general, the choice of hash

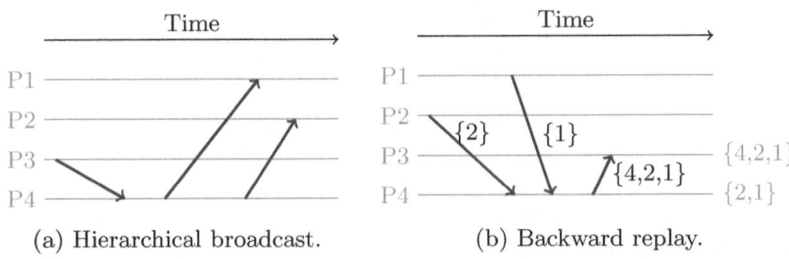

(a) Hierarchical broadcast.   (b) Backward replay.

Fig. 4. Identifying a hierarchical broadcast via backward replay

function is configurable, a feature which can be used to expand the coverage of our method. Unfortunately, MPI reduction operations can be arbitrarily complex as they support both user-defined data types and user-defined reduction operations. Correctly identifying reductions via hashes is already challenging for predefined data types such as floating point values, which is why we also store an 8-byte message prefix in addition to the hash. This enables testing against a predefined set of potential reduction operations. To support arbitrary user-defined data types, we utilize the *MPITypes* library by Rob Ross [12], which allows hashes to be calculated for arbitrary MPI messages.

4   Search for Manual Collectives

Our collectives detector searches for manual collectives in communication traces enriched with message hashes such as illustrated using a hierarchical broadcast example in Figure 4a. The actual search is performed via a backward replay of the traces. During this replay, we traverse the traces from end to beginning and reenact the recorded point-to-point communication in backward direction, that is, the roles of sender and receiver are reversed. To simplify the replay, our analyzer runs with the same number of processes as were used to trace the target application, giving a one-to-one mapping between application and analysis processes. The objective of the replay is to determine all processes that a particular message has visited on its way to the final destination and to propagate this information back to its origin. At the end, each process checks whether it acted as root (i.e., did not receive the message from anyone else) and whether the message has reached all other processes—directly or indirectly.

For this purpose, each process maintains a set of receivers for each message hash that will later contain the ranks of all processes that have received a message with this particular hash. Figure 4b shows these sets for the hash involved in a broadcast. At the beginning, all sets are empty. Whenever a process encounters a receive event during the backward replay for this hash, it adds its rank to the set and sends its own set along with the backward message. The receiver of a replay message then constructs the union of its own set with the set just received. If at the end of the replay one process has a set containing all other processes but not itself, a broadcast has happened with this process acting as the root. Sending the hash along with the message is one way of separating the traffic related to different broadcasts. In the example, processes 1 and 2 add themselves to the set once they hit their local receive events. After replaying the first two messages, the set of process 4 therefore includes $\{2, 1\}$. Before replaying the third message, process 4 adds itself to the set and sends it to process 3. The set of process 3 finally includes $\{4, 2, 1\}$, satisfying the condition for a broadcast with 3 as the root. Using this method, we can detect any broadcast irrespective of its particular implementation. The backward replay ensures that we can track every conceivable message pattern. In general, the direction of the replay depends on the nature of the collective operation. If information is spread as in the case of broadcast, we replay in backward direction. If information is concentrated as in the case of gather, we replay in forward direction. The goal is always to end up at the root. The all-to-all collective does not have a root, but can be detected by decomposing it into either its 1-to-all or all-to-1 components.

There are two major challenges arising in the context of operations such as gather, scatter, and reduce that transform messages through concatenation, split, or combine operations. To check whether a message has been created as a result of such a transformation, we need to send message hashes along with our replay messages. The checks are then performed as we go, exploiting homomorphic properties of the hash function as far as this is possible. Unfortunately, not all operands of such a transformation are necessarily stored in the trace because they might involve buffers that never appear in any communication.

Figure 5 illustrates the problem for scatter. For example, we cannot tell whether the process shown in Figure 5a is the root of a scatter operation that intends to disseminate the vector A, B, C, D, E because C is only stored locally and as such never part of a communication that causes its message hash to be recorded in the trace. To make such information available to our analyzer, we introduced a new function that a user can insert into the program:

```
void Send_To_Self(void *buf, int count, MPI_Datatype datatype);
```

A call to this function records the missing information, about a buffer that a process utilizes locally, in the trace. Calling this function for C before doing both sends, makes C available in the trace. With this information, the analyzer can infer that A,B,C,D, and E belonged together before they were scattered across the processes. Without this information, however, no positive match can be made. A similar situation is shown in Figure 5b. Here, an inner node of a scatter tree receives a message, stores a part of it locally and forwards the rest to other processes. A message containing A, B and C is received but only A and B are sent further. Only with the hash of C written to the trace, the relationship becomes visible. Nevertheless, finding all related message parts requires testing of every combination of hashes. As this would be impractical for large numbers of messages, our detector checks only adjacent messages for split and concatenation. This is accomplished by looking at their starting addresses and MPI data types.

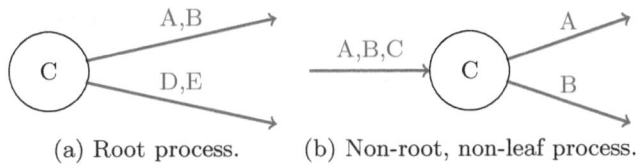

(a) Root process.   (b) Non-root, non-leaf process.

Fig. 5. Inputs and outputs of processes during scatter operations

Currently, our prototype supports the fully automatic detection of broadcast and alltoall variants composed of multiple point-to-point broadcasts. Barriers are currently recognized as alltoall with an empty payload. Moreover, our prototype is capable of semi-automatically recognizing scatter, gather, and reduce with the help of the extra information described above. Reduce is restricted to some arithmetic and logical operations on certain datatypes. Floating-point arithmetics are not supported. We believe, that our collectives detector can be upgraded to eventually support all regular MPI collective operations. Irregular collectives, however, still present serious challenges due to their high degree of flexibility, which makes it hard to formulate manageable pre- and postconditions.

5   Evaluation

To demonstrate that our method can handle even very challenging cases, we apply it first to a set of micro-benchmarks that implement different variants of broadcast. A case study with the HPL benchmark shows the potential for actual performance improvements. We conducted all experiments on the IBM Blue Gene/P *Jugene* installation located at the Jülich Supercomputing Centre.

5.1   Microbenchmarks

We start with a linear broadcast, as illustrated with solid arrows in Figure 6a. In a linear broadcast, a message visits one process after another until all processes have seen it. Thus, the communication is effectively serialized. In spite of the un-related message traffic (i.e., the noise depicted as dashed arrows), the pattern is correctly identified. Adding a redundant message in Figure 6b offers two choices for the root process (P1 and P2). Both options are reported. Repeating the same broadcast twice as in Figure 6c results in the recommendation to replace all messages involved with a single predefined broadcast. Finally, we perform a nested broadcast by passing a token from the root to each other process (Figure 6d). Whoever owns the token initiates an inner broadcast. Both types of broadcasts are reported, although replacing the outer broadcast might change the order of the inner broadcasts. The ultimate decision is left to the user. If token-passing is extended to ring communication, our algorithm will report one instance for each process involved because each process could be a potential root.

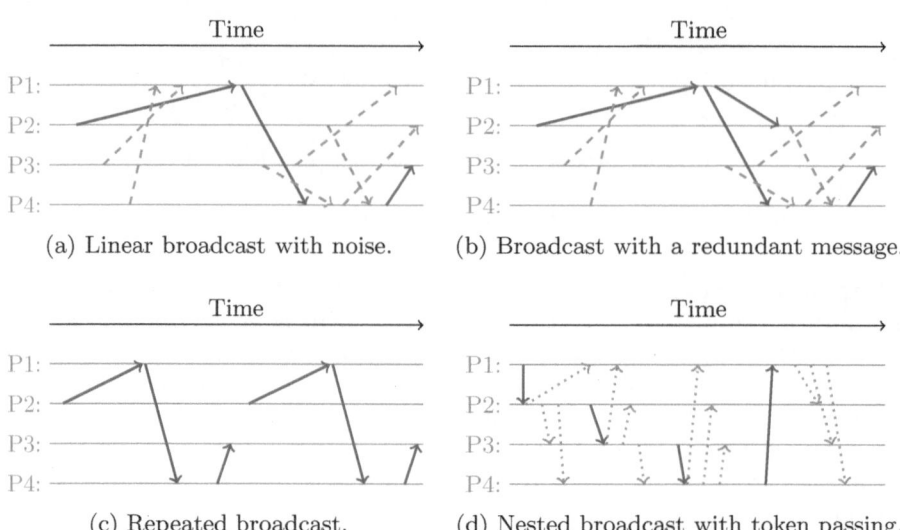

(a) Linear broadcast with noise.   (b) Broadcast with a redundant message.

(c) Repeated broadcast.   (d) Nested broadcast with token passing.

Fig. 6. Four test scenarios for broadcast. All solid arrows show messages with the same payload. Dashed arrows represent noise messages with different payloads. The dotted arrows illustrate correctly detected broadcasts.

5.2   High-Performance Linpack

The High-Performance Linpack Benchmark [1] solves a dense linear system in double precision and is used to rank the world's fastest supercomputers in the Top500 list. We selected HPL as a test candidate because it makes heavy use of collective operations implemented via point-to-point communications. Prior to running this benchmark, the user needs to select in a configuration file one out of six hand-crafted broadcast implementations.

Regardless of whether the broadcasts are implemented with blocking or non-blocking semantics in any send mode, our collectives detector correctly reported all six HPL broadcast implementations including their source-code location plus some others that are presumably used for synchronization. Figure 7a compares the HPL execution time for any of the six broadcast options with the execution time after replacing the manual variant with MPI_Bcast(). The experiments reflect the performance for 256 cores and an input problem size of 32,000. Our results not only show significant performance differences among the six manual variants, but also show that the MPI broadcast always delivers superior performance with an overall improvement of up to 9%.

Although BLongM performed worst in this test with 256 cores, we chose it for the scaling study in Figure 7b because its bandwidth-optimized implementation reveals benefits for larger core counts. The number of matrix elements per core was kept at four million. Indeed, the difference between BlongM and MPI_Bcast() is most pronounced between 2k and 16k cores. Above, BLongM plays out its own strength. Overall, MPI_Bcast() was faster in all cases and with 8k cores

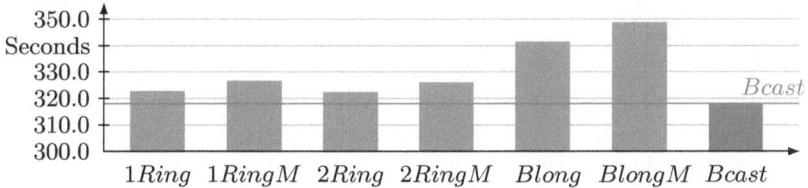

(a) Comparison between the HPL-included broadcasts and MPI_Bcast.

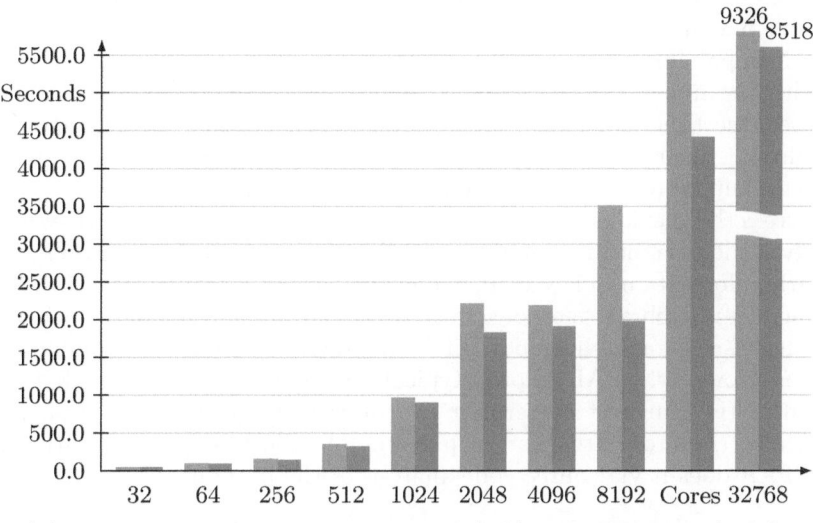

(b) Weak-scaling of HPL using BLongM (left) and MPI_Bcast (right).

Fig. 7. Total execution time of HPL with and without manual collectives

the difference was even 44%. This demonstrates that the performance advantage of predefined collectives can be substantial and that the replacement of manual collectives is often worthwhile. In addition, using MPI_Bcast() in HPL makes several hundred lines of code obsolete, reducing its code complexity.

6   Related Work

In this section, we compare our approach with alternative routes taken earlier. Preissl et al. [11] pursue an almost identical objective. They not only attempt to find collective point-to-point exchanges but also to automatically replace their occurrences in the code with equivalent predefined MPI operations. Their solution is built around ROSE, a generator for source-to-source translators. They trace the messages sent during program execution, match the structure of the message graph with predefined structures representing collective communications, find the corresponding places in the code, and carry out the substitution.

The advantage of their solution is that the user does not have to do anything except running the tool to get—in the optimal case—better code. A disadvantage, however, is that no collective communication pattern can be matched except those already thought of and stored in the tool. While this might not seem to be very important at first, one has to consider that a user can implement a collective communication in any way that suits him and his particular application. For example, since different hardware topologies support different communication topologies, the precise shape of the communication pattern might be platform dependent. Moreover, since new network topologies are being created alongside new hardware, new patterns can emerge that are not thought of yet. While it is possible to add them to such a solution, the maintenance cost will rise with the number of possibilities. Also, if the number of possible patterns increases, the time to check against all of them will grow as well. At the time of publishing their work, the tool developed by Preissl et al. was able to detect only one simple broadcast implementation.

Whereas the previous approach and ours search for manual collectives in traces with dynamic content, di Martino et al. [3] attempt to detect them relying only on static information. They define collectives using mathematical abstractions and then use algebraic methods to find them. This methodology can provide a strong basis for matching potential candidates to models of communication patterns. However, PPAR, a prototypical tool based on this method, seems restricted in the range of patterns it can interact with. Just like Preissl et al., PPAR also tries to match known patterns of collective communications, only that PPAR does it via source-code analysis of the program.

7   Conclusion and Outlook

We proposed a practical method to detect manually-implemented collective operations in MPI programs written in any language without making any assumptions about the actual communication pattern. This eliminates the detection of false negatives, which sets our approach apart from earlier work in the field. However, there are still limitations which we want to address in the future.

Our prototype needs further work to evolve into a productive tool. As collective operations involving only a subset of the processes in a given communicator are not yet recognized, we plan to extend our method to also detect those cases and to suggest the creation of sub-communicators. Moreover, our current scheme allows fully automatic detection only for collectives such as broadcast and alltoall. For more sophisticated collectives, the user needs to supply extra information by inserting function calls into the program. Future versions could extend the coverage of our detector by also reporting matches that are incomplete to a certain degree, carefully balancing false negatives with false positives. Finally, to guide the user to the most promising replacement candidates in terms of potential performance improvements, we plan to simulate the effects of a replacement using a real-time replay of modified traces [6]. This would permit the user to compare the required effort with the benefit that is likely to materialize.

References

1. HPL – A portable implementation of the high-performance Linpack benchmark for distributed-memory computers, http://netlib.org/benchmark/hpl/
2. Bernaschi, M., Iannello, G., Lauria, M.: Efficient Implementation of Reduce-scatter in MPI. In: Proceedings. 10th Euromicro Workshop on Parallel, Distributed and Network-based Processing, pp. 301–308 (2002)
3. Di Martino, B., Mazzeo, A., Mazzocca, N., Villano, U.: Parallel program analysis and restructuring by detection of point-to-point interaction patterns and their transformation into collective communication constructs. Science of Computer Programming 40, 235–263 (2001)
4. Geimer, M., Wolf, F., Wylie, B.J.N., Ábrahám, E., Becker, D., Mohr, B.: The Scalasca performance toolset architecture. Concurrency and Computation: Practice and Experience 22(6), 702–719 (2010)
5. Gorlatch, S.: Send-Receive Considered Harmful: Myths and Realities of Message Passing. ACM Transactions on Programming Languages and Systems (TOPLAS) 26, 47–56 (2004)
6. Hermanns, M.-A., Geimer, M., Wolf, F., Wylie, B.J.N.: Verifying causality between distant performance phenomena in large-scale MPI applications. In: Proc. of the 17th Euromicro International Conference on Parallel, Distributed, and Network-Based Processing (PDP), Weimar, Germany, pp. 78–84. IEEE Computer Society (February 2009)
7. Hoefler, T., Siebert, C., Lumsdaine, A.: Group Operation Assembly Language - A Flexible Way to Express Collective Communication. In: The 38th International Conference on Parallel Processing. IEEE (September 2009)
8. Hoefler, T., Siebert, C., Rehm, W.: A practically constant-time MPI Broadcast Algorithm for large-scale InfiniBand Clusters with Multicast. In: Proceedings of the 21st IEEE International Parallel & Distributed Processing Symposium, pp. 1–8. IEEE Computer Society (March 2007)
9. Kumar, S., Sabharwal, Y., Garg, R., Heidelberger, P.: Optimization of All-to-All Communication on the Blue Gene/L Supercomputer. In: Proc. of the 37th International Conference on Parallel Processing, pp. 320–329. IEEE Computer Society, Washington, DC (2008)
10. Message Passing Interface Forum. MPI: A Message-Passing Interface Standard, Version 2.2. High Performance Computing Center Stuttgart, HLRS (2009)
11. Preissl, R., Schulz, M., Kranzlmuller, D., de Supinski, B.R., Quinlan, D.J.: Transforming MPI Source code based on communication patterns. Future Generation Computer Systems 26, 147–154 (2009)
12. Ross, R., Latham, R., Gropp, W., Lusk, E., Thakur, R.: Processing MPI Datatypes Outside MPI. In: Ropo, M., Westerholm, J., Dongarra, J. (eds.) PVM/MPI. LNCS, vol. 5759, pp. 42–53. Springer, Heidelberg (2009)
13. Sanders, P., Träff, J.L.: Parallel Prefix (Scan) Algorithms for MPI. In: Mohr, B., Träff, J.L., Worringen, J., Dongarra, J. (eds.) PVM/MPI 2006. LNCS, vol. 4192, pp. 49–57. Springer, Heidelberg (2006)
14. Träff, J.L., Ripke, A., Siebert, C., Balaji, P., Thakur, R., Gropp, W.: A Pipelined Algorithm for Large, Irregular All-Gather Problems. International Journal of High Performance Compututing Applications 24, 58–68 (2010)

A Type-Based Approach
to Separating Protocol from Application Logic
A Case Study in Hybrid Computer Programming

Geoffrey C. Hulette[1], Matthew J. Sottile[2], and Allen D. Malony[1]

[1] University of Oregon, Eugene, OR
[2] Galois, Inc., Portland, OR

Abstract. Numerous programming models have been introduced to allow programmers to utilize new accelerator-based architectures. While
OpenCL and CUDA provide low-level access to accelerator programming, the task cries out for a higher-level abstraction. Of the higher-
level programming models which have emerged, few are intended to
co-exist with mainstream, general-purpose languages while supporting
tunability, composability, and transparency of implementation. In this
paper, we propose extensions to the type systems (implementable as syn-
tactically neutral annotations) of traditional, general-purpose languages
can be made which allow programmers to work at a higher level of abstraction with respect to memory, deferring much of the tedium of data
management and movement code to an automatic code generation tool.
Furthermore, our technique, based on formal term rewriting, allows for
user-defined reduction rules to optimize low-level operations and exploit
domain- and/or application-specific knowledge.

1 Introduction

Programming for hybrid architectures is a challenging task, in large part due
to the partitioned memory model they impose on programmers. Unlike a basic
SMP, devices must be set up and torn down, processing synchronized, and data
explicitly allocated on a particular device and moved around within the memory
hierarchy. Programming systems such as CUDA[1] and OpenCL[2] provide an
interface for these operations, but they are quite low-level. In particular, they
do not distinguish between the high-level computational and application logic
of a program, and the *protocol logic* related to managing heterogeneous devices.
As a result, the different types of program logic invariably become entangled,
leading to excessively complex software that is prohibitively difficult to develop,
maintain, and compose with other software. The problem we have described is
pervasive in programming for hybrid architectures; in this paper, we will focus
on the specific instance of this problem presented by GPU-based accelerators.

We present a high-level programming language called *Twig*, designed for expressing protocol logic and separating it from computational and application
logic. Twig also supports automated reasoning about composite programs that

C. Kaklamanis et al. (Eds.): Euro-Par 2012, LNCS 7484, pp. 40–51, 2012.

can, in many cases, avoid problems such as redundant memory copying. This allows Twig programs often to retain the high performance of a lower-level programming approach.

Crucially, Twig's role in the programming toolchain is to generate code in a mainstream language, such as C. The generated code is easily incorporated into the main program, which is then compiled as usual. This minimizes the complexity that Twig adds to the build process, and allows Twig code to interact easily with existing code and libraries.

Twig achieves these goals by using data types to direct the generation of code in the target language. In particular, we augment existing data types in the target language with a notion of *location*, e.g., an array of floats located on a GPU, or an integer located in main memory. In the following sections, we first present related work, and then describe Twig's code generation strategy and core semantics. Finally, we present an example demonstrating the use of located types to generate code for a GPU. In the example, we also show how Twig programs can be automatically rewritten in order to minimize data movement.

2   Related Work

Twig was inspired in part by Fig[3]. In that project, a similar formal approach was used to express bindings between different programming languages. In our experience, multi-language programming has much in common with programming hybrid systems. The overlaps include memory ownership and management, data marshaling, and managing the flow of program control across the language or device boundary. Our work builds upon the approach in Fig, and in particular aims to provide a general-purpose tool not tied to the Moby[4] programming language.

Numerous systems have been created in recent years that provide an abstraction above low-level interfaces such as OpenCL or CUDA. These include the PGI Accelerate model[5], the HMPP programming system[6], and Offload[7]. Interestingly, Offload, like Twig, uses locations encoded in the type system. While these systems provide an effective high-level abstraction, they offer little room for tuning the low-level interface to the accelerator. Twig provides a simple method for user-definable rewriting of programs, which allows architecture-, domain-, and even application-specific optimizations to be realized.

Furthermore, in large applications it is infeasible to assume that all developers of the various components will use the same high-level abstraction. This makes program composition challenging, since it may be unclear how the objects generated by independent programming systems interacts. Twig adopts a code generation approach in which a single, low-level target (such as CUDA) is used. This approach solves the composability problem, since all Twig code maps to a single "lingua franca" for programming the hybrid system.

Sequoia[8] is a language and runtime system for programming hybrid computers. It allows programmers to explicitly manage the memory hierarchy, while retaining program portability across architectures. Although Sequoia is based on C++, it is intended as a complete programming environment, not as a way to extend existing programs with hybrid computation.

Code generation approaches have had notable success in the computational science field, an exemplar being the Tensor Contraction Engine (TCE)[9]. The TCE allows computational chemists to write tensor contraction operations in a high level language, and then generates the corresponding collections of loops that implement the operations. Unlike Twig, the TCE is quite specialized, being of use only to programmers working with tensor-based computations.

3   Method Overview

In Twig, we write *rules* which express some high-level operation, such as kernel execution or copying data to a device, as a function on *types*. A Twig program is evaluated with a type given as input. The output is another type, transformed by the combined rules of the program. As a side-effect, C code is generated which performs the transformation on values in C. This basic idea is formalized in Sec. 5.

Types in Twig are based on the set provided by C, but may be augmented with additional information. For GPU programming, we augment standard data types with a *location*. The location information describes where the data is stored in memory; in this case, either in the main system memory or on the GPU. For example, we can represent an array of `ints` on the CPU with the Twig type `array(int)`. The same type located on the GPU is `gpu(array(int))`. Any standard type may be wrapped inside the `gpu` type constructor.

Note that location information is only used by Twig during evaluation and code generation. In particular, it may not be reflected in the types for the generated code. If we are generating CUDA code, for example, the generated type for both `gpu(array(int))` and `array(int)` is simply a C pointer to `int` (i.e., `int *`). In this case, the location information is erased during the code generation phase. For other target languages or APIs that have a notion of location, the information could be preserved in the target data types.

By augmenting basic data types with location information, we ensure that rules must be specific to the GPU in order to operate on GPU data. For example, a rule

```
[gpu(array(float)) -> gpu(array(int))]
```

converts an "array of floats" data type to an "array of integers" type if and only if the type describes data located on the GPU. If the type describes data located elsewhere, its type must be "converted" (i.e., the data copied to the device) with a rule such as

```
[array(float) -> gpu(array(float))]
```

This simple scheme enables Twig to reason about requirements for data motion.

It is important to understand that rules such as those given above describe transformations on *data types*, not on the data themselves. It falls to the code that is generated as a consequence of successful application of these rules to perform the promised conversion on the actual data. Code generation is described in Sec. 4.

Our scheme could be extended to support multiple GPU devices, with each device corresponding to a unique located type. In fact, we think that located types could be useful in a variety of situations; this is a topic of ongoing work.

4   Code Generation

To generate code, Twig uses an abstract, language-independent system with a small number of basic operations. The relative simplicity of the model is motivated by Twig's semantics, described in Sec. 5. It is helpful in clarifying the precise operations which Twig supports, without getting bogged down in the (potentially quite complicated) details of outputting code for a particular language.

Twig generates code in units called *blocks*. A block of code represents anything that performs some operation on a set of inputs in order to produce a set of outputs. Blocks have zero or more inputs and/or outputs. Blocks can be combined in two different ways: *sequentially*, or *in parallel*. These operations are described below.

Our current implementation of this model generates C code, although the model is general enough to generate other languages as well. Figures 1(a) and 1(b) each depict a different basic block that generates C.

4.1   Block Composition

As mentioned above, Twig provides two fundamental operations on blocks. The first is *sequential* composition, which we represent formally as addition ($+$) on blocks. Sequencing connects two blocks of code by "wiring" the outputs of the first block into the inputs of the second (see Fig. 1(c)). In C, this is done by creating uniquely-named temporary variables which are substituted into the original blocks.

The second operation is *parallel* composition, where two blocks are combined so as to execute independently of one another, but to appear as one single block (see Fig. 1(d)). We represent this operation as multiplication ($\times$).

4.2   Identity Blocks

Twig's formal semantics require the definition of a set of special *identity* blocks. An identity block I_n has n inputs and n outputs ($n > 0$). Its function is, as its name implies, to simply pass each of its inputs through, unchanged, to the corresponding output. In Twig's semantics we use I_n as a kind of "no-op." We also use I in place of I_n when the value of n is implied from the context.

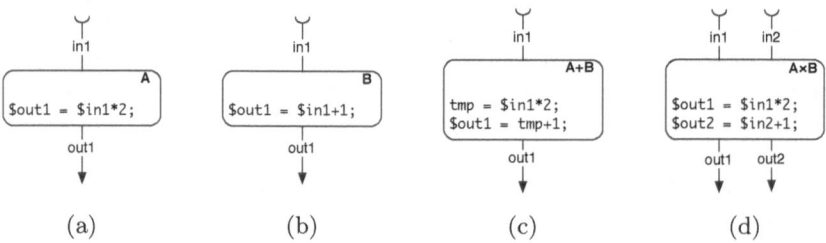

Fig. 1. Code generation using blocks. A (a) and B (b) are basic blocks. $A + B$ (c) is the sequential composition of A and B. $A \times B$ (d) is the parallel composition of A and B.

Identity blocks are subject to a few rules, which we describe here informally and only briefly. First, I_n are left- and right-identity elements under sequential composition, i.e., $I + x = x + I = x$ for all blocks x. Second, we can define parallel composition of identity blocks is defined by summing their size, i.e., $I_n \times I_m = I_{n+m}$.

5   Twig's Semantics

Twig is based on a core semantics for term rewriting called System S[10], augmented with code generation and specialized to operate on types instead of general terms. Twig uses the operators in System S to combine primitive *rules* into more complex transformations on types. These transformations are then applied to a given type, resulting in a new type and potentially generating code as a side effect. In this section we describe the Twig language.

We present an abbreviated and relatively informal description of Twig's semantics, focusing on the features used to support GPU programming. The full semantics will be described in a forthcoming paper.

Twig does not currently provide any built-in constructs for expressing general recursive expressions, including loops. We are working to address this limitation in our current work. At the moment, Twig can generate loops contained within a single generated block and, of course, Twig can also be used to generate a block of code for inclusion within a loop body.

5.1   Values

Values in Twig can be any valid *term* representing a type in the target language. Terms are tree structured data with labeled internal nodes. Examples of terms include simple values like `int` and `float`, as well as compound types like `ptr(int)`, which might represent a pointer to an integer in C.

The mapping between terms in Twig and types in the target language is a configuration option. Furthermore, the mapping need not be injective, i.e. users are free to have multiple values in Twig map to the same type in C. For example, you might use distinct Twig values `string` and `ptr(char)`, but map both to a pointer to `char` (`char *`) in C.

Twig also includes support for terms representing groups of values, i.e. *tuples*, and operations on groups. We omit these semantics here for lack of space.

5.2 Rules

The fundamental components of a Twig program are called *primitive rules*. A primitive rule describes a transformation from one term to another. For example, in C it is easy to convert an integer value to floating point, and we can write this rule in Twig as follows:

```
[int -> float]
```

The term to the left of the arrow is the input, and the term to the right is the output. In this example, the rule says that if and only if the input to the rule is the term int, then the output will be the term **float**. If the input is not int, then the output will be the special value ⊥, which can be read as "undefined" or "failure."

Primitive rules will typically generate code as a side effect of successful application. To associate a block of code with a rule, the programmer puts it immediately after the rule definition and surrounds it with braces. The brace symbols are configurable; here we use <| and |>. For example:

```
[int -> float] <| $out = (float)$in; |>
```

When the code is generated, Twig will create temporary variables for $in and $out and ensure that the various bookkeeping details, such as variable declarations, are handled. If there are multiple inputs or outputs, then the relevant placeholders are enumerated; e.g., $in1, $in2, and so on.

Note that Twig does not check the generated code for correctness – the generation procedure is essentially syntactic. This approach is similar to the strategy used tools such as SWIG[11].

5.3 Formal Semantics

In the following formal semantics, let t range over terms, m over code block expressions, and s over rule expressions, i.e., a primitive rule or a sub-expression built with operators.

Primitive Rules. A primitive rule s transforms a term t to another term t' with generated code m:

$$t \xrightarrow{s} (t', m)$$

if the application of rule s to value t succeeds. If no code is given for the rule, then m is the identity element, I (see Sec. 4). If the application of s to t fails, e.g., if t does not match the pattern in s, then

$$t \xrightarrow{s} \bot$$

Note that a code block is not emitted in this case.

Operators. Rules can be combined into more complex expressions using operators. The most useful of these is the *sequence* operator (note the distinction from the + operator for code blocks described in Sec. 4). A sequence chains the application of two rules together, providing the output of the first to the input of the second, and failing if either rule fails (see Fig. 2). With this operator, simple rules can be composed into multi-step transformations.

$$\frac{t \xrightarrow{s_1} (t', m_1) \quad t' \xrightarrow{s_2} (t'', m_2)}{t \xrightarrow{s_1;s_2} (t'', m_1 + m_2)} \qquad \frac{t \xrightarrow{s_1} \bot}{t \xrightarrow{s_1;s_2} \bot} \qquad \frac{t \xrightarrow{s_1} (t', m) \quad t' \xrightarrow{s_2} \bot}{t \xrightarrow{s_1;s_2} \bot}$$

Fig. 2. Semantics for sequence operator

Another important operator is *left-biased choice*. Choice expressions will try the first rule expression, and if it succeeds then its output is the result (see Fig. 3) of the expression. If the first rule fails (i.e., results in $\bot$), then the second rule is tried. This operator allows different paths to be taken, and different code to be generated, depending on the input type.

$$\frac{t \xrightarrow{s_1} (t', m)}{t \xrightarrow{s_1|s_2} (t', m)} \qquad \frac{t \xrightarrow{s_1} \bot \quad t \xrightarrow{s_2} (t', m)}{t \xrightarrow{s_1|s_2} (t', m)} \qquad \frac{t \xrightarrow{s_1} \bot \quad t \xrightarrow{s_2} \bot}{t \xrightarrow{s_1|s_2} \bot}$$

Fig. 3. Semantics for left-biased choice

Fig. 4 gives the formal semantics for some of Twig's other basic operators. These include constant operators and operators which discard their results.

$$\frac{}{t \xrightarrow{id} (t, I)} \qquad \frac{}{t \xrightarrow{fail} \bot} \qquad \frac{t \xrightarrow{s} (t', m)}{t \xrightarrow{?s} (t, I)} \qquad \frac{t \xrightarrow{s} \bot}{t \xrightarrow{?s} \bot} \qquad \frac{t \xrightarrow{s} (t', m)}{t \xrightarrow{\neg s} \bot} \qquad \frac{t \xrightarrow{s} \bot}{t \xrightarrow{\neg s} (t, I)}$$

Fig. 4. Semantics for basic operators

Twig also provides some special operators for tuples. These are not needed for this paper, so we omit further discussion.

Named Expressions. Twig allows rules and rule expressions to be assigned to names. The name can be used in place of the rule itself within expressions. For example:

```
intToFloat = [int -> float] { ... }
```

A Twig program is a list of such name/expression assignments. There is a special name, `main`, which designates the top-level expression for the program.

5.4 Reductions

Reductions are a mechanism provided within Twig as a way to automatically simplify expressions. Reductions can be used to exploit application or domain knowledge about primitive rules, and as such are usually developed alongside a set of rules.

As an example, consider the following two rules:

```
intToFloat = [int -> float] {
  $out = (float)$in;
}

floatToInt = [float -> int] {
  $out = (int)$in;
}
```

and the expression

```
intToFloat;floatToInt
```

We would most likely consider this conversion to be redundant and we should eliminate it wherever possible. We can tell Twig to do this with the following reduction rule:

```
reduce intToFloat;floatToInt => id
```

This statement instructs Twig to replace any subexpression `intToFloat;` `floatToInt` with the identity rule, `id`, anywhere it occurs within the program. Recall that `id` is the identity rule; it simply passes the value through unchanged.

Twig comes equipped with some standard reductions by default. These reductions rely on the meaning of Twig's combinators to normalize expressions. For example, we can replace subexpressions of the form `id;X` with `X`, where `X` represents any subexpression.

Twig's reductions are based on the theory of term rewriting; for a formal discussion see [12]. In this case, Twig's expressions constitute the terms. There are some subtleties with reductions, e.g., they must be developed carefully to avoid circular reductions.

6 Implementation

Our implementation of Twig is written in Haskell. Twig expects as input a `.twig` file containing a list of named rule expressions along with a `main` rule expression, as described in Sec. 5.3. It also expects an initial value (i.e. a term, representing a C type), which will be used as the input to the main rule expression. Twig must also be configured with a mapping from terms to C types. Currently, this mapping is provided with a simple key/value text file.

Generated code may optionally be wrapped in a C function body, with parameters corresponding to the inputs, and return value corresponding to the output. The generated code may be redirected to a separate file and included in a C program using an `#include` directive.

7   Example

Now we present an example program written in Twig. The code in Fig. 5 demonstrates how Twig is used, and how reductions can eliminate redundant memory copies.

```
copyToGPU=[array(float) -> gpu(array(float))] <|
  cudaMalloc((void **)&$out,SIZE);
  cudaMemcpy($out,$in,SIZE,cudaMemcpyHostToDevice);
|>
copyFromGPU=[gpu(array(float)) -> array(float)] <|
  $out = malloc(SIZE * sizeof(float));
  cudaMemcpy($out,$in,SIZE,cudaMemcpyDeviceToHost);
|>
kernel(k)=[gpu(array(float)) -> gpu(array(float))] <|
  $k <<<N_BLOCKS,BLOCK_SIZE>>>($in, N);
  $out = $in;
|>
runKernel(k)=copyToGPU;kernel(k);copyFromGPU
main=runKernel(<|foo|>);runKernel(<|bar|>)
reduce copyFromGPU;copyToGPU => id
```

Fig. 5. Twig code example

This example is quite simple, in the interest of brevity and clarity. We omit setup and teardown logic, and assume that the array size, block size, and other parameters are simple constants. A real application would probably pass these values around using more complex rules.

The first three rules definitions are primitives for moving data to and from the GPU (copyToGPU and copyFromGPU), and for invoking a kernel transformation on the array in GPU memory (kernel). The rules kernel and runKernel are parameterized by k, whose value is inserted directly into the generated code.

The runKernel rule will perform a single logical "function" on the GPU. Note that this rule will be semantically valid in any context where it appears, since it ensures that the data is first moved onto the GPU, the kernel is executed, and then the data is copied back. To the programmer, runKernel appears to perform a function on a local array – a considerably simpler than the abstraction presented by OpenCL or CUDA.

The main rule is the top level of the program. This example executes two kernels in sequence with two invocations of runKernel. As noted above, by design each invocation would normally result in a copy to and from the GPU – a conservative strategy. Since the data is not modified in between GPU calls, on its own this expression would generate a redundant copy in between the calls to foo and bar. To see why, we can trace the execution of the Twig program. First, variable names are substituted with the expressions they denote, so main goes from:

```
main = runKernel(foo);runKernel(bar)
```

to

```
main = copyToGPU;kernel{foo};copyFromGPU;
       copyToGPU;kernel{bar};copyFromGPU
```

Evaluating this expression on the type `array(float)` will generate the following code.

```
float *tmp01,*tmp02,*tmp03,*tmp04,*tmp05,*tmp06,*tmp07;
cudaMalloc((void **)&tmp02,SIZE);
cudaMemcpy(tmp02,tmp01,SIZE,cudaMemcpyHostToDevice);
foo <<<N_BLOCKS,BLOCK_SIZE>>> (tmp02, N);
tmp03 = tmp02;
tmp04 = malloc(SIZE * sizeof(float));
cudaMemcpy(tmp04,tmp03,SIZE,cudaMemcpyDeviceToHost);
cudaMalloc((void **)tmp05,SIZE);
cudaMemcpy(tmp05,tmp04,SIZE,cudaMemcpyHostToDevice);
bar <<<N_BLOCKS,BLOCK_SIZE>>> (tmp05,N);
tmp06 = tmp05;
tmp07 = malloc(SIZE * sizeof(float));
cudaMemcpy(tmp07,tmp06,SIZE,cudaMemcpyDeviceToHost);
```

Notice that this code, while correct, contains a redundant copy! The problem arises because we sequenced the two kernel operations, which introduces the sub-expression `copyFromGPU;copyToGPU`. This sub-expression will copy data from the GPU to main memory, and then immediately back to the device. We solve this problem using a *reduction*, as described in Sec. 5.4. The line

```
reduce copyFromGPU;copyToGPU => id
```

instructs Twig to search for the expression `copyFromGPU;copyToGPU` and replace it with `id`, the identity transformation. After the reduction step, the expanded version of `main` has the extra copies removed:

```
main = copyToGPU;kernel{foo};id;kernel{bar};copyFromGPU
```

In fact, Twig's built-in reduction rules would remove the spurious `id` as well, although this has no bearing on the meaning. Now Twig will generate this code:

```
float *tmp01,*tmp02,*tmp03,*tmp04,*tmp05;
cudaMalloc((void **)&tmp02,SIZE);
cudaMemcpy(tmp02,tmp01,SIZE,cudaMemcpyHostToDevice);
foo <<<N_BLOCKS,BLOCK_SIZE>>> (tmp02, N);
tmp03 = tmp02;
bar <<<N_BLOCKS,BLOCK_SIZE>>> (tmp03,N);
tmp04 = tmp03;
tmp05 = malloc(SIZE * sizeof(float));
cudaMemcpy(tmp05,tmp04,SIZE,cudaMemcpyDeviceToHost);
```

This code does not contain the extraneous copying. Although this example is simple, it demonstrates the power of reductions. The reduction rule given here would probably be paired with the `copyToGPU` and `copyFromGPU` rules in a module intended for consumption by domain programmers, allowing them to perform GPU operations without worrying about the design of the rules. Sophisticated users, however, could add their own rules or even application-specific reductions, enabling very powerful and customizable code generation based on domain-specific logic.

8   Future Work

We are working on expanding the Twig language with a notion of *functors*. Functors cleanly capture most cases in which users might need to generate loop constructs, allocate/free patterns, or other protocols that require a notion of *context*.

We are also investigating a number of ways in which Twig might be more closely integrated with mainstream coding practices. For example, we imagine that it may be possible for Twig code to live in the background, and express protocol logic through declarative annotations in the application code.

9   Conclusion

We have introduced the concept of separating the protocol logic inherent to hybrid systems from the computational and application logic of a program. We have demonstrated that a type-based approach can enforce this separation by making explicit in data types information related to both data location, and the representation of the data itself. By doing so, we allow the protocol logic of a program to be expressed via operations exclusively on located types. Many explicit programming chores become implicit features of the generated code, such as declaring intermediate values or reducing redundant memory movement. Finally, by adopting a code generation approach, we show that users of these higher level abstractions are not prohibited from both tuning the resultant code and composing independently developed programs that utilize standardized hybrid programming libraries like OpenCL or CUDA.

Acknowledgements. This work was supported in part by the Department of Energy Office of Science, Advanced Scientific Computing Research.

References

1. Sanders, J., Kandrot, E.: CUDA By Example: An Introduction To General-Purpose GPU Programming (July 2010)
2. Khronos OpenCL Working Group: The OpenCL Specification Version 1.0
3. Reppy, J., Song, C.: Application-specific foreign-interface generation. In: GPCE 2006, pp. 49–58 (October 2006)

4. Fisher, K., Reppy, J.: The design of a class mechanism for Moby. In: SIGPLAN 1999, pp. 37–49 (May 1999)
5. Wolfe, M.: Implementing the PGI accelerator model. In: GPGPU 2010 (2010)
6. Dolbeau, R., Bihan, S., Bodin, F.: HMPP: A hybrid multi-core parallel programming environment. In: GPGPU 2007 (2007)
7. Cooper, P., Dolinsky, U., Donaldson, A.F., Richards, A., Riley, C., Russell, G.: Offload – Automating Code Migration to Heterogeneous Multicore Systems. In: Patt, Y.N., Foglia, P., Duesterwald, E., Faraboschi, P., Martorell, X. (eds.) HiPEAC 2010. LNCS, vol. 5952, pp. 337–352. Springer, Heidelberg (2010)
8. Fatahalian, K., Knight, T., Houston, M., Erez, M., Horn, D., Leem, L., Park, H., Ren, M., Aiken, A., Dally, W., Hanrahan, P.: Sequoia: Programming the memory hierarchy. In: SC 2006 (November 2006)
9. Baumgartner, G., et al.: Synthesis of high-performance parallel programs for a class of ab initio quantum chemistry models. Proceedings of the IEEE (2005)
10. Visser, E., el Abidine Benaissa, Z.: A core language for rewriting. Electronic Notes in Theoretical Computer Science 15, 422–441 (1998)
11. Beazley, D.M.: Automated scientific software scripting with SWIG. Future Generation Computer Systems 19, 599–609 (2003)
12. Baader, F., Nipkow, T.: Term rewriting and all that. Cambridge University Press, New York (1998)

Topic 2: Performance Prediction and Evaluation

Allen D. Malony, Helen Karatza, William Knottenbelt, and Sally McKee

Topic Committee

In recent years, a range of novel methodologies and tools have been developed for the purpose of evaluation, design, and model reduction of existing and emerging parallel and distributed systems. At the same time, the coverage of the term 'performance' has constantly broadened to include reliability, robustness, energy consumption, and scalability in addition to classical performance-oriented evaluations of system functionalities. Indeed, the increasing diversification of parallel systems, from cloud computing to exascale, being fuel by technological advances, is placing greater emphasis on the methods and tools to address more comprehensive concerns. The aim of the Performance Prediction and Evaluation topic is to bring together system designers and researchers involved with the qualitative and quantitative evaluation and modeling of large-scale parallel and distributed applications and systems to focus on current critical areas of performance prediction and evaluation theory and practice.

The five papers selected for the topic area reflect the broadening perspective of parallel performance involving alternative evaluation techniques (measurement, simulation, analytical modeling), different systems components (file systems, GPUs, I/O, multicore processors), and multiple targeted metrics (execution time, energy, network latency). The two papers based on modeling methodologies looked at two different systems aspects:

The paper "Energy Consumption Modeling for Hybrid Computing" by Marowka presented analytical models based on an energy consumption metric and used the model to analyze different design options for hybrid CPU-GPU chips. They studied the joint effect of performance and energy consumption to understand their relationship, particularly with respect to greater parallelism. The paper "HPC File Systems in Wide Area Networks: Understanding the Performance of Lustre over WAN" by Aguilera et al., also looked at performance interactions, but with respect to network system design. They evaluated the performance of the Lustre file system and its networking layer, with the goal of understanding the impact that the network latency has on Lustre's performance and deriving useful "rules of thumb" to help predict Lustre performance variation.

The next two papers share a common thread of how to capture a more comprehensive evaluation of performance for predictive purposes when the environment itself is complex and difficult to study directly. While sharing this theme, the two papers look at two distinct problems areas and take two different approaches:

"Understanding I/O Performance using I/O Skeletal Applications" by Logan et al. attempts to get a handle on the causes of I/O bottlenecks in HPC applications for purposes of guiding scaling optimization. By combining an approach to generate I/O benchmark codes from a high-level description with low-level

C. Kaklamanis et al. (Eds.): Euro-Par 2012, LNCS 7484, pp. 52–53, 2012.
© Springer-Verlag Berlin Heidelberg 2012

performance characterization of I/O components, a more complete and representative picture of application I/O behavior is obtained. The methodology enables more meaningful I/O performance testing, improved prediction of I/O performance, and more flexible evaluation of new systems and I/O methods. The paper "ASK: Adaptive Sampling Kit for Performance Characterization," by Castro et al., addresses the complexity of a large optimization design space for performance tuning. Their approach measures and understands performance tradeoffs by applying multiple adaptive sampling methods and strategies with the goal of considerably reducing the cost of performance exploration. The outcome are precise models of performance, created with a small number of measures, that can be used for prediction of performance for specific features, such as memory stride accesses.

The last paper of the topic area "CRAWP: A Workload Partition Method for the Efficient Parallel Simulation of Manycores" by Jiao et al. points out the interesting aspect that the behavior of manycore execution affects the performance of parallel discrete event simuation (PDES) system used to study it. Thus, by altering how workload is partitioned in the PDES simulator, it is possible to achieve improved speed and accuracy. The authors propose an adaptive workload partition method – Core/Router-Adaptive Workload Partition (CRAWP) – that make the simulation of on-chip-network independent of the cores. Significant improvements are achieved for the simulation of the SPLASH2 benchmark applications.

Energy Consumption Modeling
for Hybrid Computing

Ami Marowka

Department of Computer Science
Bar-Ilan University, Israel
amimar2@yahoo.com

Abstract. Energy efficiency is increasingly critical for embedded systems and mobile devices, where their continuous operation is based on battery life. In order to increase energy efficiency, chip manufacturers are developing heterogeneous CMP chips.

We present analytical models based on an energy consumption metric to analyze the different performance gains and energy consumption of various architectural design choices for hybrid CPU-GPU chips. We also analyzed the power consumption implications of different processing modes and various chip configurations. The analysis shows clearly that greater parallelism is the most important factor affecting energy saving.

Keywords: Analytical model, CPU-GPU architecture, Performance, Power estimation, Energy.

1 Introduction

Energy efficiency is one of the most challenging problems confronting multi-core architecture designers. Future multi-core processors will have to manage their computing resources while maintaining their power consumption within a power budget. This constraint is forcing the microprocessor designers to develop new computer architectures that deliver better performance per watt rather than simply yielding higher sustainable performance.

Recent research shows that integrated CPU-GPU processors have the potential to deliver more energy efficient computations, which is encouraging chip manufacturers to reconsider the benefits of heterogeneous parallel computing. The integration of CPU and DSP cores on a single chip has provided an attractive solution for the mobile and embedded market segments, and a similar direction for CPU-GPU computing appears to be an obvious move. It is known that the integration of thin cores and fat cores on a single processor achieves a better performance gain per watt. For example, a study of analytical models of various heterogeneous multi-core processor configurations found that the integration of many simplified cores in a single complex core achieved greater speedup and energy efficiency when compared with homogeneous simplified cores [1]. Thus, it is generally agreed that a heterogeneous chip integrating different core architectures, such as CPU and GPU, on a single die is the most promising

C. Kaklamanis et al. (Eds.): Euro-Par 2012, LNCS 7484, pp. 54–64, 2012.

technology [2–5]. Chip manufacturers such as Intel, NIVIDIA, and AMD have already announced such architectures, i.e., Intel Sandy Bridge, AMD's Fusion APUs, and NVIDIA's Project Denver.

Despite some criticisms [6, 7] Amdahl's Law [8] is still relevant as we enter a heterogeneous multi-core computing era. Amdahl's Law is a simple analytical model that helps developers to evaluate the actual speedup that can be achieved using a parallel program. However, the future relevance of the law requires its extension by the inclusion of constraints and architectural trends demanded by modern multiprocessor chips. Here, we extend a study conducted by Woo and Lee [1] and apply it to the case of hybrid CPU-GPU multi-core processors.

We investigate how energy efficiency and scalability are affected by the power constraints imposed on modern CPU-GPU based heterogeneous processors. We present analytical models that extend Amdahl's Law by accounting for energy limitations and we analyze the three processing modes available for heterogeneous computing, i.e., symmetric, asymmetric, and simultaneous asymmetric.

The rest of this paper is organized as follows. Section 2 presents an analytical model of a symmetric multi-core processor that reformulates Amdahl's Law to capture power constraints. In Section 3 we continue by applying energy constraints to an analytical model of an asymmetric processor. In Section 4 we study how performance and power consumption are affected by simultaneous asymmetric processing. In Section 5 we compare the three analytical models. Section 6 presents related works and Section 7 concludes the paper.

2   Symmetric Processors

In this section we reformulate Amdahl's Law to capture the necessary changes imposed by power constraints. We start with the traditional definition of a symmetric multi-core processor and continue by applying energy constraints to the equations following the method of Woo and Lee [4].

2.1   Symmetric Speedup

Amdahl's law posts an upper limit on the *symmetric speedup* (*speedup$_s$*) that can be achieved by parallelization of a symmetric multi-core processor, as follows:

$$Speedup_s = \frac{1}{(1-f) + \frac{f}{c}} \tag{1}$$

where c is the number of cores, and f is the fraction of a program's execution time that is parallelizable ($0 \leq f \leq 1$).

2.2   Symmetric Performance per Watt

To model power consumption in realistic scenarios, we introduce the variable k_c to represent the fraction of power a single CPU core consumes in its idle state

$(0 \le k_c \le 1)$. In the case of a symmetric processor, one core is active during the sequential computation and consumes a power of 1, while the remaining $(c-1)$ CPU-cores consume $(c-1)k_c$. During the sequential computation period, the processor consumes a power of $1 + (n-1)k_c$. Thus, during the parallel computation time period, c CPU-cores consume c power. It requires $(1-f)$ and f/c to execute the sequential and parallel codes respectively, so the formula for the average power consumption W_s of a symmetric processor is as follows.

$$W_s = \frac{(1-f) \cdot \{1 + (c-1)k_c\} + \frac{f}{c} \cdot c}{(1-f) + \frac{f}{c}} = \frac{1 + (c-1)k_c(1-f)}{(1-f) + \frac{f}{c}} \tag{2}$$

Next, we define the *performance per watt (Perf/W)* metric to represent the amount of performance that can be obtained from 1 W of power. The *Perf* of a single CPU-core execution is 1, so the $Perf/W_s$ achievable for a symmetric processor is formulated as follows.

$$\frac{Perf}{W_s} = \frac{Speedup_s}{W_s} = \frac{1}{1 + (c-1)k_c(1-f)} \tag{3}$$

2.3   Symmetric Performance Per Joule

The definition of $Perf/W$ metric allows us to evaluate the performance achievable by a derived unit of power (watt). Power is the rate at which energy is converted, so we can define a *Performance per Joule (Perf/J)* metric where the joule is the derived unit of energy, representing the amount of performance stored in an electrical battery. The $Perf/J$ of a single CPU-core execution is 1, so the $Perf/J_s$ achievable by a symmetric processor is formulated as follows.

$$\frac{Perf}{J_s} = Speedup_s \cdot \frac{Perf}{W_s} = \frac{1}{(1-f) + \frac{f}{c}} \cdot \frac{1}{1 + (c-1)k_c(1-f)} \tag{4}$$

Figure 1 plots the $Perf/J_s$ as a function of the number of CPU-cores in a symmetric multi-core processor. It is immediately obvious that there is a huge gap between the $Perf/J_s$ obtainable when a high degree of parallelism is available $(f = 0.99)$ and that when the available parallelism is only 10% less $(f = 0.9)$. Thus, the major factor affecting the energy saving of mobile devices is the development of extremely parallel applications. When an abundance of parallelism is available $(f = 0.99)$, the $Perf/J_s$ increases linearly with the increase in the number of cores whereas with $f < 0.9$ the $Perf/J_s$ reaches it maximum at a small number of cores before decreasing slowly.

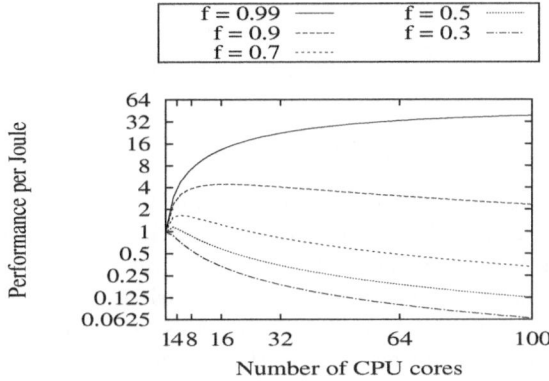

Fig. 1. Performance per joule as a function of the number of CPU-cores in a symmetric multi-core processor when $k_c = 0.3$ and various values of f

3   Asymmetric CPU-GPU Processors

We assume that a program's execution time can be composed of a time period where the program runs in parallel (f), a time period where the program runs in parallel on the CPU cores (α), and a time period where the program runs in parallel on the GPU cores ($1 - \alpha$).

To model the power consumption of an asymmetric processor we introduce another variable, k_g, to represent the fraction of power a single GPU-core consumes in its idle state ($0 \leq k_g \leq 1$). We introduce two further variables, α and β, to model the performance difference between a CPU-core and a GPU-core. The first variable represents the fraction of a program's execution time that is parallelized on the CPU-cores ($0 \leq \alpha \leq 1$), while the second variable represents a GPU core's performance normalized to that of a CPU-core ($0 \leq \beta$). For example, comparing the performance of a single core of Intel Core-i7-960 multi-core processor against the performance of a single core of a NVIDIA GTX 280 GPU processor yields values of β between 0.4 and 1.2. Moreover, recent studies such as [9] show that the GPU processor (NVIDIA GTX 280) achieves only 2.5x speedup in average compared to multi-core processor (Intel Core-i7-960).

We assume that one CPU-core in an active state consumes a power of 1 and the *power budget (PB)* of a processor is 100. Thus, $g = (PB-c)/w_g$ is the number of the GPU-cores embedded in the processor, where variable w_g represents the active GPU core's power consumption relative to that of an active CPU-core ($0 \leq w_g \leq 1$).

3.1   Asymmetric Speedup

Now, if the sequential code of the program is executed on a single CPU-core the following equation represents the theoretical achievable *asymmetric speedup* (*speedup$_a$*).

$$Speedup_a = \frac{1}{(1-f) + \frac{\alpha f}{c} + \frac{(1-\alpha)f}{g \cdot \beta}}$$ (5)

3.2   Asymmetric Performance per Watt

To model the power consumption of an asymmetric processor we assume that one core is active during the sequential computation and consumes a power of 1, while the remaining $c - 1$ idle CPU-cores consume $(c - 1)k_c$ power and g idle GPU-cores consume $g \cdot w_g \cdot k_g$ power. Thus, during the parallel computation period of the CPU-cores, c active CPU-cores consume c power and g idle GPU-cores consume $g \cdot w_g \cdot k_g$ power. During the parallel computation period of the GPU-cores, g active GPU-cores consume $g \cdot w_g$ power and c idle CPU-cores consume $c \cdot k_c$ power. Let P_s, P_c and P_g denote the power consumption during the sequential, CPU, and GPU processing phases, respectively.

$$P_s = (1 - f)\{1 + (c - 1)k_c + g \cdot w_g \cdot k_g\}$$

$$P_c = \frac{\alpha f}{c}\{c + g \cdot w_g \cdot k_g\}$$

$$P_g = \frac{(1 - \alpha)f}{g \cdot \beta}\{g \cdot w_g + c \cdot k_c\}$$

It requires $(1 - f)$ to perform the sequential computation, and $\frac{\alpha f}{c}$ and $\frac{(1-\alpha)f}{g \cdot \beta}$ to perform the parallel computations on the CPU and GPU, respectively, so the average power consumption W_a of an asymmetric processor is as follows.

$$W_a = \frac{P_s + P_c + P_g}{(1-f) + \frac{\alpha f}{c} + \frac{(1-\alpha)f}{g \cdot \beta}}$$ (6)

Consequently, $Perf/W_a$ of a an asymmetric processor is expressed as

$$\frac{Perf}{W_a} = \frac{Speedup_a}{W_a} = \frac{1}{P_s + P_c + P_g}$$ (7)

3.3   Asymmetric Performance per Joule

Based on our definition of performance per joule, the $Perf/J_a$ of a an asymmetric processor is expressed as follows.

$$\frac{Perf}{J_a} = Speedup_a \cdot \frac{Perf}{W_a} =$$ (8)

$$\frac{1}{(1-f) + \frac{\alpha f}{c} + \frac{(1-\alpha)f}{g \cdot \beta}} \cdot \frac{1}{P_s + P_c + P_g}$$

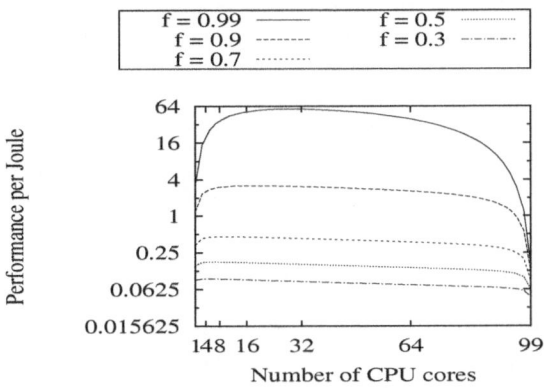

Fig. 2. Performance per joule as a function of the number of CPU-cores in an asymmetric processor when $\alpha = 0.1, k_c = 0.3, k_g = 0.2, w_g = 0.25, \beta = 0.5$ and various values of f

Figure 2 plots the $Perf/J_a$ as a function of the number of CPU-cores in an asymmetric processor. It can be observed again that high energy efficiency in a heterogeneous system is obtainable only if hybrid parallel programming models will be available for building extremely parallel programs. Such programs will need the support of runtime systems to find the optimal chip configuration for maximum battery continues operation. For example, in Figure 2 the optimal configuration (for $f = 0.99$) is achieved for 28 CPU-cores and 312 GPU-cores.

4   CPU-GPU Simultaneous Processing

In the previous analysis we assumed that a program's execution time is divided into three phases as follows: a sequential phase where one core is active, a CPU phase where the parallelized code is executed by the CPU-cores and a GPU phase where the parallelized code is executed by the GPU-cores. However, the aim of hybrid CPU-GPU computing is to divide the program while allowing the CPU and the GPU to execute their codes simultaneously.

4.1   Simultaneous Asymmetric Speedup

We conduct our analysis assuming that the CPU's execution time overlaps with the GPU's execution time. Such an overlap occurs when the CPU's execution time $\frac{\alpha f}{c}$ equals the GPU's execution time $\frac{(1-\alpha)f}{g \cdot \beta}$. Let α' denote the value of α that applies to this equality:

$$\alpha' = \frac{c}{g \cdot \beta + c}$$

Now, if the sequential code of the program is executed on a single CPU-core the following equation represents the theoretical achievable *simultaneous asymmetric speedup (speedup$_{sa}$)*.

$$Speedup_{sa} = \frac{1}{(1-f) + \frac{\alpha' f}{c}} = \frac{1}{(1-f) + \frac{f}{g \cdot \beta + c}} \tag{9}$$

4.2 Simultaneous Asymmetric Perf/W

To model the power consumption of an asymmetric processor for the case of simultaneous processing, we assume that one core is active during the sequential computation and consumes a power of 1, while the remaining $c-1$ idle CPU-cores consume $(c-1)k_c$ power and g idle GPU-cores consume $g \cdot w_g \cdot k_g$ power. During the parallel computation period, c active CPU-cores consume c power and g GPU-cores consume $g \cdot w_g$ power. It requires $(1-f)$ to execute the sequential code, and $\frac{\alpha' f}{c}$ to execute the parallel code on the CPU and GPU simultaneously, so the formula for the average power consumption W_{sa} of an asymmetric processor during simultaneous processing is as follows.

$$W_{sa} = \frac{P_s + \frac{\alpha' f}{c}\{c + g \cdot w_g\}}{(1-f) + \frac{\alpha' f}{c}} \tag{10}$$

Consequently, $Perf/W_{sa}$ of an asymmetric processor during simultaneous processing is expressed as

$$\frac{Perf}{W_{sa}} = \frac{Speedup_{sa}}{W_{sa}} = \frac{1}{P_s + \frac{\alpha' f}{c}\{c + g \cdot w_g\}} \tag{11}$$

4.3 Simultaneous Asymmetric Perf/J

Based on our definition of performance per joule, the $Perf/J_{sa}$ of an asymmetric processor in the simultaneous processing mode is expressed as follows.

$$\frac{Perf}{J_{sa}} = Speedup_{sa} \cdot \frac{Perf}{W_{sa}} = \tag{12}$$

$$\frac{1}{(1-f) + \frac{\alpha' f}{c}} \cdot \frac{1}{P_s + \frac{\alpha' f}{c}\{c + g \cdot w_g\}}$$

Figure 3 shows the $Perf/J_{sa}$ as a function of number of CPU-cores with an asymmetric processor where the CPU and the GPU are in simultaneous processing mode. As expected, a low degree of parallelism decreases significantly

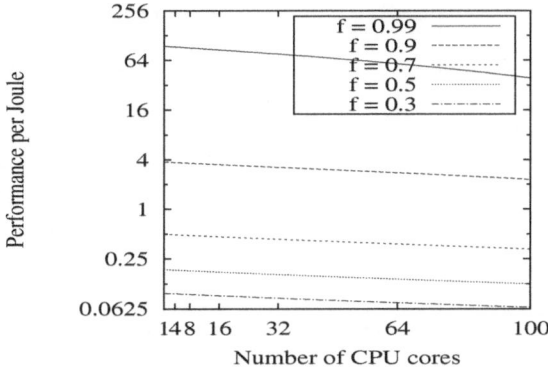

Fig. 3. Performance per joule as a function of the number of CPU-cores for an asymmetric processor in simultaneous processing mode when $\alpha = 0.1, k_c = 0.3, k_g = 0.2, w_g = 0.25, \beta = 0.5$ and various values of f

the energy efficiency. On the other hand, when an abundance of parallelism is available the energy efficiency is very high. In simultaneous processing mode, the obtainable $Perf/J_{sa}$ decreases slowly with the increase in the number of CPU-cores. This phenomenon means that it is not always necessary to support a dynamic reconfigurable processor and an associated runtime optimizer when finding the best chip configuration, because all possible chip configurations yield optimal or near-optimal configuration.

5 Synthesis

Figure 4 shows the three $Perf/J$ graphs for the analytical models investigated, i.e., symmetric (s), asymmetric (a) and simultaneous asymmetric (sa). This comparison shows that greater parallelism yields better energy efficiency and offers more chip configurations choices, while encouraging the search or better scalable software with energy saving. Simultaneous processing yields an excellent $Perf/J$ with peak performance using a chip configuration of a single CPU-core. It then decreases as the number of CPU-cores increases untill the point where all cores in the chip are CPU-cores, which is also the intersection point with the symmetric $Perf/J$. In contrast, the asymmetric processor delivers poor $Perf/J$ at extreme points where the number of CPU-cores is small or large, which requires that the dynamic configuration is identified and set for optimal chip organization.

6 Related Work

Hill and Marty [11] studied the implications of Amdahl's law on multi-core hardware and proposed the design of future chips based on the overall chip performance rather than core efficiencies. The major assumption in that model

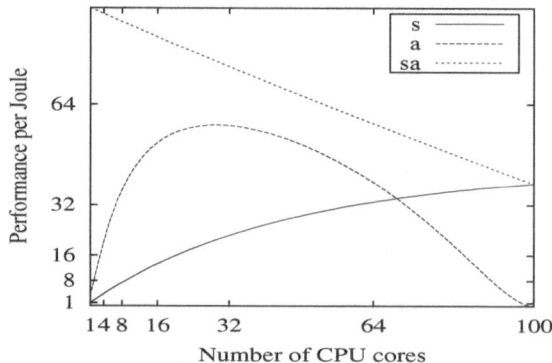

Fig. 4. Comparison between symmetric Perf/J (s), asymmetric Perf/J (a), and simultaneous asymmetric Perf/J (sa) when $\alpha = 0.1, k_c = 0.3, k_g = 0.2, w_g = 0.25, \beta = 0.5$ and $f = 0.99$

was that a chip is composed of many basic cores and their resources can be combined dynamically to create a more powerful core with higher sequential performance. Using Amdahl's law, they showed that asymmetric multi-core chips designed with one fat core and many thin cores exhibited better performance than symmetric multi-core chip designs. For example, with $f = 0.975$ (the fraction of computation that can parallelize) and $n = 256$ (Base Core Equivalents), the best asymmetric speedup was 125.0, whereas the best symmetric speedup was 51.2. Individual core resources could be dynamically combined to increase performance of the sequential component, so the performance was always improved. In our example, the speedup was increased to 186.0.

Woo and Lee [1] developed a many-core performance per energy analytical model that revisited Amdahl's Law. Using their model the authors investigated the energy efficiency of three architecture configurations. The first architecture studied contained multi-superscalar cores, the second architecture contained many simplified and energy efficient cores, and the third architecture was an asymmetric configuration of one superscalar core and many simplified energy efficient cores. The evaluation results showed that under restricted power budget conditions the asymmetric configuration usually exhibited better performance per watt. The energy consumption was reduced linearly as the performance was improved with parallelization scales. Furthermore, improving the parallelization efficiency by load balancing among processors increased the efficiency of power consumption and increased the battery life.

Sun and Chen [11] studied the scalability of multi-core processors and reached more optimistic conclusions compared with the analysis conducted by Hill and Marty [11]. The authors suggested that the fixed-size assumption of Amdahl's law was unrealistic and that the fixed-time and memory-bounded models might better reflect real world applications. They presented extensions of these models for multi-core architectures and showed that there was no upper bound on the scalability of multi-core architectures. However, the authors suggested that the major problem limiting multi-core scalability is the memory data access delay and they called for more research to resolve this memory-wall problem.

Esmaeilzadeh et al. [12] performed a systematic and comprehensive study to estimate the performance gains from the next five multi-core generations. Accurate predictions require the integration of as many factors as possible. Thus, the study included: power, frequency and area limits; device, core and multi-core scaling; chip organization; chip topologies (symmetric, asymmetric, dynamic, and fused); and benchmark profiles. They constructed models based on pessimistic and optimistic forecasts, and observations of previous works with data from 150 processors. The conclusions were not encouraging. Over five technology generations only a 7.9x average speedup was predicted with multi-core processors, while over 50% of the chip resources will be turned off due to power limitations. Neither multi-core CPUs nor many-core GPUs architectures were considered to have the potential for delivering the required performance speedup levels.

Cho and Melhem [13] studied the mutual affects of parallelization, program performance, and energy consumption. Their analytic model was applied to a machine that could turn off individual cores, while others do not make this assumption. The main prediction was that greater parallelism (a greater ratio of the parallel portion in the program) and more cores helped reduce energy use. Moreover, it was shown that is possible to reduce the processor speeds and gain further dynamic energy reductions before static energy becomes the dominant factor determining the total amount of energy used.

Hong and Kim [14] developed an integrated power and performance modeling system (IPP) for the GPU architecture. IPP is an empirical power model that aims to predict performance-per-watt and the optimal number of active cores for bandwidth-limited applications. IPP uses predicted execution times to predict power consumption. In order to predict the execution time the authors used a special-purpose GPU analytical timing model. Moreover, to obtain the power model parameters, they designed a set of synthetic micro-benchmarks that stress different architectural components in the GPU.

The evaluation of the proposed model was done by using NVIDIA GTX280 GPU. The authors show that by predicting the optimal number of active cores, they can save up to 22.09% of runtime GPU energy consumption and on average 10.99% of that for five memory bandwidth-limited benchmarks. They also calculated the power savings if a per-core power gating mechanism is employed, and the result shows an average of 25.85% in energy reduction. IPP predicts the power consumption and the execution time with an average of 8.94% error for the evaluated benchmarks GPGPU kernels. It can be used by a thread scheduler in order to manage the power system more efficiently or by the programmers to optimize program configurations.

7 Conclusions

We investigated three analytical models of symmetric, asymmetric, and simultaneous asymmetric processing. These models extended Amdahl's Law for symmetric multi-core and heterogeneous many-core processors by taking in account power constraints. The analysis of speedup and the performance per watt of various chip configurations suggests that future CMPs should be a priori designed to include one or a few fat cores alongside many efficient thin cores to support energy efficient hardware platforms. On the software side, this study shows without a doubt that increased parallelism should not be the exception, because the standard parallel programming paradigm can create energy saving applications that can be used to efficiently underpin future multi-core processor architectures.

References

1. Woo, D.H., Lee, H.S.: Extending Amdahl's Law for Energy-Efficient Computing in the Many-Core Era. IEEE Computer 38(11), 32–38 (2005)
2. Mantor, M.: Entering the Golden Age of Heterogeneous Computing. In: Performance Enhancement on Emerging Parallel Processing Platforms (2008)
3. Kogge, P., et al.: ExaScale Computing Study: Technology Challenges in Achieving Exascale Systems. DARPA, Washington, D.C (2008)
4. Fuller, S.H., Millett, L.I.: Computing Performance: Game Over or Next Level? IEEE Computer 44(1), 31–38 (2011)
5. Borkar, S.: Thousand core chips: a technology perspective. In: Proc. 44th Design Automation Conference, pp. 746–749. ACM Press (2007)
6. Gustafson, J.L.: Reevaluating Amdahl's Law. Communication of ACM 31(5), 532–533 (1988)
7. Hillis, D.: The pattern on the stone: The simple ideas that make computers work. Basic Books (1998)
8. Amdahl, G.M.: Validity of the Single-Processor Approach to Achieving Large-Scale Computing Capabilities. In: Proc. Am. Federation of Information Processing Societies Conf., pp. 483–485. AFIPS Press (1967)
9. Lee, V.W., et al.: Debunking the 100X GPU vs. CPU myth: an evaluation of throughput computing on CPU and GPU. In: Proceedings of the 37th Annual International Symposium on Computer Architecture (2010)
10. Hill, M.D., Marty, M.R.: Amdahl's Law in the Multicore Era. IEEE Computer 41(7), 33–38 (2008)
11. Sun, X.-H., Chen, Y.: Reevaluating Amdahl's law in the multicore era. Journal of Parallel and Distributed Computing 70(2), 183–188 (2010)
12. Esmaeilzadeh, H., Blem, E., Amant, R.S., Sankaralingam, K., Burger, D.C.: Dark Silicon and the End of Multicore Scaling. In: Proceeding of 38th International Symposium on Computer Architecture (ISCA), pp. 365–376 (June 2011)
13. Cho, S., Melhem, R.G.: On the Interplay of Parallelization, Program Performance, and Energy Consumption. IEEE Trans. Parallel Distrib. Syst. 21(3), 342–353 (2010)
14. Hong, S., Kim, H.: An Integrated GPU Power and Performance Model. In: Proceeding of ISCA 2010, pp. 19–23. ACM (June 2010)

HPC File Systems in Wide Area Networks: Understanding the Performance of Lustre over WAN

Alvaro Aguilera, Michael Kluge, Thomas William, and Wolfgang E. Nagel

Technische Universität Dresden, Dresden, Germany
{alvaro.aguilera,
michael.kluge,thomas.william,wolfgang.nagel}@tu-dresden.de

Abstract. Using the first commercially available 100 Gbps Ethernet technology with a link of varying length, we have evaluated the performance of the Lustre file system and its networking layer under different latency scenarios. The results led us to a better understanding of the impact that the network latency has on Lustre's performance. In particular spanning Lustre's networking layer, striped small I/O, and the parallel creation of files inside a common directory. The main contribution of this work is the derivation of useful rules of thumbs to help users and system administrators predict the variation in Lustre's performance produced as a result of changes in the latency of the I/O network.

1 Introduction

Scientific instruments create an enormous amount of data every day. For example, the NASA Earth Observation System (EOSDIS) created about 2.9 TiB data in average each day in 2010 [1]. To share this data with collaborating scientists, WAN file systems have already proven their value. The European DEISA project [2] utilizes a series of dedicated 10 Gbps links to serve a distributed GPFS file system to different HPC centers. Since about 2006, the parallel file system Luste has gained some attention while being used in WAN environments [3,4,5]. These evaluations and the consecutive use of Lustre as a production file system in the DataCapacitor project at Indiana University have demonstrated that parallel file systems can be used efficiently on networks with latencies of more than 100 milliseconds.

However, up to now, the relevant publications only describe different use cases, experiences, and tuning efforts, but none focuses on the interplay between the network latency, the Lustre tunables and the resulting performance of the file system. Advancing this understanding will certainly ease Lustre's tuning effort as well as give some hints about how to use the file system in production, e.g. how files should be striped. Our aim in this paper is to make a first step in this direction by analyzing our observations on the 100 Gpbs Ethernet testbed.

The paper is structured as follows. The next section (Section 2) introduces the testbed system itself. In Section 3, previous work on performance models,

C. Kaklamanis et al. (Eds.): Euro-Par 2012, LNCS 7484, pp. 65–76, 2012.

especially for parallel file systems, is reviewed. Within Section 4, the Lustre networking layer is evaluated and a simple performance model for this software layer is presented. After this, Section 5 deals with the performance obtained when different I/O calls are issued from a single client while Section 6 extends this work to multiple clients. Section 7 gives a conclusion and sketches future work.

2   100 GbE Testbed between Dresden and Freiberg

The 100 Gbps testbed, provided by Alcatel-Lucent, T-Systems, HP, and DDN, provided a unique resource to, on the one hand test this new technology, and on the other hand to extend our knowledge about different network services. The testbed spans the distance between the cities of Dresden and Freiberg in Saxony, Germany with a geographical distance of about 37 km and a optical cable length of about 60 km. During the test, additional boxes with optical cables have been used to extend the testbed from 60 up to 400 km. This allows us to conduct experiments using different latency configurations with a reliability not found in software-based latency injection methods. An Alcatel-Lucent 1830 photonic service switch connects both sides and can transmit 100 Gbps on a single carrier. The 7750 SR-12 service router links the optical layer and the network adapter. Both service routers (Freiberg and Dresden) have one media dependent adapter (MDA) with 100 Gbps, two adapters with 5x10 Gbps, five adapters with 2x10 Gbps, and 20x1 Gbps adapters.

HP provided 34 DL160G servers, 17 on each location, which are stocked with a ServerEngines-based 10 Gbps card, each connected to one of the 10 Gbps interfaces of the Ethernet switch. All servers are equipped with one six core Intel Westmere (Xeon 5650, 2.67 GHz) processor and 24 GiB of RAM.

Several sub-projects were scheduled on this testbed. Initial TCP tests provide the subsequent projects with a reliable base in terms of the available bandwidth and the network behavior in general. Three different parallel file systems: GPFS, Lustre, and FhGFS are installed on the HP servers to study the impact of the

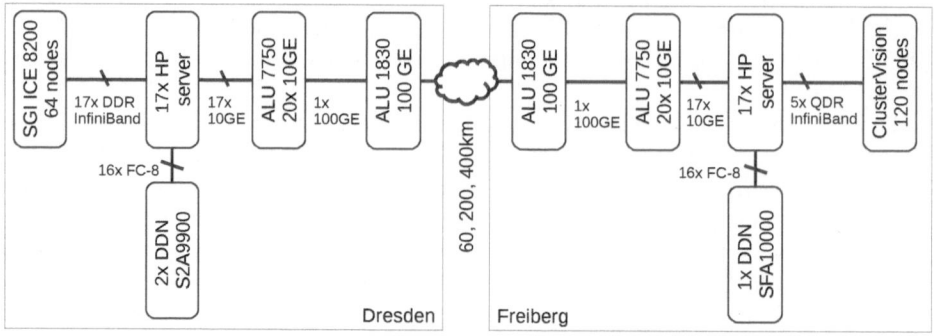

Fig. 1. 100 Gbps testbed equipment with connection cables

latency on the file system performance. An overview of the testbed setup is shown in Figure 1.

3   Related Work

The performance analysis of network file systems accessed over high latency networks such as WANs has been the focus of several studies during the last decade, and has been gaining importance as the technological trends make this use case more and more practical. The first studies describing the viability of a WAN file system in HPC context were conducted by the researchers working on the TeraGrid project, and can be found in [6]. The first published experiences using a minimally tuned Lustre file system in a WAN environment are detailed in [3,4,5,7]. More recent publications concentrate not on analyzing Lustre's raw performance over WAN, but more on its suitability for concrete use cases. An analysis of the use cases that would profit from an HPC WAN file system the most is presented in [8]. In [9], Cai et al. evaluated the suitability of a Lustre over WAN solution to sustain database operations. In [10], Rodriguez et al. describe their experience using Lustre over WAN for the CMS experiment at the LHC. Even though the modeling and simulation of storage systems have been a subject of study for at least two decades, most of the publications concentrate on modeling the individual components and not the file systems. To the best of our knowledge, there is only one publication explicitly dealing with the modeling of Lustre's performance, namely [11], in which Zhao et al. applied the idea of relative modeling to predict the performance of a Lustre file system.

4   Lustre's Networking Layer

In the first part of this paper we will discuss the performance of Lustre at its networking layer without considering the storage hardware and software components acting on lower layers. Understanding the performance of Lustre's networking layer is a first logical step in order to gain an understanding about how this parallel file system behaves when its major tunable parameters and network conditions are changed.

4.1   The LNET Protocol

Lustre Networking (LNET) is a custom networking API that leverages on the native transport protocol of the I/O network to interconnect the Metadata Server (MDS), the Object Storage Servers (OSSs), and the client systems of a Lustre cluster. It offers support for most of the network technologies used in HPC through a set of Lustre Network Drivers (LNDs) that are both available as individual kernel modules, and user space libraries. Internally, LNET uses a stateful protocol based on remote procedure calls that was derived from Sandia Portals. The bandwidths achievable by LNET during the file system

operation are determined by a combination of its own performance parameters, and the characteristics and configuration of the underlying I/O network.
In low-latency environments, the relevance of the former group of parameters is
not apparent, since Lustre achieves its maximal performance without much tuning. Their importance, however, is promptly made clear as soon as the network
latency increases.

LNET fragments all data transfers in units called Lustre RPCs, whose sizes
are always aligned to the system page size, and range from a single page (in
most cases 4096 bytes) up to one megabyte. The maximal size a Lustre RPC
may have, can be modified on a per client basis as long as the new size satisfies the conditions stated in [12]. In order to minimize the transmission and
processing overhead associated with small RPCs, Lustre tries to merge adjacent
RPCs to form RPCs of maximal size. The individual size of an RPC being transmitted is ultimately determined by a combination of the maximal RPC size, the
size of the buffers being read or written by the application, and on whether or
not the operation is immediately committed to disk (call to `fsync()`, operation
in O_DIREC mode, etc.). LNET is a stateful protocol in which every RPC being
sent must be acknowledged by the receiver. Similarly to TCP's sliding window
protocol, LNET may send more than one RPC before waiting for an acknowledgment response. These unacknowledged RPCs are normally referred to as the
RPCs in flight. Like in the previous case, the maximal number of RPCs in flight
sent by LNET during any operation is a parameter that can be defined on a per
client basis.

In the following section we explore the interplay between the size and count
of Lustre RPCs in flight, and how they affect, together with the network latency,
the LNET effective bandwidth.

4.2   Model Constraints

The performance of LNET is heavily dependent on the underlying transport protocol it relies upon, and especially on its congestion avoidance mechanisms. In
order to keep our model as simple, and as general as possible, we will constrain it
to data transmissions that are unthrottled by the transport protocol of the I/O
network. The behavior of the LNET bandwidth as a function of its in-flight data
is best exemplified in Figure 2. Our focus will lie in the unthrottled state (a) in
which changes in the RPC size or count, as well as on the network latency yield
a full effect on the LNET bandwidth. Even though this constrain is certainly
undesirable, doing otherwise would incur in excessive complexity, while simultaneously tying our model to a particular transport protocol (implementation).

4.3   Proposed Model

It is well understood that for any given network, the relation between the
bandwidth-delay product and the amount of in-flight data that is actually present
in a network segment at a given time is one of the key factors determining the
network throughput [13]. The maximal amount of in-flight data D_{max} that fits

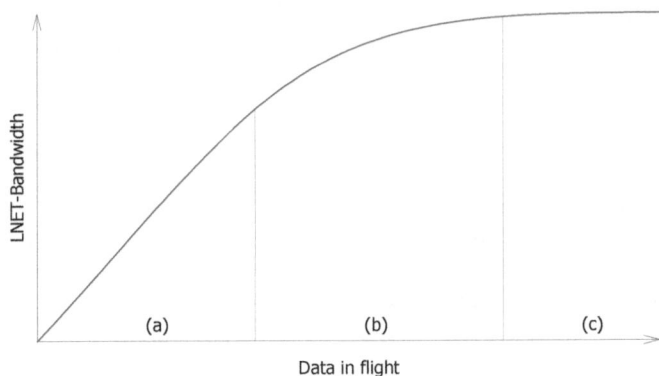

Fig. 2. Behavior of the LNET bandwidth as a function of its in-flight data. (a) un-throttled communication, (b) throttled by the transport protocol, and (c) maximal utilization reached.

inside a network path is determined by the bandwidth-delay product, defined as the round-trip time l of the network, multiplied by the network bandwidth b, plus the buffer space m of the network devices along the way (Equation 1). It is trivial to see that (omitting the buffer space m) any change in the network latency from l_0 to l_1 will immediately affect D_{max} by a factor of l_1/l_0.

$$D_{max} = l * b + m \qquad (1)$$

On the other hand, the amount of data D that LNET may put down the wire before waiting for an acknowledgment is mainly defined by the size s and number c of RPCs in flight (Equation 2). In a similar way, a change in the size and number of RPCs in flight should affect D by a factor of $(s_1c_1)/(s_0c_0)$.

$$D = s * c \qquad (2)$$

Our first modeling hypothesis will be that the LNET bandwidth will vary by the same factor, and in direct proportion to D, and by the same factor but in inverse proportion to D_{max}. This means that the expected variation in a known LNET bandwidth b_0 resulting from a change in the network latency, or in the size and number of RPCs in flight during an unthrottled communication can be calculated using Equation 3.

$$b = (\frac{s_1}{s_0} \frac{c_1}{c_0} \frac{l_0}{l_1})b_0 \qquad (3)$$

4.4   Measurement and Comparison with the Model

The benchmarking of the LNET performance was conducted on the testbed system previously introduced using the LNET-Selftest tool distributed with Lustre.

This tool allows the generation of intense LNET traffic between groups of nodes without requiring any physical I/O. The generated workload is also supposed to be similar to that produced by Lustre during real I/O operation.

Using (3) we were able to obtain a good approximation of the experimental data for changes in the network latency and RPC count. However, the bandwidth changes resulting from increasing the RPC size up from its minimum value were roughly half as big as those predicted. This difference indicates that an increase in the RPC size doesn't translate 1:1 to an increase in the LNET in-flight data. In spite of this, the model can be adapted by introducing a factor k to account for this overhead, as shown in Equation 4 .

$$b = \frac{1}{k}(\frac{s_1}{s_0}\frac{c_1}{c_0}\frac{l_0}{l_1})b_0 \qquad (4)$$

It is expected for the value of k to vary depending on each particular network configuration. For the testbed system, a value of $k = 2.27$ yielded the best results with an overall error of less than 15%. Figure 3 and 4 compare some of the predictions against the experimental data. The predictions of the model were done using the measured performance obtained with one 4 KiB RPC in flight and a network latency of 2.17 ms to extrapolate all other points.

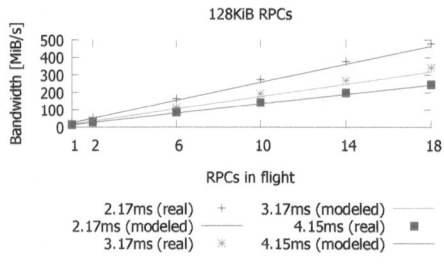

Fig. 3. Predictions of the LNET model for 128 KiB RPCs using different latency settings and RPC counts.

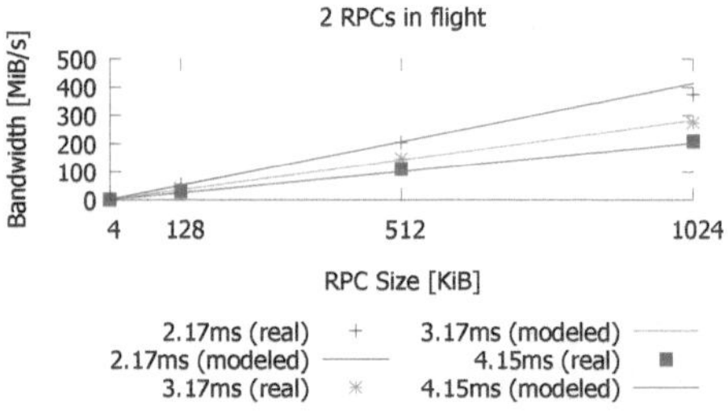

Fig. 4. Predictions of the LNET model for 2 RPCs in flight using different latency settings and RPC sizes.

5 Single Client Performance Observations

The aim of this section is to describe the impact of the latency for small file accesses and a single Lustre file system client. This data is advantageous for users that use WAN file systems similiar to home file systems, for example for compiling source code or for editing input files. The main parameters that have an impact on the performance and that can be influenced by the user are mainly the file size, the access size and for Lustre, the way the striping is done. For the striping it is worth mentioning, that the stripe size is fixed to 1 MiB as this is the native stripe size of the DDN devices. Thus, only the number of used stripes can be adjusted.

5.1   Setup and Measurements

For these tests we performed initial measurements with IOzone and a fiber length of 200 km and compared the results with results gathered at 400 km. Comparing this data with performance data collected locally (at 0 km) would have not made sense in this context as the communication locally was done via InfiniBand and any comparison would not only include the latency difference but also the difference between the protocols.

5.2   Observations and Findings

Fig. 5 shows the difference between the 200 km and the 400 km data for different I/O functions, for different file sizes and one block size. The figure shows the performance in KiB/s for the 400 km case in percent, using the performance for 200 km as 100 %. It shows that there is a noticeable performance impact only on the initial write of a file. All other functions, which reuse existing files, show only a small performance impact. This impact is due to the fact that the creation of a file needs at least one RTT. With increasing file sizes, this additional RTT has less influence on the total time of the operation.

Fig. 6 shows a more detailed performance study of the influence of the file size and the stripe count on the performance. The figure shows the differences in the latencies for the initial creation of files with different sizes and different stripe counts. Here, we subtracted the numbers gathered at 200 km from the numbers gathered at 400 km. As the next step, we normalized this time difference to the difference in the RTTs (4.14 ms $-$ 2.17 ms = 1.97 ms) between the two distances. This allows to characterize the impact of the additional distance on the performance.

Up to 1 MiB file size, all data is written to a single stripe and the number of stripes that the file can use does not influence the performance. The '1' in the figure in these cases just means that at 400 km it takes 1.97 ms longer to create a file and to write the content than it takes at 200 km. For each additional stripe used there is a penalty that is added as soon as a new stripe is used. This is due to the fact that the Lustre file system creates the objects on the storage servers for the first stripes in a sequential fashion.

This can create a large impact on file systems with a large number of stripes used by default, as the time to write the first N MiB will always be (N+1)*RTT. The +1 has to be added for the initial file creation on the metadata server, the stripes are created in an extra step. The problem with this finding is that in these cases the bandwidth is determined by the RTT, and not by the capabilities of the link.

Fig. 7 shows that there is no significant performance impact by the addition of 200 km to the distance when the same file is accessed with different block sizes or with different I/O functions. As most file I/O for these cases is rather small, this implies that the clients cache most I/O operations efficiently.

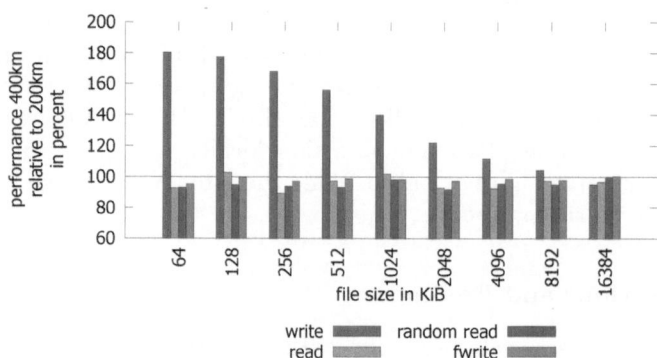

Fig. 5. Comparison of different I/O functions for 200 km and 400 km for different file sizes. A block size of 4 KiB and a stripe count of 1 was used. The performance for 400 km is given in percent relative to the performance for 200 km.

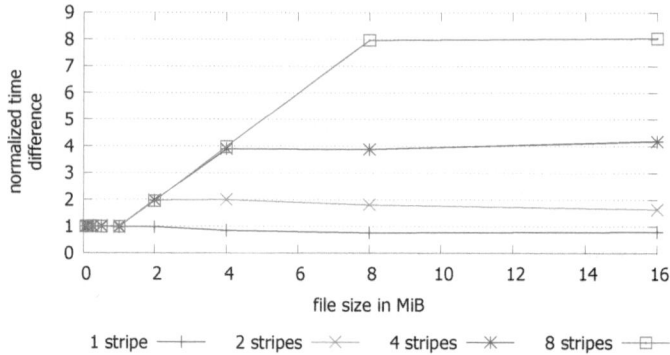

Fig. 6. Time differences for the initial file creations between the 400 km and the 200 km setup. The time difference is normalized to the difference of the RTTs for both distances.

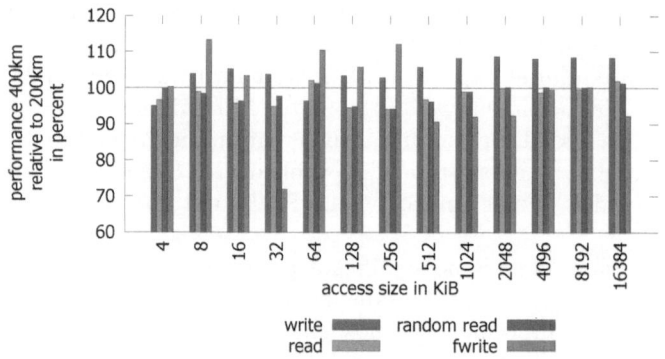

Fig. 7. Comparison of different I/O functions for 200 km and 400 km for different block sizes. A file size of 16 MiB and a stripe count of 1 was used. The performance for 400 km is given in percent relative to the performance for 200 km.

6   Performance Observations for Multiple Clients

The goal of the experiment was to determine the latency of the I/O traffic for multiple clients working in parallel, and to analyze the impact of the extension of the testbed fiber length on different file sizes. We therefore used the Lustre file system described above to generate unidirectional I/O traffic from Dresden to Freiberg. Up to 16 clients were used to measure latency and bandwidth using 8 B files for latency and 2 GiB files for bandwidth. The benchmarking tool used to generate the workload was IOR [14].

6.1   Setup and Measurements

Each process wrote its own file using POSIX I/O in O_DIRECT mode to ensure that the I/O operation is immediately committed to disk (see Fig. 8). The value *blockSize* in Fig. 9 denotes the amount of data that is written to each file per process, while *transferSize* represents the size of the payload being sent with each I/O request. This is restricted by IOR to a minimum of 8 bytes and a maximum of 2 GiB. IOR uses MPI to synchronize processes, therefore switching on *intraTestBarriers* adds a MPI_Barrier(all) between each test to ensure that there is no traffic left from previous I/O operations.

```
api=POSIX
filePerProc=1
useO_DIRECT=1
intraTestBarriers=1
repetitions=10
writeFile=1
readFile=0
```

```
RUN
  blockSize=8
  transferSize=8
  numTasks=16
RUN
  blockSize=2147483648
  . . .
```

Fig. 8. IOR configuration header

Fig. 9. I/O test with 16 processes each sending 8 B and 2 GB

The 1 GbE management links were used for the MPI communication not to disrupt the Lustre traffic. Also, each test was repeated 10 times for higher accuracy. The benchmarks were executed on the Dresden HP nodes on a mount point pointing to the Lustre in Freiberg using the SFA10K DDN storage and the 16 Freiberg HP nodes as OSTs.

In the experiments, IOR was run on one Server using one process writing to only one OST at first. Then additional servers doing the same file I/O were added up to the point where 16 processes were writing 16 files in parallel each to its own OST. This setup generated data for:

- different optical line length (60 km, 200 km, and 400 km)
- different number of parallel writes (1-16)
- different file sizes (8 B up to 2 GiB).

6.2   Observations and Findings

IOR can separately log the times for the functions involved in writing a file to disk (`open()`, `write()`, `close()`). We normalized the values to the round-trip time that was measured on the TCP layer (60 km → 0.72 ms, 200 km → 2.17 ms, 400 km → 4.14 ms) to see how many RTTs it takes to complete each of the functions and whether this depends on the number of parallel clients or not.

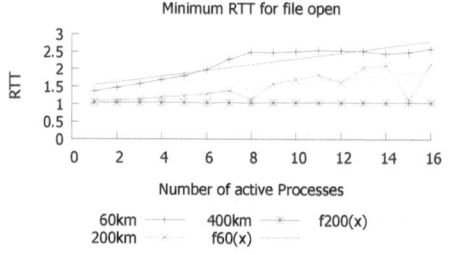

Fig. 10. Minimum RTT for file open of 8 Byte files   **Fig. 11.** Average RTT for file open of 8 Byte files

We first look at the time needed to open a file at the different line lengths and for different numbers of parallel clients. The numbers shown are for 8 Byte files but are essentially the same for 2 GiB files as well. The minimum numbers in Fig. 10 show the anticipated time of 1 RTT only for the 400 km distance. The latency of the computer hardware has a larger influence at 60 km due to the small transfer time compared to the time needed for processing. The time needed for the `open()` call is the RTT plus an overhead which in turn is a function depending on the number of parallel processes. Using a linear least squares fit we can determine the slope being 0.08 for 60 km and 0.06 for 200 km. This means that each additional process adds an overhead of 8 % of the RTT for 60 km and 6 % for 200 km to the minimum time needed to complete a `open()` call for the first process in the group that issues the open call. Additionally, the average time rises with the number of parallel processes (all processes involved and 10 repetitions) as seen in Fig. 11. This is due to the locking of the directory in which all files reside.

For the `close()` call the minimum times are nearly the same across all process counts at 1 RTT as there is no additional workload on the MDS. Again, 2 GiB files do not differ from 8 bytes files. Fig. 13 shows the transfer times for the `write()` call with minimum and maximum values as error bars. For large files the distance has no significant impact on the transfer time. As each of the processes was writing to its own OST, the number of parallel streams has no noticeable influence on the individual performance either.

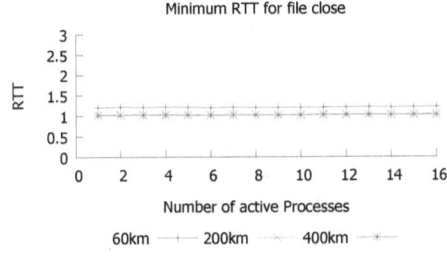

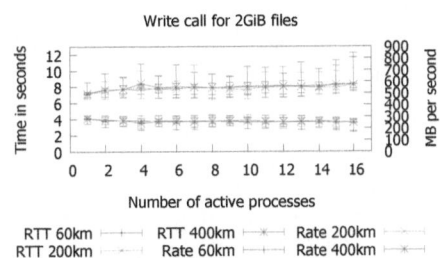

Fig. 12. Minimum RTT to close files of 8 bytes

Fig. 13. Write RTT and data rate for 2 GiB files

7   Conclusion

In this paper we explored some aspects of the performance variation exhibited by the Lustre file system when subjected to changes in the network latency. Furthermore, we used the empiric results obtained using a testbed network between the cities of Dresden and Freiberg to derive basic rules explaining the observed interaction between different performance parameters. Our findings describe the interplay between the bandwidth-delay product of the network, and the size and count of Lustre RPCs in flight for unthrottled communications, the penalty introduced by the stripe count during file creation, and the overhead encountered when concurrently opening files from multiple nodes.

There are several ways in which this work could be further improved. The first one would be to investigate whether the results are still valid for other deployments of Lustre or not. Among the other things deserving a deeper look are the relation between the RPC size and its induced overhead (expressed with the constant k), and how the congestion avoidance mechanisms acting at the transport layer of the network affect the predictions of the LNET-model.

Acknowledgments. We would like to conclude this paper by expressing our gratitude to all the sponsors and our networking group for the support they provided.

References

1. NASA. Key Science System Metrics (2010),
   http://earthdata.nasa.gov/about-eosdis/performance
2. Gentzsch, W.: DEISA, The Distributed European Infrastructure for Supercomputing Applications. In: Gentzsch, W., Grandinetti, L., Joubert, G.R. (eds.) High Performance Computing Workshop. Advances in Parallel Computing, vol. 18, pp. 141–156. IOS Press (2008)
3. Simms, S.C., Davy, M., Hammond, B., Link, M., Stewart, C., Bramley, R., Plale, B., Gannon, D., Baik, M.-H., Teige, S., Huffman, J., McMullen, R., Balog, D., Pike, G.: All in a Day's Work: Advancing Data-Intensive Research with the Data Capacitor. In: Proceedings of the 2006 ACM/IEEE Conference on Supercomputing, SC 2006. ACM, New York (2006)

4. Simms, S.C., Pike, G.G., Balog, D.: Wide Area Filesystem Performance Using Lustre on the TeraGrid. In: Proceedings of the TeraGrid 2007 Conference (2007)
5. Simms, S.C., Pike, G.G., Teige, S., Hammond, B., Ma, Y., Simms, L.L., Westneat, C., Balog, D.A.: Empowering Distributed Workflow with the Data Capacitor: Maximizing Lustre Performance Across the Wide Area Network. In: Proceedings of the 2007 Workshop on Service-Oriented Computing Performance: Aspects, Issues, and Approaches, SOCP 2007, pp. 53–58. ACM, New York (2007)
6. Andrews, P., Kovatch, P., Jordan, C.: Massive High-Performance Global File Systems for Grid Computing. In: Proceedings of the 2005 ACM/IEEE Conference on Supercomputing, SC 2005, p. 53. IEEE Computer Society, Washington, DC (2005)
7. Filizetti, J.: Lustre Performance over the InfiniBand WAN. In: Proceedings of the 2010 Lustre User Group (2010)
8. Michael, S., Simms, S., Breckenridge III, W.B., Smith, R., Link, M.: A Compelling Case for a Centralized Filesystem on the TeraGrid: Enhancing an Astrophysical Workflow with the Data Capacitor WAN as a Test Case. In: Proceedings of the 2010 TeraGrid Conference, TG 2010, pp. 13:1–13:7. ACM, New York (2010)
9. Cai, R., Curnutt, J., Gomez, E., Kaymaz, G., Kleffel, T., Schubert, K., Tafas, J.: A Scalable Distributed Datastore for BioImaging (2008), http://www.r2labs.org/pubs/BioinformaticsDatabase.ps
10. Rodriguez, J.L., Avery, P., Brody, T., Bourilkov, D., Fu, Y., Kim, B., Prescott, C., Wu, Y.: Wide Area Network Access to CMS Data Using the Lustre Filesystem. Journal of Physics: Conference Series 219(7), 072049 (2010)
11. Zhao, T., March, V., Dong, S., See, S.: Evaluation of a Performance Model of Lustre File System. In: 2010 Fifth Annual ChinaGrid Conference (ChinaGrid), pp. 191–196 (July 2010)
12. Oracle, Inc. Lustre 2.0 Operations Manual (2010), http://wiki.lustre.org/images/3/35/821-2076-10.pdf
13. Lakshman, T.V., Madhow, U.: The Performance of TCP/IP for Networks with High Bandwidth-Delay Products and Random Loss. IEEE/ACM Trans. Netw. 5, 336–350 (1997)
14. Shan, H., Shalf, J.: Using IOR to Analyze the I/O Performance for HPC Platforms. In: Cray User Group Conference (2007)

Understanding I/O Performance Using I/O Skeletal Applications

Jeremy Logan[1,2], Scott Klasky[1], Hasan Abbasi[1], Qing Liu[1],
George Ostrouchov[1], Manish Parashar[3], Norbert Podhorszki[1],
Yuan Tian[4], and Matthew Wolf[1,5]

[1] Oak Ridge National Laboratory, Oak Ridge, Tennessee, USA
[2] University of Tennessee, Knoxville, Tennessee, USA
[3] Rutgers, The State University of New Jersey, Piscataway, New Jersey, USA
[4] Auburn University, Auburn, Alabama, USA
[5] Georgia Tech., Atlanta, Georgia, USA

Abstract. We address the difficulty involved in obtaining meaningful measurements of I/O performance in HPC applications, as well as the further challenge of understanding the causes of I/O bottlenecks in these applications. The need for I/O optimization is critical given the difficulty in scaling I/O to ever increasing numbers of processing cores. To address this need, we have pioneered a new approach to the analysis of I/O performance using automatic generation of I/O benchmark codes given a high-level description of an application's I/O pattern. By combining this with low-level characterization of the performance of the various components of the underlying I/O method we are able to produce a complete picture of the I/O behavior of an application.

We compare the performance measurements obtained using Skel, the tool that implements our approach, with those of an instrumented version of the original application to show that our approach is accurate. We demonstrate the use of Skel to compare the performance of several I/O methods. Finally we show that the detailed breakdown of timing information produced by Skel provides better understanding of the reasons for the performance differences between the examined I/O methods. We conclude that our approach facilitates faster, more accurate and more meaningful I/O performance testing, allowing application I/O performance to be predicted, and new systems and I/O methods to be evaluated.

1 Introduction

Understanding and optimizing the I/O performance of high-performance scientific applications is critical for the success of many of these applications. As larger and faster platforms are introduced, the higher fidelity applications that run on these platforms produce increasingly massive data sets. At the same time, the increase in computational performance of supercomputers is greatly outpacing the I/O bandwidth of these machines [1]. To achieve reasonable I/O performance despite this growing gap, increasingly sophisticated I/O techniques are required.

C. Kaklamanis et al. (Eds.): Euro-Par 2012, LNCS 7484, pp. 77–88, 2012.

The design, implementation, and subsequent optimization of these techniques require sophisticated tools that allow the performance of I/O methods to be rapidly understood.

Meaningful measurements of I/O performance on large-scale systems are currently difficult and time consuming to acquire. One approach is to instrument a representative application with timing code to measure I/O performance. This approach presents unnecessary complexity to an I/O developer since building and running the application typically requires detailed knowledge that is unrelated to the task of I/O performance measurement.

A more manageable approach is to use *I/O kernels*, such as the FLASH-IO benchmark routine [2] and the S3D I/O kernel [3], codes that include the I/O routines from a target application but which typically have had computation and communication operations removed. While providing accuracy by performing the same I/O pattern as the application, I/O kernels have significantly shorter total execution time since they do not include the application's computational workload. Furthermore, I/O kernels may not retain all of the application's dependencies on external libraries, simplifying their use on new systems. The use of I/O kernels still presents several problems, including (i) They are seldom kept up-to-date with changes to the target application; (ii) They often retain the same cumbersome build mechanism of the target application; (iii) Different I/O kernels do not measure performance the same way; (iv) An I/O kernel is not necessarily available for every application of interest.

To address the weaknesses of current approaches, we have investigated a new approach to I/O performance measurement using *I/O skeletal applications*. These are generated codes that, like I/O kernels, include the same set of I/O operations used by an application while omitting computation and communication. In contrast to I/O kernels, I/O skeletal applications are built from the information included in a high-level I/O descriptor, and do not directly duplicate any of the application source code. Our skeletal applications share the advantages of I/O kernels, providing an accurate and concise benchmark for I/O performance. Furthermore, this approach directly addresses the four problems of I/O kernels identified above: (i) Since skeletal applications are automatically generated, they are trivial to reconstruct after a change to the application; (ii) All skeletal applications share the same relatively simple build mechanism; (iii) All skeletal applications share the same flexible measurement techniques, making it easier to perform apples to apples comparisons; (iv) Given its I/O descriptor, it is a simple process to generate an I/O skeletal application for any application.

To implement our I/O skeletal approach, we have created *Skel*. Skel generates skeletal applications based on the high-level I/O descriptors used in the Adaptive IO System (ADIOS) [4]. We have extended ADIOS to include a mechanism for timing the low-level operations performed during calls to the ADIOS API functions. Integration with this augmented ADIOS library allows more detailed measurements than are typically available from I/O kernels, enabling a user to determine, for instance, how much of the total I/O time is spent performing interprocess communication or low-level I/O operations.

In the next Section, we provide an overview of the Skel tool that implements our generative approach to creating I/O skeletal applications. In Section 3, we present some brief background of declarative I/O and ADIOS. Section 4 examines the validity of the generated skeletal applications. In Section 5, we describe the use of Skel to perform a detailed analysis of three I/O methods. We describe the related work in Section 6. Finally, Section 7 concludes the paper and presents an overview of future work.

2   The Skel System

We have implemented a tool, *Skel* [5], which supports the creation and execution of I/O skeletal applications. Skel consists of a set of python modules, each of which performs a specific task. The *skel-xml* module produces the adios-config file to be used by the skeletal application. The *skel-parameters* module creates a default parameter file based on the configuration file, which is then customized by the user. Modules *adios-config* and *skel-config* interpret the inputs, the adios-config file, which already exists for users of the ADIOS XML-based API, and the parameter file, which provides all other information needed to create a skeletal application. Based solely on data from those two input files, the *skel-source* module generates the source code, in either C or Fortran, that comprises the skeletal application. The *skel-makefile* module generates a makefile for compiling and deploying the skeletal application based on a platform specific template. Finally, the *skel-submit* module generates a submission script, also based on a template, suitable for executing the skeletal application on the target platform. Due to space limitations, we refer the reader to [5] for a more detailed discussion of the design of the Skel system.

In order to obtain more detailed timings of the ADIOS I/O methods, we have extended ADIOS with a mechanism for providing such timing information. The optional, low-level timing information is collected on each application core. We have also instrumented the I/O methods of interest with customized timing instructions using our timing mechanism. The timing functionality, though simple, is designed to be extensible, so that authors of new I/O methods can instrument their codes and extend the timer set to include arbitrary timing targets.

3   Background

Our implementation of Skel depends on the use of the declarative I/O mechanism that is present in the Adaptive IO System (ADIOS) [4]. ADIOS is a componentized I/O library designed to achieve excellent I/O performance on leadership class supercomputers. It provides a mechanism for externally describing an application's I/O requirements using an XML file that is termed the *adios-config* file. The mechanism provides a clear description of the structure of the application's I/O that is separate from the application source code.

In addition to describing the application's I/O pattern, the adios-config file allows the user to toggle between the various I/O methods[1] that are offered by ADIOS. This is particularly convenient because it allows different I/O mechanisms to be substituted without the need to change or recompile the application's source code. Such flexibility is a major strength of ADIOS since it facilitates the optimization of an application's I/O performance.

3.1 I/O Methods

In this paper we focus on three of the synchronous write methods offered by ADIOS, namely POSIX, MPI_LUSTRE and MPI_AMR [4]. We chose these methods both to show the general applicability of the Skel tool to different I/O methods, and to explore the performance of these methods over a range of scales and applications. Each of the methods takes a different approach to handling the application's I/O. In particular, the methods each produce a different number of files as shown in Fig. 1.

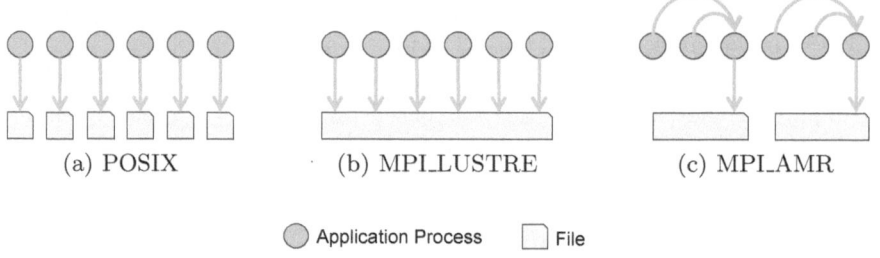

(a) POSIX   (b) MPI_LUSTRE   (c) MPI_AMR

⬤ Application Process   ☐ File

Fig. 1. Comparison of the three I/O methods

The POSIX method [6] was designed to take advantage of the concurrency of parallel file systems. With POSIX, each writing process is responsible for writing data to its own output file, as illustrated in Fig. 1a. An index file is then written by just one of the processes. So the use of the POSIX method results in the index file, along with a subdirectory containing all of the files written individually by the application processes.

The MPI_LUSTRE method, in contrast to the POSIX method, writes all data to a single file. Each process writes its data independently to an assigned location in the file as shown in Fig. 1b. This method is designed to work well with the striping scheme of the Lustre file system, thus the locations assigned to the writers are chosen to coincide with the beginning of a Lustre stripe, and the file's stripe size is carefully chosen so that each stripe accommodates the data from a single application process.

---

[1] In this paper the term *method* should be taken to mean a runtime selectable technique for performing the basic ADIOS operations (adios_open, adios_write, etc.).

Finally, the MPI_AMR method is a more sophisticated technique in which each of a subset of application processes acts as an aggregator for a group of peers. This method results in a collection of separate files, with one file corresponding to each of the aggregators. As can be seen in Fig. 1c, this results in fewer separate files than are produced by the POSIX method, but more than the single file written by MPI_LUSTRE. This strategy, in our experience, often achieves better throughput when scaled to large (>10,000) core counts.

These three methods all produce ADIOS-BP formatted files. ADIOS-BP is a structured, self-describing and resilient format for scientific data. The ADIOS-BP format was designed to provide efficient writing and reading [7,8] by allowing parallel I/O operations to be performed independently and supporting flexibility in the layout of data on the file system.

4   Validation of Skel

To understand whether an automatically generated skeletal application is an accurate predictor of application I/O performance, we compare performance measured with the skeletal apps with performance measured by applications and I/O kernels. We focus on two applications: CHIMERA [9] and GTS [10]. For both of these we compare the observed performance of our I/O skeletal application with that of the corresponding I/O kernel or application.

4.1   Test Platforms

Jaguar is a Cray supercomputer located at the Oak Ridge Leadership Computing Facility (OLCF) at Oak Ridge National Laboratory. Our experiments coincided with an upgrade of the Jaguar machine. The original configuration (Jaguar-XT5) consisted of 18688 compute nodes, each home to two Opteron 2435 "Istanbul" six-core processors and 16 GB of memory, and connected with Cray's Seastar interconnection network. The second configuration of Jaguar (Jaguar-XK6) had 9216 compute nodes, each containing a single 16 core AMD "Interlagos" processor and 32 GB of memory, and connected with the Gemini interconnect [11].

Sith is a 1280 core Linux cluster also located at the Oak Ridge National Laboratory. Sith consists of forty compute nodes connected with an InfiniBand interconnection network. Each node contains four 2.3 GHz 8 core AMD Opteron processors and 64 GB of memory [12].

All tests utilize Spider [13], the OLCF center-wide Lustre file system, which provides a total of 10.7 Petabytes of disk space. Spider consists of three separate Lustre file systems. Our experiments were performed using the widow1 file system, which offers 5 PB of storage space distributed over 672 OSTs.

4.2   CHIMERA

CHIMERA is a multi-dimensional radiation hydrodynamics code designed to study core-collapse supernovae [9]. An I/O kernel based on the CHIMERA application is the focus of this set of experiments. The comparison involved a weak scaling experiment where approximately 10 MB of data was written by each core. Each test was repeated 50 times, and the results are presented in Fig. 2. Each graph shows the average observed throughput for all of the runs that used that I/O method. The error bars represent the minimum and maximum throughput that were observed during these runs.

Parameters may be used to guide some of the ADIOS methods. The POSIX method does not take any parameters, but both MPI_LUSTRE and MPI_AMR allow method parameters to be specified. For the MPI_LUSTRE tests, we used `stripe_count=160`, `stripe_size=10485760`, `block_size=10485760`. The parameters for the MPI_AMR tests were `stripe_count=1`, `stripe_size=4194304`, `block_size=4194304`, `num_aggregators=N`, `merging_pgs=0`. We varied the number of aggregators depending on the number of cores, keeping the aggregator count fixed at one aggregator for every 4 cores.

The results of these tests are shown in Fig. 2. The graphs show the mean throughput seen during the tests, with the error bars indicating the minimum and maximum throughput observed. Also shown are the computed relative error values for the skeletal application with respect to both maximum and average throughput.[2] The relative error of the skeletal application as compared with the CHIMERA I/O kernel is less than 10% for over 93% of the cases shown.

4.3   GTS

The GTS application [10] is a first principles fusion microturbulence code that studies turbulent transport of energy. GTS uses a generalized geometry to solve the realistic geometries from many fusion reactors used today. For these tests, we looked at weak scaling performance, with each core writing approximately 55 MB of data. The method parameters used for the MPI_LUSTRE tests were `stripe_count=160`, `stripe_size=57344000`, `block_size=57344000`. For the MPI_AMR tests, we used parameters `stripe_count=1`, `stripe_size=4194304`, `block_size=4194304`, `num_aggregators=N`, `merging_pgs=0`. The number of aggregators for the MPI_AMR method varied, with one aggregator used for every 16 processor cores.

Each individual test was repeated 25 times and the results are shown in Fig. 3. Again, we have calculated the relative error for both the average and maximum throughput values. For these tests we observe that the skeletal application yields results that are within 10% of the values given by the GTS application for approximately 71% of the cases.

These results are extremely positive, particularly when we consider that skel will be most useful at larger scales with larger core counts. All cases involving 1024 or more cores, including both CHIMERA and GTS, are accurate to within 10% of the corresponding application or I/O kernel measurements.

---

[2] We have omitted a comparison of the minimum throughput as we have found that poor performance due to I/O contention with other jobs can lead to arbitrarily poor performance, making this measurement difficult to reproduce, even among separate runs of the same test.

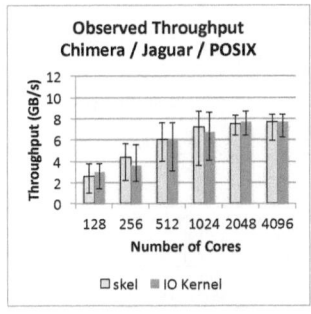
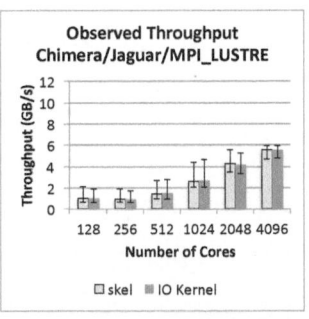
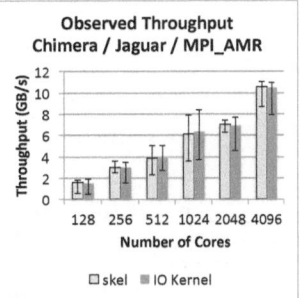

CHIMERA on Jaguar, maximum throughput % error						
Cores	128	256	512	1024	2048	4096
POSIX	1.23299306	1.16572625	-0.7565853	0.91980066	-4.3526464	0.26457144
MPI_LUSTRE	6.35053155	7.18255015	-4.5819049	-7.0988589	4.96703107	-0.9504542
MPI_AMR	-1.1075423	1.60319245	0.81126651	-6.1477011	-2.7247985	0.22539071

CHIMERA on Jaguar, average throughput % error						
Cores	128	256	512	1024	2048	4096
POSIX	-15.586524	22.9848149	1.76922079	6.63736976	-2.5146621	-0.3474594
MPI_LUSTRE	-1.1188266	-2.6544645	-9.2216744	-2.8223058	4.53511945	0.13061827
MPI_AMR	6.63407572	0.32418106	-1.2447502	-4.4013228	0.13788721	0.97436759

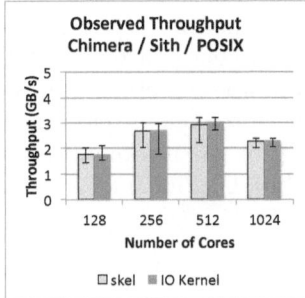
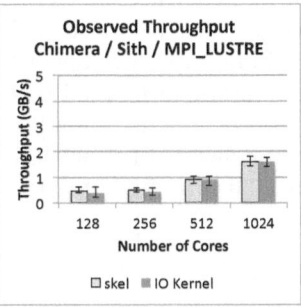
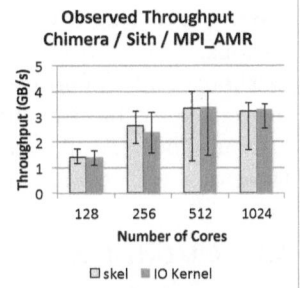

CHIMERA on Sith, maximum throughput % error					CHIMERA on Sith, average throughput % error				
Cores	128	256	512	1024	Cores	128	256	512	1024
POSIX	-4.3197237	0.88416525	-0.0444967	0.0639165	POSIX	0.20222178	-1.0747786	-3.9282669	-1.0447177
MPI_LUSTRE	-1.0537163	-0.1010961	-0.272298	2.86825714	MPI_LUSTRE	25.8798844	19.0632635	0.10644743	0.67528339
MPI_AMR	4.60492215	1.06613332	0.2709434	1.02694457	MPI_AMR	2.00347839	9.93222729	-1.0470642	-2.0402408

Fig. 2. Results of CHIMERA comparison

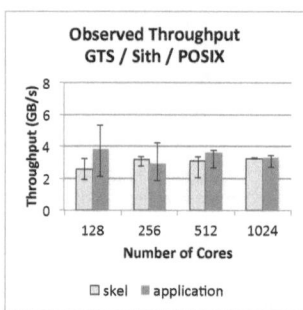 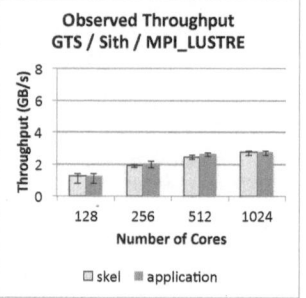 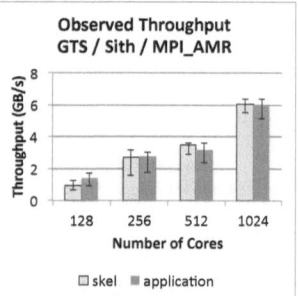

GTS on Sith, maximum throughput % error			
Cores	128	256	512	1024
POSIX	-38.483027	-19.700923	-10.558645	-3.7543821
MPI_LUSTRE	-0.2314274	-9.1780889	-5.564479	-0.8277953
MPI_AMR	-26.351491	3.7610783	0.20705444	-0.6193706

GTS on Sith, average throughput % error			
Cores	128	256	512	1024
POSIX	-31.91049	8.61128093	-13.370227	-2.184351
MPI_LUSTRE	1.36451053	-5.587962	-6.9181146	-0.8563423
MPI_AMR	-32.94111	-3.6202477	9.61607405	1.06179842

Fig. 3. Results of GTS comparison

5   Using Skel to Study I/O Performance

A common challenge in high-performance computing is determining which I/O method to use for a given situation. In general, the answer depends on a great many factors including application I/O pattern, frequency of I/O, storage hardware, file system type and configuration, network performance and system utilization to name a few. An intended use of Skel is to provide a mechanism for rapidly exploring I/O space in order to select an I/O method for arbitrary situations. This is useful to middleware developers for verifying and testing I/O methods, and to end users to assist in manual selection of I/O methods. In the future the mechanism should prove useful in guiding the autonomous selection of I/O methods. To illustrate this use of Skel, we have performed a comparison of three of the I/O methods available in ADIOS: POSIX, MPI_LUSTRE and MPI_AMR.

5.1   CHIMERA

The data from the CHIMERA skeletal application runs is shown in Fig. 4a, including two additional data points that illustrate the results of continuing these tests on the newly upgraded Jaguar-XK6. It can be seen that the POSIX method provides the highest throughput for most of the runs, with the MPI_AMR method offering better performance only for the 4096 core case. This agrees with our experience with these methods, as MPI_AMR has been observed to achieve better performance than POSIX at higher core counts. The number of cores at which it becomes advantageous to use the MPI_AMR method varies by application and platform, and also relies on the parameters used for MPI_AMR.

The CHIMERA I/O skeletal application provided us with a first glimpse at the I/O performance of Jaguar's XK6 configuration. The 8192 core and 16384 core cases were run immediately after the Jaguar-XK6 configuration became available. It appears that there may be a slight decrease in I/O performance for the POSIX and MPI_AMR methods, but the MPI_LUSTRE method exhibited truly poor performance on the new hardware. We will return to this issue in the next section.

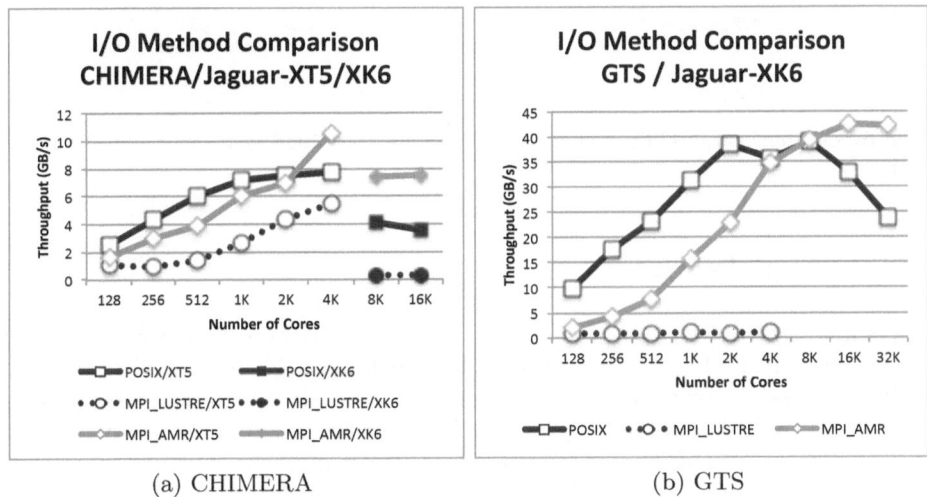

(a) CHIMERA   (b) GTS

Fig. 4. Comparison of I/O methods on Jaguar

5.2 GTS

Next we compared these same three I/O methods using a skeletal application for GTS. Each run was repeated 25 times, and the results are summarized in Fig. 4b. Again we see that the POSIX method provides the best performance among these three methods up to 8192 cores.[3] Beyond 8192 cores, however, the MPI_AMR method provides greater throughput than the other methods.

A surprising result is the particularly poor performance of the MPI_LUSTRE method. To investigate this anomaly, we examined the more detailed timing results available from Skel. We chose a single representative run of the 1024 core GTS / MPI_LUSTRE experiment, and looked at the performance for each of the cores, focusing on communication and I/O timings, and sorting by total time. The result is shown in Fig. 5. It can be seen that in both cases the times for the I/O operations are roughly equal for all of the overall slowest cores. However the communication times show dramatically different behavior. On Jaguar-XK6, a few of the cores exhibit very large communication times, with the slowest taking nearly 55 seconds. This view led to a quick diagnosis of the cause of the poor performance seen with the MPI_LUSTRE method, an expensive global collective communication operation that is not used by the other two methods.

---

[3] These results are not directly comparable to those found in [5], since those tests were performed on Jaguar-XT5, and used a different file system partition.

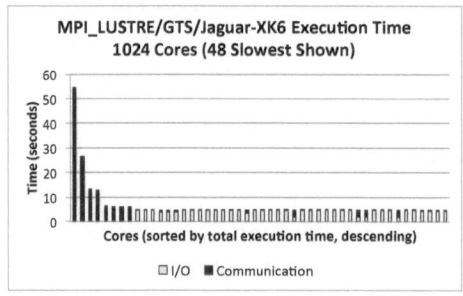 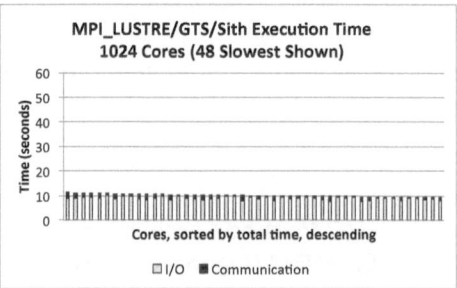

Fig. 5. Comparison of GTS execution times using MPI_LUSTRE on Jaguar-XK6 and Sith

6   Related Work

A common I/O performance measurement method is bulk I/O testing, using a tool such as IOR [14] or the NAS parallel benchmark [15,16]. As with Skel, the bulk I/O testing process is less cumbersome, since it is not necessary to deal with the complexities of an application, however the results may not provide an accurate prediction of the I/O performance that an application would obtain [17].

There are many I/O kernels that are used to benchmark I/O performance. FLASH-IO [2], MADBench2 [18] and S3D-IO are three often-used examples. Our skeletal applications are intended to provide the utility of I/O kernels, while eliminating issues such as difficulty of use, lack of availability, and outdated versions of codes.

The Darshan project [19] is examining the I/O patterns used by applications of interest. Darshan provides a lightweight library for gathering runtime event traces for later examination. The ScalaIOTrace tool [20] also addresses the measurement and analysis of I/O performance. Similar to Darshan, it works by capturing a trace of I/O activities performed by a running application. The multilevel traces may then be analyzed offline at various levels of detail.

7   Conclusion and Future Work

In this paper, we have presented our approach for the automatic generation of I/O skeletal benchmarks, as well as our tool, Skel, which implements this approach. We have examined one application and one I/O kernel and confirmed that the measurements obtained from the I/O skeletal application provide a reasonable estimate of performance. We observed a slightly better correspondence with the CHIMERA I/O kernel than with the GTS Application. In both cases, the performance predictions produced by the skeletal application improved with larger numbers of application processes.

We have conducted a performance comparison of three of the ADIOS write methods using our I/O skeletal technique. We have shown how Skel can be used to quickly measure the aggregate performance achieved by these methods. Furthermore, we have shown how the more detailed measurements provided by Skel can be leveraged to achieve a deeper understanding of the causes for the observed performance behavior.

In this paper we have focused on writing single restart files using synchronous I/O methods. In the future we will extend Skel to explore the task of writing smaller and more frequent analysis data. We will also investigate how to accurately measure the impact of asynchronous I/O. We expect that this will require generating some computation and communication operations to provide the same effect as those performed by the application. Finally, we will use Skel to investigate reading performance for very large data files.

Acknowledgment. This work was supported in part by the National Center for Computational Sciences (NCCS) at Oak Ridge National Laboratory. Support was provided by the Director, Office of Science, Office of Advanced Scientific Computing Research, of the U.S. Department of Energy under Contract No. DEAC02- 05CH11231. Additional support was provided by the SciDAC Fusion Simulation Prototype (FSP) Center for Plasma Edge Simulation (CPES) via Department of Energy Grant No. DE-FG02-06ER54857. This material is based upon work supported by the National Science Foundation under Grant No. 1003228. Support was also provided by the Remote Data Analysis and Visualization Center (RDAV) through National Science Foundation Grant No. 0906324.

References

1. Lang, S., Carns, P., Latham, R., Ross, R., Harms, K., Allcock, W.: I/O performance challenges at leadership scale. In: Proceedings of the Conference on High Performance Computing Networking, Storage and Analysis, SC 2009, pp. 40:1–40:12. ACM, New York (2009)
2. FLASH I/O benchmark routine – parallel HDF5, http://www.ucolick.org/~zingale/flash_benchmark_io/
3. Chen, J., Choudhary, A., De Supinski, B., DeVries, M., Hawkes, E., Klasky, S., Liao, W., Ma, K., Mellor-Crummey, J., Podhorszki, N., et al.: Terascale direct numerical simulations of turbulent combustion using S3D. Computational Science & Discovery 2, 015001 (2009)
4. ADIOS 1.3 user's manual, http://users.nccs.gov/~pnorbert/ADIOS-UsersManual-1.3.pdf
5. Logan, J., Klasky, S., Lofstead, J., Abbasi, H., Ethier, S., Grout, R., Ku, S.H., Liu, Q., Ma, X., Parashar, M., Podhorszki, N., Schwan, K., Wolf, M.: Skel: generative software for producing skeletal I/O applications. In: The Proceedings of D³science (2011)
6. Lofstead, J., Zheng, F., Klasky, S., Schwan, K.: Adaptable, metadata rich IO methods for portable high performance IO. In: IEEE International Symposium on Parallel Distributed Processing, IPDPS 2009, pp. 1–10 (May 2009)
7. Tian, Y., Klasky, S., Abbasi, H., Lofstead, J., Grout, R., Podhorski, N., Liu, Q., Wang, Y., Yu, W.: EDO: improving read performance for scientific applications through elastic data organization. In: Proceedings of IEEE Cluster (2011)

8. Lofstead, J., Polte, M., Gibson, G., Klasky, S., Schwan, K., Oldfield, R., Wolf, M., Liu, Q.: Six degrees of scientific data: reading patterns for extreme scale science IO. In: Proceedings of the 20th International Symposium on High Performance Distributed Computing, HPDC 2011, pp. 49–60. ACM, New York (2011)

9. Messer, O.E.B., Bruenn, S.W., Blondin, J.M., Hix, W.R., Mezzacappa, A., Dirk, C.J.: Petascale supernova simulation with CHIMERA. Journal of Physics: Conference Series 78(1), 012049 (2007)

10. Wang, W.X., Lin, Z., Tang, W.M., Lee, W.W., Ethier, S., Lewandowski, J.L.V., Rewoldt, G., Hahm, T.S., Manickam, J.: Gyro-kinetic simulation of global turbulent transport properties in tokamak experiments. Physics of Plasmas 13(9), 092505 (2006)

11. Titan configuration and timeline, http://www.olcf.ornl.gov/titan/system-configuration-timeline/

12. OLCF computing resources: Sith, http://www.olcf.ornl.gov/computing-resources/sith/

13. Shipman, G., Dillow, D., Oral, S., Wang, F.: The spider center wide file system: From concept to reality. In: Proceedings, Cray User Group (CUG) Conference, Atlanta, GA (2009)

14. IOR HPC Benchmark, http://sourceforge.net/projects/ior-sio/

15. Bailey, D., Harris, T., Saphir, W., Van Der Wijngaart, R., Woo, A., Yarrow, M.: The NAS parallel benchmarks 2.0. Technical Report NAS-95-020, NASA Ames Research Center, Tech. rep. (1995)

16. Bailey, D., Barszcz, E., Barton, J., Browning, D., Carter, R., Dagum, L., Fatoohi, R., Frederickson, P., Lasinski, T., Schreiber, R., et al.: The NAS parallel benchmarks summary and preliminary results. In: Proceedings of the 1991 ACM/IEEE Conference on Supercomputing, pp. 158–165. IEEE (1991)

17. Shan, H., Shalf, J.: Using IOR to analyze the I/O performance for HPC platforms. In: Cray Users Group Meeting (CUG) 2007, Seattle, Washington (May 2007)

18. MADbench2, http://crd.lbl.gov/~borrill/MADbench2/

19. Darshan, petascale I/O characterization tool, http://www.mcs.anl.gov/research/projects/darshan/

20. Vijayakumar, K., Mueller, F., Ma, X., Roth, P.C.: Scalable I/O tracing and analysis. In: Proceedings of the 4th Annual Workshop on Petascale Data Storage, PDSW 2009, pp. 26–31. ACM, New York (2009)

ASK: Adaptive Sampling Kit for Performance Characterization

Pablo de Oliveira Castro[1], Eric Petit[2],
Jean Christophe Beyler[3], and William Jalby[1]

[1] Exascale Computing Research,
University of Versailles - UVSQ, France
{pablo.oliveira,william.jalby}@exascale-computing.eu
[2] LRC ITACA, University of Versailles - UVSQ, France
eric.petit@uvsq.fr
[3] Intel Corporation
jean.christophe.beyler@intel.com

Abstract. Characterizing performance is essential to optimize programs and architectures. The open source Adaptive Sampling Kit (ASK) measures the performance trade-offs in large design spaces. Exhaustively sampling all points is computationally intractable. Therefore, ASK concentrates exploration in the most irregular regions of the design space through multiple adaptive sampling methods. The paper presents the ASK architecture and a set of adaptive sampling strategies, including a new approach: Hierarchical Variance Sampling. ASK's usage is demonstrated on two performance characterization problems: memory stride accesses and stencil codes. ASK builds precise models of performance with a small number of measures. It considerably reduces the cost of performance exploration. For instance, the stencil code design space, which has more than 31.10^8 points, is accurately predicted using only $1\,500$ points.

1 Introduction

Understanding architecture behavior is crucial to fine tune applications and develop more efficient hardware. An accurate performance model captures all interactions among the system's elements such as: multiple cores with an out-of-order dispatch or complex memory hierarchies. Building analytical models is increasingly difficult with the complexity growth of current architectures.

An alternative approach considers the architecture as a black box and empirically measures its performance response. The Adaptive Sampling Kit (ASK) gathers many state of the art sampling methods in a common framework simplifying the process. From the samples, engineers build a surrogate performance model to study, predict, and improve architecture and application performance on the design space. The downside of the approach is the exploration time needed to sample the design space. As the number of factors considered grows – cache levels, problem size, number of threads, thread mappings, and access patterns – the size of the design space explodes and exhaustively sampling each combination of factors becomes unfeasible.

C. Kaklamanis et al. (Eds.): Euro-Par 2012, LNCS 7484, pp. 89–101, 2012.

To mitigate the problem, the engineer must sample only a limited number of combinations. Moreover, they should be chosen with care: clustering the sampled points in a small portion of the design space biases the performance model. The two fundamental elements of a sampling pipeline are the sampling method and surrogate model.

1. The sampling method decides what combinations of the design space should be explored.
2. The surrogate model extrapolates from the sampled combinations a prediction on the full design space.

Choosing an adequate sampling strategy is not simple: for best results one must carefully consider the interaction between the sampling method and surrogate model [1]. Many implementations of sampling methods are available, but they all use different configurations and interfaces. Therefore, building and refining sampling strategies is difficult. ASK addresses this problem by gathering many state of the art sampling strategies in a common framework. Designed around a modular architecture, ASK facilitates building complex sampling pipelines. ASK also provides reporting and model validation modules to assess the quality of the sampling and find the best experimental setup for performance characterization. The paper's main contributions are:

- ASK, a common toolbox gathering state of the art sampling strategies and simple to integrate with existing measurement tools
- A new sampling strategy, Hierarchical Variance Sampling (HVS), which mitigates sampling bias by using confidence bounds
- An evaluation of the framework, and of HVS, on two representative performance characterization experiments

Section 2 discusses related works. Section 3 explains the HVS strategy. Section 4 succinctly presents ASK's architecture and usage. Finally, Section 5 evaluates ASK on two performance studies: memory stride accesses and 2D stencils.

2 Related Works

There are two kinds of sampling strategies: space filling designs and adaptive sampling. Space filling designs select a fixed number of samples with sensible statistical properties such as uniformly covering the space or avoiding clusters. For instance, Latin Hyper Cube designs [2] are built by dividing each dimension into equal sized intervals. Points are selected so the projection of the design on any dimension contains exactly one sample per interval. Maximin designs [3] maximize the minimum distance between any pair of samples; therefore spreading the samples over the entire experimental space. Finally, low discrepancy sequences [4] choose samples with low discrepancy: given an arbitrary region of the design space, the number of samples inside the region is close to proportional to its measure. By construction, the sequences uniformly distribute points in space. Space filling designs choose all points in one single draw before starting the experiment.

Adaptive sampling methods, on the contrary, iteratively adjust the sampling grid to the complexity of the design space. By observing already measured samples, they identify the most irregular regions of the design space. Further samples are drawn in priority from the irregular regions, which are harder to explore.

The definition of irregular regions changes depending on the sampling method. Variance-reduction methods prioritize exploration of regions with high variance. The rationale is irregular regions are more complex, thereby requiring more measures. Query-by-Committee methods build a committee of models trained with different parameters and compare the committee's predictions on all the candidate samples. Selected samples are the ones where the committee's models disagree the most. Adaptive Multiple Additive Regression Trees (AMART) [5] is a recent Query-by-Committee approach based on Generalized Boosted Models (GBM) [6], it selects non-clustered samples with maximal disagreement. Another recent approach by Gramacy et al. [7] combines the Tree Gaussian Process (TGP) [8] model with adaptive sampling methods [9]. For an extensive review of adaptive sampling methods please refer to Settles [10].

The Surrogate Modeling Toolbox (SUMO) [11] offers a Matlab toolbox building surrogate models for computer experiments. SUMO's execution flow is similar to ASK's: both allow configuring the model and sampling method to fully automate an experiment plan. SUMO focuses mainly on building and controlling surrogate models, offering a large set of models. It contains algorithms for optimizing model parameters, validating the models, and helping users choose a model. On the other hand, most of SUMO's adaptive methods are basic sequential sampling methods. Only a single recent approach is included, which finds trade-offs between uniformly exploring the space and concentrating on nonlinear regions of the space [12]. SUMO is open source but restricted to academic use and depends on the proprietary Matlab toolbox.

ASK specifically targets adaptive sampling for performance characterization, unlike SUMO. It includes recent state of the art approaches that were successfully applied to computer experiments [7] and performance characterization [5]. Simpson et al. [1] show one must consider different trade-offs when choosing a sampling method: affinity with the surrogate model or studied response, accuracy, or cost of predicting new samples. Therefore, ASK comes with a large set of approaches to cover different sampling scenarios including Latin Hyper Cube designs, Maximin designs, Low discrepancy designs, AMART, and TGP. Additionally, ASK includes a new approach, Hierarchical Variance Sampling (HVS).

3   Hierarchical Variance Sampling

Many adaptive learning methods are susceptible to bias because the sampler makes incorrect decisions based on an incomplete view of the design space. For instance, the sampler may ignore a region despite the fact it contains big variations because previous samplings missed the variations.

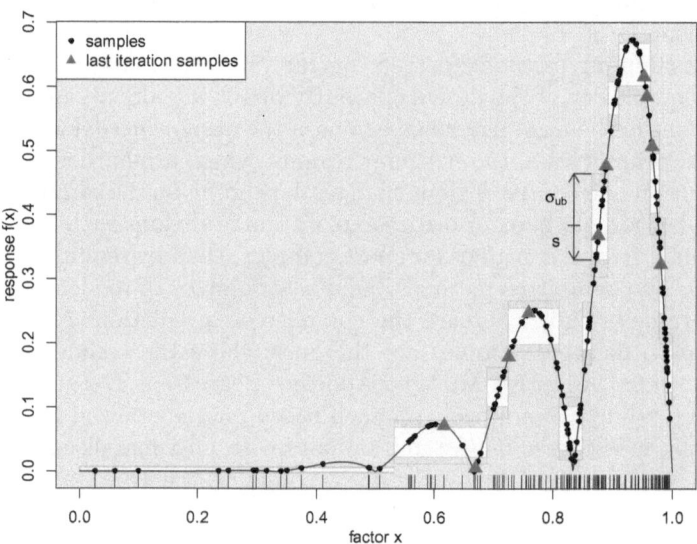

Fig. 1. HVS on a synthetic 1D benchmark after fifteen drawings of ten samples each. The true response, $f(x) = x^5|sin(6.\pi.x)|$, is the solid line. CART partitions the factor dimension into intervals, represented by the boxes horizontal extension. For each interval, the estimated standard deviation, s, is in a light color and the upper bound of the standard deviation, σ_{ub}, is dark. HVS selects more samples in the irregular regions.

To mitigate the problem, ASK includes the new Hierarchical Variance Sampling. HVS' principal concept is to reduce the bias using confidence intervals that correct the variance estimation. HVS partitions the exploration space into regions and measures the variance of each region. A statistical correction depending on the number of samples is applied to obtain an upper bound of the variance. Further samples are then selected proportionally to the upper bound and size of each region. By using a confidence upper bound on the variance, the sampler is less greedy in its exploration but is less likely to overlook interesting regions. In others words, the sampler will not completely ignore a region until the number of sampled points is enough to confidently decide the region has low variance.

HVS is similar to Dasgupta et al. [13] proposing a hierarchical approach for classification tasks using confidence bounds to reduce the sampling bias. Nevertheless, the Dasgupta et al. approach is only applicable to classification tasks with a binary or discrete response; whereas HVS deals with continuous responses, which are more appropriate for performance characterization.

To divide the design space into regions, HVS uses the Classification and Regression Trees (CART) partition algorithm [14] with the Analysis of Variance (ANOVA) splitting criteria [15] and prunes the tree to optimize cross validation error. At each step, the space is divided into two regions so the sum of the regions variance is smaller than the variance of the whole space. The result of a CART partitioning is shown in Figure 1 where each box depicts a region.

After partitioning, HVS samples the most problematic regions and ignores the ones with low variance. The sampler only knows the empiric variance s^2, which depends on previous sampling decisions; to reduce bias HVS derives an upper bound of the true variance σ^2. Assuming a close to normal region's distribution, HVS computes an upper bound of the true variance σ^2 satisfying $\sigma^2 < \frac{(n-1)s^2}{\chi^2_{1-\alpha/2,n-1}} = \sigma^2_{ub}$ with a $1-\alpha$ confidence[1]. To reduce the bias HVS uses the corrected upper bound accounting for the number of samples drawn.

For each region, Figure 1 plots the estimated standard deviation s, light colored, and upper-bound σ_{ub}, dark colored. Samples are selected proportionally to the variance upper bound multiplied by the size of the region, as shown in Figure 1. New samples, marked as triangles, are chosen inside the largest boxes. HVS selects few samples in the $[0, 0.5]$ region, which has a flat profile.

If the goal of the sampling is to reduce the absolute error of the model, then the HVS method is adequate because it concentrates on high-variance regions. On the other hand, if the goal is to reduce the relative, or percentage, error of the model it is better to concentrate on regions with high relative variance, $\frac{s^2}{\bar{x}^2}$. HVSrelative is an alternate version of HVS using relative variance with an appropriate confidence interval [16]. Section 5 evaluates the two sampling strategies, HVS and HVSrelative, in two performance studies.

4   ASK Architecture

ASK's flexibility and extensibility come from its modular architecture. When running an experiment, ASK follows the pipeline presented in Figure 2:

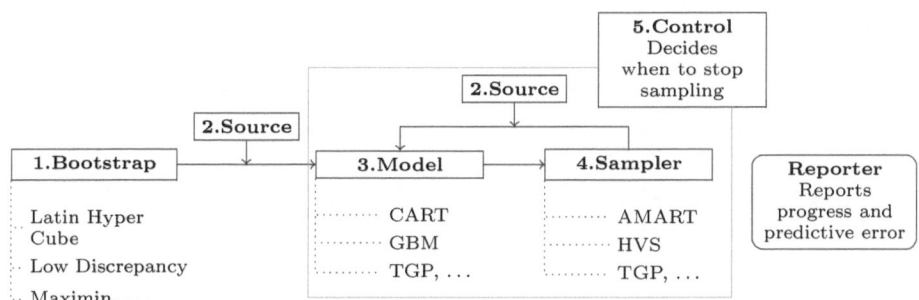

Fig. 2. ASK pipeline

---

1. A *bootstrap* module selects an initial batch of points. ASK provides standard bootstrap modules for the space filling designs described in Section 2: Latin Hyper Cube, Low Discrepancy, Maximin, and Random.
2. A *source* module, usually provided by the user, receives a list of requested points. The source module computes the actual measures for the requested factors and returns the response.
3. A *model* module builds a surrogate model for the experiment on the sampled points. Currently ASK provides CART [14], GBM [6,17], and TGP [7] models.
4. A *sampler* module iteratively selects a new set of points to measure. Some sampler modules are simple and do not depend on the surrogate model. For instance, the `random` sampler selects a random combination of factors and the `latin` sampler iteratively augments an initial Latin Hyper Cube design. Other sampler modules are more complex and base their decisions on the surrogate model.
5. A *control* module decides when the sampling process ends. ASK includes two basic strategies: stopping when it has sampled a predefined amount of points or stopping when the accuracy improvement stays under a given threshold for a number of iterations.

From the user perspective, setting up an ASK experiment is a three-step process. First, the range and type of each factor is described by writing an experiment configuration file in the JavaScript Object Notation (JSON) format. ASK accepts real, integer, or categorical factors. Then, users write a *source* wrapper around their measuring setup. The interface is straightforward: the wrapper receives a combination of factors to measure and returns their response. Finally, users choose which bootstrap, model, sampler, control, and reporter modules to execute. Module configuration is also done through the configuration file. ASK provides fallback default values if parameters are omitted from the configuration. An excerpt of a two factor configuration with the hierarchical sampler module follows:

```
"factors": [{"name": "image-size",
             "type": "integer",
             "range": {"min": 0, "max": 600}},
            {"name": "stencil-size",
             "type": "categorical",
             "values": ["small", "medium", "large"]}],
 "modules": {"sampler": {"executable": "sampler/HVS",
                         "params": {"nsamples":50}}}
```

Editing the configuration file quickly replaces any part of the ASK experiment pipeline with a different module. All the modules have clearly defined interfaces and are organized with strong separation of concerns in mind. It allows the user to quickly integrate custom made modules to the ASK pipeline. For example, by replacing `sampler/HVS` with `sampler/latin` the user replays the same experience with the same parameters but using a Latin Hyper Cube sampler instead.

5   Experimental Study

Two performance characterization experiments were conducted using ASK to achieve two different objectives. The first objective was to validate the ASK pipeline and understand on a low dimension space the behavior of each method using a synthetic microbenchmark called ai_aik. Ai_aik explores the impact of stride accesses to a same array in a single iteration. The design space is composed of 400 000 different combinations of two factors: loop trip N and k-stride. The design space is large and variable enough to challenge sampling strategies. Nonetheless, it is narrow enough to be measured exhaustively providing an exact baseline to rate the effectiveness of the sampling strategies.

The second objective was to validate the strategies on a large experimental space: 2D cross-shaped stencils of varying sizes on a parallel SMP. A wide range of scientific applications use stencils: for instance, Jacobi computation [18,19] uses 2×2 stencils and high-order finite-difference calculations [20] use 6×6 stencils. The design space is composed of five parameters: the $N \times M$ size of the matrix, the $X \times Y$ size of the stencil, and T the number of concurrent threads used. The design space size has more than 7.10^8 elements in a 8-core system and more than 31.10^8 elements in a 32-core system.

Since an exhaustive measure is computationally unfeasible, the prediction accuracy is evaluated by computing the error of each strategy on a test set of 25 600 points independently measured. The test set contains 12 800 points chosen randomly and 12 800 points distributed in a regular grid configuration over the design space. Measuring the test set takes more than twelve hours of computation on a 32-core Nehalem machine.

All studied sampling strategies use random seeds, which can slightly change the predictive error achieved by different ASK runs. Therefore, the median error, among nine different runs, is reported when comparing strategies.

Experiments ran with six of the sampling strategies included in ASK: AMART, HVS, HVSrelative, Latin Hyper Cube, TGP, and Random. All the benchmarks were compiled with ICC 12.0.0 version. The strategies were called with the following set of default parameters on both experiments:

Samples All the strategies sampled in batches of fifty points per iteration.
Bootstrapping All the strategies were bootstrapped with samples from the same Latin Hyper Cube design, except Random, which was bootstrapped with a batch of random points.
Surrogate Model The TGP strategy used tgpllm [8] model with its default parameters. The other strategies used GBM [6] with the following parameters: `ntrees=3 000, shrinkage=0.01, depth=8`.
AMART ran with a committee size of twenty as recommended by Li et al. [5].
TGP used the Active Learning-Cohn [9] sampling strategy.
HVS, HVSrelative used a confidence bound of $1 - \alpha = 0.9$.

Section 5.1 validates ASK on an exhaustive stride access experiment. Section 5.2 validates ASK on a large design space that cannot be explored exhaustively: multi-core performance of cross-shaped stencils.

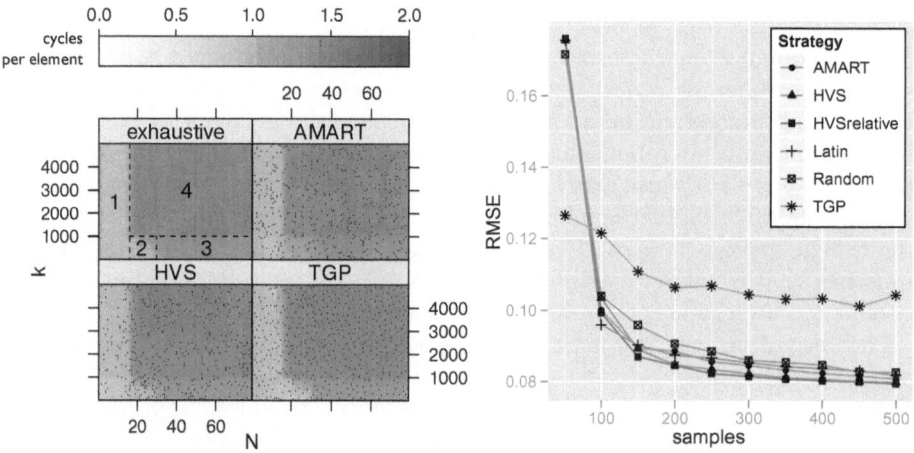

Fig. 3. Stride experiments: (*Left*) exhaustive level plot shows the true response of the studied kernel in cycles per element. AMART, HVS, and TGP respectively show the predicted response of each strategy. Black dots indicate the position of the sampled points. (*Right*) RMSE is plotted for each strategy and number of samples. The median among nine runs of each strategy was taken to remove random seed effects.

5.1   Stride Memory Effects

This section studies the stride memory accesses of the following ai_aik kernel:

```
for(i=0;i<N*256;i++) {
  res[i]=a[i]+a[i+2*k];
}
```

The baseline is an exhaustive measure of cycle per element performance on a Xeon L5609 quad-core 1.87GHz with 8GB of RAM. The measures revealed the four zones on the left side of Figure 3:

1. In Zone 1 the kernel is fastest because both i and $i + 2.k$ accesses fit in L1.
2. Zone 2 is a transition zone between Zone 1 and Zone 3: under the diagonal, the accessed elements still fit in L1.
3. In Zone 3 accesses do not fit fully in L1 anymore but the performance is still acceptable because the accesses $i + 2.k$ prefetch the data needed for the i accesses in future iterations.
4. In Zone 4, performance is the worst because accesses do not fully fit in L1 and the $2.k$ distance is too wide for efficient software prefetching.

The preceding exhaustive analysis required 400 000 measures of the ai_aik kernel. ASK ran the same experiment with different sampling strategies stopping at five hundred samples. Each method's accuracy was determined by comparing its predictions to the exhaustive baseline.

Comparing the predicted and exhaustive responses in Figure 3 shows that TGP and HVS capture the four zones in detail while AMART is less precise in

Zone 2. The diagonal effect in Zone 2 introduces high variance. Therefore, HVS concentrates its sampling, capturing the zone accurately.

The Root Mean Square Error (RMSE) is the standard metric in the literature to evaluate a model's accuracy. Figure 3, right side, shows the RMSE of each method. The methods, except TGP, are comparable in terms of convergence speed and reached accuracy. TGP scores a poor RMSE performance compared to the other methods. GBM surrogate model seems to be a better fit for this experiment.

In the experiment, ASK's five hundred point sampling successfully captures the performance features of the design space. ASK uses eight hundred times less samples than the exhaustive analysis, while preserving accuracy.

5.2   Stencil Characterization

The stencil code characterization is unfeasible with exhaustive exploration, but is possible using ASK's adaptive sampling methods. In the studied stencil code, Figure 4a, five factors are tunable – X and $Y \in \{1, 2, 4, 8, 16\}$ the horizontal and vertical sizes of the stencil, $N \in [64, 2048]$ the number of lines of the matrix, $M \in [64, 2048]$ the number of columns of the matrix, and $T \in [1, 32]$ the number of threads. The stencil was studied on two Nehalem architectures: an 8-core dual-socket Xeon E5620 at 2.40GHz with 24GB of RAM and a 32-core four-socket Xeon X7550 at 2.00GHz with 128GB of RAM. The OpenMP mapping policy was set to `Scatter`. The error was evaluated on an independent test set of 25 600 points.

TGP was not used during the second experiment because it does not handle categorical variables [8]. The computation time needed to select samples with HVS, HVSrelative, and Latin is negligible compared to the time required to measure a batch of samples. AMART is a Query-by-Committee strategy, which generates a prediction on all the candidate points, for twenty different models. In the stencil experiment the number of candidate points is in the order of billions, computing a prediction on all of them is not possible. Therefore, as suggested by Gramacy [8], ASK's AMART implementation reduces the number of candidate points to one thousand with a Latin Hyper Cube presampling.

The sampling strategies' accuracy is measured in terms of RMSE and mean relative error, in Figure 4b. Here only the 32-core results are examined because the sampling strategies' accuracy was similar for both the 8 and 32-core architectures. For RMSE, HVS outperforms all other strategies both in quality of the final model, 1.76 RMSE, and speed of convergence. For mean percentage error, AMART achieves the best final result, 8.89%, followed closely by Latin, Random, and HVSrelative. HVSrelative converges faster than the others.

Overall, HVSrelative is the best compromise between RMSE and percentage error because it achieves low final errors and converges quickly to an accurate model. Using only eight hundred samples, HVSrelative predicts the test set with a mean absolute error of 1.51 cycles and a mean relative error of 10.97%.

Figure 4c shows the performance prediction for HVS and HVSrelative on the $X \times 16$ stencils. Each square represents a unique (X, Y, T) configuration. Inside

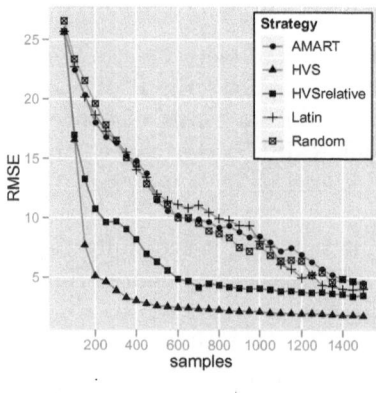

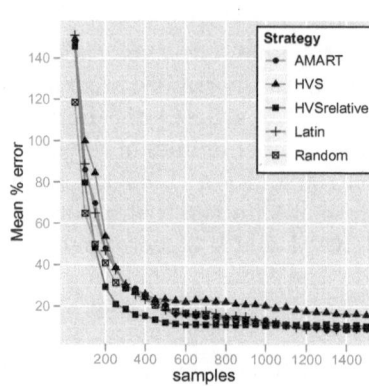

```
#pragma omp parallel for
for(i=Y; i<N-Y; i++)
  for(j=X; j<M-X; j++) {
    for(k=j-X; k<=j+X; k++)
      out[i][j] += in[i][k];
    for(l=i-Y; l<=i+Y; l++)
      out[i][j] += in[l][j];
}
```

(a) Stencil code evaluated

(b) Error curves for the exploration on 32 cores. The median among nine runs of each strategy was taken to remove random seed effects.

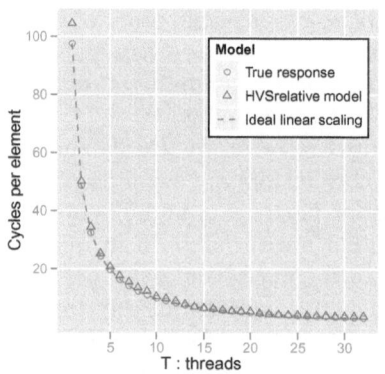

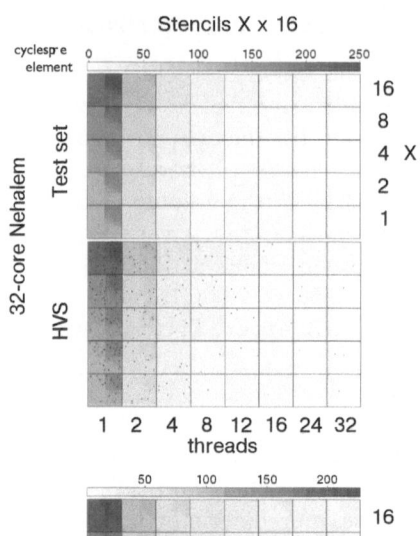

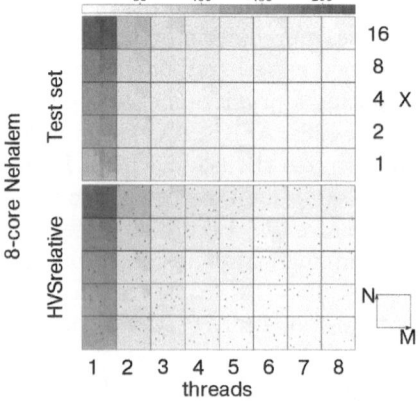

(d) Scalability for the 8×8 stencil on a $1\,000 \times 1\,000$ matrix.

(c) Stencil $X \times 16$ predicted performance vs. test set performance in cycles per element.

Fig. 4. Stencil experiments

each square the performance is plotted depending on the matrix size $N \times M$. The kernel is slowest for high Y stencils whose column order accesses stress the cache. In comparison, X stencil's size impact on performance is negligible. Performance degrades for large matrices, as shown from the darker top-right corners of each square, probably because the matrices exceed the L2 cache capacity. Therefore, choosing an adequate blocking factor should improve the performance. On ASK HVS' 32-core model the mean speed-up obtained by varying the matrix size is 1.65. Using a small matrix with a high number of threads is detrimental because the cost of synchronization predominates.

On both 8 and 32-core targets the stencil code scales linearly up to, respectively, 8 and 32 threads. As an example, the scalability of the application was studied on the 8×8 stencil on a $1\,000 \times 1\,000$ matrix. Figure 4d shows the performance per number of threads predicted by the HVSrelative strategy. The prediction follows the measured true response. The matrix sizes explored fit into the socket's 18Mb L3 cache, additionnaly the 2D stencil beneficiates from data reuse, which explains the strong scaling.

Measuring the whole design space would take centuries whereas ASK adaptive sampling took less than an hour of experiment time with a 1.76 RMSE. ASK is both efficient when dealing with narrow design spaces, as in ai_aik, and large design space, as in the stencil codes experiment.

6   Conclusion

Adaptive sampling techniques drastically reduce exploration time of large design spaces. Nevertheless, choosing the right technique can be difficult for a performance architect. ASK provides an homogeneous interface to multiple state of the art sampling strategies, making the process easier. Adding new strategies to the framework is straightforward due to ASK's modular architecture.

The new HVS strategy reduces experimental bias and is comparable, or even outperforms, other state of the art approaches in two case studies. The stencil code design space, which has more than 31.10^8 points, was accurately predicted using only $1\,500$ points. The performance characterization field could benefit from adaptive sampling techniques. Hopefully, ASK's open source release will facilitate their adoption.

Currently, users must try different models manually to find what is best suited to their experiment. Automatic model and sampling selection techniques [11,21] applied to performance experiments will be investigated in future works.

ASK will soon be released at `http://code.google.com/p/adaptive-sampling-kit`. The experimental data and benchmarks used to produce the paper results are available at the same URL.

Acknowledgments. This work has been conducted conjointly by the Exascale Computing Research lab, thanks to the support of CEA, GENCI, Intel, UVSQ, and by the LRC ITACA lab, thanks to the support of the French Ministry for Economy, Industry, and Employment throught the ITEA2 project H4H. Any opinions, findings, and conclusions or recommendations expressed in this material are those of the author(s) and do not necessarily reflect the views of the CEA, GENCI, Intel, or UVSQ.

References

1. Simpson, T., Lin, D., Chen, W.: Sampling strategies for computer experiments: design and analysis. International Journal of Reliability and Applications 2(3), 209–240 (2001)
2. Stein, M.: Large sample properties of simulations using Latin hypercube sampling. Technometrics, 143–151 (1987)
3. Johnson, M., Moore, L., Ylvisaker, D.: Minimax and maximin distance designs. Journal of Statistical Planning and Inference 26(2), 131–148 (1990)
4. Diwekar, U., Kalagnanam, J.: Efficient sampling technique for optimization under uncertainty. AIChE Journal 43(2), 440–447 (1997)
5. Li, B., Peng, L., Ramadass, B.: Accurate and efficient processor performance prediction via regression tree based modeling. Journal of Systems Architecture 55(10-12), 457–467 (2009)
6. Ridgeway, G.: Generalized Boosted Models: A guide to the gbm package. Update 1, 1 (2007)
7. Gramacy, R., Lee, H.: Adaptive design and analysis of supercomputer experiments. Technometrics 51(2), 130–145 (2009)
8. Gramacy, R.: tgp: An R package for Bayesian nonstationary, semiparametric nonlinear regression and design by treed gaussian process models. Journal of Statistical Software 19(9), 6 (2007)
9. Cohn, D.: Neural network exploration using optimal experiment design. Neural Networks 9(6), 1071–1083 (1996)
10. Settles, B.: Active Learning Literature Survey. Science 10(3), 237–304 (1995)
11. Gorissen, D., Couckuyt, I., Demeester, P., Dhaene, T., Crombecq, K.: A surrogate modeling and adaptive sampling toolbox for computer based design. The Journal of Machine Learning Research 11, 2051–2055 (2010)
12. Crombecq, K., Gorissen, D., Deschrijver, D., Dhaene, T.: A Novel Hybrid Sequential Design Strategy for Global Surrogate Modeling of Computer Experiments. SIAM Journal on Scientific Computing 33, 1948 (2011)
13. Dasgupta, S., Hsu, D.: Hierarchical sampling for active learning. In: Proceedings of the 25th International Conference on Machine learning, pp. 208–215. ACM (2008)
14. Breiman, L., Friedman, J., Olshen, R., Stone, C., Steinberg, D., Colla, P.: CART: Classification and regression trees. Wadsworth, Belmont (1983)
15. Atkinson, E., Therneau, T.: An introduction to recursive partitioning using the RPART routines. Mayo Foundation, Rochester (2000)
16. McKay, A.: Distribution of the Coefficient of Variation and the Extended t Distribution. Journal of the Royal Statistical Society 95(4), 695–698 (1932)
17. Friedman, J.: Greedy function approximation: a gradient boosting machine. Annals of Statistics, 1189–1232 (2001)
18. Datta, K., Murphy, M., Volkov, V., Williams, S., Carter, J., Oliker, L., Patterson, D., Shalf, J., Yelick, K.: Stencil computation optimization and auto-tuning on state-of-the-art multicore architectures. In: Proceedings of the 2008 ACM/IEEE Conference on Supercomputing, pp. 1–12. IEEE Press (2008)

19. Treibig, J., Wellein, G., Hager, G.: Efficient multicore-aware parallelization strategies for iterative stencil computations. J. Comput. Science 2(2), 130–137 (2011)
20. Dursun, H., Nomura, K.-i., Peng, L., Seymour, R., Wang, W., Kalia, R.K., Nakano, A., Vashishta, P.: A Multilevel Parallelization Framework for High-Order Stencil Computations. In: Sips, H., Epema, D., Lin, H.-X. (eds.) Euro-Par 2009. LNCS, vol. 5704, pp. 642–653. Springer, Heidelberg (2009)
21. Maron, O., Moore, A.: Hoeffding races: Accelerating model selection search for classification and function approximation. Robotics Institute, 263 (1993)

CRAW/P: A Workload Partition Method for the Efficient Parallel Simulation of Manycores

Shuai Jiao[1,2], Paolo Ienne[3], Xiaochun Ye[1], Da Wang[1],
Dongrui Fan[1], and Ninghui Sun[1]

[1] SKL Computer Architecture, ICT, CAS, Beijing, P.R. China
[2] Graduate University of Chinese Academy of Sciences, Beijing, P.R. China
[3] École Polytechnique Fédérale de Lausanne, Lausanne, Switzerland
{jiaoshuai,yexiaochun,wangda,fandr,snh}@ict.ac.cn,
Paolo.Ienne@epfl.ch

Abstract. This paper addresses the workload partition strategies in the simulation of manycore architectures. The key observation behind this paper is that, compared to traditional multicores, manycores feature more non-uniform memory access and unpredictable network traffic; these features degrades simulation speed and accuracy of *Parallel Discrete Event Simulators (PDES)* when one uses static workload partition schemes. Based on the observation, we propose an adaptive workload partition method: *Core/Router-Adaptive Workload Partition (CRAW/P)*. The method delivers more speedup and accuracy than static partition schemes by partitioning the simulation of on-chip-network independently from that of the cores and by synchronizing them differently. Using a PDES simulator, we evaluate the performance of CRAW/P in simulating a 256-core general purpose many-core processor. Running SPLASH2 benchmark applications, the experimental results demonstrate it can deliver speed improvement by 28%~67% over static partition scheme and reduces timing errors to <10% in very relaxed simulation (quantum size as 64).

Keywords: Parallel Simulation, Manycore, Multicore, Workload Partition.

1   Introduction

While the "manycore era" approaches, some manycore processors have already arrived [1, 2, 3, 4, 5]. Thousand-core processors are no longer infeasible, and it is likely that thousands of cores on a single die will become a commodity [6]. Simulating such parallel systems is a serious problem. Currently, the majority of simulators available are sequential [7, 8, 9] and thus run the simulation workload on a single host thread. When the number of cores increases in a target system, the simulation performance for each core goes down.

A variety of techniques have been proposed to accelerate simulation. These techniques include the following: parallel simulation [10, 11, 12, 13, 14, 15], direct execution [15, 16], and FPGA acceleration [17, 18, 19]. Among these techniques, parallel simulation speeds simulation by exploiting the parallelism inherent in the target parallel architecture. Besides, the advent of low-cost SMP computers makes

C. Kaklamanis et al. (Eds.): Euro-Par 2012, LNCS 7484, pp. 102–114, 2012.

parallel simulation very attractive. Representative works on parallel simulation are *Parallel Discrete Event Simulation (PDES)* simulators [21].

State-of-the-art PDES simulators focus on simulating multicores. These simulators partition the simulation workload in a simple static manner. Examples could be found in P-mambo [16], SlackSim [14], and Graphite [15]. In these simulators, target cores are evenly distributed among host threads. These schemes work well in simulating multicores but are inefficient in simulating manycores. The distinction lies in the architectural difference between multicores and manycores: manycores feature large scale on-chip-networks, which produce more non-uniform memory accessses (NUCA) and unpredictable network traffic.

Based on these observations, this paper proposes a partition method: core/router-adaptive workload partition (CRAW/P). The essential idea of CRAW/P is that it partitions the simulation workload adaptively, divides the simulation of on-chip-network (routers) separately from that of the cores and simulates network more strictly. The method delivers more speedup and accuracy than static partition schemes.

The main contributions of CRAW/P are 1) how to use adaptive partitions to deliver speedup and accuracy and 2) how to leverage the core/router partitioning to efficiently maintain accuracy in terms of (i) reduction of host threads that simulate the network and (ii) strict synchronization for the network. As far as we can tell, we are the first to comprehensively discuss the workload partition scheme in manycore simulation and exploit simulation speedup and accuracy saving from the division of network and cores in parallel simulating manycores.

The remainder of this paper is structured as follows. Section 2 discusses the architectural characteristics of manycores and demonstrates the observations. Section 3 discusses the details of CRAW/P. Section 4 illustrates the experimental results. Section 5 discusses some related work. Section 6 offers some concluding remarks.

2   Observation

This section explains why static workload partition scheme is far from an optimal choice for simulating manycores and demonstrates the observation that motivates the adaptive partitioning and the core/router partitioning. Some experimental examples are given using the experimental platform later described in Section 4.

Fig. 1 shows how static partition schemes are used to simulate typical multicores and manycores. The partition scheme in Fig. 1(a) represents the scheme used, for instance, by SlackSim [14] while the partition scheme in Fig. 1(b) represents what used in Graphite [15]. The static scheme may work well in these multicore simulators but it is inefficient for manycores. The key is the large-scale on-chip-network, which is widely used in modern manycore architectures to provide better scalability. A large on-chip-network enlarges the following phenomenon on manycores: non-uniform cache accesses and unpredictable network traffic.

Non-uniform cache access (NUCA) produces an obvious workload imbalance. On large scale networks, NUCA accesses might produce very non-uniform on-chip traffic due to topology and routing. The situation is worsened if hot spotting appears on the network. The non-uniform traffic results in some cores being busy running ahead due to short off-core latency, while other cores are stalling for reply. In a typical PDES simulator, simulating a busy core/router results in a large workload while simulating a

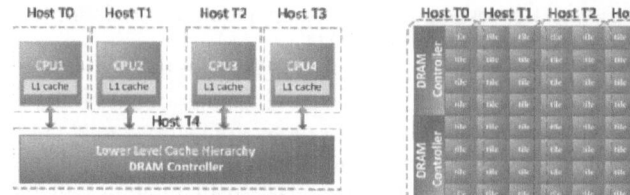

(a) A typical 4-core CMP Multi-core Processor   (b) A typical 64-core Many-core Processor

Fig. 1. A typical multicore SMP and manycore processor simulated by 5 host threads using static partition schemes. The Multi-core architecture is typical CMP architecture. Each core of a CMP has a L1 instruction and coherent data cache. The lower level cache hierarchy is made of L2 (or even L3) cache banks, which are accessed in a NUCA [21] manner. Banks can be private or shared to each core. The manycore architecture is extracted from state-of-the-art manycore processors (e.g., Intel SCC [1], Polaris [2], Tile64/Pro/Gtx [3, 4], and Godson-T [5]). These processors present common features such as distributed caches, mesh network, and DRAM controllers at the peripheral of the chip.

waiting core/router involves practically no activity. Fig. 2 shows the workload distribution when simulating a 256-core architecture running the *matrix_multiply* kernel. The arrows and circles indicate that on-chip cache-access hot spotting is the main reason of non-uniform workload distribution.

If the simulation workload is statically partitioned among host threads, the workload imbalance would slowdown simulation speed because host threads spend more time on synchronizing their local clocks. Besides, in cases of relaxed simulation (e.g., quantum method), clock skew is more likely to happen, resulting in more timing errors. Fig. 3 shows a example of static partition scheme which demonstrates the relation of simulation speed and timing error with workload imbalance. *MCPS (Million Cycles per Second)* is the measure of simulation speed: it indicates how many target cycles the simulator achieves during one wall-clock second. *ECPE (Error Cycles per Event)* is the measure of the dynamic timing error during simulation run time: it indicates the average cycle deviation of an event timestamp from the cycle it actually takes effect. Workload imbalance is measured by the *Standard Deviation (SD)* of the synchronization overhead for all host threads. As illustrated, simulation speed presents opposite variation against workload imbalance while timing error varies similarly with the imbalance. The relation is more obvious in the region of 130–150s.

Unpredictable network traffic makes parallel simulation of network quite sensitive to accuracy. The network connects so many interacting tiles that we cannot predict the on-chip traffic pattern. Besides, most modern on-chip routers achieve one-cycle latency between neighboring tiles; as a result, there is no "critical-latency" (SlackSim) for any router-to-router link. In this situation, if neighboring routers are asynchronously simulated by different threads, even a limited clock skew between host threads would easily produce timing errors.

Further examining the network traffic, we find that the traffic between core and router is less frequent and more deterministic than that between router and router. A *Router-Router (RR)* link would be quite busy even if all cores inject infrequently messages into the network because a router-router link may be shared by many interacting cores. The *Core-Router (CR)* link, however, demonstrates some

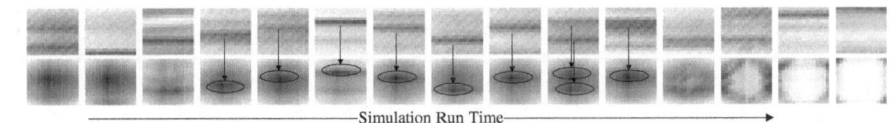

—————————————————————Simulation Run Time————————————————▶

Fig. 2. Relation between on-chip hot spotting in accessing shared cache (top) and workload distribution (bottom) during simulation. Data is collected from the simulation of a 256-tile manycore processor running *matrix-multiply*. Hot spotting is measured as the busy cycles of every L2$. Workload is measured by the time spent in simulating each tile. Each pane (16x16 mini blocks) represents a snapshot of simulation workload distribution. Each gray mini block in the pane indicates a target tile. The shade of gray indicates the workload level: darker is for heavy workload while lighter for light workload.

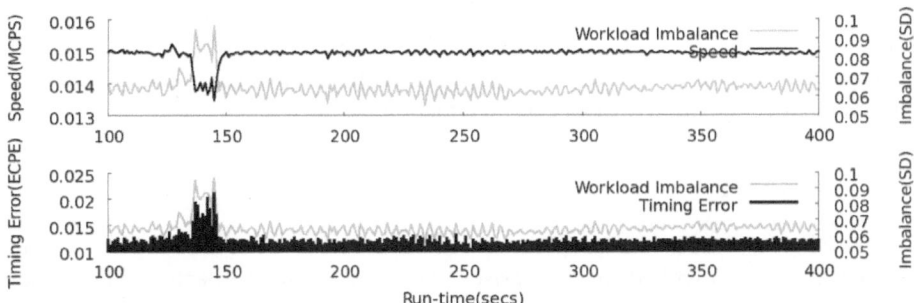

Fig. 3. Simulation speed (MCPS) and timing error (ECPE) are influenced by workload imbalance (Standard Deviation) during simulation run-time. Data is collected from a simulation section of *matrix-multiply* by 8 host threads with a relaxed quantum of 16. An obvious episode of workload imbalance is observed in the region of 130–150s.

determinism. For example, we can predict the most probable interval between two continuous Cache-Misses on a CR link. Also, we can determine the minimum interval between a Cache-Miss event and the corresponding Cache-Refill event on a CR link because the L2$ (Level 2 cache) has the minimum processing cycles for a Cache-Miss request. Although these events may happen concurrently and result in less cycles between two CR link events, the CR link still present more determinism than RR link.

To demonstrate the difference between RR link and CR link, the idle-cycle distribution of RR link and CR link is presented in fig. 4. An idle-cycle is defined as the cycles between two continuous traffic events on a specific link. For example, an idle-cycle of 4 for a CR link indicates that a link event happens 4 cycles after the previous one on that CR link. Two simulation cases are demonstrated in Fig. 4: *matrix_multiply* and *lu. Matrix_multiply* is a highly parallel application and contains little inter-core communication. Most network traffic events are Cache-Misses/Refills. *Lu* is among the applications of the SPLASH2 benchmark which contain most inter-core communication. Most of the on-chip traffic is coherence traffic between caches. Both simulations show that router-router link is busier than core-router link. For example, in *matrix_multiply*, the idle-cycles equal to 2 accounts for a proportion of

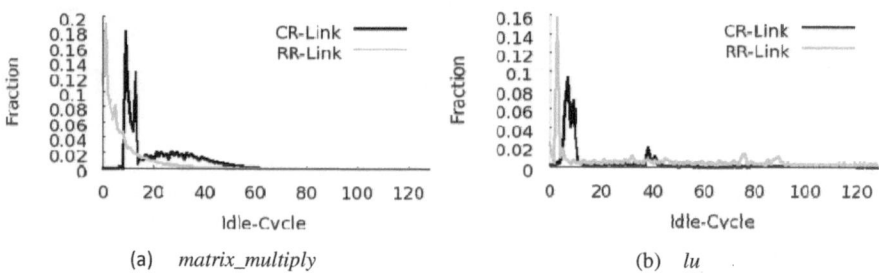

(a)   *matrix_multiply*   (b)   *lu*

Fig. 4. Idle-cycles distribution in the simulation of a 256-core architecture running *matrix_multiply* and *lu*. Idle-cycles larger than 128 is ignored.

20%. The idle-cycles for the CR link, however, mostly falls beyond 8, which is determined by the least processing cycles of Cache-Miss message in L2$ (L2$ hit latency). In *lu*, idle-cycles mainly fall in the region of 4–10, which is determined by the minimum processing cycles of Cache-Invalidation message in the core.

3   CRAW/P

This section discusses the proposed partition scheme, CRAW/P.   Two partition schemes are also described for comparison: Static and Simple-Adaptive. Fig. 5 shows examples of these three partition schemes. In these schemes, simulation of the target is distributed onto multiple host threads. Host threads run in a relaxed PDES manner. Implementation details of the used PDES method in this paper could be referred in [22].

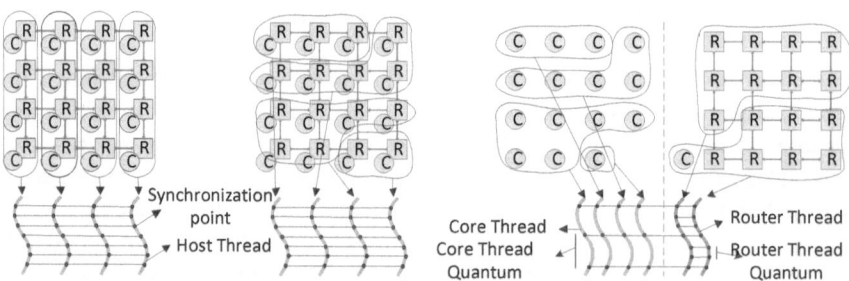

Fig. 5. Workload partitioning for Static, Simple-Adaptive, and CRAWP. Note that Static and Simple-Adaptive use one synchronization quantum among all host threads while CRAWP uses two different quantum sizes: core thread quantum (large) and router thread quantum (small).

The described Static partition scheme is exactly the scheme used in Graphite. It simply partitions the target tiles evenly onto all host threads. The Simple-Adaptive scheme is a very straightforward manner to create adaptive partitions: It tries to maintain workload balance between all host threads. As a result, host threads spend less time on synchronization, resulting in some speedup. Besides, better balance produces less clock-skew between host threads, reducing timing errors in a relaxed simulation.

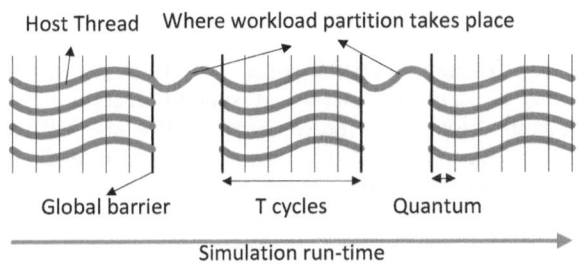

Fig. 6. Adaptive workload partition in Simple-adaptive and CRAW/P

Fig. 6 shows how adaptive partition works in a quantum based PDES simulator. During the simulation, host threads record the simulation time for each tile; all host threads barrier twice after a period of T cycles; between every barrier-pair, only one host thread runs the repartition work while other threads are simply waiting. The partitioning thread assembles the summary workload and redistributes the tiles to each thread, assuring workload balance. The idea of adaptive partition is to use the workload distribution in the near history to guide the workload partition in the near future and achieve better balance. In practical simulation, the partition interval should be set to a proper value: a big interval would result in bad partition efficiency while a small one would introduce considerable synchronization overhead. In this paper, the interval is set dynamically during simulation to ensure the partition overhead is less than a constant (e.g. 2%).

CRAW/P is essentially an adaptive partition scheme and has the same basic mechanism as Simple-Adaptive. However, it goes further; it partitions the network apart from the cores and allows the simulation of network and cores to use different synchronization strategies. To implement the network/core partition and different synchronization strategies, CRAW/P divides the host threads into two types: core thread and router thread. Core threads simulate cores (processing pipeline, L1$, and L2$) and synchronize with a coarse quantum; on the other hand, router threads simulate routers and synchronize with a fine quantum. The idea of CRAW/P could be defined through the following constraints on the simulator:

1. **Workload balance must be maintained between all host threads (core threads + router threads).** Since workload balancing is the basic mechanism to reduce synchronization overhead and clock-skew, it is a must for parallel simulation.

2. **Network must be simulated by router threads.** Since the network is quite sensitive to parallelism, the network should be simulated by as few threads as possible. Because the router is a light weight module compared to core module (functional model and core timing model), this constraint will largely reduce the number of router threads.

3. **Cores should be simulated by core threads with higher priority than by router threads.** Ideally, cores are simulated only by core threads. However, in cases of serious workload imbalance between router threads and core threads, workload balance should be achieved by migrating cores to router threads other than routers to core threads. Fig. 5(c) shows an example where a core

migrates to a router thread. This decision still obeys constraint 2 that the network must be simulated by router threads.

4. **Synchronization between router threads must be strict to increase accuracy.** Since RR link is highly-interactive, if synchronization between router threads is relaxed, the clock-skew between router threads will easily produce timing errors. So, for the sake of accuracy, router threads must synchronize very strictly. In our CRAW/P scheme, the router thread quantum is set as 1 by default. The small router thread quantum requires router threads to frequently synchronize with each other. However, the synchronization overhead is not much because the number of router thread is small.

5. **Synchronization between core threads and router threads should be relaxed to a reasonable extent to enable speedup.** The only synchronization requirement for simulating cores derives from the CR interaction. The CR links, however, are observed to be less interactive links. So, relaxing the core-core and core-router synchronization to a reasonable extent will improve the simulation speed at the expense of a moderate accuracy loss.

In summary, compared to a Static scheme, CRAW/P should present a better speedup and accuracy: Simulation speedup comes from 1) workload balancing between host threads and 2) synchronization relaxation between core threads. The accuracy improvement comes from several features: 1) it partitions the workload adaptively, achieving accuracy improvements from the workload balance; 2) it requires the network to be simulated by dedicated router threads, which limits the number of host threads that simulate the network; 3) it requires tight synchronization between router threads (small router thread quantum).

4   Results

This section presents the evaluation results that demonstrate the simulation speed and accuracy of our partition scheme compared to our two references. Section 4.1 describes the host and target configurations. Section 4.2 compares the performance of Static, Simple-Adaptive, and CRAW/P.

4.1   Experimental Setup

The multi-core host has four quad-core Intel(R) Xeon(R) E7420 CPUs running at 2.13GHZ and 128GB of DRAM. The OS is Red Hat SMP Linux with kernel version 2.6.18. Each of the experiments in this section uses the target architecture parameters summarized in Table 1 unless otherwise noted. The 256-core many-core architecture is similar with that in Fig. 1(b). The cache coherency used in the target is directory based MESI, which is similar with that of TilePro [3]. We ported representative applications from the SPLASH2 benchmark suite. Table 2 lists the problem size of these applications. All applications are threaded with 256 threads.

Table 1. Parameters of Target Architecture

Feature	Value
Clock	1GHz
L1 Cache	Private, 32 KB, 32-byte line size, 4-way associative, and LRU replacement.
L2 Cache	Shared, 128KB, 64-byte line size, 8-way associative, and LRU replacement.
Coherence	Directory based MESI.
DRAM	64GB/s = (8 Controllers * 8 GB/s each).
Interconnect	Mesh network 16x16; wormhole routing;

Table 2. Parameters of Target Applications

application	Problem size
fft	64M points
radix	256M keys
lu	1024x1024
fmm	4k
barnes	2048
cholesky	Input set tk15.O

The experiment platform is a quantum [13] based PDES simulator—QMill. QMill derives from the GAS [22] simulator, which is an accurate simulator for the Godson-T Many-core architecture [5]. We use QMill because it can conveniently simulate hundred-core general purpose manycore architectures described in fig. 1(b).

4.2   Simulation Performance

Case Study. Fig. 7 (abc) illustrates the average synchronization overhead and timing error of the three partition schemes during a simulation run. The adaptive schemes (Simple-Adaptive and CRAW/P) largely reduce the synchronization overhead from ~20% to <10%. Meanwhile, the deviation is reduced from ~23% to <5%. Note that CRAW/P is slightly worse than Simple-Adaptive in synchronization reduction. This is because of the higher partitioning overhead and extra synchronization overhead introduced by the stricter inter-router synchronization. Fig. 7 (efg) shows the simulation speed and timing error of three schemes during simulation. Simple-Adaptive improves the speed from ~0.011MCPS to >0.015 MCPS. CRAW/P

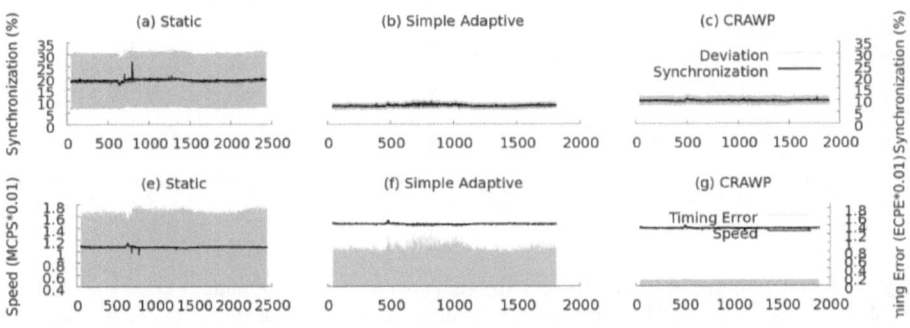

Fig. 7. The top graphs show the average synchronization overhead and the deviation (workload imbalance) for Static (a), Simple-Adaptive (b), and CRAW/P (c). The bottom graphs show the simulation speed (MCPS) and timing error (ECPE) for Static (e), Simple-Adaptive (f), and CRAW/P (g). Data are collected from the *matrix_multiply* kernel simulated by 8 host threads. For Static and Simple-Adaptive, the synchronization quantum is set to 8. For CRAW/P, the core thread quantum is 8 and the router thread quantum is 1.

improves the speed to >0.014 MCPS, which is slightly less than that achieved by Simple-Adaptive. However, CRAW/P reduces the timing error (ECPE) from ~0.016 (Static) to <0.002, which is far less than ~0.01 of Simple-Adaptive.

Simulation Speed. Fig. 8 shows the simulation speedup over the SPLASH2 benchmarks for Static, Simple-Adaptive, and CRAW/P. All applications exhibit better speed when using Simple-Adaptive or CRAW/P. The improvement of Simple-Adaptive ranges from a factor 1.37 (*choleskey_p16*) to 1.74 (*radix_p4*). The improvement of CRAW/P ranges from a factor 1.28 (*cholesky_p16*) to 1.67 (*radix_p4*).

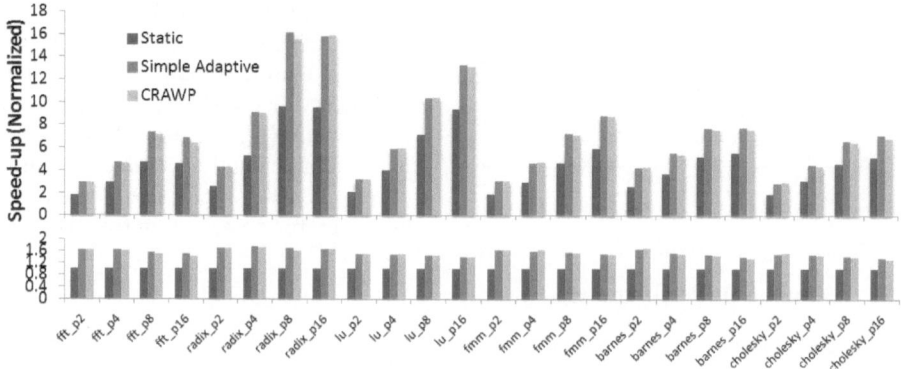

Fig. 8. Scaling of SPLASH benchmarks across various core counts using the three partition schemes. The top graph shows the speed-up normalized to a single core (sequential simulation). The bottom shows the speed improvement of Simple-Adaptive and CRAW/P compared to Static. The number of cores in the host is denoted by *_p*. In all simulations, the quantum size is 8 but in CRAW/P router thread quantum is 1. The number of router threads in all the simulations falls in the region of 1–2.3.

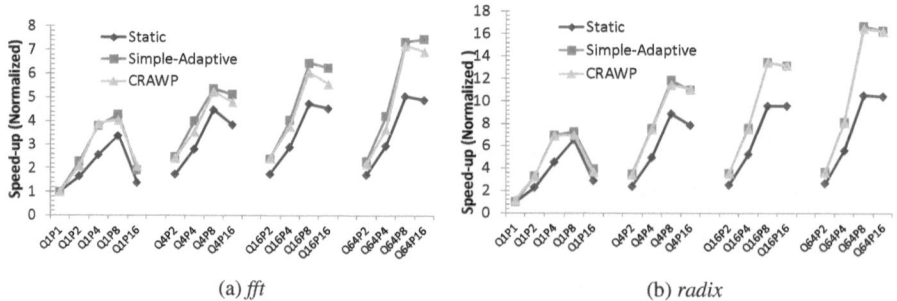

(a) *fft*   (b) *radix*

Fig. 9. Simulation speed of (a) fft and (b) radix with different quantum sizes and different numbers of host threads. The quantum size is denoted by Q*. The thread count is denoted by P*. For example, Q4P8 indicates that the simulation is run on 8 host threads with a synchronization quantum of 4.

It is notable that Simple-Adaptive behaves slightly better than CRAW/P. The difference is more obvious when more host threads are involved. That happens because CRAW/P introduces more partitioning overhead because it has to consider the workload of core and router respectively. Besides, the router threads in CRAW/P synchronize in every cycle while in Simple-Adaptive the synchronization between all host threads is relaxed. The extra synchronization slows down the simulation speed of CRAW/P but warrants better accuracy.

Another observation in is that the speed improvements of Simple-Adaptive and CRAW/P generally slowly drop as more host threads are added. The observation corresponds to the growing synchronization overhead between host threads. The benefit of workload balancing is gradually reduced as synchronization dominates the simulation overhead.

To demonstrate the effect of Simple-Adaptive and CRAW/P using different quantum, the result of *fft* (the application displaying the worst scaling) and *radix* (the application with best scaling) are selected and illustrated in Fig. 9. A general trend can be observed: adaptive partition schemes provide more improvement in cases of larger quantum sizes. The reason is the following: in quantum simulation, each host thread advances quantum cycles without synchronization, and the imbalance in each quantum is accumulated. The imbalance accumulation will be more significant as the quantum grows. A bigger workload imbalance will give more chances to adaptive partition schemes to do workload balancing.

Simulation Accuracy. Fig. 10 shows timing errors of *fft* and *radix* with different quantum sizes and different numbers of host threads. The timing error referred here is obtained by comparing the timing result with sequential simulation. The first fact to observe is that the timing error of *fft* is generally less than *radix*. This matches the application behavior of *fft* that contains a considerable sequential phase (the sequential phase is sequentially simulated and thus involves no timing error).

As illustrated, in the case of the Static scheme, the timing error grows intensely with 1) more host threads and 2) larger quantum. The simulation is particularly inaccurate with large quantum. The timing error in Q64P8 and Q64P16 exceeds 50% (51.5% in *fft* and 84.1% in *radix*), which makes the results simply worthless. Another notable fact is that timing error grows very significantly from Q16 to Q64. That's because a quantum of 64 would more easily yield clock skew of tens of cycles, which would completely cover the major part of idle-cycle distribution (cf. Fig. 4).

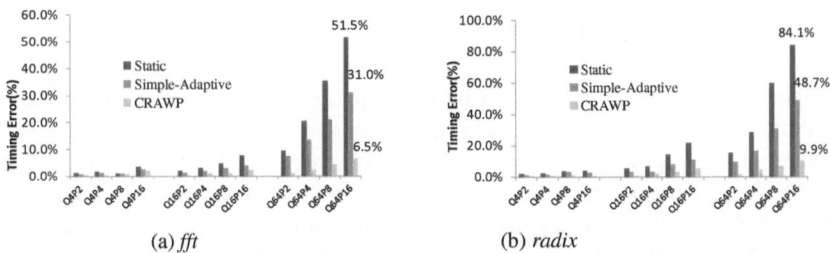

(a) *fft*   (b) *radix*

Fig. 10. Simulation error of (a) fft and (b) radix in simulation with different quanta across various numbers of host threads. Quantum size is denoted by Q*. Thread number is denoted by P*.

Simple-Adaptive does indeed reduce the timing error, but the saving is very limited. Although it shows better efficiency in cases with more host threads and larger quantum size, the timing error is still more than half that of Static. CRAW/P is very effective in improving accuracy, especially in the cases of Q64*. It reduces the timing error of *fft_Q64P16* from 51.5% to 6.5%, and reduces the timing error of *radix_Q64P16* from 84.1% to 9.9%. Overall, the experiment shows that CRAW/P behaves much better than both Simple-Adaptive and Static.

5   Related Work

Simulation is an important technique to explore new computer architectures ranging from micro-processors to parallel computers. A variety of different simulators exist, most of which are sequential. Sequential simulators run on one host thread, which limits performance. Various techniques have been studied to accelerate simulation speed including parallel simulation, direct execution, and FPGA acceleration et.

The best known parallel simulation method, PDES has been studied for decades. In conventional PDES simulators, host threads must synchronize frequently to maintain the fidelity of the simulation. Some PDES simulators adopted Quantum [13] or Slack to relax the synchronization condition.

Static workload partition is widely used in state-of-art parallel simulators. Typical examples are P-Mambo, SlackSim, and Graphite. Parallel Mambo [16] (P-Mambo) is a multi-threaded implementation of Mambo where a core based module partition is proposed to achieve high inter-scheduler parallelism. However, the evaluation only simulates a relatively small 4-core PowerPC machine.

In SlackSim, there are two types of host threads: core thread and manager thread. One dedicated thread simulates the centralized lower memory hierarchy while another set of threads (four in the paper) simulate the cores. The workload imbalance between core and memory threads can be statically avoided and the minimum L2$ access latency can be identified as safe quantum (SlackSim).

Graphite [15] uses multi-machine distributed simulation, which provides a better scalability. The tiled multicore architecture is very similar to the manycore architecture proposed in this paper. However, the workload partition is still static, with each host process simulating a set of target tiles, whose number is limited to 32.

6   Conclusion and Future Work

This paper addresses workload partitioning in manycore simulation. We discuss the architectural characteristics of manycores, present the drawbacks of a static scheme for manycore simulation, and propose an adaptive workload partition method called CRAW/P. Experimental results demonstrate that CRAW/P delivers considerable speedup (28–67%) and accuracy saving (<10% in timing error with a quantum of 64).

Further digging into the mechanisms lying behind the effects of workload imbalance and network on speed and accuracy can help us better understand manycore simulation; it can also provide future improvement opportunities. These opportunities are the focus of our future work. Another possible extension is to partition the simulation of cores and network onto different machines, seeking better performance to simulate large-scale manycore architecture containing more cores.

Acknowledgment. This work is in part supported by the National Grand Fundamental Research 973 Program of China under Grant No. 2011CB302501, the National Science Foundation for Distinguished Young Scholars of China under Grant No. 60925009, the Foundation for Innovative Research Groups of the National Natural Science Foundation of China under Grant No. 60921002, the Beijing science and technology plans under Grant No.2010B058 and the National Natural Science Foundation of China under Grant No.(61173007, 61100013 and 61100015).

References

[1] Howard, J., Dighe, S., Hoskote, Y., et al.: A 48-Core IA-32 Message-Passing Processor with DVFS in 45nm CMOS. In: Proceedings of the International Solid-State Circuits Conference, ISSCC 2010 (February 2010)

[2] Vangal, S., et al.: An 80-Tile 1.28 TFLOPS Network-on-Chip in 65nm CMOS. In: IEEE International Solid-State Circuits Conference, ISSCC 2007. Digest of Technical Papers, pp. 98–589 (2007)

[3] Bell, S., et al.: TILE64 processor: A 64-core SoC with mesh interconnect. In: Proceedings of the International Solid-State Circuits Conference, ISSCC 2008 (February 2008)

[4] The TILE-GxTM Processor Family, Tilera (2009), http://www.tilera.com/products/processors

[5] Fan, D., Zhang, H., Wang, D., et al.: High-Efficient Architecture of Godson-T Many-Core Processor. In: Proceedings of Hot Chips 23. IEEE Computer Society (2011)

[6] Kelm, J.H., Johnson, D.R., Johnson, M.R., et al.: Rigel: An Architecture and Scalable Programming. In: ISCA 2009 (2009)

[7] Burger, D., Austin, T.: The SimpleScalar tool set, version 2.0. Technical Report TR-1342, University of Wisconsin-Madison Computer Sciences Department (June 1997)

[8] Binkert, N.L., Dreslinski, R.G., Hsu, L.R., Lim, K.T., Saidi, A.G., Reinhardt, S.K.: The M5 Simulator: Modeling Networked Systems. IEEE Micro 26, 4 (2006)

[9] Magnusson, P.S., et al.: Simics: A Full System Simulation Platform. IEEE Computer 35(2), 50–58 (2002)

[10] Chidester, M.C., George, A.D.: Parallel simulation of chip-multiprocessor architectures. Proceedings of ACM Trans. Model. Comput. Simul., 176–200 (2002)

[11] Steinman, J.S.: SPEEDES: A Multiple-Synchronization Environment for Parallel Discrete-Event Simulation. International Journal in Computer Simulation 2, 251–286 (1992)

[12] Chandy, K.: Distributed Simulation: A Case Study in Design and Verification of Distributed Programs. IEEE Transactions on Software Engineering 5(5), 440–452 (1979)

[13] Mukherjee, S.S., Reinhardt, S.: Wisconsin Wind Tunnel II: A Fast, Portable Parallel Architecture Simulator. IEEE Concurrency 8(4), 12–20 (2000)

[14] Chen, J., Annavaram, M., Dubois, M.: SlackSim: A Platform for Parallel Simulations of CMPs on CMPs. SIGARCH Comput. Archit. News 37(2), 20–29 (2009)

[15] Miller, J.E.: Graphite: A distributed parallel simulator for multicores. In: HPCA 2010: The 16th IEEE International Symposium on High-Performance Computer Architecture (2010)

[16] Wang, K., Zhang, Y., Wang, H., Shen, X.: Parallelization of IBM mambo system simulator in functional modes. Operating Systems Review, 71–76 (2008)

[17] Chiou, D., Sunwoo, D.: FPGA-Accelerated Simulation Technologies (FAST): Fast, Full-System, Cycle Accurate Simulators. In: MICRO 2007: Proceedings of the 40th Annual IEEE/ACM International Symposium on Microarchitecture, pp. 249–261 (2007)

[18] Chung, E.S., Papamichael, M.K., Nurvitadhi, E., Hoe, J.C., Mai, K., Falsa, B.: ProtoFlex: Towards Scalable, Full System Multiprocessor Simulations Using FPGAs. ACM Trans. Recongurable Technol. Syst. 2(2), 1–32 (2009)

[19] Dave, N.: Implementing a functional/timing partitioned microprocessor simulator with an FPGA. In: 2nd Workshop on Architecture Research using FPGA Platforms, WARFP 2006 (February 2006)

[20] Monchiero, M., Ahn, J.H., Falcon, A., Ortega, D., Faraboschi, P.: How to simulate 1000 cores. SIGARCH Comput. Archit. News 37(2), 10–19 (2009)

[21] Dybdahl, H.: An Adaptive Shared/Private NUCA Cache Partioning Scheme for Chip Multiprocessors. In: Proc. of the Int. Symposium on High Performance Architecture (HPCA), pp. 2–12 (2007)

[22] Huiwei, L., et al.: P-GAS: Parallelizing a Cycle-Accurate Event-Driven Many-Core Processor Simulator Using Parallel Discrete Event Simulation. In: 24th ACM/IEEE/SCS Workshop on Principle of Advanced and Distributed Simulation (PADS 2010), Atlanta, USA (June 2010)

[23] Jefferson, D., Beckman, B., Wieland, F., Blume, L., Diloreto, M.: Time warp operating system. In: Proceedings of the 11th ACM Symposium on Operating System Principles, pp. 77–93 (1987)

[24] Das, S.R., Fujimoto, R., Panesar, K.S., Allison, D., Hybinette, M.: GTW: a time warp system for shared memory multiprocessors. In: Winter Simulation Conference, pp. 1332–1339 (1994)

Topic 3: Scheduling and Load Balancing

Denis Trystram, Ioannis Milis, Zhihui Du, and Uwe Schwiegelshohn

Topic Committee

More than ever, parallelism is available today at every level of computing systems, including dedicated embedded systems, basic instructions and registers, hardware accelerators, multi-core platforms, computational grids, etc. Despite of lot of efforts and nice positive results obtained during the past years, such systems are still not fully exploited. Scheduling represents the use or optimization of resources allocation in parallel and distributed systems. There are many issues to study for a better share of the load, a better reliability, a better adaptivity under computing, bandwidth or memory constraints. They are all crucial for obtaining a better use of parallel and distributed systems. It is a big challenge to study related techniques provided at both application and system levels. At the application level, the choice of the adequate computational model, the design of dynamic algorithms that are able to adapt to the particular characteristics, the mapping of applications onto the underlying computing platforms and the actual utilization of the systems are particularly relevant.

This new edition of the topic in EuroPar provides a very good coverage of the various modern perspectives. The submitted papers covered many aspects of scheduling and load balancing from theoretical foundations for modeling or analyzing new policies under specific constraints to the design of efficient and robust strategies, experimental studies, applications and development of practical tools.

This year all the submitted papers have been evaluated by four reviewers, and finally seven papers were chosen to be included into the final program. They reflect the good and necessary synergy between theoretical approaches (models, analysis) and practical realization and tools (new methods, simulation results, actual experiments, specific tuning of some applications). Problems like minimization of energy consumption, malleability for achieving the maximum possible resource utilization, on-line policies, scheduling of MapReduce jobs show how this old field always creates new problems. The objects of study evolve from year to year and reflect the new trends in scheduling showing that this classical topic remains very active and challenging.

Finally we would like to express our gratitude to all our colleagues, experts in any field of scheduling for the time and effort spent in the reviewing process. Their good job could not be achieved without the support of the organizing committee which created a good balance between a gentle pressure and the trust and freedom they gave us within the scientific topics.

Thanks also to the authors whose unfailing involvement makes EuroPar a premium forum for scheduling for parallel and distributed systems.

C. Kaklamanis et al. (Eds.): Euro-Par 2012, LNCS 7484, p. 115, 2012.
© Springer-Verlag Berlin Heidelberg 2012

Job Scheduling Using Successive Linear Programming Approximations of a Sparse Model

Stephane Chretien[1], Jean-Marc Nicod[2], Laurent Philippe[2],
Veronika Rehn-Sonigo[2], and Lamiel Toch[2]

[1] Department of Mathematics, Université de Franche-Comté, Besançon, France
[2] FEMTO-ST Institute, UMR CNRS / UFC / ENSMM / UTBM, Besançon, France

Abstract. In this paper we tackle the well-known problem of scheduling a collection of parallel jobs on a set of processors either in a cluster or in a multiprocessor computer. For the makespan objective, i.e., the completion time of the last job, this problem has been shown to be NP-Hard and several heuristics have already been proposed to minimize the execution time. We introduce a novel approach based on successive linear programming (LP) approximations of a sparse model. The idea is to relax an integer linear program and use ℓ_p norm-based operators to force the solver to find almost-integer solutions that can be assimilated to an integer solution. We consider the case where jobs are either rigid or moldable. A rigid parallel job is performed with a predefined number of processors while a moldable job can define the number of processors that it is using just before it starts its execution. We compare the scheduling approach with the classic Largest Task First list based algorithm and we show that our approach provides good results for small instances of the problem. The contributions of this paper are both the integration of mathematical methods in the scheduling world and the design of a promising approach which gives good results for scheduling problems with less than a hundred processors.

1 Introduction

Nowadays clusters of computers or large shared memory computers are widely used by many communities such as researchers, universities or industries to speed up their applications. Due to their cost these computing facilities are usually shared between several users and several parallel jobs must be run at the same time on the same platform. The problem of scheduling parallel jobs on clusters without knowing in advance the submission times of user jobs has been widely studied [20]. In this case the scheduling problem is said to be "on-line" [12]. When all characteristics of the jobs are known in advance, the scheduling problem becomes "off-line" and it has been widely studied for sequential jobs [13] and for parallel jobs [8,11].

The "off-line" problem considered here depends on the job characteristics. In the literature one distinguishes three kinds of parallel jobs. Rigid jobs [16] are performed with the number of processors originally required. Moldable jobs introduced by Turek et al. in [18] may run with different numbers of processors but cannot change their allocation after their start. Malleable jobs [9] can modify

C. Kaklamanis et al. (Eds.): Euro-Par 2012, LNCS 7484, pp. 116–127, 2012.

the number of allocated processors during their execution. The rigid job model can easily be used in most of the cases of parallel jobs. The two other models however need an interaction between the application and the scheduler to define the number of allocated processors. This is for instance the case of applications developed with the *Bulk Synchronous Parallel* (BSP) model introduced in [19] that can be run as moldable jobs. Processor virtualization however could be a solution to transparently make standard parallel applications moldable as presented in [17]. Applying virtualization to malleable jobs is probably more difficult as it would need to use virtual machine migration. For these reasons we focus on rigid and moldable jobs.

The problem of scheduling several parallel rigid and moldable jobs on homogeneous computing resources has been shown to be NP-Hard respectively in [11] and [8]. Several previous works have already tackled the issue of providing heuristics that give efficient sub-optimal solutions. In [2] static scheduling of rigid parallel jobs for minimizing the makespan is studied and in [1] for minimizing the sum of the completion time of each job. In [10], Dutot et al. consider the problem of scheduling moldable jobs with the objective of minimizing the makespan. The authors present experimental results where the well-known Largest Task First (LTF) algorithm is the best for the makespan objective.

The contribution of this paper is a novel approach for scheduling a collection of rigid or moldable jobs using successive LP approximations based on the gradient operator. To the best of our knowledge there is no existing work using this promising approach based on the sparse recovery problem in statistics domain.

The remainder of the paper is organized as follows. In Section 2 we describe the problem and the model of moldable jobs. In Section 3 we present the sparsity promoting penalization as well as linear approximation principles. Then, in Section 4 we present how to adapt this method to our scheduling problem. In Section 5 we compare our technique with the algorithm developed by Dutot et al. in [10] and show experimental results to assess the performance of our approach, and finally we conclude and give future work directions in Section 6.

2   Framework

In this section we formally define the targeted framework and the problem. We consider the problem of scheduling a collection of n independent parallel jobs. We tackle both cases of rigid and moldable jobs.

The jobs are run on a homogeneous cluster of distributed computing nodes or on a shared memory multiprocessor or multicore computer. In a cluster each node is made up of identical processors which are in turn made up of identical cores. The scheduling policy used on most clusters does not pay any attention to the exact distribution of the cores allocated on the nodes provided that the job is parallel. For this reason, in this paper, we will only consider the number of allocated cores, assimilated to processors and called Processing Elements (PEs). The results can then be applied either on clusters or on multiprocessor-multicore computers. In the remainder of the paper m denotes the number of available PEs in the execution platform.

Rigid jobs are defined by an execution time and a static number of requested PEs, i.e., the job cannot be run on neither more nor less PEs than originally requested. Each rigid job i is defined by its number of requested PEs $reqproc_i$ and its duration $reqtime_i$.

Moldable jobs can be run on a different number of PEs or cores but this number is fixed at the job execution start and cannot change during the execution. The considered moldable jobs respect the model defined in [10]. Let $reqtime_i$ be the duration of job i which requires at most $reqproc_i$ PEs. Let $t_i(n)$ be the duration of the job i if n PEs are allocated for job i. The relation between the duration of a job i and its number of allocated PEs is stated as:

$$\forall i, \ \forall n \leq reqproc_i, \ t_i(n) = \left\lceil \frac{reqproc_i}{n} \right\rceil reqtime_i$$

Given this framework our objective is to minimize the makespan of the schedule. According to the $\alpha|\beta|\gamma$ (platform | application | optimized criterion) classification of scheduling problems given by Graham in [15], the above problem is denoted by $P|\text{parallel jobs}|C_{max}$.

3   Sparsity Promoting Penalization with Successive Linearizations

The optimization method presented in the paper relays on two steps. First we formulate the problem as an integer linear program, then we relax it and apply the sparsity promoting penalization which tries to find almost integer solutions. As the sparsity promoting penalization implies to minimize a non linear objective function we use successive LP approximations to linearize it. In this section we detail the main steps of the method.

3.1   Sparsity Promoting Penalization

Recent works on the sparse recovery problem in statistics and signal processing have brought to general attention the fact that using non-differentiable penalties such as the ℓ_p norm can be an efficient ersatz to combinatorial constraints in order to promote sparsity. This approach for constructing continuous relaxations to hard combinatorial problems is a key ingredient in e.g., the new field called Compressed Sensing which originated in the work of Candès, Romberg and Tao [3]. Donoho [7] showed that finding the sparsest solution to an under-determined system of linear equations may sometimes be equivalent to finding the solution with smallest ℓ_1-norm. This discovery lead to a intense research activity in the recent years focusing on finding weaker sufficient conditions on the linear system under which it is possible to prove this equivalence. It was found in particular that for matrices satisfying certain incoherence conditions (implying that the columns of the associated matrix are almost orthogonal), the equivalence between finding the sparsest and the least ℓ_1 norm solution holds for systems with a number of unknowns to the order of exponential of the number of equations.

Other non-differentiable penalties have also been proposed in order to increase the performance of sparse recovery procedures. Candès, Wakin and Boyd proposed an iterative reweighted ℓ_1 procedure in [4]. In our setting, the standard ℓ_1 relaxation is not suitable. Indeed, as will be detailed in the sequel (e.g. equation 1 below), our constraints will always imply that the ℓ_1 is constant. A more appropriate sparsity promoting penalization in this case is the ℓ_p-quasi-norm relaxation, for $p \in (0,1)$. This corresponds to minimizing $\|x\|_p := (\sum_k x_k^p)^{1/p}$ instead of $\|x\|_1$, under the same design constraints. Such a non–convex relation was successfully implemented in, e.g. [6].

3.2   Linear and Conic Approximation

In physics and mathematics a function f is often approximated with a linear formulation at point x_0, if f is differentiable at point x_0. The gradient of a function with several parameters ($f : \mathbb{R}^n \to \mathbb{R}$), noted ∇f, is the vector whose components are equal to derivatives of f with respect to the parameters. Taylor's expansion gives

$$f(x + h) = f(x) + \langle \nabla f, h \rangle + o(h)$$

where x and h belong to $\mathbb{R}^n$, and $\langle\ \rangle$ represents the dot product.

In cases such as $x \mapsto \|x\|_p$, where f is non-differentiable, it is still possible linearize by using the appropriate generalization of the gradient, called the Clarke-subdifferential. In simple words, a non-differentiable function may have several tangents in a generalized sense and the Clarke-subdifferential, denoted by $\partial f(x)$, is the set of all such generalized tangents. The nonsmooth counterpart to Taylor's expansion is given by

$$f(x + h) = f(x) + \sup_{g \in \partial f(x)} \langle g, h \rangle + o(h)$$

.

In order to implement our ℓ_p-based relaxation, we will implement successive linearizations on a standard linear programming solver.

4   Applying the Method on the Job Scheduling Problem

In this section we apply the method on the job scheduling problem. First it implies to define a sparse representation of the problem then we apply the two steps of sparsity promoting penalization and linear approximation.

4.1   Formulation as an Integer Linear Program

In the defined framework a solution to the scheduling problem must provide at least the start time of the jobs for the rigid jobs as their duration and the number of used PEs are constants of the problem. For the moldable jobs, the duration depends upon the number of PEs that are allocated to the jobs. So the scheduled jobs are characterized by their start time and the number of allocated PEs and the duration of the job is determined as soon as this number of allocated PEs is determined. We call "configuration" of a job the number of allocated PEs. We call "position" of a job, its position determined by its start time in a discrete time scale. Finally, we call "slot" the couple (configuration, position).

Let us create a list of slots (configurations, positions) for each job. The idea is to create a vector x_i for each job i. Each component $x_{i,j}$ of the vector x_i is a binary variable which indicates whether slot j of job i is chosen or not. Then we fix a time horizon T and we let a linear program find a solution. We iteratively reduce the time horizon T until the linear program cannot find a solution any more.

The following constant values are defined to formulate the problem:

- $proc_{i,j}$: the number of PEs for the configuration j of job i
- $nconf_i$: the number of all possible configurations for job i (for rigid jobs i, $nconf_i = 1$)
- $nslots_i$: the number of all possible slots for job i
- $C_{i,s}$: the configuration index of job i used in the slot s of job i
- $run_{i,s,t}$ indicates whether in the slot s the job i is running at time t.

Then we define the binary variable $x_{i,s}$ which indicates whether slot s of job i is chosen or not. For each job i, we note x_i the vector whose components are the values $x_{i,s}$ and we define a vector x which is equal to the concatenation of the n vectors x_i of every job i, $1 \leq i \leq n$.

The problem can be formulated as an integer linear program. Since we only have to determine whether a feasible solution exists or not for a given time horizon T, we only need the constraints to be respected. That is why we set all the coefficients of the variables in the objective function to 0. The problem is stated as: *"find a feasible solution which respects the following linear constraints:"*

$$\forall 1 \leq i \leq n, \quad \sum_{s=1}^{s=nslot_i} x_{i,s} = 1 \tag{1}$$

$$\forall 1 \leq t \leq T, \sum_{i=1}^{i=n} \sum_{s=1}^{s=nslot_i} x_{i,s} \times run_{i,s,t} \times proc_{i,C_{i,s}} \leq m \tag{2}$$

Constraint 1 imposes the unicity of the chosen slot s on each job i. Constraint 2 means that at each time t, the set of all running jobs does not consume more than the m available PEs in the considered cluster.

4.2   Relaxation via Sparsity Promoting Penalization

The solution of the Integer formulation of the problem cannot be found in polynomial execution time. So we make a relaxation of it. We transform all binary variables $x_{i,s}$ into rational variables and with $0 \leq x_{i,s} \leq 1$. Since vector x – the concatenation of x_i vectors – indicates which slots are chosen, we are tempted to strongly enforce its sparsity. In fact, vector x must have exactly n "1" and many "0". Thus, we legitimately expect the binary constraints to be naturally recovered by imposing sufficient sparsity. Notice that the proposed constraints impose that the sum of the components of x is equal to one jobwise. Since the components are positive, this implies that the ℓ_1 norm is equal to one jobwise, which explains why minimizing the ℓ_1 norm for promoting sparsity is unfortunately useless in the present context. In order to overcome this difficulty, we chose to minimize the ℓ_p norm non-convex function

$$f(x) = \sum_i \|x_i\|_p \tag{3}$$

under constraints (1) and (2) for $p \in \,]0, 1[$.

4.3   Successive LP Approximation Scheme

We now apply successive LP approximation schemes to linearize the problem. Let $f_i(x_i) = \|x_i\|_p$, for all jobs i. Thus, $f = \sum_i f_i$. We use the value of each variable computed during the previous iteration. We will use the following arbitrary choice $g \in \partial f$ among all possible subgradients of f:

$$g_{i,j} = \begin{cases} x_{i,j}^{p-1} \times f_i\,(x_i)^{1-p} & \text{if } x_{i,j} \neq 0 \\ 0 & \text{otherwise.} \end{cases} \tag{4}$$

The method is implemented in Algorithm 1. It starts with any initial value e.g. the zero vector. First we compute a lower bound of the makespan at line 1, which is equal to the maximum between the duration of the longest job and $\frac{\sum_i reqproc_i \times reqtime_i}{m}$. The time horizon T is set to the makespan of the LTF list algorithm. If the linear program $\mathcal{LP}$ finds a satisfactory solution (line 9), it reduces the time horizon (line 12) until it cannot (line 23) before $maxIter$ iterations. If it does not find a satisfactory solution with $T = Listmakespan$ before $maxIter$ iterations (line), it increases the time horizon T (line 21). For a given time horizon T, it iteratively updates the objective function of the linear program (line 7) according to the subgradient-based Taylor approximation rule of the sparsity promoting penalization.

4.4   Improving the Algorithm Efficiency

During the experiment step of our work *a problem* appeared in the linear resolution. Satisfactory solutions for Algorithm 1 are only detected (at line 9) if all jobs i have their vector x_i with exactly one "1" as the algorithm is designed to find exclusively exact solutions. In fact, for a given time horizon T, the successive linear approximations manage to find a schedule for most of the jobs of the collection but it let few jobs j of the collection with fuzzy schedules. That is to say, vectors x_i contain exactly one "1" while vectors x_j do not. In this case the algorithm often continues to iterate, even if x_j is close to 1, until $maxIter$ is reached without being able to find a solution. This leads to longer computing times for the algorithm while giving inefficient solutions.

Algorithm 1. A successive LP scheme

---

1 $lb \leftarrow$ lower bound of makespan
2 $sched \leftarrow$ compute a schedule with LTF ; $listMakepsan \leftarrow makepsan(sched)$;
   $T \leftarrow listMakespan$; $end \leftarrow false$; $incT \leftarrow false$;
3 **while** $T > lb$ *and not end* **do**
4      $proc \leftarrow$ compute the configurations $(\mathcal{J})$; $run \leftarrow$ compute all possible slots $(\mathcal{J}, m, T)$;
       $iter \leftarrow 1$; $found \leftarrow false$
5      $\forall i, k, \; x_{i,k}^{(iter)} \leftarrow 0$
6   **while** $iter < maxIter$ *and not found* **do**
7   set the objective function of $\mathcal{LP}(\mathcal{J}, m, T, proc, run)$ to
           $\sum_i^{|\mathcal{J}|} f(x_i^{(iter)}) + \langle \nabla f(x_i^{(iter)}), x_i - x_i^{(iter)} \rangle$
8          $x \leftarrow$ execute $\mathcal{LP}(\mathcal{J}, m, T, proc, run)$
9   **if** $\forall i$, x_i *contains exactly one "1"* **then**
10             $sched \leftarrow$ convert into schedule $(x, proc, run)$
11             $T \leftarrow makespan(sched)$
12             $T \leftarrow T - 1$
13             $found \leftarrow true$
14   **if** $incT = true$ **then**
15                 $end \leftarrow true$
16         $\forall i, k, \; x_{i,k}^{(iter)} \leftarrow x_{i,k}$; $iter \leftarrow iter + 1$
17   **if** *not found* **then**
18   **if** $T = listMakespan$ **then**
19             $incT \leftarrow true$
20   **if** $incT = true$ **then**
21             $T \leftarrow T + 1$
22   **else**
23             $end \leftarrow true$

24 **return** $sched$

So we modify Algorithm 1 and its detection criterion at line 9 as follows: when a valid rational schedule is found we keep the exact schedule for jobs i whose x_i have exactly one "1" and we schedule the rest of the jobs for which the linear program gives fuzzy schedules with the LTF list algorithm. If a solution shorter than the time horizon T is found, then the $found$ variable is set to $true$ otherwise we continue to iterate.

5   Simulation and Results

In this section we present the results obtained on the two versions of the algorithm and we compare them to the well-known Largest Task First algorithm. We assess both cases of rigid and moldable jobs. Notice that the problem we propose to solve is nonconvex and very high dimensional. Moreover, no theoretical guarantee for convergence of the proposed iterative procedure is available and it is well known that minimizing an ℓ_p quasi-norm, $0 < p < 1$, is NP-hard already. On the other hand, various non-convex ℓ_p-based strategies have been successfully used for promoting sparsity in the literature. Despite the current lack of appropriate theoretical foundation, in most reported experiments the ℓ_p-based approach managed to reach a local solution significantly superior to the ℓ_1 minimizer for, e.g., the Compressed Sensing reconstruction problem [6]. The goal of this section is to show that such a good performance can also be observed for the studied scheduling problem. Notice that we did not optimize the computational aspects of the problem, in particular, we made no use of the very special properties of the constraint matrix. This explains why the computing time is currently much higher than what could be obtained after a careful design of the algebraic aspects of our algorithms.

5.1   Experimental Settings

Carrying out real experiments on clusters is difficult: experiments are not reproducible and may be long. Furthermore a cluster is expensive and meant to be used for calculations while experiments may monopolize it. For these reasons, we have developed a simulator of a homogeneous cluster based on a master/slave architecture. This simulator is also meant to check schedules obtained by the different algorithms. The simulator is implemented using SimGrid [5] and its MSG API. It takes a workload as input and it gives a schedule as output.

To simulate the job collection, we use synthetic workloads generated with uniform distributions. The parameters associated with a workload is the job granularity, the ratio of the duration of the longest job over the duration of the shortest one.

5.2   Assessing Performance of Algorithm 1

In a first set of experiments the simulations have been run with a ℓ_p norm where $p = 0.1$, $maxIter$ is set to 15000 in the algorithm and the machine is made up of 64 PEs. We have scheduled a collection of 60 jobs and, for each number of jobs in the collection, we performed 40 experiments to compute an average value of the ratio of the makespan over the lower bound. The results where disappointing: they were far from the optimal and very time consuming.

So we ran another set of experiments with less jobs, ℓ_p norm where $p = 0.1$, $maxIter$ set to 15000. The machine is made up of 32 PEs and the number of PEs requested by each job is uniformly chosen between 1 and 8. The granularity is set to 25. For each number of jobs in the collection we perform 20 experiments, then we remove the best and the worse results in order to reduce the deviation, and we compute an average of the ratio of the makespan over the lower bound.

Figure 1a shows the ratio of the makespan over the lower bound against the number of rigid jobs, while Figure 1b shows this ration for moldable jobs. In the figures the algorithms are noted *succ. LP approx* for our algorithm and *LIST* for the LTF implementation. The figures also show the standard deviation σ: the height of a vertical line is equal to 2σ.

We can note that for less than 25 jobs the successive LP approximation algorithm gives better results than the LTF algorithm and with more than 25 jobs the latter outperforms successive LP approximation. Note that after 40 jobs the performance ratio of LTF quickly tends toward 1.1 which means, on the one hand, that it probably finds most of the time the optimal solution and, on the other hand, that it is difficult to find better solutions. Moreover, with more than

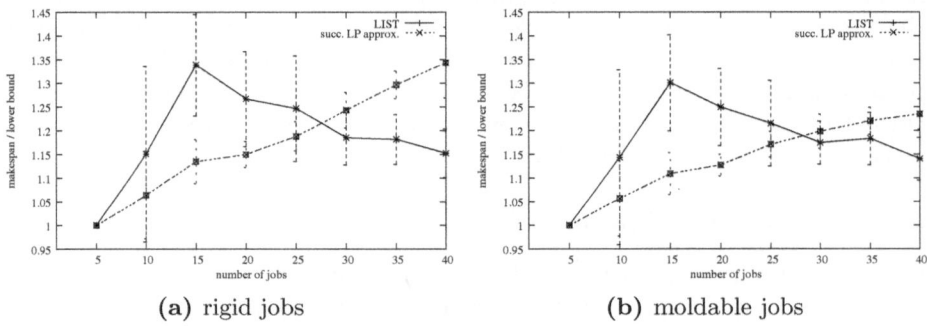

(a) rigid jobs   **(b)** moldable jobs

Fig. 1. Performance comparison 32 PEs

25 jobs, the problem becomes so complex that the successive LP approximation algorithm must increase its T to find a solution. Under this threshold the maximum gain is about 15% for 16 moldable or rigid jobs. Furthermore the standard deviation of the experiments with our new approach is less than the standard deviation of LTF. We can easily understand that for 5 jobs the optimal is found due to the experimental settings: the number of PEs that each job requires is uniformly chosen between 1 and 8. As a consequence, all jobs may start at time 0. That also explains the peak with 15 jobs which do not necessarily start at time 0. We can also notice that for rigid jobs and moldable jobs the successive LP approximation algorithm has the same behavior, that is to say, when the number of jobs increases, the ratio of the makespan over the lower bound increases.

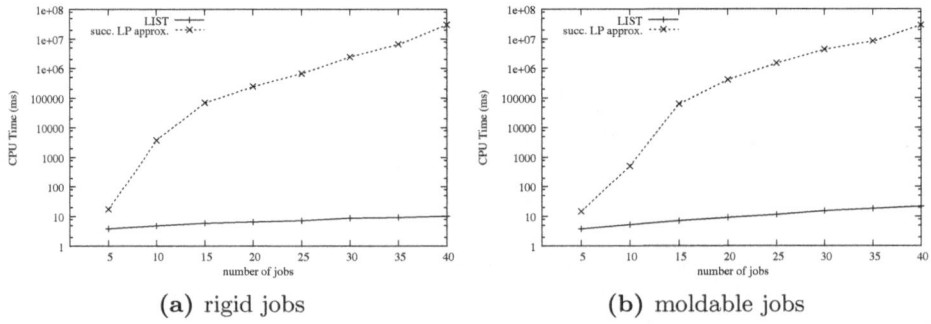

(a) rigid jobs   **(b)** moldable jobs

Fig. 2. Compute time for 32 PEs

As we can see in Figure 2a and Figure 2b, Algorithm 1 is very time consuming with both rigid jobs and moldable jobs compared to the LTF algorithm. Note however that for 15 jobs, in the case where gives the best results, the time taken by the LP approximation is not more than 1.5 minutes which is still reasonable. We assess the performance of the improved version in the following section.

5.3   Performance of the Improved Algorithm

To assess the improved algorithm, we performed experiments with two simulated machines made up of 64 and 128 PEs. The number of PEs requested by each job is uniformly chosen between 1 and 16 for the machine with 64 PEs, and between 1 and 32 for the machine with 128 PEs. Granularity is set to 25 for the machine with 64 PEs and to 10 for the machine with 128 PEs. We set $p = 0.1$ and $maxIter = 200$. For each number of jobs in the collection, we perform 40 experiments.

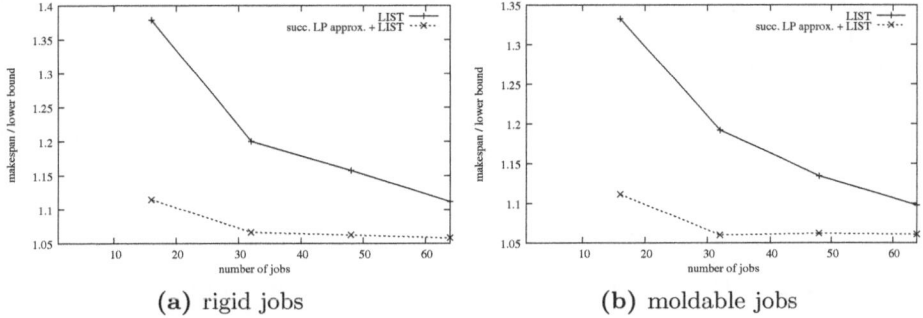

(a) rigid jobs   **(b)** moldable jobs

Fig. 3. Performance of the algorithms with 64 PEs

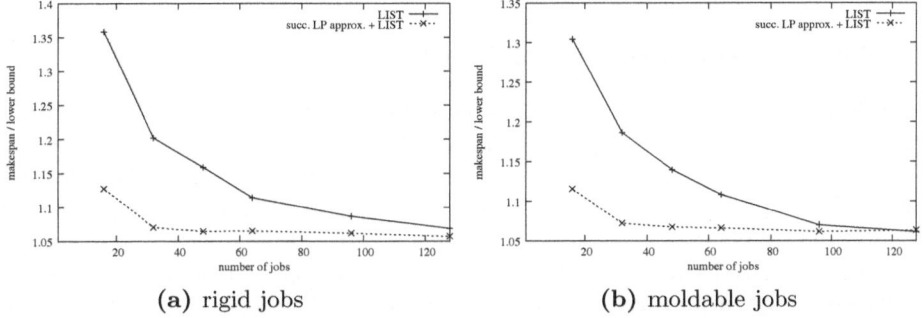

(a) rigid jobs   **(b)** moldable jobs

Fig. 4. Performance of the algorithms with 128 PEs

Figure 3a shows the ratio of the makespan over the lower bound against the number of rigid jobs in a cluster of 64 PEs, while in Figure 3b we consider scheduling moldable jobs. The performances of the new approach are better than LTF for moldable and rigid jobs, and better than the unmodified algorithm. Figure 4a and Figure 4b give the results for a machine with 128 PEs.

We can note that in the four cases the performance ratio between LTF and our approach is up to 20%. The results obtained with a 128 PEs machine show an improvement for the LTF algorithm compared to 64 case while the behavior of the new algorithm is quite similar. This is probably because our solution is very close to (if not at) the optimal solution and nothing more can be gained.

We have also recorded some statistics data after each execution of the linear program. On average with 64 PEs and 16 jobs almost 75% of jobs have exact schedules, while with 64 jobs 50% of them have exact schedules. We notice that when the number of jobs to schedule increases the number of exact schedules found by the linear program decreases. We get the same behaviour with 128 PEs: on average with 128 PEs and 16 jobs almost 80% of jobs have exact schedules, while with 128 jobs 60% of them have exact schedules.

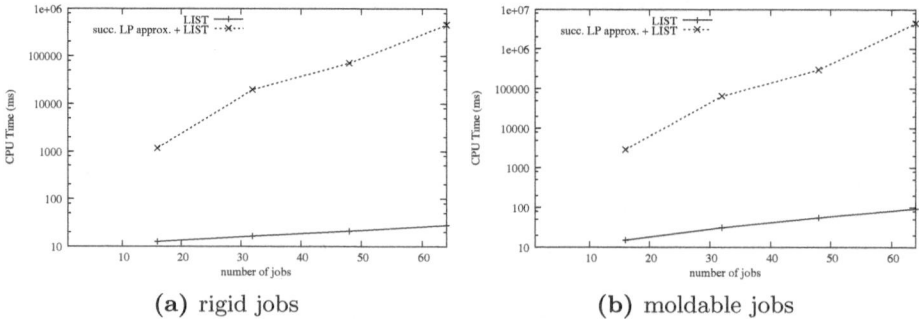

(a) rigid jobs   (b) moldable jobs

Fig. 5. CPU Time consumed to compute a schedule with 64 PEs

Figures 5a and 5b show the time spend by the two algorithms. We notice that the hybrid algorithm is less time consuming than the original algorithm but still consumes more time than the LTF algorithm.

6 Conclusion and Future Work

In this paper, we assess the use of successive linear programming approxima-tions of a sparse model for job scheduling. This method is applied on clusters to schedule rigid and moldable jobs. Experimental results show that the pure suc-cessive LP approximation only gives good performances regarding the makespan for scheduling up to dozens jobs on a machine with dozens PEs. In contrast, a variant associated with LTF gives good results for bigger instances with up to a hundred jobs on machines with up to a hundred PEs. This variant is a good alternative to the LTF algorithm and provides a significant improvement of the schedules for the range of machine size where the LTF algorithm is less efficient.

For future work we plan to implement the Split Bregman Method [14] to speed up the solving time and try other relaxations of the linear program. We also plan to use a multi-level scheduling approach for which we distinguish small jobs and large jobs. We then apply our method on different collections of jobs.

An important part of the simulations has been run thanks to the computing facilities of the *Mésocentre de Calcul de Franche-Comté* in Besançon, France.

References

1. Afrati, F.N., Bampis, E., Fishkin, A.V., Jansen, K., Kenyon, C.: Scheduling to minimize the average completion time of dedicated tasks. In: Proceedings of the 20th Conference on Foundations of Software Technology and Theoretical Computer Science, FST TCS 2000, London, UK, pp. 454–464 (2000)

2. Amoura, A.K., Bampis, E., Kenyon, C., Manoussakis, Y.: Scheduling independent multiprocessor tasks. Algorithmica 32(2), 247–261 (2002)
3. Candes, E.J., Romberg, J., Tao, T.: Robust uncertainty principles: exact signal reconstruction from highly incomplete frequency information. IEEE Transactions on Information Theory 52(2), 489–509 (2006)
4. Candes, E.J., Wakin, M.B., Boyd, S.P.: Enhancing Sparsity by Reweighted L1 Minimization. Journal of Fourier Analysis and Applications 14(5), 877–905 (2008)
5. Casanova, H., Legrand, A., Quinson, M.: Simgrid: A generic framework for large-scale distributed experiments. In: UKSIM 2008, pp. 126–131 (2008)
6. Chartrand, R., Yin, W.: Iteratively reweighted algorithms for compressive sensing. In: 33rd International Conference on Acoustics, Speech, and Signal Processing, ICASSP (2008)
7. Donoho, D.L.: Compressed sensing. IEEE Trans. Inform. Theory 52, 1289–1306 (2006)
8. Dutot, P.-F., Eyraud, L., Mounié, G., Trystram, D.: Bi-criteria algorithm for scheduling jobs on cluster platforms. In: Proceedings of the Sixteenth Annual ACM Symposium on Parallelism in Algorithms and Architectures, SPAA 2004, New York, NY, USA, pp. 125–132 (2004)
9. Dutot, P.-F., Trystram, D.: Scheduling on hierarchical clusters using malleable tasks. In: SPAA 2001, pp. 199–208 (2001)
10. Dutot, P.-F., Netto, M.A.S., Goldman, A., Kon, F.: Scheduling Moldable BSP Tasks. In: Feitelson, D.G., Frachtenberg, E., Rudolph, L., Schwiegelshohn, U. (eds.) JSSPP 2005. LNCS, vol. 3834, pp. 157–172. Springer, Heidelberg (2005)
11. Feitelson, D.G.: Job scheduling in multiprogrammed parallel systems. Research Report RC 19790 (87657). IBM T. J. Watson Research Center (1997)
12. Feitelson, D.G., Mualem, A.W.: On the definition of "on-line" in job scheduling problems. Technical report, SIGACT News (2000)
13. Garey, M.R., Johnson, D.S.: Computers and Intractability; A Guide to the Theory of NP-Completeness. W. H. Freeman & Co., New York (1990)
14. Goldstein, T., Osher, S.: The split bregman method for l1-regularized problems. SIAM J. Img. Sci. 2, 323–343 (2009)
15. Graham, R.L., et al.: Optimization and approximation in deterministic sequencing and scheduling: a survey. Ann. Discrete Math., 287–326 (1979)
16. Lublin, U., Feitelson, D.G.: The workload on parallel supercomputers: Modeling the characteristics of rigid jobs. Journal of Parallel and Distributed Computing 63, 2003 (2001)
17. Nicod, J.-M., Philippe, L., Rehn-Sonigo, V., Toch, L.: Using virtualization and job folding for batch scheduling. In: ISPDC 2011, 10th Int. Symposium on Parallel and Distributed Computing, Cluj-Napoca, Romania, pp. 39–41. IEEE Computer Society Press (July 2011)
18. Turek, J., Wolf, J.L., Yu, P.S.: Approximate algorithms scheduling parallelizable tasks. In: Proceedings of the Fourth Annual ACM Symposium on Parallel Algorithms and Architectures, SPAA 1992, pp. 323–332. ACM, New York (1992)
19. Valiant, L.G.: A bridging model for parallel computation. Commun. ACM 33, 103–111 (1990)
20. Ye, D., Zhang, G.: On-line scheduling of parallel jobs in a list. J. of Scheduling 10, 407–413 (2007)

Speed Scaling on Parallel Processors with Migration[*]

Eric Angel[1], Evripidis Bampis[2], Fadi Kacem[1], and Dimitrios Letsios[1,2]

[1] IBISC, Université d' Évry, France
{eric.angel,fadi.kacem,dimitris.letsios}@ibisc.univ-evry.fr
[2] LIP6, Université Pierre et Marie Curie, France
{Evripidis.Bampis}@lip6.fr

Abstract. We study the problem of scheduling a set of jobs with release dates, deadlines and processing requirements (works), on parallel speed-scalable processors so as to minimize the total energy consumption. We consider that both preemption and migration of jobs are allowed. We formulate the problem as a convex program and we propose a polynomial-time combinatorial algorithm which is based on a reduction to the maximum flow problem. We extend our algorithm to the multiprocessor speed scaling problem with preemption and migration where the objective is the minimization of the maximum lateness under a budget of energy.

1 Introduction

Energy consumption is a major issue in our days. Great efforts are devoted to the reduction of energy dissipation in computing environments ranging from small portable devices to large data centers. From an algorithmic point of view, new challenging optimization problems are studied, in which the energy consumption is taken into account as a constraint or as the optimization goal itself (for recent reviews see [1], [2]). This later approach has been adopted in the seminal paper of Yao et al. [11], where a set of independent jobs with release dates and deadlines have to be scheduled on a single processor so that the total energy is minimized, under the so-called *speed-scaling* model where the processor may run at variable speeds. Under this model, if the speed of a processor is s then the power consumption is s^α, where $\alpha > 1$ is a constant, and the energy consumption is the power integrated over time.

Single Processor Case. Yao et al., in [11], proposed an optimal off-line algorithm, known as the YDS algorithm, for the preemptive problem, i.e., where the execution of a job may be interrupted and resumed later on. In the same work, they initiated the study of online algorithms for the problem, introducing the Average Rate (AVR) and the Optimal Available (OA) algorithms. Bansal et al.

---

[*] Research supported by the French Agency for Research under the DEFIS program TODO, ANR-09-EMER-010, and by GDR du CNRS, RO.

C. Kaklamanis et al. (Eds.): Euro-Par 2012, LNCS 7484, pp. 128–140, 2012.

[5] proposed a new online algorithm, the BKP algorithm, which improves the competitive ratio of OA for large values of α.

Multiprocessor Case. There are two variants of the model. The first variant allows the preemption of the jobs but not their migration. We call this variant, the *non-migratory* variant. This means that a job may be interrupted and resumed later on, on the same processor, but it is not allowed to continue its execution on a different processor. In the second variant, the *migratory* variant, both the preemption and the migration of the jobs are allowed.

In [4], Albers et al. considered the multiprocessor non-migratory problem of minimizing the total energy consumption of a set of jobs with release dates and deadlines. For unit-work jobs with *agreeable* deadlines, they proposed a polynomial time algorithm. When the release dates and deadlines of jobs are arbitrary, they proved that the problem becomes NP-hard even for unit-work jobs and proposed approximation algorithms with constant approximation ratios for the off-line version of the problem. A generic reduction is given by Greiner et al. (see [9]) transforming a β-approximation algorithm for the single-processor problem to a βB_α-approximation algorithm for the multi-processor non-migratory problem, where B_α is the α-th Bell number. Furthermore, they showed that a β-approximation for multiple processors with migration yields a deterministic βB_α-approximation algorithm for multiple processors without migration.

For the migratory variant, Chen et al., in [8], initiated the study of the energy minimization speed scaling problem on multiprocessors with migration and they proposed a efficient algorithm when the jobs have arbitrary works but a common release date and deadline. In [7], Bingham and Greenstreet proposed a polynomial-time algorithm for the general problem when the release dates and deadlines of jobs are arbitrary. Their algorithm makes use of the Ellipsoid method (see [10]). Since the complexity of the Ellipsoid algorithm is high for practical applications, it was interesting to define a faster combinatorial algorithm.

When preparing a previous version of this paper, it came to our knowledge that Albers et al. [3], independently of our work, considered the same problem and proposed an optimal $O(n^2 f(n))$-time combinatorial algorithm, where n is the number of jobs and $f(n)$ is the complexity of finding a maximum flow in a graph with $O(n)$ vertices. They, also, extended the analysis of the single processor OA and AVR online algorithms to the multiprocessor case with migration.

Our Contribution and Organization of the Paper. We consider the multiprocessor migratory scheduling problem with the objective of minimizing the energy consumption. In Section 3, we give a convex programming formulation of the problem and in Section 4, we apply, the well known KKT conditions to our convex program. In this way, we obtain a set of properties that are satisfied by any optimal schedule. Then in Section 5, we propose an optimal algorithm in the case where the jobs have release dates, deadlines and the power function is of the form s^α, where $\alpha > 2$. The time complexity of our algorithm, which we call BAL, is in $O(nf(n) \log U)$, where n is the number of jobs, U is the range of all possible values of processors' speed divided by the desired accuracy and $f(|V|)$

is the complexity of computing a maximum flow in a layered graph with $O(|V|)$ vertices. Notice that our algorithm is faster than the one of Albers et al. [3] only if moderate precision is required. If full accuracy is required, our algorithm is not faster. Finally, we extend our algorithm so as to obtain an optimal algorithm for the problem of maximum lateness minimization under a budget of energy.

2   Preliminaries

Let $\mathcal{J} = \{j_1, ..., j_n\}$ be a set of jobs. Each job j_i is specified by a work w_i, a release date r_i and a deadline d_i. We define the span of a job j_i to be $span_i = [r_i, d_i]$ and we say that j_i is *alive* at time t if $t \in span_i$. We also define the density of job j_i as $den_i = w_i/(d_i - r_i)$. We assume a set of m variable-speed processors in the sense that they can all, dynamically, change their speeds and have a common speed-to-power function $P(t) = s(t)^\alpha$, where $P(t)$ is the power consumption at time t, $s(t)$ is the speed at time t and $\alpha > 2$ is a constant. Consider any interval of time $[a, b]$ and a given processor. The amount of work processed by this processor and its energy consumption, during $[a, b]$, are $\int_a^b s(t)dt$ and $\int_a^b s(t)^\alpha dt$, respectively. Hence, if the processor runs at a constant speed s, during $[a, b]$, then $s \cdot (b - a)$ units of work are executed and $s^\alpha \cdot (b - a)$ units of energy are consumed, during $[a, b]$. In our setting, preemption and migration of jobs are allowed. That is, the processing of a job may be suspended and resumed later, on the same or on different processor. Nevertheless, we do not allow parallel execution of a job which means that a job cannot be run simultaneously on two or more processors. We also assume that a continuous spectrum of speeds is available and that there is no upper bound on the speed of any processor. Our objective is to find a feasible schedule that minimizes the total energy consumed by all processors.

We define $\mathcal{T} = \{t_0, \ldots, t_L\}$ to be the set of release dates and deadlines taken in a non-decreasing order and without duplication. Clearly, $t_0 = \min_{j_i \in \mathcal{J}}\{r_i\}$ and $t_L = \max_{j_i \in \mathcal{J}}\{d_i\}$. Let $I_j = [t_{j-1}, t_j]$, for $1 \leq j \leq L$, and $\mathcal{I} = \{I_1, \ldots, I_L\}$. We denote $|I_j|$ the length of the interval I_j. Also, let $A(j)$ be the set of jobs that are alive during I_j, i.e. all the jobs j_i with $I_j \subseteq span_i$, and $a_j = |A(j)|$ be the number of jobs in $A(j)$. Given any schedule, we denote $t_{i,j}$ the total units of time that job j_i is processed during the interval I_j. As already mentioned in many other works (see [11] for example), one can show, through a simple exchange argument, that, in any optimal schedule, every job j_i is executed at a constant speed s_i and this comes from the convexity of the power function.

Next, we state a problem which is a variation of our problem that we will need throughout our analysis, we call it the *Work Assignment Problem* (or WAP) and can be described as follows: Consider a set of n jobs $\mathcal{J} = \{j_1, j_2, \ldots, j_n\}$ and a set of intervals $\mathcal{I} = \{I_1, I_2, \cdots, I_L\}$. Each job can be alive in one or more intervals in $\mathcal{I}$. During each interval I_j there are m_j available processors. Moreover, we are given a value v. Our objective is to find whether or not there is a feasible schedule that executes all jobs in $\mathcal{J}$ with constant speed v. Recall that a schedule is feasible if and only if each job is executed during its alive intervals and is executed by at most one processor at each time t. Preemption and migration of jobs are allowed. Note that the WAP is almost the feasibility scheduling problem where, given a set of jobs $\mathcal{J} = \{j_1, j_2, \ldots, j_n\}$, so that each job j_i has a processing time p_i, a release date r_i and a deadline d_i, we ask whether there exists a feasible preemptive and migratory schedule that executes each job between its release date and its deadline (according to the classical 3-field notation of Graham, this problem is denoted by $P|r_i, d_i, pmtn|-$). The $P|r_i, d_i, pmtn|-$ problem is almost the same with the WAP with the difference that, in WAP, not all intervals have the same number of available processors. Therefore, WAP is polynomially solvable by applying a variant of an algorithm for $P|r_i, d_i, pmtn|-$ (see [6]).

We also consider the problem of maximum lateness minimization given a fixed budget of energy. We are given a set of n jobs $\mathcal{J} = \{j_1, \ldots, j_n\}$, a set of m parallel homogeneous processors and a budget of energy E. Each job j_i is characterized by a release date r_i, a due date $\bar{d}_i$ and a work w_i. Given a schedule $\mathcal{S}$, the lateness of a job j_i in $\mathcal{S}$ is defined as $L_i(\mathcal{S}) = C_i(\mathcal{S}) - \bar{d}_i$, where $C_i(\mathcal{S})$ is the completion time of j_i is $\mathcal{S}$. The objective is to find a feasible schedule, where preemption and migration of jobs are allowed, with minimum $L_{max} = \max_i\{L_i\}$ whose energy consumption does not exceed a given budget E.

3   Convex Programming Formulation

In order to derive a convex program for our problem, we introduce a variable s_i and a variable $t_{i,j}$, for each $j_i \in \mathcal{J}$ and for all I_j such that $j_i \in A(j)$, to be the speed of job j_i and the total execution time of job j_i during the interval I_j, respectively. So, we propose the following convex programming formulation:

$$\min \sum_{j_i \in \mathcal{J}} w_i s_i^{\alpha-1} \tag{1}$$

$$\frac{w_i}{s_i} - \sum_{I_j:\, j_i \in A(j)} t_{i,j} = 0 \qquad\qquad j_i \in \mathcal{J} \tag{2}$$

$$\sum_{j_i \in A(j)} t_{i,j} - m \cdot |I_j| \leq 0 \qquad\qquad 1 \leq j \leq L \tag{3}$$

$$\sum_{j_i \in A(j)} t_{i,j} - a_j \cdot |I_j| \leq 0 \qquad\qquad 1 \leq j \leq L \tag{4}$$

$$t_{i,j} - |I_j| \leq 0 \qquad\qquad 1 \leq j \leq L,\ j_i \in A(j) \tag{5}$$

$$-t_{i,j} \leq 0 \qquad\qquad 1 \leq j \leq L,\ j_i \in A(j) \tag{6}$$

$$-s_i \leq 0 \qquad\qquad j_i \in \mathcal{J} \tag{7}$$

Note that the total running time and the total energy consumption of each job j_i is $\frac{w_i}{s_i}$ and $w_i s_i^{a-1}$, respectively. Then, the term (1) is the total energy consumed by all jobs which is our objective function and the constraints (2) enforce that w_i units of work must be executed for each job j_i. The constraints (3) enforce that we can use at most m processors for $|I_j|$ units of time during any interval

I_j. Also, we can use at most a_j processors operating for $|I_j|$ units of time during any interval I_j, otherwise we would have parallel execution of a job and this is expressed by (4). The constraints (5) prevent any job j_i from being executed for more than $|I_j|$ units of time during any interval $I_j \subseteq span_i$, otherwise we would have parallel execution of a job. The constraints (6) and (7) insure the positivity of the variables $t_{i,j}$ and s_i, respectively.

The above mathematical program is indeed convex because the objective function and the first constraint are convex while all the other constraints are linear. Since our problem can be written as a convex program, it can be solved in polynomial time to arbitrary precision, by applying the Ellipsoid Algorithm [10]. Nevertheless, the Ellipsoid Algorithm is not used in practice and we would like to construct a faster and less complicated combinatorial algorithm.

At this point, notice that once the speeds of the jobs are computed, by solving the convex program, a further step is needed in order to construct a feasible schedule. This can be done by solving the feasibility problem $P|r_i, d_i, pmtn|-$.

4   Structure of the Optimal Schedule

We apply the KKT conditions to our convex program so as to obtain necessary conditions for optimality of a feasible schedule. We next show that these conditions are sufficient for optimality.

The following lemma is a direct consequence of the KKT conditions applied to the convex program of our problem combined with the fact that the power function with respect to the speed is convex.

Lemma 1. *There is always an optimal schedule for our problem that satisfies the following properties:*

1. *Each job j_i is executed at a constant speed s_i.*
2. *During any interval I_j, we have that $\sum_{j_i \in A(j)} t_{i,j} = \min\{a_j, m\}|I_j|$.*
3. *If $a_j \leq m$ during an interval I_j, then $t_{i,j} = |I_j|$, for every j_i with $I_j \subseteq span_i$.*
4. *If $a_j > m$ then*
    - i. *All jobs j_i that are alive during I_j, with $0 < t_{i,j} < |I_j|$, have equal speeds.*
    - ii. *If a job j_i is not executed during an interval $I_j \subset span_i$, i.e. $t_{i,j} = 0$, then $s_i \leq s_k$ for every job j_k with $I_j \subseteq span_k$ and $t_{k,j} > 0$.*
    - iii. *If a job j_i has $t_{i,j} = |I_j|$ in an interval I_j, then $s_i \geq s_k$ for any job j_k alive during I_j with $t_{k,j} < |I_j|$.*

Proof. The Properties 1, 2 and 3 can be easily proved by applying the definition of convexity and a simple exchange argument.

Next, we focus on proving the Property 4. For this, we will use the KKT conditions whose general form can be found in the full version. In order to apply the KKT conditions, we need to associate with each constraint a dual variable. Therefore, to each set of the constraints from (2) up to (7), we associate the dual variables β_i, γ_j, δ_j, $\epsilon_{i,j}$, $\zeta_{i,j}$ and η_i, respectively.

By stationarity conditions, we have that

$$\nabla \sum_{j_i \in \mathcal{J}} w_i s_i^{\alpha-1} + \sum_{j_i \in \mathcal{J}} \beta_i \cdot \nabla \left(\frac{w_i}{s_i} - \sum_{I_j : \, j_i \in A(j)} t_{i,j} \right)$$

$$+ \sum_{j=1}^{L} \gamma_j \nabla \left(\sum_{j_i \in A(j)} t_{i,j} - m \cdot |I_j| \right) + \sum_{j=1}^{L} \delta_j \nabla \left(\sum_{j_i \in A(j)} t_{i,j} - a_j \cdot |I_j| \right)$$

$$+ \sum_{j=1}^{L} \sum_{j_i \in A(j)} \epsilon_{ij} \nabla (t_{i,j} - |I_j|) + \sum_{j=1}^{L} \sum_{j_i \in A(j)} \zeta_{ij} \nabla (-t_{i,j}) + \sum_{j_i \in \mathcal{J}} \eta_i \nabla (-s_i) = 0$$

The previous equation can be rewritten equivalently as

$$\sum_{j=1}^{L} \sum_{j_i \in A(j)} \left(-\beta_i + \gamma_j + \delta_j + \epsilon_{i,j} - \zeta_{i,j} \right) \nabla t_{i,j}$$

$$+ \sum_{j_i \in \mathcal{J}} \left((\alpha - 1) w_i s_i^{\alpha-2} - \frac{\beta_i w_i}{s_i^2} - \eta_i \right) \nabla s_i = 0 \qquad (8)$$

Furthermore, complementary slackness conditions imply that

$$\gamma_j \cdot \left(\sum_{j_i \in A(j)} t_{i,j} - m \cdot |I_j| \right) = 0 \qquad\qquad 1 \le j \le L \qquad (9)$$

$$\delta_j \cdot \left(\sum_{j_i \in A(j)} t_{i,j} - a_j \cdot |I_j| \right) = 0 \qquad\qquad 1 \le j \le L \qquad (10)$$

$$\epsilon_{ij} \cdot (t_{i,j} - |I_j|) = 0 \qquad\qquad 1 \le j \le L, \ j_i \in A(j) \qquad (11)$$

$$\zeta_{ij} \cdot (-t_{i,j}) = 0 \qquad\qquad 1 \le j \le L, \ j_i \in A(j) \qquad (12)$$

$$\eta_i \cdot (-s_i) = 0 \qquad\qquad j_i \in \mathcal{J} \qquad (13)$$

We can safely assume that there are no jobs with zero work because we may treat such jobs as if they did not exist. So, for any job j_i, it holds that $s_i > 0$ and $\sum_{I_j \subseteq span_i} t_{i,j} > 0$. Then, (13) implies that $\eta_i = 0$. We set the coefficients of the partial derivatives ∇s_i and $\nabla t_{i,j}$ equal to zero so as to satisfy the stationarity conditions. Thus, (8) gives that $\beta_i = (\alpha - 1) s_i^{\alpha}$ for each job $j_i \in \mathcal{J}$ and

$$(\alpha - 1) s_i^{\alpha} = \gamma_j + \delta_j + \epsilon_{i,j} - \zeta_{i,j} \qquad (14)$$

for each $j_i \in \mathcal{J}$ and $I_j \subseteq span_i$. Now, for each interval I_j such that $a_j > m$, because of (10), we have that $\delta_j = 0$. Next, we consider the following cases for the execution time of any job $j_i \in A(j)$:

- $0 < t_{i,j} < |I_j|$
  Complementary slackness conditions (11), (12) imply that $\epsilon_{i,j} = \zeta_{i,j} = 0$. As a result, (14) can be written as

$$(\alpha - 1) s_i^{\alpha} = \gamma_j. \qquad (15)$$

The variable γ_j is specific for each interval and thus, all such jobs have the same speed throughout the whole schedule and Property 4(i) is valid.

$-\ t_{i,j} = 0$

This means, by (11), that $\epsilon_{i,j} = 0$ and (14) is expressed as $(\alpha - 1)s_i^\alpha = \gamma_j - \zeta_{i,j}$. Thus, since $\zeta_{i,j} \geq 0$, we get that

$$(\alpha - 1)s_i^\alpha \leq \gamma_j. \tag{16}$$

$-\ t_{i,j} = |I_j|$

In this case, by (12), we get that $\zeta_{i,j} = 0$. So, (14) becomes $(\alpha - 1)s_i^\alpha = \gamma_j + \epsilon_{i,j}$. Because of dual feasibility conditions, $\epsilon_{i,j} \geq 0$. Hence, all jobs of this kind have

$$(\alpha - 1)s_i^\alpha \geq \gamma_j. \tag{17}$$

By Equations (15), (16) and (17), we get Properties 4(ii) and 4(iii).   □

Given a solution of the convex program that satisfies the KKT conditions, we derived some relations between the primal variables. Based on them, we defined some structural properties of any optimal schedule. These properties are necessary for optimality and we show that they are also sufficient because all schedules that satisfy these properties attain equal energy consumptions.

Lemma 2. *The properties of Lemma 1 are also sufficient for optimality.*

Proof. Assume for the sake of contradiction that there is a schedule A, that satisfies the properties of Lemma 1, which is not optimal and let B be an optimal schedule that also satisfies the properties (by Lemma 1 we know that the schedule B always exists). We denote E^X, s_i^X and $t_{i,j}^X$ the energy consumption, the speed of job j_i and the total execution time of job j_i during the interval I_j in schedule X, respectively. Because of our assumption, $E^A > E^B$. Let S be the set of jobs j_i with $s_i^A > s_i^B$. Clearly, there is at least one job j_k such that $s_k^A > s_k^B$, otherwise A would not consume more energy than B. So, $S \neq \emptyset$. By definition of S,

$$\sum_{j_i \in S} \sum_{I_j : j_i \in A(j)} t_{i,j}^A < \sum_{j_i \in S} \sum_{I_j : j_i \in A(j)} t_{i,j}^B.$$

Hence, there is at least one interval I_p such that

$$\sum_{j_i \in S} t_{i,p}^A < \sum_{j_i \in S} t_{i,p}^B.$$

If $a_p \leq m$, then there is at least one job j_q such that $t_{q,j}^A < t_{q,j}^B$. Due to the property 3 of Lemma 1, it should hold that $t_{q,j}^A = t_{q,j}^B = |I_j|$ which is a contradiction. So, assume that $a_p > m$. Then, the last equation gives that $t_{k,p}^A < t_{k,p}^B$ for some job $j_k \in S$. Thus, $t_{k,p}^A < |I_p|$ and $t_{k,p}^B > 0$. Both schedules have equal sum of processing times $\sum_{j_i \in I_j} t_{i,j}$ during any interval I_j. So, there must be a job $j_\ell \notin S$ such that $t_{\ell,p}^A > t_{\ell,p}^B$. Therefore, $t_{\ell,p}^A > 0$ and $t_{\ell,p}^B < |I_p|$. We conclude that $s_\ell^A \geq s_k^A > s_k^B \geq s_\ell^B$, which contradicts the fact that $j_\ell \notin S$.   □

Notice that the properties of Lemma 1 do not explain how to find an optimal schedule. The basic reason is that they do not determine the exact speed value of each job. Moreover, they do not specify exactly the structure of the optimal schedule. That is, they do not specify which job is executed by each processor at each time.

5   An Optimal Combinatorial Algorithm

In this section, we propose an optimal combinatorial algorithm for our problem. Our algorithm always constructs a schedule satisfying the properties of Lemma 1 which, as we have already showed, are necessary and sufficient for optimality.

Our algorithm is based on the notion of *critical jobs* defined below. Initially, the algorithm conjectures that all jobs are executed at the same speed and it assigns to all of them a speed which is an upper bound on the maximum speed that a job has in any optimal schedule. The key idea is to continuously decrease the speeds of jobs step by step. At each step, it assigns a speed to the critical jobs that we ignore in the subsequent steps and goes on with the remaining subset of jobs. At the end of the last step, every job has been assigned a speed. Critical jobs are recognized by finding a minimum (s,t)-cut in an (s,t)-network as we describe in the following.

Now, for each instance of the WAP, we define a graph so as to reduce our original problem to the maximum flow problem. Given an instance $< \mathcal{J}, \mathcal{I}, v >$ of the WAP, consider the graph $G = (V, E)$ that contains one node x_i for each job j_i, one node y_j for each interval I_j, a source node s and a destination node t. We introduce an edge (s, x_i) for each $j_i \in \mathcal{J}$ with capacity $\frac{w_i}{v}$, an edge (x_i, y_j) with capacity $|I_j|$ for each couple of j_i and I_j such that $j_i \in A(j)$ and an edge (y_j, t) with capacity $m_j|I_j|$ for each interval $I_j \in \mathcal{I}$. We say that this is the *corresponding graph* of $< \mathcal{J}, \mathcal{I}, v >$.

We are ready, now, to introduce the notion of criticality. Given a feasible instance for the WAP, we say that job j_c is *critical* iff for any feasible schedule and for each $I_j \subseteq span_c$, either $t_{c,j} = |I_j|$ or $\sum_{j_i \in A(j)} t_{i,j} = m_j|I_j|$. Moreover, we say that an instance $< \mathcal{J}, \mathcal{I}, v >$ of the WAP is critical iff v is the minimum speed so that the set of jobs $\mathcal{J}$ can be feasibly executed over the intervals in $\mathcal{I}$ and we refer to the speed v as the critical speed of $\mathcal{J}$ and $\mathcal{I}$.

5.1   Properties of the Work Assignment Problem

Next, we will prove some lemmas that will guide us to an optimal algorithm. Our algorithm will be based on a reduction of our problem to the maximum flow problem which is a consequence of the following theorem whose proof is omitted.

Theorem 1. *[6] There exists a feasible schedule for the work assignment problem iff the corresponding graph has maximum (s,t)-flow equal to $\sum_{i=1}^{n} \frac{w_i}{v}$.*

Based on the above theorem, we can extend the notion of criticality. Specifically, with respect to graph G that corresponds to a feasible instance of the WAP, a job j_c is critical iff, for any maximum flow, either the edge (x_c, y_j) or the edge

(y_j, t) is saturated for each path x_c, y_j, t. Recall that an edge is saturated by a flow $\mathcal{F}$ if the flow that passes through the edge according to $\mathcal{F}$ is equal to the capacity of the edge. Moreover, we say that a path is saturated if at least one of its edges is saturated.

The following lemmas that involve the notions of *critical job* and *critical instance* are important ingredients for the analysis of our algorithm. The following lemma links the concept of a critical instance with the concept of a critical job and it is omitted due to space constraints.

Lemma 3. *If $< \mathcal{J}, \mathcal{I}, v >$ is a critical instance of WAP, then there is at least one critical job $j_i \in \mathcal{J}$.*

Note that the instance $< \mathcal{J}, \mathcal{I}, v - \epsilon >$ is not feasible if $< \mathcal{J}, \mathcal{I}, v >$ is critical. Up to now, the notion of a critical job has been defined only in the context of feasible instances. We extend this notion for infeasible instances as follows: in an infeasible instance $< \mathcal{J}, \mathcal{I}, v - \epsilon >$, a job j_i is called critical if every path x_i, y_j, t is saturated by any maximum (s, t)-flow in the corresponding graph G'.

Let $< \mathcal{J}, \mathcal{I}, v >$ be a critical instance of the WAP and let G be its corresponding graph. Next, we propose a way for identifying the critical jobs of $< \mathcal{J}, \mathcal{I}, v >$ using the graph G' that corresponds to the instance $< \mathcal{J}, \mathcal{I}, v - \epsilon >$, for some sufficiently small constant $\epsilon > 0$ based on Lemmas 4 and 5 below. The value of ϵ is such that the two instances have exactly the same set of critical jobs. Moreover, the critical jobs of $< \mathcal{J}, \mathcal{I}, v - \epsilon >$ can be found by computing a minimum (s, t)-cut in the graph that corresponds to $< \mathcal{J}, \mathcal{I}, v - \epsilon >$. The proofs of Lemmas 4 and 5 can be found in the full version of the paper.

Lemma 4. *Given a critical instance $< \mathcal{J}, \mathcal{I}, v >$ of the WAP, there exists a constant $\epsilon > 0$ such that the unfeasible instance $< \mathcal{J}, \mathcal{I}, v - \epsilon >$ and the critical one have exactly the same critical jobs. The same holds for any other value ϵ' such that $0 < \epsilon' \leq \epsilon$.*

Lemma 5. *Assume that $< \mathcal{J}, \mathcal{I}, v >$ is a critical instance for the WAP and let G' be the graph that corresponds to the instance $< \mathcal{J}, \mathcal{I}, v - \epsilon >$, for any sufficiently small constant $\epsilon > 0$ in accordance with the Lemma 4. Then, any minimum (s, t)-cut $\mathcal{C}'$ of G' contains exactly:*

   i. at least one edge of every path x_i, y_j, t for any critical job j_i,
   ii. the edge (s, x_i) for each non-critical job j_i.

5.2   The BAL Algorithm

We are now ready to give a high level description of our algorithm. Initially, we will assume that the optimal schedule consumes an unbounded amount of energy and we assume that all jobs are executed with the same speed s_{UB}. This speed is such that there exists a feasible schedule that executes all jobs with the same speed. Then, we decrease the speed of all jobs up to a point where no further reduction is possible so as to obtain a feasible schedule. At this point, all jobs are assumed to be executed with the same speed, which is critical, and there is at least one job that cannot be executed with speed less than this, in any feasible schedule. The jobs that cannot be executed with speed less than the critical one form the current set of critical jobs. So, the critical job(s) is (are) assigned the critical speed and is (are) ignored after this point. That is, in what follows, the algorithm considers the subproblem in which some jobs are omitted (critical jobs), because they are already assigned the lowest speed possible (critical speed) so that they can be feasibly executed, and there are less than m processors during some intervals because these processors are dedicated to the omitted jobs.

In detail, the algorithm consists of a number of steps, where at each step a binary search is performed, in order to determine the minimum speed so as to obtain a feasible schedule for the remaining jobs, i.e. the critical speed. We denote s_{crit} the critical speed and $\mathcal{J}_{crit}$ the set of critical jobs at a given step. In order to determine s_{crit} and $\mathcal{J}_{crit}$, we perform a binary search assuming that all the remaining jobs are executed with the same speed. We know that each job will be executed with speed not less than its density. Therefore, given a set of jobs $\mathcal{J}$, we know that there does not exist a feasible schedule that executes all jobs with a speed $s < \max_{j_i \in \mathcal{J}}\{den_i\}$. Also, observe that if all jobs have speed $s = \max_j\{\frac{\sum_{j_i \in A(j)} w_i}{|I_j|}\}$, then we can construct a feasible schedule. These bounds define the search space of the binary search performed in the initial step. In the next step the critical speed of the previous step is an upper bound on the speed of all remaining jobs and a lower bound is the maximum density among them. We use these updated bounds to perform the binary search of the current step and we go on like that. A high level pseudo-code of our algorithm follows.

Algorithm 1. BAL

1: $s_{UB} = \max_j\{\frac{\sum_{j_i \in A(j)} w_i}{|I_j|}\}$, $s_{LB} = \max_{j_i \in \mathcal{J}}\{den_i\}$
2: **while** $\mathcal{J} \neq \emptyset$ **do**
3:     Find the minimum speed s_{crit} so that the instance $< \mathcal{J}, \mathcal{I}, s_{crit} >$ of the WAP problem is feasible, using binary search in the interval $[s_{LB}, s_{UB}]$, through repeated maximum flow computations.
4:     Pick a sufficiently small $\epsilon > 0$.
5:     Determine the set of critical jobs $\mathcal{J}_{crit}$ by computing a minimum (s,t)-cut in the graph G' that corresponds to the instance $< \mathcal{J}, \mathcal{I}, s_{crit} - \epsilon >$.
6:     Assign to the critical jobs speed s_{crit} and set $\mathcal{J} = \mathcal{J} \backslash \mathcal{J}_{crit}$.
7:     Update the number of available processors m_j for each interval I_j.
8:     $s_{UB} = s_{crit}$, $s_{LB} = \max_{j_i \in \mathcal{J}}\{den_i\}$
9: Apply an optimal algorithm for $P|r_i, d_i, pmtn|-$ to schedule each job j_i with processing time w_i/s_i.

Algorithm BAL produces an optimal schedule, and this holds because any schedule constructed by the algorithm satisfies the properties of Lemma 1.

Theorem 2. *Algorithm BAL produces an optimal schedule.*

Proof. First of all, it is obvious that the algorithm assigns a constant speed to every job because each job is assigned exactly one speed in one step and the Property 1 of Lemma 1 is true.

Recall that at each step of the algorithm, a set of jobs is assigned a speed and some processors during some intervals are dedicated to these jobs. Consider the k-th step. At the beginning of the step, the remaining jobs $\mathcal{J}^{(k)}$ and available intervals $\mathcal{I}^{(k)}$ form the new instance of the WAP for which the critical speed and jobs are determined. We denote $G^{(k)}$ the graph that corresponds to the instance $< \mathcal{J}^{(k)}, \mathcal{I}^{(k)}, v >$ of the WAP, where the speed v varies between $s_{UB}^{(k)}$ and $s_{LB}^{(k)}$, i.e. the bounds of the step.

Assume that the Property 2 is not true. Then, there must be an interval I_j during which $\sum_{j_i \in A(j)} t_{i,j} < \min\{a_j, m\}|I_j|$, i.e. we can decrease the speed of some job and still get a feasible schedule. Note that it cannot be the case that $\sum_{j_i \in A(j)} t_{i,j} > \min\{a_j, m\}|I_j|$ because BAL assigns speeds only if there exists a feasible schedule with respect to these speeds. So, there must be a job $j_c \in A(j)$ such that $t_{c,j} < |I_j|$ and there is an idle period during I_j such that j_c is not executed. Suppose that j_c became critical during the k-th step. Then, in the graph $G^{(k)}$, since j_c is a critical job, either the edge (x_c, y_j) or the edge (y_j, t) belongs to a minimum (s,t)-cut and as a result, for any maximum flow in $G^{(k)}$, either $f(x_c, y_j) = |I_j|$ or $f(y_j, t) = m_j^{(k)}|I_j|$ where $m_j^{(k)}$ is the number of available processors during I_j at the beginning of the k-th step. Hence, we have a contradiction on the fact that $\sum_{j_i \in A(j)} t_{i,j} < \min\{a_j, m\}|I_j|$ and $t_{c,j} < |I_j|$.

For the Property 3, we claim that during the interval I_j with $a_j \leq m$, if a job j_c becomes critical, the edge (x_c, y_j) becomes saturated by any maximum (s,t)-flow in $G^{(k)}$ (given that j_c becomes critical at the k-th step). If this was not the case, then there would be a maximum (s,t)-flow $\mathcal{F}$ in $G^{(k)}$ such that $f(x_c, y_j) < |I_j|$. Also, in $\mathcal{F}$ it holds that $f(x_i, y_j) \leq |I_j|$ for any other $j_i \in A(j)$. Hence, $f(y_j, t) < a_j|I_j| \leq m|I_j|$. So, neither the edge (x_c, y_j) nor the edge (y_j, t) becomes saturated by $\mathcal{F}$, contradicting the criticality of j_c. Therefore, the total execution time of j_c during I_j is $|I_j|$.

Next we prove the Property 4. Initially, consider two jobs j_i and j_ℓ, alive during an interval I_j, such that $0 < t_{i,j} < |I_j|$ and $0 < t_{\ell,j} < |I_j|$. We will show that the jobs are assigned equal speeds by the algorithm. For this, it suffices to show that they are assigned a speed at the end of the same step. So, assume that j_i becomes critical at the end of the k-th step. Then, either the edge (x_i, y_j) or the edge (y_j, t) belongs to a minimum (s,t)-cut $\mathcal{C}$ in $G^{(k)}$. Since $0 < t_{i,j} < |I_j|$, we know that there exists a maximum (s,t)-flow in $G^{(k)}$ such that $0 < f(x_i, y_j) < |I_j|$. So, it is the edge (y_j, t) that belongs in $\mathcal{C}$. Therefore, in $G^{(k)}$, the edge (y_j, t) is saturated by any maximum (s,t)-flow, and as a result, all the processors during the interval I_j are dedicated to the execution of some tasks at the end of the k-th step. Hence, j_ℓ cannot be assigned a speed at a step strictly greater than k. Similarly, j_i is not assigned a speed later than j_ℓ. Hence, the two jobs are assigned a speed at the same step. That is, *4(i)* is true.

Next, for the Property *4(ii)*, consider the case where $t_{i,j} = 0$ for a job j_i during an interval $I_j \subseteq span_i$ and assume that j_i becomes critical at the k-th step. Then, either y_j does not appear in $G^{(k)}$, that is no processors are available during I_j, or (y_j, t) belongs to a minimum (s, t)-cut of $G^{(k)}$. If none of these was true, then (y_j, t) would appear in $G^{(k)}$ and it would not belong to a minimum (s, t)-cut. Then, all the edges (x_ℓ, y_j) would belong to a minimum (s, t)-cut, for all j_ℓ alive during I_j that appear in $G^{(k)}$. So, (x_i, y_j) would be saturated by any maximum (s, t)-flow and we have a contradiction, since the fact that $t_{i,j} = 0$ implies that there exists a maximum (s, t)-flow with $f(x_i, y_j) = 0$. In both cases, that is if y_j does not appear in $G^{(k)}$ or (y_j, t) belongs to a minimum (s, t)-cut of $G^{(k)}$, no job executed during I_j will be assigned a speed after the k-th step. Hence, all jobs j_ℓ with $t_{\ell,j} > 0$ do not have lower speed than j_i.

Next, let j_i be a job with $t_{i,j} = |I_j|$ and assume that it is assigned a speed at the k-th step. As we have already shown, this cannot happen after a step where a job j_ℓ with $0 < t_{\ell,j} < |I_j|$ is assigned a speed because after such a step, the interval I_j is no longer considered. Also, as we showed in the previous paragraph, j_i becomes critical not after a job j_ℓ with $t_{\ell,j} = 0$. The Property 4(iii) follows.

Finally, because of Lemmas 4 and 5, BAL correctly identifies the critical jobs at each step of the algorithm. The theorem follows. □

We turn, now, our attention to the complexity of the algorithm. Because of Lemma 3 at least one job (all critical ones) is scheduled at each step of the algorithm. Therefore, there will be at most n steps. Assume that U is the range of all possible values of speeds divided by our desired accuracy. Then, the binary search needs to check $O(\log U)$ values of speed to determine the next critical speed at one step. That is, BAL performs $O(\log U)$ maximum flow calculations at each step. Thus, the overall complexity of our algorithm is $O(nf(n)\log U)$ where $f(|V|)$ is the complexity of computing a maximum flow in a graph with $|V|$ vertices.

6 Maximum Lateness with a Budget of Energy

In order to solve the problem of minimizing the maximum lateness under a budget of energy, it is sufficient to determine an upper and a lower bound on the maximum lateness of the optimal schedule and then perform a binary search within this interval. The algorithm and its optimality are given in the full version of the paper.

Theorem 3. *The multiprocessor speed scaling problem of minimizing the maximum lateness of a set of jobs under a budget of energy can be solved in polynomial time when preemption and migration are allowed.*

Acknowledgments. We thank Alexander Kononov for helpful discussions on this work.

References

1. Albers, S.: Energy Efficient Algorithms. Communications of the ACM 53(5), 86–96 (2010)

2. Albers, S.: Algorithms for Dynamic Speed Scaling. In: STACS. LIPIcs, vol. 9, pp. 1–11 (2011)
3. Albers, S., Antoniadis, A., Greiner, G.: On Multi-Processor Speed Scaling with Migration. In: SPAA, pp. 279–288 (2011)
4. Albers, S., Muller, F., Schmelzer, S.: Speed Scaling on Parallel Processors. In: SPAA, pp. 289–298 (2007)
5. Bansal, N., Kimbrel, T., Pruhs, K.: Speed Scaling to Manage Energy and Temperature. Journal of the ACM 54(1), 3 (2007)
6. Baptiste, P., Néron, E., Sourd, F.: Modèles et Algorithmes en Ordonnancement. Ellipses (2004)
7. Bingham, B., Greenstreet, M.: Energy Optimal Scheduling on Multiprocessors with Migration. In: ISPA, pp. 153–161 (2008)
8. Chen, J.J., Hsu, H.R., Chuang, K.H., Yang, C.L., Pang, A.C., Kuo, T.W.: Multiprocessor Energy Efficient Scheduling with Task Migration Considerations. In: ECRTS, pp. 101–108 (2004)
9. Greiner, G., Nonner, T., Souza, A.: The Bell is Ringing in Speed Scaled Multiprocessor Scheduling. In: SPAA, pp. 11–18 (2009)
10. Nemirovski, A., Nesterov, I., Nesterov, Y.: Interior Point Polynomial Algorithms in Convex Programming. Society for Industrial and Applied Mathematics (1994)
11. Yao, F., Demers, A., Shenker, S.: A Scheduling Model for Reduced CPU Energy. In: FOCS, pp. 374–382 (1995)

Dynamic Distributed Scheduling Algorithm
for State Space Search

Ankur Narang[1], Abhinav Srivastava[1], Ramnik Jain[1], and R.K. Shyamasundar[2]

[1] IBM India Research Laboratory, New Delhi
{annarang,abhin122,ramnjain}@in.ibm.com
[2] Tata Institute of Fundamental Research, Mumbai
shyam@tifr.res.in

Abstract. Petascale computing requires complex runtime systems that need to consider load balancing along with low time and message complexity for scheduling massive scale parallel computations. Simultaneous consideration of these objectives makes online distributed scheduling a very challenging problem. For state space search applications such as UTS, NQueens, Balanced Tree Search, SAT and others, the computations are highly irregular and data dependent. Here, prior scheduling approaches such as [16], [14], [7], HotSLAW [10], which are dominantly locality-aware work-stealing driven, could lead to low parallel efficiency and scalability along with potentially high stack memory usage.

In this paper we present a novel distributed scheduling algorithm (*LDSS*) for *multi-place*[1] parallel computations, that uses an unique combination of d-choice randomized remote (inter-place) spawns and topology-aware randomized remote work steals to reduce the overheads in the scheduler and dynamically maintain load balance across the compute nodes of the system. Our design was implemented using GASNet API[2] and POSIX threads. For the UTS (Unbalanced Tree Search) benchmark (using upto 4096 nodes of Blue Gene/P), we deliver the best parallel efficiency (92%) for $295B$ node binomial tree, better than [16] (87%) and demonstrate super-linear speedup on 1 Trillion node (largest studied so far) geometric tree along with higher tree node processing rate. We also deliver upto 40% better performance than Charm++. Further, our memory utilization is lower compared to *HotSLAW*. Moreover, for NQueens ($N = 18$), we demonstrate superior parallel efficiency (92%) as compared Charm++ (85%).

1 Introduction

State space search problems such as planning and scheduling problems in manufacturing industries and world wide web, VLSI design automation problems (routing, floorplanning, cell placement and others), N-Queens [8], Traveling Salesman problem and other discrete optimization problems are very fundamental in nature and hence frequently used in many industry application domains and systems research. Since all these problems are NP-Hard, one needs to resort to systematic but intelligent state

---

[1] Multi-place refers to a group of places. For example, with each place as an SMP(Symmetric MultiProcessor), multi-place refers to cluster of SMPs.

[2] http://gasnet.cs.berkeley.edu

C. Kaklamanis et al. (Eds.): Euro-Par 2012, LNCS 7484, pp. 141–154, 2012.

space search to find optimum solutions. The states and the transition function(s) (including constraints) between the states are defined according to the nature of the state space search problem. The objective of the state space search problem is to find a path from a start state to a desired goal state (or a path from the start to each among a set of goal states). For a lot of state space search problems, in order to search the given state space, one constructs a *search tree* where each *node* in the search tree represents the state reached during the search.

Even though it seems that state space search problems require exponential number of processors (as compared to graph algorithms such as depth-first search etc.) since their worst case time is almost always exponential, the average time complexity of heuristic search algorithms for some problems is polynomial [17] [12]. Furthermore, there are heuristic search algorithms that find suboptimal solutions for specific problems in polynomial time. In such cases, bigger problem instances can be solved using large scale parallel computing infrastructure. Many discrete optimization problems (such as robot motion planning, speech understanding, and task scheduling) require realtime solutions. For these applications, parallel processing may be the only way to obtain acceptable performance. Since the state space search involves higher irregular computation DAG, it suffers from severe load balancing problems.

Further, with the advent of petascale machines such as K-Computer [3], Jaguar [4], Blue Gene/Q [5] and others, there is an imminent demand for strong performance and scalability of large scale computations along with improved programmer productivity. Thus, there is a strong need to have efficient scheduling frameworks as part of run-time systems that can meet these performance and productivity objectives simultaneously. For handling large parallel computations, the scheduling algorithm (in the run-time system) should be designed to work in a *distributed* fashion. For the execution of dynamically unfolding irregular and data-dependent parallel computations, the on-line scheduling framework has to make decisions dynamically on where (which place and core/processor) and when (order) to schedule the computations. Further, the critical path of the scheduled computation is dependent on load balancing across the cores as well as on the computation and communication overheads. The scheduler needs to maintain appropriate trade-offs between load balancing, communication overheads and space utilization. Simultaneous consideration involving space, time, message complexity and load balance makes distributed scheduling of large scale parallel state space search applications a very challenging problem.

Distributed Scheduling for parallel computations is a well studied problem in the shared memory context starting from the pioneering research by Blumofe and Leiserson [3] on Cilk scheduling, followed by later work including [2] [1] [4] [6] amongst many others. These efforts are primarily focused on work-stealing efficiency improvement in shared-memory architectures without considering explicit affinity annotations by the programmer. With the advent of distributed memory architectures, lot of recent research on distributed scheduling looks at multi-core and many-core clusters [16] [15]. All these recent efforts primarily achieve load balancing using (locality-aware) work

---

[3] http://www.fujitsu.com/global/about/tech/k/

[4] http://www.nccs.gov/computing-resources/jaguar/

[5] http://www-03.ibm.com/systems/deepcomputing/solutions/bluegene/

stealing across the nodes in the system. Although, this strategy works well for slightly irregular computation such as UTS for geometric tree, it could result in large parallel inefficiencies when the computation is highly irregular (binomial tree for UTS). Certain other approaches such as [14] consider limited control and no data-dependencies in the parallel computation, which limits the scope of applicability of the scheduling framework.

In this paper, we address the following distributed scheduling problem.

Given:
(a) A parallel computation DAG (Fig. 1(a)) that represents a parallel multi-threaded computation. Each node in the DAG is a basic operation (instruction) such as and/or/add etc. Each edge in the DAG represents one of the following: (a) Spawn of a new thread; (b) Sequential flow of execution;or, (c) Synchronization dependency between two nodes. The DAG is a *strict* parallel computation DAG (synchronization dependency edge represents a thread waiting for the completion of a descendant thread, details in section 2).
(b) A cluster of n SMPs (refer Fig. 1(b)) as the target architecture on which to schedule the computation DAG. Each SMP also referred as *place* has fixed number(m) of processors and memory. The cluster of SMPs is referred as the *multi-place* setup.
Determine: An online schedule for the nodes of the computation DAG in a distributed fashion that ensures:
(a) good trade-off between load-balance across the nodes and communication overheads;
(b) Low space, time and message complexity for execution.

In this paper, we present the design of a novel distributed scheduling algorithm (referred as *LDSS*) that **combines** topology-aware *inter-place prioritized random* work stealing with *d-choice based randomized distributed remote spawns* to provide automatic dynamic load balancing across places. Our *LDSS* algorithm partitions the compute nodes of the target system into *disjoint groups*. By using higher priority for limited radius (within a group) work stealing as well as remote spawns across the places (as compared to farther off, outside the group) our algorithm achieves low overheads. The remote spawns happen within the group to maintain affinity, while they are enabled across the groups to improve load-balance in the system. By controlling the **rate of remote spawns** [6], **rate of remote work steals**, **granularity of work steals** and **group size** one can obtain *a balanced trade-off* point between load balancing, scheduling overheads and space utilization. Our main contributions are as follows:

– We present a novel online distributed scheduling algorithm (referred to as *LDSS*) that uses an elegant *combination of topology-aware* remote (inter-place) spawns based on randomized d-choice load balancing and remote prioritized random work steals to reduce the overheads in the scheduler and to dynamically maintain load balance across the compute nodes of the system.
– By tuning the parameters such as granularity of remote steals, remote work-steal rate, value of d in d-choice based remote spawns, compute group-size and others we obtain optimal trade-offs between load-balance and scheduling overheads

---

[6] Ratio of remote spawned threads to total spawned threads at a processor.

that results in scalable performance. The *LDSS* algorithm was implemented using GASNet API and POSIX threads to enable *asynchronous communication* across the nodes and improve *computation-communication* and *communication-communication* overlap.

– Using upto 4096 nodes of Blue Gene/P we obtained superior performance as compared to prior approaches. For the binomial tree UTS (Unbalanced Tree Search) benchmark [7], LDSS delivers: (a) Upto around 40% better performance than Charm++ [15] and [16]; (b) Best parallel efficiency (92%) for $295B$ node tree as compared to best prior work [14] [16] (87%). LDSS demonstrates superlinear speedup for 1 Trillion node geometric tree and best processing rate of around 4GNodes/s for 16Trillion node geometric tree (largest studied so far by any prior work). Further on benchmarks such as NQueens [8], LDSS demonstrates superior parallel efficiency as compared to Charm++ on Blue Gene/P, MPP architecture.

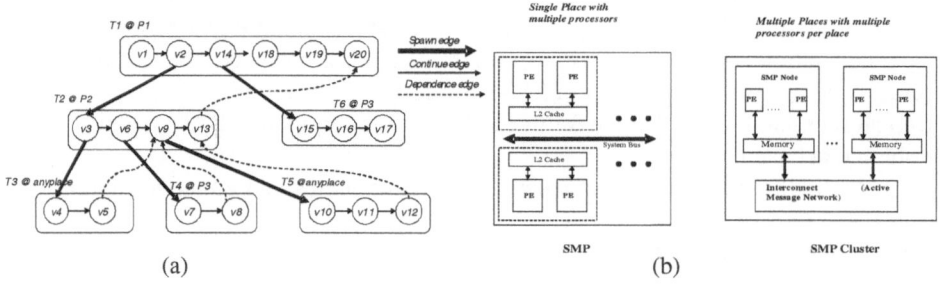

Fig. 1. (a) Computation DAG. (b) Multiple Places: Cluster of SMPs

2   System and Computation Model

The system on which the *computation DAG* is scheduled is assumed to be cluster of *SMPs* connected by an *Active Message Network* (Fig. 1(b)). Each *SMP* is a group of processors with shared memory. Each SMP is also referred to as *place* in the paper. Active Messages ((AM) [8] is a low-level lightweight RPC(remote procedure call) mechanism that supports unordered, reliable delivery of matched request/reply messages. We assume that there are n places and each place has m processors.

The parallel computation to be dynamically scheduled on the system, is assumed to be specified by the programmer in languages such as X10 and Chapel. To describe our distributed scheduling algorithm, we assume that the parallel computation has a *DAG*(directed acyclic graph) structure and consists of nodes that represent basic operations (as in a processor instruction set architecture) like *and, or, not, add* and so forth. There are edges between the nodes (basic instructions such as and/or/add etc) in the computation DAG (Fig. 1(a)) that either represent: (a) creation of new activities

---

[7] http://barista.cse.ohio-state.edu/wiki/index.php/UTS
[8] Active Messages defined by the AM-2:
  http://now.cs.berkeley.edu/AM/active_messages.html

(*spawn* edge), (*b*) sequential execution flow between the nodes within a thread/activity (*continue* edge) and (*c*) synchronization dependencies (*dependence* edge) between the nodes. In the paper, we refer to the parallel computation over nodes (basic instructions such as and/add/or) to be scheduled as the *computation DAG*. At a higher level, the parallel computation can also be viewed as a computation tree of *threads*. Each *thread* (as in multi-threaded programs) is a sequential flow of execution of instructions and consists of a set of nodes (basic operations/instructions); and it may or may not have an affinity annotation defined by the programmer. Fig. 1 shows a strict computation dag where: $v1..v20$ denote nodes, $T1...T6$ are threads and $P1..P3$ denote places).

Based on the structure of dependencies between the nodes in the computation DAG, there can be multiple types of parallel computations such as: (*a*) **Fully-strict computation**: Dependencies are only between the nodes of a thread and the nodes of its immediate parent thread; and, (*b*) **Strict computation**: Dependencies are only between the nodes of a thread and the nodes of any of its ancestor threads.

3 LDSS: Scheduling Algorithm

Our distributed scheduling algorithm, *LDSS*, attempts to achieve communication efficient load balancing across the places with low scheduling overheads. In order to achieve this goal, we make the following design choices: (*a*) **Topology Awareness:** The places in the system are clustered into small *groups* based on their distances amongst each other in the topology of the underlying target architecture; (*b*) **Two-level Work Stealing:** Our algorithm uses work-stealing at two-levels. One is intra-place randomized work stealing to achieve load balance across the processors within a place. The other is inter-place prioritized (topology-aware) randomized work stealing that provides load balance across the places in the system; (*c*) **Load Balance driven Randomized Work Pushing:** LDSS incorporates (topology-aware) work-pushing across the places (nodes) in the system. This uses the d-choice randomized load balancing algorithm to achieve low load imbalance across the groups. The rate of such *remote spawns* is automatically adjusted during the algorithm;and, (*d*) **Dedicated Communication Processor:** In order to handle inter-place spawns we assign a dedicated communication processor in each node (place). This communication processor uses GASNet API to enable asynchronous communication and improves the performance of the scheduling algorithm by enabling *computation-communication* overlap as well as *communication-communication* overlap across the places. Within a place, the online unfolding of the computation DAG happens in a depth-first manner to enable efficient space and time execution. To achieve load balancing within a place, work-stealing is enabled to allow load-balanced execution of the computation sub-graph associated with that place. The computation DAG unfolds in an online fashion in a breadth-first manner across places when the threads are pushed (remote spawns) onto remote places for better load balance. This execution strategy leads to low overall stack space requirement as compared to prior approaches which use a combination of work-first and help-first policies [10].

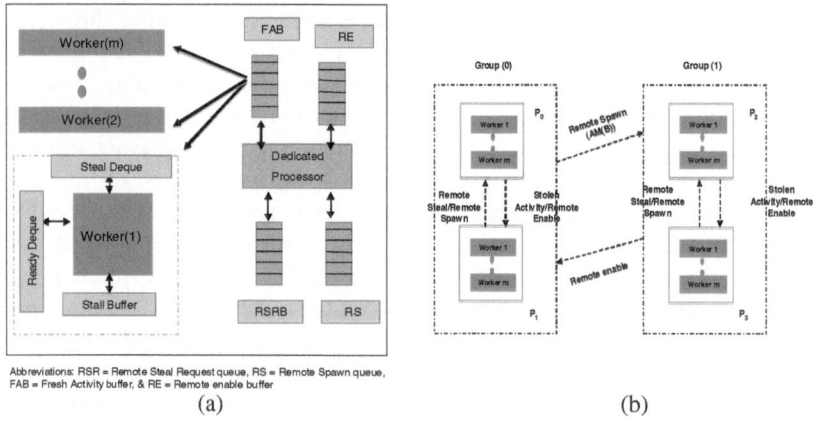

Abbreviations: RSR = Remote Steal Request queue, RS = Remote Spawn queue, FAB = Fresh Activity buffer, & RE = Remote enable buffer

(a)   (b)

Fig. 2. (a) Distributed data-structures at a Place. (b) Work Stealing & Remote Spawns

3.1 Distributed Data Structures

Each place has a dedicated communication processor (different from workers) to man-age remote communication with other places. This processor manages the following data-structures (Fig. 2(b)): (a) *Fresh Activity Buffer(FAB)* is a non-blocking FIFO data-structure. It contains threads that are remote spawned onto this place by remote nodes due to either the place annotation of the thread or the need for inter-group load bal-ancing; (b) *Remote Spawn queue* (*RS*) is a non-blocking FIFO data-structure which contains all the threads that are to be remote spawned by local processors onto remote places; (c) *Remote Enable Buffer* (*RE*) is a non-blocking deque data structure, which contains all the remote enable signals issued by local processors to remote places;and, (d) *Remote Stealing queue* (*RSRQ*) is a FIFO data-structure contains all the remote steal requests received by this particular place from other places within the same group.

Each worker (processor) at a place has the following data-structures (refer Fig. 2(b)): (a) *Ready Deque*: is a deque that contains the threads of the parallel computation that are ready to execute locally. This is accessed by the local processor only; (b) *Steal queue*: is a non-blocking deque that contains threads that are ready to be stolen by the some other processor at a local or remote place. It is accessed by other local processors or communication processor for work stealing from this processor. In helps in reducing the synchronization overheads on the local processor;and, (c) *Stall Buffer*: is a deque that contains the threads that have been stalled due to dependency on another thread that are either spawned locally or remotely in the parallel computation. This is only accessed by the local processor.

3.2 Algorithm Design

During execution, the *LDSS* algorithm is able to keep track of data and control depen-dencies in the computation DAG, by using enable signals. Flow of enable signals across the places is managed by the dedicated communication processor at each place, using the *Remote Enable* buffer. The root place receives communication from each place on the status of work at each place and based on the termination condition of the program, the root node sends termination signal to each node. This technique can be further enhanced using well-known global termination detection techniques. The actions taken by the (general) processors and the dedicated communication processor at each place, P_i, are described below.

General Processor Actions: At any step, a thread A at the rth processor (at place i), W_i^r, may perform the following actions:

1. **Spawn**

    (a) A spawns B locally: B is successfully created and starts execution whereas A is pushed into the bottom of the *Ready Deque*.

    (b) A spawns B remotely: (i) If affinity for B is for a place, P_j, the target_place = P_j. (ii) Else, if B is anyplace thread, then determine the target_place using d-choice randomized selection. (iii) Active message for B is enqueued on head of the *Remote Spawn* queue with the destination as the target_place.

2. **Terminates** (A terminates): The processor at place P_i, W_i^r, where A terminated, picks a thread from the bottom of the *Ready Deque* for execution. If none available in its *Ready Deque*, then it tries to transfer all the threads from *Steal queue* to *Ready Deque* and pick from the bottom of the deque. If *steal queue* is empty then it steals from the top of other processors' *Steal queue*. Each failed attempt to steal from another processor's *Steal queue* is followed by attempt to get the topmost thread from the *FAB* at that place. If there is no thread in the *FAB* then another victim processor is chosen from the same place. If no thread is available at that place, then enable inter place work stealing. (The communication processor helps in inter-place work-stealing by using the d-choice prioritized random selection of victim places and deciding the target_place.)

3. **Stalls** (A stalls): A thread may stall due to control or data dependencies in which case it is put in the *Stall Buffer* in a stalled state. Then same as *Terminates* (case 2) above.

4. **Enables** (A enables B): A thread, A, (after termination or otherwise) may enable a stalled thread B. If B is a local thread then the state of B changes to enabled and it is pushed onto the appropriate position of the *Ready Deque*. It B is remotely stalled then push the enable signal for that place at the bottom of the *Remote Enable* buffer.

Dedicated Communication Processor Actions: At any moment during the execution, the **dedicated communication processor** at place i will try pick up an active message from the bottom of the *Remote spawn queue*. Each failed attempt is followed by attempt to pick up an enable signal from bottom of the *Remote Enable* buffer. If there is no enable signal in *Remote Enable buffer* and inter place work stealing is enabled then it randomly non-uniformly (with priority) chooses d distinct places (with higher priority to places within its own *group*) and sends the active messages requesting workloads. On receiving reply from these places, it selects that target_place (*victim* place) as the one with the highest load (as measured in the prior time interval). If this fails, then it tries to pick up the request from the bottom of the *Remote Stealing* buffer and sends it an available thread at this place. All these operations require asynchronous and one-sided communication with other places. Hence, we implemented our *LDSS* algorithm using *GASNet.*

The dedicated communication processor also helps in maximizing *computation-communication overlap* as well as *communication-communication overlap.* When, a thread needs data from another place, it sends the request to the communication processor. The communication processor forwards that request to the place that contains that data. This goes in parallel with the computation that can be performed at the processor where the data request originated. Hence, one gets computation-communication overlap. Moreover, multiple communication requests including remote steal requests, remote enable and d-choice selection all can proceed in parallel with different places, leading to effective communication-communication overlap.

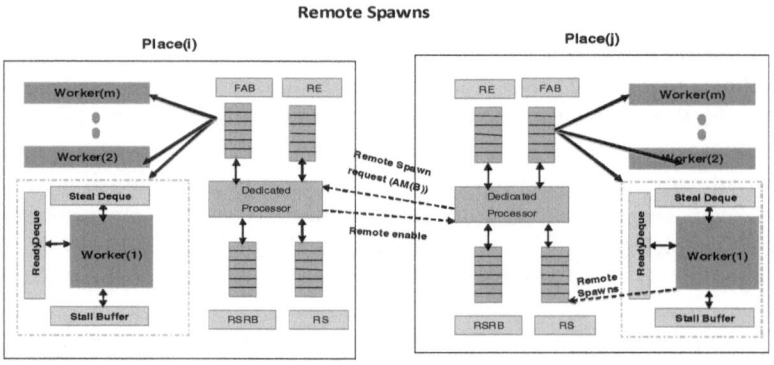

Fig. 3. Remote-spawn from Place(i) to Place(j)

Remote Spawns: Any processor that needs to spawn a thread (Fig 3), enqueues the active message for creation of that thread at the head of the *Remote Spawn* buffer. The dedicated communication processor pops the active message from the *Remote Spawn* buffer and sends it to the appropriate place asynchronously. The communication processor uses d-choice randomized load balancing for choosing the appropriate destination place. Here, random d groups are selected and the one with the lowest load is chosen as the destination. Each place maintains a load vector that contains load of d places. This load vector is updated (d each time) at periodic intervals. The **rate of remote spawns** is adjusted automatically (by considering relative load difference between this node and system average load) to reduce overheads while at the same time provide optimal trade-off between load balancing across the places and scheduling overheads. The d-choice based remote spawns result in good load balancing [11] while keeping low scheduling overheads. It is well-known [11] that pure random assignment of m balls (threads) to n bins ($m >> n$) leads to $O(\sqrt{\frac{m \log(n)}{n}})$ gap across the bins (servers, processors) while d-choice based assignment leads to $O(\ln \ln(n))$ gap across the bins. Thus, the instantaneous load-imbalance across the nodes (places) reduces in the system.

Workstealing: Each core (processor) uses *Ready Deque* (lockless queue for threads intended for local execution and *Steal queue* (synchronized queue) for threads that can be stolen. Each place in the system is associated with one and only one *group*. For inter-place work stealing higher priority is given to the groups close in the target topology as compared to the groups farther away. Once all the processors becomes idle at a place, the dedicated (communication) processor at that place (*thief*) queries a randomly selected place (*victim*) about its load. When the *victim* dedicated communication processor receives the request for worksteal from a *thief*, it randomly deques a thread from one of its local processors' *Ready deque* and sends it to thief place for its continuation.

4   Results and Analysis

Experimental Setup: We used upto 4096 compute nodes/places (with 4 cores/processors per place, total 16384 cores in the system) of Blue Gene/P (*Watson 4P* [9]) for empirical evaluation of our distributed scheduling algorithm. Each compute node (place) in *Watson 4P* is a quad-core chip with frequency of 850 MHz having 4 GB of DRAM and 32 KB of L1 instruction and data cache per core. Nodes (places) are interconnected by a 3D torus interconnect (3.4 Gbps per link in each of the six directions) apart from separate collective and global barrier networks. For efficient compute and communication overlap, *GASNet* was used since it provides asynchronous one sided message passing primitives. GASNet is a language-independent, low level networking layer that provides network-independent, high performance communication primitives tailored for implementing Parallel Global Address Space. GASNet's *DCMF* (Deep Computing Messaging Framework) conduit is the native port of GASNet to BlueGene/P architecture as it uses DCMF for the lower level communication between nodes.

Benchmarks: We implemented our distributed scheduling algorithm (*LDSS*) using pthreads (NPTL API) and GASNet as the underlying communication layer. The LDSS algorithm and benchmarks were compiled using *mpixlc_r* with optimization options -*O3*, *-qarch=450*, *-qtune=450* and *-qthreaded*. We present comparison of performance and scalability with Charm++ [15] on Blue Gene/P architecture and show that we have superior results. The benchmarks used for evaluation include:

- **Unbalanced Tree Search** (UTS): The Unbalanced Tree Search problem is to count the number of nodes in an implicitly constructed tree that is parameterized in shape, depth, size and imbalance, and,
- **NQueens**: NQueens is a backtracking search problem to place N queens on a N by N chess board so that they do not attack each other. We target at finding all solutions for N Queen problem.
  Note: UTS and NQueens are *strict* parallel computations as both of them have parent-child dependencies.

Scalability Analysis: Here, we present scalability analysis for the benchmarks. Fig. 4(a) and Fig. 4(b) demonstrate the strong scalability of *LDSS* algorithm with increasing

---

[9] http://www.research.ibm.com/bluegene

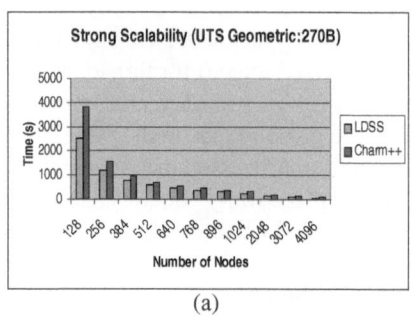 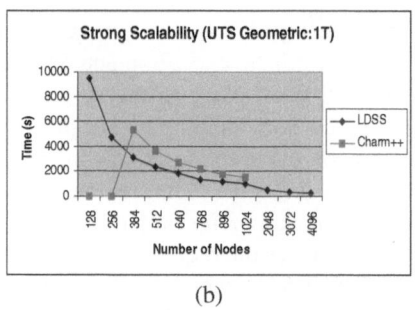

(a)   (b)

Fig. 4. UTS: (a) Strong Scalability (270B) Geometric Tree. (b) Strong Scalability (1T) Geometric Tree

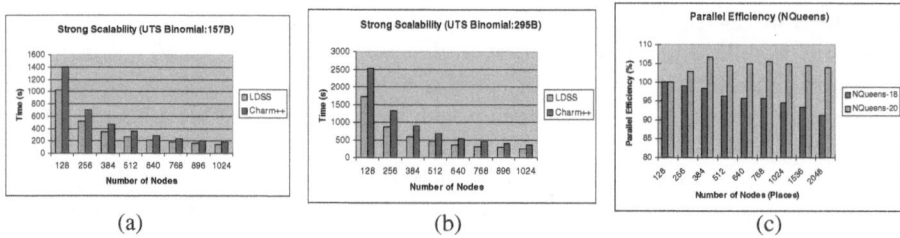

(a)   (b)   (c)

Fig. 5. UTS: (a) Strong Scalability (157B) Binomial Tree. (b) Strong Scalability (295B) Binomial Tree   (c) Strong Scalability (NQueens).

number of nodes (places) from 128 to 4096 for geometric type UTS tree with 270 Billion, and 1 Trillion tree size respectively. Here, the granularity of work-stealing was kept as 50 and the group size chosen was 8. For 270B geometric tree, the *LDSS* algorithm achieves super-inear speedup of around $36.6\times$ (from $2524s$ to $69s$) with $32\times$ increase in number of nodes. The *LDSS* algorithm has better performance as compared to Charm++ by at least 28% throughout the variation in compute nodes. For the 1T geometric tree, the *LDSS* algorithm also achieves super-linear speedup of around $37.5\times$ (from $9491s$ to $253s$) with $32\times$ increase in number of nodes. Here again, the *LDSS* algorithm has better performance ($2352s$) as compared to Charm++ ($3667s$) by around 36% at 512 compute nodes, and by 35% ($1007s$ vs $1541s$) at 1024 nodes. Charm++ gave memory error for 128 and 256 nodes (places).

On 4096 nodes, *LDSS* had a completion time of $69s$ for 270B nodes, $253s$ for 1T nodes, $993s$ for 4T nodes and $4037s$ for 16T nodes; which demonstrates better than linear ***data scalability***. For 16Trillion nodes, LDSS delivers processing rate of 3.96G Nodes/s, which is the best reported so far in the literature. Further, the parallel efficiency achieved is slightly better as compared to the best prior work [7], and this is demonstrated on the geometric tree 16Trillion tree nodes which is largest amongst the maximum sizes considered by any prior work including [14] [16] [10].

Fig. 5(a) and Fig. 5(b) demonstrate the strong scalability of *LDSS* algorithm with increasing number of nodes (places) from 128 to 1024 for binomial type UTS tree with 157 Billion, and 295 Billion tree size respectively. Here, the granularity of work-stealing was kept as 50, the group size chosen was 8, the base remote spawn rate was set at 50 and d was chosen as 3 for d-choice randomized load balancing during remote spawns. For 157B binomial tree, the *LDSS* algorithm achieves a speedup of around $7.34\times$ (from $1021s$ to $139s$) with $8\times$ increase in number of nodes, resulting in parallel efficiency of around 92%. The performance of *LDSS* is better than Charm++ by around 27% at 128 nodes and by around 22% at 1024 nodes.

For the 295B binomial tree, the *LDSS* algorithm also achieves a speedup of around $7.34\times$ (from $1715s$ to $234s$) with $8\times$ increase in number of nodes, resulting in parallel efficiency of 91.75%. The efficiency achieved is better than the best prior work [16] by around 5%. Further, for $295B$ nodes, *LDSS* has lower time than Charm++ by around 32% at 128 nodes and by around 35% at 1024 nodes.

The efficiency for binomial tree is lower than the geometric tree case since the binomial tree has larger depth and smaller breadth and hence more unbalanced as compared to the geometric tree. Due to this, the scheduling algorithm incurs larger overheads of remote spawns and work stealing in-order to achieve load balance across the compute nodes in the system. Hence, the geometric tree is able to achieve high efficiency even without remote spawns. The average (across varying number of compute nodes) single node performance of *LDSS* for Binomial tree is $1.1M$ nodes/sec as compared to $0.85M$ nodes/s for Charm++; while that for Geometric tree it is $0.96M$ nodes/s (for *LDSS*) as compared to $0.70M$ nodes/s for Charm++.

The parallel efficiency results (w.r.t. 128 nodes (places)) for NQueens benchmark are presented in Fig. 5(c). While for NQueens 20, LDSS delivers super-linear scalability and sustains parallel efficiency of 103% even at 2048 nodes; for NQueens 18 the parallel efficiency drops to 91% at 2048 nodes (places). This is due to exponential increase in size of NQueens 20 w.r.t. NQueens 18. For NQueens 18, the parallel efficiency achieved by LDSS is better than that for Charm++ (around 85%) [15].

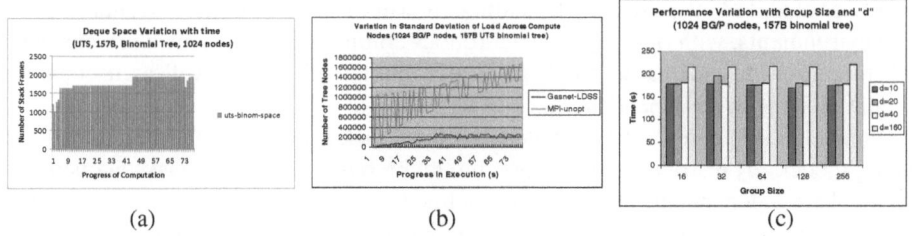

(a)   (b)   (c)

Fig. 6. UTS: (a) Space Usage (UTS). (b) Standard deviation of Load Across Compute Nodes (UTS)   (c) Performance Variation with Group Size and d value.

Fig. 6(c) illustrates the impact of change in group size on the overall performance of the *LDSS* algorithm. This variation is considered with different values of d in d-choice randomized spawns. For most group sizes (16, 64, 128, 256), as the value of d increases, the time increases while for group size 32, d value equal to 40 gives the best performance. Thus, there is an optimal combination of group size and d that gives the best performance. In general, as d increases for a given group size, the communication overheads increase leading to a larger overall time, while for some values of group size, larger d could also lead to better load balance within the system. For $d = 32$

we observed this by studying the load across the compute nodes in the system. This represents complicated trade-offs between load balance and communication overheads involved in scheduling binomial UTS trees.

Fig. 6(a) presents space usage (in number of stack frames) of the total space usage per processor (including the space used by the dedicated communication processor) as the computation progresses in the case of UTS/binomial tree with $157B$ nodes. The maximum space used is less than 2000 stack frames. This at least $3\times$ lesser than that reported by HotSLAW [10], which reports maximum space usage of around 8000 stack frames and stays above 2000 stack frames for quite sometime. This is because LDSS does not use help-first policy which leads to BFS expansion of the graph and larger usage of space, while HotSLAW uses a combination of work-first and help-first policy as does SLAW [6]. Fig. 6(b) reports the standard deviation in load across the compute nodes in the system for LDSS with tuned (*gasnet-opt*) and untuned (*MPI-unopt*) parameters [10], as the computation progresses for $157B$ binomial tree. By using GASNET and parameter tuning LDSS achieves around $8\times$ lower standard deviation (and hence better load balance) as compared to MPI implementation and untuned parameters.

5   Related Work

Distributed Scheduling for parallel computations is a well studied problem in the shared memory context starting from the pioneering research by Blumofe and Leiserson [3] on Cilk scheduling, followed by later work including [2] [1] [4] [6] amongst many others. These efforts are primarily focused on work-stealing efficiency improvement in shared-memory architectures without considering explicit affinity annotations by the programmer. With the advent of distributed memory architectures, lot of recent research on distributed scheduling looks at multi-core and many-core clusters.

Olivier et.al. [14] consider work stealing algorithms in distributed and shared memory environments, with one sided asynchronous communications. This work considers task migration on pull based mechanism and ignores affinity as well as it considers computations with no dependencies.

Dinan et.al. [7] construct distributed and local task pools for its dynamic load balancing model. [7] restricts the execution model by requiring that all tasks enqueued in task pool are independent. The model is confined to tasks that require only parent-child dependencies not other way around. Our model supports all the computations that are *strict* in nature hence allowing tasks to wait for completion of other tasks.

Saraswat et.al. [16] introduce a lifeline based global load balancing technique in *X10* which provides better load balancing for tasks as compared to random work stealing, along with global termination detection using the *finish* (X10) construct. Our algorithm considers multiple workers per place and handles data dependencies across the threads in the computation tree. We demonstrate better efficiency and performance on the binomial tree in UTS benchmark than [16].

Ravichandran et.al. [9] introduce work stealing for multi-core HPC clusters which allow multiple workers per place and two separate queues for local threads and for remote stealing, but this does not consider locality or data dependencies. Min et.al. [10]

---

[10] Rate of remote spawns, rate of workstealing, granularity of workstealing and group size.

present a task library, called *HotSLAW*, that uses *Hierarchical Victim Selection* (HVS) and *Hierarchical Chunk Selection* (HCS) to improve performance as compared to prior approaches. Our *LDSS* algorithm uses an elegant combination of two-level work stealing and remote-place (inter-group) work pushing to achieve optimal trade-offs between load balancing and scheduling overheads. Further, our space requirement is much lower than that reported by [10] for the UTS benchmark as presented in the Results section 4. Frameworks such as Scioto framework [5] and KAAPI [11] consider distributed setup but have not demonstrated results at large scale for state space search problems.

Charm++ is a C++ based parallel programming system that implements a message-driven migratable objects programming model, supported by an adaptive runtime system and work stealing [13] [15]. Charm++ supports work stealing across places [15] and uses a hierarchical mechanism [18] to migrate objects to places (processors) for load balancing. Zheng et.al. [18] consider hierarchical load balancing in Charm++. Our algorithm incorporates randomized d-choice based work pushing and prioritized inter-place work-stealing to ensure better instantaneous load balance across the places in the system. Further, on the *UTS* benchmark we demonstrate upto 40% better performance as compared to Charm++ on Blue Gene/P.

6   Conclusions and Future Work

We have addressed the challenging problem of online distributed scheduling of state space search oriented parallel computations, using a novel combination of d-choice based randomized remote spawns and topology-aware work stealing. On multi-core clusters such as Blue Gene/P (MPP architecture), our *LDSS* algorithm demonstrates superior performance and scalability (for UTS) and parallel efficiency (for NQueens benchmark) as compared to prior state-of-the-art approaches such as Charm++. For *UTS* (binomial tree) we have delivered highest parallel efficiency (close to 92%) for binomial tree (better than [16] which delivers 87%); and upto 40% better performance as compared to Charm++ [15]. In future, we plan to look into balanced allocation [11] based arguments to compute optimum trade-offs between work sharing and work stealing in large scale distributed environments.

References

1. Acar, U.A., Blelloch, G.E., Blumofe, R.D.: The data locality of work stealing. In: SPAA, New York, NY, USA, pp. 1–12 (December 2000)
2. Arora, N.S., Blumofe, R.D., Plaxton, C.G.: Thread scheduling for multiprogrammed multiprocessors. In: SPAA, Puerto Vallarta, Mexico, pp. 119–129 (1998)
3. Blumofe, R.D., Leiserson, C.E.: Scheduling multithreaded computations by work stealing. J. ACM 46(5), 720–748 (1999)
4. Blumofe, R.D., Lisiecki, P.A.: Adaptive and reliable parallel computing on networks of workstations. In: USENIX Annual Technical Conference, Anaheim, California (1997)
5. Dinan, J., Krishnamoorthy, S., Larkins, D.B., Nieplocha, J., Sadayappan, P.: A framework for global-view task parallelism. In: Proceedings of the 37th Intl. Conference on Parallel Processing, ICPP (2008)

---

[11] https://gforge.inria.fr/projects/kaapi/

6. Guo, Y., Zhao, J., Cave, V., Sarkar, V.: Slaw: A scalable localityaware adaptive work-stealing scheduler. In: IPDPS, pp. 1–12 (2010)
7. Dinan, J., Larkins, D.B., Sadayappan, P., Krishnamoorthy, S., Nieplocha, J.: Scalable work stealing. In: SC, Oregon, USA (November 2009)
8. Kalé, L.: An almost perfect heuristic or the n-queens problem. Information Processing Letters 34, 173–178 (1990)
9. Ravichandran, K., Lee, S., Pande, S.: Work Stealing for Multi-core HPC Clusters. In: Jeannot, E., Namyst, R., Roman, J. (eds.) Euro-Par 2011, Part I. LNCS, vol. 6852, pp. 205–217. Springer, Heidelberg (2011)
10. Min, S.-J., Iancu, C., Yelick, K.: Hierarchical work stealing on manycore clusters. In: PGAS 2011: Fifth Conference on Partitioned Global Address Space Programming Models. ACM (October 2011)
11. Mitzenmacher, M.: The Power of Two Choices in Randomized Load Balancing. PhD in computer science. Harvard University (1991)
12. Pearl, J.: Heuristics-Intelligent Search Strategies for Computer Problem Solving. Addison Wesley (1984)
13. Shu, W., Kale, L.: A dynamic scheduling strategy for the chare-kernel system. In: ACM/IEEE Conference on Supercomputing (Supercomputing 1989), New York, USA, pp. 389–398 (1989)
14. Oliver, S., Prins, J.: Scalable dynamic load balancing using UPC. In: ICPP, Oregon, USA (September 2008)
15. Sun, Y., Zheng, G., Jetley, P., Kale, L.V.: An adaptive framework for large-scale state space search. In: IPDPS Workshop, Alaska, USA, pp. 1798–1805 (2011)
16. Saraswat, V., Grove, D., Kambadur, P., Kodali, S., Krisnamoorthy, S.: Lifeline based global load balancing. In: PPOPP, Texas, USA (February 2011)
17. Wah, B.W., Li, G., Yu, C.F.: Multiprocessing of combinatorial search problems. IEEE Computer 18, 93–108 (1985)
18. Zheng, G., Meneses, E., Bhatele, A., Kale, L.V.: Hierarchical load balancing for charm++ applications on large supercomputers. In: 39th International Conference on Parallel Processing Workshops, ICPPW, pp. 436–444. IEEE Computer Society, Washington, DC (2010)

Using Load Information in Work-Stealing
on Distributed Systems
with Non-uniform Communication Latencies

Vladimir Janjic and Kevin Hammond

School of Computer Science, University of St Andrews, United Kingdom
{jv,kh}@cs.st-andrews.ac.uk

Abstract. We evaluate four state-of-the-art work-stealing algorithms for distributed systems with non-uniform communication latenices (Random Stealing, Hierarchical Stealing, Cluster-aware Random Stealing and Adaptive Cluster-aware Random Stealing) on a set of irregular Divide-and-Conquer (D&C) parallel applications. We also investigate the extent to which these algorithms could be improved if dynamic load information is available, and how accurate this information needs to be. We show that, for highly-irregular D&C applications, the use of load information can significantly improve application speedups, whereas there is little improvement for less irregular ones. Furthermore, we show that when load information is used, Cluster-aware Random Stealing gives the best speedups for both regular and irregular D&C applications.

1 Introduction

Work stealing [5], where idle "thieves" steal work from busy "victims", is one of the most appealing load-balancing methods for distributed systems, due to its inherently distributed and scalable nature. Several good work-stealing algorithms have been proposed and implemented for systems with non-uniform communication latencies, that is for cloud- or grid-like systems [2,4,16,17], and for high-performance clusters of multicore machines [13]. However, most of these algorithms are tailored to highly *regular* applications, such as those using simple Divide-and-Conquer (D&C) parallelism. This paper considers how work-stealing can be generalised to *irregular* parallel D&C applications, so covering a wide class of real parallel applications. In particular, we compare the effectiveness of different work-stealing approaches for such applications, and describe improvements to these approaches that provide performance benefits for "more irregular" parallel applications. This paper makes the following main research contributions:

- We compare the performance of state-of-the-art work-stealing algorithms for highly-irregular D&C applications, providing insight into whether the "best" methods for regular D&C applications also perform well for irregular ones.
- We evaluate how well these algorithms could perform if they had access to *perfect* load information, i.e. how much speedup could be improved if this

C. Kaklamanis et al. (Eds.): Euro-Par 2012, LNCS 7484, pp. 155–166, 2012.
© Springer-Verlag Berlin Heidelberg 2012

information was available. This gives insight into whether load information can improve work-stealing and also tests the limits to these improvements.

- We investigate *how accurate* this load information needs to be to provide some benefit. Since it is impossible to obtain fully-acurate instantaneous load information for real distributed systems, this gives insight into whether work-stealing can benefit from less accurate load information.

We address these three issues using high-quality simulations in the SCALES simulator [8]. The decision to use simulations is driven by our goal of considering the theoretical limits to improvements that can be obtained by using fully-accurate load information and also to quantify the extent to which those limits can be approached using more realistic partial load information. This would not be possible using a real implementation, since we cannot instantaneously communicate load information between distributed machines. The general operation of the simulation has been verified against a real distributed system [8], so we have a high degree of confidence in its ability to predict scheduling performance. We stress that, in this paper, unlike other work [16,8,2], we are not concerned with the question of how to *obtain* accurate load information. Rather, we are interested purely in the *impact* of this information on load-balancing. A comparison of different heuristics for obtaining load information can be found in [8].

2   Work-Stealing on Systems with Non-uniform Latencies

We consider distributed clusters, where each cluster contains one or more (parallel) machines (that is a cloud- or grid-like server farm). Each machine in a cluster forms a processing element (PE) that can manage its own set of independent parallel tasks (which will normally be evaluated using lightweight multi-threading) Each PE has its own *task pool*, which records the tasks that are owned by that PE. When a PE starts executing a task, it converts it into a *thread* (which we will assume is fixed to one PE). Tasks can, however, be migrated between PEs. In a work-stealing setting, whenever a PE has no tasks in its task pool, that PE becomes a *thief*. The thief sends a *steal attempt* message to its chosen *target* PE. If the target PE has more than one task in its task pool, it becomes a *victim* and returns one (or more) tasks to the thief. Otherwise, the target can either forward the steal attempt to some other target or, alternatively, a negative response can be sent to the thief, who then deals with it in an appropriate way (either by initiating another steal attempt or by delaying further stealing). The main differences between the various work-stealing algorithms that we consider lie in the way in which thieves select targets, and in the way in which targets that do not have enough tasks respond to steal attempts. This paper considers the following four state-of-the-art work-stealing algorithms:

- *Random Stealing* [4] – Targets are chosen randomly, and targets also forward steal attempts to random PEs. This is used by e.g. Cilk [3] or GUM [15].
- *Hierarchical Stealing* – The PEs are organised into a tree (based on communication latencies). A thief first attempts to steal from all of its children

(which may recursively attempt to steal from their children); only if no work is found will it ask its parent for work. This algorithm is used by the Atlas [2] and Javelin [11] runtime systems for distributed Java.

- *Cluster-Aware Random Stealing (CRS)* – Local stealing (within a cluster) and remote stealing (outside of a cluster) are done in parallel. That is, a thief will attempt to steal from a random PE within its own cluster, and, in parallel, will attempt to steal from a remote PE. Targets always forward the steal attempt to a PE in its own cluster if they have no work. A very similar algorithm is used in the Satin [18] extension to Java, which provides primitivies for divide-and-conquer and master-worker parallelism[1].
- *Adaptive Cluster-Aware Random Stealing (ACRS)* – An improvement to the CRS algorithm, where thieves prefer to steal remote tasks from clusters that are nearer to them [17].

Note that the focus of these algorithms is on the selection of potential victims. This is the main issue for distributed systems with potentially high communication latencies, so that thieves obtain work quickly. Following the usual practice in work-stealing for divide-and-conquer applications, we assume that the *oldest* task from the task pool is sent to the thief in response to the steal attempt, and that the *youngest* task from the PE's task pool is chosen for local execution. This allows locality to be preserved, while large tasks are transferred over the network. Many other factors may also influence speedups, such as task queue locking [10] and identifying termination conditions [13,14]. A comparison of several other policies for task pool management can be found in [9].

3   Irregular Divide-and-Conquer Applications

This paper focuses on the irregularity in parallelism that arises from an unbalanced task tree, i.e. where some of the tasks created by a parallel task are sequential, but where others are themselves parallel. This kind of irregularity arises in many benchmarks for load imbalance as well as in many realistic applications that deal with irregular or unbalanced data-structures. For example, the *Unbalanced Tree Search* benchmark dynamically creates highly-unbalanced trees and then processes their nodes [12]; in the *Bouncing Producer and Consumer* benchmark [6], a producer creates a set of consumer subtasks, and then nominates one of these consumers as a new producer. Unbalanced task trees arise, for example, in real applications that trace particles through a 3D scene (e.g. the *Monte Carlo Photon Transport* algorithm [7], where the input data determines how unbalanced the task tree is). Implementations of the *Min-Max* algorithm, which prune a tree of game positions, also exhibit irregularity of this kind.

We have previously introduced a formal statistical measure for the degree of irregularity for such an application [8]. Intuitively, the more unbalanced the

---

[1] In Satin, a target that has no tasks to send to the thief returns a negative response, rather than forwarding the steal attempt, as here. We have found that our version performs much better in our context.

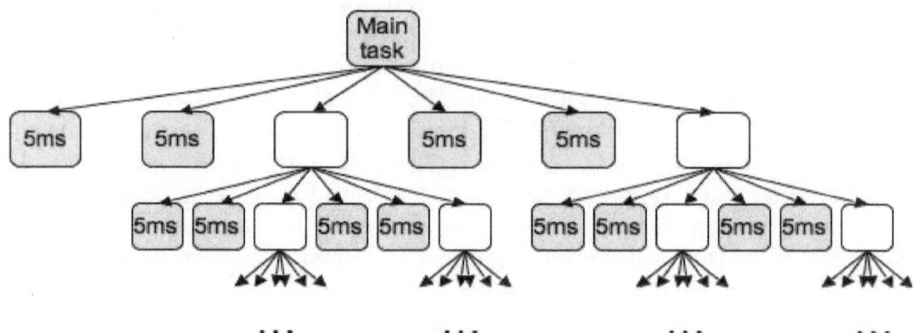

Fig. 1. Task graph for an example *DCFixedPar(6,3,5ms, T)*

task tree of such an application is, the more irregular is the application. Due to space constraints, we will use this intuitive "measure" of the irregularity of an application, without defining it rigorously here. Our main focus is on *divide-and-conquer applications with fixed parallelism*, denoted by *DCFixedPar(n,k,S,t)*. In such an application, every nested-parallel task creates n subtasks, where every k-th subtask is itself nested-parallel (and the others are sequential). Below some threshold t, all tasks are sequential with size S. Such applications are examples of irregular D&C applications, where k determines the degree of irregularity, and n determines the number of tasks. The larger k is, the more unbalanced is the task tree, and, therefore, the more irregular is the application. Figure 1 shows the task tree of an example *DCFixedPar(6,3,5ms,t)* application.

4   Using Load Information

For the work-stealing algorithms presented in Section 2, we can observe that the methods they use for selecting targets are partially (or, in the case of Hierarchical Stealing, fully) based on knowledge of the underlying network topology. They do not, however, depend on information about PE loads. Choosing targets in this way is acceptable for applications where the majority of tasks create additional parallelism (for example, *DCFixedPar(n, k, S, t)*, where k is small). During the execution of such applications a thief that steals a task will itself become a potential victim for some other thief. This means that there are a large number of potential victims in most execution phases. This makes locating victims easy, even when it is done randomly. Indeed, Nieuwpoort et al. [16] show that very good performance can be obtained using these work-stealing algorithms, with CRS giving the best speedups for simple D&C applications (those corresponding to *DCFixedPar(2,1,S,t)* applications). This is because thieves can usually obtain work locally, and remote prefetching of work (via wide-area stealing) essentially comes for free, because the latency is hidden by executing locally obtained work.

The performance of these algorithms for irregular D&C applications is, however, not well understood. In these applications, most of the tasks may be

sequential. This means that it is no longer true that almost every successful thief becomes a potential victim. Furthermore, in some execution phases the number of potential victims may be low, and high load imbalances may then exist. Locating the victims in a potentially large system can be hard if it is done randomly, and a thief may send many fruitless steal attempts over high-latency networks before it manages to locate a victim. Therefore, for irregular D&C applications, the CRS algorithm may not perform the best, and might indeed be outperformed by methods that do some kind of systematic search for targets, from closer to further ones (as with Hierarchical Stealing). In order to obtain good speedups, it may be essential to have some information about target loads, to minimise the time that thieves spend obtaining work. In order to investigate these issues, we first evaluate how basic work-stealing algorithms perform on highly irregular D&C applications, determining which of them gives the best speedups. We then investigate the extent to which these speedups could be increased if *fully accurate* load information is present and used in these algorithms; that is, if each thief, at the moment where it needs to select the stealing target, knows precisely how many tasks are in each PE's task pool. We, therefore, consider the following "perfect" work-stealing algorihms:

- *Perfect Random Stealing.* A target with non-zero load is chosen randomly.
- *Perfect Hierarchical Stealing.* A set of all PEs is organised into a tree. A thief checks the load of all of its children (where the load of a PE is the aggregate load of all of the PEs in its subtree). If a child with non-zero load exists, the steal attempt is sent to it. Otherwise, a thief tries to steal from its own parent. Whenever a target receives a steal attempt, if it has no work to send, it forwards the steal attempt using the same procedure.
- *Perfect Cluster-aware Random Stealing* (Perfect CRS). A thief attempts to steal in parallel from random local and remote PEs with non-zero load.
- *Perfect Adaptive Cluster-aware Random Stealing* (Perfect ACRS). This algorithm is similar to Perfect CRS, except that during the remote stealing, thieves prefer to steal from closer targets with non-zero load.
- *Closest-Victim Stealing* (CV). The closest target with work is chosen.
- *Highest-Loaded-Victim Stealing* (HLV). A thief steals from a target with the largest number of tasks.

Note that the last two algorithms do not have "basic" equivalents, since they depend on the presence of load information. We include them here because they represent fairly intuitive methods for selecting stealing targets in the presence of load information. We assume that a thief steals only one task at a time from a victim. While stealing more than one task may be beneficial where many tasks are sequential, for more regular D&C applications this can result in unnecessarily large amounts of work being transferred from the victim to the thief. Finally, we evaluate how the accuracy of load information relates to the performance of algorithms. In other words, we evaluate what happens if the load information is not completely accurate. This enables us to observe whether the load information needs to be fully accurate (which is impossible to obtain in the real word), or whether some approximation (which can be obtained by a heuristic) is enough.

5 Experiments

All experiments were conducted using the SCALES simulator [8], which was developed for the sole purpose of testing the performance of work-stealing algorithms on parallel systems with non-uniform communication latencies. SCALES supports several popular parallelism models, such as divide-and-conquer, data parallel and master-worker. It independently simulates the load-balancing events for each PE, such as sending/forwarding steal attempts and the transfer of tasks between PEs. It also simulates the overheads for individual load-balancing events (such as sending steal attempts, packing and unpacking of tasks, executing tasks). SCALES has been shown to accurately estimate speedups under various work-stealing algorithms for realistic runtime systems [8].

In order to keep the number of experiments manageable, we use the same simulated system in all of our experiments. A number of experiments on other simulated systems can be found in [8], which confirm the conclusions found here. Our system consists of 8 clusters of PEs, with 8 PEs in each cluster. Clusters are split into two *continents* of 4 clusters each, with an inter-continental latency of 80ms. Each continent is split into two *countries*, with an inter-country latency of 30ms. Finally, each country is split into two sites, with an inter-site latency of 10ms. In the remainder of the paper, the PE that executes the main application task is the *main* PE; the cluster containing the main PE is the *main cluster*; and all other clusters are *remote* clusters.

5.1 Performance of the Basic Algorithms

For our first set of experiments with irregular D&C applications, we focus on the *DCFixedPar(40,k,5ms,4)* applications. The size of sequential subtasks is set to *5ms* to produce an application with fine-grained tasks. Note that, as k increases, the applications become more irregular. Figure 2 shows the speedups that we obtained under the basic algorithms. We observe that the CRS and ACRS algorithms give the best speedups for more regular applications. However, as the applications become more irregular (for $k > 6$), we observe that Hierarchical Stealing starts to outperform both CRS and ACRS. The reason for this is that Hierarchical Stealing gives much more uniform work distribution than CRS, where most of the tasks are executed by PEs from the main cluster, and where PEs in the remote clusters are mostly idle. For highly irregular applications we observe that CRS and ACRS deliver poor speedups of 10-12, whereas Hierarchical Stealing still manages to deliver good speedups of 25-30. This experiment reveals two things. Firstly, it shows that the situation for irregular applications is less clear cut than for regular ones, where the CRS algorithm constantly delivers the best speedups. We can see that for less irregular applications, CRS is still the best choice. For highly-irregular ones, however, Hierarchical Stealing is better. Secondly, as the irregularity of the applications increases, speedups decrease sharply for most of the algorithms. The exception is Hierarchical Stealing, which still manages to deliver good speedups, even for highly-irregular applications. A similar situation exists for other *DCFixedPar* applications. If we increase the

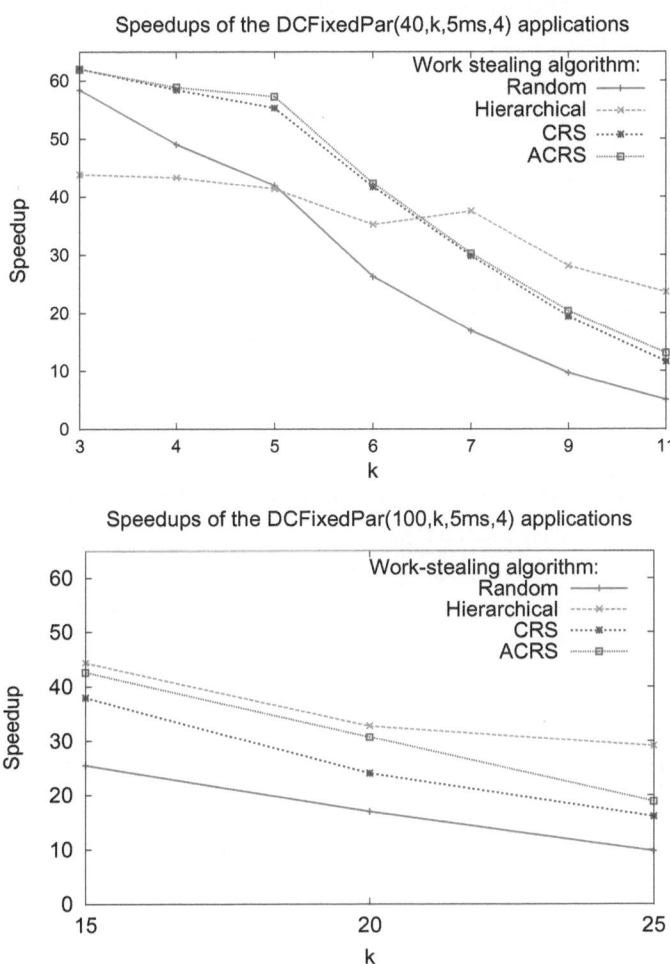

Fig. 2. Speedups under the basic algorithms for the *DCFixedPar(40,k,5ms,4)* applications (above) and *DCFixedPar(100,k,5ms,4)* applications (below)

number of subtasks from 40 to 100, as shown in the bottom of Figure 2, we obtain similar results. Since all the *DCFixedPar(100,k,5ms,4)* applications are highly irregular, Hierarchical Stealing gives the best speedups.

5.2 Performance of the Perfect Algorithms

We now consider the perfect algorithms. Figure 3 shows the corresponding speedups and relative improvements for the *DCFixedPar(40,k,5ms,4)* applications. It is clear that CRS and ACRS give the best speedups. Since all thieves know exactly where to look for work, thieves from remote clusters manage to steal a lot of work, so Perfect CRS does not suffer from the same problem as the

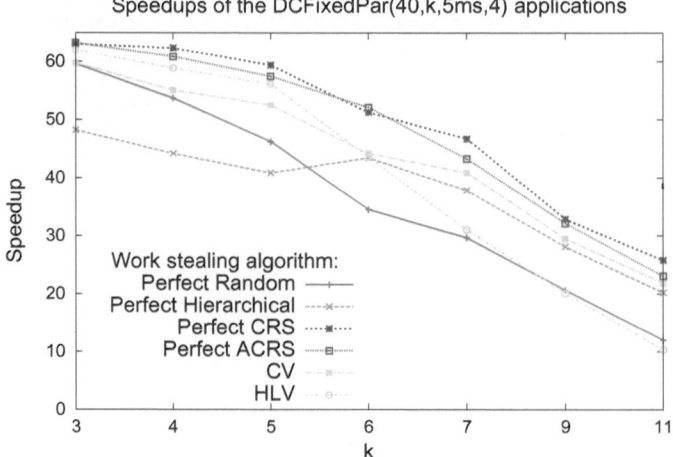

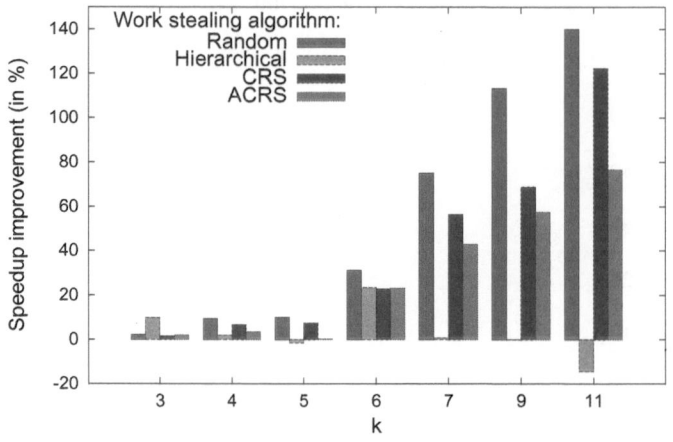

Fig. 3. Speedups (above) and relative Improvements in speedups (below) using perfect load information for the *DCFixedPar(40,k,5ms,4)* applications

basic version of CRS. CV and Hierarchical Stealing perform similarly to CRS and ACRS for more irregular applications, but are consistently worse. Random Stealing and HLV are considerably worse than these four algorithms. The bottom half of Figure 3 shows the improvement in speedup for the perfect versions. We observe small improvements for less irregular applications, but very good improvements (70%-120%) for highly-irregular applications for Random Stealing, CRS, and ACRS. For Hierarchical Stealing, we only observe very small improvements. In some cases, Hierarchical Stealing without load information is actually better than with fully accurate load information. Finally, Figure 4 shows corresponding speedups for the *DCFixedPar(100,k,5ms,4)* applications. We once again observe that CRS and ACRS give the best speedups.

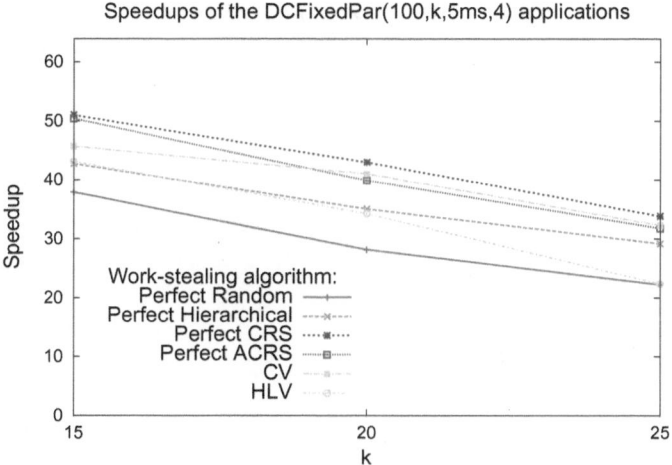

Fig. 4. Speedups under the perfect algorithms for *DCFixedPar(100,k,5ms,4)*

5.3   How Accurate Does Load Information Need to Be?

In the previous section we have seen that the use of fully-accurate load information brings significant speedup benefits when using the Random, CRS and ACRS algorithms for highly-irregular D&C applications. With fully accurate load information, CRS and ACRS give the best speedups both for less irregular and for more irregular applications. Coupled with the fact that these algorithms are also the best for regular applications, it seems that they are the algorithms of choice for work stealing on distributed systems, *provided* we can obtain a good approximation of PE loads during the execution of the application.

In this section, we will focus on the CRS algorithm, since it gives similar results to ACRS, and is easier to implement. The natural question to ask is *how accurate* load information needs to be for the load-based CRS algorithm to obtain good speedup improvements over the basic version. It is obvious that obtaining *perfect* load information (where each PE has fully accurate information about the load of all other PEs) is simply impossible in real systems. We, therefore, now investigate the extent to which an application's speedup changes when load information becomes outdated to some extent.

Let us denote the set of all PEs by $\{P_1, \ldots, P_n\}$, and the load of PE P at time t by $L(P,t)$. The load information $I(Q,t)$ that a PE Q has at time t can then be represented as a set $\{(P_i, L(P_i, t_i)|P_i \in \{P_1, \ldots, P_n\}\}$ of PE-load pairs, where the load of P_i was accurate at time t_i. For perfect information (denoted by *PI*), $t_i = t$ for all PEs, so $PI(Q,t) = \{(P_i, L(P_i,t)|P_i \in \{P_1, \ldots, P_n\}\}$. We introduce the idea of *outdated information with a delay of k time units* by

$$OI(Q,t,k) = \{(P_i, L(P_i, t - lat(Q,P_i) - k)|P_i \in \{P_1, \ldots, P_n\}\},$$

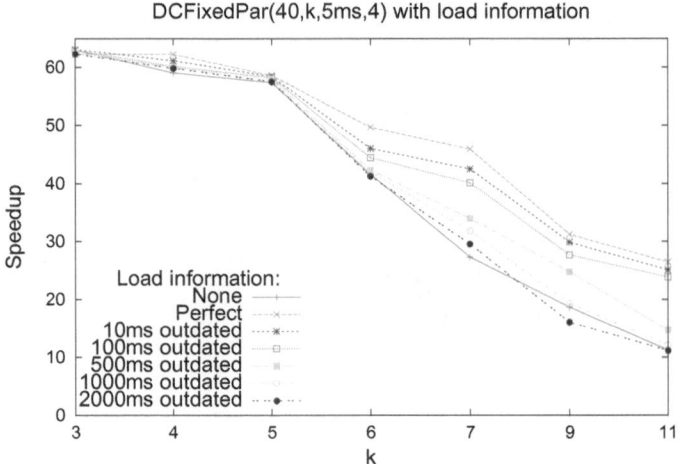

Fig. 5. Speedups with outdated load information when using CRS

where $lat(Q, P_i)$ is the communication latency between PEs Q and P_i. This represents a more realistic setup, where the age of the load information that PE Q has about PE P depends on the communication latency between P and Q, and also on the fixed delay k (in time units) of the delivery of such information. In an even more realistic setup, k could be a function, rather than a constant, so this delay could be different for different PEs.

Figure 5 shows the speedups of the *DCFixedPar(40,k,5ms,4)* applications under the CRS algorithm using outdated load information with various delays (in *ms*). As expected, speedups decrease as the load information become more outdated. However, we can still observe good speedups when the information is relatively recent (with a delay of up to 100ms). With a delay of 500ms, the information is still usable (i.e. speedups are notably better with than without load information). For large delays of 1000ms and 2000ms, the load information becomes practically unusable. Note that the load information delays in our experiments are rather high with respect to the sizes of sequential tasks. In applications with coarser-grained tasks, load information with the same delay will have less impact on the application's speedup, since PE loads will change more slowly. We therefore conclude that significant speedup improvements can be achieved for the CRS algorithm not only with perfect information, but also with relatively recent load information (which it is possible to obtain in real systems). Only if load information is completely outdated does it becomes unusable.

6   Conclusions and Future Work

In this paper, we have investigated the performance of four state-of-the-art distributed work-stealing algorithms (Random Stealing, Hierarchical Stealing,

Cluster-aware Random Stealing and Adaptive Cluster-aware Random Stealing) for irregular Divide-and-Conquer parallel applications. We have shown that, surprisingly, for highly-irregular applications Hierarchical Stealing delivers the best speedups. This differs from regular D&C applications, where previous work [16,8] has shown that CRS delivers the best speedups.

We have also investigated the speedup improvements that can be made if accurate system load is available. Our results show that perfect load information brings significant speedup benefits for highly-irregular D&C applications. Surprisingly, for less irregular ones, the availability of perfect load information is not too significant. Our results show that if some load information is available, then the CRS and ACRS algorithms deliver the best speedups for irregular D&C applications. We have, moreover, shown that in order to obtain good speedups with these algorithms, it is not necessary for the load information to be perfect: a good approximation also suffices. This clearly indicates that the CRS and ACRS algorithms are the best ones to choose for work-stealing on distributed systems with non-uniform and potentially high communication latencies.

In future, we plan to extend the CRS algorithm with mechanisms for approximating load information. Several good mechanisms for obtaining good approximations of load information already exist, e.g. Grid-GUM [1] uses a fully-distributed mechanism of load information propagation, and Atlas uses a Hierarchical mechanism [2]. We intend to consider a combination of fully-distributed and fully-centralised approaches, where load information is centralised within the low-latency networks, and distributed over high-latency ones. We also intend to investigate heuristics for estimating the number of tasks that should be transferred in one steal operation and the impact that sending more than one steal attempt has on various algorithms.

Acknowledgments. This research is supported by European Union grants SCIEnce (RII3-CT-026133), ADVANCE (IST-248828) and ParaPhrase (IST-288570), and by EPSRC grant HPC-GAP (EP/G 055181).

References

1. Al Zain, A.D., Trinder, P.W., Michaelson, G.J., Loidl, H.-W.: Managing Heterogeneity in a Grid Parallel Haskell. Scalable Computing: Practice and Experience 7(3), 9–25 (2006)
2. Baldeschwieler, J.E., Blumofe, R.D., Brewer, E.A.: ATLAS: An Infrastructure for Global Computing. In: Proc. 7th Workshop on System Support for Worldwide Applications, pp. 165–172. ACM (1996)
3. Blumofe, R.D., Joerg, C.F., Kuszmaul, B.C., Leiserson, C.E., Randall, K.H., Zhou, Y.: Cilk: An Efficient Multithreaded Runtime System. In: Proc. PPoPP 1995: ACM Symp. on Principles and Practice of Parallel Prog., pp. 207–216 (1995)
4. Blumofe, R.D., Leiserson, C.E.: Scheduling Multithreaded Computations by Work Stealing. Journal of the ACM 46(5), 720–748 (1999)
5. Burton, F.W., Sleep, M.R.: Executing Functional Programs on a Virtual Tree of Processors. In: Proc. FPCA 1981: 1981 Conf. on Functional Prog. Langs. and Comp. Arch., pp. 187–194. ACM (1981)

6. Dinan, J., Larkins, D.B., Sadayappan, P., Krishnamoorthy, S., Nieplocha, J.: Scalable Work Stealing. In: Proc. SC 2009: Conf. on High Performance Computing Networking, Storage and Analysis, pp. 1–11. ACM (2009)
7. Hammes, J., Bohm, W.: Comparing Id and Haskell in a Monte Carlo Photon Transport Code. J. Functional Programming 5, 283–316 (1995)
8. Janjic, V.: Load Balancing of Irregular Parallel Applications on Heterogeneous Computing Environments. PhD thesis, University of St Andrews (2011)
9. Janjic, V., Hammond, K.: Granularity-Aware Work-Stealing for Computationally-Uniform Grids. In: Proc. CCGrid 2010: IEEE/ACM Intl. Conf. on Cluster, Cloud and Grid Computation, pp. 123–134 (May 2010)
10. Michael, M.M., Vechev, M.T., Saraswar, V.A.: Idempotent Work Stealing. In: Proc. PPoPP 2009: 14th ACM SIGPLAN Symp. on Principles and Practice of Parallel Prog., pp. 45–54 (2009)
11. Neary, M.O., Cappello, P.: Advanced Eager Scheduling for Java-Based Adaptively Parallel Computing. In: Proc. JGI 2002: Joint ACM-ISCOPE Conference on Java Grande, pp. 56–65 (2002)
12. Olivier, S., Huan, J., Liu, J., Prins, J.F., Dinan, J., Sadayappan, P., Tseng, C.-W.: UTS: An Unbalanced Tree Search Benchmark. In: Almási, G.S., Caşcaval, C., Wu, P. (eds.) LCPC 2006. LNCS, vol. 4382, pp. 235–250. Springer, Heidelberg (2007)
13. Ravichandran, K., Lee, S., Pande, S.: Work Stealing for Multi-core HPC Clusters. In: Jeannot, E., Namyst, R., Roman, J. (eds.) Euro-Par 2011, Part I. LNCS, vol. 6852, pp. 205–217. Springer, Heidelberg (2011)
14. Saraswat, V.A., Kambadur, P., Kodali, S., Grove, D., Krishnamoorthy, S.: Lifeline-based Global Load Balancing. In: Proc. PPoPP 2011: 16th ACM Symp. on Principles and Practice of Parallel Prog., pp. 201–212 (2011)
15. Trinder, P.W., Hammond, K., Mattson Jr., J.S., Partridge, A.S., Peyton Jones, S.L.: GUM: A Portable Parallel Implementation of Haskell. In: Proc. PLDI 1996: ACM Conf. on Prog. Lang. Design and Implementation, pp. 79–88. ACM (1996)
16. Van Nieuwpoort, R.V., Kielmann, T., Bal, H.E.: Efficient Load Balancing for Wide-area Divide-and-Conquer Applications. In: Proc. PPoPP 2001: 8th ACM SIGPLAN Symp. on Principles and Practice of Parallel Prog., pp. 34–43 (2001)
17. Van Nieuwpoort, R.V., Maassen, J., Wrzesinska, G., Kielmann, T., Bal, H.E.: Adaptive Load Balancing for Divide-and-Conquer Grid Applications. Journal of Supercomputing (2004)
18. Van Nieuwpoort, R.V., Wrzesińska, G., Jacobs, C.J.H., Bal, H.E.: Satin: A High-Level and Efficient Grid Programming Model. ACM TOPLAS: Trans. on Prog. Langs. and Systems 32(3), 1–39 (2010)

Energy Efficient Frequency Scaling and Scheduling for Malleable Tasks

Peter Sanders and Jochen Speck

Department of Informatics
Karlsruhe Institute of Technology, Germany
{sanders,speck}@kit.edu

Abstract. We give an efficient algorithm for solving the following scheduling problem to optimality: Assign n jobs to m processors such that they all meet a common deadline T and energy consumption is minimized by appropriately controlling the clock frequencies of the processors. Jobs are malleable, i.e., their amount of parallelism can be flexibly adapted. In contrast to previous work on energy efficient scheduling we allow more realistic energy consumption functions including a minimum and maximum clock frequency and a linear term in energy consumption. We need certain assumptions on the speedup function of the jobs that we show to apply for a large class of practically occurring functions.

1 Introduction

In recent years the environmental impact of computing systems and their energy usage was increasingly recognized as an important issue. Additionally the energy costs of computers became a more important part of the total costs of computing systems. Hence much research effort was put into reducing the energy usage of computers. Frequency scaling and power gating (switch off entire processors if they are not needed) which were originally invented to increase battery life time of mobile devices became more and more common in PC and server systems. Theoretical research work in this area mostly considered NP-hard problems or jobs which are only sequential (see [7] and [3] for example). The well known YDS algorithm [8] which also computes an energy-optimal schedule uses a similar energy model as our work but also only considers one processor and serial jobs. This work is focused on the energy minimization problem for jobs with flexible parallelism called malleable jobs as defined in [4]. In this model we will be able to adapt previous work on makespan optimization [2], [1] and [6] to achieve polynomial time algorithms.

The main result of our work is that we can compute the optimal schedule and frequencies to minimize the energy usage for n jobs on m processors in time $\mathcal{O}(n \log(mn) \cdot \log m)$ or even in $\mathcal{O}\left(\left(\frac{n}{m} + 1\right) \log m \cdot \log(mn)\right)$ if we use all m processors for the computation. We need some assumptions about the speedup functions but show that these are fulfilled in many common cases. We also generalize the model to allow an additional linear term in energy consumption. Furthermore, we consider the problem with minimal and maximal clock frequencies,

C. Kaklamanis et al. (Eds.): Euro-Par 2012, LNCS 7484, pp. 167–178, 2012.

for which we give an ϵ-approximation. Note that this is an important step to bridge the gap between real world processors and the simple power laws usually used in theoretical results.

The rest of the paper is organized as follows: In Section 2 we describe the model. In Section 2.1 we show that the assumptions we made are commonly fulfilled. The problem for one job is solved in Section 3. In Section 4 we introduce noninteger processor numbers and the energy function for them. The algorithm is given in Section 5. The model is generalized in Section 6.

2   Model

We have a system consisting of m uniform (or physical identical) processors whose speed (processing frequency) can be adjusted independently. Since one usually has to increase the processor voltage as one increases the processing frequency and the voltage is proportional to the electrical current, it is plausible that the power consumption of a processor is proportional to the cubic frequency. For our model we set the power consumption of one processor proportional to f^α with $\alpha > 2$ and processing frequency f. Frequency changes of processors are immediate and with negligible overhead. If a processor is not used, the power consumption of that processor will be 0 during that time. In the initial machine model we have no maximum frequency and the power consumption of a processor approaches 0 if the frequency approaches 0. An enhanced model is introduced in Section 6. For the initial model we get

$$E = p \cdot f^\alpha \cdot T$$

for the energy consumption E of a job which runs on p processors with frequency f for a time T.

The n jobs we want to schedule on this machine are ready at time 0 and have a common deadline T. Job $j \in \{1, \ldots, n\}$ requires w_j clock cycles on a single processor. The jobs are malleable which means that the number of processors working on them can be changed over time. There is no penalty for changing the number of processors. However, a function $s_j(p)$ gives the speedup achieved by using p processors for job j, i.e., the amount of work done for a job j during a time interval of length t on p processors running at frequency f is $f \cdot s_j(p) \cdot t$ (we will drop the j of s_j when there is only one job or the job is clear).

For the rest of the article we assume that two restrictions are met for the speedup functions s. In Section 2.1 we argue that these conditions are met by a large class of relevant speedup functions.

The **first restriction** is that all speedup functions are concave at least between 0 and the processor number $\hat{p}$ where the maximal speedup is reached. Of course the speedup for 0 processors is 0 and the speedup for 1 processor is 1.

The **second restriction** is that the function h defined as $h(0) = 0$ and for other p as

$$h : p \mapsto \sqrt[\alpha-1]{\frac{s^\alpha(p)}{p}}$$

is monotonically increasing for all $p < \bar{p}$ for a $\bar{p} \in \mathbb{N}$ and monotonically decreasing for all $p > \bar{p}$. Additionally h should be concave on $(0, \bar{p}]$. We set $\bar{p}$ to be the smallest value for which h reaches its maximum then h is strictly increasing on $(0, \bar{p}]$ because of the concavity on $(0, \bar{p}]$. Both functions (h, s) are in general only known for integer values of p. We can see directly $\bar{p} \leq \hat{p}$.

As this work heavily depends on convexity or concavity properties of functions we restate the following definition known from calculus:

Definition 1. *A function f is said to be concave (convex) on an interval $[a, b]$ when for all $x, y \in (a, b)$ with $x < y$ the following holds:*
For each $r \in (0, 1)$ with $z = ry + (1 - r)x$ we have $f(z) \geq (1 - r)f(x) + rf(y)$ $(f(z) \leq (1 - r)f(x) + rf(y))$.

For functions which are two times continuously differentiable it is known that concavity (convexity) is equivalent to $f''(z) \leq 0$ $(f''(z) \geq 0)$ $\forall z \in (a, b)$.

2.1   Common Speedup Functions

All speedup functions considered in this section are given as closed formulas and are two times continuously differentiable except for a finite number of points. We want to show in this section that the quite technical restrictions from the model hold for many typical speedup functions and hence do not restrict the use of the results of this paper.

Linear Speedup Functions. One of the easiest types of parallel jobs are those with linear speedup between 1 and p^* processors and $s(p) = s(p^*)$ for all $p > p^*$ (even for embarrassingly parallel jobs the maximal speedup is reached for $p = m$). If we set $\bar{p} = p^*$ then h is monotonically decreasing for all $p > \bar{p}$ and $h(p) = p$ for all $p \in (0, \bar{p})$ which is monotonically increasing and concave. The speedup function is also concave between 0 and p^*. Hence these kind of jobs fulfills the restrictions.

Jobs with Amdahl's Law. Another kind of jobs one often sees are these which fulfill Amdahl's Law. If the amount of sequential work is $\frac{w}{k}$ and the amount of parallelizable work is $(1 - \frac{1}{k})w$, we get

$$s(p) = \frac{w}{\frac{w}{k} + (1 - \frac{1}{k})\frac{w}{p}} = \frac{k \cdot p}{p + k - 1}$$

as speedup function. Doing the math we get $h'(p) > 0 \Leftrightarrow p < (\alpha - 1)(k - 1)$ and thus $\bar{p} = (\alpha - 1)(k - 1)$. For $h''(p)$ we get $h''(p) < 0 \Leftrightarrow p < 2 \cdot (\alpha - 1)(k - 1)$. Hence h is concave for all $p \in (0, \bar{p})$ and monotonically increasing in $(0, \bar{p})$ and monotone decreasing in $(\bar{p}, \infty)$. The speedup function is concave for all p. Thus these kind of jobs fulfills the restrictions.

Jobs with Parallelization Overhead. A situation also very common in parallel computing is that one can parallelize all the work but one has to pay an additional overhead g depending on the number of processors with $g(p) \geq 0$ and $g'(p) \geq 0$ for all $p \geq 1$. Then we get this speedup function:

$$s(p) = \frac{w + g(1)}{\frac{w}{p} + g(p)} = \frac{p \cdot w + p \cdot g(1)}{w + p \cdot g(p)}$$

With some calculus we get:

$$h'(p) > 0 \Leftrightarrow (\alpha - 1)w > p \cdot (g(p) + \alpha p g'(p))$$
$$h''(p) < 0 \Leftrightarrow$$
$$p \cdot (g(p) + pg'(p))^2 \frac{2\alpha - 1}{\alpha - 1} < (w + pg(p))(2g(p) + 4pg'(p) + p^2 g''(p))$$

All p which fulfill $h'(p) > 0$ fulfill also $h''(p) < 0$ if $g'(p) + pg''(p) \geq 0$. To have only one local maximum of h the term $p \cdot (g(p) + \alpha p g'(p))$ has to be strictly increasing but this is the case if $2g'(p) + pg''(p) > 0$. Hence if $2g'(p) + pg''(p) > 0$ holds for an overhead g then the second restriction is fulfilled.

If $w = p^2 g'(p)$ then $s'(p) = 0$ and if the function $p \mapsto p^2 g'(p)$ is strictly increasing this only holds for only one $p = p^*$. If $2g'(p) + pg''(p) > 0$ then $p \mapsto p^2 g'(p)$ is strictly increasing and $s''(p) < 0$ for all $p \in (0, p^*]$. For $\hat{p} = p^*$ the first restriction is fulfilled.

Hence if $2g'(p) + pg''(p) > 0$ and $g(p) \geq 0$ and $g'(p) \geq 0$ hold for all $p \geq 1$ both restrictions are fulfilled. We can check this condition for some overheads:

$g(p)$	$g'(p)$	$g''(p)$	$2g'(p) + pg''(p)$
p	1	0	2
$\log p$	p^{-1}	$-p^{-2}$	p^{-1}
$p \log p$	$\log p + 1$	p^{-1}	$3 + 2\log p$
$\sqrt{p}$	$(2\sqrt{p})^{-1}$	$-(4p\sqrt{p})^{-1}$	$3 \cdot (4\sqrt{p})^{-1}$
p^2	$2p$	2	$6p$
$\log^2 p$	$2p^{-1}\log p$	$2p^{-2} \cdot (1 - \log p)$	$2p^{-1} \cdot (1 + \log p)$

So the condition is fulfilled for all of these overheads and all possible sums of these overheads.

3   The Optimal Solution for the Single-Job Case

In this section we want to compute the optimal number of processors in terms of energy usage for a job which fulfills the restriction from Section 2. Now we will give some lemmata which also will be useful for the multi-job case:

Lemma 1. *In an optimal solution for any number of processors p used during the computation of the job the processing frequency is always the same if you use p processors.*

The detailed proof will be given in the full paper.

Lemma 2. *For only one job running on the machine it is optimal to use $\bar{p}$ processors for the whole time $[0, T]$.*

Proof. If we want a job to do an amount of work w during a time interval I of length t with a constant number of processors, the operating frequency depends only on the speedup reached during I: $f = \frac{w}{s(p) \cdot t}$. Thus the energy becomes now a function only dependent on p:

$$E = \frac{w^\alpha}{t^{\alpha-1}} \cdot \frac{p}{s^\alpha(p)} = \frac{w^\alpha}{t^{\alpha-1}} \cdot \frac{1}{h^{\alpha-1}(p)}$$

Thus the energy consumption is minimized if h is maximized and it is optimal to use $\bar{p}$ processors during I (this is also true for $w = 0$).

Hence this is true for all subintervals of $[0, T]$ and thus it is optimal to use $\bar{p}$ processors all the time. □

Because of Lemma 1 the work done in an interval I for a job which runs on $\bar{p}$ processors for the whole time $[0, T]$ is proportional to the length of I. Because of Lemma 2 the algorithm to find the optimal number of processors for a job consists only of the search for the integer value $\bar{p}$ that maximizes h. It is also obvious that E is strictly decreasing on $[1, \bar{p}]$. It is possible that $\bar{p} > m$ but as h is strictly increasing in $(0, \bar{p})$, the minimal E is reached for $p = m$ in this case.

4   Energy Function for Non-integer Processor Numbers

Until this section all processor numbers were integer. In order to solve the energy-optimal scheduling problem for multiple jobs we need to handle noninteger processor numbers. For a noninteger processor number we introduce the notation $p + \tau$ for the rest of the paper where p is the integer part and $\tau \in [0, 1)$. The main goal of this Section is to compute the minimal energy consumption (and according optimal schedule) of a job with work w to be done in time T when the job can use an average number of $p + \tau$ processors. A job runs on an average processor number $p + \tau = T^{-1} \sum_{p=0}^{m} t_p p$ during time $T = \sum_{p=0}^{m} t_p$ if it runs on p processors for time t_p. We prove that for any average processor number $p + \tau$ Consider a job that runs on an average processor number $p + \tau$. We prove that it is optimal to only use p and $p + 1$ processors for the job. Then we define the the energy function E of the job to be the mapping from the average processor number $p + \tau$ to the minimal possible energy consumption with this processor number. In the end we will show some useful properties of E. All processor numbers in this section are not bigger than $\bar{p}$ otherwise you could shrink all larger processor numbers to $\bar{p}$ without using more energy (because of Lemma 2).

Lemma 3. *If the average number of processors is τ then it is optimal to run on 1 processor for time τT and not to run for time $(1 - \tau)T$.*

Proof. We consider the case of a job which runs on p processors for a time interval I_1 of length t_1 and runs on 0 processors (or does not run) for a time interval I_2 of length t_2. Let w be the total work done in I_1 (and I_2) and $E_{old} = \frac{w^\alpha}{t_1^{\alpha-1}} \cdot \frac{p}{s^\alpha(p)}$ be the energy used. Set $\tau := \frac{t_1 \cdot p}{t_1 + t_2}$. We will now use 1 processor for time $\tau(t_1 + t_2)$ and 0 processors for time $(1 - \tau)(t_1 + t_2)$ and show that this does the same work

with no more energy. The energy used to do work w with 0 and 1 processor is
$$E_{new} = \frac{w^\alpha}{(\tau \cdot (t_1 + t_2))^{\alpha - 1}} \cdot \frac{1}{s^\alpha(1)} = \frac{s^\alpha(p)}{p^\alpha} \cdot E_{old} \leq E_{old}.$$
The repeated use of this argument for all intervals where the job runs on more than one processor shows that if the average number of processors used during time T is $\tau \leq 1$ then it is optimal to use one processor during time $\tau \cdot T$ and 0 processors during time $(1 - \tau) \cdot T$.   □

Lemma 4. *If the average number of processors is $p + \tau \geq 1$ then it is optimal to run on at least 1 processor throughout $[0, T]$.*

The detailed proof will be given in the full paper.

Lemma 5. *If the average number of processors is $p + \tau \geq 1$ then it is optimal to run on $p+1$ processors for time τT and to run on p processors for time $(1-\tau)T$.*

Proof. We consider the case of a job which runs on p_1 processors for a time interval I_1 of length t_1 and runs on p_2 processors for a time interval I_2 of length t_2 (because of Lemma 4 $p_1, p_2 > 0$) w.l.o.g. $0 < p_1 < p_2 \leq \bar{p}$. Let w_i be the work done in I_i and $E_i = \frac{w_i^\alpha}{t_i^{\alpha-1}} \cdot \frac{p_i}{s^\alpha(p_i)}$ be the energy used in I_i for $i \in \{1, 2\}$. Let $w = w_1 + w_2$ be the total work done. We now compute how the work w is optimally distributed between I_1 and I_2. If we do βw work during I_1 and $(1 - \beta)w$ work during I_2 we get the following energy as a function of $\beta \in (0, 1)$:
$$E(\beta) = \frac{\beta^\alpha w^\alpha}{t_1^{\alpha-1}} \cdot \frac{p_1}{s^\alpha(p_1)} + \frac{(1 - \beta)^\alpha w^\alpha}{t_2^{\alpha-1}} \cdot \frac{p_2}{s^\alpha(p_2)}$$
We now have to find the minimum of $E(\beta)$. We do this by computing the β with $E'(\beta) = 0$. This β is a minimum because $E''(\beta) > 0$ for all $\beta \in (0, 1)$. For $A = \frac{p_1}{s^\alpha(p_1)t_1^{\alpha-1}}$ and $B = \frac{p_2}{s^\alpha(p_2)t_2^{\alpha-1}}$ the minimizing β is
$$\beta = \frac{\sqrt[\alpha-1]{B}}{\sqrt[\alpha-1]{A} + \sqrt[\alpha-1]{B}}$$
Thus the value with the optimal β for E is $E = w^\alpha \cdot (t_1 h(p_1) + t_2 h(p_2))^{-\alpha+1}$.

We set $p := \lfloor \frac{t_1 p_1 + t_2 p_2}{t_1 + t_2} \rfloor$ and $\tau := \frac{t_1 p_1 + t_2 p_2}{t_1 + t_2} - p$ then $p + \tau$ is the average number of processors used during I_1 and I_2 and $p_1 \leq p < p + 1 \leq p_2$. We now want to show that using p processors during time $(1 - \tau)(t_1 + t_2)$ and $p + 1$ processors during time $\tau \cdot (t_1 + t_2)$ is an optimal solution to do the work w during $I_1 \cup I_2$. In order to do this it is sufficient to show that $t_1 h(p_1) + t_2 h(p_2) \leq \tau \cdot (t_1 + t_2) h(p + 1) + (1 - \tau)(t_1 + t_2) h(p)$.

If we set $r := \tau \frac{t_1 + t_2}{t_1} \cdot \frac{p_2 - (p+1)}{p_2 - p_1}$ then $1 - r = (1 - \tau) \frac{t_1 + t_2}{t_1} \cdot \frac{p_2 - p}{p_2 - p_1}$ and with $s := \tau \frac{t_1 + t_2}{t_2} \cdot \frac{p+1-p_1}{p_2 - p_1}$ we get $1 - s = (1 - \tau) \frac{t_1 + t_2}{t_2} \cdot \frac{p - p_1}{p_2 - p_1}$. With this we have:
$$t_1 h(p_1) + t_2 h(p_2) = r t_1 h(p_1) + s t_2 h(p_2) + (1 - r) t_1 h(p_1) + (1 - s) t_2 h(p_2)$$
$$= \tau \cdot (t_1 + t_2) \left(\frac{p_2 - (p+1)}{p_2 - p_1} h(p_1) + \frac{p+1-p_1}{p_2 - p_1} h(p_2) \right)$$
$$+ (1 - \tau)(t_1 + t_2) \left(\frac{p_2 - p}{p_2 - p_1} h(p_1) + \frac{p - p_1}{p_2 - p_1} h(p_2) \right)$$
$$\leq \tau \cdot (t_1 + t_2) h(p + 1) + (1 - \tau)(t_1 + t_2) h(p)$$

because h is concave for $p_1, p, (p+1), p_2 \in (0, \bar{p}]$.

The repeated use of this argument for all intervals with different numbers of processors shows that if the average number of processors used during time T is $p + \tau$ with $p \geq 1$ then it is optimal to use $p+1$ processors during time $\tau \cdot T$ and p processors during time $(1 - \tau) \cdot T$.   □

With Lemma 3 and Lemma 5 we can define the energy usage for an average number of processors $p + \tau$ as the optimal energy usage of this case:

Definition 2. *A job which does work w during time T on an average number of processors $p + \tau$ with $p \in \mathbb{N}_0$ uses energy*

$$E(p + \tau) := E(p, \tau) := \frac{w^\alpha}{T^{\alpha-1}} \cdot (\tau \cdot h(p+1) + (1 - \tau)h(p))^{-\alpha+1}$$

It is immediately clear that E is a continuous function on $(0, \infty)$ and has the same values as E from Section 3 on integer p.

Lemma 6. *The function $E(p + \tau)$ as defined in Definition 2 is strictly convex on $(0, \bar{p}]$ and has its minimum at $\bar{p}$.*

Proof. The thing left to show is that $E(p+\tau)$ is strictly convex on $(0, \bar{p})$ and the minimum at $\bar{p}$. We will first show that $\frac{\partial^2 E(p,\tau)}{\partial \tau^2} = E_{\tau\tau}(p, \tau) > 0$ for all $p+1 \leq \bar{p}$.

$$E_\tau(p, \tau) = -\frac{w^\alpha}{T^{\alpha-1}} \cdot (\alpha - 1)(\tau \cdot h(p+1) + (1 - \tau)h(p))^{-\alpha}(h(p+1) - h(p))$$

$$E_{\tau\tau}(p, \tau) = \frac{w^\alpha}{T^{\alpha-1}} \cdot \alpha(\alpha - 1)(\tau \cdot h(p+1) + (1 - \tau)h(p))^{-\alpha-1}(h(p+1) - h(p))^2$$

Thus $E_{\tau\tau}(p, \tau) > 0 \Leftrightarrow h(p+1) - h(p) > 0 \Leftrightarrow p+1 \leq \bar{p}$. It remains to check that

$$\lim_{\tau \to 1} E_\tau(p, \tau) \leq \lim_{\tau \to 0} E_\tau(p+1, \tau)$$
$$\Leftrightarrow h(p+1) - h(p) \geq h(p+2) - h(p+1)$$

The last inequality is true for $p+2 \leq \bar{p}$ because h is concave and true for $p+1 = \bar{p}$ because $h(p+1) \geq h(p), h(p+2)$. Thus we have shown that E is strictly convex for $p + 1 \leq \bar{p}$ and E_τ is strictly increasing for $p + 1 \leq \bar{p}$.

The fact that E has its minimum at $\bar{p}$ directly comes from the fact that h has its maximum at $\bar{p}$.   □

Definition 3. *We define the left derivative of E for integer processor numbers as $\overrightarrow{E}(p) := \lim_{\tau \to 1} E_\tau(p - 1, \tau)$ and the right derivative as $\overleftarrow{E}(p) := \lim_{\tau \to 0} E_\tau(p, \tau)$. For noninteger $p + \tau$ the left and right derivative are the same and we define $\overrightarrow{E}(p + \tau) = \overleftarrow{E}(p + \tau) = E_\tau(p, \tau) =: E'(p + \tau)$.*

Lemma 7. *We have $\overleftarrow{E}(0) = -\infty$ and $\overrightarrow{E}(\bar{p}) \geq 0$ and $\overrightarrow{E}(p+\tau) \leq \overleftarrow{E}(p+\tau) \, \forall p + \tau \in (0, \bar{p}]$ and $\overrightarrow{E}(p+\tau), \overleftarrow{E}(p+\tau)$ are strictly increasing on $(0, \bar{p}]$.*

This lemma is obvious and needs no proof. With Lemma 7 we can define the inversion of the derivative of the energy function:

Definition 4. *We have E as in Definition 2 and the left and right derivatives as in Definition 3. Then for any $c \in (-\infty, 0]$ we define $(E')^{-1}(c) := p^* + \tau^*$ for $p^* + \tau^* \in (0, \bar{p}]$ with $\overrightarrow{E}(p^* + \tau^*) \le c \le \overleftarrow{E}(p^* + \tau^*)$.*

Lemma 8. *$(E')^{-1}$ as defined in Definition 4 is a continuous and monotonously increasing function on $(-\infty, 0]$.*

The detailed proof will be given in the full paper.

5   The Optimal Solution for the Multi-job Case

After the technical section we are now ready to prove the main theorem:

Theorem 1. *We have n jobs and for each job j an energy function E_j as in Definition 2 and the left and right derivatives as in Definition 3 and the inverse of the derivative of the energy function $(E_j')^{-1}$ as in Definition 4. If we want to minimize $\sum_j E_j(p_j + \tau_j)$ under the restriction $\sum_j(p_j + \tau_j) \le m$ then there exists a $c \in \mathbb{R}$ such that $\overrightarrow{E_j}(p_j^* + \tau_j^*) \le c \le \overleftarrow{E_j}(p_j^* + \tau_j^*)$ for all j holds for an optimal solution $(p_1^* + \tau_1^*, \ldots, p_n^* + \tau_n^*)$.*

If we have found a c such that $\sum_j(E_j')^{-1}(c) = m$ or $\sum_j(E_j')^{-1}(c) < m$ and $(E_j')^{-1}(c) = \bar{p}_j$ for all j then we have found an optimal solution.

Proof. We do the first part of the proof by contradiction. Suppose there are $i, j \in \{1, \ldots, n\}$ with $\overrightarrow{E_j}(p_j^* + \tau_j^*) > \overleftarrow{E_i}(p_i^* + \tau_i^*)$ in an optimal solution. Because $\overrightarrow{E_j}$ is continuous from the left and $\overleftarrow{E_i}$ is continuous from the right there exists a ϵ such that for all $y \in [p_j^* + \tau_j^* - \epsilon, p_j^* + \tau_j^*]$ and all $z \in [p_i^* + \tau_i^*, p_i^* + \tau_i^* + \epsilon]$ we have $\overrightarrow{E_j}(y) \ge \overleftarrow{E_i}(z) + \epsilon$. Thus we have

$$E_j(p_j^* + \tau_j^*) - E_j(p_j^* + \tau_j^* - \epsilon) \ge \inf_{y \in [p_j^* + \tau_j^* - \epsilon, p_j^* + \tau_j^*]} \overrightarrow{E_j}(y) \cdot \epsilon$$

$$\ge \sup_{z \in [p_i^* + \tau_i^*, p_i^* + \tau_i^* + \epsilon]} (\overleftarrow{E_i}(z) + \epsilon) \cdot \epsilon \ge E_i(p_i^* + \tau_i^* + \epsilon) - E_i(p_i^* + \tau_i^*) + \epsilon^2$$

$$\Leftrightarrow E_j(p_j^* + \tau_j^*) + E_i(p_i^* + \tau_i^*) \ge E_j(p_j^* + \tau_j^* - \epsilon) + E_i(p_i^* + \tau_i^* + \epsilon) + \epsilon^2$$

Hence we have shown that there exists a better solution than the optimal solution which leads to the contradiction.

In case of $\sum_j(E_j')^{-1}(c) < m$ and $(E_j')^{-1}(c) = \bar{p}_j$ for all j every job runs with its optimal number of processors. Thus no job can save energy trough running on a different number of processors. $(E_j')^{-1}(c) = \bar{p}_j$ is always the case for $c = 0$.

In case of $\sum_j(E_j')^{-1}(c) = m$ we have used all processors available. As we have $c \le 0$ and $(E_j')^{-1}(c) \le \bar{p}_j$ for all j in this case we would increase energy usage if we used less processors. As $\overrightarrow{E_j}(p_j^* + \tau_j^* + \delta) > \overleftarrow{E_i}(p_i^* + \tau_i^* - \delta)$ for all $\delta > 0$ and all i, j it is not possible to improve the solution by transferring an amount δ of processors from job i to job j. Hence the solution can not be improved and thus is optimal. □

With Theorem 1 we can now give the algorithm. Let c^* be the c of an optimal solution as in Theorem 1 and $p_j^* + \tau_j^*$ be the amount of processors of job j in the optimal solution. If we have a c such that $\sum_j (E_j')^{-1}(c) < m$ and $(E_j')^{-1}(c) < \bar{p}_j$ for at least one j then we know $c < c^*$ because all $(E_j')^{-1}$ are monotonously increasing. If we have a c such that $\sum_i (E_i')^{-1}(c) > m$ then we know $c > c^*$ also because all $(E_i')^{-1}$ are monotonously increasing. Hence we can use an interval halving technique to find the optimal $c = c^*$.

We also know in which interval to search. The maximal possible c is 0. If we use the same amount of $p_m + \tau_m = \min\{\frac{m}{n}, 1\}$ processors for each job then we know that c^* can not be smaller than $c_m = \min_i \overleftarrow{E}_i(p_m + \tau_m)$ because $p_m + \tau_m \leq \bar{p}_i$ for all i and $\sum_i (E_i')^{-1}(c_m) \leq m$.

For each job i we have a set of $2\bar{p}_i \leq 2m$ bend points. These are the points $\overleftarrow{E}_i(p)$ and $\overrightarrow{E}_i(p)$ for all $p \in \{1, \ldots, \bar{p}\}$. If we know c^* lies in an interval (c_ℓ, c_u) which contains no bend points, we can solve the problem directly. For i with $\overleftarrow{E}_i(p) \leq c_\ell < c_u \leq \overrightarrow{E}_i(p+1)$ for a certain p we have $p_i^* + t_i^* \in (p, p+1)$ (**case 1**). For i with $\overrightarrow{E}_i(p) \leq c_\ell < c_u \leq \overleftarrow{E}_i(p)$ for a certain p we have $p_i^* + t_i^* = p$ (**case 2**) thus $t_i^* = 0$. Other cases can not exist.

The computation of $(E_i')^{-1}(c)$ is done in two steps. First we search for the two adjacent bend points of c. If we are in case 2 we are done. In case 1 we know p in $E_i'(p + \tau) = c$. With some algebra we get $(\tau \cdot (h_i(p+1) - h_i(p)) + h_i(p))^{-\alpha} \cdot D_i = c$ for a positive D_i which does not depend on τ and thus we can compute τ with $\tau = \frac{\sqrt[\alpha]{c^{-1} \cdot D_i} - h_i(p)}{h_i(p+1) - h_i(p)}$ and then $(E_i')^{-1}(c) = p + \tau$.

We will now compute the exact solution if all bend points are eliminated. W.l.o.g. let the n jobs be ordered such that the jobs $1, \ldots, n_1$ are from case 1 and the jobs $n_1 + 1, \ldots, n$ are from case 2. We set $m_r = m - \sum_{i=1}^n p_i^*$ thus $\sum_{i=1}^{n_1} \tau_i^* = m_r$. For all jobs from case 1 the derivative of the energy function $E_i'(p_i^*, \tau_i^*)$ has to be the same $(= c^*)$. With some algebra we get:

$$D_1 \cdot (\tau_1^* h_1(p_1 + 1) + (1 - \tau_1^*) h_1(p_1))^{-\alpha} = \ldots =$$
$$D_{n_1} \cdot (\tau_{n_1}^* h_{n_1}(p_{n_1} + 1) + (1 - \tau_{n_1}^*) h_{n_1}(p_{n_1}))^{-\alpha} = c^*$$
$$\Leftrightarrow D_1^{\frac{-1}{\alpha}} \cdot (\tau_1^* h_1(p_1 + 1) + (1 - \tau_1^*) h_1(p_1)) = \ldots =$$
$$D_{n_1}^{\frac{-1}{\alpha}} \cdot (\tau_{n_1}^* h_{n_1}(p_{n_1} + 1) + (1 - \tau_{n_1}^*) h_{n_1}(p_{n_1})) = (c^*)^{\frac{-1}{\alpha}}$$

With $A_i = D_i^{\frac{-1}{\alpha}} \cdot (h_i(p_i + 1) - h_i(p_i))$ and $B_i = D_i^{\frac{-1}{\alpha}} h_i(p_i)$ and $G = (c^*)^{\frac{-1}{\alpha}}$ we get $A_i \tau_i^* + B_i = G$ and thus $\tau_i^* = \frac{G}{A_i} - \frac{B_i}{A_i}$ for all i of case 1. We can compute G through

$$m_r = \sum_{i=1}^{n_1} \tau_i^* = G \sum_{i=1}^{n_1} \frac{1}{A_i} - \sum_{i=1}^{n_1} \frac{B_i}{A_i}$$

and with G we can compute the τ_i^* and thus the final solution.

With these prerequisites we can now describe the algorithm:

1. Set $c_u = 0$ and $c_\ell = c_m$ and the range of bend points $[1, \bar{p}_i]$ for each job i.
2. Pick a randomly chosen bend point in (c_ℓ, c_u) and set c accordingly.

3. Check if $\sum_i (E'_i)^{-1}(c) > m$ then $c_u = c$ else $c_\ell = c$ and update the range of bend points for each job.
4. If there are bend points left in (c_ℓ, c_u) goto 2.
5. Do the exact calculation as above.

The expected number of times the loop 2-4 is executed is in $\mathcal{O}(\log(mn))$ because there are $\Theta(nm)$ bend points. Each time we have to choose one random bend point this is possible in time $\mathcal{O}(n)$. In 3 we have to invert n functions. Each can be done in time $\mathcal{O}(\log m)$ if all values of $h_i(p)$ are given. Additionally we have to update n ranges of breakpoints in 3. Each update can be done in $\mathcal{O}(1)$ with the p computed during function inversion done for the same job. The check in 4 can be done in $\mathcal{O}(n)$. The exact calculation also takes time in $\mathcal{O}(n)$.

Altogether the algorithm takes time in $\mathcal{O}(n \log(mn) \cdot \log m)$. It remains to compute the frequencies and to place the jobs onto the $m \times T$ processor $\times$ time rectangle.

We now know p_i^* and t_i^* for each job i. We reserve p_i^* processors for each job i for the whole interval $[0, T]$. For the τ_i^* and the remaining processors we can use McNaughton's wrap-around rule [5]. This can be done in time $\mathcal{O}(n)$ and each job i runs on p_i^* processors for time $(1 - \tau_i^*)T$ and on $p_i + 1$ processors for time $\tau_i^* T$ additionally any job changes its number of processors at most two times. A more detailed solution for a similar problem is given in [1]. With τ_i^* and p_i^* we can compute the work distribution between the phase with p_i^* processors and the phase with $p_i^* + 1$ processors in a similar way as we computed β in the proof of Lemma 5. Then the time, work and number of processors of both phases are known and the frequency can be computed with $w = s(p) \cdot f \cdot t$.

The parallelization is done in a similar way as we did it in [6] for another problem. If each processor does the computation of $(E'_i)^{-1}(c)$ for $\frac{n}{m}$ jobs and takes care of their list of bend points we only have to use collective operations for sum, broadcast of c and a distributed random pick. Such operations cost time $\Theta(\log m)$. In the loop 2-4 each processor computes $(E'_i)^{-1}(c)$ for $\frac{n}{m}$ jobs in $\mathcal{O}\left(\frac{n}{m} \log m\right)$ and takes part in the collective sum in time $\mathcal{O}\left(\frac{n}{m} + \log m\right)$. Thus the main part of the algorithm runs in $\mathcal{O}\left(\left(\frac{n}{m} + 1\right) \log m \cdot \log(mn)\right)$. Frequency computation can be done in time $\mathcal{O}\left(\frac{n}{m} + 1\right)$. The placement can be done with prefix sum in time $\mathcal{O}\left(\frac{n}{m} + \log m\right)$.

6   The Enhanced Model

If we have an additional linear term in the energy usage (maybe from memory or other parts which can not change their frequency) like in $E_{new} = p \cdot f^\alpha \cdot T + p \cdot \delta \cdot T$ with δ being independent of the job then $\overrightarrow{E}_{new}(p, t) = \overrightarrow{E}(p, t) + \delta T$ and $\overleftarrow{E}_{new}(p, t) = \overleftarrow{E}(p, t) + \delta T$. The optimal solution for the single job case is $\bar{p}_{new} = (E'_{new})^{-1}(0) \leq \bar{p}$ which can be noninteger. As E'_{new} is just E' with an additional constant we can use the same techniques as in the original model to get an optimal solution for the multi-job case.

Many processors have minimal and maximal operating frequencies (because of memory requirements or the length of signal paths). For this case we have to do a bit more theory first. We will restrict the presentation here to case of a maximal frequency f_G because the case of a minimal frequency is analogous.

There are two amounts of resources which are interesting. One is the absolute minimal amount of resources for our job $p_1 + \tau_1$ and the other is the minimal amount of resources $p_2 + \tau_2$ above we can use the standard techniques already introduced. Let $f_q(p + \tau)$ be the frequency for the part of the job which runs on q processors. f_q is defined on $(q - 1, q + 1)$. When we use β from the proof of Lemma 5 we can compute $f_p(p + \tau)$ and $f_{p+1}(p + \tau)$ for the optimal solution through $\beta w = s(p + 1) f_{p+1}(p + \tau) T \cdot \tau$ and $(1 - \beta) w = s(p) f_p(p + \tau) T \cdot (1 - \tau)$. Doing some algebra we get:

$$\frac{f_p(p + \tau)}{f_{p+1}(p + \tau)} = {}^{\alpha - 1}\sqrt{\frac{s(p)(p + 1)}{s(p + 1)p}} \geq 1$$

Thus $f_p(p+\tau) \geq f_{p+1}(p+\tau)$. Because the energy usage E is a continuous decreasing function of $p+\tau$ we know that f_{p+1} and f_p are continuous decreasing functions. Hence we can use the techniques for the case with unrestricted frequencies as long as $f_p(p+\tau) \leq f_G$ this gives us $p_2 + \tau_2$. The absolute minimum of resources needed for our job can be computed trough $w = s(p_1) f_G T (1 - \tau_1) + s(p_1 + 1) f_G T \tau_1$. Let p be such that $f_{p+1}(p+1) \leq f_G \leq f_p(p)$. As $f_p(p) \geq f_p(p_2 + \tau_2) \geq f_{p+1}(p_2 + \tau_2) \geq f_{p+1}(p+1)$ and $s(p) f_G T \leq w \leq s(p+1) f_G T$ we know that $p \leq p_1 + \tau_1 \leq p_2 + \tau_2 \leq p + 1$ and thus $p = p_1 = p_2$.

Now we can define the energy function $\tilde{E}$ on $(p+\tau_1, p+\tau_2)$. The frequencies on $(p+\tau_1, p+\tau_2)$ will be $\tilde{f}_p$ and $\tilde{f}_{p+1}$. We know from the proof of Lemma 5 that $E(\beta)$ is convex hence for an optimal energy usage $\tilde{f}_p(p + \tau) = f_G$ for all $\tau \in [\tau_1, \tau_2]$. Thus we can compute $\tilde{f}_{p+1}(p+\tau)$ trough $w = s(p) f_G T \cdot (1 - \tau) + s(p+1) \tilde{f}_{p+1}(p + \tau) T \tau$. With $\tilde{f}_{p+1}(p + \tau)$ we can compute $\tilde{E}(p + \tau)$. Doing some analysis we see that $\tilde{E}$ is convex on $(p+\tau_1, p+\tau_2)$. Obviously $\tilde{E}$ is continuous, $\tilde{E}(p+\tau) \geq E(p+\tau)$ and $\tilde{E}(p+\tau_2) = E(p+\tau_2)$. Hence we also have $\overleftarrow{\tilde{E}}(p+\tau_2) \leq \overrightarrow{E}(p+\tau_2)$. We also can compute $M := \overleftarrow{\tilde{E}}(p+\tau_1) = T f_G^\alpha(s(p+1) - \alpha \cdot (p+1)(s(p+1) - s(p)))s^{-1}(p+1)$ which is a function of the input values.

Let c be as in Section 5 then we get: If $c \leq \overleftarrow{\tilde{E}}(p + \tau_1)$ our job just gets $p + \tau_1$ processors because it is the minimal number so we can put this amount away and continue with the other jobs. If $c \geq \overrightarrow{\tilde{E}}(p+\tau_2)$ we can use the technique from Section 5. If we know that the optimal solution lies between the bend points for $p + \tau_1$ and $p + \tau_2$ (these are new bend points for the enhanced model) then we do not know how to compute an exact solution. But then we know the minimal derivative M (the same is true for a minimal frequency).

If we want to compute a solution which only uses an amount of ϵ more energy than the optimal solution we can do it in the following way: For each job for which the amount of processors is already known we sum up these amounts. Let the difference between these amounts and m be $\tilde{m}$. We now do interval halving on $(M, \overrightarrow{\tilde{E}}(p + \tau_2))$. For every c we invert the E' and $\tilde{E}'$ (depending on which one applies) with an maximal additive error of $\frac{\epsilon}{8n}$. Then we compare $\sum (E')^{-1}(c) + \sum (\tilde{E}')^{-1}(c)$ with $\tilde{m}$. If the remaining interval is smaller than

$\epsilon \cdot (M \cdot 2m)^{-1}$ we know that the lower end of the interval stands for a feasible solution with a maximal additive error of ϵ.

The algorithm needs time in $\mathcal{O}\big(n \log(8n\epsilon^{-1}) \cdot \log(2mM^2\epsilon^{-1})\big)$ and can be parallelized in a similar way as above.

The placement is done as in Section 5. The frequency calculations for jobs with energy function $\tilde{E}$ is clear for the others we can do it in the same way as in Section 5.

7   Conclusion

We have shown that with two restrictions it is possible to solve our energy efficient scheduling problem optimally in near linear time. The major step to solve the problem was to build continuous convex energy functions and to use some calculus on them. This is somehow surprising because many related problems are known to be NP-hard. The two restrictions do not limit the applicability too much because many classes of parallel jobs fit into these restrictions.

Acknowledgment. This work was partly supported by the German Research Foundation (DFG) as part of the Transregional Collaborative Research Center "Invasive Computing" (SFB/TR 89). We also like to thank the unknown reviewer who gave a very helpful two page review.

References

1. Blazewicz, J., Kovalyov, M.Y., Machowiak, M., Trystram, D., Weglarz, J.: Preemptable malleable task scheduling problem. IEEE Transactions on Computers 55 (2006)
2. Blazewicz, J., Machowiak, M., Weglarz, J., Kovalyov, M.Y., Trystram, D.: Scheduling malleable tasks on parallel processors to minimize the makespan: Models and algorithms for planning and scheduling problems. Annals of Operations Research 129 (2004)
3. Chen, J.-J., Kuo, T.-W.: Multiprocessor energy-efficient scheduling for real-time tasks with different power characteristics. In: International Conference on Parallel Processing, ICPP 2005 (2005)
4. Leung, J.Y.-T. (ed.): Handbook of Scheduling. CRC (2004)
5. McNaughton, R.: Scheduling with deadlines and loss functions. Management Science 6(1) (1959)
6. Sanders, P., Speck, J.: Efficient parallel scheduling of malleable tasks. In: 2011 IEEE International Parallel Distributed Processing Symposium, IPDPS (2011)
7. Yang, C.-Y., Chen, J.-J., Kuo, T.-W.: An approximation algorithm for energy-efficient scheduling on a chip multiprocessor. In: Proceedings of the Conference on Design, Automation and Test in Europe, DATE 2005, vol. 1 (2005)
8. Yao, F., Demers, A., Shenker, S.: A scheduling model for reduced cpu energy. In: Annual Symposium on Foundations of Computer Science, pp. 374–382. IEEE Computer Society (1995)

Scheduling MapReduce Jobs in HPC Clusters

Marcelo Veiga Neves, Tiago Ferreto, and César De Rose

Faculty of Informatics, PUCRS, Brazil
marcelo.neves@acad.pucrs.br, {tiago.ferreto,cesar.derose}@pucrs.br

Abstract. MapReduce (MR) has become a de facto standard for large-scale data analysis. Moreover, it has also attracted the attention of the HPC community due to its simplicity, efficiency and highly scalable parallel model. However, MR implementations present some issues that may complicate its execution in existing HPC clusters, specially concerning the job submission. While on MR there are no strict parameters required to submit a job, in a typical HPC cluster, users must specify the number of nodes and amount of time required to complete the job execution. This paper presents the MR Job Adaptor, a component to optimize the scheduling of MR jobs along with HPC jobs in an HPC cluster. Experiments performed using real-world HPC and MapReduce workloads have show that MR Job Adaptor can properly transform MR jobs to be scheduled in an HPC Cluster, minimizing the job turnaround time, and exploiting unused resources in the cluster.

1   Introduction

The MapReduce (MR) model is in increasing adoption by several researchers, including the ones that used to rely on HPC solutions [19,18]. Much of this enthusiasm is due to the highly visible cases where MR has been successfully used by companies like Google, Yahoo, and Facebook. Besides, MR provides a simpler approach to address the parallelization problem over traditional approaches, such as MPI [10].

MR implementations, such as Hadoop [20], provide a complete execution platform for MR applications, normally using a dedicated cluster in combination with an optimized distributed file system. As consequence, in order to enable the execution of regular HPC and MR jobs in a computing laboratory, two distinct clusters are required. It leads to a split in the laboratory investments, in terms of hardware and staff, to support the two models, instead of focusing in a single, large scale and powerful computing infrastructure.

We believe that users and computing laboratory administrators may benefit from using already existing HPC clusters to execute MR jobs. In order to enable it, one of the first issues that must be addressed is regarding the job submission process. While MR implementations provide a straightforward job submission process which involves the whole cluster, HPC users submit their jobs to a Resource Management System (RMS) and need to specify the number of nodes and amount of time that should be allocated for complete the job execution.

C. Kaklamanis et al. (Eds.): Euro-Par 2012, LNCS 7484, pp. 179–190, 2012.

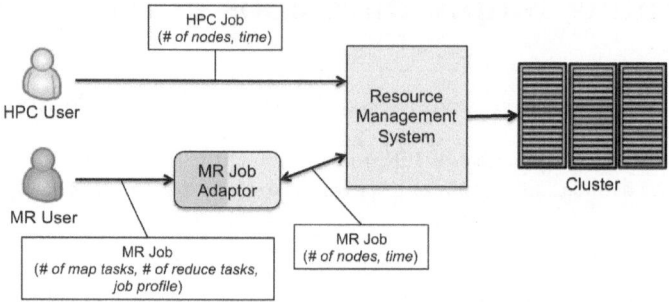

Fig. 1. Architecture of the HPC cluster with the MR Job Adaptor

Current solutions, such as Hadoop on Demand (HOD) [1] and myHadoop [13] allow one to create a virtual Hadoop cluster as a partition of a large physical cluster. However, the user must explicitly specify the number of nodes and time to be allocated as a regular HPC job. This approach may confuse typical MR users that are not used to do it, and they may, in return, always try to allocate the whole cluster for the longest time as possible. As a consequence, the turnaround time of the MR job will probably increase, since the job will be scheduled to the end of the RMS queue, which will frustrate the user again. Other solutions, such as MESOS [11] and Hamster [21], use a different approach where the responsibility for resource management is taken away from the cluster's RMS, which may conflict with the policy of use of most HPC clusters.

In order to overcome these problems, we present in this paper the MR Job adaptor, a component that converts MR jobs in order to enable their execution in HPC clusters. Figure 1 presents the MR Job adaptor and its connection with MR users and the RMS of the HPC Cluster. It receives the MR job from the user and interacts with the RMS in order to find a suitable slot to schedule the job that minimizes the resulting turnaround time of the job. We evaluated the algorithm implemented inside the MR Job Adaptor using real HPC and MR workloads and observed that it effectively decreases the turnaround time while also exploits unused resources in the cluster.

The paper is organized as follows: Section 2 provides an overview of HPC clusters and the Map Reduce model; Section 3 describes the functioning of the MR Job Adaptor for HPC Clusters; Section 4 presents the experiments performed to evaluate the MR Job Adaptor using real workloads. The conclusion and future work are presented in Section 5.

2   HPC Clusters and MapReduce

Clusters of computers, which have transformed HPC in the last decade, are still the dominant architecture in this area [22]. HPC clusters consist of a number of stand-alone computers connected by a high performance network, working together as a single computing resource, and sharing a common storage volume exported through a distributed file system.

Traditional HPC clusters typically have their resources controlled by a Resource Management System (RMS), such as PBS/TORQUE [23] or SGE [16], which enable the submission, tracking and management of jobs in the cluster. Although this approach maximizes the overall utilization of the system and enables sharing of the resources among multiple users [13], it also requires all applications to be submitted as batch jobs. The user must submit the job accompanied by the number of nodes that the parallel application should use and the maximum time that it will take to complete the job execution.

The de-facto standard for parallel programming in HPC clusters is MPI (Message Passing Interface [10]), which follows the message-passing paradigm. A parallel program using MPI consists of different processes running on the cluster and explicitly exchanging data via messages. Despite the higher complexity in the development of parallel applications using this approach, it also enables one to fine-tune the application, resulting in better performance than other high-level approaches.

In the past few years, the increase in data generation has reached rates never seen before, making it necessary to develop new technologies for storing and analyzing such a large amount of data. The processing of such large data sets, also known as big data, is normally referred as data-intensive computing [15]. Several works have already been proposed in the data-intensive computing area to address the needs of big data [17,6,9]. In this model, the data set may not fit in the main memory nor in a single disk and, therefore, a distributed storage solution is necessary.

The execution platform for data-intensive computing is typically a dedicated large-scale cluster, with the data set distributed between the cluster nodes, i.e., each node has a slice of the data set. Thus, each node is both a data and compute node, which provides scalable storage and efficient data processing (by exploiting data locality). This architecture differs from traditional HPC clusters in some ways. HPC clusters are usually shared by several users to execute different applications, through the mediation of a RMS, while in data-intensive computing the cluster is usually dedicated to process large data sets of an unique organization. HPC clusters use a shared-disk file system to share data between nodes, while in data-intensive computing each cluster node uses its own local storage (shared-nothing). Consequently, users of data-intensive computing usually adopt a dedicated cluster for their applications.

There are several frameworks for the development of data-intensive applications [6,7,12], most of them based on the MapReduce (MR) model. Hadoop [20] is currently one of the most popular open-source MapReduce implementations. Unlike typical HPC parallel programming libraries, such as MPI, MR frameworks hide much of the complexity of parallel programming from the programmer (for example, not requiring explicit data communications or application-specific logic to avoid communications). Current MR implementations allow automatic parallelization and distribution of computations on large clusters of commodity PCs, hiding the details of parallelization, fault tolerance, data distribution and load balancing [6].

The availability of several programming frameworks and the facilities to develop a parallel program has contributed to the adoption of MR by traditional HPC users. Instead of explicitly specifying the communication between processes and guaranteeing their coordination, the definition of simple map and reduce tasks seemed to be simpler in some cases. As a result, typical CPU-intensive HPC applications started being reimplemented using the MapReduce model (e.g. scientific application [19]) and, consequently, using a dedicated MR cluster.

Instead of using two clusters, one for HPC applications and another for MR applications, the RMS of the HPC cluster should also be able to schedule MR jobs. However, in order to do that, the MR job must include the number of nodes and amount of time to execute the job, which is not common in MR implementations. There are some initiatives in order to execute MR jobs in an HPC cluster. Systems such as Hadoop on Demand (HOD) [1] and myHadoop [13] allow one to create a virtual Hadoop cluster as a partition of a large physical cluster. Both systems use the TORQUE Resource Management System [23] to perform the allocation of nodes. However, the user has to specify the number of nodes and time to be allocated.

A straightforward solution would be to request the whole cluster for as long as possible to execute the job. Despite the simplicity of this approach, it can lead to longer turnaround times, since the request will probably go to the end of the RMS queue, which increases the time before the request is attended [5]. Due to the high flexibility of MR jobs (they can be executed with a variable number of nodes), the RMS could use a more intelligent approach, trying to fit the MR job in the free slots available in the RMS queue. Therefore, we propose a component called MR Job Adaptor, used to adjust the request to an HPC Cluster RMS, including number of nodes and amount of time, while ensuring that the turnaround time is minimized.

3   MapReduce Job Adaptor

This section presents the algorithm implemented inside the MapReduce Job Adaptor. The adaptor has three main goals: (i) to facilitate the execution of MR jobs in HPC clusters, (ii) to minimize the average turnaround time of MR jobs executed in an HPC cluster, and (iii) to exploit unused resources in the cluster resulted from the various shapes of HPC job requests.

MR job requests are quite different from HPC ones. They do not require any specific infrastructure parameter for submission, only straightforward application parameters such as number of map and reduce tasks. On the other hand, HPC job requests require the number of nodes and amount of time to allocate a cluster partition. The approach used by systems such as Hadoop on Demand and myHadoop to run MR jobs in HPC clusters is to ask the user how many nodes and time should be allocated for a job. However, this approach is quite cumbersome since, in general, users do not have this kind of knowledge about their MR applications for different numbers of nodes and combinations of map and reduces tasks. In practice, this causes the user to allocate the maximum allowed amount of time and resources in the cluster, which may cause longer turnaround times and waste of resources.

The proposed adaptor aims to enable the transparent execution of MR jobs in the HPC cluster, i.e., the user specifies the MR job request as he would do in a typical MR cluster and the adaptor converts it to an HPC-compatible request, which is forwarded to the Resource Management System (RMS) of the HPC cluster. Instead of always using the maximum amount of nodes and time to execute the MR job, the adaptor allocates a cluster partition which minimizes the turnaround time of the job. It does that by interacting with the RMS to get free areas (slots) in the job requests queue. Using a profile of the MR job, it estimates the job completion time for each free slot and selects the one that yields the minimum turnaround time.

This approach relies on the fact that MR jobs do not have strict requirements regarding the number of resources for execution as HPC jobs. Thus, we use the MapReduce performance model proposed by Verma et al. [24] to estimate job completion times for different number of resources. It creates a job profile comprising performance invariants from past executions and uses it as input for the time estimation. This model can be used to estimate the lower (T_J^{low}) and upper (T_J^{up}) bounds of the overall completion time of a given job J. The lower bound can be obtained as follows:

$$T_J^{low} = \frac{N_M^J + M_{avg}}{S_M^J} + \frac{N_R^J \cdot (Sh_{avg}^{typ} + R_{avg})}{S_R^J} + Sh_{avg}^1 - Sh_{avg}^{typ} \qquad (1)$$

where N_M^J is the number of map tasks, N_R^J is the number of reduce tasks, S_M^J is the number of map slots, S_R^J is the number of reduce slots and the tuple $(M_{avg}, R_{avg}, Sh_{avg}^1, Sh_{avg}^{typ})$ represents the performance invariants, for each MapReduce phase, extracted from the job profile. The equation for T_J^{up} can be written in a similar form and is detailed in Verma et al. [24]. It was reported that the average of lower and upper bounds (T_J^{avg}) is a good approximation of the job completion time, so we chose the upper bound as a conservative approach, avoiding the underestimation cases.

The algorithm implemented for the MapReduce Job Adaptor is presented in Algorithm 1. It starts by receiving the number of map and reduce tasks (Nm, Nr), and a profile p of the MR job to be executed in the cluster. It also gets information from the RMS, such as the list of free slots in the queue and maximum number of nodes and time that can be allocated in the queue. These limits in the number of nodes and time are usually imposed by HPC cluster administrators in order to enforce a fair sharing of resources between users.

Figure 2 presents an example of RMS queue of an HPC cluster with six jobs (A to F) scheduled in the queue. In this example, function $getFreeSlots()$ would return four free slots that could be used to execute the MR job. Slot 1 starts at time 10 with 25 nodes and maximum duration of 2. Slot 2 starts at time 13 with 50 nodes and maximum duration of 2. Slot 3 starts at time 16 with 25 nodes with no maximum duration, and slot 4 starts at time 17 with all cluster nodes and has also no maximum duration.

Algorithm 1 MapReduce Job Adaptor internal functioning

$(Nm, Nr) \leftarrow$ Number of map and reduce tasks of MR job
$p \leftarrow$ Job profile of MR job
$freeSlotsList \leftarrow$ getFreeSlots()
$maxNodes \leftarrow$ Maximum number of nodes allowed for allocation in the cluster
$maxTime \leftarrow$ Maximum time allowed for allocation in the cluster
$turnaround \leftarrow BigNumber$
for all $freeSlot$ in $freeSlotsList$ **do**
   $startTime \leftarrow$ getStartTime($freeSlot$)
   $slotDuration \leftarrow$ getSlotDuration($freeSlot$)
   $slotDuration \leftarrow$ MIN($slotDuration$,$maxTime$)
   $numNodes \leftarrow$ getNumberOfNodes($freeSlot$)
   $numNodes \leftarrow$ MIN($numNodes$,$maxNodes$)
   $execTime \leftarrow$ estimateJobExecutionTime(p, Nm, Nr, $numNodes$)
   $newTurnaround \leftarrow startTime + execTime - NOW$
   if $execTime <= slotDuration$ and $newTurnaround < turnaround$ **then**
     $nodes \leftarrow numNodes$
     $time \leftarrow execTime$
     $turnaround \leftarrow newTurnaround$
   end if
end for
return $(nodes, time)$

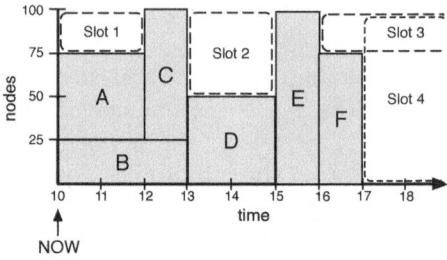

Fig. 2. Example of the queue of a Resource Management System

Variables $maxNodes$ and $maxTime$ receive the maximum number of nodes and amount of time that can be requested to the RMS. The turnaround variable, which stores the turnaround time of the best solution found by the algorithm, is initialized with a big number. After that, the algorithm starts testing each free slot in the RMS queue to verify if the MR job would fit on it while minimizing the turnaround time. Since the number of nodes can change for each slot, the execution time of the MR job needs to be estimated again. The execution time of the MR job is estimated using its parameters (number of map and reduce tasks, and job profile) and $numNodes$. The algorithm finishes with parameters $nodes$ and $time$ to be used in the job submission to the RMS which minimizes the turnaround time of the MR job. The turnaround time is calculated using the slot start time and the estimated execution time of the MR job subtracted by the current time (indicated by NOW in the algorithm).

4   Evaluation

In order to evaluate the proposed algorithm, we used a simulator based on the SimGrid toolkit [3], which provides abstractions and functionalities for the simulation of parallel and distributed systems, such as HPC clusters. We simulated a cluster of 128 nodes with 2 cores each (for the MapReduce experiments, we defined 1 map and 1 reduce task slot per node). Cluster's resources were managed by a RMS that allows users to submit jobs. We also simulated a stream of job submissions, where each job requires a number of nodes to be allocated for a particular amount of time.

The simulated RMS implements the Conservative Backfilling (CBF) [8] algorithm. The CBF algorithm enables backfilling and is a representative of the algorithms running in deployed RMS schedulers today. The main idea of CBF is that an arriving job is always inserted in the first free slot available in the schedulers queue, which offers an upper-bound to the job start time. Every time a new free slot appears, the scheduler sweeps the entire queue looking for jobs that can be brought forward without delaying the start of any other job in the queue. This means that at any time it is possible to obtain the list of available free slots in the scheduler's queue. We use this feature to provide input for the algorithm described in the previous section.

We used a naive algorithm as a baseline for comparison purposes. It consists of allocating a number of nodes, based on the number of map and reduce tasks, during a fixed amount of time, defined as the maximum allowed amount of time per request. This is the case when there is no information about the MapReduce application and the scheduler's queue state. A similar approach is used by systems such as Hadoop On Demand [1] and myHadoop [13], yet in those systems the user has to specify the required number of nodes. The algorithms were compared in terms of average job turnaround time (interval between the submission of a job and its completion) and average system utilization. The number of simulations was defined in order to provide a confidence level of 95% with an error less than 5%.

To simulate a stream of job submissions for the users of an HPC cluster, we used two different approaches. The first was to simulate a synthetic workload based on a widely used model by Lublin et al. [14], which is one of the most comprehensive and validated batch workload models in the literature. Basically, it uses two gamma distributions to model the job inter-arrival time (depending on the time of day), a two-stage uniform distribution to model the job sizes and a two-stage hyper-gamma distribution to model the runtime of jobs.

We also used real-world workload traces from the Parallel Workloads Archive [2] as input to our simulation. This archive contains log information regarding the workloads on parallel machines, such as HPC clusters. We chose traces from the San Diego Supercomputer Center SP2 (SDSC SP2), which is a well-known and widely studied workload. SDSC SP2 workload has 128 nodes and 73,496 jobs, spanning 2 years from July 1998 to December 2000.

Unfortunately, there is not yet any such workload archive publicly available for MapReduce jobs. However, recent publications [25,26,4] have reported workload

Table 1. Distribution of job sizes in Facebook workload (based on Zaharia et al. [26])

Bin	# Map Tasks	# Reduce Tasks	% Jobs at Facebook
1	1	0	39%
2	2	0	16%
3	10	3	14%
4	50	0	9%
5	100	0	6%
6	200	50	6%
7	400	0	4%
8	800	180	4%
9	2400	0	3%

characteristics for MapReduce clusters in production at Google, Facebook and Yahoo!. We used the detailed description of a Facebook workload, provided by Zaharia et al. [26], to create a synthetic MapReduce workload. This workload comes from a Hadoop cluster, in production at Facebook in October 2009, with 600 nodes running about 7,500 jobs per day.

The Facebook workload used in our experiments is distributed in 9 bins as summarized in Table 1. As can be observed, most jobs in Facebook's workload are small. However, in the original workload, jobs in the last bin range from 1,501 to 25,000 maps. We chose 2,400 maps as our representative for this bin to make it fit in the HPC cluster simulated in our experiments. The job inter-arrival times is roughly exponential with a mean of 14 seconds. We defined map and reduce tasks duration as $N(60,20)$ and $N(120,30)$ respectively, where $N(\mu, \sigma)$ is the normal distribution with a mean μ and standard deviation σ.

The first experiment performed aims to evaluated the impact of the proposed algorithm in the job performance, in terms of average turnaround time and system utilization, for an HPC cluster running a mixed workload of HPC and MR jobs. We simulated one hour of HPC job submissions (around 400 jobs, since the mean inter-arrival time in the so-called "peak hour" of the Lublin et al. model is roughly 5 seconds) mixed with one hour of MR job submissions (around 300 jobs).

Table 2 compares the results of the proposed algorithm (Adaptor) against the naive algorithm for each workload (HPC-only, MR-only and mixed HPC+MR). The proposed algorithm obtained shorter average turnaround time and improved utilization in all cases. For the MR-only workload, the use of the adaptor algorithm reduced the average turnaround time in 40%. For the mixed workload (HPC + MR), the overall average turnaround time was reduced in approximately 15%. However, the average turnaround time of the MR jobs in the mixed workload changed from 31776 (using naive algorithm) to 8616 seconds, which represent a reduction of 73%.

To evaluate the influence of MR job sizes in our algorithm, we conducted experiments for each bin in Facebook's workload. Figure 3 shows the obtained results in terms of average job turnaround time. The adaptor algorithm outperformed the naive approach regardless the job bin. However, the adaptor algorithm

Table 2. Average job turnaround time and system load for each algorithm using Lublin et al. model (HPC) and Facebook (MR) workloads.

Workload (job type)	HPC	MR		HPC + MR	
Algorithm		Naive	Adaptor	Naive	Adaptor
Avg Utilization (%)	88.9	68.5	93.7	87.5	93.3
Avg Turnaround (s)	9126	6151	3709	13680	11512

performed better for bins with smaller job sizes. This happens because small job length cause more opportunity for backfilling. We believe that it is a positive characteristic, since the first 4 bins represent approximately 80% of the jobs in Facebook workload. Moreover, similar job size distribution can be seen in workloads from Google [25] and Yahoo! [4].

In order to evaluate the adaptor algorithm with different system loads, we conducted experiments varying the inter-arrival time of job submissions. The peak hour model by Lublin et al. produces mean inter-arrival time of 5.01 seconds, which is the mean of a Gamma distribution with $\alpha = 10.23$ and $\beta = 0.49$. Thus, different HPC load characteristics were simulated varying the value of α from 4 to 60, giving inter-arrival times between approximately 2 and 30 seconds. Similarly, different inter-arrival times for MR jobs were obtained by varying the mean in the exponential distribution described earlier. The results are shown in Figure 4. In both cases, the adaptor algorithm performed better regardless of the inter-arrival time.

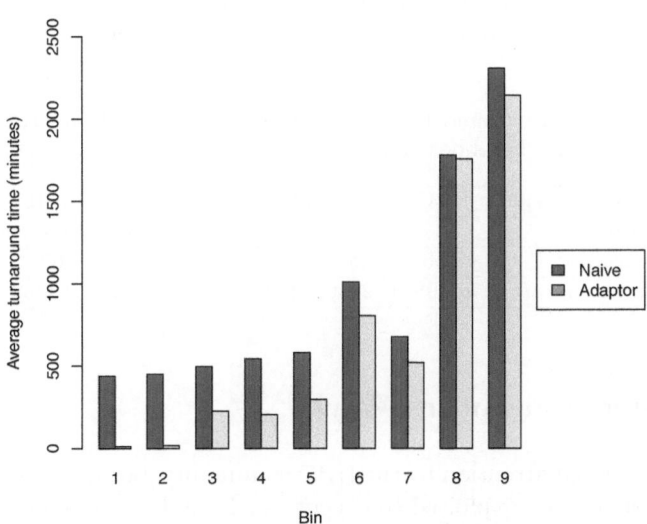

Fig. 3. Average job turnaround time for each bin in Facebook workload

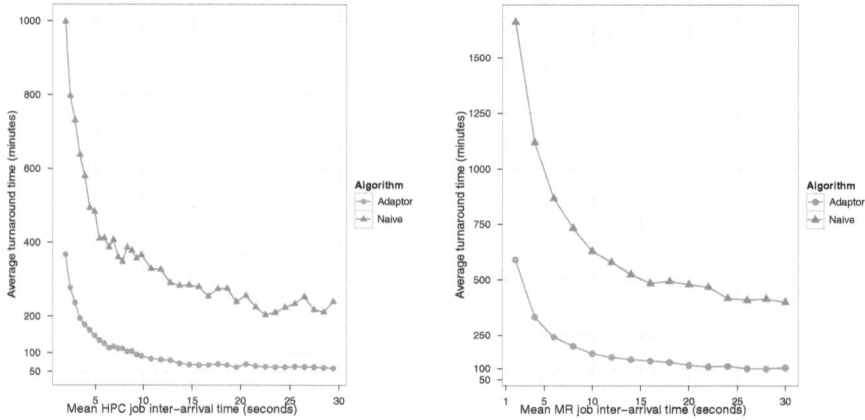

Fig. 4. Average turnaround time for a mixed workload varying (a) the mean job inter-arrival time of the HPC workload and (b) the mean job inter-arrival time of the MR workload

Finally, to evaluate the performance of the adaptor algorithm using a real-world HPC workload, we chose a day-long trace from SDSC SP2 and used it along with 1,000 MR jobs as input for our simulation. Table 3 shows the results. The adaptor algorithm performed better in all cases. In this experiment, we also observed that HPC and MR workloads are quite different. The HPC traces used to have few jobs with long running times, while MR have many jobs with short running times. This reinforces our argument that one should able to use an HPC cluster to run both HPC and MR jobs and that it can exploit unused resources.

Table 3. Average job turnaround time and system load for each algorithm using a trace from SDSC SP2 and Facebook workload

Workload (job type)	HPC	MR		HPC + MR	
Algorithm		Naive	Adaptor	Naive	Adaptor
Avg Utilization (%)	52.5	83.4	89.4	56.3	68.4
Avg Turnaround (s)	16198	22602	10269	99629	19288

5   Conclusion and Future Work

MapReduce has gained attention by the HPC community, but it is still not trivial how HPC clusters can be exploited to execute such kind of job along with HPC applications. HPC clusters present some characteristics that conflict with the MapReduce model, such as the process used to submit jobs. This paper presented the MR Job Adaptor, a module that customizes regular MR jobs for submission in HPC clusters. MR Job Adaptor estimates the execution time of the job using an MR Job Profile and tests the available slots in the Resource Management System queue in order to allocate one that results in minimal turnaround time. The experiments performed to evaluate the module demonstrated that, besides minimizing the job turnaround time, it also exploits unused resources in the cluster.

As future work, we intend to evaluate other characteristics of the MR to enhance the algorithm used by the MR Job Adaptor. We believe that using a single cluster for HPC and MR jobs can be beneficial for both users and cluster administrators.

References

1. Apache Hadoop on Demand (HOD) (2012),
   `http://hadoop.apache.org/common/docs/current/hod_scheduler.html`   (accessed on February 2012)
2. Parallel Workloads Archive (2012),
   `http://www.cs.huji.ac.il/labs/parallel/workload/` (accessed on February 2012)
3. Casanova, H.: Simgrid: A toolkit for the simulation of application scheduling. In: Proceedings of the First IEEE/ACM International Symposium on Cluster Computing and the Grid (CCGrid 2001), Brisbane, Australia (May 2001)
4. Chen, Y., Ganapathi, A., Griffith, R., Katz, R.H.: The case for evaluating mapreduce performance using workload suites. In: MASCOTS, pp. 390–399. IEEE (2011)
5. De Rose, C.A.F., Ferreto, T., Calheiros, R.N., Cirne, W., Costa, L.B., Fireman, D.: Allocation strategies for utilization of space shared resources in bag of tasks grids. Future Generation Computer Systems 24(5), 331–341 (2008)
6. Dean, J., Ghemawat, S.: Mapreduce: Simplified data processing on large clusters. Communications of the ACM 51(1), 107–113 (2008)
7. Ekanayake, J., et al.: Twister: a runtime for iterative mapreduce. In: Proceedings of the 19th ACM International Symposium on High Performance Distributed Computing, HPDC 2010, pp. 810–818. ACM, New York (2010)
8. Feitelson, D.G., Mu'alem Weil, A.: Utilization and predictability in scheduling the IBM SP2 with backfilling. In: 12th Intl. Parallel Processing Symp (IPPS), pp. 542–546 (April 1998)
9. Fox, G., et al.: Parallel data mining from multicore to cloudy grids. In: Proceedings of HPC 2008 (2011)
10. Gropp, W., Lusk, E., Skjellum, A.: Using MPI Portable Parallel Programming with the Message Passing Interface. The MIT Press (1994)
11. Hindman, B., Konwinski, A., Zaharia, M., Ghodsi, A., Joseph, A.D., Katz, R., Shenker, S., Stoica, I.: Mesos: Flexible resource sharing for the cloud. USENIX (August 2011)
12. Isard, M., et al.: Dryad: distributed data-parallel programs from sequential building blocks. In: Proceedings of EuroSys 2007 (January 2007)
13. Krishnan, S., Tatineni, M.: Myhadoop-hadoop-on-demand on traditional hpc resources. sdsc.edu (2011), `http://www.sdsc.edu/~allans/MyHadoop.pdf`
14. Lublin, U., Feitelson, D.G.: The workload on parallel supercomputers: Modeling the characteristics of rigid jobs. J. Parallel & Distributed Comput. 63(11), 1105–1122 (2003)

15. Middleton, A.: Data-intensive technologies for cloud computing. In: Handbook of Cloud Computing (January 2010)
16. Oracle: Oracle Grid Engine, previously known as Sun Grid Engine (SGE) (2012), http://www.oracle.com/technetwork/oem/grid-engine-166852.html (accessed on February 2012)
17. Schadt, E., Linderman, M., Sorenson, J.: Computational solutions to large-scale data management and analysis. Nature Reviews (January 2010)
18. Sehrish, S., et al.: Mrap: a novel mapreduce-based framework to support hpc analytics applications with access patterns. In: Proceedings of HPDC 2010, pp. 107–118 (2010), http://doi.acm.org/10.1145/1851476.1851490
19. Srirama, S., Jakovits, P.: Adapting scientific computing problems to clouds using mapreduce. Future Generation Computer Systems (January 2011)
20. Team, A.H.: Apache hadoop web site (2011), http://hadoop.apache.org (accessed on February 2012)
21. Team, A.H.: Hamster: Hadoop and mpi on the same cluster (2011), https://issues.apache.org/jira/browse/MAPREDUCE-2911 (accessed on February 2012)
22. Top 500: Top 500 Supercomputers Site (2012), http://www.top500.org (accessed on February 2012)
23. TORQUE: TORQUE Resource Manager (2012), http://www.clusterresources.com/products/torque-resource-manager.php (accessed on February 2012)
24. Verma, A., Cherkasova, L., Campbell, R.H.: Aria: automatic resource inference and allocation for mapreduce environments. In: Proceedings of ICAC 2011, pp. 235–244 (2011)
25. Wang, G., et al.: Towards synthesizing realistic workload traces for studying the hadoop ecosystem. In: MASCOTS. pp. 400–408. IEEE (2011)
26. Zaharia, M., et al.: Delay scheduling: a simple technique for achieving locality and fairness in cluster scheduling. In: Morin, C., Muller, G. (eds.) EuroSys, pp. 265–278. ACM (2010)

A Job Scheduling Approach for Multi-core Clusters Based on Virtual Malleability

Gladys Utrera[1], Siham Tabik[2], Julita Corbalan[1], and Jesús Labarta[3]

[1] Technical University of Catalonia (UPC) 08034 Barcelona, Spain
{gutrera,juli}@ac.upc.edu
[2] University of Malaga, 29071 Malaga, Spain
stabik@uma.es
[3] Barcelona Supercomputing Center (BSC) 08034 Barcelona, Spain
jesus.labarta@bsc.es

Abstract. Many commercial job scheduling strategies in multi processing systems tend to minimize waiting times of short jobs. However, long jobs cannot be left aside as their impact on the performance of the system is also determinant. In this work we propose a job scheduling strategy that maximizes resources utilization and improves the overall performance by allowing jobs to adapt to variations in the load. The experimental evaluations include both simulations and executions of real workloads. The results show that our strategy provides significant improvements over the traditional EASY backfilling policy, especially in medium to high machine loads.

Keywords: job scheduling, MPI, malleability.

1 Introduction

Modern computational clusters tend to have thousands of execution units [5]. In order to make these investments profitable, such clusters must have many users (clients). This leads to a large amount of job submissions that often exceeds the cluster capacity. Figure 1 shows a typical weekly load of the Marenostrum machine [1]. Many of these clusters are composed by nodes of multi-core processors. Multi-core processors have two or more complete computational cores integrated in the same chip. As a processing core can act as an independent processor or CPU, in this work terms core and CPU are synonyms.

A *job scheduling strategy* (JSS) is an algorithm that allocates resources to submitted jobs while applying system's administrative policies and priorities. A JSS has to deal with a wide variety of applications, from sequential to highly parallel codes, with execution times that varies from minutes to days. This scenario converts the comparison of two JSS into a difficult task. The high cost of the clusters usually makes user satisfaction the main objective for improving performance of the JSSs. For this reason, waiting times of short jobs that exceed by far their execution times are inadmissible. However, long jobs also play an important role in the performance which finally affect short jobs as well.

C. Kaklamanis et al. (Eds.): Euro-Par 2012, LNCS 7484, pp. 191–203, 2012.

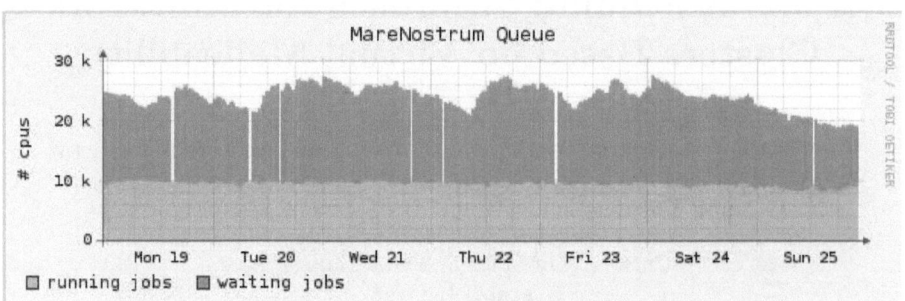

Fig. 1. Marenostrum load and wait queue during a week [1]

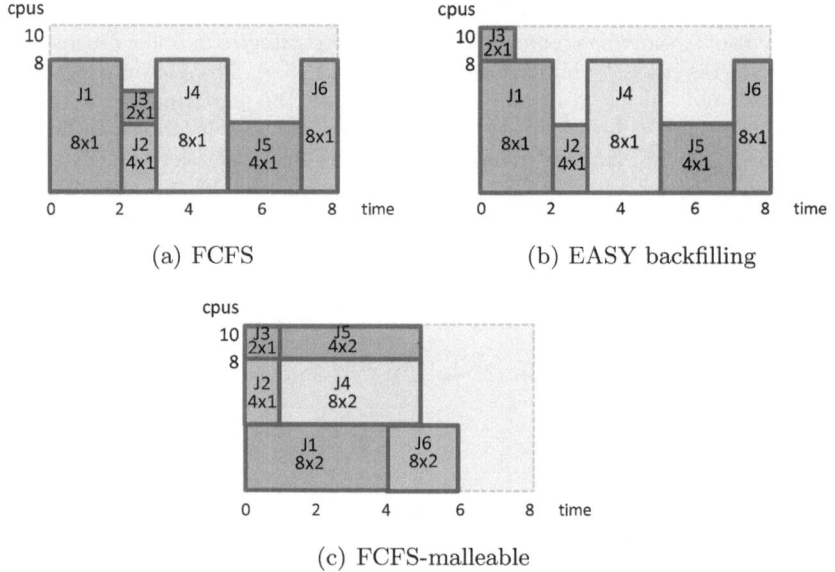

Fig. 2. Job scheduling under : (a) FCFS, (b) EASY backfilling and (c) FCFS-malleable

To quantify the performance of JSSs, this work uses three metrics, namely the *response time, slowdown* and *fragmentation*. The first two metrics depend on the waiting and execution times of jobs while the last one measures the utilization of the computing system. The following classification of job flexibility is commonly accepted in the literature [14]: *rigid* jobs, which are compiled to be run with a specific and fixed number of processes; *moldable* jobs can be executed on multiple CPU partition sizes, but once the execution starts, these sizes cannot be modified; *malleable*, if the size of the assigned partition can also be modified during the execution. Moldable and malleable jobs would increase system utilization. However, they are not the common case in production systems. This paper presents a JSS, called *First Come First Served-malleable (FCFS-malleable)*, that minimizes waiting times by maximizing system utilization no matter the jobs' execution time, nor their flexibility type. The proposed JSS is based on the idea of *Virtual Malleability* (VM)[24]. VM allows jobs to adapt to changes in the number of CPUs at runtime preserving the original number of processes. Figure 2 exemplifies two JSSs from literature, FCFS and EASY backfilling, and FCFS-malleable on a hypothetical machine of 10 CPUs where the x- and y- axis represent time and number of CPUs respectively. Each rectangle is labeled with the name of the job it represents (e.g. J1) and the number of processes per CPU (e.g., 8×1 means that each of the 8 processes is assigned to a different CPU). The jobs arrived to the wait queue in the following order: *J1, J2, J3, J4, J5, J6* with execution times equal to 2, 1, 1, 2, 2, 1 time units respectively.

Figure 2(a) shows the execution of the workload under FCFS. The execution of each job is delayed till there are enough CPUs for it. Meanwhile a group of available CPUs is not used even though there are jobs in the wait queue. For example, 2 CPUs are idle when J1 is running and 6 CPUs are idle when J5 is running which generates fragmentation.

Figure 2(b) shows how *EASY backfilling* [16] works. To alleviate the delayed execution and fragmentation problem it tries to forward jobs ahead in the queue when there are enough resources for them to be executed and provided they don't delay the first job in the queue. For example, J3 can start together with J1. However, EASY backfilling does not always find a suitable hole for a waiting job so fragmentation is not always eliminated (e.g. after J1 and J3 finish).

Figure 2(c) demonstrates how the JSS proposed in this paper behaves. As there are jobs in the wait queue, VM is applied to J1 (the oldest) so it starts its execution shrunk in 4 CPUs together with J2 and J3. Even shrinking J2 and J3, there are not enough CPUs for J4, so it has to wait and J2 and J3 can run expanded. This restriction is due to the fact that in this work VM can reduce the number of CPUs only up to half of the total assigned number. After J2 and J3 finish, J4 and J5 can start their executions shrunk. Finally, J6 starts expanded as there are no jobs in the wait queue [1].

The experimental results showed that FCFS-malleable overcomes EASY backfilling by 28% in average slowdown and 31% in average response time.

The main contributions of this work are as follows:

- A new JSS based on the concept of VM. The algorithm is easy to implement and does not require neither application recompilation nor any previous knowledge about the application.

- A study of the impact on the performance when applying VM to individual applications, taking into account intranode contention.

- Evaluations and comparison of the proposed JSS using both simulator and a runtime system.

---

[1] We assume that two processes are executed twice slower on a single CPU than on two separated CPUs. Different studies indicate that this time could be by far less then twice [22][26], since computation and communication can be overlapped.

2   Related Work

The problem of scheduling and allocating resources to jobs in parallel systems has been addressed in previous research from two perspectives. While some works focus on providing support, libraries and runtime systems, to make individual applications malleable [20,7,8,11,24,15] others attempt to integrate heuristics or new techniques to backfilling or FCFS policies [23,10,9,21].

In [20] the moldability is applied in conjunction with *folding* [18] to reduce wait times. They create as many threads as the available number of CPUs and execute the job in the assigned partition. As the load increments, the partition of the latest arrived job is reduced to a half, freeing CPUs. The folding must be done at explicit synchronizing points in the source code. This generates additional wait time and requires extra effort from programmers and system support.

In [7] the authors modify the number of used CPUs at predetermined points of execution. The evaluation was carried out using workloads of only up to 10 jobs of NAS Benchmarks, with small classes like A and B. The inclusion of dynamic malleability support using a resource manager was also studied in [8,11] They all require applications to be malleable or code modifications.

Certain levels of oversubscription improve resources utilization. The scheduling of jobs on the assigned CPUs can be done explicitly as in Gang Scheduling [13]. A static schedule of parallel communicating processes must be computed a priori, and a global context switch is used to coschedule these communicating processes. In this way, these processes have the illusion of running on a dedicated (but slower) system. These schemes usually require long time quanta to amortize the high context switch and synchronization costs, making the system less responsive for interactive and I/O-intensive applications. Furthermore, they keep the CPU idle while a process is performing I/O or is waiting for a message within its allotted time quantum. In [27] they propose an alternative to overcome this problem by matching pairs of processes: compute bound with I/O bound to share a time slice. The idea looks interesting, but they still have synchronization costs and extra effort to find, if possible, the correct matching. On the other hand, the experiments use applications with no more than 16 processes.

Implicit coscheduling [6,28] tries to overcome the drawbacks of explicit coscheduling by relying on local schedulers, so that interactive and I/O-bound jobs are properly handled. They use the communication behavior of parallel processes to make scheduling decisions. Compared to explicit coscheduling, these strategies are easier to implement on clusters, and have better scalability and reliability characteristics basing exclusively on local knowledge.

In [25] it is shown that when a job shares CPUs with itself response time and stability are improved. The fact of oversubscribing CPUs was recently re-addressed in multi-core systems [15]. Using several applications from different programming models they demonstrated that oversubscribing up to 8 tasks to each single CPU improves throughput over pure space sharing.

The most commonly used backfilling strategies are EASY backfilling and Conservative backfilling [19]. Conservative backfilling prioritizes predictability in response times and fairness, while EASY backfilling provides better response times, especially for short jobs. Many previous related works focus on making backfilling strategies more flexible by integrating moldability with them [23]. The partition size for a job is selected based on its scalability and turnaround time by applying the Downey model [10]. Results demonstrate a gain in performance over pure backfilling and pure moldability [9] on individual applications. In [21] they propose a relaxation of conservative backfilling by allowing all waiting jobs to be delayed but only to some extent which improved response time predictability and resource utilization. Bounded slowdown obtained worse results than backfilling.

The JSS presented in this paper has the following advantages over state-of-art: no predefined synchronization points are required to vary the number of CPUs; jobs are unaware of the changes in the number of assigned CPUs and consequently there is no overhead derived from data redistribution, process creation or elimination at runtime; no need for recompilation; no job classification is required; no prior knowledge of the job is needed; all the jobs are candidate to be shrunk and moldability is not required. Finally, evaluations were done using both real executions with workloads made from benchmarks with large variations in data size, number of processes and communication degree. Simulations use well-known workload traces rather than synthetic workloads with few jobs and CPUs.

3   The FCFS-Malleable Job Scheduling Strategy

First, this section describes our JSS and defines the metrics that are used for the comparisons of JSSs. After that, a brief description of the implementation of the runtime system is provided. A deeper description can be found in [26].

3.1   FCFS-Malleable Algorithm

Algorithm 1. Code executed at job arrival

1: **GetJobFromWaitQueue(J)**
2: **if** $FreeCpus \geq J.cpusRequested$ **then**
3:     $J.cpusAllocated \leftarrow J.cpusRequested$
4:     $Execute(J)$ // This function updates FreeCpus
5: **else**
6:     $listOfOrderedJobs = SortByArrivalTime(listOfRunningJobs)$
7:     // Selects as many jobs as required to execute J
8:     $listOfCandidates = SelectCandidateJobs(listOfOrderedJobs, J)$
9:     **if** $listNotEmpty(listOfCandidates)$ **then**
10:        $J.cpusAllocated = J.cpusAllocated/2;$
11:        **for** $V = JobsInList(listOfCandidates)$ **do**
12:            $V.cpusAllocated = V.cpusAllocated/2$
13:            $Shrunk(V)$ // This function updates FreeCpus and NumJobsShrunk
14:        **end for**
15:        $Execute(J)$
16:     **end if**
17: **end if**

FCFS-malleable combines FCFS with VM. Applying VM to a job consists on running it on a pool of CPUs with a size less than or equal to the number of processes. In the case the number of CPUS is smaller than the number of processes, each CPU will have binded a queue of processes belonging to the same job. The size of this queue is called *multiprogramming level* (MPL). The maximum MPL was set to 2. This maximum level was chosen based on: memory bandwidth, number of CPUs per node, number of entry points to the interconnection network.

FCFS-Malleable is an event driven algorithm executed at job arrival and at job ending. Algorithm 1 shows the code executed when a new job arrives. At that event, the algorithm evaluates whether it is possible to start a new job depending on the available CPUs. If so, it starts the job with as many CPUs as requested (lines 3-4). Otherwise, the JSS tries to free CPUs and execute the job by applying VM to some jobs that are already running, including J if necessary. When a job finishes execution, if the wait queue is not empty, Algorithm 1 is applied, otherwise, running shrunk jobs are expanded to the newly freed CPUs. Several criteria to decide which job to shrink or expand first were evaluated: the oldest first, the one with less CPU utilization, the longest first and, the shortest first. Our experiments showed that the oldest one first is the best option. Line 8 of Algorithm 1 implements this option.

Metrics Used for Evaluations. In order to quantify the performance of our technique and make comparisons with others JSSs from bibliography three metrics were used: average response time, average slowdown and fragmentation.

Response time, is the time elapsed between the job submission and termination. This metrics evidences long jobs performance and is calculated by averaging the response times of all the jobs across a workload. For example, the average response times of the example shown in Figures 2(a), 2(b), and 2(c) are equal to 4.66, 4.33 and 3.66 time units respectively.

Slowdown relates execution and wait time as it is shown in formula (1). This metrics indicates short jobs performance and is calculated by averaging the slowdown of all the jobs across a workload. It is important to note that to calculate the value of the average slowdown for the FCFS-malleable policy, the execution time in numerator of formula 1 is obtained using VM (i.e. with the overheads of running shrunk included). The average slowdown in Figures 2(a), 2(b), and 2(c) are equal to 3.5, 3.16 and 2.5 time units respectively. The utilization of the system is usually addressed as the percentage of CPUs that are busy running jobs. As we are concerned only when there are jobs in the wait queue, we will use the fragmentation concept instead (see formula (2)). In the example provided in introduction, the fragmentation values are equal to 30% for Figures 2(a) and 2(b) and 0% for 2(c).

$$Slowdown = \frac{WaitTime + ExecutionTime}{ExecutionTimeExpanded} \tag{1}$$

$$Fragmentation = \frac{\sum_{whenWaitQueueNotEmpty}^{t=start\ to\ termination} freeCPUs}{WorkloadTotalTime * TotalCPUs} \tag{2}$$

3.2   Runtime System Implementation

Let us now describe the runtime system and relevant details of implementation of the experimental framework. The runtime system is composed by a job scheduler (JS) and a runtime library. The JS receives as input a trace file and the JSS to apply. The trace file has identifications of the jobs, their arrival times and the number of requested CPUs [4]. The JS tracks information about node allocation, jobs in the wait queue, and already finished jobs.

The implementation of FCFS-malleable uses a library (VM library) that implements the concept of VM. The VM library was constructed using the *Message Passing Interface* [2], MPI, interposition mechanism. MPI was selected for being the most widely used and for its portability across shared and distributed memory architectures. The VM library is linked dynamically with jobs and communicates with the JS via TCP/IP sockets. This avoids the necessity of job recompilation. The library is in charge of CPU allocation and scheduling of processes. Process migrations are only allowed within a node and when VM is applied. Otherwise, processes remain binded to their assigned CPUs. The whole mechanism is transparent to the user. A job is said to run shrunk when is executed on a CPUs partition smaller than its number of processes. Processes belonging to the same job compete with themselves for the use of CPUs. A job is said to run expanded when is executed on a CPUs partition equal to its number of processes.

The scheduling of processes on a CPU is done by applying implicit coscheduling (see Section 2 for more details): only local knowledge (e.g. local communication events) is taken into account to make scheduling decisions. In particular Self co-scheduling [25] is applied. A running process yields the CPU and blocks immediately every time it executes a blocking operation (e.g. wait for a message that has not arrived yet). This type of scheduling promotes the overlapping of communication and computation phases.

The experiments were performed on a multi-core cluster with 10240 IBM Power PC 970MP cores at 2.3 GHz (2560 JS21 blades), 20 TB of main memory, 2510 nodes, and interconnection networks: Myrinet and Gigabit Ethernet. The operating system is Linux: SuSe Distribution. Each node has 4 cores sharing memory and each L2 cache is shared by every 2 cores.

4   Simulator

An event-driven simulator was constructed to extensively evaluate and compare JSSs. The simulator uses trace files in format of [4] as input and output. The following information about jobs is required to do the simulations: execution time, requested CPUs, requested time, CPU utilization[2]. Notice that FCFS-malleable may vary the number of CPUs of jobs at runtime. Thus, for FCFS-malleable we know only the *expanded* execution time of jobs. Next we provide a model to estimate the execution time of jobs when VM is applied to them.

---

[2] The field "CPU utilization" is used only by FCFS-malleable. In this work we refer to CPU utilization of a job to the average CPU time used by all its processes. That is the time when the CPU is doing useful work (i.e. computation).

Formula (3) arises from empirical observations. It estimates the execution time of a job when it runs isolated on different number of CPUs using the VM library. The value of MPL can vary during the execution time and is greater than 1 every time the job runs shrunk and is equal to 1 every time the job runs expanded. The parameter $CPUUtil$ is the percentage of CPU utilization when the job runs expanded. The parameter $execTime$ corresponds to the expanded execution time of the job. The parameter OV represents the overhead generated by the contention suffered when using the interconnection network. In our simulations, OV was set to random values between 0 and 1 as trace files have neither information about the communication-computation ratio nor the message sizes. We validated the proposed model by comparing results of simulations with real executions of several synthetic workloads.

$$estimatedIsolatedExecTime = \\ \sum_{t=start}^{t=termination} execTime * MPL(t) * CPUUtil \tag{3}$$

Our final model is described by formula (4).

$$estimatedExecTime = estimatedIsolatedExecTime + \\ execTime * OV \tag{4}$$

4.1   Validation of the Simulator

A synthetic workload trace was constructed to validate the simulator by applying the model in [17]. The trace was adjusted to have 150 jobs to be launched during 2 hours with average machine loads from 30% to 90%.

Table 1. Comparison of average wait times, response times and slowdowns between simulator and runtime system

	EASY backfilling						FCFS-malleable					
	Avg wait		Avg resp		Avg sld		Avg wait		Avg resp		Avg sld	
%load	S	R	S	R	S	R	S	R	S	R	S	R
30	55.0	60.0	107.0	113.0	2.1	2.1	1.6	1.3	73.5	73.3	1.2	1.4
50	69.6	74.0	121.3	125.0	6.5	6.5	13.2	13.8	93.5	93.0	3.2	3.2
70	93.0	100.0	145.2	152.0	10.7	10.3	36.8	32.0	118.3	113.0	6.0	5.7
90	175.0	162.0	228.0	214.0	19.0	16.0	118.0	111.0	220.0	198.9	10.8	9.6

In order to execute the trace generated with [17] in the runtime system, we substituted applications in the trace for real applications. Applications in the trace were matched according to their execution time and number of processes. In this way interarrival times were kept with the same characteristics as of the original trace. We used the NAS Parallel Benchmarks [3] classes A, B, C and D and number of processes varying from 1 to 128. We chose these benchmarks as they include widely used kernels. We executed the synthetic traces under FCFS-malleable and EASY backfilling JSSs both on simulator and runtime system. Table 1 provides the average waiting time, response time and slowdown obtained with the simulator (S) and with the real execution (R). The average relative error of the simulator compared to the runtime system is equal to 7%. Considering that the average gain of FCFS-malleable over EASY backfilling in the runtime system is around 30% we concluded that this error is acceptable.

5   Results and Analysis

Cleaned traces from Parallel Workload Archive [4] were used in our experiments. A cleaned trace does not contain flurries of activity by individual users which may not be representative of normal usage. Table 2 summarizes the workloads characteristics.

The columns show the names of the used workloads, total number of CPUs in the machine, number of jobs in the workload, average CPU utlization, average CPU utilization by long jobs and the ratio between the average number of requested CPUs by the machine capacity. For example, the workload in figure 1 has this ratio equal to 2. We have classified long jobs as the ones with number of processes greater than 64 and execution times greater than 8 hours and short jobs as the ones with execution times less than 10 minutes.

Table 2. Description of the workload log traces used for simulation

Workload	Cpus	Jobs	Avg CPU Util	Avg long jobs CPU Util	Req.Cpus/Cpus
CTC	430	20K-25K	57 %	70%	5.8
SDSC Blue	1152	20K-25K	23 %	70%	3.8
SDSC	128	40K-45K	66 %	90%	8.8

The CTC trace contains records from IBM SP2 located at the Cornell Theory Center. SDSC and SDSC Blue traces are from the San Diego Supercomputing Center. We now present the experimental results obtained from simulations using the workloads traces from Table 2.

5.1   Experimental Results

Figures 3(a), 3(b), 3(c) and 3(d) show the average wait time, execution time, response time and slowdown respectively for CTC, SDSC and SDSC Blue workloads under FCFS-malleable and EASY backfilling JSSs[3].

FCFS-malleable JSS obtained better average response time in all the traces, especially in trace SDSC Blue. This workload contains jobs with low CPU utilization, which leads to higher degree of overlap of communication and communication. In addition, this workload has no sequential jobs, thus all the jobs are eligible for applying VM.

---

[3] Variations of backfilling policies are used in most of the Top50 machines[12]. EASY backfilling is used as a reference for performance comparison in almost every job scheduling research. That is why we chose EASY backfilling for our comparisons.

As it was expected, average execution times are larger under FCFS-malleable due to the reduction on the number of CPUs. However, these execution times are not twice larger than the execution times in EASY backfilling.

FCFS-malleable obtains substantially better average slowdowns in CTC and SDSC Blue but not in SDSC. This means that the performance of short jobs is degraded in that workload. Analyzing this penalization we found that it was due to the strong presence of sequential jobs and the high CPU utilization of long jobs. EASY backfilling outperformed FCFS-malleable only on jobs with execution time less than 3 minutes and number of processes less than 16. EASY backfilling failed to find a suitable hole to forward long sequential jobs or with high degree of parallelism. This study can be found in [26].

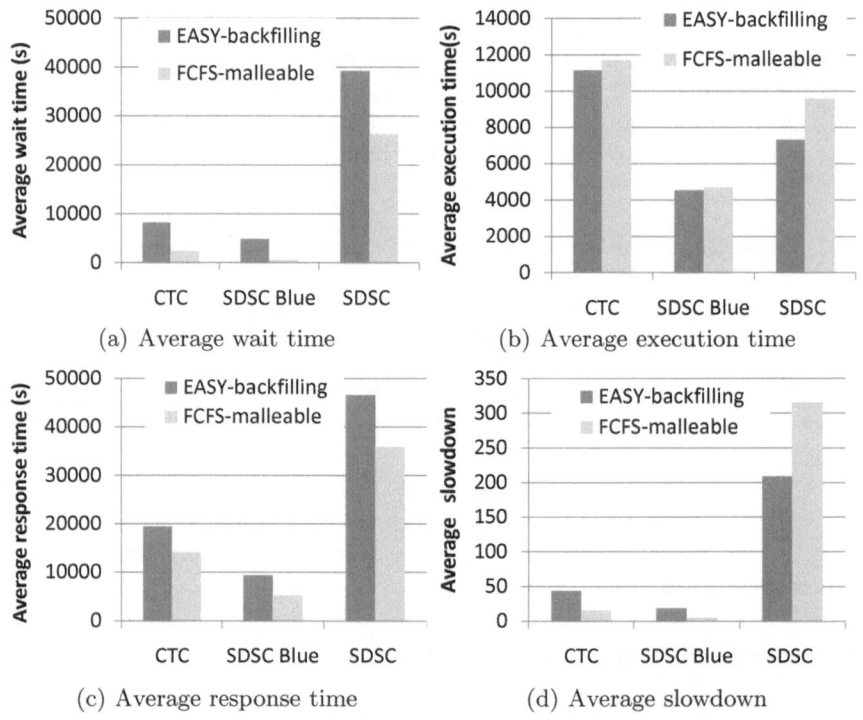

Fig. 3. Average wait, execution, response time and slowdown for CTC, SDSC and SDSC Blue under FCFS-malleable and Easy backfilling

The CPU utilization for long jobs is the highest for the SDSC workload (see Table 2). We re-simulated the SDSC workload trace varying the average value of CPU utilization of long jobs between 60% and 100%. We observed that for long jobs with CPU utilization under 90% the average slowdown for FCFS-malleable is smaller than for EASY backfilling. Due to lack of space we omitted that study here, but it can be found in [26].

Table 3. Average MPL

Workload	EASY backfilling	FCFS-malleable
CTC	0.75	0.91
SDSC Blue	0.76	0.98
SDSC	0.89	1.47

FCFS-malleable managed to eliminate fragmentation in all the workloads while EASY backfilling had fragmentation percentages from 6 for CTC to 14 for SDSC. Table 3 shows the average MPL of the three workloads for EASY backfilling and FCFS-malleable. MPL was calculated by averaging the total number of processes in the system per CPU. FCFS-malleable has average MPL below 2 (the maximum). This means that the workloads have variations so that jobs could expand from time to time decreasing in this way the average value of MPL. The value of the average MPL for the SDSC trace means that half of the CPUs run shrunk jobs all the time.

6   Conclusions and Future Work

In this work we proposed a new job scheduling strategy (JSS) for multi-core clusters: FCFS-malleable. Evaluations on the target architecture were carried out using a job scheduler and a runtime system implemented for that purpose. In addition, to extend evaluations to workloads from production systems, a simulator was constructed. Experimental results showed that FCFS-malleable outperforms EASY backfilling by 28% in average slowdown and by 31% in average response time. In addition, our JSS reduces fragmentation thanks to its capability to adapt jobs to available resources by shrinking and expanding them.

Although in this work we compete with backfilling, our JSS can be combined with it to take the most of both strategies. We are currently evaluating this approach. Memory bandwidth was not taken into account in the current study. We are working on an accurate estimation of the overhead caused by limited memory bandwidth.

Acknowledgements. This work was supported by the Ministry of Science and Technology of Spain under contracts TIN2007-60625, TIN2006-01078, TIN2010-16144 and Juan de la Cierva and the postdoctoral contract funded by the University of Malaga.

References

1. Marenostrum, http://www.bsc.es/marenostrum-support-services
2. MPI library, http://www.mcs.anl.gov/research/projects/mpi/
3. NAS Parallel Benchmarks,
   http://www.nas.nasa.gov/Resources/Software/npb.html
4. Parallel workload archive, http://www.cs.huji.ac.il/labs/parallel/workload/
5. Top500 supercomputers sites, http://www.top500.org/

6. Arpaci-Dusseau, A.C.: Implicit coscheduling: coordinated scheduling with implicit information in distributed systems. ACM Trans. Comput. Syst. 19, 283–331 (2001)
7. Buisson, J., Sonmez, O., Mohamed, H., Lammers, W., Epema, D.: Scheduling malleable applications in multicluster systems. In: Proc. of the IEEE International Conference on Cluster Computing 2007, pp. 372–381 (2007)
8. Cera, M.C., Georgiou, Y., Richard, O., Maillard, N., Navaux, P.O.A.: Supporting Malleability in Parallel Architectures with Dynamic CPUSETs Mapping and Dynamic MPI. In: Kant, K., Pemmaraju, S.V., Sivalingam, K.M., Wu, J. (eds.) ICDCN 2010. LNCS, vol. 5935, pp. 242–257. Springer, Heidelberg (2010)
9. Cirne, W., Berman, F.: Using moldability to improve the performance of supercomputer jobs. J. Parallel Distrib. Comput. 62, 1571–1601 (2002)
10. Downey, A.B.: A model for speedup of parallel programs. Technical report, University of California at Berkerley (1997)
11. El Maghraoui, K., Desell, T.J., Szymanski, B.K., Varela, C.A.: Dynamic malleability in iterative MPI applications. In: Proceedings of the Seventh IEEE International Symposium on Cluster Computing and the Grid, CCGRID 2007, pp. 591–598. IEEE Computer Society, Washington, DC (2007)
12. Ernemann, C., Krogmann, M., Lepping, J., Yahyapour, R.: Scheduling on the Top 50 Machines. In: Feitelson, D.G., Rudolph, L., Schwiegelshohn, U. (eds.) JSSPP 2004. LNCS, vol. 3277, pp. 17–46. Springer, Heidelberg (2005)
13. Feitelson, D.G., Rudolph, L.: Gang scheduling performance benefits for fine-grain synchronization. Journal of Parallel and Distributed Computing 16(4), 306–318 (1992)
14. Feitelson, D.G., Rudolph, L.: Toward Convergence in Job Schedulers for Parallel Supercomputers. In: Feitelson, D.G., Rudolph, L. (eds.) IPPS-WS 1996 and JSSPP 1996. LNCS, vol. 1162, pp. 1–26. Springer, Heidelberg (1996)
15. Iancu, C., Hofmeyr, S., Zheng, Y., Blagojevic, F.: Oversubscription on multicore processors. In: 24th International Parallel and Distributed Processing Symposium (IPDPS), pp. 1–11 (2010)
16. Lifka, D.A.: The ANL/IBM SP Scheduling System. In: Feitelson, D.G., Rudolph, L. (eds.) IPPS-WS 1995 and JSSPP 1995. LNCS, vol. 949, pp. 295–303. Springer, Heidelberg (1995)
17. Lublin, U., Feitelson, D.G.: The workload on parallel supercomputers: Modeling the characteristics of rigid jobs. Journal of Parallel and Distributed Computing 63, 2003 (2001)
18. McCann, C., Zahorjan, J.: Processor allocation policies for message-passing parallel computers. In: Proceedings of the 1994 ACM SIGMETRICS Conference on Measurement and Modeling of Computer Systems, SIGMETRICS 1994, pp. 19–32. ACM, New York (1994)
19. Mu'alem, A.W., Feitelson, D.G.: Utilization, predictability, workloads, and user runtime estimates in scheduling the ibm sp2 with backfilling. IEEE Transactions on Parallel and Distributed Systems 12(6), 529–543 (2001)
20. Padhye, J., Dowdy, L.W.: Dynamic Versus Adaptive Processor Allocation Policies for Message Passing Parallel Computers: An Empirical Comparison. In: Feitelson, D.G., Rudolph, L. (eds.) IPPS-WS 1996 and JSSPP 1996. LNCS, vol. 1162, pp. 224–243. Springer, Heidelberg (1996)
21. Sodan, A.C., Jin, W.: Backfilling with fairness and slack for parallel job scheduling. Journal of Physics: Conference Series 256(1), 012–023 (2010)

22. Subotic, V., Labarta, J., Valero, M.: Simulation environment for studying overlap of communication and computation. In: 2010 IEEE International Symposium on Performance Analysis of Systems & Software (ISPASS), White Plains, NY, pp. 115–116 (March 2010)
23. Sudarsan, R., Ribbens, C.J.: Scheduling resizable parallel applications. In: International Parallel and Distributed Processing Symposium, pp. 1–10 (2009)
24. Utrera, G., Corbalán, J., Labarta, J.: Implementing malleability on MPI jobs. In: Proceedings of the 13th International Conference on Parallel Architectures and Compilation Techniques, PACT 2004, pp. 215–224. IEEE Computer Society, Washington, DC (2004)
25. Utrera, G., Corbalán, J., Labarta, J.: Scheduling of MPI Applications: Self-co-scheduling. In: Danelutto, M., Vanneschi, M., Laforenza, D. (eds.) Euro-Par 2004. LNCS, vol. 3149, pp. 238–245. Springer, Heidelberg (2004)
26. Utrera, G., Tabik, S., Corbalán, J., Labarta, J.: A job scheduling approach to reduce waiting times. Technical report, Technical University of Catalonia, UPC-DAC-RR-2012-1 (October 2011)
27. Wiseman, Y., Feitelson, D.G.: Paired gang scheduling. IEEE Transactions on Parallel and Distributed Systems 14(6), 581–592 (2003)
28. Zhang, Y., Sivasubramaniam, A., Moreira, J., Franke, H.: A simulation-based study of scheduling mechanisms for a dynamic cluster environment. In: Proceedings of the 14th International Conference on Supercomputing, ICS 2000, pp. 100–109. ACM, New York (2000)

Topic 4: High-Performance Architecture and Compilers

Alex Veidenbaum, Nectarios Koziris, Toshinori Sato, and Avi Mendelson

Topic Committee

High-performance architecture and compilation are the foundation on which the modern computer systems are built. The two sub-topics are very strongly related and only in combination can deliver performance levels we came to expect from systems. The topic is quite broad, with sub-areas of interest ranging from multi-core and multi-threaded processors to large-scale parallel machines, and from program analysis, program transformation, automatic discovery and management of parallelism, programmer productivity tools, concurrent and sequential languages, and other compiler issues.

This year four papers were accepted after a thorough review and discussion. These papers are summarized below. We are grateful to all reviewers who helped us in this process, as we obtained at least three reviews per submitted paper.

It is clear that the remaining papers proposed interesting ideas, but this year's competition was really tough. We thank all for their submissions and hope everyone will continue to support the conference. We also thank the Euro-Par Organizing Committee for their guidance and their useful comments.

The paper "Dynamic Last-Level Cache Allocation to Reduce Area and Power Overhead in Directory Coherence Protocols" by Mario Lodde, Jose Flich, and Manuel E. Acacio proposes the reorganization of the Last Level Cache (LLC), where the storage or not of the cache blocks' data will depend on their characterization as private or shared blocks. More specifically, if a block is private (i.e. used only by one core), then the LLC will hold only its tag and any information needed by the coherence protocol. The motivation behind this proposal is the observation that a large percentage of the actions performed by the LLC concerns private blocks as they are forwarded straight to the L1 caches and do not involve the data portion of the LLC. By "eliminating" the storage of the private blocks in the LLC, the authors achieve area and power savings with a negligible impact on the performance.

The paper "A Practical Approach to DOACROSS Parallelization" by Priya Unnikrishnan, Jun Shirako, Kit Barton, Sanjay Chatterjee, Raul Silvera, and Vivek Sarkar presents a new approach for automatic parallelization of DOACROSS loops. It is based on a compiler and runtime optimization ("dependence folding") which bounds the number of synchronization variables needed to control cross-iteration dependences. Furthermore, the authors present a cost analysis for determining the profitability of parallelization, and additional techniques (unrolling, chunking) that increase granularity and reduce synchronization overhead. These characteristics render their approach practical, compared to prior similar efforts. Their approach was evaluated using 4 benchmarks on a 32-core machine. The auto-parallelization of DOACROSS loops offered

C. Kaklamanis et al. (Eds.): Euro-Par 2012, LNCS 7484, pp. 204–205, 2012.
© Springer-Verlag Berlin Heidelberg 2012

significant speedups (compared both to sequential execution and DOALL automatic parallelization) but only when cost analysis and granularity control was enabled.

The paper "Exploiting Semantics of Virtual Memory to Improve the Efficiency of the On-Chip Memory System" by Bin Li, Zhen Fang, Li Zhao, Xiaowei Jiang, Lin Li, Andrew Herdrich, Ravishankar Iyer, and Srihari Makineni proposes two hardware-based mechanisms that exploit stack memory's characteristics to optimize on-chip memory. The first mechanism reduces TLB misses by $10\% - 20\%$ by automatically creating large pages ("superpages") to host stack memory contents. The second technique treats stack accesses in a distributed shared cache in a different way than regular ones, by routing them to each core's local cache slice. The benefit of this approach is reduced interconnect power consumption by more than 14%. Both techniques are evaluated using a simulation framework and the SPEC CPU 2000 benchmarks.

Finally, the paper "From Serial Loops to Parallel Execution on Distributed Systems" by George Bosilca, Aurelien Bouteiller, Anthony Danalis, Thomas Herault, and Jack Dongarra, presents a compiler front-end for the DAGuE runtime system, to analyze annotated serial loops of tiled dense linear algebra algorithms, in order to provide symbolic information to the runtime system for the efficient execution on distributed memory machines.

Dynamic Last-Level Cache Allocation to Reduce Area and Power Overhead in Directory Coherence Protocols

Mario Lodde[1], Jose Flich[1], and Manuel E. Acacio[2]

[1] Universitat Politècnica de València, Spain
[2] Universidad de Murcia, Spain

Abstract. Last level caches (LLC) play an important role in current and future chip multiprocessors, since they constitute the last opportunity to avoid expensive off-chip accesses. In a tiled CMP, the LLC is typically shared by all cores but physically distributed along the chip, thus providing a global banked capacity memory with high associativity. The memory hierarchy is orchestrated through a directory-based coherence protocol, typically associated to the LLC banks. The LLC (and directory structure) occupies a significant chip area and has a large contribution on the global chip leakage energy. To counter measure these effects, we provide in this paper a reorganization of the LLC cache and the directory by decoupling tag and data entry allocation, and by exploiting the high percentage of private data typically found in CMP systems. Private blocks are kept in L1 caches whereas LLC area is reorganized to reduce L2 entries while still allowing directory entries for private data, thus, maximizing on-chip memory reuse. This is achieved with no performance drop in terms of execution time. Evaluation results demonstrate a negligible impact on performance while achieving 45% of area saving and 75% of static power saving. For more aggressive designs, we achieve 80% area and 82% static power savings, while impacting performance by 10%.

1 Introduction

Tiled chip multiprocessors (CMPs) have been advocated as the most effective way of organizing future many-core CMPs with dozens of processor cores [1]. These tiled architectures provide a scalable solution for managing the design complexity, and effectively using the resources available in advanced VLSI technologies. A tiled CMP is built by replicating the same tile structure on the chip surface. Each tile typically includes one (or more) processor core, one (or more) level of private caches, part (one bank) of a shared but distributed last-level cache (LLC) and a router to connect the tiles building a network-on-chip (NoC). The shared LLC is typically inclusive with respect to all of the private caches. This means that, at all times, the LLC contains a superset of the blocks in the private caches (Intels Core i7 is a good example [2] of the latter).

Private caches in these designs are kept coherent by means of a directory-based cache coherence protocol implemented in hardware. The directory structure is distributed between the LLC banks, usually included into the tags portion of every cache entry [3]. In this way, each tile keeps the sharing information of the blocks mapped to the LLC bank that it contains. This sharing information comprises two main components: the *state bits* used to codify one of the possible states that the directory can assign to the cache block, and the *sharing code*, that holds the list of current sharers.

C. Kaklamanis et al. (Eds.): Euro-Par 2012, LNCS 7484, pp. 206–218, 2012.
© Springer-Verlag Berlin Heidelberg 2012

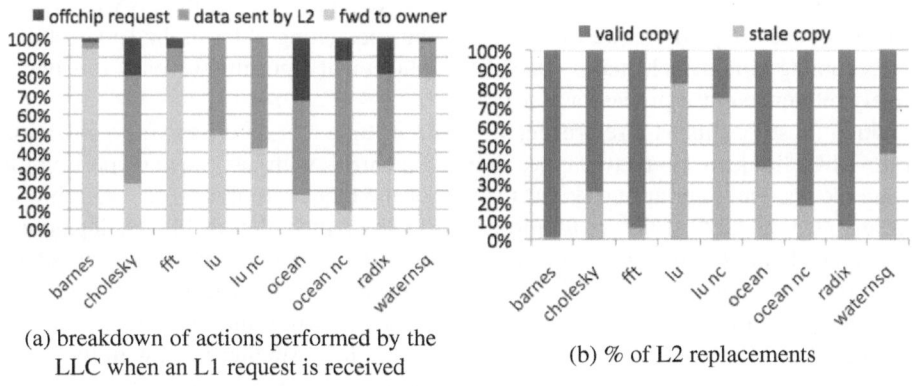

(a) breakdown of actions performed by the LLC when an L1 request is received

(b) % of L2 replacements

Fig. 1. Breakdown of actions and replacements at LLC

Last level caches (LLC) play an important role in ensuring performance since they constitute the last opportunity to avoid expensive off-chip accesses. In this way, current and future CMPs will be equipped with increasingly larger LLCs, occupying a significant fraction of the total chip area and having a large contribution on the global chip leakage energy (as an example, the size of the LLC of the Intel Core i7 can reach up to 15MB as by today [2]). In this work we focus on the design of an effective LLC for future many-core CMPs. In particular, we propose to reorganize the LLC structure to allow for the dynamic allocation of entries at this cache level depending on the specific necessities of every address.

Our work is motivated by the observation that for private blocks (memory blocks requested by a single core) the LLC contains stale data (or, more precisely, the data portion of the cache entry is not needed). While all the cached copies of a shared block have the same value, both in the private and the LLC caches, in the case of a private block (the block is owned by a single private cache) the values of the LLC and the private copies may differ (the block could be in modified state in the private cache). Therefore, the LLC copy of a private block is then probably stale, and useless until the corresponding private cache performs a *writeback* operation on the block. In that situation, the only valid copy of the block would be moved to the LLC. In this way, there are two cases in which the data portion of each LLC entry is needed: a) the block is shared by several cores, and b) a private block has been replaced by the owner private cache. In the rest of situations, keeping the data portion of the LLC entries can be seen as a waste of resources. Figure 1.(a) plots the breakdown of actions performed by the LLC of the 16-core CMP assumed in this work (details about the evaluation environment can be found in Section 3). More than 80% in some cases of the requests are forwarded to the L1 caches, thus do not involve the data portion of the LLC. Therefore, we can see that a large percentage of area and power is wasted in LLC to store private blocks.

On the other hand, LLC replacements (due to limited capacity and associativity) lead to invalidating data in private caches. Figure 1.(b) shows the percentage of LLC replacements of private blocks. As derived from the figure, a large percentage of replacements in the LLC (more than 40% in some applications) are for private blocks. Private data is only requested the first time the processor wants to operate on it and then is written and read without sending any request to the LLC, thus becoming older in the LLC set and thus becoming quickly selected by the LRU replacement algorithm (even if it is being actively referenced by a processor core).

Taking those results as a reference, we can conclude that an effective organization for the LLC should combine two types of entries: entries with just the tag's portion for private blocks, and regular entries for the rest of the blocks. In this work, we propose to reorganize the LLC structure to allow the dynamic allocation of blocks depending on the block being *private* or *shared*. In particular, we redesign the associated directory with a different (higher) associativity than the L2 data array. The tag and directory array contain information about all the cached blocks, while the data array contains only shared and replaced private blocks. This allows for a smaller LLC with the same performance as private blocks will be kept only at private caches. With our approach, we achieve large savings in static power while not hurting performance. Evaluation results demonstrate average savings of 45% in area and 75% in static power.

Our proposal can be combined with previous works that aim for reducing static power at L2 caches by dynamically powering down cache lines. This is the case of [4]. Notice that in that situation entries in L2 for private blocks (once the L1 cache writes on the block) are powered down. This means extra power saving through a line-level power gating mechanism (similar to cache decay for L1 caches [5]). These strategies are orthogonal to our approach. We provide results for the two mechanisms combined together.

The rest of the paper is organized as follows. In Section 2 we describe our proposal and its impact on area and power overhead. In Section 3, we perform a detailed analysis of performance and power savings. Then, related work is described in Section 4 and the paper is concluded in Section 5.

2 Dynamic L2 Cache Line Allocation

In this work, and without losing generality, we focus on CMPs made of 16 tiles connected through a 2D mesh NoC. Each tile includes a processor core, its private L1 cache and a bank of the L2 cache, which is shared by all the tiles. The whole L2 cache is, thus, distributed and made of 16 banks, each managing a subset of the global address space. From now on, we use the term L2 cache and LLC interchangeably.

We assume MESI states for the L1 caches. The M state is used when the block is private and has been modified; E is used when the block is private and not modified; S means read-only permissions over the block (the block is potentially shared); and I denotes a block that is invalid or not cached. Additionally, each directory entry (associated with each L2 cache entry) will have the following fields: 1) *Cache line state* field being P when the block is private, S when the block is shared with no owner, C when the block is only cached in L2, and I when the block is not cached (invalid entry). 2) *Owner* field being a pointer to the owner L1, which can provide the block to future requestors. 3) *Sharers List* field, being the list of the L1 caches which share the block.

Figure 2 shows how an L2 block switches between states depending on the requests received by L1 caches (transient states are omitted for the sake of clarity). At first, the block is not cached on chip (state I). Upon a write (GETX) or a read (GETS) request, the block is fetched from main memory and sent to the L1 requestor, which is now the owner of the block and holds a private copy (state P). At this point, a write request from another core will be forwarded to the owner, which will send the data to the requestor and invalidate its copy (the requestor will become the new owner). A read request (GETS) will also be forwarded to the owner, but in this case it will not invalidate its copy. Instead, it will provide the data to the requestor but also keep a copy of the block with read permission. However, the block state in the L2 bank will switch from P to S.

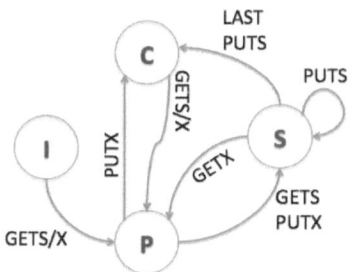

Fig. 2. Simplified FSM for the L2 cache (MESI protocol)

If the block is in P state and the owner replaces the block (PUTS or PUTX), then the only on-chip copy of the block lies in the corresponding L2 bank, which switches to state C. Further requests will be served using the L2 copy of the block.

Usually, there is a 1:1 relationship between the tag/directory and the data portions of the LLC. A different design can be chosen, with fewer data entries than tag/directory entries. If the block is private (state P in the directory), then only the tag/directory is allocated, as the only valid copy will be held by the owner L1. If, on the contrary, the block is shared or only cached in the L2 bank (state S and C in the directory), then both the tag/directory and the cache line are allocated.

We can reorganize an L2-directory bank by reducing the associativity in L2 and keeping or using that area for more tag/directory entries in the directory structure, so we break the 1:1 relationship. This will reduce replacements of blocks in P state. To do this, we keep the associativity of the tag/directory array to 16 while reducing the L2 data array associativity from 16 down to 8. This means that only the first eight ways of the tag array can store information about a block in state S or C, which data will be saved in the corresponding way of the data array, and all the 16 ways can store information about a private block. We also stress the inequality by further reducing L2 data associativity down to four and even two. Notice that this will constraint the area devoted to a shared data in L2 but will not compromise private data, as will be tracked by the directory and will be allocated in the L1 caches. With these ratios (from 1:2 to 1:8) large savings in L2 area are expected.

It has to be noted that one could think of reducing L2 size by reducing the number of sets, instead of the number of ways (see Figure 3). However, this could compromise cache capacity depending on the data set. Indeed, having less sets would lead to have less entries for shared data, while by reducing the associativity, cache capacity for shared data will not be compromised (as long as shared data does not conflict in the same set). Notice that we reduce ways as those are, expectedly, used by private blocks.

Another direction one could take is reducing in the same degree both the associativity in the L2 and in the directory (see Figure 3). However, in that case we would incur in high performance penalty (as will be seen later) as the associativity in the directory must be proportional and in line with the associativity in L1 caches. Simply, the directory will not have enough ways to avoid conflicts between both shared and private blocks.

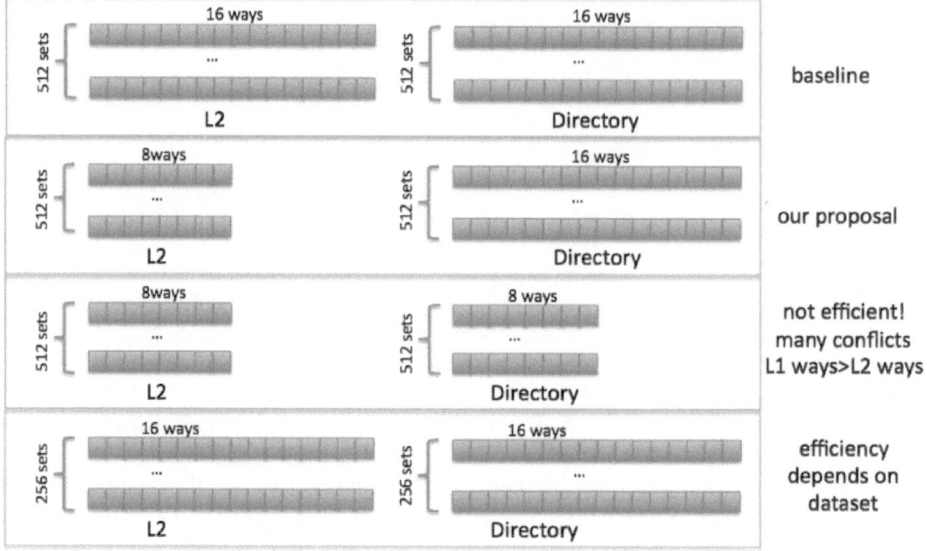

Fig. 3. Different LLC configurations by changing the number of sets and the associativity of the L2 and directory structures

Figure 4 shows an example of an L2 cache reorganized. Only the first four ways of the tag array may keep information about shared or cached blocks, which will be saved in the same way of the data array. The remaining four sets of the tag array are devoted to private blocks. However, private blocks can also be mapped in the first four ways of the tag array. In this case, the information included in the corresponding set of the data array would not be useful.[1]

Assuming an L2 cache with 4 ways and a directory with 8 ways, when a block switches from P to S or C, its entry must be moved if it doesn't lie in the first four ways. This may trigger the replacement of another block if all the first four ways are already in use, which means that the data array set is full. Thus, the automata to manage the L2 and directory needs to be slightly modified.

---

[1] With orthogonal power saving techniques as [4] these entries can be powered down. Its impact is later analyzed.

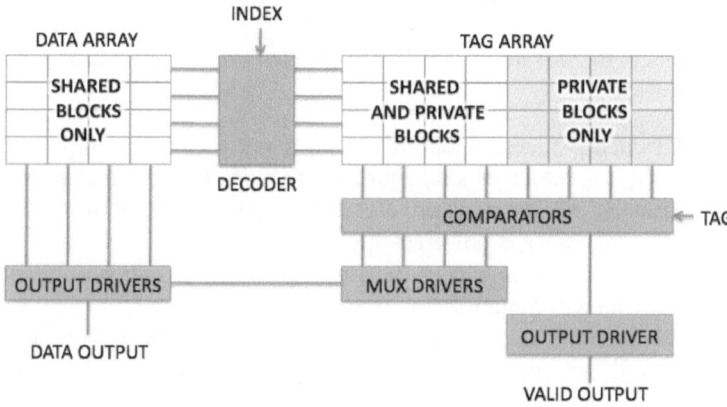

Fig. 4. Example of LLC reorganization

2.1 Replacement Policy

An LRU counter per way is used to implement the replacement policy. The counters are used in the classical way: each time the L2 receives a request for a block, all the counters with a lower value are incremented and the block counter is set to zero. When a new block is requested and all the set entries are already in use, the entry with the highest LRU counter value is selected. With the organization we propose, replacements can be triggered also when a block which is already cached must be saved in the L2 cache. As shown in Figure 2, this happens when an owner invalidates its private copy or a read request is received for a private block: the L2 state switches from P to C in the first case or from P to S in the second case. In both cases, a data entry must be allocated for the block in the L2 cache. If the first four ways are already allocated to other blocks, the replacement policy will choose the way with the highest LRU counter only between the first four entries (even though the entry with the highest LRU value could be one of the remaining ones). Notice, that LRU counters are updated for all the directory entries, thus not having two entries with the same number.

2.2 Dynamic Power Techniques

With our technique, truly shared data is promoted in the L2 data array, and private data is just tracked in the directory. However, it may happen that for a given L2-directory set more than half of the entries are for private blocks, thus not all the entries in the L2 cache will be used. In such situation, these L2 data entries are wasting energy. To mitigate this effect, our approach can be complemented with dynamic power-off strategies as [4], in which private blocks that could be allocated in the L2 cache lead to powering down the L2 entry. This can be achieved using "sleep transistors" at each cache line to eliminate the most part of the leakage current, as proposed by Kaxiras et al. [5] for L1 caches. As in [5], we use Powell's gated Vdd design [6] at cache line level, inserting a transistor between the ground and each L2 cache line to reduce the leakage current to a negligible level.

When combined, the different transitions of a block A in the L2 cache can be summarized as follows:

- When an L1 cache requests block A, which is not cached on-chip, the L2 issues an off-chip request, allocates a tag entry to the block (which can be anyone of the entries) and marks the block as private; if the tag entry is one of the first half, its corresponding data entry is powered-off.
- Subsequent write requests will cause a change of the owner but the state of the block in L2 cache will remain P.
- When the owner replaces its copy, it must be saved in the L2 cache. The same happens if the L2 receives a read request for a private block, which will become shared. If A is mapped in the first half of the tag ways, the corresponding data entry must be powered-on. If, on the contrary, A is mapped in the second half, it will trigger a potential internal swap, selecting one entry from the first half (using the LRU algorithm). The selected entry can be in one of the following states:
  - Private: this block and A are swapped, the data entry is powered on to save the write back copy of A.
  - Shared/Cached: this entry will be replaced and allocated to A.
- If a write request (in case A is in state S) or any request (in case A is in state C) is received, the block will be again treated as a private block and the data entry must be powered-off.

3   Performance Evaluation

We evaluate our proposal by using gMemNoCsim and Graphite [7] simulators. gMemNoCsim is an in-house event-driven cycle accurate cache hierarchy and NoC simulator. gMemNoCsim is embedded in Graphite, which allows execution-driven simulation of applications. Application's memory accesses are tracked by Graphite and fed into gMemNoCsim for an accurate memory coherency and on-chip network modeling. Graphite threads are blocked until memory accesses are resolved by gMemNoCSim. We implemented a two-level MESI coherence protocol, and the modifications needed to implement our block allocation and replacement policy (later we do the same for a MOESI protocol). Five different L2 designs have been evaluated and compared (block size is set to 64 bytes):

1. **L2-512-16_D-512-16**. An 512KB L2 bank with 512 sets and 16 ways. The directory also has 512 sets and 16 ways (1:1 ratio). This is the baseline for comparison.
2. **L2-512-8_D-512-16**. An 256KB L2 bank with 512 sets and 8 ways. The directory also has 512 sets but keeps 16 ways (1:2 ratio).
3. **L2-512-4_D-512-16**. An 128KB L2 bank with 512 sets and 4 ways. The directory keeps the same with 512 sets and 16 ways (1:4 ratio).
4. **L2-512-2_D-512-16**. An 64KB L2 bank with 512 sets and 2 ways. The directory keeps the same with 512 sets and 16 ways (1:8 ratio).

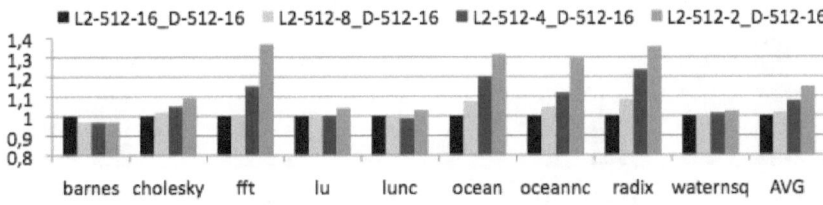

Fig. 5. Normalized execution time. MESI protocol with L2 banks with 512 sets.

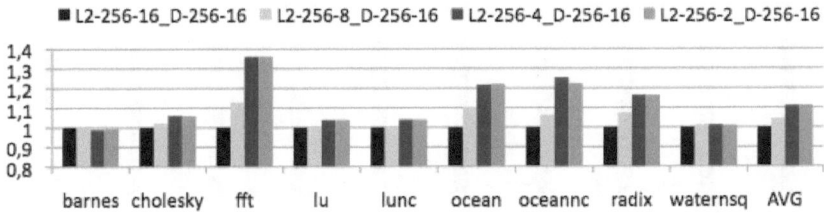

Fig. 6. Normalized execution time. MESI protocol with L2 banks with 256 sets.

In addition, we use a smaller L2 cache of 256KB as the baseline. In this case (**L2-256-16_D-256-16**), 256 sets and an associativity of 16 is used for both the L2 and the directory. Our designs on top of this baseline are **L2-256-8_D-256-16** (1:2 ratio, 128KB), **L2-256-4_D-256-16** (1:4 ratio, 64KB), and **L2-256-2_D-256-16** (1:8 ratio, 32KB).

The system is made of 16 tiles with one processor in each tile and with a private 32KB L1 data cache (with 256 sets and 4 ways). Each tile includes also the L2 bank and the associated directory. All the tiles are connected through a 2D mesh topology using the XY routing algorithm. In a first test every configuration does not incorporate any sleep transistor technology. Later we evaluate the combination of our technique with those. We ran various SPLASH-2 applications with these cache organizations.

Figure 5 shows the execution time for the different applications, normalized to the case of the first baseline L2-512-16_D-512-16. As can be seen, some applications have no impact on execution time when L2 banks are reduced. Indeed, in BARNES, CHOLESKY, LU, LUNC, and WATERNSQ, the L2 could be reduced by a factor of 8 (L2-512-2_D-512-16) with practically no impact on performance. On the other hand, other applications can be sensitive to L2 cache capacity to shared blocks. Anyway, by averaging, we can see that a good tradeoff is reducing L2 cache by half, which on average leads to only 1.7% performance decrease. Further reductions in area will tend to 7.5% performance degradation for an L2 reduction factor of 4 and to 15% for a reduction factor of 8.

For the case of smaller L2 banks (those with 256 sets), Figure 6 shows the execution time of applications. We can see similar trends with large savings (up to a factor of 8x) in area with no performance degradation, and others with some impact (up to 35%). On average, a reduction of 2x in L2 size have no large impact.

We use Cacti 5.3 [8] to compute area and leakage for the different L2-directory configurations. In Figure 7 we can see how area needs compare to the different evaluated

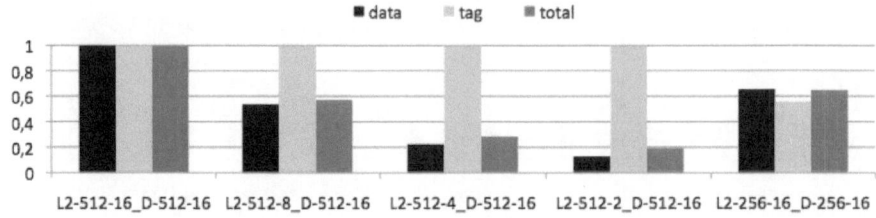

Fig. 7. Normalized LLC area occupancy

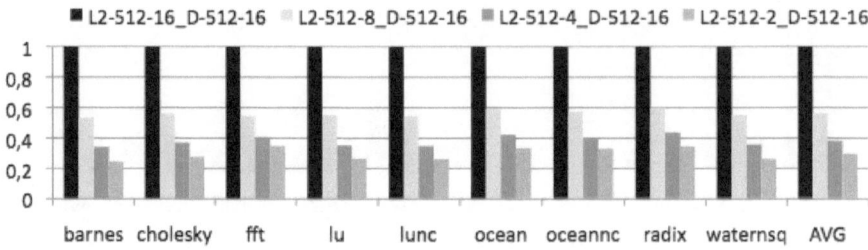

Fig. 8. Normalized L2 leakage energy. MESI protocol.

designs. Each component is normalized to the baseline design (L2-512-16_D-512-16). Tag array's area is the same for the first four designs, while in L2-256-16_D-256-16 tags take roughly half the area as the overall associativity is reduced. As far as data array is concerned, the area needs decrease with the associativity of each design. Even though L2-512-8_D-512-16 and L2-256-16_D-256-16 have the same data array size, the area of the former is lower than the area of the later, due to its lower associativity (lower number of comparators).

Figure 8 shows the leakage energy consumed by the L2 banks taking into account the entire execution of each application, and normalized to the baseline (L2-512-16_D-512-16). Leakage is reduced up to 80% due to the cache reorganization. This saving is proportional to the reduction ratio performed.

Figure 9 shows leakage savings when our proposal is combined with [6]. We compare the previous four cases and a baseline 256KB cache (L2-256-16_D-256-16). In both baselines, caches have all data entries powered on during the whole execution time, while the proposals power-on the L2 data entries only when needed, as perviously described. The average leakage energy is reduced on average by 75% (for 1:2 ratio), 83% (for 1:4 ratio) and 89% (for 1:8 ratio). Depending on the application, we achieve up to 98% in leakage energy savings (BARNES).

3.1   Benefits When Using MOESI Protocol

Another appealing protocol which can be implemented in L1 caches is MOESI. It behaves like the MESI protocol but when an owner (which has its block in state M or E) receives a forwarded GETS its state becomes O (and not S like in MESI protocol).

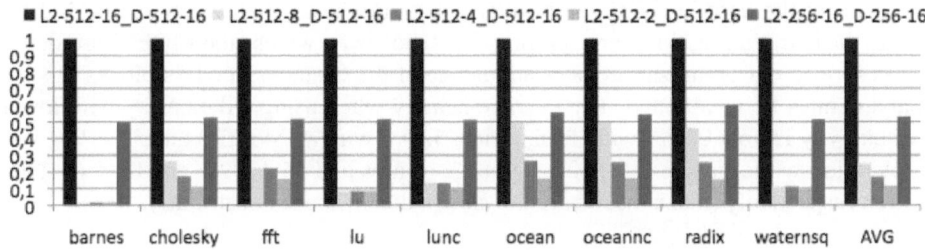

Fig. 9. Normalized L2 leakage energy (MESI) with sleep transistors

This means it remains the owner of the block with read permission, and when the L2 receives a request, it will still be forwarded to this L1. This is inefficient in a typical memory hierarchy since it takes one more step to provide the data to the requestor, but can largely benefit from our approach. Indeed, shared blocks in state O do not need to be allocated in L2 banks, thus, being all the requests forwarded to the owner. The L2 state diagram when a MOESI protocol is used in L1 caches is shown in Figure 10. A block keeps switching between states P and O until the owner invalidates its copy, and only then an L2 cache line must be allocated. In Figure 11 we compare the execution time for the different cache configurations with L1s that use a MOESI protocol (the figure shows the cases for 512 sets). However, for the sake of fairness, we use the MESI

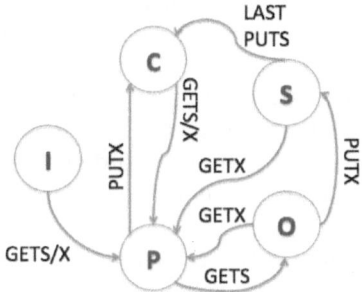

Fig. 10. Simplified FSM for the L2 cache (MOESI protocol)

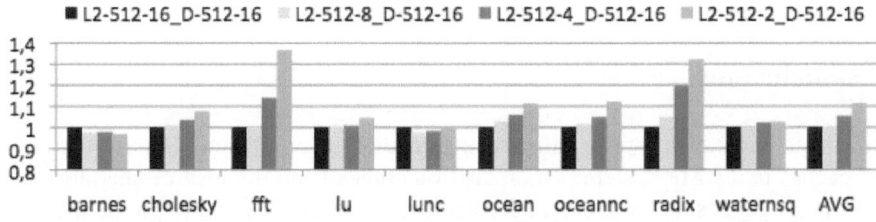

Fig. 11. Normalized execution time. MOESI (for 1:x ratio proposals) and MESI (for baseline).

protocol for the baseline configuration (which works better than when using MOESI, due to the extra indirection). For our proposals, however, we use the MOESI protocol.

As shown, our organization takes advantage of the O state since less blocks in state S and C must be invalidated. The average penalty when using a data array with lower associativity is now 0.7% (with a 1:2 ratio), 5.2% (with a 1:4 ratio) and 11.4 % (with a 1:8 ratio), while saving 43%, 72% and 81% of area and 75%, 82% and 83% of static power, respectively (these results are not shown due to space constraints).

4 Related Work

As the cache hierarchy is currently the chip component which has more impact in area and power, many efforts have been made in reducing its energy requirements. At circuit level, different techniques have been proposed to reduce cache leakage. Powell et al. [6] propose the gated-Vdd technique which powers down the L2 entries used for private blocks. Similar techniques have been proposed to reduce the supply voltage enough to reduce leakage without destroying the content of the cell, as done with drowsy [9] and superdrowsy [10] caches. The latter techniques have various drawbacks compared to the destructive technique proposed by [6], being more difficult to implement and saving less leakage since a certain voltage has to be provided to the cell to keep the data.

At higher level, alternative cache architectures have been proposed to reduce static and/or dynamic power. Savings can be achieved by modifying cache parameters like cache size and cache associativity [12, 13]. Other efforts have been made to reduce the number of cache accesses, by using snoop filters [14, 15], way predictors [16] or filter caches [17]. Various proposals turn off L1 or L2 cache ways based on different prediction techniques. As an example, Kaxiras et al [5] propose to turn off L1 cache entries which are not expected to be reused. Abella et al [11] propose a different predictor to turn off unused L2 cache entries. Li et al [4] use different policies to turn off L2 cache entries when block copies are replicated in an L1, and evaluate both conservative and destructive voltage gating techniques.

All these works always assume a 1:1 relationship between L2 entries and directory entries. In addition, our proposal is orthogonal to all these works, since in our work we simply remove the area devoted mainly to private blocks in L2. Indeed, we reduce L2 cache size by reducing its associativity, while keeping directory associativity, which is vital to keep track all the on-chip blocks, either shared or private. Indeed, we demonstrated in this paper our technique can be complemented with the one presented in [6].

5 Conclusions

In this paper we have proposed an effective method to significantly reduce the area of LLC caches in a CMP system. The reduction comes from the idea of keeping private blocks in L1 caches and not using L2 entries for such blocks. Although the idea is not new, in our approach we physically redimension the LLC cache and its associated directory and provide a 1:x approach, where different associativity degrees are used in both L2 and directory caches. The L2 cache associativity is lower in order to accommodate only shared blocks, while directory associativity is kept to keep track of all the blocks in the chip, either private or shared blocks.

Results demonstrate large savings in L2 area and static power consumption. With a MESI procotol, the L2 cache size can be halved with no impact on performance and with a rough reduction of leakage of 50%. More aggressive designs (75% of L2 size reduction) lead to low performance penalties but with very large savings in power. We also demonstrate in this paper that our technique can be complemented with power-gating mechanisms in the L2 cache.

Acknowledgement. This work has been supported by the VIRTICAL project (grant agreement n 288574) which is funded by the European Commission within the Research Programme FP7.

References

1. Tile-gx processors family, http://www.tilera.com/products/TILE-Gx.php
2. Intel Core i7 technical Specifications,
   http://www.intel.com/products/processor/corei7ee/
   specifications.htm
3. Zhang, M., Asanović, K.: Victim replication: Maximizing capacity while hiding wire delay in tiled chip multiprocessors. In: Proc. of the 32nd Int'l Symposium on Computer Architecture (ISCA-32), pp. 336–345 (2005)
4. Li, L., Kadayif, I., Tsai, Y.-F., Vijaykrishnan, N., Kandemir, M., Irwin, M.J., Sivasubramaniam, A.: Leakage energy management in cache hierarchies. In: Proceedings of the International Conference on Parallel Architectures and Compilation Techniques (2002)
5. Kaxiras, S., Hu, Z.: Cache Decay: Exploiting Generational Behavior to Reduce Cache Leakage Power. In: Proc. of the 28th Int'l Symposium on Computer Architecture (ISCA 2001) (May 2001)
6. Powell, M., Yang, S.-H., Falsafi, B., Roy, K., Vijaykumar, T.N.: Gated-Vdd: a circuit technique to reduce leakage in deep-submicron cache memories. In: Proceedings of the International Symposium on Low Power Electronics and Design, ISLPED 2000 (2000)
7. Miller, J.E., Kasture, H., Kurian, G., Gruenwald III, C., Beckmann, N., Celio, C., Eastep, J., Agarwal, A.: Graphite: A Distributed Parallel Simulator for Multicores. In: The 16th IEEE International Symposium on High-Performance Computer Architecture (HPCA) (January 2010)
8. Cacti 5 Technical Report,
   http://www.hpl.hp.com/techreports/2008/HPL-2008-20.html
9. Flautner, K., Kim, N.S., Martin, S., Blaauw, D., Mudge, T.: Drowsy caches: Simple techniques for reducing leakage power. In: Proceedings of the ACM/IEEE 29th International Symposium on Computer Architecture, ISCA 2002 (2002)
10. Kim, N.S., Flautner, K., Blaauw, D., Mudge, T.: Single-VDD and Single-VT super- drowsy techniques for low-leakage high-performance instruction caches. In: Proceedings of the ACM/IEEE International Symposium on Low Power Electronics and Design, ISLPED 2004 (2004)
11. Abella, J., Gonzales, A., Vera, X., O'Boule, M.F.P.: IATAC: A Smart Predictor to Turn-off L2 Cache Lines. ACM Transactions on Architecture and Code Optimization 2(1) (March 2005)

12. Drophso, S., Buyuktosunoglu, A., Balasubramonian, R., Albonesi, D.H., Dwarkadas, S., Semeraro, G., Magklis, G., Scott, M.L.: Integrating adaptive on-chip storage structures for reduced dynamic power. In: Proceedings of the IEEE 11th International Conference on Parallel Architectures and Compilation Techniques, PACT 2002 (2002)
13. Zhang, C., Vahid, F., Najjar, W.: A highly configurable cache architecture for embedded systems. In: Proceedings of the ACM/IEEE 30th International Symposium on Computer Architecture, ISCA 2003 (2003)
14. Moshovos, A., Memik, G., Falsafi, B., Choudhary, A.: Jetty: Filtering snoops for reduced energy consumption in SMP servers. In: Proceedings of International Symposium on High Performance Computer Architecture (HPCA 2001) (January 2001)
15. Salapura, V., Blumrich, M., Gara, A.: Design and implementation of the Blue Gene/P snoop filter. In: Proceedings of International Symposium on High Performance Computer Architecture (HPCA 2007) (February 2007)
16. Inoue, K., Ishihara, T., Muruami, K.: Way-predictive set-associative cache for high performance and low energy consumption. In: Proceedings of the ACM/IEEE International Symposium on Low Power Electronics and Design, ISLPED 1999 (1999)
17. Kin, J., Gupta, M., Mangione-Smith, W.-H.: The filter cache: An energy efficient memory structure. In: Proceedings of the ACM/IEEE 30th International Symposium on Microarchitecture, MICRO 1997 (1997)

A Practical Approach to DOACROSS Parallelization

Priya Unnikrishnan[1], Jun Shirako[2], Kit Barton[1],
Sanjay Chatterjee[2], Raul Silvera[1], and Vivek Sarkar[2]

[1] IBM Toronto Laboratory
{priyau,kbarton,rauls}@ca.ibm.com
[2] Department of Computer Science, Rice University
{js20,cs20,vs3}@rice.edu

Abstract. Loops with cross-iteration dependences (DOACROSS loops) often contain significant amounts of parallelism that can potentially be exploited on modern manycore processors. However, most production-strength compilers focus their automatic parallelization efforts on DOALL loops, and consider DOACROSS parallelism to be impractical due to the space inefficiencies and the synchronization overheads of past approaches. This paper presents a novel and *practical* approach to automatically parallelizing DOACROSS loops for execution on manycore-SMP systems. We introduce a compiler-and-runtime optimization called *dependence folding* that bounds the number of synchronization variables allocated per worker thread (processor core) to be at most the maximum depth of a loop nest being considered for automatic parallelization. Our approach has been implemented in a development version of the IBM XL Fortran V13.1 commercial parallelizing compiler and runtime system. For four benchmarks where automatic DOALL parallelization was largely ineffective (speedups of under 2×), our implementation delivered speedups of 6.5×, 9.0×, 17.3×, and 17.5× on a 32-core IBM Power7 SMP system, thereby showing that DOACROSS parallelization can be a valuable technique to complement DOALL parallelization.

1 Introduction

As hardware processors move from multicore to manycore designs, the challenge of enabling software to exploit parallelism is gaining a heightened urgency. While a number of programming models have been proposed for explicit parallelism, manual parallelization still requires a high degree of parallel programming expertise, and is often time-consuming and error-prone. It is widely believed that *automatic parallelization* can play an important role in improving the programmability of manycore-SMP systems [4] because it requires minimal or no effort by users. Furthermore, techniques for automatic parallelization can also be used in programming tools that assist in manual parallelization.

Most compilers focus on loops with no cross-iteration dependences in which all iterations can be executed completely in parallel with each other; such loops are referred to as DOALL loops. Loops with cross-iteration dependences are referred to as DOACROSS loops, and are usually serialized or, in some cases, transformed into skewed DOALL loops when practical to do so. However, Amdahl's Law dictates that it will be increasingly important to pay attention to the sequential fraction of the program, including

C. Kaklamanis et al. (Eds.): Euro-Par 2012, LNCS 7484, pp. 219–231, 2012.
© Springer-Verlag Berlin Heidelberg 2012

non-parallelized DOACROSS loops, as we move to manycore processors. Unfortunately, past approaches to DOACROSS parallelization are impractical for use in production-strength compilers due to unrealistic assumptions about *space* (*e.g.,* allocating one synchronization variable per dynamic iteration instance) or *granularity* (*e.g.,* performing synchronization operations even when their overhead exceeds the execution time of a loop iteration).

This paper presents a novel and *practical* approach to automatically parallelizing DOACROSS loops for execution on manycore-SMP systems. We introduce a compiler-and-runtime optimization called *dependence folding* that bounds the number of synchronization variables per worker thread (processor core) to be at most the maximum depth of a loop nest being considered for automatic parallelization. We also present an effective cost analysis to determine the profitability of DOACROSS parallelization, and practical techniques to increase the granularity of computation between successive synchronization operations. Our approach has been implemented in a development version of the IBM XL Fortran V13.1 commercial parallelizing compiler and runtime system. For four benchmarks where automatic DOALL parallelization was largely ineffective (speedups of under 2×), our implementation delivered speedups of 6.5× for LU, 9.0× for Poisson, 17.3× for SOR, and 17.5× for Jacobi on a 32-core IBM Power7 SMP system. Thus, DOACROSS parallelization can be a valuable technique to complement DOALL parallelism in cases where DOALL parallelization results in limited benefits [13].

The rest of the paper is organized as follows. Section 2 summarizes some of the previous work in this area. Section 3 describes our approach to DOACROSS parallelization, with details on dependence folding and runtime algorithms. Section 4 describes our methods for cost analysis and optimal grain size selection. Section 5 presents experimental results to evaluate the effectiveness of our approach. Section 6 contains our conclusions, along with suggestions for future work.

2   Previous Work

Early work on DOACROSS parallelization concentrated primarily on the synchronization mechanisms used. Cytron [2] showed how to determine a DOACROSS schedule to enforce a given set of dependences, based on the delays to be introduced for different iterations of the loop in processors that execute synchronously. Padua and Midkiff [7] focused on synchronization techniques to enforce loop carried dependences in singly-nested DOACROSS loops. They use one synchronization variable per data-dependence in the loop, and do not consider multi-dimensional loops. Wolfe [12] looked at four different synchronization mechanisms such as synchronizing at every data-dependence relationship in the loop, dividing the loop into segments of statements and pipelining the execution of the segments, inserting barriers at various points in the loop, using ordered critical sections etc. Again, only singly-nested loops were considered.

Su and Yew [10] proposed several interesting data synchronization schemes. The data-oriented scheme uses a dedicated synchronization variable for each datum involved in a dependence relationship in the loop, while the statement-oriented and process-oriented schemes have one per statement and iteration, respectively. They considered multi-dimensional loops with both a single level of parallelism and nested parallelism, but did not include any experimental results. Li [5] presents algorithms to generate synchronization code based directly on array subscripts and loop bounds using an array of event variables. This technique does not require constant data dependence distances and can target arbitrarily nested loops. Chen et al [3] proposed an algorithm for runtime parallelization of DOACROSS loops when data dependences cannot be determined at compile-time. Tang et al [8], presented synchronization schemes that can parallelize general nested loop structures with complicated cross-iteration data dependences.

Our experience in the area of automatic parallelization has led us to believe strongly that the choice of synchronization mechanism, its implementation and its tuning all have a major impact on the (im)practicality of a given approach to DOACROSS parallelization. All of the above papers [2,3,10,5,7,8,12] propose interesting techniques for synchronization, but lack quantitative measurements on the performance gains achieved, and the synchronization costs and memory requirements of the stated methods. Our synchronization mechanism uses a simple and intuitive "iteration vector" based scheme that can be easily applied to multi-dimensional loop nests. Our experimental results show that a single level of parallelism is sufficient in most cases to exploit the available resources.

There has also been some notable past work on optimizing synchronization operations. Krothapalli [11] targeted redundant synchronization elimination by removing redundant dependences in simple loops with constant dependences. Rajamony and Cox [9] used integer programming to determine the optimal solution to minimize the amount of synchronization in DOACROSS loops while retaining the parallelism that can be extracted from the loop. Chen [1] focused on increasing parallelism with statement reordering and reducing communication overhead by eliminating redundant synchronizations.

An optimal granularity of computation is required to offset the overhead of synchronization. Pan et al [14] used tiling to increase the parallelization granularity and propose a formulation for the optimal tile size. They conclude that static scheduling significantly outperforms dynamic self-scheduling by enhancing inter-tile locality. Lowenthal [6] presented a flexible runtime approach to determine the granularity for pipelined parallelization. Our work instead uses a cost-based combination of compile-time and runtime analyses to determine the granularity of work. Our results show that the accuracy of cost analysis can have a significant impact on parallel performance and scalability.

3   DOACROSS Parallelization Algorithm

Our approach to automatic DOACROSS parallelization is based on the assumption that there is one (logical) synchronization variable allocated per dynamic loop iteration. Thus, the sources and targets of inter-iteration synchronization operations can be denoted as *iteration vectors*. (Recall that iteration vector $\overrightarrow{I_v} = (I_1, I_2, I_3, ...I_n)$ represents a unique point in an n-dimensional iteration space.) The core idea is that a dependent iteration can examine the status of the synchronization variables of the iterations that it is waiting on to determine when it can start execution. Using the iteration vector as the synchronization variable interface has several advantages. It is very efficient to implement in terms of the memory required (as we will see below), simple to understand and implement, easily extensible to multidimensional loops, and does not constrain the inherent parallelism in the loop nest.

In our approach, synchronization is performed at the statement level of a given program representation. We assume that standard POST/WAIT operations can be performed on the iteration vector synchronization variables to enforce the data dependence relationships in the loop. A POST is inserted after the source statement of the dependence and the WAIT statement is inserted before the sink statement of the dependence:

1. $WAIT(w\vec{I_v})$: Causes execution to wait until the iteration specified by the iteration vector $w\vec{I_v}$ is completed. The iteration vector $w\vec{I_v}$ of WAIT is computed using the current iteration vector and the dependence distance vector $\vec{D} = (d_1, d_2, d_3, ...d_n)$ of the data dependence $w\vec{I_v} = (I_v - \vec{D}) = (I_1 - d_1, I_2 - d_2, ...I_n - d_n)$
2. $POST(p\vec{I_v})$: Indicates the completion of the iteration specified by the iteration vector $p\vec{I_v}$. The iteration vector $p\vec{I_v}$ of POST is the current iteration being executed. $p\vec{I_v} = \vec{I_v} = (I_1, I_2, ...I_n)$

3.1   Dependence Folding

With the aim of reducing synchronization overheads so as to make DOACROSS parallelization practical, our implementation folds all the loop-carried dependences in the loop into a single, conservative dependence. This leads to the insertion of at most one pair of synchronization primitives per iteration. In our experience with current hardware, the lower synchronization cost resulting from at most one synchronization per iteration far outweighs the potential loss in parallelism due to conservative approximation.

Definition 1. *A loop-carried data dependence is composed of the source statement, sink statement and the dependence distance* $\Delta = \{S_{src}\delta^* S_{sink}, \vec{D}\}$.

Consider a perfect loop nest L with n dimensions and m statements $\{S_1..S_m\}$ and k data dependences $\Delta^i = \{S_x\delta^* S_y, \vec{D^i}\}$, $i \in \{1..k\}$, $x \in \{1..m\}$ and $y \in \{1..m\}$. Each dependence vector, $\boldsymbol{D^i}$ has the form $\boldsymbol{D^i} = (d_1^i, d_2^i, ..., d_n^i)$. The single conservative dependence is computed by considering all the data dependences $\Delta^{1..k}$ in loop nest L. The source of the conservative dependence is computed as the *Lexically Latest Source* (LLS) statement across all the data dependences in the loop nest. In control flow terms, the LLS statement can be computed as follows. First compute the least common ancestor, LCA, of all source statements in the *postdominator tree* for the loop; then, find the closest ancestor of LCA in the postdominator tree that is unconditionally executed in the loop body. This statement is the LLS. Likewise, the sink of the conservative dependence is computed as the *Lexically Earliest Sink* (LES) statement across all the data dependences in the loop nest (by using the dominator tree instead of the postdominator tree). After the source and sink statements have been identified for the conservative dependence, the next step is to identify the conservative dependence distance vector $\vec{C}$. As our mechanism applies only to a single level of parallelism in the loop, it is possible to use a trivial formulation for the conservative dependence distance shown below. Assuming the outermost dimension is parallelized without loop chunking, the first dimension of $\vec{D^i}$ denotes the stride (i.e. dependence distance) along with the inter-thread loop dependence. Therefore, the first dimension of $\vec{C}$ should correspond to the maximum value of common strides in that dimension, which is the GCD value. The remaining dimensions can be conservatively computed by using $min\_vect(\vec{V_1}, \vec{V_2}, .., \vec{V_k})$, which determines the lexicographically smallest vector of $\vec{V_1}, \vec{V_2}, ..\vec{V_k}$.

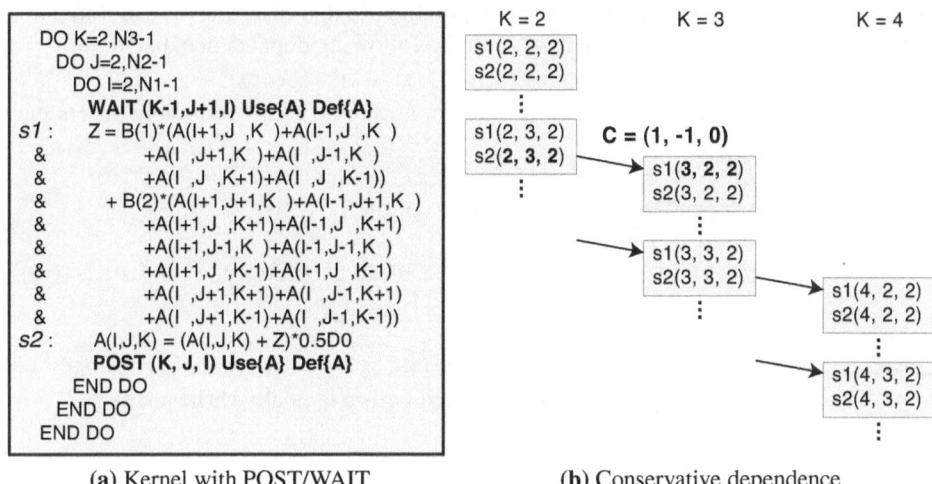

(a) Kernel with POST/WAIT   **(b)** Conservative dependence

Fig. 1. Pipelining POISSON

$$\vec{C} = \begin{pmatrix} C[1] \\ C[2..n] \end{pmatrix} = \begin{pmatrix} gcd(d_1^1, d_1^2, ...d_1^k) \\ min\_vect(D^1[2..n], ...D^k[2..n]) \end{pmatrix}$$

After insertion of POST/WAIT operations, the compiler will look for code motion opportunities to move the POST operations as early as possible in the loop, and the WAIT operations as late as possible in the loop. To ensure that such transformations do not violate any data dependences, the POST/WAIT operations are augmented with pseudo USE and DEF sets as follows:

1. Flow dependence (δ^f): A flow dependence is from a *def* of the variable to its *use*. The WAIT is inserted before the *use* and the POST is inserted after the *def*. In order to prevent the *use* of the dependence variable from moving up past the WAIT call, the variable is inserted in a pseudo DEF set for the WAIT call. Similarly, in order to prevent the *def* of the dependence variable from moving down below the POST call, the variable is inserted in a pseudo USE set for the POST call.
2. Anti dependence (δ^a): An anti dependence is from a *use* of the variable to its *def*. The WAIT is inserted before the *def* and the POST is inserted after the *use*. The dependence variable is marked as a *use* in the WAIT call to ensure that the WAIT is completed before the *def*, and as a *def* in the POST call to ensure that the POST is performed after the *use* of the variable.
3. Output dependence (δ^o): An output dependence is between two *def*'s of the same variable. The dependence variable is marked as *use* in both the WAIT and POST call to ensure that the WAIT is done before all the *defs* and POST is done after all the *defs* of that variable.

Figure 1a shows the POISSON computational kernel, which is a 3-dimensional ($400\times400\times400$) DOACROSS loop nest with the POST/WAIT synchronization primitives inserted after conservative dependence computation. In this case, there are multiple flow dependences from s2 to s1 for array A with the following dependence distances:

$$\overrightarrow{D^1} = (1,0,0), \overrightarrow{D^2} = (1,0,-1), \overrightarrow{D^3} = (1,0,1), \overrightarrow{D^4} = (1,-1,0), \overrightarrow{D^5} = (1,1,0)$$

and multiple anti dependences from s1 to s2 for array A with the following dependence distances:

$$\overrightarrow{D^6} = (1,0,0), \overrightarrow{D^7} = (1,0,1), \overrightarrow{D^8} = (1,0,-1), \overrightarrow{D^9} = (1,1,0), \overrightarrow{D^{10}} = (1,-1,0).$$

The conservative dependence are computed as follow:

$$\overrightarrow{C} = (gcd(1,1,1,1,1,1,1,1,1,1),$$
$$\qquad min\_vect((0,0),(0,-1),(0,1),(-1,0),(1,0),(0,0),(0,1),(0,-1),(1,0),(-1,0)))$$
$$= (1,-1,0)$$

Based on the conservative dependence, POST(K,J,I) is inserted after lexically last source s2 and WAIT(K-1,J+1,I) is inserted before lexically earliest sink s1.

3.2 Runtime Implementation

The compiler outlines and parameterizes the DOACROSS loops after the POST/WAIT synchronization calls to the runtime are inserted. The parallel runtime system is responsible for initializing data structures and scheduling the DOACROSS loops. The current implementation employs a static cyclic scheduling policy with a chunk size of one, where iterations are assigned to processors in a round-robin fashion. The static cyclic policy inherently brings good load balance and data locality in addition to low overhead due to static iteration mapping.

Runtime Data Structure: Let m denote the number of threads and n denote the dimension of the DOACROSS loop nest. The runtime allocates a 2-dimensional array, $sync\_vec[1 : m][1 : n]$, which is a set of $m \times n$ synchronization variables to manage the POST/WAIT synchronization on the DOACROSS loop. Given a thread with id = $thrd\_id$, the 1-dimensional sub-array $sync\_vec[thrd\_id][1 : n]$ represents the last iteration instance whose completion is ensured by the POST operation. Note that the iteration space is normalized and it is guaranteed that an iteration instance $p\overrightarrow{I_v}$ passed to a POST operation monotonically increases for each thread. A WAIT operation is blocked until when $sync\_vec[thrd\_id]$ is lexicographically larger or equal to the iteration instance $w\overrightarrow{I_v}$ passed to the WAIT.

POST/WAIT Algorithm: Algorithms 1 and 2 show the $POST(p\overrightarrow{I_v})$ and $WAIT(w\overrightarrow{I_v})$ algorithms, respectively. The boundary check for the iteration instance $p\overrightarrow{I_v}/w\overrightarrow{I_v}$ is performed at the beginning of the POST/WAIT algorithm. Note that all valid elements of $p\overrightarrow{I_v}/w\overrightarrow{I_v}$ are non-negative because of the loop normalization. The POST algorithm assigns $p\overrightarrow{I_v}$ to $sync\_vec[thrd\_id]$ for the current thread. In the implementation, this assignment is done in reverse order, i.e., starting with the innermost dimension and going outer along with appropriate memory barriers. This ensures that the intermediate state of $sync\_vec[thrd\_id]$ is always smaller than $p\overrightarrow{I_v}$. The WAIT algorithm computes the target (source) thread based on the first dimension of $w\overrightarrow{I_v}$, and waits until $sync\_vec[thrd\_id]$

Algorithm 1. POST algorithm

Input : The iteration vector of the current iteration $pI_v = (I_1, I_2, ..., I_n)$, $n =$ dimension
of loop, $m =$ number of threads

begin
    // Check for boundary conditions
    // The loops are all lower bound and bump normalized
    if $within\_boundary(pI_v[1...n])$ **then**
        $thrd\_id = mythread()$

        // Update the synchronization variable of the current thread
        $sync\_vec[thrd\_id][1..n] = pI_v[1..n]$

end

Algorithm 2. WAIT algorithm

Input : The iteration vector of the dependence source
      $wI_v = (I_1 - d_1, I_2 - d_2, ..., I_n - d_n)$, $n =$ dimension of loop, $m =$ number of
      threads, $\overrightarrow{D} = (d_1, d_2..d_n)$ is the dependence distance

begin
    if $within\_boundary(wI_v[1...n])$ **then**
        // Determine the thread executing the source iteration specified by wI_v
        // Schedule is static with chunksize=1.
        $thrd\_id = wI_v[1]\%m$

        // Block until $sync\_vec[thrd\_id]$ is lexicographically larger or equal to $wI_v[1..n]$
        while $vector\_compare(sync\_vec[thrd\_id][1..n], wI_v[1..n]) < 0$ **do**
            wait

end

---

becomes lexicographically larger or equal to $w\overrightarrow{I_v}$. This vector comparison is done starting with the outermost dimension and going inner. The order of updating and reading of the synchronization vector by the POST and WAIT calls respectively ensures that a WAIT operation will never be unblocked prematurely due to an illegal intermediate state of $sync\_vec[thrd\_id]$. The WAIT operation is relatively cheap because it only performs a read of the synchronization variable of another thread. The POST operation is very expensive because it performs a write of the synchronization variable. Because the synchronization variable in our method is an iteration vector, the number of writes is equal to the number of dimensions of the doacross loop.

4   Profitability Analysis and Grain Size Selection

Profitability and cost analysis play a major role in automatic parallelization. Excessive synchronization or insufficient granularity of computations for parallelism can result in significant performance degradation. These considerations have been studied in past work on parallelization of DOALL loops [13], and need to be extended for parallelization of DOACROSS loops. In this section, we introduce a profitability analysis to determine when it is worthwhile to parallelize a DOACROSS loop.

First, we perform a special-case check for a one-dimensional loop nest. If the POST/WAIT calls encompass the entire loop body and the conservative dependence distance equals 1, then DOACROSS parallelization cannot be profitable since the POST/WAIT calls effectively serialize the entire loop. This check does not apply if the loop nest contains $n > 1$ loops, since there may still be useful parallelism with a conservative dependence distance of 1 at the outermost level (enabled by fine-grained synchronization calls in the inner loops).

Second, we perform loop unrolling to reduce the amortized overhead of synchronization operations by increasing the granularity of computation between POST and WAIT operations. After unrolling, the lexically last POST and earliest WAIT operations are retained, and all the intervening calls to POST/WAIT are removed so as to reduce the overall synchronization overhead. Also, the iteration instance $w\vec{I_v}$ of the lexically earliest WAIT is adjusted to match the lexically last POST according to the unrolling factor.

We assume the availability of two parameters, *MinGrainSize* and *MaxLoopBodySize*, to guide our transformations for grain size selection. *MinGrainSize* imposes a lower bound on the granularity of computation to be performed between POST and WAIT operations. To compute the heuristics for the grain-size *MinGrainSize* for DOACROSS parallelization, we start by looking at the heuristics for DOALL parallelization [13]. These values were then adjusted to take into account the communication overhead. Subsequently, experimental runs were performed to further fine tune the heuristics. Similarly to determine the code-size *MaxLoopBodySize* for DOACROSS parallelization, we rely on previously calculated heuristics for the unrolling transformation. These heuristics are adjusted to prevent excessive code growth during unrolling. For the platform studied in this paper (Power7 with an XLC runtime system), it was determined that 20,000 cycles and 320 cycles are reasonable value for *MinGrainSize* and *MaxLoopBodySize* respectively. However, our approach is applicable to any other values that may be specified for these parameters.

To select the unroll factor, *UF*, for the innermost loop, we first estimate the cost of a single iteration of the loop, *LoopBodyCost*. Then, the unroll factor selected by our approach can be specified as $UF = \min(32, \lceil MaxLoopBodySize/LoopBodyCost \rceil)$, where 32 is an upper bound that is imposed on *UF* for practical reasons. If $n = 1$, an extra constraint is imposed to ensure that *UF* is less than the conservative dependence distance for the DOACROSS loop.

Finally, we perform a special form of *chunking* of the inner loops in a DOACROSS loop nest, by estimating a chunk size that we refer to as a *Runtime Granularity Factor*, *RGF*. *RGF* specifies the number of iterations of the inner loops that should be executed sequentially. This is achieved by skipping $RGF-1$ POST operations in the inner loops, so that one POST operation is performed for every *RGF* POSTs. As described in Section 3.2, the POST($p\vec{I_v}$) operation ensures that all WAIT operations whose iteration instance $w\vec{I_v}$ is lexicographically smaller or equal to $p\vec{I_v}$ can be unblocked. Therefore, it is safe to perform only the last POST operation after $RGF-1$ POSTs. To avoid a potential deadlock when the number of iterations is not an exact multiple of *RGF*, an additional POST operation is inserted at the end of each iteration of the outermost

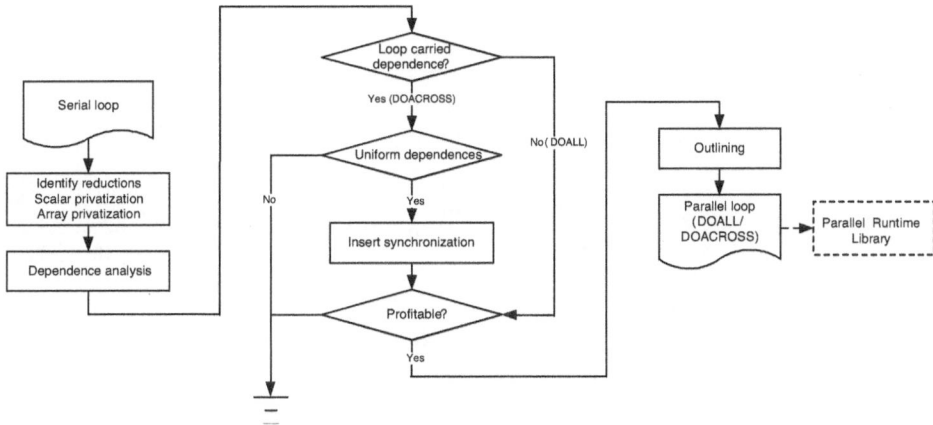

Fig. 2. Context for automatic DOACROSS parallelization in the XL Fortran and C/C++ compilers

DOACROSS loop to signal that all iteration instances included in that iteration have been completed. Note that WAIT operations, which have much smaller synchronization cost than POST operations, are always performed.

The initial value of RGF is selected at compile-time by using the formula, $RGF = MinGrainSize/(UF * LoopBodyCost)$, This value is further adjusted at runtime based on the number of threads executing the DOACROSS loop. If there are more threads, a larger value of RGF may reduce the amount of parallelism that can be exploited. Thus we adjust the RGF value selected at compile-time to a runtime value, RGF', as follows: $RGF' = 2 \times RGF/NumThreads$. Note that $RGF' = RGF$ when $NumThreads = 2$, and becomes proportionately smaller as $NumThreads$ increases, thereby balancing the trade-off between overhead and parallelism.

5   Experimental Results

This section presents results from the experiments conducted to evaluate our implementation. The experiments were performed on a Power7 system with 32-core 3.55GHz processors running Red Hat Enterprise Linux release 5.4. The measurements were done using a development version of the XL Fortran 13.1 for Linux (see Figure 2). We used 4 benchmark programs for our evaluation: Poisson, 2-dimensional LU from the NAS Parallel Benchmarks Suite (Version 3.2), SOR algorithm and 2-dimensional Jacobi computation. We manually applied array privatization for some loops in blts and buts, for which the compiler failed to automatically privatize the arrays. All these benchmarks are excellent candidates for DOACROSS parallelization. All benchmarks were compiled with option "-O5" for the sequential baseline, and "-O5 -qsmp" for the automatic parallelization enabling DOACROSS parallelization. We evaluated four experimental variants: a) **only doall** represents the speedup where the automatic DOACROSS parallelization is turned off and uses only DOALL parallelism (far left), b) **doall w/ manual skew** represents the speedup with DOALL loops including DOACROSS loops which were converted

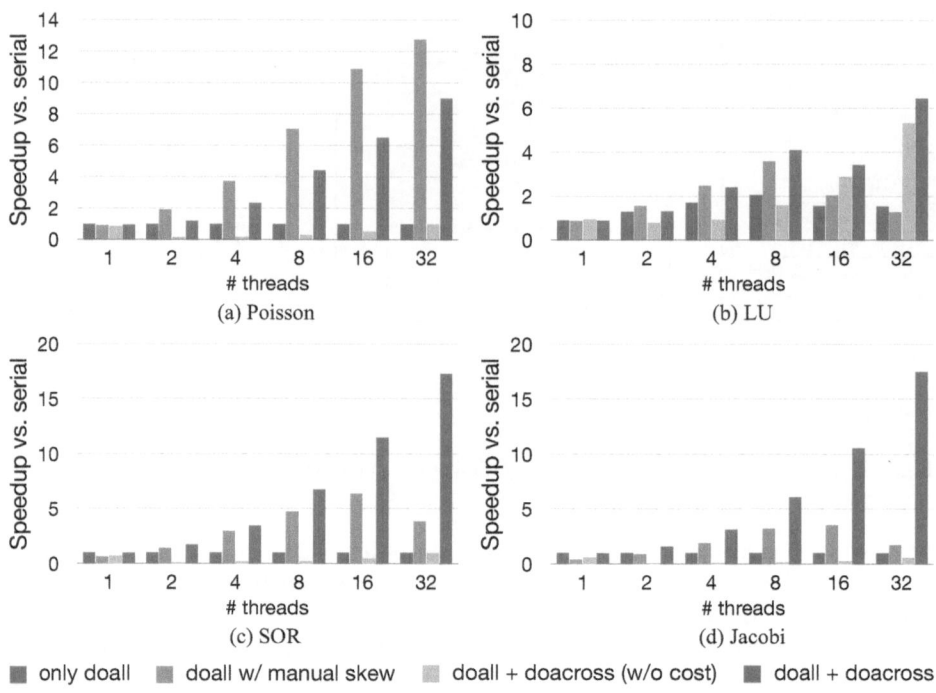

Fig. 3. Speedup related to sequential run on Power7

to DOALL loops after manual loop skewing (second left), c) **doall + doacross (w/o cost-analysis)** is the speedup where both DOALL and DOACROSS parallelization are enabled, but with the cost analysis and granularity control turned off (second right), and d) **doall + doacross** is the speedup where both DOALL and DOACROSS parallelization with cost analysis and granularity control are enabled (far right).

5.1   POISSON

We use the POISSON kernel discussed in Section 3.1. The DOACROSS loop is invoked 20 times in this experiment. Figure 3a shows the speedup of the 4 variants listed above when compared to the sequential execution.

The **only doall** case results in no speedup; **doall + doacross (w/o cost-analysis)** can results in worse performance than sequential execution because of the large synchronization overhead. **doall + doacross** delivers a speedup of up to 9.0×. Note that **doall w/ manual skew** shows better performance than **doall + doacross**. This is because the POISSON kernel is a triply nested DOACROSS loop. We manually selected the outermost and middle-nested loops for the target of loop skewing and this choice turns out to have a better granularity. On the other hand, the auto DOACROSS version inserted POST/WAIT in the innermost loop body and hence the total synchronization overhead became larger despite the granularity control. (In the future, optimizing compilers could reduce this gap by further adjusting the granularity for DOACROSS parallelization.) However, the automatic **doall + doacross** version outperforms the **doall w/ manual skew** version in the other three benchmarks whose kernel loops are doubly nested and both the manual and automatic versions use the same granularity.

5.2 LU

LU has 2 DOACROSS loops in subroutines *blts* and *buts*. Together they account for about 40% of the sequential execution time of LU. They are 2-dimensional (160×160) DOACROSS loops that are invoked 40160 times. The conservative dependence vector is $(1,0)$ for both *blts* and *buts*, and the corresponding WAIT/POST synchronizations are inserted. Although LU contains many DOALL loops, the best speedup with DOACROSS parallelism disabled is $2.1 \times$ using 8 cores. On the other hand, our DOACROSS parallelization brings more scalable performance, up to $6.5 \times$ speedup with 32 cores as shown in Figure 3b. The figure also shows the significant impact of the cost analysis and granularity control on performance. Furthermore, the DOACROSS version show even better scalability than the manual DOALL version that converted the DOACROSS loops in *blts* and *buts* into DOALL loops. It is well known that DOACROSS parallelization has better data locality and lower synchronization overhead than DOALL with loop skewing, even when they use same granularity for wavefront parallelism.

5.3 SOR and Jacobi

The kernel loop nest of SOR is a 2-dimensional 20000×10000 loop which is invoked 50 times. Note that the kernel loop of Jacobi has very similar structure as SOR, and the conservative dependence vector for both SOR and Jacobi kernels is $(1,0)$. Our framework extracted DOACROSS parallelism for both cases, and achieves up to $17.3 \times$ and $17.5 \times$ speedup for SOR and Jacobi, respectively as shown in Figures 3c and 3d. On the other hand, the best speedups when manually converting DOACROSS into DOALL by loop skewing are $6.4 \times$ for SOR and $3.5 \times$ for Jacobi. The figures also show that the granularity control is essential to obtain scalable speedup using DOACROSS parallelism.

6   Conclusions and Future Work

We presented a novel and practical approach to automatically parallelizing DOACROSS loops for execution on manycore-SMP systems, based on a compiler-and-runtime optimization called *dependence folding*. The proposed framework uses a conservative dependence vector analysis to identify suitable program points where POST/WAIT synchronization operations can be inserted. A profitability analysis is used to guide unrolling and chunking transformations to select an optimized granularity of computation for DOACROSS parallelization. Further, our runtime framework includes a lightweight and space-efficient implementation of point-to-point synchronization for DOACROSS loops.

The proposed framework has been implemented in a development version of the IBM XL Fortran V13.1 commercial parallelizing compiler and runtime system. For four benchmarks where automatic DOALL parallelization was largely ineffective (speedups of under $2\times$), our implementation delivered speedups of $6.5\times$, $9.0\times$, $17.3\times$, and $17.5\times$ on a 32-core IBM Power7 SMP system, thereby showing that DOACROSS parallelization can be a valuable technique to complement DOALL parallelization.

During the course of our work in enabling DOACROSS parallelization in the XL compilers, we encountered multiple opportunities for future work related to interactions between DOACROSS parallelization and lower-level compiler optimizations. We found cases where the DOACROSS transformation inhibited *software pipelining* (a technique for scheduling instructions to exploit instruction level parallelism in inner loops by overlapping loop iterations). In such cases, it would be desirable to extend the profitability analysis to take the impact on software pipelining into account. As another example, *predictive commoning* (an optimization to reuse computations across loop iterations by detecting indexing sequences and unrolling to avoid register copies), if performed earlier, can inhibit the detection of DOACROSS loops. A detailed study of these interactions is part of our planned future work. Other opportunities for future work include deeper analyses for synchronization overhead and parallel efficiency so as to improve accuracy of profitability analysis, and performance comparison against other existing work. As shown in the paper, POST/WAIT operations are well-suited for user annotations and the technique of dependence folding can also be adapted for the explicit parallelization using such annotations. Extensions of the proposed framework to explicit parallelization is another important direction of future work.

Acknowledgements. This work was supported in part by an IBM CAS Fellowship in 2011 and 2012.

References

1. Chen, D.K.: Compiler optimizations for parallel loops with fine-grained synchronization. PhD Thesis (1994)
2. Cytron, R.: Doacross: Beyond vectorization for multiprocessors. In: Proceedings of the 1986 International Conference for Parallel Processing, pp. 836–844 (August 1986)
3. Chen, D.-K., Torrellas, J., Yew, P.C.: An efficient algorithm for the run-time parallelization of doacross loops. In: Proc. Supercomputing 1994, pp. 518–527 (November 1994)
4. Gupta, R., Pande, S., Psarris, K., Sarkar, V.: Compilation techniques for parallel systems. Parallel Computing 25(13-14), 1741–1783 (1999)
5. Li, Z.: Compiler algorithms for event variable synchronization. In: Proceedings of the 5th International Conference on Supercomputing, Cologne, West Germany, pp. 85–95 (June 1991)
6. Lowenthal, D.K.: Accurately selecting block size at run time in pipelined parallel programs. International Journal of Parallel Programming 28(3), 245–274 (2000)
7. Midkiff, S.P., Padua, D.A.: Compiler algorithms for synchronization. IEEE Transactions on computers C 36, 1485–1495 (1987)
8. Tang, P., Yew, P., Zhu, C.: Compiler techniques for data synchronization in nested parallel loop. In: Proc. of 1990 ACM Intl. Conf. on Supercomputing, Amsterdam, pp. 177–186 (June 1990)
9. Rajamony, R., Cox, A.L.: Optimally synchronizing doacross loops on shared memory multiprocessors. In: Proc. of Intl. Conf. on Parallel Architectures and Compilation Techniques (November 1997)

10. Su, H.M., Yew, P.C.: On data synchronization for multiprocessors. In: Proc. of the 16th Annual International Symposium on Computer Architecture, Jerusalem, Israel, pp. 416–423 (April 1989)
11. Krothapalli, V.P., Sadayappan, P.: Removal of redundant dependences in doacross loops with constant dependences. IEEE Transactions on Parallel and Distributed Systems, 281–289 (July 1991)
12. Wolfe, M.: Multiprocessor synchronization for concurrent loops. IEEE Software 5(1), 34–42 (1988)
13. Zhang, G., Unnikrishnan, P., Ren, J.: Experiments with auto-parallelizing SPEC2000FP benchmarks. In: 17th Intl Workshop on Languages and Compilers for Parallel Computing (2004)
14. Pan, Z., Armstrong, B., Bae, H., Eigenmann, R.: On the interaction of tiling and automatic parallelization. In: First International Workshop on OpenMP (Wompat) (June 2005)

Exploiting Semantics of Virtual Memory to Improve the Efficiency of the On-Chip Memory System

Bin Li[1], Zhen Fang[2], Li Zhao[1], Xiaowei Jiang[1], Lin Li[1], Andrew Herdrich[1],
Ravishankar Iyer[1], and Srihari Makineni[1]

[1] Intel Corporation, Hillsboro, OR, 97124, USA
{bin.li,li.zhao,xiaowei.jiang,lin.e.li,
andrew.j.herdrich,ravishankar.iyer,srihari.makineni}@intel.com
[2] Nvidia, Austin, TX, 78717, USA
zfang@nvidia.com

Abstract. Different virtual memory regions (e.g., stack and heap) have different properties and characteristics. For example, stack data are thread-private by definition while heap data can be shared between threads. Compared with heap memory, stack memory tends to take a large number of accesses to a rather small number of pages. These facts have been largely ignored by designers. In this paper, we propose two novel designs that exploit stack memory's unique characteristics to optimize the on-chip memory system.

The first design is *Anticipatory Superpaging* - automatically create superpages for stack memory at the first page fault in a potential superpage, increasing TLB reach and reducing TLB misses. It is transparent to applications and does not require kernel to employ online analysis algorithms and page copying. The second design is *Stack-Aware Cache Placement* - stack accesses are routed to their local slices in a distributed shared cache, while non-stack accesses are still routed using cacheline interleaving. The primary benefit of this mechanism is reduced power consumption of the on-chip interconnect. Our simulation shows that the first innovation reduces TLB misses by 10% - 20%, and the second one reduces interconnect power consumption by over 14%.

1 Introduction

The concept of stack memory is universally supported in today's computer systems. By definition, it is used for local variables inside subroutines, and for private data visible only to the thread that they are attached to. The operating system (OS) supports usage of stack and heap by defining them in the virtual memory address space.

Figure 1 shows the virtual address map of a 32-bit Linux. 64-bit systems are similar. Different segments have clearly defined software semantics. For example, if the address from a user application is greater than 0x80000000, we know except for pathological cases it is a location in the stack. Virtual-to-physical address translation is performed in the memory management unit(MMU). Virtual memory semantics of memory references are not preserved after MMU, and are not exploited in the design of the memory system (e.g., cache, interconnect). For example, caches are searched using physical address.

C. Kaklamanis et al. (Eds.): Euro-Par 2012, LNCS 7484, pp. 232–245, 2012.
© Springer-Verlag Berlin Heidelberg 2012

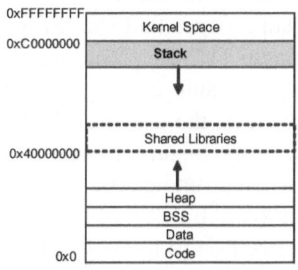

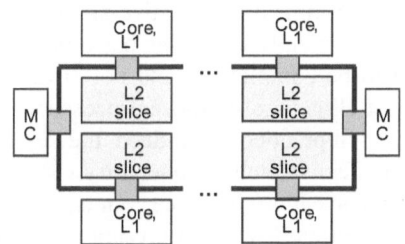

Fig. 1. A typical 32-bit virtual memory address map

Fig. 2. A CMP with L2 cache slices connected by a ring

1.1   Background: Physical Page Allocation, TLB and Superpaging

Memory for the vast majority of applications' data set is acquired through one of two dynamic allocation methods: *malloc* library call or stack growth. *malloc* triggers the *mmap* service in the OS to allocate heap memory in virtual address space. Actual physical page allocation do not happen until the first access to the memory occurs. Accessing an unmapped virtual page causes a major page fault. A physical page is then allocated by the kernel to the faulting virtual page. Compared with the explicit *malloc* method, stack growth is implicit and does not have the "registration" step. A stack pointer register simply points to the farthest location (i.e., lowest address) in a function, and page faults occur when the local variables are first referenced using relative addresses based on the value in the stack pointer. The memory allocator in the OS does not differentiate among different virtual memory regions. Whether a page fault is triggered by stack growth or heap expansion, a same page allocation algorithm is executed by the kernel.

To expedite virtual-to-physical translation, the TLB caches recent translations. Due to the performance criticality of TLB hits, its size has been limited. As a result, coverage of TLB is rather small compared with the large cache capacity in today's CMPs. TLB misses are costly events because a pagetable walk could take hundreds of clock cycles in a multi-Ghz processor. The penalty is even higher in the case of software TLB miss handling due to interferences to the instruction pipeline and instruction cache. The solution for increasing TLB coverage without slowing down TLB hits is the superpage. A superpage maps a region of multiple virtual pages to a region of physical memory. The physical pages that back a superpage must be contiguous and properly aligned. A superpage mapping occupies a single TLB entry, thus increases the reach of the TLB.

Currently, there are two ways for user applications to utilize superpages. The first is for programmers to explicitly request the OS kernel to create a superpage mapping for a virtual memory region. Imposing such an intrusive requirement upon application developers limits its popularity. An alternative approach that is transparent to applications is to leave the burden of identifying superpage candidates to the kernel. In this approach, the kernel runs a competitive algorithm, identifies superpages that should be created, and dynamically coalesces basic pages into superpages [1]. These online algorithms tend to be complicated and not always effective. As a result, although the MMU of almost all general-purpose processors support superpages, the usage of them is rather limited in practice [2]. Indeed, the key to successful usage of superpaging is two-fold: 1) an effective method to identify virtual pages that when promoted into one superpage, will receive high reference counts, and 2) to lower or avoid the cost of page copying.

In Section 2, we will present a stack-aware optimization to the page allocator that builds superpages with high efficacy and low cost. It does not entail any application programmer involvement, and does not require page copying.

1.2   Background: The Last-Level Cache and the Interconnect

Today's CMPs have increasingly large last-level caches (LLCs)[1]. Banks of the LLC have to be physically distributed on a chip due to their large geometric sizes. Figure 2 is a representative block diagram of a CMP with a number of LLC slices (banks) and two memory controllers, interconnected by a ring. There are primarily two types of logical organizations for the LLC: private caches with coherency maintained between them, and a shared cache using address-based interleaving.

In private cache architecture, the working set of each thread is attracted to its local L2 cache. Hardware-supported cache coherency ensures if a block resides in a remote L2 cache, it will be visible to the requesting core. Disadvantages of private caches include design complexity for implementing cache coherency and lower utilization of the L2 cache. Because of these disadvantages, in commercial CMPs, the predominant organization is distributed shared cache [3]. On a shared cache, memory locations are statically mapped to cache slice using physical address. With no data duplications within the LLC, it eliminates the need for maintaining cache coherency between the slices. Sharing the cache capacity among all cores also improves cache space utilization.

The address-to-cache slice mapping on a shared cache can be page interleaved or cacheline interleaved. Cacheline interleaving improves interconnect utilization efficiency, especially for data structures that are heavily shared by threads running on different cores. In this study, we assume cacheline interleaving, similar to some recent high-performance processors [3]. Requests to contiguous memory blocks are sent to different slices. Each cache slice is attached to a different access point on interconnect.

A distributed shared cache's performance relies critically on interconnect bandwidth. However, in contrast with ever-increasing core counts and cache sizes, the interconnect does not scale well. Researchers projected interconnect latencies as high as hundred cycles [4], and interconnect power up to 36% of chip power [5]. To mitigate the high interconnect traffic dictated by the distributed shared LLC, we attempt to optimize for stack memory accesses. We cache stack data in each thread's local LLC since they are private data by definition, and thus should better not be cached in remote cache.

---

[1] In this study, we model a two-level cache hierarchy. Therefore L2 cache is our last-level cache. The two terms are used interchangeably in this paper.

1.3   Overview of Innovations

We propose to exploit some unique characteristics and properties of stack memory to improve the efficiency of TLB and interconnect/LLC. Specifically, we present the following two mechanisms.

Anticipatory Superpaging: Based on an observation that stack memory tends to occupy a small set of pages with high reference counts(data to be presented in Section 2.1), we automatically build superpages in page fault handling triggered by stack references. Superpage creation is performed in anticipation of two sets of events: 1) nearby virtual pages will soon be referenced and would otherwise cause more page faults; 2) the superpage will receive high access count. We identify superpage candidates at low cost and also eliminate physical page coalescing. The primary benefit is increased TLB reach and consequently decreased TLB misses.

Stack-Aware Cache Placement: Based on stack memory's thread-private nature, we route stack accesses to the local cache slice, while non-stack accesses are still routed using the baseline block-interleaved mapping. By creating core-data affinity, we harvest the advantages of both private and shared LLCs for stack data. One thing worth noting is that depending on the system size and workload, the proposed mechanism may or may not lead to noticeable performance improvement. Rather, the major benefit is reduced traffic on the interconnect and consequently lower power consumption.

The two mechanisms are orthogonal to each other. In the next two sections, we describe both mechanisms in details. Section 2 elaborates on Anticipatory Superpaging and Section 3 discusses Stack-Aware Cache Placement.

2   Anticipatory Superpaging

2.1   Motivation

We profiled SPEC 2000 benchmarks to measure the percentages caused by stack memory in three metrics: memory references, unique pages and data TLB misses. The data are shown in Figure 3. Accesses to the stack account for about 40% of all memory references. For a few applications, e.g., *apsi, gap, mesa* and *swim*, there are actually more references to the stack than the combined sum of all other virtual memory regions. However, the total number of unique pages used by each application's stack memory is on average only 21% of the total data memory footprint, as shown in Figure 3.

An interesting observation is that although the memory footprint of the stack is small, stack accesses that miss the TLB accounts for 38% of all TLB misses. Further analysis revealed that stack pages are often evicted out of the TLB by heap page mappings. The experiment data indicate that stack memory pages have very high access density - reference count per page, as a result of high reference count and small number of unique pages. This suggests that they are good candidates for superpages.

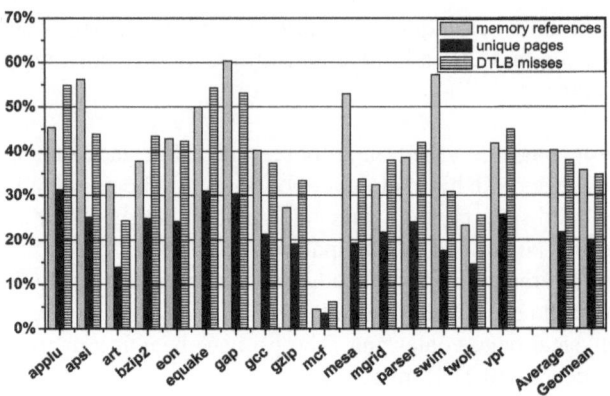

Fig. 3. Percentages of certain statistics caused by stack accesses

2.2   The Proposed Mechanism

We propose Anticipatory Superpaging to automatically create a superpage if a faulting address is believed to be a stack location. To the best of our knowledge, this is the first superpaging algorithm that possesses both of the following two features:

– Transparent to applications
– No page copying or bookkeeping overhead

The proposed flow for memory allocation is as follows. For each reference, we check whether it is a stack reference or not: (1) If the reference is non-stack reference, we perform normal memory allocation routine, that is, allocate one page. (2) If the reference is a stack reference, we create a superpage that consists of M basic pages (M is system dependent). When a stack access causes a major (i.e., "page not present") page fault, the kernel automatically allocates multiple page frames so that a superpage can be created for the faulting address, instead of just allocating one basic page.

We use 32-bit Linux to show how to quickly perform the check function described above. In most systems, shared libraries are loaded to address 0x40000000 and above by default. We can set an empirical value *TOP_OF_LIB* to around 0x50000000. As long as a faulting address belongs to range [*TOP_OF_LIB*, 0xC0000000], we build a super-page for it (0xC0000000 is the boundary between user stack and kernel space). Thus the cost for identification of stack memory address is negligible. Building a superpage that consists of M basic pages requires availability of M pages that are contiguous and aligned. With advanced page frame management such as the Buddy algorithm, these free pages can usually be found at very low cost. Since we allocate the pages upon the initial access to the to-be-created superpage, we avoid the high cost of page copying in conventional approaches.

Aggressively creating superpages for stack memory benefits applications' performance in two ways. The first is increased TLB coverage, since fewer TLB entries will be taken by stack memory. The other, less obvious, benefit is related to how TLB entries are replaced. TLB's on today's processors are usually set-associative with least-recently used (LRU), or some variant of it. A superpage competes with the basic pages "unfairly"; an access to any of the component pages of a superpage will help to push the superpage entry toward the bottom in the LRU stack. Organized as superpages, translations for stack memory have less chance of getting evicted out of the TLB. One concern maybe that useful mappings for heap data in the same set may get penalized. Our experiment results (see Section 4) reveal that this is rare with the benchmarks that we tested.

3   Stack-Aware Cache Placement

Stack-Aware Cache Placement is orthogonal to afore- described superpage/TLB work. In this section, we start with stack memory references' behavior in the cache hierarchy to motivate our cache/interconnect innovation. We then discuss the challenges in implementing the proposed idea and our design decisions.

3.1   Motivation

For stack-oriented LLC/interconnect optimizations to have meaningful benefits, a good percentage of L1 misses need to be stack references. Earlier findings by researchers suggest that *if dedicated for stack variables*, a L1 data cache would only need to be about 8KB to hold the hottest working sets [6], much smaller than a dedicated heap cache would. One's intuition might be that with a conventional L1 cache, L1 misses would have a similar mixture - dominated by heap accesses. Our experiment revealed that this intuition is not correct.

Data in Figure 4 indicate that of all cache misses from a 32KB L1D, averagely 35% are stack memory references. Even with a large, 64KB L1 data cache, stack accesses still account for 30% of all misses, not too much lower than their percentage in original programs' loads and stores (40%). Although the sequence of stack accesses in the instruction stream have strong locality, it does not translate to commensurate L1 cache hit rates. The reason is interference from non-stack references. References to the heap memory constantly cause stack data to be evicted out of the L1 cache. Most of these thread-private references then get cached in, and fetched from, remote LLC slices.

We seek to optimize the LLC organization by mapping all stack accesses to each thread's local LLC slice. The mechanism is called Stack-Aware Cache Placement. When the otherwise remote LLC hits are converted to local LLC hits, LLC latencies decrease and total ring hops drop. However, a potential downside is that a stack-intensive application could evict useful heap data out of its local slice, causing an increase in total LLC miss rates. Therefore, we need to verify that if such a scenario exists, the performance impact of increased LLC misses must not cancel off the benefits in reduced hit latencies and interconnect traffic(see Section 4.3). In general, the larger the interconnect, the more likely that create core-stack data affinity is beneficial for overall performance.

There are a few challenges in implementing this mechanism. First, it requires we be able to recognize virtual memory regions outside the processor core. Second, due to different routing rules for stack and non-stack memory, we need to make sure false

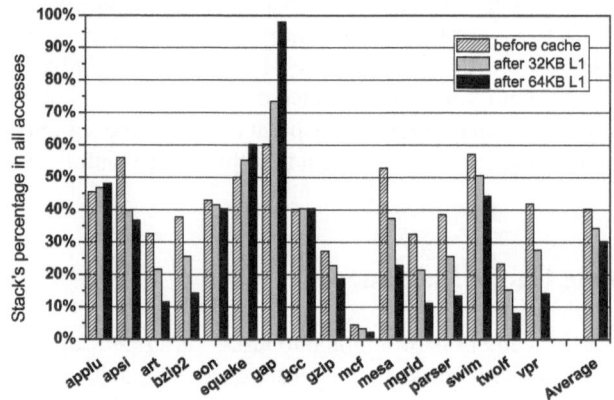

Fig. 4. Stack's percentages in the cache hierarchy

hits do not occur between memory regions. Third, we should avoid fine-grained cache coherence. In the next sections, we explain how each of these requirements is fulfilled.

3.2 Stack Reference Identification at the LLC

When a memory access misses the L1 cache, its physical address is sent out to the system. Software semantic information is lost in the virtual-to-physical translation process. We augment each pagetable and TLB entry with an extra bit for stack hint. In the page fault handler, the kernel checks the faulting virtual address. If it is a stack access, the stack hint bit is set to '1' in the pagetable entry. Otherwise set the bit to '0'. Every load/store operation acquires the stack hint bit from the TLB and carries it to the LLC if it misses the L1 cache.

Using the stack reference identification method in Section 2.2, under pathological cases we could misclassify memory references. We will explain in Section 3.5 that we do not maintain cache coherency between slices on a block-by-block basis. Therefore there is a concern on program correctness if a page misclassification does happen. Classifying a stack page as a non-stack one is harmless since that would be equivalent to degrading into the baseline interleaved mapping. Classifying a heap page as a stack page could cause program errors for multithreaded applications, because each of the threads that share the page may keep a copy in its local cache slice. To fix this issue, the virtual memory manager in the kernel only needs to increase the value of *TOP_OF_LIB* if non-stack memory (heap, shared libraries) expands above the previously set *TOP_OF_LIB*.

3.3 LLC Slice Selection

With the stack hint available for each L1 cache miss, the ring stop can select between the baseline hashing function and the local slice number as the destination LLC slice. Stack accesses are routed to their local slice and non-stack accesses are routed using the baseline hashing function.

Fig. 5. L2 cache slice selection logic

Figure 5 contains conceptual flows of the process for the baseline (Figure 5(a)) and for the proposed mechanism (Figure 5(b)). LLC access latency will not be affected since we are only adding a mux. In commercial processors, the LLC slice selection is typically performed earlier in the pipeline, e.g., as soon as a miss to the private cache is detected. In those designs the LLC slice selection logic is completely off the critical path. Based on this LLC slice selection logic, blocks that miss the L1 cache are directed to the local slice if they are stack accesses regardless of their memory block number.

3.4 Design of the Cache Tags

For a CPU with N cores and N cache slices, the slice selection logic uses $logN$ bits in the memory block address to route a request. None of the $logN$ bits is used in set indexing or tag. This is shown in the top row in Figure 6.

		$logN$ bits		
Tag	Index	Slice selection	Blk offset	**Baseline**

| 0...0 | Tag | Index | Tag | Blk offset | Stack |
| | Tag | Index | Slice selection | Blk offset | Non-stack |

False hits could occur if $logN > log(PageSize) - log(BlockSize)$

| 1 | Tag (*logN* bits longer than baseline) | Index | Blk offset | Stack |
| 0...0 | 0 | Tag | Index | Slice selection | Blk offset | Non-stack |

False hits will not occur even for large scale CMPs

Fig. 6. Avoiding false hits between stack and non-stack data

For ease of discussion, let's assume $PageSize = 4096$, $BlockSize = 64$. With stack requests mapped to their local cache slice, the tag field of each cacheline should include the baseline tag plus the $logN$ bits used for slice selection of non-stack requests, shown in the "stack" row of the middle section in Figure 6. For non-stack blocks, these extra $logN$ bits in the tag field are don't-care bits, shown as all 0's in the figure. When $logN > log(PageSize) - log(BlockSize)$, i.e., $logN > 6$ in our example, false hits can occur between stack and non-stack request/data. An incoming stack request can falsely hit a cached block with heap data, or, vise versa. The reason these can happen is that at least one low-order bit of the page number fails to be captured in the non-stack index/tag. As we have explained, false hits could occur only if $N > 64$. If $N \leq 64$, false hits are never an issue since stack and non-stack physical page numbers differ. As long as all bits of the page number are used in index/tag, stack and non-stack requests will not hit a cached block of the other region.

Table 1. Simulation Setup

Core	2.0Ghz, Out-of-order, 16 Cores.
Date TLB	4-way set associative, LRU, 4KB/page, 32/64/128/256 entries
Caches	L1: 32KB each of I and D, 4-way set-associative; L2: 0.5MBx16, 8-way set-associative
Other Cache Params.	Write-back; 64-bytes/line, LRU replacement
Hit Latencies	L1C = 2 core cycles, L2C = 12 core cycles if hit in local slice
DDR	2 Channels/MemCtrl, 64b interface, 16B/cycle @ 800Mhz, 120ns latency

Table 2. The Interconnect Setup

Ring	1.0Ghz, 18 stops: 16 core/LLC stops and 2 memory controller stops. Two uni-directional rings.
Ring Components	64-bit command/address ring and 128-bit data ring, each direction
Router Latency	2 cycles router delay, 1 cycle link propagation delay
Buffering	Data ring: 4 flit-deep at each input port; Cmd/addr ring: 2-flit deep at each input port

To cope with the stack/non-stack false hit problem in a very large-scale CMP, we add one bit in the cache tag. This bit is set to 1 for stack blocks, and 0 for non-stack blocks. In a selected set, a stack request only checks the tags of cached stack blocks, and a non-stack request only checks the tags of cached non-stack blocks. Since we separate stack and non-stack, false hits are avoided. For stack blocks, we can then use the low-order bits of the block address for cache set indexing, shown in the "Stack" row of the bottom section in Figure 6. This design solves the false hit problem.

3.5   Avoiding Cache Coherency Overhead

One key advantage of shared LLC over private LLC is elimination of the cache coherency overhead, as each memory block has a unique location in cache. The proposed mapping method would break the uniqueness property of stack block's slice number in two scenarios. The first is thread migration. The second is page swapping: the memory allocator of the kernel may decide to reallocate a page frame that currently holds stack data to a different process. In both scenarios, cached stack data in the LLC slice need to be flushed before thread migration or page reallocation can happen. To avoid duplicate caching, without turning to hardware cache coherency, the kernel should flush the involved pages out of the caches. Since swapping and thread migration are relatively rare events, performance impact of software cache flushing will be moot in practice.

4   Evaluation of the Proposed Mechanisms

In this section, we present the experiment results of both mechanisms that were proposed in Section 2 and 3. The evaluation metrics are performance and power.

4.1   Experiment Methodology

We evaluate the performance of the proposed optimizations through simulation of a multiprogrammed workload on a CMP with a distributed shared L2 cache. We use a cycle-accurate, trace-driven simulator ManySim [7] for performance modeling.

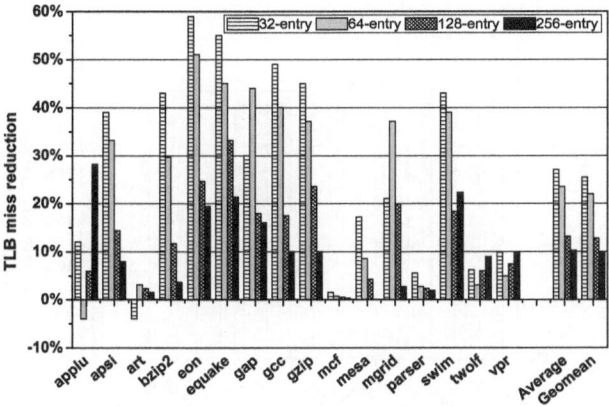

Fig. 7. TLB miss reduction when covering stack memory with superpages. Superpage size = 4

ManySim has an abstract core module to improve simulation speed, but accurately simulates the memory subsystem, including cache and TLB microarchitecture.

We model a system that is depicted in Figure 2: a 16-core CMP with a ring connecting the cores and L2 cache. Each ring stop connects one core and one slice of the L2C. Two memory controllers use dedicated ring stops. Key parameters of the simulation are presented in Table 1. The interconnect is an 18-node ring consisting of a clockwise channel and a counter-clockwise channel. Key interconnect parameters are shown in Table 2. The workload we use is a mix of 16 SPEC CPU2000 applications, each occupying a processor core. We run them so that each one at least graduates 1 billion instructions. We use Orion 2.0 [8] to estimate the power consumed by the interconnect. The power results reported by Orion 2.0 are based on an interconnect in 65 nm, 1.2V technology.

4.2   Evaluation of Anticipatory Superpaging

For the TLB performance evaluation, we model two types of processors: low-power (with 32/64 TLB entries) and high-performance (with 128/256 TLB entries). Figure 7 presents the percentage of data TLB misses that are reduced when we apply Anticipatory Superpaging. With 4KB basic pages, we assume a superpage size of 16KB. On low-power processors, superpaging for stack memory is very effective in improving TLB coverage, with an average reduction of over 20% in number of misses. Two applications (*applu* and *art*) get negative effects because superpage entries stay unfairly long in the TLB and cause basic page entries for heap memory to be constantly victimized.

With larger TLB sizes of 128 and 256 entry, the efficacy of the superpaging is lower than smaller sized TLBs, largely due to the small working set size of our benchmarks. But a few applications, such as *eon, equake, gzip* and *swim* still benefit from solid reductions in TLB misses. These results motivate us to continue this line of study. As future work, we will try applications with larger data set, such as scientific and commercial workloads to showcase the performance benefits of Anticipatory Superpaging.

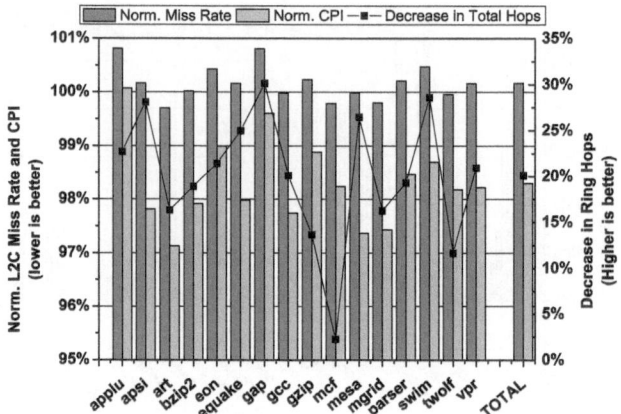

Fig. 8. Performance of Stack-Aware Cache Placement on a Multiprogrammed Workload

4.3   Evaluation of Stack-Aware Cache Placement

The main goal for Stack-Aware Cache Placement is not higher performance, but lower power and energy consumption in the interconnect. While saving power by reducing interconnect traffic is quite intuitive, the overall effect of the proposed mapping scheme on LLC miss rate, and hence application performance, is not self-evident.

Regarding the ramifications in overall performance, affinitizing stack data to the cores that they are attached to has two effects that conflict with each other. On one hand, average traversed hops in the LLC interconnect will decrease. On the other, the near-uniform distribution of LLC accesses across slices can be compromised because some applications have higher cache space occupation tendency than others. On the core-slice pair that has high cache space demand for stack data, heap accesses of all applications are penalized. The overall performance is thus the net effect of both conflicting factors. In other words, we seek to trade LLC hit rates for a more significant reward in LLC average hit latencies and interconnect switching power.

Figure 8 presents the performance results for the multiprogrammed workloads on the 16-core, 18-stop ring. Of the 16 applications, 10 exhibit higher LLC miss rates, with *applu* and *gap* being the two most impacted ones. Total LLC misses increase by 0.2% as a result. What we get in return for the cost of higher LLC miss count is that aggregated total number of traversed hops drops by 20%. Three applications, *apsi, gap* and *swim*, save around 30% in ring hop counts, due to the high percentage of stack accesses in their L1 cache miss streams. The overall effect on performance is an improvement of 1.7% in total instruction throughput, as shown in Figure 8. In fact, *applu* is the only application which sees a performance drop with a negligible 0.07% change in CPI.

Core-data affinity reduces total number of traversed ring hops and consequently switching activities in the routers and the links, and reduces dynamic power consumption. We conduct two sets of experiment to estimate the savings in power consumption of the ring. The first set of experiment consists of uni-programmed runs of the SPEC

applications. At any time only one of the 16 cores is active. In the baseline case, all
L1C misses from the active core are evenly distributed to the 16 LLC slices. In the optimized case, stack data are stored and fetched from the local LLC. Table 3 shows the
ring power when running different applications. On average, the proposed cache placement scheme saves 14% power compared with the conventional mapping. Due to the
small working sets of SPEC CPU2000, except for a few, most applications have rather
small L1 miss rates. Much of the ring's power consumption is thus due to leakage, miti-
gating the overall savings percentage. Two applications, *eon* and *equake*, hardly benefit
from the optimization at all, largely due to their small number of LLC accesses. Another interesting one is *gap*. As we have seen earlier in Figure 8, *gap*'s total hop counts
got reduced by a high 30%. However, because of its extremely low LLC accesses per
cycle, there is not much optimization opportunity on the interconnect, with a power
saving of only 2.48%. *art* demonstrates quite the opposite behavior. Its saved hop count
is below-average (Figure 8), but its power saving (43.92%) tops all the applications that
we tested, due to its high activity factor on the interconnect.

Table 3. Power Consumption of the Ring in a Uni-Program Setup

Application	eon	equake	bzip2	gap	mesa	apsi	mgrid	gzip	swim
Ring Power, Baseline(W)	0.2641	0.2642	0.2684	0.2732	0.2785	0.2936	0.3248	0.3472	0.3506
Ring Power, Optimized(W)	0.2636	0.2636	0.2654	0.2664	0.2689	0.2700	0.2915	0.3028	0.2934
Savings	0.18%	0.18%	1.10%	2.48%	3.47%	8.03%	10.27%	12.80%	16.32%
Application	gcc	parser	applu	vpr	twolf	mcf	art	Average	Geomean
Ring Power, Baseline(W)	0.3694	0.3784	0.3632	0.4113	0.5042	1.3964	1.4425		
Ring Power, Optimized(W)	0.3080	0.3125	0.2963	0.3251	0.3846	0.9259	0.8089		
Savings	16.61%	17.40%	18.42%	20.97%	23.73%	33.69%	43.92%	14.35%	6.81%

In the second experiment, we run the 16 applications simultaneously, each occupying a processor core. The aggregate traffic of the 16 applications causes significant
switching activities in the router and the links, making leakage dwarfed by dynamic
power consumption. Reduced ring activities cut total ring power from 3.58W to 2.85W,
a saving of 20.6%. Breakdown of the interconnect power is shown in Table 4.

Overall, we observe significant benefit from the two virtual memory semantic-aware
optimizations. Anticipatory Superpaging reduces TLB miss counts by 10 - 20%, while
Stack-Aware Cache Placement improves ring power consumption by over 20% on a
multiprogrammed workload despite slightly increased cache misses.

5   Related Work

5.1   Work Related to Anticipatory Superpaging

Making good use of superpaging to reduce TLB miss rates has been a perennial research
topic for the last three decades. Reservation-based physical page allocation [9] builds
superpages for heap memory with information gathered at *malloc* or *mmap* time. It is
not applicable to stack memory due to the lack of a reservation step in stack growth.
IRIX [10] and HP-UX [11] eagerly build superpages upon the initial page fault. The
major drawback of IRIX and HP-UX's superpaging mechanisms is that they require
each user application to provide superpage size hints, one for each of the segments.

Table 4. Power Consumption of the Ring in a Multi-Program Setup

	Router Power	Link Power	Total Power
Baseline	1.94W	1.64W	3.58W
Optimized	1.57W	1.28W	2.85W
Savings	19.4%	22.0%	20.6%

Recognizing limits of these static/offline superpage identification methods, Romer et al. [1] puts forth a set of dynamic, online promotion algorithms. One of the proposed algorithms is *ASAP*, in which a superpage is promoted as soon as all its component basic pages have been touched. *ASAP* is found to be suboptimal due to the high cost of page copying. Our paper can be regarded as a special form of *ASAP*. But we eliminate the cost of page copying. Similar to this paper, Cascaval et al. use different page sizes for different data categories [12]. While we classify data using virtual memory semantics, they classify data based on user applications' data structures which require offline profiling. However, offline profiling is usually not available for real-world applications.

5.2   Work Related to Stack-Aware Cache Placement

Caches in general-purpose processors are typically designed oblivious of software semantics, with some rare exceptions, e.g., building separate caches for user data versus OS data [13], and stack versus non-stack [14,6]. Based on the observation that stack data are private to each thread, Ballapuram et al. [15] propose to skip snooping for L2 accesses that belong to the stack region, assuming private caches.

Cho [16] and Jin [17] improve core-data affinity through page coloring on a distributed shared cache. Their approach assumes page-interleaving and do not exploit virtual memory semantics. We assume cacheline interleaving, which is what most commercial CMP use. But the key idea of our paper can also be achieved using page allocation proposed by Cho and Jin.

6   Conclusions

In this paper, we propose two novel mechanisms that exploit software semantics to optimize on-chip memory system components. In the first mechanism, automatically and aggressively using superpages for stack memory greatly reduces TLB misses. By creating superpages at the initial page fault of a potential superpage, we avoid the high cost of moving pages of physical memory at a later time. In the second mechanism, we force stack data, which are private in nature, to be mapped to their local cache slices. Reduced average LLC hit latencies give us slight improvement in overall application performance despite small increases in LLC miss rates, and decreased ring hop counts lead to perceivable benefits in the interconnect power consumption.

In general, we believe that there are many opportunities in microprocessor design to exploit the rich OS and programming language semantics that have been 'lost in translation' on the execution path of data movement instructions and primitives. Identifying and utilizing memory-related software semantics could lead to graceful solutions that may cost much more in software semantics-oblivious approaches.

Acknowledgement. We thank Tong Li for discussions on virtual memory management, Jim Held for his feedback on the manuscript, and the anonymous reviewers for their constructive suggestions.

References

1. Romer, T.H., Ohlrich, W.H., Karlin, A.R., Bershad, B.N.: Reducing TLB and memory overhead using online superpage promotion. In: ISCA 1995, pp. 176–187 (1995)
2. McCurdy, C., Coxa, A.L., Vetter, J.: Investigating the TLB behavior of high-end scientific applications on commodity microprocessors. In: ISPASS 2008, pp. 95–104 (2008)
3. Speight, E., Shafi, H., Zhang, L., Rajamony, R.: Adaptive mechanisms and policies for managing cache hierarchies in chip multiprocessors. In: ISCA 2005, pp. 346–356 (2005)
4. Muralimanohar, N., Balasubramonian, R., Jouppi, N.: Optimizing NUCA organizations and wiring alternatives for large caches with CACTI 6.0. In: MICRO, vol. 40, pp. 3–14 (2007)
5. Wang, H.S., Peh, L.S., Malik, S.: A power model for routers: Modeling Alpha 21364 and InfiniBand routers. IEEE Micro 23, 26–35 (2003)
6. Lee, H.H.S., Ballapuram, C.S.: Energy efficient D-TLB and data cache using semantic-aware multilateral partitioning. In: ISLPED 2003, pp. 306–311 (2003)
7. Zhao, L., Iyer, R., Moses, J., Illikkal, R., Makineni, S., Newell, D.: Exploring large-scale CMP architectures using ManySim. IEEE Micro 27, 21–33 (2007)
8. Kahng, A., Li, B., Peh, L.-S., Samadi, K.: ORION 2.0: A fast and accurate NoC power and area model for early-stage design space exploration. In: DATE 2009, pp. 423–428 (2009)
9. Navarro, J., Iyer, S., Druschel, P., Cox, A.: Practical, transparent operating system support for superpages. SIGOPS Oper. Syst. Rev. 36, 89–104 (2002)
10. Ganapathy, N., Schimmel, C.: General purpose operating system support for multiple page sizes. In: ATEC 1998 (1998)
11. Subramanian, I., Mather, C., Peterson, K., Raghunath, B.: Implementation of multiple page-size support in HP-UX. In: ATEC 1998, p. 9 (1998)
12. Cascaval, C., Duesterwald, E., Sweeney, P.F., Wisniewski, R.W.: Multiple page size modeling and optimization. In: PaCT 2005, pp. 339–349 (2005)
13. Nellans, D., Balasubramonian, R., Brunvand, E.: OS execution on multi-cores: is outsourcing worthwhile? SIGOPS Oper. Syst. Rev. 43, 104–105 (2009)
14. Huang, M., Renau, J., Yoo, S.M., Torrellas, J.: L1 data cache decomposition for energy efficiency. In: ISLPED 2001, pp. 10–15 (2001)
15. Ballapuram, C.S., Sharif, A., Lee, H.H.S.: Exploiting access semantics and program behavior to reduce snoop power in chip multiprocessors. In: ASLPED 2008 (2008)
16. Cho, S., Jin, L.: Managing distributed, shared L2 caches through OS-level page allocation. In: MICRO 2006 (2006)
17. Jin, L., Cho, S.: SOS: A software-oriented distributed shared cache management approach for chip multiprocessors. In: PaCT 2009, pp. 361–371 (2009)

From Serial Loops to Parallel Execution on Distributed Systems

George Bosilca[1], Aurelien Bouteiller[1], Anthony Danalis[1], Thomas Herault[1], and Jack Dongarra[1,2]

[1] University of Tennessee, Knoxville TN 37996, USA
[2] University of Manchester, Manchester, UK
{bosilca,bouteill,adanalis,herault,dongarra}@eecs.utk.edu

Abstract. Programmability and performance portability are two major challenges in today's dynamic environment. Algorithm designers targeting efficient algorithms should focus on designing high-level algorithms exhibiting maximum parallelism, while relying on compilers and run-time systems to discover and exploit this parallelism, delivering sustainable performance on a variety of hardware. The compiler tool presented in this paper can analyze the data flow of serial codes with imperfectly nested, affine loop-nests and if statements, commonly found in scientific applications. This tool operates as the front-end compiler for the *DAGuE* run-time system by automatically converting serial codes into the symbolic representation of their data flow. We show how the compiler analyzes the data flow, and demonstrate that scientifically important, dense linear algebra operations can benefit from this analysis, and deliver high performance on large scale platforms.

Keywords: compiler analysis, symbolic data flow, distributed computing, task scheduling.

1 Introduction and Motivation

Achieving scientific discovery through computing simulation puts such high demands on computing power that even the largest supercomputers in the world are not sufficient. Regardless of the details in the design of future high performance computers, few would disagree that a) there will be a large number of nodes; b) each node will have a significant number of processing units; c) processing units will have a non-uniform view of the memory. Moreover, computing units in a single machine have already started becoming heterogeneous, with the introduction of accelerators, like GPUs.

This creates a complex environment (the "jungle"[1]) for application and library developers. A developer, whether a domain scientist simulating physical phenomena, or a developer of a numerical library such as ScaLAPACK [4] or

---

[1] Herb Sutter, "Welcome to the Jungle", 12-29-2011,
http://herbsutter.com/2011/12/29/welcome-to-the-jungle/

C. Kaklamanis et al. (Eds.): Euro-Par 2012, LNCS 7484, pp. 246–257, 2012.
© Springer-Verlag Berlin Heidelberg 2012

PLASMA [12], is forced to compromise and accept poor performance, or waste time optimizing her code instead of making progress in her field of science. A better solution would be to rely on a run-time system that can dynamically adapt the execution to the current hardware. *DAGuE* [10], which deploys dynamic micro-task scheduling, has been shown [11] to deliver portable high performance, on heterogeneous hardware for a class of regular problems, such as those occurring in linear algebra.

Unfortunately, dynamic scheduling approaches commonly require application developers to use unfamiliar programming paradigms which hinders productivity and prevents widespread adoption. As an example, in *DAGuE*, the algorithms are represented as computation tasks decorated by symbolic expressions that describe the flow of data between tasks.

In this paper, we describe a compiler tool that automatically analyzes annotated sequential C code and generates the symbolic, problem size independent, data flow used by *DAGuE*. Through polyhedral analysis, our compiler represents the data flow of the input code as parameterized, symbolic expressions. These expressions enable each task to independently compute, at run-time, which other tasks it has dependencies with, thus defining the communication that must be performed by the system. We explain the process and the tools used to perform this translation. To the best of our knowledge, it is the first time that state-of-the-art, handcrafted software packages are outperformed by automatic data flow analysis coupled with run-time DAG scheduling on large scale distributed memory systems.

2   Related Work

Symbolic dependence analysis has been the subject of several studies [14, 18–20], mainly for the purpose of achieving powerful dependence testing, array privatization and generalized induction variable substitution, especially in the context of parallelizing compilers such as Polaris [5] and SUIF [17]. This body of work differs from the work presented in this paper in that our compiler does not focus on dependence testing, or try to statically find independent statements, in order to parallelize them. Our compiler derives symbolic parameterized expressions that describe the data flow and synchronization between tasks. Furthermore, we focus on programs that consist of loops and *if* statements, with calls to kernels that operate on whole array regions (i.e. matrix tiles), rather than operating on arrays in an element by element fashion. This abstracts away the access patterns inside the kernels, and simplifies the data flow equations enough that we can produce exact solutions using the Omega Test.

The polyhedral model [1, 3, 23], of which the Omega Test is part, has drawn a lot of attention in recent years, and newer optimization and parallelization tools, such as Pluto [6], have emerged that take advantage of it. However, unlike the work currently done within the polyhedral model, we do not use the dependence abstractions to drive code transformations, but rather export them in symbolic notation to enable our run-time to make scheduling and message exchange decisions.

In our work we harness the theoretical framework set by Feautrier [15] and Vasilache et al. [25] to compute the symbolic expressions that capture the data flow. By coupling for the first time this compiler theory with a distributed memory DAG scheduling run-time, we assert experimentally the significance of this approach in the context of high performance computing.

Finaly, Baskaran et. al [2] performed compiler assisted dynamic scheduling using compiler analysis. In their approach, the compiler generates code that scans and enumerates all vertices of the DAG at the beginning of the run-time execution. This has the same drawbacks as approaches, such as StarSS [21] and TBlas [24], that rely on pseudo-execution of the serial loops at run-time to dynamically discover dependencies between kernels. The overhead grows with the problem size and the scheduling is either centralized of replicated. In contrast, the symbolic data-flow and synchronization expressions our compiler generates can be solved at run-time by each task instance independently, in $O(1)$ time, without any regard to the location of the given instance in the DAG.

3   Compiler and Run-Time Synergy

The goal of traditional standalone parallelizing compilers is to convert a serial program into a parallel program by statically addressing all the issues involved with parallel execution. However, dynamic environments call for a run-time solution. In our toolchain, the compiler static analysis scope is reduced to producing a symbolic representation to be interpreted dynamically by the scheduler during execution. Effectively, the compiler performs static data flow analysis to convert an affine input serial program into a Direct Acyclic Graph (DAG), with program functions (kernels) as its nodes, and data dependency edges between kernels as its edges. Then, the run-time is responsible for addressing all DAG scheduling challenges, including background MPI data transfers between distributed resources [10].

To drive the scheduler decisions, the compiler needs to produce more than a boolean value regarding the existence or not of a dependency. It has to identify the exact, symbolic, dependence relations that exist in the source code. From those, it generates parameterized symbolic expressions with parameters that take distinct values for each task. The expressions are such that the run-time can evaluate them for each task T_i independently of the task's place in the DAG. Also, the evaluation of each expression costs constant time (i.e., it does not depend on the size of the DAG). The result of evaluating each symbolic expression is another task T_j, to which data must be sent, or from which data must be received[2].

---

[2] Therefore, the only parameters allowed in a symbolic expression are the parameters of the execution space of T_i, and globals used in the input code.

4   Input and Output Formats

4.1   Input Format: Annotated Sequential Code

The analysis methodology used by our compiler allows any program with regular control flow and side-effect free functions to be used as input. The current implementation focuses on codes written in C, with affine loops and array accesses. The compiler front-end is flexible enough to process production codes such as the PLASMA library [12]. PLASMA is a linear algebra library that implements tile-based dense linear algebra algorithms.

```
for (k = 0; k < A.mt; k++) {
  Insert_Task(zpotrf, A[k][k], INOUT);
  for (m = k+1; m < A.mt; m++) {
    Insert_Task(ztrsm, A[k][k], INPUT, A[m][k], INOUT);
  }
  for (m = k+1; m < A.mt; m++) {
    Insert_Task(zherk, A[m][k], INPUT, A[m][m], INOUT);

    for (n = k+1; n < m; n++) {
      Insert_Task(zgemm, A[m][k], INPUT, A[n][k], INPUT, A[m][n], INOUT);
    }
  }
}
```

Fig. 1. Cholesky factorization in PLASMA

Figure 1 shows the PLASMA code that implements the Tiled Cholesky factorization [12] (with some preprocessing and simplifications performed on the code for improving readability). The figure shows the operations that constitute the Cholesky factorization POTRF, TRSM, HERK, and GEMM. The data matrix "A" is organized in tiles, and notation such as "A[m][k]" refers to a block of data (a tile), and not a single element of the matrix. Our compiler uses a specialized parser that can process hints in the API of PLASMA. We made this choice because in the PLASMA API the following is true: a) for every matrix tile passed to a kernel as a parameter, the parameter that follows it specifies whether this tile is read, modified, or both, using the special values INPUT, OUTPUT and INOUT; b) all PLASMA kernels are side-effect free. This means that they operate only on memory pointed to by their arguments, and that memory is not aliased.

Figure 1 contains four kernels, that correspond to the aforementioned operations. In the rest of this article we will use the terms *task* and *task class*. A *task class* is a specific kernel in the application that can be executed several times, potentially with different parameters, during the life-time of the application. zpotrf and zgemm are examples of task classes in Figure 1. A *task* is a particular, and unique, instantiation of a kernel during the execution of the application, with given parameters. In the example of the figure, task class zpotrf will be instantiated as many times as the outer loop for(k) will iterate, and thus we define the task class's *execution space* to be equal to the iteration space of the loop.

4.2 Compiler Output: Job Data Flow

```
for (k = 0; k < N; k++) {
    Insert_Task( Ta, A[k][k], INOUT );
    for (m = k+1; m < N; m++) {
        Insert_Task( Tb, A[k][k], INPUT, A[m][m], INOUT );
    }
}
```

Fig. 2. Pseudocode example of input code

The compiler outputs a collection of task classes and their dependency relation in a format we refer to as the Job Data Flow (JDF). Consider the simpler input defined in Figure 2. The compiler extracts (as described in Section 5.1) data flows between T_a and T_b in a symbolic way and outputs them in the definitions of task classes T_a and T_b in the JDF. The symbolic representation of each edge is such that every task $T_a(k)$ is able to determine the tasks $T_b(k, m)$ that need to use the data defined by $T_a(k)$ and vice-versa. Consider the particular edge due to A[k][k] flowing from $T_a(k)$ to $T_b(k, m)$. In the JDF, we use the following notation to store this flow edge in task class T_a:

A[k][k] -> (k < N-1) ? A[k][k] Tb(k, (k+1)..(N-1))

Conversely, tasks of the class T_b must be able to determine which task they depend on for input. In this case the same edge has the following form:

A[k][k] <- A[k][k] Ta(k)

The full JDF that the compiler produces to represent the example code of Figure 2 is shown in Figure 3. As can be seen in the figure, in addition to the execution space and the data flow edges, there are two more elements in a JDF file. First, there is an affinity definition of the form ":A[k][k]" which signifies that the corresponding task should be run in the MPI process that owns the corresponding data element. Second, there is a BODY that consists of C-language code that the run-time will invoke in order to execute the actual kernel that constitutes the body of a task.

From interpreting that JDF output, the *DAGuE* run-time can handle distributed memory execution efficiently, the scheduler can identify which tasks must communicate with which other, without consulting a centralized entity or traversing the whole problem DAG.

```
Ta(k)
  k = 0..N-1
  : A[k][k]

  A[k][k] <- (k==0)  ? A[k][k] : A[m][m] Tb(k-1, k)
            -> (k<N-1) ? A[k][k] Tb(k, (k+1)..(N-1))
            -> A[k][k]
BODY
  Ta(A[k][k]);
END

Tb(k,m)
  k = 0..N-1
  m = k+1..N-1
  : A[m][m]

  A[k][k] <- A[k][k] Ta(k)
  A[m][m] <- (k==0) ? A[k][k] : A[m][m] Tb(k-1, m)
            -> (m==k+1) ? A[k][k] Ta(m) : A[m][m] Tb(k+1, m)
BODY
  Tb(A[k][k], A[m][m]);
END
```

Fig. 3. Example Job Description Format

5 Extracting Symbolic Data Flow and Data Exchange

5.1 Omega Relations

The Omega test [22] is the library we use for manipulating the sets of affine constraints over integer variables that arise when performing the symbolic data-flow analysis necessary when converting from sequential code to JDF. An Omega *Relation* is a mapping between two tuples, defining the execution space of the source and sink task classes, as well as the conjunction of constraints for both execution spaces. Consider the example of compiler input given in Figure 2. The iteration space of T_a is the iteration space of outer loop for(k); We denote this iteration space with the following Omega notation:

$$\{[k] : 0 <= k <= N-1\}$$

Such notation { [T] : C }, where T is a tuple, and C is a conjunction of constraints, defines the ranges of values for the elements of T for which C is true. Similarly, we define the execution space of task class T_b to be:

$$\{[k,m] : 0 <= k < N-1 \&\& k+1 <= m <= N-1\}$$

Here, the tuple has two elements, since T_b is enclosed by two loops. By examining the data-flow of the code, we can see that A[k][k] for example, will be modified (*defined*, in compiler parlance) by kernel T_a and then read (*used*, in compiler parlance) by kernel T_b. The corresponding relation due to A[k][k] flowing from $Ta(k)$ to $Tb(k,m)$ is:

$$\{[k] -> [k',m] : 0 <= k < N-1 \&\& k+1 <= m <= N-1 \&\& k == k'\}$$

In the example above, the term "[k]" represents the execution space of the source, T_a, and the term "[k',m]"[3] represents the execution space of the sink, T_b. In Omega parlance, this Relation has an *input variable* count of one and *output variable* count of two.

5.2   Interprocess Data Exchange

The symbolic data edges are associated with task classes, so that the run-time can use them to determine what messages need to be exchanged between tasks. In particular, for each task the run-time must determine the tasks that produced the input of this task and the tasks that will consume the output of this task. Therefore, the expressions stored in the JDF may contain only a) the parameters of the source task, b) symbolic and numeric constants, c) the logical constants "TRUE" and "FALSE".

Outgoing Messages. After the compiler has finished processing the input source code, it will have a collection of Omega Relations describing the data flow edges from each task class to each other task class in the code. To produce the information needed by the run-time regarding the outgoing edges of a task T_i, we need to process all Relations of flow edges that have as source the task T_i. For every parameter that appears in the execution space of a Relation's destination, we solve the equality constraints in the conjunction of constraints for this parameter. Consider, as an example, the Relation:

$$\{[k,m] \rightarrow [k'] : k' = m \text{ \&\& } 1+k = m \text{ \&\& } 1 <= m < N\}$$

which describes the flow edge from A[m][m] in T_b to A[k][k] in T_a. This edge will be stored in the JDF of T_b as:

$$A[m][m] \rightarrow ((1+k)==m) \text{ ? } A[k][k] \text{ Ta}(m)$$

This way, when the run-time is processing task $T_b(7, 8)$ for example, it can compute in $O(1)$ time that it needs to send tile A[8][8] to task $T_a(8)$. Also, when processing task $T_b(7, 11)$, the run-time can compute that A[11][11] should not be sent to any instance of T_a, since the condition $(1 + k) == m$ is not true (clearly, $1 + 7 \neq 11$).

If a destination parameter does not appear in any equality constraints in the conjunction, we determine the lower bound and upper bound of this parameter by solving the inequality constraints, and create a range of tasks that should be the receiver of this message. As an example, consider the flow edge from A[k][k] of T_a to A[k][k] of T_b which is described by the Relation:

$$\{[k] \rightarrow [k',m] : k' = k \text{ \&\& } 0 <= k < m < N\}$$

---

[3] Although both task classes share a common enclosing loop, we use different variables in the execution spaces (k and k') because the dependency could be a loop carried dependency, so we have to allow the two iteration spaces to be independent.

In order to store this edge in the JDF expression of T_a, we need to express k' and m in terms of k (and constants), since k is the only parameter in the execution space of T_a. Therefore, this edge will be translated to the following information in JDF notation:

```
A[k][k] -> (k < N-1) ? A[k][k] Tb(k, k+1..N-1)
```

since the output parameter "m" does not appear in any equality constraints. In JDF syntax, expressions with ranges signify to the run-time that a broadcast operation must be performed.

Incoming Messages. To produce the information regarding the incoming edges of a task T_i, we traverse the flow edges of all tasks searching for edges that have task T_i as the destination. For each such Relation, we compute the inverse, and then proceed with solving the inverse Relation for the output parameters, as we do for the outgoing edges.

5.3   Anti-dependence Edges

An anti-dependence edge exists between tasks T_{src} and T_{dst} if T_{src} uses a variable that T_{dst} defines, and T_{src} executes before T_{dst}[4]. In parallel execution, anti-dependence edges must be translated to synchronization edges, to avoid using wrong versions of the data. Ostensibly, anti-dependencies are not relevant in a distributed memory execution environment due to data copying. However, *DAGuE* can run on distributed memory machines, shared memory machines, or distributed memory clusters of shared memory nodes. Therefore handling anti-dependencies in a uniform and systematic way is important for preserving the semantics of the input serial algorithm.

Our compiler starts by recording all potential anti-dependencies as Omega Relations. Then, Algorithm 1 is used to minimize the number of synchronization edges by using data flow edges between tasks to eliminate the need for additional synchronization, wherever possible. This is possible because a data flow edge imposes a message exchange between tasks and therefore explicit synchronization.

6   Performance

Two metrics of performance are relevant in the context of this work. First, the performance of the compiler tool itself, and second, the performance of applications running under our system. We have tested the performance of our compiler tool by processing the dense linear algebra operations found in the PLASMA library, on hardware commonly found on average personal computers. The compilation time we have observed is in the order of 100ms when the anti-dependence minimization algorithm is not being used, and in the order of a few seconds when it is being used.

---

[4] Or more accurately, if there exists an execution path from T_{src} to T_{dst}.

Function $FinalizeAntiDependencies(I_G)$
Input: I_G, Input graph.
Result: Modifies I_G by finalizing antidependencies.
begin

    foreach *anti-dependence edge* $E_a \in I_G$ **do**

        Let G be a copy of I_G
        `/* Unless otherwise specified all nodes and edges belong   */`
        `/* to G, and all operations are done on G.                 */`
        foreach *pair of nodes* N_1, N_2 **do**

            $\mathcal{R} \leftarrow \bigcup \{ R_i : N_1 \xrightarrow{R_i} N_2 \}$
            Replace all edges from N_1 to N_2 with single edge $N_1 \xrightarrow{\mathcal{R}} N_2$

        foreach *Node* N_0 **do**

            Let $(p_1, \ldots)$ be the parameters of the task that correspond to N_0
            `/* Initiate Cycle(N0) with an empty (tautologic)      */`
            `/* Relation to self.                                  */`
            $Cycle(N_0) \leftarrow \{[p_1, \ldots] \to [p_1, \ldots]\}$

        foreach *Node* N_0 **do**

            foreach $N_0 \xrightarrow{R_0} N_1 \ldots \xrightarrow{R_{n-1}} N_0$ **do**

                `/* N0, N1...N0 is a Cycle formed following flow,   */`
                `/* and/or anti-dependence edges.                    */`
                $C \leftarrow R_0 \circ R_1 \circ \ldots \circ R_{n-1}$
                $T \leftarrow$ transitive closure of C
                $Cycle(N_0) \leftarrow Cycle(N_0) \bigcup T$

        $A \leftarrow FindTransitiveEdge(Source(E_a), \emptyset, \emptyset)$
        `/* May remove Ea if empty                              */`
        Change E_a to $(E_a - A)$ in I_G

Algorithm 1. $FinalizeAntiDependencies(I_G)$

Function $FindTransitiveEdge(N_c, T, A)$
Input: N_c, the current node in the transitive edge; T the transitive edge being built; A the union of all transitive edges found until now.
Result: Union of the transitive edges that start at N_c and end at $Sink(E_a)$
begin

    `/* Scope inlcudes the variables of FinalizeAntiDependencies()  */`
    `/* in Algorithm 1. This algorithm operates on G.               */`
    Mark N_c as visited
    $T \leftarrow Cycle(N_c) \circ T$
    foreach *Edge* $N_c \xrightarrow{R_i} N_i$ *s.t.* N_i *is not visited* **do**

        $T_{tmp} \leftarrow R_i \circ T$
        $A \leftarrow FindTransitiveEdge(N_i, T_{tmp}, A)$

    if $N_c = Sink(E_a)$ **then**
        return $A \bigcup T$
    else
        return A

Algorithm 2. $FindTransitiveEdge(N_c, T, A)$

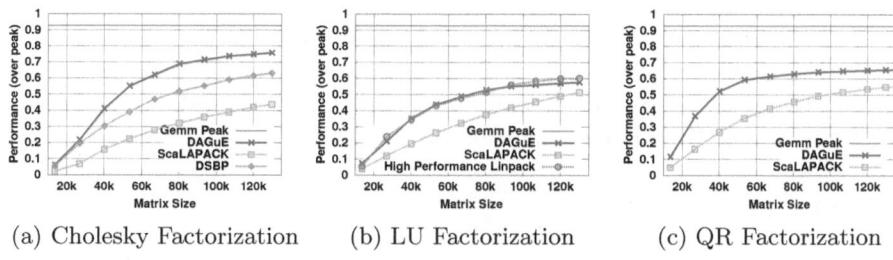

(a) Cholesky Factorization   (b) LU Factorization   (c) QR Factorization

Fig. 4. Performance comparison on the Griffon platform (on 648 cores)

The performance of the DAGuE run-time has been extensively studied in related publications [8, 9, 7]. The goal of this paper is to present the compiler front-end of the system, so we present only a summary of performance results to demonstrate that our toolchain can automatically analyze, schedule and execute non-trivial algorithms, and deliver high performance at scale. Application performance results are relevant, because the scalability achieved by our run-time is enabled by the problem size independent algebraic expressions that our compiler generates to describe inter-task dependence edges.

For the experiments we present here, we used 81 dual socket Intel Xeon L5420 quad core processors at 2.5GHz for a total of 648 cores. Each node has 16GB of memory, and is interconnected to the others by a 20Gbs Infiniband network and runs Linux 2.6.24 (Debian Sid).

The benchmarks consist of three popular dense matrix factorizations: Cholesky, LU and QR. All three operations are implemented in the ScaLAPACK numerical library [4]. Moreover, the Cholesky factorization has been implemented in a more optimized way in the DSBP software [16], using static scheduling of tasks, and a data distribution more efficient. The LU factorization with partial pivoting is also solved by the well known High Performance Linpack benchmark (HPL [13]), used to measure the performance of supercomputers.

For our comparison, we implemented these operations within *DAGuE* by using the compiler presented in this paper to generate the JDF symbolic representation from the corresponding PLASMA files. The data distribution is not generated by automatic tools, but rather chosen by the human developer. For our experiments, we have distributed the initial data following the classical 2D-block cyclic distribution used by ScaLAPACK, and used our run-time engine to schedule the operations on the distributed data. The kernels consist of the BLAS operations referenced by the sequential codes, and their implementation was the most efficient available on this machine. The same kernel implementations for ScaLAPACK, HPL, DSBP, and our engine were used on each run.

Figure 4 presents the performance measured using our system (labeled as *DAGuE*) and ScaLAPACK, and when applicable DSBP and HPL, as a function of the problem size. All data is normalized to the theoretical floating point peak of the machine. A total of 648 cores participated in the distributed run, and the data was distributed according to a 9x9 2D block-cyclic grid. Tile size was tuned to provide the best performance on each setup. As the figures illustrate, on all benchmarks and for all problem sizes, our framework outperforms ScaLAPACK, and performs as well as the state of the art, hand-tuned codes for specific problems. Our system goes from the sequential code to the parallel run automatically, with very limited human involvement, but is still able to outperform DSBP, and competes with the HPL implementation on this machine.

7 Conclusion

In this paper we presented the compiler front end of the *DAGuE* system, more precisely how the compiler extracts the Symbolic Data Flow and Data Exchanges from the input code in order to expose additional information to the run-time. We outlined JDF, *DAGuE*'s internal problem-size independent representation of task generated by the compiler and used by the run-time to make all task scheduling and communication decisions. We showed how Relations produced using the Omega test can be converted into message and synchronization requests for the run-time, and how the synchronization edges can be reduced to the minimum necessary set. Using this critical information exposed by the compiler, the run-time can take more effective decisions about inter-nodes data transfers and about how to schedule tasks in order to maximize the available parallelism not only locally but remotely. Experimental results confirm that serial codes processed by our system can match, or outperform, highly optimized, state of the art, hand tuned, distributed linear algebra codes, such as Scalapack, libSCI and HPL.

References

1. Ancourt, C., Irigoin, F.: Scanning polyhedra with do loops. In: Proceedings of ACM PPoPP 1991, Williamsburg, VA, pp. 39–50 (1991)
2. Baskaran, M.M., Vydyanathan, N., Bondhugula, U.K.R., Ramanujam, J., Rountev, A., Sadayappan, P.: Compiler-assisted dynamic scheduling for effective parallelization of loop nests on multicore processors. In: Proceedings of ACM PPoPP 2009, Raleigh, NC, pp. 219–228 (2009)
3. Bastoul, C.: Code Generation in the Polyhedral Model Is Easier Than You Think. In: Proceedings of IEEE PACT 2004, pp. 7–16. Antibes Juan-les-Pins, France (2004)
4. Blackford, L.S., Choi, J., Cleary, A., D'Azevedo, E., Demmel, J., Dhillon, I., Dongarra, J., Hammarling, S., Henry, G., Petitet, A., Stanley, K., Walker, D., Whaley, R.C.: ScaLAPACK Users' Guide. Society for Industrial and Applied Mathematics, Philadelphia (1997)
5. Blume, W., Doallo, R., Eigenmann, R., Grout, J., Hoeflinger, J., Lawrence, T., Lee, J., Padua, D., Paek, Y., Pottenger, B., Rauchwerger, L., Tu, P.: Parallel programming with polaris. IEEE Computer 29, 78–82 (1996)
6. Bondhugula, U., Hartono, A., Ramanujam, J., Sadayappan, P.: A practical automatic polyhedral parallelizer and locality optimizer. In: Proceedings of ACM PLDI 2008, Tucson, AZ, pp. 101–113 (2008)
7. Bosilca, G., Bouteiller, A., Danalis, A., Faverge, M., Haidar, H., Herault, T., Kurzak, J., Langou, J., Lemarinier, P., Ltaief, H., Luszczek, P., YarKhan, A., Dongarra, J.: Distributed-Memory Task Execution and Dependence Tracking within DAGuE and the DPLASMA Project. Tech. Rep. 232, LAWN (September 2010)

8. Bosilca, G., Bouteiller, A., Danalis, A., Faverge, M., Haidar, A., Herault, T., Kurzak, J., Langou, J., Lemarinier, P., Ltaief, H., Luszczek, P., YarKhan, A., Dongarra, J.: Flexible development of dense linear algebra algorithms on massively parallel architectures with DPLASMA. In: IEEE PDSEC 2011, Anchorage, AK (2011)
9. Bosilca, G., Bouteiller, A., Danalis, A., Herault, T., Lemarinier, P., Dongarra, J.: DAGuE: A generic distributed dag engine for high performance computing. In: HIPS 2011, Anchorage, AK (2011)
10. Bosilca, G., Bouteiller, A., Danalis, A., Herault, T., Lemarinier, P., Dongarra, J.J.: DAGuE: A generic distributed DAG engine for high performance computing. Parallel Computing (2011) (to appear), http://dx.doi.org/10.1016/j.parco.2011.10.003
11. Bosilca, G., Bouteiller, A., Hérault, T., Lemarinier, P., Saengpatsa, N.O., Tomov, S., Dongarra, J.J.: Performance portability of a gpu enabled factorization with the dague framework. In: IEEE CLUSTER, pp. 395–402 (2011)
12. Buttari, A., Langou, J., Kurzak, J., Dongarra, J.J.: A class of parallel tiled linear algebra algorithms for multicore architectures. Parallel Comput. Syst. Appl. 35, 38–53 (2009)
13. Dongarra, J.J., Luszczek, P., Petitet, A.: The LINPACK benchmark: Past, present and future. Concurrency Computat.: Pract. Exper. 15(9), 803–820 (2003)
14. van Engelen, R.A., Birch, J., Shou, Y., Walsh, B., Gallivan, K.A.: A unified framework for nonlinear dependence testing and symbolic analysis. In: Proceedings of ACM ICS 2004, Malo, France, pp. 106–115 (2004)
15. Feautrier, P.: Dataflow analysis of array and scalar references. International Journal of Parallel Programming 20, 23–53 (1991), 10.1007/BF01407931
16. Gustavson, F.G., Karlsson, L., Kågström, B.: Distributed SBP cholesky factorization algorithms with near-optimal scheduling. ACM Trans. Math. Softw. 36(2), 1–25 (2009)
17. Hall, M.W., Anderson, J.M., Amarasinghe, S.P., Murphy, B.R., Liao, S.W., Bugnion, E., Lam, M.S.: Maximizing multiprocessor performance with the SUIF compiler. IEEE Computer 29, 84–89 (1996)
18. Kyriakopoulos, K., Psarris, K.: Data dependence analysis techniques for increased accuracy and extracted parallelism. International Journal of Parallel Programming 32, 317–359 (2004)
19. Kyriakopoulos, K., Psarris, K.: Nonlinear Symbolic Analysis for Advanced Program Parallelization. IEEE Transactions on Parallel and Distributed Systems 20, 623–640 (2009)
20. Maydan, D.E., Hennessy, J.L., Lam, M.S.: Efficient and exact data dependence analysis. In: Proceedings of ACM PLDI 1991, Toronto, Ontario, pp. 1–14 (1991)
21. Perez, J.M., Badia, R.M., Labarta, J.: A dependency-aware task-based programming environment for multi-core architectures. In: Proceedings of IEEE Cluster Computing, pp. 142–151 (2008)
22. Pugh, W.: The omega test: a fast and practical integer programming algorithm for dependence analysis. In: Proceedings of the ACM/IEEE SC 1991, pp. 4–13 (1991)
23. Quilleré, F., Rajopadhye, S., Wilde, D.: Generation of efficient nested loops from polyhedra. Int. J. Parallel Program. 28, 469–498 (2000)
24. Song, F., YarKhan, A., Dongarra, J.: Dynamic task scheduling for linear algebra algorithms on distributed-memory multicore systems. In: Proceedings of ACM/IEEE SC 2009 (2009)
25. Vasilache, N., Bastoul, C., Cohen, A., Girbal, S.: Violated dependence analysis. In: Proceedings of ACM ICS 2006, Cairns, Queensland, Australia, pp. 335–344 (2006)

Topic 5: Parallel and Distributed Data Management

Domenico Talia, Alex Delis, Haimonti Dutta, and Arkady Zaslavsky

Topic Committee

The ever-increasing data volumes used to empower contemporary data–intensive applications as well as aggregations of computing systems call for novel approaches and efficient techniques in the management of geographically dispersed data. Despite recent advances, Internet-scale requirements for both applications and underlying systems require effective provisioning, staging, manipulation, continuous maintenance and monitoring of data hosted in multiple, pre-existing autonomous, distributed and often heterogeneous systems. Evidently, the notions of parallelism and concurrent execution at all levels remain key elements in attaining scalability and effective management for nearly-all modern data-intensive applications. Moreover, as underlying computing environments get transformed through the introduction of novel infrastructures, enhanced capacities and extended functionalities, new solutions are sought to cope with these changes.

In topic 5, we solicited papers in all aspects of data management (access, query, and analysis) and data-intensive applications whose central focus is weaved around the notions of concurrency, parallelism and distributed processing. Key areas that were of interest included parallel and highly-available distributed databases, data-intensive clouds, middleware solutions for processing large-scale data, distributed transaction and query processing, management of distributed data sources, Internet-scale applications, parallel and distributed information retrieval, data-intensive peer-to-peer systems efficient management of data streams, scalable web services as well as data analysis on multi-core and many-core architectures.

Each paper was reviewed by at least 3 reviewers and, finally, we were able to select 4 papers. The accepted papers discuss timely developments in the areas of clustering distributed data streams, organizing web-data using novel indexing approaches, maintaining consistent replicated data across geographically dispersed data-centers and providing fault-tolerant cache-services for search engines.

The paper entitled *"DS-Means: Distributed Data Stream Clustering"* by A. Guerrieri and A. Montresor proposes the *DS-Means* algorithm that achieves clustering of data emanating from different sources operating with minimal interaction. Instead of simply partitioning text collections across clusters of processors, paper *"3D Inverted Index with Cache Sharing for Web Search Engines"* by E. Feuerstein, G.-V. Gil-Costa, M. Marin, G. Tolosa and R. Baeza-Yates advocates the use of a 3D indexing approach which exploits the fact that data is often inherently partitioned and replicated. Paper *"Quality-of-Service for Consistency of Data Geo-Replication in Cloud Computing"* by S. Esteves, J. Nuvo Silva and L. Veiga suggests the VFC^3 approach that is a novel consistency model and

C. Kaklamanis et al. (Eds.): Euro-Par 2012, LNCS 7484, pp. 258–259, 2012.

framework capable of enforcing varying degrees of consistency in accordance to the semantics of the replicated data. Finally, paper *"A Fault-Tolerant Cache Service for Web Search Engines"* by C. Gomez-Pantoja, D. Rexachs, M. Marin and E. Luque proposes a new approach in structuring the cache of results for web-search engines; the approach is based on consistent hashing and a strategy that readily enables fault tolerance.

We would also like to take this opportunity to sincerely thank all contributing authors for their submissions, the Euro-Par 2012 Organizing Committee as well as all our referees who provided highly useful comments and whose efforts have made this topic and conference possible.

DS-Means: Distributed Data Stream Clustering

Alessio Guerrieri and Alberto Montresor[*]

University of Trento, Italy

Abstract. This paper proposes DS-MEANS, a novel algorithm for clustering distributed data streams. Given a network of computing nodes, each of them receiving its share of a distributed data stream, our goal is to obtain a common clustering under the following restrictions (i) the number of clusters is not known in advance and (ii) nodes are not allowed to share single points of their datasets, but only aggregate information. A motivating example for DS-MEANS is the decentralized detection of botnets, where a collection of independent ISPs may want to detect common threats, but are unwilling to share their precious users' data. In DS-MEANS, nodes execute a distributed version of K-MEANS on each chunk of data they receive to provide a compact representation of the data of the entire network. Later, X-MEANS is executed on this representation to obtain an estimate of the number of clusters. A number of experiments on both synthetic and real-life datasets show that our algorithm is precise, efficient and robust.

1 Introduction

Broadly speaking, *clustering* is the problem of partitioning a set of items into a collection of *clusters*, so that, given a definition of similarity, items in the same cluster are more similar to each other than they are to items in other clusters.

Most clustering algorithms assume that the items to be analyzed are available *here* and *now*, meaning that the entire data set could be easily accessed from a single machine. Sometimes both these assumptions must be relaxed, meaning that items are inherently distributed and not immediately available, for example because they are continuously generated and volatile in nature.

As a potential application area, consider the problem of detecting malicious threats like botnets, DDoS attacks, viruses, etc. [3]. In this setting, a large number of detectors, potentially belonging to different organizations (ISPs, companies, universities) collect large quantity of information about the behavior of a system (for example, by inspecting the networking traffic at routers). In order to associate detection with a (potentially immediate) reaction, data should be analyzed and clustered as it flows through the detectors. In such setting, centralized algorithms cannot be applied. There are several reason for this:

- The data to be analyzed is constituted by a continuous stream of items, and waiting for all of them to be collected in a single machine is infeasible;

---

[*] This work is supported by the Italian MIUR Project *Autonomous Security*, sponsored by the PRIN 2008 Programme.

C. Kaklamanis et al. (Eds.): Euro-Par 2012, LNCS 7484, pp. 260–271, 2012.

- Forwarding all data to one machine may be too expensive, inducing a large amount of unnecessary traffic;
- If the data is collected by different organizations, privacy issues may prohibit the gathering of the entire data set by a centralized third party.

Based on these considerations, the problem we are trying to solve is a distributed form of data stream clustering [1], where data arrives continuously at multiple nodes, without having the possibility to transmit the entire dataset to a single machine.

The main contribution of this paper is the DS-MEANS algorithm (where DS stands for Distributed Streams), a combination of various known techniques to solve this problem, without the need of previous knowledge about the number of clusters. DS-MEANS first partitions – at each node – data into chunks; then a distributed version of K-MEANS is applied on these chunks. Each time the distributed K-MEANS algorithm is executed, it returns K centroids (points equal to the average of a cluster) that are used, together with centroids obtained from previous executions, as a compact representation of the entire set of streams. Each node then locally runs X-MEANS, a clustering algorithm able to choose the number of clusters, on this representation. An aggregation protocol is executed by all nodes to reach an agreement on the number G of clusters; finally, each node runs a centralized instance of K-MEANS with $K = G$ to get the final clustering.

This approach has two important properties. Given that DS-MEANS works independently on each chunk of data, we can use it in an online setting (in which we want to look only at the last chunks of data) without the need of restarting it from scratch each time new data arrives. Also, the nodes do not directly exchange information about the single points they are trying to cluster, thus achieving a reasonable level of privacy in the presence of sensitive data.

The paper shows the good behavior of the protocol using synthetic datasets. DS-MEANS is able to reach high precision even when it starts without knowing the exact number of clusters; the amount of computation is similar to the centralized version; and finally, DS-MEANS is fault-tolerant and reaches the same level of precision with or without failures (although the computation time may increase in the presence of failures).

2   Problem Statement

We consider a distributed network of N computing nodes $p_1, \ldots, p_N$, each of them independently receiving a (possibly unbounded) stream of data items. Although the amount of points each node receives may be different, data items are homogeneous: each of them is a *point* in the same d-dimensional space $\mathbf{R}^d$, taken from a single distribution. Similarity of two points p, q is measured based on their Euclidean distance $d(p, q)$ in $\mathbf{R}^d$: the smaller their distance, the more similar two points are. We use $\mathcal{P}$ to denote the set of points received by all nodes.

Nodes may fail by crashing, meaning that they stop executing the protocol; we do not consider malicious failures, where nodes behave arbitrarily (for example spoofing other nodes with incorrect information about the points they receive).

In other words, nodes participating in the computation are trusted. Nodes are partially synchronized, e.g. through NTP.

Each node may reliably communicate with any other node; to claim a limited degree of privacy, we require that only aggregate information can be shared among nodes, without communicating individual points.

The output of our problem is a collection of G *centroids* $c_1, \ldots, c_G$, again taken from $\mathbf{R}^d$, such that (i) each centroid c_i represents a cluster $C_i \subseteq \mathcal{P}$; (ii) each point $p \in \mathcal{P}$ is assigned to the cluster represented by the closest centroid, denoted $c(p)$; (iii) the average distance

$$\frac{1}{|\mathcal{P}|} \sum_{p \in \mathcal{P}} d(p, c(p))$$

between each point and the center of its cluster is minimized.

The value of G is not known in advance; instead, the correct value must be identified based on the *Bayesian information criterion* [10]. This measure uses the log-likelihood of the dataset according to the model and its number of parameters (in our case the number of centroids multiplied by the number of dimensions) to compute the following metric:

$$\mathrm{BIC}(j) = \log M_j(S) - \frac{1}{2} k_j \log n$$

In this formula j is a model, $M_j(S)$ is the maximal likelihood of dataset S using model j, k_j is the number of parameters of the model and $n = |S|$. To compute the maximal likelihood of the dataset we used the identical spherical Gaussian assumption (see [9] for the exact formulas).

3   Background

Clustering a data set is one of the classical problems in the area of machine learning. This section provides a brief review of the clustering algorithms that are used as building blocks for our own solution.

3.1   K-Means

The most commonly known clustering algorithm is K-MEANS, based on a paper by LLoyd [6]. K-MEANS works as follows: it starts from K centroids (points representing the center of a cluster), repeatedly assigns each point to the closest centroid and recomputes the position of the centroids according to the points assigned to it. This algorithm is guaranteed to converge to a local minima of the *within-cluster sum of squares* (i.e. the average squared distance between each point and the center of its cluster), but the quality is highly dependent on the choice of the starting centroids.

Another drawback is that K needs to be initialized to the correct value of G, the actual number of clusters present in the system. If K-MEANS is given the wrong K (because such value is not known in advance), the algorithm fails to answer precisely.

3.2   X-Means

One algorithm that is able to compute a good clustering even without knowing the number of clusters is X-MEANS [9]. This algorithm uses 2-Means (K-MEANS with $K = 2$) as a subroutine and continues to divide the data set in smaller subsets using the Bayesian information criterion to decide if (and where) does the data need more clusters.

The algorithm will then return the clustering which scores the highest value based on the Bayesian information criterion. We use this algorithm mostly as a subroutine to compute the number of clusters to be created.

3.3   Distributed K-Means

The main building block of our algorithm is the distributed K-MEANS algorithm presented by Bandyopadhyay et al. [1], created to cluster datasets in P2P environments. The algorithm works as follows: each iteration of the K-MEANS algorithm is executed at the same time by all nodes in the network. Each node starts with the same centroids, updates them using its own data and computes an average between its own centroids and the centroids computed by some of its neighbors. This process is repeated until a steady state is reached.

Looking more closely to the algorithm, it appears clearly that it is based on a simple relaxation of the averaging step of K-MEANS. All nodes start from the same centroids and are able to map each of its point to the closest of them. In the centralized version of K-MEANS, we select all the points that have been mapped to a single centroid and compute their mean to get the new position of that centroid. In this distributed algorithm, each node computes the position of each centroid as the mean of the points in the neighborhood that have been mapped to it.

An interesting property to be emphasized is the fact that each node only communicates with the others using aggregate data. In the communication rounds the nodes only send their centroids to some of their neighbors. The actual points are not shared in the network and each node can avoid sharing information with the other nodes. This is highly desirable when the data is distributed between nodes representing different companies and they want to collaborate to get a good clustering without having to directly share their points with the competition.

4   The Algorithm

The original distributed K-MEANS algorithm works on static datasets and needs to know the number of clusters that better represent the data distribution in advance, or it will fail to produce a good clustering [1]. We use this algorithm as a subroutine and we create a novel distributed framework to discover the correct number of clusters in data streams.

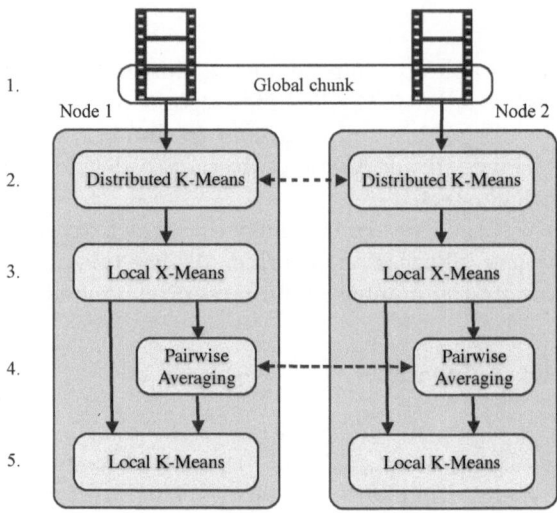

Fig. 1. Overview of DS-MEANS

We provide here an overall view of the algorithm, while implementation details are provided in the following subsections. The algorithm is divided in the following five steps, numbered (1)–(5), illustrated in Figure 1 and whose pseudo-code is outlined in Figure 2:

1. Each node p_i receives a stream of data points p, which are collected in variable C. The execution is divided into approximately synchronized *epochs* of duration Δ; at the beginning of each epoch t, a *chunk* of data denoted containing the points collected in the last Δ time units is stored in variable $chunk^i(t)$ and variable C is emptied.

2. The set of chunks, one for each node, that have been created at epoch t is called *global chunk* and is denoted $chunk(t)$:

$$chunk(t) = \{chunk^1(t), chunk^2(t), \ldots, chunk^N(t)\}$$

   The nodes execute an independent instance of the distributed K-MEANS algorithm on each global chunk $chunk(t)$, using an arbitrary value of K. Each node p_i generates a collection of centroids $centroids^i(t)$ of size K. The new centroids are added to the set $centroids^i$, which contains all centroids generated so far by node i. The original data points are discarded.

3. Each node p_i execute the centralized version of X-MEANS on $centroids^i$ to obtain the number G^i of centroids that better represent $centroids^i$.

4. The resulting number of centroids could be different at different nodes, so a pairwise averaging algorithm [5] is run on such values. All nodes will obtain the same average number $G = 1/N \sum_{i=1}^{N} G_i$.

5. Finally, each node runs a local version of K-MEANS on $centroids^i$, using K equal to the number of clusters resulting from the previous step ($K = G$). The resulting set $centroids^i_G$ is the output of the algorithm.

on receive p

(1) $\lfloor \quad C \leftarrow C \cup \{p\}$

repeat every Δ **time units**

        $t \leftarrow t + 1$

(1) $chunk^i(t) \leftarrow C$

        $C \leftarrow \emptyset$

(2) $centroids^i(t) \leftarrow \mathsf{DistKMeans}(chunk^i(t), K)$

        $centroids^i \leftarrow centroids^i \cup centroids^i(t)$

(3) $G^i \leftarrow \mathsf{XMeans}(centroids^i)$

(4) $K \leftarrow \mathsf{distributedAverage}(G^i)$

(5) $centroids^i_G \leftarrow \mathsf{KMeans}(centroids^i, K)$

Fig. 2. Algorithm executed by node p_i

DS-MEANS can be transformed into a dynamic algorithm that quickly adapts to new data arriving in the system. Assuming that the last estimate of the number of clusters is K_{old}, the framework will react as follows:

- All nodes in the network will start an instance of the distributed K-MEANS algorithm only on the new global chunk, using $K = K_{old}$.
- When the distributed K-MEANS algorithm has terminated, all nodes update their list of centroids by adding the new ones. In a "sliding window" model, they also discard the centroids generated w epochs ago, where w is the size of the sliding window (in number of epochs):

$$centroids^i \leftarrow centroids^i \cup centroids^i(t) - centroids^i(t-w)$$

- Each node will run the local X-MEANS algorithm giving it an upper bound on the number of clusters based on K_{old}.
- The pairwise aggregation step and the local K-MEANS step is repeated as usual, thus obtaining a new estimate of the number of clusters and a new clustering.

The choice to partition data into (global) chunks helps us in avoiding repeating unnecessary computations in the distributed K-MEANS step. Reusing the old number of clusters as K helps in making even less important the value of K given to DS-MEANS at startup.

Note that the nodes communicate only during two steps: for distributed K-MEANS and for pairwise averaging. During the former, only centroids are sent over the network, while in the latter the only shared value is the number of centroids. Actual points are never shared across the network, giving us a sufficient (although limited) degree of privacy.

4.1 Dividing Data into Chunks

We derive the idea of dividing data into chunks and working on each of them separately by the work of Guha et al. [4]. The easiest way to divide the data is based on the timestamp (creating a new chunk for each interval of time). Even if the nodes receive data at different rates and therefore create chunks of different size, the distributed K-MEANS algorithm makes sure that each node has similar results. We force each node to generate the same number of chunks, so that each global chunk will contain exactly one chunk from each node.

4.2   Distributed K-Means

The nodes in the network need to execute the distributed K-MEANS algorithm on each global chunk. We start an execution of the distributed K-MEANS algorithm any time a new global chunk arrives, meaning that each node has a new chunk to work on. Each execution is completely independent and chooses its own starting centroids. At the end of each execution, K new centroids are created (roughly the same in all nodes), which are added to the list of centroids computed so far. This list acts as a compact representation of all data in the network.

Each execution selects a set of starting centroids, common to all nodes, as follows: each node chooses randomly K centroids and a real number c in the range $[0, 1]$. A round-based epidemic protocol is executed to reconcile the different sets of centroids [2]. At each round, each node communicates with a random subset of neighbors and inherits both the centroids and c from the neighbor having the greatest value of c. Thanks to this epidemic protocol, $O(\log n)$ rounds are sufficient to have all nodes knowing the same set of starting centroids.

Another important part of this algorithm is the distributed termination condition. In the centralized K-MEANS, the algorithm terminates when there are no more updates to the centroids and a steady state has been reached. This approach is not viable in a decentralized setting due to the lack of a global view of the system, so we need to define a local condition to be checked by individual nodes. Each node keeps track of the last set of centroids it has generated and, at each iteration, checks if they have been changed by measuring the distance between the centroids and checking if the average change is more than a given threshold δ. A node terminates the execution and outputs its centroids when both the local centroids and the ones of the contacted neighbors have not changed in the last τ iterations.

4.3   X-Means

In this step, each node executes its own instance of X-MEANS on the list of centroids obtained from the previous step. Different runs of X-MEANS will result in different clustering and, in unlucky cases, in a number of clusters far from the correct answer. By running X-MEANS separately on each node in the network, we use redundancy to discard unlucky instances of X-MEANS.

Since we give as input to the X-MEANS algorithm a set of points $centroids^i$, each of them representing a subset of the original data points, we make each node compute the variance of each centroid in the list using its own data and we use this value as a "lower bound" on the cost of making a cluster containing that centroid.

4.4 Pairwise Averaging

The number of clusters identified by the local execution of X-MEANS can vary significantly between different nodes, while it would be desirable to obtain in all nodes a clustering with the same number of clusters. To obtain this we use a simple, epidemic-based pairwise averaging algorithm [5]. After $O(\log N)$ epidemic rounds all nodes know the average number of clusters computed by the different instances of X-MEANS.

4.5 Local K-Means

When the pairwise averaging step is over, all nodes have the same estimate of the number of clusters in the underlying data. Now each node can run the centralized K-MEANS algorithm on the local list of centroids using the value K computed in the previous steps. We can reuse some of the centroids found by the local X-MEANS as the starting centroids for the local K-MEANS, since they work on the same dataset $centroids^i$.

A more efficient implementation of this algorithm could share additional information between the two algorithms. If we wanted to compute some of the data structures commonly used to speed up clustering algorithms (like KD-Trees [8]), we could reuse them in both algorithm, thus avoiding unnecessary computations.

5 Evaluation

DS-MEANS has been tested on PeerSim [7], a P2P simulator. In this section, first the experimental framework used to evaluate DS-MEANS is presented, then the results are shown.

5.1 Experimental Framework

The input of the evaluation is a list of data points, each labeled with the correct group to which they belong. The output is a similar list, with points labeled by the clusters discovered through DS-MEANS.

We compare DS-MEANS against the centralized version of K-MEANS (instructed with the correct value of K) and the centralized version of X-MEANS. Four figures of merit are considered: *Clustering quality* is measured through the F-score, the harmonic mean of precision and recall, and through the within-cluster sum of squares. *Execution time* is measured in number of communication iterations in the simulation. Note that comparing execution time of centralized and decentralized algorithms is hardly significant, because the latter strongly depends on the underlying network. The fourth figure of merit is thus *total computational work*, measured as the number of times that the distance function (used to compare two data points) is called across the entire execution.

An artificial dataset is generated as follows. First of all, G different *mean points* are randomly chosen from a d-dimensional space, one for each of the groups in which data points are correctly subdivided. Then, each point in the dataset is created in two steps: (i) one of the groups is selected uniformly at random, and the point is labeled with it; (ii) the actual point coordinates are generated following a standard Gaussian distribution (in each of the d dimensions) centered in the corresponding mean point.

A couple of observations are in order:

- Given that mean points are independently generated, groups may overlap, inducing errors when a clustering algorithm is applied. All potential clustering algorithms are affected in the same way by this problem, so our comparison against centralized algorithms is fair.
- Data points are equally divided among nodes, and then divided in chunks. We do not ensure that a chunk will contain data from all of the original groups; this most likely occurs when the number of points in a chunk is small.

Table 1. Simulation parameters

Parameter	Symbol	Value
Network size	N	100
Groups	G	20
Data dimensionality	d	2
Threshold iterations	τ	3
Threshold variance	δ	0.5
Size of the space	M	100

Each experiment is repeated 20 times; variance is so small that it is not shown in the figures. Unless explicitly stated otherwise, our results have been obtained with the parameters listed in Table 1.

5.2   Experimental Results

Figure 3(a) compares the F-score of DS-MEANS and the two centralized clustering algorithms. We can see that the centralized K-MEANS algorithm obtains a clustering of lower quality, even when it is given the correct value of G. This is caused by the usual problem of the presence of local minima, that are better avoided by X-MEANS. DS-MEANS obtains very good F-scores not only when the correct K is used to initialize to the algorithm, but it even obtains results comparable with X-MEANS with the wrong value of K.

Figure 3(b) shows the average within-cluster sum of squares distance of DS-MEANS. As before, the experiments show results equivalent to the centralized X-MEANS.

Another interesting property of DS-MEANS is the amount of computation (measured by the number of calls to the distance function) that is needed to complete the execution. Figure 4(a) shows that our clustering algorithm is always comparable to the centralized algorithm.

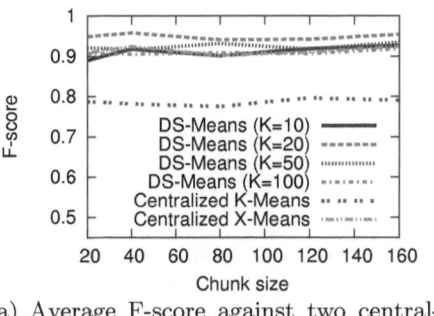

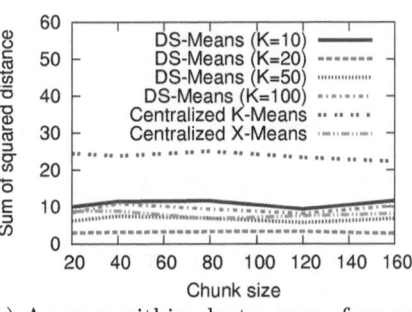

(a) Average F-score against two central- (b) Average within-cluster sum of squares
ized algorithms   against two centralized algorithms

Fig. 3. Precision of DS-MEANS using different values for K

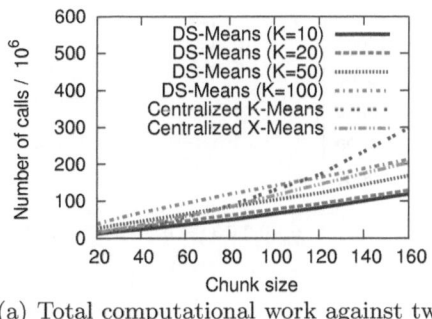

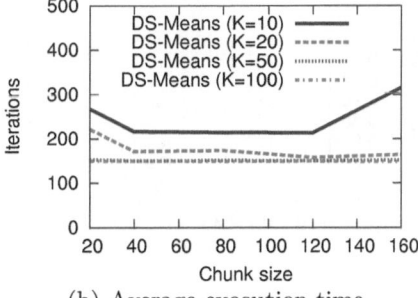

(a) Total computational work against two   (b) Average execution time
centralized algorithms

Fig. 4. Computational work and execution time of DS-MEANS using different values
for K

Figure 4(b) shows that, even if the amount of computation is larger, the
execution time (in which we only take into account the communication costs)
seems to become lower when the algorithm is given a bigger K.

While we have seen that the precision of DS-MEANS is high even when given
the incorrect number of clusters, it could be interesting to see what is the actual
number of clusters obtained from it. Figure 5(a) shows the number of clusters
found by DS-MEANS when given different values of K, when $G = 20$ different
groups have been created. The results are quite interesting: the algorithm is
closer to the correct answer when it is given a value of K bigger than necessary,
while when it is given the correct K the number of clusters it creates is bigger than necessary. This behavior is easily explained: giving a bigger K to the
distributed K-MEANS algorithm results in a bigger number of centroids generated and thus a better representation of the data. When K is small then it is
more likely that in the distributed K-MEANS algorithm more than one cluster is
mapped to a single centroid, thus creating a new point in the list of centroids
that does not correspond to any of the 20 original groups. The reason the algo-

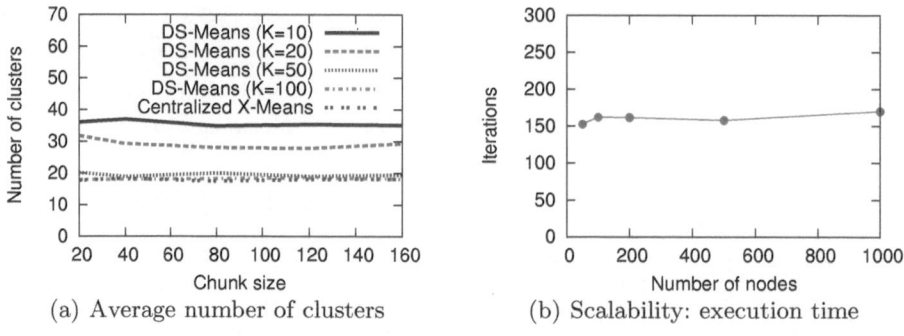

(a) Average number of clusters   (b) Scalability: execution time

Fig. 5. Average number of clusters and scalability given $G = 20$ groups and an initial value $K = 40$

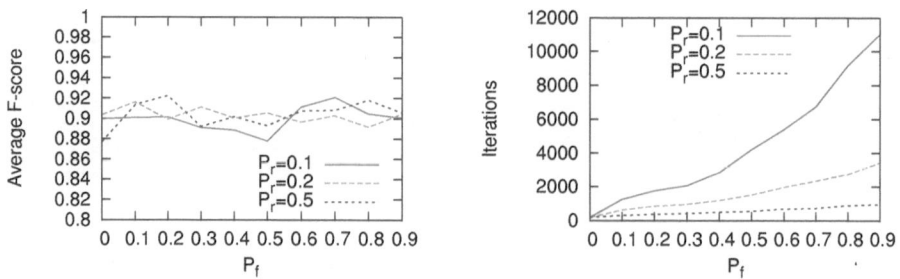

Fig. 6. Robustness of DS-MEANS with different recovery probabilities

rithm is still very precise is that these centroids, being quite far from the real distributions, will not be used when mapping the data of the entire network to the clusters.

As the next step we want to see the behavior of our distributed clustering algorithm when the network grows in size. In Figure 5(b) we see the execution time of our clustering algorithm against the number of nodes in the network. We see that the differences are very small and that we need a similar amount of communication iterations to complete the algorithm. This fact was expected since the nodes in the network only look and communicate to their local neighbors and the data is distributed homogeneously in the network.

Finally, an important aspect to be analyzed is the robustness of our approach. In our model, nodes may go down for a period of time before getting repaired and rejoining the distributed algorithm. Nodes that are down do not communicate with the rest of the network and do not work on their data. They are still able to receive a new chunk and store it until the node goes up. This model has been chosen to make sure that each node has the data on which it has to work.

Our model for robustness uses two parameters: the probability of failure (P_f) and the probability of recovery (P_r). All nodes start the computation in the up state and then, at each iteration of the simulation, each node in the up state goes down with probability P_f, while each node in the down state goes up with probability P_r.

In Figure 6(a) the F-score of our algorithm with different values of these two parameters is shown. In this simulation, while $G = 20$, the algorithm has been initialized with $K = 40$. Even with very high failure percentage and low recovery rate the algorithm is able to reach roughly the same F-score. Note that in the case in which the probability of recovery is 0.1 and the probability of failure is 0.9, on average 90% of the nodes in the network are down at all times. The algorithm is still able to reach a good clustering even in these extreme conditions, at the price of an increase in the number of communication iterations (see Figure 6(b)).

6   Conclusion

In this paper, we have presented DS-MEANS, a novel algorithm for clustering distributed data streams. Preliminary results with synthetic datasets are promising; we are now evaluating the system with realistic data for botnet detection.

References

1. Bandyopadhyay, S., Giannella, C., Maulik, U., Kargupta, H., Liu, K., Datta, S.: Clustering distributed data streams in peer-to-peer environments. Information Sciences 176(14), 1952–1985 (2006)
2. Demers, A., Greene, D., Hauser, C., Irish, W., Larson, J., Shenker, S., Sturgis, H., Swinehart, D., Terry, D.: Epidemic algorithms for replicated database maintenance. In: Proceedings of the Sixth Annual ACM Symposium on Principles of Distributed Computing, pp. 1–12. ACM (1987)
3. Gu, G., Perdisci, R., Zhang, J., Lee, W.: BotMiner: Clustering analysis of network traffic for protocol-and structure-independent botnet detection. In: Proc. of the 17th USENIX Security Conference, pp. 139–154 (2008)
4. Guha, S., Meyerson, A., Mishra, N., Motwani, R., O'Callaghan, L.: Clustering data streams: Theory and practice. IEEE Transactions on Knowledge and Data Engineering, 515–528 (2003)
5. Jelasity, M., Montresor, A., Babaoglu, O.: Gossip-based aggregation in large dynamic networks. ACM Transactions on Computer Systems (TOCS) 23(3), 219–252 (2005)
6. Lloyd, S.: Least squares quantization in PCM. IEEE Transactions on Information Theory 28(2), 129–137 (1982)
7. Montresor, A., Jelasity, M.: PeerSim: A scalable P2P simulator. In: Proc. of the 9th Int. Conference on Peer-to-Peer (P2P 2009), Seattle, WA, pp. 99–100 (September 2009)
8. Pelleg, D., Moore, A.: Accelerating exact k-means algorithms with geometric reasoning. In: Proc. of the 5th Int. Conference on Knowledge Discovery and Data Mining (KDD 1999), pp. 277–281. ACM, San Diego (1999)
9. Pelleg, D., Moore, A.W.: X-means: Extending k-means with efficient estimation of the number of clusters. In: Proc. of the 17th Int. Conference on Machine Learning (ICML 2000), pp. 727–734. Morgan Kaufmann, San Francisco (2000)
10. Schwarz, G.: Estimating the dimension of a model. The Annals of Statistics 6(2), 461–464 (1978)

3D Inverted Index
with Cache Sharing for Web Search Engines

Esteban Feuerstein[1], Veronica Gil-Costa[2,5], Mauricio Marin[4,5],
Gabriel Tolosa[3,1], and Ricardo Baeza-Yates[5]

[1] Universidad Nacional de Buenos Aires, Argentina
[2] Universidad Nacional de San Luis, Argentina
[3] Universidad Nacional de Lujan, Argentina
[4] Universidad de Santiago de Chile
[5] Yahoo! Labs Santiago, Chile

Abstract. Web search engines achieve efficient performance by partitioning and replicating the indexing data structure used to support query processing. Current practice simply partitions and replicates the text collection on the set of cluster processors and then constructs in each processor an index data structure. This paper proposes a different approach by constructing an index data structure that properly considers the fact that data is partitioned and replicated. This leads to a so-called 3D indexing strategy that outperforms current approaches. Performance is further boosted by introducing an application caching scheme devised to hold most frequently issued queries.

1   Introduction

The importance of text retrieval systems for the Web has grown dramatically during recent years due to the very rapid increase of available storage capacity. To ease the search of information, Web search engines (WSEs) are systems that index a portion of documents from the whole Web and allow to locate information through the formulation of queries. Well-known WSEs use the inverted index (or inverted file) [1] to speed up determination of relevant documents for queries.

The inverted index [1,10] is composed of a *vocabulary table* (which contains the V distinct relevant terms found in the document collection) and a set of *posting lists*. The posting list for term $c \in V$ stores the identifiers of the documents that contain the term c, along with additional data used for ranking purposes. To solve a query, one must fetch the posting lists for the query terms, compute the intersection among them, and then compute the ranking of the resulting intersection set using algorithms like BM25 or WAND [2]. Hence, the precomputed data stored in the inverted index enables the very fast computation of the top-k relevant documents for a query.

Usually these systems are expected to hold the whole index in the distributed main memory held by the processors. There are two basic types of organization for distributed inverted indexes, namely document-based partition and term-based partitions (or local and global indexes respectively). In the former, documents are evenly distributed onto P processors and an independent inverted

C. Kaklamanis et al. (Eds.): Euro-Par 2012, LNCS 7484, pp. 272–284, 2012.

index is constructed for each of the P sets of documents. Therefore answering a conjunctive query requires computing at each processor the intersection of the terms that form the query, obtaining partial results by performing the ranking to select the top-k local documents and then merging all those results.

In the term-partitioned index, a single inverted file is constructed from the whole document collection to then distribute evenly the terms with their respective posting lists onto the processors. In that way, to answer a conjunctive query one needs to determine which processor(s) hold the posting lists of the involved terms, then gather those lists in one processor and compute their intersection. Afterwards the ranking is performed over the resulting intersection set.

In our previous work [5], a 2-dimensional (2D) index was introduced. The 2D index combines document- and term-partitioning to get the "best of two worlds", i.e. to exploit the trade-off between the overhead due to the involvement of all processors in each query process as in the former, and the high communication costs required by the latter. In this paper we propose to extend the 2D index by adding processor replication, which has been widely used in conjunction with the document- and term-based approaches to improve throughput and fault tolerance [7].

We investigate the characteristics, performance and scalability of our extended 2D index called 3D index through the application of a performance evaluation framework (already used in [4]) that combines the usage of input query logs, average-cost analysis and stochastic values to compute realistic predictions of performance. The framework, based on the bulk-synchronous model of parallel computing (BSP) [12], blends empiric analysis and theoretical tools to predict the cost of real system executions which, otherwise, would be impossible to obtain in practice given the large scale of the cluster resources that would be required for experimentation.

We also improve the 3D index by adding a cache-sharing mechanism to save communication and computation overhead at the expense of an extra cost required to synchronize more processors. This new caching mechanism is also usable in simpler architectures as for example a term-partitioned index supporting replicas. The new caching mechanism constitutes an improvement to what has been proposed by Moffat et al. in [10].

The reminder of this paper is organized as follows. Section 2 presents the basic ideas behind the 3D distributed index. In Section 3 we introduce our cost estimation framework. In Section 4 we present the experimental setting. Sections 5 and 6 are devoted to show details and results of our experiments. Sections 7 and 8 respectively present related work and conclusions.

2   3D Index

The architecture we are proposing consists of the arrangement of a set of P processors in a 3D cube, formed by replicating D times each processor of a basic 2-dimensional matrix of R rows and C columns. On the row dimension, we apply the term-partition approach. On the column dimension we apply a document-partition approach. In other words, the document collection is divided into C

sub-collections, each of which is allocated into a "column" of R processors. Each processor holding a sub-collection C_i with terms belonging to R_j is replicated D times (See Figure 1).

The 2D index can be replicated in different ways. Each one may imply different search algorithms, sharing resources policies, synchronization and fault tolerance approaches, among other features. For example, by simply replicating D times the 2D index may improve the throughput roughly in a linear way as a function of the number of replicas, because the D sets of processors form D independent 2D index. Then each single 2D index replica processes a different set of queries. No communication is required between replicas, only among processors within the same replica. But it has some important drawbacks. The first one regards fault tolerance: the failure of a processor would imply the unavailability of a whole 2D index portion for a given set of queries.

The second issue regards the workload and may be explained as follows. For a fixed document collection and processor capacity, there is a minimum number of processors needed to handle the whole index on main memory. By simply maximizing the number of replicas would keep all the processors RAMs with the same load. No free memory is available for storing other cache structures. Meanwhile reducing the number of replicas and expanding the index over the C and R dimensions would allow to free RAM space in the processors that could be used for caching purposes.

A second approach consists on replicating D times each individual processor holding a portion of the 2D index. In this case queries are solved by processors of different replicas. Therefore there is communication among replicas. This approach may allow a more flexible replacement policy in case of a failure of a processor. It also allows to share the caches memories among the processors in a more intelligent way at the expense of additional synchronization costs. We explore the trade-off between these costs in Section 6.

In Figure 1 we present a schematic view of our architecture and the query flow. We propose a setting in which all the processors have a dual role of query brokers and search nodes. Each query will be introduced to the system through one of

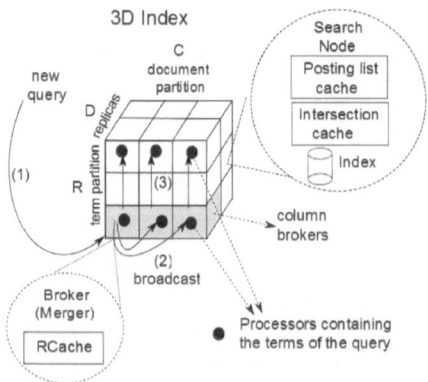

Fig. 1. 3D Index architecture

the processors that will act as its broker. That processor will then distribute the query by sending it to a random processor of each column of the same replica, the column brokers (step 2 in Figure 1).

In each column the terms may be co-resident at the same processor (row) or not. Each column broker determines which rows contain the terms involved in the query, and sends the query to the processors in each of those rows (in that same replica). Afterwards, they perform the intersection and ranking of the posting list. Each column broker that has computed and ranked the intersection sends its results to the broker. The broker merges the partial results.

As a broker, each processor keeps a cache called RCache [6]. The RCache keeps the answers for frequent queries (they are query results composed of document IDs). Each processor, as a search node, keeps the inverted index in secondary memory. Also each processor holds a posting lists cache for most frequent terms and an intersection cache for previously computed intersections. The latter will be used as a resource shared by a subset of the processors using the approaches explored in Section 6.

3   Cost Estimation Methodology

Our cost estimation framework is based on the bulk-synchronous model of parallel computing (BSP) [12], where computation is organized as a sequence of supersteps. In each of superstep the processors may perform computations on local data and/or send messages to other processors. The messages are available for processing at their destinations by the next superstep, and each superstep ends with the synchronization of the processors.

The total running cost of a BSP program is the cumulative sum of the costs of its supersteps, and the cost of each superstep is the sum of three components: computation, communication and synchronization. Computation cost is given by the maximum computation cost of a processor in the superstep, including also disk accesses. Communication cost is given by the maximum number of word-length messages sent or received by a processor during the superstep multiplied by the cost of communicating one word.

Besides, we note that just for participating in the processing of a query a processor incurs in certain overheads which must also be taken into consideration such as disk access, thread scheduling, etc. We model computation, communication and disk access overheads separately. We also separate computation cost taking into account partial costs of performing intersection, ranking and merging operations. Therefore, the total cost of a superstep is defined as:

$$\max_{p}(\max(computation + communication), disk) + Sync \qquad (1)$$

where

- p ranges over all participating processors
- $computation = processing\ overhead + intersection\ cost + ranking\ cost + merge\ cost$
- $communication = communication\ overhead + communication\ cost$
- $disk = disk\ overhead + disk\ access\ cost$

The overheads of computation, communication and disk costs of formula 1 are computed as the total number of operations (of each kind) performed in the processor. These values are then multiplied by appropriate constants corresponding to each operation. All these constants were obtained through benchmark programs run on actual hardware. Table 1 summarizes the primitive components (computation and communication costs) assigned to each query.

In our context, communication cost is by far smaller than computational cost since the sum of costly operations requiered to solve queries is much larger than the cost of sending small messages of size $|q|$ or size K among processors. Selecting the top-k results for a query requires operating on posting lists of size $x \gg K$.

Table 1. Primitive operations and concepts used in the simulation, and their values

Notation	Meaning	Cost
$t_i(x,y)$	Time required to compute the intersection of two lists of lengths x and y.	$t_i(x,y) = \min(x \log y, x+y)/6$
$t_m(x)$	Time required to merge a set of lists of total length x.	$x/6$
$t_r(x)$	Time to select the top-k results in a list of x items.	x
$I(x,y)$	Intersection length of two lists of length $x \leq y$ and constant values s and a.	$s*x*y/10^4+(s+a)*x$ according to [4]
processing overhead	Overhead due to the participation of a processor in a query.	0.1 nanoseconds
processing cost	Variable processing cost.	1 nanosecond
communication overhead	Overhead for transmitting any number of bytes	50000 nanoseconds
communication cost	Variable communication cost	50 nanoseconds (5)
disk overhead	Fixed cost for accessing disk	700 nanoseconds
disk access	Variable disk cost (per byte)	0.4 nanoseconds (0.2)

4 Experimental Setting

In this section we describe the general framework used for our experiments. We estimated the cost of different 3D index configuration by combining the variables P, C, R and D, where P is the total number of processors. We run 300,000 queries with P ranging from 32 to 2048. The number of replicas ranges from $D = 2$ to 64, and the number of columns and rows range from 1 to $P/2$ with $P = C \times R \times D$.

Queries were selected from a query log with 36,389,567 queries. We preprocessed the query log following the rules applied in [6] by removing stopwords and completely removing any query consisting only of stopwords. The resulting query trace has 16,900,873 queries. Then, we randomly selected about one million queries where 42% are unique queries, 14% occur twice and 6% occur three times. As expected, the queries' lengths are not uniformly distributed. The relative frequencies of one, two, three and four-term queries is respectively 24%,

33%, 23% and 11%. The basic index and the posting lists were built using a 1.5TB sample of the UK's web obtained from Yahoo! Search engine.

To set the values of the different primitive operation costs and constants required by our cost estimation framework, we used the same results of the benchmark programs reported in [4]. The values are expressed relative to a baseline in terms of ranking time defined as $t_r(x) = x$ (See Table 1).

To determine the communication and disk access costs we run specific benchmark programs. In both cases we used standard available technology (serial ATA disks, and a Gigabit Ethernet). Hence, both estimations can be seen as lower bounds for the real performance on a specialized and fine-tuned architecture. Benchmark programs where run on a Linux operating system with MPI.

To compute disk access time we considered the positioning of the read/write head (lseek) in response to a sequence of a random sequence of positioning depending on the length of the stream of bytes, and the time employed to read a block. We used blocks of different sizes (1 to 3000 bytes). We run six different tests combining reset and update operations. The results show that the cost of accessing x bytes can be approximated with the function $0.4x + 700$ nanoseconds.

We run benchmark programs to obtain the communication time between pairs of processors with different message sizes (in a range of 1-3000 bytes). We considered messages in the unicast and broadcast modes. We observed that the communication cost grows with the number of transmitted bytes. This cost can be approximated with the expression $50x + 50000$ nanoseconds. From these benchmark programs we established the values reported in Table 1.

5   Evaluation of the 3D Index

In this section we report the experiments performed to analyze the efficiency of our 3D index. We use a basic architecture as described in Section 2. All queries are solved within one replica of the 2D index, so no synchronization is needed among all processors at the end of every superstep, but just among those participating in the same replica. Figure 2 at left shows this approach. The black ball represents processors involved in the query process. No communication between processors of different replicas is required.

The size of cache memories used in the following experiments is reported in Section 6. We also show cost normalized to 1 to better illustrate the difference between strategies. In all cases we divide the values by the observed maximum in the respective experiment.

Figure 3 at left shows that if we allow replicas to proceed without synchronization, the best choice is to have the maximum possible number of replicas. That is natural as the whole stream of queries can be divided evenly among all replicas. In this way we obtain an optimal speed-up.

Figure 3 at right shows in the y-axis the optimal number of rows and columns selected for different values of replicas D in the x-axis. Namely, the configurations $R \times C$ with the minimum estimated cost. We observed that those configurations with many replicas (where each replica is as small as possible) it is

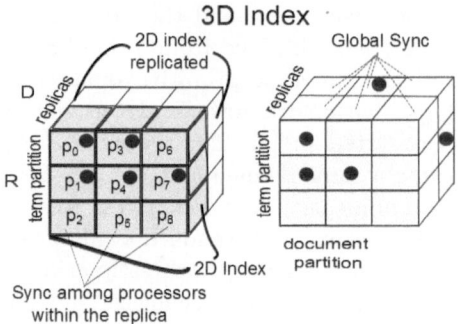

Fig. 2. Left: queries solved within a single 2D replica. Right: queries solved among all processors.

better to arrange the processors with $R = 1$ and C columns, i.e. a local index. However, configurations with a small D (and hence bigger $R \times C$) present the typical 2D index trade-off between communication and overhead costs presented among the term-partition and document-partition indexes.

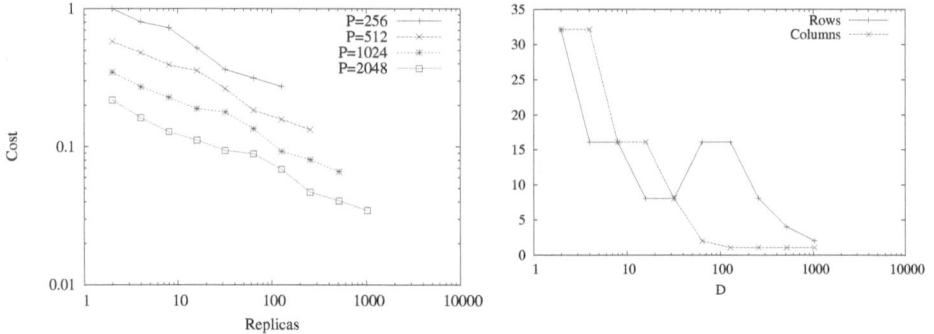

Fig. 3. Non synchronization among replicas. Left: Estimated cost. Right: Optimal number of columns C and rows R as a function of D with $P = 2048$.

Figure 4 at left shows results forcing synchronization among all P processors at the end of each superstep. In this approach queries are solved using processors belonging to different replicas. For a given query, replicas are selected in a round-robin fashion. As communication is performed among all processors a global synchronization is required. Figure 2 at right shows how a query q represented as a black ball is processed by processors belonging to different replicas.

Figure 4 at left shows the estimated cost as a function of the number of columns. The optimal value of D grows with P, but is far from tending to $D = P$. Moreover, the natural trade-off between the benefits of the global and local indexes makes that the optimal configuration can be found using a pure 2D configuration, i.e. using between 8 and 16 columns.

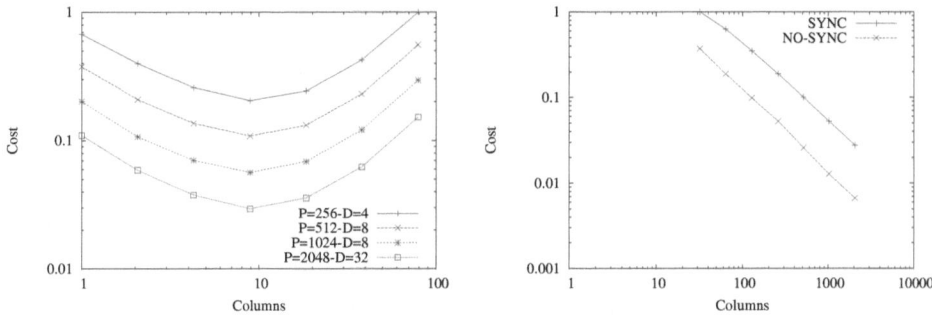

Fig. 4. Left: All processors are synchronized at the end of every superstep. Right: replica synchronization vs. global synchronization.

Figure 4 at right compares the estimated cost for different values of P when synchronization is performed only among processors within a single replica (*NO-SYNC*) versus global synchronization (*SYNC*) involving all processors. With few processors, synchronizing only processors within a replica significantly reduces the total cost. But with $P \geq 500$, the difference between both approaches is less than 1%. Moreover, the SYNC approach does not present the same fault-tolerance and workload drawbacks as described in Section 2. We conclude that, without any cache sharing mechanism (as the one introduced in Section 6), the number of replicas must be as big as possible.

5.1 Scalability of the 3D Index

In this section we study the scalability of the 3D index as the size of the document collection increases. We estimate the total cost of computing 300,000 queries for $P = 512, 1024$ and 2048. We selected the best configurations for each P. Namely, the combination $P = R \times C \times D$ which minimizes the total cost. We used $R \times C \times D$ defined by: $R = [2\ldots 64]$, $C = [4\ldots 64]$ and $D = [4\ldots 16]$. The rest of the parameters were kept unchanged, as described in Section 4. We force the synchronization of all the processors.

Figure 5 at left shows the estimated cost in the y-axis. The x-axis values represent the growth factor of posting lists. The main conclusion of this experiment is that, for a fixed value of P, the estimated cost of the system grows proportionally to the logarithm of the size of the document collection. This behavior is expected due to the fact that the running time of list intersections is logarithmic. Interestingly Figure 5 at right shows that the speed-up is constant as we increase the number of processors. Sometimes it even tends to improve as the collection size grows.

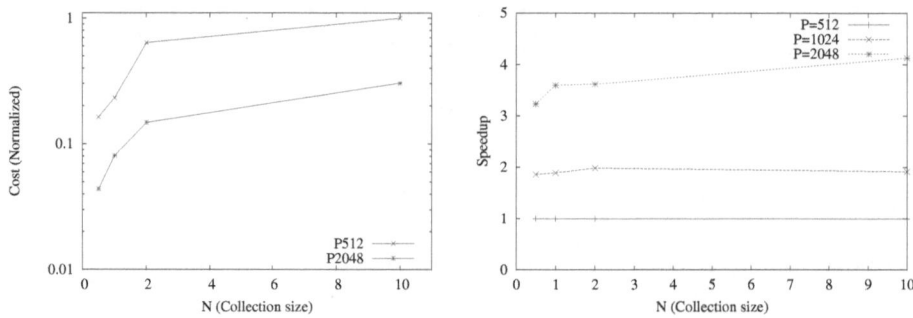

Fig. 5. Left: Estimated cost as the collection size varies, for different number of processors. Right:Speed-up as the collection size varies.

6   Pipelined Caching

In this section we propose a novel idea for managing the individual RAM memories of the processors as part of a big common cache. We propose to use a larger cache shared among processors. A different idea with the same goal was presented in [4] for a simpler index (local index with replication). The results in [4] showed that improvements in the performance can be obtained with some extra communication cost. We apply this idea to the intersection cache.

Our schema is applied to the set of processors forming a "column slice" of our 3D cube. Each column slice is a term-partitioned index with replicas for a particular sub-collection of documents (see Fig 6(a)–(c)). Within each slice the processors are partitioned in subsets (called *teams*). We use a hash function H to map each query term to one processor of the team. Different ways of constructing the teams determine different algorithms. We will explore three particular cases:

(a) One Row-many Replicas teams: Each slice is partitioned by terms in R teams. Each team has D replicas (Figure 6(a)).
(b) One Replica - many Rows teams: Each slice is divided in D teams. Each team has R processors. (Figure 6(b)).
(c) Slice-teams: All the processors in each slice form one single team (Figure 6(c)).

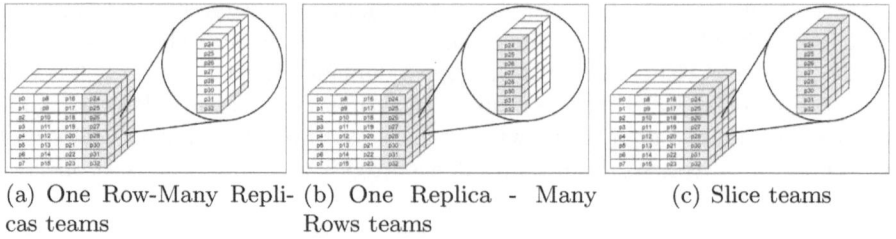

(a) One Row-Many Replicas teams

(b) One Replica - Many Rows teams

(c) Slice teams

Fig. 6. Pipelined caching

Algorithm 1 explains our caching policy. We explain the query process inside one team. Upon request of a query $t_1 \wedge t_2 \ldots \wedge t_m$, the broker directs the query to one random processor in each column of its replica. Each of these C processors will be referred to as a *column-broker*. The column-broker analyzes the query and interacts with the processors of its team to evaluate if there are, in the team's collective cache, any previously cached intersections that may be useful to compute the answer. What differentiates the three approaches presented above is the construction of the hashing function H, whose co-domain is always the team of the column-broker. For any set of terms $\{t_1 \ldots t_i\}$, $p_{H(t_1 \ldots t_i)}$ denotes the processor that, according to the hashing function H, could have in its cache the result of the intersection $t_1 \wedge t_2 \ldots \wedge t_i$.

The column-broker first looks for the whole or part of the query intersection (lines 1-4). If some intersection is found, the cached result is requested to the processor that holds it (lines 5-8). Otherwise an empty list is set (lines 9-11). This list ℓ will contain the processors identifiers that hold the remaining terms of the query (lines 13-16). If the whole query is found in cache, the retrieved list is returned (line 17). Each processor that computes an intersection caches the result and sends it to the following processor in ℓ. We use $O(t)$ to denote the processor that holds term t in any replica.

For example, for a four-term query $t_1 \wedge t_2 \wedge t_3 \wedge t_4$ the column-broker in each column looks for the intersection $t_1 \wedge t_2 \wedge t_3 \wedge t_4$ in a particular processor of the team $(p_{H(t_1 \ldots t_4)})$. If that fails, in a second step, it looks for $t_1 \wedge t_2 \wedge t_3$ on the corresponding processor $(p_{H(t_1 \ldots t_3)})$ and so on. When a hit is found the partial result of the intersection is retrieved from cache, and a schedule in ℓ is prepared to complete (by computation) the remaining intersections.

Algorithm 1. Pipelined Caching

Input: $t_1 \ldots t_m$ in increasing order of length of the posting list
1: $i \leftarrow m + 1$
2: **repeat**
3:     $i \leftarrow i - 1$
4: **until** $i = 1$ or $p_{H(t_1 \ldots t_i)}$ has $t_1 \cap \ldots \cap t_i$ in its intersection cache
5: **if** $i \neq 1$ **then**
6:     ask $p_{H(t_1 \ldots t_i)}$ the intersection $t_1 \cap \ldots \cap t_i$
7:     $\ell \leftarrow (t_1 \cap \ldots \cap t_i)$
8:     $j \leftarrow i + 1$
9: **else**
10:     $\ell \leftarrow \emptyset$
11:     $j \leftarrow 1$
12: **if** $j \leq m$ **then**
13:     Send the query $t_1 \wedge \ldots \wedge t_m$ along with ℓ to processor $p_{O(t_j)}$
14:     **while** $j < m$ **do**
15:         Processor $p_{O(t_j)}$ computes $\ell \leftarrow \ell \cap t_j$, caches ℓ in processor $p_{H(t_1 \ldots t_j)}$, makes $j \leftarrow j + 1$, and sends ℓ to $p_{O(t_j)}$
16:     Processor $p_{O(t_m)}$ computes $\ell \leftarrow \ell \cap t_m$, caches ℓ in processor $p_{H(t_1 \ldots t_m)}$, and sends ℓ to the column-broker
17: Return ℓ to the broker of the query

6.1   Results

In the following experiments we set the size of the intersection cache of each processor in terms of the number of "entries", i.e. a certain number of pre-computed intersections. We set a fixed size of $100K$ cache entries for each processor so the total size of the intersection cache is $100K \times P$ entries. This enables a natural "growing" property when adding processors to the whole system. The total space of the intersections cache in a particular team varies according to the combinations of $R \times C \times D$ and the chosen scheme. With the One Row-Many Replicas, the total size of the cache of each team is $100K \times D$. For One Replica -Many Rows it is $100K \times R$ and finally for the Slice scheme we have $100K \times R \times D$.

The RCache implemented at the broker side uses a SDC policy [3]. We made a "warm up" of the RCache with $200K$ query results. Finally, all the processors maintain caches of posting lists to reduce secondary memory access. This cache is administered with the standard LRU strategy.

In the baseline approach (2D approach) each processor cache is used and managed locally. Any information about previously computed intersections of posting lists is local to the owner of the cache memory.

Figure 7 at left shows the improvement obtained by the cache sharing model One Replica - Many Rows team. With $P = 512$, this new approach reduces the estimated cost by 40%. With the Slice approach the improvement is up to 55%.

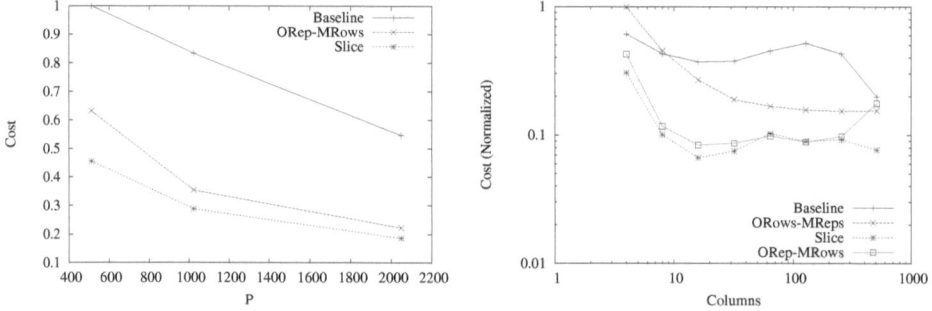

Fig. 7. Left: Normalized costs of the caching strategies. Right: Estimated cost as a function of the number of columns.

Figure 7 at right shows the estimated cost of the three cache sharing strategies and the baseline approach as a function of the number of columns for the best D value. Again the Slice approach presents very good results followed by the One Replica -Many Rows approach. The Slice allows an efficient use of the cache space since it allows to maximize the memory space shared between nodes of different replicas.

7   Related Work

The 2D index that combines document- and term-partitioning was originally proposed in [5]. Other variants of these basic schemes have been proposed in [11,13]. In this work, we use a cost estimation framework [4] based on BSP [12]. Recently the BSP model has been widely adopted by systems used in off-line computations for search engines (e.g., index construction) such as Google's Pregel [8]. In [4] the cost estimation framework was used for a simpler index (local index with replication) showing performance improvements due to the reduction in computation time and disk access. Moreover, the caches were shared by the set of replicas of each partition (in that work, a column). Within each partition all the computation was centralized at the broker of the query, which asked all its "replicas" for possible intermediate results in their caches. No "query routing" was considered there.

Also, our approach is related to the pipelined query-evaluation methodology of [10] in which partially evaluated queries are passed amongst the set of processors that host the query terms. In other words, it constructs a path or pipeline schedule through which the query "flows" to be solved. But in our case the path depends on the content of the shared cache. We combine this pipelined approach with an intersection cache distributed across the replicas of processors. The initial architecture of [10] suffered from load imbalance across the nodes of the cluster. That problem is partially solved in [9] through replication of popular terms in different processors.

8   Conclusions

In this work we have presented a 3D inveted index along a new caching algorithm for Web search engines. Replication is used as a mean to increase query throughput and to support failures. The 3D index efficiently handles replication and outperforms the 2D index which in turn outperforms well-known distributed indexes [5]. The number of rows, columns and replicas can be chosen so as to optimize the overall performance. We explored two approaches to replicate the index: SYNC and NO-SYNC. There is a trade-off. The NO-SYNC presents better performance but is less fault tolerant and viceversa.

The proposed caching algorithm is designed to make an optimized use of the distributed memory for caching posting list intersections. This significantly reduces communication of posting lists among nodes (processors). Intersections of popular terms in queries tend to remain cached, so the load of the holders of those terms is also reduced. Moreover, the hashing function of the caching algorithm can be fine-tunned to optimize a combination of hit ratio and load balance of nodes.

References

1. Baeza-Yates, R., Ribeiro-Neto, B.: Modern Information Retrieval - the concepts and technology behind search, 2nd edn. Pearson Education Ltd. (2011)
2. Broder, A.Z., Carmel, D., Herscovici, M., Soffer, A., Zien, J.Y.: Efficient query evaluation using a two-level retrieval process. In: CIKM, pp. 426–434 (2003)
3. Fagni, T., Perego, R., Silvestri, F., Orlando, S.: Boosting the performance of web search engines: Caching and prefetching query results by exploiting historical usage data. ACM Trans. Inf. Syst. 24, 51–78 (2006)
4. Feuerstein, E., Gil-Costa, V., Mizrahi, M., Marin, M.: Performance Evaluation of Improved Web Search Algorithms. In: Palma, J.M.L.M., Daydé, M., Marques, O., Lopes, J.C. (eds.) VECPAR 2010. LNCS, vol. 6449, pp. 236–250. Springer, Heidelberg (2011)
5. Feuerstein, E., Marin, M., Mizrahi, M., Gil-Costa, V., Baeza-Yates, R.: Two-Dimensional Distributed Inverted Files. In: Karlgren, J., Tarhio, J., Hyyrö, H. (eds.) SPIRE 2009. LNCS, vol. 5721, pp. 206–213. Springer, Heidelberg (2009)
6. Gan, Q., Suel, T.: Improved techniques for result caching in web search engines. In: WWW, pp. 431–440 (2009)
7. Gomez-Pantoja, C., Marin, M., Gil-Costa, V., Bonacic, C.: An Evaluation of Fault-Tolerant Query Processing for Web Search Engines. In: Jeannot, E., Namyst, R., Roman, J. (eds.) Euro-Par 2011, Part I. LNCS, vol. 6852, pp. 393–404. Springer, Heidelberg (2011)
8. Malewicz, G., Austern, M.H., Bik, A.J., Dehnert, J.C., Horn, I., Leiser, N., Czajkowski, G.: Pregel: a system for large-scale graph processing. In: PODC, p. 6 (2009)
9. Moffat, A., Webber, W., Zobel, J.: Load balancing for term-distributed parallel retrieval. In: SIGIR, pp. 348–355 (2006)
10. Moffat, A., Webber, W., Zobel, J., Baeza-Yates, R.: A pipelined architecture for distributed text query evaluation. Information Retrieval 10, 205–231 (2007)
11. Tang, C., Dwarkadas, S.: Hybrid global-local indexing for efficient peer-to-peer information retrieval. In: NSDI, p. 16 (2004)
12. Valiant, L.G.: A bridging model for multi-core computing. J. Comput. Syst. Sci. 77(1), 154–166 (2011)
13. Xi, W., Sornil, O., Luo, M., Fox, E.A.: Hybrid Partition Inverted Files: Experimental Validation. In: Agosti, M., Thanos, C. (eds.) ECDL 2002. LNCS, vol. 2458, pp. 422–431. Springer, Heidelberg (2002)

Quality-of-Service for Consistency
of Data Geo-replication in Cloud Computing*

Sérgio Esteves, João Silva, and Luís Veiga

Instituto Superior Técnico - UTL / INESC-ID Lisboa, GSD, Lisbon, Portugal
sesteves@gsd.inesc-id.pt, {joao.n.silva,luis.veiga}@inesc-id.pt

Abstract. Today we are increasingly more dependent on critical data stored in cloud data centers across the world. To deliver high-availability and augmented performance, different replication schemes are used to maintain consistency among replicas. With classical consistency models, performance is necessarily degraded, and thus most highly-scalable cloud data centers sacrifice to some extent consistency in exchange of lower latencies to end-users. More so, those cloud systems blindly allow stale data to exist for some constant period of time and disregard the semantics and importance data might have, which undoubtedly can be used to gear consistency more wisely, combining stronger and weaker levels of consistency. To tackle this inherent and well-studied trade-off between availability and consistency, we propose the use of VFC^3, a novel consistency model for replicated data across data centers with framework and library support to enforce increasing degrees of consistency for different types of data (based on their semantics). It targets cloud tabular data stores, offering rationalization of resources (especially bandwidth) and improvement of QoS (performance, latency and availability), by providing strong consistency where it matters most and relaxing on less critical classes or items of data.

1 Introduction

The great success Internet achieved during the last decade has brought along the proliferation of web applications which, with economies of scale (e.g., Google, Facebook, and Microsoft), can be served by thousands of computers in data centers to millions of users worldwide. These very dynamic applications need to achieve higher-scalability in order to provide high availability and performance. Such scalability is typically realized through the replication of data across several geographic locations (preferably close to the clients), reducing application server and database bottlenecks while also offering increased reliability and durability.

Highly-scalable cloud-like systems running around the world often comprise several levels of replication, specially among servers, clusters, inter-data centers, or even among cloud systems. More so, the advantages of geo-distributed

---

* This work was partially supported by national funds through FCT – Fundação para a Ciência e a Tecnologia, under projects PTDC/EIA-EIA/102250/2008, PTDC/EIA-EIA/108963/2008 and PEst-OE/EEI/LA0021/2011.

C. Kaklamanis et al. (Eds.): Euro-Par 2012, LNCS 7484, pp. 285–297, 2012.

micro-data centers over singular mega-data centers have been gaining significant attention (e.g., [1]), as, among other reasons, network latency is reduced to end-users and reliability is improved (e.g., in case of a fire or a natural catastrophe, or simply network outage). With this current trend that brings higher number of replicas, more and more data needs to be properly synchronized, carrying out the need of having smart schemes to manage consistency while not degrading performance.

In replication, the consistency among replicas of an object has been handled through both traditional pessimistic (lock-based) and optimistic approaches [2]. Pessimistic strategies provide better consistency but cause reduced performance, lack of availability, and do not scale well. Where optimistic approaches rely on eventual consistency, allowing some temporary divergence among the state of the replicas to favor availability and performance. Since it is not possible to fully have the best of both approaches in a distributed environment with arbitrary message loss, as stated in the CAP theorem [3], we envision that more can be done to approximate consistency from availability. One path not yet significantly explored, and that we intend to address in this work, consists of wisely and dynamically strengthen and weaken consistency in accordance to the importance of the data being replicated.

In the context of the web, replication has been extensively applied at the application and database tiers, significantly improving the performance and reducing workload. At the application level, popular in-memory caching solutions, such as memcached [4], defy developers in the following ways: they do not offer transactional consistency with the underlying database; and they offer only a key-value interface, and developers need to explicitly manage the cache consistency, namely invalidating cache data when the database changes. This management, which includes performing manual lookups and keeping the cache updated, has been a major source of programming errors.

At the database level, tuples are replicated and possibly partitioned across multiple network nodes, which can also execute queries on replicas. However, adding more database servers to a RDBMS is difficult, since the partition of database schemas, with many data dependencies and join operations, is non-trivial [5]. This is where non-relational NoSQL databases come to take place.

High-performance NoSQL data stores emerged as an appealing alternative to traditional relational databases, since they achieve higher performance, scalability, and elasticity. For example, Google has built its own NoSQL database, BigTable [6], which is used to store google entire web search system. Other solutions include Cassandra [7], Dynamo [8], PNUTS [9] and HBase [10] (which we focus on in this work).

To sum up, these several approaches to replication, existent in well-known cloud systems and components, usually treat all data at the same consistency degree and are blind w.r.t. the application and data semantics, which could and should be used to optimize performance, prioritize data, and drive consistency enforcement.

Given the current context, we propose the use of a novel consistency model with framework and programming library support that enables the definition and dynamic enforcement of multiple consistency degrees over different groups of data, within the same application, across very large scale networks of cloud data centers. This model is driven by the different semantics data might have; and the consistency levels can be automatically adjusted based on statistical information. Moreover, this framework, named VFC^3 (*Versatile Framework for Consistency in Cloud Computing*), comprises a distributed transactional in-memory cache and is intended to run on top of NoSQL high-performance databases. This way, we intend to improve QoS, rationalize resource usage (especially bandwidth), and deliver higher performance.

The remainder of this paper is organized as follows. Section 2 presents the architecture of the VFC^3 framework, and Section 3 its underlying consistency model. In Section 4, we offer details of relevant implementation aspects. Then, Section 5 presents a performance evaluation, and Section 6 reviews related work.

2   Geo-distributed Cloud Scenario and Architecture

The VFC^3 consistency model was specially designed to address very large-scale and dynamic environments (e.g., cloud computing) that need to synchronize large amounts of data (either application logic or control data) between several geographically dispersed points, while maintaining strong requirements about the quality of service and data provided. Figure 1a depicts a scenario of geo-replicated data centers where a mega data center needs to synchronize data with micro data centers scattered throughout different regions. Such micro centers replicate part of the central database, with only the more relevant data to a given corresponding region. In more detail, Figure 1b shows the constituent parts of the mega and micro data centers, where the VFC^3 middleware operates and how the interaction is carried out among data centers.

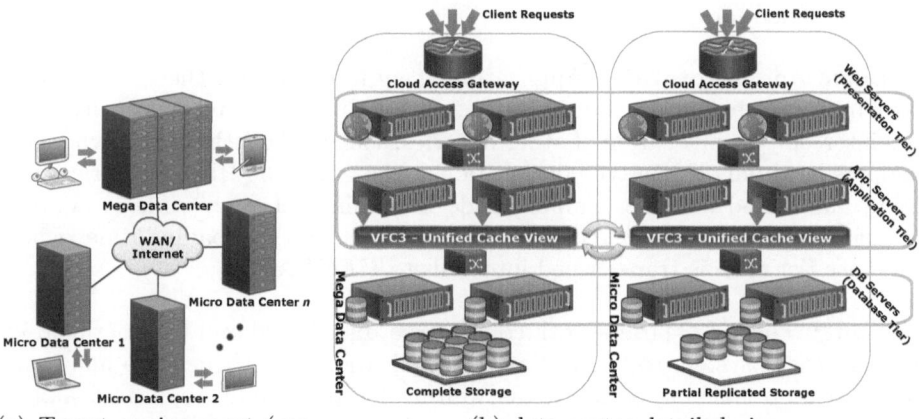

(a) Target environment (geo-distributed data centers)

(b) data center detailed view

Fig. 1. VFC^3 overall scenario and network architecture

We present an archetypal architecture (Figure 2) of the VFC^3 framework that is capable of enforcing different degrees of consistency (or conversely, bounding divergence) and runs atop very large-scale (peta-scale) databases residing in multiple data centers. The consistency constraints over the replicas are specified in accordance to the data semantics and they can be automatically adjusted at run time. Moreover, this framework comprises a distributed transactional in-memory cache system enhanced with a number of more components, described as follows.

Monitor and Control. This component analyses all requests directed to the database. It decides from where to retrieve results to queries, cache or database, and controls the workflow when an update occurs. It also collects statistics regarding access patterns to stored data in order to automatically adjust the divergence levels.

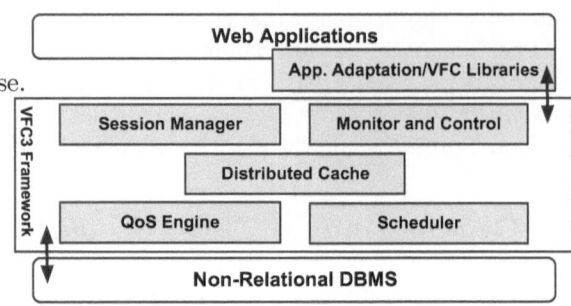

Fig. 2. VFC^3 middleware architecture

QoS Engine. It maintains data structures and control meta-data to decide when to replicate and synchronize data, obeying to consistency specifications.

Scheduler. This component verifies the time constraints over the data. When the time for data being replicated expires, the Scheduler notifies the QoS engine.

Distributed Cache. It represents an in-memory, transactional and distributed database cache for: i) temporary storing frequently accessed database items; and ii) keep tracking of which items need to be replicated.

Session Manager. It manages the configurations of the consistency constraints over the data through extended database schemas (automatically generated) defined for each application.

Application Adaptation: Applications may interact with the VFC^3 framework by explicitly invoking our libraries, but VFC^3 can also automatically adapt and intercept the invocation of other libraries, such as the HBase API, where developers need only change the declarations referencing the HBase libraries (also being automated), with remaining code unmodified. Legacy code is adapted by using annotations or pre-processor directives, during loading-time, where database tier code is transformed into calls to VFC^3 API.

Caching: The VFC^3 framework comprises a distributed and transactional in-memory cache system to be used at the application-level. It has two main purposes: i) keep tracking of the items waiting for being fully replicated and ii) temporarily store both frequently used database items and items within the same locality group (i.e., pre-fetch columns of an hot row), in order to improve read performance. Specifically, this cache stores partial database tables, with associated QoS constraints, that are very similar from the ones in the underlying database, but with tables containing many less rows. Moreover, the VFC^3 cache is completely transparent to applications; it guarantees transactional consistency with the underlying database and the data is automatically maintained and invalidated, relieving developers from a very error-prone task.

This cache can be spanned across multiple application servers within the same data center, so that it can grow in size and take advantage of the spare memory available in the commodity servers. Although the work is distributed, it still gives a logical view of a single cache. The partition of data follows an horizontal approach, meaning that rows are divided across servers and the hash of their identifiers works as key to locate the servers in which they should be stored. Hence, this cache is optimized for our target database systems, since rows constitute indexes in the multi-dimensional map and every query must contain a row identifier (apart from scans). Furthermore, each running instance of the VFC^3 cache knows all others running on neighbor nodes.

Replication: The VFC^3 framework handles the whole replication process asynchronously, supporting the two general used formats, statement-based and row/column-based. However, if the statement-based strategy is used, the element (row, column, or table), referenced in a query, with the more restrictive QoS constraints will command the replication of the statement, leading other data with less restrictive constraints to be treated at stronger consistency degrees (which can be useful for some use cases). When a maximum divergence bound, associated to an element (row, column, or table), is crossed, the changed items, and only these, within that element are broadcasted to the other interested servers. These modified items are identified through *dirty bits*.

The system constantly checks bandwidth usage of messages exchanged with other nodes. The framework can trigger data replication and synchronization on low bandwidth utilization periods, even if consistency constrains do not impose it. Replication messages, using the column-based strategy, contain the necessary indexes to identify items placement within the database map and are compressed with gzip.

3   Consistency Model

The VFC^3 consistency model is inspired on our previous work [11]. It defines three-dimensional consistency vectors (κ) that can be associated with data objects. κ bounds the maximum objects divergence; each dimension a numerical scalar defining the maximum divergence of the orthogonal constraints: time (θ), sequence (σ), value (ν).

Time: Specifies the maximum time a replica can be without being updated with is latest value. Considering $\theta(o)$ provides the time (e.g., seconds) passed since the last replica update of object o, constraint κ_θ, enforces that $\theta(o) < \kappa_\theta$ at any given time.

Sequence: Specifies the maximum number of updates that can be applied to an object without refreshing its replicas. Considering $\sigma(o)$ indicates the number of applied updates, this sequence constraint κ_σ enforces that $\sigma(o) < \kappa_\sigma$ at any given time.

Value: Specifies the maximum relative difference between replica contents or against a constant (e.g., top value). Considering $\nu(o)$ provides that difference (e.g., in percentage), this value constraint κ_ν enforces that $\nu(o) < \kappa_\nu$ at any given time. It captures the impact or importance of updates on the object's internal state.

Evaluation and Enforcement of Divergence Bounds: The evaluation of the divergence vectors σ and ν takes place every time an update request is received by the middleware. Upon such event, it is necessary to identify the affected tables/rows/columns, increment all of the associated vectors σ and verify if any σ or ν is reached when compared with the reference values (i.e., the maximum object allowed divergences). If any limit is exceeded, all updates since last replication are placed in a FIFO-like queue to be propagated and executed on other replicas. When there are multiple versions for the same mapping (table/row/column), the most recent ones are propagated first.

To evaluate the divergence vector, θ, VFC^3 uses timers (each node holding one timer per application) to check, e.g., every 1 second, if there is any object that should be synchronized (timestamp about to expire). Specifically, references to modified objects (identified by the dirty bits) are held in a list ordered ascending by time of expiration, which is the time of the last object synchronization plus θ. The Scheduler component, starting from the first element of the list, checks which objects need to be replicated. As the list is ordered, the Scheduler has only to fail one check to ignore the rest of the list; e.g., if the check on the first element fails (its timestamp has not expired yet), the Scheduler does not need to check the remaining elements of the list.

We consider 3 main events, perceived by the Monitor and Scheduler, that influence the enforcement and tuning of QoS, with their handling workflow described next.

Upon **Read Event:** 1) Try to fetch results from cache with sufficient QoS; 2) On success, return results immediately to the client application; 3) Otherwise, perform query on the database, return the respective results to the application, and store the same results in cache.

Upon **Write Event:** 1) Perform update on the database; 2) Update the cache; 3) Update θ and increment σ on table, column, or row; 4) Verify if the divergence bound, κ, is reached; 5) If so, the data is replicated to the other nodes, θ receives a new timestamp, and σ is reset.

Upon **Time Expiration Event:** 1) The data is replicated to the other interested nodes, θ is timestamped, and σ goes to 0.

Dynamic Adjustment of Consistency Guarantees: Users can specify intervals of values on the QoS vectors to let the framework automatically adjust the consistency intensity. This adjustment, performed by the QoS Engine component, is based on the observation of the frequency of read and write operations to data items during a given time frame. The general idea behind this is that many write operations, performed on different nodes over the same object (or replica), will cause conflicting accesses, and thus it is necessary to guarantee stronger consistency. Conversely, few updates, or updates concentrated only on one node, allow weakening consistency guarantees within the specified vector intervals. The frequency of read operations also contributes to tuning. Many read operations on data that is mainly written in other nodes will strengthen consistency; if the data is written in the same nodes, consistency is relaxed. Conversely, few reads, or reads concentrated on one node, will weaken consistency.

Concurrent Updates: When two or more updates occur simultaneously over the same data in different data centers, both are preserved as the all data items are versioned. We resort mostly to *last-writer-wins* rule and handlers to make data centers converge on the same values. If stronger agreement is needed in more critical (and restricted) data, rotating leases allow data centers to perform writes without contention.

4   Implementation Details

As a proof of concept, we developed, in Java, a prototype of VFC^3 to demonstrate the advantages of our consistency model when deployed as a replication middleware for high-performance databases (i.a., not supporting a full relational model). Although the framework may be adapted to other storages, our target, in the scope of this particular work, is BigTable [6] open-source Java clone, HBase [10]. This database system is a sparse, multi-dimensional sorted map, indexed by row, column (includes family and qualifier), and timestamp; the mapped values are simply an uninterpreted array of bytes. It is column-oriented and designed to scale into the petabyte, while ensuring that write and read performance remain constant. In the following we provide the more relevant details of the VFC^3 implementation.

Schema and Database Management: VFC^3 requires the registration of each application, which includes providing the schema of the required databases. For each table, row, column (and optionally sets of rows and columns), therein, it is necessary to specify the maximum object divergence, κ. Otherwise, the default κ will be used meaning no divergence at all. This schema can be built manually, specifying tables and columns, or simply introduced as the standard auto-generated XML-based schema (given by the HBase Schema Manager tool), which can be processed by VFC^3. After this, the user should create and associate divergence bounds with database objects (i.e., tables/rows/columns) through code annotations, XML specification or a UI.

W.r.t. the creation of divergence bounds, users may specify intervals of values as elements for vectors, rather than scalar constants, so that consistency constraints can be automatically adjusted within an interval. The association of κ with: i) tables is useful when the semantics of the data therein contained indicates that the data should be treated at the same consistency level (e.g., a guest list); ii) rows is beneficial to handle single records independently; iii) columns may be practical if they are supposed to hold very large values, such as media content. Furthermore, the vector ν is only applied on numeric values, and thus text or byte fields are initially precluded and supported only as byte-wise differences (e.g., number of different characters in a string). After the association of divergence bounds with schemas, a similar database is created in the domain of the VFC^3 framework for caching and tracking purposes. This database also contains the QoS attributes and a dirty bit for each cell telling if the cell was modified since the last synchronization occurred.

QoS Management: Different applications may specify overlapping QoS constraints; in this case, more restricted constraints override any others. Thus, such a scenario may happen: application1 requires $\kappa_{1,x}$ and $\kappa_{1,y}$, and application2 requires $\kappa_{2,y}$ ($\kappa_{1,y} > \kappa_{2,y}$). It could make no sense for application1 to have different consistency levels for the items x and y (as $\kappa_{1,y}$ is overridden by $\kappa_{2,y}$). To tackle this, we also allow users to define groups over items that should be handled at the same consistency level, i.e., ensuring atomicity upon serial consistency constraints, over a set or rows and/or columns, to comply with the application semantics. In the previous example, and considering application1 grouped x and y, $\kappa_{1,x}$ is thus assigned with the value of $\kappa_{1,y}$.

The QoS constraints, referring to the data consistency levels, are specified along with standard HBase XML schemas and given to the middleware with an associated application. Specifically, we introduced in the relative XSD the new element *vfc3*, which can be used inside the elements *table*, *column_family*, or *row*, to specify data requirements in relation to a table, column, or row. The vector ν is optional. Enhanced XML schemas are known by all data centers.

Library Support and API: In order to adapt HBase client applications, we provide a similar API to HBase,[1] where we only changed the implementation of some classes in order to redirect HBase calls to the VFC^3 framework, namely *Configuration.java*, *HBaseConfiguration.java*, and *HTable.java* were modified to delegate the HBase configurations to VFC^3. VFC^3 performs the management of the multiple distributed stand-alone HBase instances (without the Hadoop replication) in a transparent manner.

Cache and Metadata: The cache uses similar data structures to HBase itself, such as the *ConcurrentHashMap*, but with extensions to include metadata (the divergence bound vectors) and living in memory (albeit its state can be persisted in the underlying HBase for reliability purposes). The size of the cache and number of items to replace is configurable and new implementations of the cache eviction policy can be provided (default is LRU). Also, the types of the vector elements can be configurable.

5   Evaluation

This section presents the evaluation of the VFC^3 framework and its benefits when compared with the regular HBase/Hadoop replication scheme. All tests were

---

[1] http://hbase.apache.org/apidocs/overview-summary.html

conducted using machines with an Intel Core 2 Quad CPU Q6600 at 2.40GHz, 7825MB of RAM memory, and HDD SATA II 7200RPM 16MB. As for the network, we scattered nodes around two different and relatively distant locations and the available bandwidth was around 60Mbps (downstream and upstream). Moreover, each node/machine had a standalone HBase instance running under VFC^3.

To evaluate the performance of our replication middleware we modified and adapted the YCSB benchmark [12] to work with VFC^3, thereby only redirecting the imports of some classes. Our scenario consisted of running this benchmark, with three different workloads (95/5, 5/95, and 50/50 %updates/%reads), to measure the overall latency and throughput (operations/second) for series of 1000, 10000, 50000, and 100000 operations (reads and writes), assuming in each case that 25, 50, 75, and 100% of the data is critical and required maximum consistency. The non-critical data was associated with σ and ν constraints, meaning its replication could be postponed and not all versions of the data are required. Additionally, the case of 100% means full replication, i.e., the same as using the regular HBase replication.

The (straight) lines of figures 3a, 3c, and 3e show that the overall latency is reduced with VFC^3 (25, 50, and 75%) when compared with HBase full replication (100%): i) latency gains were almost linear (latency of single operations is nearly constant, especially for writes), e.g., under heavy updates, for 100000 ops.,

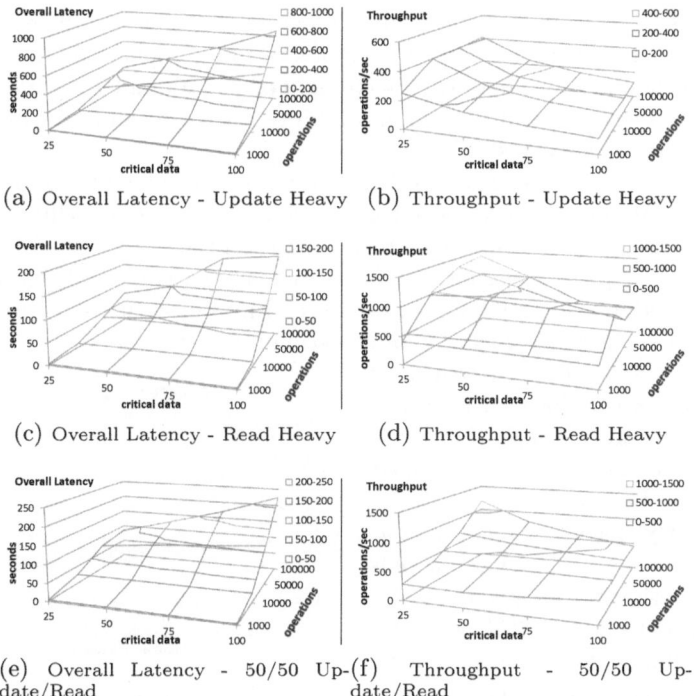

(a) Overall Latency - Update Heavy   (b) Throughput - Update Heavy

(c) Overall Latency - Read Heavy   (d) Throughput - Read Heavy

(e) Overall Latency - 50/50 Update/Read   (f) Throughput - 50/50 Update/Read

Fig. 3. VFC^3 throughput and latency performance

the gain was about 200 sec. every time critical data decreased; ii) from 10000 to 100000 ops., in all workloads, the latency increased linearly for each critical data segment; iii) for 1000 ops., the latency is not stable as the critical data increases, except for the first workload that incurs a slight improvement through the critical data axis (naturally, workloads with more updates will favor VFC^3, even for a small number of ops.); iv) the level of critical data is almost irrelevant in workloads under heavy reads, and the gains therein are mostly supported by our cache; v) the effects of the non-critical data replication were almost not noticed, since most of these data allowed version loss through vector σ (and the cache also absorbed some of these latencies); and vi) in a workload with a balanced mix of reads and writes, the average latency gain with VFC^3 is very satisfactory.

Figures 3b, 3d, and 3f show that the gains of throughput are more accentuated when critical data represents a smaller slice (25% in this case). Plus, the number of ops. only affected significantly the throughput for smaller amounts of critical data (e.g., from blue to green lines in the Read Heavy workload). The throughput for full replication was practically the lowest, comparing with other critical data levels, and almost the same in the non read heavy workloads (irrespective of the number of operations). Also, single read operations have higher latency than write operations (practically zero sec. since they are written in memory first on HBase) and that explains the instability on the Read Heavy figure.

Regarding network usage, we reduced the number of messages and also the volume of data with VFC^3. Note that only the last versions of the data were sent (like what typically happens) when σ expired; and the middleware synchronization performed compression and agglomeration of replication messages when they were inside a same small time window. For 25, 50, and 75% of critical data we saved on average about 75, 48, and 20% respectively, since we mostly relied on the σ vector (message skip).

For evaluating the cache component, we relied on different workloads (taken from the YCSB benchmark) performing 100000 operations each: a) 50%/50% reads and writes (e.g., session store recording recent actions); b) 95%/5% reads/ write mix (e.g., photo tagging); c) 100% reads; d) read latest workload (e.g., user status updates on social networks); e) read-modify-write (e.g., user database activity). The cache size was 30% of the size of each workload.

Figure 4 shows that our cache is effective and can reduce latency and save communication hops. Not surprisingly, the workload d obtained best results; i.e., since we read the most recently inserted records and have LRU as the cache eviction policy. For the other workloads, the gains were good, between 23-35% (a and b at the extremes). Note that the cache size and eviction policy can impact

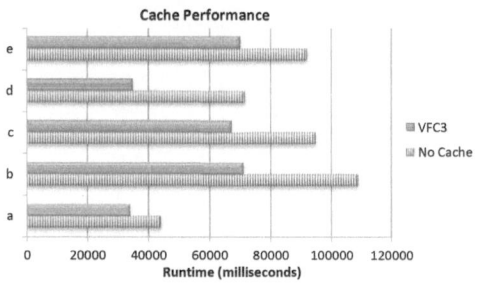

Fig. 4. Cache Performance

these results; so these parameters should be adjusted to better suit the target applications.

The average hit rate of the cache for all experimented workloads was about 51% (except for d, which had 77%), revealing that VFC^3 cache can significantly improve performance, by avoiding expensive trips to the database, for a set of typical scenarios.

6   Related Work

Many work has been done in the context of replication, transactional and consistency models. Our model is based on those already established (standard book [13]) and can be seen as an extension of them by allowing multiple levels of consistency.

In [14], a system is presented for detection-based adaptation of consistency guarantees. It takes a reactive approach to adjust consistency: upon detection of inconsistencies, this system tries to increase the current consistency levels so that they satisfy certain requirements. Contrastingly, we follow in VFC^3 a more proactive approach by trying to avoid inconsistencies from the beginning.

In [15], authors propose three metrics to cover the consistency spectrum (numerical order, order error, and staleness) which are expressed by a logical unit (conit). However, it could be difficult to associate conits with the application-specific semantics, specially in terms of granularity; whereas our work goes with a more fine-grained approach, capturing the semantics from the data itself.

In [16], authors propose creating different categories of data that are treated at different consistency levels. They demonstrate through the TPC-W benchmark that object-based data replication over e-commerce applications can definitely improve availability and performance. We go further with VFC^3, whereas we do not have such a restrictive model regarding target applications and degrees of consistency.

In [17], authors propose a system that allows developers to define consistency guarantees on the data (instead of at the transaction level) that can be automatically adjusted at runtime. Different strategies are explored to dynamically adjust consistency by gathering temporal statistics of the data. Moreover, those strategies are driven by a cost model, which associates penalty costs with different degrees of consistency. This project shares our goals of having a data-driven consistency model, however it could be difficult to associate (real) costs with transactions; and they provide only 3 consistency levels. In VFC^3, consistency degrees are not pre-categorized and hence can be used to fine-tuning and better optimize the overall system performance.

Caching at the application-level can significantly improve the performance of both web servers and underlying databases, since it can save expensive trips to the data base (e.g., [4,18]). The granularity may vary, storing partial database tables, SQL queries, entire webpages, arbitrary content, etc. The major problems: i) they usually do not provide transactional consistency with the primary data; and ii) application developers have to manually handle consistency and explicitly invalidate cache items when the underlying data changes (a very error-prone task). In VFC^3, we offer a complete solution that transparently handles consistency and data is always updated w.r.t. the local database.

Regarding cloud data stores, Amazon S3[2] and BigTable provide eventual consistency. PNUTS [9] argue that eventual guarantees do not suit well its target applications; they provide a per-record timeline consistency: all updates to a record are applied in the same order over different replicas. Conflicting records cannot exist at the same time as it is allowed by Dynamo [8].

7   Conclusion

This paper presented a novel consistency model and framework capable of enforcing different degrees of consistency, accordingly to the data semantics, for data geo-replication in cloud tabular data stores.

We implemented and evaluated a prototype architecture of the VFC^3 framework revealing promising results. It is effective on improving QoS, thereby reducing latency, bandwidth and augmenting throughput for a set of key (and typical) workloads; this, while maintaining the requirements about the quality of data provided to end-users.

To the best of our knowledge, none of the existing solutions in the areas of web-caching, database replication offer a similar flexible and data-awareness control of consistency to provide high-availability without compromising performance. Most of them only allow one level of stale data to exist, that is usually configured per application, and follow a blind approach w.r.t. the data that could require or not strong consistency guarantees; while other solutions, that share some of our goals, impose fixed consistency levels that and limited number of application classes or categories.

References

1. Church, K., Greenberg, A., Hamilton, J.: On delivering embarrassingly distributed cloud services. In: HotNets (2008), CR-ENS-GRID
2. Saito, Y., Shapiro, M.: Optimistic replication. ACM Comput. Surv. 37, 42–81 (2005)
3. Brewer, E.A.: Towards robust distributed systems (abstract). In: Proceedings of the Nineteenth Annual ACM Symposium on Principles of Distributed Computing, PODC 2000, p. 7. ACM, New York (2000)
4. Fitzpatrick, B.: Distributed caching with memcached. Linux Journal 2004, 5 (2004)
5. Coulouris, G.F., Dollimore, J.: Distributed systems: concepts and design. Addison-Wesley Longman Publishing Co., Inc., Boston (1988)
6. Chang, F., Dean, J., Ghemawat, S., Hsieh, W.C., Wallach, D.A., Burrows, M., Chandra, T., Fikes, A., Gruber, R.E.: Bigtable: a distributed storage system for structured data. In: Proceedings of the 7th USENIX Symposium on Operating Systems Design and Implementation, OSDI 2006, vol. 7, p. 15. USENIX Association, Berkeley (2006)

---

[2] http://aws.amazon.com/s3/

7. Lakshman, A., Malik, P.: Cassandra: structured storage system on a p2p network. In: Proceedings of the 28th ACM Symposium on Principles of Distributed Computing, PODC 2009, p. 5. ACM, New York (2009)
8. DeCandia, G., Hastorun, D., Jampani, M., Kakulapati, G., Lakshman, A., Pilchin, A., Sivasubramanian, S., Vosshall, P., Vogels, W.: Dynamo: amazon's highly available key-value store. In: Proceedings of Twenty-first ACM SIGOPS Symposium on Operating Systems Principles, SOSP 2007, pp. 205–220. ACM, New York (2007)
9. Cooper, B.F., Ramakrishnan, R., Srivastava, U., Silberstein, A., Bohannon, P., Jacobsen, H.A., Puz, N., Weaver, D., Yerneni, R.: Pnuts: Yahoo!'s hosted data serving platform. Proc. VLDB Endow. 1, 1277–1288 (2008)
10. George, L.: HBase: The Definitive Guide, 1st edn. O'Reilly Media (2011)
11. Veiga, L., Negrão, A., Santos, N., Ferreira, P.: Unifying divergence bounding and locality awareness in replicated systems with vector-field consistency. J. Internet Services and Applications 1, 95–115 (2010)
12. Cooper, B.F., Silberstein, A., Tam, E., Ramakrishnan, R., Sears, R.: Benchmarking cloud serving systems with ycsb. In: Proceedings of the 1st ACM Symposium on Cloud Computing, SoCC 2010, pp. 143–154. ACM, New York (2010)
13. Tanenbaum, A.S., van Steen, M.: Distributed Systems: Principles and Paradigms, 2nd edn. Prentice-Hall, Inc., Upper Saddle River (2006)
14. Lu, Y., Lu, Y., Jiang, H.: Adaptive consistency guarantees for large-scale replicated services. In: Proceedings of the 2008 International Conference on Networking, Architecture, and Storage, pp. 89–96. IEEE Computer Society, Washington, DC (2008)
15. Yu, H., Vahdat, A.: Design and evaluation of a continuous consistency model for replicated services. In: Proceedings of the 4th Conference on Symposium on Operating System Design & Implementation, OSDI 2000, p. 21. USENIX Association, Berkeley (2000)
16. Gao, L., Dahlin, M., Nayate, A., Zheng, J., Iyengar, A.: Application specific data replication for edge services. In: Proceedings of the 12th International Conference on World Wide Web, WWW 2003, pp. 449–460. ACM, New York (2003)
17. Kraska, T., Hentschel, M., Alonso, G., Kossmann, D.: Consistency rationing in the cloud: Pay only when it matters. PVLDB 2, 253–264 (2009)
18. Sivasubramanian, S., Pierre, G., van Steen, M., Alonso, G.: GlobeCBC: Content-blind result caching for dynamic web applications. Technical Report IR-CS-022, Vrije Universiteit, Amsterdam, The Netherlands (2006)

A Fault-Tolerant Cache Service
for Web Search Engines: RADIC Evaluation

Carlos Gómez-Pantoja[1,2], Dolores Rexachs[5],
Mauricio Marin[3,4], and Emilio Luque[5]

[1] Universidad Andres Bello, Santiago, Chile
[2] DCC, University of Chile, Santiago, Chile
[3] University of Santiago of Chile
[4] Yahoo! Research Latin America, Santiago, Chile
[5] University Autonoma of Barcelona, Barcelona, Spain

Abstract. Large Web search engines are constructed as a collection of services that are deployed on dedicated clusters of distributed-memory processors. In particular, efficient query throughput heavily relies on using result cache services devoted to maintaining the answers to most frequent queries. Load balancing and fault tolerance are critical to this service. A previous paper [7] described the design of a result cache service based on consistent hashing and a strategy for enabling fault tolerance. This paper goes further into implementation details and experiments related to the basic scheme to support fault-tolerance which is critical for overall performance. To this end, we evaluate the performance of the RADIC scheme [14] for fault-tolerance under demanding scenarios imposed in the caching service.

1 Introduction

Data centers for large Web search engines (WSEs) contain thousands of processors arranged in high-communicating groups called services. Usually each service is devoted to a single specialized operation related to the processing of user queries. Typically a WSE is composed by three relevant services: Front-End/Broker Service (FS), Caching Service (CS) and Index Service (IS). The FS receives queries and handles query routing; the CS keeps results for frequent queries; and the IS uses an inverted index to calculate top K results when the query results are not in the CS. The CS plays a key role in enabling high query throughput [1] as the cost of searching a query in the CS and returning the answer stored in the respective cache entry, is by far less costly in running time than computing the query answer from the IS.

The traffic generated by WSE users is not uniform neither constant, it is variable, unpredictable and follows Zipfian distributions [3]. It means that users always generate new queries and a few very popular queries can have a huge impact in performance degradation since they can cause imbalance. In addition, failing nodes can affect performance as it is necessary to distribute the load assigned to them on the remaining nodes. The service that is most exposed to imbalance situations is the CS. This is because it splits queries into disjoint sets using hash functions so that each query is allocated to only one partition. Therefore, bursty queries can overload partitions.

C. Kaklamanis et al. (Eds.): Euro-Par 2012, LNCS 7484, pp. 298–310, 2012.

The literature related to caching is extensive, but it lacks of efficient solutions for the problem relevant to this paper. A noticeable exception is the Amazon Dynamo [4] system. Almost all of previous work propose eviction algorithms, admission policies and query invalidation strategies to improve the performance of *individual* cache nodes [1,11,6]. We aim at a *distributed* system suitable for clusters of processors. Peer-to-Peer (P2P) systems using consistent hashing [8] -like Chord [15]- assume uniform key distribution, which tends to favour static assignments of keys to nodes. It is important to bear in mind that our application domain is radically different from load balancing P2P systems [16], which means that not all solutions from this domain are applicable to our setting. In our case, queries must be solved in a few tens of milliseconds and thereby there is no room for approaches based on data movement across processors [12], extra messages to locate processors [15], or the like. A fairly similar idea, though fully distributed, is proposed for P2P systems in [2]. This work does not balance the load caused by stored items or their popularity. They indirectly try to do this by making similar the range of keys that each node is responsible for. We do the opposite by using ranges of variable length as a function of node popularity. Our proposal is intended to form part of dedicated search services where homogeneous processors are not shared by other applications and no virtualization technology is used for load balancing because of its overheads.

We propose a dynamic load balancing algorithm upon consistent hashing in order to cope with imbalance in CS nodes. The balancing process is reactive in the sense that it is triggered when imbalance is detected. We also propose to mitigate the effects of node failures by using a protection system for frequent queries. For this purpose we use the RADIC approach [14]. Here, a cache p_i selects a set of pairs <query, answer> according to a criteria, and then these queries are sent to a secondary cache p_j. In case of failure of cache p_i, all of its requests are redirected to p_j (the *protector* of p_i). The protection of selected cache entries is a proactive action.

The remainder of this paper is organized as follows. Section 2 describes the system architecture. Section 3 presents our proposal. Section 4 presents experimental results using simulation, and Section 5 presents results of our RADIC implementation. Finally, Section 6 presents conclusions.

2   System Architecture

The Front-End Service (FS) comprises several replicated nodes. Each FS node receives and routes user queries, and sends back the top K results to users. After a query arrives to a FS node b_i, it asks the Caching Service (CS) to determine whether the query result has been previously stored there. A baseline CS cluster architecture is formed by an array of $P_c \times D_c$ processors (or CS nodes). A scheduling method in FS carries out the distribution of queries onto the P_c partitions. When a partition p_i has been selected, one of its D_c replicas is chosen at random to search for the query. If the query is cached, the CS node sends the query answer to b_i. Afterwards b_i sends the results to the user. If the query is not found in cache, the CS node sends a hit-miss message to b_i. At this point, b_i sends the query to the Index Service (IS).

For the IS, the standard cluster architecture is an array of $P_i \times D_i$ processors or index search nodes, where P_i indicates the level of text collection partitioning and D_i

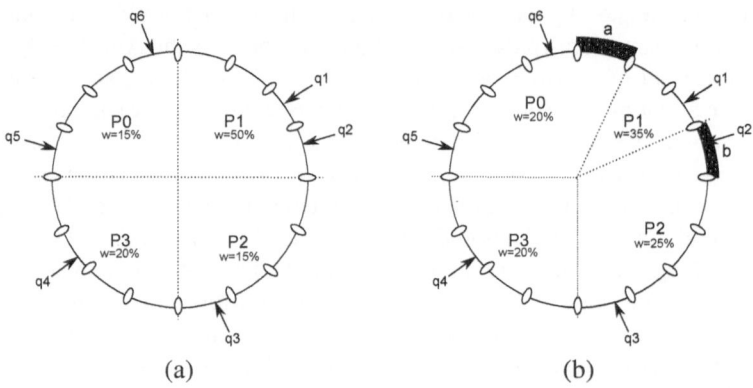

Fig. 1. (a) Consistent hashing assignment; (b) Our proposal for load balancing

the level of text collection replication. The use of this 2D array is as follows: each query is sent to all of the P_i partitions and, in parallel, a random replica in each partition determines the local top K query results. These results are then collected together by the FS to determine the global top K results for the query. Each index search node contains an inverted file which is a data structure used to efficiently map from query terms to relevant documents.

The FS and IS do not experience significant imbalance. Newly arriving queries are evenly distributed on the FS nodes. When a query is solved in the IS, all partitions work in parallel to produce the local top K results in each partition, so a query generates almost the same load in all IS partitions. Neither of these two services face serious risks of imbalance. Given the access pattern to CS partitions and the bias of user queries, the probability of significant imbalance is high in this service as we show below.

The only service in which data (inverted index and text) must be distributed before the operation is the IS. The CS populates its distributed memory with query results on-the-fly, and the FS only handles small data related to current query routing.

3   Caching Service

All memory space of CS nodes is divided into blocks (typically 4KB), and each block stores query terms and their associated top K results (HTML page). Each CS node follows an eviction policy when it is full.

The distribution of items is implemented with consistent hashing [9], which partitions the query space into P independent subsets (each subset can be seen as a non-overlapping arc in a ring (see Figure 1(a)). The idea is that each partition handles all queries (each query is assigned to one point in the ring by means of hashing) that intersect its arc. Figure 1(a) illustrates an assignment following an equidistant distribution of points (please ignore for now the small white ovals), and the assignment indicates that partition $P1$ serves queries $q1$ and $q2$.

The decision of how many partitions depends on the hit ratio we want to reach and the global set of queries (namely, the number of distincts queries in skewed distributions).

The more diverse the query space, the more partitions are needed because each node holding a partition has limited capacity. But dimensioning the number of partitions is not the only decision, also it is necessary to consider the volume of queries per unit of time that one node can accept. To cope with this later issue, replication is introduced as a form to spread the load of a particular partition; also it helps to provide fault-tolerance. This translates into D replicas for each partition. The underlying idea is to select one of the P partitions via consistent hashing, and then select one of its D replicas at random.

To reach a consistent state (all replicas of one partition), we can use an optimistic protocol [13] in each partition where replicas are guaranteed to converge after a period of time. We choose this kind of protocol instead of a protocol that provides strong consistency, because this later solution can saturate the network.

The first baseline approach we study, called *Baseline DAC* (Amazon Dynamo and Chord), can be seen as a matrix of $P \times D$ nodes. The location process works as follows: consistent hashing to select a partition, then we select one random replica. Each partition runs an optimistic consistency protocol.

The previous strategy has a configuration that can be prone to variations in the user query traffic (for example, bursty queries) and nodes failures. To control the load imbalance produced by Baseline DAC, we can use a Greedy algorithm to move replicas between partitions. The key idea is to measure the utilization at partition level each Δ units of time, and then we decide whether one partition needs more replicas taking into consideration the output of the Greedy algorithm. We call this strategy *Baseline DR* (Dynamic Reallocation). The location process is the same as Baseline DAC. Also each partition runs an optimistic protocol. We use these two strategies as a basis for comparison. More details can be found in [7].

Having fixed points in the CHR[1] is not a good policy. Our solution consists in using the same $P \times D$ matrix and a CHR at partition level (i.e., P partitions/arcs), but allowing arc lengths to dynamically change in size to split traffic between neighboring partitions.

Firstly, the ring is divided into small equally-sized buckets to discretize the range covered by each arc. Each partition covers an arc of size $1/P$ of the ring and is responsible of a disjoint set of adjacent buckets as shown in Figure 1(a). Like the strategy Baseline DR, the service utilization is reported every Δ units of time to the FS. If the efficiency[2] is less than a predefined threshold T, a load balancing algorithm that considers each partition utilization is triggered in the FS. The output of the algorithm is a set of bucket movements between neighboring partitions. This prevents all cache entries from being invalidated because only a small number of queries change of partition. Namely, those inside the buckets moved.

The threshold T used in our proposal and the DR strategy defines the maximum degree of imbalance to be tolerated. When the efficiency is below the threshold T, the balancing process is triggered.

Our proposal starts with an initial uniform distribution of the buckets as shown in Figure 1(a). Each partition handles four contiguous buckets. When measuring utilization, we detect that partition $P1$ is processing half of the system load. Figure 1(b) shows the effect of moving one bucket from $P1$ to $P0$ and one from $P1$ to $P2$ (neighbors of $P1$).

---

[1] Consistent hashing ring.

[2] The efficiency is defined as the average load divided by the maximum load.

The arc a belongs to $P0$ and arc b belongs to $P2$, meaning that $P0$ and $P2$ are now in charge of more queries which decreases the load of $P1$. An advantage of this strategy is that only little portions of cache entries are invalidated when buckets are moved. Namely, those with consistent hashing values falling in the range of the buckets moved to neighboring partitions.

The method to re-distribute the buckets is a diffusion algorithm and it is based on the Sender Initiated Diffusion (SID) algorithm presented in [17]. In this algorithm each overloaded partition distributes excess of load to neighbors with less load. The authors show that in a system with P partitions and load L unevenly distributed, the algorithm will eventually converge to load L/P in each partition and also it is stable.

Replicas associated with each partition are handled as follows. When a FS node selects partition A for query q, we apply a second hash function over the query terms to select one replica from A. This strategy increases the total number of entries available for caching different items across the replicas. This increases overall hit ratio but also node failures are expected to enhance its effects on hit ratio reduction.

To provide fault-tolerance, we use the RADIC framework [14] to efficiently replicate selected queries. It is based on two components: *Protectors* and *Observers* which we propose to use as follows.

Each partition runs a separate RADIC process. Every δ units of time all Observers send their checkpoint to the corresponding Protector. If node m_i belonging to partition A fails, all requests to m_i are re-directed to its Protector m_j allocated in the same partition. In this case, m_j processes its own queries and those originally directed to m_i. The load increment in partition A is not a severe problem due to the balancing algorithm we apply on the partitions. The imbalance generated by a faulty node m_i will be corrected decreasing the range of partition A. Performance degradation is also controlled as the most frequent entries of node m_i are likely to be already checkpointed in its Protector m_j when the failure takes place, so this node will cover the most important queries.

As said, only the most frequent queries in all nodes m_i are protected. Copying all cache entries to the Protector and doubling the number of entries in each node, is not feasible. For the purpose of comparison, to be fair to other strategies, we decrease the available cache entries in each node to make space to hold the checkpoints. From empirical evidence, we conclude that the best distribution in each CS node is 70% of memory space to hold cache entries and the remaining 30% of space to hold checkpoints.

Note that queries tend to be the same between checkpoints in any particular node, since the queries selected to be part of the checkpoint are the most accessed of that node. Hence, instead of forcing the sending of all cache entries to its Protector, only the modifications are sent to it. To this end, each node can log only modifications: priority change, entry eviction and item insertion. This decreases communication.

4   Evaluation through Simulation

To simulate the strategies described above, we have modeled and implemented discrete event simulators that are able to precisely predict a set of metrics. The methodology to build the simulator is based on the facts that (i) the major operations in our context are coarse-grained, and (ii) given our architectural design, a request in any of the nodes of a specific service takes almost the same amount of resources. The first step is the identification of the most important operations evolved in query processing (resource utilization). Then, we profile these operations and insert their costs to a discrete event simulator. We have previously used and validated this claim in [10]. In conclusion, we simulate trace-driven events, where the traces are benchmarked from real executions on a search engine using query logs of commercial WSEs.

We perform the experiments using a one-day real query log from a commercial search engine (as on April 1st, 2011). The query log comprises 68,019,311 queries. We configured the WSE as follows: (i) the FS comprises 10 replicas; (ii) the CS has $P = 20$ partitions and $D = 4$ replicas; and (iii) the IS possesses 50 partitions and 20 replicas (in order to simulate a complete inverted file loaded into RAM). Each CS node has 100,000 entries for cache. The time interval Δ to measure the partition utilization is 5 minutes and the efficiency threshold T is set to 95%. In our proposal, the service checkpoint is passed every $\delta = 10$ minutes.

Two of the most important metrics are *Average Query Response Time* and *Hit Ratio* of the CS. The first metric shows how the strategies behave under the occurrence of failures, while the second indicates the percentage of answers found in cache. We do not only show the cases of failure, but as well the results without failures for comparison purposes. We start the evaluation with 1, 2 and 3 random failures of CS nodes. Furthermore, we examine a special case, where 10% of CS nodes fail.

In all experiments, failures occur between $x = 640$ and $x = 800$. Note that the nodes where failures were injected are randomly chosen and are not re-incorporated by the service. The measurement starts with the first injected failure.

We labeled the curves in the following figures as "Baseline DAC", which is the Baseline Amazon Dynamo and Chord, and "Baseline DR", which stands for Baseline Dynamic Reallocation.

Figure 2(a) shows the average query response while no failures were triggered. Three different trends can be clearly identified in steady state (from $x = 480$): (i) the Baseline DR strategy shows an average of 98 [ms], (ii) the Baseline DAC shows an average of 95 [ms] and (iii) our approach outperforms both with an average of 83 [ms]. This presents 15.3% and 12.6% better query response time than the Baseline DAC and Baseline DR, respectively.

Figure 2(c) shows the results in the case that three failures occur. Here, as well as in Figure 2(b) and (d), it can be observed that the Baseline DR has the worst performance. The reason of this behavior is that the movements of machines between partitions is a disproportionate action, which implies that all entries of the moving nodes are lost (they are not useful for the new partition). The figures reflect the impact. On the other hand, the Baseline DAC as well as our approach present small variations in performance.

A special case is shown in Figure 2(d), in which 8 nodes are randomly chosen to stop. This is an extreme case, since 10% of CS nodes are lost. In this case, the behavior of Baseline DAC remains almost constant. While it is true that our approach experiences an increase of 9% (from 85 to 93 [ms]), it still outperforms the Baseline DAC. Moreover, our approach of dynamic load balancing helps to reduce the average query time. For this reason, a decrease of query time can be observed towards the end of Figure 2(d). Considering this metric, we have demonstrated that the proper combination

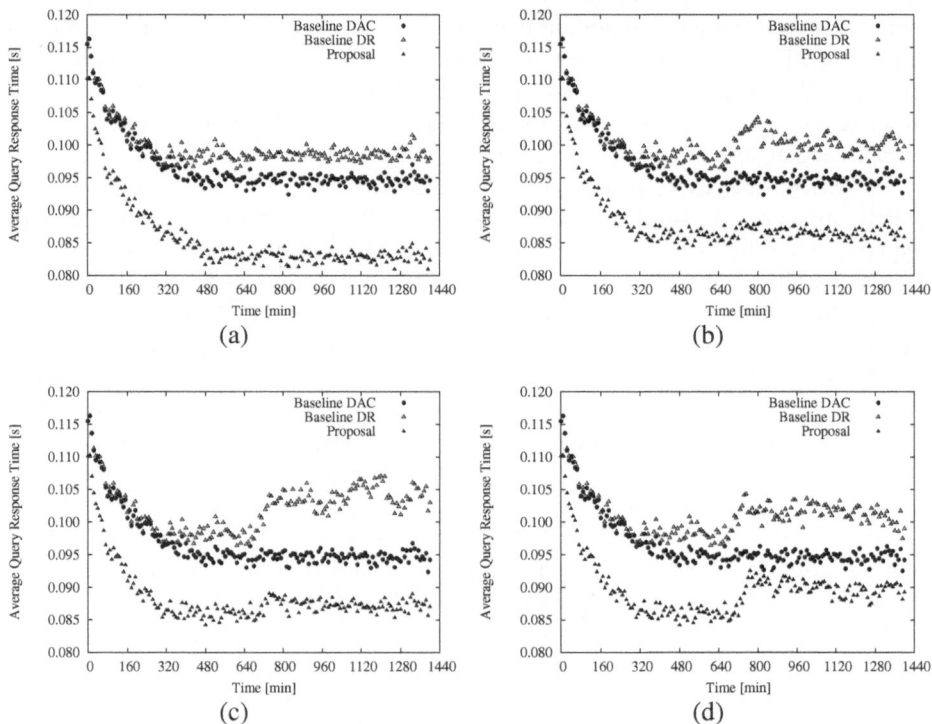

Fig. 2. Evaluation of Average Query Response Time: (a) zero, (b) one, (c) three and (d) eight failures (10%). In cases of failure, they are triggered between $x = 640$ and $x = 800$ minutes.

of dynamic load balancing and a methodology to protect valuable information (RADIC) is important to consider during the design and deployment of caching services. Table 1 summarizes the improvements that we achieved through our approach in all aforementioned cases.

The Figure 3(a) illustrates the hit ratio considering the different options. The performance of Baseline DAC and Baseline DR is similar, having a hit rate between 25% and 30% once the steady state is reached. The optimized utilization of cache entries by our strategy is another important fact. Using our strategy almost all entries show higher hit ratio, while information is only replicated for protection purpose. The proactive replication of queries helps to keep a similar hit rate in case of faults, even in situations where more than one failure occurs (Figure 3(b), (c) and (d)). Despite the failures, our proposal outperforms the other strategies in all cases and is in addition just slightly affected. Figure 3(c) shows the same behavior as before (hit ratio improved by 25% on average compared to Baseline DAC).

Figure 3(d) shows results with greater variation in all cases. This is due to imbalance issues that emerge when a large number of machines fail. The objective of reaching (and keeping) a better hit ratio compared to other strategies is accomplished, despite the high number of failures. See Table 1 for more results.

Table 1. Percentage of improvements of our Proposal against Baseline DAC and Baseline DR considering Figure 2 and 3. Values are obtained while the services were in a steady state (after failures).

	Average Query Time		Hit Ratio	
	Baseline DAC	Baseline DR	Baseline DAC	Baseline DR
Zero Failures	12.6%	15.3%	31%	46%
One Failure	8.4%	13.0%	27%	60%
Three Failures	8.4%	16.3%	32%	94%
Eigth Failures	7.2%	12.7%	37%	105%

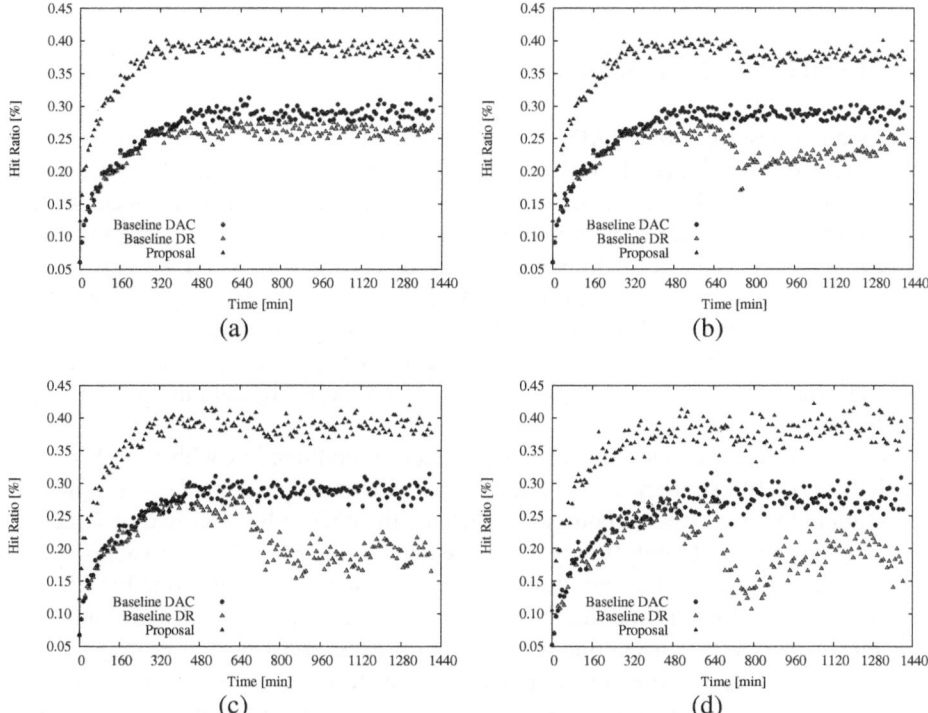

Fig. 3. Evaluation of Average Hit Ratio: (a) zero, (b) one, (c) three, and (d) eight failures (10%). In case of failures, they are triggered between $x = 640$ and $x = 800$ minutes.

4.1 Analysis

We have shown that a better organization of resources, by taking proactive actions (protect important queries) and dynamically balancing load, are important aspects to reach lower response time and higher hit ratios. We argue that the most important factor to achieve a high throughput is a suitable load balancing strategy. Nevertheless, performance grows even more when cache entries are better administered and the consistency protocol is avoided. This technique in conjunction with the protection of queries, improves the performance in all aspects as examined above through the average query time and hit ratio. Finally, this two techniques help to decrease the impact of failures in case that an organization is used, which exploits all available memory.

Also, the previous results demonstrated the benefits and limits of our proposal. At first, our strategy diminishes its performance in case of failures, but the dynamic load balancing helps to overcome this situation quickly. Moreover, it remains the best strategy. Secondly, the baseline DAC is almost not affected by failures because of the replication, but at the same time it does not utilize the resources properly, and hence this strategy does not attain the best results. Finally, there is a trade-off between replication and performance, and our proposal certainly points in the following direction: only replicate the most frequent queries and use them in case of failures following the RADIC approach.

5   RADIC Implementation

This section describes our RADIC [14] implementation and how it works in a real setting. As we mentioned above, this strategy allows us to integrate the protection of important queries to be used in case of failures. To test RADIC performance, we have implemented a C++/MPI prototype of a CS with RADIC. First of all, each processing node or processor can be seen as a container of entries with a limited capacity that follows an eviction policy when it is full. Well-known algorithms for this purpose are LRU, LFU, SDC and PDC [11]. Regardless of the strategy, in all cases important cache entries can be identified. To decide which is the next entry to be replaced, all algorithms use a priority queue. *Memcached* [5] follows the LRU policy by default (other strategies can be used).

In our service, we designed a priority queue in conjunction with a hash table to implement a LFU strategy. The choice of the LFU strategy is made to simplify the selection of the most frequent queries, which are the ones to be protected by RADIC.

Following the WSE architecture, only one node of the Front-End Service is responsible for triggering the checkpoint process (the decision is centralized at the FS side). Namely, each δ units of time the FS node sends a message to all CS nodes indicating that they must start the checkpoint process. This process consists of three stages in each node CS_i: (i) CS_i gets the most important queries from its memory, which translates into N pairs <query,answers>; (ii) CS_i sends them to its protector node using MPI; (iii) CS_i waits for the checkpoint (N pairs) from the node that it is protecting; and (iv) CS_i stores the received checkpoint in its memory (a separate area of memory). All these phases do not affect the query processing in the node (we control the concurrent access to the structures). As we mentioned above, a node only protects a node of its own partition, in this way the protection on each particion (and their replicas) is independent of other partitions.

To study the behaviour in case of failures, we did not need a fault-tolerant version of MPI, because the routing and handling of queries belongs to the FS and only this service needs to know when a node falls. Failing nodes are selected randomly and, for the sake of simplicity, we "simulate" a failure sending a MPI message to the failing nodes (to stop the processing), and then updating the state of active nodes in the FS (routing tables). Once a failure is detected in the FS, all requests to failing nodes are

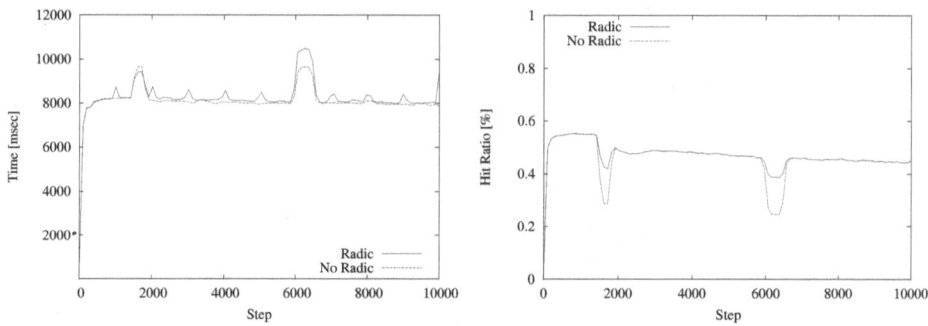

Fig. 4. Real deployments to evaluate overhead with failures: (a) running time of the complete caching service, and (b) hit ratio of a single partition of the caching service

routed to the protector of them following the RADIC algorithms/tables. As the protector maintains information regarding the failing nodes (checkpoint), it starts to process the redirected requests taking into account this information. This helps to preserve the hit ratio in case of faults.

5.1 RADIC Overheads

An important performance metric in a fault-tolerant system is the overhead imposed by the protection scheme relative to the same system without a fault tolerance strategy. Hereinafter, we evaluate the RADIC overheads through an actual implementation running in our CS service. The implementation is therefor deployed to a cluster of processors (cluster composed of 20 nodes connected by an InfiniBand switch). To effectively measure overheads and to therefor achieve a situation in which there are no queries that cause imbalance, we implemented the baseline approach (namely, the $P \times D$ matrix described in Subsection 3) with the RADIC system running in background.

We ran the complete log described in Section 4 in a caching service with P=5 partitions and D=4 replicas. We measured the time required to finish a set of queries and the resulting hit ratio (each measurement is a point in the X-axis, and only the first 10,000 points are plotted). Furthermore, we injected two failures in the system (x=1,500 and x=6,000). After the failures, nodes are re-inserted in the service after 100 y 500 steps, respectively. Note, that the implementation without RADIC has 100,000 cache entries per node, while the implementation with RADIC has 95,000 cache entries and 5,000 for checkpoints.

Figure 4(a) shows the time required to finish the queries. The overhead imposed by RADIC is 1.8% on average, what does not represent a big impact on the service. The checkpoint takes place approximately every 1,000 steps and only the top 5% of the most important queries are checkpointed. We optimized the checkpointing process by processing it through a pipeline: multiple steps are used to send the checkpoint to the Protector. The overhead and the checkpoints become important when failures occur and when queries are protected. Figure 4(b) displays the results in terms of average hit ratio inside a partition when failures occur in the same partition (D=4). It is clear that the hit ratio is less affected by failures since the implementation of RADIC allows to continue

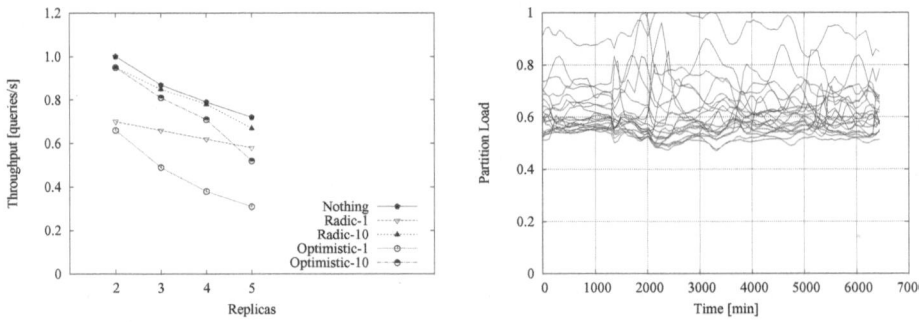

Fig. 5. (a) Evaluation of a real implementation of the Optimistic Protocol and the RADIC Proposal (Section 3) to test the scalability of one partition. (b) Load assignment to 20 partitions when we run 300 millions of queries issued during May 1-5, 2011.

the operation using the checkpointed query results in the Protector of the failing node. The average hit ratio of the partition using RADIC is 47.2% and 46% without RADIC (an increase of 2.6% in the presence of failures). We would like to remark that the results of this section were obtained with the same baseline strategy, which was extended with RADIC to expose its inherent overheads relative to the same strategy operating without RADIC.

Figure 5(a) shows results obtained with actual implementations of the consistency protocols. We ran these experiments in a cluster of processors connected by a commodity switch, taking care that each replica is allocated to a different processing node so that communication among processors takes place through the communication network. The purpose of these experiments is to evaluate the effects of extra communication required to keep consistency across the replicas of each partition at running time. The curve labeled "nothing" is the best that a protocol can do as in this case no communication bandwidth is used to replicate cache entries. They just arrive to a target replica and a search in the cache is executed. Here it is not relevant whether there is a cache hit for any query or not. Note that each replica is assumed to receive the same number of queries. This implies that as the number of replicas grows, more queries are processed in total. The remaining curves are showing the results of RADIC and the optimistic protocol for cases in which the consistency protocol is executed each 1 and 10 minutes. In both cases, the RADIC protocol outperforms the optimistic one. For 10 minute intervals the RADIC protocol achieves a performance quite similar to the optimal case. The overhead in terms of memory consumed by checkpoints is negligible because hit rate is not affected significantly keeping constant the total number of cache entries.

6   Conclusions

We conclude by referring to Figure 5(b) which shows that queries tend to produce significant imbalance when no strategy is used to load balance the amount of queries that receive each CS node.

We have shown that load balancing and protection of queries can help to improve the performance of WSEs. On the one hand, we use a load balancing algorithm to cope with user query variations and faulty nodes, this is a reactive action. Then, we move forward proposing the application of RADIC methodology to protect valuable information, which is a proactive action.

The objective of this paper was twofold: (i) analyze the resilience of our proposal, and (ii) analyze the performance and usefulness of RADIC framework. In section 4, the experimental results show that our proposal outperforms the commonly used baseline alternatives by a wide margin as shown in Figure 2 and 3 ((a) no failures; (b), (c), (d) with failures). This is because our proposal is able to significantly reduce the imbalance shown in Figure 5(b). Notice that our proposal can be easily extended to clusters with heterogeneous nodes since load balance is made considering only the utilization of processors, which can be determined by performing benchmarks on individual nodes and stablishing a relationship between incomming query traffic and utilization.

Finally, in section 5 we evidence that our RADIC implementation imposes a very low overhead to the query processing tasks, which shows its competitiveness and usefulness in the context of caching services. To the best of our knowledge, there are no previous works that addresses the protection of queries.

Acknowledgements. This research has been supported by the MICINN Spain under contract TIN2007-64974 and the MINECO (MICINN) Spain under contract TIN2011-24384, and partially funded by FONDEF project D09I1185. The first author (CG) has been supported by a Chilean PhD scholarship from CONICYT.

References

1. Baeza-Yates, R., Gionis, A., Junqueira, F., Murdock, V., Plachouras, V., Silvestri, F.: The impact of caching on search engines. In: SIGIR 2007, pp. 183–190 (2007)
2. Bienkowski, M., Korzeniowski, M., auf der Heide, F.M.: Dynamic Load Balancing in Distributed Hash Tables. In: Castro, M., van Renesse, R. (eds.) IPTPS 2005. LNCS, vol. 3640, pp. 217–225. Springer, Heidelberg (2005)
3. Breslau, L., Cue, P., Cao, P., Fan, L., Phillips, G., Shenker, S.: Web caching and zipf-like distributions: Evidence and implications. In: INFOCOM, pp. 126–134 (1999)
4. DeCandia, G., Hastorun, D., Jampani, M., Kakulapati, G., Lakshman, A., Pilchin, A., Sivasubramanian, S., Vosshall, P., Vogels, W.: Dynamo: amazon's highly available key-value store. SIGOPS 41, 205–220 (2007)
5. Fitzpatrick, B.: Distributed caching with memcached. Linux J. (2004)
6. Gan, Q., Suel, T.: Improved techniques for result caching in web search engines. In: WWW 2009, pp. 431–440 (2009)
7. Gómez-Pantoja, C., Gil-Costa, V., Rexachs, D., Marin, M., Luque, E.: A fault-tolerant cache service for web search engines. In: ISPA 2012 (to appear, 2012)
8. Karger, D., Lehman, E., Leighton, T., Panigrahy, R., Levine, M., Lewin, D.: Consistent hashing and random trees: distributed caching protocols for relieving hot spots on the world wide web. In: ACM STOC 1997, pp. 654–663 (1997)
9. Karger, D., Sherman, A., Berkheimer, A., Bogstad, B., Dhanidina, R., Iwamoto, K., Kim, B., Matkins, L., Yerushalmi, Y.: Web caching with consistent hashing. In: WWW 1999, pp. 1203–1213 (1999)
10. Marin, M., Gil-Costa, V., Gomez-Pantoja, C.: New caching techniques for web search engines. In: HPDC 2010, pp. 215–226 (2010)
11. Perego, T.F.R., Silvestri, F., Orlando, S.: Boosting the performance of web search engines: Caching and prefetching query results by exploiting historical usage data. In: ACM TOIS 2006, pp. 51–78 (2006)

12. Raiciu, C., Huici, F., Rosenblum, D.S., Handley, M.: ROAR: Increasing the flexibility and performance of distributed search. In: SIGCOMM 2009, pp. 291–302 (2009)
13. Saito, Y., Shapiro, M.: Optimistic replication. ACM Comput. Surv. 37, 42–81 (2005)
14. Santos, G., Duarte, A., Rexachs, D., Luque, E.: Providing Non-stop Service for Message-Passing Based Parallel Applications with RADIC. In: Luque, E., Margalef, T., Benítez, D. (eds.) Euro-Par 2008. LNCS, vol. 5168, pp. 58–67. Springer, Heidelberg (2008)
15. Stoica, I., Morris, R., Karger, D., Kaashoek, M.F., Balakrishnan, H.: Chord: A scalable peer-to-peer lookup service for internet applications. SIGCOMM 31, 149–160 (2001)
16. Surana, S., Godfrey, B., Lakshminarayanan, K., Karp, R., Stoica, I.: Load balancing in dynamic structured peer-to-peer systems. Perform. Eval. 63, 217–240 (2006)
17. Willebeek-LeMair, M., Reeves, A.: Strategies for dynamic load balancing on highly parallel computers. IEEE Transactions on Parallel and Distributed Systems 4(9), 979–993 (1993)

Topic 6: Grid, Cluster and Cloud Computing

Erik Elmroth, Paraskevi Fragopoulou, Artur Andrzejak, Ivona Brandic, Karim Djemame, and Paolo Romano

Topic Committee

Grid and cloud computing have changed the IT landscape in the way we access and manage IT infrastructures. Both technologies provide easy-to-use and on-demand access to large-scale infrastructures. Grid and cloud computing are major research areas with strong involvement from both academia and industry. Although significant progress has been made in the design, deployment, operation and use of such infrastructures, many key research challenges remain to achieve the goal of user-friendly, efficient, and reliable grid and cloud infrastructures. Research issues cover many areas of computer science to address the fundamental capabilities and services that are required in a heterogeneous environment, such as adaptability, scalability, reliability and security, and to support applications as diverse as ubiquitous local services, enterprise-scale virtual organizations, and internet-scale distributed supercomputing. While there are several differences, grid and cloud computing are closely related in their research issues. Both areas will greatly benefit from interactions with the many related areas of computer science, making Euro-Par an excellent venue to present results and discuss issues.

The issues to be covered include but are certainly not limited to the following: middleware; applications and platforms; interoperability and portability; aggregation and federation of grids and clouds; efficient energy usage of resources; resource/service/information discovery; resource management and scheduling; programming models, tools, and algorithms; dependability, adaptability, and scalability; security for grids and clouds; workflow management; accounting, billing and business models; automated or autonomic management of resources and applications; quality-of-service and Service-Level-Agreement.

- The paper entitled *Scalable Reed-Solomon-based Reliable Local Storage for HPC Applications on IaaS Clouds* by Leonardo Bautista Gomez, Bogdan Nicolae, Naoya Maruyama, Satoshi Matsuoka and Franck Cappello, introduces a novel persistency technique that leverages Reed-Solomon (RS) encoding to save data in a reliable fashion on IaaS Cloud computing platforms to be used for HPC applications. Compared to traditional approaches that rely on block replication, this technique demonstrates about 50% higher throughput while reducing network bandwidth and storage utilization by a factor of 2 for the same targeted reliability level. This is achieved both by modeling and real life experimentation on hundreds of nodes.
- Subsequently, the paper *Caching VM Instances for Fast VM Provisioning: A Comparative Evaluation*, by Pradipta De, Manish Gupta, Manoj Soni

C. Kaklamanis et al. (Eds.): Euro-Par 2012, LNCS 7484, pp. 311–312, 2012.
© Springer-Verlag Berlin Heidelberg 2012

and Aditya Thatte from IBM Research India, presents a method to overcome the delays in transfer and booting time for the preparation of VMs in cloud environments. Alternatively, a VM is prepared a priori, and saved in standby state in a "cache" space collocated with the compute nodes. On receiving a matching request, the VM from cache is instantly served to the user, thereby reducing service time. Based on usage data collected from an enterprise cloud, and through simulation, it is shown that a reduction of 60% in service time is achievable.

- The paper *Improving Scheduling Performance using a Q-Learning-based Leasing Policy for Clouds* by Alexander Fölling and Matthias Hofmann from TU Dortmund University, presents a reinforcement learning-based policy which controls the maximum leasing size of cloud computing resources with regard to the current resource/workload state and the balance between scheduling benefits and costs in an online adaptive fashion. Furthermore, it provides an appropriate model to evaluate such policies and presents heuristics to determine upper and lower reference values for the performance evaluation under the given model. Using event driven simulation and real workload traces, the authors were able to investigate the dynamics of the learning policy and to demonstrate the adaptivity on workload changes.

- Finally, the paper *Impact of Variable Priced Cloud Resources on Scientific Workflow Scheduling* by Simon Ostermann, Radu Prodan from the University of Innsbruck, analyzes the problem of provisioning Cloud instances to large scientific workflows that do not benefit from sufficient Grid resources as required by their computational requirements. An extension is proposed to the dynamic critical path scheduling algorithm to deal with the general resource leasing model encountered in today's commercial Clouds. The availability of the cheaper and unreliable Spot instances is analyzed and their potential to complement the unavailability of Grid resources for large workflow executions are studied. Experimental results demonstrate that Spot instances represent a 60% cheaper but equally reliable alternative to Standard instances provided that a correct user bet is made.

We would like to take the opportunity of thanking the authors who submitted a contribution, as well as the Euro-Par Organizing Committee, and the external referees with their useful comments, whose efforts have made this conference and this topic possible.

Scalable Reed-Solomon-Based Reliable Local Storage for HPC Applications on IaaS Clouds

Leonardo Bautista Gomez[1], Bogdan Nicolae[2], Naoya Maruyama[5],
Franck Cappello[2,3], and Satoshi Matsuoka[1,4]

[1] Tokyo Institute of Technology, Japan
[2] INRIA, France
[3] University of Illinois at Urbana Champaign, USA
[4] National Institute of Informatics, Japan
[5] RIKEN AICS, Japan

Abstract. With increasing interest among mainstream users to run HPC applications, Infrastructure-as-a-Service (IaaS) cloud computing platforms represent a viable alternative to the acquisition and maintenance of expensive hardware, often out of the financial capabilities of such users. Also, one of the critical needs of HPC applications is an efficient, scalable and persistent storage. Unfortunately, storage options proposed by cloud providers are not standardized and typically use a different access model. In this context, the local disks on the compute nodes can be used to save large data sets such as the data generated by Checkpoint-Restart (CR). This local storage offers high throughput and scalability but it needs to be combined with persistency techniques, such as block replication or erasure codes. One of the main challenges that such techniques face is to minimize the overhead of performance and I/O resource utilization (i.e., storage space and bandwidth), while at the same time guaranteeing high reliability of the saved data. This paper introduces a novel persistency technique that leverages Reed-Solomon (RS) encoding to save data in a reliable fashion. Compared to traditional approaches that rely on block replication, we demonstrate about 50% higher throughput while reducing network bandwidth and storage utilization by a factor of 2 for the same targeted reliability level. This is achieved both by modeling and real life experimentation on hundreds of nodes.

1 Introduction

In recent years High Performance Computing (HPC) applications have seen an increasing adoption among mainstream users, both in academia and industry. Unlike "hero" applications that are designed to run on powerful (and expensive!) supercomputers, mainstream users typically need to run medium-sized jobs that need no more than a couple of thousands of cores. For these types of jobs, Infrastructure-as-a-Service (IaaS) [4] cloud platforms present a viable alternative to purchasing dedicated resources: with thousands of virtual machines (VMs) routinely allocated by large IaaS providers [2], users can easily lease a virtual environment on the cloud for their HPC applications.

However, running HPC applications in an efficient fashion on IaaS clouds is challenging. One such open challenge is how to handle storage. Unlike supercomputing infrastructure, where storage is typically handled using a POSIX-compatible parallel file

C. Kaklamanis et al. (Eds.): Euro-Par 2012, LNCS 7484, pp. 313–324, 2012.

system (e.g., GPFS [22] or PVFS [10]), IaaS clouds feature a large variety of specialized storage solutions that are not standardized, which makes it difficult to port HPC applications. Furthermore, these storage services are often geared towards high-availability rather than high performance, not to mention that they incur costs proportional to the I/O space and bandwidth utilization. One solution to this problem is to rely on the local storage available to the VM instances. In a common IaaS configuration, local storage is plentiful (several hundreds of GB), up to an order of magnitude faster [17] and does not incur any extra operational costs. Furthermore, most HPC applications can directly take advantage of local storage or require little modifications to do so, which greatly increases scalability.

Despite its obvious advantages, local storage has a major issue: it relies on commodity components that are prone to failures [24]. Even if local disks did not fail, they would become inaccessible if their hosting compute nodes failed, effectively leading to loss of data. As a consequence, we need to deal with the reliability of local storage in order to be able to leverage it in our context. However, this invariably introduces an additional overhead, both performance-wise and resource-wise. Current cloud storage services achieve reliability and availability by replicating data, often three or more copies. However, data replication is highly space and bandwidth consuming, and it leads to an inefficient use of available resources. In this paper we explore the use of Reed-Solomon (RS) [21] based erasure encoding to address the reliability requirement for local storage in a scalable and efficient fashion. We aim to achieve a low overhead for our scheme, such that it can sustain a high I/O data access throughput and a high reliability level with minimal storage space and bandwidth utilization.

Our contributions can be summarized as follows:

– We propose a novel Reed-Solomon based encoding algorithm specifically optimized to conserve total system bandwidth in scenarios where large amounts of data are concurrently dumped to the local disks, which ultimately diminishes operational costs and frees up more bandwidth for the applications themselves. (Section 3.1)
– We introduce a formal model to compare data-replication and RS encoding analytically in order to predict the storage space and network bandwidth utilization for different levels of reliability. (Section 4)
– We show how to implement our approach in practice by integrating it into *BlobCR* [19], a distributed checkpoint-restart framework that is specifically designed to take persistent snapshots of local storage for HPC applications that are running on IaaS clouds. (Section 3.2)
– We evaluate our approach experimentally on hundreds of nodes of the Grid'5000 testbed [7], both with synthetic benchmarks and a real-life application. These experiments demonstrate significant improvement in performance and resource utilization when compared to replication for the same reliability level. (Section 5)

2   Related Work

There is a rich storage ecosystem around IaaS clouds. Cloud providers typically offer their own storage solutions, which are not standardized and expose a different access model than POSIX: key-value stores based on REST-ful access APIs [3], distributed file systems specialized for MapReduce workloads [23], database management systems [14], etc. Besides the disadvantages presented in Section 1, most of these solutions are optimized for high-availability, under the assumption that data is frequently read and only seldom updated. HPC applications on the other hand typically generate new data often and need to read it back only rarely (e.g. checkpointing). The idea of leveraging local storage for HPC applications running on IaaS clouds in a reliable fashion was introduced before by our previous work: we presented *BlobCR* [19], a checkpoint-restart framework that is able to take snapshots of local storage that survive failures. However, in order to survive failures, BlobCR relies on replication, which can lead to excessive use of storage space and bandwidth.

Different studies of block replication and erasure codes were performed before in the context of RAID systems [8], whose implementation is at hardware level, as well as for distributed data storage [25] implemented at the software level. More recently, DiskReduce [12] and Zhe Zhang et al. [26] study the feasibility of replacing three-way replication with erasure codes in cloud storage / large data centers. They are probably the closest work to the technique proposed in this paper, going in a similar direction and complementing our own approach: while they focus on space reduction and performance, in this work we focus on limiting the network bandwidth consumption as much as possible by using a novel low-communication RS encoding, which significantly decreases the application performance overhead.

RS encoding algorithms [15,11] similar to the proposal presented in our previous work create encoding groups that encompass blocks stored on different nodes. Other encoding techniques, such as bitwise XOR [16] also encode distributed blocks. Such algorithms are suitable for scenarios such as coordinated checkpoint where multiple distant processes need to reliably store data at the same time; in such context, distributed blocks of data can be encoded after synchronization. However, these algorithms cannot be used for a storage system where isolated writes will need to be performed without imposing any synchronization with other processes. Therefore, a fundamental algorithm change is necessary in order to encode isolated blocks of data.

3   Our Approach: A RS-Encoding Algorithm Proposal

In this section we introduce a novel low-communication RS encoding algorithm that guarantees high reliability with a low bandwidth consumption. The key observation leading to this choice is the widening gap between the cost of computational power and network bandwidth, for which reason we try to conserve network bandwidth at the expense of slightly higher CPU utilization.

We start with a quick overview of the RS encoding. As we can see in Figure 1(a), the RS encoding takes a data vector and encodes it by performing a matrix-vector multiplication with a distribution matrix. To recover any m failures, any sub-square matrix of the distribution matrix (including minor) must be non-singular. Thus, in practice we usually use a Cauchy or a Vandermonde matrix [20]. While using RS encoding to encode distributed data, the data vector is composed by blocks of data of distant nodes. In Figure 1(b), we can see the data to be encoded in the case of diskless checkpointing. Each process P_i holds blocks of data B_{ij} where j is the block index, going from

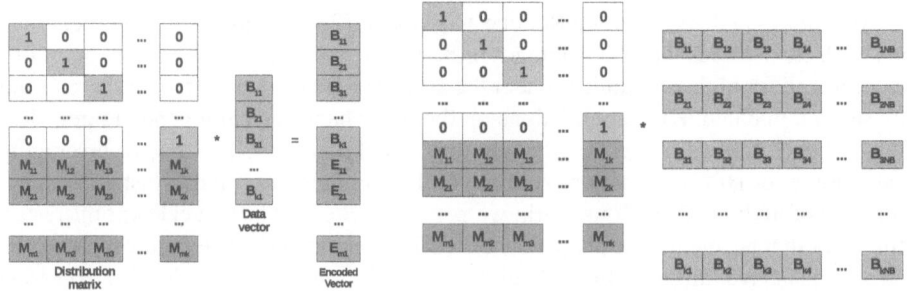

(a) Reed-Solomon encoding   (b) Reed-Solomon for distributed data

Fig. 1. Reed-Solomon encoding study

1 to NB (Number of Blocks in the checkpoint file). The first data vector will be composed of the first block of each checkpoint file, the second vector of the second block of each checkpoint file, and so on. There are several ways to implement this computation, for example, one can use MPI reductions [15], a pipeline algorithm [11] or the star algorithm proposed in our previous work [13]. However, all these algorithms require synchronizations while encoding the data. Furthermore, these algorithms produce as much (or more) communications than data replication; therefore, none of them is suitable for our purpose.

3.1 Low-Communication RS Encoding Algorithm

In an ideal setting where all m encodings generated by RS are stored on different nodes (assuming that the data blocks themselves are stored on different nodes as well), the system can tolerate m simultaneous erasures. However, this ideal scenario is costly in terms of resource utilization, as it results in the need to employ dedicated parity nodes. Thus, to leverage the available resources better, one idea is to store the encodings on the same nodes where the data itself is stored. The distributed algorithm (Algorithm 1) presented in our previous work [13,5] illustrates this idea. We propose to generate as many encodings as the group size ($m = k$), while evenly distributing the encodings among the same nodes where the data blocks are stored. Thus, one node failure will lead to two erasures and each group will be able to tolerate up to 50% of failed nodes.

In order to avoid synchronizations between processes during the encoding, we propose a novel low-communication algorithm (Algorithm 2). Instead of encoding the ith block of each one of the k nodes, we encode the first k blocks of each node and then we scatter the k blocks of original data plus the k encodings on the k nodes of the group. Notice that both algorithms encode the same amount of data and store all the original and parity data on the encoding nodes, thus offering the same level of reliability.

With respect to bandwidth consumption, Algorithm 1 needs to transfer $NB*m$ blocks in total. Since $m = k$, its communication cost is thus $Comm_{Alg_1} = NB*k$. On the other hand, for Algorithm 2 we have $Comm_{Alg_2} = \frac{NB}{k}*k*2 = NB*2$. By comparing the two

Algorithm 1. Distributed RS encoding algorithm	**Algorithm 2.** Low-communication RS encoding algorithm
1: ▷ r : the process rank	1: **for** $i \leftarrow 0..(NB/k) - 1$ **do**
2: ▷ NB : the number of blocks	2: **for** $j \leftarrow 1..k$ **do**
3: ▷ k : the group size	3: $B_{r(i*k+j)} \leftarrow$ read $(i*k+j)^{th}$ block
4: ▷ m : the number of encodings	4: **end for**
5: **for** $i \leftarrow 1..NB$ **do**	5: **for** $j \leftarrow 1..k$ **do**
6: $B_{ri} \leftarrow$ read i^{th} block	6: **for** $l \leftarrow 1..m$ **do**
7: **for** $j \leftarrow 1..m$ **do**	7: $E_{ij} \leftarrow E_{ij} + M_{jl} * B_{r(i*k+l)}$
8: $T \leftarrow M_{(r+j)r} * B_{ri}$	8: **end for**
9: Send T to P_{r+j}	9: **end for**
10: $F \leftarrow$ Recv from P_{r-j}	10: **for** $j \leftarrow 1..k$ **do**
11: $E_{ri} \leftarrow E_{ri} + F$	11: Send $B_{r(i*k+j+r)}$ and $E_{i(j+r)}$ to P_{j+r}
12: **end for**	12: Recv FB_j and FE_j from P_{j-r}
13: $writeE_{ri}$	13: Write FB_j and FE_j
14: **end for**	14: **end for**
	15: **end for**

formulas one can easily notice that the low-communication algorithm is more bandwidth friendly than the distributed algorithm. Particularly, in the case of the distributed algorithm the data transferred over the network will increase proportionally with the group size k, while the low-communication algorithm will keep it constant.

Figure 2(a) shows a performance comparison and time breakdown of both encoding algorithms measured on Tsubame2 [1], for different numbers of cores (96 and 192 cores). Although communications and computation are overlapped in both cases, the low-communication algorithm is 24% faster because of the data locality. Reducing communications not only decreases the stress on the network but also increases cache efficiency. For isolated writes on a storage system, one can design a system that implements a multi-stage striping: each chunk of data that makes up the local storage can be itself divided into the k blocks that are fed to the low-communication encoding algorithm.

3.2   Integration in Practice

In order to illustrate the benefits of the algorithm presented in the previous section in practice, we have integrated our approach into *BlobCR* [19], a distributed checkpoint-restart framework based on *BlobSeer* [18] that is specifically designed to take persistent snapshots of local storage for HPC applications that are running on IaaS clouds. BlobCR exposes local storage to the VM instances as virtual disks that can be attached to them. The initial content of the virtual disk is striped into chunks and stored in a distributed fashion among the participating storage elements. Whenever a virtual disk is attached to a VM instance, an initially empty mirror of it is created on the local disk. Reads to the virtual disk fetch any remote chunks not present in the mirror, gradually filling it on-demand. Writes to the virtual disk are always performed on the mirror. A special primitive can be used to persistently save the mirror as a new snapshot of the virtual disk that is globally shared. This is done by distributing all locally modified chunks among the storage elements, then by consolidating these changes using cloning and shadowing.

In order to provide high reliability for the chunks that make up the disk snapshots, BlobCR relies on the reliability scheme implemented in BlobSeer, which by default is replication: each chunk is stored to multiple local disks. We implemented an alternative reliability scheme based on the algorithm presented in Section 3.1 which was then integrated into BlobSeer. More precisely, instead of replicating each chunk to multiple local disks, we perform a second level of striping that splits each chunk into k small, equally sized blocks. These blocks form a group to which erasure coding is applied in order to obtain a second group of k blocks that hold parity information. Once this step has completed, we distribute the $2 * k$ blocks among a set of $2 * k$ different remote disks.

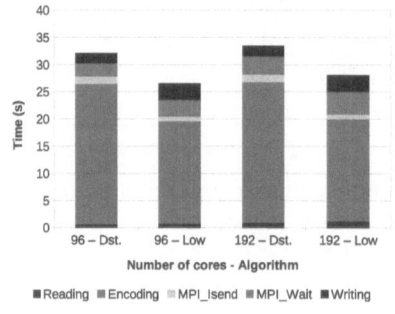

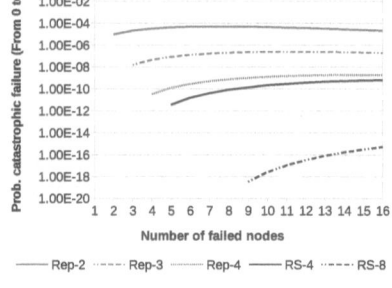

(a) Performance of Algorithm 1 vs. Algorithm 2 (1 GB/core)

(b) Reliability of replication vs. RS encoding (1000 nodes)

Fig. 2. Performance compared to previous work and reliability modeling

4   Reliability, Storage and Network Bandwidth Study

In this section we develop a model to predict the performance, storage and network bandwidth cost of both approaches (data replication and RS encoding) for comparable levels of reliability. As explained in Section 3, the distributed encoding algorithm is not suitable for storage systems, therefore we do not include it in this comparison.

First, we focus on the reliability level of both approaches. We use the reliability model presented in our previous work [6] to compute the probability of *catastrophic failures*, i.e., failures that lead to unrecoverable data loss. This will depend on the number of simultaneous erasures and the probability of those erasures to hit the replicated or parity data of a given data chunk. Figure 2(b) shows the difference of reliability for five different settings: our approach using a group size of four and eight (denoted *rs-4* and *rs-8*) vs. replication using a replication factor of two, three and four (denoted *rep-2*, *rep-3* and *rep-4* respectively). Notice that *rep-4* comes very close (without surpassing) to *rs-4* in terms of reliability only when the replication factor reaches 4. For this reason, we consider *rep-4* and *rs-4* comparable in terms of reliability for the rest of this paper.

A fair comparison between both techniques should study the storage overhead and performance overhead necessary to guarantee a comparable level of reliability. The replication technique replicates chunks of data and stores them on multiple remote disks. Similarly, the RS encoding technique generates parity data and stores it on multiple remote nodes. Let us assume that we want to reliably write a chunk of z bytes of data. We assume a replication factor of k and an RS encoding group size of k. In the replication approach we store a total of $Str_{rep} = z*k$ bytes of data. In contrast, for the RS technique we need to split the chunk in k blocks, encode them and finally store a total of $Str_{rs} = k * \frac{z}{k} * 2 = z*2$ bytes in the system. As we can see, the replication approach becomes prohibitively expensive quickly, while RS encoding is scalable in terms of storage. In addition, the amount of data transferred over the network is equal to the amount of data stored for both approaches, which means that data replication transfers more data than RS encoding for a similar level of reliability. For instance, *rep-4* transfers and stores two times more data than *rs-4*. However, we also should notice that the RS technique imposes an overhead due to the encoding work. In the next section we measure and compare the performance overhead of both approaches.

5   Experimental Evaluation

This section evaluates the benefits of our proposal both in synthetic settings and for scientific HPC applications.

5.1   Experimental Setup

The experiments were performed on Grid'5000 [7], an experimental testbed for distributed computing that federates nine sites in France. We used 100 nodes of the griffon cluster from the Nancy site, each of which is equipped with a quadcore Intel Xeon X3440 x86_64 CPU with hardware support for virtualization, local disk storage of 278 GB (access speed $\simeq$55 MB/s using SATA II AHCI driver) and 16 GB of RAM. The nodes are interconnected with Gigabit Ethernet (measured 117.5 MB/s for TCP sockets with MTU = 1500 B with a latency of $\simeq$0.1 ms).

The hypervisor running on all compute nodes is KVM 0.14.0, while the operating system is a recent Debian Sid Linux distribution. For all experiments, a 2 GB raw disk image file based on the same Debian Sid distribution was used to boot the guest operating system of the virtual machine instances that run the user application.

5.2   Synthetic Benchmarks

Our first series of experiments evaluates the performance of our approach vs. replication in two synthetic benchmarking scenarios. We compare the same five settings as in Section 4. In all five settings we aim to measure the maximal theoretical performance levels that can be achieved during checkpointing. We chose checkpointing because it is one of the tasks that stress the most the storage. More specifically, we measure the sustained throughput when checkpointing local modifications to the virtual disk that amount to 512MB. We omit running any application in parallel with the checkpointing process in

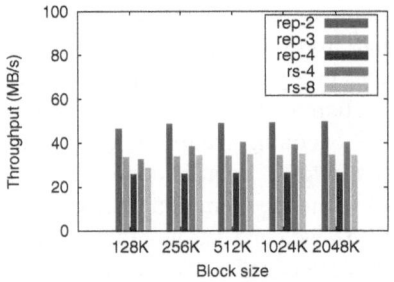

(a) Single checkpointer, variable chunk size

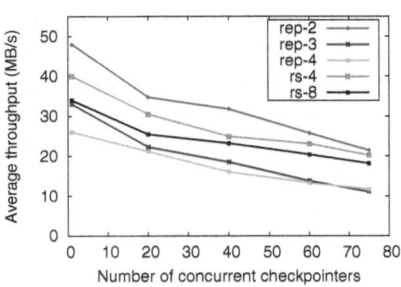

(b) Increasing nr. of concurrent checkpoint-
ers, fixed chunk size (512K)

Fig. 3. Our approach vs. replication: Sustained throughput when checkpointing 512MB worth of data (no computation running in parallel)

order to better control the experimental setting and eliminate any influence caused by the computation. Instead, the local modifications are simply randomly generated data that is written to the virtual disk before checkpointing.

The first benchmarking scenario evaluates how the chunk size impacts the sustained throughput. To this end, we deploy BlobCR on all available nodes (100) and run a single checkpointing process in each of the five settings using a variable chunk size, ranging from 128K to 2048K. The results are depicted in Figure 3(a). As expected, *rep-2*, *rep-3* and *rep-4* roughly achieve 1/2, 1/3 and 1/4 of the maximal throughput, with a slightly increasing trend as the chunk size gets larger (which is the consequence of fewer messages and thus lower overhead due to latency). This trend is observable for *rs-4* and *rs-8* too, slightly more pronounced due to the larger number of messages (i.e. 4 and 8 respectively per chunk).

When comparing our approach to replication, the advantage of lower amount of data transfer is clearly visible: for 2048K chunks, *rs-4* achieves a throughput that is 38% higher than *rep-4* while guaranteeing a comparable reliability level. At the same time, it reduces bandwidth and storage space utilization by more than 100% as predicted by our model (See Section 4). Comparing *rs-4* to *rep-2* (which consumes the same amount of bandwidth and storage space), we observe a decrease in throughput of less than 20%. This overhead is a worthwhile investment, considering that *rs-4* increases the reliability level over *rep-2* by several orders of magnitude. Even for small chunk sizes (going as low as 128KB), the advantage of our approach is clearly visible: the throughput achieved by *rs-4* is only 24% lower than *rep-2* and more than 25% higher than *rep-4*. Finally, we note better scalability for our approach: *rs-8* is already 33% faster than *rep-4*, while increasing the reliability level yet again several orders of magnitude.

So far we have analyzed our approach vs. replication for a single checkpointing process only. Our second benchmarking scenario evaluates how the five settings compare in a highly concurrent setting where an increasing number of checkpointing processes need to save the checkpointing data simultaneously and thus compete for the system bandwidth. We fix the chunk size to 512K and gradually increase the number of checkpointing processes, from one up to 75, while measuring the average sustained throughput. These results are represented in Figure 3(b).

As can be observed, all five approaches suffer a performance degradation with increasing number of checkpointing processes. In case of *rep-2*, the throughput drops from 47 MB/s to little over 20 MB/s. Although starting lower than *rep-2*, both *rs-4* and *rs-8* catch up with *rep-2* under concurrency: the average throughput drops to 20 MB/s and 19 MB/s respectively. This shows that competition for system bandwidth effectively hides the encoding overhead of our approach at larger scales, enabling it to be more scalable: we sustain virtually the same throughput as *rep-2* for the same storage space and bandwidth utilization, albeit at a much higher resilience level. Compared to *rep-3* and *rep-4*, where the throughput drops to little over 10 MB/s, we can observe an even more dramatic improvement than in the case of a single checkpointing process: we sustain a throughput almost 100% higher while keeping the same reliability level (*rs-4*) and even increasing it several levels of magnitude (*rs-8*).

5.3 Real-Life Application: CM1

In the previous section we have evaluated the throughput of our persistent storage under concurrency for an ideal environment where no application is running. However, in real life the application continues running after requesting a virtual disk snapshot to be taken, while the virtual disk snapshot is persisted in the background. This limits the amount of resources available to the persistency process: (1) there is less overall available system bandwidth, as the application generates network traffic; (2) there is less available computational power on each compute node, as the application processes run inside the virtual machine and consume CPU.

To illustrate the impact of these limitations in real life, we have chosen *CM1*: a three-dimensional, non-hydrostatic, non-linear, time-dependent numerical model suitable for idealized studies of atmospheric phenomena. This application is used to study small-scale phenomena that occur in the atmosphere of the Earth (such as hurricanes) and is representative of a large class of HPC applications that perform stencil (nearest-neighbor) computations.

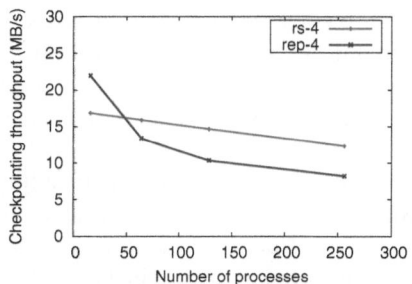 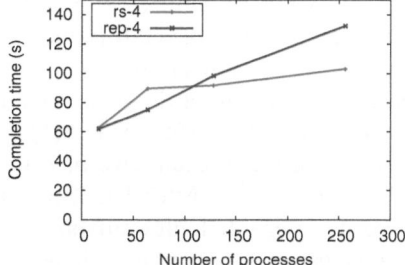

(a) Asynchronous checkpointing throughput (higher is better)

(b) Time to finish running the application (lower is better)

Fig. 4. Our approach vs. replication: Performance in the real world (CM1)

We perform a weak scalability experiment that solves an increasingly large problem (constant workload per core), starting from 16 processes up to 256 processes. As input data for CM1, we have chosen a 3D hurricane that is a version of the Bryan and Rotunno simulations [9]. The processes are distributed in pairs of four among virtual machine instances that are allotted 4 cores, such that each process has its own dedicated core. This represents the worst case scenario for our approach, as the block encoding must compete with the application processes. We compare two approaches that offer the same reliability level: *rs-4* and *rep-4*. The checkpointing frequency is fixed such that a single checkpoint is taken throughout the application run-time.

The results are depicted in Figure 4. As expected, the average checkpointing through-put (Figure 4(a)) is smaller than in the ideal case (See Section 5.2) and drops with increasing number of processes. The combined effects of both concurrent checkpoint-ing and application communication quickly saturate the system bandwidth in the case of *rep-4*, which leads to a dramatic drop in average checkpointing throughput from 22 MB/s to 8 MB/s. Our approach on the other hand is more scalable: it presents a drop from 16 MB/s to 12 MB/s. Unlike the ideal case, this time *rs-4* is significantly slower because less computation power is available for encoding. Nevertheless, it is still 50% faster than *rep-4 in the worst case* and has the potential to become even 150% faster if a dedicated core can be spared for the encoding. Taking a look at the completion times (Figure 4(b)), *rs-4* is again much more scalable: the completion time increases from 60s to 100s, which is 35% less than the increase observed in the case of *rep-4* (135s). This advantage of *rs-4* over *rep-4* can be traced back to the twice as lower bandwidth consumption, which effectively increases the bandwidth available to the application.

6   Conclusions

A large class of HPC applications can take advantage of local storage in order to im-prove performance and scalability while reducing resource utilization, however doing so raises reliability issues. In this paper we have presented a scalable Reed-Solomon based algorithm that provides a high degree of reliability for local storage at the cost of very low computational overhead and using a minimal amount of communication.

We demonstrated the benefits of our approach both theoretically through a perfor-mance and resource utilization model, as well as in practice through extensive experi-ments performed on the G5K testbed. Compared to transitional approaches that rely on replication, we show up to 50% higher throughput and 2x lower bandwidth / storage space consumption for the same reliability level, which improves overall performance of a real life HPC application (CM1) up to 35%.

Based on these results, we plan to explore in future work the issue of reliability of local storage in greater detail. In particular, we are investigating how to extend our approach to enable high availability of data under concurrent read scenarios: in this context, parity information could be used to avoid contention to the original data. This is important for a large number of HPC applications that need to share the same initial datasets between processes.

Acknowledgments. This work was supported in part by the ANR-10-01-SEGI project, the ANR-JST FP3C project and the Joint Laboratory for Petascale Computing, an INRIA-UIUC initiative. The experiments presented in this paper were carried out using the Grid'5000/ALADDIN-G5K experimental testbed, an initiative of the French Ministry of Research through the ACI GRID incentive action, INRIA, CNRS and RENATER and other contributing partners (see http://www.grid5000.fr/).

References

1. Tsubame2, http://tsubame.gsic.titech.ac.jp
2. Amazon Elastic Compute Cloud (EC2), http://aws.amazon.com/ec2/
3. Amazon Simple Storage Service (S3), http://aws.amazon.com/s3/
4. Armbrust, M., Fox, A., Griffith, R., Joseph, A., Katz, R., Konwinski, A., Lee, G., Patterson, D., Rabkin, A., Stoica, I., Zaharia, M.: A view of cloud computing. Commun. ACM 53, 50–58 (2010)
5. Bautista-Gomez, L.A., Nukada, A., Maruyama, N., Cappello, F., Matsuoka, S.: Low-overhead diskless checkpoint for hybrid computing systems. In: HiPC 2010: Proceedings of the 2010 International Conference on High Performance Computing, Goa, India, pp. 1–10 (2010)
6. Bautista-Gomez, L.A., Tsuboi, S., Komatitsch, D., Cappello, F., Maruyama, N., Matsuoka, S.: FTI: high performance fault tolerance interface for hybrid systems. In: SC 2011: 24th International Conference for High Performance Computing, Networking, Storage and Analysis, Seattle, USA, pp. 32:1–32:12 (2011)
7. Bolze, R., Cappello, F., Caron, E., Daydé, M., Desprez, F., Jeannot, E., Jégou, Y., Lanteri, S., Leduc, J., Melab, N., Mornet, G., Namyst, R., Primet, P., Quetier, B., Richard, O., Talbi, E.G., Touche, I.: Grid'5000: A large scale and highly reconfigurable experimental grid testbed. Int. J. High Perform. Comput. Appl. 20, 481–494 (2006)
8. Brown, A., Patterson, D.A.: Towards availability benchmarks: a case study of software raid systems. In: ATEC 2000: Proceedings of the USENIX Annual Technical Conference, pp. 22:1–22:15. USENIX Association, San Diego (2000)
9. Bryan, G.H., Rotunno, R.: The maximum intensity of tropical cyclones in axisymmetric numerical model simulations. Journal of the American Meteorological Society 137, 1770–1789 (2009)
10. Carns, P.H., Ligon, W.B., Ross, R.B., Thakur, R.: PVFS: A parallel file system for Linux clusters. In: Proceedings of the 4th Annual Linux Showcase and Conference, Atlanta, USA, pp. 317–327 (2000)
11. Chen, Z., Dongarra, J.: A scalable checkpoint encoding algorithm for diskless checkpointing. In: HASE 2008: Proceedings of the 11th IEEE High Assurance Systems Engineering Symposium, pp. 71–79. IEEE Computer Society, Nanjing (2008)
12. Fan, B., Tantisiriroj, W., Xiao, L., Gibson, G.: Diskreduce: Raid for data-intensive scalable computing. In: PDSW 2009: Proceedings of the 4th Annual Workshop on Petascale Data Storage, pp. 6–10. ACM, Portland (2009)
13. Gomez, L.A.B., Maruyama, N., Cappello, F., Matsuoka, S.: Distributed diskless checkpoint for large scale systems. In: CCGRID 2010: Proceedings of the 10th IEEE/ACM International Conference on Cluster, Cloud and Grid Computing, CCGRID 2010, pp. 63–72. IEEE Computer Society, Melbourne (2010)
14. Lakshman, A., Malik, P.: Cassandra: A decentralized structured storage system. SIGOPS Oper. Syst. Rev. 44, 35–40 (2010)

15. da Lu, C.: Scalable Diskless Checkpointing for Large Parallel Systems. Ph.D. thesis, Univ. of Illinois at Urbana-Champaign (2005)
16. Moody, A., Bronevetsky, G., Mohror, K., de Supinski, B.R.: Design, modeling, and evaluation of a scalable multi-level checkpointing system. In: SC 2010: Proceedings of the 23rd International Conference for High Performance Computing, Networking, Storage and Analysis, pp. 1–11. IEEE Computer Society, New Orleans (2010)
17. Nadgowda, S.J., Sion, R.: Cloud Performance Benchmark Series: Amazon EBS, S3, and EC2 Instance Local Storage. Tech. rep., Stony Brook University (2010)
18. Nicolae, B., Antoniu, G., Bougé, L., Moise, D., Carpen-Amarie, A.: BlobSeer: Next-generation data management for large scale infrastructures. J. Parallel Distrib. Comput. 71, 169–184 (2011)
19. Nicolae, B., Cappello, F.: BlobCR: Efficient Checkpoint-Restart for HPC Applications on IaaS Clouds using Virtual Disk Image Snapshots. In: SC 2011: 24th International Conference for High Performance Computing, Networking, Storage and Analysis, Seattle, USA, pp. 34:1–34:12 (2011)
20. Plank, J., Xu, L.: Optimizing cauchy reed-solomon codes for fault-tolerant network storage applications. In: Fifth IEEE International Symposium on Network Computing and Applications, NCA 2006, pp. 173–180 (July 2006)
21. Reed, I.S., Solomon, G.: Polynomial codes over certain finite fields. Journal of the Society for Industrial and Applied Mathematics 8(2), 300–304 (1960)
22. Schmuck, F., Haskin, R.: GPFS: A shared-disk file system for large computing clusters. In: FAST 2002: Proceedings of the 1st USENIX Conference on File and Storage Technologies, Monterey, USA (2002)
23. Shvachko, K., Huang, H., Radia, S., Chansler, R.: The Hadoop distributed file system. In: MSST 2010: 26th IEEE Symposium on Massive Storage Systems and Technologies, pp. 1–10 (2010)
24. Vishwanath, K.V., Nagappan, N.: Characterizing cloud computing hardware reliability. In: SoCC 2010: Proceedings of the 1st ACM Symposium on Cloud Computing, Indianapolis, USA, pp. 193–204 (2010)
25. Weatherspoon, H., Kubiatowicz, J.D.: Erasure Coding Vs. Replication: A Quantitative Comparison. In: Druschel, P., Kaashoek, M.F., Rowstron, A. (eds.) IPTPS 2002. LNCS, vol. 2429, pp. 328–338. Springer, Heidelberg (2002)
26. Zhang, Z., Deshpande, A., Ma, X., Thereska, E., Narayanan, D.: Does erasure coding have a role to play in my data center? Tech. Rep. MSR-TR-2010-52, Microsoft Research (2010)

Caching VM Instances for Fast VM Provisioning: A Comparative Evaluation

Pradipta De[1], Manish Gupta[1], Manoj Soni[2], and Aditya Thatte[1]

[1] IBM Research India, New Delhi
{pradipta.de,gmanish,adthatte}@in.ibm.com
[2] Georgia Institute of Technology, Atlanta, GA, USA
manojsoni@gatech.edu

Abstract. One of the key metrics of performance in an infrastructure cloud is the speed of provisioning a virtual machine (or a virtual appliance) on request. A VM is instantiated from an image file stored in the image repository. Since the image files are large, often GigaBytes in size, transfer of the file from the repository to a compute node running the hypervisor can take time in the order of minutes. In addition to it, booting an image file can be a time consuming process if several applications are pre-installed. Use of caching to pre-fetch items that may be requested in future is known to reduce service latency. In order to overcome the delays in transfer and booting time, we prepare a VM a priori, and save it in a standby state in a "cache" space collocated with the compute nodes. On receiving a matching request, the VM from the cache is instantly served to the user, thereby reducing service time. In this paper, we compare multiple approaches for pre-provisioning and evaluate their benefits. Based on usage data collected from an enterprise cloud, and through simulation, we show that a reduction of 60% in service time is achievable.

1 Introduction

Time to service a request for a new virtual machine in a cloud can often require several minutes. The complete workflow beginning with receiving a request till a new virtual machine is delivered to the user, follows a number of steps. First, the requested machine image template from which the VM must be instantiated is looked up in the image repository, then the image template file is copied to a compute host and the VM is then booted up. Image template files are very large in size, often in GigaBytes range. Transferring such large files over the network is time consuming. In addition to it, the boot process can be slow depending on the number of pre-installed components in the image. Due to these bottlenecks [8,19], servicing a provisioning request can take a long time.

In order to speed up virtual server provisioning, there have been approaches to expedite the transfer of the large template files using different streaming techniques [4, 18]. Caching of the template files at the compute nodes to mask the transfer latency has also been explored [8]. To reduce the boot time, one approach is to instantiate a VM, and store it in a standby mode in the cache. This saves the time to create an instance from a template and boot the VM. In essence, an inventory of readily deliverable VMs are maintained based on user request patterns.

C. Kaklamanis et al. (Eds.): Euro-Par 2012, LNCS 7484, pp. 325–336, 2012.

In this work, we compare 3 techniques for pre-provisioning VM instances. Given a fixed size cache space, each technique comes up with the composition of the inventory. In other words, for a template type, the expected number of VM instances to be requested is calculated and pre-provisioned. If a request matches a pre-provisioned VM instance, it is delivered with minimal delay to the user. The cache or VM inventory space is freed up when a VM instance is delivered to a user. The inventory is replenished periodically with new VM instances. We have analyzed the server request trace from an enterprise-wide cloud deployment to study the request pattern. We compare three approaches to highlight the benefits of each technique in maintaining the inventory or cache of VM instances for reducing the server provisioning time in cloud.

The rest of the paper is organized as follows. In Section 2, we present literature related to caching in different contexts. In Section 3, we present an overview of the cloud architecture, along with our simulation approach. Section 4 presents the pre-provisioning methodologies. In Section 5, we present detailed simulation results showing the benefits of pre-provisioning. Finally we conclude in Section 6.

2   Related Work

Caching as a technique has been extensively researched, specially in the domain of operating system and design of memory hierarchies. Our focus is more on the use of caching techniques in the context of cloud.

Web Caching and CDN: Application of caching serve content that is accessed repetitively over the Internet has been well studied [6]. Pre-fetching web resources by anticipating future trends is a challenging problem. Usage prediction methods, that apply clustering [11], neural networks [13], have proved to be of limited benefit. [6] reports that cache hit rates can rarely cross 40-50%. Instead of tackling the problem of usage prediction using standard methods, in this work we gain insight into the peculiarities of a trace to leverage the usage pattern.

In web caching, often the most common cache replacement strategy of replacing the Least-Recently-Used (LRU) item works quite well, as shown in [16]. However, in a cloud, requests from one user may arrive in bursts, therefore, the replacement policy may be inefficient if replacement is performed one item at a time. Rather, in our strategy we predict the most suitable set of image templates, as well as, the number of VM instances of each type.

Content Delivery Networks(CDN) uses caching at Internet scale. The key idea in CDN is to push content closer to the user before a request arrives. [12] proposes scheduling algorithms to push content in a timely manner to proxy cache servers, while [1] discusses a cooperative cache management scheme to maximize traffic volume served from the cache. These schemes can be useful in pushing the cloud images nearer to the user, but we are also interested in maintaining multiple instances of a template.

Caching in Cloud Context: Caching machine image templates has been studied in grid and cluster environments. In [8], Emeneker et al. show that caching of a virtual image can speed up execution of parallel jobs. However, they do not explore the pros and cons of different caching approaches.

Predictive methods are used more frequently in cloud context in order to manage resource requirement. Tackling the problem of when to scale resources by adding more VMs, or how many resources to apportion to a cloud setup has been addressed by several works. In [2], the authors develop a forecasting method to predict resource demand in a cloud by using historic data. An approach to auto-scale during flash crowds is presented in [20]. Use of cloning or de-duplication techniques to quickly add new VMs have been addressed in [10]. In our method, we deliver the server instance instantly if there is a cached VM instance matching the request to avoid the overhead of cloning.

Several works identify that fetching an image from a central repository takes up significant proportion of the service time of a request [8, 19]. A BitTorrent-like distribution system has been proposed to speed up image delivery in [4]. A similar image streaming approach, by breaking up an image in chunks, has been proposed in [18]. We are complementary to these schemes since we can leverage the speedy distribution of the image once it has been identified by our scheme.

Moka5 provides a solution for desktop virtualization, where desktop snapshots are generated at regular intervals and stored in a central repository [3, 14]. However, it assumes a large storage repository and does not present any intelligent caching mechanism to speed up the delivery of images. Similarly, Eucalyptus cloud management system mentions the use of caching without providing details [9, 15]. Even, Amazon mentions that frequent users will have the benefit of a faster turnaround time which hints at underlying caching. IBM Workload Deployer, previously known as WebSphere CloudBurst Appliance, also mentions the use of caching [5]. However, caching prepared VM instances, as opposed to image templates, distinguishes our work.

Inventory Management: A close look at our technique reveals that it is closest in design to an inventory management system. Whenever a matching request arrives, an instance of an image template is used up, and the inventory (cache) must be replenished efficiently at a minimum cost. The problem of lot sizing in inventory management [7] deals with selecting the appropriate quantities of each item to be provisioned in the inventory. Similar to inventory management, it is necessary to predict the appropriate number of each image template to be kept in the cache.

3   System Model and Assumptions

A typical cloud setup maintains a farm of compute nodes. The compute nodes are used to instantiate virtual machines from user-specified image templates, which are stored in an image repository. When a user makes a request for a new server instance, the request is first intercepted by the cloud provisioning engine. The provisioning engine checks for the available image type in the repository, and initiates a transfer of the image to a compute host. Once the image is transferred to a compute host, it is expanded, and booted to create an instance of a virtual machine. The SAN store is used for providing the user data space, similar to Amazon's Elastic Block Storage (EBS). Besides the provisioning requests, a user may also request deletion of an instance. Fig. 1 shows the different components of the cloud architecture.

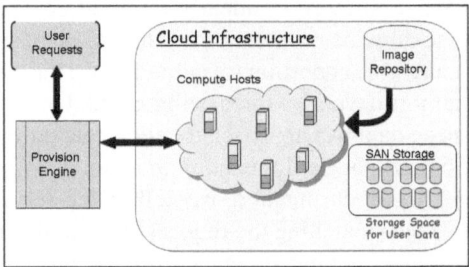

Fig. 1. Cloud architecture overview showing different functional components

3.1 Simulation Model

We model the cloud provisioning engine as a multi-server queuing system. Each server is modeled such that it handles a thread to service a request in the provisioning engine. Assuming infinite servers in the model, we can accurately compute the time to service each request, referred to as *service time*, since there is no delay in the queue. In order to model the service time of each request, we introduce a *start-event* and *finish-event* for each event type. For example, a provision request is modeled using a start and finish event for that type. Corresponding events for deletion requests are introduced.

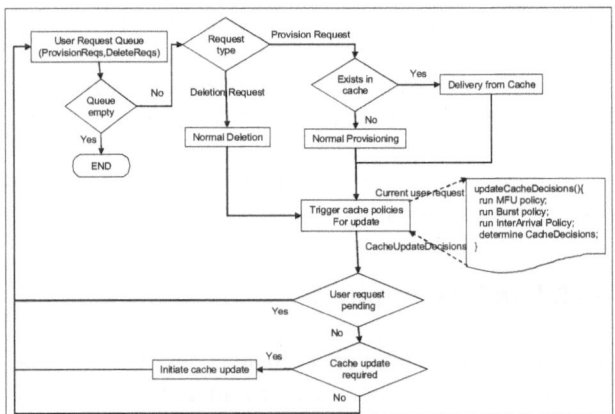

Fig. 2. Flowchart showing the steps in the simulator

In order to quantify the benefits of pre-provisioning, events denoting *cache pre-filling event* and *cache entry deletion event* are introduced. We assume that a fixed amount of space is available for pre-provisioning; therefore, we can only pre-provision a fixed number of VM instances. Before inserting a new VM instance, that are determined by the cache update policies, an instance may be deleted from the cache. The pre-fetch action never blocks a request already queued in the provisioning engine. However, once a pre-provision action is triggered, it must complete before releasing the thread. If a user request arrives while pre-fetching is in progress, then the user request must wait, thereby increasing its service time. Fig. 2 shows the flowchart for the simulator.

4   Pre-provisioning Techniques

In this section, we present the techniques for selection of the cache composition at periodic intervals. First an analytical model is explained to motivate the approaches, followed by three main approaches for selection of items to place in the cache.

4.1   Analytical Model for Pre-provisioning

The provisioning engine is modeled as a single server queue, with an additional cache entity that can store *exactly one* VM instance at a time. Request arrival for image-type-1(I_1), and image-type-2(I_2) follows Poisson distribution, with rates λ_1 and λ_2 respectively. The service time for each server request is image-type dependent, and exponentially distributed with rates, μ_1 and μ_2 respectively. A pre-provisioned instance is fetched and cached only if the cache is empty and there is no pending user request in the queue. If a request for a server instance arrives before the pre-provisioning request is complete, *the pre-provisioning request is canceled, thereby leaving the cache empty.* The cache entry is purged if there is a cache miss. On cache hit, the request is serviced instantaneously, implying zero service time.

The caching policy can be stated as follows: *Create an instance of I_1 with probability p, otherwise create an instance of I_2 with probability $(1 - p)$ whenever the system is detected to be idle. Our aim is to find the value of p such that it minimizes the average end-to-end provisioning time for the requests.*

Theorem 1. *Under the stated assumptions, the optimal policy is to set $p = 1$, otherwise set $p = 0$, if the following condition holds.*

$$\frac{\lambda_1}{\lambda_1+\lambda_2+\mu_1} > \frac{\lambda_2}{\lambda_1+\lambda_2+\mu_2}$$

Proof. Since the arrival and service processes are memory-less, due to our assumption, therefore, the evolution of the process does not depend on past history. Minimizing the expected service time is equivalent to maximizing the reduction in service time by using the cache. The reduction in expected service time can be represented as:

$$p\left(\frac{\mu_1}{\lambda_1+\lambda_2+\mu_1}\right)\left(\frac{\lambda_1}{\lambda_1+\lambda_2}\right)\frac{1}{\mu_1} + (1 - p)\left(\frac{\mu_2}{\lambda_1+\lambda_2+\mu_2}\right)\left(\frac{\lambda_2}{\lambda_1+\lambda_2}\right)\frac{1}{\mu_2}$$

The first and second terms in the expression corresponds to time reduction when I_1 and I_2 are chosen respectively. Within each block in the expression, $\frac{\mu_1}{\lambda_1+\lambda_2+\mu_1}$ corresponds to the probability that pre-provisioning I_1 completes before next request arrival; $\frac{\lambda_1}{\lambda_1+\lambda_2}$ denotes probability of arrival of I_1 request before I_2 request; and $\frac{1}{\mu_1}$ is the expected savings. The expression has the structure $pA + (1 - p)B$ where $A > 0$ and $B > 0$ are constants. Now if $A > B$ then the expression will be maximized by choosing $p = 1$ otherwise by choosing $p = 0$.   □

If the service rates for two image types are identical ($\mu_1 = \mu_2$), then I_1 would be the optimal choice for cache if $\frac{\lambda_1}{\lambda_1 + \lambda_2} > \frac{\lambda_2}{\lambda_1 + \lambda_2}$, and vice versa. Given a window of R requests from history, an image type with the highest request count within the window should be cached.

Following notations are used henceforth to explain the techniques.

$R :=$ number of user requests asking for new image instances within a given time window

$N :=$ number of image templates used to create the R requests within a given time window

$f_{ij} := 1$ if the jth request is for the template i, otherwise 0

$C :=$ Maximum number of image instances that can be kept in the pre-provisioned inventory

4.2   Techniques for Pre-provisioning

Most-Frequently-Used (MFU) strategy leverages the insight of popularity based caching from Section 4.1. Within a window of R past requests, it computes the requests for type i as $f_i = \sum_{1 \leq j \leq R} f_{ij}$. The order of importance of an image is proportional to f_i. Now, given a cache size of C, the cache is completely filled up using the following formula:

$$C_i = w_i * C, \tag{1}$$

where w_i is computed as,

$$w_i = \frac{\sum_{1 \leq j \leq R} f_{ij}}{\sum_{1 \leq i \leq N} \sum_{1 \leq j \leq R} f_{ij}} \tag{2}$$

The MFU approach implicitly assumes that the request distribution for an image type is stationary within the history window. In practice, the popularity of an image may fade over time, thereby falsifying the stationarity assumption necessary for MFU to perform effectively. If time elapsed since the last request for an image template is large, then we can assume that the likelihood of a request for that image template is low. Thus, it is required to take into account the time of arrival of a request while selecting the cache composition.

In *Most-Recently-Used(MRU)* approach, we adjust w_i for a template i by attenuating the contribution of instances whose requests are older. Values can be attenuated by applying different functions. For instance, a naive approach is to reduce the values proportional to the time elapsed since the arrival of an instance request for a specific type. Alternatively, one can assign high importance to recent image types with the assumption that image types go out of fashion very quickly. The attenuated weight, w'_i, factoring in the temporal aspect, is expressed as,

$$w'_i = \frac{\sum_{1 \leq j \leq R} A(j, f_{ij})}{\sum_{1 \leq i \leq N} \sum_{1 \leq j \leq R} A(j, f_{ij})} \tag{3}$$

where, $A(x, y)$ is the attenuation function and can be expressed as, $A(x, y) = y * exp(-x)$. The new weights, w'_i, are used in Eqn-1 to compute the number of instances of an image template to be cached.

In MFU and MRU, selection is based on the popularity of an image and the available cache size. If the cache size is large, MFU and MRU may populate the cache with a large count of VMs of a template, although in practice, the maximum request count for the template is lower than the cached count. This allows the opportunity to fill the cache more judiciously, thereby saving the resource wastage for deleting an unused cache entry during next refresh. *Burst Adjustment(BA)* technique, finds the largest burst, B_i, that an image template i has encountered in the request history of R requests, and then uses B_i to limit the number of VM instances for image template i in cache. Represented mathematically, the number of VM instances of image type i in cache, is:

$$Burst\ adjusted\ C_i = min(w'_i * C,\ B_i) \tag{4}$$

Note that the selection step in Eqn-2 or Eqn-4 computes fractional numbers. During actual provisioning, VM instances occupy integral values, derived by rounding the fractions. This leads to some VM instances, with a low fractional value, being dropped from selection while allocating in decreasing order of the count. A simple example, where the cache can store 10 VM instances, illustrates the effect of rounding. At the end of BA technique, VMs of 3 image types are to be cached with instance count [6.7, 2.6, 0.7] respectively. Rounding the values changes the allocation to [7, 3, 0] respectively, thereby discarding the third image type, thus affecting the cache hit rate.

5   Experimental Evaluation

In this section, we show the results of evaluating the pre-provisioning approaches using simulation. We explain the simulation parameters, and provide a summary of the RC2 trace data which helps in understanding the results.

5.1   Simulation Parameters

Table 1 shows the simulation parameters used in the experiments. Three key parameters are: (a) *cache size* denotes the total number of instances of VMs that can be pre-provisioned, (b) *history window* denotes the number of past requests that are taken into consideration while computing the cache composition, (c) *pool-size* denotes the number of available threads or resources that can be dedicated for computations such as deletion, pre-provisioning. *Cache-update-interval* parameter is used to trigger the computation of cache composition periodically.

Few other parameters relevant for evaluating the caching techniques are: (i) *cache entry insertion time* accounts for the time to fetch an image from repository and place it in the cache, (ii) *cache entry deletion time* accounts for the time to delete an entry from the cache, (iii) *service time on cache hit* accounts for the time to deliver a cached instance to the user request. Cache hit service time is non-zero because some user-defined configurations may need to be set up prior to delivering the VM to the user.

The MRU technique uses an attenuation function to assign higher importance to the recent requests. A *negative exponential function with a mean of 10.0* is used(refer Eqn-3). In the BA technique, we compute the burst size by clustering all requests of an image type that arrive within the *cluster size of burst*.

Table 1. Simulation Parameters

Simulation Parameter	Parameter Value
Cache Size	30 (or as mentioned)
History Window	1000 (or as mentioned)
Pool Size	100 (or as mentioned)
Cache Update Interval	15 mins
Cache Entry Insertion time	10 mins
Cache Entry Deletion time	2 mins
Servicing time on Cache Hit	2 mins
MRU Policy Parameter	10.0
Cluster Size for Burst	11 mins

5.2 RC2 Trace Summary

We collected a 1 year request log from the Research Compute Cloud (RC2), which is a cloud computing platform for use by the worldwide IBM Research community [17]. It serves on average 200 active users and 800 VM instances per month, with a user base of 700 users. 10200 requests were logged during the 1-year observation period. For each request, the time of arrival of the provision request and deletion request, as well as, the time taken to provision the request is collected. Provision time is the end-to-end time from the request arrival to the user being notified of successful deployment of the virtual machine.

1088 unique image types were requested by 743 different users over the 1-year period. Less than 10 server instances were requested for 890 image templates, with just a single request for 453 image types, making request density for an image template quite sparse. Requests for the top 15 image types constitute only 26% of the total requests serviced. Another trend in request arrival is the presence of requests for an image type arriving in bursts, which could happen when a multi-tier application is being set up with similar servers. Even if one request from this group takes longer, it will force the user to wait. Therefore, an efficient caching strategy must try to provision all the instances during a burst.

5.3 Simulation Results: Using RC2 Trace Data

We compare the cache hit ratio with a varying cache size, as shown in Fig. 3. Beyond a cache size of 20, the BA technique outperforms all other techniques. When the cache size is less than 20, then according to Eqn-4, the MRU method performs better than the BA technique; thereby the results for both of them are identical.

Results for varying history window are shown in Fig. 4 The LRU method is not impacted by a varying history window because it always replaces the least recently used entry from the cache without looking at the history. The MFU technique may degrade in performance with an increasing history window because when the history size is increased, several image types which are old are often never requested again. Therefore, giving equal importance to all requests, without taking temporal aspect into account, leads to a degraded performance for MFU. The performance improves as soon

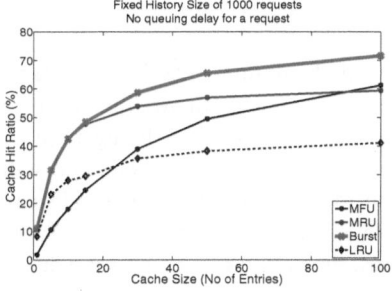

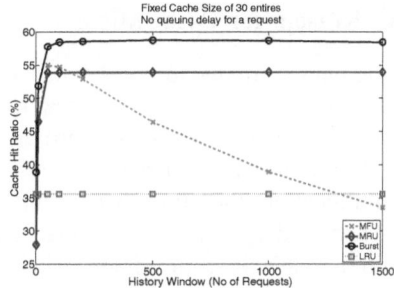

Fig. 3. Comparison of different techniques on RC2 trace data, where cache size is varied

Fig. 4. Comparison of different techniques on RC2 trace data, where history window is varied

as MRU is applied along with the MFU method. However, MRU also may end up over-allocating instances for an image type. BA reduces the number of instances of an image type to be pre-provisioned, thereby creating room to cache more image types.

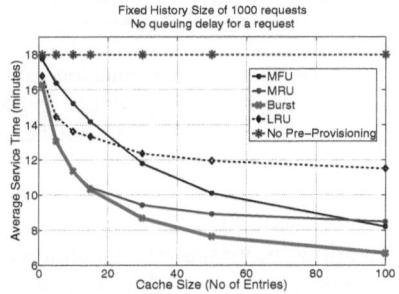

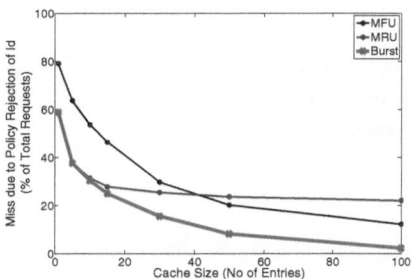

Fig. 5. Plot shows the average service time for provisioning a request

Fig. 6. Reduction in misses, due to the policy rejecting an image id, as the cache size is increased

We also report the improvement in service time with caching. *Without pre-provsioning, the average service time for a request is 18 minutes.* When pre-provisioning is applied, *the average service time can be reduced to as low as 6 minutes for some configurations,* as shown in Fig 5. The best case with a history size of 1000 requests is recorded when the cache size is 100 and burst adjustment policy is applied. *The reduction in service time is 62%. If we consider a more realistic cache size of 30 entries of an average size of 30 GB, requiring total space of approximately 1 TB, then the reduction of 51% in service time* is still significant.

5.4   Reasons for Cache Misses

Cache misses are due to several reasons, some of which are unavoidable, viz. *a request for a template is received for the first time. Choice of history window size* impacts misses since a larger history window provides a larger set of image types being requested. Third type of *miss occurs due to rounding..* Since some image types ends up with a zero allocation, therefore, it may lead to a cache miss if a request for the discarded type arrives despite the policy correctly inferring the importance of the image type. In addition to this, if request inter-arrival time is short, then although the caching decision may be accurate, the time for pre-provisioning is insufficient.

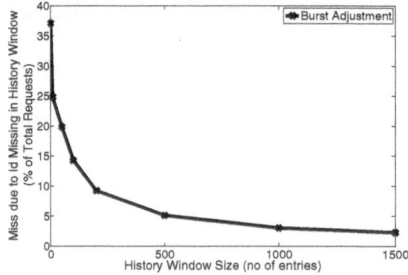

Fig. 7. Reduction in misses, due to the absence of an image type in the history window, as history window size is increased

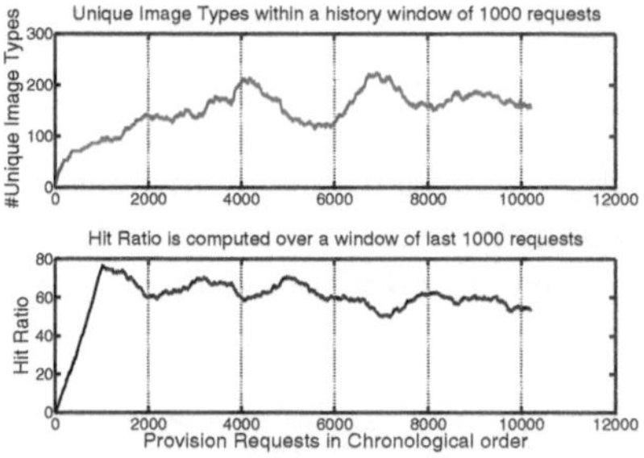

Fig. 8. Cache Hit Ratio computation over a rolling window where rolling window is 1000 entries. Graph also shows the number of unique image types in the window.

Fig. 6 shows that as the cache size is increased, it allows more space to accommodate larger number of image types. Thus while performing the integral allocation step in caching; lesser number of image types are rejected, thereby increasing the number of hits. In case of BA, since the policy trims the number of instances to be kept for each image type to the maximum size of burst observed, therefore, it helps in accommodating instances of more image types. Therefore, number of misses due to the policy rejecting an image type is lowest for the burst adjustment (BA) method.

Fig. 7 shows that as the history window size is increased, it allows the cache policy to view more image types, therefore helping the caching policy to take more image types into account while deciding the cache composition. It can be inferred from the figure that with a higher history window size the misses will reduce.

We also investigate the evolution of the cache hit ratio over a period of time. In Fig. 8, we consider a sliding window of 1000 requests, to observe the change in cache hit ratio. We also plot the number of unique image types present in the same history window. The graph confirms our observation that *as the number of unique image types increases within a history window, it leads to a drop in the cache hit ratio.*

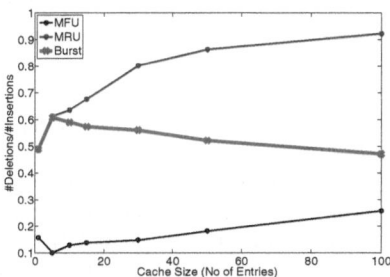
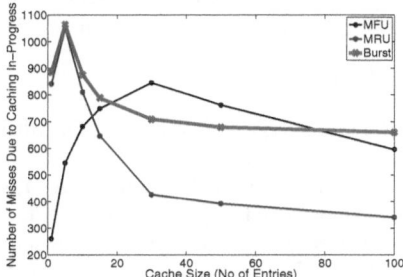

Fig. 9. Ratio of number of deletions to number of insertions in the cache. Lower ratio indicates that the time wasted for deletion was saved during an insertion.

Fig. 10. Number of misses due to caching in-progress for the RC2 traces having a total of 10210 requests

5.5 Gains Demystified

Deletion of a VM instance from the cache implies that the VM instance was provisioned unnecessarily. It wastes the time to pre-provision, as well as, time is spent in deleting it, thereby delaying the insertion of a new VM instance. In BA, since we are conservative in placing a VM instance into the cache, therefore, it reduces the number of deletions. When we compare the number of deletions, with respect to the number of insertions, per policy, we observe that this ratio is lower for the BA compared to the MRU policy, as shown in Fig 9.

Although MFU policy shows a significantly low deletion to insertion ratio, it still performs worse overall because it suffers due to the choice of image templates. Fig. 6 earlier shows that the MFU policy suffers mainly due to rejection of a number of image types when the integral allocation step is applied.

Despite an accurate prediction of the future arrival of requests by a policy, it still may not show the result, if the pre-provisioning of the instance does not complete before the next arrival. Often inter-arrival time between requests for an image type is shorter than the time to complete a pre-provisioning request. In our simulation, we assume that if the caching is in progress then it is a cache miss. Fig. 10 shows for each policy the number of requests which recorded a miss although the image instance was being cached.

6   Conclusion

Typically, it takes time in order of minutes to provision a new VM in cloud. Transfer of the large image template file from an image repository to a compute node, and booting are the main causes of delay in the provisioning workflow. We apply caching to alleviate the problem. Using request logs, we determine the image templates which will be high in demand, and also estimate the number of requests for each image type. Thus we can pre-provision VM instances by preparing and storing them in standby mode in the cache. On receiving a matching request, a cached VM is readily delivered to the user. We have compared 3 different techniques for selection of the cached VM instances.

Under specific configurations, service time to deploy a virtual machine can be reduced by 60% as compared to a no cache enabled scenario.

References

1. Borst, S., Gupta, V., Walid, A.: Distributed caching algorithms for content distribution networks. In: INFOCOM (2010)
2. Caron, E., Desprez, F., Muresan, A.: Forecasting for grid and cloud computing on-demand resources based on pattern matching. In: Proceedings of the 2010 IEEE Second International Conference on Cloud Computing Technology and Science, CLOUDCOM 2010 (2010)
3. Chandra, R., Zeldovich, N., Sapuntzakis, C., Lam, M.S.: The collective: a cache-based system management architecture. In: Proceedings of the 2nd Conference on Symposium on Networked Systems Design & Implementation, NSDI (2005)
4. Chen, Z., Zhao, Y., Miao, X., Chen, Y., Wang, Q.: Rapid provisioning of cloud infrastructure leveraging peer-to-peer networks. In: ICDCS Workshops (2009)
5. IBM Workload Deployer,
   `http://www-01.ibm.com/software/webservers/workload-deployer/`
6. Davison, B.D.: A web caching primer. IEEE Internet Computing 5 (July 2001)
7. Drexl, A., Kimms, A.: Lot sizing and scheduling – survey and extensions. European Journal of Operational Research 99(2) (1997)
8. Emeneker, W., Stanzione, D.: Efficient virtual machine caching in dynamic virtual clusters. In: SRMPDS Workshop, ICAPDS 2007 Conference (2007)
9. Eucalyptus Systems, `http://www.eucalyptus.com/`
10. Lagar-Cavilla, H.A., Whitney, J.A., Scannell, A.M., Patchin, P., Rumble, S.M., de Lara, E., Brudno, M., Satyanarayanan, M.: Snowflock: rapid virtual machine cloning for cloud computing. In: EuroSys (2009)
11. Lin, J., Huang, T., Yang, C.: Research on web cache prediction recommend mechanism based on usage pattern. In: Proceedings of the First International Workshop on Knowledge Discovery and Data Mining (2008)
12. Liran, R.C.: Scheduling algorithms for a cache pre-filling content distribution network (2002)
13. Makkar, P., Gulati, P., Sharma, A.: A novel approach for predicting user behavior for improving web performance. International Journal on Computer Science and Engineering 02(04) (2010)
14. MokaFive Desktop Management Simplified, `http://www.moka5.com/`
15. Nurmi, D., Wolski, R., Grzegorczyk, C., Obertelli, G., Soman, S., Youseff, L., Zagorodnov, D.: Eucalyptus: A technical report on an elastic utility computing archietcture linking your programs to useful systems. Tech. Rep. 2008-10, UCSB Computer Science Technical Report (October 2008)
16. Podlipnig, S., Böszörmenyi, L.: A survey of web cache replacement strategies. ACM Comput. Surv (2003)
17. Ryu, K.D., Zhang, X., Ammons, G., Bala, V., Berger, S., Da Silva, D.M., Doran, J., Franco, F., Karve, A., Lee, H., Lindeman, J.A., Mohindra, A., Oesterlin, B., Pacifici, G., Pendarakis, D., Reimer, D., Sabath, M.: Rc2-a living lab for cloud computing. In: Proceedings of the 24th International Conference on Large Installation System Administration, LISA 2010 (2010)
18. Shi, L., Banikazemi, M., Wang, Q.B.: Iceberg: An image streamer for space and time efficient provisioning of virtual machines. In: Proceedings of the 2008 International Conference on Parallel Processing - Workshops (2008)
19. Sotomayor, B., Keahey, K., Foster, I.: Combining batch execution and leasing using virtual machines. In: Proceedings of the 17th International Symposium on High Performance Distributed Computing, HPDC (2008)
20. Zhu, J., Jiang, Z., Xiao, Z.: Twinkle: A fast resource provisioning mechanism for internet services. In: INFOCOM, pp. 802–810 (2011)

Improving Scheduling Performance
Using a Q-Learning-Based Leasing Policy for Clouds

Alexander Fölling[1] and Matthias Hofmann[2]

[1] Robotics Research Institute, TU Dortmund University, 44221 Dortmund, Germany
alexander.foelling@tu-dortmund.de
[2] D-Grid GmbH, 44221 Dortmund, Germany
matthias.hofmann@d-grid-gmbh.de

Abstract. Academic data centers are commonly used to solve the major amount of scientific computing. Depending on upcoming research projects the user generated workload may change. Especially in phases of high computational demand it may be useful to temporarily extend the local site. This can be done by leasing computing resources from a cloud computing provider, e.g. Amazon EC2, to improve the service for the local user community. We present a reinforcement learning-based policy which controls the maximum leasing size with regard to the current resource/workload state and the balance between scheduling benefits and costs in an online adaptive fashion. Further, we provide an appropriate model to evaluate such policies and present heuristics to determine upper and lower reference values for the performance evaluation under the given model. Using event driven simulation and real workload traces, we are able to investigate the dynamics of the learning policy and to demonstrate the adaptivity on workload changes. By showing its performance as a ratio between costs and scheduling improvement with regard to the upper and lower reference heuristics we prove the benefit of our concept.

1 Introduction

The increase of applications in the area of HPC also increases the need for computational resources and effective management of such resources. Nowadays, the resource demand of scientific workload is justified by explicitly funded academic data centers or commercial data centers which are scaled for the average workload to balance operating costs and service quality provided to the local user community. With the new paradigm of cloud computing this rigid resource scaling can be replaced by a hybrid infrastructure which combines the advantages of local HPC resources with the potentially unlimited scalability of resources within the cloud on a pay-per-use basis [1].

In this infrastructure, academic data centers are able to temporarily extend their local computing power through cloud resources to execute waiting tasks within the cloud. This way, the data center is able to provide a better service to its local scientific community by decreasing the time researchers have to wait for their job executions in exchange for leasing fees. Besides the technical implementation, a hybrid infrastructure needs efficient cloud leasing management and scheduling algorithms which support data center administrators to gain performance benefits, e.g. decreasing wait time, and to control

C. Kaklamanis et al. (Eds.): Euro-Par 2012, LNCS 7484, pp. 337–349, 2012.

the costs for using cloud resources. Our work addresses this problem by introducing a reinforcement learning-based leasing policy. This policy adaptively steers a maximum leasing size depending on past scheduling decisions and workload characteristics in an online fashion and without any prior knowledge.

In the remainder of this paper we present the hybrid system model, the used workload traces, and the used performance metrics in Section 2 as basis for the investigation of reference heuristics in Section 3. After describing the important fundamentals in reinforcement learning (RL) (Section 4) and its use in our leasing policy (Section 5) we evaluate the performance of our approach with regard to the upper and lower reference performance in Section 6. In the subsequent Section 7 we provide some related work with reference to the technical inplementation, costs and performance in the cloud, and leasing strategies. In the end, we conclude and describe our future research perspective in Section 8.

2   System Model

As foundation for our learning policy and reference heuristics we introduce the considered system model and all assumptions or restrictions we make for simulation.

The system model can be separated into several distinct parts. The first one is a local scheduling system that describes local machines, job characteristics, and local scheduling strategies. The second part contains workload traces, criteria for their selection, and reference simulation results. The third part represents the extension of local by cloud resources and the last part describes performance metrics to evaluate the performance of a cloud leasing policy.

2.1   Local Scheduling System

As local scheduling system we assume a single academic data center with its local resource management system (LRMS). Figure 1 shows that the LRMS serves as submission point for the local user community. The submitted jobs are appended to a local job

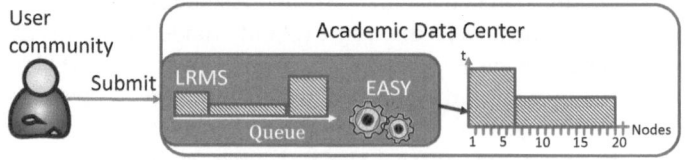

Fig. 1. Local System Model

queue. As scheduling heuristic for the LRMS we use EASY backfilling which is known to create efficient local schedules and commonly used in implemented LRMS [2]. EASY allocates the jobs to the M homogeneous resources in an online fashion. That means the jobs are submitted over time and after submission only the release time, degree of parallelism, and estimated runtime of every job are known whereas the actual runtime is unknown at scheduling decision time.

We decided to use homogeneous resources according to Fölling et al. [3,4] to reduce the complexity of the system model and focus our research on the efficient usage of cloud resources. In addition, data scheduling is beyond the scope of our paper, and hence the communication delays during the transfer of jobs between local and/or cloud nodes is neglected. In academic data centers the submitted jobs are usually scientific -sometimes parallel- rigid batch jobs which are neither moldable nor malleable and cannot be preempted during execution. Consequently, we assume the very same job model for our simulations.

2.2 Workload Traces

Table 1. Workload characteristics

Identifier	$archive$ no.	#Jobs	M	$Util$ in $\%$	$months$
KTH	8	28479	100	68.9	11
CTC	6	77199	430	66.2	11
SDSC	20	74903	1664	60.1	11

To represent the local user community in a realistic way we select three different real workload traces from Feitelson's parallel workloads archive [5] with regard to four different criteria. The utilization of the traces has to be high enough to motivate cloud leasing at all. The length of the traces must be similar to make the benefits of the model and policies comparable. We want to reach a preferably high differentiation in the amount of local resources for the different workloads. As we want to discover the system dynamics and the behaviour of policies under the hybrid model in a first attempt we try to keep the simulation overhead comparatively small.

This prohibits workloads with too many local nodes and too many submitted jobs per time. Table 1 shows the characteristics of the three workloads. The last trace was shorted from 13 to 11 months with regard to the criteria of similar length.

2.3 Extending the Resource Space

To include cloud resources for job execution in the simulation model we introduce a new local scheduling strategy which coordinates the use of cloud resources. This transfer

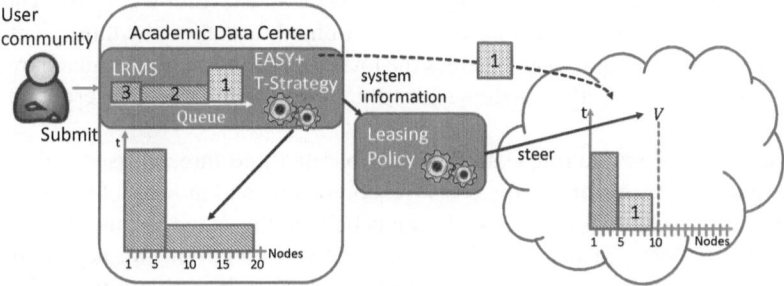

Fig. 2. Example system state with extended resource in the hybrid model

strategy is placed at the LRMS (see Figure 2) and is executed after the local scheduling algorithm EASY. It transfers waiting jobs from the local queue into the cloud if they are immediately executable within the cloud.

Whether a job is immediately executable in the cloud depends on the maximum size of the virtual cluster V which is steered by the leasing policy. A leasing policy periodically collects system state metrics from the LRMS and adjusts V for the next period respectively. V restricts the amount of cloud resources that can be used by the transfer strategy to execute jobs within the cloud. Thus, the strategy is allowed to dynamically startup only 0 to V virtual machines within a given period.

In the example a possible state of the system with $V = 10$ is shown. Therein, some jobs with different degree of parallelism in the LRMS job queue and a single job (1) can be observed, which is currently transferred by the transfer strategy to get executed in the cloud. For this jobs, five additional cloud nodes have to be leased. With the additional nodes the current value of V is not exceeded. Neither the next job within the queue (2) nor the last one (3) are immediately executable within the cloud because there is only one additional node leasable.

For completeness, during the simulation multisite execution is not allowed, hence parallel jobs cannot be executed on local and cloud resources at the same time. Thus, a single job is started on cloud resources without any wait time but causing execution costs, or it is scheduled at local resources and may wait until its execution but does not produce any costs at all.

We assume the cloud nodes as uniform to the local nodes in terms of processing speed. This model—Iosup et al. [6] call it "source-like performance"—is very optimistic as the runtime of an application is usually related to the performance decrease caused by the virtualization overhead. According to Iosup et al., the model is "[...] useful for assessing the theoretical performance of future and more mature clouds".

2.4 Performance Objective Metric

As we want to improve the quality of service for our local user community we choose their jobs' wait time as one factor for our performance objective metric. It is calculated as the sum of all wait times during simulation. The second factor represents the costs of providing a better service using cloud resources. With using more cloud resources we can expect the total wait time (twt) decreasing and costs (c) increasing while using less cloud resources may lead to a higher total wait time with lower costs. Those two contrary performance objectives have to be combined to make schedules—created by different policies—comparable. Whereas the total wait time can be calculated by adding the single wait times of all scheduled jobs the basis for the costs criterion has to be discussed in detail.

Generally, cloud computing costs can be separated into three classes which are the uptime costs for VM-instances, the costs for stored data, and the costs for data transfers into and from the cloud. As data scheduling is beyond the scope of this work (see Subsection 2.1) the costs for data transfers is neglected. In addition, we neglect the costs for storing data because we assume that cloud nodes can be directly instantiated with a preconfigured cluster-image, e.g. customized Amazon AMI [7]. Further, the image with all stored data, e.g. Amazon EBS, can be deleted when the instance is not needed anymore. This way, storage costs only appear for running virtual cluster nodes, and hence are just a scaling factor for the uptime costs. Thus, we reduce our cost factor to the resource usage for jobs running in the cloud.

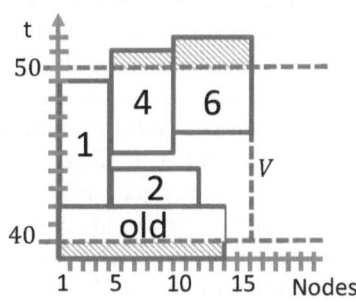

Fig. 3. Example Schedule

Job	r_j	m_j	p_j	$p_j \cdot m_j$	c
old	0	13	42	546	26
1	40	4	7	28	28
2	40	7	2	14	14
4	45	5	6	30	25
6	46	6	6	36	24

Fig. 4. Characteristics of jobs with release time r_j, parallel machines m_j, and runtime p_j

Figure 3 shows a Schedule and Figure 4 shows workload characteristics for an example cost calculation with $V = 15$ for an evaluation interval $[40, 50]$ seconds and five scheduled jobs which consume 117 cpu seconds in the cloud (The shaded areas represent costs in surrounding time intervals). Both twt and c are absolute values with different ranges and hence must be normalized before being combined. Normalizing those absolute values is only possible if the maxima are known. In this sense we are able to express each absolute value as percentual part of its corresponding maximum. To calculate the maximum for twt we simulate a whole workload with $V = 0$ and thus force the LRMS to perform local scheduling only. The resulting total wait time is refered as twt_{ref}. To calculate the maximum for c we leave V unrestricted ($V \rightarrow \infty$) and refer the result as c_{ref}. Now, we are able to normalize every absolute result pair (twt,c) of every schedule by calculating a percentage value ($twt\%$,$c\%$) with regard to the maximum values (twt_{ref},c_{ref}).

Alternative to $twt\%$ we can calculate the improvement in total waittime as $twt_{imp}\% = 100 - (twt\%)$.

Then, the proposed combination of twt and c called *balance* is calculated as *balance* = $twt_{imp}\% - c\%$. A low or even negative *balance* indicates less economic efficiency and the solution with the biggest *balance* is most effective from an economical point of view.

This *balance* metric is the central component of all following evaluations. It is used to show the lower and upper performance reference values of simple heuristics as described in the following section. Likewise, it is used within the proposed learning policy.

3 Performance Reference Evaluation

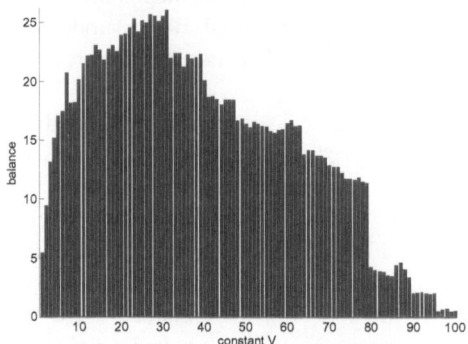

Fig. 5. *Balance* for all different constant $V \in$ [0, 100] in simulations of the KTH workload.

To evaluate the performance of our later defined learning approach we first need to determine reference *balance* values for the proposed model. Therefore, we evaluate two simple heuristics to determine an upper reference value for *balance* by an offline brute force analysis and a lower reference value for *balance* with help of a randomized online leasing policy approach. These upper and lower reference values are not upper and lower bounds in the mathematical sense. They are just reference values to ease the assessment of reached strategy results. During the brute force analysis we simulate all workloads with constant settings for V during the whole workload. As we do not know which value range is appropriate for setting V we tested all settings from 0 to M (M is the number of local nodes of the traced machine; See Section 2.2). Figure 5 exemplarily shows the results in *balance* for all different V-configurations for the KTH workload. Although the figure only shows the analysis results for the KTH workload, they are typical for the proposed model due to similar characteristics for other workloads. They show rapid *balance* increasement by extending the local resources in the first V-configurations which then reaches its zenith for a specific V. Above this V the scheduling performance may indeed increase further but this is done under disproportional increasement of the percentual costs and is thus less efficient regarding balance.

Table 2. Results for all three constant V analysis.

Identifier	Best V	$twt_{imp}\%$	$c\%$	$balance$
KTH	31	70.38	45.39	25
CTC	62	58.46	30.43	28
SDSC05	184	51.53	30.14	21.4

Table 3. Results of all workloads and 1000 simulations.

Identifier	Average of 1000	Best outlier of 1000
KTH	14	20
CTC	11	18
SDSC05	5	11

Table 2 contains statistical results for constant V evaluations and all evaluated workloads. Every row represents the best V-configuration found during brute force analysis. Beside this configuration the values for $twt_{imp}\%$, $c\%$, and the resulting balance are given. The potential in increasing the scheduling performance with regard to the upcoming costs is different between the three workloads but very high in all cases. The KTH, for example, is able to decrease its cumulative wait time by more than 70% compared to a local scheduling simulation. Nevertheless, these *balance*-results for every workload can be interpreted as a brute force calculated upper reference value as the best configuration is only known at the end of the simulation.

To get a low reference value for the learning performance we first have to define the term "online leasing policy". The underlying decision problem—which value should be set for V—is divided in predetermined evaluation steps from time t to time t' with a step length $span$. This length is a constant value for the whole simulation. Thus, the policy is periodically queried for a new setting of V during the simulation ($t \overset{span}{\rightarrow} t' \overset{span}{\rightarrow} t'' \overset{span}{\rightarrow}$ ...). The simplest way to model such an online leasing policy—and our lower reference heuristic at the same time—is letting the policy choose V for each step $t \rightarrow t'$ randomly. According to the brute force analysis, we simulate all workloads with an appropriate range for eligible V-configurations depending on the original traced machine from 0 to M. We choose the length of an evaluation step $span$ as one day arbitrarily[1]. As the policy decisions are randomized we repeat the simulations 1000 times and trace the $balance$ results for all of them. We use the average of all results in each setup as lower reference value for the evaluation of our learning approach, whereas the best outliers are of interest, too. Both information are listed in Table 3. The objective for our learning cloud leasing policy is to achieve a higher $balance$ than the random heuristic but similar to the offline calculated results of the best constant V-configuration.

4   Reinforcement Learning

Before we explain our adaptive cloud leasing policy we introduce some background on reinforcement learning (RL).

In temporal difference learning, the one-step Q-learning [8] approach is applied using two basic concepts. The first concept is the separation of a continuing—potentially infinite—learning task in distinct learning steps $t \rightarrow t'$. At the beginning of every step, the *learning agent* is in a specific state $s_t \in S$ with S as finite set of states and chooses one action a_t from a finite repertoire of actions A. At the end of a step[2] the agent senses a new state $s_{t'} \in S$ and a reward signal $r_{t'}$.

The second concept is the use of a function to store the current expected reward for each action at a given state. After each learning step this *Q-value function* is updated using the update Rule 1.

$$
Q(s_t, a_t) \leftarrow Q(s_t, a_t) + \alpha \underbrace{\left[r_{t'} + \gamma \max_{a \in A} Q(s_{t'}, a) - Q(s_t, a_t) \right]}_{sensed\ reward}
\tag{1}
$$

The update depends on the difference of the new sensed reward and the old expected one $Q(s_t, a_t)$. The new sensed reward at time t' is further separated in the reward signal $r_{t'}$, and the expected reward at time t' under the sensed state $s_{t'}$ modeled as term $\gamma \max_{a \in A} Q(s_{t'}, a)$. In this term the maximum Q-value of all actions for the new sensed state $s_{t'}$ represents the best expected reward among all actions at the new state. This term is weighted by the discount rate γ and thus the influence of the expected (future) reward on the current reward can be parameterized. Further, this whole temporal difference is weighted by the learning rate $\alpha \in [0, 1]$. This parameter is usually scaled down

---

[1] The same setting for $span$ will be used for the later learning approach either.
[2] End of a configured time interval or some special event.

from learning step to learning step to decrease the influence of the temporal difference over time and hence to assure convergence of the learning approach. In infinite learning tasks it can also be constant to keep the adaptivity of the learning algorithm stable.
The current Q-values of the actions at state s_t impact the decision which one is choosen for the next learning step.

5 Adaptive Cloud Leasing Policy

Our goal is to model a cloud leasing policy to steer the configuration of V with regard to the economical *balance* metric in an online manner. In comparison to other machine learning approaches (Genetic Programming, Artificial Neural Networks, Decision Tree Learning) RL appears to be best suitable for the underlying problem since the learning is done online without any expert knowledge about the system and with comparatively small computational demand.
To model the policy we consider each possible V-configuration as eligible action a during the reinforcement learning process. Further, we reduce the set of states S to one single state to limit the search space to $A \cdot S$. With these assumptions we can simplify the Q-value function update Rule 1 to the Rule 2.

$$Q(V_t) \leftarrow Q(V_t) + \alpha \left[r_{t'} + \gamma \max_V Q(V) - Q(V_t) \right] \tag{2}$$

In this rule, the Q-value of the configuration V_t is updated which has been choosen at the beginning of the learning step with help of the old Q-value, the learning parameters α and γ, and the a reward signal $r_{t'}$. We decided to set α and γ to 0.1 according to several Q-Learning settings of Sutton and Barto [8].

To model the immediate reward signal $r_{t'}$ we have to implement a concept to simulate all different V-configurations in parallel. At the beginning t of each cloud policy evaluation step the current schedule and the queue are cloned for every possible V-configuration. During the evaluation step $t \rightarrow t'$ each clone is simulated with its associated configuration for V and every new submitted job is forwarded to all of them. Although, all clones had to schedule the very same jobs, this could lead to different schedules at the end of the evaluation step. Consequently, each clone produces its own results in total wait time twt and costs c (See Section 2.4). With help of two additional simulations ($V = 0$ and $V \rightarrow \infty$) that are simulated during each evaluation step we are able to calculate the *balance* for each clone. A high *balance* indicates that the used configuration of V during the learning step has been more economical than that with a lower one.

To express this fact through a reward signal $r_{t'}$ we interpolate the *balance*-values of all clones analog to the interval of minimum and maximum reached *balance*. This way, the V-configuration of the best clone gets a reward of $r_{V,t'} = 1$ and that with the worst clone a reward of 0.

Having the rewards $r_{V,t'}$ for all parallel clones we are able to update all Q-values and not only that one of the choosen action V_t (See update Rule 2).

Afterwards, the policy chooses the next action by taking the one with the highest new Q-value. This behaviour is reasonable in our scenario as we are evaluating all possible actions at the same time. If there are multiple actions with the highest value, we choose the one with the smallest V-configuration. For the next evaluation step $t' \to t''$ the clone of the previous choosen action V_t serves as draft for the new clones. In a real application of the leasing policy the parallel evaluation of all possible configurations could be made during the learning step and hence the decision time and replication of datastructures tend to be zero. Even the simulation of a whole 11-month workload on a single commodity Intel(R)Core(TM)-i5 machine with 4 GB RAM is comparatively short (from 2 up to 30 minutes depending on workload and simulated machine size— KTH, CTC, or SDSC05).

Evaluating all V-configurations and choosing the best one enables us to avoid random influences in our simulations. This is advantageous to investigate the basic behaviour of the approach and to evaluate different settings for the assumed policy parameters $(span, \alpha, \gamma)$. This way we are able to achieve a good learning performance even for short traces and thus a limited amount of learning steps. However, it leads to a huge simulation complexity for traces with a realtively big amount of local nodes (e.g. bigger than 1664).

6   Policy Evaluation

As mentioned before, we evaluate our learning approach with a learning step length of one day for all three test workloads. Table 4 gives an overview on the metric results for the simulations.

Table 4. Results of the Q-Learning approach

Identifier	WT impr. in %	Costs in %	balance	Low/Up Ref.-value
KTH	46.54	22.31	24.23	14/25
CTC	46.66	23.67	22.99	11/28
SDSC05	34.44	20.69	13.75	5/21.4

The *balance* value lies between the results for constant V and the randomized leasing policy in all three setups. Every learning simulation succeeds to reach a *balance* value larger than the average and even the best random experiment of 1000 simulations. This proves the positive influence of the reinforcement learning approach regarding the economical performance of the system. However, the learning approach is not able to reach *balance* results near the upper reference value because the knowledge about the system is limited. Figure 6 shows the evolution of the overall *balance* during the 11-months simulation of the KTH workload for the best constant V-configuration ($V = 31$), the mean balance of all 1000 random policy evaluations, and the learning approach in comparison.

In the reference case the decision of the policy is always $V = 31$ and no alternative decisions are allowed. Thus, the achievable performance in *balance* only depends on the workload characteristics. There are many situations where $V = 31$ coincidentally is a good configuration which leads to a high overall *balance* $= 25$ at the end of the simulation. In contrast, the learning policy has to evaluate the different V-configurations first before it finds out appropriate configurations. However, in the last two-thirds of the simulation it is observable that the learning policy approximates the *balance* quality of the upper reference heuristic. In addition, it converges to a stable *balance* in the second half of the simulation. The learning controller is able to adapt its behaviour to the changing

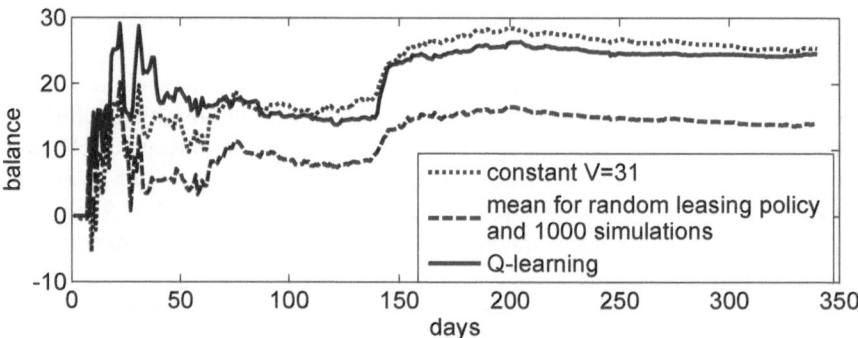

Fig. 6. Overall *balance* trace in comparison for constant $V = 31$, random policy, and Q-learning during KTH-simulation

workload and to produce similar or from time to time even better decisions than choosing a fixed configuration for the whole workload (better *balance* within the interval 30 to 50 days). Although, finding the best fixed configuration needs an offline evaluation of all possible configuration whereas the learning controller succeeds to get a high balance at the very first run without any knowledge. This is noteworthy as reinforcement learning is usually able to learn useful strategies only by training the system with a huge count of episodes (or walks) whereas the resulting knowledge of each episode serves as input for the next training episode.

In addition, learning is done while keeping the costs under one quarter of the maximum while increasing the scheduling performance by $34 - 46\%$ depending on the input workload.

7   Related Work

Chester et al. [9] investigate a system for allocating server resources to applications dynamically, thus allowing applications to automatically adapt to variable workloads. Similiar work was done by Chase et al. [10]. They develop a cluster manager that allocates servers from a common pool to multiple virtual cluster environments. The use of virtualisation technology has also been explored [11] to scale, adapt, and share resources on the same host machine or local computing infrastructure. To improve Quality of Service by reducing the job wait time, the idea of accessing additional resources from a cloud has become increasingly popular. Marshall et. al. [12] invent a cloud model by ensuring elastic provisioning of ressources and dealing with issues such as security and data privacy. Moreover, they involve and evaluate different policies to avoid under- and over-provisioning. A prominent example of exploiting clouds for scientific applications was further investigated by Deelman et al. [13]. They elaborate the costs of doing science in the cloud by simulations based on the fees of Amazon EC2 computational as well as Amazon S3 storage service. While running the same workload in different execution models they have shown that cloud computing can be applied in a cost-effective fashion especially for data-intensive applications. The tradeoff between cost and performance in the cloud context is a central aspect in various other publications such as [14], [15], [16], [17], and [18].

While our approach dynamically steers a maximum size of resources to be leased from the cloud by using a reinforcement learning algorithm, Buyya et al. [19] evaluate different so called strategy sets which compose of local scheduling and redirection strategies. Other publications show that reinforcement learning is a widely used method in the context of grid scheduling [20] and adaptive resource allocation [11], [21], [22] while reinforcement learning in the context of cloud leasing management appears as a new topic.

8 Conclusion

This paper proposes an adaptive learning-based cloud leasing policy that is evaluated in a simulated hybrid infrastructure comprising of a local and a cloud system.

A cloud leasing policy is responsible for steering the amount of additional cloud resources and tries to optimize the trade-off between extra costs and scheduling performance. This trade-off is formalised in a *balance* metric which is used for evaluation. Two heuristics that provide reference values are defined in order to examine the performance of our policy. We observe that the learning approach almost reaches the balance of the upper reference value the longer the learning is applied. The job scheduling performance of our cloud leasing policy is also compared with a scenario in which every job is computed on local resources. This comparison shows an improvement of about $34 - 46\%$ less wait time under comparatively small costs for every workload.

Using parallel evaluation of all leasing sizes during a learning step—as we do in here—leads to huge simulation complexity as the count of reasonable configurations increases with the size of the given academic data centers. In this work, relatively small reference data centers and associated workloads are used for the evaluation. However, the size of modern data centers is often much larger (K Computer with 705024 Cores)[3]. Thus, future work will focus on the incorporation of strategies into the policy which keep the number of parallel evaluations low while retaining good *balance* results. One approach in this regard is the appliance of ϵ-greedy [8] strategies to ensure explorational behaviour in a limited part of the search space instead of evaluating the whole search space at every step. This also needs further investigations in the configuration of parameters like the learning evaluation step or reinforcement learning parameters. To make the model more realistic, our future work will consider a heterogenous resource model for data centers and cloud providers and will account for realistic VM performance models rather than the mentioned "source-like performance" model.

Acknowledgment. This work was partially funded by the German Federal Ministry of Education and Research (BMBF) under grant #01IG08006.

References

1. Ostermann, S., Prodan, R., Fahringer, T.: Extending grids with cloud resource management for scientific computing. In: 10th IEEE/ACM International Conference on Grid Computing (Grid 2009), pp. 42–49. IEEE Press (2009)

---

[3] http://www.top500.org

2. Feitelson, D.G., Weil, A.M.: Utilization and predictability in scheduling the IBM SP2 with backfilling. In: Proceedings of the 12th International Parallel Processing Symposium and the 9th Symposium on Parallel and Distributed Processing, pp. 542–547. IEEE Computer Society Press (1998)

3. Fölling, A., Grimme, C., Lepping, J., Papaspyrou, A.: Robust load delegation in service grid environments. IEEE Transactions on Parallel and Distributed Systems 21(9), 1304–1316 (2010)

4. Fölling, A., Grimme, C., Lepping, J., Papaspyrou, A.: Connecting community-grids by supporting job negotiation with coevolutionary fuzzy-systems. Soft Computing - A Fusion of Foundations, Methodologies and Applications 15, 2375–2387 (2011)

5. Feitelson, D.G.: Parallel workload archive (March 2011),
   http://www.cs.huji.ac.il/labs/parallel/workload/

6. Iosup, A., Ostermann, S., Yigitbasi, N., Prodan, R., Fahringer, T., Epema, D.: Performance analysis of cloud computing services for many-tasks scientific computing. IEEE Transactions on Parallel and Distributed Systems 22, 931–945 (2011)

7. Nishimura, H., Maruyama, N., Matsuoka, S.: Virtual clusters on the fly—fast, scalable, and flexible installation. In: Proceedings of the Seventh IEEE International Symposium on Cluster Computing and the Grid (CCGrid 2007), pp. 549–556. IEEE Computer Society, Washington, DC (2007)

8. Sutton, R.S., Barto, A.G.: Reinforcement Learning - An Introduction, 4th edn. The MIT Press (1998)

9. Chester, A., Xue, J.W.J., He, L., Jarvis, S.: A system for dynamic server allocation in application server clusters. In: Proceedings of the 2008 IEEE International Symposium on Parallel and Distributed Processing with Applications, pp. 130–139. IEEE Computer Society, Washington, DC (2008)

10. Chase, J.S., Irwin, D.E., Grit, L.E., Moore, J.D., Sprenkle, S.E.: Dynamic virtual clusters in a grid site manager. In: International Symposium on High Performance Distributed Computing (HPDC 2003), pp. 90–100 (June 2003)

11. Rao, J., Bu, X., Xu, C.Z., Wang, L., Yin, G.: Vconf: a reinforcement learning approach to virtual machines auto-configuration. In: Proceedings of the 6th International Conference on Autonomic Computing (ICAC 2009), pp. 137–146. ACM, New York (2009)

12. Marshall, P., Keahey, K., Freeman, T.: Elastic site: Using clouds to elastically extend site resources. In: 10th IEEE/ACM International Conference on Cluster, Cloud and Grid Computing (CCGrid), pp. 43–52 (2010)

13. Deelman, E., Singh, G., Livny, M., Berriman, B., Good, J.: The cost of doing science on the cloud: the montage example. In: Proceedings of the 2008 ACM/IEEE Conference on Supercomputing, pp. 1–12. IEEE Press, Piscataway (2008)

14. Genaud, S., Gossa, J.: Cost-wait trade-offs in client-side resource provisioning with elastic clouds. In: 4th IEEE International Conference on Cloud Computing (CLOUD), pp. 1–8. IEEE (2011)

15. Mao, M., Li, J., Humphrey, M.: Cloud auto-scaling with deadline and budget constraints. In: 11th IEEE/ACM International Conference on Grid Computing (GRID), pp. 41–48. IEEE, Brussels (2010)

16. Ostermann, S., Iosup, A., Yigitbasi, N., Prodan, R., Fahringer, T., Epema, D.: A performance analysis of ec2 cloud computing services for scientific computing. In: Cloud Computing. LNICST, vol. 34, pp. 115–131. Springer, Heidelberg (2010)

17. Rehr, J.J., Vila, F.D., Gardner, J.P., Svec, L., Prange, M.: Scientific computing in the cloud. Computing in Science and Engineering 12, 34–43 (2010)

18. Fenn, M., Holmes, J., Nucciarone, J.: A performance and cost analysis of the amazon elastic compute cloud cluster compute instance. Research Computing and Cyberinfrastructure Group, Penn State University, Tech. Rep. (2011),
http://rcc.its.psu.edu/education/white_papers/cloud_report.pdf
19. de Assuncao, M.D., di Costanzo, A., Buyya, R.: Evaluating the cost-benefit of using cloud computing to extend the capacity of clusters. In: Proceedings of the 18th ACM International Symposium on High Performance Distributed Computing (HPDC 2009), pp. 141–150. ACM, New York (2009)
20. Zeng, B., Wei, J., Liu, H.: Dynamic grid resource scheduling model using learning agent. In: Proceedings of the 2009 IEEE International Conference on Networking, Architecture, and Storage (NAS 2009), pp. 67–73. IEEE Computer Society, Los Alamitos (2009)
21. Galstyan, A., Czajkowski, K., Lerman, K.: Resource allocation in the grid using reinforcement learning. In: International Joint Conference on Autonomous Agents and Multiagent Systems, vol. 3, pp. 1314–1315. IEEE Computer Society, Los Alamitos (2004)
22. Vengerov, D.: A reinforcement learning approach to dynamic resource allocation. Sun Microsystems Laboratories, Tech. Rep. (2005)

Impact of Variable Priced Cloud Resources on Scientific Workflow Scheduling

Simon Ostermann and Radu Prodan

Institute of Computer Science, University of Innsbruck,
Technikerstr. 21a, 6020 Innsbruck, Austria

Abstract. We analyze the problem of provisioning Cloud instances to large scientific workflows that do not benefit from sufficient Grid resources as required by their computational requirements. We propose an extension to the dynamic critical path scheduling algorithm to deal with the general resource leasing model encountered in today's commercial Clouds. We analyze the availability of the cheaper and unreliable Spot instances and study their potential to complement the unavailability of Grid resources for large workflow executions. Experimental results demonstrate that Spot instances represent a 60% cheaper but equally reliable alternative to Standard instances provided that a correct user bet is made.

Keywords: Cloud computing, Grid computing, Spot instances, Scheduling, Scientific workflows, Performance, Cost.

1 Introduction

From the rather broad amount of definitions and interpretations of the term Cloud computing, the scientific computing community mostly focuses on the Infrastructure as a Service (IaaS) interpretation characterized by leasing of computation, storage, message queues, databases, and other raw resources from specialized providers. In this context, scientific workflows emerged in the last decade as a highly successful paradigm for programming loosely-coupled high-performance computing infrastructures such as computational Grids and Clouds. In this paper, we study the possibility of extending Grid infrastructures with IaaS Cloud resources to improve the execution of large workflow applications that do not have sufficient Grid resources available for their computational demands. We design an extension to the *Dynamic Critical Path (DCP)* algorithm which demonstrated in previous work [1] better results compared to other existing heuristic strategies for most types of workflows and independent of their size, in particular when the resource availability frequently changes or when using dynamic and unreliable heterogeneous resources. We build such an environment by using a combination of a set of cluster resources offered by the Austrian Grid, complemented on-demand by "pay-as-you-go" Cloud resources offered by Amazon EC2. Besides the Standard instances rented at a fixed price per hour, Amazon gives the possibility to bet on unused resources called *Spot instances* (SIs) and rent them at variable prices with no reliability guaranteed. To support workflow execution in such hybrid environment, we extend the DCP algorithm in three directions: (1) *Rescheduling* deciding when to start the Cloud

C. Kaklamanis et al. (Eds.): Euro-Par 2012, LNCS 7484, pp. 350–362, 2012.

instances to complement the unavailability of free Grid resources; (2) *Cloud choice* determining the type and maximum amount of Cloud resources to be provisioned, and the price bet for SIs; (3) *Prescheduling* trying to minimize the impact of the scheduling overhead in case of a large number of activities. We analyze the *Spot prices* offered by Amazon EC2 and their impact to the overall workflow execution time. We find that SIs have good potential to improve the workflow execution with a significant cost reduction for longtime usage, provided that a correct price bet is made.

2   Model

2.1   Resource Model

We adopt the resource model of Amazon EC2 that offers different resources called instances of different types, where we shown the three used ones in Table 1. The processor speed of these instances is quantified with a metric called *Elastic Compute Unit (ECU)* equivalent to the speed of an Opteron 2007 processor with approximately 1.2Ghz. Additionally, there are three pricing models for renting these resource:

– *Standard instances* let the customer pay for compute capacity by the hour with no long-term commitments.
– *Spot instances (SIs)* allow customers to bid on unused Amazon EC2 capacity and run those instances for as long as their bid exceeds the current spot price. The spot price changes periodically based on supply and demand, and customers whose bids exceed/meet the spot price gain/loose access to the SIs;
– *Reserved instances* give customers the option to make a one-time payment for each instance they want to reserve for one or three years and receive in-turn a significant discount on the hourly charge.

We analyzed first the Reserved instance prices which are cheaper than the Standard instances if their usage is in the interval of $[167.3, 173.1]$ days for one year reservations, and $[252.5, 265.2]$ days for three year reservations. Since this high utilization requirement does not match our execution scenarios characterized by occasional execution of experimental workflow sets, we concentrate our work on the SIs as a cheap and more interesting alternative (see Section 4).

2.2   Application Model

We focus in this paper on scientific workflow applications that can benefit from additional Cloud resources if there are no sufficient Grid resources to support their computational requirements. We model a scientific workflow $WS = (AS, DS)$ as a set AS of

Table 1. Overview of used Amazon EC2 Linux resources and prices for the *US-East* area as of 1.2.2012; Spot prices are averaged over the last 12 months

Name	ECUs (Cores)	RAM [GB]	Archi. [bit]	I/O Performance	Disk [GB]	Standard Cost [$/h]	Spot price [$/h]	Reserved Cost [$/h]	Reservation Cost [$/j, $/3j]
m1.large	4 (2)	7.5	64	High	850	0.34	0.2124	0.12	910, 1400
c1.xlarge	20 (8)	7.0	64	High	1,690	0.68	0.3151	0.24	1820, 2800
cc1.4xlarge	33.5 (8)	23	64	10 Gigabit	1690	1.30	0.6797	0.56	4290, 6590

legacy codes called *activities* interconnected in a directed acyclic graph through control flow and data flow dependencies DS. With no loss of generality, we assume that the workflow has one initial and one final activity which has no pre- or successors. As most legacy applications do not support checkpointing, we assume that this support is not available. This restriction means that, when a SI is terminated, the intermediate results of the currently running computational activities are also lost.

To model and estimate the execution time of workflow activities on Grid/Cloud resources, we use a simple performance modeling and prediction service tuned for our pilot applications that we presented in [2]. We use the benchmarks presented in [3] to model the performance of EC2 instances.

2.3   Dynamic Critical Path Algorithm

Dynamic Critical-Path Scheduling (DCP) is a static scheduling algorithm for allocating task graphs to fully connected multiprocessors [4], which demonstrated in previous work [1] better results compared to other existing heuristic strategies for most types of workflows and independent of their size, in particular when the resource availability frequently changes or when using dynamic and unreliable heterogeneous resources. The algorithm is based on the dynamic calculation of the *critical path* (CP) in each step. The CP is defined as the set of interconnected activities from the initial to the final activity with the maximum aggregated computation and communication costs. The workflow could achieve a minimum execution time if the CP is scheduled on the resources that deliver its earliest completion time (ECT) and all other activities are executed in parallel to the CP.

Algorithm 1. DCP algorithm.

Require: $W = (AS, DS)$: workflow application, RS: resource set
Ensure: : Schedule W on RS
1: Compute *EST* **and** *LST* for all activities on all resources
2: **while** Not all activities are scheduled **do**
3:    Select $A \in AS$: $\min(LST_A - EST_A)$ and $\min(EST_A)$
4:    **if** $LST_A = EST_A$ **then**
5:      $onCP \leftarrow true$
6:    **else**
7:      $onCP \leftarrow false$
8:    **end if**
9:    $SelectResource(A, onCP, RS)$
10:    Update EST_A **and** $LST_A, \forall A \in AS$
11: **end while**

Algorithm 1 briefly outlines the DCP pseudocode that takes a workflow as input parameter and maps its activities onto the available resources for execution. First, it calculates the *earliest possible start time* (EST) and *latest possible start time* (LST) for all activities on all resources (line 1), which allows the identification of the CP. Then, the activity A with the minimum EST_A and the minimum start flexibility ($LST_A - EST_A$) is chosen for scheduling (line 3). The activities for which $LST_A = EST_A$ are on the critical path and will therefore be scheduled first. Each activity is scheduled on the resource delivering its ECT such that the LST and EST are satisfied (line 9). The *SelectResource* function (explained in Section 5.1) needs to consider if the activity is on the CP, in which case it reconsiders the already scheduled activities as the CP may have changed. This is indicated by the boolean $onCP$ variable, set to true if $LST_A = EST_A$ (lines 4–8). Depending on the scheduled activities, the EST and LST are recalculated for all activities and the algorithm repeats until all activities are scheduled (lines $10 - 11$).

3   Related Work

There are a number of important projects showing a growing interest in Cloud computing in the scientific and open source communities. The Nimbus [5] package provides a scientific Cloud middleware deployed in an informal group of four university Clouds called Science Cloud. Hadoop [6] is a toolkit for distributed computing allowing mapreduce applications to be developed and executed in a complete tool chain that supports Cloud resources and Grids. A commercial open-source implementation of a Cloud middleware compatible with EC2 is provided by Eucalyptus [7].

In [8], the impact of checkpointing when using unreliable Cloud resources is analyzed. However, scientific applications often consist of legacy codes that do not the support checkpointing mechanism needed for such an approach. Our solution shows that there is potential for using unreliable SIs even without checkpointing.

Other approaches [9] used Cloud resources to extend clusters when more requests need to be processed than the cluster can handle. A simple load balancer submits jobs either to the cluster or to the Cloud, while in our approach the Cloud instances are completely integrated into the resource pool.

In [10], several BPEL extensions for using Cloud resources in cases of peak load situations are proposed. The approach does not optimize the use of the Cloud resources, nor is well-suited for massively parallel applications, as one of the constrains of the proposed extensions is that execution servers may handle multiple requests simultaneously. In contrast, scientific applications are mostly designed to utilize dedicated resources.

The work presented in [11] is comparable to our approach, but focuses on fault tolerant scheduling that does not optimize the schedule for Cloud use and cost. Our approach is optimizing these metrics with support for cheaper SIs.

An extension of the Torque job manager to add Cloud resources to clusters is presented in [12]. Our approach could interoperate with such job manager by integrating cluster, Clouds and Grid in one single computing environment. Similar work has been done in [13] where clusters can dynamically extend their capacity with Cloud resources upon peak usage to meet given service level agreements. This gives to resource owners control over the Cloud usage and gives no transparency of the cost to the enduser, like the scheduler we proposed in this paper.

4   Spot Price Analysis

Most Cloud providers offer resources for a fixed price on an hourly basis. Since recently, Amazon EC2 introduced SI with a market-oriented dynamic pricing to better utilize the unused resources from their resource pool and we analyze them as a case study for variable prices in general. SI have a dynamic price that changes periodically based on their supply and customer's demands and bids. If the Spot price exceeds the user's bet, the instance is automatically terminated and the user does not have to pay for the last uncompleted billing interval. In the following, we analyze the Spot prices for all available instance types for the last 12 months. Since 7.2011, SI have different prices for the different availability zones of a region. Nevertheless, the data we collected until now does not show any significant difference from the overall pricing. The shown

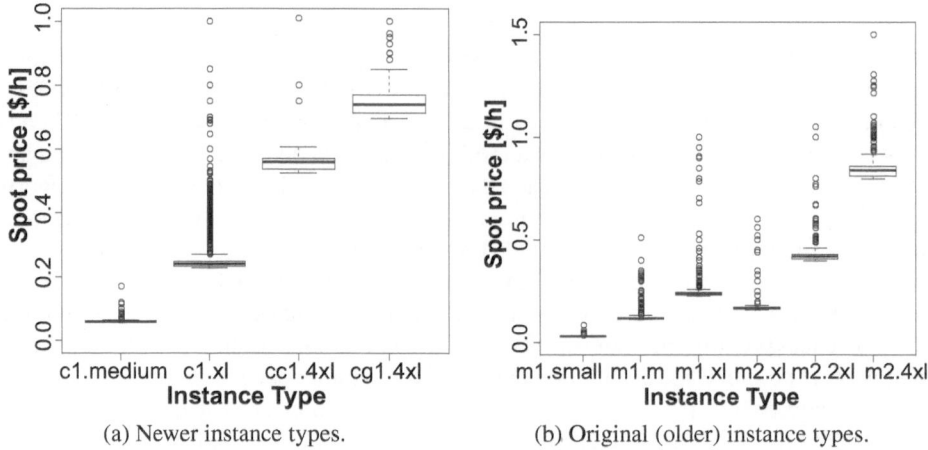

(a) Newer instance types.   (b) Original (older) instance types.

Fig. 1. EC2 Spot price analysis for the time period 22.1.2011 – 1.2.2012

methods to analyze the impact of the user bet on the reliability could be applied to other providers once they introduce similar resource models.Spot prices can be queried using the command line tools provided by Amazon EC2. Users accessing the EC2 Cloud can request Spot price information of any instance type, region, operating system and availability zone.

Figure 1 shows box-plots of the different instance types and their prices. We removed some of the outliers in Figure 1a, as the cg1.4xlarge instance had one single Spot price of $7.0 and two to four of $2.0, $2.45, $3.0, $4.0, $5.0, $6.0 and $7.0 per hour, while all other prices were lower than $1.00. We do not show the outliers in Figure 1b for the m2.4xlarge instance ($2.0 and $3.0 per hour), which only occurred once in the monitored time period. As the markers for the percentile of the dataset show, the Spot prices are stable for the m1.small, c1.medium, m2.xlarge, and m2.2xlarge instances. For the other six instance types, the price ranges are larger as shown by the visible whiskers, especially for the six xlarge types (except m2.xlarge). Not only the price fluctuations are important, but also the time periods for which the prices are valid, which we analyze in Figure 2. For a user who wants to use a SI for several hours or days, not only the maximum price to be payed is of interest, but also the average price resulting from the overall usage which approximates the cost to be payed. We analyzed the following metrics for all instance types:

– *user bet* is the price that a user will to pay for the execution of one SI for one hour;
– *average cost* is the hourly price a user has to pay for the SI if it were executed using his bet averaged over the complete timespan of our analysis;
– *maximum Spot price* is the maximum SI cost encountered still below the user bet;
– *unavailability count* represents the total number of shutdowns that occur when trying to run a SI with the user bet for the whole time period;
– *total uptime* sums the total time an instance was executed over the analyzed period with the given user bet;

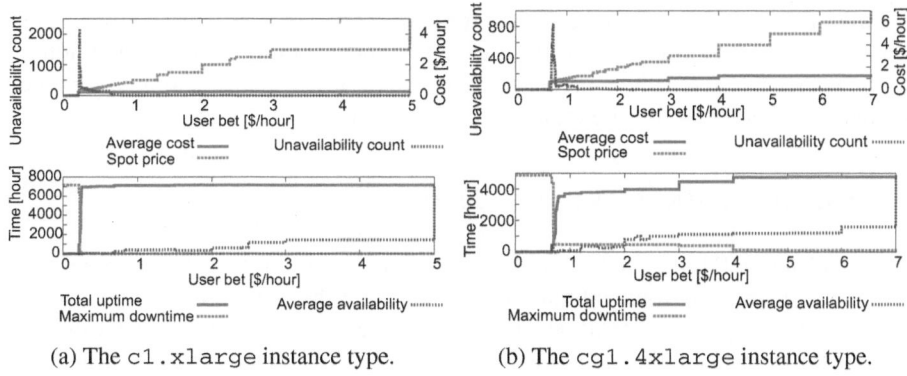

(a) The `c1.xlarge` instance type.   (b) The `cg1.4xlarge` instance type.

Fig. 2. Instance uptime and average price analysis from 22.1.2011 until 1.2.2012

- *maximum downtime* is the longest time period where the market price of a SI was above the user bet;
- *average availability* is the time between each enforced shutdown averaged across the unavailability count.

Due to space limitations, we only present the two most interesting instances: `c1.xlarge` as the mostly used instance and `cg1.4xlarge` as the most expensive and fluctuating.

Figure 2a shows that for the `c1.xlarge` instance, the user bet has an insignificant influence on the average price to be paid for its long-term usage. The top chart shows that for a low bet close to $0.22 per hour, the unavailability count can reach a peak value of 2200 which rapidly decreases when increasing the bet. We conclude that the user shall be more generous in his bet to obtain a reliable infrastructure. The maximum Spot price follows a step function showing jumps in the market prices (for example there were no prices between $1.0 and $1.36). The bottom chart illustrates that the total uptime grows as soon as the lowest market price of 0.22 is met and reaches a value of 98% at $0.252 per hour. The maximum downtime never took longer than 32 hours for a user bet over $0.24 per hour. The average availability grows to a value of 125 hours for a user bet of $0.68 per hour and raises further except for user bets between [$1.5, $2.0] per hour for which it slightly decreases as the unavailability count grows from 16 to 20.

In Figure 2b we analyze the new `cg1.4xlarge` instance type, currently only available in one of the eight regions of EC2 and since 6.4.2011. The top chart shows that the average cost increases by 14% for the analyzed period if the user bet is increased from $3.0 to $4.0 per hour. The maximum SI cost is increasing again stepwise, as in the case of the `c1.xlarge` instance. For a user bet close to $0.74 per hour, the unavailability count reaches a peak value of 833 and decreases slower than for the `c1.xlarge` instance when increasing the bet. We conclude that the user shall be more generous with this type and bet about $1.2 per hour to obtain a reliable SI. The bottom chart shows that the user bet has a higher influences on total uptime as market price fluctuations are higher. The maximum downtime for user bets up to $4 per hour is above 407 hours and reduces to 113 hours with bets over \$4. The average availability is again related to the unavailability count and fluctuates up to a bet of \$2.45 per hour where it reaches 998 hours. Nevertheless, it shows significants improvement starting from \$2.0 per hour.

The values we were able to gather from this analysis are used to estimate the recommended user bet for SI that are required to get reliable results.

5   Dynamic Critical Path for Clouds

We extend the DCP full-ahead scheduling algorithm for minimizing the workflow makespan in a dynamic environment that uses Cloud SIs if the Grid resources are not sufficient for executing large workflows. If the Spot price increases to a higher value than, the user bet, a rescheduling action is triggered to move activities from the SIs that are terminated. The Spot price is then monitored every minute using a pull mechanism until the market price is below the user bet and a new instances can be requested. A *price buffer* is added to the market price to avoid constant rescheduling when the price is fluctuating close to the user bet. If the new price is lower, the scheduler will reschedule the workflow again, compare the expected execution time and cost with the previous schedule, and use the better mapping for the activities that have not been yet submitted.

5.1   DCP-C Algorithm

The DCP-C algorithm (see Algorithm 2) is based on the original DCP presented in Section 2.3. We extend in this paper the *SelectRessource*($A_i, onCP$) function called by the main loop for mapping each activity A_i, where the *onCP* boolean variable indicates whether the activity is on the CP. For each available resource (line 3), the EST is first calculated using Algorithm 3 (lines 4–7) which considers whether the activity is on the CP or not. Afterwards, the child with the minimum $LST - EST$ difference is chosen (line 9) and a slot that fulfills its EST and LST is searched (line 10). The resource that delivers the minimum EST sum of the two activities is then chosen for scheduling.

We extended the *FindSlot* function for finding a free resource slot that satisfies the EST and LST constraints of an activity to support Cloud resources, in particular SI targeted by our work. We achieved this by adding a new resource type called *NewCloud*. If a task is mapped to a new Cloud instance, we add the startup latency of this resource (i.e. premeasured virtual machine deployment and boot times) to the estimated EST (lines 2–4), which we determine based on our previous benchmark analysis work [3]. The scheduler then looks for a resource slot that allows the execution of the activity A_i within its range $[EST, LST]$ (lines $6 - 8$) and returns an infinite start time if none is available. If the task is on the CP, the scheduler needs to consider the already mapped activities and maybe even delay some of them if the CP has been changed (lines $9 - 11$). Otherwise, the scheduler needs to check that no dependencies are violated by the new slot (line 12). Finally, the function returns the start time of the found timeslot (line 13).

In the following, we present three simple optimizations aiming to reduce the makespan of large workflows running on Grid resources complemented with on-demand dynamic SIs: rescheduling, Cloud choice, and prescheduling.

Algorithm 2. *SelectResource* function.

Require: : A_i: workflow activity; $onCP$: flag set true if $A_i \in CP$; RS: resource set;
Ensure: : Schedule A_i to one resource in RS
1: $S \leftarrow null$
2: $EST2 \leftarrow \infty$
3: **for all** $R \in RS$ **do**
4:    $EST_i \leftarrow FindSlot(A_i, R, false)$
5:    **if** $EST_i = \infty$ and $onCP$ **then**
6:       $EST_i \leftarrow FindSlot(A_i, R, true)$
7:    **end if**
8:    **if** $EST_i \neq \infty$ **then**
9:       Select $A_c \in successor(A_i)$:
        $\min(LST_c - EST_c)$
10:       $cST \leftarrow FindSlot(A_c, R, false)$
11:       **if** $cST + EST_i < EST2$ **then**
12:          $S \leftarrow R$
13:          $EST2 \leftarrow cST + EST_i$
14:       **end if**
15:    **end if**
16: **end for**
17: Schedule A_i to S

Algorithm 3. *FindSlot* function.

Require: : A: workflow activity; R: resource; $onCP$: flag set true if $A_i \in CP$;
Ensure: : start time of A_i on R;
1: Calculate EST and LST of A on R
2: **if** $R = NewCloud$ **then**
3:    $EST \leftarrow EST + CloudLatency$
4: **end if**
5: $(Start, End) \leftarrow Get\!-\!Slot(A, R, EST, LST)$
6: **if** $[Start, End] \not\subset [EST, LST]$ **then**
7:    **return** ∞
8: **end if**
9: **if** $onCP$ **then**
10:    Check if CP has changed **and** activities no longer on CP need delay
11: **end if**
12: Check if schedule does not violate dependencies
13: **return** $Start$

5.2 Rescheduling

There may appear situations during workflow execution when additional shared Grid resources become available, or when Spot prices change so that some instances may get terminated or additional ones started. This is detected by the scheduler that continuously monitors the Spot price, compares it with the user bet, and triggers a rescheduling operation to adjust the mapping of activities to the new infrastructure configuration. To perform rescheduling, the scheduler marks each activity as not scheduled except for the finished and currently running ones, and maps unexecuted subworkflow again using the proposed DCP-C algorithm. The user sets a price buffer that adds additional inertia to the mechanism to keep the number of rescheduling operations within reasonable limits.

5.3 Cloud Choice

In our model, an important task of the scheduler is to dynamically complement the Grid infrastructure with additional Cloud resources during runtime if this presents potential for improving the workflow execution. In our case, this happens when the amount of Grid resources are insufficient for executing large workflow parallel regions that need to be serialized. The scheduler detects this situation when there are no free slots on the resource an activity gets mapped to, which results in an increase in the planed starttime. In selecting a new Cloud instance, the scheduler takes two important decisions: (1) the instance type to lease from the Cloud provider, and (2) the number of such instances.

Selecting the instance type requires analysis of three important metrics: resource speed, cost, and reliability. If the total budget available for a workflow is known, the scheduler estimates the overall makespan on the Grid resources, and approximates the hourly budget required. Then, it uses this information to decide which instance type to use (representing the speed and cost parameters), including whether it should be a SI or a Standard instance (representing reliability and cost). Based on our trace data analyzed in Section 4, the scheduler will prefer SIs as long as their current market price is below the Standard price. The Spot prices are in average up to 60% lower than the regular costs (see Table 1) allowing cheaper executions with a similar reliability. To keep the complexity of the DCP-C algorithm low, this functionality is hidden in the *CloudLatency* parameter when calculating the EST on the *NewCloud* (line 3 in Algorithm 3). We quantify the speed of a Grid resource or Cloud instance in *Elastic Compute Units* (ECU), which we compute based on our previous Cloud benchmarking work [3]. Using this unit, we compare the resource speeds and predict the activity execution times for our computation-intensive workflows. When selecting the Cloud instance type, we use the maximum number of parallel activities and the total amount of available Grid resources. Since Cloud instances are only offered in bulks of cores, we use a minimum resource utilization input by the user to estimate the amount of Cloud instances needed.

5.4 Prescheduling

The DCP-C algorithm has an $O(N^3)$ complexity, where N is the number of activities, and has a non-negligible execution time for workflows with a large number of activities. To minimize the impact of this overhead on the execution time for large workflows (above 1000 activities), the scheduler does first an immediate mapping of the initial activities (most likely to be on the CP) on the fastest resources, before calling the DCP-C algorithm in a rescheduling-similar manner. With this hybrid approach we are able to obtain good scheduling results combined with a reduced scheduling overhead.

6 Evaluation

We designed and implemented the methods presented in this paper in the ASKALON environment [14] designed to support scientists in modeling, scheduling, and executing scientific workflows in Grid and Cloud infrastructures. The Cloud support is enabled through a back-end interface to the Eucalyptus middleware, compliant with the EC2 interface. To support a large number of experiments required to validate new methods, we interfaced the ASKALON enactment engine to the GroudSim [15] simulator that enables deterministic simulations of applications comprising job executions, file transfers, cost calculations, and background load on of Grid and Cloud computing infrastructures.

We used GroudSim to simulate three sites of the Austrian Grid environment and the instance types offered by Amazon EC2 (see Tables 1 and 2). We assumed a standard Amazon account which allows us to acquire a maximum of 20 instances, which results in a total of 160 cores when using the `cc1.4xlarge` instance type. We used the ASKALON environment interfaced to GroudSim to simulate two real-world scientific workflows from the material science and hydrological domains with different structure, number of activities, and computational requirements. Each workflow is characterized by a parameter x defining the "parallelization size", which determines the total number N of workflow activities. We used this parameter to simulate workflows of different sizes (small to very large), which we executed first in a pure Grid environment and then by using on-demand Cloud resources.

To model and estimate the execution time of workflow activities on Grid/Cloud resources, we use a simple performance modeling and prediction service tuned for our pilot applications that we presented in [2]. To better quan-

Table 2. Simulated Austrian Grid resources

Grid site	Location	Cores	Speed [ECU]
karwendel	Innsbruck	80	2.5
altix1.uibk	Innsbruck	16	1.5
altix1.jku	Linz	64	2.0

tify the computational performance of one ECU we use the performance benchmarks and models from previous work [3]. The average size of a file transfer in our workflows is of around 100 kilobytes, which is insignificant compared to the total computation time. Staging files into the Cloud is mostly free of charge, while staging files out of the Cloud cost around $0.12 per gigabyte, resulting in a negligible influence on the aggregated workflow cost. In all experiments, we selected from the historical data from EC2 an interval with high SI prices to study the impact of instance termination and the resulting rescheduling.

6.1   Wien2k

Wien2k is a material science workflow for performing electronic structure calculations of solids. The Wien2k workflow contains two parallel sections of size x, with sequential synchronization activities in between (see Figure 3a). The total number of activities in a Wien2k workflow is: $N_{wien2k} = 2 \cdot x + 3$.

Figure 4a shows our simulation results for scheduling and executing Wien2k workflow with different parallelization sizes. The chart first shows that using additional Cloud instances brings significant improvements in the execution time of this workflow. The faster instance types are selected by the scheduler and the more cores are added to the resource pool, the lower the execution time of the workflow gets. Using user bets of $0.35 for m1.large, $0.68 for c1.xlarge and $1.01 for cc1.4xlarge instances gives us a 99% reliability according to our analysis in Section 4 (meaning a value in the area of high availability with acceptable average prices). This means that there will be no reschedules in the execution on the SIs resulting in identical curves as when using the Standard instances. The cost for using SIs is about 60% cheaper than using the Standard instances for the same executions (see Table 1). The figure also shows that choosing lower user bets of $0.125 for m1.large, $0.25 for c1.xlarge and 0.58 $ for cc1.4xlarge instances closer to the Spot price increases the completion time due to SI unavailability and rescheduling operations (e.g. for the cc1.4xlarge instances with a parallelization size of 388). In the worse case, the same execution times as when using Grid resources only are obtained, meaning that the SIs have been paid with no benefit. Smaller workflows with $x < 200$ or 80 hours of runtime finish before any SI gets terminated which produces overlapping lines in the chart.

6.2   Invmod

Invmod [16] is a hydrological application that uses the Levenberg-Marquardt algorithm to minimize the least squares of the differences between the measured and the simulated runoff for a determined time period. The Invmod workflow displayed in Figure 3b

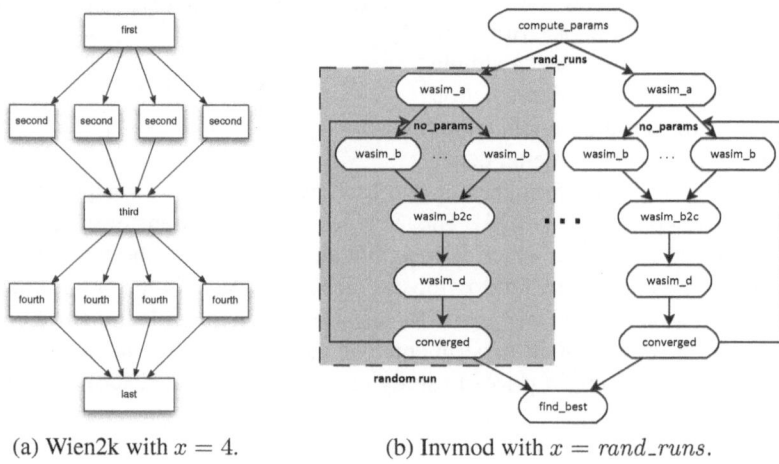

(a) Wien2k with $x = 4$.   (b) Invmod with $x = rand\_runs$.

Fig. 3. Scientific workflows

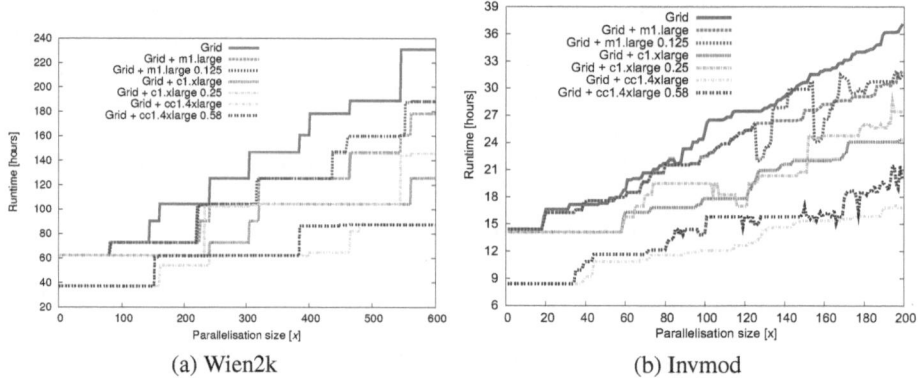

(a) Wien2k   (b) Invmod

Fig. 4. Execution times of different parallelization sizes and instance types

consists of two levels of parallelism: (1) the outermost parallel loop consists of a number of random runs x (parallelization size) that perform a local search optimization (in a sequential loop) starting from a random initial solution; (2) alternative local changes are examined for each calibrated parameter in parallel in the inner nested parallel loop. The total number of jobs in an Invmod workflow is: $N_{invmod} = 13 \cdot x + 2$.

Figure 4b shows our experiments results with the Invmod workflow. Since this workflow has a higher complexity than Wien2k, the benefits are smaller when slower m1.large instances are added to the resource pool. On the other hand, the impact of the cluster resources cc1.4xlarge is higher, as they outperform the available Grid hardware by their speed and core amount ($20 \cdot 8 = 160$ cores). Similar to the Wien2k workflow, a correct user bet for SIs can bring similar performance results with over 60% cost reduction (see Table 3). The m1.large executions show that for some parallelization sizes ($x = 140$ or $x = 150$) the performance of the workflow running on SIs that get terminated can be as low as a Grid only execution. SIs can be again detrimental when setting a too low user bid for all three instance types as compared to Standard instances. In rare cases, rescheduling triggers an improvement in execution time due to the better mappings for the remaining subworkflows obtained by the CP heuristic.

7   Conclusion

The choice of the correct Cloud instance is critical when running scientific application in the Cloud. While cheaper resources might look attractive, their slow characteristics degrade the performance in such a way that the overall execution time is not significantly im-

Table 3. Cost comparison when using Spot and Standard instances

Workflow	Size [x]	Instance Type	Cost [$]	Spot price [$]	saved [%]
wien2k	400	m1.large	439.28	167.743	61.81
wien2k	200	cc1.4xlarge	1771.2	648.818	63.37
invmod	140	c1.xlarge	171.36	64.726	62.23
invmod	160	cc1.4xlarge	385.6	154.047	60.05

proved and in the worst case even decreased. In this paper we analyzed the potential of using SIs for improving the performance of real-world scientific workflows that suffer from the insufficient Grid resources compared to their computational demand. We first collected and analyzed the characteristics of the Spot prices over a period of 12 months and found out that they are quite stable for the different instance types offered by Amazon. Furthermore, SIs offer over 99% availability when the user makes a correct bet of $0.35 for m1.large, $0.68 for c1.xlarge and $1.01 for cc1.4xlarge instances with a average price far below the standard price. Then, we extended the Dynamic Critical Path algorithm for Cloud environments using three simple extensions: rescheduling, Cloud choice, and prescheduling. We learned that SI can bring significant improvements similar to the Standard instances to the execution of scientific workflows in hybrid Grid-Cloud environment, provided that the correct user bet is made. Cost-wise, SIs can bring over 60% reduction in costs if correctly used. With the upcoming Spot price per availability zone, there will be even more room for improving the execution cost by dynamically choosing the cheapest zone for the end-user.

Acknowledgment. The research was funded by the Austrian Science Fund (FWF): TRP 72-N23 and the Standortagentur Tirol: RainCloud.

References

1. Rahma, M., Venugopal, S., Buyya, R.: A Dynamic Critical Path Algorithm for scheduling Scientific Workflow Applications on Global Grids. In: eScience, pp. 35–42. IEEE Computer Society (2007)
2. Nadeem, F., Yousaf, M., Prodan, R., Fahringer, T.: Soft benchmarks-based application performance prediction using a minimum training set. In: e-Science. IEEE Computer Society Press (2006)
3. Iosup, A., Ostermann, S., Yigitbasi, N., Prodan, R., Fahringer, T., Epema, D.: Performance Analysis of Cloud Computing Services for Many-Tasks Scientific Computing. IEEE TPDS 22(6), 931–945 (2011)

4. Kwok, Y.-K., Ahmad, I.: Dynamic Critical-Path Scheduling: An effective Technique for allocating Task Graphs to Multiprocessors. IEEE TPDS 7(5), 506–521 (1996)
5. Keahey, K., Freeman, T., Lauret, J., Olson, D.: Virtual Workspaces for Scientific Applications. In: Scientific Discovery through Advanced Computing, Boston (June 2007)
6. Apache, Apache Hadoop Project develops open-source Software for Reliable, Scalable, Distributed Computing (May 2010), http://hadoop.apache.org/
7. Nurmi, D., Wolski, R., Grzegorczyk, C., Obertelli, G., Soman, S., Youseff, L., Zagorodnov, D.: Eucalyptus: A Technical Report on an Elastic Utility Computing Architecture Linking Your Programs to Useful Systems. UCSB Computer Science. Tech. Rep. 2008-10 (2008)
8. Sangho, Y., Kondo, D., Andrzejak, A.: Reducing Costs of Spot Instances via Checkpointing in the Amazon Elastic Compute Cloud. In: CLOUD, pp. 236–243. IEEE (2010)
9. Assuncao, A.C.M., Buyya, R.: Evaluating the Cost-Benefit of using Cloud Computing to Extend the Capacity of Clusters. In: HPCC. ACM (2009)
10. Dörnemann, T., Juhnke, E., Freisleben, B.: On-demand Resource Provisioning for BPEL Workflows using Amazon's Elastic Compute Cloud. In: CCGrid, pp. 140–147. IEEE Computer Society (2009)
11. Ramakrishnan, L., Koelbel, C., Kee, Y.-S., Wolski, R., Nurmi, D., Gannon, D., Obertelli, G., YarKhan, A., Mandal, A., Huang, T.M., Thyagaraja, K., Zagorodnov, D.: Vgrads: enabling e-Science Workflows on Grids and Clouds with Fault Tolerance. In: SC. ACM (2009)
12. Marshall, P., Keahey, K., Freeman, T.: Elastic Site: Using Clouds to Elastically Extend Site Resources. In: CCGrid, pp. 43–52. IEEE (2010)
13. Blanco, C.V., Huedo, E., Montero, R.S., Llorente, I.M.: Dynamic Provision of Computing Resources from Grid Infrastructures and Cloud Providers. In: GPC Workshops, pp. 113–120. IEEE Computer Society (2009)
14. Fahringer, T., Prodan, R., Duan, R., Nerieri, F., Podlipnig, S., Qin, J., Siddiqui, M., Truong, H.L., Villazón, A., Wieczorek, M.: ASKALON: A Grid application development and computing environment. In: GRID, pp. 122–131. IEEE (2005)
15. Ostermann, S., Plankensteiner, K., Prodan, R.: Using a New Event-based Simulation Framework for Investigating Different Resource Provisioning Methods in Clouds. Scientific Programming Journal 19(2-3), 161–178 (2011)
16. Cullmann, J., Mishra, V., Peters, R.: Flow analysis with WaSiM-ETH - model parameter sensitivity at different scales. Advances in Geosciences 9, 73–77 (2006)

Topic 7: Peer to Peer Computing

Alberto Montresor, Evaggelia Pitoura, Anwitaman Datta, and Spyros Voulgaris

Topic Committee

Peer-to-peer (P2P) systems enable computers to share information and other resources with their networked peers in large-scale distributed computing environments. The resulting overlay networks are inherently decentralized, self-organizing, and self-coordinating. Well-designed P2P systems should be adaptive to peer arrivals and departures, resilient to failures, tolerant to network performance variations, and scalable to huge numbers of peers (tens of thousands to millions). As P2P research becomes more mature, new challenges emerge to support complex and heterogeneous decentralized environments for sharing and managing data, resources, and knowledge with highly dynamic and unpredictable usage patterns. This topic provides a forum for researchers to present new contributions to P2P systems, technologies, middleware, and applications that address key research issues and challenges.

This year, three papers have been accepted for publication in the peer-to-peer track. Each paper was evaluated by four referees.

The paper *ID-Replication for Structured Peer-to-Peer Systems* by Tallat Shafaat, Bilal Ahmad, and Seif Haridi from the Royal Institute of Technology (KTH), Sweden, discusses the shortcomings of existing replication schemes in DHTs and proposes a new technique called ID-Replication. ID-Replication is less sensitive to churn compared to the state-of-the-art techniques, and allows to vary the replication degree depending on the popularity of the keys.

The paper *Changing the Unchoking Policy for an Enhanced BitTorrent* by Vaggelis Atlidakis, Mema Roussopoulos and Alex Delis from the University of Athens, Greece, proposes a novel optimistic unchoking approach for BitTorrent that takes into consideration the number of peers currently interested in downloading from a client that is to be unchoked, and favors those having few peers interested in downloading data from them, to trigger the interest of additional peers.

The paper *Peer-to-Peer Multi-Class Boosting* by István Hegedüs, Róbert Busa-Fekete, Róbert Ormándi, Márk Jelasity, and Balázs Kégl from the University of Szeged, Hungary, and the University of Paris-Sud, France, deals with the problem of data mining over large-scale fully distributed databases, where each node stores only one data record. The authors extends their previous work on gossip-based machine learning by considering a well-known boosting technique.

We would like to take the opportunity of thanking the authors who submitted a contribution, as well as the Euro-Par Organizing Committee, and the external referees with their highly useful comments, whose efforts have made this conference and this topic possible.

C. Kaklamanis et al. (Eds.): Euro-Par 2012, LNCS 7484, p. 363, 2012.
© Springer-Verlag Berlin Heidelberg 2012

ID-Replication
for Structured Peer-to-Peer Systems*

Tallat M. Shafaat[1], Bilal Ahmad[1], and Seif Haridi[2]

[1] KTH - Royal Institute of Technology, Sweden
[2] Swedish Institute of Computer Science, Sweden
{tallat,bilala,haridi}@kth.se

Abstract. Structured overlay networks, like any distributed system, use replication to avoid losing data in the presence of failures. In this paper, we discuss the short-comings of existing replication schemes and propose a technique for replication, called *ID-Replication*. ID-Replication allows different replication degrees for keys in the system, thus allowing popular data to have more copies. We discuss how ID-Replication is less sensitive to churn compared to existing replication schemes, which makes ID-Replication better suited for building consistent services on top of overlays compared to other schemes. Furthermore, we show why ID-Replication is simpler to load-balance and more secure compared to successor-list replication. We evaluate our scheme in detail, and compare it with successor-list replication.

1 Introduction

Structured overlay networks provide the infrastructure used to build scalable and fault-tolerant key-value stores, e.g. Cassandra [9]. While scalability comes with using consistent hashing, fault-tolerance is achieved by replication. There are different strategies for replication in overlays, such as successor-list replication [17], using multiple hash functions, and symmetric replication [3]. Out of these, successor-list replication is the most popular and widely used in ring-based overlays. For instance, overlays including Chord [17], Pastry [14] (with a minor modification), and Cassandra [9], all use successor-list replication.

It turns out that successor-list (SL) replication has some drawbacks. SL-replication is highly sensitive to churn; hence a single node join or failure event results in updating multiple replication groups. Furthermore, the replication degree has to be constant throughout the system, restricting popular/hot data from having more replicas. Next, SL-replication is inherently difficult to load-balance. Finally, SL-replication is less secure and presents a bottleneck since there is a master replica of each replication group and all requests for that group have to go through the master replica. We discuss these issues in detail in Section 2.1.

In this paper, we propose a replication strategy called *ID-Replication*. ID-Replication does not suffer from the afore-mentioned drawbacks of SL-replication.

---

* We would like to thank Cosmin Arad, Ahmad Al-Shistawy and Niklas Ekström for their valuable discussions and feedback.

C. Kaklamanis et al. (Eds.): Euro-Par 2012, LNCS 7484, pp. 364–376, 2012.

It allows varied replication degrees in the system, and requests do not need to go through the master replica. ID-Replication gives more control to an administrator, without hampering self-management. Furthermore, ID-Replication is less sensitive to churn, thus being better suited to be used for building consistent services and in asynchronous networks where false failure detections are a norm. Since we use a generic design, ID-Replication can be used in any structured overlay network.

In this paper, we discuss the short-comings of popular existing replication schemes. We explain ID-Replication in detail and discuss the ideology behind the design decisions. We perform a thorough evaluation and compare ID-Replication to SL-replication.

2   Preliminaries

An overlay makes use of an *identifier space*, which for our purposes is defined as a set of integers $\{0, 1, \cdots, \mathcal{N} - 1\}$, where $\mathcal{N}$ is some apriori fixed, large, and globally known integer. This identifier space is perceived as a ring that wraps around at $\mathcal{N} - 1$. Each node in the system has a unique identifier from the identifier space. The *successor* of a node with identifier p is the first node found going in clockwise direction on the ring starting at p. Similarly, the *predecessor* of a node with identifier q is the first node met going in anti-clockwise direction on the ring starting at q. The *successor-list* of a node m consists of m's c immediate successors, where c is typically set to $\log_2(s)$, where s is the network size.

Each node q is responsible for storing keys between q's predecessor and q. For a replication degree of r in SL-replication, a key k is stored on the node q that is responsible for storing k, and $r - 1$ immediate successors of q. In essence, the key is stored on the responsible node q, and the first $r - 1$ members of q's successor-list (see Figure 1). In Fig 1, node 30 is responsible for storing keys $k \in (20, 30]$, and k are replicated on $\{30, 35, 40\}$, which is called the *replica group* for k. As nodes join and leave the system, the successor, predecessor and successor-lists are updated, leading to changes in the replica groups and transfer of keys between nodes.

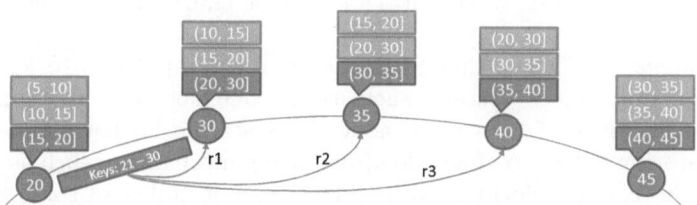

Fig. 1. Successor-list replication with replication degree 3. The replication group for $keys \in [21, 30]$ is $\{30, 35, 40\}$. Similarly, responsibility of node 35, i.e. $(30, 35]$, is replicated on 3 nodes encountered clockwise from 35, i.e. $35, 40$ and 45.

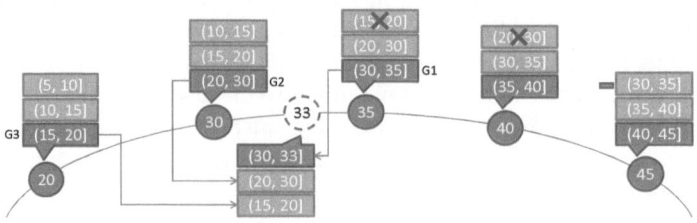

Fig. 2. A new node 33 joins in a system using successor-list replication and degree 3. 6 nodes are involved in making changes, and 4 replication groups have to be updated.

2.1 Problems with Existing Schemes

Replica Groups Affected by Churn: Churn - node joins and failures - is considered a norm in P2P systems. A desirable behaviour is that a churn event should not effect the configuration of an overlay greatly. In SL-replication, the unit of replication is a node's assigned key space, also known as the node's responsibility. For instance in Fig. 1, the key space assigned to node 30 is $(20, 30]$, which is replicated on 35 and 40. Consequently, for a replication degree of r, each node replicates r node responsibilities going anti-clockwise.

When a new node joins the overlay, it divides a node responsibility range into two ranges. Similarly, a node failure results in merger of two node responsibilities. Since each node responsibility range is replicated on r nodes, and each node replicates r ranges, a single churn event results in reconfiguration of r replication groups. Furthermore, a join event involves action on behalf of $2 \times r$ nodes, and a failure involves action on behalf of $(2 \times r) - 1$ nodes. This is shown in Figure 2 where a new node 33 joins the system in a state shown in Figure 1. Replication groups $G1$, $G2$, and $G3$ need to be updated, and nodes 20, 30, 33, 35, 40, 45 are involved in such updates.

This approach has multiple drawbacks. First, a single churn event is overly complicated, involving many nodes. Second, consistent services built on top of overlays require consistent views of replication groups. For instance, Scatter [4] and Etna [12], both require consensus whenever a replication group changes. A high number of reconfigurations for a single churn event is undesirable. Lastly, the time duration needed to stabilize for a single churn event is very high.

Load-Balancing: We argue that SL-replication is complicated to load-balance. Consider an unbalanced system, such as the one depicted in Figure 1. It is unbalanced in-terms of keys since node 30 is storing 10 keys while all other nodes are storing 5 keys. A simple load-balancing mechanism, such as [8], would move node 30 counter-clockwise to handover responsibility of some keys to 35, or move 20 clockwise so that 20 takes over responsibility of some keys from 30. Since $keys \in (20, 30]$ are replicated on 3 nodes, such a movement will reduce load from one replica node only. Hence, r node movements on the identifier ring are needed to balance the load of one key range.

Security: In SL-replication, all requests for a key k end up on the node n responsible for k. This has two drawbacks. First, it is difficult to load-balance requests since all requests for k pass through n before they can be routed to a replica. Hence, n becomes a bottleneck. Second, if n is an adversary, it can launch a malicious attack [16].

Symmetric Replication: In Symmetric Replication [3], keys are stored symmetrically on the identifier space using equivalence classes. This leads to requiring a complicated bulk operation for retrieving all keys in a given range. Node joins and failures have to use such a bulk operation to find data to be replicated.

3 ID-Replication

In this section, we describe a replication scheme for ring-based overlays, called ID-Replication. We first provide an overview of ID-Replication, give a detailed algorithmic specification, and then discuss its desirable properties.

3.1 Overview

We set out to design a replication scheme that is less sensitive to churn in terms of the number of replication groups that need to be reconfigured. In ID-Replication, we use sets of nodes, called *groups*, instead of individual nodes as the building blocks for the overlay. Instead of partitioning the identifier space amongst nodes, we partition the identifier space among groups. Thus, compared to the simple structured overlay model where nodes are responsible for key ranges, we assign responsibility ranges to groups. Consequently, groups are assigned identifiers. The idea of using groups instead of nodes can be applied to the majority of the overlays. For the sake of simplicity, we use Chord-like notation in this paper.

All nodes within a group have the same identifier as the group. To distinguish nodes within a group, each node also has a group-local identifier. The group-local identifiers of nodes only need to be unique within the group. For efficient routing, each node maintains long range links, such as fingers in Chord.

The model of ID-Replication is shown in Figure 3. There are five groups on the identifier space: $20, 30, 35, 40$ and 45. The successor of a group is the first group encountered going clockwise from that group, e.g. group 40 is the successor of group 35. Similarly, the predecessor of a group is the first group encountered going anti-clockwise, e.g. 30 is the predecessor of 35. A group is responsible for the key range from its predecessor to itself, e.g. group 35 is responsible for $keys \in (30, 35]$.

Each group is composed of a number of nodes, e.g. group 30 contains nodes $\{1, 2, 3\}$. The nodes of a group are the replicas for the keys that the group is responsible for. The size of each group is specified using two parameters: r_{min} and r_{max}. Thus, the replication degree of keys is always between r_{min} and r_{max}.

To maintain the ring under dynamism, we employ a modified version of periodic stabilization [17] that operates on groups instead of nodes. Furthermore, we use *gossiping* between nodes in a group to synchronize the view of the group among the group members.

We use two operations for reconfiguring groups: *Merge* and *Split*. When the size of a group $G1$ drops below r_{min}, we need to *merge* $G1$'s members with another group $G2$ such that the size of the merged group should be less than r_{max}. The merged group $G = G1 \cup G2$ retains the identifier of $G2$.

When the size of a group G becomes larger than r_{max}, we need to *split* it into two groups, $G1$ and $G2$, such that the size of each split group is larger than or equal to r_{min}. The identifiers of $G1$ and $G2$ are calculated in a way to increase the load-balance in the system.

A failure of a node can trigger a merge. Similarly, a new node joins an existing group, which can result in a split.

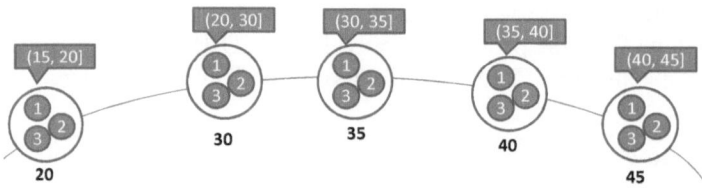

Fig. 3. A configuration of ID-Replication. Replica groups are denoted by a single identifier on the identifier space ring. Nodes in a replica group $G1$ are responsible for storing keys between $G1$'s predecessor replica group's identifier and $G1$. Nodes within a replica group are differentiated by using group-local identifiers (1, 2, and 3 in the figure).

3.2   Algorithm

We give a full specification of ID-Replication as Algorithm 1 and 2 in an *rpc*-notation. Each node stores a group-local identifier l_{id}, a group identifier id, and a set of nodes in its group, *group*. An rpc-call is denoted by '::'. For instance, $m::id$ denotes the value of id on node m.

A new node n joins the system by attempting to become a member of a group of size less than r_{max} to avoid a split operation. Ideally, n should join the lowest-sized group. Such a group can be found in a best-effort manner by a random walk, or by maintaining directories that store such information (as in [5]). If a group with less than r_{max} members is not found, n will join a group causing it to split into two.

Nodes maintain successor-lists to preserve the ring-geometry amid churn. The difference between Chord and ID-Replication successor-lists is that the lists are composed of successive groups instead of successive nodes. If all nodes in the successor group of G fail or merge with another group, G points to the next group in the successor-list (Algo 1, line 13). For ring and successor-lists maintenance, we use an algorithm similar to Chord's periodic stabilization, where nodes belonging to a group periodically stabilize the ring with nodes in their successor group (Algo 1, lines 15–27).

Algorithm 1. ID-Replication(part 1): Periodic stabilization for joins and failures

---

▷ '::' denotes a remote procedure call

1: $n.$**join**$(seed)$ ▷ Periodically retried with new seed if request fails
2:      $seed::$join_request(n)
3:
4: $n.$**join_request**(m)
5:      **if** $|group| < r_{max}$ **then**   ▷ if n's replica group has space
6:          $group := group \cup \{m\}$
7:          $m :: \langle id, group \rangle := \langle id, group \rangle$ ▷ Set joining node's id and group
8:
9: $n.$**node_failure**(f) ▷ Node f failed
10:      $group := group - \{f\}$
11:      $pred.group := pred.group - \{f\}$
12:      $succ.group := succ.group - \{f\}$
13:      **if** $succ.group == \{\}$ **then**
14:          $succ := next\_in\_successor\_groups()$
15:
    ▷ Periodically check for new successor and predecessor groups
16: $n.$**stabilize_ring**$()$
17:      $random_succ := $ select_random$(succ.group, 1)$ ▷ Select a random node
18:      $\langle x.id, x.group \rangle := rand\_succ :: pred.\langle id, group \rangle$
19:      **if** $x.id \in (id, succ.id)$ **then**
20:          $\langle succ.id, succ.group \rangle := \langle x.id, x.group \rangle$
21:      $succ.group := $ select_random$(succ.group, 1):: group$ ▷ Update my view
22:      $\forall p \in succ.group$ **do** $p :: notify(n, id, group)$
23:
24: $n.$**notify**$(src, pid, pgroup)$
25:      **if** $pid \in (pred.id, id)$ **or** $src \in pred.group$ **then**
26:          $\langle pred.id, pred.group \rangle := \langle pid, pgroup \rangle$
27:

Say the size of a group G_{size} is more than r_{min}. In such a case, even if $G_{size} - r_{min}$ nodes, called *standby* nodes, leave the group, it will neither violate the replication degree nor require a merge operation. These standby nodes can potentially become part of a group in which a node fails. Hence, standby nodes advertise themselves (Algo 2, lines 16–17) by either gossiping, or periodically updating their address information into directories (as in [5]).

Each node n periodically checks if the size of its group, G_{size}, is between r_{min} and r_{max}. If G_{size} is smaller than r_{min}, then n searches for a standby node by gossiping or contacting a directory, and tries to include it in n's group. If a standby node cannot be found, n triggers a merge operation (Algo 2, lines 7–14). A merger is required in this case to maintain a replication degree of at least r_{min}. Similarly, if G_{size} is larger than r_{max}, n initiates the split operation by dividing the group into two groups (Algo 2, lines 1–6). Furthermore, n periodically gossips with its group members to synchronize their view of the group, and can use anti-entropy to update data items.

Algorithm 2. ID-Replication(part 2): SPLIT and MERGE operations

$\triangleright$ Periodically attempt to keep $r_{min} < |group| < r_{max}$

```
1: every γ time units  and  |group| > r_max at n                    ▷ SPLIT operation
2:    peers_to_split := get_top(sort(group), r_min)   ▷ Get r_min nodes with lowest l_id
3:    ∀ p ∈ peers_to_split do  p::⟨id, group⟩ := ⟨new_key, peers_to_split⟩
4:    peers_to_retain := group − peers_to_split
5:    ∀ p ∈ peers_to_retain do  p::⟨id, group⟩ := ⟨id, peers_to_retain⟩
6: end event

7: every γ time units  and  |group| < r_min at n                        ▷ Due to failures
8:    node := search_standby_node()
9:    if node = nil then              ▷ Search failed, MERGE with successor group
10:       ⟨new_id, new_group, new_succ⟩ := ⟨succ.id, succ.group∪group, succ :: succ⟩
11:       ∀ p ∈ new_group do  p::⟨id, group, succ⟩ := ⟨new_id, new_group, new_succ⟩
12:    else                          ▷ Make the standby node part of n's group
13:       node::⟨id, group, succ, pred⟩ := ⟨id, group ∪ {node}, succ, pred⟩
14: end event

15: every δ time units at n          ▷ Periodically synch view with group-mates
16:    if index_of(n, sort(group)) > r_min then
17:       publish_as_standby_node(n)
18:    gossip_view(group)   ▷ Synchronize group view (& data) with group members
19: end event
```

3.3 Discussion

As we discuss in Section 3.1 and evaluate in Section 4, ID-Replication requires less replication group reconfigurations per churn event. This makes ID-Replication ideal for building a consistent DHT. Each replication group can be considered as a replicated state machine and operations are performed on the data in a total order within the group. To handle dynamism, we need to support the merge and split operations where the view of a group changes. For this, we can use a re-configurable replicated state machine, such as SMART [11]. Using SMART with SL-replication is both complicated and expensive as replicated state machines are implemented using Consensus. Since ID-Replication requires fewer replication group reconfigurations per churn event, it will require fewer instances of consensus. Furthermore, in an asynchronous system, false failure suspicions are very common, which will trigger much more reconfiguration requests in SL-replication.

ID-Replication allows the system user to have different replication degrees for different keys. We use two parameters, r_{min} and r_{max}, to control the replication degree. For a given range, the number of replicas is at least r_{min} and at most r_{max}. Thus, popular or critical data can have more copies than other data by setting higher values of r_{min} and r_{max} for the corresponding key range. Furthermore, requests do not need to go through a master replica. Hence, ID-Replication does not have any bottlenecks, and requests can be load-balanced across all replicas. Finally, such a design avoids the security vulnerabilities of SL-replication [16].

Owing to the design of ID-Replication, a system administrator has much more control over the system compared to SL-replication. For instance, the administrator can control how many and which machines should serve a particular key-range. This also allows the usage of specialized hardware for handling requests for certain keys. On the contrary, a node in SL-replication is responsible for replicating multiple key ranges (r key partitions anti-clockwise), making it harder to control.

Routing tables, e.g. fingers in Chord, can also be build using groups. Each routing pointer can point to a group, containing addresses of multiple nodes. Greedy routing can be done on group identifiers, and a lookup can be routed to a random node in the group. For fault-tolerance and better performance, a lookup can be routed by forwarding in parallel to all nodes in the groups at each hop, and considering only the first reply. While such a mechanism consumes more bandwidth, it (a) is more reliable as it can tolerate failure of nodes in the path, and (b) has lower latency as the lookup can exploit multiple paths. Such parallel lookup techniques have also been proposed for Chord like overlays [10].

4   Evaluation

To evaluate our work, we simulated both ID-Replication and SL-replication in Kompics [2]. The simulations were performed with an initial network size of 2000 nodes, using the King dataset [6] for message latencies between the nodes. Each experiment had the following structure: we initialized the overlay with 2000 nodes. Once the overlay converged, we subjected it to 2000 churn events (1000 joins and 1000 failures), and measured the metrics till the topology converged. The lifetimes of nodes had a poisson distribution, and each node failure was followed by a join event. We evaluated both replication schemes under various levels of churn by changing the median parameter of poisson distribution for the lifetimes. A higher median lifetime results in lower churn rate. We performed simulations for periodic stabilization periods of 30 and 60 seconds. The experiment results for both stabilization rates were the same, so we omit graphs for stabilization delay of 60 seconds due to space restrictions. We simulated 3 directories for nodes to publish and find standby nodes, and used a value of $r_{max} = 2 \times r_{min}$. Such directories can be implemented by using predefined keys, and storing information under those keys [5]. We repeated each experiment for 10 seeds and report the averages.

4.1   Replication Groups Restructured

We measured the number of replication groups that need to be reconfigured due to the churn events (see Figure 4). The x-axis shows the median lifetime used for nodes, while the y-axis depicts the number of replication groups restructured per churn event. As analyzed earlier, the figure shows that there are r number

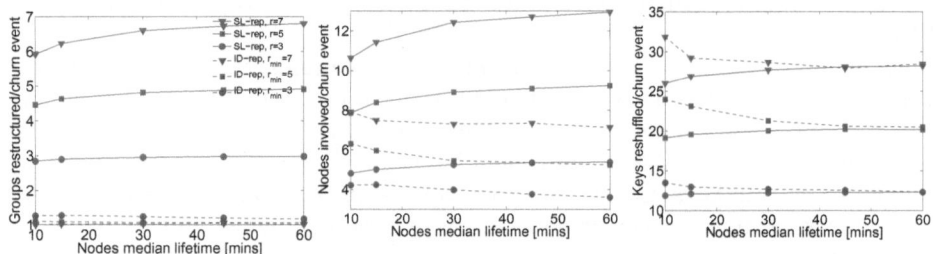

Fig. 4. Number of replica-tion groups restructured per churn event

Fig. 5. Number of nodes in-volved in updates for each churn event

Fig. 6. Number of keys transferred between nodes converge for both schemes

of reconfigurations per churn event for SL-replication, while the corresponding value stays close to one for ID-Replication. In this case, ID-Replication does not depend on the replication degree whereas SL-replication does. SL-replication has lower restructuring count at higher churn rates than lower rates. The reason being that at high churn, simultaneous node failures in a replica group can mask the cost of multiple node failures with the cost of a single node failure, since, all the failed nodes will be replaced in a single periodic stabilization round.

ID-Replication has a very low group restructuring cost and is unaffected by both r and churn rate because majority of churn events restructure only one group. Splits, merges and standby node movement restructure two groups. Since these events occur at a low frequency, the restructuring cost stays low.

4.2   Nodes Involved in Updates

Each churn event requires action on behalf of a certain number of nodes. In this experiment, we counted the number of nodes involved in the reconfiguration updates. This count is depicted in Figure 5, normalised against the number of churn events. As analyzed in Section 2.1, the count for SL-replication approaches $2 \times r$. Since ID-Replication involves only one group for a single churn event (excluding splits and merges), the number of nodes involved to handle churn stays close to r. It is noteworthy that the performance of ID-Replication improves as the mean life time of nodes increases, which is opposite to SL-replication behaviour. At lower churn rates, the number of splits and merges is reduced because of the join mechanism of ID-Replication where new nodes try to join groups with low node count. Since at low churn rates the topology changes very slowly, nodes take better decisions about which group to join. This reduces splits and merges, thus resulting in fewer nodes involved at low churn rates. Such behaviour makes ID-Replication ideal for managed systems in data-centers and cloud computing where the churn rate is low.

4.3   Keys Reshuffled

Next, we evaluate the number of keys that have to be transferred between nodes. Figure 6 shows the comparison between SL-replication and ID-Replication with-respect-to the number of keys re-shuffled per churn event. At lower churn rates, both SL-replication and ID-Replication converge to the same value. However, the two replication schemes behave differently at higher churn rates.

The reason behind SL-replication's reduced cost at high churn rate is because nodes become aware of the change in their responsibilities after each periodic stabilization step. Now, when the mean lifetime of nodes is very short it may happen that a failed node is replaced by a new node, before other nodes had the chance to detect its failure. This way the join event masks the cost associated with the failure of the node. Furthermore, a new node may fail shortly after joining (within a periodic stabilization window) without anyone noticing its join and failure, thus avoiding key re-shuffling.

The increased cost of ID-Replication at high churn is due to a high merge rate. Merge is costly in terms of keys re-shuffled as it results in transferring keys of two responsibility ranges by all members of the two groups being merged. On the other hand, the movement of a standby node requires the transfer of one responsibility range only once. When the churn rate is higher than the rate at which the standby nodes are being advertized, the number of merge operations is naturally higher. However, when the churn rate becomes comparable to the rate of publication of standby nodes, the system involves more standby node movements and thus reducing the number of merge operations. This phenomena is depicted in Figure 7, illustrating the number of splits, merges, and standby node movements per churn event. The figure shows that as the mean lifetime of nodes is increased (churn rate is reduced), the rate of standby node movements increases, which results in a decreased merge rate. Furthermore, at low churn rates, the search for lowest size group gets better results. This experiment suggests that for higher rates of churn, the rate of updating the directories with group-size and standby node information should be higher as well. It is worth noting that a lifetime of 10 minutes is considered a very high churn rate for a DHT.

4.4   Overhead of Maintaining Groups

ID-Replication maintains groups using a gossiping protocol such as Cyclon [18], which adds to the maintenance cost. Cyclon is inexpensive, especially given that the group sizes are small and the churn rates are moderate in cloud environments. We used a gossip rate equal to the periodic stabilization rate (30 seconds), and measured maintenance cost for various network sizes. Our results show that the gossiping overhead is approximately the same as periodic stabilization. Hence, using ID-Replication doubles the maintenance bandwidth requirement and the number of messages exchanged is almost 1.7 times higher. The maintenance cost is still moderate and negligible given today's interconnects.

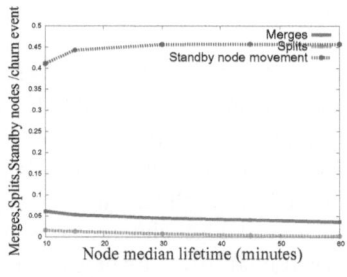

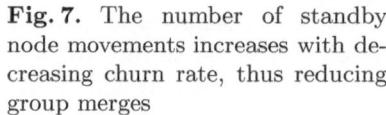

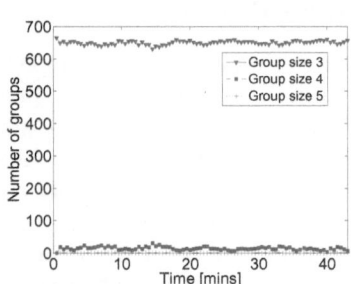

Fig. 7. The number of standby node movements increases with decreasing churn rate, thus reducing group merges

Fig. 8. Number of groups for sizes between 3 (r_{min}) and 6 (r_{max})

4.5 Evolution of Groups

Finally, we evaluated the size of groups over time, where it is desirable that the group sizes are close to r_{min}. Figure 8 shows the number of groups for each size between r_{min} and r_{max} over time for a poisson churn with mean 60 minutes. The figure confirms that most of the groups have a size of 3, which is r_{min}. We observed similar trends for other churn rates as well, which we omit here due to space constraints.

5   Related Work

Symmetric replication [3] proposes an alternative replication scheme for structured overlays. However, it requires a bulk search operation to find all data items in a key range for every join and fail event. Such a bulk operation is complex, requires extra messages, and induces a delay before the churn event is completely catered. In contrast, we do not require any such bulk operation.

Scatter [4] uses a similar scheme for achieving consistency in DHTs. Compared to our scheme, they further sub-divide the groups to differentiate between key responsibilities of each node. Furthermore, they do not evaluate or argue for the usefulness of their scheme. We provide algorithmic specification of our work, backed by design decisions and evaluation with comparison to SL-replication.

P-Grid [1] uses a notion called *structural replication*, where nodes form groups and data is replicated among nodes in these groups. Like ID-Replication, different groups can have different replication degrees. The geometry of P-Grid is a tree, while we give a solution for overlays with a ring geometry, which is the geometry of a majority of structured overlays. Compared to P-Grid, our solution uses consistent hashing [7], thus leveraging properties of consistent hashing such as self-management, load balancing, and minimized repartitioning of data under churn.

Agyaat [15] proposes to use groups of nodes, called *clouds*, to provide mutual anonymity in structured overlays. Compared to ID-replication, Agyaat maintains an R-Ring and an overlay with the clouds, which is more complicated and requires some nodes to be part of two overlays. A similar approach is taken by Narendula et al. [13], where nodes form sub-overlays with trusted nodes for better access control in P2P data management.

6 Conclusion

This paper discusses popular approaches employed for replication in structured overlay networks, including successor-list replication and symmetric replication, and outlines their drawbacks. We present the design, algorithmic specification, and evaluation of ID-Replication, a replication scheme for structured overlays that does not suffer from the afore-mentioned problems. It does not require requests to go through a particular replica. Furthermore, ID-Replication allows different replication degrees for different key ranges. This allows for using higher number of replicas for hotspots and critical data. We provide detailed evaluation of ID-Replication, and compare it with SL-replication. Our results show that ID-Replication is less sensitive to churn than SL-replication, which makes it better suited for building consistent services and for working in asynchronous networks where inaccurate failure detections are a norm.

Future Work: Due to its low sensitivity to churn, a possible step forward would be to build a consistent key-value store using ID-Replication. Each replication group can act as a replicated state machine, where operations are performed in a total order on the replicas. Since replica groups change with dynamism, we propose using a reconfigurable replicated state machine such as SMART [11].

References

1. Aberer, K., Cudré-Mauroux, P., Datta, A., Despotovic, Z., Hauswirth, M., Punceva, M., Schmidt, R.: P-Grid: a self-organizing structured P2P system. SIGMOD Record 32(3), 29–33 (2003)
2. Arad, C., Dowling, J., Haridi, S.: Developing, simulating, and deploying peer-to-peer systems using the Kompics component model. In: COMSWARE 2009 (2009)
3. Ghodsi, A., Alima, L.O., Haridi, S.: Symmetric Replication for Structured Peer-to-Peer Systems. In: Moro, G., Bergamaschi, S., Joseph, S., Morin, J.-H., Ouksel, A.M. (eds.) DBISP2P 2005 and DBISP2P 2006. LNCS, vol. 4125, pp. 74–85. Springer, Heidelberg (2007)
4. Glendenning, L., Beschastnikh, I., Krishnamurthy, A.: Scalable Consistency in Scatter. In: ACM SOSP, pp. 15–28 (2011)
5. Godfrey, B., Lakshminarayanan, K., Surana, S., Karp, R., Stoica, I.: Load balancing in dynamic structured P2P systems. In: Proceedings of the 23rd Conference of the IEEE Computer and Communications Societies (2004)
6. Gummadi, K.P., Saroiu, S., Gribble, S.D.: King: estimating latency between arbitrary internet end hosts. In: IMW 2002: Proceedings of the 2nd ACM SIGCOMM Workshop on Internet Measurment, pp. 5–18. ACM, New York (2002)
7. Karger, D., Lehman, E., Leighton, T., Panigrahy, R., Levine, M., Lewin, D.: Consistent hashing and random trees: distributed caching protocols for relieving hot spots on the world wide web. In: STOC, pp. 654–663. ACM (1997)

8. Karger, D.R., Ruhl, M.: Simple Efficient Load Balancing Algorithms for Peer-to-Peer Systems. In: Voelker, G.M., Shenker, S. (eds.) IPTPS 2004. LNCS, vol. 3279, pp. 131–140. Springer, Heidelberg (2005)
9. Lakshman, A., Malik, P.: Cassandra: a decentralized structured storage system. SIGOPS Oper. Syst. Rev. 44, 35–40 (2010)
10. Leong, B., Liskov, B., Demaine, E.D.: Epichord: Parallelizing the chord lookup algorithm with reactive routing state management. Computer Communications 29(9), 1243–1259 (2006)
11. Lorch, J.R., Adya, A., Bolosky, W.J., Chaiken, R., Douceur, J.R., Howell, J.: The smart way to migrate replicated stateful services. In: EuroSys (2006)
12. Muthitacharoen, A., Gilbert, S., Morris, R.: Etna: a Fault-tolerant Algorithm for Atomic Mutable DHT Data. Mit technical report, MIT (June 2005)
13. Narendula, R., Miklós, Z., Aberer, K.: Towards access control aware p2p data management systems. In: EDBT/ICDT Workshops, pp. 10–17 (2009)
14. Rowstron, A., Druschel, P.: Pastry: Scalable, Decentralized Object Location, and Routing for Large-Scale Peer-to-Peer Systems. In: Guerraoui, R. (ed.) Middleware 2001. LNCS, vol. 2218, pp. 329–350. Springer, Heidelberg (2001)
15. Singh, A., Gedik, B., Liu, L.: Agyaat: mutual anonymity over structured p2p networks. Internet Research 16(2), 189–212 (2006)
16. Sit, E., Morris, R.: Security Considerations for Peer-to-Peer Distributed Hash Tables. In: Druschel, P., Kaashoek, M.F., Rowstron, A. (eds.) IPTPS 2002. LNCS, vol. 2429, pp. 261–269. Springer, Heidelberg (2002)
17. Stoica, I., Morris, R., Liben-Nowell, D., Karger, D.R., Kaashoek, M.F., Dabek, F., Balakrishnan, H.: Chord: a scalable peer-to-peer lookup protocol for internet applications. IEEE/ACM Transactions on Networking (TON) 11(1), 17–32 (2003)
18. Voulgaris, S., Gavidia, D., van Steen, M.: Cyclon: Inexpensive membership management for unstructured p2p overlays. J. Network Syst. Manage. 13(2), 197–217 (2005)

Changing the Unchoking Policy
for an Enhanced Bittorrent

Vaggelis Atlidakis, Mema Roussopoulos, and Alex Delis

University of Athens, Athens, 15784, Greece
{v.atlidakis,mema,ad}@di.uoa.gr

Abstract. In this paper, we propose a novel optimistic unchoking approach for the *BitTorrent* protocol whose key objective is to improve the quality of inter-connections amongst peers. In turn, this yields enhanced data distribution without penalizing underutilized and/or idle peers. The suggested policy takes into consideration the number of peers currently interested in downloading from a client that is to be unchoked. Our conjecture is that clients having few peers interested in downloading data from them should be favored with optimistic unchoke intervals. This will enable the clients in question to receive data since they become unchoked faster and consequently, they will trigger the interest of additional peers. In contrast, clients with plenty of "interested" peers should enjoy a lower priority to be selected as "planned optimistic unchoked" as they likely have enough data to forward and have saturated their uplinks. In this context, we increase the aggregate probability that the swarm obtains a higher number of interested-in-cooperation and directly-connected peers leading to improved peer inter-connection. Experimental results indicate that our approach significantly outperforms the existing optimistic unchoking policy.

Keywords: Peer-to-peer, Content Distribution, Unchoking.

1 Introduction

Peer-to-peer applications remain of crucial importance as there is still a growing trend for exchange of large multimedia files, voice-over-*IP* and broadcasting of TV-quality programs in the World Wide Web. Content delivery networks based on the traditional client-server model were shown not to scale for large content sharing aggregations. Most of their limitations emanate from the lack of bandwidth that causes bottlenecks in light of heavy requests. In addition, quality of service at the client side inadvertently suffers when servers experience substantial loads. In contrast, highly decentralized *peer-to-peer* models do not distinguish the role of providers and consumers as peers play a dual role by being both a server and/or a client at times. The absence of a centralized authority also constitutes the foundation for scalable and adaptive applications.

Nowadays, *BitTorrent* [2] is the most popular *peer-to-peer* protocol, accounting for approximately 27-55% of all Internet traffic depending on geographical

C. Kaklamanis et al. (Eds.): Euro-Par 2012, LNCS 7484, pp. 377–388, 2012.

location, according to [1]. In the pre-*BitTorrent* era, Napster, Gnutella and Fast-Track were widely-used protocols for transferring multimedia files, such as mp3's, movies, and software. However, their centralized indexing methods and/or the lack of a *tit-for-tat* schema among peers prevented them from being an effective competitor to *BitTorrent*'s dominance.

The *BitTorrent* protocol [2] operates at three different layers: At the *swarm layer*, a peer contacts a tracker to join a swarm and receive a list of other peers to whom to connect. At the *neighborhood layer*, the core reciprocation mechanism is implemented, which forces peers to share any received data in order to receive downloading slots from counterparts. This is done locally, without any help from a centralized mechanism and constitutes the fundamental choice for the incentive policy in use. At the *data layer*, a file is viewed as a concatenation of fixed-size pieces that are requested in a rarest-first policy to ensure the highest degree of content replication. In this paper, we focus at the *neighborhood layer* and modify the neighborhood selection mechanism of the protocol known as peer unchoking; this includes regular unchoking and optimistic unchoking. Regular unchoking is the basic mechanism that implements a *tit-for-tat* schema that allocates bandwidth preferably to peers sending data and penalizes free-riders. Periodically, every peer sorts its uploaders according to the rate they provide data and allocates downloading slots only to the top-three uploaders. Peers not uploading data are excluded from this process, and therefore, they receive no reciprocation. Optimistic unchoking ensures that new peers have a chance of downloading one first piece without having sent any themselves.

The question we seek to answer in this paper is how an uploader should allocate its *optimistic unchoke interval* to downloaders to achieve the most aggregate benefit in a swarm. The existing optimistic unchoking policy uses a round-robin approach giving priority to more recently connected peers [2]. This approach guarantees at least one bootstrapping interval for any new peer, regardless of the situation (i.e., dynamics) in which it finds itself. In a set of newly connected peers, some of them may already possess data blocks, while others do not. Those who possess highly-demanded data are more likely to receive data requests, thus immediately contributing to the swarm. In contrast, peers without data on high-demand or no data at all are more likely to be underutilized. Our proposal is that clients having few peers interested in downloading data, should be favored with *optimistic unchoke intervals*. In turn, this approach enables the clients in discussion to receive data since they get unchoked and so, they may trigger the interest of additional peers. To this end, we check the number of *interested* initiated connections a client maintains and select as the planned optimistic unchoked node the one with the least number of *interested* connections. Uploading clients with few peers interested in downloading from them, receive data in order to trigger global interest and attract block requests. In the long run, the peers in question will be rewarded with additional bandwidth from others due to regular unchoking *tit-for-tat* schema and will stop being idle. As a matter of fact, more peers will participate in the distribution of data, asserting a high quality of inter-connection of peers.

We examine a number of key factors that help our approach enhance the performance of the native *BitTorrent* protocol. These include the number of peers acting as intermediates, decongestion in seeders, contribution of aggregate seeders and peers, and altruism presented by peers. The contributions of our work are:

1. enhancement of the *BitTorrent* protocol that collectively enables an increase in peer content contribution. A high number of peers now act as intermediaries as under-utilized peers have a higher priority to receive *optimistic unchoke intervals*.
2. decongestion of seeders as fewer peers remain idle and so the load on seeders eases up considerably.

Although prior related research has been carried out in a number of aspects including reciprocity mechanisms [6, 8], *tit-for-tat* schemas to discourage free riding [14], and incentives policies in [12, 7], our work is to the best of our knowledge the first effort to adopt an alternative optimistic unchoking policy. Previous research has suggested solutions regarding the modification of the *regular unchoking policy*, and has introduced techniques to encourage peers to act as uploaders and to discard idle peers. Our work, however, is the very first to modify the *optimistic unchoking policy* to encourage cooperation of peers. Our purpose is to treat underutilized uploaders as nodes that lack data to upload, rather than consider them to be selfish free-riders. It is the first time that uploaders are able to locate idle peers and "reward" them with optimistic unchoking slots; no central authority point is used to locate idle peers. Our new optimistic unchoking policy increases the number of interested-in-cooperation and directly-connected peers. In this manner, the quality of inter-connection of peers is improved and a high number of peers now act as data intermediaries, rather than remain idle. Via experimental evaluation and comparison of our protocol with the native *BitTorrent*, we show a significant increase in upload bandwidth offered by peers. We also show that a noteworthy number of peers upload more blocks than download, so we claim that our protocol modification yields an increase in altruism presented by peers.

The rest of the paper is organized as follows: Section 2 discusses the key features of our proposed enhanced *BitTorrent* scheme and Section 3 presents our main experimental results. Section 4 outlines related work while concluding remarks are found in Section 5.

2   Enhanced *BitTorrent*

In this section, we outline our proposed peer unchoking policy by first introducing the messages used by our enhanced *BitTorrent* protocol. We then introduce and analyze the ratio of interest, and the algorithms used for our unchoking policy. Finally, we give an overview of our enhanced *BitTorrent* system.

2.1   Enhanced *BitTorrent* Messages

The messages of the native *BitTorrent* protocol can be categorized into: *swarm-oriented, state-oriented* and *data-oriented* messages. To implement our enhanced *BitTorrent* protocol we use the messages of the original *BitTorrent* protocol, but we augment the **have** state-oriented message with an additional float value. The latter corresponds to the *ratio of interest* (Section 2.2) of the sender of the *have* message and helps us implement our enhanced unchoking policy.

Table 1 summarizes the *swarm-oriented* messages that are exchanged between peers and the tracker. These messages are helpful to the tracker so that it can maintain an up-to-date mapping of the dynamics of the swarm. *Swarm-oriented* messages are also helpful to peers to help them locate each other in a timely fashion. The messages in this group contain no downloadable data.

Table 1. Swarm-oriented Messages

join: A peer interested in joining a swarm sends this message to the tracker. This message contains metadata of the respective file and contact information of the sender
join_response: The tracker sends this message in response to **join**; no payload.
peerset: A peer sends this message to the tracker to request the contact information of other peers participating in the swarm; no payload.
peerset_response: The tracker sends this message in response to **peerset**. This message contains a list of listening IP–addresses and ports of peers participating in the swarm.
leave: A peer sends this message to inform the tracker that it is leaving the swarm.

The group of messages sent among cooperating peers is depicted in Table 2. We refer to these as *state-oriented* messages that help achieve cooperation among peers and implement the peer unchoking policy. All messages of Table 2 contain no downloadable data but designate when peers must exchange data or not. More specifically when peer A dispatches a **choke** message to peer B, the latter must not send any *data-oriented* messages back to A. B must receive an **unchoke** message from A in order to commence sending new *data-oriented* messages.

Furthermore, a peer will send an **unchoke** message only to remote peers that have previously sent an **interested** message. Peer A is *interested* in receiving data from peer B, if B possesses data pieces that A does not possess. **Have** and **bitfield** messages indicate the arrival of a new piece and the set of pieces possessed by a peer, respectively.

Finally, Table 3 summarizes the *data-oriented* messages that are sent between *unchoked* peers (i.e., peers that are exchanging data).

2.2   Peer Unchoking - Ratio of Interest

We define the ratio of interest RI_p of a peer p to be $RI_p = \frac{int_p}{n_p}$, where int_p is the number of interested connections p maintains, from a total of n_p initiated connections. The number of interested connections maintained by a peer may help project the number of data requests the peer in question will ultimately receive provided that data requests are received only via initiated connections marked

Table 2. State-oriented Messages

choke: Peer A sends this message to remote peer B to inform B that it is choked by A. Consequently, B must not send any *data-oriented* messages to A; no payload.
unchoke: Peer A sends this message to remote peer B to inform B that it is no longer choked by A. Consequently, B may send *data-oriented* messages to A; no payload.
interested: Peer A sends this message to remote peer B when A is interested in receiving data from B; no payload.
have: Peer A sends this message to every connected remote peer to inform it that it has received a new piece or to acknowledge the sender of a piece. The payload of this message is an integer identifying received piece, and a float corresponding to *ratio of interest* of A.
bitfield: Peer A sends this message after establishing a new connection to inform remote peer B about pieces it possesses; variable length payload that is a bitmap indicating valid blocks of A.
handshake: Peer A sends this message to establish connection with peer B. Payload includes file identifier and peer identifier of peer A.

Table 3. Data-oriented Messages

request: The sender of this message includes 3 integers denoting requested piece, block within piece and block length.
piece: The sender of this message includes an integer that is the position of requested piece, block's offset within piece and requested data block.
cancel: The sender of this message informs the recipient that it is no longer interested in a previously requested block of a piece. Payload consisting of 3 integers indicating piece index, block offset and block length.

as interested. It is evident that peers with a low ratio of interest receive few data requests and it is likely that they are underutilized and/or idle. To prevent peers from remaining idle, every time an *optimistic unchoke* is to be performed we select the peer p with the minimum RI_p to be the *planned optimistic unchoked* peer. In the long run, our optimistic unchoking policy is effective as idle peers initially unable to act as intermediaries and content replicators, will be unchoked earlier than in the native *BitTorrent* protocol where the unchoking policy is based on random choice. The peers that have saturated their uplinks will be decongested as more clients will act as content intermediaries. We anticipate that our approach will be most effective when we rotate the *planned optimistic unchoked* peer in a prioritized way, yielding the right-of-way to fresh peers and peers with minimum interest ratios. To the best of our knowledge this is the first time such a technique is suggested. Our suggested approach does not bypass the *tit-for-tat* schema, since it does not modify regular unchoking; it rather offers an alternative to improve the quality of inter-connection of peers. An improvement in the quality of inter-connection is attained as soon as an increase in the number of directly-connected and interested-in-cooperation peers is achieved. The benefit obtained from our approach is demonstrated in section 3 where we compare the unchoking policies of our enhanced *BitTorrent* and the native *BitTorrent* protocol.

2.3   Algorithms

Our enhanced *BitTorrent* protocol invokes Algorithms 1 and 2 when a client is in *leech* and *seed* state respectively. These two algorithms are invoked every

10 seconds, every time a peer disconnects from the local client, and when an unchoked peer becomes interested or uninterested. The above timing and event-driven settings are inline with the directives of the *BitTorrent* protocol [2]. As soon as these two algorithms are invoked, a "new round" starts; the number that designates a round ranges from 1 to 3.

Algorithm 1, invoked when a peer is in leech state, takes as input the set of remote *Downloaders* of the local client, the set of remote *Uploaders* to the local client and the vector RI_p denoting the ratio of interest of each remote peer p. No explicit output is produced. The effect however of the algorithm is the realization of our suggested *peer unchoking policy*. RI_p vector is updated every time a *have* message is sent from a remote peer p to the local client. Peers having sent data to the local client are sorted according to their uploading rate and the top three are kept unchoked, called *regular unchoked peers* (RU). Every third round, the remote peer with minimum RI is selected as *planned optimistic unchoked* (OU) and kept unchoked from the local client (for 30 seconds). If *planned optimistic unchoked* is a member of the regular unchoked peers, a new interested peer must be added to the regular unchoked set. Note that uninterested peers may be selected unchoked until an *interested* peer is added to the regular unchoked set. However, only four *interested* peers remain unchoked in the same round.

Algorithm 1 peer unchoking algorithm for client in *leech* state

---

Input: Uploaders, Downloaders, $RI_{p \in Downloaders}$
1: *Interested* ← {$p : \forall p \in Downloaders\ AND\ p\ interested\ in\ local\ client$}
2: **if** $round = 1$ **then**
3:     OU ← {$p : Min\{RI_p\} \forall p \in Interested$}
4:     unchoke OU
5: **end if**
6: RU ← {$p : p \in Top3\ Uploaders$}
7: **for** $p \in Interested$ **do**
8:     **if** $p \in RU$ **then**
9:        unchoke p
10:    **else**
11:       choke p
12:    **end if**
13: **end for**
14: **if** $OU \subseteq RU$ **then**
15:    **repeat**
16:       choose $p \in Downloaders$
17:       unchoke p
18:    **until** $p \in Interested$
19: **end if**

Algorithm 2, invoked when a peer is in seed state, takes as input the set of remote *Downloaders* of the local client as well as the vector RI_p. Again no explicit output is returned. Peers with pending block requests are sorted according to the time they were last unchoked (most-recently-first). Remaining peers are sorted according to their downloading rates (those displaying highest rates are given priority), and are appended to the above set of sorted peers. During two rounds (out of three), the algorithm keeps unchoked the three first peers (RU); moreover, it keeps unchoked the peer p with the minimum RI_p (OU). In the third round, the algorithm keeps unchoked the first four peers (RU).

Algorithm 2 peer unchoking algorithm for client in *seed* state

---

Input: *Downloaders*, $RI_{p \in Downloaders}$
 1: $temp1 \leftarrow \{p : \forall p \in Downloaders\ AND\ has\ pending\ requests\ OR\ recently\ unchoked\}$
 2: sort $temp1$ according to last unchoke time
 3: $temp2 \leftarrow \{p : \forall p \in Downloaders\ AND\ p \notin temp1\}$
 4: sort $temp2$ according to downloading rate
 5: **if** $round = 1, 2$ **then**
 6:    $RU \leftarrow \{p_{i=1,2,3} \in temp1 + temp2\}$
 7:    $OU \leftarrow \{p : Min\{RI_p\} \forall p \in temp1 + temp2\}$
 8:    unchoke OU
 9: **else**
10:    $RU \leftarrow \{p_{i=1,2,3,4} \in temp1 + temp2\}$
11: **end if**
12: **for** $p \in D$ **do**
13:    **if** $p \in RU$ **then**
14:       unchoke p
15:    **else**
16:       choke p
17:    **end if**
18: **end for**

2.4   Overview of Enhanced *BitTorrent*

The *initial seeder* publishes to the *tracker* the *.torrent* file including metadata describing the file to be distributed. The *initial seeder* possesses a full copy of the designated file and is the first uploader in the swarm. A *fresh* peer wishing to join the swarm must contact the tracker (*HTTP* plain text messages) to obtain the .torrent file and a *peer set* of, typically, 50 peers to whom to connect. Afterwards, the fresh peer establishes *TCP* connections with peers in its peer set. Each peer is multi-threaded and asynchronously downloads/uploads data from/to multiple counterparts, without exceeding a threshold of 40 *initiated* connections. Enhanced *BitTorrent* peers maintain bitmaps to keep track of missing and obtained data pieces; pieces are requested using the rarest-first policy. Uploaders maintain a vector of *ratios of interest* of all peers. *Optimistic unchoking* is a process that "rewards" *underutilized* and/or *idle* peers with *optimistic unchoke* slots. Fresh peers are also rewarded with optimistic unchoke intervals from our unchoking policy to acquire initial data. Our purpose is to prevent peers with low ratios of interest from being idle and to motivate them to act as *data intermediaries*. Furthermore, the *regular unchoking* policy facilitates the formation of clusters of peers with similar bandwidth. Upon completion of downloading, each peer reports its downloading statistics for the file to the tracker, and may be selfish and leave the swarm or altruistic and become an *additional* seeder.

3   Evaluation

To evaluate our enhanced *BitTorrent* protocol, we have implemented in *Python* a respective client as well as a tracker. Our implementation of both the client and the tracker run in *Windows7*, *Linux* and *MacOS*. For our experiments, we used 40 workstations, each featuring a 1 *GHz* clock and 1 *GB* memory running *GNU/Linux*. The workstations are attached to a local Ethernet network running at $100 Mps$. Our key experimental objectives were to: **a)** measure the number of directly-connected and interested-in-cooperation peers to compare the *quality* of peer inter-connections for both our enhanced and the native *BitTorrent*, **b)** examine pieces uploaded from leechers and seeders to evaluate the decongestion of seeders achieved by our enhanced *BitTorrent*, and **c)** ascertain the degree of altruism attained by leechers in our enhanced *BitTorrent*. During experimentation we used an $700MB$ test file, $512KB$ pieces were shared among peers and each peer maintained 40 initiated connections. In steady state a swarm of as many as 150 peers was formed. In all our experiments, seeders joined swarms before leechers; the former had a full copy of the file to be distributed, while the latter had no data at all.

Ratio of Interest

In this section, we examine the *ratio of interest* of peers, as defined in section 2.2. From a peer's local perspective, the ratio of interest indicates the amount of data requests a peer will receive from others. From a global perspective, the ratio of interest reflects the *quality* of inter-connection of peers. In this regard, the benefit of our approach is depicted by Figs. 1(a) and 1(b) that illustrate the ratios of interest and number of *Interested* connections maintained by peers over the duration of the experiment. In both cases, swarms are formed from 130 leechers

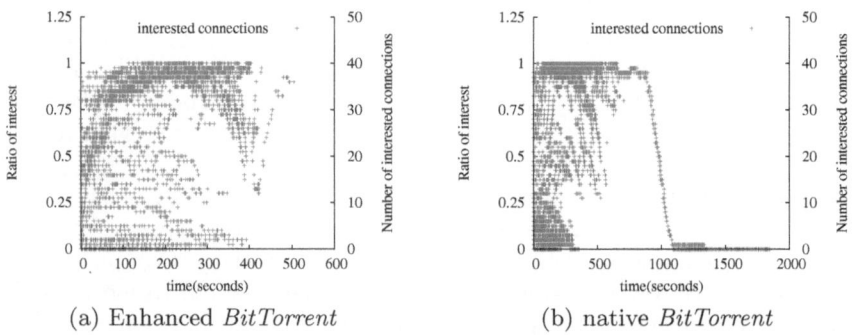

(a) Enhanced *BitTorrent*   (b) native *BitTorrent*

Fig. 1. Ratios of interest of peers and *Interested* connections under (a) our enhanced and (b) the native *BitTorrent* protocol. Initiated connections maintained per-peer are fixed at 40 and the ratio of interest is $RI \leq 1$ for both cases. The average ratio of interest is at 0.30 and 0.22 in (a) and (b) respectively.

and 15 seeders; 90% of peers join a swarm within 100 seconds. In Fig. 1(a), which corresponds to the enhanced *BitTorrent* protocol, the average ratio of interest is 0.30 per peer, while in Fig. 1(b), which corresponds to the native *BitTorrent* protocol, the average ratio of interest is 0.22 per peer. Moreover, before 500 seconds in Fig. 1(a), there is a higher coverage of *interested* connections than that of Fig. 1(b). In the first case, all peers act as intermediaries (downloading **and** uploading) and the ratio of interest is high until the completion of downloading. After completion of downloading, the ratio of interest is uniformly decreased. In the second case, there are underutilized peers with a low ratio of interest. This ratio of interest of idle peers becomes even lower and asymptotically reaches zero as soon as the majority of peers completes downloading. As a matter of fact, the enhanced *BitTorrent* displays a higher number of directly-connected and interested-in-cooperation peers than its native counterpart. An improved interconnection of peers is achieved as the new unchoking policy, as implemented by Algorithms 1 and 2, maximizes the ratio of interest and provides idle peers with data. In turn, idle peers act as additional data intermediaries and "trigger" the interest of other peers. In contrast, the unchoking policy of the native *BitTorrent* protocol has no mechanism to locate idle peers and essentially does not "prod" them to cooperate with others.

Uploading Contribution/Altruism of Leechers

In this section, we compare the uploading contribution of leechers of both protocols. We also examine the *altruism* presented by leechers that we define as the ratio: *pieces uploaded/pieces downloaded*. Figures 2(a) and 2(b) illustrate the number of pieces uploaded as a function of pieces downloaded, and the line $\epsilon : y = x$ which distinguishes between leechers with (i) *altruism* ≥ 1 and (ii) *altruism* ≤ 1. In both cases, we use swarms that consist of 15 seeders and 130 leechers. Leechers join the swarms in flash-crowds and download a fixed number of pieces to obtain a full copy of the distributed file. Although in the enhanced *BitTorrent* (Fig. 2(a)) there is a non-negligible number of peers clustered into area (i), there are only a handful of peers in the same area in the native *BitTorrent* (Fig. 2(b)). In the first case, "altruistic" leechers upload more than 2,500 pieces, but in the second case leechers can upload at most 1,300 pieces. The

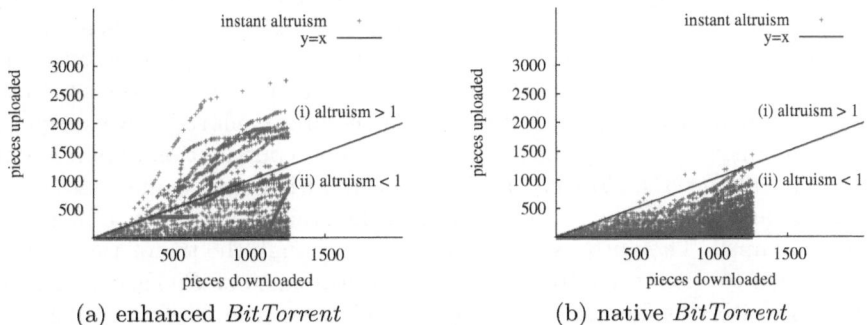

(a) enhanced *BitTorrent*   (b) native *BitTorrent*

Fig. 2. Altruism presented by leechers under (a) the enhanced and (b) the native *BitTorrent* protocol. In the first case, many leechers upload more data than they download (*altruism*> 1). In the second case, leechers display non-altruistic behavior (*altruism*< 1).

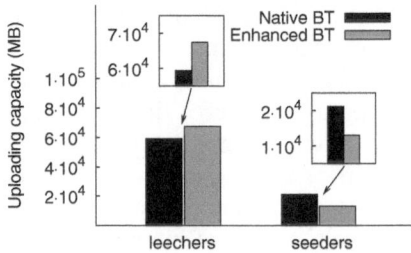

Fig. 3. Aggregate uploading contribution of leechers and seeders under the enhanced and the native *BitTorrent* protocol. Under the enhanced *BitTorrent* protocol, leechers upload 68 *GB* of data and under the native *BitTorrent* protocol leechers upload 58*GB* of data.

leechers found in the area *(i)* act more as uploaders than downloaders. These leechers decongest seeders and provide the swarm with additional uploading capacity of up to 10*GB*. As Fig. 3 shows, in the native protocol, seeders uploaded 20*GB* of data and leechers uploaded 60*GB* of data. Under the enhanced *BitTorrent*, seeders uploaded 10*GB* and leechers uploaded 70*GB*, for the respective experiment. Our approach thus achieves an increase in the contribution of leechers without involving any complex incentive policy. This is in-line with our key objective to encourage underutilized peers to act as data intermediaries, rather than penalize them. To this end, uploaders unchoke underutilized leechers in an *altruistic* manner. In turn, underutilized leechers obtain data to upload, and ultimately, provide the swarm with additional uploading capacity.

4   Related Work

A number of techniques have been suggested to improve the performance of the native *BitTorrent* [3] protocol including bartering-based approaches among peers, and incentive-based policies. In [6, 8], indirect and direct reciprocity mechanisms are examined so that peers exploit their own data contributions to obtain data from others. Our approach differs from the above efforts as we suggest an unchoking policy in which peers do not exploit their contribution to obtain data. Under our enhanced *BitTorrent* protocol, peers altruistically offer data to underutilized and/or idle counterparts. In [12], the issue of incentive compatibility was re-examined. The authors showed that even though the tit-for-tat approach was intended to discourage free-riding, the performance of *BitTorrent* has very little to do with this fact. Also through the release of the *BitTyrant* client, the conjecture of whether incentives build robustness in *BitTorrent* is evaluated. Incentives in *BitTorrent* systems are also studied in [7], where the unchoking algorithm of native *BitTorrent* is evaluated. This work shows that regular unchoking facilitates the formation of clusters of peers with similar bandwidth, which is also the case in our enhanced *BitTorrent* protocol.

A variety of mechanisms for preventing free-riding in *P2P* file-sharing systems are applied in [14, 16, 11]. Although applying mechanisms to discourage free-riding is essential to steering more peers to act as data intermediaries, it does not address the problem of locating peers with no initial data to upload. With our improved unchoking policy, uploaders immediately locate and furnish data to peers with no initial data blocks.

Neighborship consistency is defined as the ratio between the number of known nodes and the number of actual nodes within a node's area of interest and can be used to measure the quality or connectivity in *P2P* systems [5]. To compliment neighborship, in our enhanced *BitTorrent* protocol we define *ratio of interest* (Section 2.2) and use this metric to decide which peer should be selected as planned optimistic unchoked. In [15], the use of altruism in *P2P* networks is examined; altruism is defined via a parameter that reflects benefit obtained for a peer's contribution. A peer selectively decides the level of its own contribution and demands to download a specific amount of data; the amount of data a peer demands is proportional to its contribution. In our enhanced *BitTorrent* approach, a peer decides which peer to unchoke in order to maximize its *ratio of interest*. Benefit obtained from our unchoking policy is examined collectively. We increase the number of directly-connected and interested-in-cooperation peers in an attempt to build a robust swarm.

There have also been proposals for new models that essentially suggest *BitTorrent*-like protocols [4, 13, 10, 11, 9]. However, our work is the first to suggest a modification to the optimistic unchoking policy that collectively increases the number of peers acting as intermediaries and decongests seeders.

5   Conclusion

In this paper, we present the enhanced *BitTorrent* protocol whose unchoking policy better harnesses underutilized peers that have few clients interested in downloading data from them. Our proposal involves uploaders allocating optimistic unchoking slots to underutilized peers. This policy enables peers to obtain data and essentially act as content intermediaries, rather than remain idle. Experimentation with leecher and tracker prototypes shows that our approach achieves improved quality of inter-connection amongst peers compared with the native *BitTorrent* protocol. Under our enhanced *BitTorrent* protocol, the number of directly-connected and interested-in-cooperation peers increases significantly. A substantial portion of the peers in question act as data intermediaries and consequently, better peer content distribution is achieved. Moreover, our modified *BitTorrent* protocol has the effect of creating altruistic leechers who act more as uploaders than downloaders. The net result is that these altruistic leechers furnish uploading capacity that helps relieve the burden of seeders.

Acknowledgments. We would like to thank Y. Mimiyannis, Y. Kamonas, C. Christou and A. Sevastidou for their help during our experimentation as well as the anonymous reviewers for their fruitful feedback. This work was partially funded by the *iMarine EU-FP6* project.

References

[1] Internet Study. http://www.ipoque.com/en/resources/internet-studies
[2] The Bittorrent Protocol. http://www.bittorrent.org/beps
[3] B. Cohen: Incentives Build Robustness in BitTorrent. In: IPTPS. Berkeley, CA (February 2003)
[4] Chow, A.L.H., Golubchik, L., Misra, V.: BitTorrent: An Extensible Heterogeneous Model. In: INFOCOM. pp. 585–593. Rio De Janeiro, Brazil (April 2009)
[5] Jiang, J., Chiou, J., Hu, S.: Enhancing Neighborship Consistency for Peer-to-Peer Distributed Virtual Environments. In: IEEE–ICDCS Workshops. Toronto, Canada (June 2007)
[6] Landa, R., Griffin, D., Clegg, R., Mykoniati, E., Rio, M.: A Sybilproof Indirect Reciprocity Mechanism for Peer-to-Peer Networks. In: INFOCOM. pp. 343–351. Rio De Janeiro, Brazil (April 2009)
[7] Legout, A., Liogkas, N., Kohler, E., Zhang, L.: Clustering and Sharing Incentives in BitTorrent Systems. In: SIGMETRICS. pp. 301–312. San Diego, CA (June 2007)
[8] Menasché, D., Massoulié, L., Towsley, D.: Reciprocity and Barter in Peer-to-Peer Systems. In: INFOCOM. San Diego, CA (March 2010)
[9] Meulpolder, M., Epema, D.H., Sips, H.: Replication in bandwidth-symmetric BitTorrent Networks. In: 22nd IEEE Int. Parallel and Distributed Processing Symposium (IPDPS'08). pp. 1–8. Miami, FL (April 2008)
[10] M.Y., Yang, Y.: An Efficient Hybrid Peer-to-Peer System for Distributed Data Sharing. IEEE Transactions on Computers 59(9), 1158–1171 (September 2010)
[11] Peterson, R., Sirer, E.: AntFarm: Efficient Content Distribution with Managed Swarms. In: NSDI. pp. 107–122. Boston, MA (April 2009)
[12] Piatek, M., Isdal, T., Anderson, T., Krishnamurthy, A., Venkataramani, A.: Do Incentives Build Robustness in BitTorrent? In: 4th USENIX Symposium on Networked Systems Design & Implementation. Cambridge, MA (April 2007)
[13] Ren, S., Tan, E., Luo, T., Chen, S., Guo, L., Zhang, X.: TopBT: A Topology-Aware and Infrastructure-Independent BitTorrent Client. In: INFOCOM. pp. 1523–1531. San Diego, CA (March 2010)
[14] Shin, K., Reeves, D., Rhee, I.: Treat-before-Trick: Free-Riding Prevention for BitTorrent-like Peer-to-Peer Networks. In: 23rd IEEE Int. Symposium on Parallel and Distributed Processing (IPDPS'09). pp. 1–12. Rome, Italy (May 2009)
[15] Vassilakis, D.K., Vassalos, V.: An Analysis of Peer-to-peer Networks with Altruistic Peers. Peer-to-Peer Networking and Applications 2(2), 109–127 (June 2009)
[16] Yang, M., Feng, Q., Dai, Y., Zhang, Z.: A Multi-Dimensional Reputation System Combined with Trust and Incentive Mechanisms in P2P File Sharing Systems. In: IEEE–ICDCS Workshops. Toronto, Canada (June 2007)

Peer-to-Peer Multi-class Boosting[*]

István Hegedűs[1], Róbert Busa-Fekete[1,2],
Róbert Ormándi[1], Márk Jelasity[1], and Balázs Kégl[2,3]

[1] Research Group on AI, Hungarian Acad. Sci. and Univ. of Szeged, Hungary
{ihegedus,busarobi,ormandi,jelasity}@inf.u-szeged.hu
[2] Linear Accelerator Laboratory (LAL), University of Paris-Sud,
CNRS Orsay, 91898, France
[3] Computer Science Laboratory (LRI), University of Paris-Sud,
CNRS and INRIA-Saclay, 91405 Orsay, France
balazs.kegl@gmail.com

Abstract. We focus on the problem of data mining over large-scale fully distributed databases, where each node stores only one data record. We assume that a data record is never allowed to leave the node it is stored at. Possible motivations for this assumption include privacy or a lack of a centralized infrastructure. To tackle this problem, earlier we proposed the generic gossip learning framework (GoLF), but so far we have studied only basic linear algorithms. In this paper we implement the well-known boosting technique in GoLF. Boosting techniques have attracted growing attention in machine learning due to their outstanding performance in many practical applications. Here, we present an implementation of a boosting algorithm that is based on FILTERBOOST. Our main algorithmic contribution is a derivation of a pure online multi-class version of FILTERBOOST, so that it can be employed in GoLF. We also propose improvements to GoLF, with the aim of maximizing the diversity of the evolving models gossiped in the network, a feature that we show to be important. We evaluate the robustness and the convergence speed of the algorithm empirically over three benchmark databases. We compare the algorithm with the sequential ADABOOST algorithm and we test its performance in a failure scenario involving message drop and delay, and node churn.

Keywords: P2P, gossip, multi-class classification, boosting, FilterBoost.

1 Introduction

Making data analysis possible in fully distributed systems via data mining tools has been an important research direction in the past decade. Tasks such as information retrieval, recommendations, detecting spam, vandalism and intrusion require sophisticated models that are based on large amounts of data. This data is often generated in a fully distributed fashion on routers, PCs, smart phones or sensor nodes. In many cases local data cannot be collected centrally due to privacy constrains or due to the lack of computing infrastructure.

---

[*] M. Jelasity was supported by the Bolyai Scholarship of the Hungarian Academy of Sciences. This work was partially supported by the FET programme FP7-COSI-ICT of the European Commission through project QLectives (grant no.: 231200) and by the ANR-2010-COSI-002 grant of the French National Research Agency.

C. Kaklamanis et al. (Eds.): Euro-Par 2012, LNCS 7484, pp. 389–400, 2012.

In this paper we are concerned with the scenario in which there is a very large number of nodes, all of which store small amounts of data, such as personal profiles or recent sensor readings. In our previous work, we have proposed the gossip learning framework (GoLF) for data mining in such environments [19,20]. The basic idea is that models perform random walks in the network, while being improved by an arbitrary *online learning* method. Convergence can be improved significantly if nodes combine the models that pass through them, or if they use other techniques such as voting. In this framework we have so far only studied learning linear models.

In this paper we develop a boosting algorithm, which proves the viability of gossip learning also for implementing state-of-the-art machine learning algorithms. In a nutshell, a boosting algorithm constructs a classifier in an incremental fashion by adding simple classifiers (that is, weak classifiers) to a pool. The weighted vote of the classifiers in the pool determines the final classification.

Our contributions are the following. First, to enable P2P boosting via gossip, we derive a purely online multi-class boosting algorithm that can be proven to minimize a certain negative log likelihood function. We also introduce efficient multi-class weak learners to be used by the online boosting algorithm. Second, we improve GoLF to make sure that the diversity of the models in the network is preserved. This makes it meaningful to spread the current best model in the network; a technique we propose to improve local prediction performance. Finally, we perform simulation experiments where we study our algorithm under extreme message drop, message delay and node churn to prove its robustness.

2   System Model and Data Distribution

Our system model is a network of computers (peers). Each node in the network has a unique network address and can communicate with other nodes through messages if the address of the target node is locally available. We also assume that a *peer sampling service* is available that provides addresses of random available peers at any time. Here we use NEWSCAST [26] as our peer sampling service. Messages can be delayed or dropped, moreover, new nodes can join and leave the network without any warning. We assume that when a node rejoins the network it has the same state as at the time of going offline.

Regarding data distribution, we assume that the data records are distributed horizontally, that is, all the nodes store full records. At the same time, all the nodes store only very few records, perhaps only a single record. This excludes the possibility of any local statistical processing of the data. Another important assumption is that the data never leave the nodes, that is, it is not allowed to collect the data centrally due to privacy or infrastructural constraints.

3   Background and Related Work

The problem we tackle in this paper is *supervised classification* that can be formally defined as follows. We are given a training database in the form of a set of training instances. Each training instance consists of a feature vector and a corresponding class label. Let us denote this training dataset by $S = \{(\mathbf{x}_1, \mathbf{y}_1), \ldots, (\mathbf{x}_n, \mathbf{y}_n)\} \subset \mathbb{R}^d \times \{-1, +1\}^K$, where d is the *dimension* of the problem and K defines the number of classes. The goal of the classification problem is to find a function $\mathbf{f} : \mathbb{R}^d \rightarrow \{-1, +1\}^K$ that can correctly classify *any*

Algorithm 1. Skeleton of original GoLF learning protocol

1: $currentModel \leftarrow$ initModel()
2: **loop**   6: **procedure** ONRECEIVEMODEL(m)
3:     wait(Δ)   7: m.updateModel(x, y)
4:     $p \leftarrow$ selectPeer()   8: $currentModel \leftarrow m$
5:     sendModel($p, currentModel$)

---

samples, including those not in the training set, with high probability *(generalization)*. In *multi-class* classification problems—where $K > 2$—one and only one of the elements of $\mathbf{y}_i$ is $+1$, whereas in *multi-label* (or *multi-task*) classification $\mathbf{y}_i$ is arbitrary, meaning that the observation $\mathbf{x}_i$ can belong to several classes at the same time. In the former case we will denote the index of the correct class by $\ell(\mathbf{x}_i)$. In classical multi-class classification the elements of $\mathbf{f}(\mathbf{x})$ are treated as posterior scores corresponding to the labels, so the predicted label is $\widehat{\ell}(\mathbf{x}) = \arg\max_{\ell=1,\dots,K} f_\ell(\mathbf{x})$ where $f_\ell(\mathbf{x})$ is the ℓth element of $\mathbf{f}(\mathbf{x})$. The function $\mathbf{f}$ is called the *model* of the data.

As mentioned before, in this paper we focus on online boosting in GoLF. A few proposals for online boosting algorithms are known. An online version of ADABOOST [11] is introduced in [8] that requires a random subset from the training data for each boosting iteration, and the base learner is trained on this small sample of the data. The algorithm has to sample the data according to a non-uniform distribution making it inappropriate for pure online training. A gradient–based online algorithm is presented in [3], which is an extension of Friedman's gradient–based framework [12]. However, their approach is for binary classification, and it is not obvious how it can be extended to multi-class problems. Another notable online approach is Oza's online algorithm [21] whose starting point is ADABOOST.M1 [10]. However, ADABOOST.M1 requires the base learning algorithm to achieve 50% accuracy for any distribution over the training instances. This makes it impractical in multi-class classification since most of the weak learners used as a base learner do not satisfy this condition.

We also discuss work related to fully distributed P2P data mining in general. We note that we do not overview the extensive literature of parallel machine learning algorithms because they have a completely different underlying system model and motivation. We do not discuss those distributed machine learning approaches either that assume the availability of sufficient local data to build models locally (a survey can be found in [22]).

One notable and relevant research direction is gossip–based algorithms where convergence to global functions over fully distributed data is achieved through local communication. Perhaps the simplest example is gossip–based averaging [14,16], where the gossip approach is extremely robust, scalable, and efficient. However, gossip algorithms support more sophisticated algorithms that compute more complex global functions. Examples include the EM algorithm [17], LDA [2] or PageRank [13]. Numerous other P2P machine learning algorithms have also been proposed, as in [18,25]. A survey of many additional ideas can be found in [7]. This work builds on the Gossip Learning Framework (GoLF) [19,20], which offers an abstraction to implement a wide range of machine learning algorithms.

The skeleton of GoLF is shown in Alg. 1. This algorithm runs on each node. The algorithm consists of an *active loop* that runs periodically and an event handler (ONRECEIVEMODEL) which is called when a new model arrives. The models take random walks over the network by selecting a random node (line 4)

Algorithm 2. FILTERBOOST(INIT(), UPDATE($\cdot, \cdot, \cdot, \cdot$), T, C)

1: $\mathbf{f}^{(0)}(\mathbf{x}) \leftarrow 0$
2: **for** $t \leftarrow 1 \rightarrow T$ **do**
3:     $C_t \leftarrow C \log(t+1)$
4:     $\mathbf{h}^{(t)}(\cdot) \leftarrow$ INIT()
5:     **for** $t' \leftarrow 1 \rightarrow C_t$ **do**   ▷ Online base learning
6:        $(\mathbf{x}, \mathbf{y}, \mathbf{w}) \leftarrow$ FILTER($\mathbf{f}^{(t-1)}(\cdot)$)   ▷ Draw a weighted random instance
7:        $\mathbf{h}^{(t)}(\cdot) \leftarrow$ UPDATE($\mathbf{x}, \mathbf{y}, \mathbf{w}, \mathbf{h}^{(t)}(\cdot)$)
8:     $\gamma \leftarrow 0, W \leftarrow 0$
9:     **for** $t' \leftarrow 1 \rightarrow C_t$ **do**   ▷ Estimate the edge on a filtered data
10:      $(\mathbf{x}, \mathbf{y}, \mathbf{w}) \leftarrow$ FILTER($\mathbf{f}^{(t-1)}(\cdot)$)   ▷ Draw a weighted random instance
11:      $\gamma \leftarrow \gamma + \sum_\ell^K w_\ell h_\ell^{(t)}(\mathbf{x})y_\ell, \; W \leftarrow W + \sum_\ell^K w_\ell$
12:     $\gamma \leftarrow \gamma / W$ ▷ Normalize the edge
13:     $\alpha^{(t)} \leftarrow \frac{1}{2} \log \frac{1+\gamma}{1-\gamma}$
14:     $\mathbf{f}^{(t)}(\cdot) = \mathbf{f}^{(t-1)}(\cdot) + \alpha^{(t)} \mathbf{h}^{(t)}(\cdot)$
15: **return** $\mathbf{f}^{(T)}(\cdot) = \sum_{t=1}^{T} \alpha^{(t)} \mathbf{h}^{(t)}(\cdot)$
16: **procedure** FILTER($\mathbf{f}(\cdot)$)
17:     $(\mathbf{x}, \mathbf{y}) \leftarrow$ RANDOMINSTANCE()   ▷ Draw random instance
18:     **for** $\ell \leftarrow 1 \rightarrow K$ **do**
19:      $w_\ell \leftarrow \dfrac{\exp\left(f_\ell(\mathbf{x}) - f_{\ell(\mathbf{x})}(\mathbf{x})\right)}{\sum_{\ell'=1}^{K} \exp\left(f_{\ell'}(\mathbf{x}) - f_{\ell(\mathbf{x})}(\mathbf{x})\right)}$
20:     **return** $(\mathbf{x}, \mathbf{y}, \mathbf{w})$

---

and jumping there (line 5). Procedure ONRECEIVEDMODEL updates the received model using the training sample stored by the node (line 7, where x and y represent a training example and the corresponding class label, respectively). It then stores the model as the current model (line 8). In this skeleton the model is an abstract class which provides the update possibility. We note that models can also be combined [20] or they can interact through ensemble learning techniques (like voting) [19], which results in a substantial performance improvement. Regarding model interaction, additional details will be given later in relation to the boosting algorithm.

4   Multi-class Online FilterBoost

This section introduces our main contribution, a multi-class online boosting algorithm that can be applied in GoLF. We build on FILTERBOOST [5] where the main idea is to filter (sample) the training examples in each boosting iteration and to give the base learner only this smaller, filtered subset of the original training dataset, leading to fast base learning. The performance of the base classifier is also estimated on an additional random subset of the training set resulting in further improvement in speed.

Our formulation of the FILTERBOOST algorithm is given as Alg. 2. This is not yet in a form to be applied in GoLF, but the transformation is trivial as discussed in Section 6. This fully online formulation is equivalent to FILTERBOOST, except that it handles multi-class problems as well. To achieve this, while ensuring that the algorithm can still be theoretically proven to converge, our key contribution is the derivation of a new weight formula calculated in line 19. First we introduce this formula, then we explain Alg. 2 in more detail.

A boosting algorithm can be thought of as a minimization algorithm of an appropriately defined target function over the space of models. The target function is related to the classification error over the training dataset. The key idea is that we select an appropriate target function that will allow us to both derive an appropriate weight, as well as argue for convergence. Inspired by the logistic regression approach of [6], we will use the following negative log likelihood function as our target function:

$$R_{\mathrm{L}}(\mathbf{f}) = -\sum_{i=1}^{n} \ln \frac{\exp\left(f_{\ell(\mathbf{x}_i)}(\mathbf{x}_i)\right)}{\sum_{\ell'=1}^{K} \exp\left(f_{\ell'}(\mathbf{x}_i)\right)} = \sum_{i=1}^{n} \ln \left[1 + \sum_{\ell \neq \ell(\mathbf{x}_i)}^{K} \exp\left(f_\ell(\mathbf{x}_i) - f_{\ell(\mathbf{x}_i)}(\mathbf{x}_i)\right) \right]$$

(1)

Note that the FILTERBOOST algorithm returns a vector-valued classifier $\mathbf{f} : \mathbb{R}^d \to \mathbb{R}^K$. The rest of the definitions and notations were introduced in Section 3.

FILTERBOOST builds the final classifier $\mathbf{f}$ as a weighted sum of *base classifiers* $\mathbf{h}^{(t)} : \mathbb{R}^d \to \{-1, +1\}^K$ returned by a *base learner* algorithm which has to be able to handle weighted training data. The class-related weight vector assigned to $\mathbf{x}_i$ in iteration t is denoted by $\mathbf{w}_i^{(t)}$ and its ℓth element is denoted by $w_{i,\ell}^{(t)}$. It can be shown that selecting $w_{i,\ell}^{(t)}$ so that it is proportional to the output of the current strong classifier

$$w_{i,\ell}^{(t)} = \frac{\exp\left(f_\ell^{(t)}(\mathbf{x}_i) - f_{\ell(\mathbf{x}_i)}^{(t)}(\mathbf{x}_i)\right)}{\sum_{\ell'=1}^{K} \exp\left(f_{\ell'}^{(t)}(\mathbf{x}_i) - f_{\ell(\mathbf{x}_i)}^{(t)}(\mathbf{x}_i)\right)}.$$

(2)

ensures that our target function in (1) will decrease in each boosting iteration. The proof is outlined in the Appendix.

The pseudocode of FILTERBOOST is shown in Alg. 2. Here, the algorithm is implemented according to the practical suggestions given in [5]: first, the number of randomly selected instances is $C \log(t + 1)$ in the tth iteration (where C is a constant parameter), and second, in the FILTER method the instances are first randomly selected then re-weighted based on their scores given by $\mathbf{f}^{(t)}(\cdot)$. Procedure INIT() initializes the parameters of the base classifier (line 4), and UPDATE($\cdot, \cdot, \cdot, \cdot$) updates (line 7) the parameter of the base classifier using the current training instance $\mathbf{x}$ given by FILTER($\cdot$). The input parameter T is the number of iterations, and C controls the number of instances used in one boosting iteration. $\alpha^{(t)}$ is the base coefficient, $\mathbf{h}^{(t)}(\cdot)$ is the vector-valued base classifier, and $\mathbf{f}^{(T)}(\cdot)$ is the final (strong) classifier. Procedure RANDOMINSTANCE() selects a random instance from the training data $\mathbf{X}, \mathbf{Y}$.

Let us point out that there is no need to store more than one training instance anywhere during execution. Second, the algorithm does not need any global information about the training data, such as the size, so this implementation can be readily applied in a pure online environment.

5   Multi-class Online Base Learning

For the online version of FILTERBOOST, we need to propose online base learners as well. In FILTERBOOST, for theoretical reasons, the base classifiers are restricted to output discrete predictions in $\{-1, +1\}^K$ and, in addition, they have to minimize the weighted exponential loss

$$E(\mathbf{h}, \mathbf{f}^{(t)}) = \sum_{i=1}^{n} \sum_{\ell=1}^{K} w_{i,\ell}^{(t)} \exp\left(-h_\ell(\mathbf{x}_i) y_{i,\ell}\right).$$

(3)

We follow this approach and, in addition, we build on our base learning framework [15] and assume that the base classifier $\mathbf{h}(\mathbf{x})$ is vector-valued and represented as $\mathbf{h}_\Theta(\mathbf{x}) = \text{sign}(\mathbf{v}\varphi_\Theta(\mathbf{x}))$, parameterized by $\mathbf{v} \in \mathbb{R}^K$ (the *vote vector*), and $\varphi_\Theta(\mathbf{x})$: $\mathbb{R}^d \rightarrow \mathbb{R}$, a *scalar* base classifier parameterized by Θ. The coordinate-wise sign function is defined as sign : $\mathbb{R}^K \rightarrow \{-1, +1\}^K$. In this framework, learning consists of tuning Θ and $\mathbf{v}$ to minimize the weighted exponential loss (3).

Since it is hard to optimize the non-differentiable function $\mathbf{h}_\Theta$ even in batch mode, we take into account only $\widehat{\mathbf{h}}_\Theta(\mathbf{x}) = \mathbf{v}\varphi_\Theta(\mathbf{x})$. This approach is heuristic as it is hard to say anything about the relation between $E(\mathbf{h}_\Theta, \mathbf{f}^{(t)})$ and $E(\widehat{\mathbf{h}}_\Theta, \mathbf{f}^{(t)})$, but in practice this base learning approach performs quite well.

Since $\varphi_\Theta(\cdot)$ is differentiable, the stochastic gradient descent (SGD) [4] algorithm provides a convenient way to train the base learner in an online fashion. The SGD algorithm updates the parameters iteratively based on one training instance at a time. Let us denote $Q(\mathbf{x}, \mathbf{y}, \mathbf{w}, \mathbf{v}, \Theta) = \sum_{\ell=1}^{K} w_\ell \exp\left(-y_\ell v_\ell \varphi_\Theta(\mathbf{x})\right)$. Then the gradient based parameter update can be calculated as follows:

$$\Theta^{(t'+1)} \leftarrow \Theta^{(t')} + \gamma^{(t')} \nabla_\Theta Q(\mathbf{x}, \mathbf{y}, \mathbf{w}, \mathbf{v}, \Theta) \tag{4}$$

$$\mathbf{v}^{(t'+1)} \leftarrow \mathbf{v}^{(t')} + \gamma^{(t')} \nabla_\mathbf{v} Q(\mathbf{x}, \mathbf{y}, \mathbf{w}, \mathbf{v}, \Theta) \tag{5}$$

This update rule can be used in line 7 of FILTERBOOST to update the base classifier. A simple decision stump or ADALINE [27] can be easily accommodated to this multi-class base learning framework. In the following we derive the update rules for a decision stump, that is, a one-decision two-leaf decision tree having the form

$$\varphi_{j,b}(\mathbf{x}) = \begin{cases} 1 & \text{if } x^{(j)} \geq b, \\ -1 & \text{otherwise,} \end{cases} \tag{6}$$

where j is the index of the selected feature and b is the decision threshold. Since $\varphi_{j,b}(\mathbf{x})$ is not differentiable with respect to b, we decided to approximate it by the differentiable sigmoidal function, whose parameters can be tuned using SGD. The sigmoidal function can be written as

$$s_{j,\theta}(\mathbf{x}) = s_{j,(c,d)}(\mathbf{x}) = \frac{1}{1 + \exp\left(-cx^{(j)} - d\right)}.$$

where $\Theta = (c, d)$. And $\varphi_{j,b}(\cdot)$ can be approximated by $\varphi_{j,b}(\mathbf{x}) \approx 2s_{j,\theta}(\mathbf{x}) - 1$. Then the weighted exponential loss of this so-called *sigmoidal decision stump* for a single instance can be written as

$$Q_j = Q_j(\mathbf{x}, \mathbf{y}, \mathbf{w}, \mathbf{v}, \Theta) = \sum_{\ell=1}^{K} w_\ell \exp\left(-v_\ell \left(2s_{j,\theta}(\mathbf{x}) - 1\right) y_\ell\right)$$

and its partial derivatives are

$$\frac{\partial Q_j}{\partial v_\ell} = -\exp\left(-v_\ell \left(2s_{j,\theta}(\mathbf{x}) - 1\right) y_\ell\right) w_\ell \left(2s_{j,\theta}(\mathbf{x}) - 1\right) y_\ell$$

$$\frac{\partial Q_j}{\partial c} = -2 \sum_{\ell=1}^{K} \exp\left(-v_\ell \left(2s_{j,\theta}(\mathbf{x}) - 1\right) y_\ell\right) w_\ell v_\ell y_\ell x^{(j)} s_{j,\theta}(\mathbf{x}) \left(1 - s_{j,\theta}(\mathbf{x})\right)$$

$$\frac{\partial Q_j}{\partial d} = -2 \sum_{\ell=1}^{K} \exp\left(-v_\ell \left(2s_{j,\theta}(\mathbf{x}) - 1\right) y_\ell\right) w_\ell v_\ell y_\ell s_{j,\theta}(\mathbf{x}) \left(1 - s_{j,\theta}(\mathbf{x})\right)$$

The initial value of c and d were set to 1 and 0, respectively (line 4 of Alg. 2).

So far, we implicitly assumed that the index of feature j is given. To choose j, we trained sigmoidal decision stumps in parallel for each feature and we estimated the edge of each of them using the sequential training data as $\widehat{\gamma}_j = \sum_{t'=1}^{C_t} \sum_{\ell=1}^{K} w_{t',\ell} y_{t',\ell} \text{sign}\big(v_\ell^{(t')} \varphi_{j,\Theta_j^{(t')}}(\mathbf{x}_{t'})\big)$. Finally, we chose the feature with the highest edge estimate $j^* = \arg\max_j \widehat{\gamma}_j$.

In every boosting iteration we also train a *constant learner* (also known as y-intercept) and use it if its edge is higher than the edge of the best decision stump we found. The output of the constant learner does not depend on the input vector $\mathbf{x}$, that is $\varphi(\cdot) \equiv 1$, in other words it returns the vote vector $\mathbf{v}$ itself. Thus only $\mathbf{v}$ has to be learnt but this can be done easily by calculating the classwise edge $v_\ell = \sum_{t'=1}^{C_t} w_{t',\ell} y_{t',\ell}$.

6   GoLF Boosting

In order to adapt Alg. 2 to GoLF (Alg. 1), we need to define the permanent state of the FILTERBOOST model class, and we need to provide an implementation of the UPDATEMODEL method. This is rather straightforward: the model instance has to store the the actual strong learner $\mathbf{f}^{(t)}$ as well as the state of the inner part of the two for loops in Alg. 2 so that UPDATEMODEL could simulate these loops every time a new sample is processed.

This way, every model that is performing a random walk is theoretically guaranteed to converge so long as we assume that peer sampling works perfectly. However, there is a catch. Since in each iteration some nodes will receive more than one model, while others will not receive any, and since the number of models in the network is kept constant if there is no failure (since in each iteration all the nodes send exactly one model) it is clear that the *diversity* of models will decrease. That is, some models get replicated, while others "die out". Introducing failure makes things a lot worse, since we can lose models due to message loss, delay, and churn as well, which speeds up homogenization. This is a problem, because diversity is important when we want to apply techniques such as combination or voting [19,20]. Without diversity these important techniques are guaranteed not to be effective.

The effects of decreasing diversity are negligible during the timespan of a few gossip cycles, but a boosting algorithm needs a relatively large number of cycles to converge (which is not a problem, since the point of boosting is not speed, but classification quality). So we need to tackle the loss of diversity. We propose Alg. 3 to deal with this problem.

This protocol works as follows. A node sends models in an active cycle (line 4) only in two cases: it sends the last received model if there was no incoming model until 10 active cycles (line 6), otherwise it sends all of the models received since the last cycle (line 13). If there is no failure, then this protocol is guaranteed to keep the diversity of models, since all the models in the network will perform independent random walks. Due to the Poisson distribution of the number of incoming models in one cycle, the probability of bottlenecks is diminishing, and for the same reason the probability that a node does not receive messages for 10 cycles is also practically negligible.

If the network experiences message drop failures or churn, then the number of models circulating in the network will converge to a smaller value due to the

---

Algorithm 3. Diversity Preserving GoLF

1: $currentModel \leftarrow$ initModel()	13: **else**
2: $modelQueue$.add($currentModel$)	14: **for all** $m \in modelQueue$ **do**
3: $counter \leftarrow 0$	15: $p \leftarrow$ selectPeer()
4: **loop**	16: sendModel(p, m)
5: wait(Δ)	17: $modelQueue$.remove(m)
6: **if** $modelQueue$.isEmpty() **then**	18: $counter \leftarrow 0$
7: **if** $counter = 10$ **then**	
8: $p \leftarrow$ selectPeer()	19: **procedure** ONRECEIVEMODEL(m)
9: sendModel($p, currentModel$)	20: m.updateModel(x, y)
10: $counter \leftarrow 0$	21: $modelQueue$.add(m)
11: **else**	22: $currentModel \leftarrow m$
12: $counter \leftarrow counter + 1$	

---

10 cycle waiting time, and the diversity can also decrease, since after 10 cycles a model gets replicated in line 9. Interestingly, this is actually useful because if the diversity is low, it makes sense to circulate fewer models and to wait most of the time, since information is redundant anyway. Besides, with reliable communication channels that eliminate message drop (but still allow for delay), diversity can still be maintained.

Finally, note that if there is no failure, Alg. 3 has the same total message complexity as Alg. 1 except for the extremely rare messages sent in line 4. In case of failure, the message complexity decreases as a function of failure rate; however, the remaining random walks do not get slower relative to Alg. 1, so the convergence rate remains the same on average, at least if no model-combination techniques are used.

7   Experimental Results

In our experiments we examined the performance of our proposed algorithm as a function of gossip cycles, which is about the same as the number of training samples seen by any particular model. To validate the algorithm, we compared it with three baseline multi-class boosting algorithms, all using the same decision stump (DS) weak learner. The first one is the multi-class version of the well known AdaBoost [24] algorithm, the second one is the original FilterBoost [5] method implemented for a single processor, with the setting $C = 30$, and the third one is the online version of FILTERBOOST (Alg. 2). We used three multi-class classification benchmark datasets to evaluate our method, namely the CTG, the PenDigits and the Segmentaion databases. These were taken from the UCI repository [9] and have different size, number of features, class distributions and characteristics. The basic properties of the datasets can be found in Table 1.

Table 1. The main properties of the data sets, and the prediction errors of the baseline algorithms

	CTG	PenDigits	Segmentation
Training set size	1,701	7494	2100
Test set size	425	3,492	210
Number of features	21	16	19
Class labels	1325/233/143	10 classes (uniform)	7 classes (uniform)
AdaBoost (DS)	0.109347	0.060715	0.069048
FilterBoost (DS, C30)	0.094062	0.071657	0.062381

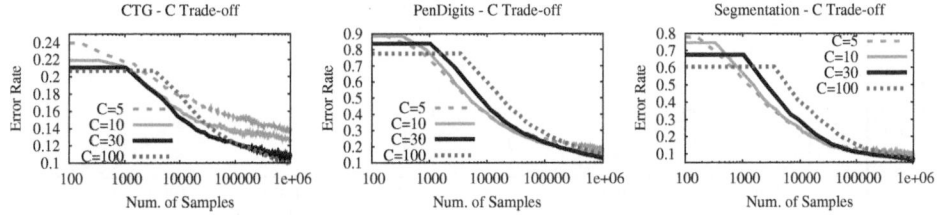

Fig. 1. The effect of parameter C in online FILTERBOOST (Alg. 2)

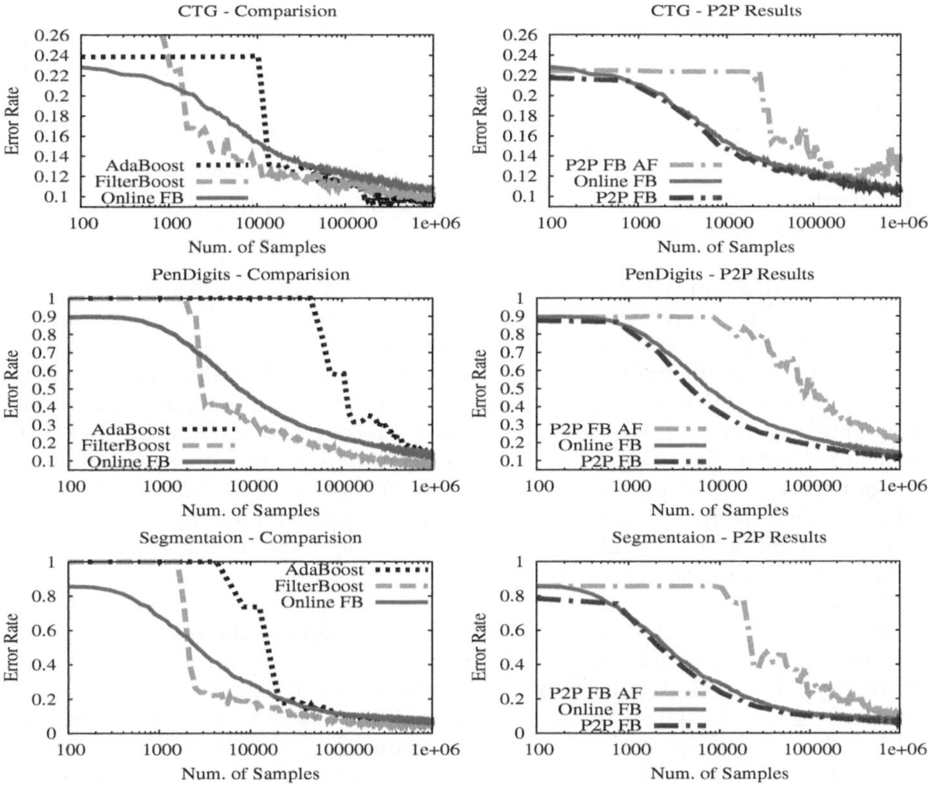

Fig. 2. Comparison of boosting algorithms (left column) and P2P simulations (right column). FB and AF stand for FilterBoost and the "all failures" scenario, respectively.

In the P2P experiments we used the PeerSim [23] simulation environment to model message *delay, drop* and peer *churn*. We used two scenarios: a perfect network without any delay, drop or churn; and a scenario with heavy failure where the message delay was drawn uniformly at random from the interval $[\Delta; 10\Delta]$, a message was dropped with a probability of 0.5 and the online/offline session lengths of peers were modeled using a real P2P bittorrent trace [1]. As our performance metric, we applied the well known *0-1 error* (or error rate), which is the proportion of test instances that were incorrectly classified.

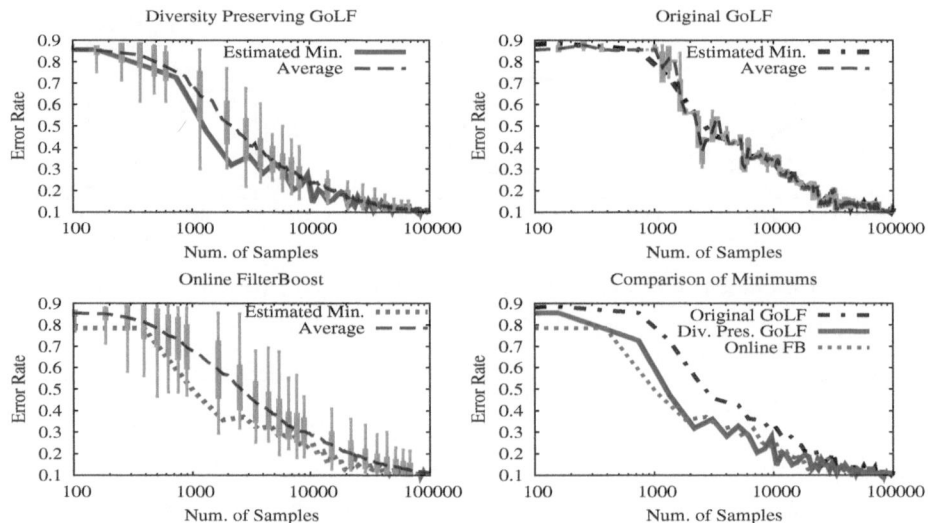

Fig. 3. The improvement due to estimating the best model based on training performance. The Segmentation dataset is shown.

Figure 1 illustrates the effect of parameter C. Larger values result in slower convergence but better eventual performance. The setting $C = 30$ represents a good tradeoff in these datasets, so from now on we fix this value.

We compared our online boosting algorithm to baseline algorithms as can be seen in Figure 2 (left hand side). The figure shows that the algorithms converge to a similar error rate, which was expected. Moreover, our online FILTERBOOST converges faster than the AdaBoost algorithm and it has almost the same convergence rate as that for the sequential FilterBoost method. Note that since two of these algorithms are not online, we had to approximate the number of (not necessarily different) training samples used in one boosting iteration. We used a lower bound to be conservative.

In our P2P evaluations of GoLF BOOSTING we used the mean error rate of 100 randomly selected nodes in the network to approximate the performance of the algorithm. Figure 2 (right hand side) shows that without failure the performance is very similar to that of our online FILTERBOOST algorithm. Moreover, in the extreme failure scenario, the algorithm still converges to the same error rate, although with a delay. This delay can be accounted for using a heuristic argument: since message delay in itself represents a slowdown of a factor of 5 on average, message drop and churn contributes approximately another factor of 2.

Finally, we demonstrate a novel way of exploiting model diversity (see Section 6): through gossip-based minimization one can spread the model with the *best training performance*, thus the best model can be made available to all nodes at all times. Figure 3 demonstrates this technique for different algorithms. We include results over the segmentation database only, the other two datasets produce similar results.

The top left plot shows results with GoLF BOOSTING. It can be seen that the best model based on training performance is not necessarily the best over the test set, but it is reasonably good, and results in a speedup of about a factor of 2. The top right plot belongs to the original GoLF implementation (Alg. 1). Due to the complete lack of diversity, the best model's performance is almost identical to the average one. The bottom left plot is a baseline experiment that represents the case with the maximal possible diversity, based on 100 completely independent runs of the online FILTERBOOST algorithm. Finally, the bottom right plot collects the most interesting curves from the other three plots allowing a better comparison.

8 Conclusions

We demonstrated that the GoLF is suitable for the implementation of multi-class boosting. The significance of this result is that boosting is a state-of-the-art machine learning technique from the point of view of the quality of the learned models, which is now available in the P2P system model with fully distributed data. To achieve this, we proposed a modification of FILTERBOOST that allows it to learn multi-class models in a purely online fashion, and we proved theoretically that the resulting algorithm optimizes a suitably defined negative log likelihood measure. Our experimental results demonstrate the robustness of the method. We also identified the lack of model diversity as a potential problem with GoLF. We provided a solution that was demonstrated to be effective in preserving the difference between the best model and the average models; this allowed us to propose spreading the best model as a way to benefit from the large number of models in the network.

References

1. Filelist. http://www.filelist.org (2005)
2. Asuncion, A.U., Smyth, P., Welling, M.: Asynchronous distributed estimation of topic models for document analysis. Statistical Methodology 8(1), 3 – 17 (2011)
3. Babenko, B., Yang, M., Belongie, S.: A family of online boosting algorithms. In: Computer Vision Workshops (ICCV Workshops). pp. 1346–1353 (2009)
4. Bottou, L.: Large-scale machine learning with stochastic gradient descent. In: Intl. Conf. on Computational Statistics. vol. 19, pp. 177–187 (2010)
5. Bradley, J., Schapire, R.: FilterBoost: Regression and classification on large datasets. In: Advances in Neural Information Processing Systems. vol. 20. The MIT Press (2008)
6. Collins, M., Schapire, R., Singer, Y.: Logistic regression, AdaBoost and Bregman distances. Machine Learning 48, 253–285 (2002)
7. Datta, S., Bhaduri, K., Giannella, C., Wolff, R., Kargupta, H.: Distributed data mining in peer-to-peer networks. IEEE Internet Comp. 10(4), 18–26 (July 2006)
8. Fan, W., Stolfo, S.J., Zhang, J.: The application of AdaBoost for distributed, scalable and on-line learning. In: Proc. 5th ACM SIGKDD Intl. Conf. on Knowledge Discovery and Data Mining. pp. 362–366 (1999)
9. Frank, A., Asuncion, A.: UCI machine learning repository (2010)
10. Freund, Y., Schapire, R.E.: Experiments with a new boosting algorithm. In: Machine Learning: Proc. Thirteenth Intl. Conf. pp. 148–156 (1996)
11. Freund, Y., Schapire, R.E.: A decision-theoretic generalization of on-line learning and an application to boosting. J. of Comp. and Syst. Sci. 55, 119–139 (1997)
12. Friedman, J.: Stochastic gradient boosting. Computational Statistics and Data Analysis 38(4), 367–378 (2002)
13. Jelasity, M., Canright, G., Engø-Monsen, K.: Asynchronous distributed power iteration with gossip-based normalization. In: Euro-Par 2007. LNCS, vol. 4641, pp. 514–525. Springer (2007)

14. Jelasity, M., Montresor, A., Babaoglu, O.: Gossip-based aggregation in large dynamic networks. ACM Trans. on Computer Systems 23(3), 219–252 (August 2005)
15. Kégl, B., Busa-Fekete, R.: Boosting products of base classifiers. In: Intl. Conf. on Machine Learning. vol. 26, pp. 497–504. Montreal, Canada (2009)
16. Kempe, D., Dobra, A., Gehrke, J.: Gossip-based computation of aggregate information. In: Proc. 44th Annual IEEE Symposium on Foundations of Computer Science (FOCS'03). pp. 482–491. IEEE Computer Society (2003)
17. Kowalczyk, W., Vlassis, N.: Newscast EM. In: 17th Advances in Neural Information Processing Systems (NIPS). pp. 713–720. MIT Press, Cambridge, MA (2005)
18. Luo, P., Xiong, H., Lü, K., Shi, Z.: Distributed classification in peer-to-peer networks. In: Proc. 13th ACM SIGKDD Intl. Conf. on Knowledge discovery and data mining (KDD'07). pp. 968–976. ACM, New York, NY, USA (2007)
19. Ormándi, R., Hegedűs, I., Jelasity, M.: Asynchronous peer-to-peer data mining with stochastic gradient descent. In: Euro-Par 2011. LNCS, vol. 6852, pp. 528–540. Springer (2011)
20. Ormándi, R., Hegedűs, I., Jelasity, M.: Efficient p2p ensemble learning with linear models on fully distributed data. CoRR abs/1109.1396 (2011)
21. Oza, N., Russell, S.: Online bagging and boosting. In: Proc. Eighth Intl. Workshop on Artificial Intelligence and Statistics (2001)
22. Park, B.H., Kargupta, H.: Distributed data mining: Algorithms, systems, and applications. In: Ye, N. (ed.) The Handbook of Data Mining. CRC Press (2003)
23. PeerSim: http://peersim.sourceforge.net/
24. Schapire, R.E., Singer, Y.: Improved boosting algorithms using confidence-rated predictions. Machine Learning 37(3), 297–336 (1999)
25. Siersdorfer, S., Sizov, S.: Automatic document organization in a P2P environment. In: Advances in Information Retrieval, LNCS, vol. 3936, pp. 265–276. Springer (2006)
26. Tölgyesi, N., Jelasity, M.: Adaptive peer sampling with newscast. In: Euro-Par 2009. LNCS, vol. 5704, pp. 523–534. Springer (2009)
27. Widrow, B., Hoff, M.E.: Adaptive Switching Circuits. In: 1960 IRE WESCON Convention Record. vol. 4, pp. 96–104 (1960)

Appendix

The second order expansion of multi-class negative log likelihood for fixed α and $\mathbf{h}(\mathbf{x}) = 0$ can be written as

$$R_{\mathrm{L}}\left(\mathbf{f}^{(t)} + \alpha\mathbf{h}\right) = \ln\left(1 + \sum_{\ell \neq \ell(\mathbf{x})}^{K} \mathcal{F}_{\ell}^{\mathbf{f}^{(t)}}(\mathbf{x})\right) - \sum_{\ell}^{K} \frac{\mathcal{F}_{\ell}^{\mathbf{f}^{(t)}}(\mathbf{x})}{\sum_{\ell'=1}^{K} \mathcal{F}_{\ell'}^{\mathbf{f}^{(t)}}(\mathbf{x})} \alpha y_{\ell} h_{\ell}(\mathbf{x})$$

$$+ \frac{1}{2} \sum_{\ell=1}^{K} \frac{\alpha y_{\ell}\left(1 + \sum_{\ell' \neq \ell} \mathcal{F}_{\ell'}^{\mathbf{f}^{(t)}}(\mathbf{x})\right) - \alpha^2 \overbrace{y_{\ell}^2 h_{\ell}^2(\mathbf{x})}^{=1} \mathcal{F}_{\ell}^{\mathbf{f}^{(t)}}(\mathbf{x})}{1 + \sum_{\ell' \neq \ell} \mathcal{F}_{\ell'}^{\mathbf{f}^{(t)}}(\mathbf{x})}$$

where $\mathcal{F}_{\ell}^{\mathbf{f}^{(t)}}(\mathbf{x}) = \exp\left(f_{\ell}^{(t)}(\mathbf{x}) - f_{\ell(\mathbf{x})}^{(t)}(\mathbf{x})\right)$. Let us note that the last term does not depend on $\mathbf{h}(\cdot)$, consequently minimizing this approximation of $R_{\mathrm{L}}\left(\mathbf{f}^{(t)} + \alpha\mathbf{h}\right)$ with respect to $\mathbf{h}(\mathbf{x})$ is equivalent to maximizing the weighted accuracy and the weight of the ℓth label is

$$w_{\ell}^{(t)} = \frac{\mathcal{F}_{\ell}^{\mathbf{f}^{(t)}}(\mathbf{x})}{\sum_{\ell'=1}^{K} \mathcal{F}_{\ell'}^{\mathbf{f}^{(t)}}(\mathbf{x})} = \frac{\exp\left(f_{\ell}^{(t)}(\mathbf{x}) - f_{\ell(\mathbf{x})}^{(t)}(\mathbf{x})\right)}{\sum_{\ell'=1}^{K} \exp\left(f_{\ell'}^{(t)}(\mathbf{x}) - f_{\ell(\mathbf{x})}^{(t)}(\mathbf{x})\right)}$$

Topic 8: Distributed Systems and Algorithms

Andrzej Goscinski, Marios Mavronicolas, Weisong Shi, and Teo Yong Meng

Topic Committee

The increasing significance of *Distributed Computing* becomes more and more crucial with the prevail of technological advances that make *Global Computing* a reality in modern world. Indeed, it is hard to imagine some application or computational activity and process that falls outside *Distributed Computing*. With the large advent of distributed systems, we are faced with the real challenges of distributed computation: How do we cope with asynchrony and failures? How (and how well) do we achieve load balancing? How do we model and analyze malicious and selfish behavior? How do we address mobility, heterogeneity and the dynamic nature of participating processes? What can we achieve in the presence of disconnecting operations that cause network partitioning?

These and many more are some of the questions that are routinely scrutinized under the light of current research in *Distributed Systems and Algorithms,* the well-known **Topic 8** of *Europar.* This *Europar* topic provides a forum for both research and development, of interest to both academia and industry, to present and discuss novel approaches to *Distributed Computing* and its relation and connection to *Parallel Processing.* The *Europar 2012* Call for Papers encouraged submission of papers accross the whole spectrum of *Distributed Systems and Algorithms,* with emphasis on several classical and currently popular subareas.

This year five papers were accepted. The paper *Towards Load Balanced Distributed Transactional Memory,* by *G. Sharma* and *C. Busch,* considers the problem of implementing transactional memory in d-dimensional mesh networks. It presents and analyzes *multibend,* a novel load balanced directory-based protocol, which is designed for the *data-flow* distributed implementation of software transactional memory. The paper *CUDA-For-Clusters: A System for Efficient Execution of CUDA Kernels on Multi-Core Clusters,* by *R. Prabhakar, G. Ramaswamy* and *M. J. Thazhuthaveetil,* presents and explores *CUDA* as a programming language for multicores and develops in this way *CUDA-For-Clusters (CFC),* a framework that transparently orchestrates execution of *CUDA* kernels on a cluster of multi-core machines. The paper *From a Store-collect Object and Ω to Efficient Asynchronous Consensus,* by *M. Raynal* and *J. Stainer,* presents an efficient algorithm to build a consensus object, which is based on an Ω failure detector (to obtain liveness) and a *store-collect* object (to maintain its safety). The paper *An Investigation into the performance of reduction algorithms under load imbalance,* by *P. Marendic, J. Lemeire, T. Haber, D. Vucinic,* and *P. Schelkens,* investigates contexts where it is not guaranteed that all processes start *reduction* at about the same time; this is a common context in practice, where significant load imbalances may occur and affect the performance of algorithms. The paper investigates the impact of such imbalances on the most commonly employed reduction algorithms and propose a new algorithm specifically adapted

C. Kaklamanis et al. (Eds.): Euro-Par 2012, LNCS 7484, pp. 401–402, 2012.
© Springer-Verlag Berlin Heidelberg 2012

for such contexts. The paper *Achieving Reliability in Master-worker Computing via Evolutionary Dynamics,* by *E. Christoforou, A. Fernández Anta, C. Georgiou, M. A. Mosteiro* and *A. Sánchez,* consider Internet-based Master-Worker Computations where a master process sends tasks, across the Internet, to worker processes; workers execute and report back some result but they are not trustworthy. To this respect, the paper models such computations using evolutionary dynamics and studies the conditions under which the master can reliably obtain tasks results. The paper develops and analyzes an algorithmic mechanism that uses reinforcement learning to provide workers with the necessary incentives to eventually become truthful.

We would like to take this opportunity to thank all authors who submitted their work to Topic 8 of *Europar 2012,* all external referees who assisted us, and all people involved in organizing the review process for their hard work.

Towards Load Balanced Distributed Transactional Memory

Gokarna Sharma and Costas Busch

Department of Computer Science, Louisiana State University
Baton Rouge, LA 70803, USA
{gokarna,busch}@csc.lsu.edu

Abstract. We consider the problem of implementing transactional memory in d-dimensional mesh networks. We present and analyze MultiBend, a novel load balanced directory-based protocol, which is designed for the *data-flow* distributed implementation of software transactional memory. It supports three basic operations, *publish*, *lookup*, and *move*, on a shared object. A pleasing aspect of MultiBend is that it is load balanced (minimizes maximum node and edge utilization) which is achieved by using paths of multiple bends in the mesh. This protocol guarantees an $\mathcal{O}(d^2 \log n)$ approximation for the load and also for the distance stretch of *move* requests, where n is the number of nodes in the network. For fixed d, both the load and the *move* stretch are optimal within a constant and a loglog factor, respectively. It also guarantees $\mathcal{O}(d^2)$ approximation for *lookup* requests which is optimal within a constant factor for fixed d. To the best of our knowledge, this is the first distributed directory protocol that is load balanced.

1 Introduction

In distributed networked systems processors are the nodes of a network which communicate through a message passing environment. We assume that there is a shared memory address space, which is equally split among the processors. Each processor has its own *cache*, where copies of objects reside. In Transactional Memory (TM) [10,9,16,8,12] a *transaction* represents a sequence of shared memory operations (i.e., reads and writes) that are all performed atomically. The individual entries at the shared memory, called *objects*, can be shared by multiple transactions on different network nodes. A transaction can either *commit* (i.e., take effect) or *abort* (i.e., have no effect at all). If a transaction aborts, it is typically restarted until it commits. When a transaction running at a processor node issues a read or write operation for a shared memory location, the data object at that location is loaded into the processor-local cache.

We consider the *data-flow distributed implementation of software transactional memory* (DTM) suggested by Herlihy and Sun [11], where transactions are immobile (i.e., running at some particular node) and shared objects are moved to those nodes that need them. In DTM, transactions can only operate on local shared objects and, if remote shared objects are required, the transaction must communicate with one or more remote processor nodes. Some distributed *cache-coherence* mechanism should ensure that shared objects remain *consistent*, i.e., writing to an object automatically *locates* and *invalidates* other cached copies of that object. A DTM protocol typically supports

C. Kaklamanis et al. (Eds.): Euro-Par 2012, LNCS 7484, pp. 403–414, 2012.
© Springer-Verlag Berlin Heidelberg 2012

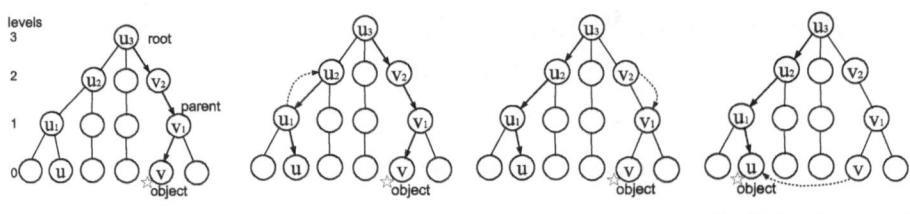

(a) Initially, owner v publishes the object

(b) The request continues up phase

(c) The request continues down phase

(d) Object is moved directly from v to u

Fig. 1. Illustration of MultiBend for a *move* request issued by node u for the object at node v, where the nodes shown are leader nodes of the respective sub-meshes

three kinds of operations: (i) *publish* operation which allows a node which created an object in its memory space to publish it so that other nodes in the network can find it; (ii) *lookup* operation, the protocol should locate the current copy of the object and move it to the requesting node's cache (*shared* access), without modifying the old copy; (iii) *move* operation, where a transaction attempts to access an object to update explicitly the DTM protocol should locate the current cached copy of the object and move it to the requesting node's cache invalidating the old copy. [3,17] also studied DTM.

Typically the performance of a DTM protocol is measured with respect to the communication cost, which is the total number of messages sent in the network. The communication cost for an operation (resp. for a set of operations) is compared to the optimal communication cost for that operation (resp. for that set of operations) to provide an approximation ratio, which is generally referred to as *stretch*. In the context of DTM, previous approaches [6,17,3,11,15] focused only on stretch bounds for various network topologies (Table 1 summarizes their properties) and they do not control the congestion. The network congestion can also affect the overall performance of the algorithm and sometimes it is a major bottleneck. We measure the network congestion as the worst node or edge utilization (the maximum number of times the object requests use any edge or node in the network while accessing the shared object).

Contributions. We present MultiBend, a DTM protocol that is suitable for d-dimensional mesh networks and is load balanced in the sense that it minimizes the congestion (maximum node and edge utilization), and at the same time maintains low stretch. Mesh networks are appealing due to their use in parallel, distributed, and high-performance computing. The low stretch is achieved through a novel labeled hierarchical directory-based approach which we first introduced in [15] for general networks and we adapted it here appropriately for the mesh network. The load balancing is achieved through an *oblivious routing* approach (e.g., [13,4,5]) for communication between different hierarchy level leader nodes. In particular, we use the oblivious routing algorithm in Busch *et al.* [5] that gives near optimal congestion while maintaining small path length stretch in the mesh networks.

For the performance analysis of MultiBend, we consider sequential and concurrent execution of requests. For the *move* operations in both the cases, MultiBend guarantees $\mathcal{O}(d^2 \log n)$ amortized stretch and $\mathcal{O}(d^2 \log n)$ approximation of the optimal congestion on any node or any edge in d-dimensional mesh networks, where n is the number

Table 1. Comparison of DTM protocols, where n, S_*, and D, respectively, are the number of nodes, stretch, and the diameter of the network kind on which they operate

Protocol	Stretch	Network Kind	Load Balanced	Runs On
Arrow [6]	$\mathcal{O}(S_{ST}) = \mathcal{O}(D)$	General	No	Spanning tree
Relay [17]	$\mathcal{O}(S_{ST}) = \mathcal{O}(D)$	General	No	Spanning tree
Combine [3]	$\mathcal{O}(S_{OT}) = \mathcal{O}(D)$	General	No	Overlay tree
Ballistic [11]	$\mathcal{O}(\log D)$	Constant-doubling	No	Hierarchical directory with independent sets
Spiral [15]	$\mathcal{O}(\log^2 n \cdot \log D)$	General	No	Hierarchical directory with sparse covers
MultiBend (this paper)	$\mathcal{O}(\log n)$ $\overline{\mathcal{O}(d^2 \log n)}$	2-D mesh d-D mesh	Yes	Hierarchical decomposition of the mesh

of nodes in the mesh. For fixed d, the *move* stretch is optimal within a loglog factor comparing to the $\Omega(\log n / \log \log n)$ lower bound by Alon *et al.* [1]; the congestion approximation is also optimal within a constant factor in light of the $\Omega(\frac{C^*}{d} \log n)$ lower bound on the approximation ratio of an oblivious algorithm due to Maggs *et al.* [13]. The communication cost of the *publish* operation is proportional to the diameter of the network (i.e., $\mathcal{O}(n)$) and it is a fixed initial cost which is only considered once and compensated by the costs of the *move* (or *lookup*) operations which are issued thereafter. Note that *lookup* operations have always $\mathcal{O}(d^2)$ stretch even when considered individually while their overall congestion is $\mathcal{O}(d^2 \log n)$ approximation in the d-dimensional mesh. To the best of our knowledge, this is the first DTM protocol that achieves low stretch in a load balanced way. It has been shown that the stretch and the congestion cannot be controlled simultaneously in general networks [5].

Techniques. For simplicity, consider an 2-dimensional $n = m \times m$ mesh network and one shared object; the general case for d-dimensional mesh is given in Section 6. (We consider transactions with only one shared object which is typical in the DTM literature [3,11,17,6,15]. A protocol for one object can be generalized to accommodate transactions with multiple objects by appropriately replicating that protocol in such a way that the replication avoids livelock of transactions.) MultiBend is a directory-based consistency protocol implemented on a hierarchy of sub-meshes (as clusters). There are $k + 1 = \mathcal{O}(\log n)$ levels such that side lengths of the sub-meshes increase by a factor of 2 between two consecutive levels. In each sub-mesh one node is chosen to act as a leader to communicate with different level sub-meshes. At the bottom level (level 0) each sub-mesh consists of individual nodes, while at the top level (level k) there is a single sub-mesh for the whole graph with a special leader node called *root*. The hierarchy forms a tree of leaders such that higher level leaders have as children the lower level leaders. Only the bottom level nodes can issue requests (*publish*, *lookup*, and *move*) for the shared object, while the nodes in higher levels are used to propagate the requests in the graph. (The difference between Spiral [15] and MultiBend is that Spiral uses sparse covers as clusters while MultiBend uses sub-meshes as clusters.)

The protocol maintains a *directory path* which is directed from the root to the bottom-level node that owns the shared object. The directory path is updated whenever the object moves from one node to another. As soon as the object is created by some bottom level node, it publishes the object by visiting its sequence of increasingly higher level leaders path towards the root, making each parent pointing to its child leader (Fig. 1a). These leader pointers correspond to path segment between the leaders and the concatenation of these path segments form the initial directory path. A *move* request from a node is served by following leader ancestors of that node, setting downward links toward it until it intersects the directory path to the owner node, and resetting the directory path it follows while descending towards the owner (Figs. 1b–1c); the directory path now points to the requesting node. As soon as the *move* reaches the owner, the object is forwarded to the requester along some shortest path in the mesh (Fig. 1d). A *lookup* operation is served similar to *move* without modifying the directory path.

In order to route the request between two consecutive leaders, we use *multi-bend* paths. In the oblivious routing algorithm of [5] they use a one-bend path between pairs of randomly selected nodes in the mesh. A one-bend path consists of two straight lines, one line in each dimension which meet at a corner where the bend occurs. Following [5], we use at most two-bend paths between leaders. The one-bend path is sufficient when the parent sub-mesh completely contains the child sub-mesh (they are at different level). There is an attribute in our algorithm where every level has actually two sub-levels with possibly the same side length sub-meshes (at least one same side length). For this a two-bend path is needed between the leaders of the same level sub-meshes.

The concatenation of the one-bend or two-bend paths form multi-bend paths. In order to obtain low congestion, every time we access the leader node of the sub-mesh we immediately replace it with another leader chosen uniformly at random among the nodes in the sub-mesh. The directory is then updated appropriately with the new leader information by updating the parent and children leaders. We note that the update cost is low in comparison to the cost of serving the requests because only the information in the nearby region needs to be updated due to the new leader. We argue that this step is necessary to control the congestion. This is because when a fixed leader is used, the node congestion on that leader is proportional to the number of requests that visit that leader. Moreover, in the fixed leader case, edge congestion can also be proportional to the number of requests as all the requests use fixed edges along the shortest path between two subsequent leaders. We also note that, using this random leader approach, if the congestion requirement on edges (or nodes) can be relaxed by the factor of κ, then leader change is only needed after every κ requests.

Outline of Paper. We proceed with network model and preliminaries in Section 2. In Section 3, we give hierarchy construction for the 2-dimensional mesh. We present MultiBend protocol in Section 4 and analyze it in Section 5. In Section 6, we extend MultiBend for the d-dimensional mesh. (Many proofs and details are omitted due to space restrictions.)

2   Network Model and Preliminaries

We begin with some necessary definitions which are adapted from [5,15]. We represent a distributed network as a d-dimensional mesh. The d-dimensional mesh $M = (V, E)$ is a d-dimensional grid of nodes (network machines) V, where $|V| = n$, with side length m_i in each dimension such that $n = \prod_{i=1}^{d} m_i$, and edges (interconnection links between machines) $E \subseteq V \times V$. Each computing node $u \in V$ is connected with each of its $2d$ neighbors (except the nodes at the boundaries of the mesh). We denote by $|E|$ the number of edges in M. A path p in M is a sequence of nodes with respective sequence of edges connecting the nodes, such that the length of the path p, denoted $\text{length}(p)$, is the number of edges it uses. A *sub-path* of p is any path obtained by a subsequence of consecutive edges in p; we may also refer to a sub-path as a *fragment* of p. Let $\text{dist}(u, v)$ denote the shortest path length (distance) between nodes u and v.

Consider a routing problem Π defined as a set of pairs of source and destination nodes. A routing algorithm for Π provides paths from every source to its respective destination. An algorithm is *oblivious* if the path choice for each pair of source destination is independent of the path choices of any other pair. The edge congestion C for any set of paths is the maximum congestion on any edge (link) of the network. Let C^* denote the optimal congestion attainable by any routing algorithm. We have symmetric definitions for node congestion. For a sub-mesh $M' \subseteq M$ (i.e., M' is any mesh that contains inside M), let $\text{out}(M')$ denote the number of edges at the boundary of M', which connect nodes in M' with nodes outside M'. For routing problem Π, we define the *boundary congestion* as follows. Consider some sub-mesh M' of the network M. Let Π' denote the messages (pairs of sources and destinations) in Π which have either their source or destination in M', but not both. All the messages in Π' will cross the boundary of M'. The paths of these messages will cause congestion at least $|\Pi'|/\text{out}(M')$. Define the boundary congestion of M' to be $B(M', \Pi) = |\Pi'|/\text{out}(M')$. For problem Π, the boundary congestion $B = \max_{M' \subseteq M} B(M', \Pi)$. Clearly, $C^* \geq B$.

We bound the stretch of the MultiBend protocol, which is the ratio of the total communication cost for a request (or for a set of requests) to the optimal cost for that operation (or for that set of requests). The congestion is the maximum number of times any node or edge is used by the object requests. We assume that M represents a network in which nodes do not crash, it implements FIFO communication between nodes (i.e. no overtaking of messages occurs), and messages are not lost. We also assume that, upon receiving a message, a node is able to perform a local computation and send a message in a single atomic step. *TM memory proxy* module [11] at each node provides interfaces both to the transactions at that node and to the proxies at other nodes on how to publish and access shared objects (details in [11,15]). The conflicts, if any, between a local transaction and a transaction running in some other node, in accessing the object, is resolved using well-known contention managers, e.g., [7,2,14].

3   Hierarchical Directory for the 2-Dimensional Mesh

We describe how to represent the 2-dimensional mesh M with equal side lengths $m = 2^k, k \geq 0$, as a hierarchy of sub-meshes (we discuss the d-dimensional case later in Section 6). We decompose M into two types of sub-meshes, type-1 and type-2 (see Fig. 2), as given below, adapting some notions from [5].

— *Type-1 sub-meshes.* There are $k + 1$ levels of type-1 sub-meshes, $\ell = 0, 1, \cdots, k$. The mesh M itself is the only level k sub-mesh. Every level ℓ sub-mesh can be partitioned into 4 sub-meshes by dividing each side by 2. Each resulting sub-mesh is a

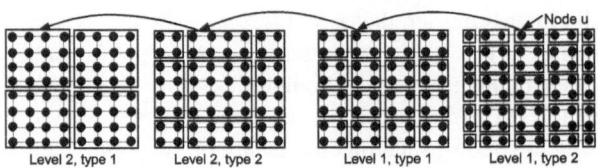

Fig. 2. Decomposition for the $2^3 \times 2^3$ 2-dimensional mesh. The arrows show the parent sub-meshes of a node u in its multi-bend path towards the root at level 3.

type-1 sub-mesh at level $\ell - 1$. According to this decomposition, at level ℓ, there are $2^{2\ell}$ sub-meshes each with side length $m_\ell = 2^{k-\ell}$. Note that the level 0 sub-meshes are the individual nodes of the mesh.

– *Type-2 sub-meshes.* There are $k - 1$ levels of type-2 sub-meshes, $\ell = 1, \cdots, k-1$. The type-2 sub-meshes at level ℓ are obtained taking the type-1 sub-meshes and shifting them by $m_\ell/2$ simultaneously in both dimensions. Some of the shifted sub-meshes may have to be truncated at the borders of M.

We assign integer sub-levels to different type sub-meshes at each level. As there are only two types of sub-meshes at any level $0 < \ell < k$, we assign sub-level 1 to type-2 and sub-level 2 to type-1 sub-meshes (Fig. 2). For levels 0 and k we have only one sub-level as there are only type-1 sub-meshes. (i, j) denotes the level i sub-level j.

We now define a hierarchy of leveled sub-meshes. The sub-mesh hierarchy $\mathcal{Z} = \{Z_0, Z_1, \ldots, Z_k\}$, is a hierarchy of $k + 1$ levels of sub-meshes such that: (i) At level k all nodes in M belong to exactly one sub-mesh, i.e., mesh M itself is the only level k sub-mesh; (ii) At level 0 each node in M is the one sub-mesh by itself; and (iii) In any level i, $1 \leq i \leq k - 1$, Z_i contains type-1 and type-2 sub-meshes of level i.

Multi-bend Paths. We define a path $p(u)$ for each node $u \in V$ which is a "multi-bend" path of u. The path $p(u)$ is built by visiting the leader nodes in all the sub-meshes that u belongs to starting from level 0 up to k. In each level, the sub-meshes are visited according to the order of their sub-levels. From an abstract point of view, the path bends (changes dimensions) multiple times while it visits sub-mesh leaders of higher levels.

In every sub-mesh X we choose a *leader* node arbitrarily at the initialization of $\mathcal{Z}$ which we denote as $\ell(X)$. If one node is the leader on many sub-level sub-meshes, we add a virtual copy node of it and create a virtual link between the virtual copy and y itself in subsequent sub-meshes. Denote the leader of sub-level (i, j) sub-mesh $X_{i,j}(u)$ as $\ell_{i,j}(u) = \ell(X_{i,j}(u))$. Since the top most Z_k consists of a single sub-level it has a unique leader which we denote $\ell_{k,0}(u) = r$ (the root). Trivially, every node $u \in V$ is a leader of its own sub-mesh at level 0, $\ell_{0,1}(u) = u$. Note that $\ell(X)$ is changed for every request by electing a new leader uniformly at random among the nodes of X. This step is necessary to minimize the congestion among the nodes and edges.

For any pair of nodes $u, v \in V$, let $s(u, v)$ denote a dimension-by-dimension (i.e., change in path from one dimension to other dimension in every bend) shortest path (an at most two-bend path) from u to v. For any set of nodes $u_1, u_2, \ldots, u_f \in V$, let $s(u_1, u_2, \ldots, u_f)$ denote the concatenation of shortest paths $s(u_1, u_2)$, $s(u_2, u_3), \ldots, s(u_{f-1}, u_f)$. Formally, the multi-bend path of node u is: $p(u) = s(u, \ell_{1,1}(u), \ell_{1,2}(u), \ldots, \ell_{k-1,1}(u), \ell_{k-1,2}(u), r)$.

We say that two multi-bend paths *intersect* if they have a common node. We also say that two multi-bend paths intersect at level i if they visit the same leader node at level i (they may intersect outside leaders but we do not consider that). Therefore,

Lemma 1 (Sharma et al. [15]). *For any two nodes $u, v \in V$, their multi-bend paths $p(u)$ and $p(v)$ intersect at level at most* $\min\{k, \lceil \log(\text{dist}(u, v)) \rceil + 1\}$.

Canonical Paths. We need later paths obtained from fragments of multi-bend paths; the fragments are formed while the object moves. These paths start at level 0 and may go up to the root. We will refer to such paths as *canonical*. As shown in the figure on the right, the newly formed path from w to v_6 is a canonical path that obtained from the fragment of $p(w)$ from w to u_2, the fragment of $p(u)$ from u_2 to v_4, and the fragment of $p(v)$ from v_4 to v_6. Formally, a canonical path q up to sub-level $(\alpha, \beta) \le (k, 1)$ is $q = s(x_{0,2}, x_{1,1}, x_{1,2}, x_{2,1}, x_{2,2}, \ldots, x_{\alpha,\beta})$, such that $x_{i,j}$'s are leader nodes along the path. A canonical path can be either *partial* when the top node is below level k (below the root), or *full* when the top node is the root. A multi-bend path $p(u)$ is a full canonical path. Any prefix of a multi-bend path is a partial canonical path. The path q up to level α is the concatenation of paths constructed by the 2 bend dimension to dimension paths in sub-meshes of (at least one) sides $2^1, 2^2, \cdots, 2^\alpha$, which sums at most $\text{length}(q) \le 2^{\alpha+3}$. Thus,

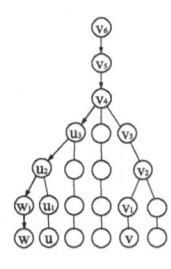

Lemma 2. *For any canonical path q up to level α,* $\text{length}(q) \le 2^{\alpha+3}$.

4   The MultiBend Algorithm

We present the MultiBend protocol (Algorithm 1) which implements a DTM for shared objects over a 2-dimensional mesh graph M. Consider some shared object ξ. The protocol guarantees that any moment of time only one node holds the shared object ξ which is the *owner* of the object. The owner is the only node who can modify the object (write the object); the other nodes can only access the object for read.

The basic idea is to maintain a *directory path* in a sub-mesh hierarchy $\mathcal{Z}$, which is a directed path from the root node r to the bottom-level node that currently owns the shared object ξ. Initially, the directory path is formed from the multi-bend path $p(v)$ of the creator node v when it issues the *publish*(ξ) operation by assigning pointers along the edges of $p(v)$ directed toward v (Fig. 1a shows hierarchy $\mathcal{Z}$ after a successful publish operation). The pseudo-code for *publish* is given in Lines 1–2 of Algorithm 1.

We define the notion of parent node before giving details of *lookup* and *move*. We denote *parent* node y of a node x in the multi-bend path $p(u)$ as $y = \text{parent}_{p(u)}(x)$, i.e., if y is the sub-level (i, j) sub-mesh leader in $p(u)$ then x is the leader of the immediate lower sub-level sub-mesh leader. Note that the leader of a level 0 sub-mesh is the node itself. Each node knows its parent in the hierarchy, except the root, whose parent is $\perp$ (null). A node might have a *link* (a downward pointer) towards one of its children (otherwise $\perp$); the *link* at the root is not $\perp$.

Lets assume a *lookup*(ξ) and a *move*(ξ) operation issued by a node u. Both operations are implemented in two phases: (i) in the *up phase*, a request message is sent from u

---

Algorithm 1. MultiBend protocol

1: **When** y **receives** $m = \langle v, up, publish \rangle$ **from** x:　　　// Publish operation
2:　set $y.link = x$; **if** y is not a root node **then send** m to $\text{parent}_{p(v)}(y)$;

3: **When** y **receives** $m = \langle u, phase, lookup \rangle$ **from** x:　　// Lookup operation
4:　**if** $m = \langle u, up, lookup \rangle$ **then**　　　　　　　　// *lookup* up phase
5:　　**if** $y.link = \perp$ **then**
6:　　　**if** $y.slink$ list is empty **then**
7:　　　　elect a leader w at sub-mesh containing $\text{parent}_{p(u)}(y)$; send m to w;
8:　　　**else** elect a leader w at sub-mesh containing first pointer of $y.slink$ list; send
　　　　　　$\langle u, down, lookup \rangle$ to w;
9:　　**else** elect a leader w at sub-mesh containing $y.link$; send $\langle u, down, lookup \rangle$ to w;
10:　**if** $m = \langle u, down, lookup \rangle$ **then**　　　　　　// *lookup* down phase
11:　　**if** y is a leaf node **then**
12:　　　send the read-only copy of ξ to u and remember u;
13:　　**else** elect a leader w at sub-mesh containing $y.link$; send m to w;

14: **When** y **receives** $m = \langle u, phase, move \rangle$ **from** x:　　// Move operation
15:　**if** $m = \langle u, up, move \rangle$ **then**　　　　　　　// *move* up phase
16:　　assign $oldlink \leftarrow y.link$ and set $y.link = x$;
17:　　add y in $slink$ list of y's special patent;
18:　　**if** $oldlink = \perp$ **then**
19:　　　elect a leader w at sub-mesh containing $\text{parent}_{p(u)}(y)$; send m to w;
20:　　**else** send $\langle u, down, move \rangle$ to $oldlink$;
21:　**if** $m = \langle u, down, move \rangle$ **then**　　　　　　// *move* down phase
22:　　**if** y is in the $slink$ list **then** erase y from $slink$;
23:　　**if** y is not a leaf node **then** $oldlink \leftarrow y.link$; $y.link \leftarrow \perp$; send m to $oldlink$;
24:　　**else** send the writable copy of ξ to u;
25:　　invalidate(ξ) from the owner node v and the read-only copies from other nodes;

26: **Leader election procedure:**
27:　select a node w in the sub-mesh containing leader z uniformly at random;
28:　copy information at old leader z to new leader w;
29:　inform the parent and child of z about the new leader w;
30:　construct a sub-path p_i from w_{i-1} to w by picking a dimension by dimension
　　　shortest path (where the sub-path is either one-bend or two-bend);

---

upward in the hierarchy $\mathcal{Z}$ along the multi-bend path $p(u)$ towards the root r until the request intersects at a node (i.e. node x) with the directory path; (ii) in the *down phase*, the request message follows the directory path from node x to the object owner; then the owner sends a copy of ξ to u (along some shortest path in M). For the *lookup* it is a read-only copy of ξ without modifying the hierarchy (see Lines 3–13 of Algorithm 1). But, for the *move* it is a writable copy invalidating the old copy of ξ and modifying the directory path (Figs. 1b−1d). In the up phase, the *move* sets the directions of the edges in the fragment of $p(u)$ between u and x to point toward u. In the down phase it deletes the downward pointers (or links) in the fragment of the directory path from x to v. Now the new directory path points toward u. When the down phase reaches v, u obtains a copy of the object (see Lines 14–25 of Algorithm 1). This process has resulted to a canonical directory path that consists of two multi-bend path fragments, a fragment of u's multi-bend path between r and x and a fragment of v's multi-bend path between x and u. Subsequent *move* operations may result into further fragmentation of the directory path into multiple (more than two) multi-bend path fragments.

Need of Special Parent. A *lookup* request may not find immediately the directory path to ξ, even if the *lookup* originates near the owner node of ξ. This is because after several *move* operations the directory path may become highly fragmented. The notion of a *special parent* node helps to avoid this situation and guarantee efficient *lookups*, such that whenever a downward link is formed at a node z the special parent of z is also informed about z holding a downward pointer. A *special-parent* node of y, denoted as sparent$_{p(u)}(y)$, at sub-level (i, j) in the multi-bend path $p(u)$ is the leader node of one of the sub-meshes $X \in Z_\eta(u)$ at level η, where $\eta = i + 4$, i.e., sparent$_{p(u)}(y)$ is some *ancestor* node of y at level η in $p(u)$. Every node knows its special parent and has *slink* (downward pointer) towards *special-child* node from its special-parent sparent$_{p(u)}(y)$ (otherwise it is $\perp$). We maintain a list of *slink* pointers if one node is the special parent for several sub-meshes. The special parent is selected in such a way that any nearby *lookup*, close to z will either reach z or its special parent. We will prove that *lookups* are always efficient using special parents (see Lines 6,8,17, and 22 of Algorithm 1).

Load Balancing. MultiBend (Algorithm 1) uses a leader election procedure (Lines 26–30 of Algorithm 1) such that *lookup* and *move* requests can be served in a load balanced way. The procedure works as follows: Let z be a leader node of the sub-mesh M' in $\mathcal{Z}$. We elect a new leader at M' by selecting a node $w \in M'$ uniformly at random. After the leader is elected, the information at old leader z is moved to new leader w and the parent and child of z are informed about the new leader w. The pointers inside M' are also updated to point to the new leader. After that, sub-path p_i from w_{i-1} (a leader of the sub-mesh that is sending a message to M') to w is formed by picking a dimension by dimension shortest path; the sub-path p_i is one-bend if sub-mesh containing w and the sub-mesh containing w_{i-1} are both type-1 sub-meshes, otherwise, p_i is of at most two bend path. If the sub-path is the two-bend path then it is picked by a random ordering of dimensions on a random node. The *lookup* uses this procedure in Lines 7,8,9, and 13 of Algorithm 1. The *move* invokes it at Line 19 of Algorithm 1.

We observe that at any time a request locks at most three nodes (level prev(i, j), (i, j), and next(i, j)) along the multi-bend path or a downward path. In concurrent situations this may be a problem; but we describe how we handle concurrent requests efficiently later in Sections 5 and 6. Note also that the special parent node doesn't need to be locked because only one specific *slink* pointer value needs to be updated.

5 Performance

We give the stretch and congestion analysis of MultiBend for sequential executions; the stretch and congestion analysis for concurrent executions is deferred to the full paper due to space limitations. Moreover, the correctness proof is omitted as it can be extended from [3,11,15].

Move Cost. We now give the analysis of MultiBend in sequential executions. Lets define a sequential execution of a set $\mathcal{E}$ of $l+1$ object ξ requests $\mathcal{E} = \{r_0, r_1, \cdots, r_l\}$, where r_0 is the initial *publish* request and the rest are the subsequent *move* requests.

For the sake of analysis, similar as in [15], we define a two-dimensional array B of size $(k+1) \times (l+1)$, where $k+1$ and $l+1$ are the number rows and columns, respectively. The $k+1$ rows of B can be denoted as $\{row_0, row_1, \cdots, row_k\}$, and the $l+1$ columns of B can be denoted as $\{col_0, col_1, \cdots, col_l\}$. All the locations of the array are initially empty ($\perp$). We fix that $[0, 0]$ is the lower left corner element and $[k, l]$ be the upper right corner element. The levels visited by each request r_i in the hierarchy $\mathcal{Z}$ while searching for the object are registered in each $col_i, 0 \leq i \leq l$. The maximum level reached by a request r_i in $\mathcal{Z}$ is called the *peak level* for that request. We have that $l \leq k$. The peak level reached by r_0 (the *publish* request) is always k and r_0 is registered at all the locations of col_0 starting from $col_0[0]$ to $col_0[k]$.

Let $A^*(\mathcal{E})$ denote the optimal cost for serving requests in $\mathcal{E}$ and $A(\mathcal{E})$ denote the total communication cost using the MultiBend. We will bound the stretch $\max_{\mathcal{E}} A(\mathcal{E})/A^*(\mathcal{E})$. For any $c, d, 0 \leq c < d \leq l$, a *valid pair* $W^j_{(c,d)}$ of two nonempty entries in $row_j, 0 \leq j \leq h$ is defined as $W^j_{(c,d)} = (row_j[c], row_j[d])$, such that $row_j[c] \neq \perp$ and $row_j[d] \neq \perp$, and $\forall e, c < e < d, row_j[e] = \perp$. In other words, $W^j_{(c,d)}$ is a pair of two subsequent non-empty entries in a row. Moreover, we denote by S_j the total count of the number of entries $row_j[i], 0 \leq i \leq l$, such that $row_j[i] \neq \perp$, and by W_j the total number of non-empty pairs ($W^j_{(c,d)}$) in it. We have that $W_j = S_j - 1$.

Theorem 1. *The move stretch of* MultiBend *is* $\mathcal{O}(\log n)$ *for sequential executions.*

Proof (sketch). Let $A^*_h(\mathcal{E})$ be the optimal communication cost for level h in the hierarchy $\mathcal{Z}$. According to the execution setup, S_h are the number of requests in $\mathcal{E}$ that reach level h, and W_h are the total number of valid pairs at that level. For any two subsequent requests that originate from nodes u and v and reach level h, $\text{dist}(u, v) \geq 2^{h-1}$ (according to Lemma 1), since otherwise their multi-bend paths would intersect at level $h-1$ or lower. Therefore $A^*_h(\mathcal{E}) \geq W_h 2^{h-1} \geq (S_h - 1)2^{h-1}$, as $W_h = S_h - 1$. Considering all the levels from 1 to k, $A^*(\mathcal{E}) \geq \max_{1 \leq h \leq k} A^*_h(\mathcal{E}) \geq \max_{1 \leq h \leq k}(S_h - 1)2^{h-1}$.

Similarly, let $A_h(\mathcal{E})$ be the total communication cost of MultiBend for all the requests in $\mathcal{E}$ that reach level h in the hierarchy $\mathcal{Z}$, while probing the shared object in their up phase. We have that $A_h(\mathcal{E}) \leq (S_h - 1)2^{h+3}$ (Lemma 2). By combining the cost for each level, $A(\mathcal{E}) = \sum_{h=1}^k A_h(\mathcal{E}) \leq \sum_{h=1}^k (S_h - 1)2^{h+3}$. We do not need to consider level 0 for $A^*(\mathcal{E})$ and $A(\mathcal{E})$ because there is no communication at that level.

Since the execution $\mathcal{E}$ is arbitrary and $\sum_{h=1}^k (S_h - 1)2^{h+3} \leq k \cdot \max_{1 \leq h \leq k}(S_h - 1)2^{h+3}$, $\max_{\mathcal{E}} A(\mathcal{E})/A^*(\mathcal{E}) \leq 16 \cdot k = \mathcal{O}(\log n)$, as $k = \lceil \log n \rceil + 1$. □

Congestion. We relate the congestion of the paths selected by MultiBend to the optimal congestion C^*. In particular, we prove the following theorem (this bound is valid for both *move* and *lookup* operations, as both do random leader change in the same way):

Theorem 2. MultiBend *achieves* $\mathcal{O}(\log n)$ *approximation on congestion w.h.p.*

Proof (sketch). Recall that every request to predecessor nodes are routed by MultiBend by selecting some paths. Precisely, these paths are the multi-bend paths. Let e denote an edge in the mesh graph M and $C(e)$ denote the load on e (the number of times the edge e is used by the paths of the requests). We bound the probability that some multi-bend path uses edge e. Consider the formation of a sub-path p_i from a sub-mesh M_1 to a sub-mesh M_2, such that $M_1 \subseteq M_2$ and e is a member of M_2. If M_1 is of type-1 then all of its sides are equal to m_ℓ, where ℓ is the level of M_1. Then the sub-path p_i uses edge e with probability at most $2/m_\ell$. Moreover, a one-bend sub-path is enough to route the request from M_1 to M_2.

Let P' be the set of paths that go from M_1 to M_2 (or vice-versa). Let $C'(e)$ denote the congestion that the messages P' cause on e. Using the similar argument as given in previous paragraph for an edge e, the upper bound in $C'(e)$, denoted as $E[C'(e)]$, is bounded by $E[C'(e)] \leq 2|P'|/m_\ell$. Moreover, from the definition of the boundary congestion $B \geq B(M_1, \Pi) \geq |P'|/\text{out}(M_1)$. Thus, $C^* \geq |P'|/\text{out}(M_1)$. Since M_1 has all sides of length m_ℓ nodes, $\text{out}(M_1) \leq 4m_\ell$. Therefore, $E[C'(e)] \leq 8C^*$. We charge this congestion to sub-mesh M_2. Between every sub-level $(i, 2)$ sub-meshes, $1 \leq i \leq k - 1$, as M_1 of sub-level $(i, 2)$ is completely contained in M_2 of sub-level $(i + 1, 2)$ and there are at most $k < \log n + 2$ levels, the expected congestion on edge e, denoted as $E[C(e)]$, is bounded by $E[C(e)] \leq 8C^*(\log n + 2)$.

According to our construction, there is one type-2 sub-mesh M_1' between every two type-1 sub-meshes M_1 and M_2 in the sub-mesh hierarchy. As the type-2 sub-mesh M_1' may not be the proper subset of M_2, the set of paths from M_1 to M_1' may go through four possible type-2 sub-meshes and they may bend at most two times before they reach to the leader node of M_2. This will increase the congestion by at most the factor of 4 between every two type-1 sub-meshes M_1 and M_2. Moreover, as we know only sub-meshes up to level $k < \log n + 2$ can contribute to the congestion on edge e and there are at most $(\log n + 2)$ levels of type-2 sub-meshes, $E[C(e)]$ increases by a constant factor only due to the type-2 sub-meshes. As every request selects its path independently of every other request (Lines 26–30 of Algorithm 1), using standard Chernoff bound, we obtain a concentration result on the congestion C. □

Publish Cost. We can prove the following theorem for any *publish* operation.

Theorem 3. *The publish operation has communication cost $\mathcal{O}(n)$.*

Lookup Cost. It can be shown that a lookup request r from w finds either the directory path to the owner v ($\text{dist}(w, v) \leq 2^i$) or a *slink* to the directory path towards v at level at most η, where $\eta = i + 4$. Therefore, we obtain:

Theorem 4. *The stretch of MultiBend is constant for a lookup operation.*

6   Extension to the d-Dimensional Mesh

We outline the alternative decomposition that has $\mathcal{O}(d^2 \log n)$ approximation for both the path stretch and the congestion in d-dimensional mesh networks. The decomposition will have type-1 sub-meshes and other shifted sub-meshes. We set $\lambda = \max\{1, m_\ell/2^{\lceil \log d + 1 \rceil}\}$, where m_ℓ is the side length of the level ℓ type-1 sub-mesh.

The type-1 sub-meshes are shifted by $(j-1)\lambda$ nodes in each dimension to get the type-j sub-meshes. According to this decomposition, there will be at most $2(d+1)$ different types of sub-meshes at any level. The hierarchy $\mathcal{Z}$ is formed similar to 2-dimensional mesh but now there will be $2d+1$ sub-levels. The multi-bend and canonical paths can also be defined similar to Section 3. We summarize the performance bounds below:

Theorem 5. *In d-dimensional mesh networks,* MultiBend *has $\mathcal{O}(d^2 \log n)$ amortized stretch for move operations and $\mathcal{O}(d^2 \log n)$ approximation on congestion w.h.p. Moreover, the publish operation has cost $\mathcal{O}(n)$ and the lookup operation has stretch $\mathcal{O}(d^2)$.*

References

1. Alon, N., Kalai, G., Ricklin, M., Stockmeyer, L.J.: Lower bounds on the competitive ratio for mobile user tracking and distributed job scheduling. Theor. Comput. Sci. 130(1), 175–201 (1994)
2. Ansari, M., Luján, M., Kotselidis, C., Jarvis, K., Kirkham, C., Watson, I.: Steal-on-abort: Improving transactional memory performance through dynamic transaction reordering. In: HiPEAC. pp. 4–18 (2009)
3. Attiya, H., Gramoli, V., Milani, A.: A provably starvation-free distributed directory protocol. In: SSS. pp. 405–419 (2010)
4. Azar, Y., Cohen, E., Fiat, A., Kaplan, H., Räcke, H.: Optimal oblivious routing in polynomial time. J. Comput. Syst. Sci. 69(3), 383–394 (2004)
5. Busch, C., Magdon-Ismail, M., Xi, J.: Optimal oblivious path selection on the mesh. IEEE Trans. Comput. 57(5), 660–671 (2008)
6. Demmer, M.J., Herlihy, M.: The arrow distributed directory protocol. In: DISC. pp. 119–133 (1998)
7. Guerraoui, R., Herlihy, M., Pochon, B.: Toward a theory of transactional contention managers. In: PODC. pp. 258–264 (2005)
8. Hammond, L., Carlstrom, B.D., Wong, V., Chen, M., Kozyrakis, C., Olukotun, K.: Transactional coherence and consistency: Simplifying parallel hardware and software. IEEE Micro 24(6), 92–103 (2004)
9. Herlihy, M., Luchangco, V., Moir, M., Scherer, III, W.N.: Software transactional memory for dynamic-sized data structures. In: PODC. pp. 92–101 (2003)
10. Herlihy, M., Moss, J.E.B.: Transactional memory: architectural support for lock-free data structures. SIGARCH Comput. Archit. News 21(2), 289–300 (1993)
11. Herlihy, M., Sun, Y.: Distributed transactional memory for metric-space networks. Distrib. Comput. 20(3), 195–208 (2007)
12. Kotselidis, C., Ansari, M., Jarvis, K., Luján, M., Kirkham, C., Watson, I.: Distm: A software transactional memory framework for clusters. In: ICPP. pp. 51–58 (2008)
13. Maggs, B., auf der Heide, F.M., Voecking, B., Westermann, M.: Exploiting locality for data management in systems of limited bandwidth. In: FOCS. pp. 284–293 (1997)
14. Sharma, G., Busch, C.: A competitive analysis for balanced transactional memory workloads. Algorithmica 63(1-2), 296–322 (2012)
15. Sharma, G., Busch, C., Srinivasagopalan, S.: Distributed transactional memory for general networks. In: IPDPS, pp. 1045–1056 (2012)
16. Shavit, N., Touitou, D.: Software transactional memory. Distrib. Comput. 10(2), 99–116 (1997)
17. Zhang, B., Ravindran, B.: Brief announcement: Relay: A cache-coherence protocol for distributed transactional memory. In: OPODIS. pp. 48–53 (2009)

CUDA-For-Clusters:
A System for Efficient Execution
of CUDA Kernels on Multi-core Clusters

Raghu Prabhakar[1,*], R. Govindarajan[2], and Matthew J. Thazhuthaveetil[2]

[1] University of California, Los Angeles
raghu@cs.ucla.edu
[2] Supercomputer Education and Research Centre,
Indian Institute of Science, Bangalore, India
{govind,mjt}@serc.iisc.ernet.in

Abstract. Rapid advancements in multi-core processor architectures coupled with low-cost, low-latency, high-bandwidth interconnects have made clusters of multi-core machines a common computing resource. Unfortunately, writing good parallel programs that efficiently utilize all the resources in such a cluster is still a major challenge. Various programming languages have been proposed as a solution to this problem, but are yet to be adopted widely to run performance-critical code mainly due to the relatively immature software framework and the effort involved in re-writing existing code in the new language. In this paper, we motivate and describe our initial study in exploring CUDA as a programming language for a cluster of multi-cores. We develop CUDA-For-Clusters (CFC), a framework that transparently orchestrates execution of CUDA kernels on a cluster of multi-core machines. The well-structured nature of a CUDA kernel, the growing popularity, support and stability of the CUDA software stack collectively make CUDA a good candidate to be considered as a programming language for a cluster. CFC uses a mixture of source-to-source compiler transformations, a work distribution run-time and a light-weight software distributed shared memory to manage parallel executions. Initial results on running several standard CUDA benchmark programs achieve impressive speedups of up to 7.5X on a cluster with 8 nodes, thereby opening up an interesting direction of research for further investigation.

Keywords: CUDA, Multi-Cores, Distributed Programming, Distributed Systems, Clusters, Software Distributed Shared Memory.

1 Introduction

Clusters of multi-core nodes have become a common HPC resource due to their scalability and attractive performance/cost ratio. Such compute clusters typically have a hierarchical design with nodes containing shared-memory multi-core

---

* The author was affiliated with the Indian Institute of Science during this work.

C. Kaklamanis et al. (Eds.): Euro-Par 2012, LNCS 7484, pp. 415–426, 2012.
© Springer-Verlag Berlin Heidelberg 2012

processors interconnected via a network infrastructure. While they provide an enormous amount of computing power, writing parallel programs to efficiently utilize all the cluster resources remains a daunting task. For example, intra-node communication between tasks scheduled on a single node is much faster than inter-node communication, hence it is desirable to structure code in a way so that most of the communication takes place locally. Interconnect networks have large bandwidth and are suitable for heavy, bursty data transfers. This task of manually orchestrating the execution of parallel tasks efficiently and managing multiple levels of parallelism is difficult. A popular programming choice is a hybrid approach [10] using multiple programming models like OpenMP[5] (intra-node) and MPI[20] (inter-node) to explicitly manage locality and parallelism. The challenge lies in writing parallel programs that can readily scale across systems with steadily increasing numbers of both cores per node and nodes in the cluster. Various programming languages and models that have been proposed as a solution to this problem ([11], [12] etc.,) are yet to be adopted widely due to the effort involved in porting applications to the new language as well as the constantly changing software stack supporting the languages.

GPGPU computation has attracted the attention of software developers and researchers off-late, and has been facilitated mainly by NVIDIA's CUDA [3] and OpenCL [4]. In particular, CUDA has become a popular language as evident from an increasing number of users [3] and benchmarks [6] [13]. However, CUDA is a shared memory programming model designed in tandem with the CUDA architecture which consists of homogeneous cores. Therefore intuitively, CUDA does not seem to fit the bill to program distributed machines. However, the semantics of CUDA enforce a structure on parallel kernels where communication between parallel *threads* is guaranteed to take place correctly only if the communicating threads are part of the same *thread block*, through some block-level *shared memory*. From a CUDA *thread*'s perspective, the *global* memory offers a relaxed consistency that guarantees coherence only across kernel invocations, and hence no communication can reliably take place through global memory within a kernel invocation. Such a structure naturally exposes data locality information that can readily benefit from the multiple levels of hardware-managed caches found in conventional CPUs. In fact, previous works such as [21] and [22] have shown the effectiveness using CUDA to program multi-core shared memory CPUs, and similar research has been performed on OpenCL as well [16]. There has also been some recent work on using OpenCL to program heterogeneous CPU/GPU clusters [17]. More recently, a compiler that implements CUDA on multi-core x86 processors has been released commercially by the Portland Group [7]. CUDA has evolved into a very mature software stack with efficient supporting tools like debuggers and profilers, making application development and deployment easy.

Considering the factors of programmability, popularity, scalability, support and expressiveness, we believe that CUDA can be used as a single language to efficiently program a cluster of multi-core machines. From a utility perspective, establishing an execution flow from CUDA to a distributed system would immediately enable many CUDA programs to achieve speedups on commodity cluster machines. In this paper, we explore this idea and describe CFC, a framework to execute CUDA kernels can be efficiently and in a scalable fashion on a cluster of multi-core machines. As the thread-level specification of a CUDA kernel is too fine grained to be profitably executed on a CPU, we employ compiler techniques described in [22] to serialize threads within a block and transform the kernel code into a *block-level* specification. The independence and granularity of thread blocks makes them an attractive schedulable unit on a CPU core. As global memory in CUDA provides only a relaxed consistency, we show it can be realized by a lightweight software distributed shared memory (DSM) that provides an abstraction of a single shared address space across the compute cluster nodes. Finally, we describe our work-partitioning runtime that distributes thread blocks across all cores in the cluster. We evaluate our framework using several standard CUDA benchmark programs from the Parboil benchmark suite [6] and the NVIDIA CUDA SDK [2] on a compute cluster with eight nodes. We achieve promising speedups ranging from 3.7X to 7.5X compared to a baseline multi-threaded execution (around 56X compared to a sequential execution). We claim that CUDA can be successfully and efficiently used to program a compute cluster and thus motivate further exploration in this area.

The rest of this paper is organized as follows: Section 2 provides the necessary background. In Section 3, we describe the CFC framework in detail. In Section 4, we describe our experimental setup and evaluate our framework. Section 5 discusses related work. In section 6 we discuss possible future directions and conclude.

2 Background

2.1 CUDA Programming Model

The CUDA programming model provides a set of extensions to the C programming language enabling programmers to execute functions on a GPU. Such functions are called *kernels*. Each kernel is executed on the GPU as a *grid* of *thread blocks*. The grid size and block size are specified by the programmer during invocation. Data transfer between the main memory and GPU DRAM is performed explicitly using CUDA APIs. Each block is scheduled to execute on one *streaming multiprocessor* (SM) on the GPU. Each SM contains a number of scalar processors (SP), a large register file and some scratch pad memory. Thread-private variables are stored in registers in each SM. Read-only GPU data that has been declared as *constant* is mapped to a different *constant* memory. Programmers can use *shared* memory - which is a low-latency, user-managed scratch pad memory - to store frequently accessed data. Shared memory data is visible to all the threads within the same block. The *syncthreads* construct provides barrier synchronization across threads within the same block.

Each thread block in a kernel grid gets scheduled independently on the SM that it is assigned to. The programmer must be aware that a race condition potentially exists if two or more thread blocks are operating on the same global memory address and at least one of them is performing a write/store operation. This is because there is no control over when the competing blocks will get scheduled. CUDA's *atomic* primitives can be used only to ensure that the accesses are serialized in some arbitrary order, but there is no mechanism to communicate globally across blocks in a single kernel invocation.

2.2   Compiler Transformations

As the per-thread code specification of a CUDA kernel is too fine grained to be scheduled profitably on a CPU, we first transform the kernel into a per-block code specification using transformations described in the MCUDA framework [22]. Logical threads within a thread block are serialized, i.e., the kernel code is executed in a loop with one iteration for each thread in the block. Loop boundaries provide implicit barrier synchronization. Hence, *__syncthreads()* is implemented using a technique called *deep fission*. The single thread loop nest is *split* into two separate loops at the point of invocation of *__syncthreads()*, thereby preserving CUDA's execution semantics. Thread-local variables that are live across such synchronization boundaries are expanded into an array so that each logical thread can maintain its state correctly. Thread-private variables are replicated selectively, avoiding unnecessary duplication while preserving each thread's instance of the variable. The end result of all transformations is a block-level specification of the CUDA kernel that can be compiled and executed on a CPU. [22] has further details on these transformations. A CUDA kernel is composed of several blocks, and is executed by calling the above function several times in a loop. The next section describes how we distribute this execution across nodes using MPI and OpenMP.

3   CUDA for Clusters (CFC)

In this section, we describe CFC in detail. Section 3.1 describes CFC's work partitioning runtime scheme. Section 3.2 describes CFC-SDSM, the Software DSM that used to realize CUDA global memory in a compute cluster.

3.1   Work Distribution

Executing a kernel involves executing the per-block code fragment for all block indices, as specified in the kernel's execution configuration. In this initial work, we employ a simple work distribution scheme that divides the set of block indices into contiguous, disjoint subsets called *block index intervals*. The number of blocks assigned to each node is determined by the number of executing nodes, which is specified as a parameter during execution. If there are more blocks than nodes (as is usually the case), each node gets assigned more than one block. For the example in Fig. 1, the set of block indices $0 - 7$ has been split into four contiguous, disjoint subsets $\{0, 1\}$, $\{2, 3\}$, $\{4, 5\}$ and $\{6, 7\}$, which are scheduled to be executed by nodes N1, N2, N3 and N4 respectively. OpenMP is

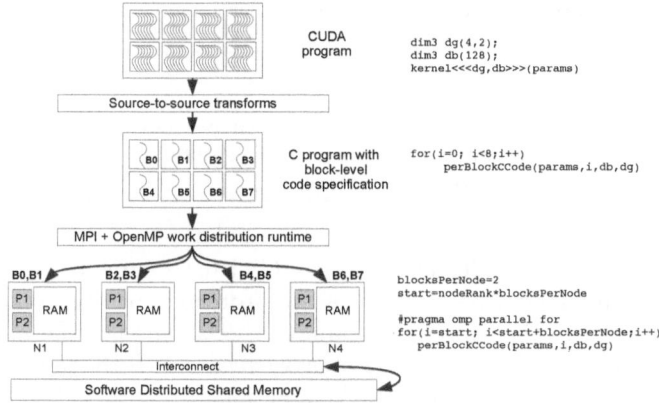

Fig. 1. Structure of the CFC framework. The pseudo-code for kernel invocation at each stage is shown on the right for clarity.

used within each node to execute the assigned work units in parallel on multiple cores. For example, in Fig. 1, within each node the assigned blocks are executed in parallel using multiple threads on cores P1 and P2. The thread blocks are thus distributed uniformly irrespective of the size of the cluster or number of cores in each cluster node.

3.2 CFC-SDSM

CFC supports CUDA kernel execution on a cluster by providing the global CUDA address space through a software abstraction layer, called CFC-SDSM. We begin by noting CUDA kernels with data races produce unpredictable results on a GPU. However, global data is coherent at kernel boundaries; all thread blocks see the same global data when a kernel commences execution. We therefore enforce a relaxed consistency semantics[8] in CFC-SDSM that ensures coherence of global data at kernel boundaries. Thus, for a data-race free CUDA program, CFC-SDSM guarantees correct execution, but provides no such guarantees for racy programs. *Constant* memory is maintained as separate local copies on every node.

As the size of objects allocated in global memory can be large, CFC-SDSM operates at page-level granularity. Table 1 describes the meta information stored by CFC-SDSM for each page of global data in its page table.

CFC-SDSM Operation CFC-SDSM treats all memory allocated using *cudaMalloc* as global data. Each allocation call typically populates several entries in the CFC-SDSM table. Every memory allocation is performed starting at a page boundary using *mmap*. At the beginning of any kernel invocation, CFC-SDSM marks every global memory page to be *read-only*. Thus, any write to a global page within the kernel results in a segmentation fault which is handled by CFC-SDSM's SIGSEGV handler. The segmentation fault handler first examines

Table 1. Structure of a CFC-SDSM page table entry

Field	Description
pageAddr	Starting address of the page.
pnum	A unique number (index) given to each page, used during synchronization.
written	1 if the corresponding page was written, else 0.
twinAddr	Starting address of the page's *twin*.

the address causing the fault. The fault could either be due to (i) a valid write access to a global memory page that is write-protected, or (ii) an illegal address caused by an error in the source program. In the latter case, the handler prints a stack trace onto standard error and aborts execution. If the fault is due to the former, the handler performs the following actions:

- Set the *written* field of the corresponding CFC-SDSM table entry to 1.
- Create a replica of the current page, called its *twin*. Store the *twin's* address in the corresponding CFC-SDSM table entry.
- Grant write access to the corresponding page and return.

In this way, at the end of the kernel's execution, each node is aware of the global pages it has modified. Note that within each node, the global memory pages and CFC-SDSM table are shared by all executing threads, and hence all cores. So, the SIGSEGV handler overhead is incurred only once for each global page in a kernel, irrespective of the number of threads/cores writing to it. Writes by a CPU thread/thread block are made visible to other CPU threads/thread blocks executing in the same node by the underlying hardware cache coherence mechanism, which holds across multiple sockets of a node. Therefore, no special treatment is needed to handle *shared* memory.

The information of global pages that have been modified within a kernel has to be communicated globally to all other nodes at kernel boundaries. To accomplish this, each node constructs a vector called *writeVector* specifying the set of global pages written by the node during the last kernel invocation. The *writeVector*s are communicated to other nodes using an all-to-all broadcast. Every node then computes the summation of all *writeVector*s. We perform this vector collection-summation operation using `MPI_Allreduce`[20]. At the end of this operation, each node knows the number of modifiers of each global page. For instance, $writeVector[p] == 0$ means that the page having $pnum = p$ has not been modified, and hence can be excluded from the synchronization operation.

Pages having *writeVector[pnum]* $== 1$ have just one modifier. For such pages, the modifying node broadcasts the up-to-date page to every other cluster node To reduce broadcast overheads, all the modified global pages at a node are grouped together in a single broadcast from that node. The actual page broadcast operation is implemented using `MPI_Bcast`.

For pages that have more than one modifier, each modifier must communicate its modifications to other cluster nodes. CFC-SDSM accomplishes this by *diff*ing the modified page with its *twin* page created by the SIGSEGV handler in each modifier node. In CFC-SDSM, each modifier node other than node 0 computes the *diff*s and sends them to node 0, which collects all the *diff*s and applies them to the page in question. *Diff*ing is an inexpensive operation that is easily performed using a bitwise *xor* operation. Node 0 then broadcasts the up-to-date page to every other node. The coherence operation ends with each node receiving the modified pages and updating the respective pages locally.

We show in section 4 that centralizing the *diff*ing process at node 0 does not cause much of a performance bottleneck mainly because the number of pages with multiple modifiers is relatively less. For pages with multiple modifiers, CFC-SDSM assumes that the nodes modified disjoint chunks of the page. If multiple nodes have modified overlapping regions in a global page the program has a data race, and under CUDA semantics the results are unpredictable. CFC-SDSM does not guarantee correctness for such programs.

3.3 Lazy Update

Broadcasting every modified page to every other node creates a high volume of network traffic, which is unnecessary most of the times. We therefore implement a *lazy update* optimization in CFC-SDSM where modified pages are sent to nodes *lazily* on demand. CFC-SDSM uses lazy update if the total number of modified pages across all nodes exceeds a certain threshold. We have found that a threshold of 2048 works reasonably well for many benchmarks (see section 4). In lazy update, global data is updated only on node 0 and no broadcast is performed. Instead, in each node n, read permission is set for all pages p that were modified only by n (since the copy of page p is up-to-date in node n), and the write permission is reset as usual. If a page p has been modified by some other node(s), node n's copy of page p is stale. Hence, CFC-SDSM *invalidates* p by removing all access rights to p in n. Pages which have not been modified by any node are left untouched (with read-only access rights). At the same time, on node 0, a *server thread* is forked to receive and service lazy update requests from other nodes. In subsequent kernel executions, if a node tries to read from an invalidated page (i.e. a page modified by some other node in the previous kernel call), a request is sent to the daemon on node 0 with the required page's *pnum*. In section 4, we show that the *lazy update* scheme offers appreciable performance gains for a benchmark with a large number of global pages.

4   Performance Evaluation

In this section, we evaluate CFC using several representative benchmarks from standard benchmark suites.

4.1 Experimental Setup

For this study, we performed all experiments on an eight-node cluster, where each node is running Debian Lenny Linux. Nodes are interconnected by a high-bandwidth Infiniband network. Each node is comprised of two quad-core Intel Xeon processors running at 2.83GHz, thereby having eight cores.

Compiler Framework. Fig. 2 shows the structure the CFC compiler framework. We use optimization level O3 in all our experiments.

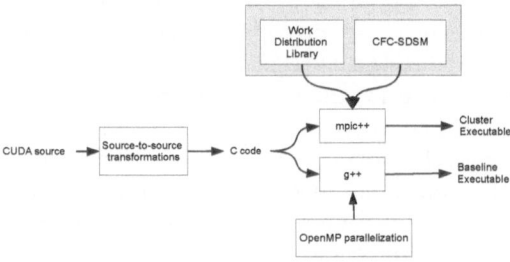

Fig. 2. Structure of the compiler framework

Benchmarks. We used five benchmark applications and one kernel. Four are from the Parboil Benchmark suite [6]. *Blackscholes* and the *Scan* kernel are applications from the NVIDIA CUDA SDK[2]. The benchmarks are from different computing disciplines, and are representative of present day workloads, all of which have mature CUDA implementations in standard benchmark suites. Table 2 briefly describes each benchmark.

Table 2. Benchmarks and description

Benchmark	Description
cp	Coulombic potential computation over one plane in a 3D grid, 100000 atoms
mri-fhd	$F^H d$ computation using in 3D MRI reconstruction, 40 iterations
tpacf	Two point angular correlation function
blackscholes	Call and put prices using Black-Scholes formula, 50000000 options, 20 iterations
scan	Parallel prefix sum, 25600 integers, 1000 iterations
mri-q	Q computation in 3D MRI reconstruction, 40 iterations

Performance Metrics. In all our experiments, we keep the number of threads equal to the number of cores on each node (eight threads per node in our cluster). We haven't explored variable number of threads per node. We define speedup of an *n node* execution as:

$$speedup = \frac{t_{baseline}}{t_{CLUSTER}} \qquad (1)$$

, where $t_{baseline}$ represents the baseline multi-threaded execution time on one node, and $t_{CLUSTER}$ represents execution time in the CFC framework on n nodes. The baseline uses a single node and hence requires only OpenMP (and not MPI). The baseline can only gain because of this, thereby ensuring fairness in comparison. Observe that the speedup is computed for a cluster of n nodes (i.e., $8n$ cores) relative to performance on one node (i.e., 8 cores). In effect, for $n = 8$, the maximum obtainable speedup would be 8. Each benchmark has been run 10 times, and the median value is reported.

4.2   Results

Table 3 shows the number of pages of global memory as well as the number of modified pages. Our benchmark set has a mixture of large and small working sets along with varying percentages of modified global data, thus covering a range of GPGPU behavior suitable for studying an implementation such as ours. Benchmark speedups are shown in Fig. 3. Fig. 3(a) shows speedups with the lazy

Table 3. Number of pages of global memory declared and modified in each benchmark

Benchmark	Global pages	Modified	% Unmodified
Cp	1024	1024	0
Mri-fhd	1298	510	60.7
Tpacf	1220	8	99.3
BlackScholes	244145	97658	60
Mri-q	1286	508	60.49
Scan	50	25	50

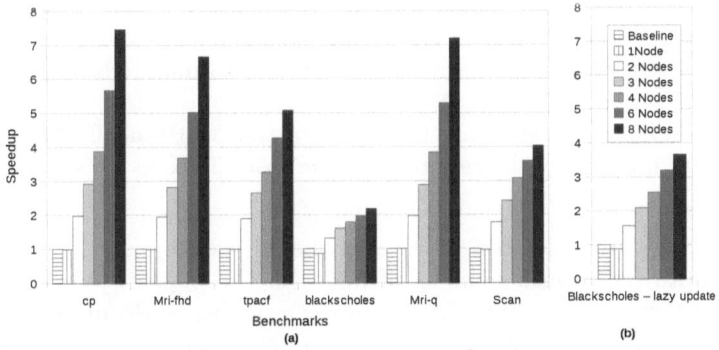

Fig. 3. Comparison of execution times of various benchmark applications on our system. (a) shows normalized speedups on a cluster with 8 nodes without lazy update. (b) shows the performance of *BlackScholes* with the *lazy update* optimization.

update optimization disabled for all the benchmarks, while 3(b) shows speedups for the *BlackScholes* benchmark when the lazy update optimization is enabled. As we have set a threshold of at-least 2048 global pages to trigger CFC-SDSM to operate in lazy mode, only blackscholes triggers this operation. Any number of pages less than this can easily be handled by CFC-SDSM in the normal mode, and the experimental results demonstrate it. Hence we study the lazy update effect only on *Blackscholes*. We make the following observations:

- Our implementation has low runtime overhead. Observe the speedups for $n = 1$, i.e., the second bar. In almost all cases, this value is close to the baseline. *BlackScholes* slows down by about 14% due to its large global data working set.

- The *Cp* benchmark shows very high speedups in spite of having a high percentage of global data pages being modified. *Cp* is a large benchmark with lots of computations that can utilize many nodes efficiently.
- The *Scan* benchmark illustrates the effect of a CUDA kernel design on its performance on a cluster. Originally, the *Scan* kernel is small where only 512 elements are processed per kernel. Spreading such a small kernel's execution over many nodes was an overkill and provided marginal performance gains comparable to *Blackscholes* in Fig. 3(a). However, after the kernel was modified (*coarsened* or *fattened*) to processes 25600 elements per kernel, we achieve the speedups shown in 3(a).
- The *BlackScholes* benchmark shows scalability, but low speedups. Due to the large volume of network traffic it generates, this benchmark benefits from lazy update. On a cluster with 8 nodes, we obtain a speedup of 3.7X with lazy update, compared to 2.17X without lazy update. This suggests that the performance gained by reducing interconnect traffic compensates for the overheads incurred by creating the daemon thread. We have observed that for this application, invalidated pages are never read in any node.
- We can observe network overhead specifically only in *BlackScholes* where we've overloaded CFC-SDSM with many pages. The lazy update scheme seems to work pretty well even for large memory sizes. We would like to explore potential problems when we scale this to hundreds of nodes in the future.
- Across the benchmarks, our runtime approach to extend CUDA programs to clusters has achieved speedups ranging from 3.7X to 7.5X on an 8 node cluster.

In summary, we are able to achieve appreciable speedup and a good scaling efficiency (upto 95%) with number of nodes in the cluster. While an 8-node cluster is not a very big cluster, it serves as a reasonable platform to demonstrate the effectiveness of CFC. Future work will deal with studying larger clusters and problems arising from that.

5   Related Work

We briefly discuss a few previous works related to programming models, using CUDA on non-GPU platforms and software DSMs. The *Partitioned Global Address Space* family of languages (Chapel[11], X10[12] etc.) aims to combine the advantages of both message-passing and shared-memory models. Intel's Concurrent collections [1] is another shared memory programming model that aims to abstract the description of parallel tasks.

Previous works like [7], [14] and [22] use either compiler techniques or binary translation to execute kernels on x86 CPUs. In all the works mentioned here, CUDA kernels have been executed on single shared-memory hardware.

Various kinds of software DSMs have been suggested in literature like [9], [15], [18], and [19], to name a few. CFC-SDSM differs from the above works in the sense that locks need not be acquired and released explicitly by the programmer.

All global memory data is 'locked' just before kernel execution and 'released' immediately after, by definition. Also, synchronization operation proceeds either eagerly or lazily, depending on the total size of global memory allocated. This makes our DSM very lightweight and simple.

6   Conclusions and Future Work

In this paper, we have presented an initial study in exploring CUDA as a language to program clusters of multi-core machines. We have implemented CFC, a framework that uses a mixture of compiler transformations, work distribution runtime and a lightweight software DSM to collectively implement CUDA's semantics on a multi-core cluster. We have evaluated our implementation by running six standard CUDA benchmark applications to show that there are indeed promising gains that can be achieved.

Many interesting directions can be pursued in the future. One direction could be towards optimizing network usage by building a static communication cost estimation model or tracking global memory access patterns that can be used by the runtime to schedule blocks across nodes appropriately. Another interesting and useful extension to this work would be to consider GPUs on multiple nodes as well, along with multi-cores. Automatic compile-time kernel coarsening and automatic kernel execution configuration tuning are other interesting areas.

References

1. Intel concurrent collections for c++, http://software.intel.com/en-us/articles/intel-concurrent-collections-for-cc/
2. Nvidia cuda c sdk, http://developer.download.nvidia.com/compute/cuda/sdk
3. Nvidia cuda zone, http://www.nvidia.com/cuda
4. Opencl overview, http://www.khronos.org/developers/library/overview/opencl_overview.pdf
5. Openmp specifications, version 3.0, http://openmp.org/wp/openmp-specifications/
6. The parboil benchmark suite, http://impact.crhc.illinois.edu/parboil.php
7. The portland group, http://www.pgroup.com
8. Adve, S.V., Gharachorloo, K.: Shared memory consistency models: A tutorial. IEEE Computer 29, 66–76 (1995)
9. Amza, C., Cox, A.L., Dwarkadas, S., Keleher, P., Lu, H., Rajamony, R., Yu, W., Zwaenepoel, W.: Treadmarks: Shared memory computing on networks of workstations. Computer 29(2), 18–28 (1996)
10. Cappello, F., Etiemble, D.: Mpi versus mpi+openmp on ibm sp for the nas benchmarks. In: Proceedings of the 2000 ACM/IEEE Conference on Supercomputing (CDROM), Supercomputing 2000. IEEE Computer Society, Washington, DC (2000)
11. Chamberlain, B., Callahan, D., Zima, H.: Parallel programmability and the chapel language. Int. J. High Perform. Comput. Appl. 21(3), 291–312 (2007)

12. Charles, P., Grothoff, C., Saraswat, V., Donawa, C., Kielstra, A., Ebcioglu, K., von Praun, C., Sarkar, V.: X10: an object-oriented approach to non-uniform cluster computing. In: OOPSLA 2005: Proceedings of the 20th Annual ACM SIGPLAN Conference on Object-oriented Programming, Systems, Languages, and Applications, pp. 519–538. ACM, New York (2005)
13. Danalis, A., Marin, G., McCurdy, C., Meredith, J.S., Roth, P.C., Spafford, K., Tipparaju, V., Vetter, J.S.: The scalable heterogeneous computing (shoc) benchmark suite. In: Proceedings of the 3rd Workshop on General-Purpose Computation on Graphics Processing Units, GPGPU 2010, pp. 63–74. ACM, New York (2010)
14. Diamos, G.F., Kerr, A.R., Yalamanchili, S., Clark, N.: Ocelot: a dynamic optimization framework for bulk-synchronous applications in heterogeneous systems. In: PACT 2010: Proceedings of the 19th International Conference on Parallel Architectures and Compilation Techniques, pp. 353–364. ACM, New York (2010)
15. Gelado, I., Stone, J.E., Cabezas, J., Patel, S., Navarro, N., Hwu, W.M.W.: An asymmetric distributed shared memory model for heterogeneous parallel systems. SIGARCH Comput. Archit. News 38(1), 347–358 (2010)
16. Gummaraju, J., Morichetti, L., Houston, M., Sander, B., Gaster, B.R., Zheng, B.: Twin peaks: a software platform for heterogeneous computing on general-purpose and graphics processors. In: Proceedings of the 19th International Conference on Parallel Architectures and Compilation Techniques, PACT 2010, pp. 205–216. ACM, New York (2010)
17. Kim, J., Seo, S., Lee, J., Nah, J., Jo, G., Lee, J.: Opencl as a programming model for gpu clusters. In: LCPC 2011: Proceedings of the 24th International Workshop on Languages and Compilers for Parallel Computing, (2011)
18. Li, K., Hudak, P.: Memory coherence in shared virtual memory systems. ACM Trans. Comput. Syst. 7(4), 321–359 (1989)
19. Manoj, N.P., Manjunath, K.V., Govindarajan, R.: Cas-dsm: a compiler assisted software distributed shared memory. Int. J. Parallel Program. 32(2), 77–122 (2004)
20. Snir, M., Otto, S., Huss-Lederman, S., Walker, D., Dongarra, J.: MPI-The Complete Reference, Volume 1: The MPI Core. MIT Press, Cambridge (1998)
21. Stratton, J.A., Grover, V., Marathe, J., Aarts, B., Murphy, M., Hu, Z., Hwu, W.M.W.: Efficient compilation of fine-grained spmd-threaded programs for multicore cpus. In: CGO 2010: Proceedings of the 8th Annual IEEE/ACM International Symposium on Code Generation and Optimization, pp. 111–119. ACM, New York (2010)
22. Stratton, J.A., Stone, S.S., Hwu, W.-M.W.: Mcuda: An efficient implementation of cuda kernels for multi-core cpus, pp. 16–30 (2008)

From a Store-Collect Object and Ω
to Efficient Asynchronous Consensus

Michel Raynal[1,2] and Julien Stainer[2]

[1] Institut Universitaire de France,
[2] IRISA, Université de Rennes 1, France
{raynal,Julien.Stainer}@irisa.fr

Abstract. This paper presents an efficient algorithm that build a consensus object. This algorithm is based on an Ω failure detector (to obtain consensus liveness) and a store-collect object (to maintain its safety). A store-collect object provides the processes with two operations, a store operation which allows the invoking process to deposit a new value while discarding the previous value it has deposited and a collect operation that returns to the invoking process a set of pairs (i, val) where val is the last value deposited by the process p_i. A store-collect object has no sequential specification.

While store-collect objects have been used as base objects to design wait-free constructions of more sophisticated objects (such as snapshot or renaming objects), as far as we know, they have not been explicitly used to built consensus objects. The proposed store-collect-based algorithm, which is round-based, has several noteworthy features. First it uses a single store-collect object (and not an object per round). Second, during a round, a process invokes at most once the store operation and the value val it deposits is a simple pair $\langle r, v \rangle$ where r is a round number and v a proposed value. Third, a process is directed to skip rounds according to its view of the current global state (thereby saving useless computation rounds). Finally, the algorithm benefits from the adaptive wait-free implementations that have been proposed for store-collect objects, namely, the number of shared memory accesses involved in a collect operation is $O(k)$ where k is the number of processes that have invoked the store operation. This makes the proposed algorithm particularly efficient and interesting for multiprocess programs made up of asynchronous crash-prone processes that run on top of multicore architectures.

Keywords: Asynchronous shared memory system, Building block, Concurrent object, Consensus, Distributed algorithm, Eventual leader, Failure detector, Fault-tolerance, Modularity, Multicore systems, Process crash, Store-collect object.

1 Introduction

1.1 On the Implementation of Consensus Objects

Consensus object and its universality An implementation of any object (or service) is *wait-free* if the crash of any number of processes does not prevent the other processes from terminating their operation invocations on the constructed object [13]. It has been shown by M. Herlihy [13] that *consensus objects* are universal when one has to design

C. Kaklamanis et al. (Eds.): Euro-Par 2012, LNCS 7484, pp. 427–438, 2012.

wait-free implementation of any object (or service) defined from a sequential specification. This means that, as soon as we are provided with consensus objects and atomic read/write registers, it is possible to design algorithms (called *universal constructions*) that build wait-free implementations of any concurrent object defined by a sequential specification. Such implementations are said to be *linearizable* [14].

A *consensus* object is a one-shot object that provides the processes with a single operation denoted propose() (*one-shot* means that a process invokes at most once the operation propose() on a consensus object). When a process invokes propose(v), we say that it "proposes v". A consensus object allows processes to *agree* (if the processes of a multiprocess program do not have to agree in one way or another, they are independent and do not constitute a distributed computation). More specifically a consensus object is defined as follows. Each process is assumed to propose a value and has to decide a value in such a way that the following properties are satisfied: each correct process which invokes propose() decides a value (wait-free termination), a decided value is a proposed value (validity) and no two processes decide different values (agreement).

Consensus impossibility and ways to circumvent it. While consensus objects are fundamental objects for the design and the implementation of crash-prone distributed systems, the bad news is that they cannot be wait-free implemented in asynchronous systems. *Wait-free* means here "whatever the number of process crashes". Moreover, this impossibility is independent of the underlying communication medium, which means that it holds for both read/write shared memory systems [18] and asynchronous message-passing systems [8]).

Several approaches to circumvent this impossibility have been investigated in the context of read/write shared memory systems. One consists in enriching the system model by providing the processes with registers stronger (from a computability point of view) than read/write atomic registers. This approach has given rise to the notion of *consensus number* introduced and developed by Herlihy [13]. An object X has consensus number n if n is the largest integer such that it is possible to wait-free implement n-process consensus objects from atomic read/write registers and objects X. If X allows to wait-free implement n-process consensus for any value of $n > 0$, the consensus number of the object X is $+\infty$. It is shown it in [13] that there are objects (such as compare&swap or LL/SC registers) whose consensus number is $+\infty$.

Another approach consists in enriching the base read/write system with a failure detector [4]. Intuitively, a failure detector can be seen as a distributed module that provides each process with information on failures. According to the type and the quality of this information, several failure detectors can be defined. Failure detectors have initially been proposed for message-passing systems before being used in shared memory systems [17]. One of the most important result associated with the failure detector-based approach is the proof that the failure detector denoted Ω is the one that captures the minimal information on failures that allows processes to wait-free implement a consensus object despite asynchrony and process crashes [3]. A failure detector Ω is characterized by the following behavioral property: after a finite but unknown and arbitrary long period, Ω provides forever the processes with the same (non-crashed) leader.

Modular approach: on the liveness side. Implementations of consensus objects have to ensure that a single among the proposed values is decided (safety) and that each process that proposes a value and does not crash eventually decides despite the behavior of the other processes (wait-freedom).

Interestingly, when considering round-based algorithms (i.e., algorithms in which the processes execute asynchronously a sequence of rounds), the safety and wait-freedom properties of a consensus object can be ensured by different means, i.e., by different object types. More precisely, the eventual leadership property provided by Ω can be used to ensure that at least one process will terminate (thereby entailing the termination of the other processes). Hence, an Ω failure detector constitutes a liveness building block on which implementations of consensus objects can rely in order to obtain the wait-freedom property.

Modular approach: on the safety side. To our knowledge three types of read/write-based objects that ensure the safety properties of a consensus object have been proposed.

The first (which is given the name $alpha$ in [12,21]) has been proposed by Lamport in [16] in the context of message-passing systems and adapted to the read/write shared memory model by Lamport & Gafni [10]. An alpha object is a round-based abstraction which has a single operation denoted deposit(). *Round-based* means that the specification of deposit() involves the round number which is passed as a parameter. This operation takes also a value as input parameter and returns a proposed value or a default value $\perp$ indicating that the current invocation is aborted. An alpha object is implemented with an array of n shared single-writer/multi-reader registers where n is the total number of processes. Each register contains two round numbers plus a proposed value or the default value $\perp$.

The second object, denoted *adopt-commit* has been introduced by Gafni [9]. It is a round-free object: its specification does not depend on round numbers, and its implementation requires two arrays of size n asynchronously accessed by each process one after the other. As adopt-commit objects are round-free, each round of an adopt-commit-based consensus algorithm requires its own adopt-commit object and, when it executes a round, a process accesses only the corresponding adopt-commit object.

A third object that has used to ensure the safety property of a consensus object is the *weak set* object type proposed by Delporte & Fauconnier [5]. This object is a set from which values are never withdrawn. Similarly to adopt-commit objects, these sets are round-free objects but, differently from them, during each round a process is required to access three distinct sets (the ones associated to the previous, the current and the next rounds).

1.2   Content of the Paper

Step complexity and number of objects. The step complexity (number of shared memory accesses) involved by each invocation of an operation on an alpha, adopt-commit or weak set object is $O(n)$.

On another side, the consensus algorithms based on adopt-commit or weak set objects requires one such object per round and, due the very nature of Ω, the number of rounds that have to be executed before processes decide is finite but can can be arbitrary large. This means that the number of adopt-commit or weak set objects used in an execution cannot be bounded and, consequently, these objects have to be dynamically created. (Due to the distributed nature of the computation and the possibility of failures, such dynamic object creations are much more difficult to manage than iterative or recursive objects creation in sequential or parallel failure-free computing.) Interestingly, alpha-based consensus algorithms (e.g., [10,11,16]) needs a single alpha object.

A question. Hence, the question: Is it possible to design a consensus object from Ω (for the consensus liveness part) and (for the consensus safety part) an object such that (a) a single instance of this object is necessary (as in alpha-based consensus) and (b) whose step complexity of each operation is *adaptive* (i.e. depends on the number of processes that have invoked operations and not on the total number of processes)? The paper answers positively the previous question. To that end it considers *store-collect* objects.

Store-collect object. Such an object, which can be seen as an array with an entry per process, provides processes with two operations denoted store() and collect(). The first operation allows a process to deposit a new value in the store-collect object, this new value overwriting the value it has previously deposited. The second operation is an asynchronous read of the last values deposited by each process. A *store-collect* object has no sequential specification.

While a *store-collect* object has a trivial wait-free implementation based on an array of size n with operations whose step complexity is $O(n)$, in a very interesting way, efficient adaptive wait-free implementations have been proposed. As an example, when considering the implementation described in [2], the step complexity of each invocation of collect() is $O(k)$ where $k, 1 \leq k \leq n$, is the number of processes that have previously invoked the operation store()) and the step complexity of each invocation of store() by a process is $O(1)$ but for its first invocation which can be up to $O(k)$.

A variant of a store-collect object is one in which the operations store() and collect() are merged to obtain a single operation denoted store_collect() (whose effect is similar to store() followed by collect()). A wait-free implementation of such a variant is described in [6] where it is shown that in some concurrency patterns the step complexity of store_collect() is $O(1)$.

Content of the paper. The paper presents an algorithm that wait-free implements a consensus object from Ω (building block for wait-free termination) and a single store-collect object (building block for consensus safety).

When compared to consensus algorithms that needs an unbounded number of adopt-commit or weak set objects, the proposed algorithm (similarly to alpha-based consensus algorithms [10,12,16]) needs a single base (store-collect) object. Moreover, when compared to alpha-based algorithms, it has several noteworthy features. (a) A better step complexity (measured as the number of accesses to the shared memory) during each round. (b) The fact that an entry of the store-collect object has only two components (a round number plus a proposed value) while an entry of an alpha object has three components (two round numbers plus a proposed value). And (c) the fact that the next round executed by a process is dynamically computed from its current view of the global state and not a priori defined from a predetermined sequence (thereby allowing a process to skip useless computation rounds). It is important to notice that, while the algorithm is relatively simple, the proof that a single value is decided is not trivial.

Hence, the paper presents a new consensus algorithm suited to shared memory systems which, from an efficiency point of view, compares favorably with existing algorithms. It is important to notice that, with the advent of multicore architectures, the design of such efficient fault-tolerant algorithms become a real challenge.

Roadmap. The paper is made up of 4 sections. Section 2 presents the computation model (base read/write registers, store-collect object and Ω), and the consensus object. Then, Section 3 describes, discusses and proves correct an efficient algorithm that builds a consensus object from Ω and a single store-collect object as underlying building blocks. Finally, Section 4 concludes the paper.

2 Computation Model

2.1 Crash-Prone Asynchronous Processes

The system is made up of a set Π of n sequential processes denoted $p_1, \ldots, p_n$. The integer i is the index of the process p_i. The processes are asynchronous which means that each process proceeds at its own speed which can vary arbitrarily. The execution of a sequential process is represented by a sequence of steps which are accesses to its local memory or to the shared memory (see below).

Any number of processes may crash. A crash is a premature halt. After it has crashed (if ever it does) a process executes no more step. It is only assumed that a process that does not crash eventually executes its next step as defined by the code of its algorithm. Given an execution, a process that crashes is said to be *faulty*, otherwise it is *correct*.

2.2 Cooperation Model

¿From a notational point of view, the names of the objects shared by the processes are denoted with upper case letters (e.g., DEC) while the name of a local variable of a process p_i is denoted with lower case letters with i as a subscript (e.g., set_i).

Cooperation objects: an atomic register and a store-collect object. The processes cooperate through an atomic multi-writer/multi-reader register denoted DEC (initialized to the default value $\perp$) and a single store-collect object denoted MEM. Such an object contains a set of pairs (i, v) where i is a process index and v a value. For any i, this set contains at most one pair $(i, -)$. Initially, a store-collect object is empty.

The operation store() *and* collect(). As indicated in the introduction, such an object has two operations denoted store() and collect(). A process p_i invokes MEM.store(val) to deposit the value val, i.e., the pair (i, val) is added to the store-collect and overwrites the previous pair $(i, -)$ (if any)[1]. Hence, when (i, val) belongs to the store-collect object, val is the last value stored by the process p_i.

---

[1] In the algorithm proposed in Section 3 a value val is a pair made up of a round number r and a proposed value v. To prevent confusion, the notation $(-, -)$ is used for a pair written into a store-collect object, while the notation $\langle -, - \rangle$ is used for a pair val.

A process invokes MEM.collect() to obtain a value of the store-collect object. The set that is returned is called a *view* and contains the latest pairs deposited by the processes that have invoked MEM.store().

Partial order on the views. To define precisely the notion of "latest" pairs returned in a view, we use the following partial order relation on views. Let $view1$ and $view2$ be two views. The notation $view1 \leq view2$ means that, for for every i such that $(i, v1) \in view1$, we have $(i, v2) \in view2$, where the invocation MEM.store($v2$) by p_i is issued after (or is) its invocation MEM.store($v1$).

Properties of the operations store() *and* collect() The invocations of these operations satisfy the following properties.

- Validity. Let col be an invocation of collect() that returns the set $view$. For any $(i, v) \in view$, there is an invocation store(v) issued by the process p_i that has started before the invocation col terminates.
  This property means that a collect() operation can neither read from the future, nor output values that have never been deposited.
- Partial order consistency. Let col1 and col2 be two invocations of the operation collect() that return the views $view1$ and $view2$, respectively. If col1 terminates before col2 starts, then $view1 \leq view2$.
  This property expresses the mutual consistency of non-concurrent invocations of the operation collect(): an invocation of collect() cannot obtain values older than the values obtained by a previous invocation of collect(). On the contrary, there is no constraint on the views returned by concurrent invocations of collect() (hence the name *partial order* for that consistency property).
- Freshness. Let st and col be invocations of store(v) and collect() issued by p_i and p_j, respectively, such that st has terminated before col starts. The view returned by p_j contains a pair (i, v') such that v' is v or a value deposited by p_i after v.
  This property expresses the fact that the views returned by the invocations of collect() are *up to date* in the sense that, as soon as a value has been deposited, it cannot be ignored by future invocations of collect(). If store(v) is executed by a process p_i, the pair (i, v) must appear in a returned view (provided there are enough invocations of collect()) unless v has been overwritten by a more recent invocation of store() issued by p_i.
- Wait-free termination. Any invocation of an operation by a process that does not crash terminates.

It is easy to see from these properties that a store-collect object has no sequential specification (two invocations of collect() which obtain incomparable views cannot be ordered).

Wait-free implementations of store-collect objects. Such implementations are described in several papers (see Chapter 7 of [21] for a survey). The implementations described in [1,2] are based on atomic read/write registers. As noticed in the introduction, they are adaptive to the number k of processes that have invoked the operation store(). Let the *step complexity* of an operation be the maximum number of shared memory accesses it can issue. When considering the implementation presented in [2], the step complexity of an invocation of collect() or of the first invocation of store() by a process is $O(k)$ and the step complexity of the other invocations of store() by the same process is $O(1)$.

Fast store-collect object Such an object, introduced in [6], is a store-collect object where the store() and the collect() operations are merged into a single operation denoted store_collect(). This object is particularly interesting when a process invokes repeatedly store() followed by collect() without executing other steps in between, which is exactly what the store-collect-based consensus algorithm presented in Section 3 does.

An implementation of such a store-collect object is presented in [6], where the step complexity of an invocation of store_collect() converges to $O(1)$ when, after some time, a single process invokes that operation[2].

2.3 The Failure Detector Ω

This failure detector, which has been informally defined in the Introduction, has been proposed and investigated in [3]. It provides each process p_i with a read-only variable denoted leader$_i$ that always contains a process index. The set of these variables satisfies the following property.

- Eventual leadership. There is a finite time τ after which the local variables leader$_i$ of all the correct processes contain the same process index and this index is the index of a correct process.

It is important to notice that, before time τ, there is an anarchy period during which the variables leader$_i$ can have arbitrary values (e.g, there no common leader and crashed processes can be leaders). Moreover, τ can be arbitrarily large and is never explicitly known by the processes.

As already indicated, Ω is the weakest failure detector that allows a consensus object to be wait-free implemented [3]. Moreover, as consensus cannot be solved in a pure asynchronous read/write system prone to process crashes, it follows that such a system has to be enriched with time-related behavioral assumptions in order Ω can be built. Examples of such assumptions and associated Ω algorithms are described in [7].

Notation The previous read/write system model enriched with the additional computability power provided by Ω is denoted $\mathcal{ASM}[\Omega]$.

3 The Store-Collect-Based Consensus Algorithm

This section presents and proves correct an algorithm that implements the operation propose() of a consensus object $CONS$. As previously announced, this construction is based on a store-collect object to ensure the consensus safety properties and a failure detector Ω to guarantee its wait-free termination property.

---

[2] As we will see, this is exactly what does occur in the proposed algorithm after Ω elects forever the same correct process.

3.1   Description of the Algorithm

Internal representation of the consensus object The two base objects used in the algorithm have been introduced in Section 2.2. The aim of the atomic register DEC is to contain the decided value. The aim of the store-collect object MEM is to guarantee that no two different values are decided.

The algorithm implementing the operation propose(). Algorithm 1 is a round-based asynchronous algorithm. A process p_i invokes $CONS$.propose(v_i) where v_i is the value it proposes. Its invocation terminates when it executes the statement return(DEC) where DEC contains the value it decides (line 17).

The local variable r_i contains the current round number of p_i while est_i contains its current estimate of the decision value (these local variables are initialized at line 1). A process executes a while loop (lines 2-16) until it decides (or crashes). Moreover, it executes the loop body (lines 4-14) only if it is currently considered as a leader by Ω (predicate of line 3).

When it is considered as a leader, p_i does the following. First it stores its current local state $\langle r_i, est_i \rangle$ into the store-collect object MEM and then reads its current content by invoking MEM.collect() (line 4). (Let us observe that line 4 can be replaced by the single statement $mem_i \leftarrow MEM$.store_collect($\langle r_i, est_i \rangle$) if one wants to use a fast store-collect object instead of a more general store-collect object.) Let us notice that line 4 is the only line where p_i accesses the store-collect object, i.e., the part of the shared memory related to the consensus safety property. All the other statements executed by p_i in a round (but the write into DEC if it decides) are local statements.

---

operation $CONS$.propose(v_i) **is**
(1) $r_i \leftarrow 1$; $est_i \leftarrow v_i$;
(2)   **while** $(DEC = \bot)$ **do**
(3)   **if** $(\texttt{leader}_i = i)$ **then**
(4) MEM.store($\langle r_i, est_i \rangle$); $view_i \leftarrow MEM$.collect();
(5) $mem_i \leftarrow \{ \langle r, v \rangle \mid (-, \langle r, v \rangle) \in view_i \}$;
(6) $rmax_i \leftarrow \max\{r \mid \langle r, - \rangle \in mem_i\}$;
(7)   **if** $(r_i = rmax_i)$
(8)   **then** $set_i \leftarrow \{v \mid \langle r, v \rangle \in mem_i$ where $r \in \{rmax_i, rmax_i - 1\}\}$;
(9)   **if** $(r_i > 1) \wedge (set_i = \{est_i\})$
(10)   **then** $DEC \leftarrow est_i$
(11)   **else** $r_i \quad \leftarrow r_i + 1$
(12)   **end if**
(13)   **else** $est_i \leftarrow v$ such that $\langle rmax_i, v \rangle \in mem_i$; $r_i \leftarrow rmax_i$
(14)   **end if**
(15)   **end if**
(16)   **end while**;
(17)   return(DEC)
end operation.

Algorithm 1: The store/collect-based consensus operation propose()

Then, p_i stores into mem_i the pairs $\langle r, v \rangle$ contained in the view $view_i$ it has obtained (line 5) and computes the greatest round $rmax_i$ that, from its point of view, has ever been attained (line 6). Its behavior depends then on the fact that it is or not late with respect to $rmax_i$.

- If it is late ($r_i < rmax_i$), p_i jumps to the round $rmax_i$ and adopts as new estimate a value that is associated with $rmax_i$ in the view it has previously obtained (line 13).
- If it is "on time" from a round number point of view ($r_i = rmax_i$), p_i checks if it can write a value into DEC and decide. To that end, it executes lines 8-12. It first computes the set set_i of the values that are registered in the store-collect object with a round number equal to $rmax_i$ or $rmax_i - 1$, i.e., the values registered by the processes that (from p_i's point of view) have attained one of the last two rounds. If p_i has passed the first round ($r_i > 1$) and its set set_i contains only the value kept in est_i, it writes it into DEC (line 10) just before deciding at line 17. If it cannot decide, p_i proceeds to the next round without modifying its estimate est_i (line 11).

Hence, the base principle on which rests this algorithm is pretty simple to state. (It is worth noticing that this principle is encountered in other algorithms that solve other problems such as termination detection of distributed computations). This principle can be stated as follows: processes execute asynchronous rounds (observation periods) until a process sees two consecutive rounds in which "nothing which is relevant has changed".

3.2   Discussion

A particular case   It is easy to see that, when all processes propose the same value, no process decides in more than two rounds whatever the pattern failure and the behavior of Ω. Similarly, only two rounds are needed when Ω elects a correct common leader from the very beginning. In that sense, the algorithm is optimal from a "round number" point of view [15].

On the management of round numbers   In adopt-commit-based or alpha-based consensus algorithms, the processes that execute rounds do execute a predetermined sequence of rounds[3]. Differently, the proposed algorithm allows a process p_i that executes rounds to jump from its current round r_i to the round $rmax_i$ which can be arbitrarily large (line 13). These jumps make the algorithm more efficient. More specifically, let us consider a time τ of an execution such that (a) up to time τ, when a process executes line 9, the decision predicate is never satisfied, (b) processes have executed rounds and mr is the last round that has been attained at time τ, (c) from time τ, Ω elects the same correct leader p_ℓ at any process p_i, and (d) p_ℓ starts participating at time τ. It follows from the algorithm that p_ℓ executes the first round during which it updates r_ℓ to mr, and then (according to the values in the store-collect MEM) at most either the rounds mr and $mr + 1$ or the rounds mr, $mr + 1$ and $mr + 2$. As the sequence of rounds is not predetermined, p_ℓ saves at least $mr - 2$ rounds.

---

[3] In an adopt-commit-based algorithm each process that executes rounds does execute the predetermined sequence of rounds numbered 1, 2, etc., while, in an alpha-based algorithm each process p_i that executes rounds does execute the predetermined sequence of rounds numbered $i, i + n, i + 2n$, etc.

3.3 Proof of the Algorithm

This proof is based only on the properties of Ω and the store-collect object MEM. It does not require MEM to be built from atomic registers (they can be regular registers only). Due to page limitation, the missing proofs can be found in [22].

Lemma 1. *If a process invokes first MEM.store($\langle r, -\rangle$) and later MEM.store($\langle r', -\rangle$), we have $r' > r$.*

Lemma 2. *Let $r > 1$. If a process p_i invokes MEM.store($\langle r, v\rangle$) at time τ, then there is a process p_j that has invoked MEM.store($\langle r - 1, v\rangle$) at a time $\tau' < \tau$.*

Lemma 3. *A decided value is a proposed value.*

Lemma 4. *No two processes decide different values.*

Proof. As a decided value is a value that has been written into DEC and a process writes at most once into DEC, the proof consists in showing that distinct processes do not write different values into DEC.

Preliminary definitions. Let $view_i^r$ be the value of $view_i$ obtained by p_i during round r. Let $\tau(i, r, b, st)$ and $\tau(i, r, e, st)$ be the time instants at which process p_i starts and terminates, respectively, the invocation of the operation store() during round r. $\tau(i, r, b, c\ell)$ and $\tau(i, r, e, c\ell)$ have the same meaning when considering the invocation of the operation collect().

Let r be the first round during which processes write into DEC, p_i one of these processes and v the value it writes. Let us observe that, due to line 9, $r > 1$; hence $r - 1$ exists. We claim that, for any w such that $(-, \langle r, w\rangle)$ is returned by an invocation of collect() we have $w = v$ (Claim C1). It follows (a) from this claim that no process can decide a value different from v at round r and (b) from this claim, Lemma 1 and Lemma 2 that no process ever writes $\langle r', w\rangle$ with $r' > r$ and $w \neq v$. Consequently, no value different from v can be decided which proves the consensus agreement property.

Proof of the claim C1. Let w be any value such that $(-, \langle r, w\rangle)$ is returned by an invocation of collect(). To prove the claim (i.e., $w = v$), let us consider the following definition given for each value w.

1. Let $\tau(k_w, r_w, e, c\ell)$ be the first time instant at which a process (let p_{k_w} denote this process) returns from an invocation of collect() (let r_w denote the corresponding round) and the view it obtains is such that $(-, \langle r, w\rangle) \in view_{k_w}^{r_w}$.
2. Let j_w be a process index such that $(j_w, \langle r, w\rangle) \in view_{k_w}^{r_w}$ (hence p_{j_w} invokes store($\langle r, w\rangle$)).

We claim (Claim C2, proved in [22]) that (a) p_{j_w} executes round $r - 1$, and during that round both (b) invokes store($\langle r - 1, w\rangle$) and (c) executes line 11 (i.e., $r_{j_w} \leftarrow r_{j_w} + 1$).

To prove the claim C1, let us consider any process p_i that writes into DEC at round r (the first round during which processes write into DEC). This process obtained $view_i^r$ when it invoked collect() at round r. Considering any value w and its associated process p_{j_w} as previously defined, we analyze the different cases which can occur according to value r' such that $(j_w, \langle r', v'\rangle) \in view_i^r$ or the fact that no pair $(j_w, -)$ belongs to $view_i^r$.

- $(j_w, \langle r', -\rangle)$ is such that $r' > r$. This case is not possible because otherwise we would have $rmax_i \geq r' > r$ when p_i executes round r and it would consequently execute line 13 and not line 10 (the line at which it writes into DEC).
- $(j_w, \langle r', v'\rangle)$ is such that $r' = r$. In that case, it follows from line 8 and the predicate evaluated by p_i at line 9 that we necessarily have $v' = v$. Moreover, as p_{j_w} writes at most once in a round (Lemma 1), it follows from the definition of j_w (see Item 2 above) that $v' = w$. Hence, $w = v$.
- $(j_w, \langle r', v'\rangle)$ is such that $r' = r - 1$. In that case, it follows from Item (b) of Claim C2 that p_{j_w} has invoked store($\langle r - 1, w\rangle$). Then the proof is the same as in the previous case, and we have $w = v$.
- $(j_w, \langle r', v'\rangle)$ is such that $r' < r - 1$ or there is no pair $(j_w, -)$ in $view_i^r$.
  It then follows from Item (a) of Claim C2 that p_{j_w} executes the round $r - 1$ and we have then $\tau(i, r, b, c\ell) < \tau(j_w, r-1, e, st)$ (otherwise the freshness property of the store-collect object would be violated). According to the sequential code executed by p_{j_w} and p_i we consequently have

$$\tau(i, r, e, st) < \tau(i, r, b, c\ell) < \tau(j_w, r-1, e, st) < \tau(j_w, r-1, b, c\ell).$$

It then follows from the previous line, the freshness property of the store-collect object and the fact that p_i does not write the store-collect object after it has written into DEC, that $(i, \langle r, v\rangle) \in view_{j_w}^{r-1}$. Consequently, p_{j_w} reads $\langle r, -\rangle$ during round $r - 1$, it executes line 13 which contradicts Item (c) of Claim C2 (which states that p_{j_w} executes line 11 during round $r - 1$). Hence, this case cannot appear, which concludes the proof of Claim C1.

$\square_{Lemma\ 4}$

Lemma 5. *Let assume that the eventual leader elected by Ω participates. Any correct process decides a value.*

The next theorem follows from Lemma 3, Lemma 4 and Lemma 5.

Theorem 1. *Let assume that the eventual leader elected by Ω participates. Algorithm 1 is a wait-free implementation of a consensus object in the system model $\mathcal{ASM}[\Omega]$.*

4   Conclusion

This paper was motivated by the use of store-collect objects to build a consensus object. It has presented such an algorithm based on a single store-collect object in which a value stored by a process is a simple pair made up of a round number and a proposed value. Due to the fact that it uses a single store-collect object, the algorithm is practically interesting. Moreover, as it can benefit from the adaptive wait-free implementations that have been proposed for store-collect objects and it directs processes to skip rounds (thereby saving "useless" computation), this consensus algorithm is also particularly efficient and relevant for practical implementations. These features, together with its simplicity, make it attractive for multiprocess programs made up of asynchronous crash-prone processes that run on top of multicore architectures.

Acknowledgments. This work has been supported by the French ANR project DIS-PLEXITY devoted to computability and complexity in distributed computing.

References

1. Afek Y., Stupp G., Touitou D., Long-lived adaptive collect with applications. *Proc. 40th IEEE Symposium on Foundations of Computer Science Computing (FOCS'99)*, IEEE Computer Press, pp. 262-272, 1999.
2. Attiya H., Fouren A. and Gafni E., An adaptive collect algorithm with applications. *Distributed Computing*, 15(2):87-96, 2002.
3. Chandra T., Hadzilacos V. and Toueg S., The weakest failure detector for solving consensus. *Journal of the ACM*, 43(4):685-722, 1996.
4. Chandra T. and Toueg S., Unreliable failure detectors for reliable distributed systems. *Journal of the ACM*, 43(2):225-267, 1996.
5. Delporte-Gallet C. and Fauconnier H., Two consensus algorithms with atomic registers and failure detector Ω. *Proc. 10th Int'l Conference on Distributed Computing and Networking (ICDCN'09)*, Springer Verlag #5408, pp. 251-262, 2009.
6. Englert B. and Gafni E., Fast collect in the absence of contention. *Proc. IEEE Int'l Conference on Distributed Computing Systems (ICDCS'02)*, IEEE Press, pp. 537-543, 2002.
7. Fernández A., Jiménez E., Raynal M. and Trédan G., A timing assumption and two t-resilient protocols for implementing an eventual leader service in asynchronous shared-memory systems. *Algorithmica*, 56(4):550-576, 2010.
8. Fischer M.J., Lynch N.A. and Paterson M.S., Impossibility of distributed consensus with one faulty process. *Journal of the ACM*, 32(2):374-382, 1985.
9. Gafni E., Round-by-round fault detectors: unifying synchrony and asynchrony. *Proc. 17th ACM Symp. on Principles of Distr. Computing (PODC'98)*, ACM Press, pp. 143-152, 1998.
10. Gafni E. and Lamport L., Disk Paxos. *Distributed Computing*, 16(1):1-20, 2003.
11. Guerraoui R. and Raynal M., The information structure of indulgent consensus. *IEEE Transactions on Computers*. 53(4):453-466, 2004.
12. Guerraoui R. and Raynal M., The alpha of indulgent consensus. *The Computer Journal*, 50(1):53-67, 2007.
13. Herlihy M.P., Wait-free synchronization. *ACM Transactions on Programming Languages and Systems*, 13(1):124-149, 1991.
14. Herlihy M.P. and Wing J.L., Linearizability: a correctness condition for concurrent objects. *ACM Transactions on Programming Languages and Systems*, 12(3):463-492, 1990.
15. Keidar I. and Rajsbaum S., On the cost of fault-tolerant consensus when there are no faults. *ACM SIGACT News, Distributed Computing Column*, 32(2):45-63, 2001.
16. Lamport L., The part-time parliament. *ACM Trans. on Comp. Syst.*, 16(2):133-169, 1998.
17. Lo W.-K. and Hadzilacos V., Using failure detectors to solve consensus in asynchronous shared memory systems. *Proc. 8th Int'l Workshop on Distributed Algorithms (WDAG'94)*, Springer Verlag #857, pp. 280-295, 1994.
18. Loui M. and Abu-Amara H., Memory requirements for for agreement among Unreliable Asynchronous processes. *Advances in Computing Research*, 4:163-183, JAI Press Inc., 1987.
19. Mostéfaoui A. and Raynal M., Leader-based consensus. *Parallel Processing Letters*, 11(1):95-107, 2001.
20. Raynal M., Communication and agreement abstractions for fault-tolerant asynchronous distributed systems. *Morgan & Claypool Publishers*, 251 p., 2010 (ISBN 978-1-60845-293-4).
21. Raynal M., Concurrent programming: algorithms, principles and foundations. *To appear*, Springer, 420 pages, 2012.
22. Raynal M. and Stainer J., ¿From a store-collect object and Ω to efficient asynchronous consensus. *Tech Report #1987*, 19 pages, IRISA/INRIA, Université de Rennes 1 (France), 2011.

An Investigation into the Performance of Reduction Algorithms under Load Imbalance

Petar Marendić[1,2], Jan Lemeire[1,2], Tom Haber[3],
Dean Vučinić[1,2], and Peter Schelkens[1,2]

[1] Vrije Universiteit Brussel (VUB), ETRO Dept.,
Pleinlaan 2, B-1050 Brussels, Belgium
petar.marendic@vub.ac.be
[2] Interdisciplinary Institute for Broadband Technology (IBBT), FMI Dept., Gaston
Crommenlaan 8 (box 102), B-9050 Ghent
[3] EDM, UHasselt, Diepenbeek
tom.haber@uhasselt.ac.be

Abstract. Today, most reduction algorithms are optimized for balanced
workloads; they assume all processes will start the reduction at about the
same time. However, in practice this is not always the case and significant
load imbalances may occur and affect the performance of said algorithms.
In this paper we investigate the impact of such imbalances on the most
commonly employed reduction algorithms and propose a new algorithm
specifically adapted to the presented context. Firstly, we analyze the
optimistic case where we have a priori knowledge of all imbalances and
propose a near-optimal solution. In the general case, where we do not
have any foreknowledge of the imbalances, we propose a dynamically
rebalanced tree reduction algorithm. We show experimentally that this
algorithm performs better than the default OpenMPI and MVAPICH2
implementations.

Keywords: MPI, imbalance, collective, reduction, process skew, benchmarking.

1 Introduction

The reduction algorithm - *extracting a global feature from distributed data* such
as the sum of all values - is a common and important communication operation.
However, it has two downsides which degrade the performance of a parallel program. Firstly, all of the reduction algorithms scale superlinearly as a function of
the number of processors, as shown later on in the text. Secondly, any reduction
operation breaks the independence of process execution, as it requires a global
process synchronization. Unless of course if the reduction could be performed in
the background, i.e. asynchronously - a case we will not consider here. Reduction
algorithms are vulnerable to load imbalances in the sense that if one process is
delayed before starting the reduction, the execution of part of the reduction will
also be delayed. One can however change the order in which process subresults

C. Kaklamanis et al. (Eds.): Euro-Par 2012, LNCS 7484, pp. 439–450, 2012.

are combined so that this concomitant delay is significantly reduced. In the extreme case where the imbalance dwarfs the combination times, this delay can be effectively eliminated.

1.1 Performance Cost Model

For analytical evaluation and comparison of various reduction algorithms we will employ a simple flat model as defined by [1] wherein there are p participating processes and each process has an input vector size of n bytes. We denote the local computation cost for one binary operation on two vector bytes as $\gamma[sB^{-1}]$. Communication time is modeled as $\alpha + n\beta$, where α is per message latency and $\beta[sB^{-1}]$ per byte transfer time.

We further assume that any process can send and receive one message at the same time, so that p parallel processes can send p messages in parallel.

The next section discusses the state-of-the art on reduction algorithms and the effect of load imbalances. In section 3 we present a static load balancing algorithm, while in section 4 we propose a new dynamic load balancing algorithm. Section 5 presents the experimental results.

2 Reduction

By definition, a reduction operation combines elements of a data vector residing on each of the participating processes by application of a specified reduction operation (e.g. maximum or sum), and returns the combined values in the output vector of a distinguished process (the root).

All reduction operators are required to be associative, but not necessarily commutative. However it is always beneficial to know whether a particular operator is commutative as there are faster ways of performing a reduction in that case.

One interesting case where a non-commutative operator arises is in the image compositing step of a distributed raytracing algorithm. In such an algorithm global data is distributed across processes and each process generates an image of its share of that data. To produce the final image, a composition needs to be performed on the produced images. This composition is a complex reduction step using the so-called 'over' operator and needs to happen in the correct back-to-front order.

The reduction that we've thus far been talking about is actually known as all-to-one reduction [2] since the end result is sent to one distinguished process. Variants of the reduction operation are the allreduce and reduce-scatter

2.1 Related Work

The simplest implementation of an all-to-one reduce is to have all processes send their local result to the root and the root combine these subresults in the next step. This approach is known as *Linear Reduction Algorithm*. It usually results in a bottleneck at the root process. Using our cost model, the complexity of this algorithm can be expressed as:

$$T(n,p) = (p-1)(\alpha + n\beta + n\gamma) \tag{1}$$

One straightforward way to eliminate this bottleneck is to employ a divide and conquer strategy that will order all participating processes in a binary tree and where at each step half of the processes will finish their work. This *Binary Tree Reduction* algorithm is efficient for small message sizes but suffers from suboptimal load balance, as the execution time is:

$$T(n, p) = \lceil \log_2 p \rceil (\alpha + n\beta + n\gamma) \tag{2}$$

Another variation on this idea is *Binomial Tree Reduction* where a binomial tree structure is used instead of a binary tree. This structure has the advantage of producing less contention on root nodes over that of the binary tree.

Other well known algorithms are *Direct-Send* and *Scatter-Gather*. In direct-send, every process is responsible for $\frac{1}{p}$th of the data vector and scatters the remaining chunks to their corresponding processes. In the second stage, once all processes have reduced the chunks they had received, a gather operation is performed on these subresults to form the result vector at the root process. This approach will result with maximal utilization of active computational resources, and with only a single communication step. However, it will also generate $p \times (p-1)$ messages to be exchanged among all participating processes. In a communication network where each of the participating processes are connected by network links, this will likely generate link contention as multiple processes will simultaneously be sending messages to the same process [3]. The execution time for direct-send is:

$$T(n, p) = p \times (p-1)(\alpha + \frac{n}{p}\beta) + n\gamma \tag{3}$$

It should be noted that this only states the time to perform a Reduce-Scatter, i.e. having the result vector scattered across participating processes. To implement an All-to-One reduction we need to follow up the reduce-scatter step by a gather to the root, which is typically performed with a binomial tree algorithm. Reduce-scatter can be also performed by other well known methods, such as *Binary Swap* or *Radix-k* algorithms. Another well-performing algorithm of this type is Rabenseifner's algorithm which was shown to perform well with longer message sizes [4,1,5].

$$T(n, p) = 2 \log_2 p\alpha + n\beta + (1 - \frac{1}{p})(n\beta + n\gamma), \; where \; p = 2^x, x \in \mathbb{N} \tag{4}$$

This algorithm is considerably more efficient than the binary tree reduction when the complexity of the reduction operation is significant.

As far as we know, no work has been done on analyzing and optimizing reduction algorithms under load imbalances. We will show in the following chapter that this leaves many real world scenarios unaccounted for.

2.2 Load Imbalances

We can identify three sources of imbalances:

- (Type 1) imbalances in the phase that precedes reduction
- (Type 2) imbalances in the amount of data that is sent at each step
- (Type 3) imbalances in the completion time of the combination operation.

The distributed raytracing algorithm we previously mentioned is a nice example of type 1, 2 and 3 imbalance occurrences. As it is typical for applications of this sort to generate images in the multi megapixel range, compression schemes are often employed to reduce the amount of data to be sent across the network. The time to combine these images using the *over operator* is a linear function of their size, where the effective size of the image is measured in non-black, that is relevant only pixels. Since this size varies across processes, the time to combine such images will vary as well.

3   Static Load Balancing under Perfect Knowledge

Here we assume perfect knowledge of the load imbalances and the time reduction phases will be finished. We analyze which reduction scheme gives the minimal completion time. For the communication and combination step we assume one of the following two performance models. The one of Fig. 1 is based on three parameters σ, τ and ψ, while the one of Fig. 2 is a simplification in which $\tau = 0$. We assume that the three parameters are constant during the total reduction. The parameters incorporate α, β and γ discussed before. Parameter σ denotes the time which is consumed on the sending process. Parameter τ denotes the time in which the sending process has already started the communication, but the receiving process does not yet have to participate, in the sense that no cycles are consumed. This happens if the message has not arrived yet or part of the receiving is performed in the background. These cycles can be used for other computations, so during τ, the receiving process might be busy with other things. We then assume that the receiving process is ready after τ. After that, it consumed ψ cycles to finish the communication and combination phase. When the receiving process would not be ready after τ, the phase will start when ready and still consume ψ. Not that ψ includes the receiving and combination phase.

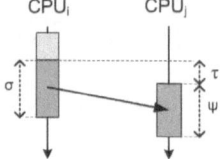 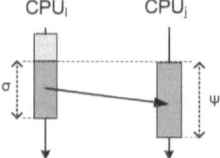

Fig. 1. Performance model of a communi- **Fig. 2.** Simplified performance model of a cation and combination phase   communication and combination phase

Under these assumptions, we propose the following algorithm. The algorithm is executed by each process when it starts the reduction.

Algorithm 1. The static optimized reduction algorithm

While step S2 has not been performed, do the following

  C check if another process has sent its subresults to you

      S1 if so, receive it and accumulate it with own subresult
      S2 if not, send own subresult to the first process that will be executing check
          C

Under the assumption of 'perfect knowledge' we know during step S2 which process will be first to be ready to receive a message. Secondly, in step C we assume that we can test without cost that there is a message on the way. We neglect the fact that the test for an incoming message (a 'MPI_Probe call') could give a negative answer while process just has posted a message which is not yet detectable by the receiving process.

The algorithm gives the optimal reduction when using the simplified performance model.

Lemma 1. *The static optimized reduction algorithm (Alg. 1) gives the minimal completion time under the simplified performance model.*

Proof. At S2, a process has to decide to whom sending its subresults. By sending it to the first process ready to receive (step C), say process 2 at t_2, the receiving and combination will be ready first, at $t_2 + \psi$. By sending it to another process, ready at t_3 with $t_2 < t_3$, the merge step will only be ready later, at $t_3 + \psi$. This would not give an advantage. Process 2 could start sending its subresults at t_2 to process 4, which will finish at $t_4 + \psi$. But this is not faster than any other process that would merge with process 4. The earliest that this can finish is $t_4 + \psi$. In this way we have proven that no other choice of receiving process can complete the reduction earlier.

Alg. 1 is, however, not always optimal for the first communication model. In some very specific cases, an alternative merge order gives a better completion time. Consider the case shown in Fig. 3. P1 sends its data to P2 (the first one to finish next) and P3 merges with P4. A better merge order is shown in Fig. 4. Here P1 communicates with P3 instead and P2 with P4. The first message (P1 to P3) arrives later but the second one (P2 to P4) arrives earlier than in the first scheme. Due to the configuration of the imbalances in P5, P6, P7 and P8, this gives rise to a merge order which finishes τ earlier.

Hence, the given algorithm is suboptimal. Nonetheless, it will be optimal in most cases. Only in exceptional fine-tuned cases, alternative schemes will exist. Moreover, the difference with the optimal completion time will be small because we expect τ to be quite small. Concluding, in most cases, the algorithm will be optimal. In the exceptional, suboptimal cases the algorithm will be approximately optimal.

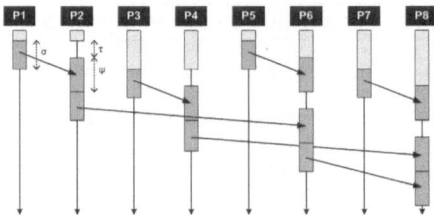

Fig. 3. Case in which the static load balancing algorithm is not optimal

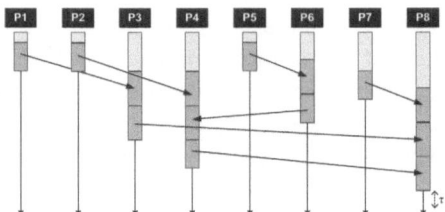

Fig. 4. Alternative merge order which completes faster than the static load balancing algorithm

It must be stressed that this algorithm is impractical, since the knowledge on when processes will be ready will not be present (it is difficult to predict and would in general lead to too much overhead to communicate). On the other hand, the algorithm gives a hint of how the optimal reduction would have been. Every solution can be compared to it.

4 Dynamic Load Balancing

Our initial idea was to take a regular binary tree reduction algorithm and augment it by installing a timeout period at each node of the tree. Should at any time a node time out waiting on data from its children, it would delegate these busy child nodes to its parent, reducing the number of steps their messages will eventually have to take on their way to the root. This process would continue until the root node received contribution from all nodes. Benchmarking however showed that this algorithm lacked robustness, as it was hard to pick a proper timeout value for varying vector sizes, process numbers and operator complexities.

We therefore turned our attention to a more deterministic algorithm that although tree-based was capable of dynamically reconfiguring its structure to minimize the effect an imbalance might have. The algorithm allows neighbours that are ready to start combining their subresults. The processes are ordered in a linear sequence and will send their local subresult to their right neighbour when finished, as described by Algorithm 2. Since the right neighbour might already have terminated, first a handshake is performed by sending a *completion* message and waiting for the *handshake* message. Once a process is finished, its left neighbour should be redirected to its 'current' right neighbour. This is illustrated by the example run shown in Fig. 5.

Algorithm 2. Local reduction algorithm

S1 Initialize left and right neighbours according to the neighbours in the predefined linear ordering of the processes. The processes at the head or tail do not have left or right neighbour respectively.
S2 Send *completion* message to right neighbour.
S3 Wait for incoming messages.

    S3.1 On receipt of *completion* message, initiate *handshake*.
    S3.2 On receipt of *redirect* message, change right to new node.

S4 Complete *handshake*, exchange *data* and perform reduction. Change left to sender's left neighbour.
S5 If data was received goto 3
S6 Wait for message from right neighbour and *redirect* to left.

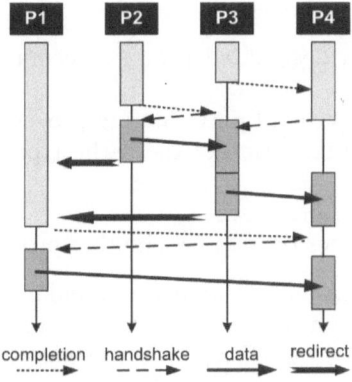

Fig. 5. Example run of the local reduction algorithm

5   Experimental Results

We devised an experiment with the primary purpose of ordering several well known algorithms in terms of their performance under various conditions of load imbalance. Our benchmarking scheme was as follows: before running the actual timed benchmark for a given algorithm, we perform a single warm up run; second, we synchronize the participating processes with a single call to MPI::Barrier; third, we sleep each process for *delay* milliseconds, where *delay* is a time offset that we distribute individually across participating processes using one of the schemes enumerated below; then, we run the algorithm k times for each predefined operator, where k is a parameter to our benchmark program, random shuffling the data vector between each iteration. Finally, we repeat this process r times, each time random generating a new data vector and a new time offset for each participating process.

5.1   Completion Time

In ideal conditions, where no load imbalances and process skew are present, the completion time of a reduction operation should typically be measured from its initiation till its termination at the root process, as this best reflects the time a given user would have to wait before the results of a reduction operation become available to him.

However, in real-life applications it is rarely the case that all participating processes begin the reduction at the same time, even when they have been explicitly synchronized with a call to MPI::Barrier [6]. To compound matters, we explicitly introduce process skew ensuring that participating processes will be initiating and completing their share of the reduction operation at different time instances.

With this in mind, and having resolved to report only a single number as the elapsed time of a reduction operation, several different schemes of reducing the initiation and termination times at each process present themselves. [7] and [6] enumerate the following approaches

T_1 the time needed at a designated process (termination - initiation time at root)
T_2 maximum of the times on all participating processes
T_3 the time between the initiation of the earliest process and the termination of the last
T_4 the average time of all processes
T_5 minimimum of the times on all participating processes

One can however also take into account the imbalance time of a given process, and treat both this time and the reduction time as one unit of interest, as would be the case in image rendering where the imbalance time is the time required to generate an image of the local data and the reduction time, the time to perform image compositing that results with an image of global data. Thus we decided to report the mean time T_1 plus the imbalance time at root process, across r iterations.

5.2   Benchmark Parameters

To study the performance of reduction algorithms we developed a test suite that can create workloads controlled by the following parameters:

− Statistical imbalance distributions:
  • a Gaussian distribution with mean m and standard deviation s.
  • a Gamma distribution with mean $m=\theta.k$, where θ is the scale parameter and k the shape parameter. It is frequently used as a probability model for waiting times.

- data vectors sizes. No imbalances were generated (Type 2).
- reduction operator complexity. No imbalances were generated (Type 3), completion time was a function of vector size.

We decided to take into consideration imbalances of type 1 only, as it is our opinion that they are most representative of real life scenarios.

We included the following algorithms into our analysis:

1. Binary Tree Reduction
2. OpenMPI's and MPICH's default implementation
3. All-to-All followed by local reductions and final gather to root
4. Our local reduction algorithm

The experiments were performed on a cluster machine available in the Intel Exascience lab, Leuven. It is a 32 node cluster with 12 cores per node (2 x Xeon X5660 processor with $\gamma = 1.79\ 10^{-10}$s defined as $2/R_{peak}$ with R_{peak} the theoretical maximum floating point rate of the processing cores) and QDR Infiniband interconnect (measured MPI bandwidth of $\sim$ 3500MB/s, $\alpha = 3.13$ μs, $\beta = 0.00263\mu$s)

5.3   Performance of Default Implementation

Our tests have shown that the default implementation of the reduction algorithm under MVAPICH2, when tested without any load imbalances, is consistently slower than our All-to-All reduce implementation. Even though we don't report the timings here, we have run the same battery of tests on OpenMPI as well, but the ranking of the tested algorithms was unchanged. In addition, we checked the performance of the default implemented Reduce-Scatter algorithm and have confirmed that it too is faster than the default Reduce algorithm. This leads us to believe that the default implementation is in fact a binomial tree reduction algorithm that performs well for small vector sizes only. This is a rather surprising revelation considering that significantly better algorithms have been published more than 5 years ago [4,1,5].

5.4   Impact of Data Size

For small vector sizes, the default implementation was regularly the fastest, scaling very well with increasing number of processors which is indicative of binomial tree reduction. On the other hand, for big and very big vectors the All-to-All (and Reduce-Scatter) algorithms outperform everyone else thanks to their linear γ factor scaling (Eq. 4). We should point out that All-to-All reduce can only be applied if the data vector can be sliced - something that is not always feasible with custom user defined data. In such case the only recourse is to revert to the binomial tree algorithm. Our benchmarks have shown that for reasonably big vectors the local reduce algorithm is of approximately the same performance as the default implementation and considering its superior performance under load imbalance we can confidently state that in this case it is the more robust algorithm of the two.

5.5   Impact of Reduction Operator's Complexity

Reading the relevant literature, one cannot escape the sentiment that little research has been performed in evaluating the performance of reduction algorithms with operators of varying complexity. We decided therefore to include into our tests two operators: std::plus and a special operator that is two orders of magnitude slower. Additionally, for sake of generality we assumed that both of these operators are non-commutative. The significant disparity in complexity of these two operators made it immediately apparent how inadequate the default implementation is, as it was completely outperformed by All-to-All (and Reduce-Scatter) implementations with execution times up to 6 times slower and with significantly worse scaling as visible in Fig. 7. It was however consistently faster than our local reduce algorithm in test runs without imbalance.

5.6   Impact of Load Imbalances

To test the impact of load imbalances, we identified two interesting cases:

1. there is a single slow process.
2. imbalances are distributed according to a gamma distribution for which $k=2$ and $\theta=0.5$, where the 90th percentile was 2.7 the mean imbalance.

In the first case where one of the processes is experiencing a slowdown, our local reduction algorithm proves itself as the best performer, often exhibiting flat scaling due to its ability to hide communication and computation overheads behind the incurred imbalance (see Fig. 8). The improvement we were able to achieve with our algorithm is dependent on the ratio of imbalance time and the time to reduce two vectors as is visible from Table 1. However, when the imbalances are distributed according to a gamma distribution law the local reduction algorithm only remains competitive up to and including vector size of 4MB (see Fig. 9). For a 40MB vector, the All-to-All implementation was the fastest.

Table 1. Speedup obviously depends on the imbalance time. Here, the time to reduce two vectors was 4 ms. All runtimes are reported in seconds.

128 processors with a 1024000 elements vector				
Imbalances in ms				
Algorithm	10	20	30	40	60
LocalReduce	0.0466108	0.0458913	0.0466785	0.050597	0.0695058
Default	0.0502286	0.0599159	0.0696198	0.079698	0.0996538
Speedup	1.08	1.32	1.49	1.56	1.43

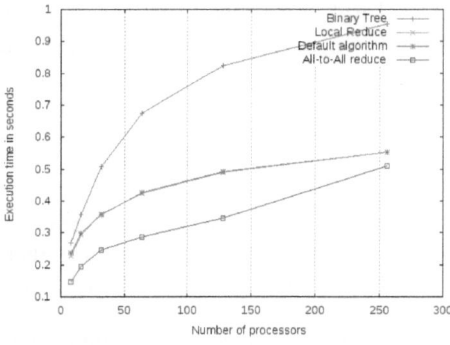
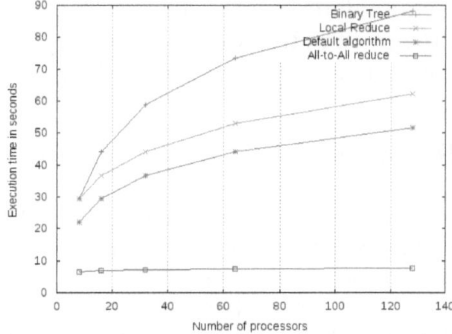

Fig. 6. Performance result for a 2^{20} elements vector using operator std::plus

Fig. 7. Performance result for a 2^{20} elements vector using special slow operator

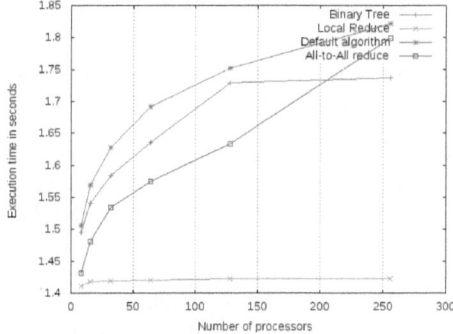
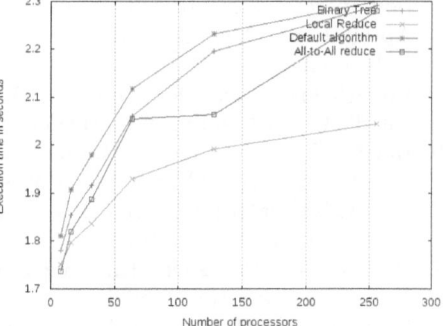

Fig. 8. Performance result for a 2^{20} elements vector using operator std::plus with an imbalance of 120ms at one node

Fig. 9. Performance result for a 2^{20} elements vector using operator std::plus with gamma distributed imbalances

6 Conclusions

We can establish two important conclusions: when designing reduction algorithms one should take into account operators significantly more complex than std::plus, as our tests have confirmed that algorithms well suited for cheap operations do not necessarily perform as well when expensive operations come into play; secondly, load imbalances do impact the performance of state-of-the art reduction algorithms and there are ways, as we have shown, to mitigate this.

The next step should be to investigate what benefits could be achieved by using the ideas here presented in a real world scenario such as a distributed raytracing algorithm that exhibits all three identified types of imbalances and verifying whether the results obtained with these synthetic tests are indeed relevant.

Acknowledgments. This work is funded by Intel and by the Institute for the Promotion of Innovation through Science and Technology in Flanders (IWT), in the context of the Exascience lab.

References

1. Rabenseifner, R., Träff, J.L.: More Efficient Reduction Algorithms for Non-Power-of-Two Number of Processors in Message-Passing Parallel Systems. In: Kranzlmüller, D., Kacsuk, P., Dongarra, J. (eds.) EuroPVM/MPI 2004. LNCS, vol. 3241, pp. 36–46. Springer, Heidelberg (2004)
2. Kumar, V., Grama, A., Gupta, A., Karypis, G.: Introduction to Parallel Computing. Benjamin/Cummings, Redwood City (1994)
3. Yu, H., Wang, C., Ma, K.L.: Massively parallel volume rendering using 2-3 swap image compositing. In: Proceedings of IEEE/ACM Supercomputing 2008 Conference, SC (2008)
4. Thakur, R., Rabenseifner, R., Gropp, W.: Optimization of collective communication operations in MPICH. International Journal of High Performance Computing Applications 19(1), 49–66 (2005)
5. Rabenseifner, R.: Optimization of Collective Reduction Operations. In: Bubak, M., van Albada, G.D., Sloot, P.M.A., Dongarra, J. (eds.) ICCS 2004, Part I. LNCS, vol. 3036, pp. 1–9. Springer, Heidelberg (2004)
6. Hoefler, T., Schneider, T., Lumsdaine, A.: Accurately measuring overhead, communication time and progression of blocking and nonblocking collective operations at massive scale. International Journal of Parallel, Emergent and Distributed Systems 25(4), 241–258 (2010)
7. Worsch, T., Reussner, R., Augustin, W.: On Benchmarking Collective MPI Operations. In: Kranzlmüller, D., Kacsuk, P., Dongarra, J., Volkert, J. (eds.) PVM/MPI 2002. LNCS, vol. 2474, pp. 271–279. Springer, Heidelberg (2002)

Achieving Reliability in Master-Worker Computing via Evolutionary Dynamics

Evgenia Christoforou[1], Antonio Fernández Anta[2], Chryssis Georgiou[1],
Miguel A. Mosteiro[3], and Angel (Anxo) Sánchez[4]

[1] University of Cyprus, Nicosia, Cyprus
[2] Institute IMDEA Network & Univ. Rey Juan Carlos, Madrid, Spain
[3] Rutgers University, Piscataway, NJ, USA & Univ. Rey Juan Carlos, Madrid, Spain
[4] Universidad Carlos III de Madrid, Madrid, Spain & BIFI Institute, Zaragoza, Spain

Abstract. This work considers Internet-based task computations in which a master process assigns tasks, over the Internet, to rational workers and collect their responses. The objective is for the master to obtain the correct task outcomes. For this purpose we formulate and study the dynamics of evolution of Internet-based master-worker computations through reinforcement learning.

1 Introduction

Motivation: As an alternative to expensive supercomputing parallel machines, Internet is a feasible computational platform for processing complex computational jobs. Several Internet-based applications operate on top of this global computation infrastructure. Examples are volunteer-based "@home" projects [2] such as SETI and profit-seeking computation platforms such as Amazon's Mechanical Turk.

Although the potential is great, the use of Internet-based computing is limited by the untrustworthy nature of the platform's components [2]. In SETI, for example, there is a machine, call it the *master*, that sends tasks, across the Internet, to volunteers' computers, call them *workers*, that execute and report back some result. However, these workers may not be trustworthy and it might be at their best interest to report incorrect results; that is, workers, or their owners, can be viewed as *rational* [1,14]. In SETI, the master attempts to minimize the impact of these bogus results by assigning the same task to several workers and comparing their outcomes (i.e., redundant task allocation is employed [2]).

Prior work [8,9,18] has shown that it is possible to design algorithmic mechanisms with reward/punish schemes so that the master can reliably obtain correct task results. We view these mechanisms as one-shot in the following sense: In a round, the master sends a task to be computed to a collection of workers, and the mechanism, using auditing and reward/punish schemes guarantees (with high probability) that the master gets the correct task result. For another task to be computed, the process is repeated (with the same or different collection of workers) but without taking advantage of the knowledge gained.

C. Kaklamanis et al. (Eds.): Euro-Par 2012, LNCS 7484, pp. 451–463, 2012.

Given a long running computation (such as SETI-like master-worker computations), it can be the case that the best interests, and hence the behavior of the workers, might change over time. So, one wonders: Would it be possible to design a mechanism for performing many tasks, over the course of a possibly infinite computation, that could positively exploit the repeated interaction between a master and the same collection of workers?

Our Approach: In this work we provide a positive answer to the above question. To do so, we introduce the concept of *evolutionary dynamics* under the biological and social perspective and relate them to Internet-based master-worker task computing. More specifically, we employ *reinforcement learning* [4, 15] to model how system entities or learners interact with the environment to decide upon a strategy, and use their experience to select or avoid actions according to the consequences observed. Positive payoffs increase the probability of the strategy just chosen, and negative payoffs reduce this probability. Payoffs are seen as parameterizations of players' responses to their experiences. Empirical evidence [3, 5] suggests that reinforcement learning is more plausible with players that have information only on the payoffs they receive; they do not have knowledge of the strategies involved. This model of learning fits nicely to our master-worker computation problem: the workers have no information about the master and the other workers' strategies and they don't know the set of strategies that led to the payoff they receive. The workers have only information about the strategies they choose at each round of the evolution of the system and their own received payoffs. The master also has minimal information about the workers and their intentions (to be truthful or not). Thus, we employ reinforcement learning for both the master and the workers in an attempt to build a reliable computational platform.

Our Contributions

1. We *initiate* the study of the evolutionary dynamics of Internet-based master-worker computations through reinforcement learning.
2. We develop and analyze a mechanism based on reinforcement learning to be used by the master and the workers. In particular, in each round, the master allocates a task to the workers and decides whether to audit or not their responses with a certain probability p_A. Depending on whether it audits or not, it applies a different reward/punish scheme, and adjusts the probability p_A for the next round (a.k.a. the next task execution). Similarly, in a round, each worker i decides whether it will truthfully compute and report the correct task result, or it will report an incorrect result, with a certain probability p_{Ci}. Depending on the success or not of its decision, measured by the increase or the decrease of the worker's *utility*, the worker adjusts probability p_{Ci} for the next round.
3. We show necessary and sufficient conditions under which the mechanism ensures *eventual correctness*, that is, we show the conditions under which, after some finite number of rounds, the master obtains the correct task result in every round, with minimal auditing, while keeping the workers satisfied (w.r.t. their utility). Eventual correctness can be viewed as a form of *Evolutionary*

Stable Strategy [6,10] as studied in Evolutionary Game Theory: even if a "mutant" worker decides to change its strategy to cheating, it will soon be brought back to an honest strategy.

4. Finally, we show that our mechanism, when adhering to the above-mentioned conditions, reaches eventual correctness quickly. In particular, we show analytically, probabilistic bounds on the convergence time, as well as bounds on the expected convergence time. Our analysis is complemented with simulations.

Background and Related Work: Evolutionary dynamics were first studied in evolutionary biology, as a tool to studying the mathematical principles according to which life is evolving. Many fields were inspired by the principles of evolutionary dynamics; our work is inspired by the dynamics of evolution as a mean to model workers' adaptation to a truthful behavior.

The dynamics of evolution have mainly been studied under the principles of Evolutionary Game Theory (EGT) [11]. In EGT the concept of evolutionarily stable strategy (ESS) is used [6,10]. A strategy is called evolutionary stable if, when the whole population is using this strategy, any group of invaders (mutants) using a different strategy will eventually die off over multiple generations (evolutionary rounds). It is shown [10] that an ESS is a Nash Equilibrium, but the reverse is not true.

While evolution operates on the global distribution of strategies within a given population, reinforcement learning [15] operates on the individual level of distribution over strategies of each member of the population. There are several models of reinforcement learning. A well-known model is the Bush and Mosteller's model [4]. This is an aspiration-based reinforcement learning model where negative effects on the probability distribution over strategies are possible, and learning does not fade with time. The player's adapt by comparing their experience with an *aspiration* level. In our work we adapt this reinforcement learning model and we consider a simple aspiration scheme where aspiration is fixed by the workers and does not change during the evolutionary process.

Phelps, McBurney and Parsons [13] discusses the concept of Evolutionary Mechanism Design. The evolutionary mechanism has a continues interaction and feedback from the current mechanism, as opposed to classical mechanism design [12] than when the mechanism is introduced in the system, it remains in the same Nash Equilibrium forever. In some way, our mechanism can be seen as an evolutionary mechanism, since the probability of auditing of the master and the probability of cheating of the workers, change, which is similar to changing the mechanism.

An extended account on related work (discussing applications of game theory to distributed computing, the concept of combinatorial agencies, the BAR model, etc.) can be found in [17].

2   Model and Definitions

Master-Worker Framework: We consider a distributed system consisting of a master processor that assigns, over the Internet, computational tasks to a set of n workers (w.l.o.g., we assume that n is odd). In particular, the computation is broken into rounds, and in each round the master sends a task to the workers to compute and return the task result. The master, based on the workers' replies, must decide on the value it believes is the correct outcome of the task in the same round. The tasks considered in this work are assumed to have a unique solution; although such limitation reduces the scope of application of the presented mechanism [16], there are plenty of computations where the correct solution is unique: e.g., any mathematical function.

Following Abraham et al. [1], and Shneidman and Parkes [14], we assume that workers are *rational*, that is, they are selfish in a game-theoretic sense and their aim is to maximize their benefit (utility) under the assumption that other workers do the same. In the context of this paper, a worker is *honest* in a round, when it truthfully computes and returns the task result, and it *cheats* when it returns some incorrect value. So, a worker decides to be honest or cheat depending on which strategy maximizes its utility. We denote by p_{Ci}^r the probability of a worker i cheating in round r. This probability is not fixed, the worker adjusts it over the course of the computation.

While it is assumed that workers make their decision individually and with no coordination, it is assumed that all the workers that cheat in a round return the same incorrect value (as done, for example, in [7] and [8]). This yields a worst case scenario (and hence analysis) for the master with respect to obtaining the correct result using mechanisms where the result is the outcome of voting; it subsumes models where cheaters do not necessarily return the same answer. (In some sense, this can be seen as a cost-free, weak form of collusion.)

Auditing, Payoffs, Rewards and Aspiration: To "persuade" workers to be honest, the master employs, when necessary, *auditing* and *reward/punish* schemes. The master, in a round, might decide to audit the response of the workers (at a cost). In this work, auditing means that the master computes the task by itself, and checks which workers have been honest. We denote by $p_{\mathcal{A}}$ the probability of the master auditing the responses of the workers. The master can change this auditing probability over the course of the computation. However, unless otherwise stated, we assume that there is a value $p_{\mathcal{A}}^{min} > 0$ so that at all times $p_{\mathcal{A}} \geq p_{\mathcal{A}}^{min}$.

Furthermore, the master can reward and punish workers, which can be used (possibly combined with auditing) to encourage workers to be honest. When the master audits, it can accurately reward and punish workers. When the master does not audit, it decides on the majority of the received replies, and it rewards only the majority. We refer to this as the $\mathcal{R}_m$ reward scheme.

The payoff parameters considered in this work are detailed in Table 1. Note that the first letter of the parameter's name identifies whose parameter it is. M stands for master and W for worker. Then, the second letter gives the type of parameter. P stands for punishment, C for cost, and B for benefit. Observe that there are different parameters for the reward WB_y to a worker and the cost MC_y of this reward to the master. This models the fact that the cost to the master might be different from the benefit for a worker.

Table 1. Payoffs. The parameters are non-negative.

$WP_{\mathcal{C}}$	worker's punishment for being caught cheating
$WC_{\mathcal{T}}$	worker's cost for computing the task
$WB_{\mathcal{Y}}$	worker's benefit from master's acceptance
$MP_{\mathcal{W}}$	master's punishment for accepting a wrong answer
$MC_{\mathcal{Y}}$	master's cost for accepting the worker's answer
$MC_{\mathcal{A}}$	master's cost for auditing worker's answers
$MB_{\mathcal{R}}$	master's benefit from accepting the right answer

We assume that, in every round, a worker i has an *aspiration* a_i, that is, the minimum benefit it expects to obtain in a round. In order to motivate the worker to participate in the computation, the master must ensure that $WB_{\mathcal{Y}} \geq a_i$; in other words, the worker has the potential of its aspiration to be covered. We assume that the master knows the aspirations. This information can be included, for example, in a contract the master and the worker agree on, prior to the start of the computation.

Note that, among the parameters involved, we assume that the master has the freedom of choosing $WB_{\mathcal{Y}}$ and $WP_{\mathcal{C}}$; by tuning these parameters and choosing n, the master can achieve the goal of eventual correctness. All other parameters can either be fixed because they are system parameters or may also be chosen by the master (except the aspiration, which is a parameter set by each worker).

Eventual Correctness: The goal of the master is to eventually obtain a reliable computational platform. In other words, after some finite number of rounds, the system must guarantee that the master obtains the correct task results in every round with probability 1. We call such property *eventual correctness*.

3   Algorithmic Mechanism

We now detail the algorithms run by the Master and the workers.

Master's Algorithm: The master's algorithm begins by choosing the initial probability of auditing. After that, at each round, the master sends a task to all workers and, when all answers are received (a reliable network is assumed), the master audits the answers with probability $p_{\mathcal{A}}$. In the case the answers are not audited, the master accepts the value contained in the majority of answers and continues to the next round with the same probability of auditing. In the case the answers are audited, the value $p_{\mathcal{A}}$ of the next round is reinforced (i.e., modified according to the outcome of the round). Then, the master rewards/penalizes the workers appropriately. The master initially has scarce or no information about the environment (e.g., workers initial p_C). The initial probability of auditing will be set according to the information the master possesses. For example if it has no information about the environment, a safe approach may be to initially set $p_{\mathcal{A}} = 0.5$.

Observe that, when the answers are not audited, the master has no information about the number of cheaters in the round. Thus, the probability $p_{\mathcal{A}}$ remains

Algorithm 1 Master's Algorithm	**Algorithm 2** Algorithm for Worker i
$p_{\mathcal{A}} \leftarrow x$, where $x \in [p_{\mathcal{A}}^{min}, 1]$	$p_{Ci} \leftarrow y$, where $y \in [0, 1]$
for $r \leftarrow 1$ **to** ∞ **do**	**for** $r \leftarrow 1$ **to** ∞ **do**
send a task T to all workers in W	**receive** a task T from the master
upon receiving all answers **do**	set $S_i \leftarrow -1$ with probability p_{Ci}, and
audit the answers with probability $p_{\mathcal{A}}$	$S_i \leftarrow 1$ otherwise
if the answers were not audited **then**	**if** $S_i = 1$
accept the majority	**then** $\sigma \leftarrow compute(T)$
else	**else** $\sigma \leftarrow$ arbitrary solution
$p'_{\mathcal{A}} \leftarrow p_{\mathcal{A}} + \alpha_m(cheaters(r)/n - \tau)$	**send** response σ to the master
$p_{\mathcal{A}} \leftarrow \min\{1, \max\{p_{\mathcal{A}}^{min}, p'_{\mathcal{A}}\}\}$	get payoff Π_i
$\forall i \in W$: pay/charge Π_i to worker i	$p'_{Ci} \leftarrow p_{Ci} - \alpha_w(\Pi_i - a_i)S_i$
	$p_{Ci} \leftarrow \max\{0, \min\{1, p'_{Ci}\}\}$

the same as in the previous round. When the answers are audited, the master can safely extract the number of cheaters. Then, the master adapts the auditing probability $p_{\mathcal{A}}$ according to this number. (We denote by $cheaters(r)$ the number of cheaters in round r.) Observe that the algorithm guarantees $p_{\mathcal{A}} \geq p_{\mathcal{A}}^{min}$. This, combined with the property $p_{\mathcal{A}}^{min} > 0$ will prevent the system to fall in a permanent set of "bad" states where $p_{\mathcal{A}} = 0$ and $p_C > 0$. A discount factor, which we call *tolerance* and denote by τ, expresses the master's tolerable ratio of cheaters (typically, we will assume $\tau = 1/2$). Hence, if the proportion of cheaters is larger than τ, $p_{\mathcal{A}}$ will be increased, and otherwise, $p_{\mathcal{A}}$ will be decreased. The amount by which $p_{\mathcal{A}}$ changes depends on the difference between these values, modulated by a *learning rate* α_m. This latter value determines to what extent the newly acquired information will override the old information. (For example, if $\alpha_m = 0$ the master will never adjust $p_{\mathcal{A}}$.)

Workers' Algorithm: The workers' algorithm begins with each worker i deciding an initial probability of cheating p_{Ci}. At each round, each worker receives a task from the master and, with probability $1 - p_{Ci}$ calculates the task, and replies to the master with the correct answer. If the worker decides to cheat, it fabricates an answer, and sends the incorrect response to the master. We use a flag S_i to model the decision of a worker i to cheat or not. After receiving its payoff (detailed in the analysis section), each worker i changes its p_{Ci} according to the payoff Π_i received, the chosen strategy S_i, and its aspiration a_i. Observe that the workers' algorithm guarantees $0 \leq p_{Ci} \leq 1$. The workers have a learning rate α_w. We assume that all workers have the same learning rate, that is, they learn in the same manner (see also the discussion in [15]; the learning rate is called step-size there); note that our analysis can be adjusted to accommodate also workers with different learning rates.

4 Analysis

We now analyze the mechanism, which is composed of the Master's and the workers' algorithms presented in the previous section. We first model the evolution of the mechanism as a Markov Chain, and then we prove necessary and sufficient conditions for achieving eventual correctness. Then, we provide analytical evidence that convergence to eventual correctness can be reached rather quickly. Observe in Algorithms 1 and 2 that there are a number of variables that may change in each round. We will denote the value of a variable X after a round r with a superindex r as X^r.

4.1 The Mechanism as a Markov Chain

We analyze the evolution of the master-workers system as a Markov chain. To do so, we first define the set of states and the transition functions:

Let the state of the Markov chain be given by the vector of probabilities $(p_{\mathcal{A}}, p_{C1}, p_{C2}, \ldots, p_{Cn})$. Then, the state after round r is $(p_{\mathcal{A}}^r, p_{C1}^r, p_{C2}^r, \ldots, p_{Cn}^r)$. Observe from Algorithms 1 and 2 that any state $(p_{\mathcal{A}}, p_{C1}, p_{C2}, \ldots, p_{Cn})$ in which $p_{\mathcal{A}} \in [p_{\mathcal{A}}^{min}, 1]$, and $p_{Ci} \in [0, 1]$ for each worker i, is a possible initial state of the Markov chain. The workers' decisions, the number of cheaters, and the payoffs in round r are the stochastic outcome of the probabilities used in round r. Then, restricted to $p_{\mathcal{A}}^r \in [p_{\mathcal{A}}^{min}, 1]$ and $p_{Ci}^r \in [0, 1]$, we can describe the transition function of the Markov chain in detail. For each subset of workers $F \subseteq W$, $P(F) = \prod_{j \in F} p_{Cj}^{r-1} \prod_{k \notin F} (1 - p_{Ck}^{r-1})$ is the probability that the set of cheaters is exactly F in round r. Then, we have the following.

- With probability $p_{\mathcal{A}}^{r-1} \cdot P(F)$, the master audits when the set of cheaters is F, and then, (0) the master updates $p_{\mathcal{A}}$ as $p_{\mathcal{A}}^r = p_{\mathcal{A}}^{r-1} + \alpha_m(|F|/n - \tau)$, and (1) each worker $i \in F$ updates p_{Ci} as $p_{Ci}^r = p_{Ci}^{r-1} - \alpha_w(a_i + WP_C)$, (2) each worker $i \notin F$ updates p_{Ci} as $p_{Ci}^r = p_{Ci}^{r-1} + \alpha_w(a_i - (WB_Y - WC_T))$.
- With probability $(1 - p_{\mathcal{A}}^{r-1})P(F)$, the master does not audit when F is the set of cheaters. Then, the master does not change $p_{\mathcal{A}}$ and the workers update p_{Ci} as follows. For each $i \in F$, (3) if $|F| > n/2$ then $p_{Ci}^r = p_{Ci}^{r-1} + \alpha_w(WB_Y - a_i)$, (4) if $|F| < n/2$ then $p_{Ci}^r = p_{Ci}^{r-1} - \alpha_w \cdot a_i$, and for each $i \notin F$, (5) if $|F| > n/2$ then $p_{Ci}^r = p_{Ci}^{r-1} + \alpha_w(a_i + WC_T)$, (6) if $|F| < n/2$ then $p_{Ci}^r = p_{Ci}^{r-1} + \alpha_w(a_i - (WB_Y - WC_T))$.

The following terminology will be used throughout. Let a *covered worker* be one that is paid at least its aspiration a_i and the computing cost WC_T. In any given round r, let an *honest worker* be one for which $p_C^{r-1} = 0$. Let an *honest state* be one where the *majority* of workers are honest. Let an *honest set* be any set of honest states. We refer to the opposite cases as *uncovered worker*, *cheater worker* ($p_C^{r-1} = 1$), *cheat state*, and *cheat set* respectively.

4.2 Conditions for Eventual Correctness

We show the conditions under which the system can guarantee eventual correctness. We begin with some terminology. Let a set of states S be called *closed* if, once the chain is in any state $s \in S$, it will not move to any state $s' \notin S$. (A singleton closed set is called an *absorbing* state.) For any given set of states S, we say that the chain *reaches* (resp. *leaves*) the set S if the chain reaches some state $s \in S$ (resp. reaches some state $s \notin S$).

In order to show eventual correctness, we must show eventual convergence to a closed honest set. Thus, we need to show (i) that there exists at least one such closed honest set, (ii) that all closed sets are honest, and (iii) that one honest closed set is reachable from any initial state. *Omitted proofs are given in [17].*

Lemma 1. *Consider any set of workers $Z \subseteq W$ such that $\forall i \in Z : WB_{\mathcal{Y}} \geq a_i$. If $|Z| > n/2$, then the set of states*

$$S = \{(p_{\mathcal{A}}, p_{C1}, \ldots, p_{Cn}) | (p_{\mathcal{A}} = 0) \wedge (\forall w \in Z : p_{Cw} = 1)\},$$

is a closed cheat set.

Given (ii) above, the necessity of $p_A^{min} > 0$ is motivated by the above lemma. Hence, $p_A > 0$ is assumed for the rest of the analysis.

Lemma 2. *If there exists a set of workers $Z \subseteq W$ such that $|Z| > n/2$ and $\forall i \in Z : WB_{\mathcal{Y}} < a_i + WC_{\mathcal{T}}$, then no honest set is closed.*

Given (i) above, the necessity of a covered majority is motivated by Lemma 2. Hence, in the remainder we assume that the majority of workers are covered.

Lemma 3. *Consider any set of workers $Z \subseteq W$ such that $\forall i \in Z : WB_{\mathcal{Y}} \geq a_i + WC_{\mathcal{T}}$ and $\forall j \notin Z : WB_{\mathcal{Y}} < a_j + WC_{\mathcal{T}}$. If $|Z| > n/2$, then the set of states*

$$S = \{(p_{\mathcal{A}}, p_{C1}, \ldots, p_{Cn}) | \forall w \in Z : p_{Cw} = 0\},$$

is a closed set.

Hence Lemma 3 proves (i) above. We continue with the proof of the other properties.

Lemma 4. *Consider any set of workers $Z \subseteq W$ such that $\forall i \in Z : WB_{\mathcal{Y}} \geq a_i + WC_{\mathcal{T}}$ and $\forall j \notin Z : WB_{\mathcal{Y}} < a_j + WC_{\mathcal{T}}$. Then, for any set of states*

$$S = \{ (p_{\mathcal{A}}, p_{C1}, \ldots, p_{Cn}) | \exists Y \subseteq W : (|Y| > n/2) \wedge (\forall w \in Y : p_{Cw} = 0) \wedge (Z \not\subseteq Y) \},$$

S is not a closed set.

Lemma 5. *Consider any set of workers $Z \subseteq W$ such that $\forall i \in Z : WB_{\mathcal{Y}} \geq a_i + WC_{\mathcal{T}}$ and $\forall j \notin Z : WB_{\mathcal{Y}} < a_j + WC_{\mathcal{T}}$. If $|Z| > n/2$ and $p_{\mathcal{A}} > 0$, then for any set of states*

$$S = \{(p_{\mathcal{A}}, p_{C1}, \ldots, p_{Cn}) | \exists Y \subseteq W : (|Y| > n/2) \wedge (\forall w \in Y : p_{Cw} > 0)\},$$

S is not a closed set.

Together, Lemma 4 and 5 prove (ii), and also (iii) because, if only honest sets are closed, then there is a way of going from non-honest sets to one of them. Lemmas 3–5 give the overall result:

Theorem 1. *If $p_A > 0$ then, in order to guarantee with positive probability that, after some finite number of rounds, the system achieves eventual correctness, it is* **necessary** *and* **sufficient** *to set* $\boxed{WB_y \geq a_i + WC_T}$ *for all $i \in Z$ in some set $Z \subseteq W$ such that $|Z| > n/2$.*

The above theorem shows that there is a positive probability of reaching some state after which correctness can be guaranteed, as long as for a chosen majority of workers, the payment is enough to cover their aspiration and cost of performing the task.

Remark: From Algorithm 1 it is easy to see that once the closed set $S = \{(p_A, p_{C1}, \ldots, p_{Cn}) | \forall w \in Z : p_{Cw} = 0\}$ is reached, eventually $p_A = p_A^{min}$ and stays such forever.

4.3 Convergence Time

Theorem 1 shows necessary and sufficient conditions to achieve eventual correctness. However, in order to have a practical system, it is necessary to bound the time taken to achieve it, which we call the *convergence time*. In other words, starting from any initial state, we want to compute the number of rounds that takes to the Markov chain to reach an honest closed set. In this section, we show bounds on the convergence time. *Omitted proofs are given in [17].*

Expected Convergence Time: Let C be the set of all covered workers. We assume, as required by Theorem 1, that $|C| > n/2$. From transitions (1) and (2) in the Markov chain definition, it can be seen that it is enough to have a consecutive sequence of $1/(\alpha_w \min\{WB_y - a_i - WC_T, WP_C + a_i\}), \forall i \in C$, audits to enforce $p_C = 0$ for all covered workers. Which gives the following upper bound on the convergence time.

Theorem 2. *The expected convergence time is at most $\rho/(p_A^{min})^\rho$, where $\rho = 1/(\alpha_w \min_{i \in C}\{WB_y - a_i - WC_T, WP_C + a_i\})$ and C is the set of covered workers.*

The upper bound shown in Theorem 2 may be too pessimistic for certain values of the parameters. The following theorem provides a tighter bound under certain conditions.

Theorem 3. *Let us define, for each worker i, $dec_i \triangleq \alpha_w \min\{WP_C + a_i, WB_y - WC_T - a_i\}$, and $inc_i \triangleq \alpha_w \max\{WB_y - a_i, WC_T + a_i\}$. Let C be the set of covered workers. If $p_A^{min} = \max_{i \in C}\{inc_i/(inc_i + dec_i)\} + \varepsilon$, for some $0 < \varepsilon < 1 - \max_{i \in C}\{inc_i/(inc_i + dec_i)\}$, the expected convergence time is $1/(\varepsilon \min_{i \in C} dec_i)$.*

The following corollary is derived from the previous theorem for a suitable scenario.

Corollary 1. *If $WP_C + a_i \geq WB_y - WC_T - a_i$ and $WB_y - a_i \leq WC_T + a_i$, $\forall i \in C$, and if*

$$p_A^{min} = \frac{WC_T + \max_{i \in C} a_i}{WB_y} + \varepsilon,$$

where C is the set of covered workers and $0 < \varepsilon < 1 - (WC_T + \max_{i \in C} a_i)/WB_y$, then the expected convergence time is ρ/ε, where $\rho = 1/(\alpha_w(WB_y - WC_T - \max_{i \in C} a_i))$.

Probabilistic Bound on the Number of Rounds for Convergence: We show now that, under certain conditions on the parameters of the system, it is possible to bound the probability to achieve convergence and the number of rounds to do so. Assume that $p_A^0 > 0$. Since p_A is not changed unless the master audits, we have the following.

Lemma 6. *Let $p_A^0 = p > 0$. Then, the master audits in the first $\rho = \ln(1/\varepsilon_1)/p$ rounds with probability at least $1 - \varepsilon_1$, for any $\varepsilon_1 \in (0, 1)$.*

Let us assume that the system parameters are such that, for all workers i, $\alpha_w(WP_C + a_i) \in [0, 1]$ and $\alpha_w(WB_y - WC_T - a_i) \in (0, 1]$ (all workers are covered). Let us define $dec\_cheater \triangleq \alpha_w \min_i\{WP_C + a_i\}$ and $dec\_honest \triangleq \alpha_w \min_i\{WB_y - WC_T - a_i\}$. From transitions (1) and (2) we derive the following lemma.

Lemma 7. *Let r be a round in which the master audits, and F be the set of cheaters in round r. Then,*

$$p_{Ci}^r \leq 1 - \alpha_w(WP_C + a_i) \leq 1 - dec\_cheater, \forall i \in F$$
$$p_{Cj}^r \leq 1 - \alpha_w(WB_y - WC_T - a_j) \leq 1 - dec\_honest, \forall j \notin F$$

Denoting the sum of all cheating probabilities before a round r as $P^{r-1} \triangleq \sum_i p_{Ci}^{r-1}$.

Lemma 8. *Let r be a round in which the master audits such that $P^{r-1} > n/3$. If $dec\_cheater \geq dec\_honest$ and $dec\_cheater + 3 \cdot dec\_honest \geq 8/3$, then $P^r \leq n/3$ with probability at least $1 - \exp(-n/96)$.*

Let us now define $dec_i \triangleq \alpha_w \min\{a_i, WB_y - WC_T - a_i\}$. Let, $dec \triangleq \min_i dec_i$. Assume $WP_C \geq 0$ and $a_i \geq 0$, for all workers.

Lemma 9. *Consider a round r such that $P^{r-1} \leq n/3$. Then, with probability at least $1 - \exp(-n/36)$ each worker i has $p_{Ci}^r \leq \max\{0, p_{Ci}^{r-1} - dec\}$, and hence $P^r \leq n/3$.*

Lemmas 6–9 lead to the following result:

Theorem 4. *Assume $\alpha_w(WP_C + a_i) \in [0, 1]$ and $\alpha_w(WB_y - WC_T - a_i) \in (0, 1]$ for all workers i. (Observe that all workers are covered.) Let $dec\_cheater \triangleq \alpha_w \min_i\{WP_C + a_i\}$, $dec\_honest \triangleq \alpha_w \min_i\{WB_y - WC_T - a_i\}$, and $dec \triangleq \alpha_w \min_i\{a_i, WB_y - WC_T - a_i\}$. If $p_A^0 = p > 0$, $dec\_cheater \geq dec\_honest$ and $dec\_cheater + 3 \cdot dec\_honest \geq 8/3$, then eventual convergence is reached in at most $\ln(1/\varepsilon_1)/p + 1/dec$ rounds, with probability at least $(1 - \varepsilon_1)(1 - \exp(-n/96))(1 - \exp(-n/36))^{1/dec}$, for any $\varepsilon_1 \in (0, 1)$.*

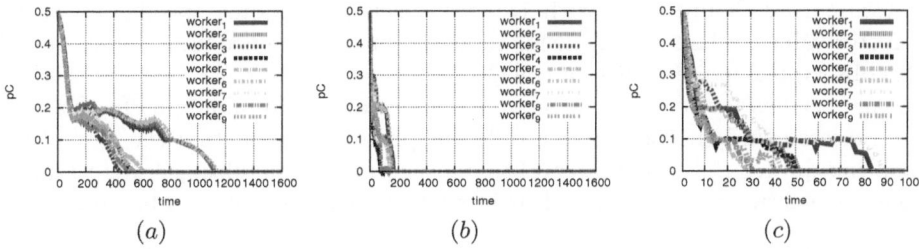

Fig. 1. Cheating probability for the workers as a function of time (number of rounds) for parameters $WP_\mathcal{C} = 0$, $WC_\mathcal{T} = 0.1$ and $a_i = 0.1$. (a) $\alpha = 0.01$, $WB_\mathcal{Y} = 1$; (b) $\alpha = 0.1$, $WB_\mathcal{Y} = 1$; (c) $\alpha = 0.1$, $WB_\mathcal{Y} = 2$.

5   Simulations

In this section we complement the theoretical analysis with simulations. Our analytical upper bounds on convergence time correspond to worst case scenarios. Here we present simulations for a variety of parameter combinations likely to occur in practice. We have created our own simulation setup by implementing our mechanism; technical details can be found in [17]. Each depicted plot value represents the average over 10 executions of the implementation.

We choose sensible parameter values, likely to be encountered in real applications. In particular, the number of workers has been set to nine (providing majority). Nine workers seems like an appropriate workforce, compared to Seti-like systems using three workers. The initial cheating probability of each worker i is not known, and therefore we have set it at $p_{Ci} = 0.5$ as a reasonable assumption. Similarly, we have set $p_A = 0.5$ as the master's initial probability of auditing. The minimum probability of cheating is set to be $p_A^{min} = 0.01$ and tolerance $\tau = 0.5$, hence the master will not tolerate a majority of cheaters.

The payoffs for the workers are set using $WB_\mathcal{Y} \in \{1, 2\}$ as our normalizing parameter and we take in analogy $WP_\mathcal{C} = 0$ and $WC_\mathcal{T} = 0.1$ as realistic values to explore the effects of these choices. The aspiration is a parameter defined by the workers in an idiosyncratic manner; for simplicity, here we consider all workers having the same aspiration level $a_i = 0.1$. The values of the aspiration and $WC_\mathcal{T}$ satisfy the necessary conditions of Theorem 1 and hence eventual convergence is reached. Finally, we consider the same learning rate for the master and the workers, i.e., $\alpha = \alpha_m = \alpha_w$. The learning rate, as discussed for example in [15] (called step-size there), for practical reasons it can be set to a small constant value; experimentally we notice that high values make the learning unstable. So we consider $\alpha \in \{0.1, 0.01\}$. A rich account of our results, on several scenarios under different parameter values (providing as well more intuition on system parameters e.g., tolerance) can be found in [17].

Figure 1 shows that convergence can be reached very quickly (in a few hundred rounds) even if no punishment is given to the workers caught cheating, and the number of workers and $WB_\mathcal{Y}$ are small. We also notice that a slightly higher value of α can make the convergence time shorter.

Comparing Figures 1(b) with 1(c) we observe that for a specific set of parameter values, a larger WB_y leads to a shorter convergence time. Interestingly, this observation points out to a trade-off between convergence time and the cost the master has for reaching faster convergence and maintaining it. In this way, the master could choose between different protocols estimating the cost of the auditing during the whole interval to convergence: less auditing leads to larger convergence times, so it is not clear in principle what is going to be optimal.

6   Conclusions

This work applies reinforcement learning techniques to formulate the evolution of Internet-based master-worker computations. The mechanism developed is presented and analyzed. In particular we show that under necessary and sufficient conditions, the master reaches a state after which the correct task result is received at each round, with minimal cost. In addition we show that such state can be reached quickly. The convergence analysis is complemented with simulations; our simulation results suggest that when having a positive reinforcement learning (i.e., $WP_C = 0$) the master can reach fast convergence, while applying negative reinforcement learning (i.e., $WP_C = \{1, 2\}$) provides even faster convergence (see [17]). In fact, we may conclude that applying only negative reinforcement is enough to have fast convergence.

Acknowledgments. This work is supported by the Cyprus Research Promotion Foundation grant ΤΠΕ/ΠΛΗΡΟ/0609(ΒΕ)/05, NSF grants CCF-0937829, CCF-1114930, Comunidad de Madrid grant S2009TIC-1692, Spanish MOSAICO and RESINEE grants and MICINN grant TEC2011-29688-C02-01, and National Natural Science Foundation of China grant 61020106002. We thank Carlos Diuk for useful discussions.

References

[1] Abraham, I., Dolev, D., Goden, R., Halpern, J.Y.: Distributed computing meets game theory: Robust mechanisms for rational secret sharing and multiparty computation. In: Proc. of PODC 2006, pp. 53–62 (2006)

[2] Anderson, D.: BOINC: A system for public-resource computing and storage. In: Proc. of GRID 2004, pp. 4–10 (2004)

[3] Bendor, J., Mookherjee, D., Ray, D.: Aspiration-based reinforcement learning in repeated interaction games: An overview. International Game Theory Review 3(2-3), 159–174 (2001)

[4] Bush, R.R., Mosteller, F.: Stochastic Models for Learning. Wiley (1955)

[5] Camerer, C.F.: Behavioral game theory: Experiments in strategic interaction. Roundtable Series in Behavioral Economics (2003)

[6] Easley, D., Kleinberg, J.: Networks, Crowds, and Markets: Reasoning About a Highly Connected World. Cambridge University Press (2010)

[7] Fernández, A., Georgiou, C., Lopez, L., Santos, A.: Reliably executing tasks in the presence of untrusted processors. In: Proc. of SRDS 2006, pp. 39–50 (2006)

[8] Fernández Anta, A., Georgiou, C., Mosteiro, M.A.: Designing mechanisms for reliable Internet-based computing. In: Proc. of NCA 2008, pp. 315–324 (2008)

[9] Fernández Anta, A., Georgiou, C., Mosteiro, M.A.: Algorithmic Mechanisms for Internet-based Master-Worker Computing with Untrusted and Selfish Workers. In: Proc. of IPDPS 2010, pp. 1–11 (2010)

[10] Gintis, M.C.: Game Theory Evolving. Princeton University Press (2000)

[11] Maynard Smith, J.: Evolution and the Theory of Games. Cambridge U. Press (1982)

[12] Nisan, N., Ronen, A.: Algorithmic mechanism design. Games and Economic Behavior 35, 166–196 (2001)

[13] Phelps, S., McBurney, P., Parsons, S.: Evolutionary mechanism design: A review. Journal of Autonomous Agents and Multi-Agent Systems (2010)

[14] Shneidman, J., Parkes, D.C.: Rationality and Self-interest in P2P Networks. In: Kaashoek, F., Stoica, I. (eds.) IPTPS 2003. LNCS, vol. 2735, pp. 139–148. Springer, Heidelberg (2003)

[15] Szepesvári, C.: Algorithms for Reinforcement Learning. Synthesis Lectures on Artificial Intelligence and Machine Learning. Morgan & Claypool Publishers (2010)

[16] Taufer, M., Anderson, D., Cicotti, P., Brooks, C.L.: Homogeneous redundancy: a technique to ensure integrity of molecular simulation results using public computing. In: Proc. of IPDPS 2005 (2005)

[17] Technical report of this work, TR-12-02, Dept. of Computer Science, University of Cyprus (February 2012), http://www.cs.ucy.ac.cy/~chryssis/EvolMW-TR.pdf

[18] Yurkewych, M., Levine, B.N., Rosenberg, A.L.: On the cost-ineffectiveness of redundancy in commercial P2P computing. In: Proc. of CCS 2005, pp. 280–288 (2005)

Topic 9: Parallel and Distributed Programming

Sergei Gorlatch, Rizos Sakellariou, Marco Danelutto, and Thilo Kielmann

Topic Committee

This topic provides a forum for the presentation of the latest research results and practical experience in parallel and distributed programming in general, except for work specifically targeting multicore and manycore architectures, which has matured to becoming a Euro-Par topic of its own.

The challenge addressed by the topic is how to produce correct, portable parallel software with predictable performance on existing and emerging parallel and distributed architectures. This requires advanced algorithms, realistic modeling, efficient design tools, high-level programming abstractions, high-performance implementations, and experimental evaluation. Related to these central needs, it is also important to address methods for reusability, performance prediction, large-scale deployment, self-adaptivity, and fault-tolerance. Given the rich history in this field, practical applicability of proposed methods, models, algorithms, or techniques is a key requirement for timely research.

Each submission was reviewed by at least four reviewers and, finally, we were able to select 7 high-quality papers, one of them as distinguished paper. The presented research spans the broad scope, ranging from low-level issues like transactional access to shared memory and dynamic thread mapping, over algorithmic methods for partitioning and fault-tolerance, all the way up to scalable collective operations and pipelined MapReduce.

We are proud of the ambitious scientific program that we managed to assemble for this topic. Of course, this was only possible by combining the efforts of many people. We would like to take the opportunity to thank the authors who submitted their contributions, the external referees who have made the efficient selection process possible, and the conference organizers for a perfectly organized and very pleasant cooperation.

C. Kaklamanis et al. (Eds.): Euro-Par 2012, LNCS 7484, p. 464, 2012.
© Springer-Verlag Berlin Heidelberg 2012

Dynamic Thread Mapping Based on Machine Learning for Transactional Memory Applications

Márcio Castro[1], Luís Fabrício Wanderley Góes[2],
Luiz Gustavo Fernandes[3], and Jean-François Méhaut[1]

[1] INRIA - CEA - LIG Laboratory - Grenoble University,
ZIRST 51, avenue Jean Kuntzmann, 38330 Montbonnot Saint Martin, France
{marcio.castro,jean-francois.mehaut}@imag.fr
[2] Department of Computer Science - Pontifical Catholic University of Minas Gerais,
Av. Dom José Gaspar, 500, Belo Horizonte, MG, Brazil
lfwgoes@pucminas.br
[3] PPGCC - Pontifical Catholic University of Rio Grande do Sul,
Av. Ipiranga, 6681 - Prédio 32, Porto Alegre, RS, Brazil
luiz.fernandes@pucrs.br

Abstract. Thread mapping is an appealing approach to efficiently exploit the potential of modern chip-multiprocessors. However, efficient thread mapping relies upon matching the behavior of an application with system characteristics. In particular, Software Transactional Memory (STM) introduces another dimension due to its runtime system support. In this work, we propose a dynamic thread mapping approach to automatically infer a suitable thread mapping strategy for transactional memory applications composed of multiple execution phases with potentially different transactional behavior in each phase. At runtime, it profiles the application at specific periods and consults a decision tree generated by a Machine Learning algorithm to decide if the current thread mapping strategy should be switched to a more adequate one. We implemented this approach in a state-of-the-art STM system, making it transparent to the user. Our results show that the proposed dynamic approach presents performance improvements up to 31% compared to the best static solution.

Keywords: transactional memory, dynamic thread mapping, machine learning.

1 Introduction

Thread mapping is an appealing approach to efficiently exploit the potential of modern chip-multiprocessors by making better use of cores and memory hierarchy. It allows multithreaded applications to amortize memory latency and/or reduce memory contention. However, efficient thread mapping relies upon matching the behavior of an application with system characteristics.

Software Transactional Memory (STM) appears as a promising concurrency control mechanism for those modern chip-multiprocessors. It allows programmers to write parallel code as transactions, which are guaranteed to execute atomically and in isolation regardless of eventual data races [3,9]. At runtime, transactions are executed speculatively and the STM runtime system continuously keeps track of concurrent accesses and detects conflicts. Conflicts are then solved by re-executing conflicting transactions.

C. Kaklamanis et al. (Eds.): Euro-Par 2012, LNCS 7484, pp. 465–476, 2012.

However, due to its runtime support, applications can behave differently depending on the characteristics of the underlying STM system. Thus, the prediction of a suitable thread mapping strategy for a specific application/STM system becomes a daunting task.

Our previous work focused on a machine learning-based approach to statically infer a suitable thread mapping strategy for transactional memory applications [2]. This means that the predicted thread mapping strategy is applied once at the beginning and does not change during the execution of the application. We demonstrated that this approach improved the performance of all STAMP applications [10], since most of the transactions within each application usually have very similar behavior.

We have constantly seen efforts for a wider adoption of Transactional Memory (TM). For instance, the latest version of the GNU Compiler Collection (GCC 4.7) now supports TM primitives and new BlueGene/Q processors have hardware support for TM. Moreover, Intel recently released details of the Transactional Synchronization Extensions (TSX) for the future multicore processor code-named "Haswell". Thus, it is expected that more complex applications will make use of TM in a near future. In those cases, static thread mapping will no longer improve the performance of those applications, emerging the necessity of dynamic or adaptive approaches.

In this paper, we propose a dynamic approach to do efficient thread mapping on STM applications composed of more diverse workloads. These workloads may go through different execution phases, each phase with potentially different transactional characteristics. At runtime, we gather useful information from the application, STM system and platform at specific periods. At the end of each profiling period, we rely on a decision tree previously generated by a Machine Learning (ML) algorithm to decide if the current thread mapping strategy should be switched to a more adequate one. This dynamic approach was implemented within TinySTM [5] as a module, so the core of TinySTM remains unchanged and it is transparent to the user. Our results show that the dynamic approach is up to 31% better than the best static thread mapping for those applications.

The rest of this paper is organized as follows. Section 2 further describes STM and our previous work. In Section 3, we propose our dynamic thread mapping mechanism. Section 4 evaluates our dynamic thread mapping on several applications. Finally, Section 5 presents related work and Section 6 concludes.

2   Background

2.1   Software Transactional Memory

Transactional Memory is an alternative synchronization solution to the classic mechanisms such as locks and mutexes [9]. It removes from the programmer the burden of correct synchronization of threads on data races and provides an efficient model for extracting parallelism from the applications.

Transactions are portions of code that are executed atomically and with isolation. Concurrent transactions *commit* successfully if their accesses to shared data did not conflict with each other; otherwise some of the conflicting transactions will *abort* and none of their actions will become visible to other threads. Conflicts can be detected during the execution of transactions when the TM system uses an *eager conflict detection policy* whereas they are only detected at commit-time when the system uses a *lazy conflict detection policy*.

When a transaction aborts, the runtime system *rollbacks* some of the conflicting transactions. The choice among the conflicting transactions is done according to the *conflict resolution policies* implemented in the runtime system. Two common alternatives are to squash one of the conflicting transactions immediately (*suicide* strategy) or to wait for a time interval before restarting the conflicting transaction (*backoff* strategy).

Transactional Memory can be software-only, hardware-only or hybrid. In this work we are interested in STM since hardware and hybrid solutions are not yet available in commercial processors. This allows us to carry out experiments in current platforms without relying on simulations.

2.2   Static Thread Mapping Based on Machine Learning

Our previous work proposed a machine learning-based approach to predict a suitable thread mapping for TM applications [2]. It was composed of the following steps (Figure 1). Firstly, we profiled several TM applications from the STAMP benchmark suite [10] considering characteristics from the application, STM system and platform to build a set of input instances. Then, a Decision Tree Learning method (ID3) [11] was fed with these input instances and *trained*. The ID3 algorithm outputted a decision tree (predictor) capable of infering a thread mapping strategy for new unobserved instances.

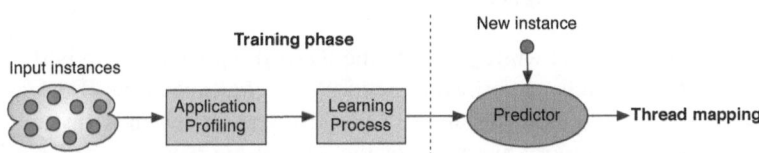

Fig. 1. Overview of our machine learning-based approach

We evaluated the performance of all TM applications from STAMP when applying the predicted thread mapping strategies *statically*. This means that the strategy is applied at the beginning and remains unchanged during the whole execution of the application. Our results showed that our approach usually makes correct predictions [2]. However, a deeper analysis of STAMP applications revealed that most of them do not have multiple phases with different transactional characteristics. For the upcoming complex workloads, there may not exist a single best thread mapping strategy that delivers the best performance for all phases. In the next section we present our solution to tackle this problem, which employs a dynamic approach adapted for more complex applications.

3   Dynamic Thread Mapping for Transactional Memory

As we previously stated, we will naturally face more complex TM applications due to the wider adoption of TM. These applications will probably have multiple execution phases with a different transactional behavior in each phase. Thus, we need a more dynamic thread mapping approach able to identify these different phases and switch to a more adequate thread mapping strategy during the execution. In the following sections, we explain the basic concepts of our dynamic approach as well as its implementation within a state-of-the-art STM system.

3.1 Proposed Approach

Our dynamic thread mapping approach is based on the fact that the performance of a TM application is not only governed by its characteristics but also by the characteristics of the TM system and platform. Those characteristics must be taken into account to choose a thread mapping strategy adapted to behavior of the workload. Thus, we consider the following criteria that have an important impact on the performance of TM applications:

- **Transactional time ratio:** fraction of the time spent inside transactions to the total execution time;
- **Abort ratio:** fraction of the number of aborts to the number of transactions issued (aborted + committed);
- **Conflict detection policy:** eager or lazy;
- **Conflict resolution policy:** suicide or backoff;
- **Last-level cache miss ratio:** fraction of the number of cache misses to the number of accesses on the last-level cache.

We considered these criteria while profiling the STAMP applications to build a *thread mapping predictor* as briefly described in Section 2.2. We trained the ID3 learning algorithm with two sets of input instances. The difference between them comes from the complexity of the memory hierarchy of the underlying platform. The predictor is represented in Figure 2.

The subtree on the left considers a single level of shared L2 caches whereas the subtree on the right considers a more complex memory hierarchy with two levels of shared caches (L2 and L3). Internal nodes represent our criteria (rectangles). Leaves represent the thread mapping strategy to be applied (rounded rectangles).

The predictor chooses a thread mapping strategy among four possible configurations: *scatter, compact, round-robin* and *linux*. *Scatter* distributes threads across different processors avoiding cache sharing between cores in order to reduce memory contention. In contrast, *compact* places threads on sibling cores that share all levels of the cache hierarchy. The *round-robin* strategy is an intermediate solution in which threads share higher levels of cache (*i.e.*, L3) but not the lower ones (*i.e.*, L2). Finally, *linux* is the default scheduling strategy implemented by the operating system.

Since most of the considered characteristics can vary during the execution of applications composed of several phases, they need to be profiled at runtime. We thus use profiling to gather the information needed by the predictor at specific periods. We specify two periods: the profiling period and the interval between profilings. These values are specified by the number of committed transactions instead of time. This guarantees that our measures occur when transactions are being executed. We use a hill-climbing

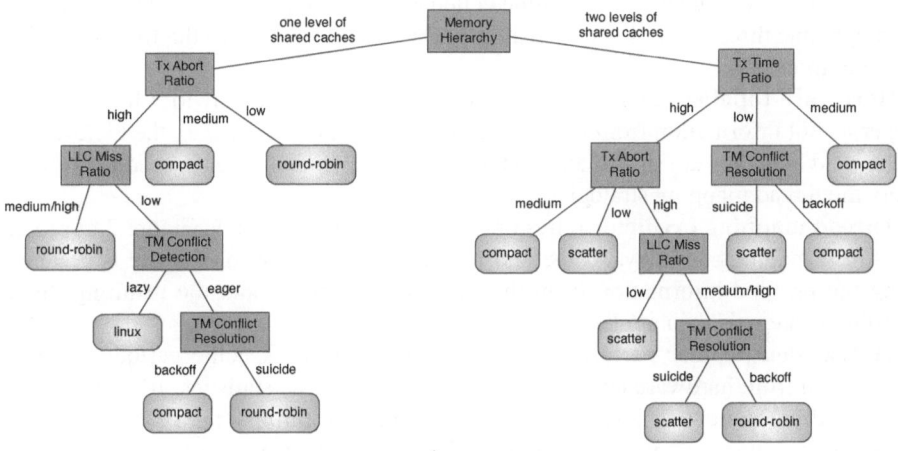

Fig. 2. Thread mapping predictor based on machine learning

strategy to adapt those values during the execution. We start with short periods and we double them each time the predicted thread mapping strategy was not changed. This is done until a maximum interval size is reached. When the thread mapping strategy is changed due to a phase transition, we reset them to their initial values and the hill-climbing strategy is restarted.

3.2 Implementation

For our solution to be transparent to users, we decided to implement it within a STM system. We chose TinySTM [5] among other STM systems because it is lightweight, efficient and its implementation has a modular structure that can be easily extended with new features. Figure 3 shows the organization schema of TinySTM and as well as our dynamic thread mapping module and its main components.

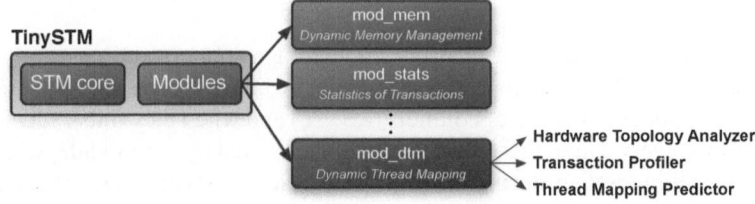

Fig. 3. Implementation of our dynamic thread mapping in TinySTM

Basically, TinySTM is composed of a STM core in which most of the STM code is implemented, and some additional modules. These modules implement basic features such as the dynamic memory management (mod_mem) and transaction statistics (`mod_stats`). We added a new module called `mod_dtm` that extends TinySTM to perform dynamic thread mapping transparently. Our module combines the following three main components:

Hardware topology analyzer uses the Hardware Locality (hwloc) library [1] to gather useful information from the underlying platform topology (*i.e.*, the hierarchy of caches and how they are shared among the cores). Such information is used to correctly apply the thread mapping strategies.

Thread mapping predictor relies on the decision tree shown in Figure 2 to predict the thread mapping strategy. At the end of each profiling period, the tree is traversed using the profiled information from the *transaction profiler* and the resulting thread mapping strategy is then applied.

Transaction profiler performs runtime profiling during specific periods to gather information from hardware counters and transactional basic statistics. Its pseudo-code is depicted in Figure 4. The *cache miss ratio* is obtained through the Performance Application Programming Interface (PAPI) [12] to access hardware counters. We maintain two counters to calculate the *abort ratio* (named **Aborts** and **Commits**). The *transactional time ratio* is an approximation obtained by measuring the time spent inside and outside transactions.

```
// on transaction start
if is profiling period then
    if first tx in this period then
        StartPapi(LLCAccess, LLCMiss);
        ProfileTime ← GetClock();
    end
    TxTime ← GetClock();
end

// on transaction abort
if is profiling period then
    Aborts ← Aborts + 1;
end
```

```
// on transaction commit
if is profiling period then
    TxTime ← GetClock() − TxTime;
    TotalTxTime ← TotalTxTime + TxTime;
    Commits ← Commits + 1;
    if last tx in this period then
        StopPapi(LLCAccess, LLCMiss);
        ProfileTime ← GetClock() − ProfileTime;
        TotalNonTxTime ← ProfileTime − TotalTxTime;
        ThreadMapping ← TMPredictor();
        ResetAllCounters();
    end
end
```

Fig. 4. Transaction profiler pseudo-codes

TinySTM allows the inclusion of user-defined extensions. In our case, we instrumented three basic TM operations that are called when transactions start (`start`), when they are rollbacked in case of conflicts (`abort`) and when they finish successfully (`commit`). Thus, every call to these operations is intercepted by our module, which executes the *transaction profiler* during the profiling periods and calls the *thread mapping predictor* to switch the thread mapping strategy when necessary.

When a TM application is executed, only one thread among all concurrent running threads is chosen to be the *transaction profiler*. The reason for that is threefold: (i) it considerably reduces the intrusiveness on the overall system, so the behavior of the application is not changed; (ii) we do not need to use extra synchronization mechanisms to guarantee reliable measures among concurrent threads; and (iii) most workloads of current TM applications are uniformly distributed among the threads. However, our implementation can be adapted to gather information from all threads. This may be necessary for non-SPMD applications, where different threads execute different flows of control.

4 Experimental Evaluation

In this section, we demonstrate that our dynamic thread mapping can benefit from applications composed of multiple execution phases with potentially different transactional behavior on each one. First, we describe our experimental setup as well as the set of characteristics we considered to create TM applications composed of multiple phases. Afterwards, we compare our performance gains with static solutions. Finally, we present a deeper analysis of our mechanism.

4.1 Experimental Setup

Since most of the transactions within each STAMP application usually have very similar behavior, they are not suitable for the evaluation of our dynamic thread mapping approach. For this reason, we used EigenBench [8] to create new TM applications with different phases. This micro-benchmark allows a thorough exploitation of the orthogonal space of TM applications characteristics.

Varying all possible orthogonal TM characteristics involves a high-dimensional search space [8]. Thus, we decided to vary 4 out of 8 orthogonal characteristics that govern the behavior of TM applications. We used the first three (transaction length, contention and density) to create a set of workloads (Table 1). Since we assume two possible discrete values for each one, we can create a total of 2^3 distinct workloads (named $W_1, W_2, \ldots, W_8$) by combining those values. It is important to mention that these values were obtained after an empirical study based on several experiments with different configurations of TinySTM (conflict detection and resolution policies) and EigenBench parameters. The fourth orthogonal characteristic is *concurrency* and it is further discussed in Section 4.3.

Table 1. TM orthogonal characteristics used to compose our set of workloads

Characteristic	Definition	Values
Tx Length	number of shared accesses per transaction	short (≤ 64) long (≥ 128)
Contention	probability of conflict	low-conflicting ($< 30\%$) contentious ($\geq 30\%$)
Density	fraction of the time spent inside transactions to the total execution time	sparse ($< 80\%$) dense ($\geq 80\%$)
Concurrency	number of concurrent threads/cores	$2 - 16$

We conducted our experiments on a multi-core platform based on four six-core 2.66GHz Intel Xeon X7460 processors and 64 GB of RAM running Linux 2.6.32. Each processor has 16MB of shared L3 cache and each group of two cores shares a L2 cache (3MB). TinySTM and all applications were compiled with GCC 4.4.5 using -O3. All results in the following sections are based on arithmetic means of 30 runs.

4.2   Dynamic Thread Mapping vs. Static Thread Mapping

Our first set of experiments explores the effectiveness of our dynamic thread mapping in comparison to the thread mapping strategies individually. We derived a set of applications from the 8 distinct workloads discussed in Section 4.1. We fixed the number of phases to 3, thus each application will be composed of three workloads. Therefore, all possible applications composed of three distinct workloads is determined by the number of k-combinations from a given set of n elements, $i.e.$, $C_k^n = C_3^8$, which results in 56 applications (named $A_1, A_2, \ldots, A_{56}$). Thus, the set of applications can be represented as follows: $A_1 = \{W_1, W_2, W_3\}$, $A_2 = \{W_1, W_2, W_4\}$, $\ldots$, $A_{56} = \{W_5, W_6, W_7\}$. Phases (workloads) are parallelized using Pthreads and there is no synchronization barrier between phases, $i.e.$, threads may not be computing the same workload at the same time.

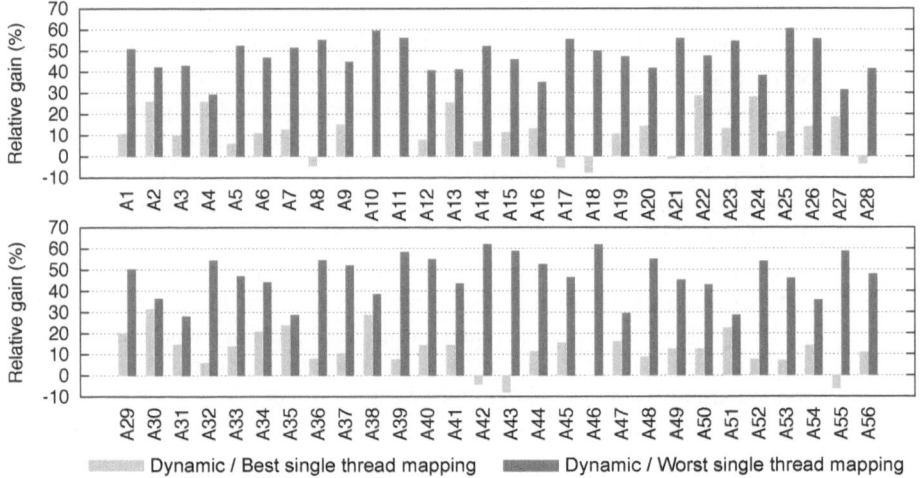

Fig. 5. Relative gains of our dynamic thread mapping compared to the best and worst single thread mappings. We considered applications composed of 3 phases (A_1 to A_{56}).

We ran all the applications with each one of the static thread mappings (*compact, round-robin* and *scatter*), the Linux default scheduling strategy and our dynamic approach. Figure 5 presents the relative gains of our dynamic thread mapping when compared to the **best** and **worst** single thread mappings. The relative gain is given by $1 - \overline{x}_d \div \overline{x}_s$, where $\overline{x}_d$ and $\overline{x}_s$ are mean execution times of 30 executions using the dynamic and the best/worst single thread mapping, respectively. Thus, positive values mean performance gains whereas negative values mean performance losses. All applications were executed with 4 threads and TinySTM was configured with lazy conflict detection and backoff conflict resolution.

We can draw at least two important conclusions from these results. Firstly, the thread mapping strategy had an important impact on the performance. This can be easily observed when comparing the relative gains between the best and worst single thread mappings. Secondly, our dynamic thread mapping usually improved the performance of the applications by switching to an adequate thread mapping strategy in each phase. We achieved performance gains up to 31% and 62%, when comparing to the best and worst single thread mappings respectively. However, our dynamic thread mapping did not deliver performance improvements on 3 applications and presented some performance losses in 8 applications when comparing with the best single thread mapping strategy. In the case of A_{10}, A_{11} and A_{46}, a single thread mapping strategy (*compact*) was best for all phases, thus we cannot expect performance improvements by using our dynamic approach. The performance losses were due to wrong decisions of the predictor, which did not select the best thread mapping strategy on all phases. The maximum performance loss was about 8% (A_{43}). One reason for that may come from the characteristics that we take into account in training phase and profiling. We leave the discussion of other possible characteristics to enrich the predictions to future work.

4.3 Varying Concurrency

Our second set of experiments focuses on the performance impacts of the thread mapping strategies when varying the number of threads. We selected 4 interesting cases. Cases 1 and 2 are applications that presented a single best thread mapping strategy for all thread counts. Cases 3 and 4 are applications whose the best single thread mapping varied according to the number of threads.

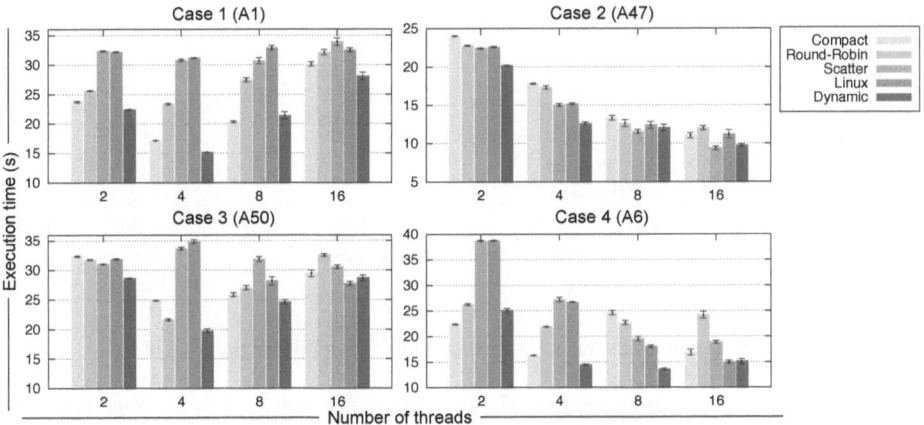

Fig. 6. Execution times when varying the number of threads

Figure 6 compares the execution times of the four single thread mapping strategies with our dynamic thread mapping mechanism. Results represent mean execution times of 30 executions with 95% confidence intervals. We do not consider more than 16 threads for two reasons: (i) placing threads on different cores when all available cores are used does not impact the overall performance because the applications tend to communicate uniformly, and (ii) most of our workloads did not scale beyond 16 threads.

In Case 1, the best single thread mapping for all thread counts was *compact* whereas in Case 2 it was *scatter*. In both cases our dynamic thread mapping presented lower execution times for most of the thread counts. Case 3 represents a scenario in which the best single thread mapping strategy relied on the number of threads (*scatter*, *round-robin*, *compact* and *linux* with 2, 4, 8 and 16 threads respectively). In case 4, we observed that *compact* was best for low thread counts whereas *linux* was best for high thread counts. In both cases 3 and 4, our dynamic thread mapping usually resulted in better results than single thread mappings.

4.4 Dynamic Thread Mapping in Action

In order to observe how our dynamic thread mapping reacts when it encounters several different phases, we created a single application composed of all the 8 distinct workloads. We then executed this application with our dynamic thread mapping while tracing the information obtained by the *transaction profiler* at the end of each profiling period. Figure 7 shows the variance of the profiled metrics during the execution with 4 threads. Vertical bars represent the intervals in which each thread mapping strategy was applied.

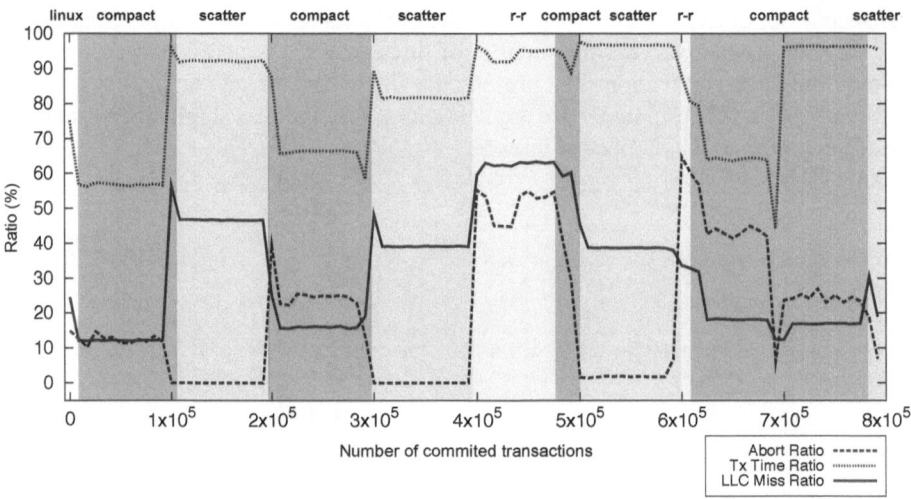

Fig. 7. Profiled metrics during the execution of an application with 8 phases

At the beginning, our dynamic thread mapping mechanism applies *linux* as its default strategy and profiles some transactions. After the first profiling period, the predictor decided to apply *compact* and did not switch to another strategy until it reached a different phase near 1×10^5. At this point, the predictor switched to *scatter*. Overall, the predictor detected more than 8 phases due to the variance of some profiled metrics but it still detected correctly the 8 main phase changes, reacting by applying a suitable thread mapping strategy for each phase. We can also observe that the variance of the profiled metrics confirms the fact that the 8 workloads have distinct characteristics.

5 Related Work

Thread Mapping. In [15], the authors presented a process mapping strategy for MPI applications. The strategy used a graph partitioning algorithm to generate an appropriate process mapping for an application. The proposed strategy was then compared with *compact* and *scatter*. In [4], two thread mapping algorithms were proposed. These algorithms relied on memory traces extracted from benchmarks to find data sharing patterns between threads. These patterns were extracted by running the workloads on a simulator. The proposed approach was compared to *compact*, *scatter* and other strategies. In [7], the authors proposed a dynamic thread mapping strategy for regular data parallel applications implemented with OpenMP. The strategy considered the machine description and the application characteristics to map threads to processors and it was evaluated using simulations. Contrary to these works, our mechanism relies on machine learning to predict the thread mapping strategy without simulations. Instead, we use hardware counters and software libraries to gather information about the platform and applications.

Machine Learning. In [6], authors proposed a ML-based compiler model that accurately predicts the best partitioning of data-parallel OpenCL tasks. Static analysis was used to extract code features from OpenCL programs. These features were used to feed a ML algorithm which was responsible for predicting the best task partitioning among GPUs and CPUs. In [13], the authors proposed a two-staged parallelization approach combining profiling-driven parallelism detection and ML-based mapping to generate OpenMP annotated parallel programs. In this method, first they used profiling to identify portions of code that can be parallelized. Afterwards, they applied a previously trained ML-based prediction mechanism to each parallel loop candidate in order to select a scheduling policy from the four options implemented by OpenMP (*cyclic*, *dynamic*, *guided* or *static*). In [14], the authors proposed a ML-based approach to do thread mapping on parallel applications developed with OpenMP. The proposed solution was capable of predicting the ideal number of threads and the scheduling policy for an application. This approach was compared with the default OpenMP runtime through experiments on a Cell platform. In contrast to those works, we target a different domain of applications, *i.e.*, STM applications. These applications can be more sensitive to thread mapping due to their complex memory access patterns and effects of the underlying STM system.

6 Conclusion

In this paper, we proposed a dynamic thread mapping approach based on Machine Learning for TM applications. We focused on TM applications composed of multiple execution phases with potentially different transactional behavior in each phase. We defined and implemented this mechanism in a state-of-art STM system, making it transparent to the user. To the best of our knowledge, our work is the first to implement dynamic thread mapping for TM applications.

Our results showed that there is not a single thread mapping strategy adapted for all those complex applications. Instead, we could deliver a solution capable of detecting phase changes during the execution of the applications and then predicting a suitable thread mapping strategy adapted for each phase. We achieved performance improvements up to 31% in comparison to the best single strategy.

As future work, we aim at extending our predictor to consider a broader range of STM conflict detection and resolution policies. Additionally, we intend to consider more orthogonal TM characteristics to build even more diverse applications. Consequently, we can extend the evaluation of our approach over more diverse scenarios. Finally, we plan to use other machine learning algorithms to build new thread mapping predictors and compare their performances.

References

1. Broquedis, F., Clet-Ortega, J., Moreaud, S., Goglin, B., Mercier, G., Thibault, S.: hwloc: a Generic Framework for Managing Hardware Affinities in HPC Applications. In: PDP, pp. 180–186. IEEE Computer Society, Pisa (2010)
2. Castro, M., Góes, L.F.W., Ribeiro, C.P., Cole, M., Cintra, M., Méhaut, J.F.: A Machine Learning-Based Approach for Thread Mapping on Transactional Memory Applications. In: HiPC. IEEE Computer Society, Bangalore (2011)
3. Castro, M., Georgiev, K., Marangonzova-Martin, V., Méhaut, J.F., Fernandes, L.G., Santana, M.: Analysis and Tracing of Applications Based on Software Transactional Memory on Multicore Architectures. In: PDP, pp. 199–206. IEEE Computer Society, Aya Napa (2011)
4. Diener, M., Madruga, F., Rodrigues, E., Alves, M., Schneider, J., Navaux, P., Heiss, H.U.: Evaluating Thread Placement Based on Memory Access Patterns for Multi-core Processors. In: HPCC, pp. 491–496. IEEE Computer Society, Melbourne (2010)
5. Felber, P., Fetzer, C., Riegel, T.: Dynamic Performance Tuning of Word-Based Software Transactional Memory. In: PPoPP, pp. 237–246. ACM, NY (2008)
6. Grewe, D., O'Boyle, M.F.P.: A Static Task Partitioning Approach for Heterogeneous Systems Using OpenCL. In: Knoop, J. (ed.) CC 2011. LNCS, vol. 6601, pp. 286–305. Springer, Heidelberg (2011)
7. Hong, S., Narayanan, S.H.K., Kandemir, M., Özturk, O.: Process Variation Aware Thread Mapping for Chip Multiprocessors. In: DATE, pp. 821–826. IEEE Computer Society, Nice (2009)
8. Hong, S., Oguntebi, T., Casper, J., Bronson, N., Kozyrakis, C., Olukotun, K.: Eigenbench: A Simple Exploration Tool for Orthogonal TM Characteristics. In: IISWC, pp. 1–11. IEEE Computer Society, Atlanta (2010)
9. Larus, J., Rajwar, R.: Transactional Memory. Morgan & Claypool (2006)
10. Minh, C.C., Chung, J., Kozyrakis, C., Olukotun, K.: STAMP: Stanford Transactional Applications for Multi-Processing. In: IISWC, pp. 35–46. IEEE Computer Society, Seattle (2008)
11. Quinlan, J.R.: Induction of Decision Trees. Machine Learning 1, 81–106 (1986)
12. Terpstra, D., Jagode, H., You, H., Dongarra, J.: Collecting Performance Data with PAPI-C. In: Parallel Tools Workshop, pp. 157–173. Springer, Berlin (2010)
13. Tournavitis, G., Wang, Z., Franke, B., O'Boyle, M.F.: Towards a Holistic Approach to Auto-Parallelization: Integrating Profile-Driven Parallelism Detection and Machine-Learning Based Mapping. ACM SIGPLAN Not. 44, 177–187 (2009)
14. Wang, Z., O'Boyle, M.F.: Mapping Parallelism to Multi-cores: A Machine Learning Based Approach. ACM SIGPLAN Not. 44, 75–84 (2009)
15. Zhang, J., Zhai, J., Chen, W., Zheng, W.: Process Mapping for MPI Collective Communications. In: Sips, H., Epema, D., Lin, H.-X. (eds.) Euro-Par 2009. LNCS, vol. 5704, pp. 81–92. Springer, Heidelberg (2009)

A Checkpoint-on-Failure Protocol
for Algorithm-Based Recovery in Standard MPI

Wesley Bland, Peng Du, Aurelien Bouteiller, Thomas Herault,
George Bosilca, and Jack Dongarra

Innovative Computing Laboratory, University of Tennessee
1122 Volunteer Blvd., Knoxville, TN 37996-3450, USA
{bland,du,bouteill,herault,bosilca,dongarra}@eecs.utk.edu

Abstract. Most predictions of Exascale machines picture billion way
parallelism, encompassing not only millions of cores, but also tens of
thousands of nodes. Even considering extremely optimistic advances in
hardware reliability, probabilistic amplification entails that failures will
be unavoidable. Consequently, software fault tolerance is paramount to
maintain future scientific productivity. Two major problems hinder ubiquitous adoption of fault tolerance techniques: 1) traditional checkpoint
based approaches incur a steep overhead on failure free operations and
2) the dominant programming paradigm for parallel applications (the
MPI standard) offers extremely limited support of software-level fault
tolerance approaches. In this paper, we present an approach that relies
exclusively on the features of a high quality implementation, as defined by
the current MPI standard, to enable algorithmic based recovery, without
incurring the overhead of customary periodic checkpointing. The validity
and performance of this approach are evaluated on large scale systems,
using the QR factorization as an example.

1   Introduction

The insatiable processing power needs of domain science has pushed High Performance Computing (HPC) systems to feature a significant performance increase
over the years, even outpacing "Moore's law" expectations. Leading HPC systems, whose architectural history is listed in the Top500 [1] ranking, illustrate the
massive parallelism that has been embraced in the recent years; current number
1 – the K-computer – has half a million cores, and even with the advent of GPU
accelerators, it requires no less than 73,000 cores for the Tsubame 2.0 system
(#5) to breach the Petaflop barrier. Indeed, the International Exascale Software
Project, a group created to evaluate the challenges on the path toward Exascale,
has published a public report outlining that a massive increase in scale will be
necessary when considering probable advances in chip technology, memory and
interconnect speeds, as well as limitations in power consumption and thermal
envelope [6]. According to these projections, as early as 2014, billion way parallel machines, encompassing millions of cores, and tens of thousands of nodes,

---

[1] www.top500.org

C. Kaklamanis et al. (Eds.): Euro-Par 2012, LNCS 7484, pp. 477–488, 2012.

will be necessary to achieve the desired level of performance. Even considering extremely optimistic advances in hardware reliability, probabilistic amplification entails that failures will be unavoidable, becoming common events. Hence, fault tolerance is paramount to maintain scientific productivity.

Already, for Petaflop scale systems the issue has become pivotal. On one hand, the capacity type workload, composed of a large amount of medium to small scale jobs, which often represent the bulk of the activity on many HPC systems, is traditionally left unprotected from failures, resulting in diminished throughput due to failures. On the other hand, selected capability applications, whose significance is motivating the construction of supercomputing systems, are protected against failures by ad-hoc, application-specific approaches, at the cost of straining engineering efforts, translating into high software development expenditures. Traditional approaches based on periodic checkpointing and rollback recovery, incurs a steep overhead, as much as 25% [15], on failure-free operations. Forward recovery techniques, most notably Algorithm-Based Fault Tolerant techniques (ABFT), are using mathematical properties to reconstruct failure-damaged data and do exhibit significantly lower overheads [13]. However, and this is a major issue preventing their wide adoption, the resiliency support ABFT demands from the MPI library largely exceeds the specifications of the MPI standard [16] and has proven to be an unrealistic requirement, considering that only a handful of MPI implementations provide it.

The current MPI-2 standard leaves open an optional behavior regarding failures to qualify as a "high quality implementation." According to this specification, when using the MPI_ERRORS_RETURN error handler, the MPI library should return control to the user when it detects a failure. In this paper, we propose the idea of Checkpoint-on-Failure (CoF) as a minimal impact feature to enable MPI libraries to support forward recovery strategies. Despite the default application-wide abort action that all notable MPI implementations undergo in case of a failure, we demonstrate that an implementation that enables CoF is simple and yet effectively supports ABFT recovery strategies that completely avoid costly periodic checkpointing.

The remainder of this paper is organized as follows. The next section presents typical fault tolerant approaches and related works to discuss their requirements and limitations. Then in Section 3 we present the CoF approach, and the minimal support required from the MPI implementation. Section 4 presents a practical use case: the ABFT QR algorithm and how it has been modified to fit the proposed paradigm. Section 5 presents an experimental evaluation of the implementation, followed by our conclusions.

2   Background and Related Work

Message passing is the dominant form of communication used in parallel applications, and MPI is the most popular library used to implement it. In this context, the primary form of fault tolerance today is rollback recovery with periodical checkpoints to disk. While this method is effective in allowing applications to recover from failures using a previously saved state, it causes serious scalability concerns [2]. Moreover, periodic checkpointing requires precise heuristics for fault frequency to minimize the number of superfluous, expensive protective actions [17,10,14,4,1]. In contrast, the work presented here focuses on enabling *forward recovery*. Checkpoint actions are taken only *after* a failure is detected; hence the checkpoint interval is optimal by definition, as there will be one checkpoint interval per effective fault.

Forward recovery leverages algorithms properties to complete operations despite failures. In Naturally Fault Tolerant applications, the algorithm can compute the solution while totally ignoring the contributions of failed processes. In ABFT applications, a recovery phase is necessary, but failure damaged data can be reconstructed only by applying mathematical operations on the remaining dataset [12]. A recoverable dataset is usually created by initially computing redundant data, dispatched so as to avoid unrecoverable loss of information from failures. At each iteration, the algorithm applies the necessary mathematical transformations to update the redundant data (at the expense of more communication and computation). Despite great scalability and low overhead [13,8], the adoption of such algorithms has been hindered by the requirement that the support environment must continue to consistently deliver communications, even after being crippled by failures.

The current MPI Standard (MPI-2.2, [16]) does not provide significant help to deal with the required type of behavior. Section 2.8 states in the first paragraph: *"MPI does not provide mechanisms for dealing with failures in the communication system. [...] Whenever possible, such failures will be reflected as errors in the relevant communication call. Similarly, MPI itself provides no mechanisms for handling processor failures."* Failures, be they due to a broken link or a dead process, are considered resource errors. Later, in the same section: *"This document does not specify the state of a computation after an erroneous MPI call has occurred. The desired behavior is that a relevant error code be returned, and the effect of the error be localized to the greatest possible extent."* So, for the current standard, process or communication failures are to be handled as errors, and the behavior of the MPI application, after an error has been returned, is left unspecified by the standard. However, the standard does not prevent implementations from going beyond its requirements, and on the contrary, encourages high-quality implementations *to return* errors, once a failure is detected. Unfortunately, most of the implementations of the MPI Standard have taken the path of considering process failures as unrecoverable errors, and the processes of the application are most often killed by the runtime system when a failure hits any of them, leaving no opportunity for the user to mitigate the impact of failures.

Some efforts have been undertaken to enable ABFT support in MPI. FT-MPI [9] was an MPI-1 implementation which proposed to change the MPI semantic to enable repairing communicators, thus re-enabling communications for applications damaged by failures. This approach has proven successful and applications have been implemented using FT-MPI. However, these modifications were not adopted by the MPI standardization body, and the resulting lack of portability undermined user adoption for this fault tolerant solution.

In [11], the authors discuss alternative or slightly modified interpretations of the MPI standard that enable some forms of fault tolerance. One core idea is that process failures happening in another MPI world, connected only through an inter-communicator, should not prevent the continuation of normal operations. The complexity of this approach, for both the implementation and users, has prevented these ideas from having a practical impact. In the CoF approach, the only requirement from the MPI implementation is that it does not forcibly kill the living processes without returning control. No stronger support from the MPI stack is required, and the state of the library is left undefined. This simplicity has enabled us to actually implement our proposition, and then experimentally support and evaluate a real ABFT application. Similarly, little effort would be required to extend MPICH-2 to support CoF (see Section 7 of the Readme[2]).

3   Enabling Algorithm-Based Fault Tolerance in MPI

3.1   The Checkpoint-on-Failure Protocol

In this paper, we advocate that an extremely efficient form of fault tolerance can be implemented, strictly based on the MPI standard, for applications capable of taking advantage of forward recovery. ABFT methods are an example of forward recovery algorithms, capable of restoring missing data from redundant information located on other processes. This forward recovery step requires communication between processes, and we acknowledge that, in light of the current standard, requiring the MPI implementation to maintain service after failures is too demanding. However, a high-quality MPI library should at least allow the application to regain control following a process failure. We note that this control gives the application the opportunity to save its state and exit gracefully, rather than the usual behavior of being aborted by the MPI implementation.

Algorithm 1. The Checkpoint-on-Failure Protocol

1. MPI returns an error on surviving processes 2. Surviving processes checkpoint 3. Surviving processes exit 4. A new MPI application is started 5. Processes load from checkpoint (if any) 6. Processes enter ABFT dataset recovery 7. Application resumes	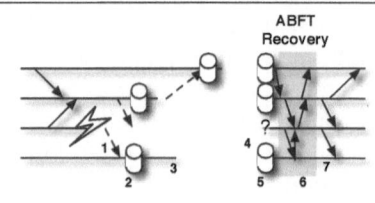

Based on these observations, we propose a new approach for supporting ABFT applications, called Checkpoint-on-Failure (CoF). Algorithm 1 presents the steps

---

[2] http://www.mcs.anl.gov/research/projects/mpich2/documentation/
files/mpich2-1.4.1-README.txt

involved in the CoF method. In the associated explanatory figure, horizontal lines represent the execution of processes in two successive MPI applications. When a failure eliminates a process, other processes are notified and regain control from ongoing MPI calls (1). Surviving processes assume the MPI library is dysfunctional and do not further call MPI operations (in particular, they do not yet undergo ABFT recovery). Instead, they checkpoint their current state independently and abort (2, 3). When all processes exited, the job is usually terminated, but the user (or a managing script, batch scheduler, runtime support system, etc.) can launch a new MPI application (4), which reloads processes from checkpoint (5). In the new application, the MPI library is functional and communications possible; the ABFT recovery procedure is called to restore the data of the process(es) that could not be restarted from checkpoint (6). When the global state has been repaired by the ABFT procedure, the application is ready to resume normal execution.

Compared to periodic checkpointing, in CoF, a process pays the cost of creating a checkpoint only when a failure, or multiple simultaneous failures have happened, hence an optimal number of checkpoints during the run (and no checkpoint overhead on failure-free executions). Moreover, in periodic checkpointing, a process is protected only when its checkpoint is stored on safe, remote storage, while in CoF, local checkpoints are sufficient: the forward recovery algorithm reconstructs datasets of processes which cannot restart from checkpoint. Of course, CoF also exhibits the same overhead as the standard ABFT approach: the application might need to do extra computation, even in the absence of failures, to maintain internal redundancy (whose degree varies with the maximum number of simultaneous failures) used to recover data damaged by failures. However, ABFT techniques often demonstrate excellent scalability; for example, the overhead on failure-free execution of the ABFT QR operation (used as an example in Section 4) is inversely proportional to the number of processes.

3.2   MPI Requirements for Checkpoint-on-Failure

Returning Control over Failures: In most MPI implementations, MPI_ERRORS_ABORT is the default (and often, only functional) error handler. However, the MPI standard also defines the MPI_ERRORS_RETURN handler. To support CoF, the MPI library should never deadlock because of failures, but invoke the error handler, at least on processes doing direct communications with the failed process. The handler takes care of cleaning up at the library level and returns control to the application.

Termination after Checkpoint: A process that detects a failure ceases to use MPI. It only checkpoints on some storage and exits without calling MPI_Finalize. Exiting without calling MPI_Finalize is an error from the MPI perspective, hence the failure cascades and MPI eventually returns with a failure notification on every process, which triggers their own checkpoint procedure and termination.

3.3   Open MPI Implementation

Open MPI is an MPI 2.2 implementation architected such that it contains two main levels, the runtime (ORTE) and the MPI library (OMPI). As with most MPI library implementations, the default behavior of Open MPI is to abort after a process failure. This policy was implemented in the runtime system, preventing any kind of decision from the MPI layer or the user-level. The major change requested by the CoF protocol was to make the runtime system resilient, and leave the decision in case of failure to the MPI library policy, and ultimately to the user application.

Failure Resilient Runtime: The ORTE runtime layer provides an out-of-band communication mechanism (OOB) that relays messages based on a routing policy. Node failures not only impact the MPI communications, but also disrupt routing at the OOB level. The default routing policy in the Open MPI runtime has been amended to allow for self-healing behaviors; this effort is not entirely necessary, but it avoids the significant downtime imposed by a complete redeployment of the parallel job with resubmission in queues. The underlying OOB topology is automatically updated to route around failed processes. In some routing topologies, such as a star, this is a trivial operation and only requires excluding the failed process from the routing tables. For more elaborate topologies, such as a binomial tree, the healing operation involves computing the closest neighbors in the direction of the failed process and reconnecting the topology through them. The repaired topology is not rebalanced, resulting in degraded performance but complete functionality after failures. Although in-flight messages that were currently "hopping" through the failed processes are lost, other in-flight messages are safely routed on the repaired topology. Thanks to self-healing topologies, the runtime remains responsive, even when MPI processes leave.

Failure Notification: The runtime has been augmented with a failure detection service. To track the status of the failures, an incarnation number has been included in the process names. Following a failure, the name of the failed process (including the incarnation number) is broadcasted over the OOB topology. By including this incarnation number, we can identify transient process failures, prevent duplicate detections, and track message status. ORTE processes monitor the health of their neighbors in the OOB routing topology. Detection of other processes rely on a failure resilient broadcast that overlays on the OOB topology. This algorithm has a low probability of creating a bi-partition of the routing topology, hence ensuring a high accuracy of the failure detector. However, the underlying OOB routing algorithm has a significant influence on failure detection and propagation time, as the experiments will show. On each node, the ORTE runtime layer forwards failure notifications to the MPI layer, which has been modified to invoke the appropriate MPI error handler.

4   Example: The QR Factorization

In this section, we propose to illustrate the applicability of CoF by considering a representative routine of a widely used class of algorithms: dense linear factorizations. The QR factorization is a cornerstone building block in many applications, including solving $Ax = b$ when matrices are ill-conditioned, computing eigenvalues, least square problems, or solving sparse systems through the GMRES iterative method. For an $M \times N$ matrix A, the QR factorization produces Q and R, such that $A = QR$ and Q is an $M \times M$ orthogonal matrix and R is an $M \times N$ upper triangular matrix. The most commonly used implementation of the QR algorithm on a distributed memory machine comes from the ScaLAPACK linear algebra library [7], based on the block QR algorithm. It uses a 2D block-cyclic distribution for load balance, and is rich in level 3 BLAS operations, thereby achieving high performance.

4.1   ABFT QR Factorization

In the context of FT-MPI, the ScaLAPACK QR algorithm has been rendered fault tolerant through an ABFT method in previous works [8]. This ABFT algorithm protects both the left (Q) and right (R) factors from fail-stop failures at any time during the execution. At the time of failure, every surviving process is notified by FT-MPI. FT-MPI then spawns a replacement process that takes the same grid coordinates in the $P \times Q$ block-cyclic distribution. Missing checksums are recovered from duplicates, a reduction collective communication recovers missing data blocks in the right factor from checksums. The left factor is protected by the Q-parallel panel checksum, it is either directly recovered from checksum, or by recomputing the panels in the current Q-wide section (see [8]). Although this algorithm is fault tolerant, it requires continued service from the MPI library after failures – which is a stringent requirement that can be waived with CoF.

4.2   Checkpoint-on-Failure QR

Checkpoint Procedure: Compared to a regular ABFT algorithm, CoF requires a different checkpoint procedure. System-level checkpointing is not applicable, as it would result in restoring the state of the broken MPI library upon restart. Instead, a custom MPI error handler invokes an algorithm specific checkpoint procedure, which simply dumps the matrices and the value of important loop indices into a file.

State Restoration: A ScaLAPACK program has a deep call stack, layering functions from multiple software packages, such as PBLAS, BLACS, LAPACK and BLAS. In the FT-MPI version of the algorithm, regardless of when the failure is detected, the current iteration of the algorithm must be completed before entering the recovery procedure. This ensures an identical call stack on every process and a complete update of the checksums. In the case of the CoF protocol, failures interrupt the algorithm immediately, the current iteration cannot be completed due to lack of communication capabilities. This results in potentially diverging call stacks and incomplete updates of checksums. However, because failure notification happens only in MPI, lower level, local procedures (BLAS, LAPACK) are never interrupted.

To resolve the call stack issue, every process restarted from checkpoint undergoes a "dry run" phase. This operation mimics the loop nests of the QR algorithm down to the PBLAS level, without actually applying modifications to or exchanging data. When the same loop indices as before the failure are reached, the matrix content is loaded from the checkpoint; the state is then similar to that of the FT-MPI based ABFT QR after a failure. The regular recovery procedure can be applied: the current iteration of the factorization is completed to update all checksums and the dataset is rebuilt using the ABFT reduction.

5   Performance Discussion

In this section, we use our Open MPI and ABFT QR implementations to evaluate the performance of the CoF protocol. We use two test platforms. The first machine, "Dancer," is a 16-node cluster. All nodes are equipped with two 2.27GHz quad-core Intel E5520 CPUs, with a 20GB/s Infiniband interconnect. Solid State Drive disks are used as the checkpoint storage media. The second system is the Kraken supercomputer. Kraken is a Cray XT5 machine, with 9,408 compute nodes. Each node has two Istanbul 2.6 GHz six-core AMD Opteron processors, 16 GB of memory, and are connected through the SeaStar2+ interconnect. The scalable cluster file system "Lustre" is used to store checkpoints.

5.1   MPI Library Overhead

One of the concerns with fault tolerance is the amount of overhead introduced by the fault tolerance management additions. Our implementation of fault detection and notification is mostly implemented in the non-critical ORTE runtime. Typical HPC systems feature a separated service network (usually Ethernet based) and a performance interconnect, hence health monitoring traffic, which happens on the OOB service network, is physically separated from the MPI communications, leaving no opportunity for network jitter. Changes to MPI functions are minimal: the same condition that used to trigger unconditional abort has been repurposed to trigger error handlers. As expected, no impact on MPI bandwidth or latency was measured (Infiniband and Portals results not shown for lack of space). The memory usage of the MPI library is slightly increased, as the incarnation number doubles the size of process names; however, this is negligible in typical deployments.

5.2   Failure Detection

According to the requirement specified in Section 3.2, only in-band failure detection is required to enable CoF. Processes detecting a failure checkpoint then exit, cascading the failure to processes communicating with them. However, no recovery action (in particular checkpointing) can take place before a failure has been notified. Thanks to asynchronous failure propagation in the runtime, responsiveness can be greatly improved, with a high probability for the next MPI call to detect the failures, regardless of communication pattern or checkpoint duration.

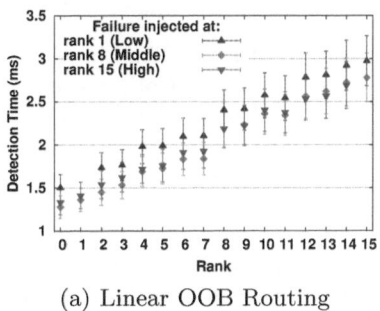
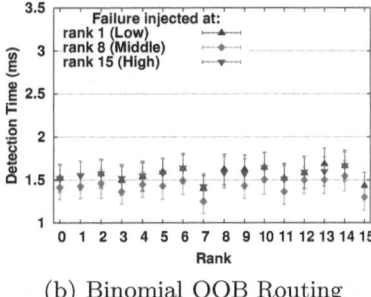

(a) Linear OOB Routing   (b) Binomial OOB Routing

Fig. 1. Failure detection time, sorted by process rank, depending on the OOB overlay network used for failure propagation

We designed a micro-benchmark to measure failure detection time as experienced by MPI processes. The benchmark code synchronizes with an MPI_Barrier, stores the reference date, injects a failure at a specific rank, and enters a ring algorithm until the MPI error handler stores the detection date. The OOB routing topology used by the ORTE runtime introduces a non-uniform distance to the failed process, hence failure detection time experienced by a process may vary with the position of the failed process in the topology, and the OOB topology. Figure 1(a) and 1(b) present the case of the linear and binomial OOB topologies, respectively. The curves "Low, Middle, High" present the behavior for failures happening at different positions in the OOB topology. On the horizontal axis is the rank of the detecting process, on the vertical axis is the detection time it experienced. The experiment uses 16 nodes, with one process per node, MPI over Infiniband, OOB over Ethernet, an average of 20 runs, and the MPI barrier latency is four orders of magnitude lower than measured values.

In the linear topology (Figure 1(a)) every runtime process is connected to the *mpirun* process. For a higher rank, failure detection time increases linearly because it is notified by the *mpirun* process only after the notification has been sent to all lower ranks. This issue is bound to increase with scale. The binomial tree topology (Figure 1(b)) exhibits a similar best failure detection time. However, this more scalable topology has a low output degree and eliminates most contentions on outgoing messages, resulting in a more stable, lower average detection time, regardless of the failure position. Overall, failure detection time is on the order of milliseconds, a much smaller figure than typical checkpoint time.

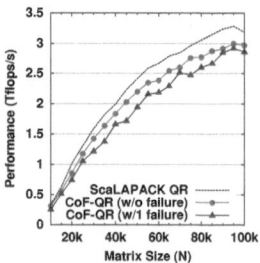
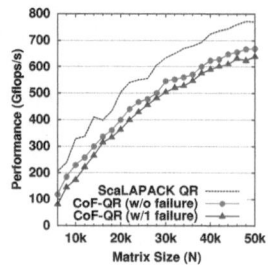
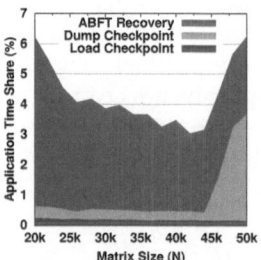

Fig. 2. ABFT QR and one CoF recovery on Kraken (Lustre)

Fig. 3. ABFT QR and one CoF recovery on Dancer (local SSD)

Fig. 4. Time breakdown of one CoF recovery on Dancer (local SSD)

5.3 Checkpoint-on-Failure QR Performance

Supercomputer Performance: Figure 2 presents the performance on the Kraken supercomputer. The process grid is 24×24 and the block size is 100. The CoF-QR (no failure) presents the performance of the CoF QR implementation, in a fault-free execution; it is noteworthy, that when there are no failures, the performance is exactly identical to the performance of the unmodified FT-QR implementation. The CoF-QR (with failure) curves present the performance when a failure is injected after the first step of the PDLARFB kernel. The performance of the non-fault tolerant ScaLAPACK QR is also presented for reference.

Without failures, the performance overhead compared to the regular ScaLAPACK is caused by the extra computation to maintain the checksums inherent to the ABFT algorithm [8]; this extra computation is unchanged between CoF-QR and FT-QR. Only on runs where a failure happened do the CoF protocols undergoe the supplementary overhead of storing and reloading checkpoints. However, the performance of the CoF-QR remains very close to the no-failure case. For instance, at matrix size N=100,000, CoF-QR still achieves 2.86 Tflop/s after recovering from a failure, which is 90% of the performance of the non-fault tolerant ScaLAPACK QR. This demonstrates that the CoF protocol enables efficient, practical recovery schemes on supercomputers.

Impact of Local Checkpoint Storage: Figure 3 presents the performance of the CoF-QR implementation on the Dancer cluster with a 8 × 16 process grid. Although a smaller test platform, the Dancer cluster features local storage on nodes and a variety of performance analysis tools unavailable on Kraken. As expected (see [8]), the ABFT method has a higher relative cost on this smaller machine. Compared to the Kraken platform, the relative cost of CoF failure recovery is smaller on Dancer. The CoF protocol incurs disk accesses to store and load checkpoints when a failure hits, hence the recovery overhead depends on I/O performance. By breaking down the relative cost of each recovery step in CoF, Figure 4 shows that checkpoint saving and loading only take a small percentage of the total run-time, thanks to the availability of solid state disks on every node.

Since checkpoint reloading immediately follows checkpointing, the OS cache satisfy most disk access, resulting in high I/O performance. For matrices larger than N=44,000, the memory usage on each node is high and decrease the available space for disk cache, explaining the decline in I/O performance and the higher cost of checkpoint management. Overall, the presence of fast local storage can be leveraged by the CoF protocol to speedup recovery (unlike periodic checkpointing, which depends on remote storage by construction). Nonetheless, as demonstrated by the efficiency on Kraken, while this is a valuable optimization, it is not a mandatory requirement for satisfactory performance.

6   Concluding Remarks

In this paper, we presented an original scheme to enable forward recovery using only features of the current MPI standard. Rollback recovery, which relies on periodic checkpointing has a variety of issues. The ideal period of checkpoint, a critical parameter, is particularly hard to assess. Too short a period wastes time and resources on unnecessary Input/Output. Overestimating the period results in dramatically increasing the lost computation when returning to the distant last successful checkpoint. Although Checkpoint-on-Failure involves checkpointing, it takes checkpoint images at optimal times by design: only after a failure has been detected. This small modification enables the deployment of ABFT techniques, without requiring a complex, unlikely to be available MPI implementation that itself survives failures. The MPI library needs only to provide the feature set of a high quality implementation of the MPI standard: the MPI communications may be dysfunctional after a failure, but the library must return control to the application instead of aborting brutally.

We demonstrated, by providing such an implementation in Open MPI, that this feature set can be easily integrated without noticeable impact on communication performance. We then converted an existing ABFT QR algorithm to the CoF protocol. Beyond this example, the CoF protocol is applicable on a large range of applications that already feature an ABFT version (LLT, LU [5], CG [3], etc.). Many master-slave and iterative methods enjoy an extremely inexpensive forward recovery strategy where the damaged domains are simply discarded, and can also benefit from the CoF protocol.

The performance on the Kraken supercomputer reaches 90% of the non-fault tolerant algorithm, even when including the cost of recovering from a failure (a figure similar to regular, non-compliant MPI ABFT). In addition, on a platform featuring node local storage, the CoF protocol can leverage low overhead checkpoints (unlike rollback recovery that requires remote storage).

The MPI standardization body, the MPI Forum, is currently considering the addition of new MPI constructs, functions and semantics to support fault-tolerant applications[3]. While these additions may decrease the cost of recovery, they are likely to increase the failure-free overhead on fault tolerant application

---

[3] https://svn.mpi-forum.org/trac/mpi-forum-web/wiki/FaultToleranceWikiPage

performance. It is therefore paramount to compare the cost of the CoF protocol with prospective candidates to standardization on a wide, realistic range of applications, especially those that feature a low computational intensity.

References

1. Cappello, F., Casanova, H., Robert, Y.: Preventive migration vs. preventive checkpointing for extreme scale supercomputers. PPL 21(2), 111–132 (2011)
2. Cappello, F., Geist, A., Gropp, B., Kalé, L.V., Kramer, B., Snir, M.: Toward exascale resilience. IJHPCA 23(4), 374–388 (2009)
3. Chen, Z., Fagg, G.E., Gabriel, E., Langou, J., Angskun, T., Bosilca, G., Dongarra, J.: Fault tolerant high performance computing by a coding approach. In: Proceedings of the Tenth ACM SIGPLAN Symposium on Principles and Practice of Parallel Programming, PPoPP 2005, pp. 213–223. ACM, New York (2005)
4. Daly, J.T.: A higher order estimate of the optimum checkpoint interval for restart dumps. Future Gener. Comput. Syst. 22, 303–312 (2006)
5. Davies, T., Karlsson, C., Liu, H., Ding, C., Chen, Z.: High Performance Linpack Benchmark: A Fault Tolerant Implementation without Checkpointing. In: Proceedings of the 25th ACM International Conference on Supercomputing (ICS 2011). ACM (2011)
6. Dongarra, J., Beckman, P., et al.: The international exascale software roadmap. IJHPCA 25(11), 3–60 (2011)
7. Dongarra, J.J., Blackford, L.S., Choi, J., et al.: ScaLAPACK user's guide. Society for Industrial and Applied Mathematics, Philadelphia (1997)
8. Du, P., Bouteiller, A., Bosilca, G., Herault, T., Dongarra, J.: Algorithm-based Fault Tolerance for Dense Matrix Factorizations. In: 17th ACM SIGPLAN Symposium on Principles and Practice of Parallel Programming. ACM (2012)
9. Fagg, G.E., Dongarra, J.: FT-MPI: Fault Tolerant MPI, Supporting Dynamic Applications in a Dynamic World. In: Dongarra, J., Kacsuk, P., Podhorszki, N. (eds.) PVM/MPI 2000. LNCS, vol. 1908, p. 346. Springer, Heidelberg (2000)
10. Gelenbe, E.: On the optimum checkpoint interval. JoACM 26, 259–270 (1979)
11. Gropp, W., Lusk, E.: Fault tolerance in message passing interface programs. Int. J. High Perform. Comput. Appl. 18, 363–372 (2004)
12. Huang, K.H., Abraham, J.A.: Algorithm-based fault tolerance for matrix operations. IEEE Transactions on Computers 100(6), 518–528 (1984)
13. Luk, F.T., Park, H.: An analysis of algorithm-based fault tolerance techniques. Journal of Parallel and Distributed Computing 5(2), 172–184 (1988)
14. Plank, J.S., Thomason, M.G.: Processor allocation and checkpoint interval selection in cluster computing systems. JPDC 61, 1590 (2001)
15. Schroeder, B., Gibson, G.A.: Understanding Failures in Petascale Computers. SciDAC, Journal of Physics: Conference Series 78 (2007)
16. The MPI Forum. MPI: A Message-Passing Interface Standard, Version 2.2. Technical report (2009)
17. Young, J.W.: A first order approximation to the optimum checkpoint interval. Commun. ACM 17, 530–531 (1974)

Hierarchical Partitioning Algorithm for Scientific Computing on Highly Heterogeneous CPU + GPU Clusters

David Clarke[1], Aleksandar Ilic[2], Alexey Lastovetsky[1], and Leonel Sousa[2]

[1] School of Computer Science and Informatics, University College Dublin, Belfield, Dublin 4, Ireland
[2] INESC-ID, IST/Technical University of Lisbon, Rua Alves Redol, 9, 1000-029 Lisbon, Portugal

Abstract. Hierarchical level of heterogeneity exists in many modern high performance clusters in the form of heterogeneity between computing nodes, and within a node with the addition of specialized accelerators, such as GPUs. To achieve high performance of scientific applications on these platforms it is necessary to perform load balancing. In this paper we present a hierarchical matrix partitioning algorithm based on realistic performance models at each level of hierarchy. To minimise the total execution time of the application it iteratively partitions a matrix between nodes and partitions these sub-matrices between the devices in a node. This is a self-adaptive algorithm that dynamically builds the performance models at run-time and it employs an algorithm to minimise the total volume of communication. This algorithm allows scientific applications to perform load balanced matrix operations with nested parallelism on hierarchical heterogeneous platforms. To show the effectiveness of the algorithm we applied it to a fundamental operation in scientific parallel computing, matrix multiplication. Large scale experiments on a heterogeneous multi-cluster site incorporating multicore CPUs and GPU nodes show that the presented algorithm outperforms current state of the art approaches and successfully load balance very large problems.

Keywords: parallel applications, heterogeneous platforms, GPU, data partitioning algorithms, functional performance models, matrix multiplication.

1 Introduction

In this paper we present a matrix partitioning algorithm for load balancing parallel applications running on highly heterogeneous hierarchical platforms. The target platform is a dedicated heterogeneous distributed memory platform with multi level hierarchy. More specifically, we focus on a platform with two levels of hierarchy. At the top level is a distributed memory cluster of heterogeneous nodes, and at the lower level, each node consists of a number of devices which may be a combination of multicore CPUs and specialized accelerators/co-processors (GPUs). We refer to both nodes and devices collectively as processing

C. Kaklamanis et al. (Eds.): Euro-Par 2012, LNCS 7484, pp. 489–501, 2012.

elements. The applications we target perform matrix operations and are characterised by discretely divisible computational workloads where the computations can be split into independent units, such that each computational unit requires the same amount of computational work. In addition, computational workload is directly proportional to the size of data and dependent on data locality. High performance of these applications can be achieved on heterogeneous platforms by performing load balancing at each level of hierarchy. Load balancing ensures that all processors complete their work within the same time. This requirement is satisfied by partitioning the computational workload unevenly between processing elements, at each level of hierarchy, with respect to the performance of that element.

In order to achieve load balancing on our target platform, the partitioning algorithm must be designed to take into account both the hierarchy and the high level of heterogeneity of the platform. In contrast to the traditional, CPU-only distributed memory systems, highly heterogeneous environments employ devices which have fundamental architectural differences. The ratio of performance differences between devices may be orders of magnitude more than the ratio between traditional heterogeneous platforms; moreover this ratio can vary greatly with a change in problem size. For example, accelerators need to physically load and offload portions of data on which computations are performed in order to ensure high performance and full execution control, and the executable problem size is limited by the available device memory. Finally, architectural differences impose new collaborative programming challenges, where it becomes necessary to use different programming models, vendor-specific tools and libraries in order to approach the per-device peak performance. However, even if some of the already existing collaborative execution environments are used (such as OpenCL, StarPU [1] or CHPS [11]), the problem of efficient cross-device problem partitioning and load balancing still remains.

The work proposed herein takes into account this complex heterogeneity by using realistic performance models of the employed devices and nodes. The model of each device or node is constructed by measuring the real performance of the application when it runs on that device or node. Thus, they are capable of intrinsically encapsulating all the above-mentioned architectural and performance diversities. Traditional partitioning algorithms define the performance of each processor by a single number. We refer to this simplistic model of processor performance as a constant performance model (CPM). The functional performance model (FPM), proposed in [16], is a more realistic model of processor performance, where processor speed is a function of problem size. Partitioning algorithms which use these FPMs always achieve better load balancing than traditional CPM-based algorithms.

The main contribution of this work is a new hierarchical matrix partitioning algorithm, based on functional performance models, which performs load balancing at both the node and device levels. This algorithm performs a one to one mapping of computational workload and data to nodes and a one to one mapping of workload to devices. The device level partitioning is performed on each node by sub-partitioning workload assigned to that node. In contrast to the some state of the art approaches, this algorithm does not require any a priori information about the platform, instead all required performance information is found by performing real benchmarks of the core computational kernel of an application.

To the best of our knowledge this is the first work that targets large scale partitioning problems for hierarchical and highly heterogeneous distributed systems. To show the effectiveness of the proposed algorithm we applied it to parallel matrix multiplication, which is representative of the class of computationally intensive parallel scientific applications that we target. Experiments on 3 interconnected computing clusters, using a total of 90 CPU+GPU heterogeneous nodes, showed that, for a wide range of problem sizes, the application based on FPM-based partitioning outperformed applications based on CPM algorithms.

The rest of the paper is organized as follows. In Section 2, we discuss related work. In Section 3, we propose hierarchical partitioning algorithm for highly heterogeneous CPU+GPU clusters. The experimental results are presented in Section 4. Finally, concluding remarks are given in Section 5.

2   Related Work

Divisible load theory(DLT), surveyed in [21], defines a class of applications characterised by workload that can be divided into discrete parts for parallel computation. The applications we target belong to this class. Scheduling and work stealing algorithms [7,3,20], often used in DLT, move workload between processing elements, during execution of the application, to achieve load balancing. However, on distributed memory platforms, such an approach can incur a high cost of data migration with applications where data locality is important. Moreover we are not aware of any dynamic-scheduling/work-stealing matrix multiplication application for highly heterogeneous distributed memory platforms.

A different class of load balancing algorithms are *partitioning algorithms*, also known as *predicting-the-future*; so called because they rely on performance models as input to predict the future execution characteristics of the application. The global workload is partitioned between the available processing elements. Traditional partitioning algorithms [2,8,10,14,18,19] model processor performance by a single positive number and partition workload proportionally. We refer to these simplistic models as constant performance models (CPM).

The partitioning algorithm proposed in this paper predicts future performance by using more realistic functional performance models (FPM) [16]. This algorithm is designed to be self-adaptable [17], making it suitable for applications for which each run is considered to be unique because of a change of input parameters or execution on a unique subset of hardware. This is achieved by dynamically building partial estimates of the full speed functions to the required degree of accuracy. It has been shown in [5] that applications using partitioning based on FPMs can outperform applications based on CPMs. In [12], we investigated the potentials of hierarchical divisible load scheduling on our target platform using the master-worker paradigm. Experiments on a network of off-the-shelf heterogeneous desktops (CPU + GPU), shows the benefit of using realistic performance models to load balance and efficiently overlap computations and communications at the GPU device level. In this paper, we focus on load balancing with respect to computational performance of processing elements, and to this end, we do not measure the interconnect speed between each pair of processing elements; instead we arrange elements such that the communication volume is minimised [2].

Several scientific studies have already dealt with the problems investigated herein, but only partially. For example, MAGMA [9] is a library for matrix algebra for GPU and multicore which uses scheduling for load balancing, but only on a single node. In terms of the target platform, [15,13] consider homogeneous multi-GPU cluster systems without CPUs, whereas [6] is designed for a homogeneous hierarchical platform.

3 Hierarchical Matrix Partitioning Algorithm

A typical computationally intensive parallel scientific application performs the same iterative core computation on a set of data. The general scheme of such an application can be summarised as follows: (i) all data is partitioned over processing elements, (ii) some independent calculations are carried out in parallel, and (iii) some synchronisation takes place. High performance on a distributed memory, hierarchical heterogeneous platform, for such an application, is achieved by partitioning workload in proportion to the speed of the processing elements. The speed of a processing element is best represented by a continuous function of problem size [5]. These FPMs are built empirically for each application on each processing element.

Building these speed functions for the full range of potential problem sizes can be expensive. To reduce this cost and allow the parallel application to be self adaptable to new platforms we make two optimisations: (i) many computationally intensive scientific applications repeat the same core computational kernel many times on different data; to find the performance of this application for a given problem size it is only necessary to benchmark one representative iteration of the kernel; (ii) partial estimates of the speed functions may be built at application run-time to a sufficient level of accuracy to achieve load balancing [17].

Our target platform is a two level hierarchical distributed platform with q nodes, $Q_1, \ldots, Q_q$, where a node Q_i has p_i devices, $P_{i1}, \ldots, P_{ip_i}$. The problem to be solved by this algorithm is to partition a matrix between these nodes and devices with respect to the performance of each of these processing elements. The proposed partitioning algorithm is iterative and converges towards an optimum distribution which balances the workload. It consists of two iterative algorithms, *inter-node partitioning algorithm (INPA)* and *inter-device partitioning algorithm (IDPA)*. The IDPA algorithm is nested inside the INPA algorithm.

Without loss of generality we will work with square $N \times N$ matrices. We introduce a blocking factor b to allow optimised libraries to achieve their peak performance as well as reducing the number of communications. For simplicity we assume N to be a multiple of b, hence there is a total of W computational units to be distributed, where $W = (N/b) \times (N/b)$.

The INPA partitions the total matrix into q sub-matrices to be processed on each heterogeneous computing node. The sub-matrix owned by node Q_i has an area equal to $w_i \times b \times b$, where $w_1 + \ldots + w_q = W$. The *Geometric partitioning algorithm* (GPA) uses experimentally built speed functions to calculate a load balanced distribution $w_1, \ldots, w_q$. The shape and ordering of these sub-matrices is calculated by the *communication minimising algorithm* (CMA). The CMA uses column-based 2D arrangement of nodes and outputs the heights bm_i and widths bn_i for each of the q nodes, such that $m_i \times n_i = w_i$, $bm = b \times m$ and $bn = b \times n$ (Fig. 1(a)). This two dimensional partitioning algorithm uses a column-based arrangement of processors. The values of m_i and n_i are chosen so that the column widths sum up to N and heights of sub-matrices in a column sum to N.

The IDPA iteratively measures, on each device, the time of execution of the application specific core computational kernel with a given size while converging to a load balanced inter-device partitioning. It returns the kernel execution time of the last iteration to the INPA. IDPA calls the GPA to partition the sub-matrix owned by Q_i into vertical slices of width d_{ij}, such that $d_{i1} + \ldots + d_{ip} = bn_i$ (Fig. 1(b)) to be processed on each device within a Q_i node. Device P_{ij} will be responsible for doing matrix operations on $bm_i \times d_{ij}$ matrix elements.

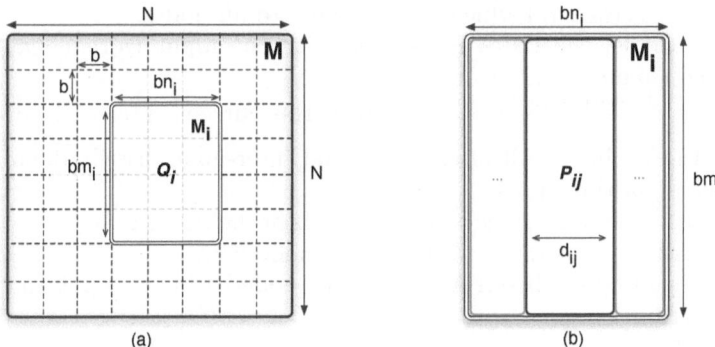

(a)   (b)

Fig. 1. Two level matrix partitioning scheme: (a) two dimensional partitioning between the nodes; (b) one dimensional partitioning between devices in a node

We now present an outline of a parallel application using the proposed hierarchical partitioning algorithm. The partitioning is executed immediately before execution of the parallel algorithm. The outline is followed by a detailed description of the individual algorithms.

INPA$\Big($IN: $N, b, q, p_1, \ldots, p_q$ OUT: $\{m_i, n_i, d_{i1}, \ldots, d_{ip}\}_{i=1}^q\Big)$ {
   WHILE inter-node imbalance
     CMA$\Big($IN: $w_1, \ldots, w_q$ OUT: $(m_1, n_1), \ldots, (m_q, n_q)\Big)$;
    On each node i (IDPA):
      WHILE inter-device imbalance
       On each device j: **kernel**$\Big($IN: bm_i, bn_i, d_{ij} OUT: $t_{ij}\Big)$;
       GPA$\Big($IN: $p_i, bn_i, p_i\text{FPMs}$ OUT: $d_{i1}, \ldots, d_{iq}\Big)$;
      END WHILE
    GPA$\Big($IN: $q, W, q\text{FPMs}$ OUT: $w_1, \ldots, w_q\Big)$;
   END WHILE
}
Parallel application$\Big($IN: $\{m_i, n_i, d_{i1}, \ldots, d_{ip}\}_{i=1}^q, \cdots\Big)$

Inter-Node Partitioning Algorithm (INPA)

Run in parallel on all nodes with distributed memory. Inputs: square matrix size N, number of nodes q, number devices in each node $p_1, \ldots, p_q$ and block size b.

1. To add initial small point to the model, each node, in parallel, invokes the IDPA with an input $(p_i, bm_i = 1, bn_i = 1)$. This algorithm returns a time which is sent to the head node.
2. The head node calculates speeds from these times as $s_i(1) = 1/t_i(1)$ and adds the first point, $(1, s(1))$, to the model of each node.
3. The head node then computes the initial homogeneous distribution by dividing the total number of blocks, W, between processors $w_i = W/q$.
4. The CMA is passed $w_1, \ldots, w_q$ and returns the inter-node distributions $(m_1, n_1), \ldots, (m_q, n_q)$ which are scattered to all nodes.
5. On each node, the IDPA is invoked with the input (p_i, bm_i, bn_i) and the returned time t_i is sent to the head node.
6. IF $\max\limits_{1 \le i,j \le q} \left| \frac{t_i(w_i) - t_j(w_i)}{t_i(w_i)} \right| \le \varepsilon_1$ THEN the current inter-node distribution solves the problem. All inter-device and inter-node distributions are saved and the algorithm stops;
   ELSE the head node calculates the speeds of the nodes as $s_i(w_i) = w_i/t_i(w_i)$ and adds the point $(w_i, s_i(w_i))$ to each node-FPM.
7. On the head node, the GPA is given the node-FPMs as input and returns a new distribution $w_1, \ldots, w_q$
8. GOTO 4

Inter-Device Partitioning Algorithm (IDPA)

This algorithm is run on a node with p devices. The input parameters are p and the sub-matrix sizes bm, bn. It computes the device distribution $d_1, \cdots, d_p$ and returns the time of last benchmark.

1. To add an initial small point to each device model, the *kernel* with parameters $(bm, bn, 1)$ is run in parallel on each device and its execution time is measured. The speed is computed as $s_j(1) = 1/t_j(1)$ and the point $(1, s_j(1))$ is added to each device model.

2. The initial homogeneous distribution $d_j = bn/p$, for all $1 \leq j \leq p$ is set.
3. In parallel on each device, the time $t_j(d_j)$ to execute the kernel with parameters (bm, bn, d_j) is measured.
4. IF $\max\limits_{1 \leq i,j \leq p} \left| \frac{t_i(d_i) - t_j(d_j)}{t_i(d_i)} \right| \leq \varepsilon_2$ THEN the current distribution of computations over devices solves the problem. This distribution $d_1, \cdots, d_p$ is saved and $\max\limits_{1 \leq j \leq p} t_j(d_j)$ is returned;
   ELSE the speeds $s_j(d_j) = d_j/t_j(d_j)$ are computed and the point $(d_j, s_j(d_j))$ is added to each device-FPM.
5. The GPA takes bn and device-FPMs as input and returns a new distribution $d_1, \ldots, d_p$.
6. GOTO 3

Geometric Partitioning Algorithm (GPA)

The geometric partitioning algorithm presented in [16] can be summarised as follows. To distribute n computational units between p processing elements, load balancing is achieved when all elements execute their work within the same time: $t_1(x_1) \approx t_2(x_2) \approx \ldots \approx t_p(x_p)$. This can be expressed as:

$$\begin{cases} \frac{x_1}{s_1(x_1)} \approx \frac{x_2}{s_2(x_2)} \approx \ldots \approx \frac{x_p}{s_p(x_p)} \\ x_1 + x_2 + \ldots + x_p = n \end{cases} \tag{1}$$

The solution of these equations, $x_1, \ldots, x_p$, can be represented geometrically by intersection of the speed functions with a line passing through the origin of the coordinate system. Any such line represents an optimum distribution for a particular problem size. Therefore, the space of solutions of the partitioning problem consists of all such lines. The two outer bounds of the solution space are selected as the starting point of algorithm. The upper line represents the optimal distribution for some problem size $n_u < n$, while the lower line gives the solution for $n_l > n$. The region between two lines is iteratively bisected. The bisection line gives the optimum distribution for the problem size n_m. If $n_m < n$, then bisection line becomes the new upper bound, else it becomes the new lower bound. The algorithm iteratively progresses until converging to an integer solution to the problem.

Communication Minimising Algorithm (CMA)

This algorithm is specific to communication pattern of application and the topology of the communication network. It takes as input the number of computational units, w_i, to assign to each processing element and arranges them in such away, (m_i, n_i), as to minimise the communication cost. For example, for matrix multiplication, $\mathbf{A} \times \mathbf{B} = \mathbf{C}$, the total volume of data exchange is minimised by minimising the sum of the half perimeters $H = \sum_{i=1}^{q}(m_i + n_i)$. A column-based restriction of this problem is solved by an algorithm presented in [2].

4   Experimental Results

To demonstrate the effectiveness of the proposed algorithm we used parallel matrix multiplication as the application. This application is hierarchical and uses nested parallelism. At the inter-node level it uses a heterogeneous modification of the two-dimensional blocked matrix multiplication [4], upon which ScaLAPACK is based. At the inter-device level it uses one-dimensional sliced matrix multiplication. It can be summarised as follows: to perform the matrix multiplication $C = A \times B$, square dense matrices A, B and C are partitioned into sub-matrices A', B', C' (Fig. 2(a)), according to the output of the INPA. The algorithm has N/b iterations, within each iteration, nodes with sub-matrix A' that forms part of the pivot column will send their part horizontally and nodes with sub-matrix B' that forms part of the pivot blocks from the pivot row will broadcast their part vertically. All nodes will receive into a buffer $A_{(b)}$ of size $bm_i \times b$ and $B_{(b)}$ of size $b \times bn_i$. Then on each node Q_i with devices P_{ij}, for $0 \leq j < p_i$, device P_{ij} will do the matrix operation $C'_j = C'_j + A_{(b)} \times B_{(b)j}$ where sub-matrix C'_j is of size $bm_i \times d_{ij}$ and sub-matrix B'_j is of size $b \times d_{ij}$ (Fig. 2(b)). Therefore the kernel that is benchmarked for this application is the dgemm operation $C'_j = C'_j + A_{(b)} \times B_{(b)j}$.

The Grid'5000 experimental testbed proved to be an ideal platform to test our application. We used 90 dedicated nodes from 3 clusters from the Grenoble site. 12 of these nodes from the Adonis cluster included NVIDIA Tesla GPUs. The remaining nodes where approximately homogeneous. In order to increase the impact of our experiments we chose to utilise only some of the CPU cores on some machines (Table 1). Such an approach is not unrealistic since it is possible to book individual CPU cores on this platform. For the local *dgemm* routine we used high performance vendor-provided BLAS libraries, namely Intel MKL for CPU and CUBLAS for GPU devices. Open MPI was used for inter-node communication and OpenMP for inter-device parallelism. The GPU execution time includes the time to transfer data to the GPU. For these experiments, an out of core algorithm is not used when the GPU memory is exhausted. All nodes are interconnected by a high speed InfiniBand network which reduces the impact of communication on the total execution time, for $N = 1.5 \times 10^5$ all communications (including wait time due to any load imbalance) took 6% of total execution time. The full functional performance models of nodes, Fig. 3, illustrate the range of heterogeneity of our platform.

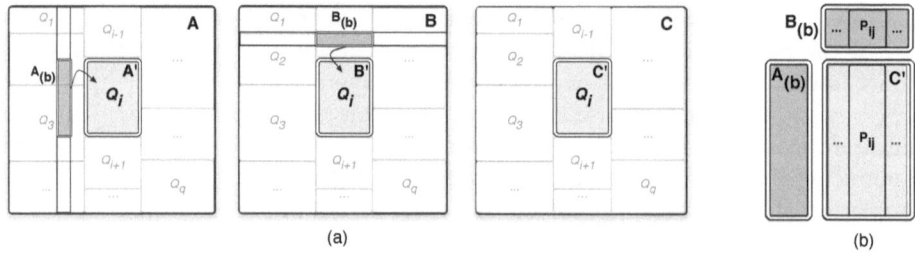

(a)   (b)

Fig. 2. Parallel matrix multiplication algorithm: (a) two-dimensional blocked matrix multiplication between the nodes; (b) one-dimensional matrix multiplication within a node

Table 1. Experimental hardware setup using 90 nodes from three clusters of the Grenoble site from Grid'5000. All nodes have 8 CPU cores, however, to increase heterogeneity only some of the CPU cores are utilised as tabulated below. One GPU was used with each node from the Adonis cluster, 10 nodes have Tesla T10 GPU and 2 nodes have Tesla C2050 GPU, and an CPU core was devoted to control execution on the GPU. As an example, we can read from the table that two Adonis nodes used only 1 GPU and 6 Edel nodes used just 1 CPU core. All nodes are connected with InfiniBand 20G & 40G.

Cores:	0	1	2	3	4	5	6	7	8	Nodes	CPU Cores	GPUs	Hardware
Adonis	2	1	1	1	1	1	2	3	0	12	48	12	2.27/2.4GHz Xeon, 24GB
Edel		0	6	4	4	4	8	8	8	8	50	250	0	2.27GHz Xeon, 24GB
Genepi	0	3	3	3	3	4	4	4	4	28	134	0	2.5GHz Xeon, 8GB
Total										90	432	12	

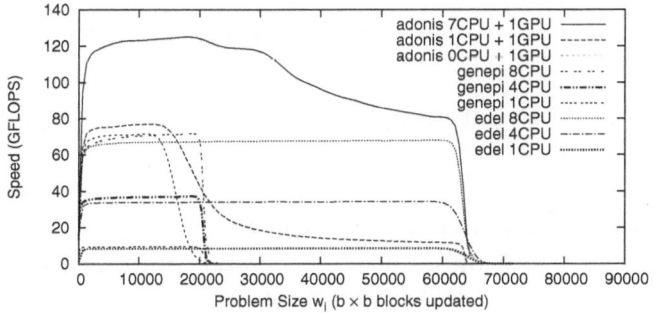

Fig. 3. Full functional performance models for a number of nodes from Grid'5000 Grenoble site. Problem size is in number of $b \times b$ blocks of matrix C updated by a node. For each data point in the node model it was necessary to build device models, find the optimum inter-device distribution and then measure the execution time of the kernel with this distribution.

Before commencing full scale experiments it was necessary to find an appropriate block size b. A large value of b allows the optimised BLAS libraries to achieve their peak performance as well as reducing the number of communications, while a small value of b allows fine grained load balancing between nodes. We conducted a serious of experiments, using one Adonis node with 7 CPU cores + 1GPU, for a range of problem sizes and a range of values of b. The IDPA was used to find the optimum distribution between CPU cores and GPU. As shown in Fig. 4, a value of $b = 128$ achieves near-peak performance, especially as N increases, while still allowing reasonably fine grained inter-node load balancing. For all subsequent experiments we used $b = 128$.

In order to demonstrate the effectiveness of the proposed FPM-based partitioning algorithm we compare it against 3 other partitioning algorithms. All four

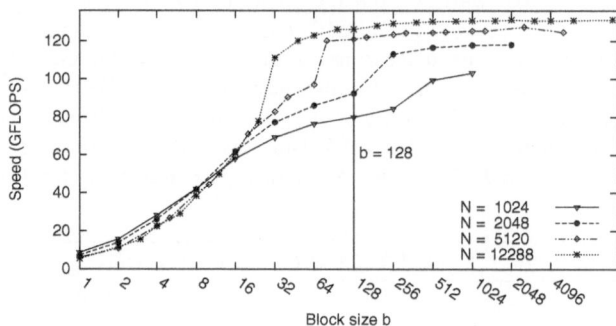

Fig. 4. Overall node performance obtained for different ranges of block and problem sizes when running optimal distribution between 7 CPU cores and a GPU

algorithms invoke the *communication minimisation algorithm* and are applied to an identical parallel matrix multiplication application. They differ on how load balancing decisions are made.

- **Multiple-CPM Partitioning** uses the same algorithm as proposed above, with step 7 of the INPA and step 5 of the IDPA replaced with $w_i = W \times \frac{s_i}{\sum_q s_i}$ and $d_j = bn \times \frac{s_j}{\sum_p s_j}$ respectively, where s_i and s_j are constants. This is equivalent to the approach used in [8,19,18].
- **Single-CPM Partitioning** does one iteration of the above multiple-CPM partitioning algorithm. This is equivalent to the approach used in [10,2].
- **Homogeneous Partitioning** uses an even distribution between all nodes: $w_1 = w_2 = \cdots = w_q$ and between devices in a node: $d_{i1} = d_{i2} = \cdots = d_{ip_i}$.

Fig. 5 shows the speed achieved by the parallel matrix multiplication application when the four different algorithms are applied. It is worth emphasizeing that the performance results related to the execution on GPU devices take into account the time to transfer the workload to/from the GPU. The speed of the application with the *homogeneous distribution* is governed by the speed of the slowest processor (a node from Edel cluster with 1CPU core). The *Single-CPM* and *multiple-CPM* partitioning algorithms are able to load balance for N up to 60000 and 75000 respectivly, however this is only because the speed functions in these regions are horozontal. In general, for a full range of problem sizes, the simplistic algorithms are unable to converge to a balanced solution. By chance, for $N = 124032$, the multiple-CPM algorithm found a reasonably good partitioning after many iterations, but in general this is not the case. Meanwhile the *FPM-based partitioning* algorithm reliably found good partitioning for matrix multiplication involving in excess of 0.5TB of data.

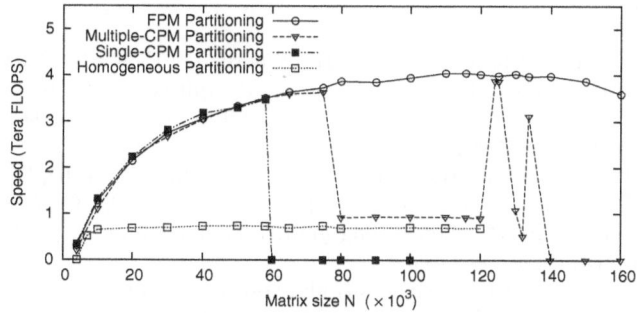

Fig. 5. Absolute speed for a parallel matrix multiplication application based on four partitioning algorithms. Using 90 heterogeneous nodes consisting of 432 CPU cores and 12 GPUs from 3 dedicated clusters.

5   Conclusions

In this paper a novel hierarchical partitioning algorithm for highly heterogeneous (CPU+GPU) clusters was presented. The algorithm load balances an application run on a hierarchical platform by optimally partitioning the workloads at both levels of hierarchy, i.e. nodes and processing devices. The presented approach is based on realistic functional performance models of processing elements which are obtained empirically in order to capture the high level of platform's heterogeneity. The efficiency of the proposed algorithm was tested in a real system consisting of 90 highly heterogeneous nodes in 3 computing clusters and compared to similar approaches for a parallel matrix multiplication case. The results show that the presented algorithm was not only capable of minimising the overall communication volume in such a complex environment, but it was also capable of providing efficient load balancing decisions for very large problem sizes where similar approaches were not able to find the adequate balancing solutions. Future work will include an out of core device kernel for when the memory limit of a device is reached; a communication efficient inter-device partitioning and multi-GPU experimental results.

Acknowledgments. This publication has emanated from research conducted with the financial support of Science Foundation Ireland under Grant Number 08/IN.1/I2054. This work was supported by FCT through the PIDDAC Program funds (INESC-ID multiannual funding) and a fellowship SFRH/BD/44568/2008. Experiments were carried out on Grid'5000 developed under the INRIA ALADDIN development action with support from CNRS, RENATER and several Universities as well as other funding bodies (see https://www.grid5000.fr). This work was also partially supported by the STSM COST Action IC0805.

References

1. Augonnet, C., Thibault, S., Namyst, R., Wacrenier, P.-A.: STARPU: A Unified Platform for Task Scheduling on Heterogeneous Multicore Architectures. In: Sips, H., Epema, D., Lin, H.-X. (eds.) Euro-Par 2009. LNCS, vol. 5704, pp. 863–874. Springer, Heidelberg (2009)
2. Beaumont, O., Boudet, V., Rastello, F., Robert, Y.: Matrix Multiplication on Heterogeneous Platforms. IEEE Trans. Parallel Distrib. Syst. 12(10), 1033–1051 (2001)
3. Blumofe, R., Leiserson, C.: Scheduling multithreaded computations by work stealing. JACM 46(5), 720–748 (1999)
4. Choi, J.: A new parallel matrix multiplication algorithm on distributed-memory concurrent computers. Concurrency: Practice and Experience 10(8), 655–670 (1998)
5. Clarke, D., Lastovetsky, A., Rychkov, V.: Column-Based Matrix Partitioning for Parallel Matrix Multiplication on Heterogeneous Processors Based on Functional Performance Models. In: Alexander, M., D'Ambra, P., Belloum, A., Bosilca, G., Cannataro, M., Danelutto, M., Di Martino, B., Gerndt, M., Jeannot, E., Namyst, R., Roman, J., Scott, S.L., Traff, J.L., Vallée, G., Weidendorfer, J. (eds.) Euro-Par 2011, Part I. LNCS, vol. 7155, pp. 450–459. Springer, Heidelberg (2012)
6. Dongarra, J., Faverge, M., Herault, T., Langou, J., Robert, Y.: Hierarchical qr factorization algorithms for multi-core cluster systems. Arxiv preprint arXiv:1110.1553 (2011)
7. Drozdowski, M., Lawenda, M.: On Optimum Multi-installment Divisible Load Processing in Heterogeneous Distributed Systems. In: Cunha, J.C., Medeiros, P.D. (eds.) Euro-Par 2005. LNCS, vol. 3648, pp. 231–240. Springer, Heidelberg (2005)
8. Galindo, I., Almeida, F., Badía-Contelles, J.M.: Dynamic Load Balancing on Dedicated Heterogeneous Systems. In: Lastovetsky, A., Kechadi, T., Dongarra, J. (eds.) EuroPVM/MPI 2008. LNCS, vol. 5205, pp. 64–74. Springer, Heidelberg (2008)
9. Horton, M., Tomov, S., Dongarra, J.: A class of hybrid lapack algorithms for multicore and gpu architectures. In: SAAHPC, pp. 150–158 (2011)
10. Hummel, S., Schmidt, J., Uma, R.N., Wein, J.: Load-sharing in heterogeneous systems via weighted factoring. In: SPAA 1996, pp. 318–328. ACM (1996)
11. Ilic, A., Sousa, L.: Collaborative execution environment for heterogeneous parallel systems. In: IPDPS Workshops and Phd Forum (IPDPSW), pp. 1–8 (2010)
12. Ilic, A., Sousa, L.: On realistic divisible load scheduling in highly heterogeneous distributed systems. In: PDP 2012, Garching, Germany (2012)
13. Jacobsen, D.A., Thibault, J.C., Senocak, I.: An MPI-CUDA Implementation for Massively Parallel Incompressible Flow Computations on Multi-GPU Clusters. In: AIAA Aerospace Sciences Meeting Proceedings (2010)
14. Kalinov, A., Lastovetsky, A.: Heterogeneous Distribution of Computations while Solving Linear Algebra Problems on Networks of Heterogeneous Computers. In: Sloot, P.M.A., Hoekstra, A.G., Bubak, M., Hertzberger, B. (eds.) HPCN-Europe 1999. LNCS, vol. 1593, pp. 191–200. Springer, Heidelberg (1999)
15. Kindratenko, V.V., et al.: GPU clusters for high-performance computing. In: CLUSTER, pp. 1–8 (2009)
16. Lastovetsky, A., Reddy, R.: Data Partitioning with a Functional Performance Model of Heterogeneous Processors. Int. J. High Perform. Comput. Appl. 21(1), 76–90 (2007)
17. Lastovetsky, A., Reddy, R., Rychkov, V., Clarke, D.: Design and implementation of self-adaptable parallel algorithms for scientific computing on highly heterogeneous HPC platforms. Arxiv preprint arXiv:1109.3074 (2011)

18. Legrand, A., Renard, H., Robert, Y., Vivien, F.: Mapping and load-balancing iterative computations. IEEE Transactions on Parallel and Distributed Systems 15(6), 546–558 (2004)
19. Martínez, J., Garzón, E., Plaza, A., García, I.: Automatic tuning of iterative computation on heterogeneous multiprocessors with ADITHE. J. Supercomput. (2009)
20. Quintin, J.-N., Wagner, F.: Hierarchical Work-Stealing. In: D'Ambra, P., Guarracino, M., Talia, D. (eds.) Euro-Par 2010, Part I. LNCS, vol. 6271, pp. 217–229. Springer, Heidelberg (2010)
21. Veeravalli, B., Ghose, D., Robertazzi, T.G.: Divisible load theory: A new paradigm for load scheduling in distributed systems. Cluster Computing 6, 7–17 (2003)

Encapsulated Synchronization and Load-Balance in Heterogeneous Programming

Yuri Torres, Arturo Gonzalez-Escribano, and Diego Llanos

Departamento de Informatica, Universidad de Valladolid
{yuri.torres,arturo,diego}@infor.uva.es

Abstract. Programming models and techniques to exploit parallelism in accelerators, such as GPUs, are different from those used in traditional parallel models for shared- or distributed-memory systems. It is a challenge to blend different programming models to coordinate and exploit devices with very different characteristics and computation powers. This paper presents a new extensible framework model to encapsulate run-time decisions related to data partition, granularity, load balance, synchronization, and communication for systems including assorted GPUs. Thus, the main parallel code becomes independent of them, using internal topology and system information to transparently adapt the computation to the system. The programmer can develop specific functions for each architecture, or use existent specialized library functions for different CPU-core or GPU architectures. The high-level coordination is expressed using a programming model built on top of message-passing, providing portability across distributed- or shared-memory systems. We show with an example how to produce a parallel code that can be used to efficiently run on systems ranging from a Beowulf cluster to a machine with mixed GPUs. Our experimental results show how the run-time system, guided by hints about the computational-power ratios of different devices, can automatically part and distribute large computations across heterogeneous systems, improving the overall performance.

1 Introduction

Currently, heterogeneous systems provide computing power using mixed types of devices and architectures, such as CPU-cores, GPUs or FPGAs [6,10]. General-Purpose Programming for devices such as GPUs (GP-GPU) has been simplified by the introduction of higher level data parallel languages, such as CUDA or OpenCL. However, to obtain efficient codes the programmer needs knowledge about the underlying target architecture, and how it relates to the programming model. The intrinsic complexity of the code generation for heterogeneous systems increases every time we add any different hardware device. Thus, it is an important goal to devise abstractions and tools that allow the programmer to blend the different programming models involved, also simplifying the tasks of data-distribution and device coordination across an heterogeneous system.

In previous works we presented Hitmap [3,5], a library to support both data and task parallelism in distributed-memory environments, through manipulation

C. Kaklamanis et al. (Eds.): Euro-Par 2012, LNCS 7484, pp. 502–513, 2012.

and mapping of hierarchical tiling arrays Hitmap features an extensible plug-in system that allows the programmer to choose among different data-partition and distribution techniques, or easily program and reuse new ones. It provides functionalities for tile communication, allowing to build complex and scalable communication patterns in terms of the results of the mapping functions.

This paper presents a new framework model to encapsulate run-time decisions related to data partition, granularity, load balance, synchronization, and communication for heterogeneous systems. It introduces a new abstraction layer in the conceptual structure of Hitmap. More precisely, in this work we present the following contributions:

(1) We propose a new plug-ins layer to encapsulate the decisions related to map tile computations to specific accelerator devices. We discuss how load-balancing techniques relate to the different plug-ins layers.

(2) We introduce a high-level API that selects the proper kernel for a given device, and hides all details of synchronization and communication between logical processes and accelerators.

(3) We discuss an implementation of this framework model currently supporting distributed-memory clusters of multicore CPUs and NVIDIA GPUs.

(4) We show with an example how to produce a single parallel code that adapts the computation to efficiently run on systems ranging from a Beowulf cluster to a machine with mixed GPUs. Our experimental results show how the run-time system can automatically part and distribute large computations across very different devices, improving the performance of a homogeneous approach.

The rest of the paper is organized as follows. Section 2 discusses some previous approaches and their limitations. Section 3 introduces our conceptual approach. Section 4 describes the architecture of our solution and the design problems faced. Section 5 shows a case study. In section 6 we present a performance evaluation of the case study with a load-balancing strategy in different scenarios. Finally, section 7 discusses some conclusions and future work.

2   Related Work

Several research groups are working in the problem of simplifying heterogeneous programming without sacrificing hardware accelerators performance.

Quintana-Ortí et al. [11] presented the FLAME programming model. It focus on programming dense linear algebra operations on complex platforms, including multicore processors, and hardware accelerators such as GPUs, and Cell B.E. FLAME abstracts the target accelerator architecture. It divides the parallelism in two levels, the first one considering each accelerator device as a computation unit (coarse-grain parallelism), and the second one considering each hardware accelerator as a set of multiple cores (fine-grain parallelism). They rely on the BLAS library to exploit this second level. Besides the limited application domain, global configuration parameters are fixed, while it has been shown that it is important to adapt them to the particular thread memory access pattern [14].

MCUDA [13] is a framework to mix CPU and GPU programming. In MCUDA it is mandatory to define kernels for all available devices. No data distribution policy is provided, and the toolkit can not make any assumption about the relative performance of the supported devices. Introducing any of these features would involve a redesign of the framework. Other works [8,15] try to exploit at the same time CPU and GPU devices, attempting to obtain good load balancing with the help of heuristics. Data structures partition and manipulation is not abstracted and they do not support flexible mechanisms to add new partition and layout policies. Finally, papers as [9,2] use MPI and CUDA parallel programming model in order to exploit all GPUs devices in heterogeneous systems. However, the authors do not abstract the use of both models and the target underlying hardware details.

Chapel [1], a PGAS language, proposes a transparent plug-in system for domain partitions in generic systems. The PGAS approach tries to hide the communication issues to the programmer. Thus, efficient aggregated communications can not be directly expressed, and most of the times can not be automatically derived from generic codes. Contributors to Chapel are currently working in prototyped layouts to generate array allocations, data transfers, and parallel operations in CUDA. However, they do not offer a different layer for accelerator partition policies, or synchronization among different CPU and GPU devices.

3   Conceptual Approach

Heterogeneous systems can be built with very different hardware devices (CPU-cores, accelerators) in several nodes interconnected in a distributed environment. Portable codes for such systems should implement parallel algorithms abstracting them from the mapping activities that adapt the computation to the platform. Thus, the programming model should encapsulate the mapping techniques and the CPU/accelerator synchronization with appropriate abstractions.

We propose a programming framework based on: (1) Several layers of plug-in modules that encapsulate the mapping functions; and (2) functionalities to build the coordination (synchronization and communication) structures of the algorithms, which are transparently adapted at run-time in terms of the results of the mapping functions.

Hitmap [3,5] is a parallel programming library where partition policies are implemented through a set of plug-ins with a common interface. The programmer may select, or change the chosen plug-in in the program code, using only its name. Hitmap automatically associates logical processes to processing nodes. The data-partition plug-in interface returns an object containing information about which parts of the data are mapped to the logical process, taking into account the neighbourhood relationships of a virtual topology. Coordination patterns are built with high-level point-to-point or collective tile communications, using the results stored in the map objects. If partition details change, the communication structure will reflect the changes automatically. Thus, the coordination among processes may be programmed in an abstract form, independently of the target system topology.

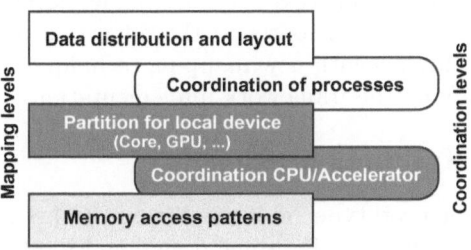

Fig. 1. Mapping/Coordination levels. White boxes show the original Hitmap approach.

Our work extends the Hitmap approach. Figure 1 shows the different mapping levels of the original Hitmap, and our proposed extension. Hitmap has a single level of data-partition and layout. It is designed to encapsulate coarse-grain mapping techniques, appropriate for distributed-memory nodes.

We propose to add a second, middle-grain partitioning level that allows to adapt the local part of data to the specific characteristics and architecture of the actual device associated to the logical process by the virtual topology. The programmer naturally introduces a third level of mapping inside the kernel code by implementing specific thread-level memory access patterns.

The second-level mapping plug-ins use information about the device and the global memory access pattern of the kernel, to generate domain partitions that exploit locality, maximum occupancy, coalescence, or other device properties that affects performance. The result is an object encapsulating information about a partition of the local computation in a grid of blocks. The same abstraction can be used for techniques of very different nature: CPU-cores, GPUs, or other kind of accelerator. Finally, the coordination, data movement between the CPU and accelerators, and kernel launch, can be automatized by a run-time system, using the second-level partition results. Padding can be automatically added to tiles if needed to properly align data to the memory banks of the particular device, alleviating memory bottlenecks, and improving cache use.

This approach can be used together with techniques to automatically generate kernels for different architectures from common specifications (see e.g. [12,4]), avoiding the current need to supply optimized kernels for all the architectures that compose the target heterogeneous system, By encapsulating the CPU/accelerator coordination in a transparent system, we also allow to integrate as kernels libraries specifically optimized for a given architecture, such as CUBLASfor GPUs. We also promote the abstraction of hierarchical tiles to specific programming languages for accelerators (in this work we use CUDA as a proof of concept, doing this exercise for more generic languages such as OpenCL is straightforward). Thus, we introduce a common array abstraction, simplifying the porting of code between CPU cores and accelerators.

This conceptual framework has been implemented adding new functionalities to the Hitmap library, without modifying the original structure. This imposes a minimal impact on the original Hitmap codes. The original Hitmap code takes care of the coordination of processes in the higher level. The new extension takes care of adapting the local parts to the device automatically assigned to the logical process. Plug-ins with new mapping techniques may be included and tested without modifying the framework implementation.

4   Design and Implementation

We have developed a prototype implementation of this framework extending Hitmap. In this section we describe some design and implementation considerations, and problems solved.

The original Hitmap library was written in C language. Nevertheless, it has and object-oriented design, and future releases could provide a neater C++ interface. Hitmap is designed to manipulate hierarchical tiling arrays. The *Hit-Shape* class implements tile domains. A shape object represents a subspace of array indexes defined as an n-dimensional rectangular parallelotope. Its limits are determined by n *Signature* objects. Each *Signature* is a tuple of three integer numbers (begin, end, and stride), representing the indexes in a domain axis.

Hitmap defines an API for data-partition modules, named *Layout* plug-ins. It defines a wrapper function that links the main code with the chosen plug-in. The Layout plug-ins receive as parameters: (1) a virtual topology object (*HitTopology*), (2) a domain to be mapped (a HitShape object), and (3) optional parameters for the specific technique. They return a *HitLayout* object containing a local domain (another HitShape), information about neighbor relations, and other mapping details. These objects are used as parameters in the constructors of *HitComm* objects that express tile communications across logical processes.

Partitions. We follow the same approach for the new second-level *Partition* plug-ins. The wrapper function is similar, but also selects different implementations of the same plug-in name depending on the architecture of the target device. Our current wrapper differences between CPU-cores, and different Nvidia's CUDA supported architectures.

The Hitmap initialization function gathers information about the particular system devices and builds an internal physical topology object. The virtual topology constructors attach each logical process to one device, storing its information. The Partition plug-ins receive as parameters: (1) the attached device data; (2) a HitLayout object with information of the local domain to be mapped. Optional parameters indicating the memory-access patters of the low-level threads can be supplied. The result is a new *HitPartition* containing information about block shapes, grid sizes, and information to generate tile paddings if needed.

As example, we have implemented a trivial partition plug-in. The CPU-cores implementation simply creates a grid with one element containing the full local shape. The GPU implementations split the local domain in rectangular blocks with 1×512 threads. This is appropriate for computations that access data linearly in both Nvidia's architectures [14]. For specific grid sizes an extra block is added at the end of each row to alleviate GPU memory contention effects. More sophisticated policies can be integrated as new plug-ins.

Assigning Several Logical Processes to the Same Device. We have introduced a new technique in the virtual topology modules of Hitmap. It allows to assign more than one logical process to the same device. It has two purposes. As a potential load-balancing technique (see section 6), and to transparently use accelerators to perform large computations whose data do not directly fit in the accelerator global memory. Thus, the full computation is done in smaller parts, coordinated by the Hitmap upper-level communication structures.

Kernel Definition and Launch. We provide a macro function to define with a common interface the function headers of different kernel versions for different architectures. The following example shows the headers of two implementations (one for CPU-cores, another for pre-Fermi GPUs) of the same kernel:

```
hit_kernelDefinition( CORE, mmult, HitTile_float *A, HitTile_float *B, HitTile_float *C ) {
hit_kernelDefinition( GPU_R1, mmult, HitTile_float *A, HitTile_float *B, HitTile_float *C ) {
```

We have developed one function that transparently do the coordination with the assigned device. It receives as parameters the kernel name, a partition object, and the kernel parameters, indicating which ones are inputs and outputs. See the example in Fig. 2. This function deals with linking issues of kernels written if specific languages. For example, the launch of a kernel for an Nvidia GPU needs a special syntax and the launching code has to be compiled with the CUDA compiler. We use internal wrapper functions with different implementations for different architectures. Each implementation is compiled with the proper tools before linking. A selection mechanism checks at run-time the assigned device architecture and calls the appropriate implementation for the local device.

For CPU-cores the wrapper simply calls the proper C function passing the indicated arguments. For accelerators the process is more complex, and involves communication between the main system memory and the device memory. We have implemented the synchronization with Nvidia's GPUs with the following stages: (1) Move to the GPU memory the input tiles (the data and the tile handler structure). Padding restrictions expressed in the HitPartition object are applied to the memory allocation in this step. (2) Launch the kernel, using the grid parameters from the Partition object, passing the pointers to the new tile handlers in GPU memory. (3) Copy data from output GPU-memory tiles to the CPU, eliminating padding if needed. Finally, a mutual exclusion mechanisms has been added in the kernel launch function to allow several processes assigned to the same device to coordinate themselves for the use of the device.

These abstractions completely encapsulate the synchronization and coordination between CPU and different devices, such as cores and accelerators. The same primitive call automatically invokes CPU-core functions written in plain C language, or launches CUDA kernels.

Running the Programs. Hitmap programs are started like any MPI program, using the *mpiexec* command. The MPI hosts file is used to select the machines where the processes are started. Processes in the same machine are automatically attached to CPU-cores or GPU devices. If data do not fit into the memory of an accelerator device, more MPI processes are required to obtain a finer partition.

5   Case Study

In this section we show with an example how Hitmap abstractions lead to codes which are independent of the encapsulated mapping techniques. We have chosen the Cannon's algorithm for matrix multiplication (see e.g. [7]). It is a task-parallel algorithm focused on reducing local memory usage for distributed systems. Thus, it shows the interaction of different levels of parallelism.

In Cannon's algorithm the available processes are organized in a perfect square topology to generate neighbor relations. Each matrix A, B and C is divided into rectangular blocks, distributing them across processes. It starts with an initial communication stage to relocate A and B blocks in a circular shift ($A_{ij} = A_{i(j-i)}$, and $B_{ij} = B_{(i-j)j}$). On each step, every process multiply its local blocks of A and B, accumulating the partial results in the local block of C. It then sends the used block of A to the leftward process, and the used block of B to the upward process, both in a circular shift. There are as many communication-computation steps as the square-root of the number of total processes.

Figure 2 shows the Cannon's matrix multiplication algorithm implemented with the Hitmap library for heterogeneous systems. We use float base elements. The code is the same used in previous versions of Hitmap for distributed-memory systems except lines 40–41 (that encapsulate the low-level partition for the assigned device), and lines 47 and 50, that encapsulates the coordination between the CPU and the accelerators.

Lines 3–6 declare the full domain of the three matrices with a global-view approach. Memory is not yet allocated. Line 9 builds a virtual topology enforcing a perfect square of processes, as required by the algorithm. Lines 12–14 create layout objects that distribute the matrices domains across the virtual topology. The layout plug-in modules used are different for the three matrices. Figure 3 shows a diagram of the resulting layouts. Matrix B uses a classical block data partition, with evenly sized parts. Matrices C and A use a load-balancing plug-in technique. The rows dimension is split unevenly according to a *Balance factor*, decided in terms of the relative computing power of the device types as recorded in the low-level topology description. Currently, it is experimentally determined.

In lines 17–21 each logical process creates and allocates the local part of the matrices. Thanks to the *maxShape* padding function, n and m do not need to be exact multiples of the number of processes in a given axis. Lines 24–26 read in parallel the tiles of the input matrices. The C matrix is initialized with 0 values.

Lines 29–32 perform the initial relocating stage prescribed by the Cannon's algorithm, shifting A and B tiles. Lines 35–37 build the shifting communication pattern that will be used between the computation stages. The layout objects and the tiles provide all the information needed to internally find neighbors and build MPI derived data types to optimize the communications. Thus, communications are adapted to the partition transparently. For this example we choose

```
1   void cannonsMM( int n, int m, int p ) {
2       /* 1. DECLARE FULL MATRICES WITHOUT MEMORY */
3       HitTile_double A, B, C;
4       hit_tileDomain( &A, float, 2, n, m );
5       hit_tileDomain( &B, float, 2, m, p );
6       hit_tileDomain( &C, float, 2, n, p );
7
8       /* 2. CREATE VIRTUAL TOPOLOGY */
9       HitTopology topo = hit_topology( plug_topSquare );
10
11      /* 3. COMPUTE PARTITIONS */
12      HitLayout layC = hit_layout( plug_layBlocksLB, topo, C, 0 );
13      HitLayout layA = hit_layoutWrap( plug_layBlocksLB, topo, A, 0 );
14      HitLayout layB = hit_layoutWrap( plug_layBlocks, topo, B );
15
16      /* 4. CREATE AND ALLOCATE TILES */
17      HitTile_double tileA, tileB, tileC;
18      hit_tileSelectNoBoundary( &tileA, &A, hit_layMaxShape(layA,1) );
19      hit_tileSelectNoBoundary( &tileB, &B, hit_layMaxShape(layB,0) );
20      hit_tileSelect( &tileC, &C, hit_layShape(layC) );
21      hit_tileAlloc( &tileA ); hit_tileAlloc( &tileB ); hit_tileAlloc( &tileC );
22
23      /* 5. INITIALIZE MATRICES */
24      hit_tileFileRead( &tileA, "matrixA.dat" );
25      hit_tileFileRead( &tileB, "matrixB.dat" );
26      float aux=0; hit_tileFill( &tileC, &aux );
27
28      /* 6. INITIAL ALIGNMENT PHASE */
29      HitComm commRow = hit_comShiftDim( layA, 1, -hit_layRank(layA,0), &tileA );
30      HitComm commCol = hit_comShiftDim( layB, 0, -hit_layRank(layB,1), &tileB );
31      hit_comDo( commRow ); hit_comDo( commCol );
32      hit_comFree( commRow ); hit_comFree( commCol );
33
34      /* 7. REUSABLE COMM PATTERN */
35      HitPattern shift = hit_pattern( HIT_PAT_UNORDERED );
36      hit_patternAdd( &shift, hit_comShiftDim( layA, 1, 1, &tileA ) );
37      hit_patternAdd( &shift, hit_comShiftDim( layB, 0, 1, &tileB ) );
38
39      /* 8. COMPUTE DEVICE PARTITION USING ACCESS PATTERN INFO */
40      HitPartition parts = hit_partition( plug_partBlocks, hit_layShape(layC),
41                          2, hit_shape( 2, ALL, THIS ), hit_shape( 2, THIS, ALL ) );
42
43      /* 9. DO COMPUTATION */
44      int loopIndex;
45      int loopLimit = max( hit_layNumActives(layA,0), hit_layNumActives(layB,1) );
46      for (loopIndex = 0; loopIndex < loopLimit-1; loopIndex++) {
47          hit_kernelLaunch( mmult, parts, 3, IN, tileA, IN, tileB, INOUT, tileC );
48          hit_patternDo( shift );
49      }
50      hit_kernelLaunch( mmult, parts, 3, IN, tileA, IN, tileB, INOUT, tileC );
51
52      /* 11. WRITE RESULT */
53      hit_tileFileWrite( &tileC, "matrixC.dat" );
54
55      /* 12. FREE RESOURCES */
56      hit_partitionFree( parts );
57      hit_layFree( layA ); hit_layFree( layB ); hit_layFree( layC );
58      hit_patternFree( &shift );
59      hit_topFree( topo );
60  }
```

Fig. 2. Heterogeneous Hitmap implementation of Cannon's matrix multiplication

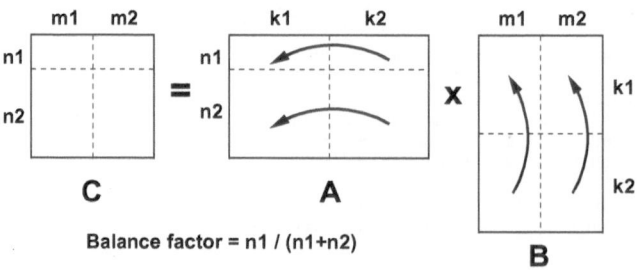

Fig. 3. Load balancing layout scheme in the Cannon's matrix multiplication example

synchronous communication to avoid the need of double buffers, exploiting our full system memory to do larger computations.

Lines 40–41 generate a partition object tailored to the device assigned to the logical process. Line 41 is a shape expression that represents the global memory access pattern; indicating, in relative coordinates, which elements are accessed by a thread. Lines 44–50 implement the main loop of the algorithm. The computation stage of the last iteration has been unrolled to avoid the last unneeded communication stage. The computation is launched by the *hit_kernelLaunch* primitive, independently of the actual device. The shifting communication pattern is activated by the *hit_patterDo* primitive. Line 53 writes the output matrix tiles to a file in parallel. Lines 56–59 free all the Hitmap resources before finishing.

6   Experimental Work

We have designed experimental work to show that: (1) Our new abstractions do not impose a significant overhead on the computation; and (2) this framework allows to easily exploit different devices to obtain performance benefits.

In order to show the efficiency of the Hitmap codes, we have manually developed and optimized reference codes for matrix multiplication: (a) A direct MPI implementation of the Cannon's algorithm (see [7]); and (b) a direct CUDA implementation that may also split and multiply the matrices block by block if they are too big to fit in the GPU device memory.

For our experiments we have used two different platforms. The first one is a Beowulf cluster with up to 18 dual-core PC computers. The second one is an Intel(R) Core(TM) i7 CPU 960, 3.20GHz with active hyper-threading. This system has two GPUs: a GeForce 8500 GT, and a GerForce 9600 GT, both managed by the CUDA driver included in the 4.0 toolkit. From now on, we identify the different available devices in this machine with the following letters: (A) GeForce 9600 GT; (B) GeForce 8500 GT; (C) Cores of the CPU.

We select several matrix sizes: $N = M = 2048, 8192, 12288$. The first size is small enough to allocate the three matrices in any of the devices of both systems. The second size cannot be fully allocated on the second GPU (device B) of the mixed CPU-GPU machine. The last size cannot be allocated in any of the GPUs.

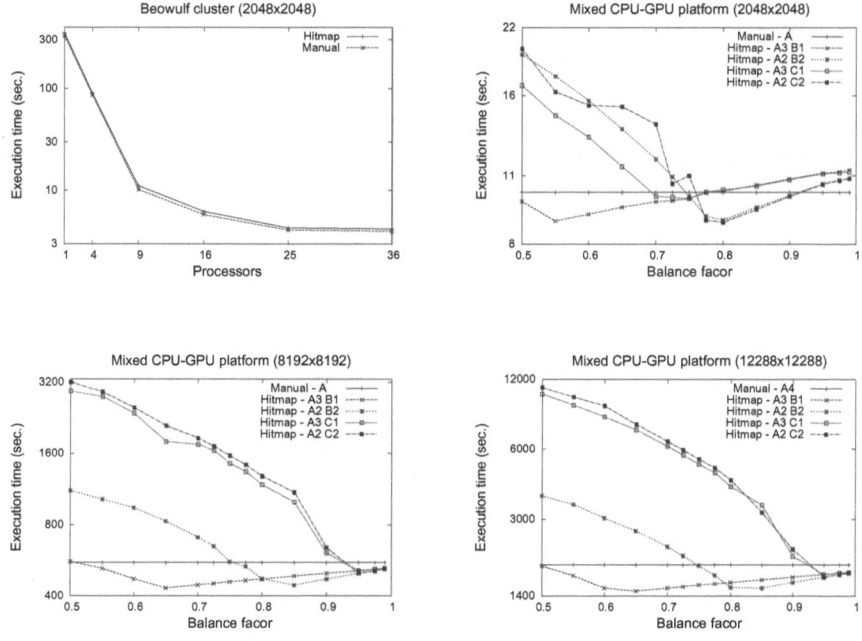

Fig. 4. Experimental work results

We also test using modified sizes (e.g. $N = M = 2039, 2057$), that our padding mechanisms do not impose a significant performance effect on the results.

Figure 4 presents execution times obtained in different scenarios. Notice that all y-axis are in logarithmic scale. The experiments in the Beowulf cluster show, even for the small matrix size, that Hitmap implementations have the same scalability and overall performance than the manually optimized MPI code. A minimal Hitmap performance overhead is observed in all our experimental work.

In the more heterogeneous machine the best performance results are obtained for a small number of processes. Remind that Cannon's algorithm forces more synchronization stages when the number of processes increase. Thus, for Cannon's algorithm, more MPI processes lead to bigger communication overhead, while reducing the computation load of each task. In this machine, our experiments show the best results for four MPI processes. We show results for the following scenarios. *Reference code:* (A) Manually developed CUDA code, executing the whole computation with only one kernel launch in device A, the fastest GPU; (A4) For matrices that do not fit in the GPU device memory, the reference code parting the matrices in four even parts and executing the computation in several kernel launches. *Hitmap code:* Changing the topology module we can easily experiment with different assignments of devices to logical processes. (A3-B1) Mixed GPUs: 3 processes mapped to device A, and 1 process to device B; (A2-B2) Mixed GPUs: 2 processes mapped to device A, and 2 process to device B; (A3-C1) GPU and core: 3 processes mapped to device A, and 1 process to one CPU-core; (A2-C2) GPU and cores: 2 processes mapped to device A, and 2 processes, each one mapped to a different CPU-core. For all the experiments with GPUs and Hitmap we have manipulated the partition plug-in to experiment with different load-balance factors, between 0.5 and 0.975.

Consider the execution time of the reference CUDA code (A and A4). The results show that it is always possible to improve these performance results with the Hitmap code, exploiting heterogeneity with more than one device. The results for the small matrix size are more unstable, and impacted by the kernel initialization times, including the communication between CPU and GPU. However, as the matrix size increases, the results are more stable, and show exactly the same trends. We obtain performance improvements of up to 10% for the small matrices, and a consistent best improvement of 20.5% for medium and big input data sizes. Traces of the executions show that the MPI communication times are always less than 10% of the total execution time for the small matrix size. And their impact quickly decreases as the data input size grows.

On the left part of the plots (load-balance factor 0.5), the load is evenly distributed, not taking into account the different computing powers of the devices. The critical path is dominated by the slower devices. As the load-balance factor grows, the balance is improved proportionally reducing the total execution time. After the optimum balance point is found, an increase of the factor leads to too few computation on the slower devices. Thus, the critical path is dominated by the fastest device, proportionally reducing performance again.

An important question is: Is it possible to predict the best load-balance factor for a given set of devices? Profiling tests with simple benchmarks show that the relative computing power between devices A and B is approximately $r = 3.826$; and between device A and a core (device C) it is $r = 14.153$. In order to assign to each device a computation proportional to its relative computing power, the load-balance factor may be calculated as $LB = r/(r+1)$ for the A2 scenarios, and $LB = (r-1)/(r+1)$ for the A3 scenarios. The experimental results show that, for big enough matrices, this estimation is always a little lower than the value that leads to the best performance: 10% in both A3 scenarios, 2% and 6% on A2 scenarios. A more sophisticated model, taking into account the synchronization stages, is needed to automatically predict the best factor in the Layout plug-ins.

7   Conclusion

In this paper we present a new framework for heterogeneous programming. It encapsulates the mapping techniques into plug-ins at two different layers of abstraction: one related to logical processes coordination, and another related to adapting the computations to the inherent parallelism and architecture details of the actual device associated to each logical process. We propose a high-level API that transparently deals with all the details of communication and synchronization between logical processes and accelerator devices, such as GPUs.

This framework allows to generate codes which are transparently adapted to heterogeneous systems with mixed types of accelerator devices.

Current on-going work involves: (1) Introducing in the framework more sophisticated mapping policies that better exploit CPU-cores and GPU architecture information; and (2) test the applicability of these techniques to more types of programs, including well-know benchmarks and real applications.

References

1. Chamberlain, B., Deitz, S., Iten, D., Choi, S.E.: User-defined distributions and layouts in Chapel: Philosophy and framework. In: 2nd USENIX Workshop on Hot Topics in Parallelism (June 2010)
2. Kui Chen, Q., Kang Zhang, J.: A stream processor cluster architecture model with the hybrid technology of mpi and cuda. In: ICISE 2009, pp. 86–89 (December 2009)
3. de Blas Cartón, C., Gonzalez-Escribano, A., Llanos, D.R.: Effortless and Efficient Distributed Data-Partitioning in Linear Algebra. In: HPCC 2011, pp. 89–97. IEEE (September 2010)
4. Farooqui, N., Kerr, A., Diamos, G.F., Yalamanchili, S., Schwan, K.: A framework for dynamically instrumenting GPU compute applications within GPU Ocelot. In: GPGPU, p. 9 (2011)
5. Fresno, J., Gonzalez-Escribano, A., Llanos, D.R.: Automatic Data Partitioning Applied to Multigrid PDE Solvers. In: PDP 2011, pp. 239–246. IEEE (February 2011)
6. Gelado, I., Stone, J.E., Cabezas, J., Patel, S., Navarro, N., Mei, W.: An asymmetric distributed shared memory model for heterogeneous parallel systems. In: ASPLOS 2010, pp. 347–358. ACM, New York (2010)
7. Grama, A., Gupta, A., Karypis, G., Kumar, V.: Introduction to Parallel Computing, 2nd edn. Addison Wesley (2003)
8. Hong, C., Chen, D., Chen, W., Zheng, W., Lin, H.: MapCG: writing parallel program portable between CPU and GPU. In: PACT 2010, pp. 217–226. ACM, New York (2010)
9. Karunadasa, N., Ranasinghe, D.: Accelerating high performance applications with cuda and mpi. In: ICIIS 2009, pp. 331–336 (December 2009)
10. Luk, C.K., Hong, S., Kim, H.: Qilin: Exploiting parallelism on heterogeneous multiprocessors with adaptive mapping. In: MICRO-42, pp. 45–55 (December 2009)
11. Quintana-Ortí, G., Igual, F.D., Quintana-Ortí, E.S., van de Geijn, R.A.: Solving dense linear systems on platforms with multiple hardware accelerators. In: PPoPP 2009, pp. 121–130. ACM, New York (2009)
12. Singh, S.: Computing without processors. Commun. ACM 54, 46–54 (2011)
13. Stratton, J.A., Stone, S.S., Hwu, W.-M.W.: MCUDA: An Efficient Implementation of CUDA Kernels for Multi-core CPUs. In: Amaral, J.N. (ed.) LCPC 2008. LNCS, vol. 5335, pp. 16–30. Springer, Heidelberg (2008)
14. Torres, Y., Gonzalez-Escribano, A., Llanos, D.R.: Using Fermi architecture knowledge to speed up CUDA and OpenCL programs. In: Proc. ISPA 2012, Leganes, Madrid, Spain (2012)
15. Yao, P., An, H., Xu, M., Liu, G., Li, X., Wang, Y., Han, W.: CuHMMer: A loadbalanced CPU-GPU cooperative bioinformatics application. In: HPCS 2010, pp. 24–30 (July 2010)

Transactional Access to Shared Memory in StarSs, a Task Based Programming Model

Rahulkumar Gayatri[1,2], Rosa M. Badia[1,3],
Eduard Ayguade[1,2], Mikel Luján[4], and Ian Watson[4]

[1] Barcelona Supercomputing Center, Barcelona, Spain
{rgayatri,rosa.m.badia,eduard.ayguade}@bsc.es
[2] Universitat Politècnica de Catalunya, Spain
[3] Artificial Intelligence Research Institute (IIIA),
Spanish National Research Council (CSIC), Spain
[4] University of Manchester, UK
{mikel.lujan@manchester.ac.uk,watson@cs.man.ac.uk}

Abstract. With an increase in the number of processors on a single chip, programming environments which facilitate the exploitation of parallelism on multicore architectures have become a necessity. StarSs is a task-based programming model that enables a flexible and high level programming. Although task synchronization in StarSs is based on data flow and dependency analysis, some applications (e.g. *reductions*) require *locks* to access shared data.

Transactional Memory is an alternative to lock-based synchronization for controlling access to shared data. In this paper we explore the idea of integrating a lightweight Software Transactional Memory (STM) library, TinySTM , into an implementation of StarSs (SMPSs). The SMPSs runtime and the compiler have been modified to include and use calls to the STM library. We evaluated this approach on four applications and observe better performance in applications with high lock contention.

1   Introduction

Over the past decade, single-core processors ran into three walls, namely ILP (Instruction Level Parallelism), power and memory. The ensuing stalemate led to the trend of placing multiple slower processors on a single chip. But achieving good performance on these architectures is hard. It often requires programmers to rewrite the code or implement algorithms anew. In multi-core programming, the programmer's efforts are directed towards hardware details, such as movement of data between processors and synchronization, than on the details of the algorithm. Every new architecture additionally comes with its associated SDK, which raises the issue of portability. Hence, what is needed now are programming environments, i.e. sets of compilers, runtimes and communication libraries, that make multi-core programming easier while achieving maximum performance. The effectiveness of such a programming model can be evaluated using the following measures:

C. Kaklamanis et al. (Eds.): Euro-Par 2012, LNCS 7484, pp. 514–525, 2012.
© Springer-Verlag Berlin Heidelberg 2012

- Performance of an application using the programming model versus the native SDK
- Level of complexity exposed to the programmer.
  - Increase in number of lines compared to sequential program
  - Use of specific API calls
- Ease of portability of applications.

OpenMP [5] is a widely used programming model for share memory architectures. It supports multi-platform multiprocessor programming in C, C++, and Fortran. Cilk [7] is a similar programming model developed at MIT. Both OpenMP and Cilk support task-based and loop-based parallelism but neither performs task-based data dependence analysis. Magma [4] is a programming language designed to investigate algebraic, geometric and combinatorial structures. It is not intended for general-purpose programming since its structure is preconditioned for linear algebra problems.

StarSs[9] is a programming environment for parallel architectures such as Symmetric Multiprocessors (SMP), the Cell Broadband Engine (Cell B./E.), Graphical Processing Units (GPU) and clusters. An application written with this programming model, can be executed on any of the architectures mentioned above with no change to the code, effectively achieving portability. In this paper we focus on SMPSs [10], the implementations of StarSs for SMP.

SMPSs allows programmers to write sequential applications, while the runtime exploits the inherent concurrency and schedules tasks to different cores of an SMP. We will have more to say on this topic in Section 2. In order to protect the atomicity of shared memory locations, SMPSs uses *locks*. But locks suffer from the traditional drawbacks of:

- Deadlock - two tasks trying to lock two different objects, each getting access to one and waiting for the other one to be released.
- Livelock - similar to deadlock, except that the state of a livelocked process changes constantly, although without progressing.
- Priority Inversion - a high-priority thread blocked by a low-priority thread.

Software Transactional Memory (STM) is an alternative method to lock-based synchronization for accessing shared-memory locations. To this end a program wraps operations (i.e., reads and writes) in a transaction and STM guarantees that either all the operations in the transaction occur or none. It is a non-blocking approach where a transaction tentatively updates shared memory. If successful it makes the changes permanent and visible to other transactions, else the transaction aborts and restarts [8]. This opportunistic strategy helps us in avoiding problems arising from locks.

There are many TM systems available which allow programmers to access and modify data through transactions. The Intel C++ STM compiler provides extensions to its C++ compiler with support for STM language extensions [1]. RSTM is a set of STM systems available from the Rochester Synchronization Group. It consists of different library implementations and a smart-pointer API for relatively transparent access to STM, and requires no compiler changes. TinySTM

[6] is a word-based STM implementation of the LSA algorithm, available from the University of Neuchatel. In this paper we explore the idea of integrating TinySTM into SMPSs, as a replacement to *locks* for synchronizing simultaneous access to critical memory locations. The rest of the paper is organized as follows: Section 2 explains the basic framework of SMPSs, Section 3 discusses STM and TinySTM, Section 4 presents our idea of integrating TinySTM in SMPSs, Section 5 evaluates and characterizes the performance of our idea. section 6 presents the conclusions and section 7 discusses the future work that we intent to do.

2   SMPSs

SMP Superscalar (SMPSs) consists of a source-to-source compiler and a runtime library. The programmer annotates the sequential code and marks tasks or units of computation using *pragmas* provided by the SMPSs compiler. During execution, the SMPSs runtime analyzes the data accesses of these tasks, but does not immediately perform the corresponding computation. Instead it builds a Task Dependency Graph(TDG), where each node represents a task instance, and edges denote dependencies between tasks. SMPSs uses the TDG to schedule tasks to cores. Independent tasks, i.e. tasks without incoming edges can execute in parallel.

2.1   SMPSs Syntax

As mentioned previously, the programmer typically annotates the functions using pragmas and declares tasks:

```
1    #pragma css task [clauses]
2 function definition / function declaration
```

Listing 1.1. Syntax of a Task Declaration

The clauses indicate the type of access that a task performs for each parameter. Every task parameter must appear in one of the clause, along with its dimensions. SMPSs supports the following clauses in the task pragma:

```
1 The list of main clauses is the following:
2     input ([list of parameters])
3     output ([list of parameters])
4     inout ([list of parameters])
5     reduction ([list of parameters])
```

The runtime builds a TDG based on the directionality of the parameters. Shown below is an example of a task pragma:

```
1 #pragma css task input(A[NB][NB],B[NB][NB],NB) inout(C[NB][NB])
2 void matmul(float  *A, float *B, float *C, unsigned long NB)
```

2.2   The *Reduction* Clause

Although the SMPSs runtime only schedules independent tasks for parallel execution, the programming model supports a *reduction*-clause which allows parallel updates to a specified memory regions. The runtime does not insert an edge in the TDG in this case. Responsibility falls on the programmer to access shared memory in the critical section using *lock* and *unlock* pragmas provided by SMPSs, for example:

```
1 #pragma css task input (n, j, a[n]) inout (results) reduction (results)
2 void nqueens_ser_task(int n, int j, char *a, int *results)
3 {
4    ....
5    #pragma css mutex lock (results)
6        *results = *results + local_sols;
7    #pragma css mutex unlock (results)
8    .....
9 }
```

Listing 1.2. Example of reduction

In the above Listing 1.2, the reduction applies to the variable *results*, which implies the latter can be updated simultaneously by different tasks. Hence the atomicity of the updates need to be guaranteed by lock and unlock pragmas.

3   Software Transactional Memory

Software Transactional Memory (STM) is an optimistic approach to manage concurrent accesses to shared memory locations. When two different transactions simultaneously try to update the same memory location, STM detects a conflict and allows only one of the transactions to complete successfully. The other transaction is either delayed or aborted. The delaying or aborting of the transaction is also called rollback and the transaction is called the conflicting transaction. The idea was first implemented by Shavit and Touitou [12]. STM simplifies the implementation of shared memory access since each transaction can now be viewed as an isolated series of operations. Every transaction is composed of 4 basic steps:

1. Start of a transaction.
2. Load values from memory into the current transactional environment.
3. Store values back to memory.
4. Commit the results.

Different STM libraries check for conflicts at different steps, depending on their implementation and design. Conflicts can be handled in various ways. The decision of which transaction should be allowed to complete and which should roll back, depends on the contention manager being used.

STM places limitations on the kind of operations that can be executed in a transaction. Only operations that can be rolled back should be employed, whereas for example I/O operations (like *printf("")* in C) cannot be included in a transaction. The use of STM implies a performance degradation due to the overhead incurred in the roll back of transactions.

3.1   TinySTM

TinySTM[6] is a word based STM library based on the *atomic_ops* library[3]. It implements a single version, word-based variant of the Lazy Snapshot Algorithm(LSA)[11]. Like most STM implementations TinySTM uses a shared array of locks to manage concurrent memory accesses. It maps addresses to locks, via a hash function, and uses a shared counter to maintain the timestamp validity of memory locations in transactions. It has three strategies to access memory:

1. WRITE_BACK_ETL - locks are acquired during encounter time.
2. WRITE_BACK_CTL - locks are only acquired during commit.
3. WRITE_THROUGH - directly updates memory and maintains an undo log for rollbacks.

In order to decide which transaction should roll back in case of a conflict, TinySTM has several built in contention managers:

1. CM_SUICIDE - Abort the transaction that detects the conflict.
2. CM_DELAY - Similar to CM_SUICIDE, but the rolled back transaction waits until the contended lock has been released.
3. CM_BACKOFF - Like CM_SUICIDE, but delay restarting the transaction for a random time.
4. CM_AGGRESSIVE - Kill the other transaction.

We use WRITE_BACK_CTL to access memory with TinySTM in SMPSs.

This choice is in line with our future work, where we plan to introduce the speculative execution of tasks in SMPSs. Tasks will execute speculatively, but the committing of the results is postponed to later stages. For handling conflicts we use the CM_DELAY contention manager. CM_DELAY restarts the rolled back transaction when the contended lock is released. As such we avoid the possibility of a transaction being rolled back repeatedly because of the same contention.

4   Integrating TinySTM in SMPSs

In order to incorporate TinySTM library calls in SMPSs applications, the library in question has to be initialized and threads have to be made as transactional threads. The initialization of TinySTM library (*stm_init*) takes place in the main thread of SMPSs and each SMPSs thread registers itself as a transactional thread (*stm_init_thread*). In order to replace *locks* with STM, the operations executed between lock and unlock pragmas must be wrapped in a transaction. The memory locations accessed in this region must be updated using STM calls. An SMP thread finally must commit the result of these operations and make them permanent. For example, the code of Listing 1.2 is transformed into :

```
1 #pragma css task input (n, j, a[n]) inout (results) reduction (results)
2 void nqueens_ser_task(int n, int j, char *a, int *results)
3 {
4   ....
5   sigjmp_buf* jump = stm_start(NULL); //start the transaction
6   if(jump != NULL)
7     setjmp(*jump, 0); //save stack context
8   int buffer += stm_load_int(results) + local_sols;
9   stm_store_int(results, buffer);
10  stm_commit(); //commit transaction
11  ...
12 }
```

Listing 1.3. Implementation of Listing 1.2 with TinySTM library calls

As soon as a transaction starts (line 5 of Listing 1.3), the stack context is saved using a call to *setjmp* (line 7 of Listing 1.3). Critical memory locations are loaded into the current transactional context via calls to TinySTM (line 8 of Listing 1.3). The SMP thread performs the updates and at the end, stores the results and commits the transaction (lines 9 and 10). If it detects a conflict while commit-ting the results, then it performs a *longjmp* and the execution is restarted from the location of the associated *setjmp*. In Listing 1.3., we inserted the transac-tional calls to the TinySTM library manually. But our objective is to implement transactional access to shared memory locations in SMPSs, without burdening the programmer with these implementation details. TinySTM calls then have to be performed from the SMPSs runtime or inserted by the compiler. Instead of adding new pragmas for this purpose, we decided to modify the implementation of the existing lock and unlock pragmas.

When The SMPSs compiler processes the lock and unlock pragmas, it replaces them with runtime calls to *css_lock* and *css_unlock* respectively. css_lock locks the parameter passed to the lock pragma while css_unlock unlocks it. The basic idea is to start a transaction when a lock pragma is encountered and to commit the results with the corresponding unlock pragma. Its implementation in the SMPSs library is troublesome, due to the use of stack calls of *setjmp* and *longjmp* by TinySTM. This restricts the start and end of a transaction to the same stack frame.

As the locking pragmas map to different functions in the SMPSs runtime, the stack context changes between the occurrence of a lock and its associated unlock. If we modify the runtime such that a transaction is started from css_lock and the results committed in css_unlock, conflicting transaction will not be rolled back correctly.

Instead the SMPSs compiler was modified to insert transactional calls to the TinySTM library. A transaction is started when a thread acquires a lock and the address to be locked is loaded into the transactional environment. The compiler creates a local variable and assigns it the memory address passed to the lock pragma. All updates are performed on this local variable. When the thread re-leases the lock for the memory location, i.e. at the location of an unlock pragma, the value of the local variable is stored back into the memory address and the transaction is committed. If another thread, and hence a different transaction, has modified the shared memory location since the time it was loaded into the local variable, the transaction is rolled back and restarted.

Below we show how the compiler transforms lock and unlock pragmas for Listing 1.2:

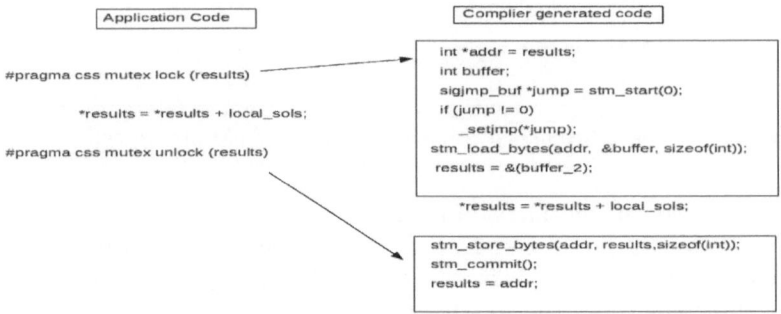

Fig. 1. Compiler generated code of Listing 1.2

The local variable is private to the thread and its scope is the task scope, i.e., this variable is alive only within this instance of the task.

We observed that in some applications it is more efficient to load multiple shared memory addresses into a single transaction rather than to generate a separate transaction for every address. Therefore we extended the *lock* pragma to accept more than one address and load them into a single transactional context. An example of multiple memory locations passed to a single lock pragma is shown in Listing 1.4.

5   Results

To evaluate the performance of our implementation, we executed 4 benchmark programs and the results are compared between SMPSs applications using locks and STM. The applications chosen are:

- NQueens - of placing chess queens on a n*n board such that no queens attack each other. The problem size of the results presented is a chess board of size 14*14 . The problem can have more than one unique solution. The critical section in this application was when updating the number of unique solutions that the problem has.
- Gmeans - a data mining algorithm to find clusters in a dataset. It returns the number of Gaussian distributions and their centers contained in the dataset. The atomicity in this application is required while updating the centers of clusters in the data set.The problem size is a data set of 10 million points of 10 dimensions each. It was observed that the application was not scaling with more than 8 threads.

- Matrix Multiplication - In this application the values of resultant matrix are simultaneously updated by different tasks. Hence, while storing the results a lock needs to be acquired on element of the matrix. The dimensions of the matrix are (128*16) * (128*16).
- Specfem3D - the algorithm simulates seismic wave propagation in sedimentary basins or any other regional geological model. In this application, locks are used in two different stages, once while localizing the data in tasks from global vectors and again while summing the values from each tasks in the global mesh. The global mesh is accessed both directly and indirectly which leads to conflicting accesses of the same position some times.

5.1   Performance Evaluation

The performance evaluation is shown below:

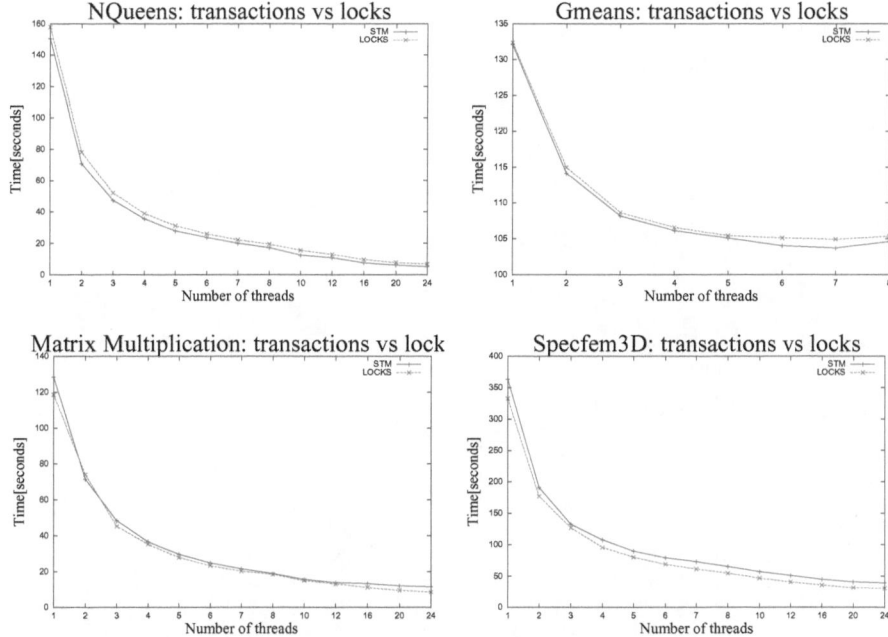

Fig. 2. Performance comparison between SMPSs examples using Transactions and Locks

The above mentioned applications were executed on an Intel(R) Xeon(R) E7450@2.40GHz machine with 24 cores. Thread affinity was controlled by assigning one thread to each core. From Fig.2., we observe that in Nqueens and Gmeans applications STM performs better than locks, whereas in matrix multiplication example, the execution timings are nearly similar. In Specfem3D, even though locks perform better than STM, we observe that the STM version scales. We hope that with further optimizations and better hardware support, STM will have higher performance.

5.2   Performance Characterization

While using STM, with increasing number of threads there is a higher probability of transactions conflicting with each other. Since more threads try to simultaneously update the shared memory locations, more conflicts occur resulting in more rollbacks. Hence we analyzed the behavior of above mentioned applications with increasing number of threads. We used Paraver[2], a flexible visualization tool to analyze characteristics of transactions generated while running the applications. SMPSs runtime generates performance trace-files if *tracing* flag is enabled during compilation. These traces can be analyzed using paraver. Transaction specific events such as time spent in executing operations in a transaction, time spent in commit and rollback were added to paraver tracefile. Shown below is analysis of time spent by applications in different phases of transaction when executed with varying number of threads.

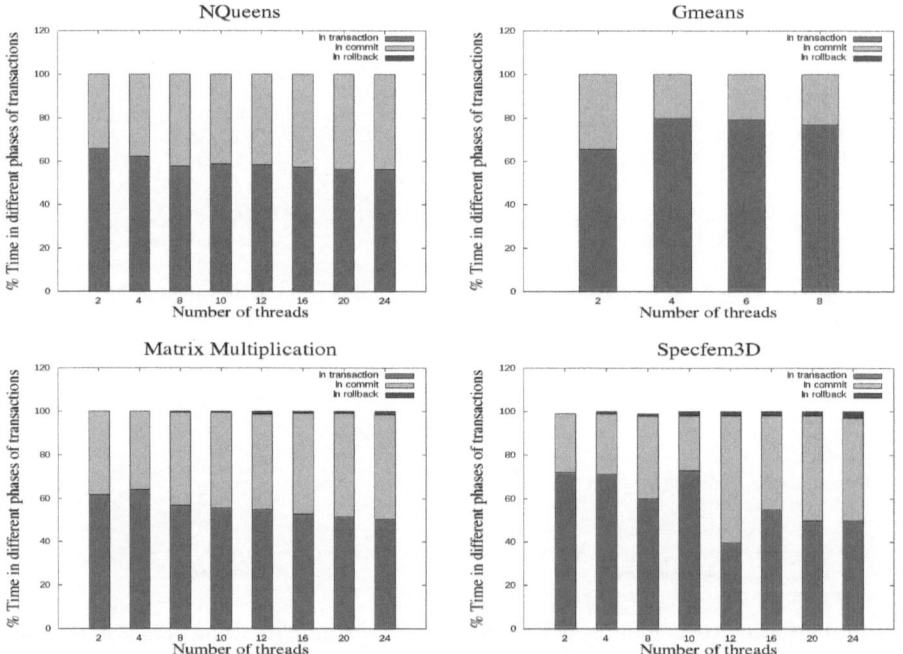

Fig. 3. Time spent in different phases of transaction

From Fig.3., we can observe that in NQueens and Gmeans applications, time spent in rollback is minimal. This is the main reason for their better performance when compared to their lock based implementations. We can also observe that in Gmeans application, threads spend longer time in commit compared to operations executed in transactions. The reason is, since every point in the data set is of multiple dimensions, while committing the results of updated centers the value of each dimension has to be committed. This leads to higher amount of time being spent in commit. We can also observe that matmul and specfem3D spend significantly more time in rollback, compared to the other two benchmarks. As rollback contributes to run-time overhead and thus to the overall execution time, the performance decreases accordingly. However, the rollback overhead does not increase linearly with the number of threads. The potential for contention, and hence rollback, of an application is bounded by the amount of parallelism it admits. The latter is a characteristic of the TDG of an application, not of the number of resources (like the number of threads) with which one chooses to execute. This observation gives us optimism that, with further optimizations and better hardware support, we can further improve the performance of our parallel programming model using STM.

As mentioned earlier sometimes it is more efficient to update multiple shared memory locations in a single transaction compared to generating one transaction for every single address. The trade-off is, to create multiple smaller transactions and thus spend more time in start and commit of transactions versus longer transactions and hence longer time in rollbacks in case of a conflict. In Specfem3D instead of updating 3 different memory locations separately we updated all 3 in a single transaction. Shown below is a snippet of the code.

```
1  //3 Transactions
2  #pragma css mutex lock((temp_x))
3  *(temp_x) += sum_terms[elem][k][j][i][X];
4  #pragma css mutex unlock((temp_x))
5
6  #pragma css mutex lock((temp_y))
7  *(temp_y) += sum_terms[elem][k][j][i][Y];
8  #pragma css mutex unlock((temp_y))
9
10 #pragma css mutex lock((temp_z))
11 *(temp_z) += sum_terms[elem][k][j][i][Z];
12 #pragma css mutex unlock((temp_z))
13
14 //1 big transaction
15 #pragma css mutex lock(temp_x,temp_y,temp_z)
16 *(temp_x) += sum_terms[elem][k][j][i][X];
17 *(temp_y) += sum_terms[elem][k][j][i][Y];
18 *(temp_z) += sum_terms[elem][k][j][i][Z];
19 #pragma css mutex unlock(temp_x,temp_y,temp_z)
```

Listing 1.4. Specfem3D

We observed that it was more optimal to update three different shared memory locations in a single transaction compared to creating a transaction for each of them. Shown in Fig.4., is how in Specfem3D these two approaches affected the number of rollbacks:

While in this case longer transactions performed better, sometimes they can degrade the performance due to longer rollback time.

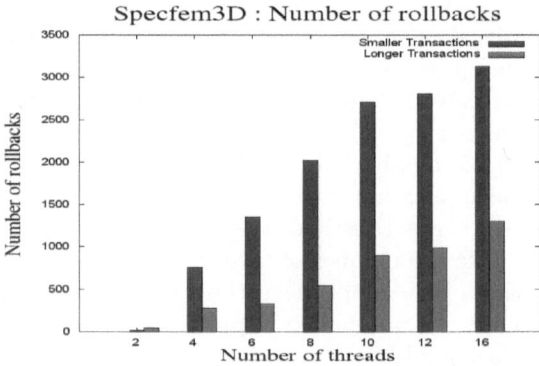

Fig. 4. Specfem3D : Number of rollbacks

6   Conclusion

To keep pace with Moore's law, the trend in the processor industry is to place multiple processors on a single chip. To completely utilize this power, the need of the hour is of programming models offering an easier way to exploit parallelism. StarSs is one such programming model for widely used multicore architectures. It uses *lock* based synchronization during simultaneous updates of shared memory. STM is an alternative shared memory synchronization technique. It is optimistic in nature and simplifies concept of concurrent access to shared memory. We integrate TinySTM, a lightweight STM library with SMPSs (one implementation of StarSs) and replace locks with transactions in SMPSs applications. The results were optimistic with higher performance in applications with high lock contention.

7   Future Work

We plan to use STM to speculatively execute tasks. SMPSs provides programmers with synchronization constructs such as *barriers* and *wait-on*(wait for a particular variable or memory location to be updated before continuing execution). Such constructs lead to problems of load balancing. Hence we plan to introduce speculation using STM. In cases where there is a control dependency between tasks and not data dependency, we can use STM to speculate and execute tasks but postpone the committing of their results till the dependency is resolved.

Acknowledgements. We thankfully acknowledge the support of the European Commission through the TERAFLUX project (FP7-249013) and the HiPEAC-2 Network of Excellence (FP7/ICT 217068), the support of the Spanish Ministry of Education (TIN2007-60625, CSD2007-00050 and FI program) and the Generalitat de Catalunya (2009- SGR-980).

References

1. http://software.intel.com/en-us/articles/intel-c-stm-compiler-prototype-edition-20/
2. http://www.bsc.es/media/1364.pdf
3. http://www.hpl.hp.com/research/linux/atomicops/
4. magma.maths.usyd.edu.au/magma/pdf/intro.pdf
5. Duran, A., Perez, J.M., Ayguadé, E., Badia, R.M., Labarta, J.: Extending the OpenMP Tasking Model to Allow Dependent Tasks. In: Eigenmann, R., de Supinski, B.R. (eds.) IWOMP 2008. LNCS, vol. 5004, pp. 111–122. Springer, Heidelberg (2008)
6. Felber, P., Fetzer, C., Riegel, T.: Dynamic performance tuning of word-based software transactional memory. In: PPoPP 2008, Salt Lake City, Utah, USA. ACM (2008)
7. Frigo, M., Leiserson, C.E., Randall, K.H.: The implementation of the cilk-5 multithreaded language. In: Proceedings of the 1998 ACM SIGPLAN Conference on Programming Language Design and Implementation, PLDI 1998, Montreal, Canada. ACM (1998)
8. Harris, T., Larus, J., Rajwar, R.: Transactional Memory, 2nd edn. Morgan and Claypool Publishers (2010)
9. Perez, J.M., Badia, R.M., Labarta, J.: A dependency-aware task-based programming environment for multi-core architectures. In: IEEE Int. Conference on Cluster Computing, pp. 142–151 (September 2008)
10. Perez, J.M., Badia, R.M., Labarta, J.: Handling task dependencies under strided and aliased references. In: Proceedings of the 24th ACM International Conference on Supercomputing, ICS 2010, pp. 263–274. ACM, New York (2010)
11. Riegel, T., Felber, P., Fetzer, C.: A lazy snapshot algorithm with eager validation. In: Proceedings of the 20th International Symposium on Distributed Computing
12. Shavit, N., Touitou, D.: Software transactional memory. In: 14th ACM Symposium on the Principles of Distributed Computing, Ottawa, Ontario, Canada. ACM (1995)

On-the-Fly Task Execution for Speeding Up Pipelined MapReduce

Diana Moise[1], Gabriel Antoniu[1], and Luc Bougé[2]

[1] INRIA Rennes - Bretagne Atlantique / IRISA, France
[2] ENS Cachan - Brittany / IRISA, France

Abstract. The MapReduce programming model is widely acclaimed as a key solution to designing data-intensive applications. However, many of the computations that fit this model cannot be expressed as a single MapReduce execution, but require a more complex design. Such applications consisting of multiple jobs chained into a long-running execution are called pipeline MapReduce applications. Standard MapReduce frameworks are not optimized for the specific requirements of pipeline applications, yielding performance issues. In order to optimize the execution on pipelined MapReduce, we propose a mechanism for creating map tasks along the pipeline, as soon as their input data becomes available. We implemented our approach in the Hadoop MapReduce framework. The benefits of our dynamic task scheduling are twofold: reducing job-completion time and increasing cluster utilization by involving more resources in the computation. Experimental evaluation performed on the Grid'5000 testbed, shows that our approach delivers performance gains between 9% and 32%.

Keywords: MapReduce, pipeline MapReduce applications, intermediate data management, task scheduling, Hadoop, HDFS.

1 Introduction

The MapReduce abstraction has revolutionized the data-intensive community and has rapidly spread to various research and production areas. Google introduced MapReduce [8] as a solution to the need to process datasets up to multiple terabytes in size on a daily basis. The goal of the MapReduce programming model is to provide an abstraction that enables users to perform computations on large amounts of data.

The MapReduce abstraction is inspired by the "map" and "reduce" primitives commonly used in functional programming. When designing an application using the MapReduce paradigm, the user has to specify two functions: *map* and *reduce* that are executed in parallel on multiple machines. Applications that can be modeled by the means of MapReduce, mostly consist of two computations: the "map" step, that applies a filter on the input data, selecting only the data that satisfies a given condition, and the "reduce" step, that collects and aggregates all the data produced by the first phase. The MapReduce model exposes a simple interface, that can be easily manipulated by users without any experience with parallel and distributed systems. However, the interface is versatile enough so that it can be employed to suit a wide range of data-intensive applications. These are the main reasons for which MapReduce has known an increasing popularity ever since it was introduced.

C. Kaklamanis et al. (Eds.): Euro-Par 2012, LNCS 7484, pp. 526–537, 2012.

An open-source implementation of Google's abstraction was provided by Yahoo! through the Hadoop [5] project. This framework is considered the reference MapReduce implementation and is currently heavily used for various purposes and on several infrastructures. The MapReduce paradigm has also been adopted by the cloud computing community as a support to those cloud-based applications that are data-intensive. Cloud providers support MapReduce computations so as to take advantage of the huge processing and storage capabilities the cloud holds, but at the same time, to provide the user with a clean and easy-to-use interface. Amazon released ElasticMapReduce [2], a web service that enables users to easily and cost-effectively process large amounts of data. The service consists of a hosted Hadoop framework running on Amazon's Elastic Compute Cloud (EC2) [1]. Amazon's Simple Storage Service (S3) [3] serves as storage layer for Hadoop. AzureMapReduce [9] is an implementation of the MapReduce programming model, based on the infrastructure services the Azure cloud [6] offers. Azure's infrastructure services are built to provide scalability, high throughput and data availability. These features are used by the AzureMapReduce runtime as mechanisms for achieving fault tolerance and efficient data processing at large scale.

MapReduce is used to model a wide variety of applications, belonging to numerous domains such as analytics (data processing), image processing, machine learning, bioinformatics, astrophysics, etc. There are many scenarios in which designing an application with MapReduce requires the users to employ several MapReduce processing. These applications that consist of multiple MapReduce jobs chained into a long-running execution, are called *pipeline MapReduce applications*. In this paper, we study the characteristics of pipeline MapReduce applications, and we focus on optimizing their execution. Existing MapReduce frameworks manage pipeline MapReduce applications as a sequence of MapReduce jobs. Whether they are employed directly by users or through higher-level tools, MapReduce frameworks are not optimized for executing pipeline applications. A major drawback comes from the fact that the jobs in the pipeline have to be executed sequentially: a job cannot start until all the input data it processes has been generated by the previous job in the pipeline.

In order to speed up the execution of pipelined MapReduce, we propose a new mechanism for creating "map" tasks along the pipeline, as soon as their input data becomes available. Our approach allows successive jobs in the pipeline to overlap the execution of "reduce" tasks with that of "map" tasks. In this manner, by dynamically creating and scheduling tasks, the framework is able to complete the execution of the pipeline faster. In addition, our approach ensures a more efficient cluster utilization, with respect to the amount of resources that are involved in the computation. We implemented the proposed mechanisms in the Hadoop MapReduce framework [5] and evaluated the benefits of our approach through extensive experimental evaluation.

In section 2 we present an overview of pipelined MapReduce as well as the scenarios in which this type of processing is employed. Section 3 introduces the mechanisms we propose and shows their implementation in Hadoop. Section 4 is dedicated to the experiments we performed; we detail the environmental setup and the scenarios we selected for execution in order to measure the impact of our approach. Section 5 summarizes the contributions of this work and presents directions for future work.

2   Pipeline MapReduce Applications: Overview and Related Work

Many of the computations that fit the MapReduce model, cannot be expressed as a single MapReduce execution, but require a more complex design. These applications that consist of multiple MapReduce jobs chained into a long-running execution, are called *pipeline MapReduce applications*. Each stage in the pipeline is a MapReduce job (with 2 phases, "map" and "reduce"), and the output data produced by one stage is fed as input to the next stage in the pipeline. Usually, pipeline MapReduce applications are long-running tasks that generate large amounts of *intermediate* data (the data produced between stages). This type of data is transferred between stages and has different characteristics from the meaningful data (the input and output of an application). While the input and output data are expected to be persistent and are likely to be read multiple times (during and after the execution of the application), intermediate data is *transient* data that is usually *written once*, by one stage, and *read once*, by the next stage.

However, there are few scenarios in which users directly design their application as a pipeline of MapReduce jobs. Most of the use cases of MapReduce pipelines come from applications that *translate* into a chain of MapReduce jobs. One of the drawbacks of the extreme simplicity of the MapReduce model is that it cannot be straightforwardly used in more complex scenarios. For instance, in order to use MapReduce for higher-level computations (for example, the operations performed in the database domain) one has to deal with issues like multi-stage execution plan, branching data-flows, etc. The trend of using MapReduce for database-like operations led to the development of high-level query languages that are executed as MapReduce jobs, such as Hive [14], Pig [12], and Sawzall [13]. Pig is a distributed infrastructure for performing high-level analysis on large data sets. The Pig platform consists of a high-level query language called *PigLatin* and the framework for running computations expressed in PigLatin. PigLatin programs comprise SQL-like high-level constructs for manipulating data that are interleaved with MapReduce-style processing. The Pig framework compiles these programs into a pipeline of MapReduce jobs that are executed within the Hadoop environment.

The scenarios in which users actually devise their applications as MapReduce pipelines, involve binary data whose format does not fit the high-level structures of the aforementioned frameworks. In order to facilitate the design of pipeline MapReduce applications, Cloudera recently released Crunch [4], a tool that generates a pipeline of MapReduce jobs and manages their execution. While there are several frameworks that generate pipeline MapReduce applications, few works focus on optimizing the actual execution of this type of applications. In [11], the authors propose a tool for estimating the progress of MapReduce pipelines generated by Pig queries. The Hadoop Online Prototype (HOP) [7] is a modified version of the Hadoop MapReduce framework that supports online aggregation, allowing users to get snapshots from a job as it is being computed. HOP employs pipelining of data between MapReduce jobs, i.e., the reduce tasks of one job can optionally pipeline their output directly to the map tasks of the next job. However, by circumventing the storing of data in a distributed file system (DFS) between the jobs, fault tolerance cannot be guaranteed by the system. Furthermore, as the computation of the reduce function from the previous job and the map function of the next job cannot be overlapped, the jobs in the pipeline are executed sequentially.

3 Introducing Dynamic Scheduling of Map Tasks in Hadoop

3.1 Motivation

In a pipeline of MapReduce applications, the intermediate data generated between the stages represents the output data of one stage and the input data for the next stage. The intermediate data is produced by one job and consumed by the next job in the pipeline. When running this kind of applications in a dedicated framework, the intermediate data is usually stored in the distributed file system that also stores the user input data and the output result. This approach ensures intermediate data availability, and thus, provides fault tolerance, a very important factor when executing pipeline applications. However, using MapReduce frameworks to execute pipeline applications raises performance issues, since MapReduce frameworks are not optimized for the specific features of intermediate data. The main performance issue comes from the fact that the jobs in the pipeline have to be executed sequentially: a job cannot start until all the input data it processes has been generated by the job in the previous stage of the pipeline. Consequently, the framework runs only one job at a time, which results in inefficient cluster utilization and basically, a waste of resources.

3.2 Executing Pipeline MapReduce Applications with Hadoop

The Hadoop project provides an open-source implementation of Google's MapReduce paradigm through the Hadoop MapReduce framework [5,15]. The framework was designed following Google's architectural model and has become the reference MapReduce implementation. The architecture is tailored in a master-slave manner, consisting of a single master *jobtracker* and multiple slave *tasktrackers*. The jobtracker's main role is to act as the task *scheduler* of the system, by assigning work to the tasktrackers. Each tasktracker disposes of a number of available *slots* for running tasks. Every active map or reduce task takes up one slot, thus a tasktracker usually executes several tasks simultaneously. When dispatching "map" tasks to tasktrackers, the jobtracker strives at keeping the computation as close to the data as possible. This technique is enabled by the data-layout information previously acquired by the jobtracker. If the work cannot be hosted on the actual node where the data resides, priority is given to nodes closer to the data (belonging to the same network rack). The jobtracker first schedules "map" tasks, as the reducers must wait for the "map" execution to generate the intermediate data.

Hadoop executes the jobs of a pipeline MapReduce application in a sequential manner. Each job in the pipeline consists of a "map" and a "reduce" phase. The "map" computation is executed by Hadoop tasktrackers only when all the data it processes is available in the underlying DFS. Thus, the mappers are scheduled to run only after all the reducers from the preceding job have completed their execution. This scenario is also representative for a Pig processing: the jobs in the logical plan generated by the Pig framework are submitted to Hadoop sequentially. In consequence, at each step of the pipeline, at most the "map" and "reduce" tasks of the same job are being executed. Running the mappers and the reducers of a single job involves only a part of the cluster nodes. The rest of the computational and storage cluster capabilities remains idle.

3.3   Our Approach

In order to speed-up the execution of pipeline MapReduce applications, and also to improve cluster utilization, we propose an optimized Hadoop MapReduce framework, in which the scheduling is done in a *dynamic* manner. For a better understanding of our approach, we first detail the process through which "map" and "reduce" tasks are created and scheduled in the original Hadoop MapReduce framework.

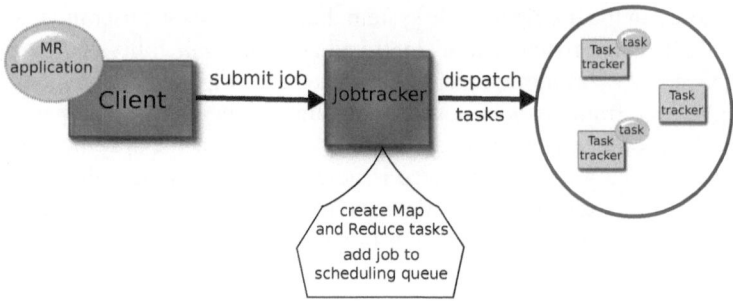

Fig. 1. Job submission process in Hadoop

Figure 1 displays the job submission process. The first step is for the user to specify the "map" and "reduce" computations of the application. The Hadoop client generates all the job-related information (input and output directories, data placement, etc.) and then submits the job for execution to the jobtracker. On the jobtracker's side, the list of "map" and "reduce" tasks for the submitted job is created. The number of "map" tasks is equal to the number of chunks in the input data, while the number of "reduce" tasks is computed by taking into account various factors, such as the cluster capacity, the user specification, etc. The list of tasks is added to the *job queue* that holds the jobs to be scheduled for execution on tasktrackers. In the Hadoop MapReduce framework, the "map" and "reduce" tasks are created by the jobtracker when the job is submitted for execution. When they are created, the "map" tasks require to know the location of the chunks they will work on.

In the context of multiple jobs executed in a pipeline, the jobs are submitted by the client to the jobtracker sequentially, as the chunk-location information is available only when the previous job completes. Our approach is based on the remark that a "map" task is created for a single input chunk. It only needs to be aware of this very chunk location. Furthermore, when it is created, the only information that the "map" task requires, is the list of nodes that store the data in its associated chunk. We modified the Hadoop MapReduce framework to create "map" tasks *dynamically*, that is, as soon as a chunk is available for processing. This approach can bring substantial benefits to the execution of pipeline MapReduce applications. Since the execution of a job can start as soon as the first chunk of data is generated by the previous job, the total runtime is significantly reduced. Additionally, the tasks belonging to several jobs in the pipeline can be executed at the same time, which leads to a more efficient cluster utilization.

The modifications and extensions of the Hadoop MapReduce framework that we propose, are further presented and summarized on Figure 2.

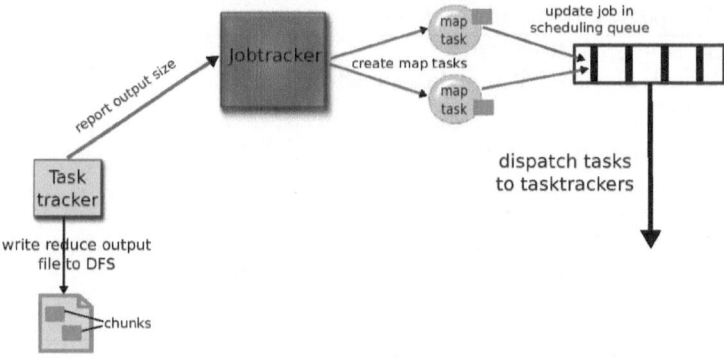

Fig. 2. Dynamic creation of "map" tasks

Algorithm 1. Report output size (**on tasktracker**)

1: **procedure** COMMITTASK
2:     $(size, files) \leftarrow tasktracker.writeReduceOutputData()$
3:     $jobtracker.transmitOutputInfo(size, files)$
4: **end procedure**

Job-Submission Process

Client side. On the client side, we modified the submission process between the Hadoop client and the jobtracker. Instead of waiting for the execution to complete, the client launches a *job monitor* that reports the execution progress to the user. With this approach, a pipeline MapReduce application employs a single Hadoop client to run the application. The client submits all the jobs in the pipeline from the beginning, instead of submitting them sequentially, i.e., the modified Hadoop client submits the whole set of jobs $job_1...job_n$ for execution.

Jobtracker side. The job-submission protocol is similar to the one displayed on Figure 1. However, at submission time, only the input data for job_1 is available in the DFS. Regarding $job_2...job_n$, the input data has to be generated throughout the pipeline. Thus, the jobtracker creates the set of "map" and "reduce" tasks only for job_1. For the rest of the jobs, the jobtracker creates only "reduce" tasks, while "map" tasks will be created along the pipeline, as the data is being generated.

Job Scheduling

Tasktracker side: For a job_i in the pipeline, the data produced by the job's "reduce" phase ($reduce_i$) represents the input data of job_{i+1}'s "map" task (map_{i+1}). When $reduce_i$ is completed, the tasktracker writes the output data to the backend storage. We modified the tasktracker code to notify the jobtracker whenever it successfully completes the execution of a "reduce" function: the tasktracker informs the jobtracker about the size of the data produced by the "reduce" task.

Algorithm 2. Update job (**on jobtracker**)

```
1:  procedure TRANSMITOUTPUTINFO(size, files)
2:      invoke updateJob(size, files) on taskscheduler
3:  end procedure

4:  procedure UPDATEJOB(size, files)
5:      for all job ∈ jobQueue do
6:          dir ← job.getInputDirectory()
7:          if dir = getDirectory(files) then
8:              if writtenBytes.contains(dir) = False then
9:                  writtenBytes.put(dir, size)
10:             else
11:                 allBytes ← writtenBytes.get(dir)
12:                 writtenBytes.put(dir, allBytes + size)
13:             end if
14:             allBytes ← writtenBytes.get(dir)
15:             if allBytes ≥ CHUNK_SIZE then
16:                 b ← job.createMapsForSplits(files)
17:                 writtenBytes.put(dir, allBytes − b)
18:             else
19:                 job.addToPending(files)
20:             end if
21:         end if
22:     end for
23: end procedure
```

Jobtracker side: In our modified framework, the jobtracker keeps track of the output data generated by reducers in the DFS. This information is important for the scheduling of the jobs in the pipeline, as the output directory of job_i is the input directory of job_{i+1}. Each time data is produced in job_i's output directory, the jobtracker checks to see if it can create new "map" tasks for job_{i+1}. If the data accumulated in job_{i+1}'s input directory is at least of the size of a chunk, the jobtracker creates "map" tasks for the newly generated data. For each new chunk, the jobtracker creates a "map" task to process it. All the "map" tasks are added to the scheduling queue and then dispatched to idle tasktrackers for execution.

The modifications on the tasktracker side are described in Algorithm 1. We extended the code with a primitive that sends to the jobtracker the information about the "reduce" output data: the files written to the DFS and the total size of the data. Algorithm 2 shows the process of updating a job with information received from tasktrackers. The algorithm is integrated in the jobtracker code, mainly in the scheduling phase. The jobtracker also plays the role of *task scheduler*. It keeps a list of data written to the input directory of each job. For each received update, the jobtracker checks if the data in the job's input directory reaches at least a chunk in size (64 MB default). If it is the case, "map" tasks will be created, one per each new data chunk. Otherwise, the job's information is stored for subsequent processing. The mechanism of creating "map" tasks is presented in Algorithm 3, executed by the jobtracker, and integrated into the job

---

Algorithm 3. Create map tasks (**on job**)

1: **procedure** ADDTOPENDING($files$)
2:     $pendingFiles.addAll(files)$
3: **end procedure**

4: **function** CREATEMAPSFORSPLITS($files$) **returns** splitBytes
5:     $pendingFiles.addAll(files)$
6:     $splits \leftarrow getSplits(pendingFiles)$
7:     $pendingFiles.clear()$
8:     $newSplits \leftarrow splits.length$
9:     $jobtracker.addWaitingMaps(newSplits)$
10:     **for** $i \in [1..newSplits]$ **do**
11:         $maps[numMapTasks + i] \leftarrow newMapTask(splits[i])$
12:     **end for**
13:     $numMapTasks \leftarrow numMapTasks + newSplits$
14:     $notifyAllReduceTasks(numMapTasks)$
15:     **for all** $s \in splits$ **do**
16:         $splitBytes \leftarrow splitBytes + s.getLength()$
17:     **end for**
18:     **return** $splitBytes$
19: **end function**

---

code. We extended the code so that each job holds the list of files that were generated so far in the job's input directory. When the jobtracker computes that at least a chunk of input data has been generated, new "map" tasks are created for the job. The data in the files is split into chunks. A "map" task is created for each chunk and the newly launched tasks are added to the scheduling queue. The jobtracker also informs the "reduce" tasks that the number of "map" tasks has changed. The reducers need to be aware of the number of mappers of the same job, as they have to transfer their assigned part of the output data from all the mappers to their local disk.

4   Evaluation

We validate the proposed approach through a series of experiments that compare the original Hadoop framework with our modified version, when running pipeline applications.

4.1   Environmental Setup

The experiments were carried out on the Grid'5000 [10] testbed. The Grid'5000 project is a widely-distributed infrastructure devoted to providing an experimental platform for the research community. The platform is spread over 10 geographical sites located through on French territory and 1 in Luxembourg. For our experiments, we employed nodes from the Orsay cluster of the Grid'5000. The nodes are outfitted with dual-core x86_64 CPUs and 2 GB of RAM. Intra-cluster communication is done through a 1 Gbps

Ethernet network. We performed an initial test at a small scale, i.e., 20 nodes, in order to assess the impact of our approach. The second set of tests involves 50 nodes belonging to the Orsay cluster.

4.2   Results

The evaluation presented here focuses on assessing the performance gains of the optimized MapReduce framework we propose, over the original one. To this end, we developed a benchmark that creates a pipeline of n MapReduce jobs and submits them to Hadoop for execution. Each job in the pipeline simulates a load that parses key-value pairs from the input data and outputs 90% of them as final result. In this manner, we manage to obtain a long-running application that generates a large amount of data, allowing our dynamic scheduling mechanism to optimize the execution of the pipeline. The computation itself is not relevant in this case, as our goal is to create a scenario in which enough data chunks are generated along the pipeline so that "map" tasks can be dynamically created. We run this type of application first with the original Hadoop framework, then with our optimized version of Hadoop. In both cases, we measure the pipeline completion-time and compare the results.

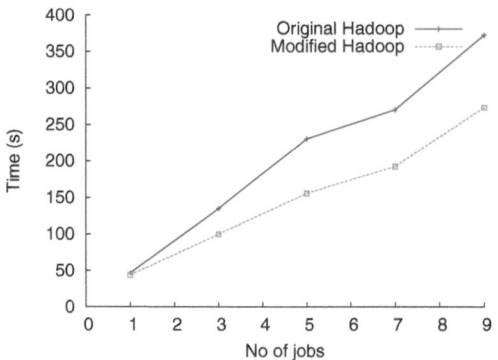

Fig. 3. Completion time for short-running pipeline applications

Short-Running Pipeline Applications

In a first set of experiments, we run the benchmark in a small setup involving 20 nodes, on top of which HDFS and Hadoop MapReduce are deployed as follows: a dedicated machine is allocated for each centralized entity (namenode, jobtracker), a node serves as the Hadoop client that submits the jobs, and the rest of 17 nodes represent both datanodes and tasktrackers. At each step, we keep the same deployment setup and we increase the number of jobs in the pipeline to be executed. The first test consists in running a single job, while the last one runs a pipeline of 9 MapReduce jobs. The application's input data, i.e., job_1's input data, consists of 5 data chunks (a total of 320 MB). Job_i keeps 90% of the input data it received from job_{i-1}. In the case of the 9-job pipeline, this data-generation mechanism leads to a total of 2 GB of data produced throughout the pipeline, out of which 1.6 GB account for intermediate data.

Figure 3 shows the execution time of pipeline applications consisting of an increasing number of jobs (from 1 to 9), in two scenarios: when running on top of the original Hadoop, and with the pipeline-optimized version we proposed. In the first case, the client sequentially submits the jobs in the pipeline to Hadoop's jobtracker, i.e., waits for the completion of job_i before submitting job_{i+1}. When using our version of Hadoop, the client submits all the jobs in the pipeline from the beginning, and then waits for the completion of the whole application. As expected, the completion time in both cases increases proportionally to the number of jobs to be executed. However, our framework manages to run the jobs faster, as it creates and schedules "map" tasks as soon as a chunk of data is generated during the execution. This mechanism speeds-up the execution of the entire pipeline, and also exhibits a more efficient cluster utilization. Compared to the original Hadoop, we obtain a performance gain between 26% and 32%.

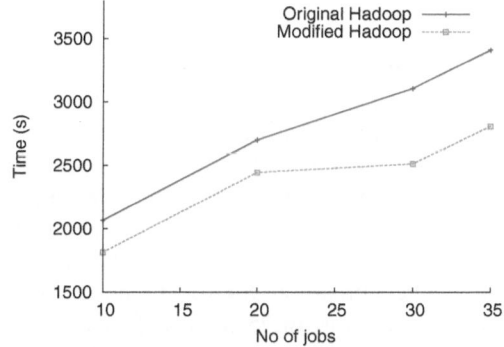

Fig. 4. Completion time for long-running pipeline applications

Long-Running Pipeline Applications

The first experiment we presented was focused on pipeline applications that consist of a small up to a medium number of jobs (1 to 9). Due to the long-running nature of pipeline applications and considering the significant size of the intermediate data our benchmark generates, we performed experiments with larger applications and larger datasets in a different setup, including 50 nodes. HDFS and Hadoop MapReduce are deployed as for the previous experiment, employing thus 47 tasktrackers. The size of the input data for each pipeline application amounts to 2.4 GB (40 data chunks). We vary the number of jobs to be executed in each pipeline, from 10 to 35. For the longest-running application, the generated data add up to a total of 24.4 GB.

The results for this setup are displayed on Figure 4. Consistently with the previous results, our approach proves to be more efficient for long-running applications as well. The performance gains vary between 9% and 19% in this scenario. The benefits of our optimized framework have a smaller impact in this case, because of the data size involved in the experiment. Since more chunks are used as input, and substantially more chunks are being generated throughout the pipeline, a large part of the tasktrackers is involved in the current computation, leaving a smaller number of resources available for dynamically running created "map" tasks.

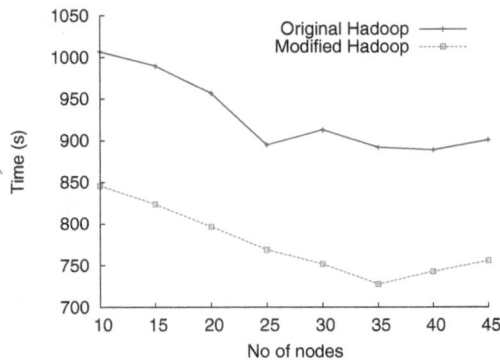

Fig. 5. Impact of deployment setup on performance

Scaling Out

In the context of pipeline applications, the number of nodes involved in the Hadoop deployment can have a substantial impact on completion time. Furthermore, considering our approach of dynamic scheduling "map" tasks, the scale of the deployment is an important factor to take into account. Thus, we performed an experiment in which we vary the number of nodes employed by the Hadoop framework. At each step, we increase the number of nodes used for the deployment, such that the number of task-trackers that execute "map" and "reduce" tasks is varied from 10 to 45. In each setup, we run the aforementioned benchmark with a fixed number of 7 jobs in the pipeline. The input data is also fixed, consisting of 25 chunks of data, i.e., 1.5 GB.

Figure 5 shows the completion time of the 7-job pipeline when running with both original Hadoop and modified Hadoop, while increasing the deployment setup. As the previous experiments also showed, our improved framework manages to execute the jobs faster than the original Hadoop. In both cases, as more nodes are added to the deployment, the application is executed faster, as more tasktrackers can be used for running the jobs. However, increasing the number of nodes yields performance gains up to a point, which corresponds to 25 tasktrackers for the original Hadoop. This number is strongly related to the number of chunks in the input data, since the jobtracker schedules a tasktracker to run the "map" computation on each chunk. For the modified Hadoop, the point after which expanding the deployment does not prove to be profitable any longer, is higher than for the original Hadoop. The reason for this behavior lies in the scheduling approach of both frameworks: in original Hadoop, the scheduling of jobs is done sequentially, while in modified Hadoop, the "map" tasks of each job are scheduled as soon as the data is generated. The completion time starts to increase in both cases after a certain point, as the overhead of launching and managing a larger number of tasktrackers overcomes the advantage of having more nodes in the deployment.

5   Conclusions

In this paper we address a special class of MapReduce applications, i.e., applications that consist of multiple jobs executed in a pipeline. In this context, we focus on improving the performance of the Hadoop MapReduce framework when executing pipelines. Our proposal consists mainly of a new mechanism for creating tasks along the pipeline, as soon as their input data become available. This dynamic task scheduling leads to an improved performance of the framework, in terms of job completion time. In addition, our approach ensures a more efficient cluster utilization, with respect to the amount of resources involved in the computation. The approach presented in this paper can be further extended so as to allow the overlapping of several jobs in the pipeline. However, this aspect would require careful tuning of the scheduling of tasks in MapReduce frameworks. Deciding whether to execute reducers for the current job or to start mappers for the next jobs is a crucial aspect that requires complex metrics. As future direction, we also plan to validate the proposed approach through experiments with higher-level frameworks in the context of pipelined MapReduce, such as Pig.

Acknowledgments. This work was supported in part by the Agence Nationale de la Recherche (ANR) under Contract ANR-10-SEGI-01 (MapReduce Project). The experiments presented in this paper were carried out using the Grid'5000 testbed, an initiative of the French Ministry of Research through the ACI GRID incentive action, INRIA, CNRS and RENATER and other contributing partners (see http://www.grid5000.org/).

References

1. Amazon Elastic Compute Cloud (EC2), http://aws.amazon.com/ec2/
2. Amazon Elastic MapReduce, http://aws.amazon.com/elasticmapreduce/
3. Amazon Simple Storage Service (S3), http://aws.amazon.com/s3/
4. Crunch, https://github.com/cloudera/crunch
5. The Hadoop MapReduce Framework, http://hadoop.apache.org/mapreduce/
6. The Windows Azure Platform, http://www.microsoft.com/windowsazure/
7. Condie, T., Conway, N., Alvaro, P., et al.: Mapreduce online. In: Procs. of NSDI 2010, Berkeley, CA, USA, p. 21. USENIX Association (2010)
8. Dean, J., Ghemawat, S.: MapReduce: simplified data processing on large clusters. Communications of the ACM 51(1), 107–113 (2008)
9. Gunarathne, T., Wu, T.-L., Qiu, J., Fox, G.: MapReduce in the Clouds for Science. In: Procs. of CLOUDCOM 2010, Washington, DC, pp. 565–572 (2010)
10. Jégou, Y., Lantéri, S., Leduc, J., et al.: Grid'5000: a large scale and highly reconfigurable experimental Grid testbed.. Intl. Journal of HPC Applications 20(4), 481–494 (2006)
11. Morton, K., Friesen, A., Balazinska, M., Grossman, D.: Estimating the progress of MapReduce pipelines. In: Procs. of ICDE, pp. 681–684. IEEE (2010)
12. Olston, C., Reed, B., Srivastava, U., et al.: Pig Latin: a not-so-foreign language for data processing. In: Procss of SIGMOD 2008, pp. 1099–1110. ACM, NY (2008)
13. Pike, R., Dorward, S., Griesemer, R., Quinlan, S.: Interpreting the data: Parallel analysis with Sawzall. Scientific Programming Journal 13, 277–298 (2005)
14. Thusoo, A., Sarma, J.S., Jain, N., et al.: Hive: A warehousing solution over a MapReduce framework. In: Procs. of VLDB 2009, pp. 1626–1629 (2009)
15. White, T.: Hadoop: The Definitive Guide. O'Reilly Media, Inc. (2009)

Assessing the Performance and Scalability of a Novel Multilevel K-Nomial Allgather on CORE-*Direct* Systems

Joshua S. Ladd, Manjunath Gorentla Venkata,
Richard Graham, and Pavel Shamis

Computer Science and Mathematics Division
Oak Ridge National Laboratory
One Bethel Valley Road
Oak Ridge, TN 37831, USA
laddjs@ornl.gov

Abstract. In this paper, we propose a novel allgather algorithm, Reindexed Recursive K-ing (RRK), which leverages flexibility in the algorithm's tree topology and ability to make asynchronous progress coupled with Core-Direct communication offload capability to optimize the MPI_Allgather for Core-Direct enabled systems. In particular, the RRK introduces a reindexing scheme which ensures contiguous data transfers while adding only a single additional send and receive operation for any radix, k, or communicator size, N. This allows us to improve algorithm scalability by avoiding the use of a scatter/gather elements (SGE) list on InfiniBand networks. The implementations of the RRK algorithm and its evaluation shows that it performs and scales well on Core-Direct systems for a wide range of message sizes and various communicator configurations.

Keywords: Collectives, MPI, Allgather, CORE-Direct.

1   Introduction

The MPI_Allgather operation, in which each rank must share a message with all other ranks, falls within the class of all-to-all collective communications. The allgather is an important operation employed in such varied applications as parallel matrix multiplication and video compression algorithms. While much work has been done on developing good allgather algorithms for a broad range of networks and data sizes, several caveats exist that leave many questions open for exploration on how to best achieve good performance for an arbitrary communicator size, message length, and communication substrate. In particular, several implementors have recommended the use of logarithmic algorithms for latency sensitive small message operations and linear algorithms for bandwidth bound large message operations. For the current class of logarithmic algorithms one must circumvent one of two obstacles: dealing with the non power-of-two case, or dealing with packing and unpacking noncontiguous data.

C. Kaklamanis et al. (Eds.): Euro-Par 2012, LNCS 7484, pp. 538–549, 2012.

In this paper, we focus on optimizing the MPI_Allgather collective operation by 1) introducing a novel logarithmic allgather algorithm 2) making the algorithm hierarchy aware, and 3) leveraging Mellanox Technologies' CORE-*Direct* capability which provides the capability to fully offload a linked list of network operations to the host channel adapter (HCA).

The novel algorithm introduced is the Reindexed Recursive K-ing (RRK) algorithm. The novelty is two-fold; first, so far as we are aware, no one has implemented a general k-nomial allgather algorithm, a feature we believe could be leveraged on emerging networks, and second, the way in which the algorithm handles the non power-of-k case requires no additional steps and need not make concessions for noncontiguous data. It's possible to implement recursive k-ing on CORE-*Direct* systems by leveraging scatter-gather hardware via scatter/gather lists (SGE) and using this to handle the non power-of-k case, however, this limits the scalability of the algorithm due to limitations on the maximum size of an SGE list. It's with this constraint in mind that we propose an alternative algorithmic approach to handle the generic non power-of-k case.

All algorithms presented herein have been implemented within the *Cheetah* framework [6] in order to design hierarchy aware implementations. We compare the non power-of-2 performance to Open MPI's Bruck implementation [4]. We investigate the cost and benefit of pipelining fragments for the three-level allgather when messages are too large to fit in preregistered library buffers. We also revisit the comparison between logarithmic versus linear algorithms for large messages this time on CORE-*Direct* enabled systems by comparing the performance of the one level zero-copy RRK allgather to the performance of the one level, zero-copy neighbor exchange.

The rest of this paper is organized as follows. In Section 2 we give describe the background and related works, in Section 3 we give an overview of the design and implementation of the algorithms. In section 4 we present results and finally, in Section 5 we discuss our results and give concluding remarks.

2   Background

All-to-all collectives have been extensively studied, a survey of all-to-all collectives on various networks with various communication modes can be found in [5], [8]. Some of the earliest work on optimizing the allgather for modern message passing systems was done by Bruck et al. [2]. From this work emerged the well-known and oft implemented "Bruck" algorithm. Träff et al. have studied the allgather extensively; they proposed to optimize this operation for large SMP systems by implementing a shared-memory aware algorithm as well as introducing a logarithmic-linear hybrid allgather [13]. Benson et al. [1] have studied both recursive doubling and dissemination based allgather algorithms. The latter algorithm must deal with noncontiguous data and the former handles the non power-of-two case with with a modified recursive doubling algorithm that can require up to $2 * \lfloor log(n) \rfloor$ steps. Kandalla et al. have explored a multi-leader algorithm thereby extending the notion of hierarchy to more effectively utilize modern high-speed interconnects. Sur et al. implemented Remote Direct Memory Access (RDMA) based allgather algorithms for InfiniBand (IB) systems to implement allgather for IB based clusters [12]. Hierarchical algorithms have been proposed by several authors, including Sanders, Träff, and Kandalla [10] [13] [11]. Chen et al. [3] studied several common allgather algorithms on terascale Linux clusters with fast ethernet and proposed a linear time neighbor exchange algorithm for the large message allgather.

3   Algorithm Design

In this section, we describe the allgather algorithms designed and implemented. Throughout this paper, with the exception of the neighbor exchange, the algorithms under examination implement a general k-nomial exchange pattern. Recursive k-ing, as we have dubbed it, is a logarithmic algorithm which has $log_k(N)$ steps, where on each step each rank performs a pair-wise exchange with $(k - 1)$ peers.

3.1   Reindexed Recursive K-ing (RRK)

Recursive k-ing is a generalization of the well known recursive doubling algorithm. Whereas recursive doubling's communication pattern is encoded in a binomial tree, as the name suggests, recursive k-ing's pattern is encoded in a k-nomial tree. To begin the description, first, consider the case where the number of ranks, N, is a power of the radix, k, i.e. $N = k^p$. Then the algorithm consists of p steps where on each step every rank communicates in pair-wise fashion with $(k - 1)$ peers, namely with those at distances $k^i, 2k^i, 3k^i, ..., (k - 1)k^i$ away. In a recursive k-ing allgather of input size s bytes, at step i, each rank sends and receives $s * k^i$ bytes to and from each of the $(k - 1)$ peers.

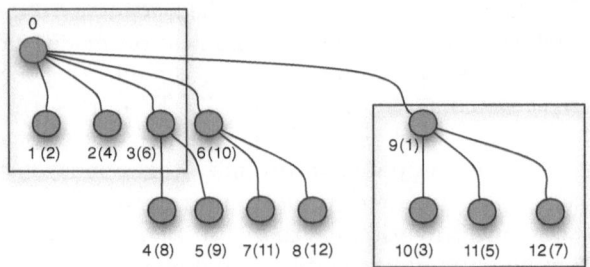

Fig. 1. An incomplete 3-nomial tree consisting of 13 nodes. R is contained in the right hand box and its mapping into T_0 is contained in the left box. The relabeled node values appear in parentheses next to the original labels.

For the non power-of-k case we introduce a reindexing scheme which ensures contiguous data transfers during the communication phase. Consider N processes where $k^p < N < k^{p+1}$ and divide the processes into complete or *full* subtrees of size k^p. There are $m < k$ such subtrees, denoted here as: $T_0, T_1, ..., T_{m-1}$. There exists a single, incomplete, remainder subtree of size $N - mk^p < k^p$, denoted here as R. We define a set of $N - mk^p$ proxies within the complete subtree T_0 by performing the following regrouping. Starting at the root of T_0, we redistribute the nodes of the tree in T_0 and R so that, rank 0 remains the root of T_0, rank 1 is the root of subtree R, rank 2 is the first leaf in T_0, rank 3 is the first leaf in R, and so on. We continue this way, redistributing in round robin fashion between T_0 and R until R is a subtree whose nodes are physically composed of the first $N - mk^p$ odd ranks, and T_0 is a subtree whose first $N - mk^p$ nodes are composed of the first $N - mk^p$ even ranks. Starting with rank $N - mk^p$ we sequentially fill out each subtree, T_i. The first $N - mk^p$ nodes in T_0 are natural proxies for the nodes in R since elements of T_0 would be connected to the nodes of R in this way at step $p + 1$ if $N = k^{p+1}$. In the first step of the algorithm, all nodes in R send their data to their naturally connected proxy in T_0. It follows by induction on p that each full subtree, T_i, may apply recursive k-ing independent of any other subtree, T_j. In this way, after p steps, every member of subtree T_i has all of the data contributed by all nodes in T_i and the data is contiguous. On step $p + 1$, each node in T_i has a unique communication partner in subtree T_j for all $j \neq i$. Each node performs $(m - 1) < (k - 1)$ pairwise exchanges, at the conclusion of which all ranks within $T_0 \cup T_1 \cup ... \cup T_{m-1}$ have all of the data. In the last step, the proxy nodes in T_0 send the full message back to the extra nodes in R. This algorithm has $p + 3$ distinct communication phases, and requires, at most, $(p + 1)(k - 1)$ send/receives all but one set being pair-wise. Figure 1 shows the positions of the ranks in an incomplete 3-nomial tree before and after regrouping. This scheme is appealing since, by construction, it ensures contiguous data by pairing proxy ranks with their consecutive counterparts who have been placed logically in the incomplete remainder tree R.

3.2   Small Message Algorithms

For small messages, defined as those where the received message can fit entirely within a single library buffer, we have implemented a hierarchical method. For an n-level hierarchy, we gather the data up $n - 1$ distinct k-nomial trees, the radices can be chosen independently allowing one to choose an optimal radix for each communication substrate. The same reindexing is employed by the gather operation to deal with any extra ranks. The RRK allgather is executed at level n. Subsequent to this, the data is broadcast down the $n - 1$ levels and copied from the library buffer back into the user's space. If the buffer space is exhausted, then we fall back on a pipelined fragmentation scheme.

RRK Implementations. We have implemented two separate RRK algorithms for the purposes of this study; the first is the CORE-*Direct*-RRK or *offloaded*

RRK. A complete overview of CORE-*Direct*'s capabilities is beyond the scope of this paper and we refer the reader to [7] for a complete overview of the CORE-*Direct* queue structure. For the small message CORE-*Direct*-RRK we employ remote direct memory access (RDMA) calls with sender-side "put" with immediate data to transfer the data into remote pre-registered buffers. The immediate data is used to generate a completion queue entry which is used by the host channel adapter (HCA) for collective communication management. Processes that participate in this collective create a list of network tasks which is posted to a management queue (MQ), which in turn posts to the appropriate queue pair (QP). After the post, the CORE-*Direct* HCA progresses the collective completely freeing the CPU for other work save for polling on collective completion. We also implement a point-to-point RRK (p2p-RRK) which involves the same communication patterns, but implements non-blocking send/receive semantics atop Open MPI's point-to-point layer to achieve data transfers.

3.3   Large Message Algorithms

For CORE-*Direct* systems we have implemented a zero-copy CORE-*Direct*-RRK algorithm that does not copy data into library buffers but instead works entirely in the user's buffer space. This implementation employs send/receive semantics instead of RDMA. In order to ensure that the receiver's buffer is ready at any given step, we implement a rendezvous protocol via ready-to-receive messages (RTR). After a receiver posts a receive, it must inform the sender that it is "ready-to-receive" by sending an RTR message to a buffer that was pre-posted by the sender. This coordinated exchange happens on each exchange. Furthermore, user buffers must be registered with the HCA in order to pin the memory and perform various address translations. Messages must be quite large in order to amortize the cost of the rendezvous and registration.

For comparison, we have implemented the linearly scaling zero copy CORE-*Direct* neighbor exchange (NX) algorithm whose data pattern was first described in [3]. In this algorithm, processes perform pairwise exchange alternating between left and right-most neighbors. We again implement a rendezvous protocol via RTR messages to ensure buffers are ready to receive.

4   Results

In this section, we assess the performance of the small message RRK implementations including: the single level RRK implementations without fragmentation, the hierarchical RRK with fragmentation enabled, and the zero copy large message RRK and NX algorithms.

4.1   System Configuration

For all experiments conducted herein, we used the *Hyperion* system, which is a 1,152 node test-bed for experimenting with high-performance computing technologies installed at Lawrence Livermore National Laboratory. The partition that we used for our experiments is comprised of 64 nodes where each node contains two sockets each, with four Intel Xeon 2.5 GHz processor cores. Each node contains10 GB of memory and a CORE-*Direct* HCA.

Each Hyperion node runs an instance of CHAOS, an Open Mosix based operating system, with kernel version 2.6.18. For latency measurements, each node was configured with eight MPI process and the performance was measured on up to 512 processes. For overlap experiments, each node was configured with one process per node with 64 nodes active.

4.2   Single Level Performance

In this experiment, we ran both the p2p-RRK and the CORE-*Direct*-RRK together over a range of data sizes and radices up to 128 processors on Hyperion. The choice of data sizes presented is a measured balance between the size of library buffers used and the size of the final message which is a product of the number of cores and input message size. We considered 1 byte message sizes for small data, 128 bytes and 512 bytes for medium message size and 1024 bytes for medium-large message size. We intentionally did not go beyond this because we did not want to engage the fragmentation logic. We will look at larger message sizes as well as the effects of fragmentation later in the paper. In each experiment and for each of the two implementations, we considered a range of different radices namely, $r = \{2, 3, 4\}$.

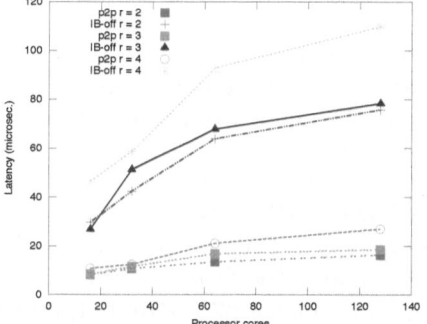 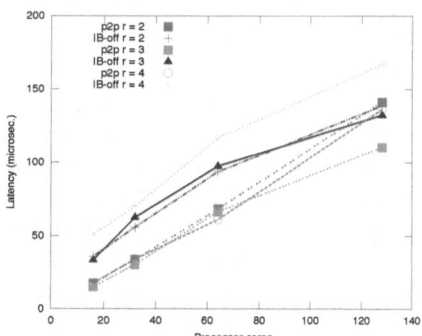

Fig. 2. Performance versus the radix for the single level RRK allgather implementations. On the left, latency is shown for the single byte allgather as the number of cores increases. On the right, the 128 byte allgather latency is presented.

Figure 2 shows the results of both the 1 byte and 128 byte experiments. At 128 processors the p2p-RRK with radix 2 is the superior performer with an average latency of 16.44 μ-sec. It outperforms the radix 3 and 4 cases of the same implementation by 13% and 64% respectively. The radix 2 CORE-*Direct*-RRK performs the best amongst the three CORE-*Direct*-RRK curves with a latency of 75.87 μ-sec. It outperforms the radix 3 and 4 cases of the same implementation

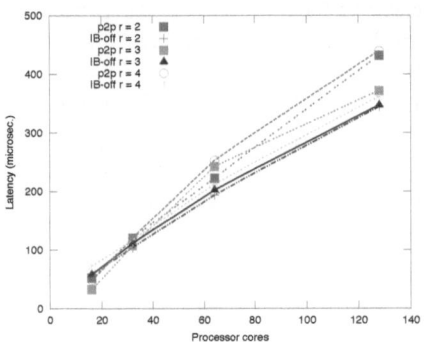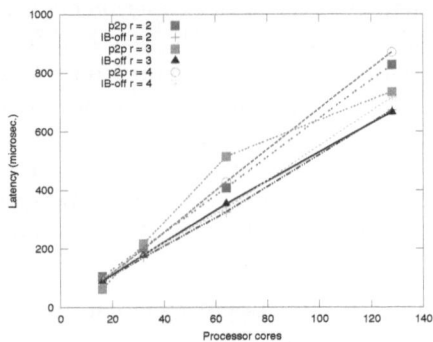

Fig. 3. Performance versus the radix for the single level RRK allgather implementations. On the left, latency is shown for the 512 byte allgather as the number of cores increases. On the right, the 1024 byte allgather latency is presented.

by 3.4% and 44% respectively. However, the radix 2 p2p-RRK outperforms the CORE-*Direct*-RRK by 361.4%.

Figure 2 also shows the 128 byte allgather results. At 128 ranks, the radix 3 p2p-RRK achieve the best performance with an average latency of 110 μ-sec. The second best performer is the radix 3 CORE-*Direct*-RRK with an average latency of 132.5 μ-sec. In this case, the radix 3 p2p-RRK outperforms the radix 2 and 4 p2p-RRK by 27% and 23% respectively. The radix 3 CORE-*Direct*-RRK outperforms the radix 2 and 4 CORE-*Direct*-RRK by 4.8% and 26% respectively. The radix 3 p2p-RRK outperforms the radix 3 CORE-*Direct*-RRK by about 20%. It is interesting to note that 128 is neither a power of 3 or 4, and, as a result, these experiments engage the extra rank logic for radix 3 and 4.

Figure 3 shows the latency as a function of the number of processor cores for the 512 byte and 1024 byte allgather. In this experiment, we see quite a different trend than for the small data. For 512 bytes at 128 cores, the best performing algorithms are the radix 2 and radix 3 CORE-*Direct*-RRK algorithms which have latencies of 344.5 μ-sec and 347 μ sec respectively. They outperform the best performing p2p-RRK, radix 3, by about 8 % and outperform the radix 2 RRK (recursive doubling) by 25%. Interestingly, the radix 3 p2p-RRK outperforms recursive doubling by 16.2%. The same trend persists for the 1024 byte allgather. There is nearly a tie for top performer between radix 3 and radix 2 CORE-*Direct*-RRK beating the radix 3 p2p-RRK by 24%. Again, we see the radix 3 p2p-RRK outperform recursive doubling (radix 2 p2p-RRK) by 12.5%.

4.3   Single Level Non Power-of-2 Performance

As seen in the previous section, the performance for non power-of-k ranks can be quite good. To determine how effective our reindexing strategy is we compare the performance of the radix 2 p2p-RRK against Open MPI's non power-of-two

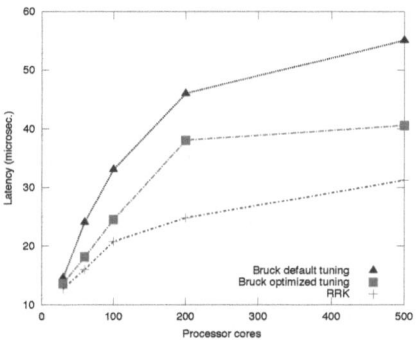 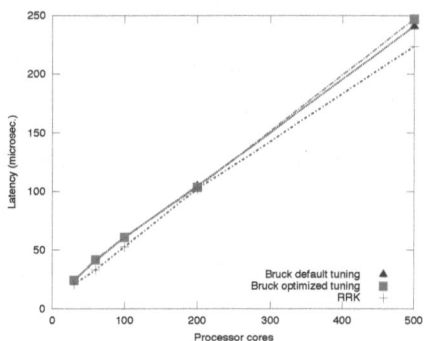

Fig. 4. Performance as a function of the number of cores. On the left, 1 byte p2p-RRK versus the optimized and default Bruck algorithm implemented in Open MPI's Tuned collectives for non power-of-two ranks. On the right, the same comparison is made at 64 bytes. The ranks considered are 30, 60, 100, 200, 500.

default allgather algorithm[1] which is a Bruck algorithm also implemented with non-blocking send/receives atop Open MPI's point-to-point layer. We chose a collection of non power-of-two communicator sizes, namely 30, 60, 100, 200, and 500 and measured the latency of both algorithms as a function of communicator size.

It has been noted by multiple authors that pair-wise exchange should provide better performance on TCP networks due to the fact that TCP can optimize a pair-wise exchange by piggybacking data on ACK packets [1],[3]. We observed a similar pair-wise exchange optimization in Open MPI's byte transfer layer (BTL). The OpenIB BTL implements internal flow-control algorithms in order to control network load. Some parts of this algorithm utilize a piggyback credit exchange mechanism to optimize network utilization. OpenIB BTL's default tuning of this mechanism is not optimized for unidirectional communication patterns. As a result, allgather algorithms based on this pattern, e.g. a Bruck allgather, may exhaust internal resources and cause substantial performance degradation. In order to improve the Bruck allgather performance we disabled the OpenIB BTL communication protocols that use piggyback credit mechanisms, thereby allowing us to better account for actual algorithmic differences. In figure 4, the curve labeled "Bruck default" is the default performance observed with no tuning, the curve labeled "Bruck optimized" is the observed performance with piggbacking optimizations disabled. As evidenced in the figure, the radix 2 p2p-RRK outperforms the optimized Bruck algorithm by nearly 30% for a 1 byte allgather on 500 cores. For a 64 byte message size, the disparity in performance decreases, however, the radix 2 p2p-RRK still outperforms the optimized Bruck algorithm by 10.3%. Beyond this data size, fragmentation is engaged and the comparison ceases to be fair.

---

[1] Recursive doubling is Open MPI's default allgather algorithm for power-of-two communicators.

4.4   Overlap Performance of CORE-*Direct* Implementations

To measure communication computation overlap, we extended the COMB bench-mark suite developed by Lawry et al. [9], which studies asynchronous point-to-point operations. The extensions allow one to use the benchmark suite to study asynchronous collective operations, and to study overlap using a post-work-wait, and a post-work-test-loop method.

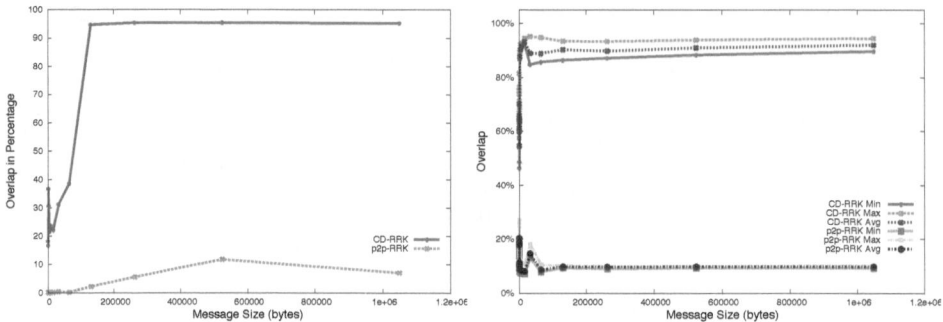

Fig. 5. Overlap measurements for the one level CORE-*Direct*-RRK algorithms. The poll method results are seen of the left and the wait method results are on the right.

With the post-work-test-loop method, each process posts the collective opera-tion, and iterates over a series of work loops followed by a completion test, exiting the work loop on test completion. Total completion time and total time spent in the work loop are measured, using a CPU cycle counter to compute computation cycles available. The work loop is a loop over the "nop" asm instruction, with the loop range count used to control work-time.

In the post-work-wait method, each process posts the collective operation and executes a pre-defined amount of work, and then waits for collective completion. The test searches for the work-loop size that results in allgather operation latency similar to that of the no-work loop latency measurement, starting at a size that corresponds to about 10% of the no-work allgather, incrementing the work-loop size. We consider the time available for computation, as time of the largest work-loop that does not increase the allgather operation latency.

Figure 5 shows the CPU computation time available for the nonblocking CORE-*Direct*-RKK and p2p-RKK allgather algorithms, as a function of mes-sage size for the post-work-wait benchmark test executed across 64 processes. The average, minimum, and maximum values are displayed for the HCA and the CPU managed algorithms. As expected, the HCA managed allgather makes available more CPU cycles. In a 128 bytes-per-process allgather operation, on average, 58% of algorithm completion time is available for CPU work with the CORE-*Direct*-RRK algorithm, whereas the CPU managed p2p-RRK algorithm makes, on average, only 10% of the algorithm time available for application CPU work. For large messages, the differences are much greater. As expected, the HCA managed CORE-*Direct*-RRK algorithm makes available many more CPU cycles. For a 2 megabytes-per-process allgather operation, on average, 95% of algorithm completion time is available for CPU work with the CORE-*Direct*-RRK algorithm, whereas for the CPU managed p2p-RRK algorithm, on average, only 9.5% of the time is available for application work.

4.5   Hierarchical Performance versus Fragment Size

One option for dealing with large messages is to fragment the input data. Conventional wisdom dictates that the fewer the fragments, the better the performance. For large allgathers, fragmenting the data over multiple library buffers is necessary in our hierarchical algorithm. In these experiments we consider a three-level p2p-RRK and a three-level CORE-*Direct*-RRK made of shared memory socket and uniform memory access (UMA) subgroups below, and either a p2p or CORE-*Direct* subgroup on top. In these experiments we vary the size of the internal library buffers to determine an optimal size. Small buffers consume fewer resources but require more fragments, large buffers result in a larger footprint but will result in less fragmentation. In figure 6, on the left, the latency as a function of message size on 512 cores for the three-level p2p-RRK allgather is shown. It's no surprise that the larger the buffers, the better the performance, due to the fact that larger buffers simply means fewer fragments. However, the plot on the right in figure 6 tells a more interesting story, this shows the results for the three-level CORE-*Direct*-RRK where we see a clear range of data sizes where 16KB buffers provides superior performance over the large 64KB buffer, by as much as 36.3% in the case of a 512 byte allgather. At 512 bytes on 512 cores, the final message size is 262,144 bytes which results in about 7 fragments for the 64KB buffers and 17 fragments for the 16KB buffers.

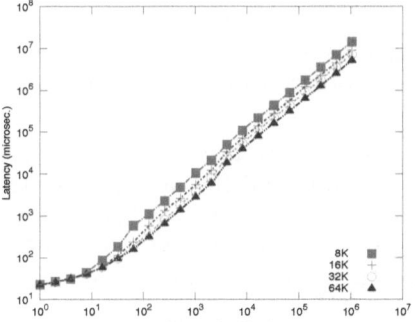 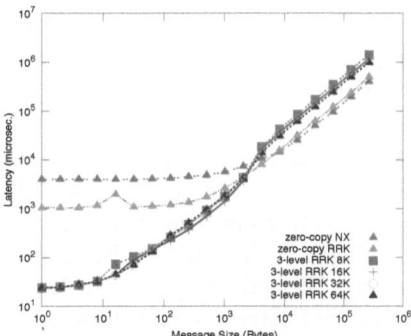

Fig. 6. On the left, performance versus message size for the three-level p2p-RRK with various fragment sizes. On the right, performance versus message size for the three-level CORE-*Direct*-RRK with various fragment sizes. Also depicted on the right, the performance of the single-level zero-copy CORE-*Direct*-NX and zero-copy CORE-*Direct*-RRK algorithm. In both plots, the number of cores employed is 512.

These observations can be explained as follows; in the three-level hierarchy we much broadcast the result of each fragment back down and memcpy the result out to the user. The CORE-*Direct* implementation allows us to overlap network operations with CPU operations such as progressing broadcasts in shared memory and making system calls to copy data back to the user. In this particular experiment, the pipeline depth was set to 5 fragments so both 64K and 16KB buffers have an opportunity to fill the pipeline.

4.6   Zero Copy Performance

Figure 6 also contains the performance results of the zero copy CORE-*Direct* algorithms. In this experiment, we compared only recursive doubling (k=2) with the NX algorithm. At 512 cores the zero copy RRK outperforms the linearly scaling NX up to message size 1024 bytes. After 1024 bytes, the NX algorithm outperform the radix 2 RRK where at the largest sized measured, 200KB (100MB final message size), the NX algorithm outperform the radix 2 RRK by 24%. For the 200KB message the zero copy NX algorithm outperforms the best three-level implementation by 145%.

5   Conclusion

We have presented a novel logarithmic allgather algorithm and have assessed its performance and scalability on CORE-*Direct* enabled systems. The algorithm is appealing since it handles the non power-of-k communicator by adding only a single additional send and receive and ensures contiguous data transfers. This strategy allows us to avoid using SGE lists for non contiguous data, which, while effective, necessarily limits scalability. We have demonstrated the efficacy of RRK's reindexing by comparing it against Open MPI's default non power-of-two allgather. We have examined a range of algorithm parameters including the radix and fragment size and have demonstrated excellent overlap characteristics. We have demonstrated how to leverage overlap in the three-level hierarchical implementation by pipelining smaller fragments allowing the CPU to perform memcpys while the HCA makes simultaneous progress on network operations.

Acknowledgements. This research is sponsored by the Office of Advanced Scientific Computing Research's FASTOS program and the Math/CS Institute EASI; U.S. Department of Energy, and performed at ORNL, which is managed by UT-Battelle, LLC under Contract No. DE-AC05-00OR22725. The HPC Advisory Council (http://www.hpcadvisorycouncil.com) provided computational resource for testing and data gathering.

References

1. Benson, G.D., Chu, C.-W., Huang, Q., Caglar, S.G.: A Comparison of MPICH Allgather Algorithms on Switched Networks. In: Dongarra, J., Laforenza, D., Orlando, S. (eds.) EuroPVM/MPI 2003. LNCS, vol. 2840, pp. 335–343. Springer, Heidelberg (2003)

2. Bruck, J., Member, S., Tien Ho, C., Kipnis, S., Upfal, E., Member, S., Weathersby, D.: Efficient algorithms for all-to-all communications in multi-port message-passing systems. In: IEEE Transactions on Parallel and Distributed Systems, pp. 298–309 (1997)
3. Chen, J., Zhang, L., Zhang, Y., Yuan, W.: Performance evaluation of allgather algorithms on terascale linux cluster with fast ethernet. In: Proceedings. Eighth International Conference on High-Performance Computing in Asia-Pacific Region, pp. 6–442 (July 2005)
4. Fagg, G., Bosilca, G., Pješivac-Grbović, J., Angskun, T., Dongarra, J.: Tuned: An open mpi collective communications component. In: Distributed and Parallel Systems, pp. 65–72. Springer, US (2007)
5. Fraigniaud, P., Lazard, E.: Methods and problems of communication in usual networks. Discrete Applied Mathematics 53, 79–133 (1994)
6. Graham, R., Venkata, M.G., Ladd, J., Shamis, P., Rabinovitz, I., Filipov, V., Shainer, G.: Cheetah: A framework for scalable hierarchical collective operations. In: CCGRID 2011 (2011)
7. Graham, R.L., Poole, S., Shamis, P., Bloch, G., Bloch, N., Chapman, H., Kagan, M., Shahar, A., Rabinovitz, I., Shainer, G.: Connectx-2 infiniband management queues: First investigation of the new support for network offloaded collective operations. In: CCGRID, pp. 53–62 (2010)
8. Hedetniemi, S.M., Hedetniemi, S.T., Liestman, A.L.: A survey of gossiping and broadcasting in communication networks. Networks (1988)
9. Lawry, W., Wilson, C., Maccabe, A., Brightwell, R.: Comb: a portable benchmark suite for assessing mpi overlap. In: 2002 IEEE International Conference on Cluster Computing, pp. 472–475 (2002)
10. Sanders, P., Träff, J.L.: The Hierarchical Factor Algorithm for All-to-All Communication. In: Monien, B., Feldmann, R.L. (eds.) Euro-Par 2002. LNCS, vol. 2400, p. 799. Springer, Heidelberg (2002)
11. Sur, S., Bondhugula, U.K.R., Mamidala, A.R., Jin, H.-W., Panda, D.K.: High Performance RDMA Based All-to-All Broadcast for InfiniBand Clusters. In: Bader, D.A., Parashar, M., Sridhar, V., Prasanna, V.K. (eds.) HiPC 2005. LNCS, vol. 3769, pp. 148–157. Springer, Heidelberg (2005)
12. Sur, S., Jin, H.-W., Panda, D.K.: Efficient and scalable all-to-all personalized exchange for infiniband-based clusters. In: Proceedings of the 2004 International Conference on Parallel Processing, ICPP 2004, pp. 275–282. IEEE Computer Society (2004)
13. Träff, J.L.: Efficient Allgather for Regular SMP-Clusters. In: Mohr, B., Träff, J.L., Worringen, J., Dongarra, J. (eds.) PVM/MPI 2006. LNCS, vol. 4192, pp. 58–65. Springer, Heidelberg (2006)

Topic 10: Parallel Numerical Algorithms

Iain Duff, Efstratios Gallopoulos, Daniela di Serafino, and Bora Ucar

Topic Committee

The solution of large-scale problems in Computational Science and Engineering relies on the availability of accurate, robust and efficient numerical algorithms and software that are able to exploit the power offered by modern computer architectures. Such algorithms and software provide building blocks for prototyping and developing novel applications, and for improving existing ones, by relieving the developers from details concerning numerical methods as well as their implementation in new computing environments.

From the papers submitted to this year's Europar, the topic of Parallel Numerical Algorithms involving these themes attracted submissions from Europe, Asia and the Africa. Each paper received at least four reviews and finally three were selected for presentation following extensive discussions between members of EUROPAR's Program Committee.

Donfack, Grigori and Khabou present an algorithm for dense LU factorization that is suitable for machines with multiple levels of parallelism and describe its implementation on a cluster of multicore processors based on MPI and Pthreads. Korch describes data-parallel implementations of ODE solvers, specifically explicit Adams-Bashforth methods. He examines locality and scalabiity and shows how the careful use of pipelining can improve the locality of memory references. Cotronis, Konstantinidis, Louka and Missirlis describe special local SOR methods optimized for GPUs.

Based on these interesting papers, this session provides a forum for the discussion of recent developments in the design and implementation of numerical methods on modern parallel architectures such as multicores and GPU systems.

It is appropriate, at this time, to thank the authors who submitted papers to the session and congratulate those whose papers were accepted. We are especially grateful to the referees who provided us with carefully written and informative reviews. Finally, we thank the Conference Organizers for providing the opportunity to the participants to present and discuss the state-of-the-art in Parallel Processing on the beautiful island of Rhodes. In this Olympic year, our authors took "hic Rhodus, hic salta" to heart as all papers demonstrate implementations of numerical algorithms on advanced computer systems.

C. Kaklamanis et al. (Eds.): Euro-Par 2012, LNCS 7484, p. 550, 2012.
© Springer-Verlag Berlin Heidelberg 2012

Avoiding Communication through a Multilevel LU Factorization

Simplice Donfack, Laura Grigori, and Amal Khabou

INRIA Saclay-Ile de France,
Laboratoire de Recherche en Informatique, Université Paris-Sud
simplice.donfack@lri.fr, {laura.grigori,amal.khabou}@inria.fr

Abstract. Due to the evolution of massively parallel computers towards deeper levels of parallelism and memory hierarchy, and due to the exponentially increasing ratio of the time required to transfer data, either through the memory hierarchy or between different compute units, to the time required to compute floating point operations, the algorithms are confronted with two challenges. They need not only to be able to exploit multiple levels of parallelism, but also to reduce the communication between the compute units at each level of the hierarchy of parallelism and between the different levels of the memory hierarchy.

In this paper we present an algorithm for performing the LU factorization of dense matrices that is suitable for computer systems with two levels of parallelism. This algorithm is able to minimize both the volume of communication and the number of messages transferred at every level of the two-level hierarchy of parallelism. We present its implementation for a cluster of multicore processors based on MPI and Pthreads. We show that this implementation leads to a better performance than routines implementing the LU factorization in well-known numerical libraries. For matrices that are tall and skinny, that is they have many more rows than columns, our algorithm outperforms the corresponding algorithm from ScaLAPACK by a factor of 4.5 on a cluster of 32 nodes, each node having two quad-core Intel Xeon EMT64 processors.

Keywords: LU factorization, communication avoiding algorithms, multiple levels of parallelism.

1   Introduction

Due to the evolution of massively parallel computers towards deeper levels of parallelism and memory hierarchy, and due to an exponentially increasing ratio of the time necessary to transfer data, either through the memory hierarchy or between different compute units, to the time required to perform floating point operations, the algorithms are confronted with two challenges. They need to be able to exploit multiple levels of parallelism, and they also need to minimize communication and synchronization at each level of the hierarchy of parallelism and memory. The particularity of a computer system with multiple levels of parallelism is that a compute unit at a given level can be formed by several smaller

C. Kaklamanis et al. (Eds.): Euro-Par 2012, LNCS 7484, pp. 551–562, 2012.

compute units connected together. A machine formed by nodes of multicore processors can be seen as an example of a machine with two levels of parallelism. An approach to exploit such an architecture for an existing algorithm consists of identifying functions which are executed sequentially on a node, and then replacing them by a call to their multithreaded version. This approach, based mainly on combining MPI and threads, is easy to implement and has been used in many applications, but can have several drawbacks. It can lead to more communication and synchronization between MPI processes or between threads, load imbalance, and in general a simple adaptation of an existing algorithm can cause an important degradation of the overall performance on this type of architectures. For such a computer system, there are two levels of communication: inter-node communication, that is communication between two or more nodes, and intra-node communication, that is communication performed inside each node. For both types, an algorithm that takes into account at the design level the two levels of parallelism can be able to reduce communication at every level.

Motivated by the increased cost of communication with respect to the cost of computation [11], a new class of algorithms has been introduced in the recent years for dense linear algebra, referred to as communication avoiding algorithms. These algorithms, first proposed for dense LU factorization (CALU) [12] and QR factorization (CAQR) [6], allow to minimize communication on a computer system with one level of parallelism (or between two levels of fast and slow memory) and are as stable as classic algorithms as for example implemented in LAPACK [2] and ScaLAPACK [3]. They were shown to lead to good performance on distributed memory machines [6, 12], on multicore processors [7], and on grids [1]. In the distributed version of CALU and CAQR, blocks of the input matrix are distributed among processors, and data is communicated via MPI messages during the factorization. In the approach used for multicore processors [7], operations on a block are performed as tasks, which are scheduled statically or dynamically to the available cores. However, none of these algorithms has addressed the more realistic model of today's hierarchical parallel computers.

In this paper we introduce an algorithm for performing the LU factorization of a dense matrix that is suitable for computer systems with two levels of parallelism, and that can be further extended to multiple levels of parallelism. We refer to this algorithm as multilevel CALU. It can be seen as a generalization of CALU [13, 12]. At each iteration of the initial 1-level CALU algorithm, a block column, referred to as a panel, is factored and then the trailing matrix is updated. A classic algorithm as Gaussian elimination with partial pivoting (GEPP) is not able to minimize the number of messages exchanged during the factorization. This is because of partial pivoting, which requires to permute the element of maximum magnitude to the diagonal position at each step of the panel factorization. To minimize communication, CALU uses tournament pivoting, a different strategy shown to be very stable in practice [12]. With this strategy, the panel factorization is performed in two steps. In the first step, tournament pivoting uses a reduction operation to select a set of pivot rows from different blocks of the panel distributed among different processors or different cores, where GEPP is the operator used at each node of the reduction tree. In the second step, these pivot rows are permuted to the top of the panel, and then the LU factorization without pivoting of the panel is performed.

Multilevel CALU uses the same approach as CALU, and it is based on tournament pivoting and an optimal distribution of the input matrix among compute units to reduce communication at the first level of the hierarchy. However, each building block of CALU is itself a recursive function that allows to be optimal at the next level of the hierarchy of parallelism. For the panel factorization, at each node of the reduction tree of tournament pivoting, CALU is used instead of GEPP to select pivot rows, based on an optimal layout adapted to the current level of parallelism. We present this algorithm in section 2. We also model the performance of our approach by computing the number of floating-point operations, the volume of communication, and the number of messages exchanged on a computer system with two levels of parallelism. We show that our approach is optimal at every level of the hierarchy and attains the lower bounds on communication of the LU factorization (modulo polylogarithmic factors). The lower bounds on communication for the multiplication of two dense matrices were introduced in [14, 15] and were shown to apply to LU factorization in [6]. We discuss how these bounds can be used in the case of two levels of parallelism. Due to the multiple calls to CALU, multilevel CALU performs additional flops compared to 1-level CALU. It is known in the literature that in some cases, these extra flops can degrade the performance of a recursive algorithm (see for example in [8]). However, for two levels of parallelism, the choice of the optimal layout at each level of the hierarchy allows to keep the extra flops as a lower order term. Furthermore, multilevel CALU may also change the stability of the 1-level algorithm. We argue in section 3 through numerical experiments that 2-level CALU specifically studied here is stable in practice. We also show that multilevel CALU is up to 4.5 times faster than the corresponding routine PDGETRF from ScaLAPACK tested in multithreaded mode on a cluster of multicore processors.

2   CALU for Multiple Levels of Parallelism

In this section we introduce a multilevel communication avoiding LU factorization, presented in Algorithm 1, that is suitable for a hierarchical computer system with L levels of parallelism. Each compute unit at a given level i is formed by P_{i+1} compute units of level $i + 1$. Correspondingly, the memory associated with a compute unit at level i is formed by the sum of the memories associated with the P_{i+1} compute units of level $i + 1$. Level 1 is the first level and level L is the last level of the hierarchy of parallelism. Later in this section we model the communication cost of the algorithm for a computer system with two levels of parallelism.

The goal of multilevel CALU is to minimize communication at every level of a hierarchical system. It is based on a recursive approach, where at every level of the recursion optimal parameters are chosen, such as optimal layout

and distribution of the matrix over compute units, optimal reduction tree for
tournament pivoting. Algorithm 1 receives as input the matrix A of size $m \times n$,
the number of levels of parallelism in the hierarchy L, and the number of compute
units P_1 that will be used at the first level of the hierarchy and that are organized
as a two-dimensional grid of compute units of size $P_1 = P_{r_1} \times P_{c_1}$. The input
matrix A is partitioned into blocks of size $b_1 \times b_1$,

$$
A = \begin{pmatrix}
A_{11} & A_{12} & \dots & A_{1N} \\
A_{21} & A_{22} & \dots & A_{2N} \\
\vdots & \vdots & & \vdots \\
A_{M1} & A_{M2} & \dots & A_{MN}
\end{pmatrix},
$$

where $M = m/b_1$ and $N = n/b_1$. The block size b_1 and the dimension of the grid
$P_{r_1} \times P_{c_1}$ are chosen such that the communication at the top level of the hierarchy
is minimized, by following the same approach as for the 1-level CALU algorithm
[12]. That is, the blocks of the matrix are distributed among the P_1 compute units
using a two-dimensional block cyclic distribution over the two-dimensional grid
of compute units $P_1 = P_{r_1} \times P_{c_1}$ (we will discuss later in this section the values
of P_{r_1} and P_{c_1}) and tournament pivoting uses a binary reduction tree. At each
step of the factorization, a block of b_1 columns (panel) of L is factored, a block
of b_1 rows of U is computed, and then the trailing submatrix is updated. Each
of these steps is performed by calling recursively functions that will be able to
minimize communication at the next levels of the hierarchy of parallelism. Note
that Algorithms 1 and 2 do not detail the communication performed during
the factorization, which is triggered by the distribution of the data. By abuse
of notation, the permutation matrices need to be considered as extended by
identity matrices to the desired dimensions. For simplicity, we also consider that
the number of processors are powers of 2.

We describe in more detail now the panel factorization (line 8 of Algorithm 1)
computed by using mTSLU, described in Algorithm 2. It is a multilevel version
of TSLU, the panel factorization used in the 1-level CALU algorithm [13]. Let
B denote the first panel of size $m \times b_1$, which is partitioned into P_{r_i} blocks.
As in TSLU, mTSLU is performed in two steps. In the first step, a set of b_1
pivot rows are selected by using tournament pivoting. In the second step, these
rows are permuted into the diagonal positions of the panel, and then the LU
factorization with no pivoting of the panel is computed. Tournament pivoting
uses a reduction operation, where at the leaves of the tree b_1 candidate pivot rows
are selected from each block B_I of B. Then a tournament is performed among
the P_{r_i} sets of candidate pivot rows to select the final pivot rows that will be
used for the factorization of the panel. At each node of the reduction tree, a new
set of candidate pivot rows is selected from the sets of candidate pivot rows of
the children nodes in the reduction tree. The initial 1-level TSLU uses GEPP
to select a set of candidate pivot rows. However this means that at the second
level of parallelism, the compute units involved in one GEPP factorization will
need to exchange $O(b_1)$ messages for each call to GEPP due to partial pivoting,
and hence the number of messages will not be minimized at the second level

of the hierarchy of parallelism. Differently from 1-level TSLU, multilevel TSLU
selects a set of rows by calling multilevel CALU, hence being able to minimize
communication at the next levels of parallelism. That is, at each phase of the
reduction operation every compute unit from the first level calls multilevel CALU
on its blocks with adapted parameters and data layout. At the last level of the
recursion, 1-level CALU is called (referred to in the algorithms as CALU).

Once the panel factorization is performed, the trailing submatrix is updated
using a multilevel solve for a triangular system of equations (referred to as dtrsm)
and a multilevel algorithm for multiplying two matrices (referred to as dgemm).
We do not detail here these algorithms, but one should use recursive versions of
Cannon [4], or SUMMA [10], or a cache oblivious approach [9] if the transfer of
data across different levels of the memory hierarchy is to be minimized as well
(with appropriate data storages).

Algorithm 1 mCALU: multilevel communication avoiding LU factorization

1: **Input:** $m \times n$ matrix A, level of parallelism i in the hierarchy, block size b_i, number
   of compute units $P_i = P_{r_i} \times P_{c_i}$
2: **if** $i == L$ **then**
3:   $[\Pi_i, L_i, U_i] = CALU(A, b_i, P_i)$
4: **else**
5:   $M = m/b_i, N = n/b_i$
6:   **for** $K = 1$ to N **do**
7:     $[\Pi_{KK}, L_{K:M,K}, U_{KK}] = mTSLU(A_{K:M,K}, i, b_i, P_{r_i})$
8:     /* Apply permutation and compute block row of U */
9:     $A_{K:M,:} = \Pi_{KK} A_{K:M,:}$
10:     **for each** compute unit at level i owning a block $A_{K,J}, J = K+1$ to N **do**
        in parallel
11:       $U_{K,J} = L_{KK}^{-1} A_{K,J}$ /* call multilevel dtrsm on P_{i+1} compute units */
12:     **end for**
13:     /* Update the trailing submatrix */
14:     **for each** compute unit at level i owning a block $A_{I,J}$ of the trailing submatrix,
        $I, J = K+1$ to M, N **do in parallel**
15:       $A_{I,J} = A_{I,J} - L_{I,K} U_{K,J}$ /* call multilevel dgemm on P_{i+1} compute units */
16:     **end for**
17:   **end for**
18: **end if**

2.1 Performance Model

In this section we present a performance model of the 2-level CALU factorization
of a matrix of size $n \times n$ in terms of the number of floating-point operations
(#flops), the volume of communication (#words moved), and the number of
messages exchanged (#messages) during the factorization. Let $P_1 = P_{r1} \times P_{c1}$ be
the number of processors and b_1 be the block size at the first level of parallelism.
Each compute unit at the first level is formed by $P_2 = P_{r2} \times P_{c2}$ compute units at
the second level of parallelism. The total number of compute units at the second

---

Algorithm 2 mTSLU: multilevel panel factorization

1: **Input:** matrix B, level of parallelism i in the hierarchy, block size b_i, number of compute units P_{r_i}
2: Partition B on P_{r_i} blocks /* Here $B = (B_1^T, B_2^T, ..., B_{P_{r_i}}^T)^T$ */
3: /*Each compute unit owns a block B_I*/
4: **for** each block B_I **do in parallel**
5:     $[\Pi_I, L_I, U_I] = mCALU(B_I, i+1, b_{i+1}, P_{i+1})$
6:     Let B_I be formed by the pivot rows, $B_I = (\Pi_I B_I)(1 : b_i, :)$
7: **end for**
8: **for** $level = 1$ to $log_2(P_{r_i})$ **do**
9:     **for** each block B_I **do in parallel**
10:         **if** $((I - 1) \mod 2^{level-1} == 0)$ **then**
11:             $[\Pi_I, L_I, U_I] = mCALU([B_I; B_{I+2^{level-1}}], i+1, b_{i+1}, P_{i+1})$
12:             Let B_I be formed by the pivot rows, $B_I = (\Pi_I[B_I; B_{I+2^{level-1}}])(1 : b, :)$
13:         **end if**
14:     **end for**
15: **end for**
16:     Let Π_{KK} be the permutation performed for this panel
17: /* Compute block column of L */
18: **for** each block B_I **do in parallel**
19:     $L_I = B_I U_1(1 : b_i, :)^{-1}$ /* using multilevel dtrsm */
20: **end for**
21: **end if**

---

level is $P = P_1 \cdot P_2$. Let b_2 the block size at the second level of parallelism. We note CALU(m, n, P, b) the routine that performs 1-level CALU on a matrix of size $m \times n$ with P processors and a panel of size b.

We first consider the arithmetic cost of 2-level CALU. It is formed by the factorization of the panel, the computation of a block row of U, and the update of the trailing matrix, at each step k of the algorithm. To factorize the panel k of size b_1, we perform 1-level CALU on each block of the reduction tree, using a grid of P_2 smaller compute units and a panel of size b_2. The number of flops performed to factor the k-th panel is,

$$\#\text{flops}(CALU(\frac{n_k}{P_{r1}}, b_1, P_2, b_2)) + \log P_{r1} \cdot \#\text{flops}(CALU(2b_1, b_1, P_2, b_2)),$$

where n_k denotes the number of columns of the k-th panel. To perform the rank-b_1 update, first the input matrix of size $n \times n$ is divided into P_1 blocks of size $\frac{n}{P_{r1}} \times \frac{n}{P_{c1}}$. Then each block is further divided among P_2 compute units. Hence each processor from level two computes a rank-b_1 update on a block of size $\frac{n}{P_{r1} \times P_{r2}} \times \frac{n}{P_{c1} \times P_{c2}}$. It is then possible to estimate the flops count of this step as a rank-b_1 update of a matrix of size $n \times n$ distributed into $P_{r1}.P_{r2} \times P_{c1}.P_{c2}$ processors. The same reasoning holds for the arithmetic cost of the computation of block row of U.

We estimate now the communication cost at each level of parallelism. At the first level we consider the communication between the P_1 compute units. This corresponds to the communication cost of the initial 1-level CALU algorithm, which is presented in detail in [12]. The size of the memory of one compute unit at the first level is formed by the sum of the sizes of the memories of the compute units at the second level. We consider here that this size is of $O(n^2/P_1)$, that is each node stores a part of the input and output matrices, and this is sufficient for determining a lower bound on the volume of communication that needs to be performed during our algorithm. However, the number of messages that are transferred at this level depends on the maximum size of data that can be transferred from one compute unit to another compute unit in one single message. We consider here the case when the size of one single message is of the order of n^2/P_1, which is realistic if shared memory is used at the second level of parallelism. However, if the size of one single message is smaller, and it can be as small as n^2/P when distributed memory is used at the second level of parallelism, the number of messages and the lower bounds presented in this section need to be adjusted for the given memory size.

At the second level we consider in addition the communication between the P_2 smaller compute units inside each compute unit of the first level. We note that we consider Cannon's matrix-matrix multiplication algorithm [4] in our model. Here we detail the communication cost of the factorization of a panel k at the second level of parallelism. Inside each node we first distribute the data on a grid of P_2 processors, then we apply 1-level CALU using P_2 processors and a panel of size b_2:

$$\begin{aligned}
\#\text{messages}_k &= \#\text{messages}(CALU(\tfrac{n_k}{P_{r1}}, b_1, P_2, b_2)) \\
&\quad + \log P_{r1} \cdot \#\text{messages}(CALU(2b_1, b_1, P_2, b_2)) + \log P_2 \times (1 + \log P_{r1}), \\
\#\text{words}_k &= \#\text{words}(CALU(\tfrac{n_k}{P_{r1}}, b_1, P_2, b_2)) \\
&\quad + \log P_{r1} \cdot \#\text{words}(CALU(2b_1, b_1, P_2, b_2)) + b_1^2 \log P_2 \times (1 + \log P_{r1}).
\end{aligned}$$

We estimate now the performance model for a square matrix using an optimal layout, that is we choose values of P_{ri}, P_{ci}, and b_i at each level of the hierarchy that allow to attain the lower bounds on communication. By following the same approach as in [12], for two levels of parallelism these parameters can be written as $P_{r1} = P_{c1} = \sqrt{P_1}$, $P_{r2} = P_{c2} = \sqrt{P_2}$, $b_1 = \frac{n}{\sqrt{P_1}} \log^{-2} P_1$, and $b_2 = \frac{b_1}{\sqrt{P_2}} log^{-2} P_2 = \frac{n}{\sqrt{P_1 P_2}} log^{-2} P_1 log^{-2} P_2$. We note that $P = P_1 \cdot P_2$. Table 1 presents the performance model of 2-level CALU. It shows that 2-level CALU attains the lower bounds on communication of dense LU, modulo polylogarithmic factors, at each level of parallelism.

2.2   Implementation on a Cluster of Multicore Processors

In the following we describe the specific implementation of these algorithms on a cluster of nodes of multicore processors. At the top level, we have used a static distribution of the data on the nodes, more specifically the matrix is distributed using a two-dimensional block cyclic partitioning on a two-dimensional grid of processors. This is similar to the distribution used in 1-level CALU [13], and hence the communication between nodes is performed as in the 1-level CALU

Table 1. Performance estimation of parallel (binary tree based) 2-level CALU with optimal layout. The matrix factored is of size $n \times n$. Some lower-order terms are omitted.

	Communication cost at the first level of parallelism	Lower bound	Memory size
# messages	$O(\sqrt{P_1} \log^3 P_1)$	$\Omega(\sqrt{P_1})$	$O(\frac{n^2}{P_1})$
# words	$O(\frac{n^2}{\sqrt{P_1}} \log P_1)$	$\Omega(\frac{n^2}{\sqrt{P_1}})$	$O(\frac{n^2}{P_1})$
	Communication cost at the second level of parallelism		
# messages	$O(\sqrt{P} \log^6 P_1 + \sqrt{P} \log^3 P_1 \log^3 P_2)$	$\Omega(\sqrt{P})$	$O(\frac{n^2}{P})$
# words	$O(\frac{n^2}{\sqrt{P}} \log^3 P_1 \log P_2)$	$\Omega(\frac{n^2}{\sqrt{P}})$	$O(\frac{n^2}{P})$
	Arithmetic cost of 2-level CALU		
# flops	$\frac{1}{P} \frac{2n^3}{3} + \frac{3n^3}{2P \log^2 P} + \frac{5n^3}{6P \log^3 P}$	$\frac{1}{P} \frac{2n^3}{3}$	

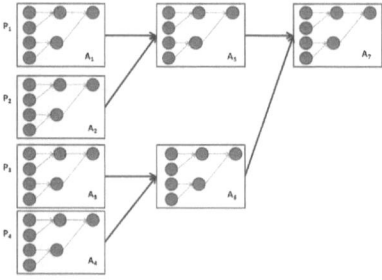

Fig. 1. Multilevel TSLU on a computer system with two levels of parallelism

factorization. For each node, the blocks are again decomposed using a two-dimensional layout, and the computation of each block is associated with a task. A dynamic scheduler is used to schedule the tasks to the available threads as described in [7].

Figure 1 shows an example of execution of multilevel TSLU algorithm on a cluster of 4 nodes, each node being formed by a 4 cores processor. The white squares represent the nodes and the blue circles represent the cores inside a node. We consider that a binary tree is used at each level of the hierarchy of parallelism. The figure shows two levels of reduction tree. The red lines represent communication between different nodes during one panel factorization, and the blue lines represent synchronization between the different cores during one step of a smaller panel factorization performed at the second level of parallelism.

3   Experimental Results

In this section we discuss first the stability of 2-level CALU, and then we evaluate its performance on a cluster of multicore processors.

3.1 Stability

It was shown in [12] that CALU is as stable as GEPP in practice. Since 2-level CALU is based on a recursive call to CALU, its stability can be different from 1-level CALU. We present here a set of experiments performed on random matrices and a set of special matrices that were also used in [12] for discussing the stability of 1-level CALU, where a detailed description of these matrices can be found. The size of the test matrices is varying from 1024 to 8192. We use different combinations of the number of processors P_1 and P_2 and of the panel sizes b_1 and b_2. We study both the stability of the LU decomposition and of the linear solver, in terms of growth factor and three different backward errors, the normwise backward error, the componentwise backward error, and $\|PA - LU\|/\|A\|$. For all the test matrices, the worst growth factor obtained is smaller than 10^2. Figure 2 shows the ratios of 2-level CALU's backward errors to those of GEPP. For almost all test matrices, 2-level CALU's normwise backward error is at most $3\times$ larger that GEPP's normwise backward error. However, for special matrices the other two backward errors of 2-level CALU can be larger by a factor of the order of 10^2 than the corresponding backward errors of GEPP. We note that for several test matrices, at most two steps of iterative refinement are used to attain the machine epsilon. These experiments show that 2-level CALU exhibits a good stability, however further investigation is required if more than two levels of parallelism are to be used.

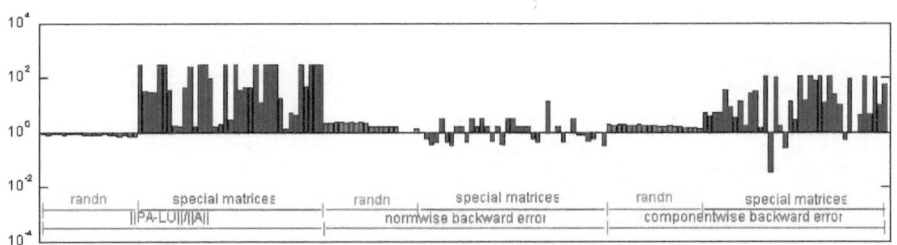

Fig. 2. The ratios of 2-level CALU's backward errors to GEPP's backward errors

3.2 Performance of 2-Level CALU

We perform our experiments on a cluster composed of 32 nodes based on Intel Xeon EMT64 processors running on Linux, each node is a two-socket, quad-core processor. Each core has a frequency of 2.50 GHz. The cluster is part of Grid 5000 [5]. Both ScaLAPACK and multilevel CALU are linked with BLAS from MKL 10.1. We compare the performance of our algorithm with the corresponding routine from ScaLAPACK. Our algorithm is implemented using MPI and Pthreads. In the experiments, P refers to the number of nodes (which corresponds to the number of compute units P_1 at the first level of parallelism described in section 2), T refers to the number of threads per node (which corresponds to the number of compute units P_2 at the second level of parallelism described in section 2), and b_1 and b_2 are the block sizes used respectively at the first and the second

level of the hierarchy. The choice of the block size at each level depends on the
architecture, the number of levels in the hierarchy, and the input matrix size. In
our experiments, we empirically tune the block sizes b_1 and b_2. This tuning is
simple because we only have two levels of parallelism, but it should be replaced
by an automatic approach for multiple levels of parallelism.

At the second level of parallelism, 2-level CALU uses a grid $T = T_r \times T_c$ of
threads, where T_r is the number of threads on the vertical dimension working
on the panel, and T_c is the number of threads on the horizontal dimension. Here
we evaluate the performance of three different parametric choices, $T = 8 \times 1$,
$T = 4 \times 2$, and $T = 2 \times 4$. We note that ScaLAPACK is executed over P MPI
processes, and in each node we call multithreaded BLAS routines with T threads.

Figure 3 shows the performance of 2-level CALU and GEPP as implemented
in ScaLAPACK for tall and skinny matrices with varying number of rows. We
observe that 2-level CALU is scalable and faster than ScaLAPACK. For a matrix
of size $10^6 \times 1200$, 2-level CALU is twice faster than ScaLAPACK. Furthermore,
an important loss of performance is observed for ScaLAPACK when $m = 10^4$,
while all the variants of 2-level CALU lead to good performance.

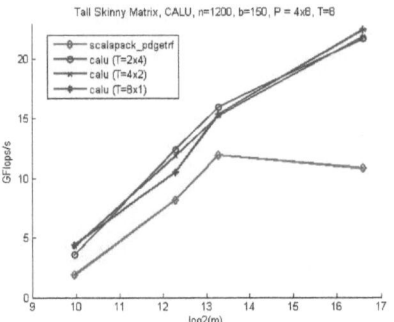
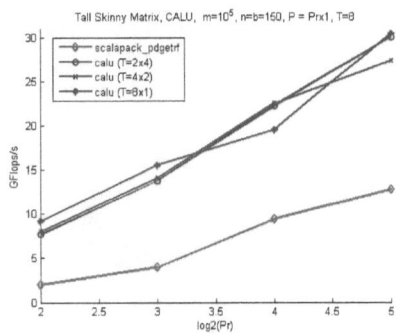

Fig. 3. Performance of 2-level CALU and ScaLAPACK on $P = 4 \times 8$ nodes, for matrices with n=1200, m varying from 10^3 to 10^6, $b_1 = 150$, and $b_2 = MIN(b_1, 100)$

Fig. 4. Performance of 2-level CALU and ScaLAPACK on $P = P_r \times 1$ nodes, for matrices with $n = b_1 = 150$, $m = 10^5$, and $b_2 = MIN(b_1, 100)$

Figure 4 shows the performance of 2-level CALU on tall and skinny matrices
with varying number of processors working on the panel factorization. Since
the matrix has only one panel, recursive TSLU is called at the top level of the
hierarchy and the adapted multithreaded CALU is called at the second level.
We observe that for $P_r = 4$, 2-level CALU is 4.5 times faster than ScaLAPACK.
When $P_r > 4$, ScaLAPACK's performance is at most 10 GFlops/s, while 2-level
CALU's performance is up to 30 GFlops/s, that is twice faster. We note that for
$P_r = 8$, the variant $T = 8 \times 1$ is slightly better than the others. This shows the
importance of determining a good layout at runtime.

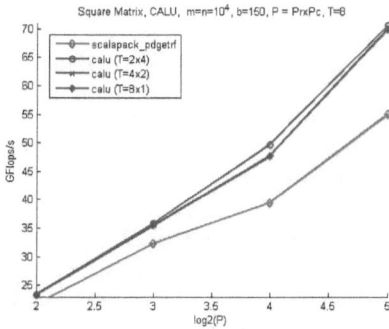

Fig. 5. Performance of 2-level CALU and ScaLAPACK on $P = P_r \times P_c$ nodes, for matrices with $m = n = 10^4$, $b_1 = 150$, and $b_2 = MIN(b_1, 100)$

Figure 5 shows the performance of 2-level CALU on a square matrix when the number of processors varies. For each value of P, we use the same layout for ScaLAPACK and at the top level of parallelism of 2-level CALU. The layout at the second level is one of the three variants discussed previously. We observe that all the variants of 2-level CALU are faster than ScaLAPACK. The variant $T = 2 \times 4$ is usually better than the others. This behavior has also been observed for multithreaded 1-level CALU [7]. We recall that the matrix is partitioned into $T_r \times N/b$ blocks. Thus increasing T_r increases the number of tasks and the scheduling overhead, and this impacts the performance of the entire algorithm.

4   Conclusion

In this paper we have introduced a communication avoiding LU factorization adapted for a computer system with two levels of parallelism, which minimizes communication at each level of the hierarchy of parallelism in terms of both volume of data and number of messages exchanged during the decomposition. On a cluster of multicore processors, that is a machine with two levels of parallelism based on a combination of both distributed and shared memories, our experiments show that our algorithm is faster than the corresponding routine from ScaLAPACK. On tall and skinny matrices, a loss of performance is observed for ScaLAPACK when the number of rows increases, while 2-level CALU shows an improving speedup. Our performance model shows that 2-level CALU increases the number of flops, words, and messages just by a polylogarithmic factor.

As future work, we plan to model and to evaluate the performance of our algorithm for multiple levels of parallelism, and to extend the same approach to other factorizations as QR. It will also be important to evaluate the usage of autotuning for computing the optimal layout and the optimal block size at each level of parallelism.

References

[1] Agullo, E., Coti, C., Dongarra, J., Herault, T., Langem, J.: QR factorization of tall and skinny matrices in a grid computing environment. In: Parallel Distributed Processing Symposium (IPDPS), pp. 1–11. IEEE (2010)

[2] Anderson, E., Bai, Z., Bischof, C., Blackford, S., Demmel, J., Dongarra, J., Du Croz, J., Greenbaum, A., Hammarling, S., McKenney, A., Sorensen, D.: LAPACK Users' Guide. SIAM, Philadelphia (1999)

[3] Blackford, L.S., Choi, J., Cleary, A., D'Azevedo, E., Demmel, J., Dhillon, I., Dongarra, J., Hammarling, S., Henry, G., Petitet, A., Stanley, K., Walker, D., Whaley, R.C.: Scalapack: A linear algebra library for message-passing computers. In: SIAM Conference on Parallel Processing (1997)

[4] Cannon, L.E.: A cellular computer to implement the Kalman filter algorithm. PhD thesis, Montana State University (1969)

[5] Cappello, F., Desprez, F., Dayde, M., Jeannot, E., Jegou, Y., Lanteri, S., Melab, N., Namyst, R., Primet, P.V.B., Richard, O., et al.: Grid5000: a nation wide experimental grid testbed. International Journal on High Performance Computing Applications 20(4), 481–494 (2006)

[6] Demmel, J., Grigori, L., Hoemmen, M., Langou, J.: Communication-optimal parallel and sequential QR and LU factorizations. Technical Report UCB/EECS-2008-89, University of California Berkeley, EECS Department, LAWN #204 (2008)

[7] Donfack, S., Grigori, L., Gupta, A.K.: Adapting communication-avoiding LU and QR factorizations to multicore architectures. In: IEEE International Parallel and Distributed Processing Symposium (IPDPS). IEEE (2010)

[8] Elmroth, E., Gustavson, F.: New Serial and Parallel Recursive QR Factorization Algorithms for SMP Systems. In: Kågström, B., Elmroth, E., Waśniewski, J., Dongarra, J. (eds.) PARA 1998. LNCS, vol. 1541, pp. 120–128. Springer, Heidelberg (1998)

[9] Frigo, M., Leiserson, C.E., Prokop, H., Ramachandran, S.: Cache-oblivious algorithms. In: 40th Annual Symposium on Foundations of Computer Science, pp. 285–297 (1999)

[10] Van De Geijn, R.A., Watts, J.: SUMMA: Scalable Universal Matrix Multiplication Algorithm. Concurrency Practice and Experience 9(4), 255–274 (1997)

[11] Graham, S.L., Snir, M., Patterson, C.A.: Getting up to speed: The future of supercomputing. National Academies Press (2005)

[12] Grigori, L., Demmel, J., Xiang, H.: CALU: A communication optimal LU factorization algorithm. SIAM Journal on Matrix Analysis and Applications 32, 1317–1350 (2011)

[13] Grigori, L., Demmel, J.W., Xiang, H.: Communication avoiding Gaussian elimination. In: Proceedings of the 2008 ACM/IEEE Conference on Supercomputing, p. 29. IEEE Press (2008)

[14] Hong, J.-W., Kung, H.T.: I/O complexity: The red-blue pebble game. In: Proceedings of the Thirteenth Annual ACM Symposium on Theory of Computing. ACM (1981)

[15] Irony, D., Toledo, S., Tiskin, A.: Communication lower bounds for distributed-memory matrix multiplication. Journal of Parallel and Distributed Computing 64(9), 1017–1026 (2004)

Locality Improvement of Data-Parallel Adams–Bashforth Methods through Block-Based Pipelining of Time Steps

Matthias Korch

University of Bayreuth
Applied Computer Science 2
korch@uni-bayreuth.de

Abstract. Adams–Bashforth methods are a well-known class of explicit linear multi-step methods for the solution of initial value problems of ordinary differential equations. This article discusses different data-parallel implementation variants with different loop structures and communication patterns and compares the resulting locality and scalability. In particular, pipelining of time steps is employed to improve the locality of memory references. The comparison is based on detailed runtime experiments performed on parallel computer systems with different architectures, including the two supercomputer systems JUROPA and HLRB II.

1   Introduction

Many time-dependent processes can be modeled by initial value problems (IVPs) of ordinary differential equations (ODEs):

$$\mathbf{y}'(t) = \mathbf{f}(t, \mathbf{y}(t)), \quad \mathbf{y}(t_0) = \mathbf{y}_0, \tag{1}$$

where $\mathbf{y}(t) \in \mathbb{R}^n$ is the solution function to be computed for the interval $t \in [t_0, t_e]$, $\mathbf{y}_0$ is the given *initial value*, i.e., the initial state of the process to be simulated at time t_0, and $\mathbf{f} : \mathbb{R} \times \mathbb{R}^n \to \mathbb{R}^n$ is the given *right-hand-side function*, which describes the rates of change of the process to be simulated.

This article considers Adams–Bashforth (AB) methods [2,5,6] on an equidistant grid. At each time step $\kappa = 0, 1, 2, \ldots$, these methods compute an approximation $\mathbf{y}_{\kappa+1}$ to the solution function at time $t_{\kappa+1}$, $\mathbf{y}(t_{\kappa+1})$, using function results of the last k preceding time steps and weights $\beta_1, \ldots, \beta_k$ according to the scheme

$$\mathbf{y}_{\kappa+1} = \mathbf{y}_\kappa + h \sum_{l=1}^{k} \beta_l \mathbf{f}(t_{\kappa-l+1}, \mathbf{y}_{\kappa-l+1}). \tag{2}$$

AB methods belong to the class of *explicit linear k-step* or, more generally, *multi-step* methods and are suitable for *nonstiff* IVPs (see [2,5,6] for a discussion of stiffness and explicit vs. implicit methods).

Many parallel IVP solution methods have been proposed. An overview of the fundamental work and further references can be found in [1]. Recent work on

C. Kaklamanis et al. (Eds.): Euro-Par 2012, LNCS 7484, pp. 563–576, 2012.
© Springer-Verlag Berlin Heidelberg 2012

parallel ODE methods includes variants of iterated Runge–Kutta methods [4] and peer two-step methods [12]. Many of the parallel ODE methods proposed concentrate on *parallelism across the method*, i.e., they provide a small number of independent coarse-grained computational tasks inherent in the computational structure of the method, for example, independent stages. Examples are Parallel Adams–Bashforth (PAB) and Parallel Adams–Moulton (PAM) methods [13], which belong to the class of *general linear methods* [2]. Different parallel execution schemes for these methods are investigated and discussed in [11].

This article focuses on the *parallelism across the ODE system* available in classical k-step AB methods (2), i.e., the computation of the components $y_{\kappa+1,1}, \ldots, y_{\kappa+1,n}$ of $\mathbf{y}_{\kappa+1}$ is distributed across the processing elements and performed in data-parallel style. Different loop structures for (2) and corresponding communication patterns for ODE systems with arbitrary coupling and for ODE systems with a special coupling structure called *limited access distance* are described, and the influence on locality and scalability is discussed. Double-precision implementations have been written in C for shared and distributed address space using POSIX Threads (Pthreads) and MPI, respectively. Starting point were Pthread implementations [10], which have, for this article, been optimized for NUMA (*non-uniform memory access*) architectures and been complemented by MPI implementations. Scalability and locality have been investigated using runtime experiments on several computer systems with different architectures, including the two supercomputer systems HLRB II and JUROPA.

2   Parallel Implementation of General AB Solvers

2.1   Possible Loop Structures

Equation (2) leads to a doubly nested, fully permutable loop structure, since one iteration over the summands $\beta_l \mathbf{f}(t_{\kappa-l+1}, \mathbf{y}_{\kappa-l+1})$ for $l = 1, \ldots, k$ and one iteration over the system dimension $j = 1, \ldots, n$ is required. It is sufficient to compute one evaluation of the right-hand-side function $\mathbf{f}(t_\kappa, \mathbf{y}_\kappa)$ per time step if the function results of the previous $k - 1$ time steps are kept in memory. These considerations lead to the following three loop structures:

j–l. The j-loop over the large system dimension is chosen as outer loop. The l-loop to compute the j-th component of the vector $\mathbf{y}_{\kappa+1}$, $y_{\kappa+1,j} = y_{\kappa,j} + h \sum_{l=1}^{k} \beta_l F_{l,j}$ with $F_{l,j} := f_j(t_{\kappa-l+1}, \mathbf{y}_{\kappa-l+1})$ runs inside the j-loop. This results in a high temporal locality for this vector component since its storage location is reused in the partial sum updates.

l–j. The l-loop over the k steps of the AB method is chosen as outer loop. In each iteration of the l-loop, a j-loop over the system dimension is executed, which accesses the two vectors $\mathbf{F}_l := \mathbf{f}(t_{\kappa-l+1}, \mathbf{y}_{\kappa-l+1})$ and $\mathbf{y}_{\kappa+1}$. This results in a high spatial locality since cache lines of these two vectors can be reused for subsequent vector components once they have been loaded into the cache. Moreover, this loop structure can benefit from hardware prefetching, and its access pattern is easily predictable by the hardware prefetcher.

Tiling. Since the two loops are fully permutable, they can also be tiled to create an additional working space that fits in the cache. This leads to a triply nested loop structure, where the outermost loop (j-loop) iterates over the system dimension with stride B (*tile size* or *block size*). Inside the j-loop runs the l-loop, which iterates over the k steps. The innermost loop (jj-loop) again iterates over the system dimension and accesses the components $j, \ldots, j + B - 1$ of the two vectors $\mathbf{F}_l$ and $\mathbf{y}_{\kappa+1}$. Thus, a high spatial locality results from the innermost loop iterating over two vectors with stride 1, but also a high temporal locality results from the reuse of a block of size B of the vector $\mathbf{y}_{\kappa+1}$ in successive iterations of the l-loop.

2.2   Parallelization

In data-parallel implementations, the system dimension $1, \ldots, n$ is partitioned among the processing elements. For highest spatial locality, a blockwise distribution is appropriate, such that each of the p processing elements is assigned a block of n/p consecutive components. As data structures, $k - 1$ function results and the two approximation vectors $\mathbf{y}_\kappa$ and $\mathbf{y}_{\kappa+1}$ have to be stored in memory. In a sequential implementation, the function results can be stored in a $(k - 1) \times n$ 2D array that is used in a cyclic fashion such that $\mathbf{f}(t_{m+k-1}, \mathbf{y}_{m+k-1})$ overwrites $\mathbf{f}(t_m, \mathbf{y}_m)$. Similarly, $\mathbf{y}_\kappa$ and $\mathbf{y}_{\kappa+1}$ can be stored in two 1D arrays of size n, the pointers to which are swapped at each time step.

The 2D array holding the function results can be distributed to the processing elements such that each processing element stores locally a partition of size $(k-1) \times n/p$ of this 2D array. This is necessary in an MPI implementation because of the separate address spaces. But even for shared address space, a distributed storage of the function results in separate, thread-local $(k - 1) \times n/p$ 2D arrays often is preferable, because it ensures that all function results associated with a thread can be stored in local memory of the processing element on which the thread is executed. This is particularly important on NUMA architectures. Though the "first touch" policy applied by modern operating systems to support NUMA will also move memory pages of shared arrays to local memory of the thread that first writes to the page, shared data structures may lead to sharing of memory pages and thus to remote memory accesses at the borders of the data ranges of the threads. At a finer grained level, sharing of cache lines may decrease performance even on UMA (*uniform memory access*) architectures.

For the two vectors $\mathbf{y}_\kappa$ and $\mathbf{y}_{\kappa+1}$, distributed storage is not always possible or preferable, because the function $\mathbf{f}$ has to be evaluated for $\mathbf{y}_\kappa$ and, in the general case, this function evaluation may use all components of $\mathbf{y}_\kappa$. Therefore, the general shared-address-space implementations considered in this article implement these two vectors as shared data structures. One barrier operation is required per time step to prevent that threads start the function evaluation before all other threads have computed their share of the argument vector. The MPI implementations require a replicated storage of the argument vector. Since each of the MPI processes computes n/p components of the argument vector, it must be gathered by all processes using a multibroadcast operation (`MPI_Allgatherv()`). The

```
for (i = k - 1; i < steps; i++)
{
  MPI_Allgatherv(Y_cur + first_elem, num_elems, MPI_DOUBLE,
                 Y_arg, counts, offsets, MPI_DOUBLE, MPI_COMM_WORLD);

  for (j = first_elem; j <= last_elem; j += B)
  {
    for (jj = j; jj < j + B; jj++) Y_new[jj] = b[k - 1] * F[i % (k - 1)][jj];

    for (l = 1; l < k - 1; l++)
      for (jj = j; jj < j + B; jj++) Y_new[jj] += b[j] * F[(i - 1) % (k - 1)][jj];

    for (jj = j; jj < j + B; jj++) F[i % (k - 1)][jj] = f(jj, t0 + i * h, Y_arg);
    for (jj = j; jj < j + B; jj++) Y_new[jj] += b[0] * F[i % (k - 1)][jj];
    for (jj = j; jj < j + B; jj++) Y_new[jj] = Y_cur[jj] + h * Y_new[jj];
  }

  swap_vectors(&Y_new, &Y_cur);
}
```

Listing 1. General parallel MPI implementation with tiled loop structure

implementations considered in this article therefore store in each process n/p components of $\mathbf{y}_\kappa$ and $\mathbf{y}_{\kappa+1}$ and one additional array of size n, in which $\mathbf{y}_\kappa$ is gathered and which is then used as argument vector for the function evaluation.

Listing 1 shows a code fragment of a general parallel implementation of one time step of an AB method with tiled loop structure that uses MPI as programming environment. The loop structures j–l and l–j can be interpreted as special cases of the tiled loop structure using $B = 1$ and $B = n$, respectively.

3 Reducing Parallel Overhead through Specialization

There are many sparse ODE systems where the components of $\mathbf{y}_\kappa$ accessed by a component function $f_j(t_\kappa, \mathbf{y}_\kappa)$ are located nearby the index j. Examples are ODE systems resulting from a spatial discretization of PDE systems by the method of lines. This property is measured by the *access distance* $d(\mathbf{f})$, which is the smallest value b, such that all component functions $f_j(t_\kappa, \mathbf{y}_\kappa)$ access only the components $\{y_{\kappa,j-b}, \ldots, y_{\kappa,,j+b}\}$. We say $d(\mathbf{f})$ is *limited* if $d(\mathbf{f}) \ll n$.

For ODE systems with limited access distance, data-parallel implementations with a blockwise data distribution only need to exchange $d(\mathbf{f})$ components of $\mathbf{y}_\kappa$ at the left and at the right border of their data range. Using MPI as programming environment, the expensive multibroadcast operation MPI_Allgatherv() (cf. Fig. 1) can be replaced by non-blocking single transfer operations (MPI_Isend() and MPI_Irecv()), thus potentially overlapping communication with computations. Replicated storage of the argument vector for the function evaluation using an array of size n is no longer required. Instead, the two arrays holding the local parts of $\mathbf{y}_\kappa$ and $\mathbf{y}_{\kappa+1}$ of size n/p are enlarged by $d(\mathbf{f})$ at each border to store the data received from the neighbor processes (*ghost cells*).

In shared-address-space implementations, it is desirable to avoid the high costs of the global barriers (cf. Fig. 2). If the ODE system has a limited access

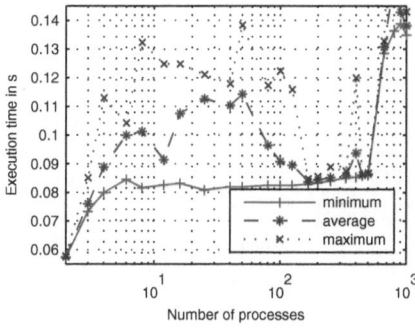
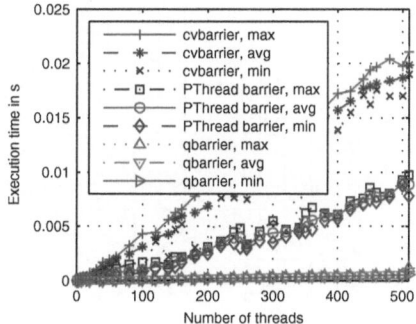

Fig. 1. Execution time of the communication operation `MPI_Allgatherv()` for $8 \cdot 10^6$ vector elements on HLRB II

Fig. 2. Comparison of the execution time of barrier operations on HLRB II (barrier based on condition variables, Pthread barrier, barrier based on busy waiting)

distance, only data from neighbor threads are required for the function evaluation. Hence, it is sufficient to use locks for synchronization between neighbors. Similar to the way communication and computation could be overlapped in an MPI implementation, no waiting times for acquiring the locks occur if all threads process the ODE components synchronously at the same speed.

While shared data structures are needed in general implementations where the function evaluation may access all components of its argument vector, implementations specialized in a limited access distance can store the vectors $\mathbf{y}_\kappa$ and $\mathbf{y}_{\kappa+1}$ in a distributed fashion similar to the MPI implementations, thus avoiding page and cache line sharing. In this case, the data required from neighbor threads have to be copied to ghost cells, which consumes CPU time.

4   Pipelining of Time Steps

The loop structures considered in Section 2.1 allow that the evaluation of the right-hand-side function $\mathbf{f}(t_\kappa, \mathbf{y}_\kappa)$ accesses all components of $\mathbf{y}_\kappa$. Next, a pipeline-like loop structure covering several time steps of the AB method is described that can be used for ODE systems with limited access distance to increase locality. A similar approach has been proposed for the stages of embedded Runge–Kutta methods [8], the corrector steps of iterated Runge–Kutta methods [7], and the micro-steps of extrapolation methods [9].

The pipeline-like loop structure is based on a subdivision of all n-vectors into $n_B = \lceil n/B \rceil$ blocks of size B. This subdivision is similar to the subdivision for loop tiling, but, for the pipelining scheme to work, the block size must be larger than the access distance, i.e., $B \geq d(\mathbf{f})$, and the number of blocks, n_B, must be at least as large as the pipeline length L.

Given this subdivision, the function evaluation of a block $J \in \{1, \ldots, n_B\}$ of $\mathbf{y}_\kappa$ uses only components of the blocks $J-1$, J, and $J+1$ of $\mathbf{y}_\kappa$ if these blocks

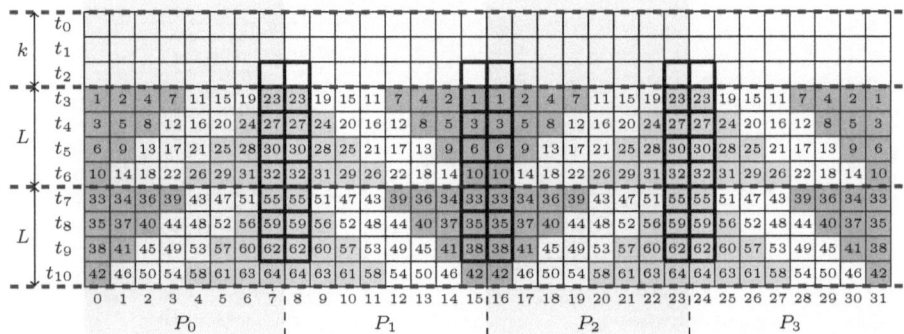

Fig. 3. Illustration of the pipeline-like processing of time steps in a data-parallel implementation for $n_B = 32$, $p = 4$, $k = 3$, and $L = 4$. Blocks that need to be exchanged between processing elements are highlighted by thick borders.

exist, i.e., if $1 < J < n_B$. Now, due to the dependence pattern of the blocks, a sequence of L successive time steps can be computed in a single sweep over the system dimension using the block computation order displayed in Fig. 3. In a data-parallel implementation, neighbor processing elements process their pipelines in opposite directions so that during the initialization and the finalization of the pipelines all data that have to be received from neighbor processing elements are available before they are needed for a function evaluation.

Data-parallel implementations of the pipeline-like loop structure can use the same distributed data structures and the same efficient communication patterns as the specialized implementations optimized for limited access distance described in Section 3.

A modification of the computation order to support ODE systems where the access distance is limited only in a cyclic sense as it occurs in discretized PDE systems with periodic boundary conditions is possible but not discussed here.

5   Storage and Working Spaces

Sequential implementations store $k - 1$ function results and the two vectors $\mathbf{y}_\kappa$ and $\mathbf{y}_{\kappa+1}$, which amounts to a total storage space of

$$S_{\text{seq}}(k, n) = (k + 1)n. \tag{3}$$

All parallel implementations store n/p components of the $k - 1$ function results per processing element. Differences in the storage space between the parallel implementations result from how $\mathbf{y}_\kappa$ and $\mathbf{y}_{\kappa+1}$ are handled. The general and specialized shared-address-space implementations with shared storage of $\mathbf{y}_\kappa$ and $\mathbf{y}_{\kappa+1}$ require the same storage space as sequential implementations:

$$S_{\text{sas,ys}}(p, k, n) = p \left[(k - 1)\frac{n}{p} \right] + 2n = (k + 1)n. \tag{4}$$

General MPI implementations store n/p components of $\mathbf{y}_\kappa$ and $\mathbf{y}_{\kappa+1}$ in local arrays, the pointers to which are swapped at every time step. Additionally, one array of size n to collect $\mathbf{y}_\kappa$ is used, which amounts to a total storage space of

$$S_{\text{das,mbcast}}(p, k, n) = p\left[(k-1)\frac{n}{p} + 2\frac{n}{p} + n\right] = (p + k + 1)n. \tag{5}$$

Specialized implementations for distributed address space which use neighbor-to-neighbor communication and specialized implementations for shared address space which store $\mathbf{y}_\kappa$ and $\mathbf{y}_{\kappa+1}$ in a distributed fashion, need additional storage space, compared with the sequential implementations, only for the ghost cells:

$$S_{\text{das,single}}(p, k, n, d(\mathbf{f})) = S_{\text{sas,yd}}(p, k, n) = p\left[(k-1)\frac{n}{p} + 2\left(\frac{n}{p} + 2d(\mathbf{f})\right)\right]$$
$$= (k+1)n + 4d(\mathbf{f})p. \tag{6}$$

Since all implementations iterate over all their data structures during one time step, the storage space they require per processing element constitutes the most significant working space of their loop structures. If the ODE system is large so that not all data used by a processing element per time step fits in the cache, the tiled or the pipeline-like loop structure can be expected to be more efficient, because they create additional smaller working spaces that allow temporal reuse of cache data.

The working space created by loop tiling, i.e., the size of one *tile*, consists of $k - 1$ blocks of size B of the function results, one block of size B of $\mathbf{y}_{\kappa+1}$ computed in the current time step, and one block of size $B + 2d(\mathbf{f})$ of $\mathbf{y}_\kappa$ for which the function is evaluated:

$$W_{\text{tile}}(k, d(\mathbf{f}), B) = (k+1)B + 2d(\mathbf{f}). \tag{7}$$

The most important working space created by the pipeline-like loop structure using pipeline length L is the working space of one pipelining step, i.e., the computation of one diagonal consisting of L blocks. This working space can be viewed as being built up of L loop tiling working spaces, but components within the access distance of the function evaluation partially overlap. The resulting size of the working space is

$$W_{\text{pipe}}(k, d(\mathbf{f}), B, L) = L(k+1)B + 4d(\mathbf{f}). \tag{8}$$

6   Experimental Results and Discussion

6.1   Experimental Setup

In this article, we present selected experimental results measured on three computer systems. Sequential jobs needed for empirical search of optimal block sizes and pipeline lengths were run on a small cluster system consisting of 32 2-way AMD Opteron DP 246 nodes with 64 KB L1 data cache and 1024 KB L2 cache.

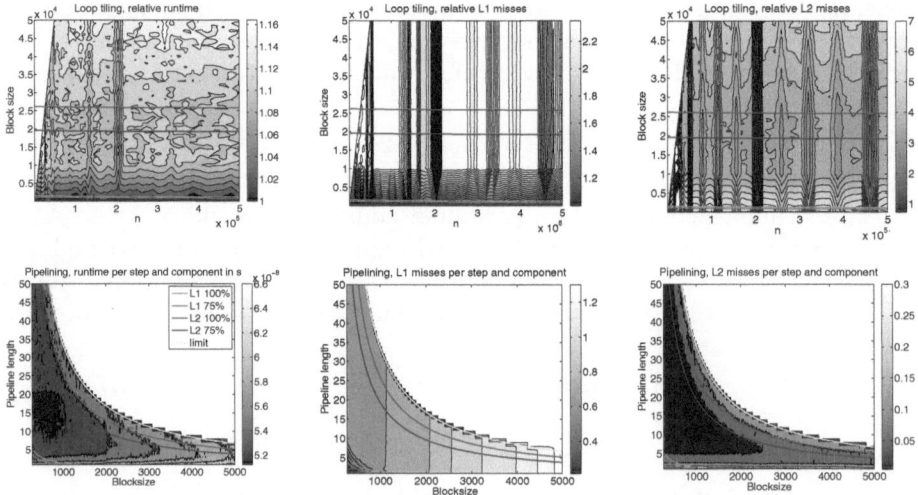

Fig. 4. Sequential runtime and locality behavior of the tiled and the pipeline-like loop structur on AMD Opteron DP 246. $k = 4$. Test problem: BRUSS2D-MIX. For the tiled loop structure, the problem size is varied. For the pipeline-like loop structure, results for problem size $N = 130$ ($n = 33\,800$) are shown. Dark color indicates better behavior.

To compare the sequential and parallel implementations, results measured on the two supercomputer systems JUROPA and HLRB II are shown. JUROPA (Jülich Supercomputing Centre (JSC)) consists of $2\,208$ compute nodes equipped with two quad-core Intel Xeon X5570 (Nehalem-EP) processors running at $2.93\,$GHz and interconnected by an Infiniband QDR network. The L1 data cache has a size of $32\,$KB; the L2 cache size is $256\,$KB. The L3 cache is shared between the four cores and has a size of $8\,$MB. HLRB 2 (Leibniz Supercomputing Centre (LRZ) Munich) is an SGI Altix 4700 system based on dual-core Itanium 2 9040 (Montecito) processors running at $1.6\,$GHz with a total number of $9\,728$ CPU cores. The system is interconnected by an SGI NUMAlink 4 network and is divided into 19 shared-memory partitions containing 512 cores. The L1 data cache has a size of $16\,$KB, but does not store floating point data. The L2 and the L3 cache have a size of $256\,$KB and $9\,$MB, respectively. All cache levels are non-shared.

The test problems considered are BRUSS2D-MIX (2D Brusselator reaction-diffusion equation [1,5]), which is derived from a first order 2D PDE system with two variables using an $N \times N$ grid and which has a system size of $n = 2N^2$ and an access distance of $d(\mathbf{f}) = 2N$, and STRING (mechanical vibration of a string [5]), which is derived from a second order 1D PDE system with one variable and which has a system size of $n = 2N$ and an access distance of $d(\mathbf{f}) = 3$.

6.2   Choosing Blocksize and Pipeline Length

For the tiled loop structure, the runtime depends on the block size B. The upper part of Fig. 4 illustrates the influence of the block size on the sequential runtime

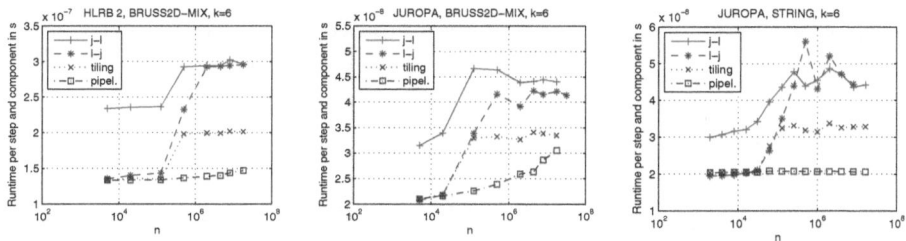

Fig. 5. Comparison of the normalized runtime of the sequential implementations

on an AMD Opteron DP 246 processor for the test problem BRUSS2D-MIX, $k = 4$ and varying system sizes between $n = 200$ and $n = 500\,000$. For this example, small block sizes up to ≈ 1000 deliver the smallest execution time. But the block size should not be too small, i.e., $B \gtrsim 100$, so that it spans several cache lines. This observation conforms with the working space model (7), which suggests maximum block sizes between 1630 ($n = 200$) and 1238 ($n = 500\,000$) for a tile to fit in the L1 cache. In fact, Fig. 4 shows that in the ranges of the block size with best execution times the smallest numbers of L1 misses occur.

The execution time of the pipeline-like loop structure is influenced by two parameters: the block size and the pipeline length. The lower part of Fig. 4 illustrates this influence for the test problem BRUSS2D-MIX with problem size $N = 130$ ($n = 33\,800$) and $k = 4$ on an AMD Opteron DP 246 processor. Though according to (8) it is possible to fit the working space of a pipelining step in the L1 cache using pipeline length 5 or smaller, best performance, in this example, is obtained for a pipeline length between about 10 and 20 and block sizes up to 1000. Generally, an area of good performance is framed by the working space model (8) applied to the size of the L2 cache, but within this area neither the block size nor the pipeline length should be chosen too large.

6.3   Influence of the Working Spaces on Sequential Performance

Figure 5 compares the sequential implementations on one processor core of HLRB II and JUROPA using normalized runtime, i.e., the execution time per step and component. For the loop tiling and the pipelining implementations, a set of block sizes and pipeline lengths were precomputed using their working space models, and the runtime of the best parameter choice is shown. Since the function evaluation costs per component for the two test problems are independent of the system size, an increase in the normalized runtime is usually caused by working spaces of loops growing larger than a cache level. For small system sizes, where all data structures used in a time step fit in the cache, general implementations can obtain a good performance. For larger system sizes, loop tiling or pipelining is required for best performance. Pipelining performs best in the range of system sizes where the pipelining working space (8) fits in the cache but the overall working space of a time step (3) is too large to fit in the cache.

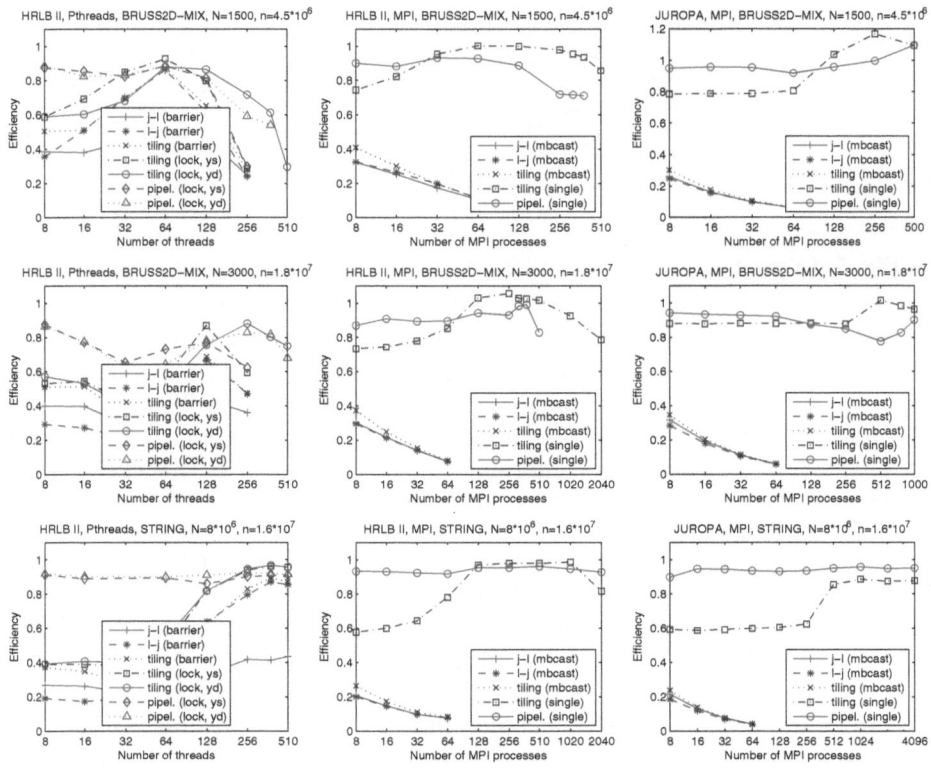

Fig. 6. Strong scalability of the Pthread and MPI implementations

6.4 Parallel Performance on Different Architectures

To investigate the strong scalability of the parallel implementations, Fig. 6 shows selected experimental results measured on HLRB II and JUROPA. The comparison is based on efficiency, i.e., execution time of the fastest sequential implementation divided by the execution time of the parallel implementation and the number of processing elements. Thus, an efficiency of 1 is optimal. A set of block sizes and pipeline lengths is precomputed and the best efficiency is shown, similarly to the comparison of the sequential implementations.

The general MPI implementations, which need to use `MPI_Allgatherv()`, do not scale. The MPI pipelining implementation obtains a very high, nearly constant efficiency (measured for STRING up to 4096 cores on JUROPA). A loop tiling implementation that exploits limited access distance by using single transfer operations is not as good as the pipelining implementation, in particular for small numbers of processor cores, but it catches up with or even outperforms the pipelining implementation as the number of processing elements is increased and the amount of data processed by each processor core per time step decreases.

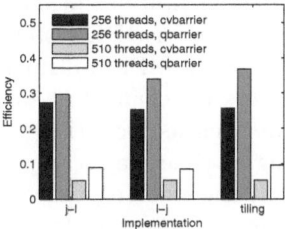 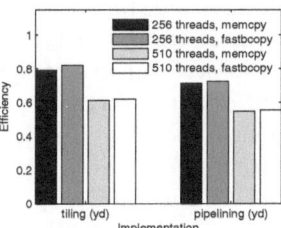 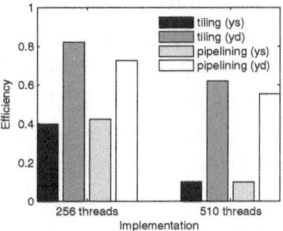

Fig. 7. Influences on the scalability of Pthread implementations on HLRB II. Test problem: BRUSS2D-MIX with $N = 2000$ ($n = 8 \cdot 10^6$). $k = 6$. Left: Barrier operations. Middle: Memory copy operations. Right: Distributed vs. shared storage.

The Pthread implementations could be investigated on HLRB II using up to 510 threads. To improve the performance of the general implementations, which need barrier operations, a barrier operation based on busy waiting [3] was used (cf. Fig. 7 (left)). In contrast to the general MPI implementations, the general Pthread implementations benefit from a parallel execution. Depending on the test problem and the problem size, even several hundred threads can be run efficiently. Using 256 threads and $k = 6$, speedups between 92 (j–l) and 121 (loop tiling) have been measured for BRUSS2D-MIX and $N = 3000$ and between 62 (j–l) and 72 (loop tiling) for BRUSS2D-MIX and $N = 1500$. Using 510 threads, for STRING and $N = 8 \cdot 10^6$, the general implementations even reached speedups between 222 (j–l) and 445 (loop tiling).

The specialized Pthread implementations are even more efficient than the general Pthread implementations but do not reach the performance of the specialized MPI implementations. In the example cases shown in Fig. 6 speedups between 346 and 383 have been measured for BRUSS2D-MIX and $N = 3000$ and between 464 and 489 for STRING using 510 threads. At least for BRUSS2D-MIX, the implementations with distributed storage of y_κ and $y_{\kappa+1}$ are more efficient than those with shared storage of these vectors (cf. Fig. 7 (right)). To further improve these implementations, which have to copy data from neighbor threads, the `memcpy()` operation from the standard C library was replaced by the faster (internal) `fastbcopy()` operation of the SGI MPT library. This, however, increased efficiency only marginally (cf. Fig. 7 (middle)). As for the MPI implementations, pipelining is most efficient for smaller numbers of processing elements, where the amount of data processed by each processing element is larger than the cache.

7   Conclusions

Data-parallel implementations of Adams–Bashforth methods can be used efficiently on hundreds and thousands of processing elements if the ODE system is large enough. Since general MPI implementations require the use of multibroadcasts, only specialized MPI implementations which exploit the specific structure of the ODE system can reach high speedups. Pthread implementations also can obtain significant speedups from a data-parallel execution, even for ODE systems with arbitrary coupling, but the performance of the specialized MPI implementations is higher. If the amount of data processed per processing element at each time step exceeds the cache size, locality optimizations such as loop tiling or pipelining are required for best performance. Pipelining of time steps, as proposed in this paper, outperforms standard loop tiling if the working space of a pipelining steps fits in the cache. Efficient block sizes and pipeline lengths can be chosen using a working space model. In future work, the most efficient implementation, block size and pipeline length could be chosen automatically.

Acknowledgments. We thank the JSC and the LRZ Munich for providing access to their supercomputer systems. This work was supported by the German Research Foundation (DFG) [grant numbers RA 524/17-1 and RA 524/17-2].

References

1. Burrage, K.: Parallel and Sequential Methods for Ordinary Differential Equations. Oxford University Press, New York (1995)
2. Butcher, J.C.: Numerical methods for ordinary differential equations, 2nd edn. John Wiley & Sons, Chichester (2008)
3. Chen, J., Watson III, W.: Software barrier performance on dual quad-core Opterons. In: Proceedings of the 2008 International Conference on Networking, Architecture, and Storage, pp. 303–309. IEEE Computer Society (2008)
4. Cong, N.H., Xuan, L.N.: Twostep-by-twostep PIRK-type PC methods with continuous output formulas. J. Comput. Appl. Math. 221, 165–173 (2008)
5. Hairer, E., Nørsett, S.P., Wanner, G.: Solving Ordinary Differential Equations I: Nonstiff Problems, 2nd rev. edn. Springer, Berlin (2000)
6. Hairer, E., Wanner, G.: Solving Ordinary Differential Equations II: Stiff and Differential-Algebraic Problems, 2nd rev. edn. Springer, Berlin (2002)
7. Korch, M., Rauber, T.W.: Locality Optimized Shared-Memory Implementations of Iterated Runge-Kutta Methods. In: Kermarrec, A.-M., Bougé, L., Priol, T. (eds.) Euro-Par 2007. LNCS, vol. 4641, pp. 737–747. Springer, Heidelberg (2007)
8. Korch, M., Rauber, T.: Parallel low-storage Runge-Kutta solvers for ODE systems with limited access distance. Int. J. High Perf. Comput. Appl. 25(2), 236–255 (2011)
9. Korch, M., Rauber, T., Scholtes, C.: Scalability and locality of extrapolation methods on large parallel systems. Concurrency Computat.: Pract. Exper. 23(15), 1789–1815 (2011)
10. Ley, K.: Parallele Implementierung und Analyse eines expliziten Adams-Verfahrens. Bachelor's thesis, University of Bayreuth (November 2010)
11. Rauber, T.W., Rünger, G.: Execution Schemes for Parallel Adams Methods. In: Danelutto, M., Vanneschi, M., Laforenza, D. (eds.) Euro-Par 2004. LNCS, vol. 3149, pp. 708–717. Springer, Heidelberg (2004)
12. Schmitt, B.A., Weiner, R., Jebens, S.: Parameter optimization for explicit parallel peer two-step methods. Appl. Numer. Math. 59, 769–782 (2008)
13. van der Houwen, P.J., Messina, E.: Parallel Adams methods. J. Comput. Appl. Math. 101, 153–165 (1999)

Parallel SOR for Solving the Convection Diffusion Equation Using GPUs with CUDA

Yiannis Cotronis, Elias Konstantinidis,
Maria A. Louka, and Nikolaos M. Missirlis

Department of Informatics and Telecommunications,
University of Athens,
Panepistimiopolis, 15784, Athens, Greece
{cotronis,ekondis,mlouka,nmis}@di.uoa.gr

Abstract. In this paper we study a parallel form of the SOR method for the numerical solution of the Convection Diffusion equation suitable for GPUs using CUDA. To exploit the parallelism offered by GPUs we consider the fine grain parallelism model. This is achieved by considering the local relaxation version of SOR. More specifically, we use SOR with red black ordering with two sets of parameters ω_{ij} and ω'_{ij}. The parameter ω_{ij} is associated with each red (i+j even) grid point (ij), whereas the parameter ω'_{ij} is associated with each black (i+j odd) grid point (ij). The use of a parameter for each grid point avoids the global communication required in the adaptive determination of the best value of ω and also increases the convergence rate of the SOR method [3]. We present our strategy and the results of our effort to exploit the computational capabilities of GPUs under the CUDA environment. Additionally, a program for the CPU was developed as a performance reference. Significant performance improvement was achieved with the three developed GPU kernel variations which proved to have different pros and cons.

Keywords: Iterative methods, SOR, R/B SOR, GPU computing, CUDA.

Subject classification: AMS(MOS), 65F10, 65N20, CR:5.13.

1 Introduction

Traditionally, conventional processors have been used to solve computational problems. Modern graphics processors (GPUs) have become coprocessors with significantly more computational power than general purpose processors. Their large computational potential has turned them to a special challenge for solving general-purpose problems with large computational burden. Thus, application programming environments have been developed like the proprietary CUDA (Compute Unified Development Architecture) by NVidia [16,12] and the OpenCL (Open Computing Language) [20] which is supported by many hardware vendors, including NVidia.

CUDA environment is rapidly evolving and a constantly increasing number of researchers is adopting it in order to exploit GPU capabilities. It provides an elegant way for writing GPU parallel programs, by using a kind of extended C

C. Kaklamanis et al. (Eds.): Euro-Par 2012, LNCS 7484, pp. 575–586, 2012.

language, without involving other graphics APIs. In this paper we use GPUs for the numerical solution of Partial Differential equations. In particular, we consider the solution of the second order convection diffusion equation

$$\Delta u - f(x,y)\frac{\partial u}{\partial x} - g(x,y)\frac{\partial u}{\partial y} = 0 \tag{1}$$

on a domain $\Omega = \{(x,y)\}|0 \leq x \leq 1, 0 \leq y \leq 1\}$, where $u = u(x,y)$ is prescribed on the boundary $\partial\Omega$. The discretization of (1) on a rectangular grid $M_1 \times M_2 = N$ unknowns within Ω leads to

$$u_{ij} = \ell_{ij}u_{i-1,j} + r_{ij}u_{i+1,j} + t_{ij}u_{i,j+1} + b_{ij}u_{i,j-1}, \tag{2}$$
$$i = 1,2,\ldots,M_1 \ , \ j = 1,2,\ldots,M_2$$

with

$$\ell_{ij} = \frac{k^2}{2(k^2+h^2)}(1+\frac{1}{2}hf_{ij}) \ , \ r_{ij} = \frac{k^2}{2(k^2+h^2)}(1-\frac{1}{2}hf_{ij})$$

$$\tag{3}$$

$$t_{ij} = \frac{h^2}{2(k^2+h^2)}(1-\frac{1}{2}kg_{ij}) \ , \ b_{ij} = \frac{h^2}{2(k^2+h^2)}(1+\frac{1}{2}kg_{ij}),$$

where $h = 1/(M_1+1)$, $k = 1/(M_2+1)$, $f_{ij} = f(ih,jk)$ and $g_{ij} = g(ih,jk)$. For a particular ordering of the grid points (2) yield a large, sparse, linear system of equations of order N of the form

$$Au = b. \tag{4}$$

The Successive Overrelaxation (SOR) iterative method, which is given by the form

$$u_{ij}^{(n+1)} = (1-\omega)u_{i,j}^{(n)} + \omega(\ell_{ij}u_{i-1,j}^{(n+1)} + r_{ij}u_{i+1,j}^{(n)} + t_{ij}u_{i,j+1}^{(n)} + b_{ij}u_{i,j-1}^{(n+1)}) \tag{5}$$

is an important solver for large linear systems [14], [15]. It is also a robust smoother as well as an efficient solver of the coarsest grid equations in the multigrid method. However, the SOR method is essentially sequential in its original form. Several parallel versions of the SOR method have been studied by coloring the grid points [1], [13].

In order to use a parallel form of the SOR method with fine grain parallelism we have to color the grid points red-black [1], [13] so that sets of points of the same color can be computed in parallel. However, the parameter ω which accelerates the rate of convergence of SOR is computed adaptively in terms of $u^{(n+1)}$ and $u^{(n)}$ [7]. This computation requires global communication between the processors for each iteration. To overcome this problem local relaxation methods are used [4], [5], [11]. In these methods each point in the grid has its own relaxation parameter ω_{ij} which is determined in terms of the local coefficients of the PDE. In [2], [4], [5], the local SOR (LSOR) with different formulas for the optimum values of the relaxation parameters was studied numerically and compared with the classic SOR method for the 5-point stencil. It was found that LSOR possesses better convergence rate than SOR using only local communication. However, the first theoretical results about the convergence of LSOR were presented in [11] under the assumption that the coefficient matrix is symmetric and positive definite. Following a similar approach but using two different sets of parameters ω_{ij} and ω'_{ij} for the red and black points, respectively, it was proved in [3] that the local Modified SOR method (LMSOR) possesses a better rate of convergence than LSOR for the 5-point stencil. This comparison was carried out in case the eigenvalues of the Jacobi matrix possesses either real (real case) or imaginary (imaginary case) eigenvalues.

The SOR method has been implemented on GPUs as applied to medical analysis [6] as well as to computational fluid dynamics [9] problems.

Our contribution is to explore the LMSOR method for the 5-point stencil exploiting the computational capabilities of GPUs under the CUDA environment. In our study we used three different techniques to find the one that best exploits the capabilities of the GPU.

The remainder of the paper is organized as follows. In section 2 we present a general description of the LMSOR method. In section 3 we present its implementation in GPUs, in section 4 we present our performance results and finally, in section 5, we state our remarks and conclusions.

2   The Local Modified SOR Method

The LSOR method was introduced by Ehrlich [4], [5] and Botta and Veldman [2] in an attempt to further increase the rate of convergence of SOR. The idea is based on letting the relaxation factor ω vary from equation to equation. This means that each equation of (2) has its own relaxation parameter denoted by ω_{ij}. Kuo et. al [11] combined LSOR with red black ordering and showed that is suitable for parallel implementation on mesh connected processor arrays. In [3] we generalized LSOR by letting two different sets of parameters $\omega_{ij}, \omega'_{ij}$ to be used for the red ($i+j$ even) and black ($i+j$ odd) points, respectively. An application of our method to (2) can be written as follows:

$$u_{ij}^{(n+1)} = (1 - \omega_{ij})u_{ij}^{(n)} + \omega_{ij}J_{ij}u_{ij}^{(n)}, \quad \text{red points} \tag{6}$$

$$u_{ij}^{(n+1)} = (1 - \omega'_{ij})u_{ij}^{(n)} + \omega'_{ij}J_{ij}u_{ij}^{(n+1)}, \quad \text{black points} \tag{7}$$

where
$$J_{ij}u_{ij}^{(n)} = l_{ij}u_{i-1,j}^{(n)} + r_{ij}u_{i+1,j}^{(n)} + t_{ij}u_{i,j+1}^{(n)} + b_{ij}u_{i,j-1}^{(n)} \tag{8}$$

and J_{ij} is called the local Jacobi operator. The parameters $\omega_{ij}, \omega'_{ij}$ are called local relaxation parameters and (6)–(8) will be referred to as the local Modified SOR (LMSOR) method. Note that if $\omega_{ij} = \omega'_{ij}$, then (6), (7) reduce to the LSOR method studied in [11]. Moreover, if $\omega_{ij} = \omega'_{ij} = \omega$ (6), (7) degenerate into the classical SOR method with red black ordering. Using Fourier analysis, Boukas and Missirlis [3] proved that the optimum values of the local relaxation parameters $\omega_{1,i,j}$ and $\omega_{2,i,j}$ for the LMSOR method in case the eigenvalues μ_{ij} of the local Jacobi operator J_{ij} are all real or all imaginary are the following.

Case 1: μ_{ij} are real. This case applies when $\ell_{ij}r_{ij} \geq 0$ and $t_{ij}b_{ij} \geq 0$. The optimum values of the LMSOR parameters are given by

$$\omega_{1,i,j} = \frac{2}{1 - \overline{\mu}_{ij}\underline{\mu}_{ij} + \sqrt{(1 - \overline{\mu}_{ij}^2)(1 - \underline{\mu}_{ij}^2)}}$$

and $\qquad\qquad\qquad\qquad\qquad\qquad\qquad\qquad\qquad\qquad\qquad$ (9)

$$\omega_{2,i,j} = \frac{2}{1 + \overline{\mu}_{ij}\underline{\mu}_{ij} + \sqrt{(1 - \overline{\mu}_{ij}^2)(1 - \underline{\mu}_{ij}^2)}}$$

where

$$\overline{\mu}_{ij} = 2\left(\sqrt{\ell_{ij}r_{ij}}\cos\pi h + \sqrt{t_{ij}b_{ij}}\cos\pi k\right) \tag{10}$$

and

$$\underline{\mu}_{ij} = 2\left(\sqrt{\ell_{ij}r_{ij}}\cos\frac{\pi(1-h)}{2} + \sqrt{t_{ij}b_{ij}}\cos\frac{\pi(1-k)}{2}\right). \tag{11}$$

Note that μ_{ij} is the spectral radius of the local Jacobi operator J_{ij} where

$$\mu_{ij} = 2\left(\sqrt{\ell_{ij}r_{ij}}\cos\frac{k_1\pi}{M_1 + 1} + \sqrt{t_{ij}b_{ij}}\cos\frac{k_2\pi}{M_2 + 1}\right), \tag{12}$$

with $k_1 = 1, 2, \ldots, M_1$, $k_2 = 1, 2, \ldots, M_2$, for periodic boundary conditions.

Case 2: μ_{ij} are imaginary. This case applies when $\ell_{ij}r_{ij} \leq 0$ and $t_{ij}b_{ij} \leq 0$. The optimum values of the LMSOR parameters are given by

$$\omega_{1,i,j} = \frac{2}{1 - \overline{\mu}_{ij}\underline{\mu}_{ij} + \sqrt{(1 + \overline{\mu}_{ij}^2)(1 + \underline{\mu}_{ij}^2)}}$$

and $\qquad\qquad\qquad\qquad\qquad\qquad\qquad\qquad\qquad\qquad\qquad$ (13)

$$\omega_{2,i,j} = \frac{2}{1 + \overline{\mu}_{ij}\underline{\mu}_{ij} + \sqrt{(1 + \overline{\mu}_{ij}^2)(1 + \underline{\mu}_{ij}^2)}}$$

where $\overline{\mu}_{ij}$ and $\underline{\mu}_{ij}$ are computed by (10) and (11), respectively.

3   Parallel Implementation

Implementations for the LMSOR method were developed for both the CPU and GPU. The CPU version is a single threaded program, used as a performance reference for our experiments. In contrast, the GPU version is a massively parallel program. The speedup that is observed between the sequential CPU program and the parallel GPU program is a sequential versus parallel program comparison, although not in the strict sense due to the GPU architectural differences.

As the solution of the Laplace equation using the red/black SOR method is memory bound [10], this problem can also be characterized as memory bound. Inspecting the computations by (6), (7) and (8) we note that, in case of good cache behavior, each element needs to access roughly 8 elements (accesses to $u_{i,j}$ count as 3, 2 reads and 1 write) per iteration (either red or black). The same computation reveals also that 11 floating point operations are needed per element. Thus, the ratio of floating point operations per accessed elements is $11/8$, and considering the use of double precision arithmetic, the ratio of floating point operations per byte accesses is $11/(8 \times 8) \approx 0.17$. This ratio is quite low as the GPUs are able to handle much more compute operations than memory access operations [8]. For instance, the NVidia GTX480 has a double precision peak performance of 168 GFlops and memory throughput of 177 GB/sec. Thus, a balanced algorithm should perform at least ≈ 1 double precision floating point operation per byte accessed. It should be noted that double precision operations were applied in all developed programs of this work.

In the implementation of the LMSOR method, the parameters ω_{ij} and ω'_{ij} are precomputed on the CPU for two of the three developed kernels.

As our program implements a red/black ordering, it is beneficial to apply reordering by color strategy into separate matrices in order to optimize performance by coalescing [10], [17]. Points in a mesh are split into two different matrices, one for the red points and one for the black. This strategy can improve bandwidth utilization by improving locality and coalescing of memory accesses and mostly, by utilizing all points contained in a memory segment, which is not possible with a natural interleaved red/black ordering.

Moreover, our program utilizes 6 matrices during the computation procedure (u, ω, l, r, t and b) as formulae (6), (7) and (8) indicate, all of which feature a red/black ordering. In contrast, the solution of the Laplace equation with R/B SOR requires accessing on a single matrix [10]. As the reordering strategy can be applied on every red-black ordered matrix it is possible to apply it on all 6 matrices. This factor raises the importance of the use of point reordering by color strategy.

In order to alleviate the high memory bandwidth requirements set by the program, an alternative approach will be used. Some read-only matrices, having their elements computed during the program initialization, can be eliminated by replacing accesses on them with computations in the GPU kernel. As the GPU has very high instruction throughput capability this trade-off can be beneficial. In summary, LMSOR was implemented in three variations as three different kernels. Each variation differs by the amount of redundant computations it performs iteratively. All kernels employ the reordering by color strategy as it is expected to be beneficial. These kernels are:

Kernel #1 - No Redundant Computations. All values required in (6), (7) and (8) reside in matrices situated in the GPU device memory. Beyond employing the reordering by color strategy, this kernel is the natural outcome implementation as no extra computations are performed. Thus, the 6 aforementioned matrices are required in this scheme and about 8 element accesses per computed element. As previously shown, the ratio of floating point operations per byte accessed is 0.17, which is particularly low.

Kernel #2 - Redundant Computations of $l_{i,j}$, $r_{i,j}$, $t_{i,j}$, $b_{i,j}$. The values of the two matrices $f_{i,j}$ and $g_{i,j}$ multiplied by h, are precomputed and stored in two matrices in the device memory. Thus, instead of 4 matrices for $l_{i,j}$, $r_{i,j}$, $t_{i,j}$ and $b_{i,j}$ we just need to keep 2 matrices only in device memory. Memory requirements are lower since only 4 matrices are required to reside in device memory (for $u_{i,j}$, $\omega_{i,j}$, $f_{i,j}$ and $g_{i,j}$). However, it comes at a cost of extra operations needed to recompute the required terms for the formula on every iteration. In this case, each element requires 6 accesses and at least $11+4 = 15$ floating point operations, as formula (3) indicates. Now, the ratio is about $15/(6 \times 8) = 0.31$ flops per byte, which is a more balanced ratio but still less than 1.0.

Kernel #3 - Redundant Computations of All Terms. In this implementation, recomputation is applied to the extreme point that all terms, excluding $u_{i,j}$, are recomputed in flight. The type of f and g functions is passed as a parameter to the kernel and all required terms are recomputed on every iteration. In this case only 3 accesses per computed element are required. The required flops are dependent on the selected $f(x,y)$ and $g(x,y)$ functions. A rough estimate is that at least $15 + 30 = 45$ flops are required plus the extra flops for the computation of $f(x,y)$ and $g(x,y)$. An approximation of the least ratio value is $45/(3 \times 8) \approx 1.9$, which clearly exceeds 1. Thus, this kernel is compute bound, as opposed to the previous kernels.

During computation only $u_{i,j}$ terms are accessed from memory and all other terms are recomputed as required. Thus, this variation has the least memory requirements of all kernels, as it requires only 1 matrix residing in device memory. The performance of the GPU version relies on the global memory cache present on Fermi GPU devices. As it has been shown [10] the global memory cache can offer the potential of high performance without the need to utilize special memory types (i.e. shared memory or texture memory). Additionally, it accomodates previously non-coalesced memory accesses with spacial locality. Therefore, the application is not expected to run efficiently on older hardware, i.e. GT-200 based GPUs. On such architectures, an alternative approach should have been chosen utilizing texture memory or shared memory of the device.

It should be noted that all implementations perform convergence checking on every iteration which raises the execution overhead. Convergence checking in the GPU kernel is implemented as a reduction of all computed maximum values. On a production environment convergence checking should be avoided, at least on most iterations, in order to attain peak performance.

The CPU version is a fairly straightforward implementation without employing any sophisticated access patterns. Elements are processed sequentially in rows and no cache blocking has been employed.

4   Performance Results

In order to test our theoretical results we considered the numerical solution of
(1) with $u = 0$ on the boundary of the unit square. The initial vector was chosen
as $u^{(0)}(x, y) = xy(1 - x)(1 - y)$. The solution of the problem above is zero. For
the purpose of comparison we considered the application of LMSOR method
with red black ordering, on CPU and GPU. In all cases the iterative process was
terminated when the criterion $||u^{(n)}||_\infty \leq 10^{-6}$ was satisfied. Various functions
for the coefficients $f(x, y)$ and $g(x, y)$ were chosen such that the eigenvalues μ_{ij} to
be either real or imaginary. The type of eigenvalues for each case is indicated by
the tuple (# real, # imaginary) in the second row of each table. The coefficients
used in each problem are:

1. $f(x, y) = Re(2x - 10)^3$, $g(x, y) = Re(2y - 10)^3$
2. $f(x, y) = Re(2x - 10)$, $g(x, y) = Re(2y - 10)$
3. $f(x, y) = g(x, y) = Re \cdot 10^4$

where the Reynold operator $Re = 10^m$, $m = 0, 1, 2, 3$ and 4.

All experiments were performed on a Linux environment. The CPU implementation was compiled with GCC version 4.4.4 on a 64bit environment, with
all essential optimization flags enabled (-O2 -fomit-frame-pointer -ftree-vectorize
-msse2 -msse -funroll-loops -fassociative-math -fno-signed-zeros -fno-trapping-
math -fno-signaling-nans). The GPU implementation was compiled using CUDA
Toolkit version 4.1 and GCC version 4.1.2 on a 64bit environment. The graphics
driver version was 295.53. The parameter "–use_fast_math" had been used.

The hardware used for the experimental runs was an AMD Opteron 6180 SE
(2.5GHz), for the CPU executions. For the GPU executions, a Nvidia GTX-480
and a Tesla C2050 [19] were used. Both GPUs are Fermi architecture based, featuring global memory cache which is essential for the performance of our kernel.

Three different series of experimental runs were performed, each investigating
the GPU and CPU implementations from a different aspect. The first series
of runs were performed in order to determine the most efficient out of the 3
developed kernels applying the LMSOR method. The second series of runs was
performed in order to compare the GPU version with the CPU version, for the
three problems, in terms of performance, on various Re values. The fluctuation
of Re values causes a varying number of required iterations to meet convergence.
The last series of runs was performed to measure the performance of the GPU
kernel and the CPU on one specific problem, on a wider range of mesh sizes
where the CPU version execution is heavily time-consuming.

In the results that follow, two different time measurements were carried out.
The first, referred as *computation time*, is the net computation time, without
extra overheads like the PCI-Express data transfer time overhead and, in case
of GPU kernels, the element reordering time overhead. The second, referred as
execution time, includes all the aforementioned overhead times. The function
used to measure time is the *gettimeofday()* function, which is available on Linux
platform.

4.1 Three Kernel Comparison

All kernels, of both methods, were executed in solving the three aforementioned problems, on mesh size $h = k = \frac{1}{\sqrt{N}+1}$ where $\sqrt{N} = M_1 = M_2 = \{402, 2002\}$. The GPU used in this experiment was the GTX480. The results of the executions are depicted on table 1. Large matrices are more important, as the GPUs are optimized for massive parallelism and therefore suited for large array processing. Thus, it is sensible to focus on the case where $\sqrt{N} = 2002$.

Table 1. Kernel comparison in LMSOR execution on GTX480, for $\sqrt{N} = \{402, 2002\}$

f,g	Experimental results	$\sqrt{N} = 402$			$\sqrt{N} = 2002$		
		#1	#2	#3	#1	#2	#3
1	(R,I)	(0,161604)			(0,4008004)		
	Iterations	412	412	412	1998	1998	1998
	Computation time (secs)	0.0561	0.0503	0.1489	3.6108	2.9294	13.7794
	Total execution time (secs)	0.0613	0.0541	0.1507	3.6636	2.9711	13.7940
	Comp. time/iteration (msecs)	0.1361	0.1220	0.3614	1.8072	1.4662	6.8966
2	(R,I)	(161604,0)			(4008004,0)		
	Iterations	554	554	554	2704	2704	2704
	Computation time (secs)	0.0752	0.0670	0.1889	4.9431	4.0572	17.4341
	Total execution time (secs)	0.0812	0.0702	0.1905	4.9960	4.0987	17.4497
	Comp. time/iteration (msecs)	0.1358	0.1209	0.3409	1.8281	1.5004	6.4475
3	(R,I)	(0,161604)			(0,4008004)		
	Iterations	1015	1015	1015	2018	2018	2018
	Computation time (secs)	0.1372	0.1223	0.3429	3.6871	3.0273	12.9215
	Total execution time (secs)	0.1418	0.1255	0.3442	3.7400	3.0668	12.9357
	Comp. time/iteration (msecs)	0.1352	0.1204	0.3378	1.8271	1.5002	6.4031

As it is obvious from the results, kernel #3 presents the worst performance. Kernel #2 seems to be the best performing of all. Although, it executes more operations per computed element, it actually performs better ($\approx 22\%$) than the first one, revealing the memory throughput bottleneck in this program. On the C2050 the improvement of kernel #2 was even more notable ($25 - 39\%$) due to its higher double precision operation throughput.

The advantage of kernel #3 is its limited memory access requirements. Kernel #1 makes use of 6 $\sqrt{N} \times \sqrt{N}$ matrices, one for each mesh. Kernel #2 makes use of 4 matrices of the same order and kernel #3 makes use of just 1 matrix of the same order. This makes it suitable for solving a large problem when memory size is a critical limitation.

Due to the different memory access requirements of the 3 kernels, inspecting the effective bandwidth can lead to misleading conclusions about the performance of each kernel. As can be seen on table 2, some profiling data were captured during one iteration of computation of red elements, for $\sqrt{N} = 4002$. Bytes accessed by kernel were extrapolated by using the first two performance counters. For kernel #1 the achieved effective bandwidth was estimated to be almost 148GB/sec, computing about 2250 elements/sec. Kernel #2 achieved calculating near 2800 elements by utilizing about 10GB/sec less bandwidth. In contrast, kernel #3 suffers by low occupancy and instruction execution pressure.

Table 2. Profiling on one iteration of red elements calculation on the GTX480, for $\sqrt{N} = 4002$

	kernel #1	kernel #2	kernel #3
fb_subp0_read_sectors	7204214	5102010	2013049
fb_subp0_write_sectors	1001454	1001464	1102356
gputime	3549.952	2871.936	12671.392
registers/thread	25	25	63
occupancy	0.667	0.667	0.333
Bytes accessed (extrapolated)	525162752	390622336	199385920
Bandwidth (GB/sec)	147.94	136.01	15.74
MegaElements/sec	2253.55	2785.58	631.34

Each kernel is characterized by different memory bandwidth requirements and thus, it cannot be used as a direct comparison measure. Thus, pure bandwidth does not expose the actual performance of these kernels.

4.2 CPU - GPU Comparison

In this series of executions the GPU kernel #2, and the CPU program were compared, for both methods, in executions for various Re values. Matrix order $\sqrt{N}$ was kept constant ($\sqrt{N} = 1002$) and the program was executed for $Re=\{1000.0, 10000.0, 100000.0\}$. Results are depicted in table 3. It is worth to note that the GPU version is constantly achieving an over $\times 50$ speed-up over the single threaded CPU version. The GPU shows a stable performance behavior by computing elements at a rate of less than half a millisecond per iteration.

Table 3. Kernel comparison in LMSOR execution on GTX480, for $\sqrt{N} = 1002$, for various values of Re, for the three problems, * indicates no convergence after 20000 iterations

f,g	Experimental results	$Re = 1000.0$		$Re = 10000.0$		$Re = 100000.0$	
		CPU	GPU	CPU	GPU	CPU	GPU
1	(R,I)	(0,1004004)		(0,1004004)		(0,1004004)	
	Iterations	2620	2620	5394	5394	6243	6243
	Computation time (secs)	56.7055	1.1146	118.1573	2.2959	133.6569	2.6551
	Total execution time (secs)	56.7055	1.1269	118.1573	2.3082	133.6569	2.6674
	Comp. time/iteration (msecs)	21.6433	0.4254	21.9053	0.4256	21.4091	0.4253
	Computation speedup	1.0000	50.8773	1.0000	51.4650	1.0000	50.3394
2	(R,I)	(0,1004004)		(0,1004004)		(0,1004004)	
	Iterations	1003	1003	1112	1112	3170	3170
	Computation time (secs)	21.0700	0.4266	23.9960	0.4728	69.4001	1.3464
	Total execution time (secs)	21.0700	0.4369	23.9960	0.4831	69.4001	1.3568
	Comp. time/iteration (msecs)	21.0070	0.4253	21.5791	0.4252	21.8928	0.4247
	Computation speedup	1.0000	49.3880	1.0000	50.7524	1.0000	51.5448
3	(R,I)	(0,1004004)		(0,1004004)		(0,1004004)	
	Iterations	5514	5514	6271	6271	7034	7034
	Computation time (secs)	118.6613	2.3459	135.8946	2.6648	154.8432	2.9918
	Total execution time (secs)	118.6613	2.3562	135.8946	2.6751	154.8432	3.0027
	Comp. time/iteration (msecs)	21.5200	0.4254	21.6703	0.4249	22.0135	0.4253
	Computation speedup	1.000	50.5827	1.0000	50.9970	1.0000	51.7567

4.3   CPU - GPU Scalability

The CPU and GPU versions were executed for a wider range of mesh sizes with $\sqrt{N} = \{402, 1002, 2002, 3002, 4002\}$, for the 2nd problem and $Re = 10.0$. The results are depicted on table 4.

The speed-up observed is further increased as $\sqrt{N}$ obtains higher values. For mesh size with $\sqrt{N} = 4002$, the speed-up exceeds $\times 110$. The GTX-480 needs just 31.16 seconds to execute 5406 iterations on that mesh which is near 150 milliseconds per iteration. This rate reaches to 2.8 Giga elements computed per second. These numbers include the time required for checking of convergence criterion.

The rate of computations of elements per second and the speed-up observed for the GPU computation times can be summarized on figure 1.

The C2050, although targeted to HPC environments it lacks the high bandwidth of the GTX480. Additionally, as the Tesla ECC protections was enabled, the memory bandwidth was further stressed roughly by 20% [18]. Thus, the performance results are lower on C2050 than on GTX-480, which does not feature ECC memories. The CPU version achieves about 25 MegaElements/sec which corresponds to $8 \times 8 \times 25 = 1600$ MB/sec bandwidth. This straightforward CPU implementation, features strided accesses (reading red or black elements) that avoid vectorization and data are used inefficiently as only half of them read in a cache line are actually used in computations.

Table 4. Various executions for the 2nd problem, (a) on CPU AMD Opteron 6180 SE, (b) on GPU NVidia GTX480 (kernel #2) and (c) on GPU NVidia Tesla C2050 (kernel #2), for mesh sizes with $\sqrt{N} = \{402, 1002, 2002, 3002, 4002\}$ and Re=10.0

Matrix $\sqrt{N} \times \sqrt{N}$	Total Iterations	(R,I)	Model	Execution time	Computation time	Mega Elements computed per second	Computation Speed-up
402 × 402	554	(161604,0)	(a)	1.54	1.54	57.55	1.00
			(b)	0.07	0.07	1329.55	23.10
			(c)	0.14	0.12	721.78	12.54
1002 × 1002	1384	(1004004,0)	(a)	28.19	28.19	49.10	1.00
			(b)	0.60	0.59	2354.79	47.96
			(c)	0.91	0.89	1555.39	31.68
2002 × 2002	2704	(4008004,0)	(a)	256.50	256.50	42.17	1.00
			(b)	4.10	4.06	2665.86	63.22
			(c)	6.02	5.91	1831.57	43.44
3002 × 3002	4055	(9012004,0)	(a)	1402.17	1402.17	26.03	1.00
			(b)	13.35	13.27	2750.96	105.69
			(c)	20.43	20.05	1820.49	84.55
4002 × 4002	5406	(16016004,0)	(a)	3473.73	3473.73	24.90	1.00
			(b)	31.30	31.16	2776.04	111.49
			(c)	46.71	46.27	1869.18	75.07

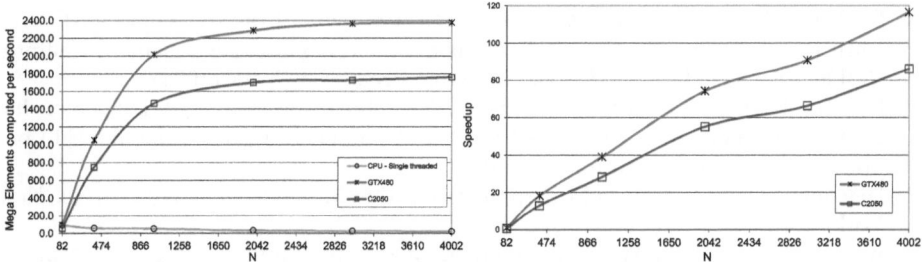

Fig. 1. Mega Elements computed per second on CPU & GPUs (left) and Computation speed-up of GPUs over CPU (right) for different matrices

5   Remarks and Conclusions

GPU is a suitable platform for massive parallel computations like those provided by the red/black ordering of iterative methods in solving systems of linear equations. In order to achieve memory coalescing, the locality of accesses must be ensured. Thereafter, the high memory bandwidth of the GPU can be exploited and attain high performance.

GPU recomputation can be beneficial in cases where memory accessing becomes a bottleneck. Instead of keeping the processing units idle, one strategy is to recompute data in order to avoid multiple memory accesses. This is a tradeoff and in many cases when a kernel is bandwidth limited, compute resources can be traded for less demand in memory bandwidth. It is applicable when a few operations at most are required for recomputation, so that computation does not turn to a bottleneck. It can provide a performance speed-up and moreover, it can release portions of device memory, allowing to solve larger problems.

Even in cases where recomputation is applied to the extreme, although performance is worsened, there can be other benefits. Recomputation leaves more available memory for other uses and thus a bigger problem is allowed to be solved. The size of the problem that is to be solved can determine the appropriate kernel to be used.

Acknowledgments. We would like to acknowledge the kind permission of the Innovative Computing Laboratory at the University of Tennessee to use their NVidia Tesla C2050 installation for the purpose of this work.

References

1. Adams, L.M., Leveque, R.J., Young, D.: Analysis of the SOR iteration for the 9-point Laplacian. SIAM J. Num. Anal. 9, 1156–1180 (1988)
2. Botta, E.F., Veldman, A.E.P.: On local relaxation methods and their application to convection-diffusion equations. J. Comput. Phys. 48, 127–149 (1981)
3. Boukas, L.A., Missirlis, N.M.: The Parallel Local Modified SOR for Nonsymmetric Linear Systems. Intern. J. Computer Math. 68, 153–174 (1998)

4. Ehrlich, L.W.: An Ad-Hoc SOR Method. J. Comput. Phys. 42, 31–45 (1981)
5. Ehrlich, L.W.: The Ad-Hoc SOR method: A local relaxation scheme, in elliptic Problem Solvers II, pp. 257–269. Academic Press, New York (1984)
6. Ha, L., Króger, J., Joshi, S., Silva, C.T.: Multiscale Unbiased Diffeomorphic Atlas Construction on Multi-GPUs. GPU Computing Gems. Emerald Edition, pp. 771–791. Morgan Kaufmann (2011)
7. Hageman, L.A., Young, D.M.: Applied Iterative Methods. Academic Press, New York (1981)
8. Kirk, D.B., Hwu, W.W.: Programming Massively Parallel Processors. Morgan Kaufmann (2009)
9. Komatsu, K., Soga, T., Egawa, R., Takizawa, H., Kobayashi, H., Takahashi, S., Sasaki, D., Nakahashi, K.: Parallel Processing of the Building-Cube Method on the GPU Platform. In: Computers & Fluids Special Issue "22nd International Conference on Parallel Computational Fluid Dynamics", vol. 45(1), pp. 122–128 (2011)
10. Konstantinidis, E., Cotronis, Y.: Accelerating the Red/Black SOR Method Using GPUs with CUDA. In: Wyrzykowski, R., Dongarra, J., Karczewski, K., Waśniewski, J. (eds.) PPAM 2011, Part I. LNCS, vol. 7203, pp. 589–598. Springer, Heidelberg (2012)
11. Kuo, C.-C.J., Levy, B.C., Musicus, B.R.: A local relaxation method for solving elliptic PDE's on mesh-connected arrays. SIAM J. Sci. Statist. Comput. 8, 530–573 (1987)
12. Nickolls, J., Buck, I., Garland, M., Skadron, K.: Scalable Parallel Programming with CUDA. In: ACM SIGGRAPH 2008 Classes, vol. 16, pp. 1–14 (2008)
13. Ortega, J.M., Voight, R.G.: Solution of Partial Differential Equations on Vector and Parallel Computers. SIAM, Philadelphia (1985)
14. Varga, R.S.: Matrix Iterative Analysis. Prentice-Hall, Englewood (1962)
15. Young, D.M.: Iterative Solution of Large Linear Systems. Academic Press, New York (1971)
16. NVidia CUDA Reference Manual v. 4.0, NVidia (2011)
17. NVidia CUDA C Best Practices Guide Version 4.0, NVidia (2011)
18. Tuning CUDA Applications for Fermi, NVidia (2011)
19. Tesla C2050 And Tesla C2070 Computing Processor Board, NVidia (2011)
20. The OpenCL Specification, Khronos group (2009)

Topic 11: Multicore and Manycore Programming

Eduard Ayguade, Dionisios Pnevmatikatos, Rudolf Eigenmann,
Mikel Luján, and Sabri Pllana

Topic Committee

Modern multicore and manycore systems enjoy the benefits of technology scaling and promise impressive performance. However, harvesting this potential is not straightforward. While multicore and manycore processors alleviate several problems that are related to single-core processors – known as memory-, power-, or instruction-level parallelism-wall – they raise the issue of the programmability and programming effort. This topic focuses on novel solutions for multicore and manycore programmability and efficient programming in the context of general-purpose systems.

The wall calls for new parallel programming methods and tools.

The quality of submissions was very high. Papers have been selected based on the recommendations of four reviewers. The seven accepted papers address a wide range of issues related to the multicore and manycore programming.

The paper "Efficient Support for In-Place Metadata in Transactional Memory" by Ricardo J. Dias, João M. S. Lourenço, Tiago Vale addresses the management of transactional memory metadata in Java virtual machines. The proposed approach extends the DeuceSTM storing the metadata information in-place achieving better scalability.

The paper "Folding of Tagged Single Assignment Values for Memory-Efficient Parallelism" by Dragos Sbirlea, Kath Knobe, Vivek Sarkar discusses how using concurrent collections the compiler can allow in place update optimization in a dynamic single assignment programming abstraction.

The paper "High-Level Support for Pipeline Parallelism on Manycore Architectures" by Siegfried Benkner, Enes Bajrovic, Erich Marth, Martin Sandrieser, Raymond Namyst, Samuel Thibault describes how a component based parallel execution model can be used to efficiently exploit pipeline parallelism in heterogenous multicore architectures.

The paper "Node.Scala: Implicit Parallel Programming for High-Performance Web Services" by Daniele Bonetta, Danilo Ansaloni, Achille Peternier, Cesare Pautasso, Walter Binder presents a improved event-driven programming framework in web services. This framework allows better scalability and thought put in multicore machines.

The paper "Task-parallel Programming on NUMA Architectures" by Christian Terboven, Dirk Schmidl, Tim Cramer, Dieter an Mey evaluates two programming strategies for OpenMP task-level parallelization on NUMA architectures, and the related programming practices.

The paper "Speeding Up OpenMP Tasking" by Spiros N. Agathos, Nikolaos D. Kallimanis, Vassilios V. Dimakopoulos describes and evaluates an OpenMP

C. Kaklamanis et al. (Eds.): Euro-Par 2012, LNCS 7484, pp. 587–588, 2012.
© Springer-Verlag Berlin Heidelberg 2012

tasking implementation with optimized queues for fast synchronization and work distribution.

The paper "An Efficient Unbounded Lock-Free Queue for Multi-Core Systems" by Marco Aldinucci, Marco Danelutto, Peter Kilpatrick, Massimiliano Meneghin, Massimo Torquati addresses the problem of single-producer/single-cinsumer proposing and evaluating two unbounded, wait-free queues.

We are grateful to all authors for submitting their high-quality papers to this topic and to reviewers for their efforts to evaluate submitted papers. Furthermore, we would like to acknowledge the encouragement and support of conference chairs Christos Kaklamanis, Theodore Papatheodorou, and Paul Spirakis.

Efficient Support for In-Place Metadata in Transactional Memory*

Ricardo J. Dias, Tiago M. Vale, and João M. Lourenço

Departamento de Informática and CITI
Universidade Nova de Lisboa, Portugal
{ricardo.dias,t.vale}@campus.fct.unl.pt,
joao.lourenco@fct.unl.pt

Abstract. Implementations of Software Transactional Memory (STM) algorithms associate metadata with the memory locations accessed during a transaction's lifetime. This metadata may be stored either in-place, by wrapping every memory cell in a container that includes the memory cell itself and the corresponding metadata; or out-place (also called external), by resorting to a mapping function that associates the memory cell address with an external table entry containing the corresponding metadata. The implementation techniques for these two approaches are very different and each STM framework is usually biased towards one of them, only allowing the efficient implementation of STM algorithms following that approach, hence inhibiting the fair comparison with STM algorithms falling into the other. In this paper we introduce a technique to implement in-place metadata that does not wrap memory cells, thus overcoming the bias by allowing STM algorithms to directly access the transactional metadata. The proposed technique is available as an extension to the DeuceSTM framework, and enables the efficient implementation of a wide range of STM algorithms and their fair (unbiased) comparison in a common STM infrastructure. We illustrate the benefits of our approach by analyzing its impact in two popular TM algorithms with two different transactional workloads, TL2 and multi-versioning, with bias to out-place and in-place respectively.

1 Introduction

Software Transactional Memory (STM) algorithms differ in the used read strategies (visible or invisible), update strategies (direct or deferred), conflict resolution policies (contention management), progress guarantees (blocking or non-blocking), consistency guarantees (opacity or snapshot isolation), and interaction with non-transactional code (weak or strong isolation), among others. Some STM frameworks (e.g., DSTM2 [7] and DeuceSTM [8]) aim at allowing the implementation

---

* This research was partially supported by the EU COST Action IC1001 (Euro-TM), the Portuguese national research projects RepComp (PTDC/EIA-EIA/108963/2008), Synergy-VM (PTDC/EIA-EIA/113613/2009), and the research grant SFRH/BD/41765/2007.

C. Kaklamanis et al. (Eds.): Euro-Par 2012, LNCS 7484, pp. 589–600, 2012.

and comparison of different STM algorithms using a unique transactional interface, and are frequently used for experimenting with new algorithms.

STM algorithms manage information per transaction (frequently referred to as a *transaction descriptor*), and per memory location (or object reference) accessed within that transaction. The transaction descriptor is typically stored in a thread-local memory space and maintains the information required to validate and commit the transaction. The per memory location information depends on the nature of the STM algorithm, which we will henceforth refer to as *metadata*, and may be composed by e.g. locks, timestamps or version lists. Metadata is stored either "near" each memory location (*in-place* strategy), or in an external mapping table that associates the metadata with the corresponding memory location (*out-place* or *external* strategy).

STM libraries targeting imperative languages, such as C, frequently use an out-place strategy, while those targeting object-oriented languages bias towards the in-place strategy. The out-place strategy is implemented by using a table-like data-structure that efficiently maps memory references to its metadata. Storing the metadata in a pre-allocated table avoids the overhead of dynamic memory allocation, but incurs in overhead for evaluating the location-metadata mapping function and has limitations imposed by the size of the table. The in-place strategy is usually implemented by using the *decorator* design pattern [6] that is used to extend the functionality of an original class by wrapping it in a *decorator* class, which also contains the required metadata. This technique allows the direct access to the object metadata without significant overhead, but is very intrusive to the application code, which must be rewritten to use the decorator classes. This *decorator* pattern based technique also incurs in two other problems: some additional overhead for non-transactional code, and multiple difficulties to cope with primitive and array types. Riegel et al. [10] briefly describe the tradeoffs of using in-place *versus* out-place strategies.

DeuceSTM is among the most efficient STM frameworks for the Java programming language and provides a well defined interface that is used to implement several STM algorithms. On the application developer's side, a memory transaction is defined by adding the annotation @Atomic to a Java method, and the framework automatically instruments the application's bytecode by injecting callbacks to the STM algorithm, intercepting the read and write memory accesses. The injected callbacks provide the referenced memory address as argument, limiting the range of viable STM algorithms to be used by forcing an out-place strategy.

This paper describes the adaptation and extension of DeuceSTM to support the in-place metadata strategy without making use of the decorator pattern. Our new approach complies to the following properties:

Efficiency. Our extension does not rely on an auxiliary mapping table, thus providing fast direct access to the transactional metadata; transactional code avoids the extra memory dereference imposed by the decorator pattern; no performance overhead is introduced for non-transactional code, as it is oblivious to the presence of metadata in objects; primitive types are fully supported, even in transactional code; and we propose a solution for supporting transactional N-dimensional arrays with a negligible overhead for non-transactional code.

Flexibility. Our extension supports both the original out-place and the new in-place strategies simultaneously, hence imposing no restrictions on the nature of the algorithms and their implementations.

Transparency. Our extension automatically identifies, creates and initializes all the necessary additional metadata fields in objects; non-transactional code is oblivious to the presence of metadata in objects, hence no source code changes are required, although it does some light transformation on the non-transactional bytecode; the new transactional array types (that support metadata for individual cells) are compatible with the standard arrays, hence not requiring pre-/post-processing of the arrays when invoking standard or third-party non-transactional libraries.

Compatibility. Our extension is fully backwards compatible and the already existing implementations of STM algorithms are executed with no changes and with null or negligible performance overhead.

In the remainder of this paper, we describe the DeuceSTM framework and the usage of out-place strategy in §2. In §3 we describe the properties of in-place strategy, and its implementation as an extension to DeuceSTM. We evaluate our implementation with some benchmarks in §4, and discuss the related work in §5. We finish with some concluding remarks in §6.

2   DeuceSTM and the Out-Place Strategy

Algorithms such as TL2 [4] or LSA [11] use an out-place strategy by resorting to a very fast hashing function and storing a single lock in each table entry. However, due to performance issues, the mapping table does not avoid hash collisions and thus two memory locations may be mapped to the same table entry, resulting in the false sharing of a lock for two different memory locations.

The out-place strategy fits well to algorithms whose metadata information does not depend on the memory locations, such as locks and timestamps, but is unfitting for algorithms that need to store location-dependent metadata information, e.g., multi-version based algorithms. The out-place implementations for these algorithms require a mapping table with collision lists, which impose a significant and unacceptable performance overhead.

DeuceSTM provides the STM algorithms with a unique identifier for an object field, composed by a reference to the object and the field's logical offset within that object. This unique identifier can then be used by the STM algorithms as a key to any map implementation that associate the object fields with the transactional metadata. Likewise for array types, the unique identifier of an array's cell is composed by the array reference and the index of that cell. It is worthwhile to mention that DeuceSTM relies heavily on bytecode instrumentation to provide a transparent transactional interface to application developers, which are not aware of how the STM algorithms are implemented nor of the strategy being used to store the transactional metadata.

DeuceSTM is an extensible STM framework that may be used to compare different STM algorithm implementations. However, it is not fair to compare an algorithm that fits very well to the out-place strategy with another algorithm that does not. In the concrete case of DeuceSTM, the framework only supports an out-place strategy, therefore being inappropriate for e.g. multi-version oriented STM algorithms. We have extended DeuceSTM to, in addition to the out-place strategy, also support an efficient in-place strategy, while keeping the same transparent transactional interface to the applications.

3   Support for In-Place Strategy

The unique identifier of an object's field is composed by the object reference and the field's logical offset. DeuceSTM computes that logical offset at compile time, and for every field f in every class C an extra static field f^o is added to that class, whose value represents the logical offset of f in class C. No extra fields are added for array cells, as the logical offset of each cell corresponds to its index. When there is a read or write memory access (within a memory transaction) to a field f of an object O, or to the array element $A[i]$, the run-time passes the pair (O, f^o) or (A, i) respectively as the argument to the callback function. The STM algorithm shall not differentiate between field and array accesses. In DeuceSTM, if the algorithm needs to e.g. associate a lock with a field, it has to store the lock in an external table indexed by the hash value of the pair (O, f^o).

In our approach for extending DeuceSTM to support an in-place strategy, we replace the previous pair of arguments to callback functions (O, f^o) with a new metadata object f^m, whose class is specified by the STM algorithm's programmer. We guarantee that there is a unique metadata object f^m for each field f of each object O, and hence the use of f^m to identify an object's field is equivalent to the pair (O, f^o). The same applies to arrays where we ensure that there is a unique metadata object a^m for each position of an array A.

3.1   Implementation

Although the implementation of the support for in-place metadata objects differs considerably for class fields and array elements, a common interface is used to interact with the STM algorithm implementation. This common interface is supported by a well defined hierarchy of metadata classes, illustrated in Figure 1, where the rounded rectangle classes are defined by the STM algorithm developer.

All metadata classes associated with class fields extend directly from the top class TxField. For array elements, we created specialized metadata classes for each primitive type in Java, the TxArr*Field classes, where * ranges over the Java primitive types[1]. All the TxArr*Field classes extend from TxField, providing the STM algorithm with a simple and uniform interface for callback functions, which shall be extended to include the support of new STM algorithms.

---

[1] **int, long, float, double, short, char, byte, boolean**, and Object.

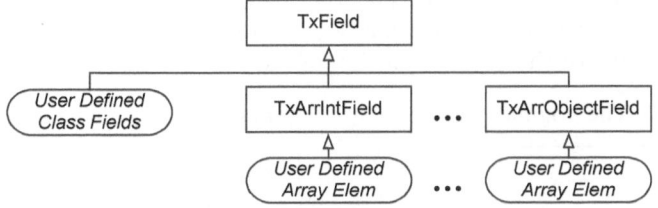

Fig. 1. Metadata classes hierarchy

The newly defined metadata classes need to be registered in our framework to enable its use by the instrumentation process, using a Java annotation in the class that implements the STM algorithm, as exemplified in Listing 1.1.

Listing 1.1. Declaration of the STM algorithm specific metadata

```
@InPlaceMetadata(
    fieldObjectClass="TL2Field",
    fieldIntClass="TL2Field",
    ...
    arrayObjectClass="TL2ArrObjectField",
    arrayIntClass="TL2ArrIntField",
    ...
)
final public class TL2Context implements ContextMetadata {
    ...
}
```

The STM algorithm must implement a well defined interface that includes a callback function for the read and write operations on each Java type. These functions always receive an instance of the super class TxField, but each one knows precisely which metadata subclass was actually used to instantiate the metadata object.

Lets now see where and how the metadata objects are stored, and how they are used on invocation of the callback functions. We will explain separately the management of metadata objects for class fields and for array elements.

Class Fields. During the execution of a transaction, there must be a metadata object f^m for each accessed field f of object O. A very efficient way to implement this metadata object f^m is by making it accessible by a single dereference operation from object O. Therefore, for each declared field in a class C, we add an additional metadata field of the appropriate type. The general rule can be described as: given a class C that has a set of declared fields $\{f_1, \ldots, f_n\}$, we add a metadata object field for each of the initial fields, such that the class ends with the set of fields $\{f_1, \ldots, f_n, f_{1+n}^m, \ldots, f_{n+n}^m\}$ where the field f_k is associated with the metadata field f_{k+n}^m for any $k \leq n$. In Listings 1.2 and 1.3 we show a concrete example of the transformation of a class with two fields.

Listing 1.2. The original class

```
class C {
    int a;
    Object b;
}
```

$\Longrightarrow$

Listing 1.3. The transformed class

```
class C {
    int a;
    Object b;
    TxField a_metadata;
    TxField b_metadata;
}
```

Each metadata field is instantiated at the constructor of the class where the field is declared. This ensures that whenever a new instance of a class is created, the corresponding metadata objects are also new and unique.

Opposed to the approach based in the *decorator* pattern, where primitive types must be replaced with their object equivalents (e.g., an **int** field is replaced by an `Integer` object), our transformation approach keeps the primitive type fields untouched, simplifying the interaction with non-transactional code, limiting the code instrumentation and avoiding autoboxing and its overhead.

Array Elements. The structure of an array is very strict, with each array cell containing a single value of a well defined type, and no other information can be added to those elements. The common approach to overcome this limitation is to change the array to an array of objects that wrap the original value and the additional information. This transformation has strong implications in the remaining of the application code, as code statements expecting the original array type or array element will now have to be rewritten to receive the new array type or wrapping class respectively. This problem is even more complex if the arrays with wrapped elements were to be manipulated by non-instrumented libraries, such as the JDK libraries.

The solution we propose is also based on changing the type of the array to be manipulated by the instrumented application code, but strongly limiting the implications for the remaining non-instrumented code. We keep all the values in the original array, and have a sibling second array, only manipulated by the instrumented code, that contains the additional information and references to the original array. The type of the declaration of the base array is changed to the type of the corresponding sibling array (`TxArr*Field`), as shown in Figure 2. This Figure also illustrates the general structure of the sibling `TxArr*Field` arrays (in this case, a `TxArrIntField` array). Each cell of the sibling array has the metadata information required by the STM algorithm, its own position/index in the array, and a reference to the original array where the data is stored (i.e., where the reads and updates take place). This scheme allows the sibling array to keep a metadata object for each element of the original array, while maintaining the original array always updated and compatible with non-transactional legacy code.

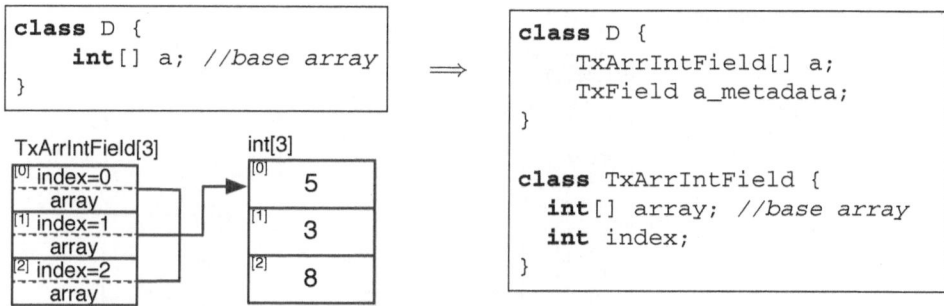

Fig. 2. Memory structure of a `TxArrIntField` array

Non-transactional methods that have arrays as parameters are also instrumented to replace the array type by the corresponding sibling `TxArr*Field`. The value of an array element is then obtained by dereferencing the pointer to the original array kept in the sibling, as illustrated in Listings 1.4 and 1.5. When passing an array as argument to an uninstrumented method (e.g., from the JDK library), we can just pass the original array instance. Although the instrumentation of non-transactional code adds a dereference operation when accessing an array, we do avoid the autoboxing of primitive types that would impose an increased overhead.

Listing 1.4. Access to an array cell

Listing 1.5. Access to an array cell from the transformed array

```
void foo(int[] a) {
    // ...
    t = a[i];
}
```

$\implies$

```
void foo(TxArrIntField[] a) {
    // ...
    t = a[0].array[i];
}
```

Multi-dimensional arrays. The special case of multi-dimensional arrays is tackled using the `TxArrObjectField` class, which has a different implementation from the other specialized metadata array classes. This class has the additional field `nextDim`, which may be null in the case of a uni-dimensional reference type array, or may hold the reference of the next array dimension by pointing to another array of type `TxArr*Field`. Once again, the original multi-dimensional array is always up to date and can be safely used by non-transactional code.

Figure 3 depicts the memory structure of a bi-dimensional array of integers. Each element of the first dimension of the sibling array has a reference to the original integer matrix. The elements of the second dimension of the sibling array have a reference to the second dimension of the matrix array.

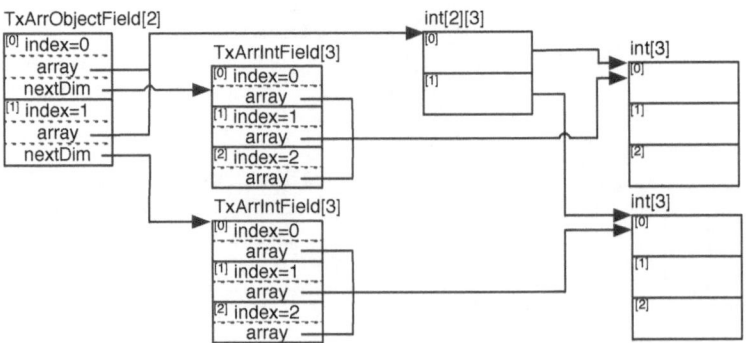

Fig. 3. Memory structure of a multi-dimensional `TxArrIntField` array

4   Performance Evaluation

We evaluated our approach in two dimensions: the performance overhead resulting from the introduction of metadata associated with object fields, and the performance improvements achieved by implementing a multi-versioning STM algorithm (JVSTM [3]) using our extension (with in-place metadata), when compared to an equivalent implementation in the original DeuceSTM (with out-place metadata). To measure the transactional throughput we used the vanilla micro-benchmarks available in the DeuceSTM framework. No changes were necessary to execute the benchmarks on our extension of DeuceSTM with in-place metadata, as all the necessary bytecode transformations were performed automatically.

The benchmarks were executed in a computer with four AMD Opteron 6168 12-Core processors @ 1.9 GHz with 12x512 KB of L2 cache and 128 GB of RAM, running Red Hat Enterprise Linux Server Release 6.2 with Linux 2.6.32 x86_64.

To evaluate the overhead of our extension, we compared the performance of the TL2 algorithm as provided by the original DeuceSTM distribution, with another implementation of TL2 using the new interface of our modified DeuceSTM. The original DeuceSTM interface for callback functions provide a pair with the object reference and the field logical offset. The new interface provides a reference to the field metadata (`TxField`) object. Despite using the in-place metadata feature, the new implementation of TL2 resembles the original one as much as possible and still uses an external table to map memory references to locks. By comparing these two similar implementations, we can measure the overhead introduced by the management of the metadata object fields and sibling arrays.

Figure 4 depicts the overhead of our extension with respect to the original DeuceSTM for two data structures: a Red-Black Tree and a Skip List. The former only uses metadata objects for class fields, while the latter also make use of metadata arrays. We executed each data structure with two different workloads: a *read-only* workload, and a *read-write* workload with an average of 10% of update operations. The overhead is in percentage and is relative to the out-place implementation of TL2 in the original DeuceSTM.

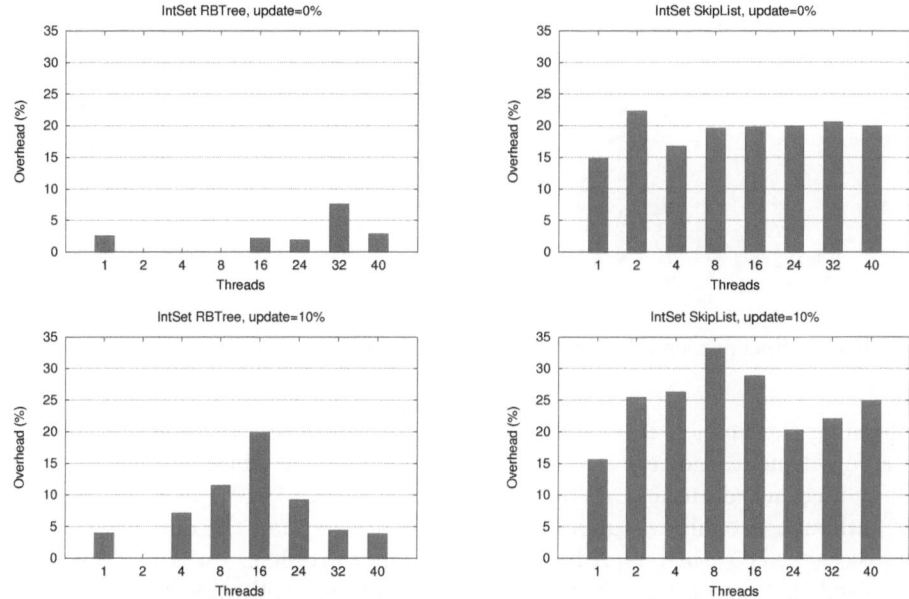

Fig. 4. Overhead measure of the usage of metadata objects relative to out-place TL2

In the Red-Black tree benchmark, the use of metadata objects in class fields in a read-only workload (top-left chart) has a negligible overhead. In a read-write workload (bottom-left chart) there is an average overhead of 10% with respect to the out-place version. This overhead results from the additional allocations necessary to initialize metadata objects. For instance, when adding a new node to the tree, we need to allocate additional metadata objects for the value, color, left, right and parent fields. In the case of the Skip List benchmark, each node contains an array of nodes. In the read-only workload (top-right chart), there is an average overhead of 20% with respect to the out-place strategy that uses the original arrays declared in the program. Although no new nodes are allocated, there is a performance penalty to pay for the additional dereference imposed by the support of in-place metadata for arrays. In the read-write workload (bottom-right chart), which allocates new nodes, we get a slightly higher overhead averaging 25%.

From this analysis we conclude that our in-place strategy is a viable option for implementing algorithms biased to in-place transactional metadata. To stress this fact, we implemented two versions of the JVSTM algorithm as proposed in [3], one in the original DeuceSTM framework using the out-place strategy (JVSTM-Out), and another in the extended DeuceSTM using the in-place strategy (JVSTM-In). The JVSTM-Out implementation uses an open concurrent hash table to map each accessed memory location to its list of versions. Hence, for each memory location accessed, we perform a search in the hash table to find the respective version list. If none is found, a new one is created and added to the hash map. The JVSTM-In implementation uses metadata objects

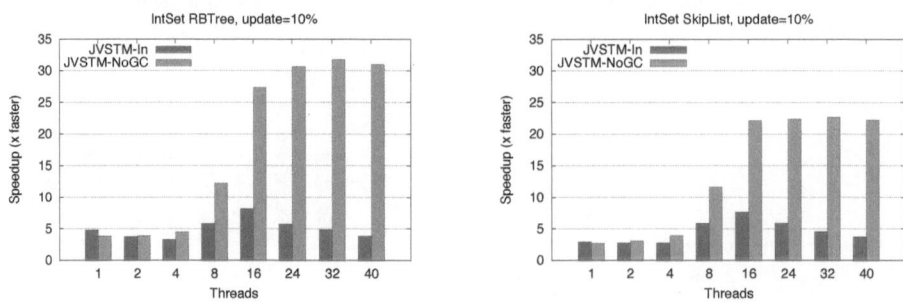

Fig. 5. Speedup of JVSTM-In and JVSTM-NoGC relative to JVSTM-Out

containing the version lists, thus when there is an access to an object field or an array element, a direct reference to its version list is obtained from the corresponding metadata field. JVSTM-In algorithm follows the specification in [3], and although it exhibits a much better performance than the out-place version (in average it is 5× faster), it has some scalability problems due to: i) the global lock used in the commit phase, and ii) the garbage collection mechanism for *vboxes* (used by JVSTM to wrap the object and its list of versions). Hence we implemented an optimized variant of the JVSTM-In (JVSTM-NoGC), which replaces the global lock in the commit phase with separate locks for each memory locations in the transaction write-set, and which eliminates the *vbox* garbage collection mechanism by imposing an upper bound on the size for the list of versions for any memory location. In this new algorithm, transactions accessing an old version that is not available anymore are aborted and restarted.

Figure 5 depicts the speedup results when comparing the JVSTM-In and JVSTM-NoGC implementations with respect to the JVSTM-Out implementation. JVSTM-In is in average 5× faster than JVSTM-Out on both benchmarks. JVSTM-NoGC is in average 33× faster in the Red-Black Tree, and 23× faster in the Skip List. These results prove that our strategy to support in-place metadata in DeuceSTM gave it leverage to implement algorithms that need direct access to transactional metadata, thus enabling the fair comparison of a wide range of STM algorithms, including those that could not be implemented efficiently in the original DeuceSTM.

5   Related Work

Several STM algorithms were developed in the last few years, and comparing their performance always requires a great implementation effort while using the same transactional interface and programming language. Some STM frameworks address this problem and provide a uniform transactional interface front-end and a flexible run-time back-end, which is normally biased towards one of the in-place or out-place strategies.

DSTM2 [7] is a flexible STM framework for the Java language which permits the use of different synchronization techniques and recovery strategies as framework plug-ins. DSTM2 creates transactional objects using the factory pattern, and new factories can be implemented to test different properties of STM algorithms. DSTM2 only allows to implement algorithms using the in-place strategy.

DeuceSTM [8], which is the base of our work, is one of the most efficient STM frameworks available for the Java programming language. It provides a well defined interface that allows to implement several STM algorithms, and relies in Java bytecode instrumentation to intercept transaction limits and transactional memory accesses and invoke developer-defined callback functions. DeuceSTM has a strong bias towards the out-place approach.

STM algorithms such as TL2 [4], LSA [11] and SwissTM [5] are usually implemented using an out-place strategy, thus viable for use in DeuceSTM. Others such as JVSTM [3] and SMV [9] are better fit for the in-place strategy and impracticable for DeuceSTM. Our extension of DeuceSTM overcomes this limitation and allows the efficient implementation of algorithms using any of those strategies, enabling their fair comparison.

Anjo et al. [2] developed a specialized transactional array targeting specifically the JVSTM framework, achieving considerable performance improvements in read-dominant workloads that use arrays. Our approach when extending DeuceSTM aimed at providing an efficient implementation for transactional arrays that is backwards compatible, where no autoboxing is required and whose values are kept in their original primitive format and are accessible to both transactional and non-transactional code.

All the static optimizations proposed by Afek et al. [1] are orthogonal to our work and can also be applied to algorithms using the new in-place strategy, thus increasing the overall performance.

6   Concluding Remarks

To the best of our knowledge, the extension of DeuceSTM as described in this paper creates the first Java STM framework providing a balanced support of both in-place and out-place strategies. This is achieved by a transformation process of the program bytecode that adds new metadata objects for each class field, and that includes a customized solution for N-dimensional arrays that is fully backwards compatible with primitive type arrays. The creation or structural modification of arrays are not supported outside instrumented code, which is oblivious to TxArr*Field and metadata.

In the current state of the proposed extension every field is subjected to this transformation, hence there will be a considerable increase in the application's memory footprint. For example, for the Red-Black Tree benchmark with 50 000 elements in a read-only workload, the GNU time command reported 196 MB of memory used when using the out-place version of the TL2 algorithm, and 248 MB when using the in-place version. This memory overhead can be minimized by doing code analysis to discover the fields that are not accessed within transactions, and skipping the creation and initialization of the metadata associated with those fields, which will never be needed.

We evaluated our system by measuring the overhead introduced by our new in-place interface with respect to the TL2 algorithm implementation provided in DeuceSTM distribution package as reference. Although we can observe a light slowdown in our new implementation of arrays, we would like to reinforce that our solution has no limitations whatsoever concerning the type of the array elements, the number of dimensions, fits equally algorithms biased towards in-place or out-place strategies, and all bytecode transformations are done automatically requiring no changes to the source code. We also evaluated the effectiveness of the new in-place interface by comparing the performance of two multi-version STM implementations: one using the newly proposed in-place strategy, and another using an out-place strategy resorting to an external mapping table. The version using the new in-place strategy was in average 5× faster than the one using the out-place strategy. The optimized version of the JVSTM algorithm using the in-place strategy was in average 33× faster than the out-place version.

References

1. Afek, Y., Korland, G., Zilberstein, A.: Lowering STM Overhead with Static Analysis. In: Cooper, K., Mellor-Crummey, J., Sarkar, V. (eds.) LCPC 2010. LNCS, vol. 6548, pp. 31–45. Springer, Heidelberg (2011)
2. Anjo, I., Cachopo, J.: Lightweight Transactional Arrays for Read-Dominated Workloads. In: Xiang, Y., Cuzzocrea, A., Hobbs, M., Zhou, W. (eds.) ICA3PP 2011, Part II. LNCS, vol. 7017, pp. 1–13. Springer, Heidelberg (2011)
3. Cachopo, J., Rito-Silva, A.: Versioned boxes as the basis for memory transactions. Sci. Comput. Program. 63, 172–185 (2006)
4. Dice, D., Shalev, O., Shavit, N.N.: Transactional Locking II. In: Dolev, S. (ed.) DISC 2006. LNCS, vol. 4167, pp. 194–208. Springer, Heidelberg (2006)
5. Dragojević, A., Guerraoui, R., Kapalka, M.: Stretching transactional memory. In: Proc. Int. Conf. on Programming Language Design and Implementation, pp. 155–165. ACM (2009)
6. Gamma, E., Helm, R., Johnson, R., Vlissides, J.: Design Patterns: Elements of Reusable Object-Oriented Software. Addison-Wesley Professional (1994)
7. Herlihy, M., Luchangco, V., Moir, M.: A flexible framework for implementing software transactional memory. In: Proc. 21st Conference on Object-Oriented Programming Systems, Languages, and Applications, pp. 253–262. ACM (2006)
8. Korland, G., Shavit, N., Felber, P.: Noninvasive concurrency with Java STM. In: Proc. MultiProg 2010: Programmability Issues for Heterogeneous Multicores (2010)
9. Perelman, D., Byshevsky, A., Litmanovich, O., Keidar, I.: SMV: Selective Multi-Versioning STM. In: Peleg, D. (ed.) DISC 2011. LNCS, vol. 6950, pp. 125–140. Springer, Heidelberg (2011)
10. Riegel, T., Brum, D.B.D.: Making object-based STM practical in unmanaged environments. In: Proc. of the 3rd Workshop on Transactional Computing (2008)
11. Riegel, T., Felber, P., Fetzer, C.: A Lazy Snapshot Algorithm with Eager Validation. In: Dolev, S. (ed.) DISC 2006. LNCS, vol. 4167, pp. 284–298. Springer, Heidelberg (2006)

Folding of Tagged Single Assignment Values for Memory-Efficient Parallelism

Dragoş Sbîrlea[1], Kathleen Knobe[2], and Vivek Sarkar[1]

[1] Department of Computer Science, Rice University
{dragos,vsarkar}@rice.edu
[2] Intel Corporation
kath.knobe@intel.com

Abstract. The dynamic-single-assignment property for shared data accesses can establish data race freedom and determinism in parallel programs. However, memory management is a well known challenge in making dynamic-single-assignment practical, especially when objects can be accessed through tags that can be computed by any step.

In this paper, we propose a new memory management approach based on user-specified *folding functions* that map logical dynamic-single - assignment (DSA) tags into dynamic-multiple-assignment (DMA) tags. We also compare folding with *get-counts*, an approach in which the user specifies a reference count for each single-assignment value. The context for our work is parallel programming models in which shared data accesses are coordinated by put/get operations on tagged DSA data structures. These models include dataflow programs with I-structures, functional subsets of parallel programs based on tuple spaces (notably, Linda), and programs written in the Concurrent Collections (CnC) coordination language. Our conclusion, based on experimental evaluation of five CnC programs, is that folding and get-counts can offer significant memory efficiency improvements, and that folding can handle cases that the get-counts cannot.

1 Introduction

The multicore revolution has increased the urgency for developing programming models that deliver scalable parallelism with minimal effort by programmers. The use of shared data structures by parallel tasks has proved to be a two-edged sword in pursuing this goal. On the one hand, a shared address space can reduce the semantic gap between a sequential program and its parallel version. On the other, uncoordinated accesses to shared data structures are a notorious source of bugs that arise from data races and other sources of nondeterminism leading to the *programmability wall*.

One approach to addressing the drawbacks of shared data structures is to enforce a *dynamic-single-assignment property* for shared data accesses, since it in turn can establish data race freedom and determinism in parallel programs. Thus, the context for our work is parallel programming models for multicore and many-core processors in which all shared data accesses are performed through put/get

C. Kaklamanis et al. (Eds.): Euro-Par 2012, LNCS 7484, pp. 601–613, 2012.

operations on dynamic-single-assignment data structures indexed using associative tags (keys). These models include dataflow programs with I-structures [1], functional subsets of parallel programs based on tuple spaces (notably, Linda [7]), and programs written in the Concurrent Collections (CnC) coordination language [3].

However, past experiences with implementations of functional languages have shown that memory management can be challenging with the dynamic-single-assignment property. It becomes even more challenging when objects can be accessed through user-computable tags, since standard reference-based garbage collection cannot be applied in that case. In this paper, we propose a new memory management approach based on user-specified *folding functions* that map logical dynamic-single-assignment (DSA) tags into dynamic-multiple-assignment (DMA) tags. We also compare folding with *get-counts*, an approach in which the user supplies a function that maps tags to integers indicating the number of gets that will occur on the item Both approaches are *fail-safe* i.e., an exception is thrown if the program performs accesses that are inconsistent with the folding functions or get-counts.

There has been a lot of past work focused on converting a multiple-assignment program to dynamic single assignment form so as to simplify program optimization and transformation. An early paper [6] described several applications of dynamic single assignment, such as conversion of a program to a set of recurrence equations, scalar expansion, array expansion [5], program verification and parallel program construction. In contrast, folding addresses the dual problem of converting a dynamic single assignment program to multiple-assignment form with reduced memory requirements. Based on the well known challenges in transforming static single assignment form to multiple assignment form [2], it is natural to expect that translating out of dynamic single assignment form will be a challenging problem too, especially when the original non-DSA program is unavailable. To the best of our knowledge, this paper is the first to propose a user-specified "folding" approach to address this problem.

In summary, this paper includes the following contributions:

- *Basic folding* (Section 2.1), a novel memory management technique for accesses to associative dynamic-single-assignment data structures (item collections). This technique relies on user-specified folding functions with fail-safe checks for correctness at runtime.
- *Update-in-place memory reuse* (Section 2.2), an extension that allows the user to specify *GetForUpdate* operations that allow an input item to be rewritten as an output. This approach can be used both with folding functions and get-counts, and includes fail-safe checks as well.
- *Extended folding* (Section 2.5), an extension to basic folding for items that are written but never read.
- A *design and implementation* (Section 3) of the above folding and get-count techniques for the CnC model.
- *Empirical results* (Section 4) that show that folding and get-counts can offer significant improvements in memory efficiency over the baseline version without these techniques.

2   Folding of Dynamic Single Assignment Values

2.1   Basic Folding

The intuition behind folding is as follows: if we know that two values have non-overlapping lifetimes, we can assign them to the same physical storage thereby reducing the maximum memory requirement for the application. Following the terminology used in the CnC model, we refer to the associative dynamic-single-assignment (DSA) data structures assumed in this work as *item collections*, to keys as *tags*, values as *items*, and computational tasks as *steps*. The two operations supported by item collections are *put(tag, item)* and *get(tag)*. The DSA property requires that dynamically at most one *put()* operation be performed for a given tag. Further, each *get()* operation is assumed to be blocking i.e., it only returns a value after a *put()* operation has been performed with that tag.

Definition 1 (Folding function). *A folding function f transforms a logical tag t_1 to a physical tag, $f(t_1)$. Thus, the logical* put(t_1, i_1) *operation is transformed into a physical* put(t_1, f(t_1), i_1) *operation, where $f(t_1)$ is the physical location used to store the item and the original tag t_1 is stored as an auxiliary value. Likewise, the logical* get(t_1) *operation is transformed into a physical* get(t_1, f(t_1)) *operation.*

Thus, the folding function maps DSA tags to dynamic multiple assignment (DMA) tags which are associative indices into a physical store. When a new item i_2 is mapped to the same physical store location as a previous item i_1 (because $f(t_1) = f(t_2)$), the space of i_1 is freed. Example executions of a program that computes the n-th Fibonacci element are in Figures 1 and 2 (without and with folding, respectively). Item n can fold over item $n-2$. The folding function used is: $fold(n) = (n+1)\%2 + 1$.

This use of a folding function is called basic folding. As discussed later in Section 2.3, a runtime error may be thrown if the folding function is specified incorrectly, but a *get()* operation will never return an incorrect logical value.

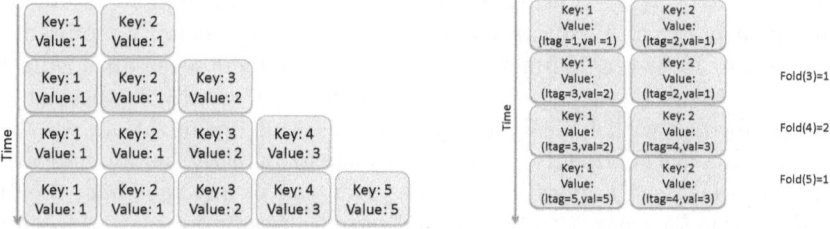

Fig. 1. Item collection content for a baseline execution of Fibonacci

Fig. 2. Item collection content for a folding execution of Fibonacci

We now identify the conditions under which folding is legal. As an example, consider the following sequence of logical *get()* and *put()* operations: "*put(t₁, i₁)*;"... *get(t₁); put(t₂, i₂); get(t₁)*". In this case, it would be illegal to fold items i_1 and i_2 on the same location because they have interfering live ranges [11]. To ensure safety for folding two items, they must have disjoint lifetimes in any possible schedule of the program.

Definition 2 (Item lifetime). *The lifetime of an item in a program execution is the interval between the execution point at which the item is produced by a put() operation and the execution point of the last get() operation performed on the item. If there are no get() operations, the lifetime begins and ends at the put().*

Definition 3 (Legal program). *A legal program is one that always completes execution with all get() operations having successfully completed, for all possible schedules.*

Definition 4 (Correct folding transformation). *A folding transformation for a legal program P specified by folding function f is correct if, for every input I, an execution of P with input I and folding function f is also legal (no blocked gets()) and results in the same result for each get() operation as the original execution of program P without folding.*

Theorem 1 (Folding correctness requirement). *For a folding transformation of a legal program to be correct, the folding function must not fold together any two items whose lifetimes may overlap.* [Proof omitted due to space limitations.]

2.2   Folding with Update-in-Place Memory Reuse

Basic folding ensures that memory can be reclaimed after the end of a computational step that performs the last logical *get()* operation on an item. However, many steps have the following computational structure: "$i_1 = get(t_1)$; allocate i_2; $i_2.set(G(i_1))$; put(t_2, i_2);". With basic folding, both i_1 and i_2 will be assumed to be simultaneously live and will contribute to the maximum memory requirement for the program. However, if function G can be implemented as an *update-in-place* function, then i_1's storage can be reused for i_2 if get(t_1) is the last get operation performed with logical tag t_1. To enable this optimization, we allow the user to use a *getForUpdate()* operation instead of *get()*, as an indication that this is the last *get()* operation for the given tag in any schedule, thereby making it possible for item i_1 to be updated in place to obtain item i_2. Figure 3 is an example. As with the folding function, the correctness of a *getForUpdate()* operation will also be checked at runtime so as to guarantee fail-safe behavior (see Section 2.3).

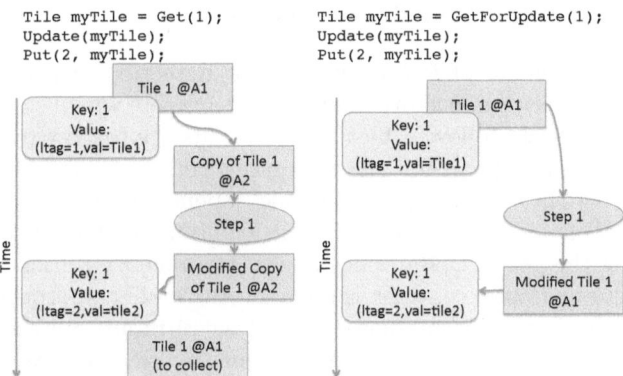

Fig. 3. Left: With a get() call, the item memory is copied before being returned to the step, which can modify it and put() it with some other tag. This leaves the old item memory to be collected when an item folds over its entry in the store. Right: With getForUpdate, the copy is not performed and no memory will need to be collected, as it is reused by the new item.

2.3   Error Detection

The folding error detection mechanisms are based on the assumption that the original program is legal (Definition 3) without the folding optimization. We define any behavior of a legal program in the presence of folding that differs from the behavior of a non-folded execution as an error.

For example, a *get()* that returns an incorrect value would constitute an error. This could happen, if the content of the physical store location corresponding to a particular tag is returned without checking that the logical tag of the item in that location corresponds to the logical tag of the item we are trying to get. If the item in the store does not have the same logical tag, we need to wait for it to be produced. However, if the item was previously produced and some other item was erroneously folded over it, we will never find the item. Without an error checking mechanism, the program may finish with blocked steps instead of the correct non-folded behavior.

To enable detection of such errors, we define a debug mode for folding, in which a boolean flag is stored for each tag that is *put()* during execution. Using this flag, we can differentiate between items that are not present in the physical store because some other item was folded over them and items that have not been produced as yet. A *get()* performed on previously overwritten items should throw an exception reporting an incorrect folding function, but a *get()* should block until the item is produced if that is not the case. In debug mode the system also detects dynamic single assignment violations (on every put, if the boolean flag for that logical tag was previously *put()*, we report an exception) with or without the presence of folding.

2.4   Programmability Benefits of Folding

To illustrate the benefits of folding with error detection, consider a common technique used by performance-oriented C programmers where storage is reused instead of calling free() followed by malloc(). This approach can be especially error-prone for parallel programs, because the overlap in lifetime between the initial and subsequent values may be schedule-dependent. With folding, a similar reuse of memory could be achieved in a fail-safe manner by folding the two logical items and using the getForUpdate mechanism for memory reuse.

As a concrete example, consider the classic two-buffer approach used by iterative algorithms in which one buffer is used as an input and the other as the output, and their roles are swapped in each sequential iteration. With our folding approach, the programmer can think in terms of allocating a new DSA output buffer in each iteration, and a folding function can effectively perform the swap. This approach was used in our implementation of a Routing simulation application (see Section 4) where the routing tables for one iteration are built using the routing tables of the previous one, and a folding function was specified as follows:

```
public final Object fold(point tag) {
    int i, j, k; //i: node id, j: iteration id; k: repetition #
    i = p.get(0); j = p.get(1); k = p.get(2);
    return new point(i, j%2, k);
}
```

2.5   Extended Folding: Folding with Ordering

Items with empty lifetimes pose an interesting research challenge for folding. Consider a program that expects to produce and consume items in order as follows: "*Step1: [put(t_1, i_1)] Step2:[get(t_1);put(t_2, i_2)] Step3:[get(t_2); put(t_3, i_3)] Step4:[get(t_3)]* ". In such a case, it might seem reasonable to fold t_1, t_2, and t_3 to the same physical location. However, if (say) get(t_2) is not performed for some reason, there is no way (if using blocking-get synchronization only) to ensure that put(t_2, i_2) completes before put(t_3, i_3), thereby making the folding incorrect (because get(t_3) may never find t_3 as it has been folded over).

This is an instance of the more general problem caused by optional *get()* calls but in this particular case there is a way to solve the problem. We propose an extension to folding that allows folding of items that may never be consumed. Such items can appear when control dependent gets are used, for example with short-circuit boolean operations such as "get(t_1) && get(t_2)". We observe that items that are never read have an empty lifetime and can be optimized away from the physical store. However, this may not be known at the time of the *put()* operation, but may be known when a subsequent *put()* is performed on the same physical location.

We can express this by allowing the presence of an additional user function that acts like a "compare age" operation. If an item that is being put maps to a physical location where another item resides and should be declared dead, the function returns *true* ("newer"), and the new item is stored. Otherwise, if the new item is known to never be read, it returns returns *false* ("older"), the incoming item is not stored and the old item is retained.

To perform the age comparison, the function needs two parameters: the tag of the item being put currently and the tag of the old item that exists in the location in the physical store where the new item would be inserted. The programmer has to identify if the tag of the current item in the item collection means that all of the steps that could access the incoming item have executed and did not access the incoming item. If this is the case, then the incoming item can safely be discarded. The Rician Denoising benchmark (see Section 4) uses this extension.

3   Implementation

We have implemented folding as an extension to the Habanero Java CnC runtime [3]. The Java key-value data structure used to implement item collections is now indexed by DMA tags instead of DSA tags. When an item is put() with DSA tag t_1 its corresponding DMA location in the store is determined by identifying $pt_1 = f(t_1)$, where f is the folding function. Then, the physical store is accessed to see if there is any entry at that physical location. If there is none, we create it, and label it with the logical tag t_1. If there is, we need to hold a lock on the physical store location while the following operations are performed. First, we update the logical tag of the physical store entry to the logical tag of the item that has just been put. Then, we go through the list of steps waiting on that particular physical store location and, for each step that is waiting for the current item mark it as ready for execution. The marked marked continue their execution by performing a get() that will succeed because the desired item is already in the physical store.

When a get() on item with DSA tag t_1 is performed, its DMA tag is determined by identifying $pt_1 = f(t_1)$. If the entry does not exist, it is created, inserted in the physical store and the step is added to its list of waiting steps. If the entry does not correspond to the logical tag of the item, it registers itself to wait also. Compared to a non-folding execution, the only extras step needed for insertion is the application of the folding function (which does not need synchronization and has minimal overhead). The bigger overhead is in the *put()* , where the list of waiting steps has to be checked linearly to unblock only the steps that are waiting for the new item and this happens while holding the lock. We chose to have the overhead in the *put()* and not *get()* as the *get()* is usually performed multiple times on a single item and our approach leads to less contention.

Both the get-counts and folding policies only remove items from item collections, so that there is no object reference pointing to them; the Java garbage collection subsequently reclaims the memory.

4   Results

The following results were obtained on a 16 core Xeon system with 16GB RAM, running Habanero Java implementation of Concurrent Collections [3] on a 64 bit Java 1.6, using 16 workers for the work-stealing CnC runtime and Java default garbage collection mechanism. In this section we compare the performance and memory footprint of the following CnC memory management policies:

1. *Baseline*: non-collecting CnC (items are never removed from item collections) leading to memory leaks, but also no folding overhead.
2. *Get-counts*: memory management in which the user specifies a reference count for selected items, the count is decremented on every get() operation on a specified item, and the item is freed when the count becomes zero.
3. *Folding*: the folding runtime described in Section 3. We used the ordering extension described in Section 2.5 as needed and the tables contain the "Ordered" specifier where this happened.

For each policy used, we obtained the following measurements:

1. Execution Time - We performed thirty repetitions of the program in the same JVM instance, and reported the average, as advocated in [8].
2. Memory at end - the program footprint after the CnC graph finishes execution. With this metric, get-counts has an advantage because it removes items immediately, where as folding waits for the birth of another item, so at the end folding usually has more live items. In contrast, folding saves some work by taking a lazy approach to freeing items.
3. Items at end - similar to the previous metric, but expressed in items.

We evaluated the impact of folding and get-counts on the following applications:

1. *Microbenchmark* showing the difference in scalability between get-counts and folding with the number of reads per item.
2. *N-body simulation* for performance analysis.
3. *Routing simulation* as an application in which get-counts might lead to leaks because items have a number of accesses unknown at creation time, but folding works without needing the Ordered extension.
4. *Rician denoising* as example of an application in which folding with ordering can safely be used, but get-counts leads to leaks because some items have data-dependent accesses whose number is unknown.
5. *Cholesky factorization* as an example of memory reuse via the getForUpdate optimization.

Microbenchmark: Scalability with read/write ratio This benchmark varies the reads to write ratio to analyze the performance of the two collection mechanisms. Because folding performs most of the synchronization on *put()* as opposed to get-counts, which performs most of the synchronization on *get()*, we checked if the best performing policy might be get-counts for low read/write ratio. However, as shown in Figure 4, the folding version runs faster than both get-counts and baseline CnC even for a ratio of 1. Some applications may have a read/write ratio lower than one; performance for this case is analysed later using the Rician Denoising application.

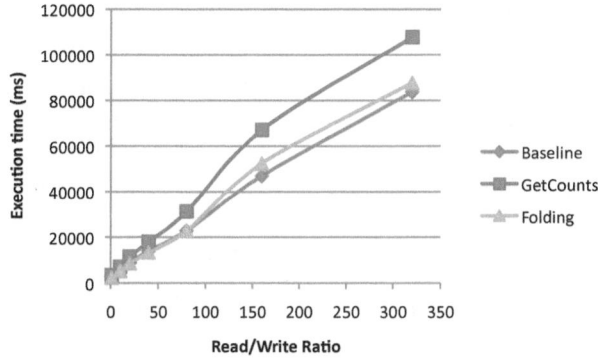

Fig. 4. Performance with read/write ratio (16 core Xeon)

N-Body Simulation. We implemented the $O(N^2)$ algorithm for N-body simulation with both get-counts and folding and the results are shown in Table 1. The
folding policy performs well because this benchmark has a small step granularity,
large number of items and thus more contention on the item collections. The fact
that folding has less synchronization of gets (in this application there are 10 gets
per item) leads to a consistent (1.3×) performance improvement compared to
get-counts. The get-counts footprint is smaller because folding can only reduce
the footprint to the maximum footprint of the program during its execution, and
in this case, that footprint is 20 items, which is also the maximum theoretical
footprint for get-counts.

Table 1. Experimental results for
NBody (5 bodies, 100000 timesteps)

CnC policy	Time (s)	Memory at end (bytes)	(items)
Baseline	16.9	277.0 MB	1,000,005
Get-Counts	18.1	3.6 KB	10
Folding	**13.0**	7.0 KB	20

Table 2. Experimental results for
Routing, with reliable links

CnC policy	Time (s)	Memory at end (bytes)	(items)
Baseline	21	61.0MB	102000
Get-Counts	25	10.7KB	1000
Folding	21	1.3MB	2000

Routing Simulation The routing simulation benchmark has unknown number of
gets on each item, making it a challenge for the get-counts approach. It simulates
the convergence of min-distance routing protocols such as IS-IS [10] and OPSF [9].
As links might go down, when a routing table is being built, we cannot know how
many gets will be performed on that node. In such cases, the get-count will never
reach zero and the item will become a memory leak. To see how the number of
leaked items varies with the chance of links failing we varied the chance of a message
not getting through from 0 to 10%, as shown in Figure 5: at only 1% failure rate
half the items are leaked. Even in the absence of link failure, folding shows a 16%
performance improvement over get-counts (Table 2).

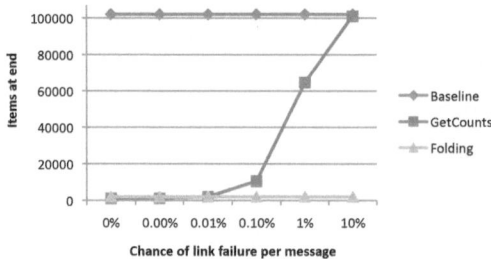

Fig. 5. Relation of link fail rate and memory leaks

Table 3. Performance comparison for Rician Denoising: image size 2560*1280, tile size 128*64. Shortcircuit reductions DISABLED (top) and ENABLED (bottom)

Shortcircuit operators	CnC policy	Time (s)	(MB)	Memory at end (Items in each collection)					
				Image	Gradient	Image× Gradient	Factor	Convergence Status
Enabled	Baseline	6.5	2888	10800	10400	10400	10400	10400
	Get-Counts	**2.6**	740	800	0	0	0	9897
	Folding (Ordered)	**2.6**	800	1200	800	800	800	800
Disabled	Baseline	8.1	2888	108006	10400	10400	10400	10400
	Get-Counts	3.7	720	800	0	0	0	0
	Folding	**3.5**	830	1200	800	800	800	800

Rician Denoising Rician (Poisson) denoising is an image processing application. Its global convergence check is a reduction on the convergence status of all tiles and it is sped up using a short-circuit evaluation: if a single tile changes significantly we do not need to wait for the convergence condition of all the other tiles to be evaluated, we immediately know we will need an additional iteration and can start spawning the corresponding steps.

The results (Table 3) show the performance of get-counts and folding: folding offers the best performance. Furthermore, get-counts leads to leaks of items from the ConvergenceStatus item collection in which the operands of the short-circuit operators are stored (the cause of the leaks is the unknown number of gets): 95% of items stored in that item collection are leaked, totaling 20MB. However, without short-circuit operators, get-counts collects more items because at the end of the program, all the items stored in the item collections that store intermediate results (Gradient, ImageTimesGradient, etc) can be collected. This does not affect the actual high water-mark of the program which is the same in both folding and get-counts executions.

Cholesky Factorization Cholesky factorization is a numerical application whose input is a symmetrical positive-definite matrix and output a lower-triangular matrix. One possible CnC implementation was previously described and benchmarked in [4] and the results were encouraging.

Table 4 shows that the proposed update-in-place optimization, if applied on either get-counts or folding, can lead to a large performance increase. Using get-ForUpdate leads to a performance improvement between 10% and 20% for both collecting policies. Baseline CnC cannot safely apply this optimization without additional programmer input to ensure that whenever GetForUpdate is called, the item accessed is indeed dead. To work around this, we manually added this call only when such accesses are safe.

Table 4. Performance comparison for Cholesky factorization (125*125 tiles)

Input Size	CnC Policy	Without update-in-place			With update-in-place		
		Time (s)	Item collection memory (MB at end)	(items at end)	Time (s)	Item collection memory (MB at end)	(items at end)
2000	Baseline	**0.9**	142.2	952	0.8	33.7	952
	Get-Counts	**0.9**	33.7	272	0.8	33.7	272
	Folding	**0.9**	33.7	272	**0.7**	33.7	272
4000	Baseline	7.3	1008.5	6512	5.6	133.2	6512
	Get-Counts	**6.2**	133.2	1056	5.6	133.2	1056
	Folding	**6.2**	133.2	1056	**5.0**	133.2	1056
6000	Baseline	26.6	2680.2	20776	19.5	298.4	20776
	Get-Counts	22.3	298.4	2352	19.2	298.4	2352
	Folding	**21.6**	298.4	2352	**19.1**	298.4	2352

Memory High-watermark comparison Table 5 shows the maximum number of live items during the execution of the benchmarks. This metric shows, in the schedules and with the parallelism actually used during execution, what is the maximum number of items that were live - the memory "high-water mark" of the program. To obtain these values we used atomic counters that tracked the number of stored items. The results show that maximum live items number

Table 5. Maximum number of items live during execution

Benchmark		Baseline	Get-Counts	Folding
Nbody		1,000,005	19	20
Routing		102000	1100	2000
RicianDenoising	Image	10800	800	800
	Gradient	10400	27	800
	Image × Gradient	10400	26	800
	ConvergenceStatus	10400	9897	800
Cholesky (6000)		20776	2352	2352

is lower than the bound identified by folding. However, in the future, as the number of processors grows, more tasks will run concurrently and the number of live items will increase.

5 Conclusions and Future Work

In this paper, we introduced a new memory management approach based on user-specified *folding functions* that map logical dynamic-single-assignment (DSA) tags into dynamic-multiple-assignment (DMA) tags, while preserving semantic guarantees of data race freedom and determinism. Our approach is applicable to parallel programming models in which shared data accesses are coordinated by put/get operations on tagged DSA data structures. These models include dataflow programs with I-structures, functional subsets of parallel programs based on tuple spaces (notably, Linda), and programs written in the Intel Concurrent Collections (CnC) coordination language. Our conclusion, based on experimental evaluation of five CnC programs, is that folding can offer significant memory efficiency improvements, and that folding can handle cases that get-counts (an alternative approach to user-specified memory management) cannot. An interesting direction for future work is automatic generation of folding functions. In many of the benchmarks that we studied, it is possible to use static analysis of get and put function parameters to identify candidates for folding.

Acknowledgments. We are grateful to the Intel Concurrent Collection team, in particular Frank Schlimbach, James Brodman and Ryan Newton (now at Indiana University), for proposing the Get-Counts idea and for having stimulating discussions. We thank Shams Imam for his debugging help and thorough feedback and the reviewers for their helpful comments.

References

1. Arvind, Nikhil, R.S., Pingali, K.K.: I-structures: data structures for parallel computing. ACM Trans. Program. Lang. Syst. 11 (October 1989)
2. Boissinot, B., Darte, A., Rastello, F., de Dinechin, B.D., Guillon, C.: Revisiting out-of-ssa translation for correctness, code quality and efficiency. In: CGO 2009, Washington, DC, USA, pp. 114–125 (2009)
3. Budimlic, Z., Burke, M., Cavè, V., Knobe, K., Lowney, G., Newton, R., Palsberg, J., Peixotto, D., Sarkar, V., Schlimbach, F., Tasirlar, S.: Concurrent collections. Scientific Programming (2010)
4. Chandramowlishwaran, A., Knobe, K., Vuduc, R.: Performance evaluation of concurrent collections on high-performance multicore systems. In: IPDPS (2010)
5. Feautrier, P.: Array expansion. In: Proceedings of the 2nd International Conference on Supercomputing, ICS, pp. 429–441. ACM, New York (1988)
6. Feautrier, P.: Dataflow analysis of array and scalar references. International Journal of Parallel Programming 20(1), 23–51 (1991)
7. Gelernter, D.: Generative communication in linda. ACM Trans. Program. Lang. Syst. 7, 80–112 (1985)

8. Georges, A., Buytaert, D., Eeckhout, L.: Statistically rigorous java performance evaluation. In: Proceedings of the 22nd Annual ACM SIGPLAN Conference on Object-oriented Programming Systems and Applications (2007)
9. Moy, J.: OSPF Version 2. RFC 2178, Obsoleted by RFC 2328 (July 1997)
10. Oran, D.: Osi is-is intra-domain routing protocol. RFC 1142 (February 1990)
11. Torczon, L., Cooper, K.: Engineering A Compiler, 2nd edn. Morgan Kaufmann Publishers Inc., San Francisco (2011)

High-Level Support for Pipeline Parallelism on Many-Core Architectures

Siegfried Benkner[1], Enes Bajrovic[1], Erich Marth[1], Martin Sandrieser[1],
Raymond Namyst[2], and Samuel Thibault[2]

[1] Research Group Scientific Computing, University of Vienna, Austria
[2] University of Bordeaux, LaBRI-INRIA Bordeaux Sud-Ouest, Talence, France

Abstract. With the increasing architectural diversity of many-core architectures the challenges of parallel programming and code portability will sharply rise. The EU project PEPPHER addresses these issues with a component-based approach to application development on top of a task-parallel execution model. Central to this approach are multi-architectural components which encapsulate different implementation variants of application functionality tailored for different core types. An intelligent runtime system selects and dynamically schedules component implementation variants for efficient parallel execution on heterogeneous many-core architectures. On top of this model we have developed language, compiler and runtime support for a specific class of applications that can be expressed using the pipeline pattern. We propose C/C++ language annotations for specifying pipeline patterns and describe the associated compilation and runtime infrastructure. Experimental results indicate that with our high-level approach performance comparable to manual parallelization can be achieved.

1 Introduction

With the shift towards heterogeneous many-core architectures combining different types of execution units like conventional CPU cores, GPUs and other accelerators, the challenges of parallel programming will sharply rise. For an efficient utilization of such architectures usually different programming models and APIs, tailored for the different types of execution units, have to be combined within a single parallel application. Available technologies like Intel TBB [1], CUDA [2], Cell SDK [3], and OpenCL [4] are characterized by a low level of abstraction, forcing programmers to take into account a myriad of architecture details, usually beyond the capabilities of average users. Several recent research projects including ParLab [5], PetaBricks [6], Elastic Computing [7], ENCORE [8] and others, are addressing these challenges by proposing higher-level programming support for emerging many-core systems.

The European research project PEPPHER [9] targets programmability and performance portability for single-node heterogeneous many-core architectures by means of a component-based approach in combination with a task-parallel execution model. In this paper we present our contributions towards high-level support for pipelined applications within the PEPPHER framework. Section 2

C. Kaklamanis et al. (Eds.): Euro-Par 2012, LNCS 7484, pp. 614–625, 2012.

outlines the PEPPHER approach and describes the proposed language features for realizing pipelined C/C++ applications. Section 3 describes our source-to-source transformation framework as well as the coordination and runtime support for pipelining. Experimental results for two real-world applications and a comparison to TBB are presented in Section 4. The paper closes with a discussion of related work and future directions.

2   High-Level Programming Support

Since there exists no parallel programming model that covers all types of parallel applications and architectures, we argue that a programming framework for heterogeneous parallel architectures should support the use of different programming models and APIs within an application.

2.1   The PEPPHER Component Model

Within the PEPPHER model performance-critical parts of an application are realized by means of multi-architectural components that encapsulate, behind an interface, different implementation variants of a function[1] tailored for different execution units of a heterogeneous many-core system.

Component implementation variants are usually written by expert programmers using different programming APIs (e.g., CUDA, OpenCL) or are taken from optimized vendor-supplied libraries. Non-expert programmers may then construct applications at a high level of abstraction by invoking component functionality using conventional interfaces and source code annotations to delineate asynchronous (or synchronous) component calls. With this approach, a sequential program spawns component calls, which are then scheduled for task-parallel execution by the runtime system. A source-to-source compiler transforms annotated component calls such that they are registered with the runtime system and generates corresponding glue-code.

The compiler and runtime system make use of rich component meta-data, usually supplied by expert programmers via external XML descriptors. Besides information about the data read and written by components, meta-data includes information about resource requirements, possible target platforms, and performance relevant parameters [10]. The runtime system, built on top of the StarPU [11] heterogeneous runtime system, relies on a representation of the program as a directed acyclic graph (DAG) where nodes represent component calls (tasks) and edges represent data dependences. The runtime system dynamically schedules component calls to the available execution units of a heterogeneous many-core architecture such that (1) independent component calls execute in parallel on different execution units and (2) the "best" implementation variants for a given architecture are selected based on historical performance information captured in performance models.

---

[1] These functions must be pure, i.e. they must not access global data, they must be stateless, and they are to be executed non-preemptively.

2.2   Language Support for Expressing Pipeline Patterns

A pipeline consists of several inter-connected stages, where a stream of data flowing through the pipeline is processed at every stage. Data entering a stage via input port(s) is consumed, processed and emitted at the output port(s) accordingly. Usually, stages are connected via buffer structures from which data is fed into stages. While buffered pipelines require additional memory, they allow to decouple stages and mitigate relative performance differences. In our approach buffers between stages are generated automatically, but we provide language features for the user to control certain aspects of buffering.

The pipeline pattern has the potential of exploiting two levels of parallelism: inter-stage parallelism, if different stages execute on different cores, and, intra-stage parallelism, if a stage is itself parallel and executes on, e.g., a GPU.

In our framework, pipelines may be constructed from while loops where the loop body comprises two or more calls to multi-architectural components as considered within the PEPPHER framework. The *pipeline* pragma indicates that the subsequent *while* loop represents a pipeline. Each stage of the pipeline corresponds to a single component call statement within the loop body.

```
1  #pragma pph pipeline
2  while(data.size != 0) {
3      func1(iFile,data);              // connect func1 to func2 via data
4    #pragma pph stage replicate(4) // replicate stage 4 times
5      func2(data,cdata);              // connect func2 to func3 via cdata
6      func3(cdata,oFile);
7  }
```

Listing 1.1. Example of a pipeline directive

2.3   Stage Replication and Stage Merging

Provided application logic permits, stage replication aims to increase the potential parallelism of pipelined applications by creating multiple replicas of a stage that may then be executed in parallel. Stages can be replicated using the *stage* pragma with the *replicate(N)* clause, specifying that a stage should be replicated N times (see Listing 1.1). As a consequence, multiple stage instances will be generated by the compiler to enable processing of different data packets in parallel, if enough execution units are available. While replication is a suitable technique for increasing pipeline throughput by replicating stages with (relative) high execution times, the programmer has to be aware that the order in which data-packets are processed may change (unless priority ordering is used), resulting in unpredictable application behavior. Moreover, sizes of connected in- and output buffers may have to be adapted as well. Also, stage replication might result in a performance degradation if not enough execution units are available to execute all stage replicas in parallel.

The *stage* pragma may also be used to merge multiple stages into a single stage (see, e.g., Listing 1.4). This allows the programmer to manually increase the granularity of stages, if the involved individual component calls do not exhibit enough computational density to mandate processing within a separate stage. Note that stages can only be merged if for all involved stages compatible component implementation variants are available. The interface of the resulting single stage is automatically generated by the compiler, describing all input and output ports of the merged stage.

2.4 Buffer Management

With our framework buffers are automatically generated in between pipeline stages. These buffers temporarily store data packets, generated by the source stage(s), and consumed by the target stage(s). Depending on the type of application, different order guarantees and sizes for buffers may be required. Therefore we provide the buffer clause for specifying the order guarantee, including *priority*, *random*, and *fifo*, as well as the size of buffers (see Listing 1.2). Global buffer settings may be specified by using the *with buffer* clause together with the *pipeline* pragma. Local buffer settings for individual stages may be specified with the *buffer for port* clause within the *stage* pragma. In Listing 1.2 the buffer for port *cdata* in the second stage is changed to *RANDOM* and buffer size to 8, while for all other buffers *FIFO* ordering is used.

```
1  #pragma pph pipeline with buffer(FIFO)
2  while(data.size != 0) {
3      func1(iFile,data);
4      #pragma pph stage buffer for port(cdata,RANDOM,8)
5      func2(data,cdata);
6      func3(cdata,oFile);
7  }
```

Listing 1.2. Influencing Buffer Management

3 Implementation

We have developed a prototype source-to-source compiler that transforms C/C++ applications with the described annotations into C++ with calls to a pipeline coordination layer that utilizes the StarPU [11] heterogeneous runtime system for scheduling stages for parallel execution onto the execution units of a heterogeneous many-core system comprising CPUs as well as GPUs. The source-to-source compiler has been implemented using the ROSE compiler infrastructure [12].

3.1 Source-to-Source Transformation

After the usual front-end processing phases an abstract syntax tree is constructed. Pipeline constructs are then further processed to determine the structure of the pipeline (stage interconnection) by analyzing the data types of objects passed between pipeline stages. For each stage interconnection (port) corresponding buffer structures, as specified globally or locally, are generated.

The generated target code contains calls to the pipeline coordination layer which comprises various classes for coordinating the execution of pipeline stages

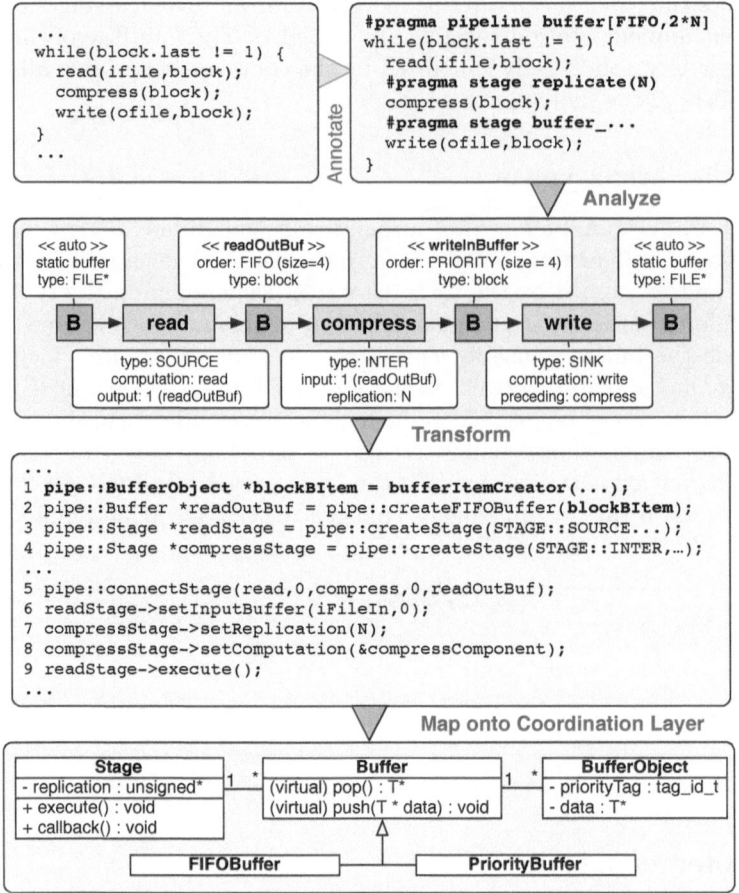

Fig. 1. Overview of the transformation process

on top of the StarPU runtime system. The *Stage* class encapsulates information on the stage functionality (component), directly connected buffers and stages for each port, the number of instances of a stage to be processed in parallel (replication count), and the position of the stage within the pipeline (*source*, *inter*, or *sink*). Each stage owns a local coordination mechanism, described later, which slightly differs depending on the position of a stage. The abstract *Buffer* class generalizes the buffer access interface and is used to derive concrete buffer implementations like FIFO buffers or priority buffers. Each buffer stores information about connected stages, internal storage layout, and the type of data packets.

Figure 1 gives an overview for a three-stage data compression pipeline example. First buffer objects are generated (line 1) for holding data packets and meta-data such as priority tags, creation dates, and other information. Next, all inter-stage buffers are generated to hold the previously generated buffer objects, setting the default data type for each buffer, as shown in line 2. Class instances for the *read* and *compress* stages are generated in line 3 and 4. The stages are further configured by specifying connected buffers, replication counts, and the stage computations, as shown in line 5 to 8. After all stages have been created and configured, stages are posted for execution to the runtime, initiating the actual execution of the pipeline, as shown in line 9.

3.2 Task-Based Heterogeneous Runtime

The StarPU runtime system [11] utilized in our framework is based on an abstraction of the underlying heterogeneous many-core architecture as a set of *workers*, each representing either a general purpose computing core, or an accelerator (e.g., a GPU). The runtime system is responsible for selecting suitable component implementation variants for pipeline stages and for scheduling their execution to workers in a performance- and resource-efficient way, according to a specific scheduling policy. StarPU manages data transfers between workers, ensures memory coherency, and provides support for different scheduling strategies which may be selected at runtime.

Besides the well-known EAGER scheduling policy, StarPU also features the Heterogeneous Earliest Finish Time (HEFT) [13] policy. The HEFT policy considers inter-component data dependencies, and schedules components to workers taking into account the current system load, available component implementation variants, and historical execution profiles, with the goal of minimizing overall execution time by favoring implementations variants with the lowest expected execution time.

3.3 Coordination

The coordination layer controls the execution of a pipelined application by deciding when to post stage component calls to the runtime system. We utilize a local coordination strategy where each stage is at runtime controlled by a corresponding stage object (an instance of the *Stage* class). The *Stage* class provides two methods for coordinating the execution of a pipelined application: the method *execute()* for posting a stage for execution to the runtime system and the method *callback()* for transferring control back to a stage object after its associated component has finished execution on a worker. In the following we outline a coordination scenario for the code shown in Figure 1.

First the runtime system is initialized and the required stage and buffer objects are instantiated. Next, the execution of all stages is initiated by invoking the *execute()* method on each stage object. This method posts a stage for execution to the runtime system provided its input buffer(s) and free slots in its output buffer are available. In our scenario, initially only the *read* stage is posted to the runtime while execution of all other stages is postponed since no input data is yet available. The runtime system then selects a suitable component implementation variant for the *read* stage and schedules it for execution onto a worker. When the *read* stage finishes execution on the selected worker, the runtime system invokes its callback method. Within the callback, first all connected buffers of the *read* stage are updated and then the *execute()* method of the connected *compress* stage is called, which results in posting the *compress* stage to the runtime system. Finally, if a new data packet to be processed and a free output buffer slot are available, the *read* stage re-posts itself for execution to the runtime system. The runtime can now schedule the second instance of the *read* stage and the first instance of the *compress* stage for parallel execution on different workers. With this scheme, stages coordinate themselves only in combination with their immediate neighbors, but without a central coordinator.

4   Experimental Evaluation

An initial evaluation has been performed with two real world codes, a data compression application and a face detection application. We compare our approach to Threading Building Blocks (TBB) as well as to existing parallel implementations on two different architectures. Architecture A represents a homogeneous system with two Intel Xeon X7560 (8 cores, 2.26 GHz) running RHEL 5. Architecture B is a heterogeneous system with two Intel Xeon X5550 (4 cores, 2.67 GHz), one Nvidia GeForce GTX 480 (480 Cores, 1.5GB, 1.40GHz), and one Nvidia GeForce GTX 285 (240 Cores, 1GB, 1.48GHz), CUDA 4.0 and RHEL 5.6. The performance numbers shown are average execution times over ten runs.

4.1   BZIP2 Compression

BZIP2 is an open-source lossless data compression tool, based on the Burrows-Wheeler-Transformation [14]. Compared to other compression techniques, BZIP2 is embarrassingly parallel, compressing data at the granularity of blocks with a fixed size. Since there are currently no heterogeneous implementations available, BZIP2 cannot utilize the GPUs of architecture B.

Listing 1.3 outlines the implementation using our language constructs. We implement a three stage pipeline with *FIFO* and *PRIORITY* buffers and utilize the replication feature for the middle stage.

```
1  unsigned int N = get_max_cpu_cores();
2  #pragma pph pipeline with buffer(FIFO,N*2)
3  while(b->iSize == bs) {
4    readBlock(ifile,b);
5    #pragma pph stage replicate(N)
6    compress(b);
7    #pragma pph stage buffer for port(b,PRIORITY)
8    writeCompressedBlock(ofile,b);
9  }
```

Listing 1.3. BZIP2 Compression Pipeline

For comparison, we have implemented bzip2 using the pipeline pattern of TBB to create a pipeline [15] with three stages (read, compress, write) using the utility functions provided by the bzip2 library [16]. Since in TBB stages cannot be replicated explicitly, we have set the number of alive objects to twice the number of cores available. This allows TBB to schedule stages in parallel if

Table 1. bzip2 Performance Results (execution times in s)

# Cores	Architecture A					Architecture B			
	1	2	4	8	16	1	2	4	8
TBB	248.19	133.18	64.99	32.63	19.58	201.42	103.43	52.11	28.73
pbzip2	268.88	128.20	65.86	33.75	20.16	207.57	106.52	55.28	31.83
PEPPHER	288.04	143.42	76.27	37.46	19.99	208.88	107.17	61.24	44.71

possible. For correct compression in time, we have enabled order preservation. Moreover, we measured an existing code from the pbzip2 project [17], which represents a manually implemented pipeline with three stages (read, compress, and write) and priority buffers.

Table 1 shows the measured (average) execution times (wall-clock times) for the bzip2 benchmark for compressing a file with size of 1 GB. As can be seen, our high-level approach delivers performance results which are very similar to the other two approaches. However, the programming effort with our approach as well as the total lines of code required is significantly reduced.

4.2 OpenCV Image Processing

The Open Source Computer Vision (OpenCV) library provides extensive support for the implementation of real time computer vision applications [18]. Originally developed for homogeneous architectures, current releases offer built-in support for GPUs based on CUDA. Using OpenCV we have implemented a face detection application in a pipelined manner, where for the detection stage two different implementation variants, for CPUs and GPUs, are provided.

```
1  unsigned int N = get_max_execution_units();
2  #pragma pph pipeline with buffer(PRIORITY,N*2)
3  while(inputstream >> file) {
4     readImage(file,image);
5     #pragma pph stage replicate(N) {
6        resizeAndColorConvert(image);
7        detectFace(image,outimage);
8     }
9     writeFaceDetectedImage(file,outimage);
10 }
```

Listing 1.4. OpenCV Image Processing Pipeline

Listing 1.4 sketches our implementation as a three-stage pipeline with priority buffers. The pipelines exits once 32 images have been processed. The middle stage, which merges two component calls, is replicated according to the available number of execution units within the system.

For comparison, a pipelined TBB version [15] has been implemented in a similar way. Again, instead of stage replication, we have set the number of alive data packets to twice the number of execution units, enabling TBB to schedule stages in parallel.

Table 2. OpenCV Performance Results (execution times in secs)

Architecture A				
Image Size	VGA	SVGA	XGA	QXGA
TBB (1 Core)	15.61	23.51	41.84	170.58
PEPPHER (1 Core)	12.40	17.85	30.72	140.86
TBB (16 Core)	1.26	1.92	3.39	13.60
PEPPHER (16 Cores)	1.16	1.72	2.91	12.33
Architecture B				
TBB (1 Core)	12.75	20.07	35.15	145.68
PEPPHER (1 Core)	9.62	14.33	24.94	111.45
PEPPHER (1 Core + 1 GPU)	3.94	5.91	10.35	46.30
PEPPHER (1 Core + 2 GPUs)	2.95	2.72	6.53	30.81
TBB (8 Cores)	1.47	2.29	4.13	17.4
PEPPHER (8 Cores)	1.18	1.78	3.58	13.69
PEPPHER (7 Cores + 1 GPU)	1.13	1.63	2.91	11.89
PEPPHER (6 Cores + 2 GPUs)	0.94	1.40	2.44	10.71

Table 2 shows performance results for the face detection code in comparison to the TBB version using different image resolutions including VGA(640x480), SVGA(800x600), XGA(1024x768), and QXGA(2048x1536). As can be seen, on architecture A our high-level approach outperforms the TBB version by about 20%. On architecture B, when using only the CPU cores, we get similar results. As opposed to TBB, however, our approach can take advantage of the GPUs by utilizing the GPU implementation variant for the (merged) middle pipeline stage. Since for each GPU an additional CPU core is required, the number of usable general purpose cores is reduced accordingly. As can be seen, using one CPU core and one GPU (GTX 480) the execution time is reduced by a factor of up to 3.14 compared to the TBB version using one CPU core. Using a second GPU results in only modest further performance improvements. Again the results show that our high-level approach to pipelining has the potential to outperform TBB, while significantly improving programmability. Moreover, based on our concept of multi-architectural components together with a versatile heterogeneous runtime system, we can take advantage of a heterogeneous CPU/GPU-based architecture without modifying the high-level application code.

5   Related Work

Over the last two decades design patterns as well as skeletons had a significant impact on software development in general as well as on parallel and distributed programming [19–22].

Intel Threading Building Blocks (TBB) provides direct support for pipeline patterns. As opposed to our work, TBB targets only homogeneous architectures and does not support stage implementation variants.

Thies at al. [23] propose StreamIt, a domain specific language (DSL) for designing stream-based applications where in combination with the underlying compiler pipeline specific optimizations [24] are applied. However, support for heterogeneous architectures is not addressed in this work.

Schaefer at al. [25, 26], propose a language-based approach for engineering parallel application using tunable patterns. Although different stage implementations are supported within the pipeline pattern, support for heterogeneous architectures has not been addressed.

Suleman at al. [27], propose a feedback-directed approach with integrated support for tuning. In combination with online-monitoring, pipelined applications are optimized on a coarse-grain level. Neither different stage implementation variants, nor optimizations for heterogeneous architectures are supported.

There have been several proposals for extending C or Fortran in order to support programming of heterogeneous systems comprised of CPUs and GPUs including OmpSS [28], PGI Accelerate [29], HMPP [30], and OpenACC[31]. These approaches are based on directives for specifying regions of code in Fortran or C programs that can be offloaded from a host CPU to an attached GPU by the compiler. None of these approaches, however, supports pipelining or other parallel patterns.

6 Conclusion and Future Work

We have presented high-level language annotations for C/C++ for developing pipelined applications on heterogeneous many-core architectures without having to deal with complex low-level architectural issues. We provided an overview of the associated implementation framework which is currently being developed within the European PEPPHER project. Our work relies on a component-based approach where pipeline stages correspond to multi-architectural components that encapsulate different implementation variants optimized by expert programmers for different execution units of a heterogeneous many-core system. A source-to-source compiler translates pipelined applications to an object-oriented coordination layer which is built on top of a heterogeneous task-based runtime system. The task-based runtime system attempts to schedule the best component implementation variants for parallel execution on the free execution units of a many-core system such that overall performance is optimized.

Our approach enables to run the same high-level application code without changes on homogeneous and heterogeneous multi-core architectures, delegating to the runtime system task scheduling and implementation variant selection. Experimental results on two different architectures are encouraging and indicate that a performance comparable to manual parallelization can be achieved, despite the considerably higher level of abstraction provided by our language features for the pipeline pattern.

For future work we plan to extend our framework with autotuning capabilities to, for example, determine replication counts and buffer sizes for pipeline stages. Furthermore, we plan to experiment with different runtime scheduling strategies as supported by StarPU and to provide support for other parallel patterns.

Acknowledgment. This work was supported by the European Commission as part of the FP7 Project PEPPHER under grant 248481.

References

1. Intel, Threading Building Blocks (2009), http://threadingbuildingblocks.org
2. Nvidia, C.: Compute Unified Device Architecture Programming Guide. NVIDIA, Santa Clara (2007)
3. Kahle, J.A., Day, M.N., Hofstee, H.P., Johns, C.R., Maeurer, T.R., Shippy, D.J.: Introduction to the Cell Multiprocessor. IBM Journal of Research and Development 49(4-5), 589–604 (2005)
4. Munshi, A. (ed.): OpenCL 1.0 Specification. Khronos OpenCL Working Group (2011)
5. Pan, H., Hindman, B., Asanović, K.: Composing Parallel Software Efficiently with Lithe. In: Proceedings of the 2010 ACM SIGPLAN Conference on Programming Language Design and Implementation, PLDI 2010, pp. 376–387. ACM, New York (2010)
6. Ansel, J., Chan, C.P., Wong, Y.L., Olszewski, M., Zhao, Q., Edelman, A., Amarasinghe, S.P.: PetaBricks: A Language and Compiler for Algorithmic Choice. In: Proceedings of the 2009 ACM SIGPLAN Conference on Programming Language Design and Implementation, PLDI 2009, pp. 38–49 (2009)
7. Wernsing, J.R., Stitt, G.: Elastic Computing: A Framework for Transparent, Portable, and Adaptive Multi-core Heterogeneous Computing. In: Proceedings of the ACM SIGPLAN/SIGBED 2010 Conference on Languages, Compilers, and Tools for Embedded Systems (LCTES), pp. 115–124. ACM (2010)
8. Vandierendonck, H., Pratikakis, P., Nikolopoulos, D.S.: Parallel Programming of General-Purpose Programs using Task-based Programming Models. In: Proceedings of the 3rd USENIX Conference on Hot Topics in Parallelism, HotPar 2011, Berkeley, CA, USA, p. 13 (2011)
9. Benkner, S., Pllana, S., Traff, J., Tsigas, P., Dolinsky, U., Augonnet, C., Bachmayer, B., Kessler, C., Moloney, D., Osipov, V.: PEPPHER: Efficient and Productive Usage of Hybrid Computing Systems. IEEE Micro 31(5), 28–41 (2011)
10. Sandrieser, M., Benkner, S., Pllana, S.: Using explicit platform descriptions to support programming of heterogeneous many-core systems. Parallel Computing 38(12), 52–65 (2012),
    http://www.sciencedirect.com/science/article/pii/S0167819111001396
11. Augonnet, C., Thibault, S., Namyst, R., Wacrenier, P.-A.: StarPU: A Unified Platform for Task Scheduling on Heterogeneous Multicore Architectures. Concurrency and Computation: Practice and Experience (23), 187–198 (2011)
12. Quinlan, D.: ROSE: Compiler Support for Object-Oriented Frameworks. Parallel Processing Letters 49 (2005)
13. Topcuoglu, H., Hariri, S., Wu, M.-Y.: Performance-Effective and Low-Complexity Task Scheduling for Heterogeneous Computing. IEEE Transactions on Parallel and Distributed Systems 13(3) (March 2002)

14. Burrows, M.: A Block-Sorting Lossless Data Compression Algorithm. Research Report 124, Digital Systems Research Center (1994)
15. Intel, Intel Threading Building Blocks - Pipeline Documentation, http://threadingbuildingblocks.org/files/documentation/a00150.html
16. Seward, J.: BZIP2 Library Utility Function Documentation (September 2011), http://bzip.org/1.0.5/bzip2-manual-1.0.5.html#util-fns
17. Gilchrist, J.: Parallel Data Compression with bzip2. In: Proceedings of the 16th IASTED International Conference on Parallel and Distributed Computing and Systems, vol. 16, pp. 559–564 (2004)
18. Gary, B.: Learning openCV: Computer Vision with the openCV Library. O'Reilly, USA (2008)
19. Benoit, A., Robert, Y.: Mapping Pipeline Skeletons onto Heterogeneous Platforms. In: Shi, Y., van Albada, G.D., Dongarra, J., Sloot, P.M.A. (eds.) ICCS 2007, Part I. LNCS, vol. 4487, pp. 591–598. Springer, Heidelberg (2007)
20. Cole, M.: Bringing Skeletons out of the Closet: A Pragmatic Manifesto for Skeletal Parallel Programming. Parallel Computing (2004)
21. Mattson, T., Sanders, B., Massingill, B.: Patterns for Parallel Programming. Addison-Wesley (2005)
22. Pop, A., Cohen, A.: A Stream-Computing Extension to OpenMP. In: Proceedings of the 6th International Conference on High Performance and Embedded Architectures and Compilers. ACM (2011)
23. Thies, W., Karczmarek, M., Amarasinghe, S.: StreamIt: A Language for Streaming Applications. In: Horspool, R.N. (ed.) CC 2002. LNCS, vol. 2304, pp. 179–196. Springer, Heidelberg (2002)
24. Sermulins, J., Thies, W., Rabbah, R., Amarasinghe, S.: Cache Aware Optimization of Stream Programs. ACM SIGPLAN Notices 40(7) (2005)
25. Schaefer, C., Pankratius, V., Tichy, W.: Engineering Parallel Applications with Tunable Architectures. In: ICSE 2010: Proceedings of the 32nd ACM/IEEE International Conference on Software Engineering, vol. 1 (May 2010)
26. Otto, F., Schaefer, C.A., Dempe, M., Tichy, W.F.: A Language-Based Tuning Mechanism for Task and Pipeline Parallelism. In: D'Ambra, P., Guarracino, M., Talia, D. (eds.) Euro-Par 2010, Part II. LNCS, vol. 6272, pp. 328–340. Springer, Heidelberg (2010)
27. Suleman, M., Qureshi, M., Khubaib, Patt, Y.: Feedback-Directed Pipeline Parallelism. In: PACT 2010: Proceedings of the 19th International Conference on Parallel Architectures and Compilation Techniques (2010)
28. Ayguade, E., Badia, R.M., Cabrera, D., Duran, A., Gonzalez, M., Igual, F., Jimenez, D., Labarta, J., Martorell, X., Mayo, R., Perez, J.M., Quintana-Ortí, E.S.: A Proposal to Extend the OpenMP Tasking Model for Heterogeneous Architectures. In: Müller, M.S., de Supinski, B.R., Chapman, B.M. (eds.) IWOMP 2009. LNCS, vol. 5568, pp. 154–167. Springer, Heidelberg (2009)
29. Wolfe, M.: Implementing the PGI Accelerator Model. In: GPGPU 2010: Proceedings of the 3rd Workshop on General-Purpose Computation on Graphics Processing Units. ACM (March 2010)
30. Bodin, F., Bihan, S.: Heterogeneous Multicore Parallel Programming for Graphics Processing Units. Scientific Programming 17, 325–335 (2009)
31. OpenACC. Directives for Accelerators, http://www.openacc-standard.org/

Node.Scala: Implicit Parallel Programming for High-Performance Web Services

Daniele Bonetta, Danilo Ansaloni, Achille Peternier,
Cesare Pautasso, and Walter Binder

University of Lugano (USI), Switzerland
Faculty of Informatics
{name.surname}@usi.ch

Abstract. Event-driven programming frameworks such as Node.JS have recently emerged as a promising option for Web service development. Such frameworks feature a simple programming model with implicit parallelism and asynchronous I/O. The benefits of the event-based programming model in terms of concurrency management need to be balanced against its limitations in terms of scalability on multicore architectures and against the impossibility of sharing a common memory space between multiple Node.JS processes. In this paper we present Node.Scala, an event-based programming framework for the JVM which overcomes the limitations of current event-driven frameworks. Node.Scala introduces safe stateful programming for event-based services. The programming model of Node.Scala allows threads to safely share state in a standard event-based programming model. The runtime system of Node.Scala automatically parallelizes and synchronizes state access to guarantee correctness. Experiments show that services developed in Node.Scala yield linear scalability and high throughput when deployed on multicore machines.

1 Introduction

Services published on the Web need to guarantee high throughput and acceptable communication latency while facing very intense workloads. To handle high peaks of concurrent client connections, several engineering and research efforts have focused on Web server design [2]. Of the proposed solutions, event-driven servers [3,12] have proven to be very scalable, as they are able to handle concurrent requests with a simple and efficient runtime architecture [9,7]. Servers of this class are based on the ability offered by modern operating systems to communicate asynchronously (through mechanisms such as Linux's `epoll`), and on the possibility to treat such requests as streams of events. In event-driven servers each I/O-based task is considered an *event*. Successive events are enqueued for sequential processing (in an *event queue*), and processed in an infinite *event-loop*. The event-loop allows the server to process concurrent connections nondeterministically by automatically partitioning the time slots assigned to the processing of each request, thus augmenting the number of concurrent requests handled by

C. Kaklamanis et al. (Eds.): Euro-Par 2012, LNCS 7484, pp. 626–637, 2012.

the server through time-sharing. In this way, request processing is overlapped with I/O-bound operations, maximizing throughput and guaranteeing fairness between clients. Thanks to the event-loop model, servers can process thousands of concurrent requests using a very limited number of processes (usually, one process per core on multicore machines).

The performance of event-driven architectures has promoted programming models for Web service development that rely (explicitly or implicitly) on event-loops. Examples of such programming models include libraries (e.g., Python Twisted [6] or Java NIO [1]), and language-level integrations such as Node.JS [14]. Node.JS is a programming framework for the development of Web services using the JavaScript language and Google's V8 JavaScript Engine. In Node.JS the event-loop is hidden behind a convenient programming abstraction, which allows the developer to treat event-driven programming as a set of callback function invocations, taking advantage of the functional nature of the JavaScript language. Since the event-loop is run by a single thread, while all I/O-bound operations are carried out by the OS, the developer only writes the sequential code to be executed for each event within each callback, without worrying about concurrency issues.

Despite of the high performance of the V8 Engine, frameworks like Node.JS still present some limitations preventing them from exploiting modern multicore machines. For example, long running callbacks may block the entire service due to the single-threaded, sequential event loop architecture. We have overcome these limitations with the design of Node.Scala, a programming framework for the development of scalable Web services which takes full advantage of modern multicores. In more detail, our work makes the following contributions:

- We introduce Node.Scala, an event-loop-based framework targeting the Scala language and the JVM. Node.Scala features automatic parallelization of concurrent request processing, automatic synchronization of stateful request processing, and allows the developer to use both blocking and non-blocking programming styles. Node.Scala can be used to build HTTP-based services, including RESTful Web services [4].
- We describe the design of the Node.Scala runtime, which features multiple event-loops which have been safely parallelized.
- We illustrate the performance of Node.Scala with a set of benchmark results obtained with both stateless and stateful Web services.

The rest of this paper is structured as follows. In Section 2 we further discuss the motivations underlying Node.Scala and provide background information on event-loop frameworks. Section 3 presents the programming model of Node.Scala. Section 4 presents the parallel runtime system of Node.Scala. Section 5 presents an evaluation of the performance of Node.Scala-based Web services. Section 6 discusses related work, while Section 7 concludes.

2   Background and Motivation

Despite of being very scalable in terms of handling concurrent connections, event-driven frameworks like Node.JS are limited by their runtime system at least in two aspects, namely (1) the impossibility of sharing a common memory space among processes, and (2) the difficulty of building high throughput services using blocking method calls.

Concerning the first limitation, common event-based programming frameworks are not designed to express thread-level parallelism, thus the only way of exploiting multiple cores is by replicating the service process. This approach forces the developer to adopt parallelization strategies based on master-worker patterns (e.g., WebWorkers[1]), which however require a share-nothing architecture to preserve the semantics of the event-loop. Whenever multiple processes need to share state (e.g., to implement a stateful Web service), the data needs to be stored into a database or in an external repository providing the necessary concurrency control.

Concerning the second limitation, event-based programming requires the developer to deeply understand the event-loop runtime architecture and to write services with non-blocking mechanisms so as to break down long-running operations into multiple processing steps. Unfortunately, such mechanisms usually involve the adoption of programming techniques (e.g., nested callbacks, closures) which increase the complexity of developing even simple services. Moreover, while non-blocking techniques help increasing the throughput of a service, they also increase the latency of the responses. As a consequence, services need to be developed by carefully balancing blocking and non-blocking operations.

As an example, consider Fig. 1. The two code snippets in the figure correspond to two different implementations of a simple Node.JS Web service for calculating the n-th Fibonacci sequence number. The code in Fig. 1 (a) implements the Fibonacci function using the recursive algorithm by Leonardo da Pisa, while the one in Fig. 1 (c) implements the same algorithm exploiting non-blocking programming (using a hybrid synchronous/asynchronous algorithm). If a request is issued for a Fibonacci number which is greater than a fixed threshold (over which the standard recursive algorithm is known to block the event-loop for too long), the result is calculated using the non-blocking algorithm (`fiboA`). Otherwise, the blocking recursive algorithm (`fiboS`) is used. The non-blocking implementation does not use the stack for the entire recursion. Instead, it generates a series of nested events (through the `nextTick` runtime function call), each one corresponding to a single recursive function invocation. This fragments the execution flow of each request, as control is returned to the event-loop which can accept other incoming requests.

A comparison of the performance of the two services is given in Fig. 1 (b). For each implementation, a mixed workload of "light" (20th Fibonacci number) and "heavy" (35th Fibonacci number) requests is executed; the workload amounts to 100 requests per second. For the machine used in the experiment, the threshold

---

[1] http://dev.w3.org/html5/workers/

```
1   function fiboS(n) {
2     if(n<2) return n
3     else return fiboS(n−1)+fiboS(n−2)
4   }

5   http.createServer(function(req,res){
6     var n = req.query
7     res.end('result: '+fiboS(n))
8   }).listen(8080)
```

(a) Blocking Version

Heavy(%)	Bk (msg/s)	NBk (msg/s)
0.00	100.0	100.0
0.05	24.4	94.2
0.10	9.8	88.9
0.15	5.1	83.6

(b) Performance Comparison

```
1   function fiboA(n,done) {
2     if(n<2) done(n)
3     else process.nextTick(function() {
4       fiboA(n−1, function(num1) {
5         process.nextTick(function() {
6           if(n>threshold)
7             fiboA(n−2,function(num2){
8               done(num1+num2)
9             })
10            else done(num1+fiboS(n−2))
11        })})})
12  }

13  http.createServer(function(req,res){
14    var n = req.query
15    if(n>threshold)
16      fiboA(n, function(value) {
17        res.end('result: '+value) })
18    else res.end('result: '+fiboS(n))
19  }).listen(8080)
```

(c) Non-blocking Version

Fig. 1. Blocking (Bk) vs. Non-blocking (NBk) Fibonacci Web Service in Node.JS

has been set to 30. Experiments have been performed with different percentages of heavy requests (up to 15%). Results show a notable difference between the two implementations. The blocking implementation achieves only low throughput compared to the non-blocking one, even with a low percentage of "heavy" requests. The reason is the sequential event-loop architecture: calling the fiboS function with values higher than the threshold keeps the event-loop blocked, thus preventing it from processing other clients' requests. This aspect, coupled with the impossibility of sharing a global memory space among different processes, constitutes a significant limitation for the development of high-throughput Web services using Node.JS.

3   The Programming Model of Node.Scala

The programming model of Node.Scala is similar to the one of Node.JS, as it features an implicit parallel programming model based on asynchronous callback invocations for the Scala language. However, blocking methods can be invoked without blocking the service, and concurrent requests running on different threads can safely share state. The goal is to let developers write services using the same assumptions (single-process event-loop) made on the Node.JS platform, while automatically and safely carrying out the parallelization to fully exploit multicore machines. This has the effect of freeing the developer from dealing with the issues identified in the previous section, while keeping all the benefits of the asynchronous programming model with implicit parallelism, overlapping I/O- and CPU-bound operations, and lock-free synchronization.

An example of a Node.Scala Web service (Fig. 2) similar to the one computing the n-th Fibonacci sequence number (Section 2) makes use of the two distinguishing features of Node.Scala, i.e., global stateful objects and blocking

```
1    def fiboS(n: Int): Int = n match {
2        case 0 | 1 => n
3        case _ => fiboS(n-1) + fiboS(n-2)
4    }
5    val cache = new NsHashMap[Int,Int]()
6    val server = new NsHttpServer(8080)
7    server.start( connection => // 1st callback
8    {
9        val n = connection.req.query("n").asInstanceOf[Int]
10       if( cache.contains(n) )
11           connection.res.end("result: " + cache.get(n) )
12       else
13           server.nextTick( => // 2nd callback
14           {
15               val result = fiboS( n )
16               cache.put (n, result)
17               connection.res.end("result: " + result )
18           })
19   })
```

Fig. 2. Simple Stateful Web Service in Node.Scala

synchronous calls. The stateful object (`cache`, of type `NsHashMap`) is used as a cache to store the values of previously computed requests. To perform the computation, a simple blocking function call (`fiboS`) is used. The algorithm used is the Scala-equivalent version of the recursive implementation from Fig. 1 (a). The service makes also use of two callback functions. As in Node.JS, the first callback represents the main entry point for the service, that is, the callback function that will be triggered for every new client's connection. The callback is passed as an argument to the `start()` method (implemented in the `NsHttpServer` class). The second callback used in the example is the argument to the `nextTick` method, which registers the callback to perform the actual calculation and to update the cache.

Each callback is invoked by the Node.Scala runtime whenever the corresponding data is available. For instance, as a consequence of a client connection, an HTTP request, or a filesystem access, the runtime system *emits* an event, which is put into the event-queue (i.e., into the list of all pending events to be processed). The event will then be taken from the queue by one of the threads running the event-loop, which will invoke the corresponding callback function with the received data passed as an argument. In this way, when a new client request is received, the runtime calls the first user-defined callback function passing the `connection` object as argument. The object (created by the runtime) can be accessed by all other nested callbacks, and holds all the details of the incoming request (`connection.req`), as well as the runtime object for generating the answer (`connection.res`).

The service is stateful because the first callback uses an object with global scoping, `cache`, which is not local to a specific client request (to a specific callback), but is global and thus shared among all parallel threads running the event loop. Node.Scala enables services to safely share state through a specific library of common Scala data structures, which are used by the runtime system to automatically synchronize multiple concurrent callbacks accessing the same shared data structure. The details of the runtime mechanisms allowing such safe implicit parallel processing are described in Section 4.

The second callback calls a synchronous method. In common event-loop frameworks such a blocking call would result in a temporary interruption of the event-loop, as discussed in Section 2. The parallel runtime system of Node.Scala overcomes this limitation using its architecture based on parallel event-loops. Therefore, blocking synchronous calls do not have a negative impact on Node.Scala service performance as they would have in traditional frameworks. Consequently, programmers can focus on developing the service business logic without having to employ complex non-blocking programming techniques to achieve scalability.

4   System Architecture

In this section we describe the system architecture (Fig. 3), focusing on the constructs that allow Node.Scala to safely parallelize request processing. Node.Scala uses a single JVM process with multiple threads to execute a Web service, granting shared memory access to the threads running the parallel event-loops. As illustrated in Fig. 3 (b), the request processing pipeline consists of tree stages: (1) handling, (2) processing, and (3) completion.

Request Handling. Incoming HTTP connections are handled by a dedicated server thread, which pre-processes the request header and emits an event to the parallel event-loop to notify a new request. All the operations performed by the HTTP server thread are implemented using the Java New I/O (NIO) API for asynchronous I/O data processing. Since each event-loop thread has a dedicated event-queue, the HTTP server thread adopts the *join-the-shortest-queue* policy to select which queue to fill.

Request Processing. Multiple event-loop threads concurrently process events generated by incoming requests. In particular, each event-loop thread removes an event from its local event queue, accesses the callback table associated with that event type, and executes the registered callbacks. New events generated by the execution of a callback are inserted into the local event queue of the processing thread. This mechanism ensures that all the events generated by a specific request are processed sequentially, according to the event-driven programming model. The callback table is automatically updated each time the execution flow encounters the declaration of a new callback function (see lines 7 and 13 in Fig. 2).

Request Completion. Responses are buffered using the **end** method. Once all events generated by a request are processed, the system replies to the client using the HTTP server thread, which also performs some post-processing tasks (e.g., generating the correct HTTP response headers and eventually closing the socket connection).

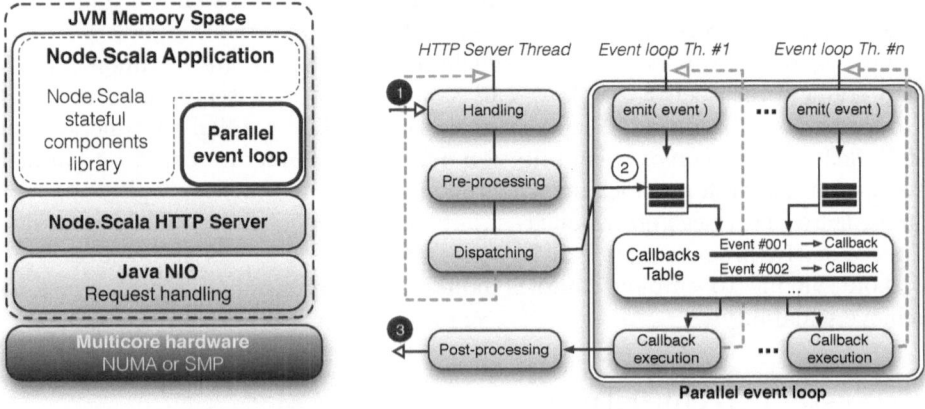

(a) High-level Architecture   (b) The Parallel Event-loop

Fig. 3. Overview of Node.Scala

4.1 Thread Safety

Node.Scala Web services are automatically guaranteed to be thread safe. To
this end, the runtime distinguishes between three types of requests: *stateful exclusive, stateful non-exclusive,* and *stateless.* This classification depends on the
type of accesses to global variables[2]. If the processing of a request can trigger
the execution of a callback that writes to at least one global variable, the request is considered stateful exclusive. Similarly, if the processing of a request
can result in at least one read access to a global variable, the request is considered stateful non-exclusive. All other requests are considered stateless. As a
consequence, a stateful exclusive request cannot be processed in parallel with
other stateful requests. Instead, multiple stateful non-exclusive requests can be
executed in parallel as long as no stateful exclusive requests are being processed.
Finally, stateless requests can be executed in parallel with any other stateless
and stateful request.

To perform this classification, Node.Scala intercepts class loading by means
of the `java.lang.Instrument` API and performs load-time analysis of the byte-
codes of each callback. Each user-defined callback is parsed by Node.Scala
to track accesses to global variables. To speedup the analysis, methods of
classes from the Node.Scala library are marked with two custom annotations:
`@exclusive` and `@nonexclusive`.

Each time the analysis classifies a new callback as stateful exclusive or stateful
non-exclusive, its bytecode is manipulated to inject all read (i.e., `ReadLock`) and
write (i.e., `WriteLock`) locks necessary to ensure thread safety[3]. Lock acquisition
instructions are injected at the beginning of the body of a callback, while lock

---

[2] Accesses to final values are not considered for the classification of requests.

[3] The semantics of `ReadLock` and `WriteLock` are defined in the documentation of the
standard Java class library.

release operations are injected at the end. Therefore, the entire body is guarded by the necessary locks. This mechanism allows the event-loop thread to try to acquire all necessary locks at once. In case of failure, the event-loop thread can delay the execution of the callback and process events generated by different requests without breaking the programming model. After a predefined number of failed attempts, the event-loop thread blocks waiting for all the locks to avoid starvation. To prevent deadlocks, we associate a unique ID to each lock and we sort the order of the inserted lock acquisition and release instructions accordingly.

In the worst-case scenario (i.e., all callbacks always require the acquisition of the same set of exclusive locks) only a single event-loop thread can execute a single request at any given time. In this case, the performance of the service is comparable to the one of single-process, event-based frameworks that make use of sockets to communicate between different processes (e.g., Node.JS). In all the other cases, Node.Scala can effectively and safely parallelize the execution of callbacks, taking advantage of all available cores to increase throughput, as illustrated in the following section.

5   Performance Evaluation

To assess the performance of the Node.Scala runtime, we have implemented a Web service similar to the one presented in Fig. 2. Instead of the simple Fibonacci function, we used the entire set of CPU-bound benchmarks of the SciMark 2.0[4] suite, a well-known collection of scientific computing workloads. The service has been implemented using only blocking function calls, while both stateless and stateful services performance have been evaluated.

The machine hosting the service is a Dell PowerEdge M915 with four AMD Opteron 6282 SE 2.6 GHz CPUs and 128 GB RAM. Each CPU consists of 8 dual-thread modules, for a total of 32 modules and 64 hardware threads. Since threads on the same module share access to some functional units (e.g., early pipeline stages and the FPUs), the throughput of Node.Scala is expected to scale linearly until 32 event-loop threads. The system runs Ubuntu GNU/Linux 11.10 64-bit, kernel 3.0.0-15, and Oracle's JDK 1.7.0_2 Hotspot Server VM (64-bit).

The runtime performance of Node.Scala is measured using a separate machine, connected with a dedicated gigabit network connection. We use httperf-0.9.0[5] to generate high amounts of HTTP requests and compute statistics about throughput and latency of responses. For each experiment, we report average values of five tests with a minimum duration of one minute and a timeout of 5 seconds. Requests not processed within the timeout are dropped by the client and not considered for the computation of the throughput.

5.1   Stateless Services

To evaluate the performance of the Node.Scala runtime with stateless requests (i.e., with callbacks neither modifying nor accessing any global state), we have

---

[4] http://math.nist.gov/scimark2/
[5] http://code.google.com/p/httperf/

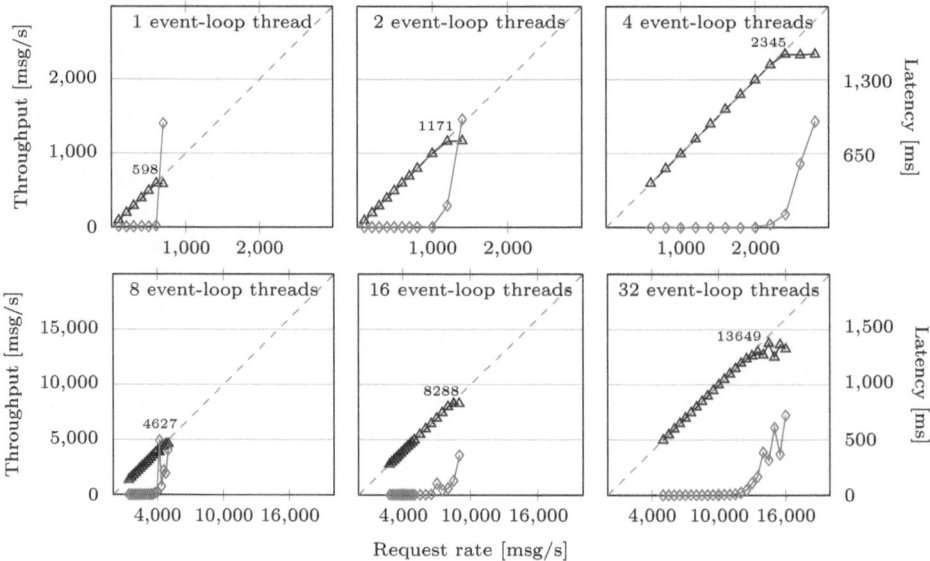

Fig. 4. Stateless service: throughput (—▲—) and latency (—◇—) depending on the arrival rate and the number of event loop threads. The dashed reference line (- - -) indicates linear scalability.

disabled the caching mechanism in the evaluated service. Therefore, the service is a pure-functional implementation of the SciMark benchmark suite.

Fig. 4 illustrates the variation of throughput and latency of responses depending on the request rate and on the number of event-loop threads. The experiment with a single event-loop thread resembles the configuration of common single-threaded event-driven frameworks for Web services, such as Node.JS. In this case, the throughput matches the request rate until a value of 600 requests per second. During this interval, the latency remains below 10ms. Afterwards, the system saturates because the single event-loop thread cannot process more requests per unit time. As a consequence, the throughput curve flattens and the latency rapidly increases to more than one second.

Experiments with larger amounts of event-loop threads follow a similar behavior: the latency remains small as long as the system is not saturated, and it rapidly increases afterwards. The peak throughput measured at the saturation point scales almost linearly with the number of event-loop threads, up to a value of 13600 msg/s with 32 threads. This confirms the ability of Node.Scala to take advantage of all available CPU cores to improve the throughput of stateless Web services. Our experiments also confirm that the parallel runtime of Node.Scala allows the developer to use blocking function calls without any performance degradation.

5.2   Stateful Services

To evaluate the performance of stateful services, we enabled the caching mechanism of the Node.Scala service used for the evaluation, and we have tested it with

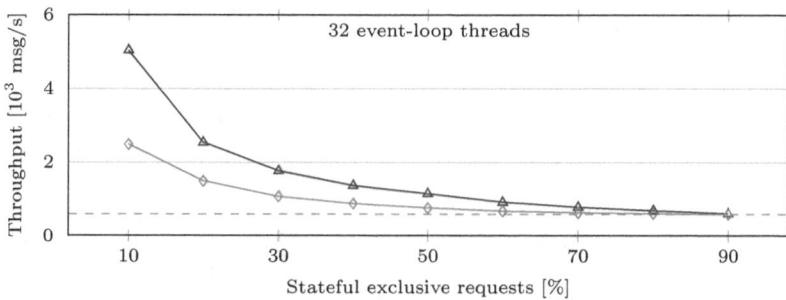

Fig. 5. Stateful services: throughput of SciMarkSf1 (——▲——) and SciMarkSf2 (——◇——) depending on the percentage of stateful exclusive requests. The reference line (- - -) refers to the throughput achievable using a single event loop thread.

two different workloads. The first one (called `SciMarkSf1`) makes an extensive use of the caching mechanism, forcing the runtime to execute either exclusive or non-exclusive callbacks. The second one, (called `SciMarkSf2`) uses the caching mechanism only to store new data. Therefore, the second workload requires the runtime to process both exclusive and stateless callbacks.

The goal of both workloads is to assess the performance of the service in the worse possible cases, i.e., when the service is intensively using a single common shared object.

Fig. 5 reports the peak throughput of the two considered Web services, executed with 32 event-loop threads, depending on the amount of stateful exclusive requests. We do not report the values for corner cases, that is, 0% and 100%, because they are equivalent to the peak throughput presented in Fig. 4 for the cases with 32, respectively 1, event-loop threads. As reference, we plot a line corresponding to the performance with a single event-loop thread. When the number of stateful exclusive requests is high, performance is comparable to those of traditional, single-threaded, event-driven programming frameworks. However, when this number is smaller, Node.Scala can effectively take advantage of available cores to achieve better throughput.

6   Related Work

Web server architectures can be roughly classified into three categories [12]: thread-based, event-based, and hybrid [5,8]. The runtime of Node.Scala lies in the latter category, as it uses both event-loops and threads. A similar approach is represented by the SEDA architecture [16]. Both SEDA systems and the Node.Scala runtime feature multiple event queues and multiple threads. However, Node.Scala features a programming framework built on top of its runtime architecture which allows to develop stateful services, while SEDA's focus is only at the runtime level and does not handle state. There are several examples of event-based Web servers [11], as well as thread-based servers [15]. A long-running debate (e.g., [10,15]) comparing the merits of the two approaches has been summarized in [12]. In the same paper, an exhaustive evaluation shows that event-based servers yield higher throughput (in the order of 18%) compared to thread-based servers under certain circumstances. A previous attempt to parallelize event-based services has been presented in [17]. The approach proposed to manually annotate callbacks with color-based annotations, and then to schedule callbacks for parallel execution according to their color. In Node.Scala no manual intervention from the developer is needed to parallelize the service since callbacks do not have to be annotated. Akka[6] is a JVM framework for developing scalable Web services using the Actor model. Like Node.Scala, Akka supports HTTP and REST, as well as Java NIO. Differently, Node.Scala features a library to share state among different client requests, while Akka relies on Software Transactional Memory. Out of the realm of the JVM, event-based programming is implemented in several frameworks and languages. For instance, Ruby's EventMachine[7] allows services to be developed using the Reactor event-loop pattern [13].

7   Conclusion

In this paper we presented Node.Scala, a programming framework and a runtime system for the development of high-throughput Web services in Scala. Node.Scala features an event-based programming model with implicit parallelism and safe state management. Node.Scala developers have to deal neither with abstractions such as parallel processes or threads, nor with synchronization primitives such as locks and barriers. Instead, the developer can focus on the service business logic, while the Node.Scala runtime takes care of the parallel processing of concurrent requests. Services built with Node.Scala do not suffer from limitations of single-threaded event-based frameworks like long-running blocking methods and lack of support for shared memory. Thanks to the parallel event-loop architecture of Node.Scala, services leverage current shared-memory multicore machines with both stateless (i.e., purely functional) services and stateful ones. Stateless services exhibit controlled latency and linear scalability up to saturation. In stateful scenarios the parallel runtime system allows Node.Scala services to exploit a shared memory space and thus obtain better performance compared to other single-process solutions.

Our ongoing research focuses on extending the Node.Scala library with additional objects from the Scala standard library. To this end, we are experimenting with bytecode analysis techniques to automatically annotate Scala types with the @exclusive/@nonexclusive annotations used by the Node.Scala runtime to protect callback invocations. Finally, we are also consolidating the Node.Scala approach by generalizing its runtime system in order to port the Node.Scala parallel event-loop system to other JVM-based functional programming languages such as Groovy, Clojure, and Rhino JavaScript.

---

[6] http://akka.io/
[7] http://rubyeventmachine.com/

Acknowledgment. This work is partially funded by the Swiss National Science Foundation with the SOSOA project (SINERGIA grant nr. CRSI22-127386).

References

1. Bahi, J., Couturier, R., Laiymani, D., Mazouzi, K.: Java and Asynchronous Iterative Applications: Large Scale Experiments. In: Proc. of the IEEE International Parallel and Distributed Processing Symposium (IPDPS), pp. 1–7 (2007)
2. Cardellini, V., Casalicchio, E., Colajanni, M., Yu, P.S.: The State of the Art in Locally Distributed Web-Server Systems. ACM Comput. Surv. 34, 263–311 (2002)
3. Dabek, F., Zeldovich, N., Kaashoek, F., Mazières, D., Morris, R.: Event-Driven Programming for Robust Software. In: Proc. of the 10th ACM SIGOPS European Workshop (EW), pp. 186–189 (2002)
4. Fielding, R.T.: Architectural Styles and the Design of Network-Based Software Architectures. Ph.D. thesis, UCI, Irvine (2000)
5. Haller, P., Vetta, A.: Actors That Unify Threads and Events. In: Murphy, A.L., Ryan, M. (eds.) COORDINATION 2007. LNCS, vol. 4467, pp. 171–190. Springer, Heidelberg (2007)
6. Kinder, K.: Event-Driven Programming with Twisted and Python. Linux J. (2005)
7. Li, P., Wohlstadter, E.: Object-Relational Event Middleware for Web Applications. In: Proc. of the Conference of the Center for Advanced Studies on Collaborative Research (CASCON), pp. 215–228 (2011)
8. Li, P., Zdancewic, S.: A Language-based Approach to Unifying Events and Threads. CIS Department University of Pennsylvania (April 2006)
9. Li, Z., Levy, D., Chen, S., Zic, J.: Auto-Tune Design and Evaluation on Staged Event-Driven Architecture. In: Proc. of the 1st Workshop on MOdel Driven Development for Middleware (MODDM), pp. 1–6 (2006)
10. Ousterhout, J.: Why Threads are a Bad Idea (for Most Purposes). In: USENIX Winter Technical Conference (1996)
11. Pai, V.S., Druschel, P., Zwaenepoel, W.: Flash: an Efficient and Portable Web Server. In: Proc. of the USENIX Annual Technical Conference (USENIX), p. 15 (1999)
12. Pariag, D., Brecht, T., Harji, A., Buhr, P., Shukla, A., Cheriton, D.R.: Comparing the Performance of Web Server Architectures. In: Proc. of the 2nd ACM SIGOPS European Conference on Computer Systems (EuroSys), pp. 231–243 (2007)
13. Schmidt, D.C., Rohnert, H., Stal, M., Schultz, D.: Pattern-Oriented Software Architecture: Patterns for Concurrent and Networked Objects, 2nd edn. Wiley (2000)
14. Tilkov, S., Vinoski, S.: Node.js: Using JavaScript to Build High-Performance Network Programs. IEEE Internet Computing 14(6), 80–83 (2010)
15. Von Behren, R., Condit, J., Brewer, E.: Why Events Are a Bad Idea (for High-Concurrency Servers). In: Proc. of the 9th Conference on Hot Topics in Operating Systems, vol. 9, p. 4 (2003)
16. Welsh, M., Culler, D., Brewer, E.: SEDA: an Architecture for Well-Conditioned, Scalable Internet Services. In: Proc. of the ACM Symposium on Operating Systems Principles (SOSP), pp. 230–243 (2001)
17. Zeldovich, N., Yip, E., Dabek, F., Morris, R.T., Mazires, D., Kaashoek, F.: Multiprocessor Support for Event-Driven Programs. In: Proc. of the USENIX Annual Technical Conference (USENIX), pp. 239–252 (2003)

Task-Parallel Programming on NUMA Architectures*

Christian Terboven, Dirk Schmidl, Tim Cramer, and Dieter an Mey

JARA, RWTH Aachen University, Germany
Center for Computing and Communication
{terboven,schmidl,cramer,anmey}@rz.rwth-aachen.de

Abstract. The multicore era has led to a renaissance of shared memory parallel programming models. Moreover, the introduction of task-level parallelization raises the level of abstraction compared to thread-centric expression of parallelism. However, tasks might exhibit poor performance on NUMA systems if locality cannot be controlled and non-local data is accessed.

This work investigates various approaches to express task-parallelism using the OpenMP tasking model, from a programmer's point of view. We describe and compare task creation strategies and devise methods to preserve locality on NUMA architectures while optimizing the degree of parallelism. Our proposals are evaluated on reasonably large NUMA systems with both important application kernels as well as real-world simulation codes.

1 Introduction

Recent multi-core architectures and the availability of cost-efficient two- and quad-socket compute nodes with large memory led to an increasing interest in shared memory programming models, both in combination with MPI or as the sole source of parallelism. The increasing number of cores imply a non-uniform memory access (NUMA) to provide appropriate memory bandwidth, even on commodity x86 architectures. In a NUMA architecture, the memory is partitioned and the latency and bandwidth of a memory access depend on the distance to the core from which the access occurs. The thread-centric expression of parallelism, like worksharing in OpenMP, works fine on such machines for well-structured code and evenly balanced algorithms. However, this has been found unsuitable to be applied to certain types of codes, such as recursive algorithms, unbounded loops, or irregular problems in general. Task-level parallelism provides solutions for these applications and promises to provide a high level abstraction for the programmer.

While threads can be bound to cores, or to a subset of the machine in general, recent parallelization paradigms embracing tasks do not offer means to control by which thread and on which core tasks are executed. Thus, they might exhibit poor performance on NUMA systems if tasks are executed on a NUMA node that does not contain the data being consumed during execution, and non-local data has to be accessed. As OpenMP [11] has become the de-facto standard for shared memory parallelization in

---

* Parts of this work were funded by the German Federal Ministry of Research and Education (BMBF) under Grant No. 01IH11006.

C. Kaklamanis et al. (Eds.): Euro-Par 2012, LNCS 7484, pp. 638–649, 2012.

HPC applications, we concentrate on the OpenMP tasking model in this work. By observing how current implementations work and execute tasks, we derive strategies for task-parallel programming that take the data allocation on NUMA architectures into account. We show that our patterns are successful for real-world applications by comparing task-parallel and thread-centric implementations on current NUMA systems.

This paper is structured as follows: Chap. 2 summarizes related work. Chap. 3 first emphasizes our expectations on the tasking model compared to a thread-centric view of parallelism and then explains the various patterns to express parallelism with tasks. In Chap. 4 we discuss our observations on the behavior of task-parallel kernels on NUMA architectures. In Chap. 5 we report our findings applying the presented patterns to real-world applications. Chap. 6 contains our conclusions and advice to programmers.

2   Related Work

Tasking [2] has been introduced in OpenMP 3.0 to allow for the expression and exploitation of unstructured parallelism. In order to extend the OpenMP standard, a reference implementation has to be provided for every new feature, which in the case of tasking was done at the Barcelona Supercomputing Center [15]. It was shown early on that OpenMP tasking is able to deliver comparable performance to OpenMP worksharing implementations [3]. While we also compare task-parallel implementations to worksharing, we additionally focus on the programmer's view of how to express task-parallelism especially on NUMA architectures. We introduce patterns for data setup/initialization as well as the actual computation and carry our findings from kernels to real-world application codes.

Several articles deal with the efficient scheduling of OpenMP tasks on multi-core multi-socket (NUMA) machines [10,4]. The main challenge is to reflect the system's memory hierarchy in the execution of the OpenMP tasks, while little or no knowledge is present of how task are being executed inside the application. Furthermore, task-stealing has to be applied in order to perform load balancing, which means the assignment of tasks from an overutilized thread to an underutilized thread. However, if tasks are moved to a different NUMA node, data of 'stolen' tasks remain on the NUMA node of the initialization, which then leads to remote memory accesses during task execution, as the Linux operating system does not perform any auto-migration of memory pages. In our work, we exploit knowledge of the implementation internals to present task generation patterns that allow task scheduling while maintaining data locality.

3   Patterns for Task-Parallelism

Two performance-critical aspects of shared memory parallel programming are load balancing and data locality. While the execution of iterations in loop-level parallelization can be controlled by schedule clauses and the threads in the team executing the worksharing construct can be bound to cores, the behavior of tasks is much less predetermined by the OpenMP specification. It is specified that a task may be picked up by any thread of the current team and that `tied` tasks may be suspended at so-called task scheduling points, `untied` tasks at every point in time. These restrictions allow

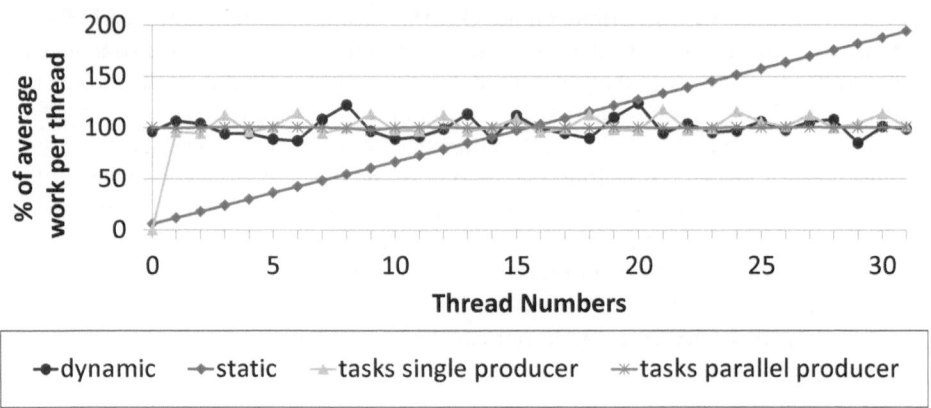

Fig. 1. Distribution of loop iterations to threads with linearly increasing load

OpenMP implementors to schedule tasks in many different ways. For example, executing all tasks by the creating thread immediately after encountering the task construct would fullfil the specification, but result in sequential execution. Pushing tasks to multiple task queues and applying work stealing approaches is a similarly valid way to handle tasks. The user has no direct influence on the scheduling decisions of the OpenMP runtime, but the scheduling has significant influence on the efficiency of a task-parallel program.

In general, there are two different patterns of task creation:

- *single-producer multiple-executors*: This pattern is popular for that it often requires little changes to code and data structures. The `single` construct ensures that a code region is executed by one thread only and thus avoids data races. The thread executing the `single` construct is responsible for creating all tasks of appropriate *task chunk size (tcs)* and all data necessary for the computation inside the tasks can be packed up at creation time using the `firstprivate` clause. The implicit barrier at the end of the `single` construct waits for the termination of all tasks.
- *parallel-producer multiple-executors*: A parallel OpenMP `for` worksharing construct loops over the outer iteration space with an increment specified as *task chunk size (tcs)*. In every iteration a task is spawned, performing the iteration over a range of size *tcs*. Thus, all threads of the team executing the worksharing construct create multiple tasks in parallel. The implicit barrier at the end of the `for` construct waits for the termination of all tasks. This pattern can also be expressed without any worksharing construct at all, as the content of a parallel region is executed by all threads of the corresponding team and thus a task construct encountered by all threads creates multiple tasks. Then the synchronization is performed at the end of the parallel region, or by appropriate task synchronization contructs or an explicit `barrier`.

Load Balancing. In order to investigate the load balancing capabilities of tasks, we created a simple test program and used the Intel C/C++ Compiler version 12.1.2. In this program, a loop over an array with 128,000 elements has been parallelized with a `for` worksharing construct and also with two task-parallel approaches for a direct comparison. In every iteration an array element is filled with a constant value. To investigate the load balancing behavior, every array is an array itself, and the length of the inner arrays is increasing linearly. In our experiments we record which thread performed the work on the elements of the outer array, resulting in a mapping of work to threads. Fig. 1 shows the distribution of iterations for the parallel `for` workshare variant with `static` and `dynamic` loop schedules as well as for the task-parallel executions. All measurements were carried out using 32 threads on the system described in Chap. 4, consisting of four NUMA nodes with 32 physical cores in total. As expected, with a `static` schedule the work is distributed unevenly over the threads, resulting in load imbalance. The linearly increasing load per iteration is the 'worst case' situation for this schedule. It assigns $\frac{128,000 \; its}{32 \; threads} = 4000$ iterations to every thread in which the first iterations are computationally much less expensive than the last chunk of iterations. The `dynamic` schedule distributes the load much better over all threads, which then execute between 87% and 120% of the average work. In the task-parallel single-producer version, the first thread executes nearly no work, since it is responsible for creating all the tasks. But the distribution of work to the other 31 threads is as good as in the `dynamic` schedule variant. The parallel-producer scheme reaches a nearly even distribution of work over all threads, close to the optimal load balancing.

Data Affinity. We also aim to understand the behavior of worksharing and task constructs with regard to data locality, taking the same array of arrays as described above for our experiment setup. The data has been distributed among the NUMA nodes using a chunk size of 4000 elements of the outer array, meaning the first 4000 elements (including the inner arrays) reside on the NUMA node that thread 0 has been bound to. The next chunk is on the NUMA node of thread 1, and so on. Again we recorded for every thread the number of iterations it worked locally and remotely, the averages work (number of updates of an inner array element) are shown in Table 1. For the `static` schedule only local accesses occur since the distribution of array elements among the threads is exactly the same for both the initialization and the iteration phases. The `dynamic` schedule and the task-parallel single-producer scheme, which both show good load balancing, lead to only about 3% of local accesses, because for both the distribution of iterations or task, respectively, to threads is undeterministic and obviously is not the same for the initialization and iteration phases. The parallel-producer scheme achieves a local access rate of about 80%. Note that due to the uneven distribution of data over NUMA nodes, 'perfect' data locality would imply weak load balancing. The parallel-producer pattern delivers the best compromise between load balancing and locality.

We also investigated how to employ tasks in the initialization of a sparse matrix structure, as it appears for example in CG-type solvers, comparing to the common practice of initializing the data in a parallel loop over the matrix rows [14]. We compared four different initialization strategies: using just one thread (serial), a `static` schedule with a `for` workshare construct, the `numactl` tool to enforce a round robin page distribution over NUMA nodes when the actual initialization is performed by one thread

Table 1. Average percentage of local and remote data accesses

	local iterations	remote iterations
`for` worksharing with `static` schedule	100%	0%
`for` worksharing with `dynamic` schedule	3.06%	96.94%
tasks-parallel single-producer	3.10%	96,90%
tasks-parallel parallel-producer	79.51%	20.49%

only, and finally tasks to initialize the data row-wise. Again, for the tasking variants we distinguish between the single- and the parallel-producer patterns. Fig. 2 shows that the `static` schedule, the round robin and the parallel-producer strategies result in a regular page distribution while for a serial initialization the complete memory is located on one NUMA node. However, the results also show that the single-producer pattern leads to an irregular distribution of pages, in which most are allocated on the 'single' node. This would lead to a serious performance degradation. The reason is that the initialization tasks are computationally very cheap and short-lived and the other threads on NUMA node 0 - besides the one performing the task creation - execute tasks at a fast enough pace. Task stealing does not occur in a noteworthy amount from other NUMA nodes.

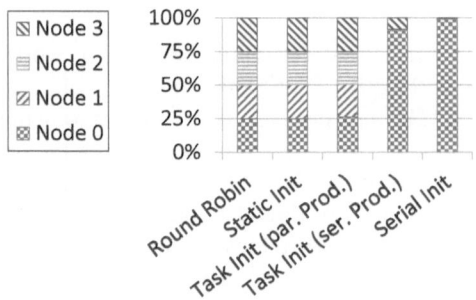

Fig. 2. Page distribution over the NUMA nodes after the matrix initialization

Multiple Levels of Parallelism. Composability of software components is not well-supported in OpenMP, i.e. worksharing constructs may not be nested within the dynamic extend of one single parallel region. Tasks can be nested inside other tasks and a worksharing construct as well, this particularly opens the opportunity for the parallel-producer pattern. The nesting of parallel regions has been supported by OpenMP early on, but only few application success stories have been reported [1]. Furthermore, nested parallel regions introduce several problems in general and on NUMA architectures in particular: (1) the thread teams for the inner parallel region are not guaranteed to be the same for two consecutive calls [5], thus data affinity cannot be maintained; and (2) the end of the inner parallel region always implies a barrier. These problems do not occur with nested tasks, especially with the techniques to create tasks we discussed so far. In Chap. 5 we show that for the FIRE code an implementation with nested tasks clearly outperforms an implementation with nested parallel regions.

Summing Up. We have shown that both task-parallel implementations perform the load balancing as well as OpenMP's `for` workshare with the `dynamic` schedule, but the parallel-producer pattern provides significantly better data locality. Tasks created in a thread bound to a particular NUMA node are picked up for execution on the same NUMA node, so that data locality is maintained if the same pattern is used during data initialization and the actual computation. This observation complies with the schedule strategies expressed in articles on OpenMP task implementations, as outlined in Chap. 2. Furthermore, one can expect the parallel-producer pattern to scale better than the single-producer in case of many small tasks, since the task creation occurs in parallel instead of being serialized. This will be analyzed in Chap. 4.

4   Task Behavior on NUMA Architectures

In this chapter we examine the behavior of two kernels, which both employ tasks, on a NUMA architecture. All measurements in this chapter have been performed on a bullx s6010 compute node, equipped with four Intel Xeon X7550 processors running at 2.0 GHz, offering 32 physical cores and 64 logical cores with hyper-threading, and 256 GB of main memory. The Intel Quickpath Interconnect (QPI) used to connect the four sockets with each other and with I/O facilities creates a system topology with four NUMA domains, with every NUMA node being separated from any other by just one hop. The system is running Scientific Linux 6.1.

4.1   STREAM

The first set of experiments has been carried out with the STREAM [9] benchmark. We picked this particular kernel to investigate effects of task-parallel implementations on NUMA architectures for two reasons: (1) if the data initialization is not done in the right way, the performance will be degraded significantly, as the computation performed in the individual kernels is memory-bound; and (2) the naive worksharing implementation delivers optimal performance if the same loop schedule is used during the data initialization and the actual computation. For the sake of brevity we only discuss results from the triad (daxpy) operation, since they are consistent with the other ones.

In Fig. 3 we compare the original parallel STREAM implementation referred to as *workshare: static-init for-loop* to several task-parallel versions. The original parallel version employs an OpenMP `for` worksharing construct with a `static` schedule both during data initialization and the actual computation, meaning that for t threads the arrays are divided into t parts of approximately equal size. Given four NUMA nodes in the system and a *scatter* thread binding, meaning threads are spread as far apart as possible, $\frac{t}{4}$ threads will be bound to each NUMA node. This results in an even data distribution over all NUMA nodes in the system. And as the computation is performed in the same manner, the number of remote accesses should be minimal. The arrays

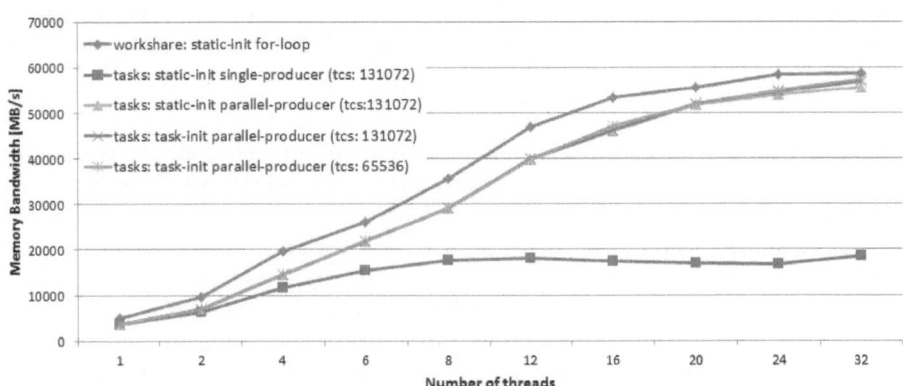

Fig. 3. STREAM triad operation: worksharing vs. task-parallel variants

have a dimension of $256, 435, 456$ double elements, which results in 1.96 GB of memory consumption per array, or 5.87 GB of total kernel size in the triad operation. Although the system offers 64 GB of memory per NUMA node, this kernel size is much larger than the cumulated cache size and thus we achieve reliable measurements of the memory bandwidth of the system. We implemented the following task-parallel variants, in all three the task chunk size (tcs) refers to the number of iterations grouped together in a single task:

- *tasks: static-init single-producer*: The data initialization is performed in the same way as in the original parallel version. The generation of tasks is performed by one thread only within a single construct (*single-producer* pattern).
- *tasks: static-init parallel-producer*: Again the data initialization is performed in the same way as in the original parallel version, but now the creation of tasks is performed in parallel (*parallel-producer multiple-executors* pattern).
- *tasks: task-init parallel-producer*: In this version both the data initialization and the computation is performed task-parallel by applying the same pattern to both code regions.

The results in Fig. 3 show that the worksharing version outperforms the best task-parallel version by just 3 %. The two task-parallel variants employing the parallel-producer pattern deliver approximately the same performance, as both distribute the data in an optimal fashion over the NUMA nodes. The single-producer tasking version clearly suffers from two effects: (1) the runtime cannot maintain data affinity, as all task are created from a single NUMA node and the work-stealing will just pick arbitrary tasks from the queue; and (2) the single thread responsible for creating the tasks cannot completely keep the other threads executing the tasks busy. The parallel-producer pattern has to be used in the data initialization and the computation so that the OpenMP runtime is able to maintain data affinity. If only one thread creates all the tasks, the runtime's task-stealing mechanism cannot take the data distribution into account during the 'stealing' and thus the performance on NUMA systems obviously suffers. As with the results discussed in the previous chapter, the task chunk size does not have a significant influence on the performance as long as enough tasks are spawned to generate enough parallelism and as long as the work per task is computationally expensive enough compared to the task creation and scheduling overhead.

4.2 Sparse-Matrix-Vector-Multiplication in a CG-Method

While STREAM served our purpose as a simple benchmark indicating fine differences in the memory access pattern, the Sparse-Matrix-Vector-Multiplication (SMXV) in a CG-Method [8] much more resembles a real-world compute kernel as part of many PDE solvers. Depending on the problem the matrix for the system of linear equations can be very irregular. In this case the sparse matrix vector product is a typical example of the importance of adequate load balancing. Especially in cases where the optimal work distribution cannot be calculated in advance, we expect task-parallel implementations to help avoiding performance issues. On the one hand the programmer has to ensure that a sufficient number of tasks is used to avoid load imbalance, on the other hand too many tasks introduce additional overhead. In our CG implementation all vector operations and the dot-product are parallelized with OpenMP `for` constructs. Only the SMXV is parallelized with tasks. The work is distributed by chunks of rows and the chunk size is the same for each task, calculated as

$$chunk\_size(tasks) = \begin{cases} \lfloor N/tasks \rfloor, & \text{if } N\%tasks = 0 \\ \lfloor N/tasks \rfloor + 1, & \text{otherwise} \end{cases} \tag{1}$$

where N is the dimension of the square matrix and $tasks$ the number of tasks. The matrix used here represents a computational fluid dynamics problem (Fluorem/HV15R) and is taken from the University of Florida Sparse Matrix Collection [6]. The dimension is $N = 2,017,169$ and the number of nonzero values is $nnz = 283,073,458$, which results in a memory footprint of approximately 3.2 GB. As shown in Fig. 4 the sparsity pattern is slightly unbalanced regarding a `static` distribution.

Fig. 5 shows the SMXV performance when executing 1000 CG iterations. It compares the effect of different initialization strategies for both tasking patterns introduced in Chap. 3. The page distribution after the initialization of the sparse matrix correlates with the performance results of this experiment (see Fig. 2). As shown in Fig. 5(a), we reach a peak performance of 10 GFLOPS for the given workload. It also shows that using a `static` schedule for the data initialization is much better than using serial or a serial-producer task initialization. However, the performance for the `static` schedule decreases for more than 256 tasks while the random initialization still works well for 8192 tasks, which translates to a chunk size of 247 rows. It is obvious that the number of tasks is very important to reach the best performance. If too few tasks are used the load imbalance decreases the performance slightly. The overhead for the use of too many tasks dominates the runtime if chunks consists of only a few rows. Fig. 5(b) shows that for the parallel-producer pattern the peak performance reaches almost 13 GFLOPS. The performance of the round-robin initialization (10 GFLOPS) is comparable to the performance of the single-producer case, but the performance decline only occurs for

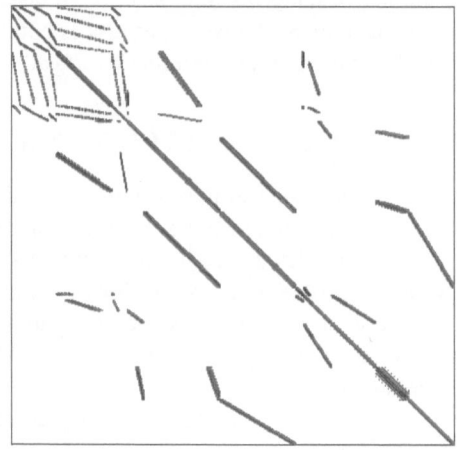

Fig. 4. Sparsity pattern of the matrix used in the CG-method

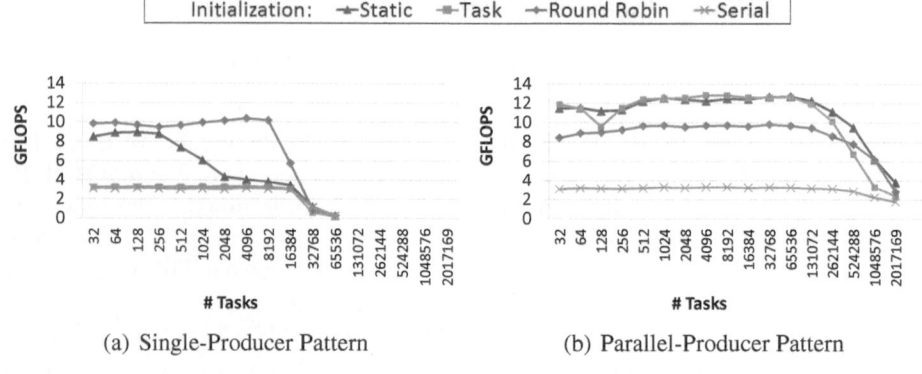

(a) Single-Producer Pattern

(b) Parallel-Producer Pattern

Fig. 5. Performance of the SMXV kernel within the CG-method

more than 100,000 tasks and is not that significant. This proves that the task creation overhead in the OpenMP runtime is 'parallelized' in the parallel-producer pattern. The fact that results for the `static` schedule and the parallel-producer variants are better shows that the programmer has a much better influence on data locality by using this pattern.

5   Application Case Studies

In order to prove the applicability of the patterns and strategies we discussed so far, we employed them to two real-world application codes. In this chapter we show that our tasking implementation of TrajSearch reaches the same performane as the state of the art worksharing implementation, and for the FIRE code it event outperforms the corresponding worksharing variant. For all performance experiments in this chapter we use a Bull BCS system consisting of four bullx s6010 system as described in Chap. 4. The four systems are equipped with Bull's proprietary BCS cards providing a cache-coherent and high performant interconnect, creating a 128 core system with 16 NUMA nodes.

5.1 Trajectory Search

TrajSearch is a code to investigate turbulences which occur during combustion. It is a post-processing code for dissipation element analysis developed by Peters and Wang [12] from the Institute for Combustion Technology[1] at the RWTH Aachen University. It decomposes a highly resolved 3D turbulent flow field obtained by Direct Numerical Simulation (DNS) into non-arbitrary, space-filling and non-overlapping geometrical elements called 'dissipation elements'. Starting from every grid point in the direction of ascending and descending gradient of an underlaying diffusion controlled scalar field, the local maximum and minimum point are found. A dissipation element is defined as a volume from which all trajectories reach the same minimum and maximum point.

Every trajectory can be investigated independently from the others in parallel. A version of this code was parallelized with traditional worksharing and a different version has been parallelized with tasks. Fig. 6 (left) shows the performance results of tests done with both versions on the NUMA system. Due to the long execution time, we restricted our experiments to at least 16 threads. The `static` tests with the `for` worksharing construct perform slightly worse than the `dynamic` workshare and the task-parallel version. This is because the time for a single search for a trajectory is not constant, it depends on the length of the trajectory which is unknown a priori. This leads to some load imbalance and thus to a performance penalty when a `static` schedule is used. The `dynamic` parallel `for` loop and the tasking versions perform better, since the load is distributed among the threads more evenly. In conclusion it can be seen, that the load balancing capabilities of tasks for this application are as good as when a `for` worksharing loop with `dynamic` schedule is used.

5.2 FIRE

The Flexible Image Retrieval Engine (FIRE) [7] was developed at the Human Language Technology and Pattern Recognition Group[2] of RWTH Aachen University. The retrieval engine takes a set of query images and for each query image it returns a number of similar images from an image database. The similarity is derived from comparing various image features. The existing parallelization of the FIRE code uses OpenMP [13] on two nested levels. On the outer level all query images are processed in parallel and on the inner level the comparison of one query image to the database images is also done in parallel.

We re-implemented the parallelization using OpenMP tasks. For every query image one task is created. Inside these tasks for every comparison of the query image to one

---

[1] http://www.itv.rwth-aachen.de
[2] http://www-i6.informatik.rwth-aachen.de

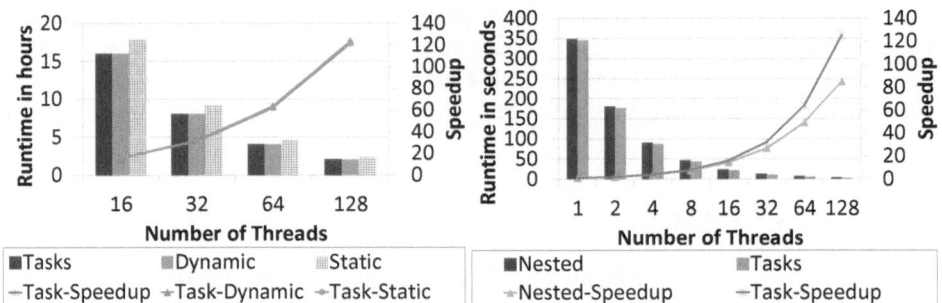

Fig. 6. Runtime and speedup of two application codes, TrajSearch (left) and FIRE (right). A comparison of tasking versions with a parallel for loop using different schedules for TrajSearch and with a version applying nested parallel regions for FIRE.

element of the database another task is created. Both parallel versions express the same amount of parallelism. Our test dataset comparing both versions processes 18 query images in a database consisting of 1000 images. The measured runtime and speedup on the 16-socket machine are shown in Fig. 6 (right). For the nested parallel regions the best combination of threads at the outer and inner regions has been used. E.g. the value for 16 threads is the minimum runtime for 1:16, 2:8, 4:4, 8:2 and 16:1 threads used at the outer:inner parallel regions.

Both versions of the code deliver nearly the same serial runtime, so the overhead of the OpenMP constructs is in the same order of magnitude. With more threads the tasking version outperforms the nested parallel region. For 128 threads the tasking version reaches a nearly linear speedup of 127 on 128 threads, whereas the nested parallel region only reaches a speedup of 85.

6 Conclusion

The introduction of task-level parallelism in OpenMP raised the level of abstraction compared to thread-centric worksharing models, by delegating the responsibility of distributing the work among the threads to the runtime. On hierarchical NUMA architectures, tasks exhibit poor performance if remote data is accessed frequently, that means if the runtime cannot maintain data locality when selecting a thread to execute a given task. As we have shown, if thread binding is used and the task-parallelism is expressed using an appropriate pattern during both the data setup/initialization as well as during the actual computation, modern OpenMP runtimes, like the one from Intel we used, can maintain data affinity and thus achieve performance on par with state-of-the-art worksharing implementations.

Furthermore, with the real-world application use cases we have shown that tasking implementations may outperform the comparable worksharing implementations. This is particularly true for situations in which the load is not evenly balanced and a `dynamic` scheduling scheme is employed, in which case tasks may offer an even finer load balancing plus the ability to maintain data locality by applying the patterns presented above. This also extends to cases in which the worksharing approach is limited, such as when OpenMP parallel regions have to be nested.

References

1. an Mey, D., Sarholz, S., Terboven, C.: Nested Parallelization with OpenMP. International Journal of Parallel Programming 35, 459–476 (2007), 10.1007/s10766-007-0054-1
2. Ayguadé, E., Copty, N., Duran, A., Hoeflinger, J., Lin, Y., Massaioli, F., Teruel, X., Unnikrishnan, P., Zhang, G.: The Design of OpenMP Tasks. IEEE Transactions on Parallel and Distributed Systems 20(3), 404–418 (2009)
3. Ayguadé, E., Duran, A., Hoeflinger, J., Massaioli, F., Teruel, X.: An Experimental Evaluation of the New OpenMP Tasking Model. In: Adve, V., Garzarán, M.J., Petersen, P. (eds.) LCPC 2007. LNCS, vol. 5234, pp. 63–77. Springer, Heidelberg (2008)
4. Broquedis, F., Furmento, N., Goglin, B., Wacrenier, P.-A., Namyst, R.: ForestGOMP: An Efficient OpenMP Environment for NUMA Architectures. International Journal of Parallel Programming 38, 418–439 (2010), doi:10.1007/s10766-010-0136-3
5. Terboven, C., an Mey, D., Schmidl, D., Jin, H., Wagner, M.: Data and Thread Affinity in OpenMP Programs. In: Proceedings of the 2008 Workshop on Memory Access on Future Processors: a Solved Problem?, MAW 2008, pp. 377–384. ACM (2008)
6. Davis, T.A.: University of Florida Sparse Matrix Collection. NA Digest 92 (1994)
7. Deselaers, T., Keysers, D., Ney, H.: Features for Image Retrieval - a quantitative comparison. Information Retrieval 11(2), 77–107 (2008)
8. Hestenes, M.R., Stiefel, E.: Methods of Conjugate Gradients for Solving Linear Systems. Journal of Research of the National Bureau of Standards 49(6), 409–436 (1952)
9. McCalpin, J.: STREAM: Sustainable Memory Bandwidth in High Performance Computers
10. Olivier, S.L., Porterfield, A.K., Wheeler, K.B., Prins, J.F.: Scheduling task parallelism on multi-socket multicore systems. In: Proceedings of the 1st International Workshop on Runtime and Operating Systems for Supercomputers, ROSS 2011, pp. 49–56. ACM, New York (2011)
11. OpenMP ARB. OpenMP Application Program Interface, v. 3.1, http://www.openmp.org
12. Peters, N., Wang, L.: Dissipation element analysis of scalar fields in turbulence. C. R. Mechanique 334, 493–506 (2006)
13. Terboven, C., Deselaers, T., Bischof, C., Ney, H.: Shared-Memory Parallelization for Content-based Image Retrieval. In: ECCV 2006 Workshop on Computation Intensive Methods for Computer Vision (CIMCV), Graz, Austria (May 2006)
14. Terboven, C., Spiegel, A., an Mey, D., Gross, S., Reichelt, V.: Parallelization of the C++ Navier-Stokes Solver DROPS with OpenMP. In: Joubert, G.R., Nagel, W.E., Peters, F.J., Plata, O.G., Tirado, P., Zapata, E.L. (eds.) PARCO. John von Neumann Institute for Computing Series, vol. 33, pp. 431–438. Central Institute for Applied Mathematics, Jülich (2005)
15. Teruel, X., Martorell, X., Duran, A., Ferrer, R., Ayguadé, E.: Support for OpenMP tasks in Nanos v4. In: Lyons, K.A., Couturier, C. (eds.) Proceedings of the 2007 Conference of the Centre for Advanced Studies on Collaborative Research, pp. 256–259. IBM (October 2007)

Speeding Up OpenMP Tasking*

Spiros N. Agathos**, Nikolaos D. Kallimanis***, and Vassilios V. Dimakopoulos

Department of Computer Science, University of Ioannina
P.O. Box 1186, Ioannina, Greece, GR-45110
{sagathos,nkallima,dimako}@cs.uoi.gr

Abstract. In this work we present a highly efficient implementation of OpenMP tasks. It is based on a runtime infrastructure architected for data locality, a crucial prerequisite for exploiting the NUMA nature of modern multicore multiprocessors. In addition, we employ fast work-stealing structures, based on a novel, efficient and fair blocking algorithm. Synthetic benchmarks show up to a 6-fold increase in throughput (tasks completed per second), while for a task-based OpenMP application suite we measured up to 87% reduction in execution times, as compared to other OpenMP implementations.

1 Introduction

Parallel computing is quickly becoming synonymous with mainstream computing. Multicore processors have conquered not only the desktop but also the hand-held devices market (e.g. smartphones) while many-core systems are well under way. Still, although highly advanced and sophisticated hardware is at the disposal of everybody, programming it efficiently is a prerequisite to achieving actual performance improvements.

OpenMP [13] is nowadays one of the most widely used paradigms for harnessing multicore hardware. Its popularity stems from the fact that it is a directive-based system which does not change the base language (C/C++/Fortran), making it quite accessible to mainstream programmers. Its simple and intuitive structure facilitates incremental parallelization of sequential applications, while at the same time producing actual speedups with relatively small effort.

The power and expressiveness of OpenMP has increased substantially with the recent addition of tasking facilities. In particular V3.0 of the specifications include directives that allow the creation of a task out of a given code block. Upon creation, tasks include a snapshot of their data environment, since their execution may be deferred for a later time or when task synchronization/scheduling directives are met. Tasking is already supported by many commercial and non commercial compilers (e.g. [2,4,16]).

Most of these implementations rely on sophisticated runtime libraries that provide each participating thread with private and/or shared queues to store tasks pending for

---

* This work has been supported in part by the General Secretariat for Research and Technology and the European Commission (ERDF) through the Artemisia SMECY project (grant 100230).
** S.N. Agathos is supported by the Greek State Scholarships Foundation (IKY).
*** N.D. Kallimanis is supported by the Empirikion Foundation.

C. Kaklamanis et al. (Eds.): Euro-Par 2012, LNCS 7484, pp. 650–661, 2012.

execution. Work-stealing [3] is usually employed for task scheduling, whereby idle threads, with no local tasks to execute, try to "steal" tasks from other thread queues. Work-stealing is a widely studied and deployed scheduling strategy, well known for its load balancing capabilities. Efficient implementation of the work-stealing algorithm and its related data structures is hence crucial for the performance of an OpenMP tasking system. The associated overheads for enqueing, dequeuing and stealing tasks can easily become performance bottlenecks limiting system's scalability as the number of cores keeps increasing.

In this work we present a high-performance tasking infrastructure built in the runtime system of the OMPi OpenMP/C compiler [6]. Support for tasking was recently added to OMPi [1], including an initial functional, albeit non-optimized, general tasking layer in its runtime library. Here we present a complete redesign of OMPi's tasking system, engineered to take advantage of modern multicore multiprocessors. The deep cache hierarchies and private memory channels of recent multicore CPUs make such systems behave with pronounced non-uniform memory access (NUMA) characteristics. To exploit these architectures our runtime system is organized in such a way as to maximize local operations and minimize remote accesses which may have detrimental performance effects. This organization is coupled with a work-stealing system which is based on an efficient blocking algorithm that emphasizes operation combining and thread cooperation in order to reduce synchronization overheads.

We have tested our system exhaustively. Using a synthetic benchmark we reveal a very significant— up to 6x—increase in attainable throughput (tasks completed per second), as compared to other OpenMP compilers, thus enjoying scalability under high task loads. At the same time applications from the BOTS tasking suite [8] experience reduced execution times (up to 87%), again in comparison to the rest of the available OpenMP systems.

The rest of the paper is organized as follows: in Section 2 we present OMPi and the way it handles tasking. The organization of its optimized runtime system is presented in detail. A key part, namely the work-stealing subsystem, is discussed separately in Section 3. Section 4 is devoted to the experiments we performed in order to assess the performance of our implementation and finally Section 5 concludes this work.

2   Tasking in the OMPi Compiler

OMPi [6] is an experimental, lightweight OpenMP V3.0 infrastructure for C. It consists of a source-to-source compiler and a runtime library. The compiler takes as input C code with OpenMP pragmas and outputs multithreaded C code augmented with calls to its runtime library, ready to be compiled by any standard C compiler.

Upon encountering an OpenMP `task` construct, the compiler uses *outlining* to move the code residing within the `task` region to a new function. Because each task is a block of code that may be executed asynchronously at a later time, its data environment must be captured at the time of task *creation*. Thus the compiler inserts code which allocates the required memory space, copies the relevant (firstprivate) variables and places a call to the runtime system to create the task using the outlined function and the captured data environment.

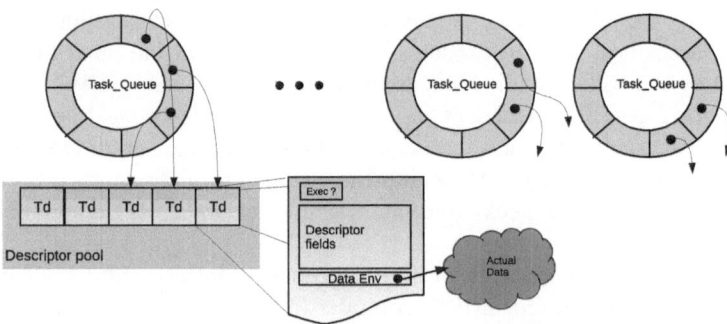

Fig. 1. Task queues organization. Each thread owns a circular queue (TASK_QUEUE) where pointers to task descriptors (Td) are inserted. Each Td carries bookkeeping information, a special flag (Exec) and a pointer to the task data.

If there exists an if clause whose condition evaluates to false or the runtime system cannot (or selects not to) create the task, then the task must be executed immediately. To optimize this case, the compiler produces a second copy of the task code (called *fast path*), this time *inlined*. Local variables are declared to capture directly the data environment and are used within the task code. In this manner, the task is executed with almost no overheads. Depending on the runtime conditions, either the normal (outlined) or the fast (inlined) path is executed.

2.1 Optimized Runtime

Our runtime organization is based on distributed task queues, one for each OpenMP thread as shown in Fig. 1. These are circular queues (TASK_QUEUES) of fixed size which is user-controlled as one of OMPi's environment variables. When a thread meets a new task region then it has the choice of executing it immediately or submitting it for deferred execution. OMPi follows the second approach, that is our runtime uses a *breadth-first* task creation policy, and the new task is stored in the thread's local TASK_QUEUE. Whenever a thread is idle and decides to execute a task, then it dequeues a deferred task from its TASK_QUEUE. If a thread's TASK_QUEUE is empty then this thread becomes a thief and traverses other threads queues in order to steal tasks. The manipulation of a TASK_QUEUE is a crucial synchronization point in OMPi, since multiple threads may concurrently access it. OMPi utilizes a highly efficient work-stealing algorithm described in the next section.

If a thread tries to store a new task in its queue and there is no space, the thread enters *throttling mode*. In throttling mode newly created tasks are executed immediately and hence the task creation policy changes to *depth-first*. In addition, as described above, throttled threads utilize the fast execution path. While in throttling mode all descendant tasks are executed immediately in the context of parent task, favoring data locality. Notice that a suspended parent task never enters the TASK_QUEUE hence it can never be stolen by any other thread. This is to say that in OMPi all tasks are *tied*.

A thread's entrance in throttling mode is one of the runtime objectives. However, a thread operating in throttling mode does not produce deferred tasks, which results in a

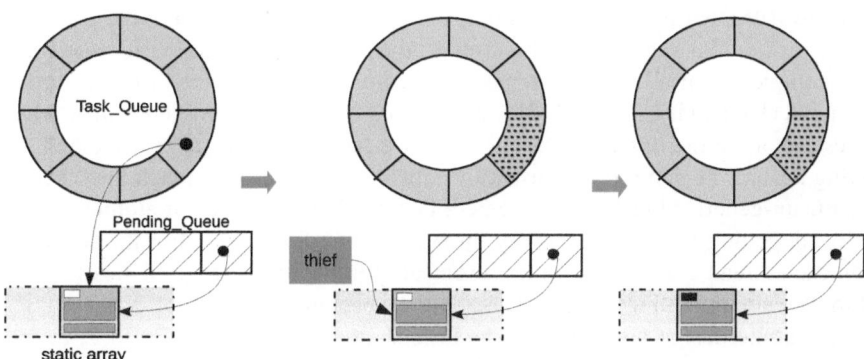

Fig. 2. Pending, executing (stolen) and finished task. When a task is pending for execution then corresponding entries in TASK_QUEUE and PENDING_QUEUE point to its Td. Upon dequeing, only the link in TASK_QUEUE is removed, freeing one slot. When the task is finished, the executing thread sets the Exec flag to announce that the descriptor can be recycled.

reduction of available parallelism. To strike a balance, before a throttled thread executes a new task, it checks its TASK_QUEUE free space. If the queue has become at least 30% empty then throttling is disabled and task creation policy returns to breadth-first.

As shown in Fig. 1 each entry in the TASK_QUEUE is a pointer to a task descriptor (Td), which stores all the runtime information related to the task execution as well as the task data environment. The descriptor is obtained out of the thread's descriptor pool. This pool contains an array of pre-allocated descriptors (in order to speed up the allocation process) and a dynamic overflow list for the case the array becomes empty. Whenever a task finishes its execution, the corresponding Td is returned to a descriptor pool, recycled for future reuse. A task created by a thread might be stolen and executed by another thread in its team. When the task finishes and the descriptor must be recycled, a decision has to be made as to which pool the descriptor should return to. If it enters the pool of the thread that executed the task, severe memory consumption is possible in cases where only few threads create a big number of tasks while the rest execute them. On the other hand, this option is a local operation, enjoying lack of contention. Memory consumption is reduced if the descriptor is put back to the task creator's pool, and this is what OMPi does. Notice though that synchronization needs arise since threads that stole tasks from the same thread may try to store to its descriptor pool concurrently.

In order to avoid the aforementioned synchronization overheads, we have used a garbage-collecting strategy, shown in Fig. 2. Each thread t maintains a private set of pointers (PENDING_QUEUE) to the task descriptors it has created and are either stored for deferred execution or are currently executing. When a task is dequeued for execution (e.g. because a thief stole it), the Td pointer is removed form TASK_QUEUE but remains intact in PENDING_QUEUE. The descriptor contains a special flag ('Exec' in Fig. 1). When the task completes its execution, the executing thread sets this flag to announce that the descriptor can now be recycled. On specific occasions thread t traverses its PENDING_QUEUE to find Tds that represent executed tasks and returns them to its pool for future use.

The PENDING_QUEUE plays a central role in the implementation of the taskwait and barrier constructs, too. Whenever a task meets a taskwait, it must wait until the completion of all tasks it created (it is actually then that the task execution and stealing mechanism is triggered). This completion condition is fulfilled simply when all the descriptors in the thread's PENDING_QUEUE have been flagged as executed. Upon meeting a barrier, a thread must wait until: (i) all its siblings reach the barrier and (ii) all team-generated tasks are executed. For the first condition an atomic counter is employed, getting increased by every thread reaching the barrier. For the second condition each thread contiguously executes/steals pending tasks from all TASK_QUEUE's within its team until all team PENDING_QUEUEs become empty.

Our runtime design aims at using as little shared data as possible, so as to reduce atomic operations and minimize thread synchronization. It is worth noting that in our tasking system thread synchronization occurs only in two cases. The first is during the unavoidable barrier construct and the second is during the work-stealing operations, as described in the next section, for which a very fast algorithm is employed. All data structures (e.g. Td's in the descriptors pool) are cache line-size aligned so as to eliminate false sharing phenomena and avoid triggering coherency protocol actions, which deteriorate the performance, especially in NUMA platforms.

While in OMPi each thread owns a public task queue where it stores newly created tasks, other compilers use different organizations. In IBM XL compilers [16], a shared task pool is associated with each parallel region where new tasks are put in the end of the queue and threads pick up tasks from the front. In contrast, in OpenUH [4] each OpenMP thread retains two task queues. The first queue is private and used for keeping tied tasks, while the second is public and used to store newly created and untied tasks. In Nanos [17], two types of queues are used. Here, a team of threads has a shared queue for newly created and untied tasks. Furthermore, each thread owns a private local queue used for tied tasks. A detailed comparison of many other queue organization alternatives has been performed by Korch and Rauber [11].

3   A Fast Work-Stealing Algorithm

The work-stealing mechanism is a crucial component of an OpenMP runtime and should thus be designed in a way to be efficient and scalable in cases of high contention. A number of workstealing algorithms with various characteristics has been proposed, such as Intel TBB's AP/SP and Lazy Binary Splitting [14,15] which are targeting taks generated by do-all loops. Cilk's workstealing infrastructure [3] is another well-known example; however Cilk's runtime is not directly applicable to OpenMP since OpenMP allows barriers among team threads. The initial implementation of OMPi tasks [1] utilized a lock-free workstealing algorithm based on [5].

In many applications task creation is unbalanced and it is a very common phenomenon few threads to produce many tasks and all other threads to consume them. In such cases contention could be lowered if threads cooperated instead of competed for obtaining the next tasks to execute. In our OpenMP tasking runtime each thread maintains (owns) a TASK_QUEUE, as explained above. A TASK_QUEUE is a shared object similar to the shared queue [12] supporting two operations: OwnerEnqueue and

`Dequeue` for inserting and removing tasks, correspondingly. `OwnerEnqueue`(q, t) inserts a new task t in queue q in case there is enough free space and returns true; otherwise, `OwnerEnqueue` fails and returns false. In contrast to `Enqueue` of a conventional shared queue, `OwnerEnqueue` is executed only by the thread that owns q. `Dequeue` is executed by any thread and removes the most early inserted task of q. Recently, Fatourou and Kallimanis [10] presented `CC-Synch`, an object which is able to implement (simulate) any shared object very efficiently. For example, to implement a shared queue, it is enough to use one instance of `CC-Synch` and to supply the sequential code for the `Enqueue` and `Dequeue` operations. `CC-Synch` supports only one operation called `ApplyOp`(*sfunc*, *arg*, *th_id*); *sfunc* is the serial code of the operation, *arg* is the argument of the operation and *th_id* is the id of the thread that executes the operation.

In [10], it is shown that `CC-Synch` significantly outperforms the state-of-the-art synchronization techniques. This is a result of the efficient implementation of the *combining* technique whereby, one thread (the *combiner*) holds a coarse lock, and additionally to the application of its own operation, serves the operations of all other active threads. Whenever a thread executes an operation using a conventional synchronization technique (such as spin-locks), it causes cache misses by fetching part of a shared object's state to the local processor cache in order to apply its operation. In the combining technique, only the combiner fetches parts of object's state and applies the operations of all active threads. Therefore, a lot of cache misses are avoided and the communication overheads among processors are much lower.

Using `CC-Synch` to implement an operation that is executed only by a single thread in any point of time is rather expensive. Thus, in our work-stealing queue implementation, we designed `OwnerEnqueue` (which is is executed only by the owner of the work-stealing queue) in a way that it does not make calls to `ApplyOp`. Thus, we avoid making the expensive calls of `CC-Synch`, wherever possible. It is noticeable that `CC-Synch` is better suited for cache-coherent NUMA machines, which constitute the majority of modern multicore multiprocessors.

We now give more details for our work-stealing implementation. Our work-stealing task queue (Fig. 3) consists of (i) a shared array of pointers to `TASK` structs, which is called `TASK_QUEUE`, (ii) a shared integer *Top* which points to the topmost element of the queue, (iii) a shared integer *Bottom* which points to the bottommost element of the queue, and (iv) an instance of `CC-Synch`. Since the `OwnerEnqueue` operation is executed only by the owner of the queue, its design is simplified. Whenever a thread p executes an `OwnerEnqueue` operation, it firstly executes a `read` on *Bottom* and after that a `read` on *Top*. If there exists free space, p inserts the new task and increases *Top* by one; otherwise, `OwnerEnqueue` returns false. Since p is the owner of the work-stealing queue and `OwnerEnqueue` is executed only by the owner, p is the only thread that modifies the shared variable *Top*. Therefore, no special care is needed while modifying *Top*. Whenever p wants to execute a `Dequeue` operation, it first checks if at least one element exists in the queue and in that case increases *Bottom* by one. Many threads may access *Bottom* simultaneously, since any thread is able to execute `Dequeue` in any `TASK_QUEUE`. We implement `Dequeue` using an instance of the `CC-Synch`

```
typedef struct WSQueue {
   int Bottom, Top;
   TASK *QArray[m];
   an instance of CC-Synch synchronization technique;
} WSQueue;

bool OwnerEnqueue(WSQueue *1, TASK *arg, int pid) {
   int top = 1->top, bottom = 1->bottom;
   int new_top = (top + 1) % TASKQUEUESIZE;

   if (new_top == bottom) return false;
   else {
      1->QArray[top] = arg;
      1->top = new_top;
      return true;
   }
}

TASK *Dequeue(WSQueue *1, int pid) { // Serial code for Dequeue, the concurrent
   void *ret;                        // version is implemented using CC-Synch.

   if (1->bottom == 1->top) ret = NULL;
   else {
      ret = 1->QArray[bottom]
      1->bottom = (1->bottom + 1) % TASKQUEUESIZE;
   }
   return ret;
}
```

Fig. 3. Pseudocode for the work-stealing queue implementation

synchronization queue. Since CC-Synch is a synchronization technique that serves operations with FIFO order, threads that execute Dequeue operations are also served with a FIFO order. Thus, our implementation satisfies strong fairness properties.

4   Performance Evaluation

In this section we evaluate the efficiency of our OpenMP tasking implementation. A synthetic producer/consumer benchmark was used to measure the task creation and the task execution throughput. Furthermore, the Barcelona OpenMP Tasks suite (BOTS) [8] was utilized in order to test our system in a broad range of task applications. All experiments were run on a 16-core machine equipped with two 8-core AMD Opteron 6128 CPUs running at 2.0GHz and with a total of 16GB RAM. The system runs Debian Squeeze based on Linux kernel 2.6.32.5. We compare the performance of our compiler with GNU GCC (version 4.4.5-8), Intel ICC (version 12.1.0) and Oracle SunStudio SUNCC (version 12.2). For reference the initial unoptimized implementation of OMPi in [1] is also included, labeled as 'OLD'.

In [7], it is shown that choosing the appropriate limits to enable and disable task cut-off is not an easy task. When dealing with task cut-off, it is required to have good knowledge of application's behavior for a specified input size, and of the runtime's tasking implementation. We thus chose to deactivate all manual cut-off techniques in all our benchmarks and let the OpenMP implementation operate under its default settings. As far as OMPi and OLD compilers are concerned, we used the default values for the size of TASK_QUEUEs which is 24.

```
main()                                    do_random_work()
{                                         {
  #pragma omp parallel num_threads(nthr)    volatile long i;
    if(omp_get_thread_num() < nprod) {
      for (int i=0;i<16E6/nprod);i++)        for (i=0;i<RandomRange(0,maxload);i++)
        #pragma omp task                        ;
        do_random_work();                 }
    }
}
```

Fig. 4. Code for synthetic microbenchmark

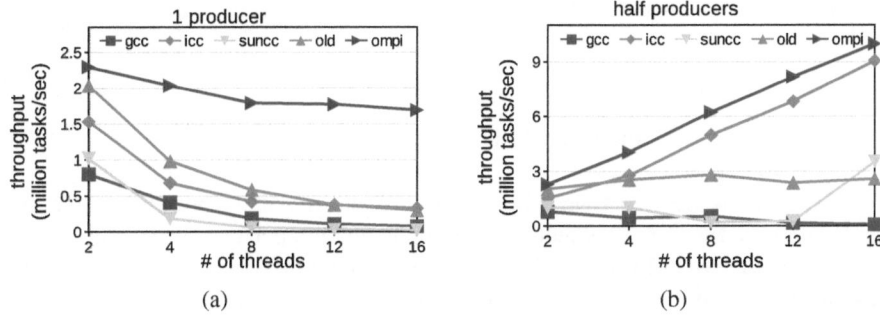

Fig. 5. Synthetic benchmark, `maxload=128`

We used GNU GCC with the "-O3" flag as a back-end compiler for OMPi. The corresponding flags for GCC, ICC and SUNCC were "-O3 -fopenmp", "-fast -openmp" and "-fast -xopenmp=parallel". We experimented with a lot of other flag combinations for all compilers but we didn't notice significant performance differences. All experiments were executed twelve times each, then the best and worst runs were discarded; from the ten remaining executions average values were calculated and reported.

4.1 Synthetic Benchmark

In order to evaluate the performance of OMPi, a synthetic benchmark with a controllable number of task producers and task consumers was used, as shown in Fig. 4. In this benchmark, a parallel region is created and a specified number of threads (equal to `nthr`) is created. Only `nprod` threads become producers and are allowed to create tasks. The rest of threads simply reach the end of parallel region and become consumers (executors) of the created tasks. Each run of the specified benchmark creates 16×10^6 tasks, the creation of which is equally assigned to producer threads. Each task consists of a dummy loop used to simulate workload that a task may have to execute in a way similar to [12,9,10]. The number of iterations is a random number between 0 and `maxload`, a variable controlling the task granularity. Iterator variable i is annotated as volatile in order to avoid compiler code elimination optimizations. This benchmark aims to stress the runtime's ability to create, steal and execute tasks.

We run several tests for different values of `nthr`, `nprod` and `maxload`. In Figs. 5–6 we present each implementation's throughput, measured as the number of tasks completed per second. For Fig. 5(a) we employed one producer and `nthr`−1 consumers.

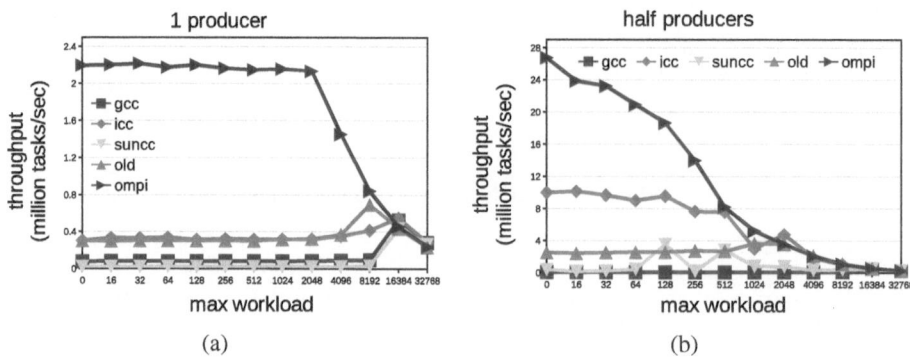

Fig. 6. Synthetic benchmark, nthr=16

In this experiment maxload was chosen to be equal to 128, representing fine-grain work. Some lock-free shared objects show unrealistic high performance when choosing a maxload value equal to 0, thus it is a common benchmarking strategy [12,9,10] to choose a small value for maxload, but not equal to 0. In this experiment, as more threads try to steal from the task queue of the producer, task throughput decreases. This is the result of extra synchronization overhead added, since more threads compete to get shared access to the same TASK_QUEUE. Due to the combining technique in our work-stealing implementation, OMPi outperforms all other compilers even in cases with very high contention and has the best scalability among them. Specifically, OMPi exhibits up to 5 times higher task throughput (at 16 threads) compared to ICC which is ranked as second best. The original OMPi implementation performs well only when 2 threads are used but its throughput quickly decreases. In Fig. 5(b), we study the behavior for different nthr values when nprod=nthr/2, while maxload is still equal to 128. The results are similar, confirming OMPi's superiority.

In Fig. 6(a), the performance results for different values of maxload and for a total of 16 threads (one of which produces tasks) are displayed. In this benchmark, our run-time exhibits higher throughput when compared to all other compilers for almost any maxload value. For values of 8192 or less, the work that each task executes is quite small and is overwhelmed by the contention that the work-stealing part induces. Since OMPi exploits the combining technique in its work-stealing queue, the synchronization overheads between threads are vastly minimized and the performance advances a lot. We achieved a little more that 6 times better performance compared to ICC and even better compared to GCC and SUNCC when application produces fine-grain tasks. When the task's granularity becomes coarser (maxload values greater than 8192), synchronization overheads between threads are not a bottleneck anymore and all compilers tend to exhibit similar behavior. Similar observations can be made with the results in Fig. 6(b), where 8 out of the 16 threads produce tasks for different values of maxload. For maxload values between 0 and 256 our new runtime achieves from 2.6 to 1.8 times higher throughput than the second best (ICC).

4.2   Performance of the BOTS Application Suite

The Barcelona OpenMP Tasks Suite (BOTS) v.1.1.1 was used for evaluating our tasking environment's efficiency in a wide range of tasking scenarios. Due to space limitations we present detailed results for the Fib, NQueens and Floorplan applications, while a brief discussion is made for Alignment, FFT, Health, Sort, SparseLU and Strassen. In order for every compiler to have full scheduling opportunities, we run both the tied and the untied task versions of the applications (while OMPi always utilizes tied tasks). We report the best execution times observed, although there were no significant performance differences as noted also in [8].

The Fib application computes the nth Fibonacci number using a recursive paralellization producing a very large number of fine-grain tasks. In Fig. 7, execution time results for the the 40th Fibonacci number are shown. Since it was a very common phenomenon for OMPi to outperform some compilers by a factor of ten or more, a logarithmic scale is used in y-axis. OMPi appears to be from 4 to 8 times faster than ICC and 20 to 80 times faster than the original (OLD) implementation. Since Fib exploits nested task parallelization which creates a deep tree of small tasks, it is a common phenomenon some threads to fill their queues. OMPi has a significant performance advantage by leveraging the new work-stealing implementation and the fast execution path produced by the compiler; task load is quickly balanced between threads, and the application delves into throttling mode. Moreover, OMPi, along with ICC, scales up with the number of threads.

NQueens calculates all the solutions of the n-queens chessboard problem. It uses a backtracking search algorithm with pruning that creates unbalanced tasks. Similarly to Fib, Nqueens exploits nested task parallelization which creates a deep tree of tasks. In the NQueens benchmark displayed in Fig. 8, for an input of 14 queens we get similar results to Fib and OMPi gives the best times. OMPi is up to 2 times faster than OLD and up to 3 times faster than ICC(not shown clearly in the logarithmic scale).

Floorplan calculates the optimal floor plan distribution of a number of cells. Tasks are hierarchically generated for each branch of the solution space. This application induces many data synchronizations and comes with a very irregular and aggressive pruning mechanism, which results in a heavily unbalanced task tree. Fig. 9 displays results of the application when the input.20 file is used; ICC is not included here because the application could not compile properly with this compiler. OMPi achieves the fastest times and our original implementation follows. Since Floorplan generates deep nested tasks, OMPi performs well due to the the work-stealing implementation along with the efficient fast path execution. SUNCC cannot exhibit speed-up, while GCC experiences significant slow-down when more threads are used.

Results from the rest of BOTS applications are given in Table 1, for the case of 16 threads. In this table we included results from OMPi when using ICC as back-end compiler, which in many situations produces faster code for the sequential part of the application. In FFT, SparseLU, Strassen and Alignment applications OMPi with ICC as backend proves to be faster, while performing second best only in two applications with very small margins (3% in Sort and 0.2% in Health). ICC has the best behavior in Health application, while our OLD system is the fastest as far as the Sort application is concerned. Thus, OMPi proves to perform consistently well in many different application

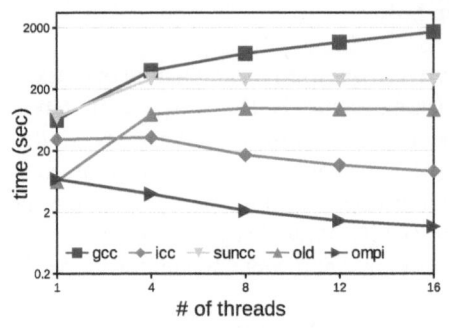

Fig. 7. Fibonacci

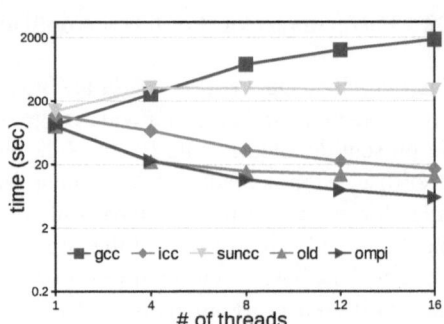

Fig. 8. Nqueens

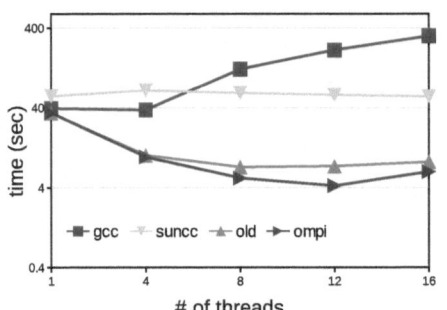

Fig. 9. Floorplan

Table 1. Execution time (sec) of BOTS using 16 threads

Compiler	FFT	Health	Sort	SpLU	Str.	Align.
GCC	17.571	141.85	2.007	1.679	24.602	1.576
ICC	2.086	**4.778**	0.621	1.676	20.641	1.338
SUNCC	2.473	15.694	0.652	1.835	21.619	1.218
OLD	2.086	7.114	**0.591**	1.766	21.589	1.587
OMPi	1.918	5.327	0.610	1.668	22.368	1.604
OMPi_ICC	**1.889**	4.787	0.621	**1.667**	**20.524**	**0.957**

scenarios, and especially when it uses an efficient back-end compiler, giving it a serious performance advantage. In general, ICC and SUNCC perform quite well with few exceptions. The version of GCC we had available does not perform up to par.

5   Conclusion

We present a highly optimized implementation of OpenMP tasking in the context of the OMPi compiler. The implementation is based on a carefully designed runtime system that emphasizes locality and operation combining while minimizing remote accesses which have detrimental performance effects in modern NUMA multicore multiprocessors. As a result, our system exhibits excellent scalability for high task loads and impressive improvement in actual application execution times, where OMPi was shown to offer competitive performance in comparison to other OpenMP implementations.

Currently we are working on analyzing the performance impact of the different portions of our runtime system and optimizing OMPi even more for some corner cases. We also work on supporting the recently released V3.1 of the OpenMP specifications [13] which offer even more opportunities for fast execution through the new `mergeable` and `final` clauses. Our preliminary experiences confirm the performance potential.

References

1. Agathos, S.N., Hadjidoukas, P.E., Dimakopoulos, V.V.: Design and Implementation of OpenMP Tasks in the OMPi Compiler. In: Proc. PCI 2011, 15th Panhellenic Conference on Informatics, pp. 265–269. IEEE, Kastoria (2011)
2. Ayguadé, E., Duran, A., Hoeflinger, J., Massaioli, F., Teruel, X.: An Experimental Evaluation of the New OpenMP Tasking Model. In: Adve, V., Garzarán, M.J., Petersen, P. (eds.) LCPC 2007. LNCS, vol. 5234, pp. 63–77. Springer, Heidelberg (2008)
3. Blumofe, R.D., Joerg, C.F., Kuszmaul, B.C., Leiserson, C.E., Randall, K.H., Zhou, Y.: Cilk: An Efficient Multithreaded Runtime System. J. Parallel Distrib. Comput. 37(1), 55–69 (1996)
4. Addison, C., LaGrone, J., Huang, L., Chapman, B.: OpenMP 3.0 tasking implementation in OpenUH. In: Proc. Open64 Workshop in Conjunction with the Int'l Symposium on Code Generation and Optimization, Seattle, USA (March 2009)
5. Chase, D., Lev, Y.: Dynamic circular work-stealing deque. In: Proc. SPAA 2005, 17th Annual ACM Symposium on Parallelism in Algorithms and Architectures, pp. 21–28. ACM, Las Vegas (2005)
6. Dimakopoulos, V.V., Leontiadis, E., Tzoumas, G.: A portable C compiler for OpenMP V.2.0. In: Proc. EWOMP 2003, 5th European Workshop on OpenMP, Aachen, Germany, pp. 5–11 (September 2003)
7. Duran, A., Corbalán, J., Ayguadé, E.: Evaluation of OpenMP Task Scheduling Strategies. In: Eigenmann, R., de Supinski, B.R. (eds.) IWOMP 2008. LNCS, vol. 5004, pp. 100–110. Springer, Heidelberg (2008)
8. Duran, A., Teruel, X., Ferrer, R., Martorell, X., Ayguadé, E.: Barcelona OpenMP Tasks Suite: A Set of Benchmarks Targeting the Exploitation of Task Parallelism in OpenMP. In: Proc. ICPP 2009, 38th Int'l Conference on Parallel Processing, Vienna, Austria, pp. 124–131 (September 2009)
9. Fatourou, P., Kallimanis, N.D.: A highly-efficient wait-free universal construction. In: Proc. SPAA 2011, Proceedings of the 23rd ACM Symposium on Parallelism in Algorithms and Architectures, pp. 325–334. ACM, San Jose (2011)
10. Fatourou, P., Kallimanis, N.D.: Revisiting the combining synchronization technique. In: Proc. PPoPP 2012, 17th ACM SIGPLAN Symposium on Principles and Practice of Parallel Programming, pp. 257–266. ACM, New Orleans (2012)
11. Korch, M., Rauber, T.: A comparison of task pools for dynamic load balancing of irregular algorithms: Research Articles. Concurr. Comput.: Pract. Exper. 16(1), 1–47 (2003)
12. Michael, M.M., Scott, M.L.: Simple, fast, and practical non-blocking and blocking concurrent queue algorithms. In: Proc. PODC 1996, 15th Annual ACM Symposium on Principles of Distributed Computing, pp. 267–275. ACM, Philadelphia (1996)
13. OpenMP ARB: OpenMP Application Program Interface V3.1 (July 2011)
14. Reinders, J.: Intel threading building blocks, 1st edn. O'Reilly & Associates, Inc., Sebastopol (2007)
15. Tzannes, A., Caragea, G.C., Barua, R., Vishkin, U.: Lazy binary-splitting: a run-time adaptive work-stealing scheduler. In: Proc. PPoPP 2010, 15th ACM SIGPLAN Symposium on Principles and Practice of Parallel Programming, pp. 179–190. ACM, Bangalore (2010)
16. Teruel, X., Unnikrishnan, P., Martorell, X., Ayguade, E., Silvera, R., Zhang, G., Tiotto, E.: OpenMP tasks in IBM XL compilers. In: Proc. CASCON 2008, 2008 Conference of the Center for Advanced Studies on Collaborative Research, Ontario, Canada, pp. 207–221 (October 2008)
17. Teruel, X., Martorell, X., Duran, A., Ferrer, R., Ayguadé, E.: Support for OpenMP tasks in Nanos v4. In: CASCON, pp. 256–259 (2007)

An Efficient Unbounded Lock-Free Queue for Multi-core Systems

Marco Aldinucci[1], Marco Danelutto[2], Peter Kilpatrick[3],
Massimiliano Meneghin[4], and Massimo Torquati[2]

[1] Computer Science Department, University of Torino, Italy
`aldinuc@di.unito.it`
[2] Computer Science Department, University of Pisa, Italy
`{marcod,torquati}@di.unipi.it`
[3] Computer Science Department, Queen's University Belfast, UK
`p.kilpatrick@qub.ac.uk`
[4] IBM Dublin Research Lab, Ireland
`massimiliano_meneghin@ie.ibm.com`

Abstract. The use of efficient synchronization mechanisms is crucial for implementing fine grained parallel programs on modern shared cache multi-core architectures. In this paper we study this problem by considering Single-Producer/Single-Consumer (SPSC) coordination using unbounded queues. A novel unbounded SPSC algorithm capable of reducing the row synchronization latency and speeding up Producer-Consumer coordination is presented. The algorithm has been extensively tested on a shared-cache multi-core platform and a sketch proof of correctness is presented. The queues proposed have been used as basic building blocks to implement the FastFlow parallel framework, which has been demonstrated to offer very good performance for fine-grain parallel applications.

Keywords: Lock-free algorithms, wait-free algorithms, bounded and unbounded SPSC queues, cache-coherent multi-cores.

1 Introduction

In modern shared cache multi-core architectures the efficiency of synchronization mechanisms is the cornerstone of performance and speedup of fine-grained parallel applications. For example, concurrent data structures in multi-threaded applications require synchronization mechanisms which enforce the correctness of concurrent updates. They typically involve various sources of overhead which have an increasingly significant effect on performance with increasing parallelism degree and decreasing synchronization granularity.

In this respect, mutual exclusion using lock/unlock, is widely considered excessively demanding for high-frequency synchronisations [1]. Among other methods, *lock-free* algorithms for concurrent data structures are the most frequently targeted. These algorithms have been devised by way of a hardware-implemented class of atomic operations — so-called CAS, because of its paradigmatic member *Compare-and-Swap* — in order to avoid an explicit consensus that would

C. Kaklamanis et al. (Eds.): Euro-Par 2012, LNCS 7484, pp. 662–673, 2012.
© Springer-Verlag Berlin Heidelberg 2012

increase the overhead for data accesses [2,3,4,5,6]. Unfortunately, CAS operations are not inexpensive since they might fail to swap operands when executed and may be re-executed many times, thus introducing other sources of potential overhead, especially under high contention [1]. Furthermore, without explicit consensus among parallel entities, the problem of correct memory management arises for dynamic concurrent data structures because of the complexity in tracking which chunk of memory is really in use at a given time. In general, lock-free dynamic concurrent data structures that use CAS operations should be supported by safe memory reclamation techniques in programming environments without automatic garbage collection [7].

In this work we study the synchronization problem for the simplest concurrent data structure: the Single-Producer/Single-Consumer (SPSC) queue. SPSC queues are widely used in many application scenarios: their efficiency can boost performance in terms of both latency and scalability to a non-negligible degree. In particular, SPSC-based synchronisations are used both in the implementation of high-level and completely general models of computation based on streams of tasks [8], and in a number of parallel frameworks as basic building blocks [9,10,11].

SPSC queues can be classified in two main families: *bounded* and *unbounded*. Bounded SPSC queues, typically implemented on top of a circular buffer, are used to limit memory usage and avoid the overhead of dynamic memory allocation. Unbounded queues are mostly preferred to avoid deadlock issues without introducing heavy communication protocols in the case of complex streaming networks, i.e. graph with multiple nested cycles. Bounded SPSC queues have been studied extensively since the emergence of the first wait-free algorithm presented by Lamport in the late 1970s [12]. More recently, some research work [13,14] revisited the Lamport queue, introducing a number of cache optimizations. On the other hand, unbounded SPSC queues, which are not any less relevant, have attracted less attention, resulting in quite a gap between the two SPSC families.

With the aim of filling this gap, we introduce and analyze here a novel algorithm for unbounded lock-free SPSC FIFO queues which minimizes the use of dynamic memory allocation. Furthermore, we provide a new implementation for the widely used dynamic list-based SPSC queue, along with proof sketches of correctness for both algorithms. Their performance is evaluated on synthetic benchmarks and on a simple yet relevant microkernel. The performance and the benefits deriving from the use of our SPSC queue when programming complete and complex application have already been assessed in [15,16].

The paper is organized as follows: Section. 2 provides the relevant background and related work discussing the reference implementations of the SPSC queue for shared-cache multi-cores. Section 3 introduces the list-based unbounded algorithm (dSPSC), while in Sec. 4 a novel algorithm for the unbounded queue (uSPSC) is presented, together with a proof sketch of its correctness. Performance results are discussed in Sec. 5. Section 6 summarizes the contribution of the work.

2 Producer-Consumer Coordination Using SPSC Queues: Background and Related Work

Producer-Consumer coordination is typically implemented by means of a FIFO queue, often realized with a circular buffer. Lamport proved that, under the *Sequential Consistency* (SC) memory model [12], a SPSC buffer can be implemented using only read and write operations [17]. Lamport's circular buffer is a *wait-free* algorithm, i.e. it is guaranteed to complete after a finite number of steps, regardless of the timing behavior of other operations. Another important class of algorithms are the *lock-free* algorithms, which enforce a weaker property than wait-free: they guarantee that at any time at least one process will make progress, although fairness cannot be assumed.

Lamport's circular buffer algorithm is no longer correct if the SC requirement is relaxed. This happens, for example, in all architectures where write-to-write memory ordering ($W \to W$ using the notation of [18]) is relaxed, i.e. two distinct writes at different memory locations may be executed out of program order (as in the *Weak Ordering* memory model [18]). A few modifications to the basic Lamport algorithm allow correct execution even under weakly ordered memory consistency models; they have been presented first and proved formally correct by Higham and Kavalsh [19]. The idea behind the Higham and Kavalsh queue basically consists in tightly coupling control and data information into a single buffer operation by extending the data domain with a new value called BOTTOM, which cannot be inserted into the queue. The BOTTOM value can be used to denote an empty cell, and then used to check if the queue is empty or full without directly comparing the indexes of the queue's *head* and *tail*.

Ten years later Giacomoni et al. [13] followed a similar line by proposing the same basic algorithm and studying its behavior in cache-coherent multiprocessor systems. As a matter of fact, Lamport's queue results in heavy cache invalidation/update traffic because both producer and consumer share both head and tail indexes[1]. This can be avoided, as already noted in [19], by using a BOTTOM value that makes it possible for the producer to write and read only the tail and for the consumer to write and read only the head indexes. Since this technique applies nicely to data pointers where NULL is the BOTTOM value, Giacomoni et al. proved that on weakly ordered memory model, a Write Memory Barrier (WMB) is actually required to enforce completion of the data write by the producer before the data pointer is passed to the consumer[2]. Figure 1 presents an implementation of the SPSC algorithm proposed in [13] which may be regarded as the reference algorithm for bounded SPSC queues.

Avoiding cache-line thrashing due to false-sharing is a critical aspect in shared-cache multiprocessors and thus much research has been focused on trying to minimize this effect. In [13] the authors present a *cache slipping* technique suitable for avoiding false sharing on true dependencies (i.e. pointers stored within

---

[1] The producer updates the tail index, the consumer updates the head index, and both the producer and the consumer read both head and tail indexes.

[2] WMB is also referred to as store-fence.

```
1 bool push(void* data) {                    10 bool pop(void** data) {
2   if (buf[pwrite]==NULL) {                  11   if (buf[pread]==NULL)
3     WMB(); // write−memory−barrier          12     return false;
4     buf[pwrite] = data;                     13   *data = buf[pread];
5     pwrite+=(pwrite+1>=size)?(1−size):1;    14   buf[pread]=NULL;
6     return true;                            15   pread+=(pread+1>=size)?(1−size):1;
7   }                                         16   return true;
8   return false;                             17 }
9 }
```

Fig. 1. SPSC circular buffer implementation as proposed in [13], where `buf` is an array of size `size` initialized to `NULL` values

queue cells) and for enforcing partial filling of the queues in such a way that producer and consumer operate on different cache lines.

A different approach for optimizing cache usage, named *cache line protection*, has been proposed in MCRingBuffer [14]. The producer and consumer thread update private copies of the head and tail indexes for several iterations before updating a shared copy. Furthermore, MCRingBuffer performs batch update of control variables, thus reducing the frequency of writing the shared control variables to main memory. A variation of the MCRingBuffer approach is used in the Liberty Queue [20]. The Liberty Queue shifts most of the overhead to the consumer end of the queue. Such customization is useful in situations where the producer is expected to be slower than the consumer.

Unbounded SPSC queues have not benefited from a similar optimization effort and, to the best of our knowledge, have been approached only through the more general and more demanding CAS-based Multiple-Producer/Multiple-Consumer (MPMC) queues.

3   Basic Unbounded List-Based Wait-Free SPSC Queue

A way to design a SPSC queue is to use as a starting point the well-known two-lock Multi-Producer/Multi-Consumer (MPMC) queue described by Michael and Scott (MS) [6]. The MS queue is based on a dynamically linked list of `Node`(s) data structure, using head and tail pointers which (both) initially point to a dummy `Node` (i.e. containing NULL values). The `Node` structure contains the actual user value and a *next* pointer. Concurrency between multiple producers is managed by a lock for enqueue operations and symmetrically consumers use a different lock for dequeue operations.

Inspired by the MS queue, we propose a new unbounded SPSC queue whose algorithm is sketched in Fig. 2 (where lines §2.10 and §2.23 can be safely ignored here at the moment as they introduce a further optimization that is described later in this section)[3]. The `push` method allocates a new `Node` data structure, fills it and then adjusts the tail pointer to point to the current `Node`. The `pop` method gets the current head `Node`, places the data values into the application buffer, adjusts the head pointer and, before exiting, deallocates the head `Node`.

---

[3] We use the §M.N notation to reference line N from the pseudo-code in Fig. M.

```
1  struct Node {                        14     tail −>next = n; tail = n;
2    void*          data;               15     return true;
3    struct Node* next;                 16  }
4  };
5  Node* head,* tail;                   18  bool pop(void** data) {
6  SPSC cache;                          19     if (!head−>next) return false;
                                        20     Node* n = head;
8  bool push(void* data) {              21     *data = (head−>next)−>data;
9    Node* n;                           22     head = head−>next;
10   if (!cache.pop(&n))                23     if (!cache.push(n)) free(n);
11       n = (Node*)malloc(sizeof(Node)); 24    return true;
12   n−>data = data; n−>next = NULL;    25  }
13   WMB(); // write−memory−barrier
```

Fig. 2. Unbounded list-based dSPSC queue implementation with Node(s) caching. The list is initialized with a dummy Node.

In general, one of the main problems with the list-based implementation of queues is the overhead associated with dynamic memory allocation/deallocation of Node structures. To mitigate the overhead, it is common to use a data structure as cache, where elements are kept for future fast reuse, instead of being deallocated [21].

For a more tailored optimization, the specific allocation pattern can be taken into account: the producer only allocates while the consumer only frees nodes. To take advantage of this pattern, we add a bounded wait-free SPSC queue implementing a Node cache, which is used to sustain a "return" path from consumer to producer of Node structures that can be reused.

The introduced optimization clearly moves allocation overhead outside the critical path at the steady state. The resulting algorithm, called *dSPSC*, is shown in Fig. 2; line §2.10 and line §2.23 introduce the proposed cache optimization.

Along with some standard definitions we now provide, for the presented dSPSC, a sketch proof of FIFO queue correctness and the lock-freedom property.

Definition 1 (Correctness). *Assuming that simple memory read and write operations are atomic, a SPSC queue is defined as* correct *if it always exhibits FIFO behavior for any interleaving of push and pop operations.*

Note that the condition that simple memory reads and writes are atomic is typically satisfied in any modern general-purpose processor for aligned memory word loads and stores.

Theorem 1 (dSPSC). *Under a weak consistency memory model, the dSPSC queue defines a correct lock-free SPSC FIFO queue if a lock-free allocator is used.*

Proof (Sketch). In a sequentially consistent model, correctness of the dSPSC derives trivially from correctness of the two-lock MS queue (where the two locks have been removed as there is no concurrency between producers or between consumers) and of the bounded SPSC queue used for the Node cache [13].

Moving to a weak memory model, the bounded SPSC queue is still correct ([13]) while the memory barrier at line §2.13 guarantees correctness regarding the dynamic linked list management. Indeed, all changes to the structure of Node n as well as to the memory pointed by data have to be committed to memory before the node itself can be visible to the consumer (i.e. before the tail is set to point at the new node). It is trivial to see that, similar to what happens with the SPSC queue in [13], no other store fence is required inside the push and pop methods under weakly ordered memory model.

Concerning the lock-free property, the strategy that is used by dSPSC for the memory management is lock-free because the allocator is lock-free by hypothesis and the SPSC used as a cache is lock-free by construction. As the rest of the dSPSC algorithm does not present any statement where producer or consumer can block, progress is guaranteed.   □

4   Fast Unbounded Lock-Free SPSC Queue

The SPSC algorithm [19,13], described in Sec. 2, is extremely fast (see Sec. 5) but implements a bounded queue. The dSPSC algorithm, presented in Sec 3, is lock-free and realizes an unbounded queue, but pays for the flexibility achieved via a list-based implementation with decreased spatial locality in cache behavior. However, the two approaches can be combined in a new lock-free algorithm for the unbounded SPSC (called *uSPSC*) inheriting the best features of both. The new algorithm is sketched in Fig. 3. The basic idea underpinning uSPSC is the nesting of the two queues. A pool of SPSC bounded queues (called *buffers* from now on) is linked together into a list as a dSPSC queue. The implementation of the pool of buffers aims to minimize the impact of dynamic memory allocation/deallocation by using a fixed-size SPSC queue as a freelist as in the list-based dSPSC queue.

The unbounded queue uses two pointers: buf_w which points to the writer's buffer (i.e. the "tail" pointer), and buf_r which points to the reader's buffer (i.e. the "head" pointer). Initially both buf_w and buf_r point to the same buffer.

The push method works as follows: the producer first checks whether the current buffer is not full (line §3.4), and then pushes the data. If the current buffer is full, it asks the pool for a new buffer (line §3.5), adjusts the buf_w pointer and pushes the data into the new buffer.

The pop method, called by the consumer, first checks whether the current buffer is not empty and if so pops data from the queue. If the current buffer is empty, there are two possibilities: a) there are no items to consume, i.e. the unbounded queue is really empty; b) the current buffer is empty (i.e. the one pointed by buf_r), but there may be some items in the next buffer.

If the buffer is empty for the consumer, it switches to a new buffer releasing the current one to be recycled by the buffer pool (lines §3.14–§3.16). From the consumer viewpoint, the queue is really empty when the current buffer is empty and both the read and write pointers (buf_r and buf_w, respectively) point to the same buffer. If the read and writer queue pointers differ, the consumer has to re-check the current queue emptiness because in the meantime (i.e. between the execution of instructions §3.11 and §3.12) the producer could have written

```
1  int  size  = N; //SPSC size         22  struct Pool {
                                        23    dSPSC inuse;
3  bool push(void* data) {              24    SPSC cache;
4    if (buf_w->full())
5      buf_w = pool.next_w();           26    SPSC* next_w() {
6    buf_w->push(data);                 27      SPSC* buf;
7    return true;                       28      if (!cache.pop(&buf))
8  }                                    29        buf = allocateSPSC(size);
                                        30      inuse.push(buf);
10 bool  pop(void** data) {             31      return buf;
11   if (buf_r->empty()) {              32    }
12     if (buf_r == buf_w) return false; 33   SPSC* next_r() {
13     if (buf_r->empty()) {            34      SPSC* buf;
14       SPSC* tmp = pool.next_r();     35      return (inuse.pop(&buf)? buf : NULL);
15       pool.release(buf_r);           36    }
16       buf_r = tmp;                   37    void release(SPSC* buf) {
17     }                                38      buf->reset(); // reset pread and pwrite
18   }                                  39      if (!cache.push(buf)) deallocateSPSC(buf);
19   return buf_r->pop(data);           40    }
20 }                                    41 }
```

Fig. 3. Unbounded lock-free uSPSC queue implementation

some new elements into the current buffer before switching to a new one. This is the most subtle condition whose occurrence must be proved to be impossible since, if the consumer switches to the next buffer while the previous one is not really empty, a data loss will occur. In the next section we prove that the if condition at line §3.13 is sufficient to ensure correct execution.

Theorem 2 (uSPSC). *The uSPSC unbound queue given in Fig. 3 is correct (according to Def. 1) on architectures with Weak Ordering consistency model (and therefore with any stricter ordering).*

Proof (Sketch). The SPSC and the dSPSC queues used as building blocks have been proven to be correct under Weak Ordering (WO) consistency and stricter models (e.g. Total Store Ordering) in [19,13] and Theorem 1, respectively.

We distinguish four cases with respect to the values of buf_r and buf_w used by producer and consumer (respectively): buf_r and buf_w are 1) equal or 2) different throughout execution of a push/pop pair; 3) they are different and become equal; or 4) they are equal and become different. In case 1) uSPSC is correct because of the correctness of the underlying SPSC. In case 2) uSPSC is correct because the producer and consumer work on different SPSC buffers and correctness follows from the correctness of the underlying dSPSC. In case 3 the consumer catches up with the producer: correctness follows because, when buf_w was assigned (line §3.5), that assignment will have been preceded by the issue of a WMB within the dSPSC push that commits all values of the previous buffer (allowing them all to be read by the consumer before it advances buf_r).

Case 4 is more subtle: here the two buffers are equal and become different when the producer observes that buf_w is full and prompts a move to a new write buffer. The concern is that, because of WO, in the case where the SPSC buffer size = 1 the consumer may see the buffer as empty and release it *before* a write to it has been committed, thus causing data loss. We prove that this cannot happen and the FIFO ordering is preserved.

Under WO model the consumer may be aware that the producer has changed the write buffer only after a synchronization point that enforces program order between two store operations has been traversed. In our algorithm the synchronization point is the WMB. In fact, the new value of buf_w (line §3.5) and the value that is written to the buffer might appear in memory in any order or not at all. Thus it might be thought possible that the reading buffer buf_r could still be perceived as empty while a new writing buffer has already been started (thus buf_r $\neq$ buf_w); the condition at line §3.13 could therefore evaluate to true even if the previous buffer is not actually empty. This condition could lead to data loss because the consumer might overtake and abandon a buffer still holding a valid value. In the uSPSC this subtle case can never arise however, because, in order to change the write buffer, a push operation in the dSPSC queue is called (§3.30) thus enforcing a WMB, which commits all previous writes to memory, and so the if condition at line §3.13 is evaluated to true only if the consumer buffer is really empty. FIFO ordering is trivially enforced by the FIFO ordering of both nested queues dSPSC and SPSC queues.   □

It is worth noticing that, regardless of the implementation of the pool used in the uSPSC queue, if $size > 1$ (line §3.1) the two conditions buf_r incorrectly perceived empty and buf_r $\neq$ buf_w, cannot hold together as at least two push and one WMB must occur to make an empty queue become a full queue.

Corollary 1 (lock-free). *The uSPSC queue is lock-free provided a lock-free allocator is used.*

Proof (Sketch). The SPSC queue, and the dSPSC queue coupled with a lock-free allocator, are lock-free. Suppose we use a lock-free allocator in the allocateSPSC and in deallocateSPSC. As the push and pop methods contain no cycle nor can they block on any non lock-free function, progress is assured.   □

Enhancing the queues to wait-freedom property. It can be demonstrated that the SPSC queue proposed in [13] as well as the dSPSC when a wait-free allocator is used, are both wait-free. The uSPSC is wait-free if a wait-free dSPSC queue is used and if a wait-free allocator is used in the pool: in fact both the push and pop methods complete in a bounded number of steps.

5   Experiments

All experiments reported in this section have been conducted on an Intel workstation with 4 eight-core double context Xeon E7-4820 @2.0GHz with 18MB L3 shared cache, 256K L2, and 24 GBytes of main memory with Linux x86_64. Some of the tests presented have been executed also on a different architecture and results can be found in [22]. Similar results to those presented in this paper have been obtained on the AMD Opteron platform.

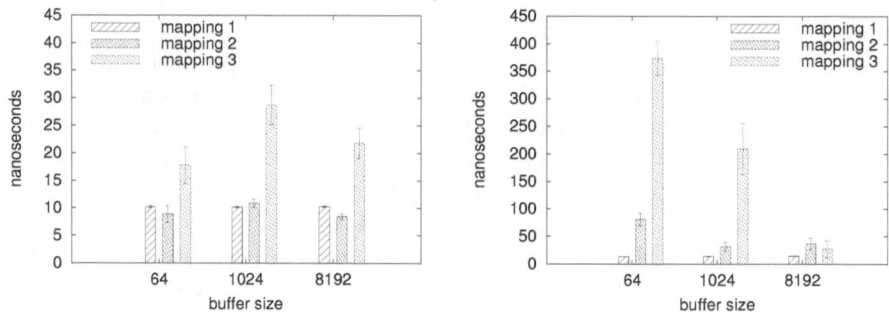

Fig. 4. Bounded SPSC (left) and unbounded dSPSC (right) average latency time in nanoseconds varying the internal buffer size and cache size respectively

The first test is a two-stage pipeline in which the first stage (P) pushes 1M tasks (a task is just a memory pointer) into a FIFO queue and the second stage (C) pops tasks from the queue and checks for correct values. Neither additional memory operations nor additional computation is executed. With this simple test we are able to measure the raw performance of a single push/pop operation by computing the average value of 100 runs, varying the buffer size for the bounded SPSC queue and the cache size for the dSPSC queue. We tested three distinct cases that differ in terms of the physical mapping of the two threads corresponding to the two stages of the pipeline. The first and the second stage of the pipeline are pinned: i) on the same physical core but on different HW contexts (`mapping1`); on the same CPU but on different physical cores (`mapping2`); on two cores of two distinct CPUs (`mapping3`).

Figure 4 reports the values obtained by running the first benchmark for the SPSC queue and the dynamic list-based dSPSC queue, varying the buffer size and the internal cache size, respectively. Fig. 5 (left) reports the values obtained by running the same benchmark using the unbounded uSPSC queue.

The bounded SPSC queue is almost insensitive to buffer size in all cases. It takes on average 8–12 ns corresponding to almost 16–24 cycles per push/pop operation with standard deviation less than 1.5 ns when the producer and the consumer are on the same CPU, and takes on average 16–36 ns if the producer and the consumer are on separate CPUs. The dSPSC queue is instead quite sensitive to the internal cache size on the tested architecture. The best values for the dSPSC queue range from 14 to 36 ns with a standard deviation that ranges from 0.5 to 11 ns. Such values are obtained with sufficiently large cache size (8192 slots). As expected, the bigger the internal cache, the better the performance obtained. As a reference, the MS queue implementation is one order of magnitude slower, going from 110–190 ns on sibling cores to 430–490 ns on non-sibling cores. The uSPSC queue (Fig. 5 left) is more sensitive to the internal buffer size in the case where the producer and the consumer are pinned to separate CPUs and when the internal buffer is small. The values obtained for the uSPSC

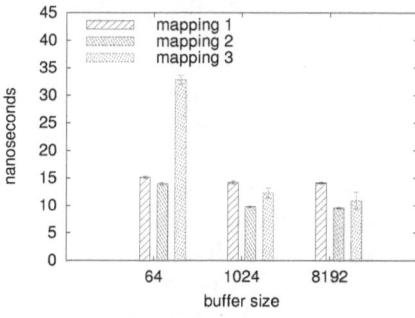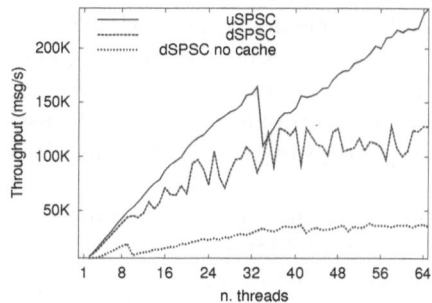

Fig. 5. Unbounded uSPSC average latency time in nanoseconds (left) varying the buffer size of the internal SPSC queues (pool cache size set to 32). Throughput in msgs/s (right) running the ring microkernel when using the dSPSC queue without any cache, the dSPSC with a cache size of 2048 and the uSPSC queue with an internal buffer size of 2048 elements (pool cache size set to 32).

are very good if compared with those obtained for the dSPSC queue, and are almost the same (or better) if compared with the bounded SPSC queue when using sufficiently large buffer size. It takes on average 9.7–14 ns per push/pop operation with standard deviation less than 1.5 ns when the internal buffer size is greater than or equal to 1024. The dSPSC queue is slower than the uSPSC version in all cases. If the producer and the consumer for the dSPSC queue are not pinned on the same core the dSPSC queue is more than 10 times slower than the uSPSC queue. Instead, when the producer and the consumer are pinned on the same core the performance is much better for the dSPSC queue (although always worse than the uSPSC one) because they work in lock step as they share the same ALUs and so dynamic memory management is reduced.

It is worth noting that caching strategies for the dSPSC queue implementation significantly improve performance but are not sufficient to obtain optimal figures like those obtained in the uSPSC implementations.

To test scalability of the queues we used a simple synthetic microkernel. We consider N threads linked into a ring using an unbounded queue (dSPSC and uSPSC). The first thread emits a number of messages which flow around the ring. The message is just a pointer obtained from dynamic allocation of a small segment of memory. The other threads accept messages, perform basic integrity verification, copy the input message into a new dynamically allocated buffer, free the input message and pass the new pointer to the next thread. When all messages return to the first thread, the program terminates. Each thread is statically pinned to a core whose id is the same as the thread id. For the architectures considered, the core ids are linear so core 0 and 32 as well as core 0 and 1 are on the same physical core and on the same CPU, respectively, whereas core 0 and 8 are on different CPUs. In Fig. 5 (right), we present the performance in messages per second (msgs/s) obtained while varying the number of threads of the ring. Three queue implementations were tested: the dSPSC queue without using any internal cache (the basic algorithm); the dSPSC queue with a cache size of 32K elements (i.e. with a SPSC queue of size 32K); and the uSPSC queue using a 32K internal SPSC queue and a cache size of 32 elements.

The uSPSC queue implementation obtains the best performance reaching a maximum throughput of ~250K msgs/s, whereas the dSPSC reaches a maximum throughput of ~128K msgs/s when using an internal cache of Node(s), and ~37K msgs/s when no cache is used. For this test, the MS queue implementations (not shown in the graph) obtain almost the same speedup as the dSPSC queue without internal cache. The uSPSC queue scales almost linearly up to 32 cores, then the performance drops due to the fact that on core 0 we have 2 threads in separate contexts (the first and the last one of the ring) producing a bottleneck. Adding more threads in the ring, the bottleneck is slowly absorbed by the increasing throughput thus reaching an optimal final 32X improvement.

In this section we have shown only synthetic benchmarks in order to present evidence of the distinctive performance of the uSPSC implementations. The simple tests shown here prove the effectiveness of the uSPSC queue with respect to the dSPSC implementation, and prove also how a fast implementation of a cache of references inside the dSPSC queue leads to much higher throughput. Since the uSPSC queue is used in the FastFlow framework, more performance figures on real-world applications can be found in [15,16].

6   Conclusions

In this paper we studied several possible implementations of fast lock-free Single-Producer/Single-Consumer (SPSC) queues for shared cache multi-core platforms, starting from the well-known Lamport circular buffer algorithm. A new implementation, called dSPSC, of the widely used dynamic list-based algorithm has been proposed. Moreover, a novel unbounded lock-free SPSC queue algorithm called uSPSC has been introduced together with a sketch proof of its correctness and several performance assessments.

The uSPSC queue algorithm and implementation are able to minimize dynamic memory allocation/deallocation and increase cache locality thus obtaining very good performance figures on modern shared cache multi-core platforms. Our uSPSC implementation has been used as a foundation for a skeleton based parallel programming framework (FastFlow [9]) that has been demonstrated to be more efficient than other state-of-the-art programming environments, including OpenMP and Cilk, on significant fine-grain parallel applications.

References

1. Orozco, D.A., Garcia, E., Khan, R., Livingston, K., Gao, G.R.: Toward high-throughput algorithms on many-core architectures. TACO 8(4), 49 (2012)
2. Moir, M., Nussbaum, D., Shalev, O., Shavit, N.: Using elimination to implement scalable and lock-free FIFO queues. In: Proc. of the 7th ACM Symposium on Parallelism in Algorithms and Architectures, pp. 253–262 (2005)

3. Ladan-Mozes, E., Shavit, N.: An optimistic approach to lock-free FIFO queues. Distributed Computing 20(5), 323–341 (2008)
4. Prakash, S., Lee, Y.H., Johnson, T.: A nonblocking algorithm for shared queues using compare-and-swap. IEEE Trans. Comput. 43(5), 548–559 (1994)
5. Tsigas, P., Zhang, Y.: A simple, fast and scalable non-blocking concurrent fifo queue for shared memory multiprocessor systems. In: Proc. of the 13th ACM Symposium on Parallel Algorithms and Architectures (SPAA), pp. 134–143 (2001)
6. Michael, M.M., Scott, M.L.: Nonblocking algorithms and preemption-safe locking on multiprogrammed shared memory multiprocessors. Journal of Parallel and Distributed Computing 51(1), 1–26 (1998)
7. Michael, M.M.: Hazard pointers: Safe memory reclamation for lock-free objects. IEEE Trans. Parallel Distrib. Syst. 15(6), 491–504 (2004)
8. Kahn, G.: The semantics of simple language for parallel programming. In: IFIP Congress, pp. 471–475 (1974)
9. FastFlow framework: website (2009), http://mc-fastflow.sourceforge.net/
10. Thies, W., Karczmarek, M., Amarasinghe, S.: StreamIt: A Language for Streaming Applications. In: Horspool, R.N. (ed.) CC 2002. LNCS, vol. 2304, pp. 179–196. Springer, Heidelberg (2002)
11. Reinders, J.: Intel Threading Building Blocks: Outfitting C++ for Multi-core Processor Parallelism. O'Reilly (2007)
12. Lamport, L.: How to make a multiprocessor computer that correctly executes multiprocess programs. IEEE Trans. Comput. 28(9), 690–691 (1979)
13. Giacomoni, J., Moseley, T., Vachharajani, M.: Fastforward for efficient pipeline parallelism: a cache-optimized concurrent lock-free queue. In: Proc. of the 13th ACM SIGPLAN Symposium on Principles and Practice of Parallel Programming (PPoPP), pp. 43–52 (2008)
14. Lee, P.P.C., Bu, T., Chandranmenon, G.P.: A lock-free, cache-efficient multi-core synchronization mechanism for line-rate network traffic monitoring. In: Proc. of the 24th Intl. Parallel and Distributed Processing Symposium, IPDPS (2010)
15. Aldinucci, M., Ruggieri, S., Torquati, M.: Porting Decision Tree Algorithms to Multicore Using FastFlow. In: Balcázar, J.L., Bonchi, F., Gionis, A., Sebag, M. (eds.) ECML PKDD 2010, Part I. LNCS, vol. 6321, pp. 7–23. Springer, Heidelberg (2010)
16. Aldinucci, M., Danelutto, M., Meneghin, M., Kilpatrick, P., Torquati, M.: Efficient streaming applications on multi-core with FastFlow: the biosequence alignment test-bed. In: Parallel Computing: From Multicores and GPU's to Petascale. Advances in Parallel Computing, vol. 19, pp. 273–280. IOS Press (2009)
17. Lamport, L.: Concurrent reading and writing. CACM 20(11), 806–811 (1977)
18. Adve, S.V., Gharachorloo, K.: Shared memory consistency models: A tutorial. IEEE Computer 29, 66–76 (1995)
19. Higham, L., Kawash, J.: Critical sections and producer/consumer queues in weak memory systems. In: Proc of the Intl. Symposium on Parallel Architectures, Algorithms and Networks (ISPAN), pp. 56–63. IEEE (1997)
20. Jablin, T.B., Zhang, Y., Jablin, J.A., Huang, J., Kim, H., August, D.I.: Liberty queues for epic architectures. In: Proc. of the 8th Workshop on Explicitly Parallel Instruction Computer Architectures and Compiler Technology, EPIC (2010)
21. Hendler, D., Shavit, N.: Work dealing. In: Proc. of the 4th ACM Symposium on Parallel Algorithms and Architectures (SPAA), pp. 164–172 (2002)
22. Torquati, M.: Single-producer/single-consumer queues on shared cache multi-core systems. Technical Report TR-10-20, Computer Science Dept., University of Pisa, Italy (2010), http://compass2.di.unipi.it/TR/Files/TR-10-20.pdf.gz

Topic 12: Theory and Algorithms
for Parallel Computation

Geppino Pucci, Christos Zaroliagis,
Kieran T. Herley, and Henning Meyerhenke

Topic Committee

Parallelism permeates all levels of current computing systems, from single CPU machines, to large server farms, to geographically dispersed "volunteers" who collaborate over the Internet. The effective use of parallelism depends crucially on the availability of faithful, yet tractable, models of computation for algorithm design and analysis, and on efficient strategies for solving key computational problems on prominent classes of computing platforms. No less important are good models of the way the different components/subsystems of a platform are interconnected. With the development of new genres of computing platforms, such as multicore parallel machines, desktop grids, clouds, and hybrid GPU/CPU-based systems, new models and paradigms are needed that will allow parallel programming to advance into mainstream computing. Topic 12 focuses on contributions providing new results on foundational issues regarding parallelism in computing, and/or proposing improved approaches to the solution of specific algorithmic problems.

This year, papers submitted to Topic 12 covered a considerable amount of subjects indicated in the call for papers, among the others, communication complexity issues on various computational models, parallel algorithms and data structures for combinatorial optimization problems, and finally parallelization of loop and finite automata computations. Submissions indicated a significant interest of the parallel computing community towards developing new sound and solid methods for parallel problem solving as well as towards investigating the limitations of parallelism.

Among all submissions, two high-quality papers were selected for presentation at the conference. The first paper, *A Lower Bound Technique for Communication on BSP with Application to the FFT*, by Gianfranco Bilardi, Michele Scquizzato, and Francesco Silvestri, focuses on the *Bulk Synchronous Parallel* (BSP) model of computation and provides a general technique to derive lower bounds on the communication complexity (i.e., the sum of the degrees of all supersteps) of algorithms implementing DAG computations, based on the switching capabilities of the underlying DAG topology. The authors show the worth of their approach by providing a novel tight lower bound for the important case of the FFT DAG. The second paper, *A fast parallel algorithm for minimum-cost small integral flows*, by Andrzej Lingas and Mia Persson, presents an interesting reduction of the problem of computing the minimum-cost integral flow problem on a capacitated, weighted directed network to the problem of performing multiple tests of simple multi-variable polynomials over a finite field of characteristic two for

C. Kaklamanis et al. (Eds.): Euro-Par 2012, LNCS 7484, pp. 674–675, 2012.

non-identity with zero. The proposed reduction is extremely efficient when the value of the maximum flow is small, and indeed yields an RNC^2 algorithm for (sub-)logarithmic values of the flow.

A Lower Bound Technique for Communication on BSP with Application to the FFT*

Gianfranco Bilardi, Michele Scquizzato, and Francesco Silvestri

Department of Information Engineering, University of Padova, Italy
{bilardi,scquizza,silvest1}@dei.unipd.it

Abstract. *Communication complexity* is defined, within the *Bulk Synchronous Parallel* (BSP) model of computation, as the sum of the degrees of all the supersteps. A lower bound to the communication complexity is derived for a given class of DAG computations in terms of the *switching potential* of a DAG, that is, the number of permutations that the DAG can realize when viewed as a switching network. The proposed technique yields a novel and tight lower bound for the FFT graph.

1 Introduction

A substantial fraction of the time and energy cost of a parallel algorithm is due to the exchange of information between processing and storage elements. As in all endeavors where performance is pursued, it is important to be able to evaluate the distance from optimality of a proposed solution.

In this paper, we consider the *Bulk Synchronous Parallel* (BSP) model of computation [23]. We develop a lower bound technique for a metric, called *communication complexity*, which captures a relevant component of the cost of BSP computations. This technique applies to a class of computations that can be modeled in terms of a *Directed Acyclic Graph* (DAG), whose vertices represent operations (of both input/output and processing type) and whose arcs represent data dependencies. The same DAG computation can be performed in many different ways, depending on the superstep and the processing element chosen for the execution of an operation and the way (routing path and schedule of the message along such a path) in which a value is routed from the processor that computes it to a processor that utilizes it. Our proposed technique further assumes that each operation is executed only once, but the case where repetitions are allowed is also of interest.

The complexity of communication of DAGs on various models of computation has received considerable attention. Lower bounds are often established through adaptations of the techniques of Hong and Kung [13] for hierarchical memory, or by critical path arguments, such as those in [1]. For applications of these and other techniques see [18,2,12,9,5,14,3] as well as [19] and references therein.

---

* This work was supported, in part, by MIUR-PRIN Project *AlgoDEEP*, by PAT-INFN Project *AuroraScience*, by the University of Padova Projects *STPD08JA32* and *CPDA099949*, and by the IBM *Visiting Scientist Program*.

C. Kaklamanis et al. (Eds.): Euro-Par 2012, LNCS 7484, pp. 676–687, 2012.

The resulting bounds are often tight, but not in all cases. A notable example is the computation of an n-input FFT DAG on a BSP with p processors: when inputs are initially evenly distributed among the processors, an adaptation of the *dominator set* technique of [13] yields a lower bound of[1] $\Omega\left(\frac{n\log n}{p\log(n\log n/p)}\right)$ to the communication complexity, which does not match the best known upper bounds when $p = \omega(n/\log n)$. As this example indicates, communication lower bounds deserve further exploration.

The main contribution of this paper is the *switching potential* technique, to obtain communication lower bounds for DAG computations in the BSP model. The communication complexity of a BSP computation is defined as the sum of the degrees of all its supersteps. The proposed technique applies to DAGs with n input nodes where all nodes, except for inputs and outputs, have out-degree equal to the in-degree. Such a graph can be viewed as a *switching network* [19], whose switching potential $\gamma(n)$ is defined as the number of different permutations that it can realize. We show that, for executions of a DAG *without recomputation* of its nodes, the BSP communication complexity satisfies a suitable lower bound expressed in terms of the switching potential. As a corollary of this general result, we obtain a tight bound for the communication complexity of the FFT on the BSP. The bound has the form $\Omega\left(\frac{n\log n}{p\log(n/p)}\right)$ and matches an upper bound of [23] for any $p = O(n)$.[2] A similar bound was derived earlier in [12], for the special class of algorithms performing exclusively supersteps of degree $\Theta(n/p)$.

Our FFT lower bound has the same form as the lower bound derived for the communication complexity of the FFT in the LPRAM model, by Aggarwal, Chandra, and Snir [1]. In addition to being developed for a different model, the argument of [1] follows a different route: the lower bound is first established for sorting, then claimed (by analogy) for permutation networks, and finally adapted to the FFT network, by exploiting the property that, as shown in [24], the cascade of three FFT networks has the topology of a full permutation network. Finally, we observe that while when recomputation is not allowed our FFT lower bound improves on the dominator-set result mentioned above, the latter remains of interest when recomputation is allowed.

In addition to the well known general motivations for lower bound techniques, we stress that striving for tight bounds for the whole range of model's parameters has special interest in the study of so-called *oblivious* algorithms, whose specification does not refer to such parameters, but are designed with the goal of achieving (near) optimality for all values of the parameters. Notable examples are *cache-oblivious algorithms* [11], *multicore-oblivious algorithms* [10] and, closer to the scenario of this paper, *network-oblivious algorithms* [8,7], where algorithms are designed and analyzed on a BSP-like model. In fact, many BSP algorithms are only defined or analyzed for a number of processors p that is sufficiently small with respect to the input size n. For the analysis of the FFT

---

[1] We denote by $\log n$ the logarithm in base two, and by $\ln n$ the natural logarithm.

[2] When $p = \Omega(n)$ a suitable adaptation of our argument gives an $\Omega(\log n)$ bound, which is also tight.

DAG, it is often assumed $p^2 \leq n$, where the complexity is $\Theta(n/p)$. Our results allow for the removal of such restrictions.

The rest of the paper is organized as follows. Section 2 introduces the concept of switching DAG and its switching potential. Then, it formulates the envelope game, a convenient framework for studying the communication occurring when evaluating a DAG. Section 3 briefly reviews the BSP model and develops a relationship between the switching potential of a DAG and its communication complexity on BSP, in the form of a mathematical program. The latter is analyzed in Section 4 and the results are applied to the FFT DAG. Finally, in Section 5 we draw some conclusions and discuss future work.

2   The Switching Potential of Computation DAGs

A *computation DAG* $G = (V, E)$ is a directed acyclic graph where nodes represent *operations* and arcs represent *data dependencies*. More specifically, an arc $(u, v) \in E$ indicates that the value produced by the operation associated with u is one of the operands of the operation associated with v, and we say that u is a *predecessor* of v and v a *successor* of u. The number of predecessors of a node v is called its *in-degree* and denoted $\delta_{in}(v)$, while the number of its successors is called its *out-degree* and denoted $\delta_{out}(v)$. A node v is called an *input* if $\delta_{in}(v) = 0$ and an *output* if $\delta_{out}(v) = 0$. We denote by V_{in} and V_{out} the set of input and output nodes, respectively. The remaining nodes are said to be *internal* and their set is denoted by V_{int}.

For many models of computation, the execution of an algorithm on a particular input can be naturally described by a computation DAG. Of particular interest is the case when this DAG is the same for all inputs of the same size n, and can then be denoted as $G(n)$. In fact, a number of graph-theoretic properties of $G(n)$ can be related to processing, storage, and communication requirements of the underlying algorithm, as well as to its amount of parallelism. In this context, we introduce one such property, the switching potential, defined for a class of relevant computation DAGs.

Definition 1. *A switching DAG $G = (V, E)$ is a computation DAG where for any internal node $v \in V_{int}$ we have $\delta_{out}(v) = \delta_{in}(v)$. We refer to $n = |V_{in}|$ as to the* input size *of G and introduce the* switching size *of G defined as*

$$N = \sum_{v \in V_{in}} \delta_{out}(v) = \sum_{v \in V_{out}} \delta_{in}(v),$$

where the equality between the two summations is easily established.

It is not difficult to see that if, for any internal node of G, a one-to-one relation R is established between the incoming arcs and the outgoing arcs, then a set $\mathcal{R}$ of N arc-disjoint paths naturally arises, where paths are formed by the arcs that belong to the same equivalence class of R^*, the transitive closure of R. Let us now number the arcs incident upon input nodes from 1 to N (in some arbitrarily chosen order) and do the same for the arcs incident upon the output nodes. Then, to the above set $\mathcal{R}$ there corresponds a permutation $\rho = (\rho(1), \rho(2), \ldots, \rho(N))$ of $(1, 2, \ldots, N)$, where $\rho(j)$ is the (number of the) last arc of the (unique) path in $\mathcal{R}$ whose first arc is numbered j. In terms of these concepts, we now introduce a key property of switching DAGs.

Definition 2. *Given a switching DAG $G = (V, E)$, consider the set Γ of all permutations corresponding to one or more sets of N arc-disjoint paths. The switching potential of G is defined as the number $\gamma = |\Gamma|$ of such permutations.*

Intuitively, if the internal nodes are viewed as switches, then items initially positioned on the input nodes can travel without conflicts on arc-disjoint paths and reach the output nodes. Indeed, in the special case where $\delta_{out}(v) = 1$ for all input nodes and $\delta_{in}(v) = 1$ for all output nodes, one has $N = n = |V_{in}| = |V_{out}|$ and the switching DAG can be viewed as a switching network in the traditional sense. Furthermore, if $\gamma = n!$ (all permutations can be realized), then the switching network is said to be a *permutation* (or, *rearrangeable*) *network*.

Next, we define the *envelope game*, to be played on a switching DAG G, based on a given one-to-one relation R between the incoming arcs and the outgoing arcs of each internal node. The game is subject to the following rules.

1. A set of N distinguishable *envelopes* is given, with exactly $\delta_{out}(v)$ envelopes placed on each input node v.
2. The set of envelopes remains invariant during the game and at any stage each envelope is at exactly one node of G.
3. One elementary move consists in moving one envelope along an arc, that is, from one node u to one node v, such that $(u, v) \in E$.
4. No envelope can be moved from a node v before all $\delta_{in}(v)$ envelopes that must be placed on v have actually been placed.
5. The game is completed when all envelopes have reached an output node.

Speaking rather informally, it is easy to see that from the orchestration of the envelope game on a given model of computation one can immediately derive a schedule without recomputation for evaluating a DAG G, on the same model, and viceversa. We just need to imagine that each envelope carries a (rewritable) card where, when a node of G is computed, its result is written on the card of each envelope currently at that node. It is also intuitive that, if nodes u and v of arc (u, v) are processed at different sites, then moving the envelope from u to v will result in some communication. It ought to be observed that given two arcs (u, v) and (u, w) with the same origin, if both v and w are processed at sites different from that of u, then two envelope moves will contribute to communication. This may result in an overcounting, in the case when v and w are processed at the same site, as just one of the two envelopes would be sufficient here, since they carry the same information. However, this overcounting is bounded from above by the maximum out-degree of any node, $\Delta = \max_{v \in V} \delta_{out}(v)$, which is a small constant for many interesting DAGs. The reverse process, of obtaining an execution of the envelope game from an evaluation of the DAG, is also straightforward, with an increase in communication upper bounded by Δ.

While the preceding considerations can be made precise only after having specified a model of computation, they do convey a useful intuition, which will be made rigorous for BSP in the next section, but could prove valuable on other models as well.

In the next section, we show that executing a switching DAG on BSP requires an amount of communication bounded from below by a certain function of its switching potential. This result is of interest, since several relevant computation DAGs are switching DAGs. Examples include the DAGs of networks of switches, of networks of comparators (e.g., for sorting or merging), the DAGs modeling computations of bounded-degree networks (as defined, e.g., in [17]), the DAGs of several stencil computations, and others.

3   Switching Potential and Communication on BSP

The Bulk Synchronous Parallel (BSP) model was introduced by Valiant [23] as a "bridging model" for general-purpose parallel computing, providing an abstraction of both parallel hardware and software. It has been widely studied (see, e.g., [20] and references therein) together with a number of variants (such as D-BSP [22,6], BSP* [4], E-BSP [15], and BSPRAM [21]).

The architectural component of the model consists of p processing elements $P_1, P_2, \ldots, P_p$, each equipped with unbounded local memory, interconnected by a communication medium. The execution of a BSP algorithm consists of a sequence of phases, called *supersteps*: in one superstep, each processor can perform operations on data residing in its local memory, send messages and, at the end, execute a global synchronization instruction. A message sent during a superstep becomes visible to the receiver only at the beginning of the next superstep. The running time of the i-th superstep is expressed in terms of two parameters g and ℓ as $T_i = w_i + h_i g + \ell$, where w_i is the maximum number of local operations performed by any processor and h_i is the maximum number of messages sent or received by any processor (i.e., the i-th superstep performs an h_i-relation). Intuitively, $1/g$ can be interpreted as the available bandwidth per processor, while ℓ as an upper bound on the time required for global barrier synchronization. The *running time* $T_{\mathcal{A}}$ of a BSP algorithm $\mathcal{A}$ is the sum of the times of its supersteps and can be expressed as $W_{\mathcal{A}} + H_{\mathcal{A}} g + S_{\mathcal{A}} \ell$, where $S_{\mathcal{A}}$ is the number of supersteps, $W_{\mathcal{A}} = \sum_{i=1}^{S_{\mathcal{A}}} w_i$ is the *local computation complexity* and $H_{\mathcal{A}} = \sum_{i=1}^{S_{\mathcal{A}}} h_i$ is the *communication complexity*. In this paper, we study the latter metric, which often represents the dominant component.

We focus on algorithms whose execution can be described by a computation DAG $G(n)$ solely determined by the input size n. The lower bounds are derived under the assumption that $G(n)$ is a switching DAG and that each node of the DAG (operation) is executed only once (no recomputation). In particular, we analyze the envelope game on G. In any given execution of such a game, a given node of G is assigned to a unique BSP processor. If $(u, v) \in E$ is an arc with u assigned to processor P and v assigned to processor $P' \neq P$, then the envelope must be routed from P to P', possibly through intermediate processors. A key observation is that a BSP execution of the envelope game corresponding to a given relation R on arcs (intuitively, a setting of the switches) can be adapted to any other relation R' without changing the number of superstep of the sources and destinations sent at each superstep. Simply, the messages will carry different envelopes. We now introduce the critical quantity that we analyze.

Definition 3. *Consider the execution of a switching DAG $G(n)$ on the BSP. The* distribution potential *at superstep j, denoted $\eta_j(n,p)$, is defined as the number of different distributions of the N envelopes across the p processors that result at the end of the j-th superstep, when relation R is varied in all possible ways. (The order of the envelopes within a processor is irrelevant.)*

Intuitively, two tradeoffs are captured by the lower bound argument developed below. First, the communication complexity h of a given superstep is bounded from below in terms of the growth of the distribution potential in that superstep. Second, the distribution potential after the last superstep is bounded from below by the switching potential of the DAG.

At the beginning of the computation (after the 0-th superstep), $\eta_0(n,p) = 1$, since the only achievable distribution of envelopes among processors is the one corresponding to the input distribution protocol. Denote by U the maximum number of envelopes held by any processor at the end of the algorithm. If the algorithm completes in K supersteps, then $\eta_K(n,p) \geq \gamma(n)/(U!)^{N/U}$, where $(U!)^{N/U}$ is a corrective term due to the definition of $\eta_K(n,p)$. Let $o_i \leq U$ be the number of envelopes stored at the end of the algorithm in the i-th processor; then, there are at most $\Pi_{i=1}^{p}(o_i!) \leq (U!)^{N/U}$ envelope permutations differing only on the output values held by the same processor which yield the same distribution of the envelopes among processors. We denote the number of envelopes held by the i-th processor after the j-th superstep by $t_{i,j}$, for each $i \in [p]$ and $j \in [K]$, where $[x]$ denotes the set $\{1, 2, \ldots, x\}$. Clearly, by the rules of the envelope game, we have $\sum_{i=1}^{p} t_{i,j} = N$ and $t_{i,j} \geq 0$. (The latter equation would not necessarily hold if the envelope game were extended in order to allow for recomputation.)

Now consider a processor $i \in [p]$ and a superstep $j \in [K]$. The $t_{i,j}$ envelopes held by processor i after the j-th superstep are of two kinds: the $s_{i,j}$ envelopes that will be sent by i to some other processors during the subsequent superstep, and the other $r_{i,j} = t_{i,j} - s_{i,j}$ remaining envelopes. (The quantities $t_{i,j}$, $s_{i,j}$, and $r_{i,j}$ are all functions of n and p, although this dependence is not made explicit in the notation, for better readability. For the same reason, when clarity is not compromised, we will write η_j in place of $\eta_j(n,p)$.) Thus, there are $\binom{t_{i,j}}{s_{i,j}}$ choices of the set of envelopes to send and (given a fixed schedule of the algorithm, i.e., a fixed pattern of communication) these envelopes can be sent in at most $s_{i,j}!$ different ways to the other processors. Hence, at each superstep j each processor i has at most

$$\binom{t_{i,j}}{s_{i,j}} s_{i,j}! = \binom{r_{i,j} + s_{i,j}}{s_{i,j}} s_{i,j}! = \frac{(s_{i,j} + r_{i,j})!}{r_{i,j}!}$$

communications choices. Then, $\eta_j/\eta_{j-1} \leq \Pi_{i=1}^{p}(s_{i,j} + r_{i,j})!/r_{i,j}!$.

Assembling the above observations, we conclude that the communication complexity H of any algorithm for G is no smaller than the value an optimal solution to the following mathematical program.

$$H \geq \min \sum_{j=1}^{K} \max_{i} s_{i,j}$$

$$\text{s.t.} \prod_{j=1}^{K} \prod_{i=1}^{p} \frac{(s_{i,j} + r_{i,j})!}{r_{i,j}!} \geq \gamma(n)/(U!)^{N/U}$$

$$\sum_{i=1}^{p} (s_{i,j} + r_{i,j}) = N \qquad \forall j \in [K]$$

$$r_{i,j}, s_{i,j} \geq 0 \qquad \forall i \in [p], \forall j \in [K].$$

4   Solving the Mathematical Program

We relax the above system by observing that, for each $j \in [K]$,

$$\prod_{i=1}^{p} \frac{(s_{i,j} + r_{i,j})!}{r_{i,j}!} \leq \prod_{i=1}^{p} (s_{i,j} + r_{i,j})^{s_{i,j}}.$$

The relaxation will enable us to exploit the following lemma.

Lemma 1. *Let q and N be two positive integer values. Then, an optimal solution of the following mathematical program*

$$\max \prod_{i=1}^{q} (a_i + b_i)^{a_i}$$

$$\text{s.t.} \sum_{i=1}^{q} (a_i + b_i) = N$$

$$a_i, b_i \geq 0 \qquad \forall i \in [q]$$

must satisfy $b_i A = a_i (N - A)$ for each $i \in [q]$, where $A = \sum_{i=1}^{q} a_i$.

Proof. When $A = 0$ or there is just one $a_i \neq 0$ the lemma is straightforward. It is also easy to see that, in an optimal solution, $b_i = 0$ whenever $a_i = 0$. Hence, in the following we assume that $a_i \neq 0$ for each $i \in [q]$.

Let $A > 0$. We first study the case $q = 2$, and then use this as a building-block for determining the solution to the general case. Consider an optimal solution (a_1, b_1, a_2, b_2), and suppose a_1 and a_2 are given. Consider the first derivative in b_1 of the objective function. The constraint of the system imposes

$$b_1 + b_2 = N - (a_1 + a_2) = N - A, \tag{1}$$

hence we have

$$\frac{d}{db_1}(a_1 + b_1)^{a_1}(a_2 + b_2)^{a_2} = \frac{d}{db_1}(a_1 + b_1)^{a_1}(N - a_1 - b_1)^{a_2}$$

$$= a_1(a_1 + b_1)^{a_1-1}(N - a_1 - b_1)^{a_2} - (a_1 + b_1)^{a_1}a_2(N - a_1 - b_1)^{a_2-1}.$$

Since $a_1, a_2 > 0$, we have that the derivative is non-negative when $a_1(N - a_1 - b_1) \geq a_2(b_1 + a_1)$, that is, using Equation 1,

$$b_1 \leq \frac{a_1(N - a_1 - a_2)}{a_1 + a_2} = a_1\left(\frac{N - A}{A}\right).$$

Since the above derivative is first non-negative and then non-positive, the point b_1 where the value of the derivative is zero is unique, and thus must satisfy

$$b_1 = a_1\left(\frac{N - A}{A}\right) \quad \text{and} \quad b_2 = a_2\left(\frac{N - A}{A}\right).$$

We now turn our attention to the situation when q is arbitrary. Let (a, b) be an optimal solution, with $a = a_1, a_2, \ldots, a_q$ and $b = b_1, b_2, \ldots, b_q$, and a is given. We claim that $b_i = a_i\left(\frac{N-A}{A}\right)$ for each $i \in [q]$. In fact, suppose this is not true. Then, there must exist an optimal solution $(a, \bar{b}) \neq (a, b)$ and a pair of indices h, k such that $\bar{b}_h/a_h \neq \bar{b}_k/a_k$. We can prove this by contradiction. In fact, if $\bar{b}_h/a_h = \bar{b}_k/a_k$ for each $h \in [q]$, we have the following system of equations with q variables $\bar{b}_1, \bar{b}_2, \ldots, \bar{b}_q$ and q constraints:

$$\begin{cases} \dfrac{\bar{b}_h}{a_h} = \dfrac{\bar{b}_{h+1}}{a_{h+1}} & \forall h \in [q-1] \\ \sum_{j=1}^{q}(\bar{b}_j + a_j) = N. \end{cases}$$

To derive its unique solution, we can rewrite the last constraint as $\sum_{j=1}^{q}(\bar{b}_j/a_j + 1)a_j = N$. By the first $q-1$ constraints we have $\bar{b}_h/a_h = \bar{b}_k/a_k$ for each $h, k \in [q]$, and thus

$$\sum_{j=1}^{q}\left(\frac{\bar{b}_h}{a_h} + 1\right)a_j = N \quad \forall h \in [q],$$

that is

$$\left(\frac{\bar{b}_h}{a_h} + 1\right)A = N \quad \forall h \in [q],$$

which implies

$$\bar{b}_h = a_h\left(\frac{N - A}{A}\right) = b_h \quad \forall h \in [q],$$

a contradiction. Therefore, we have shown that there exists a pair $(\bar{b}_h, \bar{b}_k)$ such that $\bar{b}_h/a_h \neq \bar{b}_k/a_k$. However, as seen for the case $q = 2$, we can always find a pair $(\bar{b}_h, \bar{b}_k)$ such that $(\bar{b}_h + a_h)^{a_h}(\bar{b}_k + a_k)^{a_k} > (\tilde{b}_h + a_h)^{a_h}(\tilde{b}_k + a_k)^{a_k}$, thus contradicting the optimality of $(\bar{b}, v)$ and (u, v). This pair is

$$\tilde{b}_h = a_h\left(\frac{\bar{b}_h + \bar{b}_k}{a_h + a_k}\right) \quad \text{and} \quad \tilde{b}_k = a_k\left(\frac{\bar{b}_k + \bar{b}_h}{a_k + a_h}\right),$$

which is the solution of the system

$$
\begin{cases}
\dfrac{\tilde{b}_h}{a_h} = \dfrac{\tilde{b}_k}{a_k} \\
\tilde{b}_h + \tilde{b}_k = \bar{b}_h + \bar{b}_k.
\end{cases}
$$

It remains to check that the mathematical program is not unbounded. Observe that the objective function is real-valued and continuous on a domain which is non-empty, closed, and bounded. By the classical Weierstrass theorem, such a function admits a maximum, and this must be achieved at (a, b). □

We are now ready to prove the main result of the paper; we will establish the desired lower bound for the FFT graph as a corollary. The lower bound in the theorem exhibits a tradeoff between the communication and the maximum number U of envelopes held by a processor at the end of the algorithm: indeed, as U increases the number of envelope permutations differing only on the output values stored in the same processor increases as well, and thus the required communication complexity may decrease.

Theorem 1. *Let $x^\star$ be the value of an optimal solution of the mathematical program of the previous section. Then,*

$$
x^\star \geq \frac{\ln(\gamma(n)/(U!)^{N/U})}{p \ln(eN/p)}.
$$

Proof. We consider only supersteps where at least one message is sent over the network. (Supersteps without communication do not increase the number of envelope distributions.) We use the following notation: $s_j = \max_i s_{i,j}$ and $S_j = \sum_{i=1}^{p} s_{i,j}$. By setting $a_i = s_{i,j}$, $b_i = r_{i,j}$, and $q = p$ in Lemma 1, we have that, for a given superstep j,

$$
\prod_{i=1}^{p}(s_{i,j} + r_{i,j})^{s_{i,j}} \leq \prod_{i=1}^{p}\left(s_{i,j} + s_{i,j}\left(\frac{N - S_j}{S_j}\right)\right)^{s_{i,j}}
$$

$$
= \prod_{i=1}^{p} s_{i,j}^{s_{i,j}} \left(\frac{N}{S_j}\right)^{s_{i,j}} = \left(\frac{N}{S_j}\right)^{S_j} \prod_{i=1}^{p} s_{i,j}^{s_{i,j}}.
$$

We partition the values of the index j into three sets K_1, K_2, and K_3 as follows: $j \in K_1$ iff $s_j > N/p$, $j \in K_2$ iff $N/(ep) < s_j \leq N/p$, $j \in K_3$ iff $s_j \leq N/(ep)$, where e is the base of the natural logarithm. For simplicity, we assume $p \leq N/e$. If $j \in K_1$, we have

$$
\left(\frac{N}{S_j}\right)^{S_j} \prod_{i=1}^{p} s_{i,j}^{s_{i,j}} \leq e^{N/e} s_j^{N},
$$

because function $(N/x)^x$ is increasing in x until $x < N/e$ and $x = N/e$ is the maximum of the function. The constraints on the problem implies $\sum_{i=1}^{p} s_{i,j} \leq N$, and then $\prod_{i=1}^p s_{i,j}^{s_{i,j}}$ is maximized when N/s_j values $s_{i,j}$ are set to s_j and the remaining ones to zero. On the other hand, when $j \in K_2$, we have

$$\left(\frac{N}{S_j}\right)^{S_j} \prod_{i=1}^p s_{i,j}^{s_{i,j}} \leq \left(\frac{N}{S_j}\right)^{S_j} s_j^{ps_j} \leq e^{N/e} s_j^{ps_j},$$

since, $s_{i,j} \leq s_j$. Finally, when $j \in K_3$ we have

$$\left(\frac{N}{S_j}\right)^{S_j} \prod_{i=1}^p s_{i,j}^{s_{i,j}} \leq \left(\frac{N}{S_j}\right)^{S_j} s_j^{ps_j} \leq \left(\frac{N}{ps_j}\right)^{ps_j} s_j^{ps_j} = \left(\frac{N}{p}\right)^{ps_j},$$

since the function $(N/x)^x$ is increasing in x until $x < N/e$ and the maximum value is $(N/(ps_j))^{ps_j}$ when $S_j \leq ps_j \leq N/e$.

Therefore, the first constraint of the minimization problem that we are studying can be relaxed as follows:

$$\prod_{j \in K_1} e^{N/e} s_j^N \prod_{j \in K_2} e^{N/e} s_j^{ps_j} \prod_{j \in K_3} \left(\frac{N}{p}\right)^{ps_j} \geq \frac{\gamma(n)}{(U!)^{N/U}}.$$

By taking the natural logarithm of both sides we have

$$\sum_{j \in K_1} \left(\frac{N}{e} + N \ln s_j\right) + \sum_{j \in K_2} \left(\frac{N}{e} + ps_j \ln s_j\right) + \sum_{j \in K_3} ps_j \ln(N/p) \geq \ln\left(\frac{\gamma(n)}{(U!)^{N/U}}\right).$$

Since $s_j > N/(ep)$ if $j \in K_1 \cup K_2$, we get

$$\sum_{j \in K_1} N \ln(es_j) + \sum_{j \in K_2} ps_j \ln(es_j) + \sum_{j \in K_3} ps_j \ln(N/p) \geq \ln\left(\frac{\gamma(n)}{(U!)^{N/U}}\right).$$

Let $\widehat{K}_i = \sum_{j \in K_i} s_j$. By the concavity of $\ln(es_j)$ and the convexity of $s_j \ln(es_j)$, we have that the first two summations are maximized when $s_j = N/p$ for each $j \in K_1 \cup K_2$, $|K_1| = \widehat{K}_1/(N/p)$ and $|K_2| = \widehat{K}_2/(N/p)$. Then we get

$$p\widehat{K}_1 \ln(eN/p) + p\widehat{K}_2 \ln(eN/p) + p\widehat{K}_3 \ln(N/p) \geq \ln\left(\frac{\gamma(n)}{(U!)^{N/U}}\right),$$

which yields the sought lower bound to the minimum solution of the problem:

$$\min \sum_{j=1}^K s_j \geq \widehat{K}_1 + \widehat{K}_2 + \widehat{K}_3 \geq \frac{\ln(\gamma(n)/(U!)^{N/U})}{p \ln(eN/p)}.$$

$\square$

Corollary 1. *Let $\mathcal{A}$ be any algorithm computing an n-input FFT DAG on a BSP with p processors, and let U be the maximum number of envelopes held by any processor at the end of the algorithm. If $U \leq N/2 = n$ and recomputation is not allowed, then the communication complexity of the algorithm is*

$$H_{\mathcal{A}}(n,p) = \Omega\left(\frac{n \log n}{p \log(n/p)}\right).$$

Proof (Sketch). The FFT DAG has $n(\log n + 1)$ nodes and can produce at the output nodes $\gamma(n) = 2^{n(\log n+1)}$ distinct permutations of the $N = 2n$ envelopes. Hence, by Theorem 1, we get $H_{\mathcal{A}}(n,p) \geq (n/p)\log(2n/U^2)/\log(n/p)$. Since an FFT DAG can perform any cyclic shift of a vector, an $\Omega(U)$ lower bound follows by an argument based on the information flow of cyclic shifts [19, Lemma 10.5.2]. Therefore, we have

$$H_{\mathcal{A}}(n,p) = \Omega \left(\frac{n \log(2n/U^2)}{p \log(n/p)} + U \right).$$

We have that $U \geq 2n/p$. In this range, the above bound is minimized by setting $U = 2n/p$, yielding the stated bound.   □

Several similar results can be obtained, for example for the Beneš permutation network, and for the bitonic and the AKS sorting networks.

5   Conclusions

In this paper, we have studied some aspects of the communication complexity of parallel algorithms. We have developed a new technique for deriving lower bounds on communication complexity for computations that can be represented by a certain kind of DAGs. We have demonstrated the power of this technique on the FFT DAG for which, assuming non-recomputation, the derived lower bound is tight for any possible values of parameters n and p, thus improving previous work.

It is natural to wonder whether our main lower bound holds (asymptotically) when recomputation of intermediate values is allowed. (Re-execution of operations is known, for instance, to enhance some simulations among networks [16].) While it is not difficult to see that our lower bound holds when each node of the DAG can be recomputed $O(1)$ times, in the general case (an adaptation of) our technique yields, for the FFT DAG, the same bound as that of the dominator-set result mentioned in the introduction. We feel that settling this question might shed new light on the role of recomputation in I/O- and communication-efficient computing, which is not yet fully understood.

Acknowledgments. The authors would like to thank Andrea Pietracaprina and Geppino Pucci for insightful discussions.

References

1. Aggarwal, A., Chandra, A.K., Snir, M.: Communication complexity of PRAMs. Theor. Comp. Sci. 71, 3–28 (1990)
2. Aggarwal, A., Vitter, J.S.: The input/output complexity of sorting and related problems. Comm. ACM 31(9), 1116–1127 (1988)

3. Ballard, G., Demmel, J., Holtz, O., Schwartz, O.: Graph expansion and communication costs of fast matrix multiplication. In: Proc. 23rd SPAA, pp. 1–12. ACM (2011)
4. Bäumker, A., Dittrich, W., Meyer auf der Heide, F.: Truly efficient parallel algorithms: 1-optimal multisearch for an extension of the BSP model. Theor. Comp. Sci. 203(2), 175–203 (1998)
5. Bilardi, G., Pietracaprina, A., D'Alberto, P.: On the Space and Access Complexity of Computation DAGs. In: Brandes, U., Wagner, D. (eds.) WG 2000. LNCS, vol. 1928, pp. 47–58. Springer, Heidelberg (2000)
6. Bilardi, G., Pietracaprina, A., Pucci, G.: Decomposable BSP: A bandwidth-latency model for parallel and hierarchical computation. In: Handbook of Parallel Computing: Models, Algorithms and Applications, pp. 277–315. CRC Press (2007)
7. Bilardi, G., Pietracaprina, A., Pucci, G., Scquizzato, M., Silvestri, F.: Network-oblivious algorithms (to be submitted, 2012)
8. Bilardi, G., Pietracaprina, A., Pucci, G., Silvestri, F.: Network-oblivious algorithms. In: Proc. 21st IPDPS, pp. 1–10. IEEE (2007)
9. Bilardi, G., Preparata, F.: Processor-time tradeoffs under bounded-speed message propagation: Part II, lower bounds. Theor. Comp. Syst. 32(5), 531–559 (1999)
10. Chowdhury, R.A., Silvestri, F., Blakeley, B., Ramachandran, V.: Oblivious algorithms for multicores and network of processor. In: Proc. 24th IPDPS, pp. 1–12. IEEE (2010)
11. Frigo, M., Leiserson, C.E., Prokop, H., Ramachandran, S.: Cache-oblivious algorithms. ACM Trans. Algorithms 8(1), 4:1–4:22 (2012)
12. Goodrich, M.T.: Communication-efficient parallel sorting. SIAM J. Computing 29(2), 416–432 (1999)
13. Hong, J.W., Kung, H.T.: I/O complexity: The red-blue pebble game. In: Proc. 13th STOC, pp. 326–333. ACM (1981)
14. Irony, D., Toledo, S., Tiskin, A.: Communication lower bounds for distributed-memory matrix multiplication. J. Par. & Distr. Comp. 64(9), 1017–1026 (2004)
15. Juurlink, B.H.H., Wijshoff, H.A.G.: A quantitative comparison of parallel computation models. ACM Trans. Comput. Syst. 16(3), 271–318 (1998)
16. Koch, R.R., Leighton, F.T., Maggs, B.M., Rao, S.B., Rosenberg, A.L., Schwabe, E.J.: Work-preserving emulations of fixed-connection networks. J. ACM 44(1), 104–147 (1997)
17. Leighton, F.T.: Introduction to Parallel Algorithms and Architectures: Arrays, Trees, Hypercubes. Morgan Kaufmann Publishers (1992)
18. Papadimitriou, C.H., Ullman, J.D.: A communication-time tradeoff. SIAM J. Computing 16(4), 639–646 (1987)
19. Savage, J.E.: Models of Computation: Exploring the Power of Computing. Addison-Wesley (1998)
20. Tiskin, A.: BSP (bulk synchronous parallelism). In: Encyclopedia of Parallel Computing, pp. 192–199. Springer (2011)
21. Tiskin, A.: The bulk-synchronous parallel random access machine. Theor. Comp. Sci. 196(1-2), 109–130 (1998)
22. de la Torre, P., Kruskal, C.P.: Submachine Locality in the Bulk Synchronous Setting. In: Fraigniaud, P., Mignotte, A., Robert, Y., Bougé, L. (eds.) Euro-Par 1996. LNCS, vol. 1124, pp. 352–358. Springer, Heidelberg (1996)
23. Valiant, L.G.: A bridging model for parallel computation. Comm. ACM 33(8), 103–111 (1990)
24. Wu, C.L., Feng, T.Y.: The universality of the shuffle-exchange network. Trans. Computers 30, 324–332 (1981)

A Fast Parallel Algorithm
for Minimum-Cost Small Integral Flows

Andrzej Lingas[1] and Mia Persson[2]

[1] Department of Computer Science, Lund University, 22100 Lund, Sweden
Andrzej.Lingas@cs.lth.se
[2] Department of Computer Science, Malmö University, 205 06 Malmö, Sweden
mia.persson@mah.se

Abstract. We present a new approach to the minimum-cost integral flow problem for small values of the flow. It reduces the problem to the tests of simple multi-variable polynomials over a finite field of characteristic two for non-identity with zero. In effect, we show that a minimum-cost flow of value k in a network with n vertices, a sink and a source, integral edge capacities and positive integral edge costs polynomially bounded in n can be found by a randomized PRAM, with errors of exponentially small probability in n, running in $O(k \log(kn) + \log^2(kn))$ time and using $2^k (kn)^{O(1)}$ processors. Thus, in particular, for the minimum-cost flow of value $O(\log n)$, we obtain an RNC^2 algorithm.

1 Introduction

The maximum network flow problem is a well known fundamental problem in algorithms and optimization with plenty of important applications [1, 7, 8, 16]. It is known to be P-complete even in its integral version provided that the edge capacities are exponentially large in the size of the network [13]. The minimum-cost flow problem is a well known important generalization of the maximum flow problem [1, 8, 10, 16]. The objective is to compute a maximum flow of minimum cost in a directed graph where each edge is assigned a cost. For a flow f in a directed graph (V, E), the cost of f is simply $\sum_{e \in E} f(e) cost(e)$.

The prospects for designing a fast and processor efficient parallel algorithm, in particular an NC algorithm [19], for maximum integral flow or minimum-cost integral flow are small. The fastest known parallel implementations of general maximum flow and/or minimum-cost flow algorithms achieve solely a moderate speed up and still run in $\Omega(n^\alpha)$ time, where α is a positive constant, see [2, 12].

The situation changes when the edge capacities or the supply of flow as well as edge costs are substantially bounded. For example, if the edge capacities and edge costs are bounded by a polynomials in n, both problems admit RNC algorithms. Then, the maximum integer flow problem admits even an RNC^2 algorithm [17, 18, 20] while the minimum-cost integer flow problem admits an RNC^3 algorithm [18]. At the heart of the aforementioned RNC solutions is the randomized method of detecting a perfect matching by randomly testing Edmonds' multi-variable polynomials for non-identity with zero [6, 14, 17, 20].

C. Kaklamanis et al. (Eds.): Euro-Par 2012, LNCS 7484, pp. 688–699, 2012.

When the flow supply is relatively small, e.g., logarithmic in the size of the network or a poly-logarithmic one, then just an NC implementation of the basic phase in the standard Ford-Fulkerson method [1, 7–9, 16] yields an NC algorithm (NC^3 when the supply is logarithmic) for maximum integer flow that can be extended to an NC algorithm for minimum-cost integer flow (when edge costs are polynomially bounded). The number of processors used corresponds to that required by a shortest path computation.

In this paper, we present a new approach to the minimum-cost integral flow problem for a small value k of the flow. We directly associate a simple polynomial over a finite field with the corresponding problem of the existence of k mutually vertex disjoint paths of bounded total length, connecting two sets of k terminals in a directed graph. By using the idea of monomial cancellation, the latter problem reduces to testing the polynomial over a finite field of characteristic two for non-identity with zero. We combine the DeMillo-Lipton-Schwartz-Zippel lemma [5, 21] on probabilistic verification of polynomial identities with parallel dynamic programming to perform the test efficiently in parallel. Additionally, we use the isolation lemma to construct the minimum-cost flow [17, 20].

In effect, we infer that a minimum-cost flow of value k in a network with n vertices, a sink and a source, integral edge capacities and positive integral edge costs polynomially bounded in n can be found by a randomized PRAM, with errors of exponentially small probability in n, running in $O(k \log(kn) + \log^2(kn))$ time and using $2^k (kn)^{O(1)}$ processors. Thus, in particular, for the minimum-cost flow of value $O(\log n)$, we obtain an RNC^2 algorithm.

Related Work. For the RNC algorithms for the related problem of minimum-cost perfect matching see [14, 17, 20]. For the comparison of time and substantial processor complexities of prior RNC algorithms for the minimum-cost flow see page 7 in [18]. The fastest of the reported algorithms is not an RNC^2 one even when the flow supply and thus the edge capacities are logarithmic in the size of the network. The idea of associating a polynomial over a finite field to the sought structure has been already used by Edmonds to detect matching [6] and then in several papers presenting RNC algorithms for perfect matching construction [14, 17, 20]. It appears in several recent papers that also exploit the idea of monomial cancellation [3, 4, 15, 22].

Organization. In the next section, we comment briefly on the basic notation and the model of parallel computation used in the paper. In Section 3, we derive our fast randomized parallel method for detecting the existence of k mutually vertex disjoint paths of bounded total length connecting two sets of k terminals in a directed graph. In Section 4, we generalize the method to include edge costs which enables us to replace the total length bound with the total cost one. In section 5, we show a straightforward reduction of the minimum-cost integer flow problem parametrized by the flow value to the corresponding disjoint paths problem which enables us to derive our main result on detecting minimum-cost small flows in parallel.

2   Terminology

For a natural number n, we let $[n]$ denote the set of natural numbers in the interval $[1, n]$. The cardinality of a set A will be denoted by $|A|$.

We assume the standard definitions of *flow* and *flow value* in a network (directed graph) with integral edge capacities, a distinguished source vertex s and a distinguished sink vertex t (e.g., see [7]) .

For the definitions of parallel random access machines (PRAM), the classes NC and RNC and the corresponding notions of NC and RNC algorithms, the reader is referred to [19].

The *characteristic* of a ring or a field is the minimum number of 1 in a sum that yields 0. A finite field with q elements is often denoted by F_q.

3   Connecting Vertex-Disjoint Paths

It is well known that the maximum integral network flow problem with bounded edge capacities corresponds to a disjoint path problem (cf. [7]). In Section 5, we provide an efficient parallel reduction of the minimum-cost integral flow problem parametrized by the flow value to a parametrized disjoint path problem. This section is devoted to a derivation of a fast randomized parallel method for the decision version of the parametrized path problem.

Let $L = (V, E)$ be a network in a form of a directed graph with n vertices, among them a distinguished set $X = \{x_1, ..., x_k\}$ of k source vertices and a disjoint distinguished set $Y = \{y_1, ..., y_k\}$ of k sink vertices.

A *walk* in L is a sequence of vertices $v_1, v_2, ..., v_l$ of L such that for $j = 1, ..., l-1$, $(v_j, v_{j+1}) \in E$, v_1 is in X, $v_2, ..., v_{l-1}$ are in $V \setminus (X \cup Y)$, v_l is in Y. The length of the walk is $l-1$. In other words, a walk is just a (not necessarily simple) path starting from a vertex in X, having intermediate vertices in $V \setminus (X \cup Y)$, and ending at a vertex in Y.

A *proper set S of walks* in L is a set W of k walks of total length $\leq k(n-1)$, each with a distinct start vertex in X and a distinct end vertex in Y.

A *signature* of S is the pair (i, j) that is smallest in lexicographic order such that the two walks that start at x_i and x_j respectively intersect.

Note that walks in S are pairwise vertex disjoint iff the signature of S is not defined.

We define the transformation ϕ on S as follows. If S has the signature (i, j) then ϕ switches the suffix of the walk starting at x_i with that of the walk starting at x_j at the first intersection vertex of these two walks. See Fig. 1. Otherwise, if the signature of S is not defined then ϕ is an identity on S.

Observe that if the signature of S is defined then $\phi(S)$ has the same signature as S and $\phi(S) \neq S$. The first observation is immediate. To show the second one it is sufficient to note that $\phi(S) = S$ holds iff ϕ transforms the two walks which yield the signature of S onto themselves. The latter is however impossible since they have different start vertices and different end vertices. Note also that the walks in $\phi(S)$ have the same total length as those in S.

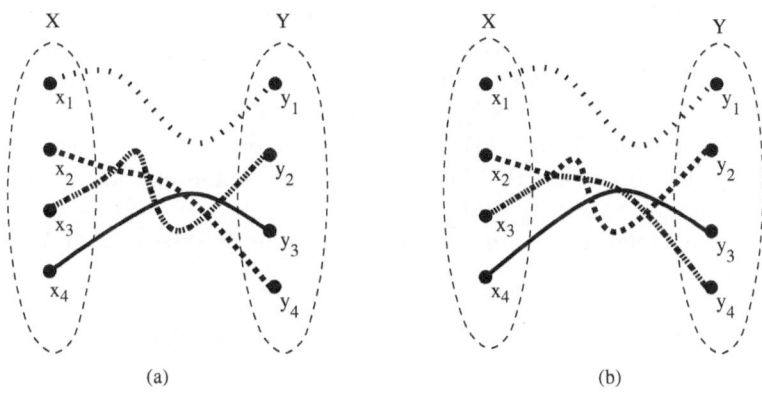

Fig. 1. An example of a proper set S of walks and the companion proper set $\phi(S)$ of walks

It follows that ϕ is an involution on sets of proper walks of total length l, i.e., $\phi(\phi(S)) = S$ holds for any proper set S of walks of total length l.

For the network L and $l \in [k(n-1)]$, let $F_{L,l}$ be the family of all proper sets of k walks of total length l in L. Assign a distinct variable x_e to each edge e in L. For a walk $W \in F_{L,l}$, let M_W be the monomial, where x_e has multiplicity equal to the number of occurrences of e in W. Next, let $Q_{L,l}$ denote the polynomial $\sum_{S \in F_{L,l}} \prod_{W \in S} M_W$.

Lemma 1. *For the network L and $l \in [k(n-1)]$, there is a proper set of k mutually vertex-disjoint walks of total length l in L iff $Q_{L,l}$ is not identical to zero over a field of characteristic two.*

Proof. $F_{L,l}$ can be partitioned into the family $F_{L,l}^1$ of sets S of walks such that $\phi(S) = S$ and the family $F_{L,l}^2$ of sets S of walks such that $\phi(S) \neq S$. The polynomial $\sum_{S \in F_L^2} \prod_{W \in S} M_W$ is identical to zero over a field of characteristic two since for each $S \in F_L^2$ the monomials $\prod_{W \in S} M_W$ and $\prod_{W \in \phi(S)} M_W$ contain equal multiplicities of the same variables and $\phi(\phi(S)) = S$ so S and $\phi(S)$ can be paired. On the other hand, since each set S of walks in $F_{L,l}^1$ consist of mutually vertex-disjoint walks, the monomials in the polynomial $\sum_{S \in F_{L,l}^1} \prod_{W \in S} M_W$ are in one-to one correspondence with S and therefore are unique.

$\square$

To warm up, we prove the following lemma on sequential evaluation of Q_L.

Lemma 2. *$Q_{L,l}$ can be evaluated for a given assignment of values over a field $F_{2^{O(\log n)}}$ of characteristic two in $O(k^2 n^3 + 2^k k^3 n^2)$ time.*

Proof. For $B \subset Y$, $l \in [(n-1)|B|]$, we consider the family $W_l(B)$ of all sets S consisting of $|B|$ walks connecting $|B|$ distinct sources in $\{x_1, ..., x_{|B|}\}$ with the $|B|$ distinct sinks in B so that the total length of the walks is exactly l. Next, we define the polynomial $Q_l(B)$ as $\sum_{S \in W_l(B)} \prod_{W \in S} M_W$. Note that $Q_{L,l} = Q_l(Y)$.

On the other hand, for $l \in [k(n-1)]$, $x \in X$ and $z \in V \setminus X$, we consider the set $W_l(x, z)$ of walks of length l in L that start at x and end at z. Let $Q_l(x, z)$ be the polynomial $\sum_{W \in W_l(x,z)} M_W$.

We have the following recurrence for a nonempty subset B of Y and $l \in [|B|(n-1)]$:

$$Q_l(B) = \sum_{y \in B} \sum_{q \in [|B|(n-1)-|B|+1]} Q_{l-q}(B \setminus \{y\}) Q_q(x_{|B|}, y).$$

Next, we have also the following recurrence for $x \in X$, $z \in V \setminus X$, and $q \in [k(n-1)]$:

$$Q_q(x, z) = \sum_{u \in V \setminus (X \cup Y) \& (u,z) \in E} Q_{q-1}(x, u) x_{(u,z)}.$$

We have also $Q_1(x, z) = x_{(x,z)}$ if $(x, z) \in E$, and otherwise $Q_1(x, z) = 0$. Consequently, we can evaluate all the polynomials $Q_q(x, z)$ by the second recurrence in $O(k^2 n^3)$ time.

Now, by using the first recurrence and setting $Q_l(\emptyset)$ to 1 in the field, we can evaluate all the polynomials $Q_l(B)$ in the increasing order of the cardinalities of B in $O(2^k k^3 n^2)$ time.

$\square$

We can partially parallelize the sequential evaluation of $Q_{L,l}$ in order to obtain the following lemma.

Lemma 3. $Q_{L,l}$ can be evaluated for a given assignment of values over a field $F_{2^{O(\log n)}}$ of characteristic two in $O(k \log n + \log^2 n)$ time by a CREW PRAM using $O(kn^4 + 2^k k^3 n^2)$ processors.

Proof. We generalize the definition of the set $W_q(x, z)$ and the corresponding polynomial $Q_q(x, z)$ to include arbitrary start vertex $x \in V \setminus Y$, requiring $z \in V \setminus X$ as previously. Then, we can evaluate $Q_q(x, z) = \sum_{W \in W_l(x,z)} M_W$ for $x \in V \setminus Y$ and $z \in V \setminus X$, for $q \in [k(n-1)]$ by the following standard doubling recurrence for $q \geq 2$:

$$Q_q(x, z) = \sum_{y \in V \setminus (X \cup Y)} Q_{\lceil q/2 \rceil}(x, y) Q_{\lfloor q/2 \rfloor}(y, z)$$

At the bottom of the recursion, we have $Q_1(x, z) = x_{(x,z)}$ if $(x, z) \in E$, otherwise $Q_1(x, z) = 0$. It follows that all $Q_q(x, z)$ for $x \in V \setminus Y$, $z \in V \setminus X$, and $q \in [k(n-1)]$ can be evaluated in a bottom-up manner in $O(\log^2 n)$ time by a CREW PRAM using $O(kn^4)$ processors.

Recall the first recurrence from the proof of Lemma 2. When the polynomials $Q_q(x_i, y_j)$ for $x_i \in X$ and $y_i \in Y$ are evaluated, we can evaluate in turn the polynomials $Q_l(B)$, where $B \subset Y$, $l \in [|B|(n-1)]$ in k phases in the increasing order of the cardinalities of B by this recurrence. It can be done in $O(k(\log k + \log n)) = O(k \log n)$ time by a CREW PRAM using $2^k k^3 n^2$ processors.

By $Q_{L,l} = Q_l(Y)$, $k \le n$, we conclude that $Q_{L,l}$ can be evaluated in $O(k \log n + \log^2 n)$ time by a CREW PRAM using $O(kn^4 + 2^k k^3 n^2)$ processors.

$\square$

The following lemma on polynomial identities verification has been shown independently by DeMillo and Lipton, Schwartz, and Zippel.

Lemma 4. *[5, 21] Let $Q(x_1, x_2, ..., x_m)$ be a nonzero polynomial of degree d over a field of size r. Then, for f_1, f_2, ...,f_m chosen independently and uniformly at random from the field, the probability that $Q(f_1, f_2, ..., f_m)$ is not equal to zero is at least $1 - \frac{d}{r}$.*

Note that the polynomial $Q_{L,l}$ is of degree l not larger than $k(n-1) \le n^2$. We can use Lemma 4 with a field $F_{2^{c \log n}}$ of characteristic two to obtain a randomized test of the polynomial $Q_{L,l}$ for not being identical to zero with one side errors. For sufficiently large constant c, the one side errors are of probability not larger than a constant smaller than 1. By performing $O(n)$ such independent tests, the probability of one side errors can be decreased to exponentially small in n one.

By Lemma 3, the series of the tests can be performed in $O(k \log n + \log^2 n)$ time by a PRAM using $O(kn^4 + 2^k k^3 n^3)$ processors. By Lemma 1, these tests verify if there is a proper set of mutually vertex-disjoint walks of total length l in the network L. The latter in turn is equivalent to the existence of k mutually vertex-disjoint paths of total length l connecting X with Y in L by the definition of a proper set of walks in L. Hence, observing that each walk can be trivially pruned to a simple directed path with the same endpoints, we obtain our main result.

Theorem 1. *The problem of whether or not there is a set of k mutually vertex-disjoint simple directed paths of total length l connecting X with Y in the network L can be decided by a randomized CREW PRAM, with one-sided errors of exponentially small probability in n, running in $O(k \log n + \log^2 n)$ time and using $O(kn^4 + 2^k k^3 n^3)$ processors.*

4   Vertex-Disjoint Connecting Paths of Bounded Cost

In this section, we shall consider a more general situation where there are a positive integer C and a cost function c assigning to each of the m edges e in the network L a cost $c(e) \in [C]$. The cost of a walk or a path is simply the sum of the costs of the edges forming it (the cost of an edge is counted the number of times it appears on the walk or path). We would like to detect a proper set of k walks in L that achieves the minimum cost.

For this reason, we consider the following generalization of the polynomial $Q_{L,l}$. For $U \in [mC]$, let $H_{L,U}$ be the set of all proper sets of walks in the edge-costed network L that have total cost equal to U. Next, for a walk W in L, as previously, let M_W be the monomial which is the product of x_e over the occurrences of edges e on W. The polynomial $CQ_{L,U}$ is defined by $\sum_{S \in H_{L,U}} \prod_{W \in S} M_W$.

By using the proof method of Lemma 1, we obtain the following counterpart of this lemma for $CQ_{L,U}$.

Lemma 5. *For the edge-costed network L, there is a proper set of k mutually vertex-disjoint walks of total cost U in L iff $CQ_{L,U}$ is not identical to zero over a field of characteristic two.*

Next, we obtain the following counterpart of Lemma 3 for $CQ_{L,U}$.

Lemma 6. *$CQ_{L,U}$ can be evaluated for a given assignment f of values over a field $F_{2^{O(\log n)}}$ of characteristic two in $O(k \log(Cn) + \log^2(Cn))$ time by a PRAM using $O(kC^4n^8 + 2^k k^3 C^2 n^4)$ processors.*

Proof. The proof reduces to that of Lemma 3. We replace each directed edge e of cost $c(e) \in [C]$ in the network L by a directed path of length $c(e)$ introducing $c(1) - 1$ additional vertices. With each edge on such a path, we associate a variable. We assign $f(x_e)$ to the variable associated with the first edge on the path replacing e, and just 1 of the field to the variables associated with the remaining edges on the path.

The resulting network L' is of size $O(Cn^2)$. Let $H_{L',U}$ be the family of all proper sets of k walks of total cost U in the network L'. We can evaluate the polynomial $Q_{L',U} = \sum_{S \in H_{L',U}} \prod_{W \in S} M_W$ in parallel analogously as $Q_{L,l}$ in the proof of Lemma 3. It remains to observe that the value of $CQ_{L,U}$ under the assignment f is equal to that of $Q_{L',U}$ under the aforementioned assignment. □

Now, we are ready to derive our main result in this section.

Theorem 2. *The minimum cost of a set of k mutually vertex-disjoint simple directed paths connecting X with Y in the network L with edge costs in $[C]$ can be computed by a randomized CREW PRAM, with errors of exponentially small probability in n, running in $O(k \log(Cn) + \log^2(Cn))$ time and using $O(kC^5n^{10} + 2^k k^3 C^3 n^7)$ processors.*

Proof. The minimum cost of the sought set of vertex-disjoint paths is in $[Cn^2]$. Hence, by Lemma 5, it is sufficient to test the polynomials $CQ_{L,U}$ for non-identity with zero for all $U \in [Cn^2]$ in parallel. By applying Lemmata 4 and 6 in a manner analogous to the proof of Theorem 1, we conclude that it can be done by a randomized CREW PRAM, with one-sided errors of exponentially small in n probability, running in $O(k \log(Cn) + \log^2(Cn))$ time and using $O(Cn^2 \times (kC^4n^8 + 2^k k^3 C^2 n^5))$ processors. □

5   Finding Vertex-Disjoint Connecting Paths

A straightforward approach of extending our randomized parallel method for deciding if there is a proper set of k mutually vertex-disjoint walks (of a given total cost) between two sets of vertices of cardinality k to include the finding variant could be roughly as follows. In parallel, for each k-tuple of respective neighbors of the k start vertices in X, replace the set of start vertices by the k-tuple and apply our method recursively to the resulting network. If the test is positive, the first edges on the walks are known, and we can iterate the method. The problem with this approach is that its recursive depth is proportional to the maximum length of a walk in the resulting set of mutually vertex-disjoint walks between X and Y.

Also, it is not clear how one could implement a straightforward divide-and-conquer approach of guessing intermediate vertices in order to find a set of k mutually-vertex disjoint walks of a given cost efficiently in parallel.

We need more advanced methods to obtain a very fast parallelization of the finding variant. We shall modify the edge cost in the network L in order to use the so called *isolation lemma* in a manner analogous to the RNC method of finding a perfect matching given in [20].

Lemma 7. *(The isolation lemma [20]). Let F be a family of subsets of a set with q elements and let r be a non-negative integer. Suppose that each element s of the set is independently assigned a weight $w(s)$ uniformly at random from $[r]$, and the weight of a subset S in F is defined as $w(S) = \sum_{x \in S} w(x)$. Then, the probability that there is a unique set in F of minimum weight is at least $1 - \frac{q}{r}$.*

Corollary 1. *For each of the m edges e in the network L, modify its cost $c(e)$ to $c'(e) = c(e)rm + w(e)$, where the weight $w(e)$ is drawn uniformly at random from $[r]$. Then, the probability that there is a unique minimum-cost set of mutually vertex-disjoint paths connecting X with Y in the edge weighted network L is at least $1 - \frac{m}{r}$.*

Proof. To use the isolation lemma, let the underlying set to consist of all edges in the network L. Next, note that a set of mutually vertex-disjoint paths connecting X with Y achieving a minimum cost consists of simple paths and thus it can be identified with the set of edges on the paths. Let P be the family of all sets of mutually vertex-disjoint simple paths connecting X with Y in the network L. By the setting of new costs $c'(e)$, solely those sets in P that achieved the minimum cost, say D, under the original costs $c(e)$ can achieve a minimum cost under the new costs $c'(e)$. So, we can set F to the aforementioned sub-family of P, and define the weight of a set of k paths in F as the sum of the weights $w(e)$ of the edges e on the paths in this set in order to use the isolation lemma. By the isolation lemma, there is a unique set S in F that achieves the minimum weight $w(S)$ with the probability at least $1 - \frac{m}{r}$. The corollary follows since each set S in F has the cost $c'(S)$ equal to $D + w(S)$. □

Throughout the rest of this section, we shall assume that each of the m edges e in the network L is assigned the cost $c'(e)$ as in Corollary 1 and that $r \in [n^{O(1)}]$.

Suppose that we know the minimum cost of a set of k mutually vertex-disjoint paths connecting X with Y in the network L with the edge costs indicated, and such a minimum-cost set is unique. Then, it is sufficient to show that we can test quickly in parallel if the network L with an arbitrary edge removed still contains a set of k mutually vertex-disjoint paths connecting X with Y that achieves the minimum cost. By performing the test for each edge of L in parallel, we can determine the set of edges forming the unique minimum-cost set of k mutually vertex-disjoint paths connecting X with Y.

To carry out these tests, we need to generalize the polynomial $CQ_{L,U}$ to a polynomial $CP_{L,e,U}$, where e is an edge in L and U is a cost constraint from $[mr(mC+1)] = [Cn^{O(1)}]$. Let $H_{L,e,U}$ be the family of all proper sets of k walks in the network L with the edge e removed that have total cost equal to U. (In the total cost of a set of walks, we count the cost of an edge the number of times equal to the sum of the multiplicities of the edge in the walks.)

As in the definition of $Q_{L,l}$ assign a distinct variable x_e to each edge e in L, and for a walk $W \in H_{L,e,W}$, let M_W be the monomial, where x_e has multiplicity equal to the number of occurrences of e in W. The polynomial $CP_{L,e,U}$ is defined by $\sum_{S \in H_{L,e,W}} \prod_{W \in S} M_W$.

By using the proof method of Lemma 1, we obtain the following counterpart of this lemma for $CP_{L,e,U}$.

Lemma 8. *For the edge-costed network L with m edges, edge e, and $U \in [Cn^{O(1)}]$, there is a proper set of k mutually vertex-disjoint walks of total cost U in the network L with the edge e removed iff $CP_{L,e,U}$ is not identical to zero over a field of characteristic two.*

Next, we obtain the counterpart of Lemma 3 for $CP_{L,e,U}$ following the proof of Lemma 6.

Lemma 9. *$CP_{L,e,U}$ can be evaluated for a given assignment of values over a field $F_{2^{O(\log n)}}$ of characteristic two in $O(k \log(Cn) + \log^2(Cn))$ time by a PRAM using $2^k(kCn)^{O(1)}$ processors.*

Now, we are ready to derive our main result in this section.

Theorem 3. *There is a randomized PRAM returning almost certainly (i.e., with probability at least $1 - \frac{1}{n^\alpha}$, where $\alpha \geq 1$) a minimum-cost set of k mutually vertex-disjoint paths connecting X with Y in the network L with the original edge costs in $[C]$ (iff such a set exists) in $O(k \log(Cn) + \log^2(Cn))$ time using $2^k(kCn)^{O(1)}$ processors.*

Proof. We set r to, say, n^2m, and specify the new edge costs $c'(e)$ in the network L drawing the weights $w(e)$ uniformly at random from $[r]$ as in Corollary 1. Next, for each $U \in [mr(mC+1)] = [Cn^{O(1)}]$, we proceed in parallel as follows. For each edge e of the network L, we test the polynomial $CP_{L,e,U}$ for the non-identity with zero by using Lemma 4 and Lemma 9 (we can perform a linear in n number of such tests in parallel in order to decrease the probability of the one-sided error to an exponentially small one). Next, we verify if the edges that passed the test positively yield a set of k mutually vertex-disjoint paths connecting X with Y. For example, it can be done by checking for each endpoint of the edges outside $X \cup Y$ if it is shared by exactly two of the edges, and then computing and examining the transitive closure of the graph induced by the edges (see [19]). If so, we save the resulting set of paths of total (new) cost U. By Corollary 1, there is a $U \in [Cn^{O(1)}]$ for which the above procedure will find such a set of paths that achieves the minimum (original) cost with probability at least $1 - \frac{1}{n}$. $\square$

6   Minimum-Cost Logarithmic Integral Flow is in RNC^2

The following lemma is a straightforward generalization of a folklore reduction of maximum integral flow to a corresponding disjoint connecting path problem (for instance cf. [7]) to include minimum-cost integral flow. We shall call a flow proper, if it ships each flow unit along a simple path from the source to the sink.

Lemma 10. *The problem of whether or not there is a proper integral flow of value k and cost D from a distinguished source vertex s to a distinguished sink vertex t in a directed network with n vertices. integral edge capacities and edge costs in $[C]$ can be (many-one) reduced to that of whether or not there is set of k mutually vertex-disjoint simple directed paths of total cost D^*, where $\lfloor \frac{D^*}{kn} \rfloor = D$, connecting two distinguished sets of k vertices in a directed network on $O(kn^2)$ vertices in $O(\log k + \log n)$ time by a CREW PRAM using $O(kn^2 + k^2 n)$ processors.*

Proof. Let $K = (V, E)$ be the directed network with integral edge capacities, edges costs in $[C]$ and the distinguished source vertex s and sink vertex t. Since we are interested in a flow of value k, we can assume w.l.o.g that all edge capacities do not exceed k.

We form a directed network K^* on the basis of the network K as follows.

Let $v \in V$. Next, let $E_{in}(v)$ be the set of edges in K incoming into v, and let $E_{out}(v)$ be the set of edges in K leaving v. For each $e \in E_{in}(v)$ and $i \in [capacity(e)]$, we create the vertex $v_{in}(e, i)$. Analogously, for each $e' \in E_{out}(v)$ and $i' \in [capacity(e')]$, we create the vertex $v_{out}(e', i')$. Furthermore, we direct an edge from each vertex $v_{in}(e, i)$ to each vertex $v_{out}(e', i')$. To each such an edge, we assign the cost 1. Also, for each edge $e = (v, w)$ of K, we direct an edge from $v_{out}(e, i)$ to $w_{in}(e, i)$ for $i \in [capacity(e)]$. To each such an edge, we assign the cost $c(e)kn$. See Fig. 2.   Let X' be the set of vertices of the form $s_{out}(...)$, and let Y' denote the set of vertices of the form $t_{in}(...)$. Create an additional set X of k vertices and from each vertex in X direct an edge to each vertex in X'. Symmetrically, create another additional set Y of k vertices and from each vertex in Y' direct an edge to each vertex in Y.

It is easy to observe that there is a proper integral flow of value k and cost D from s to t in the network K iff there is a set of k mutually vertex-disjoint simple paths of total cost D^* connecting X with Y in the network K^*, such that $\lfloor \frac{D^*}{kn} \rfloor = D$.

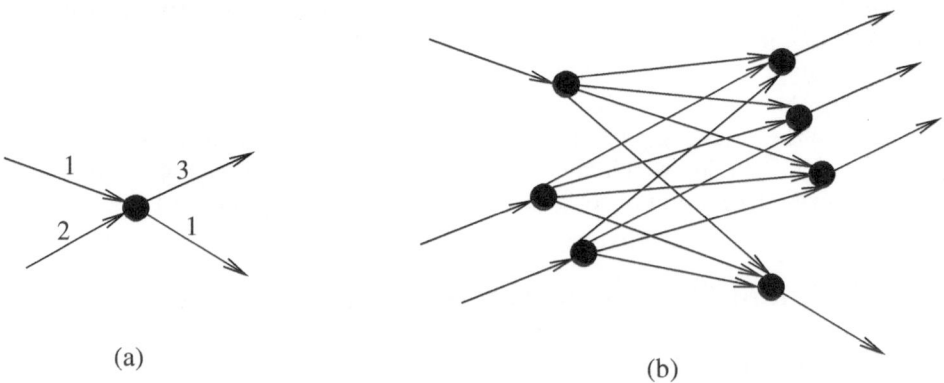

(a)   (b)

Fig. 2. An example of a vertex of the network K and the corresponding part of the network K^*

Now it is sufficient to note that the construction of K^*, X and Y on the basis of K easily implemented by a CREW PRAM in $O(\log k + \log n)$-time using $O(kn^2 + k^2n)$ processors, where n is the number of vertices in K.   □

By combining Theorem 1 with Lemma 10, we obtain our first main result.

Theorem 4. *The minimum cost of a flow of value k in a network with n vertices, a sink and a source, integral edge capacities and positive integral edge costs in $[C]$ can be found by a randomized PRAM, with errors of exponentially small probability in n, running in $O(k \log(Ckn) + \log^2(Ckn)$ time and using $O(k^{11}C^5n^{20} + 2^k k^{10}C^3n^{14})$ processors.*

By combining in turn Theorem 3 with the finding variant of Lemma 10 using exactly the same reduction, we obtain our second main result.

Theorem 5. *There is a randomized PRAM algorithm returning almost certainly a minimum-cost flow of value k (iff a flow of value k exists) in a network with n vertices, a sink and a source, integral edge capacities and edge costs in $[n^{O(1)}]$, in $O(k \log(kn) + \log^2(kn))$ time using $2^k(kn)^{O(1)}$ processors.*

Corollary 2. *The problem of finding a minimum-cost flow of value $O(\log n)$ in a network with n vertices, a sink and a source, and integral edge capacities bounded polynomially in n admits an RNC^2 algorithm.*

7   Final Remarks

We have resented a new approach to the minimum-cost integral flow problem. In particular, it yields an RNC^2 algorithm when the flow supply is (at most) logarithmic in the size of the network.

All our results can be extended to to include undirected networks by a straightforward reduction.

References

1. Ahuja, R.K., Magnanti, T.L., Orlin, J.B.: Network Flows: Theory, Algorithms, and Applications. Prentice-Hall, Inc. (1993)
2. Anderson, R.J., Setubal, J.C.: On the Parallel Implementation of Goldberg's Maximum Flow Algorithm. In: Proc. Annual ACM Symposium on Parallel Algorithms and Architectures (SPAA), pp. 168–177 (1992)
3. Björklund, A.: Determinant sums for undirected Hamiltonicity. In: Proc. 51th IEEE Symposium on Foundations of Computer Science (FOCS), pp. 173–182 (2010)
4. Björklund, A., Husfeldt, T., Taslaman, N.: Shortest Cycle Through Specified Elements. In: Proc. Annual ACM-SIAM Symposium on Discrete Algorithms, SODA (2012)
5. DeMillo, R.A., Lipton, R.J.: A probabilistic remark on algebraic program testing. Information Processing Letters 7, 193–195 (1978)
6. Edmonds, J.: Systems of distinct representatives and linear algebra. J. Res. Nat. Bur. Standards Sect. 71B(4), 241–245 (1967)
7. Even, S.: Graph Algorithms. Computer Science Press Inc. (1979)
8. Edmonds, J., Karp, R.M.: Theoretical improvements in algorithmic efficiency for network flow problems. Journal of the ACM 19(2), 248–264 (1972)
9. Ford Jr., L.R., Fulkerson, D.R.: Maximal flow through a network. Can. J. Math. 8, 399–404
10. Fulkerson, D.R.: An out-of-Kilter method for minimal cost flow problems. SIAM J. Appl. Math. 9, 18–27
11. Galil, Z., Pan, V.: Improved processor bounds for algebraic and combinatorial problems in RNC. In: Proc. 26th IEEE Symposium on Foundations of Computer Science (FOCS), pp. 490–495 (1985)
12. Goldberg, A.: Parallel Algorithms for Network Flow Problems. In: Reif, J.H. (ed.) Synthesis of Parallel Algorithms. Morgan-Kauffman (1993)
13. Goldschlager, L.M., Shaw, R.A., Staples, J.: The Maximum Flow Problem is Log Space Complete for P. Theoretical Computer Science 21(1), 105–111 (1982)
14. Karp, R., Upfal, E., Wigderson, A.: Constructing a perfect matching is in random NC. Combinatorica 6(1), 35–48 (1986)
15. Koutis, I.: Faster Algebraic Algorithms for Path and Packing Problems. In: Aceto, L., Damgård, I., Goldberg, L.A., Halldórsson, M.M., Ingólfsdóttir, A., Walukiewicz, I. (eds.) ICALP 2008, Part I. LNCS, vol. 5125, pp. 575–586. Springer, Heidelberg (2008)
16. Lawler, E.L.: Combinatorial optimization: Networks and matroids. Holt, Rinehart and Winston, New York
17. Mulmuley, K., Vazirani, U.V., Vazirani, V.V.: Matching is as Easy as Matrix Inversion. Combinatorica 7(1), 105–113 (1987)
18. Orlin, J.B., Stein, C.: Parallel Algorithms for the Assignment and Minimum-Cost Flow Problems. Operations Research Letters 14, 181–186 (1993)
19. Reif, J.H.: Synthesis of Parallel Algorithms. Morgan-Kauffman (1993)
20. Vazirani, V.V.: Parallel Graph Matching. In: Reif, J.H. (ed.) Synthesis of Parallel Algorithms. Morgan-Kauffman (1993)
21. Schwartz, J.T.: Fast probabilistic algorithms for verification of polynomial identities. Journal of the ACM 27(4), 701–717 (1980)
22. Williams, R.: Finding paths of length k in $O^*(2^k)$. Information Processing Letters 109, 301–338 (2009)

Topic 13: High Performance Network and Communication

Chris Develder, Emmanouel Varvarigos,
Admela Jukan, and Dimitra Simeonidou

Topic Committee

This topic on *High-Performance Network and Communication* is devoted to communication issues in scalable compute and storage systems, such as parallel computers, networks of workstations, and clusters. All aspects of communication in modern systems were solicited, including advances in the design, implementation, and evaluation of interconnection networks, network interfaces, system and storage area networks, on-chip interconnects, communication protocols, routing and communication algorithms, and communication aspects of parallel and distributed algorithms.

This year we selected 4 papers and one of the papers was selected for the best paper session. The authors of the paper entitled *"Topology Configuration in Hybrid EPS/OCS Interconnects"* consider a hybrid interconnection network for future high performance computing (HPC) and data center systems, combining Electronic Packet Switching (EPS) and Optical Circuit Switching (OCS). They present heuristic algorithms to map the tasks of parallel HPC applications onto physical processors to allow efficient communication over the hybrid EPS/OPS network. The article entitled *"Towards an Efficient Fat-tree like Topology"* proposes extensions of the fat-tree topology and analyzes the way the routing algorithm affects the complexity of the switch. The authors not only look into the impact of topology and routing on performance (i.e., throughput, latency etc.), but also their influence on switch cost (e.g., in terms of number of switching elements required). In *"An adaptive, scalable, and portable technique for speeding up MPI-based applications"*, the authors present a portable optimization of MPI called PRAcTICaL-MPI (Portable Adaptive Compression Library - MPI). PRAcTICaL-MPI enhances the performance and scalability of MPI applications by applying run-time lossless compression, in a transparent way for applications, to reduce the data volume exchanged among processes, selecting the most appropriate compression algorithm at run-time. Finally, the authors of the paper *"Cost-effective Contention Avoidance in a CMP with Shared Memory Controllers"* study, for large chip multiprocessors (CMP), the cause and effect of network congestion due to traffic local to the applications, and traffic caused by memory access. They present a mechanism to reduce head-of-line blocking in the switches, hence efficiently reducing network congestion, increasing network performance, and evening out performance differences between CMP applications.

C. Kaklamanis et al. (Eds.): Euro-Par 2012, LNCS 7484, p. 700, 2012.
© Springer-Verlag Berlin Heidelberg 2012

Topology Configuration
in Hybrid EPS/OCS Interconnects

Konstantinos Christodoulopoulos[1], Marco Ruffini[1], Donal O'Mahony[1],
and Kostas Katrinis[2]

[1] School of Computer Science and Statistics, Trinity College Dublin, Ireland
[2] IBM Research, Ireland
christok@tcd.ie

Abstract. We consider a hybrid Electronic Packet Switched (EPS) and Optical
Circuit Switched (OCS) interconnection network (IN) for future HPC and DC
systems. Given the point-to-point communication graph of an application, we
present a heuristic algorithm that partitions logical parallel tasks to compute resources and configures the (re-configurable) optical part of the hybrid IN to efficiently serve point-to-point communication. We measure the performance of a
hybrid IN employing the proposed algorithm using real workloads, as well as
extrapolated traffic, and compare it against application mapping on conventional fixed, electronic-only INs based on toroidal topologies.

Keywords: Reconfigurable interconnection networks, optical circuit switching,
communication graph, application mapping, partitioning, topology configuration.

1   Introduction

High Performance Computing (HPC) systems and datacenters (DCs) are being built
with ever-increasing numbers of processors. Currently systems with tens of thousands
of servers have already been reported to be in operation, while their scale is expected
to grow to the order of millions of cores towards Exascale [1]. To obtain high system
efficiency, computation and communication performance need to be balanced. Given
the aggressive increase in compute density – thanks to the increasing number of
cores/node and the growing deployment of accelerators – it is of paramount importance to avoid having the interconnection network (IN) become a bottleneck [1-2];
instead, IN technologies and system software need to grow in hand with the evolution
in compute density to enable next generation HPC and DC systems.

Flagship supercomputers typically employ regular topologies of electronic switches, such as hypercubes and toroidal structures. For instance, the Cray XT5 [3] utilizes
a 3D torus topology, while the K supercomputer [4] employs a 6D torus (although not
all dimensions are complete). Such low-degree regular topologies are adopted due to
their inherent ability to scale linearly with the number of compute nodes. Still, these
sparse topologies tend –for specific applications– to aggravate the mapping of application communication to the underlying IN [5-10]. At the far end, many HPC clusters
and DCs adopt indirect routing IN, such as fat-trees [11] that provide for

C. Kaklamanis et al. (Eds.): Euro-Par 2012, LNCS 7484, pp. 701–715, 2012.
© Springer-Verlag Berlin Heidelberg 2012

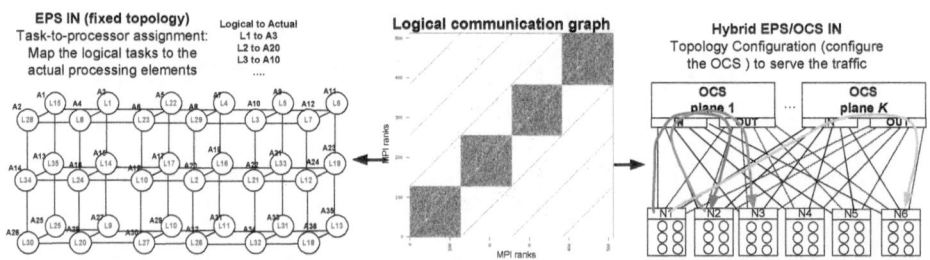

Fig. 1. Topology-aware mapping on fixed electronic-only IN (left) and topology configuration in reconfigurable OCS IN (right)

full-bisection bandwidth and thus simplify the mapping of applications. Though, this comes at a cost that scales super-linearly with the size and thus raises scalability concerns when considered at extreme scale, especially if the investment is not justified due to poor utilization [12]. Oversubscription [13] is often proposed as a means of controlling the capital and power-consumption cost. Still, the inflexibility of placing network capacity once and for all at design time remains, similar to the case of low-degree regular topologies.

To close the gap between constant-degree, fixed INs and costly, full-bisection INs, past work [14-17] has proposed building low-degree, reconfigurable INs that allow for on-demand bandwidth allocation wherever is needed. Maintaining a low-degree IN reduces its contribution to the capital cost of the system, while these designs have been shown to fit well the communication patterns exhibited by specific classes of HPC applications [14] and DC workloads [16-18]. The proposed reconfigurable IN architecture constitutes hybrids of two switching technologies, namely using both electronic packet-switches (EPS part) and optical circuit-switches (OCS part). The OCS part, typically implemented with commodity MEMS-based switches, handles high-rate, long-lived point-to-point flows, whereby the EPS part serves low-rate signaling, short-lived flows and collectives (many-to-many communication).

In this paper, we focus on traditional HPC applications that exhibit static point-to-point logical communication graphs. We call static an application that follows specific communication patterns so that its logical communication graph at intermediate phases and the aggregated graph at the end of its execution have well defined and consistent structures irrespective of the input. This definition is very close to the definition given in [14], where a static application is considered to have known communication pattern at compilation time. It must be noted here that logical point-to-point communication graphs are influenced only by application-level communication, i.e. point-to-point communication between logical entities, and thus do not depend on machine or IN characteristics. Work on identifying and classifying the point-to-point communication graphs of static HPC applications include [18] and [19]. Finally, stream computing applications [15] can be also included in this category of static applications.

Given the IN of an HPC system, mapping the tasks of a parallel application onto physical processors to allow efficient communication becomes critical to performance. This problem is usually referred to as *topology-aware mapping* and has been shown to be NP-hard [5]. The most efficient known approach [6] is to adapt the application source code to optimize task-to-task communication against a specific IN topology. Albeit efficient, this approach is cumbersome, while also being application- and system-specific, thus precluding re-use across machines or applications. More practical alternatives [7-10] do not require any changes to application code and thus trade-off efficiency for being application-agnostic. In particular, specialized system software takes as input the logical point-to-point communication graph of a static application and performs task-to-processor assignment taking into account the IN topology.

Among the various advantages [14-17] brought by reconfigurable INs, the ability to look at task partitioning/mapping from a different angle remains unexplored: instead of performing a sophisticated task-to-processor assignment on a fixed topology as in [5-10], the logical tasks can be partitioned and the topology of the (reconfigurable) IN that interconnects them can be configured for optimized communication and thus improved application performance. This comprises the topic of this paper. Fig. 1 contrasts these two different approaches.

Our methods apply and improve use cases that are relevant in mapping of static applications on fixed INs [5-10]. The logical point-to-point communication graph of a static application is identified at compilation time and/or the application is executed once to profile and capture its communication graph. Using this input, an optimized OCS topology configuration is computed and used to speedup subsequent executions of the application. We assume that the OCS network is configured once, that is, it is configured at the outset of application execution and remains the same throughout its execution. Although the reconfigurable network could dynamically adapt its topology to support more efficiently the different application phases, we will not consider such cases here. References [16-17] examine the dynamic reconfiguration of the OCS network to follow the traffic variations of DC applications, considering only single-hop transmissions and mostly single-layer optical networks, features that we plan to extend in our future work. Note that dynamic reconfiguration can be viewed as a sequence of well defined static instances, whereby each one can be solved by the method proposed here.

For the purpose of evaluation, we profile several static HPC application kernels run on an HPC cluster using IPM [20] and derive their logical point-to-point communication graphs. Using *hop-bytes*[1] [8-10] as the performance metric and for various architectural choices of the hybrid target EPS/OCS IN, the level at which optical interconnection occurs and the number of optical ports available, we evaluate the performance of the proposed joint partitioning and topology-configuration heuristic and compare it against application mapping on conventional fixed, electronic-only INs that are based on toroidal topologies. We also develop simple models for estimating the capital cost of the hybrid EPS/OCS vs. a torus-like electronic-only IN.

2   System Model

We consider a generic multi-rack system architecture in which processing elements are multi-core processor chips. A given number of chips is mounted to a (compute)

---

[1]   The hop-bytes metric is defined as the weighted sum of traversed hops. Weights correspond to message sizes.

node. In turn, a number of nodes comprises a mid-plane, and a set of mid-planes is installed in a rack.

We assume an interconnection network (IN) adhering to a hybrid architecture comprising both an Electronic Packet Switched (EPS) and an Optical Circuit Switched (OCS) network [14]. The OCS network is typically implemented with one or more Micro Electro-Mechanical Systems (MEMS) optical switches (crossbar). MEMS are layer-0 switches that establish an end-to-end optical connection by reflecting a light beam from an input to an output port. The signal is switched transparently without performing any processing on switched data. MEMS-based switches exhibit circuit setup times that are typically in the order of tens of milliseconds and thus would result in prohibitively high per-packet switching overhead, should they be operated as packet switches. Therefore, it makes sense to use MEMS as switching elements of high-rate, long-lived, point-to-point flows. Flows at core- or chip-level do not currently have such characteristics. Moreover, since MEMS-switches support solely point-to-point communication, creating a chip-level network would require a high-degree OCS topology, which is prohibitive in terms of scalability. Therefore, all-optical switching is applied to aggregated traffic at a higher interconnect level that we hereafter refer to as the *optical aggregation level*. Given current cores/node figures and typical byte/flops ratios, setting the optical aggregation at mid-plane or rack level maximizes the utilization of optical circuits and reduces the frequency of reconfiguration of OCS-switches (due to aggregation). Still, as compute density packed into a single node keeps increasing, so will do the number of compute tasks and bandwidth per node [1], justifying the placement of optical aggregation at the node level. To allow our work to capture this trend and thus be future-proof, we apply an abstraction to the assumed system model to enable us to carry parametric studies.

Specifically, we logically cluster processing elements (cores) together to form groups that we refer to as *logical clusters* (LCs). The processing elements comprising an LC are interconnected via an electronic packet switched (EPS) network (to be referred to as the first-level of EPS network or EPS-edge) that serves both intra-LC communication, as well as aggregating LC traffic destined to distant LCs. At the optical aggregation level, M EPS (bidirectional) ports are connected to the second-level of the EPS network (EPS-core) and also K (bidirectional) ports are connected to the OCS network. At the optical aggregation level the traffic towards distant LCs can be served by two parallel networks: either the EPS-core or the OCS network. The OCS network handles persistent point-to-point, high-rate inter-LC flows, while the EPS-core network handles lower bandwidth and collective communications, as well as bursty flows. INs with two levels of hierarchy are also found in fixed topology EPS systems [10], and in the hybrid EPS/OCS IN [14-17]. While the above abstraction enables us to evaluate our approach against various choices of placing the optical aggregation level, it still allows a straightforward mapping of our approach to real systems. To showcase this, we assume an HPC cluster with both levels of the EPS network being implemented using Ethernet. The optical aggregation point is then either a multi-port Ethernet NIC (node-level optical aggregation) or an Ethernet top-of-rack switch (rack or mid-plane level optical aggregation), whereby optical transceivers are used to carry packets to/from the OCS network. The transmit (resp. receive) side of each transceiver is connected to an input (resp. output) port of the MEMS-switch.

Let T denote the number of processing elements that are grouped together to form a logical cluster (LC) and N denote the number of LCs of the system. In the hybrid EPS/OCS network model we adopt, each LC is connected to K parallel OCS planes at the optical aggregation level [14]. Fig. 2 depicts the architectural diagram of the system model we adopt in this paper. Since, each parallel OCS plane is implemented with an NxN crossbar switch, K NxN crossbar switches are required in total. A large MEMS crossbar switch can be configured to create smaller NxN crossbar switches, and so the total number of switches required can be less than K. Currently 320x320 MEMS crossbar switches are available by a number of vendors, while there are prototypes with 1000 in/out ports. Note that our goal is to create a hybrid EPS/OCS network with $K \ll N$ to keep the cost as low as possible and aim at massive scale out.

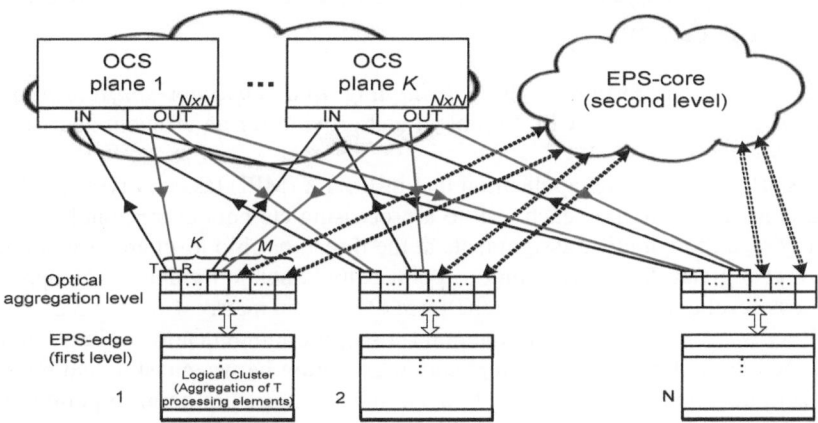

Fig. 2. Reference hybrid EPS/OCS IN consisting of N logical clusters (LCs). At the optical aggregation level, M and K bidirectional EPS ports are connected to the EPS-core and the OCS network, respectively. At each of the K EPS ports accessing the OCS network, the transmit (T) side of the transceiver is connected to an IN (input) port and the receive (R) side to an OUT (output) port of the OCS switch.

In this paper we will focus mainly on the OCS part of the hybrid IN. The method that we propose in the next section takes as input the logical point-to-point communication graph of parallel application tasks, given e.g. in the form of a traffic matrix, and a specific OCS architecture as defined by the related K and T parameters. We assume that the rest of the traffic that is not point-to-point (e.g. collectives) is routed over the EPS-core part of the hybrid IN. The goal of the proposed method is to serve the point-to-point transmissions in an efficient way. To do so, we partition the tasks to form logical clusters (LCs) and also derive the configuration of the OCS network to optimize the communication between LCs. The partitioning process divides the communication into inter- and intra-LCs communication, similar to the hierarchical mapping problem examined in [10]. Intra-LC communication that is served by the first-level, EPS-edge is considered to be *cheap* and is neglected, while inter-LC communication, which is routed over the OCS network, is the traffic that is optimized by our topology-configuration algorithms. In the remainder of this paper we will call this problem the **Partitioning and Topology-Configuration with Bounded Connectivity** (PTCBC) problem.

Given the adopted OCS IN with K parallel OCS planes, an application that after the grouping to LCs has an inter-LC communication graph with directed connectivity degree less than K, can be served over the OCS network with direct point-to-point connections. Otherwise, we resort to multi-hop transmissions. Multi-hop refers to OCS-based communication that involves electronic processing of packets at the EPS-edge, beyond the source/destination LCs, at intermediate LC *hops*. Multi-hop increases the effective bandwidth between LCs at the expense of increased latency as well as increased congestion and average network load.

3   Partitioning and Topology Configuration with Bounded Connectivity

In this section we formulate the **Partitioning and Topology-configuration with Bounded Connectivity** (PTCBC) problem and provide for an efficient heuristic algorithm to solve it.

We start with a parallel application that utilizes Z (MPI) tasks. A typical allocation would consist of assigning each task to a processing element corresponding to a processor core, although other assignments at the thread or chip level are also applicable. We are also given the logical point-to-point communication graph of the application (MPI task level communication) in the form of a traffic matrix Λ of size $Z \times Z$. Element Λ_{nm} ($1 \leq n,m \leq Z$) corresponds to the point-to-point communication volume (in bytes) exchanged throughout the entire application execution between task n and m; thus Λ has a zero diagonal. The remaining transmissions that are not point-to-point (e.g. collectives) are routed over the EPS-core part of the hybrid IN and are not considered here. Tasks are first partitioned into logical clusters (LCs, see Section 2 for the definition of LCs). In particular, we assume that T tasks are grouped together to form the LC (one of the LCs may contain less than T elements, if Z mod $T > 0$). Let N stand for the number of LCs formed after clustering the Z tasks; then $N = \lceil Z / T \rceil$.

By partitioning into logical clusters (LCs) we transform Λ into a new traffic matrix λ of size $N \times N$, with each element λ^{ij} ($1 \leq i,j \leq N$) corresponding to the point-to-point communication volume (in bytes) between LC_i and LC_j (i.e. λ corresponds to the inter-LC communication graph). Note that unlike Λ being part of the problem input, the traffic matrix λ is conditioned on the clustering method employed and as such is an intermediate output.

Inter-LC traffic, as described by traffic matrix λ, is served by the reconfigurable OCS interconnection network (IN). The LCs are mapped to the physical compute resources to form compute aggregations (racks, mid-planes or nodes), by putting tasks that belong to the same LC together in the same compute aggregation without considering the position of the aggregation in the IN topology, since the OCS IN will be configured around these aggregations. As outlined in Section 2, the OCS network we consider consists of K parallel OCS planes. There are two different versions of the problem: one that assumes unidirectional and one that assumes bidirectional connections over the OCS network. This is related to the type/configuration of aggregation switches, the related transceivers used at the optical aggregation level and their connection to the MEMS switch(es). Given that the unidirectional case is more generic and that Ethernet commodity switches can support this operating mode, we will focus on unidirectional connections, although our algorithms can be applied with minor changes to bidirectional connections. Note that if traffic matrix λ includes both directions of communication for an LC pair, the solution will include connections for both directions, although they may utilize different paths.

We start with a fully connected graph $G=(V,E)$, with V being the set of vertices that correspond to the logical clusters (LCs) and E the set of candidate links that connect every pair in V. The links included in E are not actually functioning but are the candidate links that can be established in the OCS network. In this context, we will say that we establish an optical link (i,j) when we configure the OCS network to establish a circuit connection between LC_i and LC_j. Since we are considering directed connections, the order of i and j is important. The number of links that can be established is constrained by the number K of parallel OCS planes. We let $L \subseteq E$ be the set of links that are chosen to be established and $G'=(V,L)$ the related graph, with connectivity degree less or equal to K. G' is the graph of the constructed OCS network.

The Partitioning and Topology-Configuration with Bounded Connectivity (PTCBC) problem is defined as follows. We are given the traffic matrix Λ, the desired number T of tasks per LC and the number K of parallel OCS planes, and we seek to identify a "good" partitioning of tasks to LCs (i.e. find the λ matrix), and a "good" configuration of the OCS network that interconnects the LCs (i.e. identify the set L of optical links to be established). The goodness of a joint partitioning/topology-configuration solution is measured against its ability to minimize a communication objective, namely *average hop-bytes*. The average hop-bytes metric is defined as the sum of path lengths (measured in hop-count) taken by the messages, weighted by the respective message size [8-10]. The purpose of the average hop-bytes metric is to capture approximately the average load on the network. In early works in the fields of graph embedding and VLSI design, emphasis was placed on the maximum dilation metric, which is the longest path (resp. the longest wire in a circuit). Ref. [8-10] argue that reducing the longest path (maximum dilation) is not as critical as reducing the average hops across all message sizes, as captured by the hop-bytes metric. Our algorithmic formulations are general and can be used to optimize other performance metric, such as the maximum dilation or link congestion, but the hop-bytes was chosen as the most appropriate metric for our comparisons.

If the optical aggregation level is placed directly at the processing elements, then partitioning to logical clusters becomes obsolete. This is a special case of the PTCBC problem with $T=1$ and thus $\lambda=\Lambda$. In this case, the problem is reduced to the topology-configuration problem with bounded connectivity degree, which we proved to be NP-hard by a reduction to the circular arrangement problem [21] (the proof is omitted due to space limitations). This case covers also the version of the problem where we are given directly the traffic matrix λ capturing the traffic that will be routed over the OCS network. Since the general problem described above with $T \geq 1$ includes as a special case an NP-hard problem, the general problem is also NP-hard. The optimal solution of PTCBC is bound to give at least as good performance results as the application mapping on hierarchical electronic-only fixed INs [10], as long as the connectivity degree of the hybrid network is not less than that of the fixed IN. For example, assuming that the optimal architecture to serve the traffic is a 3D torus, the optimal

PTCBC solution would be to configure the OCS network to form a 3D torus and obtain exactly the same performance as the fixed IN, assuming $K \geq 6$ (directed) planes.

3.1   Heuristic Algorithm

The PTCBC problem is computationally difficult, thus, in what follows we present a heuristic algorithm to solve it. We decompose the problem by first solving the partitioning problem (P) and then the topology-configuration with bounded connectivity degree (TCBC) problem. The proposed heuristic involves 3 phases that are described in the following paragraphs.

Partitioning into Logical Clusters
In the first phase we use spectral clustering to partition the traffic matrix Λ of size ZxZ and form logical clusters (LCs) of size T each. We obtain thus a new traffic matrix λ of size NxN that captures inter-LC traffic in bytes. We use the 2-way spectral clustering algorithm and apply it recursively [22] until we obtain partitions of the target cardinality (T).

Demand Ordering and Simulated Annealing
The sequential algorithm to be described next, establishes optical links based on the demands of traffic matrix λ, by serving the demands one-by-one in some particular order. The number of optical links emanating/terminated at an optical aggregation point is constrained due to bounded connectivity. Due to this constraint, not all demands in λ can be served by a single optical circuit and some have to be served by multi-hop transmissions. Demands that are served earlier have higher probability of finding free resources to establish an optical link, as opposed to demands that are served later. Thus, different orderings result in different topology-configuration solutions with different costs (hop-bytes). In this work, we employ the Highest Demand First (HDF) ordering policy: we order the demands based on communication volume, and serve the demand with the highest communication volume (in bytes) first. According to this policy, heavy demands that have high communication sizes are served first. Since the performance metric of interest is hop-bytes, serving the demands with the highest volume by single-hop transmissions is a rational way to minimize the specific metric.

A number of other policies can be easily defined, based on other parameters of the traffic matrix λ, e.g. based on the total load from a source to all destinations, the connectivity bound, etc. However, since the performance depends on many parameters, it is quite difficult to come up with a very good ordering policy. Thus, to find good orderings, we use the Simulated Annealing (SA) meta-heuristic. Specifically, we start with an HDF ordering and calculate its cost by sequentially serving the demands, using the heuristic algorithm described in the following paragraph (the cost function is the "fitness function" in the SA terminology). For a particular ordering $((s_1,d_1), ..., (s_k,d_k))$ of the demands, we define its neighbor as the ordering where (s_i,d_i) is interchanged with (s_j,d_j) for some i and j. Note that $k \leq N(N-1)$, depending on the number of non-zero point-to-point flows. To generate a random neighbor, we choose the pivots i and j with uniform probability among the demands. We use this neighbor generation procedure and the sequential demand heuristic as the fitness function in a typical SA iteration process.

Sequential Demand Serving Heuristic Algorithm

The algorithm described here serves sequentially the traffic between the LCs described in λ, by configuring the OCS network to establish optical links to connect two LCs. To keep track of the configuration of the OCS network we use two integer vectors to hold the number of outgoing and incoming connections that are established for each LC. In particular, we denote by $O = [O_i] = (O_1, O_2, ..., O_N)$, the vector of size N (N is the number of LCs) where each element O_i corresponds to the number of established outgoing optical connections from LC_i. Similarly, we denote by $I = [I_i] = (I_1, I_2, ..., I_N)$ the vector that keeps track of the incoming connections. We also keep a set L of established optical links in the form of (i,j) LC pairs.

The demands are served sequentially in the order specified in the previous phase. For a demand, e.g. (s,d), we establish an optical link between s and d as long as the maximum connectivity degree bound is not exceeded. In particular, we check if s has free outgoing ports ($O_s \leq K$) and also if d has free incoming ports ($I_d \leq K$). If both constraints are satisfied, we establish a direct optical connection between s and d by cross-connecting the respective optical switch ports. Subsequently, we increase the O_s and I_d elements, update the set L of established links by appending (s,d) to it and continue with serving the next demand. If the connectivity constraint for s and/or d is not met, we move to the next demand. Once we have finished with all demands, the links to be established are given by the set L. We use the graph formed by LCs as vertices and L as edges as input to Johnson's algorithm and compute the all-pairs shortest paths. The set of computed shortest paths is used to calculate the average hop-bytes of the solution. To improve performance, we additionally use Simulated Annealing (SA): the sequential heuristic of this phase is run multiple times for different orderings. For each ordering, a new set of established optical links and thus set of (shortest) paths is found, yielding a new average hop-bytes value. We keep the solution that produces the lowest hop-bytes value. By controlling the number of SA iterations, we trade-off optimality for computation time.

The heuristic algorithm presented above is of polynomial complexity. The spectral heuristic used in the first phase is generally considered efficient for solving partitioning problems. The second phase involves the ordering of N^2 elements, while the third phase involves the execution of Johnson's algorithm that takes $O(|V|^2 log|V| + |V||L|) = (N^2 log N + N^2 K)$ time, which is faster than Floyd-Warshall's algorithm for sparse networks as the one considered here ($K << N$). Last, if SA is used, the number of iterations drives the times that the third phase is executed.

4   Performance Results

We performed experiments to estimate the performance of the proposed PTCBC algorithm in a hybrid EPS/OCS IN. We considered a number of design choices of the hybrid IN, in particular different levels of optical aggregation and various numbers of optical planes. We also compare the results obtained in the hybrid IN to a torus-like electronic-only IN.

To estimate the performance of the hybrid IN, we implemented the heuristic algorithm that solves the PTCBC problem. To obtain comparison results for the electronic-only IN we used a hierarchical mapping heuristic algorithm that takes as input the logical communication graph of an application, partitions the logical entities into logical clusters (LCs) and then assigns the LCs to compute resources of a fixed IN topology to minimize the hop-bytes. The heuristic used follows the approach in [10]. It first performs a partitioning using spectral clustering, and then it assigns the created LCs to compute resources. It starts with an initial assignment and improves over that using the SA pivoting process.

To evaluate the efficacy of our approach with pragmatic traffic input, we captured the logical communication graphs of two kernels of representative HPC applications. In particular, we installed SuperLU and FFTW applications on a cluster located at IBM Dublin. SuperLU performs LU factorization of a sparse matrix and FFTW performs forward and inverse Fast Fourier transformations. We used the IPM monitoring tool [20] to capture the MPI point-to-point traffic that is generated by these applications. The applications were run on 240 up to 1920 ranks. Fig. 3 presents the point-to-point communication graphs obtained after a single run of each of the tested applications on 480 ranks. Various executions were performed to verify that the communication graphs of these applications exhibit similar patterns irrespective of the input and thus these applications fall within the static category, as described in Section 1. The captured logical communication graphs were used as input for evaluation of our approach, as well as for performing hierarchical mapping on the electronic-only IN. Given the lack of a widely-accepted method to scale these graphs, we extrapolated to higher scales by producing synthetic traffic matrices that are isomorphic to the captured ones, and to generate traffic for large problem executions that we were not able to perform on our cluster.

We assumed that the hierarchy of the system consists of 12 cores per node and 40 nodes per rack (total of 480 cores per rack), driven by the specifications of the IBM cluster. We report results for the case where the OCS network is either connected directly to compute nodes (where $T=12$) or to rack switches (where $T=480$), and for $K=4,6$ and 8 parallel OCS planes. For comparison purposes we also estimated the performance of electronic-only 2D-, 3D, and 4D-torus INs, (thus having the same connectivity degrees as the corresponding OCS networks). For both hybrid and electronic-only INs we used 100 SA iterations.

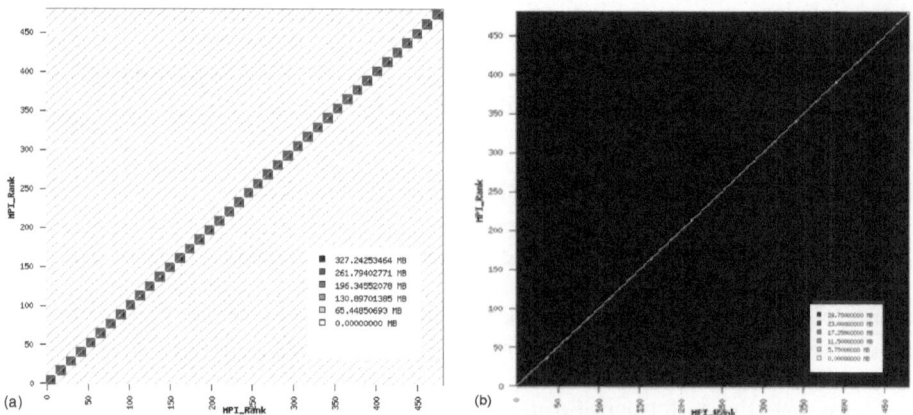

Fig. 3. Logical communication graphs of SuperLU and FFTW kernels on 480 mpi-ranks

Tables 1 and 2 report results for SuperLU, assuming that the OCS network is connected directly to compute nodes; also, that a torus EPS network is connected directly to compute nodes. More specifically, Table 1 reports the results using the communication graphs that were captured by executing SuperLU to factorize the webbase-1M matrix taken from [23], on 240, 480, 960, 1920 MPI-ranks (N=20,40,80 and 160). Note that we used Mbytes to measure the volume of data and thus the values reported in the tables are measured in hop-Mbytes. From Table 1 we can observe that the hybrid IN exhibits lower hop-Mbytes than the electronic-only IN. Even a hybrid IN with K=4 parallel optical planes has lower hop-Mbytes for almost all problem instances examined than a 4D electronic-only IN, which has double connectivity degree. Note that the improvement that we obtain when using the hybrid IN does not come from the highest capacity supported by the OCS network. The hop-bytes metric used here does not consider the capacity of the underlying networks. Instead, the improvement comes from the configurability of the OCS part of the hybrid IN that was exploited to serve efficiently the communication graph of the application. This confirms our expectation that performing static configuration improves communication performance.

Table 2 reports results obtained when using our custom built traffic generator. In particular, we generated traffic matrices to emulate the execution of SuperLU on N=20, 40, 80, 160 and 320 nodes. Fig. 4 presents the relative hop-bytes improvement brought by our approach applied in a hybrid IN over a 3D-torus electronic-only IN. Captured traffic (indicated by webbase matrix used as input) and synthetically generated traffic results are depicted in Fig. 4. By comparing results up to 160 nodes we observe that the improvement we got using the synthetically generated traffic is in-line with the one obtained with real input. As expected, the improvement is higher as we increase the number of parallel OCS planes. A 3D torus network cannot be fully constructed on a low number of nodes (up to a few tens of nodes, e.g. 40), so the improvement obtained by the hybrid IN as opposed to the 3D torus networks for problems of that size is partially explained by the 3D torus deficiency. For K=6 (equal connectivity degree to the 3D torus), we observe that the improvement is approximately 24% when executing SuperLU on 80 nodes, and increases to 30% and 38% for 160 and 320 nodes, respectively.

Table 1. Performance for communication graphs captured by running SuperLU using the "webbase-1M" matrix as input

	Hybrid IN hop-Mbytes			Electronic-only IN hop-Mbytes		
Input	K=4	K=6	K=8	2D	3D	4D
webbase-1M (N=20 nodes)	5235	4046	3730	6973	5604	5604
webbase-1M (N=40 nodes)	12703	9944	8683	13537	13272	12724
webbase-1M (N=80 nodes)	16164	12622	11136	23514	16784	16481
webbase-1M (N=160 nodes)	37868	27381	23861	60685	41362	36965

Table 2. Performance results for synthetically generated communication graphs for the SuperLU application

	Hybrid IN hop-Mbytes			Electronic-only IN hop-Mbytes		
Size	K=4	K=6	K=8	2D	3D	4D
N=20 nodes (1/2 rack)	5221	4170	3869	7178	5594	5594
N=40 nodes (1 rack)	12915	10436	9172	14620	13438	13438
N=80 nodes (2 racks)	32289	26006	23400	62141	33142	33142
N=160 nodes (4 racks)	85586	64414	56549	148767	89936	77521
N=320 nodes (8 racks)	284455	172890	146990	392423	274562	178720

Fig. 5 presents the improvement over a 3D electronic-only IN topology for the case where the OCS network is connected to Top of Rack switches (ToR). Note that in this set of experiments we neglected the performance of the network that interconnects nodes within a rack, and we only consider inter-LC (or in this particular case inter-rack) traffic. We report results using synthetically generated traffic to emulate the execution of SuperLU from 8 up to 128 racks. Note that for small scale problem instances, the 3D torus cannot be created and the comparison results are thus unfair. However, for larger scale-out instances, the comparison indicates that the hybrid IN performs up to 35% better than the electronic-only IN.

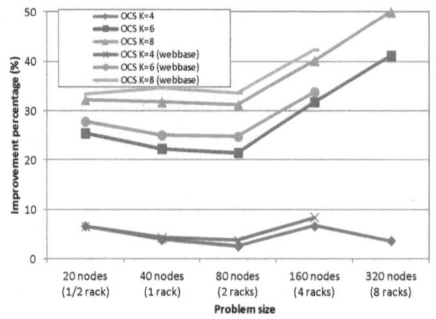 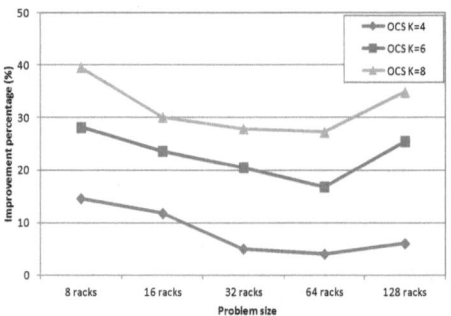

Fig. 4. Hop-bytes improvement of the hybrid IN over the 3D torus electronic-only IN (node-level aggregation)

Fig. 5. Hop-bytes improvement of the hybrid IN over the 3D torus electronic-only IN (rack-level aggregation)

Similar improvement was observed for FFTW (not reported here due to space limitations).

Capital Cost Comparison

In addition to performance evaluation, we also created simple models to estimate the cost of the hybrid IN and compare it to alternative electronic-only solutions. We let C_{OCS} be the cost of an OCS port, and C_{EPS} and C_{TR} be the cost of the EPS port and the EPS plug-in transceiver, respectively. Assuming that optical aggregation is performed at the rack level, the cost of the electronic edge (Top-of-rack switch), the electronic part of the core network and the optical part of the hybrid IN is $C_H = N \cdot K \cdot (C_{OCS} + C_{EPS} + C_{TR}) + M \cdot (C_{EPS} + C_{TR}) + N \cdot R \cdot (C_{EPS} + C_{TR})$, where N is the number of racks, K is the number of parallel OCS planes, M is the number of EPS ports of the second level, and R is the number of electronic ports of edge that are connected to the compute nodes (assuming R compute nodes per rack). The cost of the electronic-only IN where the racks are connected in an X-D torus topology and the compute nodes are directly connected to the top-of-rack switches is $C_E = N \cdot 2 \cdot X \cdot (C_{EPS} + C_{TR}) + N \cdot R \cdot (C_{EPS} + C_{TR})$, where X is equal to 2, 3, or 4 for a 2-, 3-, or 4-D torus topology, respectively.

Due to extreme price volatility in an evolving technology domain, any specific price trend assumption would be speculative and may not survive over time. Instead, we conducted a cost comparison that is parametric to R_{OCS}. We define R_{OCS} as the relative cost of an OCS port over the cost of an EPS port plus the cost of an electro-optical transceiver, i.e. $R_{OCS}=C_{OCS}/(C_{EPS}+C_{TR})$. In this comparison we assume $R=40$ compute nodes per rack and $M=1$ port for the second level of the EPS network in the hybrid IN system.

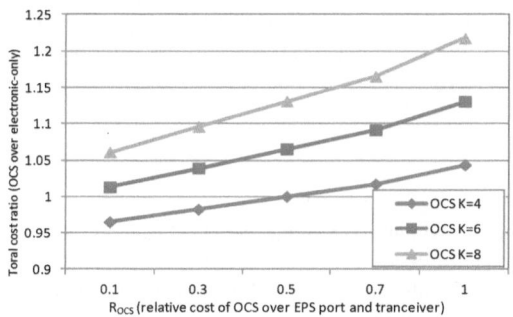

Fig. 6. Ratio of the cost of the hybrid IN over the cost of the electronic-only network that interconnects the racks in a 3D torus topology

We found R_{OCS} to be currently in the [0.5-0.6] interval, given the cost of an OCS port C_{OCS} to be around \$500 [16-17] and 40Gbps EPS technology. The price of OCS switches will tend to fall rapidly, as high as 80% [24], as vendors pursue widely deploying OCS in DC, indicating R_{OCS} reduction to around 0.15 (without considering in this calculation the reduction in the cost of the electronics). A benefit brought by the proposed hybrid IN that is not factored in the cost model is that it is future-proof, since the OCS network is agnostic to protocols, modulation formats and data rates. Thus, to increase the capacity of the hybrid IN system we only need to upgrade the electronic-edge that accesses the OCS network. This also indicates that the cost of the OCS port and that of the hybrid network will almost certainly not increase, while the cost of the electronic-only IN might increase rapidly, especially if we move to a newer technology with higher bandwidth (e.g. 100Gbps). Lastly, note that the optical technology, due to its transparent nature, consumes much less energy compared to active electronic switching, a cost-saving factor not captured in this model.

5   Conclusions

We presented a method to partition the logical tasks and identify the topology configuration of the (reconfigurable part) of a hybrid EPS/OCS interconnection network (IN) to efficiently serve the point-to-point traffic produced by a parallel application with known logical communication graph. The method presented is general and can be used in many different settings, irrespective of the level at which the optical network is deployed and the number of parallel optical planes. We used the proposed algorithm to estimate the performance of the target hybrid IN and compare it against application mapping on conventional fixed, electronic-only INs that are based on toroidal topologies. Our results indicated that the hybrid IN can exhibit better average hop-bytes performance than electronic-only INs based on toroidal topologies with higher connectivity degrees. Subject to the relative costs of the electronic and optical ports, these performance improvements can come at a slightly higher but comparable cost, while the data rate agnostic and transparent nature of optical technology ensures better upgradability and lower power consumption for the hybrid IN.

Acknowledgements. This work has been partially supported by Industrial Development Agency (IDA) Ireland and the Irish Research Council for Science, Engineering and Technology (IRCSET).

References

1. Geist, A.: Paving the roadmap to exascale. SciDAC Review, Special Issue on Information Technology in the Next Decade 16, 52–59 (2010)
2. Brightwell, R., et al.: Challenges for High-Performance Networking for Exascale Computing. In: ILCCN 2010 (2010) (invited paper)
3. Cray Inc. Cray XT Specifications (2009),
   http://www.cray.com/Products/XT/Specifications.aspx
4. Ajima, Y., Sumimoto, S., Shimizu, T.: Tofu: A 6d mesh/torus interconnect for exascale computers. Computer 42, 36–40 (2009)
5. Bokhari, S.H.: On the Mapping Problem. IEEE Trans. Computers 30(3), 207–214 (1981)
6. Fitch, B.G., Rayshubskiy, A., Eleftheriou, M., Ward, T., Giampapa, M., Pitman, M.C.: Blue matter: Approaching the limits of concurrency for classical molecular dynamics. In: Supercomputing (2006)
7. Bhanot, G., Gara, A., Heidelberger, P., Lawless, E., Sexton, J.C., Walkup, R.: Optimizing task layout on the Blue Gene/L supercomputer. IBM Journal of Research and Development (2005)
8. Bhatele, A., Gupta, G.R., Kale, L.V., Chung, I.-H.: Automated Mapping of Regular Communication Graphs on Mesh Interconnects. Computer Science Research and Tech Reports (2010)
9. Bhatele, A., Kale, L.V.: Heuristic-Based Techniques for Mapping Irregular Communication Graphs to Mesh Topologies. In: HPLC 2011 (2011)
10. Chung, I.-H., Lee, C.-R., Zhou, J., Chung, Y.C.: Hierarchical Mapping for HPC Applications. In: International Parallel & Distributed Processing Workchop, IPDPSW (2011)
11. Al-Fares, M., Loukissas, A., Vahdat, A.: A Scalable, Commodity Data Center Network Architecture. In: SIGCOMM (2008)
12. Benson, T., Akella, A., Maltz, D.A.: Network traffic characteristics of data centers in the wild. In: Conference on Internet measurement (IMC), pp. 267–280 (2010)
13. Greenberg, A., et al.: VL2: A Scalable and Flexible Data Center Network. In: SIGCOMM 2009 (2009)
14. Barker, K.J., et al.: On the Feasibility of Optical Circuit Switching for High Performance Computing Systems. In: Supercomputing (2005)
15. Schares, L., et al.: A Reconfigurable Interconnect Fabric with Optical Circuit Switch and Software Optimizer for Stream Computing Systems. In: Optical Fiber Communications, OFC (2009)

16. Farrington, N., et al.: Helios: a hybrid electrical/optical switch architecture for modular data centers. In: SIGCOMM 2010 (2010)
17. Wang, G., et al.: c-Through: Part-time Optics in Data Centers. In: SIGCOMM 2010 (2010)
18. Asanovic, K., et al.: The Landscape of Parallel Computing Research: A View from Berkeley. Technical report, Berkeley (2006)
19. Kamil, S., Oliker, L., Pinar, A., Shalf, J.: Communication Requirements and Interconnect Optimization for High-End Scientific Applications. Transactions on Parallel and Distributed Systems (2009)
20. Integrated Performance Monitoring (IPM), http://ipm-hpc.sourceforge.net/
21. Liberatore, V.: Circular Arrangements. In: Widmayer, P., Triguero, F., Morales, R., Hennessy, M., Eidenbenz, S., Conejo, R. (eds.) ICALP 2002. LNCS, vol. 2380, p. 1054. Springer, Heidelberg (2002)
22. Shi, J., Malik, J.: Normalized Cuts and Image Segmentation. IEEE Trans. on Pattern Analysis and Machine Intelligence 22(8) (2000)
23. http://www.cise.ufl.edu/research/sparse/matrices/
24. http://www.lightreading.com/document.asp?doc_id=213809&f_src =lightreading_gnesws

Towards an Efficient Fat–Tree like Topology

D. Bermúdez Garzón[1], C. Gómez[2], M.E. Gómez[1], P. López[1], and J. Duato[1,*]

[1] DISCA Department, Universitat Politècnica de Valencia,
Camino de Vera, 14, 46071–Valencia, Spain
dieberg1@posgrado.upv.es
[2] Dept. of Computing Systems, University of Castilla-La Mancha, Spain

Abstract. Topology and routing algorithm are two key design parameters for interconnection networks. They highly define the performance achieved by the network, but also its complexity and cost. Many of the commodity interconnects for clusters are based on the fat–tree topology, which allows both a rich interconnection among the nodes of the network and the use of adaptive routing. In this paper, we analyze how the routing algorithm affects the complexity of the switch, and considering this, we also propose and analyze some extensions of the fat–tree topology to take advantage of the available hardware resources. We analyze not only the impact on performance of these extensions but also their influence over switch complexity, analyzing its cost.

Keywords: Regular indirect topologies, fat–trees, adaptive and deterministic routing.

1 Introduction

Cluster machines have become very popular to build high performance computers and data centers in the last years due to their excellent cost–performance ratio. These machines use commodity computers linked by a high–performance interconnection network, which plays a critical role to achieve a high performance. Two of the main design issues of interconnection networks are topology and routing [1]. In deterministic routing schemes, packets traverse a fixed, predetermined path between their source and their destination, while in adaptive routing schemes, packets may use one of the available different alternative paths from their source to their destination. An adaptive routing algorithm is composed of the routing and selection functions [2]. For each packet, the routing function supplies the set of available routing options, while the selection function [3] selects one of them. This selection function usually takes into account the status of the network. Adaptive routing usually helps in balancing network traffic, thus allowing the network to obtain a higher throughput. However, with adaptive routing, in–order packet delivery can not be ensured, which is mandatory for some applications. This is the case, for example, for certain cache coherence protocols, some communication libraries and network technologies. On the other hand, deterministic routing algorithms usually do a very poor job balancing traffic among the network links, due to the

* This work was supported by the Spanish MICINN, Consolider Programme and Plan E funds, as well as European Commission FEDER funds, under Grants CSD2006-00046 and TIN2009-14475-C04.

C. Kaklamanis et al. (Eds.): Euro-Par 2012, LNCS 7484, pp. 716–728, 2012.

lack of path diversity, but they are easier to be implemented. Moreover, deterministic routing guarantees in–order packet delivery by design.

Concerning topology, cluster–based machines usually choose either regular direct networks (tori and meshes) or, more frequently, multistage indirect networks (MINs). In particular, fat–trees [4] have raised in popularity in the past few years, since most of the commonly–used interconnect technologies provide support for this topology. Moreover, some of the most powerful machines ever built implement a fat–tree topology, such as, the CM-5, the Cray BlackWidow machine, or the recent number 1 machine in the Top500 list, Tianhe-1A [5].

Routing and topology, besides impacting network performance, also highly define the hardware cost and complexity of the interconnection network. In this paper, we analyze several proposals of topologies and routing algorithms analyzing not only their performance but also estimating the resources needed to implement them, with the aim of taking the highest advantage of the available hardware resources.

2   Motivation

Interconnection networks are often designed only for performance. However, the increase in complexity of interconnection network negatively affects its cost and power consumption. A balanced design should consider both performance and complexity, trying to obtain a good tradeoff between them.

In this paper, we focus on routing in the fat–tree topology and how the routing algorithm can be used to simplify the switch complexity and even modify the topology. Routing in fat–trees has two different subpaths (an upwards subpath and a downwards one). In the commonly–used routing algorithm [6], the upwards path is fully adaptive and the downwards one is deterministic. The unique downwards subpath is determined by the path selected in the upwards one. So, in fat–trees, unlike other topologies, the decisions made in the upwards subpath can be critical, having a strong impact on network performance. In [7], the DESTRO mechanism was proposed to effectively select the upwards path for each packet in a fat–tree, in order to highly reduce the Head–of–Line (HoL) blocking effect [8] and in this way improving performance. This mechanism can be implemented in a very simple way, as only a component of the destination identifier is used for routing, both in the up and down subpaths. In [9], the RUFT topology was proposed which is a simplification of the fat–tree topology taking advantage of the nice properties of DESTRO. RUFT allows to reduce the hardware resources to the half while providing similar performance.

Besides obtaining different levels of performance, fat-trees (with adaptive routing), DESTRO and RUFT require switches with different complexity. Therefore, a fair comparison among them should consider both performance and complexity issues. On the other hand, as the RUFT topology use simpler switches than fat–trees, some extensions with a complexity level similar to the fat–tree can be devised. We propose two extensions. The first one (RUFT-PL) substitutes each original link by two physical parallel links, allowing the use of some adaptivity in the network and at the expense of increasing switch complexity. The second one (RUFT-DB) substitutes each original link by another one with double width, therefore duplicating link bandwidth. The new proposed

extensions of the RUFT topology are able to increase the throughput of the interconnection network by 2.5x compared to the fat–tree while having a similar or even lower
cost.

The rest of the paper is organized as follows. Section 3 revisits the fat–tree topology and presents the notation and assumptions used in the following sections. It also
describes the adaptive routing algorithm commonly–used in fat–trees and DESTRO
analyzing the switch requirements. Section 4 presents RUFT analyzing also its switch
requirements. Section 5 presents the RUFT topology extensions proposed in this paper.
Section 6 provides several evaluation results for different configurations of topology
and routing algorithm. Finally, some conclusions are drawn.

3   Fat–Tree Topology

The fat-tree topology is based on a complete tree that gets thicker near the root. The
arity of the switches increases as we go nearer to the root, which makes the physical implementation unfeasible. For this reason, some alternative implementations have
been proposed in order to use switches with fixed arity. In particular, the k–ary n–tree
[6] is a parametric family of regular multistage topologies. The number of stages is n
and k is the arity or the number of links of a switch that connects to the previous or to the
next stage (i.e., the switch degree is $2k$). A k–ary n–tree is able to connect $N = k^n$ processing nodes using nk^{n-1} switches. Each processing node is represented as a n–tuple
$\{0, 1, ..., k-1\}^n$, and each switch is defined as a pair $\langle s, o \rangle$, where s is the stage where
the switch is located at, $s \in \{0..n-1\}$, and o is a $(n-1)$-tuple $\{0, 1, ..., k-1\}^{n-1}$ which
identifies the switch inside the stage. Two switches $\langle s, o_{n-2}, ..., o_1, o_0 \rangle$ and $\langle s', o'_{n-2},$
$..., o'_1, o'_0 \rangle$ are connected by an edge if $s' = s + 1$ and $o_i = o'_i$ for all $i \neq s$. On the
other hand, there is an edge between the switch $\langle 0, o_{n-2}, ..., o_1, o_0 \rangle$ and the processing
node $p_{n-1}, ..., p_1, p_0$ if $o_i = p_{i+1}$ for all $i \in \{n-2, ..., 1, 0\}$. This edge is labeled with
p_0. In what follows, we will assume that descending links are labeled from 0 to $k - 1$,
and ascending links from k to $2k - 1$.

3.1   Adaptive Routing in Fat–Trees

In k–ary n–trees, minimal routing from a source to a destination can be accomplished
by sending packets upwards to one of the nearest common ancestors of the source and
destination nodes and then, from there, downwards to destination. When crossing stages
in the upwards direction several paths are possible, thus providing adaptive routing. In
fact, each switch can select any of its k up output ports. Once a nearest common ancestor
has been reached, the packet is turned around and sent downwards to its destination as
just a single path is available. The stage up to which the packet must be forwarded
is obtained by comparing the source and destination components beginning from the
most significant one. The first pair of components that differs indicates the last stage
to forward up the packet. Once in that stage, the descending path is deterministic. At
each stage, the descending link to choose is indicated by the component corresponding
to that stage in the destination n–tuple. In the example, from stage i, the packet must be
forwarded through the p'_i link; from stage $i - 1$ through link p'_{i-1}, and so on.

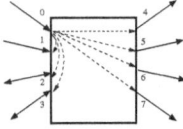

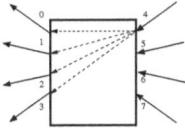

(a) Requested ports in the upwards direction by port 0.

(b) Requested ports in the downwards direction by port 4.

Fig. 1. Ports that can be requested in a 4–ary n–tree using adaptive routing

Switch complexity of fat–trees can be easily computed considering that each switch has k bidirectional input and output ports, leading to $2k \times 2k = 4k^2$ switching elements. However, this rationale does not account for the actual requirements of switching activity of the routing algorithm. As can be observed, in the upwards subpath, at each switch, the k input ports can forward packets through either any of the up k output ports, if the packet continues in its upwards subpath, or any of its down output ports if the packet starts its downwards subpath. On the other hand, in the downwards subpath, there are k input ports that can only request k down output ports, since once a packet has started its downwards subpath, the packet must continue going downwards. Figures 1a and 1b show the output ports that can be requested in the upwards and downwards directions, respectively, in the switches of a 4–ay n–tree.

A common way of implementing switches is by using as many multiplexers as the number of required output ports. Each multiplexer has a number of inputs equal to the number of input ports that can request the corresponding output port. In the switch we are considering, ports in the upwards direction require k multiplexers with k inputs each one or a $k \times k = k^2$ complexity. On the other hand, ports in the downwards direction require k multiplexers with $2k$ inputs each one or a $k \times 2k = 2k^2$ complexity. Total switch complexity can be easily obtained as the sum of the upwards and downwards directions complexities, leading to a the switch complexity of $3k^2$ switching elements.

3.2 DESTRO Routing in Fat–Trees

Contrary to the previously presented routing algorithm, DESTRO [7] is deterministic, that is, in both subpaths there is only one path for each source–destination pair. We selected this algorithm due to its good results and also because it is able to highly reduce the switch complexity. The high performance is due to the appropriate selection of packet upwards subpaths which distribute destinations in a very effective way to highly reduce the HoL blocking effect. The packet downwards subpaths are determined by the upwards subpath followed by the packet and with DESTRO the interferences among different destinations in the packet downwards subpaths are completely eliminated. All the packets destined to a particular node are kept inside the same sub-tree (See Figure 2a), and have a unique and exclusive down path. This is performed by using the destination identifier to select one of the multiple available upwards subpaths. In DESTRO, the output port for routing a packet in a particular switch is given both by the destination identifier and the stage where the switch is located. In particular, it considers the component of the packet destination corresponding to that stage (i.e., a switch located at

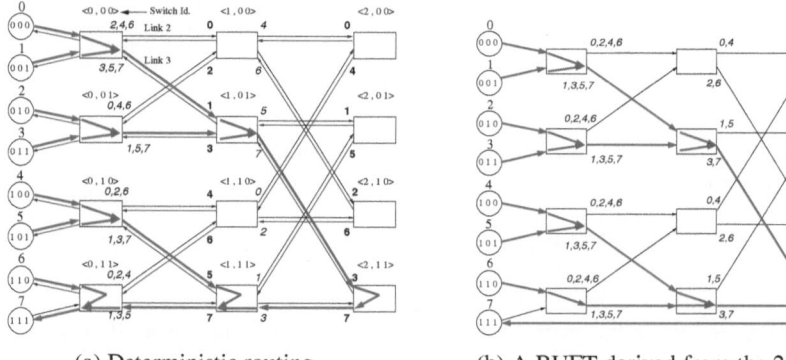

(a) Deterministic routing.   (b) A RUFT derived from the 2–ary 3–tree using DESTRO.

Fig. 2. Deterministic routing in a 2–ary 3–tree

stage s considers the sth component of the destination identifier, that is p_s). Therefore, at the switch $\langle s, o_{n-2}, ..., o_1, o_0 \rangle$, the selected output port for a packet with destination $p_{n-1}, ..., p_1, p_0$ will be $k + p_s$.

Figure 2a shows the destination node distribution in the ascending and descending links of a 2–ary 3–tree using DESTRO. In the first stage, the least significant component of the packet destination identifier (the least significant bit in this example) is used to select the ascending output port. At the second stage, the destinations of all packets that reach a given switch have the same least significant component. Hence, the component to consider in the selection of the up output port in this stage is the next one in the destination address. For instance switch 4 is only reached by packets destined to nodes 0, 2, 4 and 6. These nodes have the same least significant component, which is 0. Of them, only packets destined to nodes 4 ($\langle 100 \rangle$) and 6 ($\langle 110 \rangle$) must be forwarded upwards. Packets destined to node 4 will select the first up link, and packets destined to node 6 the another one. Following this mechanism in all the upwards stages, finally, packets destined to a particular destination reach the same switch at the last stage and have a unique down subpath. Figure 2a highlights all the paths to node 7 and how all of them share the same downwards subpath.

By using DESTRO, the switch complexity can be highly reduced. The upwards switch activity is the same as in the fat-tree with adaptive routing. That is, each input port in the upwards subpath can request either any of the up output port, or any of the down output ports. However, in the down subpath, each link, input port and output port is used exclusively by packets sent to a unique destination. As a consequence, a given input port will always request the same output port, since all the packets that arrive to a particular input port, in their downwards subpath, are destined to the same node and are always forwarded to the same output port. This allows a noticeable reduction in switch complexity. Using multiplexers to implement switches, ports in the upwards direction require k multiplexers, each with k inputs, or a $k \times k = k^2$ complexity, and ports in the downwards direction require k multiplexers with $k+1$ inputs (the k upwards ports plus the unique downwards one) or $k \times (k+1) = k^2 + k$ switching elements. Therefore the required switch complexity required by DESTRO is $2k^2 + k$.

4   RUFT

The RUFT topology [9] is a simplification of the fat–tree topology obtained by taking advantage of the nice properties of DESTRO. In particular, since there is no switching activity in the downwards subpath, the switches are simplified by making them unidirectional. Therefore, the whole downwards subpath is transformed in links that connect the last stage to the different processing nodes; (see Figure 2b). Notice also that, as the topology is unidirectional, there is not chance to start the downwards subpath before reaching the last stage and, therefore, the paths are longer since all the packets must reach the last stage, contrary to the fat–tree topology where depending on the source–destination pair, different number of stages must be traversed.

Using RUFT, the switch complexity corresponds to the switch complexity of a unidirectional switch of k input ports and k output ports, where any of the k input ports can request any of the k output ports, so the switch complexity is k^2. As it can be seen, the switch complexity has been reduced more than twice when comparing it with DESTRO and three times when comparing with the fat-tree with adaptive routing.

5   Extensions of the RUFT Topology

In this section, taking into account the difference in switch complexity among RUFT, DESTRO, and fat-trees, we propose two different enhancements of the RUFT topology that, by using more complex switches, improve its performance. The idea is to obtain a new topology with a switch complexity similar to DESTRO and fat-trees, which will allow us to perform a fair comparison among all the feasible choices.

Both of them, as RUFT, are unidirectional MINs. Therefore, all packets will traverse the same number of hops (the number of network stages) regardless of their source–destination pair. The first proposal (which will be referred to as RUFT-Parallel Links or RUFT-PL) uses switches with a number of ports equal to the number of ports of the fat–tree switches. As fat-tree ports are bidirectional, RUFT-PL have twice number of ports than RUFT. Therefore, each pair of switches are connected by two parallel links. On the other hand, the number of network links is also equal to the number of network links of the fat–tree topology. The second proposal (which will be referred to as RUFT-Double Bandwidth or RUFT-DB) uses switches with the same number of ports as the RUFT topology, but these ports and their associated links have double width (thus double link bandwidth) than the ones of RUFT and fat–tree topologies. The idea is to make the network bandwidth equal to the one of the fat–tree, while maintaining the number of ports equal to the one of the RUFT switches.

5.1   RUFT with Parallel Links

RUFT-PL uses the same number of switches as the fat–tree topology, as the RUFT topology does, but they have the same number of ports as the fat–tree switches. Our proposal pursues to implement the RUFT topology but without reducing the switch complexity. As fat–tree switches have bidirectional ports, RUFT-PL switches can double the number of ports of RUFT switches, leading to a complexity of $2k \times 2k$ switching elements.

We propose to use the available k additional ports to have two parallel links connecting each pair of switches of the original RUFT topology. These two parallel links provides additional routing flexibility that can be exploited in different ways.

A first approach is to distribute different destinations among them and, in this way, the remaining HoL blocking that still appears in the upwards subpaths of the RUFT topology is reduced even more. To do so, we propose that, at a given switch, the pair of output channels (i.e. output ports) to be used will be given by the destination component corresponding to the stage of the switch (as RUFT and DESTRO do) but the parallel channel of the port that will be finally used will be given by the least significant bit of the next destination component. Notice that we use the next destination component because it will be used in the next stage to select the output port (the pair of output ports in RUFT-PL). Therefore, packets that will take different output ports in the next switch will be forwarded through different channels, thus reducing interferences among them.

Formally, at switch $\langle s, o_{n-2}, ..., o_1, o_0 \rangle$, the selected pair of output ports for a packet with destination node $p_{n-1}, ..., p_1, p_0$ will be $k + p_s$, as in RUFT, and from this output port, the parallel channel to be used will be given by $p_{s+1} \bmod 2$ to select the least significant bit of the next destination node component. Figure 3 shows how the destination nodes are distributed among the output links of switches in RUFT–PL, for $k = 2$ and $n = 3$. Notice that the least significant bit of the next destination node component is used to decide the parallel link to be used by a given destination at a particular set of output ports. As can be seen, thanks to classifying destination nodes in this way, each input port of a given switch only requests a particular set of output ports. This is better shown in Figure 4a, where the output ports requested by the two parallel channels of input port 0 are shown. Figure 3 suggests that the last stage is not required since packets with different destinations already arrive through different channels. However, this is only true for $k = 2$, if $k > 2$ is used, the last stage is required. In Figure 4b another example for $k = 4$ is shown. In this case, a given input port only requests the two channels of two different output ports.

In RUFT-PL, each input port can only request the k output channels associated to $k/2$ output ports, whose implementation require k multiplexers with $2k$ inputs, which leads to a complexity of $k \times 2k = 2k^2$, which is even smaller than the one required by DESTRO. We will refer to this way of using parallel links as RUFT–PL–C, given that packets are classified in different channels of the parallel link according to its destination. The way RUFT–PL–C selects parallel output channels may not be the best option under some non-uniform traffic patterns. For these traffic patterns, in most of the switches only one of the two parallel channels will be actually used. For this reason, we have also evaluated another version of RUFT–PL that selects the link of the parallel channel following other criteria, without classifying packets. We will refer to this approach as RUFT–PL–NC. One possible criterion can be selecting the channel with more free buffer. The key point of RUFT–PL–NC is that it can always use both parallel channels per port regardless what kind of traffic is used. However, in this case, the switch complexity is increased since each input channel can ask for any of the up output ports, so the switch complexity is $2k \times 2k = 4k^2$ switching elements, which is higher than the complexity of the fat–tree switch with adaptive routing. Table 1 summarizes the switch complexities of the different proposals.

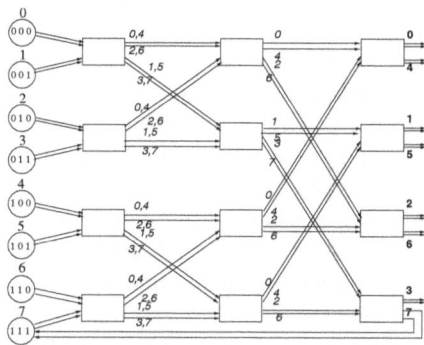

Fig. 3. A 2-ary 3-tree RUFT–PL

Table 1. Switch complexity comparison

Routing	Switch complexity
Adaptive	$3k^2$
DESTRO	$2k^2 + k$
RUFT	k^2
RUFT DB	$2k^2$
RUFT PL C	$2k^2$
RUFT PL NC	$4k^2$

5.2 RUFT with Double Bandwidth

RUFT–DB is a slight modification made on the RUFT topology, where we keep the same number of ports per switch, but doubles the bandwidth of all links connecting the switches by making them wider. On the other hand, the size of the buffers of the input and output are also doubled. The idea is to have the same bandwidth as the one of the fat–tree topology and also the same buffer resources. That is, RUFT–DB is a RUFT topology where buffers and link bandwidth are doubled. Regarding switch complexity, the number of required switching elements are the same as in RUFT, but as each one has twice the width of RUFT, the required switch complexity measured in RUFT switching elements is $2k^2$.

6 Evaluation

6.1 Simulation Environment

In order to compare the performance of the different topologies and routing algorithms analyzed in this paper, we have developed an event-driven simulator. The simulator models a virtual cut-through network composed by a set of switches. Each switch has a routing control unit that applies the routing algorithm and configures its internal crossbar. Links allow pipelined transmission and are characterized by the delay required by a flit to reach the edge of the link (fly time) and the interval among consecutive flit transmissions (link delay). Routing, crossbar, fly and link delays has been assumed equal to one clock cycle. In the case of RUFT, RUFT-DB and RUFT–PL, an increased fly time has been considered in the long links that connect the last stage to the processing nodes. This time is modelled as the product of one fly time by the number of network stages. Three packet sizes have been considered (8B, 64B and 4KB). Several synthetic traffic were considered: uniform, bit-reversal and hotspot with a traffic concentration of 5% and 15%.

6.2 Performance Results

In Figure 5a we compare the different analyzed configurations with uniform traffic and 64B packets in 2–ary 4–tree topologies. As can be seen, fat-tree with adaptive routing (referred to as FTA), DESTRO and RUFT obtain similar throughputs, being DESTRO the best of them and RUFT the worst one since its switches have half the number of ports than the fat–trees switches (FTA and DESTRO). It also has half number of links (unidirectional links are used) than FTA and DESTRO. On the other hand, concerning the three RUFT extensions proposed in this paper, both RUFT–PL configurations and RUFT–DB are able to double the throughput obtained by the first three configurations. In particular, RUFT–PL–NC is the one that achieves the highest throughput, whereas the other two configurations obtain similar throughputs, having RUFT–DB a lower network latency except near the saturation point. The effect of a greater link bandwidth of RUFT–DB compensates the partial adaptivity provided by RUFT–PL. Only at very high loads, the additional routing freedom is worth. Concerning the two versions of RUFT–PL, classifying packets between the two parallel links does not seem a good idea. The explanation is that although by classifying packets HoL effect is further reduced, it is also true that the number of routing options is limited to only one versus two in the non-classify option. Indeed, traffic in RUFT is already classified by design, and the additional distribution offered by splitting destinations among the two links of the parallel channel does not compensate the reduction in routing flexibility, as only one of the parallel links can be used for a given destination. Finally, notice though, that the RUFT extensions require more complex switches than RUFT. An additional comment can be stated regarding the fact that configurations that allow multiple paths among source–destination pairs (i.e., those ones that provide adaptive routing), which include not only FTA but also RUFT-PL-NC, may introduce out-of-order delivery of packets.

Influence of Network Size. In this section, we analyze the influence of network size on the performance of the different analyzed configurations. Uniform traffic and 64B packets were considered. Figures 5a to 5e show results for k–ary n–trees with different number of stages and arities, increasing accordingly the network size. As expected, as the number of stages increases, the latency of the different configurations also increases, since more stages must the crossed by packets. Indeed, the performance is slightly worse for the configurations with higher arity (k=4) because larger switches mix more destinations in each link, thus contributing to increase the HoL blocking effect. Anyway, the relative behavior of the different analyzed configurations remain the same. Notice that RUFT–PL–C gets slightly worse as the number of stages increases. Again, the less flexibility for routing is amplified by the fact of traversing more stages (i.e., more routings are performed to reach a given destination). The configuration that achieves the highest throughput is RUFT–PL–NC, and RUFT–DB is the one that obtains the lowest latency before saturation due to the fact that data is transmited twice faster through links and switches.

Impact of Packet Size. In this section, we analyze how the behavior of the different analyzed configurations are affected by the packet size. We have tested several network sizes with different packet sizes, figure 6a and 6b shows the results for a 2–ary 6–tree with a packet size of 8 bytes and 4 Kbytes. As expected, latency increases with the packet size and so does network throughput as the contribution of the transmission time of packets is by far more important than the one that depends on the number of hops. The relative positions of the different configurations are the same regardless

(a) Requested ports in a switch, $k = 2$.   (b) Requested ports in a switch, $k = 4$.

Fig. 4. Output ports that can be requested by an input port in RUFT switches

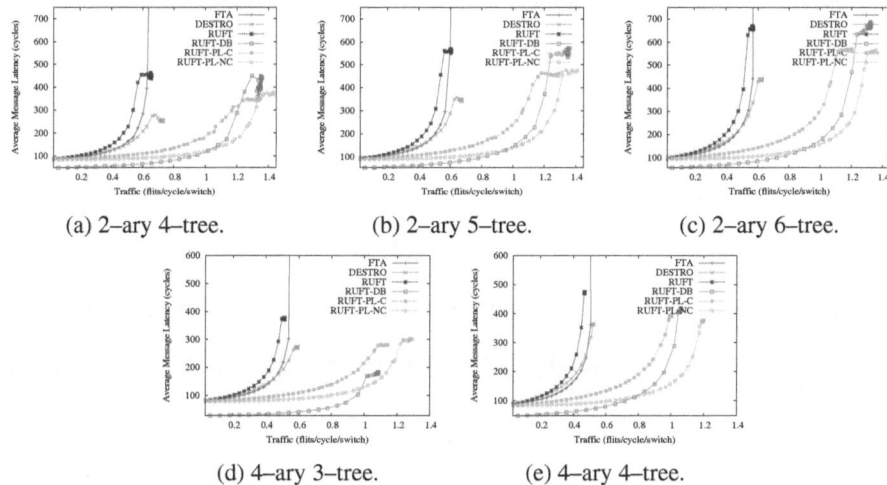

(a) 2–ary 4–tree.   (b) 2–ary 5–tree.   (c) 2–ary 6–tree.

(d) 4–ary 3–tree.   (e) 4–ary 4–tree.

Fig. 5. Network latency versus throughput for different network sizes with uniform traffic and a packet size of 64B

of the packet size, with some advantage of RUFT–DB with large packets, taking full advantage of the increased link bandwidth. Anyway, the best configuration is always RUFT–PL–NC.

Effect of Traffic Pattern. Figure 7 shows the results for 2–ary 5–tree topologies with a packet size of 64 bytes with other traffic patterns. For the bit–reversal traffic pattern, RUFT and RUFT–PL–C obtains the worst results. In this case, the paths provided by RUFT collide among different source–destination pairs and the additional channels that the PL extension supplies are useless in most cases as packets are always sent to the same node and the channel of the parallel link are selected according to the destination node. FTA is able to improve performance thanks to the use of alternative paths. The additional routing flexibility of RUFT–PL–NC also helps in improving network throughput. Finally, once more, the doubled channel bandwidth allows RUFT-DB to double the network throughput reached by RUFT.

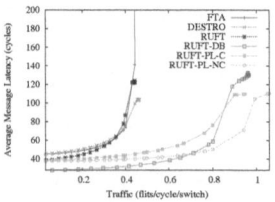
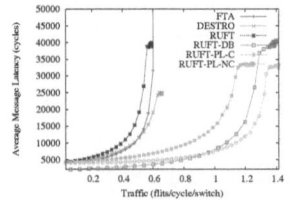

(a) 2–ary 6–tree, Packet Size 8B.   (b) 2–ary 6–tree, Packet Size 4KB.

Fig. 6. Network latency versus throughput for two network sizes and different packet sizes with uniform traffic

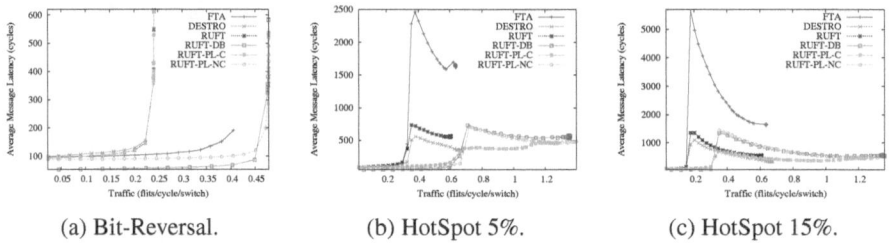

(a) Bit-Reversal.   (b) HotSpot 5%.   (c) HotSpot 15%.

Fig. 7. Network latency versus throughput with different traffic patterns in 2–ary 5–tree networks and a packet size of 64B

When considering hotspot traffic (see Figures 7b and 7c), we can observe that the higher the severity of the hot-spot, the lower the achieved network throughput. Notice the unusual shape of the plots. Once the paths that reach the hot-spot are saturated, only traffic destined to other nodes move through the network, until the network is completely saturated. The final saturation point reached is the same obtained for the uniform traffic pattern (see Figure 5b). The results also confirm RUFT–PL–NC and RUFT–DB as the best routing algorithms.

Effect of Routing and Fly Times. In this section we analyze how the obtained conclusions may change if the path setup time was higher than we considered. Path setup time is given by the sum of routing, switch and fly times. We have analyzed the impact of a higher routing time, which should impact all configurations and fly time, which should especially impact the long RUFT links of the last stage.

Figure 8a shows the results of a increased routing time in a 2-ary 5-tree with uniform traffic. As it can be seen, the higher average distance of fat-trees leads to a higher latency values of FTA and DESTRO. Moreover, the throughput of all the configurations is strongly reduced, as packets spend most of its time at the routing control units. Figure 8b shows the effect of a high fly time. As expected, all the RUFT configurations are strongly penalized, increasing their average latency. On the other hand, RUFT–DB becomes the best configuration for the full range of traffic. When traversing a link takes a long time, transmiting more information per time unit allows amortizing this time.

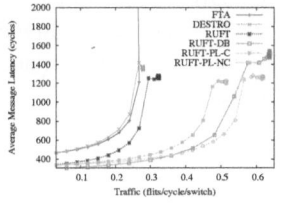

(a) Routing 50 cycles, Fly 1 cycle.

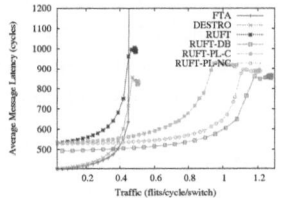

(b) Routing 1 cycle, Fly 50 cycles.

Fig. 8. Network latency versus throughput with uniform traffic in 2–ary 5–tree networks and a packet size of 64B

Performance/Cost Evaluation. This section analyzes and compares the hardware cost of each configuration evaluated in the paper. In particular, we consider for each configuration, the number of links, the number of switches and the number of switching elements required. In particular, the number of switching elements is the most appropriate measure to compare the different configurations since it considers both, the number of switches and its degree. Moreover, we define two different figures of merit that take into account both, performance and cost of a given configuration. The first one is the throughput per switching element. The higher the value of this parameter, the better performance/cost ratio has the configuration. To account for the latency, we also obtain the product of the latency by the number of switching elements. Although this value could be obtained for every value of traffic analyzed, for the sake of shortness, only results for very low load traffic are shown. In this case, the higher the value, the worse the configuration.

A subset of the results are shown in Table 2. As it can be seen, the cheapest topology is RUFT, as it has the lowest number of links and switching elements. RUFT uses half the number of links and switching elements of the RUFT–PL–C configurations and even less compared to the others. Nevertheless, RUFT does not reach the best throughput. RUFT–PL–NC is able to double the throughput obtained by FTA, DESTRO and RUFT, and outperforms the other two RUFT extensions proposed in this paper by 6%. If we consider both performance and cost, as shown by the defined figures of merit, results confirm that the best configurations are RUFT–PL–NC and RUFT–DB, in this order,

Table 2. Performance-Cost on different network sizes with several algorithms and uniform traffic

	Topology	Links	Switching Elements	Throughput	Throughput/ Switch Element	Low Load Latency	Base Lat./ Switch Element
2–ary 5–tree	FTA	320	960	0,6401	0,000667	94,4	90643
	DESTRO	320	800	0,6746	0,000843	95,4	76313
	RUFT	192	320	0,6103	0,001907	91,7	29349
	RUFT-DB	384	640	1,3611	0,002127	52,9	33835
	RUFT-PL-C	384	640	1,2892	0,002014	90,0	57623
	RUFT-PL-NC	384	1280	1,4107	0,001102	89,0	113950
4–ary 4–tree	FTA	2048	12288	0,5076	0,000041	90,6	1113818
	DESTRO	2048	9216	0,5250	0,000057	92,3	850411
	RUFT	1280	4096	0,4712	0,000115	87,2	357389
	RUFT-DB	2560	8192	1,0618	0,000130	48,6	398076
	RUFT-PL-C	2560	8192	1,0033	0,000122	85,2	697777
	RUFT-PL-NC	2560	16384	1,1993	0,000073	84,0	1376813

closely followed by RUFT. RUFT-PL-NC and RUFT-DB are able to increase by 3x the performance obtained by FTA, and by nearly 2.5x the DESTRO one. On the other hand, when the latency is considered, the one that obtains the lowest performance–cost is by far RUFT–DB, due to the fact that packet are sent faster.

7 Conclusions

In this paper, we analyze and compare several topologies and routing algorithms for fat-tree-related (fat-tree and RUFT) topologies. As RUFT topology requires less network resources than fat-trees, in order to perform a more fair comparison, we propose some extensions to the RUFT topology that tries to match the resources committed in the fat-tree. The final goal is to obtain a good tradeoff between performance and required resources. In particular, RUFT–DB uses links that are twice as wide as RUFT ones, therefore increasing network bandwidth, also doubling the required number of switching elements. The RUFT–PL–C and RUFT-PL–NC proposals also enhances the network by aggregating two links connecting switches, therefore increasing routing flexibility but at the cost of a higher number of switching elements. RUFT–PL–C tries to reduce the number of resources needed by restricting routing, statically classifying destinations among the aggregated links, also reducing the HoL blocking effect.

We evaluated all the proposals for different network sizes and network loads, obtaining that the RUFT–PL–NC approach, which adds routing flexibility to RUFT, is the one that obtains the best network throughput while the RUFT–DB one, which adds channel bandwidth helps in reducing packet latency. Most important, when we combine cost (measured in number of network resources) and performance, results show that RUFT–DB, RUFT–PL-NC and RUFT are very cost–efective configurations, obtaining the best throughput per switching element values. When latency–number of switching elements is consider, RUFT–DB is by far the best option due to its increased channel bandwidth.

References

1. Dally, W., Towles, B.: Principles and Practices of Interconnection Networks. Morgan Kaufmann (2003)
2. Duato, J., Yalamanchili, S., Ni, L.: Interconnection Networks. An Engineering Aproach. Morgan Kaufmann (2004)
3. Gilabert, F., Gómez, M.E., López, P., Duato, J.: On the influence of the selection function on the performance of fat-trees, pp. 864–873 (2006)
4. Leiserson, C.E.: Fat-trees: Universal networks for hardware-efficient supercomputing. In: ICPP, pp. 393–402 (1985)
5. Tianhe-1a, http://www.nscc-tj.gov.cn/en/
6. Petrini, F., Vanneschi, M.: k–ary n–trees: High performance networks for massively parallel architecture. IEEE Micro 15 (1995)
7. Gómez, C., Gilabert, F., Gómez, M.E., López, P., Duato, J.: Deterministic versus adaptive routing in fat-trees. In: 21th Int. Parallel and Distributed Processing Symposium (April 2007)
8. Nachiondo, T., Flich, J., Duato, J.: Buffer management strategies to reduce hol blocking. IEEE Trans. Parallel Distrib. Syst. 21(6), 739–753 (2010)
9. Gómez, C., Gilabert, F., Gómez, M.E., López, P., Duato, J.: Ruft: Simplifying the fat-tree topology (December 2008)

An Adaptive, Scalable, and Portable Technique for Speeding Up MPI-Based Applications

Rosa Filgueira[1], Malcolm Atkinson[1], Alberto Nuñez[2], and Javier Fernández[3]

[1] University of Edinburgh, School of Informatics, Edinburgh EH8 9AB, U.K.
{rosa.filgueira,mpa}@ed.ac.uk
[2] University Complutense de Madrid, Dept. Sistemas Informáticos y Computación,
28040 Madrid, Spain
alberto.nunez@pd.ucm.es
[3] University Carlos III de Madrid, Dept. Arquitectura de Computadores,
30 28911 Leganés, Spain
jfernand@arcos.inf.uc3m.es

Abstract. This paper presents a portable optimization for MPI communications, called *PRAcTICaL-MPI* (Portable Adaptive Compression Library- MPI). *PRAcTICaL-MPI* reduces the data volume exchanged among processes by using lossless compression and offers two main advantages. Firstly, it is independent of the MPI implementation and the application used. Secondly, it allows for turning the compression on and off and selecting the most appropriate compression algorithm at runtime, depending on the characteristics of each message and on network performance.

We have validated *PRAcTICaL-MPI* in different MPI implementations and HPC clusters. The evaluation shows that compressing MPI messages with the best algorithm and only when it is worthwhile, we obtain a great reduction in the overall execution time for many of the scenarios considered.

Keywords: MPI Library, Parallel techniques, High-Performance Computing, Compression algorithms, Adaptive systems, Portable optimizations.

1 Introduction

Parallel computation on cluster architectures has become the most common solution for developing High-Performance Computing applications. The Message Passing Interface (MPI) standard [1] is one of the most commonly used communication middleware frameworks on clusters.Several implementations of MPI are available, like MPICH [2], XT-MPI, OPENMPI [3], and LAM [4].

The current trend in High-Performance Computing is to use multicore clusters in order to increase computation capability, thus allowing an increase in the number of processes per application. Despite the fact that networks used in multicore clusters are fast and have low latency, the number of transferred messages may cause a bottleneck in the communication system, as communication-intensive,

C. Kaklamanis et al. (Eds.): Euro-Par 2012, LNCS 7484, pp. 729–740, 2012.

parallel MPI applications spend a significant amount of their total execution time exchanging messages between processes. This problem may lead to poor performance and scalability in many cases.

In this paper, we present a portable optimization of MPI called *PRAcTICaL-MPI* (Portable Adaptive Compression Library-MPI), which is fully transparent both to applications and MPI implementations. The main goal of *PRAcTICaL-MPI* is to enhance the performance and scalability of MPI-based applications and to reduce the volume of communications by applying run-time lossless compression in a transparent way for applications and MPI implementations. *PRAcTICaL-MPI* is capable of using the following compression algorithms: RLE [5], Huffman [6], Rice [7], FPC [8], and LZO [9]. Furthermore, the technique presented applies the Run-time Adaptive Strategy (RAS) developed in [10] to select the most appropriate compression algorithm to be used dynamically for each message exchange, and the size threshold form which a benefit is achieved by using data compression.

We have implemented *PRAcTICaL-MPI* by using the standard MPI profiling interface (PMPI) with the lowest possible overhead. The major contributions of *PRAcTICaL-MPI* can be summarised by looking the following properties:

- Transparency: *PRAcTICaL-MPI* uses the standard MPI profiling interface (PMPI), allowing transparent data compression for different applications and MPI implementations.
- Portability: *PRAcTICal-MPI* can be run by any MPI implementation that supports PMP, and is hence fully portable.
- Scalability: Since *PRAcTICaL-MPI* applies run-time compression to reduce the volume of messages transferred, the execution time of the application is reduced, thus enhancing the performance and scalability of MPI-based applications.

The remainder of this paper is structured as follows: Section 2 discusses related work. Section 3 summarises our Run-time-Adaptive Strategy. Section 4 introduces the *PRAcTICaL-MPI* architecture in detail. Section 5 presents an extensive evaluation of *PRAcTICaL-MPI*I in several scenarios. Finally, Section 6 presents conclusions and a discussion of potential future work.

2   Related Work

Two main background techniques and existing contributions are reviewed in this section: The standard MPI profiling interface, and the most popular works to extend MPI with compression capabilities.

2.1   PMPI: Standard MPI Profiling Interface

The MPI Forum defined [1] the MPI profiling interface (PMPI) as a mechanism for application developers to obtain high-level performance information about the behavior of both the application algorithm and the parallel system. Note that PMPI is part of the MPI standard specification.

The main concern of PMPI [11] is to provide a mechanism by which the developers of profiling (and other) tools can collect performance information they require without access to the underlying implementation. The mechanism is based on each MPI-routine having a corresponding PMPI-routine with identical syntax and functionality, so that it can be used to intercept all MPI calls and change their functionality. Therefore, tools can create wrappers for any MPI_routine and then insert them "between" the MPI library and the application. This is a powerful feature that is exploited in many applications and tools, such as the performance visualization tool Jumpshot [12]. Note that one of the major features of PMPI is that it allows selective replacement of MPI routines at link time without the need to re-compile or re-link the MPI implementation.

2.2   Adding Compression to MPI

The use of compression within MPI is not new, although it has been only used in specific ways for very few special cases. Major examples of such approaches include cMPI, PACX-MPI, COMPASSION, MiMPI, CoMPI, Adaptive-CoMPI.

PACX-MPI (PArallel Computer eXtension to MPI) [13,14] is an on-going project of the HLRS, Stuttgart. It enables an MPI application to run on a meta-computer consisting of several, possibly heterogeneous machines, each of which may itself be massively parallel. Compression is used for TCP message exchange among different systems in order to increase bandwidth, but a fixed compression algorithm is used and compression is not used for messages within single sub-system. *cMPI* [15,16] has similar goals o those of *PACX-MPI*, namely to enhance the performance of inter-cluster communication with a software-based data compression layer. Compression is added to all communication, so it does not offer any flexibility as to how to configure when and how to use compression.

COMPASSION [17] is a parallel I/O run-time system which includes chunking and compression for irregular applications. The LZO algorithm is used for fast compression and decompression, but again it is only used for the I/O part of irregular application.

MiMPI [18] is a prototype of a multithread implementation of MPI with thread-safe semantics that adds run-time compression of messages sent among nodes. Although the compression algorithm can be changed (providing more flexibility), the use of compression is global for all processes pertaining to an MPI application.

CoMPI [19] was the first work in which a compression library was fully integrated into MPICH. *CoMPI* is based on run-time compression of the MPI messages exchanged among applications. The user can choose the compression algorithm from a pool of algorithms, and all the communications will be compressed with the same algorithm. The problem with this approach is that the user can not always select the most suitable compression algorithm, and compression is always turned on by default.

Adaptive-CoMPI [10] allows for turning the compression on and off. It also selects the most appropriate compression algorithm at run-time. Although

Adaptive-CoMPI is independent of the application, it is dependent to the MPI implementation.

3   The Run-time-Adaptive Strategy

For the *Adaptive-CoMPI* technique [10] we developed two strategies, the *Run Time Strategy* (RAS) and the *Guided Strategy* (GS), to decide whether to apply compression or not on a message-by-message basis, as well as to decide which compression algorithm should be applied. The GS strategy makes these decisions by analysing the structure of the messages off-line. Once this selection process has been completed, the decisions are applied to the next executions of the same application with the same input parameters. In contrast to this, the RAS strategy makes these decisions at run-time, while the application is being executed. Because the GS strategy is not completely independent of the application, the RAS strategy has been chosen to be implemented in *PRAcTICaL-MPI*. With this in mind, we describe the main features of the RAS strategy in more detail.

As we explained RAS decides at run-time per message whether to compress a message before sending it or not, and which compression algorithm to apply. To make these two decisions, there are some cases in which RAS has to estimate the speedup, we will describe which ones these are late. To calculate this speedup, some network and compression information is needed. In order to provide RAS with this information, we have developed two modules:

- The Network Behavior module estimates the latency and bandwidth in order to predict the time needed to send a message, generating a network-behavior heuristics file for each installation.
- The Compression Behavior module selects the best compression algorithm depending on the message datatype and its redundancy level. Also, this model estimates the time needed to compress and decompress a message with different compression algorithms. Furthermore, it generates a compression-behavior heuristics file for each installation. This file is used to decide which algorithm to choose in order to compress a message depending on the message features.

These two modules have to be generated once per cluster, in order to obtain the heuristics files. Furthermore, the Network Behavior model also needs to be updated when there is a change in the network topology, to capture the new situation.

RAS uses length and datatype of the message, and the location of the processes to decide whether to compress or not and which compression algorithm to apply. RAS deactivates the compression when the processes involved in the communication are located in the same node. In other cases, when the processes are located in different nodes, RAS distinguishes between four kinds of datatype: Integer, floating-point, double precision floating-point, and "others" datatypes. The strategy analyses the four kinds of datatype separately and makes different decisions for each datatype. To choose the most appropriate algorithm for each datatype, RAS consults the compression-behavior heuristics taking the message features into account. Moreover, it builds a compression window for each datatype, with two adaptive thresholds that state from which minimum size to which maximum size a benefit can be achieved by compressing the data as shown in figure 1. Thus, RAS only estimates the speedup to send a message compressed when the size of the message is between both thresholds. To calculate the speedup, the information provided by the network-behavior and compression-behavior heuristics are used.

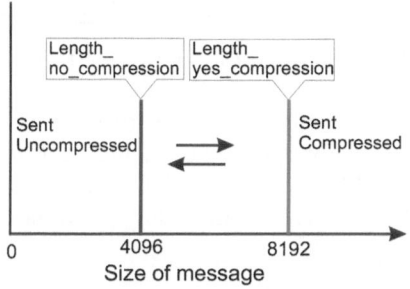

Fig. 1. Window compression

4   PRAcTICaL-MPI

The *PRAcTICaL-MPI* technique, is an optimization of MPI communications that exploits the MPI profiling interface (PMPI) to apply run-time lossless compression (and decompression), thus reducing the volume of communications. As Figure 2 shows, PMPI intercepts the MPI calls and wraps the *PRAcTICaL-MPI* technique around the actual MPI library invocation. *PRAcTICaL-MPI* is portable in the sense that it can be used with any MPI implementation, not just a with a specific MPI implementation. Besides, *PRAcTICaL-MPI* is transparent both to applications and MPI implementations, because it can be applied without changing their source code in any way.

We have built a library called *Practical*, where the most common routines of point-to-point and collective communications are wrapped inside a *PRAcTICaL-MPI* layer : MPI_Send, MPI_Isend, MPI_Bcast, MPI_Recv, MPI_Irecv, MPI_Wait, MPI_Waitall, MPI_Scatter, MPI_Gather. If we want to apply *PRAcTICaL-MPI* to another MPI communication, we only have to add a new wrapper to the respective routine in *Practical* library. The only requirement that *PRAcTICaL-MPI* makes is that the user needs to relink their applications with the *Practical* library to include our adaptive compression functionality.

Different compression algorithms are used depending on the specific characteristics of each communication. All compression algorithms have been included in a single library called Compression-Library. To include more compression

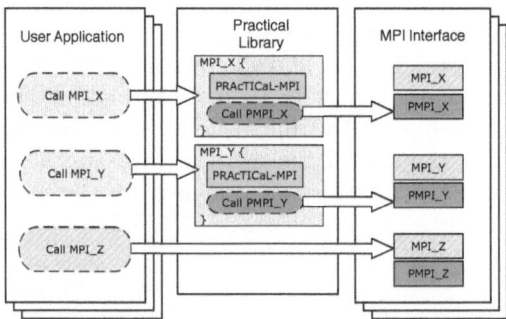

Fig. 2. *PRAcTICaL-MPI* architecture

algorithms, we only have to replace this library with a new version. Therefore, *PRAcTICaL-MPI* can be easily updated to include new compression algorithms. Currently, the compression library includes: RLE, Huffman, Rice8, Rice16, Rice32, rice8s, rice16s, rice32s, LZ, LZ77, LZ_f LZ77 Fast, Shannon-Fano, LZO, and FPC.

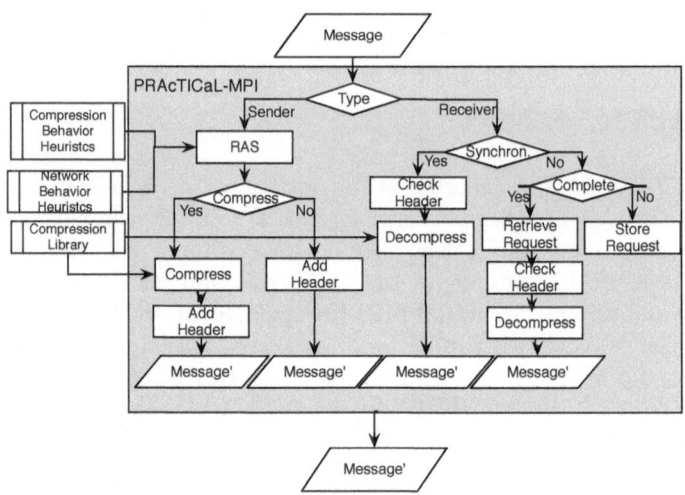

Fig. 3. *PRAcTICaL-MPI* schema

Figure 3 shows the internal workings of *PRAcTICaL-MPI* in more detail. The first step of the process is to identify which kind of operation has to be performed by the process that executes the MPI routine. If the process has to send data to other process, it is classified as a "sender". Otherwise, it is classified as a "receiver". For example, all the processes that execute a MPI_Send routine have to send data, so all these processes are classified as "senders". In the case of the MPI_Bcast routine, only the root process has to send data, and the others have to receive data. Therefore, only the root process is classified as "sender", and the rest of processes as "receivers". The reason for this classification is that *PRAcTICaL-MPI* takes different actions in each case. In the case of the "sender" type, *PRAcTICaL-MPI* tries to compress the message with the best algorithm possible. In the case of a "receive" operation, *PRAcTICaL-MPI* decompresses the message in case it was sent compressed.

More specifically, the actions performed by *PRAcTICaL-MPI* can be described as follows:

- Send Actions: Firstly, *PRAcTICaL-MPI* applies the RAS strategy to select the appropriate compression algorithm for the message depending on the location of the node and its datatype. Note that if the two processes involved in the communication are located in the same node, the message is sent without compression. Secondly, RAS compares the size of the message with the two adaptive thresholds corresponding to the datatype of the message. As a result of this operation, the decision to compress the message or not is taken. Thirdly, in case RAS decides to compres the data, the data is compressed and also the size of the compressed message is checked. If the size of the compressed data is larger than the original data, the original message is sent without compression. Otherwise, it is sent compressed. Finally, the method adds a header to the message in order to notify the receiver whether the message has to be decompressed and which decompression algorithm has to be used after receiving it.
- Receive Actions: The decompression operation is performed in two different places depending on whether message passing is synchronous or asynchronous. For asynchronous communication, such as MPI_Irecv, the decompression is performed only after message transfer is complete. Therefore, for the asynchronous receive routines, *PRAcTICaL-MPI* only stores the request pointer of the operation in a global table. Once reception has been completed, probably during the execution of MPI_Wait or MPI_Waitall, the request pointer is retrieved from the global table, and after this the decompression is performed. On the other hand, for synchronous communication, message decompression is performed when the receiver has received the complete message. To decompress a message, *PRAcTICaL-MPI* checks the header of the message in order to know whether the message has to be decompressed and which algorithm has to be employed. Finally, it applies the decompression algorithm indicated by the sender.

The ways in which the *PRAcTICaL-MPI* technique is applied depend on the characteristics of each routine. For example, in case of MPI_Send, first *PRAcTICaL-MPI* is applied to compress the data, and PMPI_Send is called after that. On the other hand, for MPI_Recv, the data is received with the PMPI_Recv routine first, and *PRAcTICaL-MPI* is applied to decompress the data after that.

5   Evaluation

We evaluate our approach using the BIPS3D application with different input meshes representing different semiconductor devices. We compare the performance of *PRAcTICaL-MPI* with the MPICH2.3 and XT-MPI distributions. The experiments were conducted using two different High-Performance Clusters called HECToR and EDDIE. We start with an overview of the BIPS3D application in section 5.1. Section 5.2 describes the HPC clusters used in our evaluation. The evaluation results themselves are presented in section 5.3.

5.1   The BIPS3D Application

BIPS3D is a 3-dimensional simulator of BJT and HBT bipolar devices described in [20]. The goal of the 3D simulation is to relate electrical characteristics of the device to its physical and geometrical parameters. The basic equations to be solved are Poisson equations and models describing electron and hole continuity in a stationary state.

Finite element methods are applied in order to discretize the Poisson equation, hole and electron continuity equations by using tetrahedral elements. The result is an unstructured mesh. In this work, we have used three different meshes, as described later.

Using the METIS library [21], the meshes are divided into sub-domains, in such a manner that one sub-domain corresponds to one process. The next step is decoupling the Poisson equation from the hole and electron continuity equations. They are linearized using the Newton method. Then we construct the part corresponding to the associated linear system for each sub-domain in a parallel manner. Each system is solved using domain decomposition methods. Finally, the results are written to a file.

For our evaluation BIPS3D has been executed using three different meshes: mesh1 (47200 nodes), mesh2 (732563 nodes) and mesh3 (289648 nodes). BIPS3D associates a data structure with each node of a mesh. The contents of these data structures constitute the data written to disk during the I/O phase. The number of elements that this structure has for each mesh entry is given by the *load* parameter. This means that, given a mesh and a load, the amount of data written to file is calculated as the product of the number of mesh elements and the load. In this work, we have evaluated our method using two different loads, 100 and 500.

5.2   HPC Clusters and MPI Implementations

We have performed our experiments on two different High-Performance Clusters in order to demonstrate how PRAcTICal-MPI adapts adapts itself to each architecture. In each cluster, a different MPI implementation is used. The main features of the clusters and MPI implementations used for our evaluation are:

1. HECToR is a Cray XT6 machine with contains 1856 nodes. Each node consists of two 12 core 2.1 GHz AMD opteron processors with 32 Gbytes of memory. The network used is Gemini interconnection. The MPI implementation used to perform our evaluation in this architecture is XT MPI 3.0.

2. EDDIE consists of 130 IBM dx360M2 iDataPlex servers with two Intel Westmere E5620 quad core processors and 24 GB of RAM, all connected through
Gigabit ethernet. MPICH2.3 is the MPI implementation used for our experiments on EDDIE. We chose this implementation as it is one of the most
popular MPI implementations.

5.3   Evaluation Results

We studied the performance of *PRAcTICaL-MPI* technique using the BIPS3D
application and two different clusters, HECToR and EDDIE. Figures 4 and 5
show the overall speedup achieved using *PRAcTICaL-MPI* for *mesh*1, *mesh*2,
*mesh*3 with two loads 100 and 500, and with 8, 16, 32, 64 and 128 processes,
respectively.

Each speedup shown in these diagrams is calculated by comparing the original MPI implementation (MPICH2.3 in Figure 4 and XT MPI 3.0 in Figure
5) with the same MPI implementation wrapped with *PRAcTICaL-MPI*. Then,
equation 1 is applied to these values. Values greater than one imply a reduction
of the overall execution time using *PRAcTICaL-MPI*.

$$Speedup = \frac{Execution\_time\_MPI\_Implementation}{Execution\_time\_MPI\_Implementation\_with\_PRAcTICaL} \qquad (1)$$

In general, the speedups achieved in 90% of the scenarios showed in Figures 4
and 5 are greater than or equal to one. These results are due to *PRAcTICaL-
MPI* applying run-time compression to reduce the volume of the messages with
the best algorithm per message, thus reducing execution time. Moreover, it deactivates the compression when it is not worth while applying any compression.
The original MPI distribution performs better only in 10% of all cases, but even
in those cases, the loss is nearly one in all of them.

The difference between the speedups achieved in the two scenarios is due to
the cluster architecture, i.e. network speed and number of cores per node. On
one hand, the EDDIE cluster (Figure 4) uses a Gigabit ethernet. This network
is slower than the Gemini network, used in HECToR (Figure 5). Due to the fact
that the network in HECToR is very fast, the compression is deactivated more
often, because is less worthwhile sending the message compressed and decom-
pressing it later than sending the message without compression. This behavior
can be observed in Figure 5(a) for a load of 500 and 8, 16, and 32 processes. In
these cases, the speedup is nearly one, because the compression is deactivated.
On the other hand, the cluster architecture affects also the results, too. EDDIE
has 8 cores per node, and HECToR has 12 cores per node. When the 12-core
architecture is used, the number of processes in the same node increases, and
therefore the number of communications between different nodes is lower than in
the 8-core architecture. This means that in HECToR, compression is deactivated
more times than in EDDIE. Therefore, we can observe how *PRAcTICaL-MPI*
is able to adapt to different architectures at run-time.

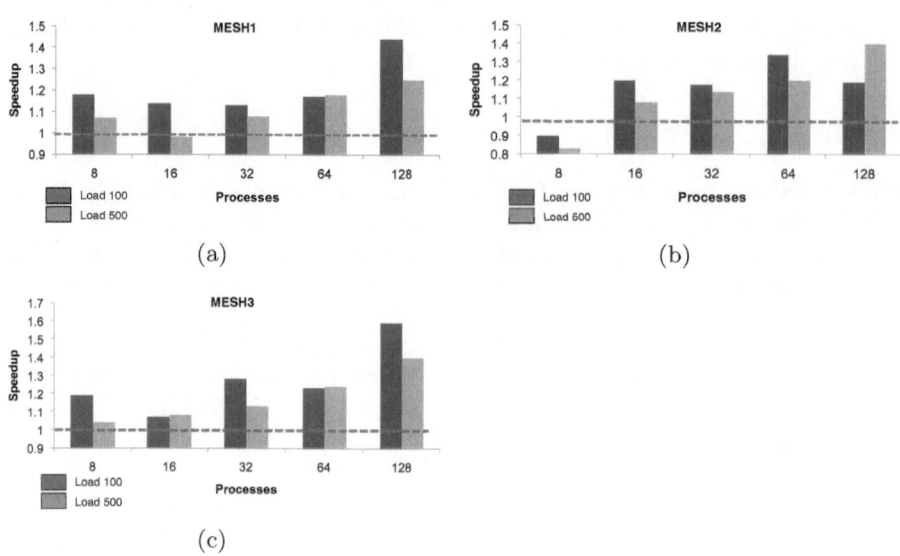

Fig. 4. Execution time improvement of BIPS3D on the EDDIE cluster: (a) Mesh1 (b) Mesh2 (c) Mesh3

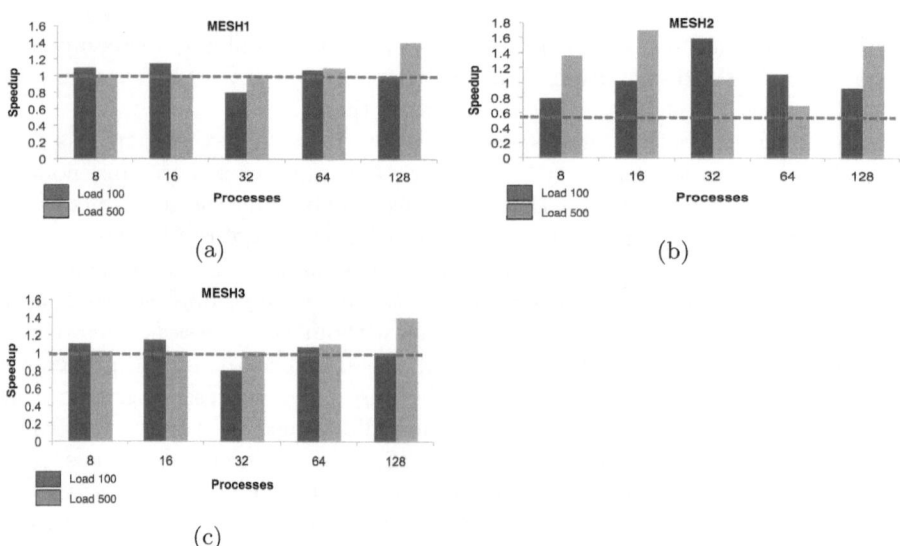

Fig. 5. Execution time improvement of BIPS3D in HECToR cluster: (a) Mesh1 (b) Mesh2 (c) Mesh3

Finally, we can notice that, the greater the number of processes, the bigger the application speedup achieved by *PRAcTICAL-MPI*. This behavior is due to the increasing number of communications. Therefore, the improvement of the communication performance has a bigger impact on the overall application performance. Thus, we can conclude that overal scalability is enhanced with *PRAcTICAL-MPI*.

6   Conclusions and Future Work

In this paper we have presented a portable optimization of MPI communications, called *PRAcTICAL-MPI*. The main goal of *PRAcTICaL-MPI* is to enhance the performance and scalability of MPI-based applications reducing the volume of communications by applying adaptive run-time lossless compression. Furthermore, *PRAcTICaL-MPI* is fully portable and transparent for both applications and MPI Implementations.

The evaluation results show that *PRAcTICAL-MPI* improves the speedup of BIPS3D for most of the scenarios considered, because the volume of communications is reduced by using the best compression algorithm per message. It also demonstrates that, even when compression is deactivated, application performance speedup is close to one. Furthermore, the run-time performance gain is bigger in most of the cases when more processes are employed, which increases scalability, and illustrates that our method will be most useful when utilised for massively parallel systems.

In future work, we want to evaluate the performance of *PRAcTICAL-MPI* technique with new compression algorithms like Snappy or PFOR. Furthermore, we want to apply *PRAcTICAL-MPI* to more MPI routines, such as collective IO, non-contiguous communications.

Acknowledgments. This work has been performed by using the facilities of HECToR, the UKs national high performance computing service, which is provided by UoE HPCx Ltd at the University of Edinburgh, Cray Inc and NAG Ltd, and funded by the Office of Science and Technology through EPSRCs High End Computing Programme.

References

1. Message Passing Interface Forum, MPI: A message-passing interface standard. International Journal of Supercomputer Applications 8, 165–414 (1994)
2. Gropp, W., Lusk, E., Doss, N., Skjellum, A.: A high-performance, portable implementation of the MPI message passing interface standard. Parallel Computing 22(6), 789–828 (1996)
3. Gabriel, E., Fagg, G.E., Bosilca, G., Angskun, T., Dongarra, J.J., Squyres, J.M., Sahay, V., Kambadur, P., Barrett, B., Lumsdaine, A., Castain, R.H., Daniel, D.J., Graham, R.L., Woodall, T.S.: Open MPI: Goals, Concept, and Design of a Next Generation MPI Implementation. In: Kranzlmüller, D., Kacsuk, P., Dongarra, J. (eds.) EuroPVM/MPI 2004. LNCS, vol. 3241, pp. 97–104. Springer, Heidelberg (2004)

4. Burns, G., Daoud, R., Vaigl, J.: LAM: An open cluster environment for MPI. In: Proceedings of Supercomputing Symposium 1994 (1994)
5. Zigon, R.: Run length encoding. Dr. Dobb's Journal of Software Tools 14(2) (February 1989)
6. Knuth, D.E.: Dynamic huffman coding. J. Algorithms 6(2), 163–180 (1985)
7. Salvatore Coco, D.G., D'Arrigo, V.: A Rice-based Lossless Data Compression System For Space. In: Proceedings of the 2000 IEEE Nordic Signal Processing Symposium, pp. 133–142 (2000)
8. Burtscher, M., Ratanaworabhan, P.: FPC: A High-Speed Compressor for Double-Precision Floating-Point Data. IEEE Transactions on Computers 58(1), 18–31 (2009)
9. Oberhumer, M.F.X.J.: Lzo real-time data compression library (2005)
10. Filgueira, R., Carretero, J., Singh, D.E., Calderon, A., Garcia, F.: Adpative-compi: Enhancing mpi based applications performance and scalability by using adaptive compression. International Journal of High Performance Computing and Applications (April 2010)
11. Schulz, M., de Supinski, B.R.: A flexible and dynamic infrastructure for mpi tool interoperability. In: Proceedings of the 2006 International Conference on Parallel Processing, ICPP 2006, pp. 193–202. IEEE Computer Society, Washington, DC (2006), http://dx.doi.org/10.1109/ICPP.2006.6
12. Zaki, O., Lusk, E., Swider, D.: Toward scalable performance visualization with Jumpshot. High Performance Computing Applications 13, 277–288 (1999)
13. Balkanski, D., Trams, M., Rehm, W.: Heterogeneous Computing With MPICH/Madeleine and PACX MPI: A Critical Comparison (2003)
14. Keller, M.L.R.: Using PACX-MPI in metacomputing applications. In: 18th Symposium Simulationstechnique, Erlangen, September 12-15 (2005)
15. Ratanaworabhan, P., Ke, J., Burtscher, M.: Fast Lossless Compression of Scientific Floating-Point Data. In: DCC 2006: Proceedings of the Data Compression Conference, pp. 133–142. IEEE Computer Society, Washington, DC (2006)
16. Ke, J., Burtscher, M., Speight, E.: Runtime Compression of MPI Messages to Improve the Performance and Scalability of Parallel Applications. In: SC 2004: Proceedings of the 2004 ACM/IEEE Conference on Supercomputing, p. 59. IEEE Computer Society, Washington, DC (2004)
17. Carretero, J., No, J., Park, S.-S., Choudhary, A., Chen, P.: COMPASSION: a Parallel I/O Runtime System Including Chunking and Compression for Irregular Applications. In: Sloot, P., Bubak, M., Hertzberger, B. (eds.) HPCN-Europe 1998. LNCS, vol. 1401, pp. 668–677. Springer, Heidelberg (1998)
18. Garíca, F., Galderón, A., Carretero, J.: MiMPI: A Multithread-Safe Implementation of MPI. In: Margalef, T., Dongarra, J., Luque, E. (eds.) PVM/MPI 1999. LNCS, vol. 1697, pp. 207–214. Springer, Heidelberg (1999)
19. Filgueira, R., Singh, D.E., Calderón, A., Carretero, J.: CoMPI: Enhancing MPI Based Applications Performance and Scalability Using Run-Time Compression. In: Ropo, M., Westerholm, J., Dongarra, J. (eds.) PVM/MPI. LNCS, vol. 5759, pp. 207–218. Springer, Heidelberg (2009)
20. Loureiro, A., González, J., Pena, T.F.: A parallel 3D semiconductor device simulator for gradual heterojunction bipolar transistors. Int. Journal of Numerical Modelling: Electronic Networks, Devices and Fields 16, 53–66 (2003)
21. Karypis, G., Kumar, V.: Metis a software package for partitioning unstructured graphs, partitioning meshes, and computing fill-reducing orderings of sparse matrices. Tech. Rep. (1998)

Cost-Effective Contention Avoidance
in a CMP with Shared Memory Controllers[*]

Samuel Rodrigo[1], Frank Olaf Sem-Jacobsen[1], Hervé Tatenguem[3],
Tor Skeie[1,2], and Davide Bertozzi[3]

[1] Simula Research Laboratory, Norway
srodrigo@simula.no
[2] Dept. of Informatics, University of Oslo
tskeie@simula.no
[3] Dept. Engineering, University of Ferrara
ttnhrv@unife.it

Abstract. Efficient CMP utilisation requires virtualisation. This forces
multiple applications to contend for the same network resources and
memory bandwidth. In this paper we study the cause and effect of network congestion with respect to traffic local to the applications, and
traffic caused by memory access. This reveals that applications close to
the memory controller suffer because of congestion caused by memory
controller traffic from other applications. We present a simple mechanism
to reduce head-of-line blocking in the switches, which efficiently reduces
network congestion, increases network performance, and evens out the
performance differences between the CMP applications.

1 Introduction

The access to the off-chip memory in large chip multiprocessors (CMPs) based on
a switched interconnect (NoC) consumes a significant portion of the bandwidth
in the on-chip network. Furthermore, this traffic is targeted towards specific areas of the chip where the memory controllers are connected. Previous studies [1]
show that the placement of these memory controller connections have a significant impact on the network load, but the authors do not study how this impacts
the performance of the applications themselves and the complex interaction between the *local traffic* (caused by cache coherency protocols) and the *memory
controller traffic*. The initial intuitive understanding of the effect of memory
controller access point placement on application performance is that the applications located closest to the memory controller access points will experience
better performance compared to applications allocated further away [4]. However, we show that applications that reside close to the memory controller might
be more severely affected by the interplay between local traffic and memory
controller traffic.

[*] This work has been supported by the project NaNoC (grant agreement no. 248972)
which is funded by the European Commission within the Research Programme FP7.

C. Kaklamanis et al. (Eds.): Euro-Par 2012, LNCS 7484, pp. 741–752, 2012.

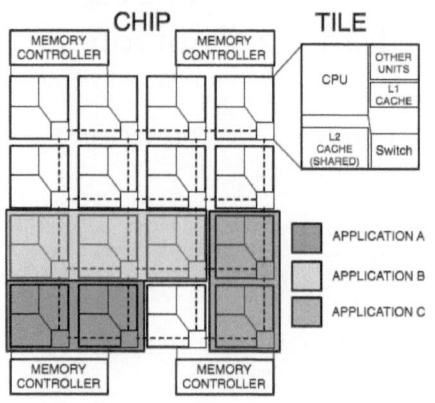
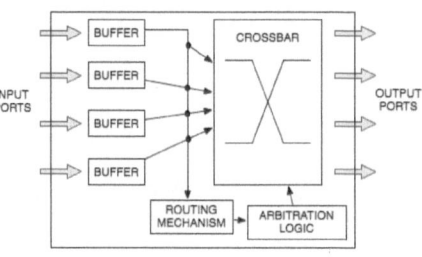

(a) CMP tile-based design with dynamic application domains

(b) Basic switch architecture

Fig. 1. CMP, core, and switch layout

To ease development, the tiles in a CMP are usually homogeneous, with a structure as displayed in Figure 1a. Every tile has a private level I and (usually) a shared level II cache, together with the processing core and a switch to access the on chip network. Off-chip memory (DRAM) is accessed through one or more memory controllers connected to the on-chip network, usually at the edge of the chip, through one or more ports. The network on chip carries cache coherency traffic between the level I and level II caches, and memory access traffic to and from the memory controller. The two traffic types may or may not be divided into two virtual networks (using virtual channels). The interaction between these two traffic types is the core of the issue we study in this paper. Local traffic from one application (cache coherency traffic) should not interfere with the local traffic from other applications, as application isolation is a core concept of CMP virtualisation [20]. Most applications will, however, be affected by the memory controller traffic from other applications.

In this paper we study how and to what extent the local and memory controller traffic contribute to network congestion and how this affects application performance. Based on this study we present a mechanism to reduce head-of-line blocking, and thus network congestion, both with and without virtual channels.

The structure of this paper is as follows: Section 2 presents the NoC background and the related work. In Section 3 we describe the congestion problem in on-chip networks and the causes behind it, and we present our congestion control solution in Section 4. Next, in Section 5, we detail the evaluation scenario and the results obtained, and finally in Section 6, we present some conclusions and future work.

2 NoC Background and Related Work

There is significant ongoing research to study how application mapping and basic properties of virtualisation is related to network performance. In [20], the importance of traffic isolation and contiguous application mapping is presented to demonstrate the foundations of virtualisation. Das et al. [4] study application mapping mechanisms, and show that memory intensive applications should be located close to the memory controller, but the authors do not study the impact of application traffic and memory controller traffic on the mapping. To simplify such mapping problems, Abts et al. [1] study alternative memory controller placements by moving the memory controller access points towards the centre of the chip. This breaks the regularity of the chip, both in design and routing, so further study is required before these strategies may realistically be employed. Finally, Sánchez et al. [18] describe how different NoC topologies can impact application performance, but do not consider the location of the application relative to other applications and the memory controllers in the CMP.

There are some solutions in [5, 9, 10] that attempt to reduce the negative effect of shared resources through quality of service (QoS) based on priority schemes. Although all these solutions can alleviate network congestion by prioritising different traffic types, their objective is to differentiate the traffic and they do not focus on the congestion problem itself. As a consequence, there may be congestion within each traffic class for unpredictable traffic patterns.

A number of solutions for CMP NoCs are presented in [8, 11, 12, 22]. The authors describe mechanisms that collect congestion information from the neighbouring nodes through the routing process and buffer ingress/egress monitoring. The idea is to offer an alternative path to route around a congested area of the chip. However, this assumption will impact negatively creating more congested resources, as it is impossible to avoid the congested region if all the congested traffic has the same target (the memory controller). Van den Brand et al. [2] and Thottethodi et al. [19] rely on a central controller to gather congestion information from the network. Whereas the former uses a guaranteed service traffic class to propagate congestion notifications to the sources, the latter uses a separate control network for this purpose. Neither solution offers great scalability because of the centralisation.

There is still an ongoing field of research on many aspects related to congestion management in NoCs for CMPs, but we have found that the basic congestion problem in a virtualised CMP is not well understood. Therefore, our objective with this paper is to present a study on how congestion problems arise in the event of many concurrent applications with shared resources (memory controllers). This work serves as a motivation and guide in the search for cost-effective resource management solutions, and we present a solution to deal with the congestion problems in these scenarios.

3   NoC Congestion

In this section we describe the concept of network on chip contention and how this leads to network congestion. Furthermore, we examine the relationship between the local traffic and memory controller traffic in order to determine how this will affect application performance based on its location on the chip relative to the memory controller access points.

3.1   NoC Contention

Whenever a NoC packet enters a switch, it is buffered, and the header information is read to determine the output port for the packet from the switch (see Figure 1b). The packet must then wait until it reaches the head of the queue and the appropriate output port is available (i.e. not receiving a packet from another port in the switch and not blocked by flow control). NoCs employ *flow control* to ensure that packets are never dropped by guaranteeing that there is buffer space available in the next hop switch before forwarding the packet across the link [6]. Since multiple packets in different input ports in the switch may have the same output port, a given packet might have to wait for several scheduling rounds (output port contention) before it is allowed access to the switch crossbar and can continue. During this time there may be other packets in the same buffer with different output ports that are available. However, these packets cannot proceed because they are blocked by the first packet in the queue. This is known as *head-of-line blocking.* ·

Together with head-of-line blocking, the flow control mechanism causes congestion trees to build up in the network. Whenever a packet is blocked, it will block packets upstream, gradually expanding the congestion tree branches from the tree root through this back-pressure mechanism. The root is the switch without enough capacity to forward all incoming packets (the place where the packets are first blocked). For the memory controller traffic, congestion tree roots will typically be the cores that are the memory controller access points. Output port contention and head-of-line blocking combined with the back-pressure caused by the flow control mechanism leads to network congestion at high network load, which has a significant impact on the performance of the NoC.

3.2   Application Performance Relative to Memory Controller Location

When using virtualisation to support multiple concurrent applications on a CMP, the two traffic types (local traffic and memory controller traffic) may or may not be separated into two different virtual networks using virtual channels. If all the traffic runs on the same virtual channel (i.e. one virtual channel) it is obvious that applications that suffer congested transit memory controller traffic will experience congestion in the local traffic as well. However, using two virtual networks allows a separation of the traffic which reduces the interaction between the two traffic types.

With two virtual channels, each channel is typically guaranteed 50% of the physical channel bandwidth. Consequently, as long as neither of the two traffic types have a demand greater than 50% of the channel bandwidth, there is no significant interaction between the traffic. However, most applications have a larger amount of local traffic than memory controller traffic. Thus, the applications that are located far away from the memory controller and have little transit memory controller traffic, the application is free to use more than 50% of the bandwidth for local traffic. For the applications located closer to the memory controller the amount of transit memory controller traffic increases drastically, which reduces the effective local traffic down to max 50% and may introduce congestion problems for the local application traffic. Consequently, applications located closer to the memory controller will exhibit worse performance. This contradicts previous studies which concluded that applications close to the memory controller had better performance [3, 4], and we clearly see this effect in the evaluation section.

This discussion has shown that even though a large degree of traffic isolation can be achieved using virtual channels, there is still interaction which can adversely affect application performance as we will see in the evaluation section (Section 5). We will also see that not separating the traffic has even more adverse effects on application performance. Efficient resource management in terms of congestion control is therefore required, both to increase overall efficiency of the chip and fairness between the running applications.

4   NoC Congestion Control

For congestion management, we propose HACS (Head-of-line Avoidance Congestion Skip-ahead), a head-of-line blocking observation mechanism that allows buffered packets to bypass the packet that is at the head of the queue. The core mechanism is presented in Figure 2. Note that this mechanism is supported under virtual cut-through packet switching. Whenever a packet is stalled for a given time period at the head of a buffer, HACS will search further back in the queue for the first packet that is routed to a free output port, because of a different destination, and let this skip to the head of the queue. This effectively reduces head-of-line blocking with the result of reduced congestion.

We now discuss the implementation of HACS in xpipesLite [6]. All packets are assumed to be 4 flits long by padding shorter ones and by splitting longer messages into multiple packets. The network guarantees in-order delivery of packets headed to the same destination. An arbiter is instantiated for each output port to perform round robin arbitration among all inputs with valid asserted and presenting a head flit. The switch implements the LBDR mechanism [17].

Assume two packets "A" and "B" are stored in an input buffer (see Figure 2), and let the arbiter of the output port requested by "A" be stalled (blocked), thus preventing packet forwarding. In the HACS switch, a timer is activated upon snooping such a stall condition. If the stall signal changes during the countdown, the timer will be reset till the generation of the next stall. If the stall is still high at the time-out, the control logic shifts the read pointer to the head

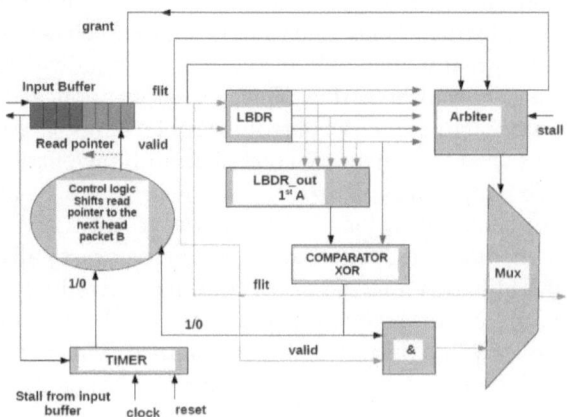

Fig. 2. Switch architecture

flit of the second packet. Before computing the destination of "B", the LBDR routing logic saves the destination of "A" in backup registers (Lbdr-out-A in the figure) for further comparison with the target output of packet "B". A XOR comparator compares the two target ports required by "A" and "B". If they are different and the port requested by "B" is available, the stall goes down and "B" is forwarded. If not, the read pointer shifts once again to packet "A" till the stall is deasserted. If the stall signal is removed while performing the target port comparison, the valid signal is driven low to avoid sampling "B" before "A", thus preserving the order on each output port. In this unfortunate and very unlikely case, the switch experiments one (1) cycle overhead before being able to forward "A". HACS can also be applied to each virtual layer of a network with virtual channel support. As previously illustrated, 2 VCs may be considered for memory controller traffic separation. In practical terms, we followed the strategy proposed in [7]. Essentially, the HACS switch is replicated twice, while placing a demux in front of each input port and a mux with associated arbiter after each output port. The link is enhanced with a virtual channel identifier and with a flow control signal for each virtual channel. This is done to exploit logic synthesis optimisations for the sake of area efficiency.

5   Evaluation

In this section we first describe the simulation environment we used for the evaluations, followed by a discussion of the results obtained, including the results obtained with HACS and its implementation costs after synthesis.

5.1   System Configuration

Our simulation framework is a combination of tools chosen to simulate a CMP system as closely as possible. Multi2sim [21] is a simulation framework for heterogeneous computing that allows one or more applications to run on top of it in CMP-like scenarios. It is able to model a complete memory hierarchy system integrated into the CMP and its connection to the respective processor cores. We combined Multi2sim with a cycle-accurate flit-level network-on-chip simulator called gNoCsim (developed by *Universidad Politécnica de Valencia*, and being used in the NaNoC project [15] by different partners). gNoCsim is able to simulate the network between all the resources in the chip; caches, memory controllers, and processor cores.

For the evaluation process, we modelled a CMP that resembles current chip configurations like in Figure 1a. This configuration implements a tile-based system, and each tile is composed of a processor core, a private L1 cache, a bank of a L2 shared cache, a memory directory bank to be used with the directory-based MOESI cache coherency protocol, and different configurations of memory controllers. Each memory controller is connected to the main memory with 2 channels (each memory controller has two access points). A detailed overview of the chip configuration is shown in Table 1.

Table 1. CMP configuration

Parameter	Configuration	Parameter	Configuration
Core	x86	Topology	10 × 10 2-D mesh
L1 cache	16 KBytes Instructions 16 KBytes Data Total 32 KBytes per core 2 cycles latency 2-way associativity 64 bytes block size	Routing mechanism	LBDR + SR
L2 cache	256 Kbytes per core 20 cycles latency 4-way associativity 64 bytes block size	Packet switching	Virtual cut-through (VCTlite) [16]
Main memory	1 Gbyte total 200 cycles latency	Buffer queue size	12 flits
Coherence protocol	MOESI CMP, directory-based	Flit-size	8 bytes

A 10 × 10 2-D regular mesh topology was used for the CMP system. The LBDR [17] mechanism was used for the routing purposes allowing for routing-contained application domains in combination with the Segment-Based Routing algorithm (SR) [13]. Virtual networks are used for different levels of traffic of the memory hierarchy system, implemented as multiple virtual channels (a total of two virtual channels are used) except for Figure 4b were no virtual channels are used.

For the evaluations we used a collection of applications from the SPLASH-2 benchmark with the default parameters defined in [14]. The applications are statically mapped to the chip when the experiment is set up. Applications are mapped to completely fill the chip, giving a fair share of cores to each application. Every batch consists of a single application type from the benchmark suite rather than being composed of a collection of mixed applications. This regularity makes it significantly easier to generate relevant statistics and spot trends in the results, such as to get averaged results for the execution time comparison. Running a mix of applications will introduce spikes in the communication, but this will be evened out by the number of applications over time, so the conclusions will still be the same. See Figure 3 for an example of the mapping of 32 concurrent applications with 4 memory controllers.

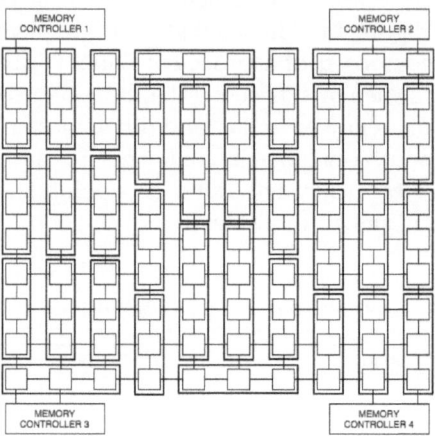

Fig. 3. 32 concurrent applications mapped on the system

5.2   Results

We have evaluated several combinations of number of concurrent applications and memory controllers for a 10×10 mesh. Specifically, we have evaluated 6, 8, and 12 applications with one memory controller, 12 and 16 applications with two memory controllers, and 32 applications with four memory controllers. Due to space constraints we report the results for 12 applications with one memory controller and 32 applications with four memory controllers. The general trend from the results is that network performance decreases and unfairness (the difference in runtime based on application location, with two virtual channels) increases as the number of concurrent applications increases for a given number of memory controllers.

We have plotted the execution time distribution for 12 applications (ocean workload) with a single memory controller both with (Figure 4a) and without (Figure 4b) virtual channels. The memory controller is located in the uppermost corner. The figure with virtual channels clearly shows how the threads that are located closer to the memory controller have a longer execution time (as much as 7.5% longer than when running alone) than the thread is located farther away. The picture is more chaotic without the use of virtual channels. There is no clear unfairness, however, the overall increase in execution time (8.1%) is larger than with virtual channels.

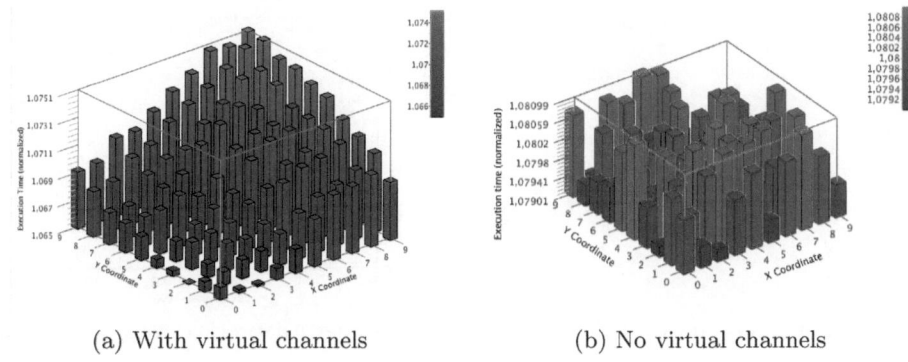

(a) With virtual channels (b) No virtual channels

Fig. 4. Execution time distribution, 1 memory controller, ocean workload

In Figure 5a we display the mean squared error between injected and accepted traffic for different applications in the scenario with 32 concurrent applications with 4 memory controllers, with and without HACS implemented. In the figure, HACS2 is allowed to skip ahead the second packet, while HACS3 may skip ahead the second or third packet. The figure shows the impact of performance degradation due to congestion for the different applications. HACS2 and HACS3 are able to reduce the penalty to only 2.5% in average. Note that there is negligible difference between HACS2 and HACS3. The performance degradation is significantly worse with fewer memory controllers.

Figure 5b shows network throughput as a function of time for the *ocean* workload. The uppermost plot is the injected traffic, and the bottom plot is the accepted traffic without congestion control, a clear case of a congested network. HACS2 and HACS3 solutions almost remove all the congestion, handling almost all the injected traffic. The second to bottom line is HACS without virtual channels. This still increases averaged network throughput by around 40%, although the result is poorer than with virtual channels. The designer has to assess the trade-off between performance and implementation costs.

Summarising, the objective of these evaluation cases was to reproduce scenarios that try to reflect current chip configurations, and realistically illustrate the effect of multiple simultaneous applications. The cost/performance trade-off depends on how much resources are available (in our case, the amount of memory controllers) and there is a need for congestion management strategies that can alleviate the problem with minimal impact on the design of the chip. In the next section we evaluate the cost of the congestion control mechanism we have developed.

5.3   Hardware Breakdown

This subsection characterises area and critical path delay overhead of the HACS switch compared to a baseline one taken from the xpipesLite NoC library [6]. The reference switch implements input buffering, stall/go flow control and virtual cut-through switching.

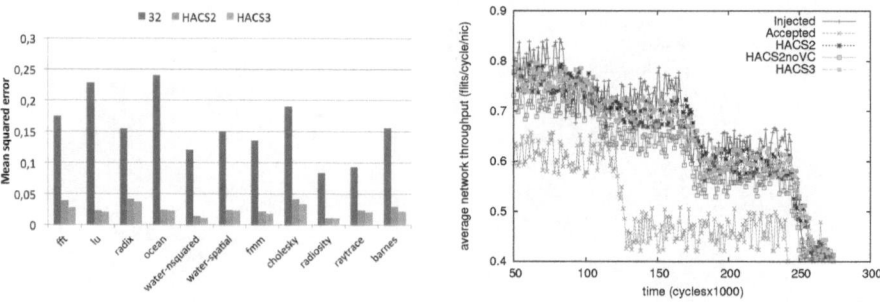

(a) MSE between injected and accepted traffic

(b) Averaged network throughput, ocean workload

Fig. 5. Results for 32 concurrent applications with 4 memory controllers

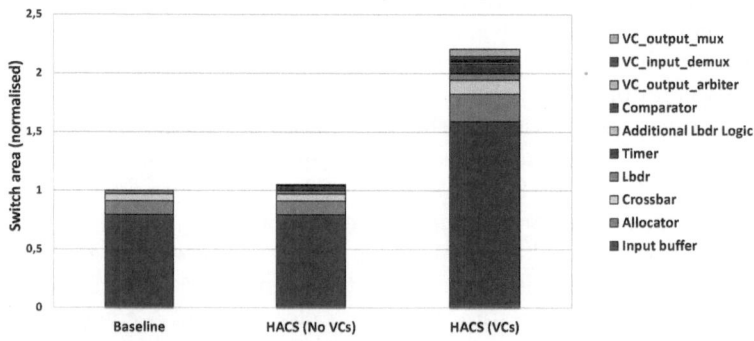

Fig. 6. Switch area at 500 MHz

Baseline and HACS switches were synthesised for a target speed of 500 MHz in a 65nm industrial technology library. Normalised post-synthesis area results are illustrated in Figure 6. The area of the HACS switch without virtual channels is about 5.34% larger compared to the baseline switch. The implementation with virtual channels results in 2.09x the area of the virtual channel-less switch. Observe in the figure that the baseline input buffer features approximately the same area of the input buffer with the new logic to shift the read and write pointers, thus denoting the marginal impact on control logic. On the other hand, most of the area overhead is due to the timer inserted in the switch and is about 4.09%. This could be improved in future solutions by using buffer thresholds instead of a timer.

After synthesis, the critical path of the new switch without VCs was proved to be degraded by less than 1% with respect to the baseline one. The virtual channel implementation contributes an additional 3% of critical path degradation, associated with the arbiters in the switch output ports selecting which virtual channel to move forward.

6   Conclusion

We have studied the effects of shared memory access in a CMP with multiple concurrent applications. It is often assumed that network congestion is not an issue for CMP systems because of the abundant bandwidth in the network on chip. Our evaluations show that network congestion may indeed be a problem when multiple applications access shared memory through the memory controllers available on a typical CMP, and we developed a simple solution, HACS, to remove this congestion.

We have observed some effects from our experimental results. First, the hotspots formed by the memory controller traffic lead to network congestion, which can degrade the performance of applications by 15% in average in our scenarios. Second, if there is a high degree of local traffic (which is often the case), the applications allocated close to the memory controller will have less bandwidth for local traffic than applications located further away. The applications closest to the memory controllers are therefore penalised and have longer execution times compared to the others. Further work includes evaluating a wide variety of network controller configurations and workloads.

References

1. Abts, D., Enright Jerger, N.D., Kim, J., Gibson, D., Lipasti, M.H.: Achieving predictable performance through better memory controller placement in many-core CMPs. ACM SIGARCH Computer Architecture News 37(3), 451 (2009), http://portal.acm.org/citation.cfm?doid=1555815.1555810
2. van den Brand, J., Ciordas, C., Goossens, K., Basten, T.: Congestion-Controlled Best-Effort Communication for Networks-on-Chip. In: 2007 Design, Automation & Test in Europe Conference & Exhibition, pp. 1–6 (April 2007), http://ieeexplore.ieee.org/lpdocs/epic03/wrapper.htm?arnumber=4211925
3. Chen, G., Li, F., Son, S.W., Kandemir, M.: Application mapping for chip multi-processors. In: Proceedings of the 45th Annual Conference on Design Automation - DAC 2008, p. 620 (2008), http://portal.acm.org/citation.cfm?doid=1391469.1391628
4. Das, R., Mutlu, O., Kumar, A., Azimi, M.: Application-to-core mapping policies to reduce interference in on-chip networks. Tech. rep., SAFARI Technical Report No. 2011 (2011), http://www.ece.cmu.edu/ omutlu/pub/interference-aware-noc-mapping-TR-SAFARI-2011-001.pdf
5. Das, R., Mutlu, O., Moscibroda, T., Das, C.R.: Application-aware prioritiza-tion mechanisms for on-chip networks. In: Proceedings of the 42nd Annual IEEE/ACM International Symposium on Microarchitecture - Micro-42, p. 280 (2009), http://portal.acm.org/citation.cfm?doid=1669112.1669150
6. Flich, J., Bertozzi, D.: Designing Network On-Chip Architectures in the Nanoscale Era. Chapman & Hall/CRC (2010)
7. Gilabert, F., Gómez, M.E., Medardoni, S., Bertozzi, D.: Improved utilization of noc channel bandwidth by switch replication for cost-effective multi-processor systems-on-chip. In: Proceedings of the 2010 Fourth ACM/IEEE International Sympo-sium on Networks-on-Chip, NOCS 2010, pp. 165–172. IEEE Computer Society, Washington, DC (2010), http://dx.doi.org/10.1109/NOCS.2010.25
8. Gratz, P., Grot, B., Keckler, S.W.: Regional congestion awareness for load balance in networks-on-chip. In: HPCA, pp. 203–214. IEEE Computer Society (2008)

9. Grot, B., Keckler, S.W., Mutlu, O.: Preemptive virtual clock: a flexible, efficient, and cost-effective QOS scheme for networks-on-chip. In: Proceedings of the 42nd Annual IEEE/ACM International Symposium on Microarchitecture, pp. 268–279. ACM (2009), http://portal.acm.org/citation.cfm?id=1669149

10. Iyer, R., Zhao, L., Guo, F., Illikkal, R., Makineni, S., Newell, D., Solihin, Y., Hsu, L., Reinhardt, S.: QoS policies and architecture for cache/memory in CMP platforms. ACM SIGMETRICS Performance Evaluation Review 35(1), 25 (2007), http://portal.acm.org/citation.cfm?doid=1269899.1254886

11. Li, M., Zeng, Q.-A., Jone, W.-B.: DyXY: a proximity congestion-aware deadlock-free dynamic routing method for network on chip. In: Proceedings of the 43rd Annual Design Automation Conference, DAC 2006, pp. 849–852. ACM, New York (2006), http://doi.acm.org/10.1145/1146909.1147125

12. Marescaux, T., Rangevall, A., Nollet, V., Bartic, A., Corporaal, H.: Distributed congestion control for packet switched networks on chip. In: Proceedings of the International Conference of Parallel Computing: Current Future Issues of High-End Computing, vol. 33, pp. 761–768. Citeseer (2005), http://citeseerx.ist.psu.edu/viewdoc/download?doi=10.1.1.89.1586&rep=rep1&type=pdf

13. Mejía, A., Flich, J., Duato, J., Reinemo, S.A., Skeie, T.: Segment-based routing: An efficient fault-tolerant routing algorithm for meshes and tori. In: International Parallel and Distributed Processing Symposium, p. 84 (2006)

14. Multi2sim Wiki: SPLASH–2 execution commands., http://www.multi2sim.org/wiki/index.php5/SPLASH2_Execution_Commands

15. NaNoC: NaNoC design platform, http://www.nanoc-project.eu

16. Roca, S., Flich, J., Silla, F., Duato, J.: VCTlite: Towards an efficient implementation of virtual cut-through switching in on-chip networks. In: International Conference on High Performance Computing (HiPC), pp. 1–12 (2010)

17. Rodrigo, S., Flich, J., Roca, A., Medardoni, S., Bertozzi, D., Camacho, J., Silla, F., Duato, J.: Addressing manufacturing challenges with cost-efficient fault tolerant routing. In: NOCS 2010: Proceedings of the 4th ACM/IEEE International Symposium on Networks-on-Chip, pp. 25–32 (2010)

18. Sanchez, D., Michelogiannakis, G., Kozyrakis, C.: An analysis of on-chip interconnection networks for large-scale chip multiprocessors. ACM Transactions on Architecture and Code Optimization (TACO) 7(1), 4 (2010), http://portal.acm.org/citation.cfm?id=1736069

19. Thottethodi, M., Lebeck, A., Mukherjee, S.: Self-tuned congestion control for multiprocessor networks. In: The Seventh International Symposium on High-Performance Computer Architecture, HPCA, pp. 107–118. IEEE (2001), http://ieeexplore.ieee.org/xpls/abs_all.jsp?arnumber=903256

20. Triviño, F., Sánchez, J.L., Alfaro, F.J., Flich, J.: Virtualizing network-on-chip resources in chip-multiprocessors. Microprocessors and Microsystems 35(2), 230–245 (2011), http://linkinghub.elsevier.com/retrieve/pii/S0141933110000712

21. Ubal, R., Sahuquillo, J., Petit, S., López, P.: Multi2Sim: A Simulation Framework to Evaluate Multicore-Multithreaded Processors. In: Proc. of the 19th Int'l Symposium on Computer Architecture and High Performance Computing (2007)

22. Wu, D., Al-Hashimi, B.M., Schmitz, M.T.: Improving routing efficiency for network-on-chip through contention-aware input selection. In: Proceedings of the 2006 Asia and South Pacific Design Automation Conference, ASP-DAC 2006, pp. 36–41. IEEE Press, Piscataway (2006), http://dx.doi.org/10.1145/1118299.1118310

Topic 14: Mobile and Ubiquitous Computing

Paolo Santi, Sotiris Nikoletseas, Cecilia Mascolo, and Thiemo Voigt

Topic Committee

The tremendous advances in wireless networks, mobile computing, and sensor networks, along with the rapid growth of small, portable and powerful computing devices, offers more and more opportunities for pervasive computing and communications. This topic deals with cutting-edge research in various aspects related to the theory and practice of mobile computing or wireless and mobile networking. These aspects include architectures, algorithms, networks, protocols, modeling and performance issues, data management, and novel applications and services. The aim of this topic is to bring together computer scientists and engineers from both academia and industry working in this exciting and emerging area of pervasive computing and communications, to share their ideas and results with their peers.

After careful selection, two papers have been selected for this topic. The first paper falls within the emerging area of wireless sensor networks, and proposes a new clustering algorithm for improving energy efficiency when gathering data from the sensor network by means of a mobile collector. The main idea of the clustering algorithm is borrowed from the image processing field, and is based on the notion of watershed transformation which is used to select clusterheads within the network. After clusterhead selection, a mobile sink node periodically visits the clusterhead nodes to collect data. The proposed clustering method is shown by means of simulations to significantly outperform existing approaches in terms of extended network lifetime. The second paper lies within the realm of mobile network modeling, which is also perceived as a very important topic within the mobile computing and networking community. In particular, the authors of the paper analyze for the first time a property of a mobile network called "liveness", which can be informally defined as absence/presence of relatively long disconnection periods during the network lifetime. To analyze "liveness", the authors perform several simulations using different mobility models, including both synthetic models and GPS traces collected from real-world experiments. The analysis discloses interesting insights that might turn useful for the design of mobile networking protocols, such as that the "liveness" property of a network does not depend on the speed of nodes, but on other parameters such as node density.

C. Kaklamanis et al. (Eds.): Euro-Par 2012, LNCS 7484, p. 753, 2012.
© Springer-Verlag Berlin Heidelberg 2012

Watershed-Based Clustering
for Energy Efficient Data Gathering
in Wireless Sensor Networks with Mobile Collector[*]

Charalampos Konstantopoulos[1], Basilis Mamalis[2],
Grammati Pantziou[2], and Vasileios Thanasias[2]

[1] Department of Informatics, University of Piraeus, Greece
[2] Department of Informatics, Technological Educational Institute of Athens, Greece
konstant@unipi.gr, {vmamalis,pantziou,cs061110}@teiath.gr

Abstract. This paper presents a clustering protocol combined with a mobile sink (MS) solution for efficient data gathering in wireless sensor networks (WSNs). The main insight for the cluster creation method is drawn from image processing field and namely from the watershed transformation which is widely used for image segmentation. The proposed algorithm creates multi-hop clusters whose clusterheads (CHs) as well as all cluster members near the CHs have high energy reserves. As these are exactly the nodes most burdened with relaying of data from other cluster members, the higher levels of available energy at these nodes prolong the network lifetime eventually. After cluster creation, a MS periodically visits each CH and collects the data from cluster members already gathered at the CH. Simulation results show the higher performance of the proposed scheme in comparison to other competent approaches from the literature.

Keywords: mobile sink, wireless sensor networks, watershed transformation, node clustering, data gathering.

1 Introduction

The interest in the use of WSNs has grown enormously during the last decade, pointing out the crucial need for efficient and reliable routing and data gathering protocols in corresponding application environments. Energy efficiency is one of the main design goals in a WSN, towards the above direction. Moreover, the appropriate minimization of nodes energy consumption as well as the uniform energy depletion of all nodes, are critical parameters in order to increase the time the network is fully operational. In typical WSNs a main reason of energy depletion concerns the need for transmitting the sensed data from the sensor nodes (SNs) to remote sinks. These data

---

[*] This research has been co-financed by the European Union (European Social Fund – ESF) and Greek national funds through the Operational Program "Education and Lifelong Learning" of the National Strategic Reference Framework (NSRF) – Research Funding Program: Archimedes III. Investing in knowledge society through the European Social Fund.

C. Kaklamanis et al. (Eds.): Euro-Par 2012, LNCS 7484, pp. 754–766, 2012.

are typically relayed using ad hoc multi-hop routes in the WSN. A side-effect of this approach is that the SNs located closer to the sink are heavily used to relay data from all network nodes; hence, their energy is consumed faster, leading to a non-uniform depletion of energy in the WSN [6]. This results in network disconnections and limited network lifetime.

Several protocols have been proposed so far for efficient data gathering in WSNs taking also into account the above problem in order to increase the lifetime of the WSN. The most promising of them involve the mobility of the sink, based on the key idea of changing progressively the neighbors of the sink so that the energy consumption for data relaying is balanced throughout the network [6]. The MS may visit each SN and gather its data [8] (single-hop communication) or may visit only some locations and the SNs send their data to the MS through multi-hop communication [2][3][9-11]. The delay in data gathering is minimized appropriately in the latter case, however special attention has to be given in the increased energy consumption due to the multi-hop communication.

A solution in between is to have the SNs send first their data to a certain number of intermediate nodes (building direct or indirect hierarchical clustering structures) which buffer the received data and send them to the MS when it comes within their transmission range or when they receive a query from the MS asking for the buffered data [4][5][12-16]. Most of these approaches naturally strike the balance between the data gathering delay and the energy consumption overhead, whereas also, they are usually highly effective in applications where there are restrictions (e.g. isolated urban areas or building blocks) with regard to the sites that can be visited by the MS. However, all previous works have not faced yet effectively the problem of the energy-holes caused around the intermediate data-relaying nodes, especially near the sinks.

The protocols proposed so far for sink mobility can also be distinguished according to the nature of the mobility itself; namely, to 'controlled' and 'uncontrolled' sink mobility protocols. In uncontrolled sink mobility protocols the sink is sent to gather data through the network at moments and along routes (e.g. random or explicitly fixed) that are out of the control of the network [3][4][13][15-16]. On the other hand, several protocols adopt controlled sink mobility [2][5][9-12][14]. These protocols determine the movement of the sink by taking into account some crucial parameters (e.g. the residual energy), either statically or dynamically, and they have led to remarkable improvements, especially in network lifetime.

In this paper (as opposed to our previous work of [4]) we focus on application environments where all the sites within the sensor-covered area are accessible by the MS without any topological restrictions. Specifically, we propose a hybrid approach that combines distributed node clustering and controlled sink mobility. Our approach first involves the building of a suitable number of energy-rich multi-hop clusters and then the MS is programmed to visit the elected CHs and gather the sensed data. In this way, we achieve a satisfactory balance between the data gathering delay and the energy consumption caused by the multi-hop communication. However, clustering introduces an additional problem, which relates with the faster depletion of CHs neighborhoods, since they tend to deplete their energy faster due to their data relay overhead. To overcome the above problem and achieve as more uniform nodes energy depletion as possible, we introduce a novel energy efficient clustering algorithm, whose basic idea comes from the watershed transform used for image segmentation [18]. The main goal of our scheme is to form suitable clusters with not only high-energy CHs, but also having energy-rich neighborhoods. Specifically, the cluster formation is made in such a way that the energy increases progressively as getting closer to the CH, thus resulting to suitable, smoothly grown, energy-rich hills around the CHs. Furthermore, we have performed a number of simulated experiments, which show the high performance of our data gathering scheme, with regard to both energy consumption (and the desired uniform energy depletion of all nodes) and the network lifetime. Moreover, our protocol is shown to have considerably better behavior, according to both the above measures, when compared to the corresponding schemes of [2] and [3], which are relevant and competent works in the literature.

The rest of the paper is organized as follows. In section 2, the detailed description of our watershed-based node-clustering algorithm is given. In section 3, the proposed data gathering protocol is presented. Section 4 outlines the experimental results, whereas section 5 concludes the paper.

2   The Node Clustering Scheme

Our approach borrows some ideas from the watershed transform [18], which is a well known method of mathematical morphology for image segmentation. The watershed transform is usually applied to a gradient image, i.e. the original image after applying edge detection operators. The gradient image is considered as a topographic relief and watershed segmentation amounts to extracting significant catchment basins in this relief. Each catchment basin is associated with a local lowest point (the bottom of the basin) and it is essentially the set of all points at which when a drop of water starts flowing will end up at the lowest point of the basin. Watersheds are the limits of adjacent catchment basins. Several implementations of the watershed algorithm can be found in the literature [18]. A significant part of them is based on topological distance, like the 'hill climbing' algorithm [1]. Moreover, since it is a computationally intensive task, several parallel and distributed solutions have also been proposed in the literature [1][19].

In our work, we follow the basic guidelines of the related algorithm given in [1] by making the necessary modifications. Specifically, we consider the SNs as 'pixels', and the residual energy of each node as the 'gray-level' of the corresponding pixel. We also regard as 'neighbors' of each SN, the nodes within its transmission range. Moreover, by substituting each min operator with the max one in the watershed transformation, we follow a complementary approach and determine the mountains of the topological relief instead of the catchment basins (Fig. 1). The mountain peaks clearly correspond to energy-rich regions of the network and thus each of these mountains defines a cluster in the network.

Naturally, our clustering algorithm leads to a partition of the WSN into multi-hop clusters where the elected CHs (local maxima) are the SNs with the higher residual energy. In determining the clusters, the distance between each pair of nodes is also considered in order to avoid long-distance and hence costly communication between cluster members. Importantly also, the residual energy of SNs is progressively increasing as we are getting closer to the CH and this effectively deals with the increased energy consumption observed around CHs due to many-to-one communication pattern taking place inside each cluster. The detailed presentation of our clustering algorithm directly follows.

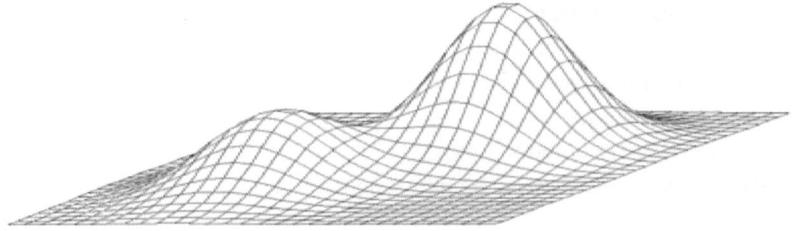

Fig. 1. Mountains and local maxima with the 'hill climbing' algorithm

---

Algorithm *CH_ELECTION*

Initialization Phase

1. each node *v*
2.   broadcasts a message *Evaluate_Msg(v.Node_ID,v.E$_{residual}$)* at a fixed power level
3.   starts a "Wait Timer" *T1*
4.   while *T1* has not expired
5.   each time *v* receives an *Evaluate_Msg(u.Node_ID,u.E$_{residual}$)* from a node *u*
6.   if *v.E$_{residual}$* > *u.E$_{residual}$*
7.   *u* is added to *LE_N(v)*
8.   else
9.   if *v.E$_{residual}$* < *u.E$_{residual}$*
10.   *u* is added to *GE_N(v)*
11.   else
12.   *u* is added to *EQ_N(v)*

CH Election Phase I

13. for each node *v*
14.   if *GE_N(v)* is non empty
15.   *v* is attached to the node *u* ∈ GE_N(v) that maximizes
        (u.E$_{residual}$ - v.E$_{residual}$)/dist(u,v)
16.   *v* starts a "Wait Timer" *T2*
17.   *v* waits for a *CH_Announce_Msg* from *u* *(flooding neighbor)*
        until *T2* expires
18.   else
19.   if *EQ_N(v)* is non empty
20.   *v* is attached to the node *u* ∈ *EQ_N(v)* that minimizes *dist(u,v)*
21.   *v* starts a "Wait Timer" *T2*
22.   *v* waits for a *CH_Announce_Msg* from *u* *(flooding neighbor)*
            until *T2* expires
23.   *else*

24.   *v* broadcasts a *CH_Announce_Msg(v.Node_ID,*
                v.Location) to its attached nodes
25. each node *v* that receives *a CH_Announce_Msg* from the
    node *u* it is attached to
26.   sets *u* as *v*'s parent node
27.   forwards the message to each node attached to *v*

CH Election Phase II

28. for each node *v* that has not received a *CH_Announce_Msg* from the
    node *u* it is attached to
29.   if *u ∉ EQ_N(v)* and there is no *u∈ EQ_N(v)* attached to *v*
30.   *v* starts a "Wait Timer" *T3*
31.   *v* waits for a *CH_Announce_Msg* from *u (flooding neighbor)*
            until *T3* expires
32.   else
33.   *v* sets *v.#attached* equal to the number of the nodes attached to *v*
34.   *v* sets *v.new_Node_ID* equal to the concatenation of
            the numbers *v.#attached* and *v.Node_ID*
35.   *v* participates to a leader election algorithm to elect as a leader the
            node *u* with the maximum *new_Node_ID*
36. each elected leader *w*
37.   broadcasts a *CH_Announce_Msg(w.Node_ID, w.Location)* to its attached
        nodes and to the node it is attached to
38. each node *v* that receives a *CH_Announce_Msg* from a node *u*
39.   sets *u* as *v*'s parent node
40.   forwards the message to each node attached to *v* and
        to the node that *v* is attached to

During an initialization phase, each node *v* broadcasts at a fixed power level an evaluate message announcing its residual energy (*Evaluate_Msg(v.Node_ID, v.E_{residual})*). Each node *v* waits until it receives all the messages sent by its neighboring nodes and builds the sets *LE_N(v)*, *GE_N(v)* and *EQ_N(v)* of neighbors that have less, more and equal residual energy to *v*, respectively (lines 1-12). Next, in the beginning of the CH election phase I, each node *v* is attached to an appropriate neighbor *u*, called flooding neighbor, as follows:

- Each node *v* with nonempty *GE_N(v)* is attached to the neighbour *u ∈ GE_N(v)* that maximizes the ratio *(u.E_{residual} - v.E_{residual})/dist(u,v)*. In the sequel, *v* sets a timer and waits for a *CH_Announce_Msg(w. Node_ID,w.Location)* from *u* (flooding neighbor) which announces the ID and the location (coordinates) of the CH *w* it will be associated with (lines 14-17).

- If the set *GE_N(v)* of *v* is empty and *EQ_N(v)* is non-empty then *v* is attached to the node *u* in *EQ_N(v)* that minimizes *dist(u,v)*. Then *v* waits for a *CH_ Announce _Msg(w.Node_ID,w.Location)* from *u* (flooding neighbor) which announces the ID and the location of the CH *w* it will be associated with (lines 18-22).

In the case that both sets *GE_N(v)* and *EQ_N(v)* are empty, *v* has the largest residual energy from all its neighbors and becomes a CH. Therefore, it broadcasts a *CH_Announce_Msg* to its attached nodes announcing its decision to become a CH and its location (lines 23-24). When a node *v* receives a *CH_Announce_Msg (w.Node_ID, w.Location)* from the node *u* it is attached to, it sets *u* as its parent node in the path towards its CH *w*, it stores the *ID* and the location of its CH *w* and forwards the message to each node attached to *v* (lines 25-27). Note that so far (CH Election Phase I) a node becomes a CH if it has the highest residual energy in its neighborhood (a local energy-maximum, resulting to empty *GE_N* and *EQ_N* sets). The CH announcing messages travel along decreasing energy paths and the nodes of each path are associated with the CH in the head of the path. Therefore, the CH has more residual energy than any other node in its cluster.

The rest of the algorithm (CH Election Phase II) addresses the case of neighboring nodes with equal residual energy which are attached one to each other i.e., none of them is attached to a higher energy node (flooding neighbor) and therefore, they have not received a *CH_Announce_Msg* and have not been associated with a CH. As a consequence, any lower residual energy node that is attached to the above equal energy nodes has also not received a CH_Announce_Msg. If a node *v* has not received a *CH_Announce_Msg* from the node *u* it is attached to, we first consider the case that $u \notin EQ\_N(v)$ and there is no $u \in EQ\_N(v)$ attached to *v*. In this case *v* starts a timer T3 and waits for a *CH_Announce_Msg* from *u* (flooding neighbor). Otherwise, *v* participates in a *leader election* algorithm that runs among all neighboring nodes that have the same residual energy with *v* (lines 28-35).

Each node *v* that participates in the leader election algorithm sets *v.#attached* equal to the number of the nodes attached to *v* and *v.new_Node_ID* equal to the concatenation of the numbers *v.#attached* and *v.Node_ID*. The elected leader is finally the node with the maximum *new_Node_ID*, i.e. one of the nodes with the largest number of attached nodes to it among the equal energy nodes that participate in the leader election process. Thus, the traffic load in the neighborhood of the CH is shared among as many SNs as possible. Then, each elected leader *w* broadcasts a *CH_Announce_Msg* to its attached nodes and to the node it is attached to announcing its election as a CH and its location. When a node *v* receives the first *CH_Announce_Msg (w.Node_ID, w.Location)* from a node *u*, it sets *u* as its parent in the path towards its CH *w*, it stores the ID and the location of *w* and forwards the received message to each node attached to *v* and to the node *v* is attached to (lines 36-40). Any other *CH_Announce_Msg* that may be received by a node *v* after the first *CH_Announce_Msg* is ignored.

A detailed example of the execution of the algorithm is given in Fig. 2. Note that CHs 9 and 15 were formed during the execution of the CH election phase I while CH 39 was formed during the execution of the CH election phase II. In this phase, nodes 34, 35, 39 and 40 were all candidates CHs and node 39 was the winning node since it has more attached nodes to it.

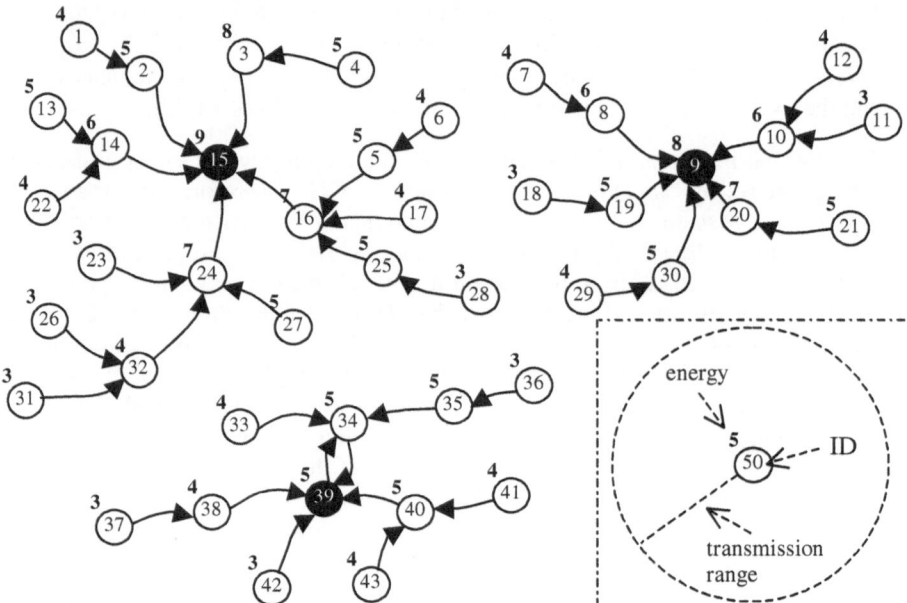

Fig. 2. A detailed execution example of the cluster formation algorithm

3   The Data Gathering Protocol

As also discussed in the previous sections, the use of a MS offers a provably efficient and reliable solution with regard to the sink-neighborhood problem. Additionally by the use of node clustering, we can achieve a quite satisfactory balance between the data gathering delay and the energy consumption overhead and also guarantee efficient aggregation procedures. Following these directions in our work, and using the watershed-based clustering scheme described in section 2, we can develop a quite simple data gathering protocol that guarantees both uniform energy depletion of all the SNs and increased network lifetime. Once the clustering hierarchy has been established, the MS has only to compute an optimal route for visiting the elected CHs, and then it can efficiently proceed to periodic data gathering through simple data packet protocols.

Furthermore, the main goal of our combined approach is to always keep the SNs organized in appropriate energy-rich clusters, where both the CHs and their neighbors can effectively afford the extra data relaying overhead caused by the multi-hop communication pattern, and thus avoid the non-uniform faster energy depletion of some nodes (energy holes). The exact time of the need for re-clustering can easily be determined by having the MS gathering information concerning the average residual energy of the visited clusters. After reclustering, the MS has to be informed (through the old CHs) with respect to the new CHs locations and then compute the new corresponding optimal route. The proposed protocol consists of two basic phases: the setup phase and the data gathering phase. Their description directly follows:

A. Setup Phase

1. Initial clustering is performed through the watershed-based clustering routine presented in the previous section.

2. The MS makes an initial walk across the network following a predefined route that covers the whole sensors area, and gathers the initial CHs locations.

3. Based on the known CHs locations, an optimal route is then computed by the MS, through one of the known TSP solutions in the literature [17].

Alternatively, based also in the assumption that all the SNs initially have the same energy, we could predefine the initial role of each SN (CH or cluster member), and embed/program the information in each SN before its deployment, building a suitable initial static/offline clustering hierarchy.

B. Data Gathering Phase

The MS periodically visits the elected CHs following the optimal route previously computed. Specifically, during each round, the MS acts as follows:

1. The MS moves to the next CH location according to its scheduled path.

2. When reaching a CH, the MS gathers the buffered data from that CH.

3. The CH may also send to the MS a 'below_ threshold' message, if the decrease of the average residual energy of its cluster compared to the average residual energy of that cluster at the time of last reclustering, is higher than a threshold. Each CH can easily keep track of the average residual energy of its cluster by periodically collecting the necessary information from all its members.

4. After visiting all CHs, the MS returns to the central base station, where it delivers the gathered data.

5. Before the next round, if the MS realizes that it has received a 'below_threshold' message from at least one CH, reclustering is decided. In that case, the following actions should also be taken by the MS:

   — It first notifies all the SNs that they have to perform reclustering, by broadcasting a message, either directly at the desired (high) power level or by flooding.

   — It then waits at the central base station until the end of reclustering (i.e. for a predefined estimated time).

   — Immediately after, it performs the next data gathering round, following the old route (thus visiting again the old CHs locations).

— When reaching each old CH, it gathers (along with the buffered sensed data) the information that this CH has collected with regard to the locations of the new CHs, as it was sent from its cluster members that have been elected as CHs in the new clustering structure (if any).

— When it returns to the central base station, it computes the new optimal path based on the new CHs locations.

When reclustering has to be performed, the following steps should be taken:

1. The watershed-based clustering routine is executed, with the additional requirement that each sensor should also keep (not overwrite) during that execution, the necessary information with regard to the previous clustering structure (its previous flooding neighbour and its previous CH id).

2. At the end of the clustering formation, each elected CH transmits its location (including its ID and its coordinates) to its previous CH (through the multi-hop routing path of the previous clustering structure). Thus, at the end of the reclustering procedure, the old CHs will keep the locations of the new CHs.

4   Simulation Results

The clustering-based data gathering protocol proposed in the previous section, has been further evaluated through a sufficient number of simulated experiments, and according to a number of suitable realistic parameters, in order that a reliable comparison with other approaches in the literature is feasible. All the experiments have been performed using the Castalia simulator, which is based on the OMNeT++ platform and has been developed especially to simulate realistic wireless node behaviour and wireless channel and radio models [7]. Specifically, in our simulation tests, we compare our protocol with the distributed protocols proposed in [2] and [3], which also consider the residual energy as the basic criterion for determining the routing paths from sensors to the MS. We have run experiments for varying number of nodes (600 to 1600 nodes), which are deployed uniformly at random, within a square area of side L = 1000m. The maximum transmission range R of the sensors nodes is equal to 45m and their initial energy is set to 500 Joules. The power consumption for each transmission depends on the target distance and varies form 29.04mW to 57.42mW (4.3m-45m). The power consumption for reception and sleep mode is 62mW and 0.016 mW, respectively.

The results are summarized in figures 3, 4, and 5. As a general notice, our data gathering protocol is shown to behave considerably better in terms of all the referred performance metrics. Less average energy consumption is achieved in total, whereas also the energy depletion of all nodes is sufficiently uniform; as opposed to the other two protocols. This naturally results in significant increase of the network lifetime in all testing cases. In [2] and [3] the routing trees have to be re-built every time the MS moves to another site, and they normally extent to the whole network area, thus leading to quite long routing paths. As an additional result, the nodes that are close to the MS have to relay quite larger amounts of data and their energy depletion can not be controlled and balanced in such a satisfactory way as in our protocol. On the other hand, in our protocol each SN always sends its sensed data only to its CH through a relatively short multi-hop path. So, the energy of the nodes that are close to the CHs deplete in much more controlled and balanced way, thus eventually leading to an almost uniform total energy depletion scheme.

More concretely, as shown in Fig. 3, the network lifetime achieved by our protocol is higher for all the numbers of sensor within the terrain. The corresponding differences are over 5% in almost all the experiments (from 5% to 20%, approximately), except the case of very small number of SNs (600) where the other two protocols behave almost equivalently well, due to the fact that the routing trees that have to be built are quite small and the total communication overhead is appropriately restricted. Moreover, in our protocol the network lifetime remains almost the same as the number of sensors increase. This happens due to the stable behavior of our clustering structure, which keeps both the average energy consumption almost constant, as well as the variance of the residual energy very low and almost constant too.

In Fig. 4 the average residual energy is shown for all the three protocols (for 1400 SNs, where the other two protocols present their most stable behavior), as it decreases with the time. Staring at Fig. 4, it can be easily observed that the decrease of the average residual energy is almost linear for all the three protocols, as it was naturally expected due to their structural properties. Also, in our protocol the average residual energy decreases with slower rate, which means that the average energy consumption is lower than in the other two protocols. Furthermore, in Fig. 5 the variance of the residual energy is presented, again for 1400 nodes. As it can be seen, the variance for our protocol is very low, almost constant during the whole network lifetime, and much lower than in the other two protocols. Specifically, the variance for our protocol ranges from 0.5 to 1.5 during the whole network lifetime, whereas for the other two protocols raises up to 12.5 at the end of the network lifetime. The latter means that in our protocol the energy of all nodes depletes in a much more uniform way than in the other two protocols. This also explains the fact that the differences in the network lifetime of the three protocols (Fig. 3) are quite high, and even more significant than the differences in the average energy consumption (Fig. 4).

It must also be noted that our clustering scheme, due to its multi-hop structure, strikes the balance between the number of clusters formed (i.e. the number of locations periodically visited by the MS) and the average number of member-nodes and communication hops within each cluster. As a consequence, the total data gathering delay is appropriately kept in quite low levels, acceptable in most of the known WSNs applications.

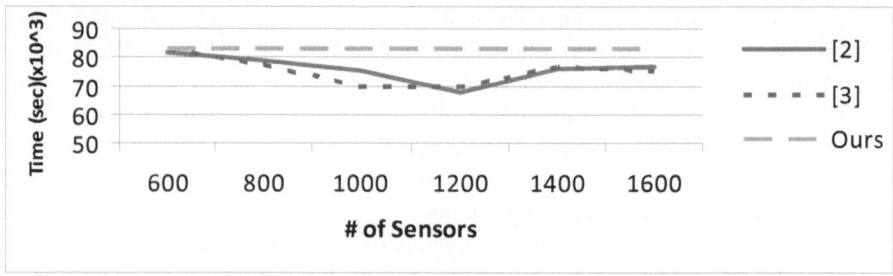

Fig. 3. Network lifetime for varying # of SNs

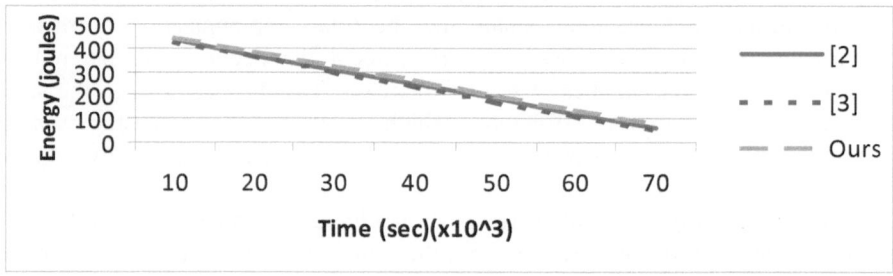

Fig. 4. Average residual energy for 1400 nodes

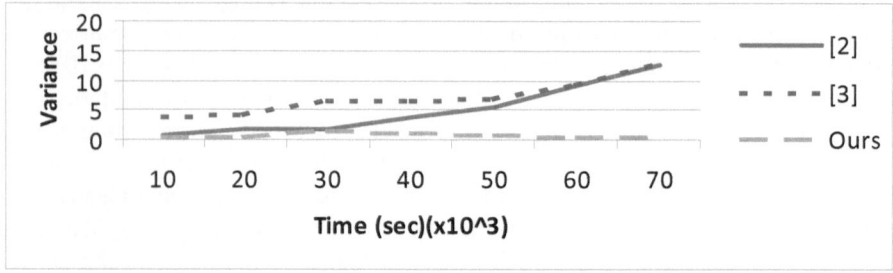

Fig. 5. Variance of the residual energy for 1400 nodes

5   Conclusion

A highly efficient data gathering protocol for WSNs that combines controlled sink mobility with a novel clustering scheme is presented throughout the paper. The proposed clustering scheme forms multi-hop clusters with not only high-energy CHs but with energy-rich CH neighborhoods too, and thus, it finally achieves uniform energy depletion of all SNs. As a consequence, the energy sink-hole problem is effectively handled and this, in turn, leads to highly improved network lifetime. In a number of simulation tests, our protocol is shown to have considerably better behavior (in terms of network lifetime and nodes residual energy) than the protocols of [2] and [3]. In a future work, we plan to generalize our approach by using more than one sinks (multiple mobile sinks [20-21]).

References

1. Galilée, B., Mamalet, F., Renaudin, M., Coulon, P.: Parallel Asynchronous Watershed Algorithm. IEEE Transactions on Parallel and Distributed Systems 18(1), 44–56 (2007)
2. Basagni, S., Carosi, A., Melachrinoudis, E., Petrioli, C., Wang, Z.M.: Controlled sink mobility for prolonging WSNs lifetime. Wireless Networks 14(6), 831–858 (2008)
3. Ammari, H., Das, S.: Promoting Heterogeneity, Mobility, and Energy-Aware Voronoi Diagram in Wireless Sensor Networks. IEEE Transactions on Parallel and Distributed Systems 19(7), 995–1008 (2008)
4. Konstantopoulos, C., Pantziou, G., Gavalas, D., Mpitziopoulos, A., Mamalis, B.: A Rendezvous-Based Approach for Energy-Efficient Sensory Data Collection from Mobile Sinks. IEEE Transactions on Parallel and Distributed Systems 23(5), 809–817 (2012)
5. Tirta, Y., Li, Z., Lu, Y.H., Bagchi, S.: Efficient Collection of Sensor Data in Remote Fields Using Mobile Collectors. In: Proc. IEEE ICCCN Conference, pp. 515–520 (2004)
6. Li, X., Nayak, A., Stojmenovic, I.: Sink Mobility in Wireless Sensor Networks. In: Wireless Sensor and Actuator Networks, ch. 6, pp. 153–184. Wiley (2010)
7. Castalia: WSNs simulator, http://castalia.npc.nicta.com.au/
8. Sugihara, R., Gupta, R.: Optimal Speed Control of Mobile Node for Data Collection in Sensor Networks. IEEE Transactions on Mobile Computing 9(1), 127–139 (2010)
9. Luo, J., Hubaux, J.P.: Joint Mobility and Routing for Lifetime Elongation in Wireless Sensor Networks. In: Proc. IEEE INFOCOM 2005, pp. 1735–1746 (2005)
10. Demirbas, M., Soysal, O., Tosun, A.Ş.: Data Salmon: A Greedy Mobile Basestation Protocol for Efficient Data Collection in Wireless Sensor Networks. In: Aspnes, J., Scheideler, C., Arora, A., Madden, S. (eds.) DCOSS 2007. LNCS, vol. 4549, pp. 267–280. Springer, Heidelberg (2007)
11. Vincze, Z., Vass, D., Vida, R., Vidacs, A., Telcs, A.: Adaptive Sink Mobility in Event-driven Densely Deployed Wireless Sensor Networks. Ad Hoc & Sensor Wireless Networks 3(2-3), 255–284 (2007)
12. Ma, M., Yang, Y.: SenCar: An Energy-Efficient Data Gathering Mechanism for Large-Scale Multihop Sensor Networks. IEEE Transactions on Parallel and Distributed Systems 18(10), 1476–1488 (2007)
13. Xing, G., Wang, T., Jia, W., Li, M.: Rendezvous Design Algorithms for Wireless Sensor Networks with a Mobile Base Station. In: Proc. ACM MobiHoc, pp. 231–239 (2008)
14. Rao, J., Biswas, S.: Network-assisted Sink Navigation for Distributed Data Gathering: Stability and Delay-energy Trade-offs. Computer Communications 33, 160–175 (2010)
15. Hamida, E., Chelius, G.: Strategies for Data Dissemination to Mobile Sinks in Wireless Sensor Networks. Wireless Communications 15(6), 31–37 (2008)
16. Chatzigiannakis, I., Kinalis, A., Nikoletseas, S.: Efficient Data Propagation Strategies in Wireless Sensor Networks using a Single Mobile Sink. Computer Communications 31, 896–914 (2008)
17. Helsgaun, K.: An Effective Implementation of the Lin-Kernighan Traveling Salesman Heuristic. European Journal of Operational Research 126(1), 106–130 (2000)

18. Soille, P.: Morphological Image Analysis, Principles and Applications. Springer (2004)
19. Moga, A., Cramariuc, B., Gabbouj, M.: Parallel watershed transformation algorithms for image segmentation. Parallel Computing 24, 1981–2001 (1998)
20. Chatzigiannakis, I., Kinalis, A., Nikoletseas, S., Rolim, J.: Fast and energy efficient sensor data collection by multiple mobile sinks. In: Proc. of MOBIWAC Conf., pp. 25–32 (2007)
21. Basagni, S., Carosi, A., Petrioli, C., Phillips, C.: Coordinated and controlled mobility of multiple sinks for maximizing the lifetime of Wireless Sensor Networks. Wireless Networks 17(3), 759–778 (2011)

Distribution of Liveness Property Connectivity Interval in Selected Mobility Models of Wireless Ad Hoc Networks

Jerzy Brzeziński, Michał Kalewski, Marcin Kosiba, and Marek Libuda

Institute of Computing Science
Poznań University of Technology
Piotrowo 2, 60–965 Poznań, Poland
Michal.Kalewski@cs.put.poznan.pl

Abstract. The *ad hoc network liveness property* disallows permanent partitioning to occur by requiring (informally) that from each time moment *reliable direct connectivity* must emerge between some nodes from every (non-empty) subset of hosts and its complementary set within some finite, but unknown, *connectivity time interval I*. An analysis of the connectivity interval is important because its finite values legitimise the liveness property assumption. Moreover, since the connectivity interval demonstrates a crucial factor of message dissemination time in ad hoc networks, its distribution significantly affects the efficiency of all protocols based on the liveness property. Therefore, in this paper, we present the distribution of the connectivity interval determined experimentally by simulation of several entity and group mobility models and real-life GPS traces of mobile nodes. We also conduct a statistical analysis of received results and show how the connectivity interval correlates with other network parameters.

1 Introduction

Mobile ad hoc networks (MANETs) [1,2] are composed of autonomous and mobile hosts (or communication devices) which communicate through wireless links. The distance from a transmitting device at which the radio signal strength remains above the minimal usable level is called the *transmission* (or *wireless*) *range* of that host. Therefore, each pair of such devices, whose distance is less than their transmission range, can communicate directly with each other—a message sent by any host may be received by all hosts in its vicinity. Hosts can come and go or appear in new places. As such, the resulting network topology may change all the time and can get partitioned and reconnected in a highly unpredictable manner.

The highly dynamic network topologies with partitioning and limited resources are the reasons why heuristic group communication and broadcast protocols with only probabilistic guarantees have been mainly proposed for the use in ad hoc networks (e.g. [4]). On the other hand, if it can be assumed that a group of collaborating nodes in an ad hoc network can be partitioned and that partitions

C. Kaklamanis et al. (Eds.): Euro-Par 2012, LNCS 7484, pp. 767–778, 2012.

heal eventually, it is possible to develop deterministic dissemination protocols used subsequently to develop more complex distributed algorithms like consensus or coherency protocols. The liveness property disallows permanent partitioning to occur by requiring (informally) that from each time moment *reliable direct connectivity* must emerge between some nodes from every (non-empty) subset of hosts and its complementary set within some finite, but unknown, *connectivity time interval I*.

In this context, an analysis of the connectivity interval is important because its finite values legitimise the liveness property assumption. Moreover, since the connectivity interval demonstrates a crucial factor of message dissemination time in ad hoc networks, its distribution significantly affects the efficiency of all protocols based on the liveness property. Therefore, in this paper, we present the distribution of the connectivity interval determined experimentally by simulation of several entity (Random Walk, Random Waypoint, Random Direction, Chiang Model, Haas Model, Gauss-Markov Model) and group (Exponential Correlated Random Mobility, Column Model, Nomadic Community Model) mobility models and real-life GPS traces of mobile nodes. We also conduct a statistical analysis of received results and show how the connectivity interval correlates with other network parameters like partition sizes or the average number of neighbouring nodes.

The paper has the following structure. First, following [10,11], the formal model of ad hoc systems with the liveness property is described in Section 2. A short review of mobility models used in our study is presented in Section 3. Section 4 describes simulation environment that has been used to perform all of our test. In Section 5, we make a statistical analysis of received results, and the paper is, finally, shortly concluded in Section 6.

2   Ad Hoc Network Liveness Property

In this paper, the topology of the distributed ad hoc system is modelled by an undirected *connectivity graph* $\mathcal{G} = (\mathcal{V}, \mathcal{E})$, where $\mathcal{V}$ is a set of all nodes, $p_1, p_2, \ldots, p_n$, and $\mathcal{E} \subseteq \mathcal{V} \times \mathcal{V}$ is a set of links (p_i, p_j) between *neighbouring* nodes p_i, p_j, i.e. nodes that are within transmission range of each other. (Note that (p_i, p_j) and (p_j, p_i) denote the same link, since links are always bidirectional.) The set $\mathcal{E}$ changes with time, and thus the graph $\mathcal{G}$ can get disconnected and reconnected. Disconnection fragments the graph into isolated sub-graphs called *components* (or *partitions* of the network), such that there is a path in $\mathcal{E}$ for any two nodes in the same component, but there is no path in $\mathcal{E}$ for any two nodes in different components.

It is presumed that the system is composed of $N = |\mathcal{V}|$ uniquely identified nodes and each node is aware of the number of all nodes in the $\mathcal{V}$ set (that is of N). The nodes communicate with each other only by sending messages (*message passing*). Any node, at any time can initiate the dissemination of message m, and all nodes that are neighbours of the sender, at least for the duration of a message transmission, can receive the message. More formally, the links can be described using the concept of a *dynamic set* function [9]. Let $\mathcal{E}'$ be a product set of $\mathcal{V}$: $\mathcal{E}' = \mathcal{V} \times \mathcal{V}$, and $\Gamma(\mathcal{E}')$ be the set of all subsets (power set) of $\mathcal{E}'$: $\Gamma(\mathcal{E}') = \{\mathcal{A} \mid \mathcal{A} \subseteq \mathcal{E}'\}$. Then, the dynamic set $\mathcal{E}_i$ of node p_i is defined as follows:

Definition 1. *The **dynamic set** $\mathcal{E}_i$ of node p_i in some time interval $T = [t_1, t_2]$ is a function:*

$$\mathcal{E}_i : T \to \Gamma(\mathcal{E}')$$

such that $\forall t \in T : \mathcal{E}_i(t)$ is a set of all links of p_i at time t.

Let δ be the maximum message transmission time between neighbouring nodes. Then, we define *direct connectivity* as follows ([11,10]):

Definition 2. *Let $T = [t,\ t + B]$, where $B \gg \delta$ is an application-specified parameter. Then, two operative nodes p_i and p_j are said to be **directly connected** at t iff:*

$$\forall \tau \in T\ (\ (p_i,\ p_j) \in \mathcal{E}_i(\tau)\ ).$$

It is assumed that channels between directly connected hosts are *reliable channels* which do not alter and lose, duplicate or create messages.

2.1   Network Liveness Property

Let $\mathcal{P}$ be a non-empty subset of $\mathcal{V}$ at some time t, and $\overline{\mathcal{P}}$ be its complementary set in $\mathcal{V}$ ($\overline{\mathcal{P}}$ contains all nodes that are not in $\mathcal{P}$). Then, the *network liveness property* is specified as follows ([11,10]):

Definition 3. *A distributed ad hoc system that was initiated at t_0 satisfies the **network liveness property**, iff:*

$$\forall t \geqslant t_0\ \forall \mathcal{P}\ \exists I \geqslant B\ (\ I \neq \infty\ \wedge\ \exists\{p_i,\ p_j\}\ (\ p_i \in \mathcal{P}\ \wedge\ p_j \in \overline{\mathcal{P}}\ \wedge$$
$$(\ \exists\{t_1, t_2\}\ (\ (t \leqslant t_1 < t_2 \leqslant t + I)\ \wedge\ (t_2 - t_1 \geqslant B)\ \wedge$$
$$(\forall t_c \in [t_1,\ t_2]\ ((p_i,\ p_j) \in \mathcal{E}_i(t_c)))))))).$$

Informally, the network liveness requirement disallows permanent partitioning to occur by requiring that reliable direct connectivity must emerge between some nodes of every $\mathcal{P}$ and $\overline{\mathcal{P}}$ within some finite, but unknown, **connectivity time interval** I.

3   Mobility Models

To facilitate research on the performance of numerous already existing and newly proposed protocols in the field of ad hoc networking, many synthetic mobility models (in two-dimensional space) have been proposed [3,8]. The literature categorises them as being either *entity* or *group models*.

Entity models are used as a tool to model the behaviour of individual mobile nodes, treated as autonomous, independent entities. On the other hand, the key assumption behind the group models is that individual nodes influence each other's movement to some degree. Therefore, group models have become helpful in simulating the motion patterns of a group as a whole.

3.1 Entity Mobility Models

Random Walk. In the Random Walk model, a mobile node randomly chooses its velocity, that is its speed and direction, from the predefined interval of $[v_{min}, v_{max}]$ and $[0, 2\Pi]$, respectively. The new values of these two parameters are calculated each time the node moves by some constant distance d or after some constant time interval $\triangle t$. Upon reaching the area boundary, the node "bounces" off it at an angle equal to the hitting angle, and moves along until the next calculation occurs. The probabilistic variant of the model known as the Chiang Model [5] makes the node's trajectory more linear and deterministic.

Random Waypoint. In this model, at each step a node first stops for some constant *pause time*. Then, the node randomly picks a point within the simulation area and starts moving toward it with a constant, but randomly selected speed that is uniformly distributed between $[v_{min}, v_{max}]$.

Random Direction. The Random Direction model is a modification of the Random Waypoint model. The only difference is that, instead of choosing a point, the node chooses direction (angle) from the $[0, 2\pi]$ range and travels along this direction until it reaches the area boundary.

The main drawback of the above models is that they generate unpredictable motion patterns. In particular, they allow some unrealistic movements, such as sharp turns or sudden stops, to occur. In order to eliminate these undesirable effects, other entity models allow to limit the level of randomness by making new steps more or less dependent on the previous ones.

Haas Model. The model Haas Model assumes that the movement of each node is characterised by the vector of speed and direction $\boldsymbol{v} = (v, \theta)$, and that the node's position is updated each $\triangle t$ time interval, according to the formulas:

$$v(t + \triangle t) = min\,[\,max\,[\,v(t) + \triangle v, 0\,], v_{max}\,]$$
$$\theta(t + \triangle t) = \theta(t) + \triangle\theta,$$

where v_{max} is a simulation constant that denotes the maximum speed, $\triangle v$ is within $[-A_{max} * \triangle t, A_{max} * \triangle t]$, A_{max} is constant maximum node's acceleration, $\triangle\theta$ is taken from the range $[-\omega * \triangle t, \omega * \triangle t]$, and ω represents the maximum angular acceleration. Parameters $\triangle v$ and $\triangle\theta$ are uniformly distributed. This movement pattern defines a Markov stochastic process, since the new position and speed at time $t + \triangle t$ depend only on their previous values at time t.

Gauss-Markov Model. In the Gauss-Markov Model, motion of a single mobile node is modelled in the form of a Gauss-Markov stochastic process, and formally is defined by the following equations:

$$v(t + \triangle t) = \alpha v(t) + (1 - \alpha)\bar{v} + \sqrt{1 - \alpha^2} V$$
$$\theta(t + \triangle t) = \alpha\theta(t) + (1 - \alpha)\bar{\theta} + \sqrt{1 - \alpha^2} D,$$

where v and θ represent the node speed and direction at timeslot t, $\bar{v}$ and $\bar{\theta}$ are constants for asymptotic speed and direction mean as $t \to \infty$, whereas random variables V and D are speed and direction random variables with a Gaussian distribution. The level of randomness is controlled by the normalised α parameter representing the preset memory level. At one extreme, if α is equal to 0, the model reduces to the Random Walk model, because the velocity in a current timeslot does not depend on its previous value at all. On the other hand, if α is 1, the random factor disappears and the velocity becomes effectively constant. For any other value of α the model has some degree of memory, which makes the node's trajectory more or less linear.

3.2 Group Mobility Models

Exponential Correlated Random Mobility. In the Exponential Correlated Random Mobility model, a new position (of a group or a single node) $Pos(t+\triangle t)$ is updated after each timestep $\triangle t$, and is given by the formula:

$$Pos(t + \triangle t) = Pos(t)e^{-\frac{1}{\tau}} + (\sigma\sqrt{1 - (e^{-\frac{1}{\tau}})^2})r,$$

where τ parameter ($\tau > 0$) controls how much two consecutive positions differ (the smaller τ the greater change), and r is a Gaussian random variable with variance σ. The model has not become popular because modelling any realistic motion pattern with its use is difficult.

Column Model. The Column Model is meant to describe a group of nodes that form a line heading in a given direction like a column. Individual nodes are allowed to deviate slightly from their reference positions (determined by the column structure) according to some entity model. The Column Model is well suited for searching and scanning applications (for instance, in a rescue team).

Nomadic Community. Sometimes, the group nodes are focused around some *reference point* (e.g. the leader node) and collectively travel from one location to another. In such settings, the Nomadic Community model is useful. In this model, the group (treated as an entity) moves randomly, because the reference point is the source of randomness. Within a group, individual nodes are free to diverge from the reference point up to some predefined maximum distance.

4 Simulator

Most available simulators provide few implementations of mobility models and usually have poor support for the creation of complex mobility models [6]. Because of that, we have decided to create a mobility model centered simulator which would facilitate fast and effortless mobility model implementations and simulations. Our simulator named MANETSim [6] was implemented in *Haskell*— a purely functional programming language. By using specific features of the language, we were able to create a *Domain Specific Language* (DSL) to describe mobility models. It enables the creation of expressive implementations which closely resemble pseudocode while retaining a high level of functionality. In order to use MANETSim as a part of another simulation environment, we have developed a communications protocol and an interoperability module for the $OMNeT++$[1] network simulation framework. We used MANETSim to analyse the liveness property in the context of different mobility models.

4.1 Measured Metrics

In order to precisely describe the measured metrics, we first introduce a definition of a *partitions set*:

Definition 4. *The **partitions set** $\mathcal{Q}$ at time t in the network is composed of all sets $\mathcal{Q} \subseteq \mathcal{V}$ such that:*

$$\mathcal{Q} \neq \emptyset \wedge \forall p_i \in \mathcal{Q} \; \exists p_j \; (p_j \in \mathcal{E}_i(t)) \wedge$$
$$\nexists \mathcal{Q}' \subset \mathcal{V} \; (\mathcal{Q}' \neq \emptyset \wedge \mathcal{Q} \cap \mathcal{Q}' \neq \emptyset \wedge \forall p_k \in \mathcal{Q}' \; \exists p_l \; (p_l \in \mathcal{E}_k(t))).$$

We also denote T_s to be the length of a simulation step, and $\text{Pos}_{n-1}(p_i)$ to be the position vector of node $p_i \in \mathcal{V}$ in the $n-1$-st time step. Thus, the momentary speed of node p_i is be expressed as:

$$V_n(p_i) = \frac{|\text{Pos}_n(p_i) - \text{Pos}_{n-1}(p_i)|}{T_s}.$$

Based on the above specification, the metrics, which are measured by MANET-Sim, are as follows:

- number of links (neighbours) of each $p_i \in \mathcal{V}$: $\frac{1}{2}|\mathcal{E}_i(t_n)|$ (since links are always bidirectional);
- momentary speed of each $p_i \in \mathcal{V}$: $V_n(p_i)$;
- size of each partition $\mathcal{Q} \in \mathcal{Q}$: $|\mathcal{Q}|$;
- **value of the liveness property connectivity interval I**, as defined in Section 2.1.

4.2 Simulation Parameters

The parameters and its values which were used in our simulation study are as follows:

- **number of repetitions of each test**: *10*;
- **simulation duration**: *6000 s*;
- **simulation transient period duration** (measurements taken during this period are ignored): *1000 s*;
- **simulated area size**: *1000 m× 1000 m*;
- **number of nodes**: *50*;
- **wireless range of each host**: *[30 m, 50 m, 80 m, 100 m, 150 m, 200 m, 250 m, 300 m]*;
- **frequency of node position updates**: *4 Hz*;

---

[1] http://www.omnetpp.org/

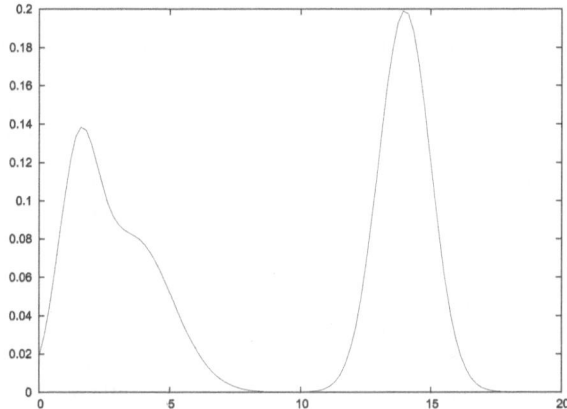

Fig. 1. The Γ node speed distribution. The X axis represents speed in m/s whereas the Y axis represents probability.

- **average node speed:** *[1 m/s, 5 m/s, 10 m/s, 15 m/s, Γ]*
- **node pause time:** *[none, Γ]*;
- *B* **parameter** (for how long do two nodes have to be in a wireless range for them to be considered connected, as specified in Section 2): *[0.5 s, 1.0 s, 2.0 s, 5.0 s, 10.0 s]*;
- **node interrupt condition** (when will a node change direction, speed etc.): *[collision, 30 m distance, 100 m distance, 30 s time, 100 s time]*;
- **Hass Model:** (model specific parameters) A_{max}=0.9 m/s², ω=10 deg/s;
- **Gauss-Markov Model:** (model specific parameter) α =*[0.1, 0.3, 0.5, 0.7, 0.9]*.

As a speed distribution we have also used Γ distribution, which was found by analysing GPS traces available freely on the Internet[2] [6]. The Γ distribution is shown in Figure 1, and Γ pause distribution is a uniform distribution over the range of *[10 s, 180 s]*.

Most of above values were based on or follow suggestions present in the articles describing the mobility model and the liveness property. A more detailed explanation of the chosen values can be found in [6].

5 Simulation Analysis

To determine the distribution of the connectivity interval *I*, we have performed simulation tests with the use of the MANETSim simulator and all the unity and group mobility models mentioned in Section 3, along with each combination of the common and model specific parameters described in Section 4.2. Based on the information from the simulator, we were able to calculate values of the connectivity time *I*, and assess how they correlate with other network parameters.

---

[2] http://www.openstreetmap.org/ and http://www.gpsies.com

Table 1. Percentage of simulation tests for which the value of the connectivity time interval was finite within simulation time

Mobility Model	Tests with Finite Value of I
Random Walk	98.69%
Chiang Model	99.90%
Random Waypoint	100.00%
Random Direction	97.49%
Haas Model	99.50%
Gauss-Markov Model	91.57%
Exponential Correlated Mobility	99.49%
Column Model	99.69%
Nomadic Community	100.00%

Table 1 shows the percentage of simulation tests, for which the value of the connectivity time interval was finite within simulation time, for all considered mobility models. As it can be seen, all these results are above 91% and in case of two models (Random Waypoint and Nomadic Community) all the simulation had finite values of this parameter.

We begin our study with analysing the distribution of the connectivity interval I parameter among mobility models. Even though the values of I varies between different mobility models, the shape of the distribution is similar amongst them. This is illustrated by Figure 2, where similarities between the distribution for the Random Direction mobility model can be seen (Figures 2(a), 2(b)) and Chiang Model (Figures 2(c), 2(d)). The *same similarity* can be observed between *all* of the analysed mobility models [6]. But despite this, the distributions differ in a statistically significant way—Wilcoxon test at $\alpha = 0.05$. The distribution of I is almost exponential (as can be seen on the logarithmic plots: 2(d), 2(b)), which means that the smallest values of I are the most probable ones. That in turn means, that most of the partitions in the network exist only for a (relatively) short time (under 5 minutes in our simulations).

The value of I is not independent of other network parameters such as average node speed or the number of links. To illustrate those dependencies, we use the scatter plot depicted in Figure 3, where the following values are shown: **coverage**—percent of the area covered by a node's wireless range; **I-upper**—value which is greater than 90% of the observed I values; **neighbours**—average number of nodes to which a node has connectivity; **I-average**—average value of the observed I values; **partitions size**—average size of a partition and **speed**—average speed of nodes.

It can be seen on the basis of Figure 3 that an increase in value of partition size, neighbours number and coverage is connected with a decrease of the I-upper and I-average values. This tendency is not symmetrical as there are observations where small partition sizes with both small and large values of I-upper. Speed has a minimal impact on both values of I-average and I-upper, which means that while analysing protocols based on the liveness property it does not suffice to vary

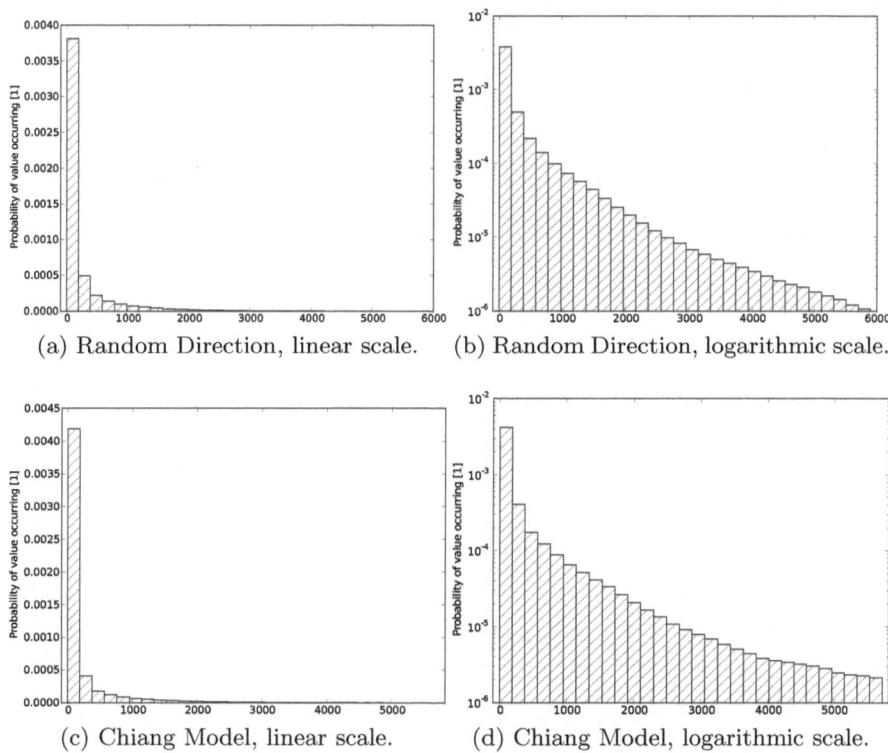

(a) Random Direction, linear scale.   (b) Random Direction, logarithmic scale.

(c) Chiang Model, linear scale.   (d) Chiang Model, logarithmic scale.

Fig. 2. Distributions of the liveness property connectivity interval I

node speed. This is contrary to common practice in MANET studies where the node speed is usually the only variable parameter of the mobility model [7]. It is worth noting that observations which do not exhibit finite values of connectivity interval I also have low coverage and a small number of neighbours.

Statistical parameters of the analysed mobility models are presented in Table 2. The results have been categorised according to the mobility model and speed distribution used: average value (denoted as *single*) and Γ distribution. To compare Γ distribution and average values results, we have used the Wilcoxon sign test. For all mobility models, the difference was found to have been statistically significant at $\alpha = 0.05$. The difference between those two types of distributions is not clear because, while most entity models with the Γ distribution had higher values of I, there were also less observations without a finite value of the connectivity interval among that group.

Finally, Table 3 depicts the percentage of observations for which I values were greater than those in the I_{cutoff} column—this can also be viewed as an approximation of the probability that a network would have not the connectivity interval for the given value of I_{cutoff}.

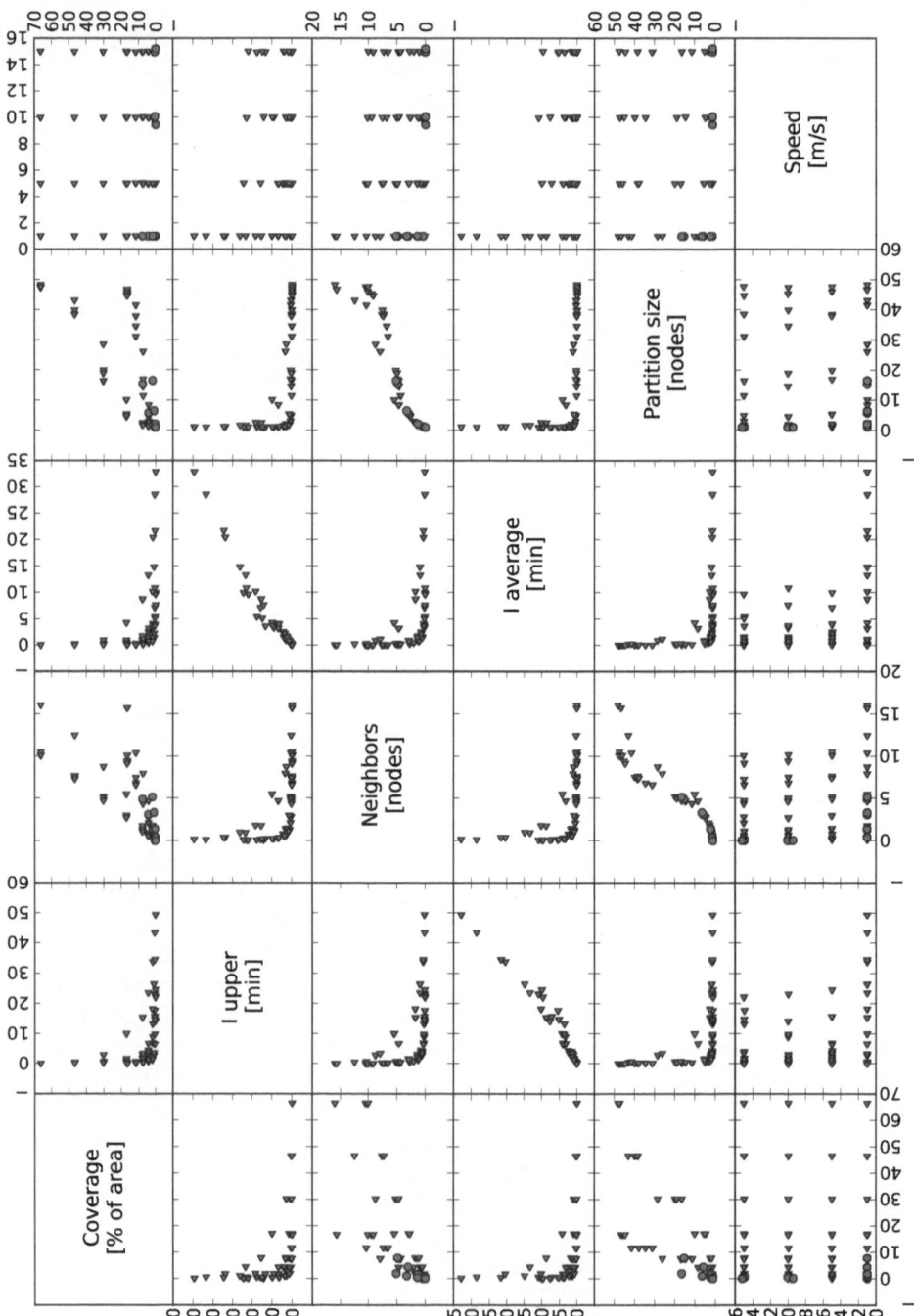

Fig. 3. Scatter plot for the Random Direction mobility model. Green triangles mark measurements with a finite value of the connectivity interval, red dots mark measurements without a finite value of the connectivity interval.

Table 2. Statistical parameters of I for all simulation experiments

Mobility Model	Speed Dist.	I						
		$E(I)$	σ_I	I_9	$min(I)$	$max(I)$	kurtosis	skew
Random Walk	single	258.74	488.07	711.00	0.25	5993.25	14.77	3.54
Random Walk	Γ	348.94	628.07	1038.75	0.25	5920.75	11.21	3.11
Chiang Model	single	205.25	388.87	519.75	0.25	5727.75	16.84	3.77
Chiang Model	Γ	120.63	214.81	217.25	0.25	5456.50	23.88	4.43
Random Waypoint	single	170.51	275.28	452.00	0.25	5903.00	17.94	3.56
Random Waypoint	Γ	188.25	357.55	362.50	0.25	4478.00	26.18	4.70
Random Direction	single	216.99	392.36	571.25	0.25	5903.75	14.98	3.53
Random Direction	Γ	384.47	571.04	959.00	0.25	5858.25	11.62	3.07
Haas Model	single	103.89	172.20	191.25	0.25	5959.50	24.41	4.45
Haas Model	Γ	91.95	191.63	153.00	0.25	5959.50	25.73	4.83
Gauss-Markov	single	235.35	371.13	687.75	0.25	5773.75	8.44	2.65
Gauss-Markov	Γ	385.00	474.05	1026.00	0.25	5309.00	3.29	1.72
Exponential Correlated	single	338.52	495.75	864.25	0.25	5860.50	10.17	2.95
Column Model	single	189.97	379.32	379.25	0.25	5699.50	28.50	4.88
Nomadic Community	single	197.06	311.80	507.75	0.25	5978.25	12.71	3.21

Table 3. Percentage of observations for which I values were greater than those in the I_{cutoff} column

Mobility Models	Value of I_{cutoff} [s]						
	5	10	20	50	100	200	400
All	98.0%	97.0%	96.0%	87.0%	74.0%	56.0%	38.0%
Entity	97.0%	96.0%	94.0%	83.0%	70.0%	54.0%	38.0%
Group	99.0%	99.0%	99.0%	96.0%	83.0%	61.0%	37.0%
	700	1000	2000	3000	4000	5000	6000
All	26.0%	19.0%	9.0%	5.0%	3.0%	2.0%	2.0%
Entity	27.0%	21.0%	10.0%	6.0%	4.0%	3.0%	3.0%
Group	22.0%	15.0%	7.0%	3.0%	1.0%	0.4%	0.2%

6   Conclusions

In this paper, we have presented a theoretical model of ad hoc networks with the liveness property, defined with the concept of dynamic sets, and several entity and group mobility models. We used these concepts to determine the distribution of the liveness property connectivity interval by simulation tests which build on our implementation of a mobility model centered simulator. The obtained results have shown that for all considered mobility models the probability that a network will have a finite value of the connectivity interval is very high, and that there is a strong correlation between the average number of neighbours and the value of I parameter. Other correlations were also considered, and we have observed that generally, an increase in values of partition size and coverage also is connected with a decrease of I value, while the speed of nodes have a minimal impact on the value. The distribution of I in our results is almost exponential, which indicates that the smallest values of the parameter are most probable. Consequently, it should be expected that most of the partitions in a network will exist for only a relatively short time.

Acknowledgment. The research presented in this paper has been partially supported by the European Union within the European Regional Development Fund program no. POIG.01.03.01–00–008/08.

References

1. Aggelou, G.: Mobile Ad Hoc Networks: From Wireless LANs to 4G Networks, 1st edn. McGraw-Hill (November 2004)
2. Brzeziński, J., Kalewski, M., Libuda, M.: A short survey of basic algorithmic problems in distributed ad hoc systems. Pro Dialog (Polish Information Processing Society Journal) 21, 29–46 (2006)
3. Camp, T., Boleng, J., Davies, V.: A survey of mobility models for ad hoc network research. Wireless Communications and Mobile Computing 2(5), 483–502 (2002)
4. Chandra, R., Ramasubramanian, V., Birman, K.P.: Anonymous gossip: Improving multicast reliability in mobile ad-hoc networks. In: Proceedings of the 21st International Conference on Distributed Computing Systems (ICDCS 2001), pp. 275–283. IEEE Computer Society (April 2001)
5. Chiang, C.-C.: Wireless network multicasting. PhD thesis, University of California, Los Angeles, CA, Chair-Gerla, Mario (1998)
6. Kosiba, M.: The liveness property of wireless ad hoc networks and its analysis based on selected mobility models (in Polish). Master's thesis, Poznań University of Technology (August 2009)
7. Kurkowski, S., Camp, T., Colagrosso, M.: Manet simulation studies: The incredibles. SIGMOBILE Mob. Comput. Commun. Rev. 9(4), 50–61 (2005)
8. Lin, G., Noubir, G., Rajaraman, R.: Mobility models for ad hoc network simulation. In: IEEE INFOCOM 2004, Conference of the IEEE Communications Society, Hong Kong (March 2004)
9. Liu, S., McDermid, J.A.: Dynamic sets and their application in VDM. In: Proceedings of the 1993 ACM/SIGAPP Symposium on Applied Computing (SAC 1993), pp. 187–192. ACM Press (February 1993)
10. Vollset, E.W.: Design and Evaluation of Crash Tolerant Protocols for Mobile Ad-hoc Networks. PhD thesis, University of Newcastle Upon Tyne (September 2005)
11. Vollset, E.W., Ezhilchelvan, P.D.: Design and performance-study of crash-tolerant protocols for broadcasting and supporting consensus in MANETs. In: Proceedings of the 24th IEEE Symposium on Reliable Distributed Systems (SRDS 2005), pp. 166–178. IEEE Computer Society (October 2005)

Topic 15: High Performance and Scientific Applications

Thomas Ludwig, Costas Bekas, Alice Koniges, and Kengo Nakajima

Topic Committee

Many fields of science and engineering are characterized by an increasing demand for computational resources. Coupled with important algorithmic advances, high performance computing allows for the gain of new results and insights by utilizing powerful computers and big storage systems. Workflows in science and engineering produce huge amounts of data through numerical simulations and derive new knowledge by a subsequent analysis of this data. Progress in these fields depends on the availability of HPC environments, from medium sized teraflops systems up to leading petaflops systems.

The High Performance and Scientific Applications Topic highlights recent progress in the use of high performance parallel computing and puts an emphasis on success stories, advances of the state-of-the-art and lessons learned in the development and deployment of novel scientific and engineering applications.

Today's complex research issues have to be mapped onto complex compute and storage environments: We find powerful computers being composed of hundreds and thousands of nodes which themselves are shared memory parallel computers with many processor cores and sometimes additional accelerator hardware. Storage hardware consists at least of high volume disk systems but could also comprise tape libraries. With the advent of the Exascale era a way to increased resource usage might enforce hardware-software co-design.

The papers accepted for this workshop characterize typical steps on the way to exploit maximum system performance. Four papers focus on an optimal adaptation of the software system onto the available hardware system. They deal with the layout of data structures in memory and the mapping of software components onto hardware in multi-core environments. One contribution analyses how data volumes could be reduced when a certain loss of quality is acceptable. Finally, we include an interesting report on how quality of data can be assessed and how it influences subsequent post-processing. We invite you to read this collection of papers and get inspired by the results of our colleagues.

Kunaseth et al. present in "Memory-Access Optimization of Parallel Molecular Dynamics Simulation via Dynamic Data Reordering", a novel data-reordering scheme aimed to optimize runtime memory access in the context of large scale molecular dynamics simulations.

Reiter et al. present in "On Analyzing Quality of Data Influences on Performance of Finite Elements driven Computational Simulations" a thorough analysis of mechanisms with which data quality can dramatically influence the results as well as the performance of scientific simulations.

C. Kaklamanis et al. (Eds.): Euro-Par 2012, LNCS 7484, pp. 779–780, 2012.

Malakar et al. demonstrate in "Performance Evaluation and Optimization of Nested High Resolution Weather Simulations" a significant reduction in run time of complex climate simulations by exploiting a careful combination of compiler optimizations coupled with overlapping computation and communication.

Fietz et al. present in "Optimized Hybrid Parallel Lattice Boltzmann Fluid Flow Simulations on Complex Geometries" an optimized hybrid parallelization strategy, that is capable of solving large-scale fluid flow problems on complex computational domains.

Aktulga et al. quantitatively show in "Topology-aware Mappings for Large-Scale Eigenvalue Problems" that topology-aware mapping of processes to physical processors can have a significant impact on the efficiency of high-performance computing applications, with a particular view to modern large-scale multi-core architectures.

Finally, Iverson et al. developed in "Fast and Effective Lossy Compression Algorithms for Scientific Datasets" effective and efficient algorithms for compressing scientific simulation data computed on structured and unstructured grids.

Memory-Access Optimization of Parallel Molecular Dynamics Simulation via Dynamic Data Reordering

Manaschai Kunaseth, Ken-ichi Nomura, Hikmet Dursun, Rajiv K. Kalia, Aiichiro Nakano, and Priya Vashishta

University of Southern California, Los Angeles, CA 90089, USA
{kunaseth,knomura,hdursun,rkalia,anakano,priyav}@usc.edu

Abstract. Dynamic irregular applications such as molecular dynamics (MD) simulation often suffer considerable performance deterioration during execution. To address this problem, an optimal data-reordering schedule has been developed for runtime memory-access optimization of MD simulations on parallel computers. Analysis of the memory-access penalty during MD simulations shows that the performance improvement from computation and data reordering degrades gradually as data translation lookaside buffer misses increase. We have also found correlations between the performance degradation with physical properties such as the simulated temperature, as well as with computational parameters such as the spatial-decomposition granularity. Based on a performance model and pre-profiling of data fragmentation behaviors, we have developed an optimal runtime data-reordering schedule, thereby archiving speedup of 1.35, 1.36 and 1.28, respectively, for MD simulations of silica at temperatures 300 K, 3,000 K and 6,000 K.

Keywords: Data reordering, memory-access optimization, data fragmentation, performance degradation, molecular dynamics.

1 Introduction

Molecular dynamics (MD) simulation is widely used to study material properties at the atomistic level [2,7,9,10]. One of the major problems on improving performance of MD simulations is to maintain data locality. Since the memory-access pattern in MD simulation is highly non-uniform and unpredictable, the locality optimization problem is challenging.

To address the locality issue, a commonly used method is data reordering, which organizes data of irregular memory-access patterns in memory according to a certain locality metric [3,4,11,13]. However, a further challenge arises from the dynamic, irregular nature of MD computations. Mellor-Crummey *et al.* showed that data reordering and computation restructuring enhance data locality, resulting in the reduction of cache and TLB misses and accordingly considerable performance improvement of MD simulations [6]. The study also suggested the necessity of repeated runtime reorderings, for which the remaining

C. Kaklamanis et al. (Eds.): Euro-Par 2012, LNCS 7484, pp. 781–792, 2012.

problem is how often the reordering should be performed in order to achieve the optimal overall performance. In addition, as the data reordering and computation restructuring incur computational overhead, such reordering cost should be considered to find the optimal reordering frequency.

Runtime data behaviors of MD simulations are closely related to both physical properties of the system being simulated (*e.g.* temperature, diffusion rate, and atomic configuration) and computational parameters (*e.g.* spatial decomposition granularity in parallel MD). Therefore, understanding how these quantities affect the deterioration of data locality is a prerequisite for designing the optimal runtime data-reordering schedule.

To address these challenges, we first introduce a *data fragmentation ratio* metric that quantifies the locality of atom data arrays during MD simulation. Then, we perform an analysis on how MD simulations with different physical/computational characteristics (*e.g.* temperature, diffusion rate, and granularity) impact the data fragmentation ratio and performance deterioration of the system. Based on the data fragmentation analysis, we finally design and evaluate an memory optimization scheme that features an optimal runtime data-reordering schedule during MD simulation.

2   Parallel Molecular Dynamics

Molecular dynamics simulation follows the phase-space trajectories of an N-atom system, where force fields describing the atomic force laws between atoms are spatial derivatives of a potential energy function $E(\mathbf{r}^N)$ ($\mathbf{r}^N = \{\mathbf{r}^1, \mathbf{r}^2, ..., \mathbf{r}^N\}$ is the positions of all atoms). Positions and velocities of all atoms are updated at each MD step by numerically integrating coupled ordinary differential equations. The dominant computation of MD simulation is the evaluation of $E(\mathbf{r}^N)$, which, in the program considered in this paper, consists of two-body $E_2(\mathbf{r}_i, \mathbf{r}_j)$ and three-body $E_3(\mathbf{r}_i, \mathbf{r}_j, \mathbf{r}_k)$ terms [7].

Figure 1(a) shows a schematic of the computational kernel of MD, which employs a linked-list cell method to compute interatomic interactions in $O(N)$ time. Periodic boundary condition is applied to the system in three Cartesian dimensions. Here, a simulation domain is divided into small cubical cells, and the linked-list data structure is used to organize atomic data (*e.g.* coordinates, velocities and atom type) in each cell. Traversing through the linked list, one retrieves the information of all atoms belonging to a cell, and thereby computes interatomic interactions. Thus, the dimensions of the cells are usually determined by the cutoff radius r_c of the interatomic interaction.

The MD program considered in this paper employs spatial decomposition at an outermost level of hierarchical parallelization [8]. Here, the physical system is partitioned into subsystems, and atoms residing in different subsystems are assigned to different compute nodes. When the atomic coordinates are updated according to the time-integration algorithm, some resident atoms may have moved out of the subsystem boundary, and such atoms are migrated to proper nodes.

MD is an archetype of irregular memory-access pattern applications, and this paper addresses two distinct sources of irregularity: (1) physically induced disorder such as atom diffusion or flow; and (2) algorithmically induced disorder such as atom migration between spatial decomposition domains in parallel MD.

3   Data Fragmentation in Parallel Molecular Dynamics

High memory-access penalty is a major obstacle of irregular memory accessing applications, which needs to be mitigated. Our MD application performs data and computation orderings by organizing atom data contiguously in memory in the same order as the linked lists of the cells access them. The reordering algorithm performed in $O(N)$ is shown in Fig. 1(b). Though the spatial locality after data reordering is retained for a while, data get disordered as simulation progresses due to atom movement among cells. How the benefit from data reordering deteriorates as simulation evolves at runtime essentially affects the performance of MD simulations. To quantify the level of fragmentation, we first define a fragmentation measurement ratio metric.

Let $C(i, j, k)$ be a linked-list cell with cell indices i, j, and k in the x, y, and z directions, respectively. Each linked-list cell contains indices of atoms whose coordinates are within their cell dimensions. Before data reordering, atom data are likely scattered in memory as illustrated in Fig. 1(c), and data reordering improves the data locality as shown in Fig. 1(d), where the atom data of the same physical cell volume reside continuously. Fragmentation in the atom data array could occur as follows. Suppose that the a-th atom in linked-list cell C moves to another cell C'. Consequently, the memory block of C' becomes partially fragmented because the a-th atom is not contiguous in memory space with other atoms in C' such that their distances in the memory space exceed the page size. Therefore, any atom moving out of its original linked-list cell introduces fragmentation in memory, which likely causes data translation lookaside buffer (DTLB) misses. To quantify the degree of fragmentation, we thus define a *data fragmentation ratio* as $f = N_{fragment}/N$, where $N_{fragment}$ is the number of atoms whose positions have moved out of the originally ordered cells and N is the total number of atoms. The data fragmentation ratio is between 0 (*i.e.*, all atoms in all cells reside continuously in memory—fully ordered state, see Fig. 1(d)) and 1 (*i.e.*, no atom resides in the original cell—fully disordered state, see Fig. 1(c)). The data fragmentation ratio will be used to quantify fragmentation extensively throughout this paper. Note that the page size used in all experiments is 4 KB.

Data fragmentation in memory is dictated by the dynamics of atoms, and thus understanding the factors that control the atom dynamics would provide an insight on how to prevent the fragmentation. One of the physical factors that are directly related to the dynamics is the temperature, since high temperature drives atoms to move faster and more freely. Among the computational factors, migration of atom data from one node to another across a subsystem boundary in parallel MD also causes fragmentation. The granularity of spatial decomposition

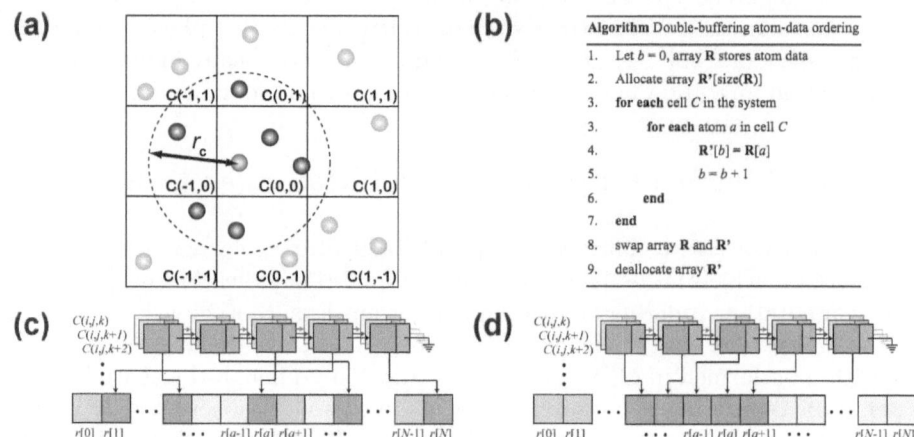

Fig. 1. (a) 2D schematic of the linked-list cell method. The center cell $C(0,0)$ is surrounded by eight neighbor cells. The cell dimensions are often chosen to be the cutoff radius (represented by the two-heads arrow) of interatomic interaction. Only force exerted from atoms within the cutoff radius are computed. (b) Pseudocode of the data reordering algorithm. (c) Schematic of memory layout for atom data in a disordered state and (d) a fully ordered state, where $C(i,j,k)$ is the linked list for the cell with indices i, j, and k in the x, y, and z directions, respectively, $r[a]$ is the data associated with the a-th atom.

(*i.e.* the number of atoms per compute node) is related to the subsystem surface-to-volume ratio, and hence is likely to control the degree of fragmentation via the amount of atom migrations. In the following subsections, we measure and analyze the effects of temperature and granularity on the data fragmentation.

3.1   Temperature Induced Fragmentation

In this subsection, we study the effect of temperature on the data fragmentation during MD simulations on two systems: Silica material [1] and combustion of aluminum nanoparticles [12]. The silica simulations involving 98,304 atoms and 8,000 linked-list cells ($20 \times 20 \times 20$ in x, y, and z directions, respectively) are performed for 3,000 MD time steps using a time discretization unit of 2 femtoseconds. The simulations are performed on a dual quadcore Intel Xeon E5410 2.33 GHz (Harpertown) at the High Performance Computing and Communication (USC-HPCC) facility of the University of Southern California. Temperature is a major physical factor that enhances the diffusion of atoms and hence the disordering of data arrays. To examine the effect of atomic diffusions on the data fragmentation ratio, we thus perform a set of simulations with three different initial temperatures, 300, 3,000, and 6,000 Kelvin (K), starting from the same initial atomic configuration—an amorphous structure with a uniform density across the system. Each dataset represents a distinct phase of silica—solid

(300 K), highly viscous liquid (3,000 K) [5], and low-viscosity liquid (6,000 K) (note that the melting temperature of silica is 1,800 K). Here, data ordering is performed only once at the beginning of the simulation, so that the atom data are fully ordered (Fig. 1(d)) initially, and subsequently we measure the data fragmentation ratio as a function of MD steps without further data reordering.

In order to study the influence of temperature on data fragmentation, Fig. 2(a) plots the data fragmentation ratio as a function of MD steps. Simulations at higher temperatures exhibit larger fragmentation ratios. The fragmentation of 300 K dataset fluctuates in the first 500 steps, then rises to 0.15 after 1,000 steps, and remains nearly constant throughout the simulation thereafter. At 3,000 K, the fragmentation ratio quickly increases to 0.24 in the first 200 steps, and then increases linearly with time (0.033 per 1,000 steps), reaching 0.36 after 3,000 MD steps. In contrast, the fragmentation ratio at 6,000 K rapidly rises to 0.73 in the first 500 steps, then continues to increase with a slightly slower rate until the data gets almost fully disordered at 0.93 after 3,000 MD steps.

In order to understand the physical origin of the correlation between the fragmentation ratio and temperature in Fig. 2(a), we measure the mean squared displacement (MSD) of atoms in each dataset as a function of time. MSD is used to identify the phase (e.g. solid vs. liquid) of the material and to calculate the diffusion rate of atoms in the system, which may be used to quantify the dynamics of atoms during the simulation. Figure 2(b) shows the MSD of 300 K, 3,000 K, and 6,000 K datasets over 6 picoseconds (3,000 MD steps). The result shows that at 300 K, silica remains in solid phase (MSD remains constant). In contrast, 3,000 K silica is melted and becomes a highly viscous liquid with a small diffusion coefficient of 2.31×10^{-5} cm^2/s (the diffusion coefficient is obtained from the linear slope of the MSD curve). Similarly, MSD of 6,000 K dataset shows that the system is completely melted with a much larger diffusion coefficient of 4.17×10^{-4} cm^2/s. Because atoms are not diffusing in 300 K dataset, 83% of the atoms ($f = 0.17$) remain in the same linked-list cell throughout the simulation.

Only atoms close to the cell boundaries move back and forth between the original and the neighbor cells due to thermal vibration. However, atoms in 3,000 K dataset are melted, and their slow diffusion from their original cells causes the gradually increasing fragmentation ratio. For 6,000 K system, atoms diffuse approximately 18 times faster than 3,000 K system, resulting in the rapid rise of fragmentation ratio. Only 3.2% of the atoms remain in the same cell after 3,000 steps. These results clearly show that diffusion is a major physical factor that contributes to data fragmentation in MD simulations.

Although the silica experiment shows that the temperature significantly affects data fragmentation, we also found that its effect on the fragmentation ratio is sensitive to the material and phenomenon being simulated. To demonstrate this point, the second experiment simulates flash heating of an aluminum nanoparticle surrounded by oxygen environment using a reactive interatomic potential [12], which involves 15,101,533 atoms on 1,024 processors of dual core AMD Opteron 270 2.0 GHz (Italy) at the USC-HPCC facility. Similar to the first simulation, the atom data arrays are ordered only once at the beginning

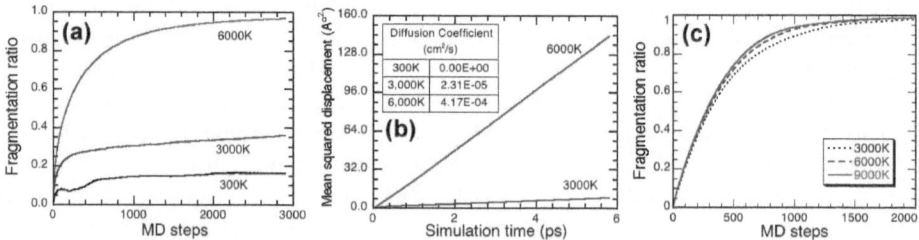

Fig. 2. (a) Time variation of data fragmentation ratio over 3,000 MD steps. Three datasets of different initial temperatures (300 K, 3,000 K, and 6,000 K) are plotted. (b) Mean squared displacement (MSD) of 300 K, 3,000 K, and 6,000 K datasets. The MSD of 300 K dataset remains constant at $0.23\,\text{\AA}^{-2}$ (too small to be seen in the figure). The inset shows the diffusion coefficients at the three temperatures. (c) Data fragmentation ratio during aluminum combustion simulation over 2,000 MD steps. The dataset involves 15 million atoms on 1,024 processors.

of the simulation. Then, the fragmentation ratio of 3,000 K, 6,000 K and 9,000 K datasets are measured at each step. Figure 2(c) shows the data fragmentation ratio as a function of MD steps. We see that the fragmentation ratios of all datasets rapidly rise to above 90% after 1,000 MD steps, and continue to increase to over 98% after 2,000 MD steps. Since the aluminum nanoparticles are at considerably high temperatures (far above the melting temperature of aluminum $\sim$ 930 K) and are surrounded by oxygen atoms in the gas phase, the atoms are highly reactive and move very rapidly. This accelerates the fragmentation to proceed very quickly regardless of the temperature. In such applications, data reordering is indispensable and is required to be performed more often in order to maintain good spatial locality. These results indicate that fragmentation is highly system dependent, so that data reordering needs to be performed dynamically at runtime, adapting to the systems behaviors.

3.2 Granularity Induced Fragmentation

Parallel MD using spatial decomposition introduces computational artifacts such as spatial subsystem (or domain) boundaries, which also contribute to the fragmentation of atom data arrays. In this subsection, we study the influence of granularity (*i.e.* the number of atoms per compute node) on data fragmentation. Atoms near a domain-boundary surface tend to migrate among domains regardless of physical simulation conditions. This inter-domain atomic migration triggers rearrangements of data arrays, including the deletion of the migrated atom data from the original array, compression of data array after the removal of the migrated atoms, and their appending to the array in the destination domain. These newly migrated atoms in the destination domain cause data fragmentation, since they do not reside continuously in memory.

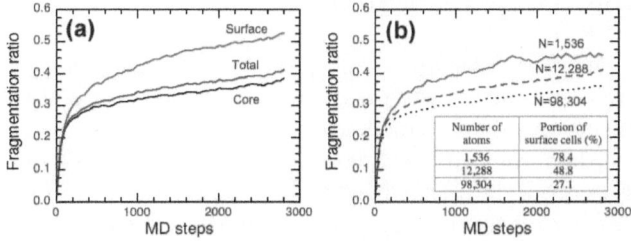

Fig. 3. (a) Time variation of the data fragmentation ratio of 12,288-atom silica at 3,000 K over 3,000 MD steps, for surface cells, core cells, and both combined. (b) Time variation of the data fragmentation ratio of 3,000 K silica varying granularities over 3,000 MD steps. Inset table shows computational parameters of silica datasets.

To confirm the expected high fragmentation ratio at the domain boundaries as explained above, the data fragmentation ratios of the surface cells (*i.e.* the outermost layer cells that share at least one facets with the inter-domain surfaces) and core cells (*i.e.* non-surface cells deep inside each domain) are measured separately. Here, we consider a 12,288-atom silica dataset initially at 3,000 K temperature similar to the dataset used in the first experiment of section 3.1 but with a reduced domain size (their dimensions are reduced by half in all three directions, so that the domain volume is one-eighth of that in section 3.1). This dataset consists of 1,000 cells in total (10 × 10 × 10), of which 488 cells (48.8%) are surface cells. Figure 3(a) clearly shows that the data fragmentation ratio of the surface cells is larger than that of the core cells. In the first 100 MD steps, the fragmentation ratios of the two groups are almost identical. Then, the fragmentation ratio of the surface cells begins to increase at higher rate reaching 0.53 after 3,000 MD steps, whereas the fragmentation ratio of the core cells is only 0.39. Thus, the atom data of the surface cells is 14% more fragmented that that in the core cells, yielding a total fragmentation ratio of 0.42 for the entire system. This result confirms that computational artifacts such as domain boundary indeed induce additional fragmentation.

The domain-boundary induced fragmentation also implies that systems with smaller granularities will have more fragmentation due to their larger surface-to-volume ratios (*i.e.*, larger portions of linked-list cells are domain-boundary cells). To test this hypothesis, we perform MD simulations for 3,000 K silica system with three different granularities—98,304, 12,288, and 1,536 atoms, over 3,000 MD steps. Figure 3(b) shows that datasets with smaller granularities indeed have larger fragmentation ratios. After 3,000 MD steps, the data array of the smallest granularity (1,536 atoms) dataset is 9.7% more fragmented than that in the largest granularity (98,304 atoms) dataset. Also, the 12,288 atoms dataset is 5.2% more fragmented compare to the largest dataset. The figure also shows large fluctuation for the fragmentation ratio for $N = 1,536$ dataset, due to less statistics inherent for the small system size.

4   Performance Measurements

In this section, we first establish a correlation between the data fragmentation ratio and the performance of the program. In a fragmented dataset, atom data are likely to reside in different memory pages, which causes a large number of DTLB misses, when they are fetched. The reordering algorithm explained in section 3 clusters atom data that need to be fetched and computed in relatively proximate times (*e.g.* atoms in the same cells) in the same page, thereby reducing DTLB misses tremendously. However, as the simulation progresses, atoms that are once ordered possibly move out of their original cells. It is thus expected that the increase of DTLB misses caused by data fragmentation is a major source of performance degradation in MD.

To confirm the correlation between the number of DTLB misses and the data fragmentation ratio, we perform a test using the same datasets from the first experiment of section 3.1. In addition to the fragmentation ratio measured in section 3.1, we here measure the DTLB miss rate as a function of MD steps. We use the Intel VTune Performance Analyzer to monitor DTLB miss events during MD simulation on Intel Core i7 920 2.67 GHz (Nehalem) processor. We measure the number of DTLB misses after data ordering is performed, then normalize it by the original number of DTLB misses without data ordering. We find that the initial DTLB miss rates at all temperatures are approximately 0.07 ($\ll 1$) right after the reordering is performed (namely, the reordering reduces DTLB misses by 93%). The great improvement highlights a significant role of data reordering in reducing DTLB misses for irregular memory-access applications. We also monitor the number of DTLB misses as a function of MD time steps and observe distinct profiles at different temperatures (Fig. 4(a)). The DTLB misses ratio at 300 K and 3,000 K saturates at 0.08 and 0.13, respectively. In contrast, the 6,000 K simulation exhibits a continuous increase of the number of DTLB misses. The DTLB miss rate reaches 0.62 at 3,000 MD steps after the initial data reordering.

Figure 4(b) shows the relation between the data fragmentation ratio and the DTLB misses rate at the three temperatures. The DTLB misses rate remains relatively small when the fragmentation ratio is small, while it rapidly increases when the fragmentation ratio increases above 0.8. This result clearly demonstrates a strong correlation between the DTLB miss rate and the data fragmentation ration. Namely, in MD simulations where atoms are moving extensively, the DTLB miss rate increases rapidly as its fragmentation ratio does.

To show that DTLB misses are the source of performance deterioration during simulation, we measure the running time of the same datasets as in the DTLB miss measurement. Figure 4(c) shows the running time of each MD step for 3,000 MD steps after ordering data at the first step. At 6,000 K, the running time gradually increases over time, and after 3,000 MD steps, the average running time per step increases by 8%. To study how much more performance degradation occurs after the initial data ordering, we extend the execution of 6,000 K dataset to 20,000 MD steps. The result shows an increase of the running time per step by 21%. This result indicates that without data reordering, the performance

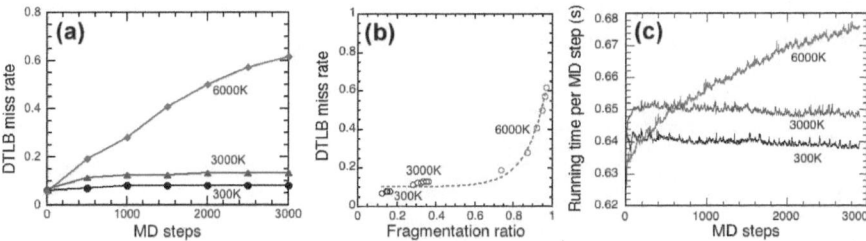

Fig. 4. (a) Time variation of the DTLB miss rate in 300, 3,000 and 6,000 K dataset over 3,000 MD steps. (b) Relation between the DTLB miss rate and the data fragmentation ratio in 300, 3,000 and 6,000 K datasets. (c) Running time per MD step as a function of MD steps at temperatures 300, 3,000, and 6,000 K for 98,304 silica atoms.

continues to degrade, and the running time per step continues to increase. On the other hand, the running time rapidly increases at the first 100 MD steps but then remains almost constant at 300 K and 3,000 K. These performance behaviors are akin to the fragmentation ratio and DTLB misses profiles in Figs. 2(a) and 4(a), respectively. These results thus confirm that data fragmentation in memory is indeed the source of performance deterioration during runtime through the increase of DTLB misses.

In Fig. 4(c), we also observe that 6,000 K simulation initially executes fastest compare to the other simulations (3.1% faster than 300 K and 2.5% faster than 3,000 K), which cannot be explained by the difference in the DTLB miss rate. One possible reason for this discrepancy is that higher temperature systems have sparser atomic distributions (*i.e.* lower atomic number densities), such that there are less number of atom interactions within the interaction range r_c, and hence less number of floating-point operations. To test this hypothesis, we measure the atom-distance distribution for all datasets. The results show that the number of atom pairs within distance r_c of 6,000 K system is approximately 3.0% and 1.4% less than those of 300 K and 3,000 K systems, respectively. As a result, 6,000 K simulation has the least computational load resulting in the fastest running time at the beginning of the simulation, which however is rapidly offset by the increase in DTLB misses.

In addition to the atom ordering discussed above, the cell ordering in the memory space also affects the locality. To address this issue, we measure the performance of three different cell ordering methods: 1) sequential ordering; 2) Hilbert-curve ordering; and 3) Morton-curve ordering [6]. The performance measurements are performed with 300 K temperature silica systems with grain size ranging from 12,288 - 331,776 atoms with reordering period of 20 MD steps on Intel Core i7 920 2.67 GHz (Nehalem) processor. The performance comparisons of all ordering methods are shown in Table 1. The results do not exhibit significant improvement due to different ordering methods. For example, the running time of Hilbert curve ordering is at most 1.76% less than that of the sequential ordering at largest granularity (331,776 atoms), while the running time of Morton curve ordering is less than 1% different for all granularities. One possible

Table 1. Runtime result with different ordering methods for silica systems at temperature 300 K. Reordering costs shown are obtained from sequential ordering.

Number of atoms	System dimensions (cells)	Average running time per MD step (ms)			Reordering cost (ms)
		Sequential	Morton	Hilbert	
12,288	10×10×10	78.48	78.57	78.45	0.66±0.02
98,304	20×20×20	639.9	640.1	637.1	8.41±0.07
331,776	30×30×30	2125.7	2117.4	2088.8	28.21±0.10

reason of the insensitivity of the performance on the cell-ordering method is the small number of references across intra-node cell boundaries. For simplicity, we employ the sequential ordering in the following.

5   Reordering Frequency Optimization

To minimize the performance degradation due to data fragmentation, we propose to repeat data reordering periodically during MD simulation. Specifically, we reorder atom arrays after every N_{rp} MD steps (*i.e.*, the reordering period N_{rp} is the number of MD steps between successive reordering operations). Though such data reordering is beneficial and often improves overall application performance, ordering arrays itself introduces an additional computational cost. Therefore, we here develop a dynamic data-reordering schedule based on a performance model and the runtime measurement in section 4. To do so, we first introduce a model that accounts for the reordering overhead. Let t_{cost} be the data reordering cost (*i.e.* the time to reorder arrays), N_{total} is the total simulation steps, and $t(n)$ is the running time at step n after ordering. The total running time, τ, as a function of reordering period, N_{rp}, is then written as

$$\tau(N_{rp}) = \left\lfloor \frac{N_{total}}{N_{rp}} \right\rfloor \left(\sum_{n=1}^{N_{rp}} t(n) + t_{cost} \right) + \sum_{n=1}^{\mathrm{mod}(N_{total}, N_{rp})} t(n) \qquad (1)$$

The optimal reordering period is then determined as the one that minimizes the total running time $N_{rp}^* = \mathrm{argmin}(\tau(N_{rp}))$. To find the ordering cost parameters in Eq. (1), we measure the reordering cost with different grain sizes. The results are summarized in Table 1.

Figure 5(a) shows the total running times for 3,000 MD steps as a function of the reordering period estimated from Eq. (1) with the measured reordering costs in Table 1. We obtain the optimal reordering period (which minimizes the total running time) as 69, 5, and 3 steps for 300 K, 3,000 K and 6,000 K datasets, respectively (see the arrows in Fig. 5(a)). With the optimally scheduled reordering thus determined, the overall performance is estimated to be improved by a factor of 1.35, 1.36 and 1.28 at 300 K, 3,000 K and 6,000 K, respectively.

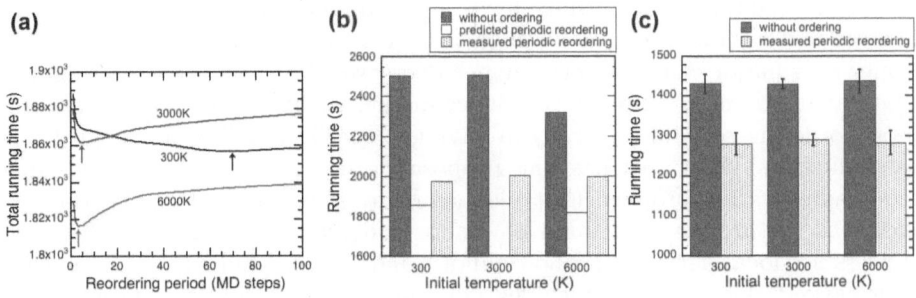

Fig. 5. (a) Total run time after 3,000 MD steps as a function of reordering period. The optimal period at 300 K, 3,000 K and 6,000 K are 69, 5 and 3 steps, respectively. (b) Comparison of total run time of silica MD over 3,000 steps achieved by model prediction and actual measurement of periodic reordering compared with that of the original code without ordering. (c) Total run time over 1,000 steps of parallel runs.

To verify this model prediction, we measure the running time of MD simulations that implement data reordering with the optimal reordering schedule. Here, MD simulations of silica containing 98,304 atoms at temperatures 300 K, 3,000 K, and 6,000 K are executed using periodic reordering at every 69, 5, and 3 steps, respectively. The results show that the measured speedups in actual executions are 1.27, 1.25, and 1.17, respectively, with 6.4%, 7.6%, and 10.0% error from the estimated values. The running time of the system without ordering, the optimized running time estimated from the model, and the measured running time with optimally scheduled reordering are compared in Fig. 5(b). To confirm that this performance benefit carries over to parallel runs, we performed a performance benchmark of silica MD with the same initial temperatures. The benchmark is executed on 192 cores of dual hexcore Intel Xeon X5650 2.66 GHz (Westmere) using 192,000 atoms per core (36 million atoms total). Figure 5(c) shows that speedups of 1.12, 1.11, and 1.12 are obtained from 300 K, 3,000 K, and 6,000 K runs, respectively. These figures shows a substantial effect of run-time data reordering on the performance of MD simulations. It should be noted that the state-of-the-art MD simulations are run up to 10^{12} time steps [2], for which the 10^3-step pre-profiling run for constructing the performance model, Eq. (1), can be amortized by updating the model every 10^6 steps.

6   Conclusions

We have developed an optimal data-reordering schedule for molecular dynamics simulations on parallel computers. Our analysis has identified physical/computational conditions such as a high temperature and a small granularity, which considerably accelerate data fragmentation in the memory space, thereby causing continuous performance degradation throughout the simulations at runtime. Our profiling results have revealed that the degree of data fragmentation correlates with the number of DTLB misses and have identified the former as a major cause of performance decrease. Based on the data fragmentation analysis and a simple performance model, we have developed an optimal data-reordering schedule, thereby archiving a speedup of 1.36 for 3,000 K silica simulation. This paper has thus proposed a practical solution to a dynamic data-fragmentation problem that plagues many scientific and engineering applications. Future research could be focused on other physical properties such as pressure, local density, and non-uniform mechanical loadings (such as shear deformation). Also, performance degradation in the light of other performance metrics such as cache misses could be explored. This work was partially supported by DOE-BES/SciDAC and NSF-CDI/PetaApps.

References

1. Chen, Y.C., Nomura, K., Kalia, R.K., Nakano, A., Vashishta, P.: Void deformation and breakup in shearing silica glass. Phys. Rev. Lett. 103(3) (2009)
2. Dror, R.O., Jensen, M., Borhani, D.W., Shaw, D.E.: Molecular dynamics and computational methods exploring atomic resolution physiology on a femtosecond to millisecond timescale using molecular dynamics simulations. J. Gen. Physiol. 135, 555–562 (2010)
3. Han, H., Tseng, C.W.: Exploiting locality for irregular scientific codes. IEEE Trans. Par. Dist. Sys. 17(7), 606–618 (2006)
4. Hu, Y.C., Cox, A., Zwaenepoel, W.: Improving fine-grained irregular shared-memory benchmarks by data reordering. In: Supercomputing (2000)
5. Kushima, A., Lin, X., Li, J., Eapen, J., Mauro, J.C., Qian, X.F., Diep, P., Yip, S.: Computing the viscosity of supercooled liquids. J. Chem. Phys. 130(22), 224504 (2009)
6. Mellor-Crummey, J., Whalley, D., Kennedy, K.: Improving memory hierarchy performance for irregular applications using data and computation reorderings. Int'l J. Par. Prog. 29(3), 217–247 (2001)
7. Nomura, K., Dursun, H., Seymour, R., Wang, W., Kalia, R.K., Nakano, A., Vashishta, P., Shimojo, F., Yang, L.H.: A metascalable computing framework for large spatiotemporal-scale atomistic simulations. In: IPDPS (2009)
8. Peng, L., Kunaseth, M., Dursun, H., Nomura, K., Wang, W., Kalia, R.K., Nakano, A., Vashishta, P.: A scalable hierarchical parallelization framework for molecular dynamics simulation on multicore clusters. In: PDPTA (2009)
9. Phillips, J.C., Zheng, G., Kumar, S., Kale', L.V.: NAMD: Biomolecular simulations on thousands of processors. In: Supercomputing (2002)
10. Shaw, D.E.: A fast, scalable method for the parallel evaluation of distance-limited pairwise particle interactions. J. Comp. Chem. 26(13), 1318–1328 (2005)
11. Singh, J.P., Hennessy, J.L., Gupta, A.: Implications of hierarchical N-body methods for multiprocessor architectures. ACM Trans. Comput. Sys. 13(2), 141–202 (1995)
12. Wang, W.Q., Clark, R., Nakano, A., Kalia, R.K., Vashishta, P.: Fast reaction mechanism of a core-shell nanoparticle in oxygen. Appl. Phys. Lett. 95(26) (2009)
13. Yao, Z.H., Wang, H.S., Liu, G.R., Cheng, M.: Improved neighbor list algorithm in molecular simulations using cell decomposition and data sorting method. Comput. Phys. Commun. 161(1-2), 27–35 (2004)

On Analyzing Quality of Data Influences on Performance of Finite Elements Driven Computational Simulations

Michael Reiter[1], Hong-Linh Truong[2], Schahram Dustdar[2],
Dimka Karastoyanova[1], Robert Krause[3], Frank Leymann[1], and Dieter Pahr[4]

[1] Institute of Architecture of Application Systems, Universität Stuttgart
{reiter,karastoyanova,leymann}@iaas.uni-stuttgart.de
[2] Distributed Systems Group, Vienna University of Technology
{truong,dustdar}@infosys.tuwien.ac
[3] Institute of Applied Mechanics (CE), Universität Stuttgart
krause@mechbau.uni-stuttgart.de
[4] Institute of Lightweight Design and Structural Biomechanics
Vienna University of Technology
pahr@ilsb.tuwien.ac.at

Abstract. For multi-scale simulations, the quality of the input data as well as the quality of algorithms and computing environments will strongly impact the intermediate results, the final outcome, and the performance of the simulation. To date, little attention has been paid on understanding the impact of quality of data (QoD) on such multi-scale simulations. In this paper, we present a critical analysis of how QoD influences the results and performance of basic simulation building blocks for multi-scale simulations. We analyze the impact of QoD for Finite Element Method (FEM) based simulation building blocks, and study the dependencies between the QoD of input data and results as well as the performance of the simulation. We devise and implement novel QoD metrics for data intensive, FEM-based simulations and show experiments with real-world applications by demonstrating how QoD metrics can be efficiently used to control and tune the execution of FEM-based simulation at runtime.

1   Introduction

For complex multi-scale simulations, e.g. to investigate structural changes within a human bone after a fracture of the arm, a common approach to perform scientific simulations is to transform the partial differential equations (PDEs) by means of the FEM to a system of linear or nonlinear matrix equations that must be solved. In such multi-scale simulations, FEM algorithms can be used at different scales, such as the skeleton, the bone structure and the bone cell simulations. Therefore, FEM algorithms play an important role in computational science. Because of their importance, understanding quality of these algorithms have attracted several research projects. However, most of them focus on subareas, for instance, optimizing matrix solver [1] or other performance aspects [2].

C. Kaklamanis et al. (Eds.): Euro-Par 2012, LNCS 7484, pp. 793–804, 2012.
© Springer-Verlag Berlin Heidelberg 2012

FEM algorithms, in general, consume or produce input data (e.g. to describe the simulation object), intermediate results (e.g. results for intermediate time steps), the final result, other output data (e.g. status data), and internal used data (e.g. FEM grid or matrix). These types of data have large volume and some of them have complex data structures.

We argue that the quality of the input data, together with, the quality of algorithms and computing environments will strongly impact the intermediate results, the final simulation output, as well as the performance and storage provisioning of the simulation. This is particular applicable to multi-scale simulations in which the output data in one scale will be used as input data on the other scale. Not to mention that even in a single scale simulation, there are various steps, in which different types of data are processed and produced. Quality of data (QoD) can strongly affect the selection and operation of algorithms as well as computing and storage resources in these steps and in the data exchange among these steps and among scale-specific simulations. While several research papers have discussed possible computing environments and algorithms in detail for FEM, little attention has been paid on understanding the impact of QoD on the performance and resource/storage provisioning in phases of FEM-based simulations. Because FEM-based simulations in computational science are typical long running and produce large amount of data, and are expensive in terms of time and money [3,4]. Detecting poor data quality and able to understanding the impact of QoD in FEM-based simulations could potentially save time and money. In particular, being able to understand QoD influences on performance and resource provisioning for FEM-based simulations can also help to develop better strategies for pay-per-use resources in cloud computing environments.

In this paper, we focus on understanding major QoD metrics for FEM-based simulations and on analyzing the dependencies among QoD metrics to the quality of according intermediate results, performance, storage needs, and the QoD of the simulation as a whole. Fundamentally, we focus our work on general FEM-based simulation steps that can be considered as a basic building block for multi-scale simulations. We have developed several QoD metrics and analyzed trade offs between different QoD findings and simulation execution time. We present our experiments with a real world simulation. To the best of our knowledge, this is the first attempt to analyze the impact of FEM QoD metrics on simulations.

The rest of this paper is organized as follows: Section 2 discuss the background of our work. Section 3 presents and defines QoD metrics. Section 4 presents our prototype and experiments. Section 5 presents related work, followed by the conclusions and future work follows in Section 6.

2   Quality of Data Implications in FEM Based Simulations

2.1   Identifying Important Types of Data in FEM-Based Simulations

Different tools and frameworks exist to execute FEM based simulations. Some of them, e.g. Ansys[1], use pre-implemented functions and have a strong FEM focus

---

[1] http://www.ansys.com/

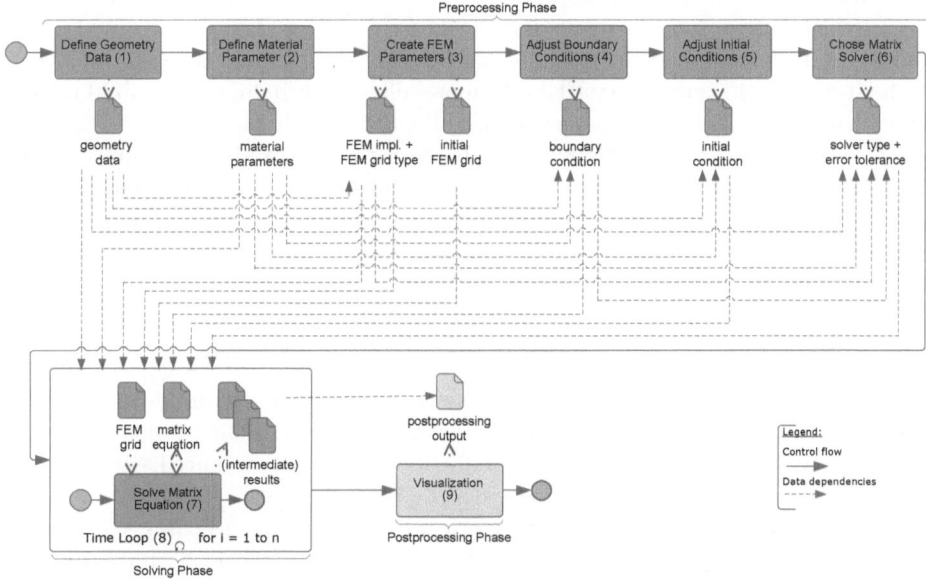

Fig. 1. Steps in a basic FEM-based simulation building block and important types of data and important data dependencies

while others could be part of a global technical computing environment, such as MatLab[2]. Physics specific FEM frameworks, like PANDAS[3], permit to create simulations in a very flexible way.

A FEM-based multi-scale simulation can have millions of basic simulation building blocks, e.g. a human skeleton has approximately 206 bones, each bone has thousands to millions of cells, where each cell can be simulated by a FEM-based simulation building block. Nevertheless a FEM-based simulation building block has a common procedure that can be divided into three phases: (i) *preprocessing*, (ii) *solving*, and (iii) *postprocessing*. The three phases can be divided again into different steps. Figure 1 illustrates the main steps for transient problems, in which the preprocessing phase includes steps 1 to 6, steps 7 and 8 belongs to the solving phase, and step 9 is in the postprocessing phase. In the preprocessing phase all relevant input data are collected, based on that an initial FEM grid and matrix equation can be generated. Matrix equations will be adapted and solved in the solving phase. The results are processed (e.g. visualized) in the postprocessing phase. In this paper, we focus on QoD metrics for the FEM-based simulation building block, in particular for the preprocessing and the solving phases.

From our study, we determine five relevant *classes of data* in the preprocessing and solving phases: (i) *PDE driven input data* (input data), (ii) *FEM driven*

---

[2] http://www.mathworks.com/products/matlab/
[3] http://www.mechbau.uni-stuttgart.de/pandas/index.php

input data (input data), (iii) *internal/temporary FEM data* (output/input data), (iv) *intermediate results* (output data), and (v) *final result* (output data). PDE and FEM driven input data are generated during the preprocessing phase. Most of the internal and temporary FEM data as well as intermediate and final results are computed at the solving phase. Generally, the QoD in these classes is critical for the quality of the simulation result, the performance, and the amount of data. Several types of data that can influence the QoD of FEM based simulations, such as the *geometry, material parameter, FEM interpolation, FEM grid type, boundary condition, initial condition, matrix solver type, matrix error tolerance, matrix equation, FEM grid,* and *time step*[4].

2.2 Defining and Evaluating Quality of Data Metrics

In general, there can be several QoD metrics. However, [6] pointed out that the term QoD is used in different domains with several meanings and must be regarded specific to the application domain. QoD can be measured with the aid of metrics [6]. A metric must be provided by one or more well defined measurement method. Furthermore, several meta information are necessary: (i) where the measurement is taken, (ii) what data are included, (iii) the measurement device, and (iv) the scale on which results are reported. To evaluate QoD, we distinguish between QoD verification and validation with respect to the FEM phases. Related to the IEEE-STD-610 definition of validation and verification in software products [7] QoD verification and validation is defined as follow:

Definition 1 (Quality Of Data Verification). *The process of evaluating quality of data during or at the end of the preprocessing phase to determine whether it satisfies specified quality of data demands.*

Definition 2 (Quality Of Data Validation). *The process of evaluating quality of data during or at the end of the solving phase to determine whether it satisfies specified quality of data demands.*

The QoD Verification definitions means that verification makes sure that the simulation (i) fulfills the specification of the simulation object, (ii) is derived from the specification of the simulation object that the simulation fulfills the specification of the PDE, and (iii) is derived from the specification of the PDE that the simulation fulfills the specification of the FEM; or for short: computes the problem right. The QoD Validation definitions means that validation make sure that the simulation actually fulfills the simulation intention; or for short: computes the right problem.

2.3 Identifying Factors Influencing Quality of Data

Each step has different influences to simulation characteristics like QoD, data quantity, or performance. In Figure 1, in step 1, the dimension (e.g. 3 D) or

---

[4] Due to the lack of space, we provide a supplement report to document important types of data and their possible influences as well as possible FEM-based specific QoD metrics in [5].

complexity (e.g. smoothness) of the simulation object – described by *geometry data* – has strong influences to the size of the result data and to the performance. The *material parameters* in step 2 approximate the material behavior within the PDE based on a given physics. Changes to material parameter influences the behavior of the PDE. In step 3, data that describe the *FEM grid type*, the *FEM implementation*, and an *initial FEM grid* are involved. Hence, interpolation functions (e.g. power series) affect the performance and the result quality. The size, complexity, and type of the FEM grid strongly influence the performance, the data quantity, and the QoD. *Boundary condition* and *initial condition* in step 4 and 5 could be defined with several accuracies and numerical characteristics which affect the QoD and the performance. A matrix *solver type* and, in some cases, an *error tolerance* must be chosen in step 6. Each solver has different characteristics in relation to error behavior, QoD, and performance.

Based on a *FEM grid* a *matrix equation* (step 7) could be solved with different precisions which influence the performance and the result quality. Additional internal or temporary data, such as the Jacobian determinant, influence the quality of the *(intermediate) result*. In step 8 the increment of *time step* δt_i influences the accuracy, the performance, and the data quantity.

Furthermore, there are relevant constrains between the steps regarding to QoD. Figure 1 depict essential data dependencies that need to be observed. Considering the important role of quality of data, we present novel QoD metrics for FEM based simulations and discuss the influences.

3   QoD Metrics for FEM-Based Simulations

QoD metrics are determined based on our evaluation important types of data that influence FEM based simulations detailed in [5]. We distinguish between *basic* and *constrained* input data as characteristics of types of data. As shown in Figure 1, basic input data have no strong dependencies to other input data and are typically data to describe the PDE. Constrained input data are essentially dependent on basic input data and can have relations to other constrained input (or combined output/input) data.

We consider QoD as a tuple of characteristics and goodness [8]. A characteristics of data will be analyzed without any simulation context, while the goodness of the characteristics of data will be evaluated with respect to the specific simulation context. To simplify the wording, we use the term QoD metric result in the following as a synonym for the goodness value of the QoD tuple. Regarding to [6], a QoD metric result will have a value calculated by a well-defined quality objective. In this paper, we use for the quality objective an interval from 0 to 1. Hence, we defined the QoD metric result for FEM-based simulations as follows:

Definition 3 (Quality of Data / Goodness). *The Quality of Simulation Data / Goodness (or QoD metric result) is an objective represented by a value $Q \in [0, 1]$ that determines the quality of input and output data of a FEM based simulation with both limits $0 = bad$ and $1 = good$. $Q \in (0, 1)$ determines a quality between bad and good.*

Table 1. Selected Quality of Data metrics for FEM-based simulations

QoD Metrics	Description
Geometry Accuracy	Metric related to Figure 1 step 1 to verify geometry data of the PDE
Material Parameter Accuracy	Metric related to Figure 1 step 2 to verify material parameters of the PDE
Interpolation Accuracy	Metric related to Figure 1 step 3 to verify interpolation functions of FEM implementation data
FEM Grid Type Adequacy	Metric related to Figure 1 step 3 to verify the FEM grid type
FEM Grid Accuracy	Metric related to Figure 1 step 3 to verify the fineness of the (initial) FEM grid
Boundary Condition Accuracy	Metric related to Figure 1 step 4 to verify the accuracy of the boundary condition of the PDE
Initial Condition Accuracy	Metric related to Figure 1 step 5 to verify the accuracy of the initial condition of the PDE
Matrix Solver Accuracy	Metric related to Figure 1 step 6 to verify the accurate selection of a matrix solver type for the FEM
Matrix Solver Error	Metric related to Figure 1 step 6 to verify the maximal error tolerance for a matrix solver type for the FEM
FEM Element Condition	Metric related to Figure 1 step 7 to validate the condition of an element within a FEM grid
Matrix Equation Condition	Metric related to Figure 1 step 7 to validate the numerical condition of a matrix within a matrix equation
Vector Condition	Metric related to Figure 1 step 7 to validate the numerical condition of a vector within a matrix equation
Time Discretization Accuracy	Metric related to Figure 1 step 8 to validate the accuracy of a time step

Based on this definition a value $Q_{QoDmetric}$ determines the QoD metric result with respect to the specific *QoDmetric*. Corresponding to the influencing factors, Table 1 presents a list of QoD metrics we have developed and their associated data as well as data types described in Section 2 (see our supplement report [5] for detailed explanation).

We introduce in detail three implementations of QoD metrics: Material Parameter Accuracy (Q_{MPA}), Matrix Solver Accuracy (Q_{MSA}), and Vector Condition (Q_{VC}). Q_{MPA} (Figure 1 step 2) and Q_{MSA} (step 6) verify the quality in the preprocessing phase. Q_{MPA} is based on the quality of PDE driven basic input data and will be used if the implemented parameter approximates a known parameter. Q_{MSA} is premised on the quality of FEM driven constrained input data and makes a statement about the expected accuracy of the numerical solution before the solving phase starts. With this knowledge a time and money consuming simulation can be adopted or aborted if a poor QoD is expected. Q_{VC} (step 7) of both vectors **b** and **x** concerns the solving phase and determines the quality of constrained internal/temporary FEM data within a matrix equation $\mathbf{Ax} = \mathbf{b}$. In contrast to Q_{MSA}, Q_{VC} validates and makes a statement about the real condition of the numerical solution at the solving phase. Nevertheless, even as the solving phase a simulation can be adopted or aborted to save time and money.

3.1 Material Parameter Accuracy

Most simulations run with estimated material parameters or average values. The QoD metric Q_{MPA} helps to verify implications of inaccurate material parameters. The correctness of the material parameter to describe the phenomenological behavior of the material based on a given physics depends on the accurate description of all relevant parameter. We define the error rate of all implemented material parameter: Given a region bm that describes the model of the simulation object and an interval $[t_0, t_n]$ that describes the simulation time period. Given a function $rMP(x, t)$ that describe the phenomenological behavior of the real material and a function $mMP(x, t)$ that describe the phenomenological behavior of the material in the implemented FEM model. The characteristic of material parameter accuracy Q_{MPA} without respect to the specific simulation context is defined by $min\left(\left|\frac{mMP(x,t)}{rMP(x,t)}\right|, \left|\frac{rMP(x,t)}{mMP(x,t)}\right|\right)$ for all $x \in bm$ and $t \in [t_0, t_n]$. Based on this definition the error rate of a specific material parameter j can be defined in the same manner.

3.2 Matrix Solver Accuracy

A matrix solver implements numerical methods to solve matrix equations and approximates (in most cases) the exact solution of the matrix equation. Q_{MSA} helps to verify implications of numerical problems. x_e is the exact solution of a test matrix equation $\mathbf{Ax} = \mathbf{b}$ and x_m is the numerical solution by using matrix solver m. We define the characteristics of the matrix solver accuracy Q_{MSA} by $\|x_e - x_m\|$ with $\|\|$ is a appropriate norm. If it is not possible to determine the exact solution x_e objectively a domain expert can determine the characteristics of Q_{MSA} subjectively.

3.3 Vector Condition

To solve a matrix equation $\mathbf{Ax} = \mathbf{b}$ with numerical methods, the condition of the vector $\mathbf{b}$ (and of $\mathbf{A}$) influences the solving performance and the quality of the solution $\mathbf{x}$. Q_{VC} helps to validate implications of numerical problems. If the difference regarding the least absolute value and the maximum absolute value of $\mathbf{b}$ is "too big" numerical errors can be estimated. b_l is the least absolute value with $b_l \neq 0$ and b_m the maximum absolute value of $\mathbf{b}$. The characteristics of the vector condition without respect to the specific simulation context is defined by $Q_{VC} = \frac{b_l}{b_m}$. We define the vector condition for the vector $\mathbf{x}$ in the same way. For this, we replace vector $\mathbf{b}$ and the values b_l and b_m by vector $\mathbf{x}$ and the values x_l and x_m.

4   Evaluating the Influence of Quality of Data Metrics

4.1   QoD Evaluation Framework

In order to measure, monitor, and evaluate QoD metrics for FEM-based simulations, we utilize our extensible QoD Evaluation Framework for workflows
developed in [8]. Generally, with this framework, we can determine QoD in a
very flexible way: (i) platform and language independent metrics and interpretations can be invoked, (ii) separate metrics as well as metrics that are included
in comprehensive algorithms (e.g. a solver that include algorithms to determine
the Jacobian determinant) can be used, and (iii) objective (automatically determined by a computer) and subjective (manual determined by a human) QoD
determination is supported. During runtime, relevant data is needed for the determination of QoD is passed to the QoD Evaluation Framework by values or
by references. This approach enables us to shift data to QoD metrics and interpretations or shift QoD metrics and interpretations to data to improve the
performance.

Conceptually, Figure 2 describes how we utilize the QoD Evaluation Framework for understanding the dependencies among QoD for inputs, intermediate results, and final outcomes, as well as the influences of QoD on the performance
and resource provisioning of the simulation. All necessary information about available QoD metrics and interpretations are stored into `Metric Definition`. Implementations of QoD metrics and interpretations can be found in `Software-based`
`Evaluator` for automatic QoD determination and in `Human-based Evaluator` for
manual QoD determination. Both kinds of evaluators and the data provisioning
are managed by `Manager`. Furthermore, the `Manager` has all meta information
described in Section 2.2. For QoD determination the `Manager` searches `Metric`
`Definition` for an appropriated QoD metric and a corresponding QoD `Evaluator`
and passes information about data (in XML messages) to the `Evaluator`. The
`Evaluator` analyzes the specified data about their characteristics and returns XML-
based `Metric Result` – including values of QoD metrics – to the `Manager`. The
metric results will be analyzed and displayed together with, e.g., performance, failures, or storage informations.

4.2   Experiments

To analyze the influence of QoD metrics and goodness we used simulations developed atop the PANDAS framework. PANDAS was designed for simulations
of multiphasic materials [9]. We used it for two reasons: It represents a typical
FEM based simulation framework that can be used within a workflow environment (e.g. to perform multi scale simulations) and FEM-based basic simulation
building blocks can be structured into the steps shown in Figure 1 [10].

We implement two simple but well proved simulations [11] simulating the same
problem but having fundamental different characteristics in terms of the elasticity of the boundary: "fluid-saturated elastic column in an impermeable rigid tub"
(EC) and "rigid slab on a fluid-saturated elastic half space" (RS). Furthermore,
two solvers (GMRES and BiCG) with different numerical behaviors are utilized.

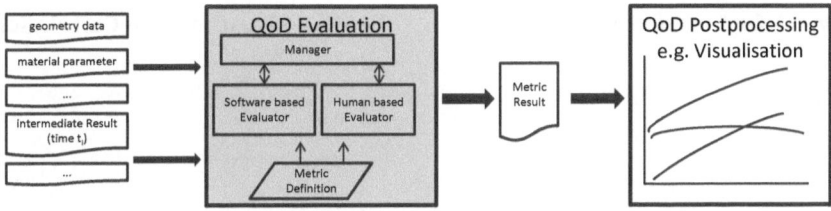

Fig. 2. Components and steps in evaluating QoD using the QoD Evaluation framework

To analyze the impact of QoD the presented QoD metrics Material Parameter Accuracy (Q_{MPA}), Matrix Solver Accuracy (Q_{MSA}), and Vector Condition (Q_{VC}) as well as performance measurement metrics (time in hour) are used. The goodness of QoD metrics with respect to the simulation context is defined based on domain expert's knowledge (due to the lack of space, the exact setting of the experiments can be found in [5]). We execute the experiments on a four CPU machine because it is sufficient to calculate our QoD metrics for FEM-based basic simulation building blocks, in particular for the preprocessing and the solving phase shown in Figure 1, of multi-scale simulations. Furthermore, it clarifies the analysis of the influence of QoD metrics and performance. Nevertheless, in practice we could use our approach to control multi-scale simulations with the aid of QoD findings within a multiple of computational resources when a large of basic simulation building blocks are employed.

To demonstrate the applicability and strength of our novel approach we have performed different correlations. We present three examples in detail: (i) the correlation between the goodness of Q_{MPA} and Q_{VC}, (ii) the correlation between the goodness of Q_{MPA} and the fault behavior, and (iii) the correlation between the goodness of Q_{MSA} and the performance. Several other correlations can be implemented. Examples are the correlation between the goodness of a QoD metric and storage provisioning or the correlation between the goodness of QoD metrics determined in different scales with a multi-scale simulation. Due the lack of space we present only selected findings in the following.

QoD-QoD Correlation: The QoD-QoD correlation (Figure 3 left) describes the relation between the goodness of Q_{MPA} (material parameter accuracy) and Q_{VC} (vector condition). Although the EC and the RS simulations are similar only with the exception of the flexibility of the boundary, the results of the Q_{VC} differ considerably. The RS simulation is less sensitive to material parameter changes than the EC simulation. As shown in Figure 1 the Q_{VC} of matrix data outputted in step 7 (vector **x**) depends on Q_{MPA} in step 2. In the RS simulation a poor QoD within material parameter data do not influence the limited or good QoD of the vector condition. In contrast, at the EC simulation the Q_{VC} is overall poor and tends to have critical values when $Q_{MPA} \leq 0.7$. Based on this QoD-QoD correlation findings we concluded that: (i) the RS simulation has an uncritical numerical behavior at the vector **x** with respect to a not well accurate material parameters, (ii) the EC simulation has overall a poorer numerical behavior in

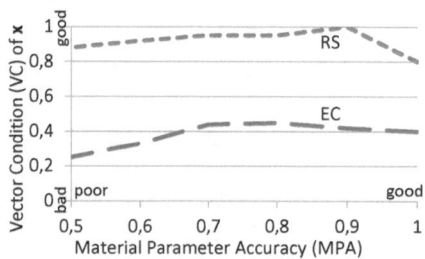
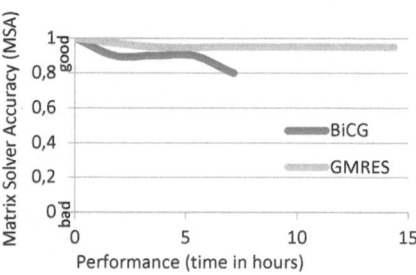

Fig. 3. Results of EC and RS simulations relating to parameter accuracy (MPA) and vector condition (VC) (left) and results of EC simulations relating to matrix solver accuracy (MSA) and performance (right)

solving the vector $\mathbf{x}$, (iii) the EC simulation shows an critical numerical behavior at the vector $\mathbf{x}$ with respect to a not well accurate material parameters. As a consequence of the findings, our tool indicates that EC simulation should use correct material parameters as well as suitable matrix solvers that can handle a poor vector condition.

QoD-Fault Correlation: The QoD-Fault correlation describes the relation between the goodness of Q_{MPA} and the fault behavior (termination of the simulation framework) by executing the EC simulation. When $Q_{MPA} \leq 0.3$ the simulation framework PANDAS frequently terminates the calculation during the solving phase with error messages (no convergence). Overall we observed that in one of five simulations PANDAS terminated at the initial time step δt_0, in three of five simulations PANDAS terminated before the time step δt_{5000}, and in one of five simulations ended without error. Based on this knowledge a simulation should be adapted at the preprocessing phase with material parameters that have a sufficient QoD value in order to avoid simulation failures.

QoD-Performance Correlation: The QoD-Performance correlation describes the relation between Q_{MSA} and runtime of the EC simulation in hours. Figure 3 (right) shows that the used solvers have a different behavior: for 5000 time steps, the BiCG matrix solver took a half of the execution time (7 hours and 10 minutes) that the GMRES solver (14 hours and 21 minutes) required. However, for longer running simulations the GMRES solver provides results with a good QoD (0.95) but the BiCG solver calculates results that differs slightly from a proven result and the achieved QoD is limited to 0.8. Hence, our QoD framework indicates that (i) the GMRES solver provides reliable results, (ii) for simulations that run with a relative small number of time steps, the BiCG solver produces good quality of results in a fast way. By using our framework, scientists can find out the maximum number of time steps on which the Q_{MSA} is useful for their specific simulations.

5   Related Work

This section presents related work in the fields of QoD influences for computational simulations. Batini et al. present general concepts, methodologies, and techniques as well as a process to determine QoD [6]. We adopt it for the field of FEM based computational simulations. Hey et al. pointed out the principle need to observe QoD in scientific applications [3]. Heber and Gray concertize this need for FEM driven computational simulations [4] and have implemented a runtime environment for FEM based simulation [4,12]. But those implementations do not support any QoD evaluation.

The principles of the FEM are characterized by [13]. Those principles includes simulation dependent QoD aspects, e.g., basic approaches for error minimizing in the overall FEM and for several FEM steps. But the authors do not focus on simulation independent QoD metrics. Several approaches were already created for error minimizing in FEM. Lots of papers discuss special aspects in one step in detail. An example is description of the implementation of the concept of geometric multigrid algorithms and hierarchical local grid refinement [1]. Those kinds of paper depict in each case not the whole simulation (preprocessing, solving, and postprocessing phase). Nevertheless, we can use those approaches within specific QoD metrics.

Beyond that, performance analysis of FEM based simulation is well studied as well as optimization, e.g., in [2,14]. Our work is different as we focus primly on understanding QoD and only effected by this on research questions like performance or data quantity. In existing FEM tools or frameworks, such as Ansys, MatLab, or PANDAS QoD checking has not been provided. To our best knowledge, there is no work to define QoD metrics for FEM driven computational simulations in general and to study how QoD metrics impact on simulations.

6   Conclusions and Future Work

This paper analyzes common steps in FEM based computational simulations and proposes novel QoD metrics for the evaluation of the QoD and performance of FEM-based simulations. We have investigated how such metrics could provide benefit for understanding the inter-dependencies among QoD of inputs, intermediate data, and outputs as well as the performance needs of entire simulations. By monitoring and analyzing QoD influences with two real world simulation examples, we have shows that different types of QoD analyses could be very useful for understanding and steering FEM-based simulations.

We concentrate on basic simulation building blocks that can be used for general FEM-based multi-scale simulations. Therefore, new metrics and analysis methods must be investigated for data exchanged among different building blocks within and across specific simulation scales as well as for specific FEM approaches. In our future work, we focus on steering multi scale simulations and adapting resource provisioning based on QoD influences.

Acknowledgments. The work presented in this paper has partly been funded by the DFG Cluster of Excellence Simulation Technology (http://www.simtech.uni-stuttgart.de) (EXC310).

References

1. Bastian, P., Blatt, M., Dedner, A., Engwer, C., Kloefkorn, R., Ohlberger, M., Sander, O.: A generic grid interface for parallel and adaptive scientific computing. part i: abstract framework. Computing (2) (2008)
2. Chen, P., Zheng, D., Sun, S., Yuan, M.: High performance sparse static solver in finite element analyses with loop-unrolling. Adv. Eng. Softw. (4) (2003)
3. Hey, A., Tansley, S., Tolle, K.M.: The fourth paradigm: data-intensive scientific discovery. Redmond, Wash.: Microsoft Research (2009)
4. Heber, G., Gray, J.: Supporting finite element analysis with a relational database backend, part i: There is life beyond files. CoRR (2007)
5. Reiter, M., Truong, H.L.: Supplement report for quality of data implications. Technical report (2012), http://www.iaas.uni-stuttgart.de/institut/mitarbeiter/reiter/Report/FemQoD.pdf
6. Batini, C., Scannapieco, M.: Data Quality: Concepts, Methodologies and Techniques. Data-Centric Systems and Applications. Springer (2006)
7. IEEE: IEEE standard computer dictionary: A compilation of IEEE standard computer glossaries, IEEE std 610-1990 (1990)
8. Reiter, M., Breitenbuecher, U., Dustdar, S., Karastoyanova, D., Leymann, F., Truong, H.L.: A novel framework for monitoring and analyzing quality of data in simulation workflows. In: 7th IEEE e-Science International Conference (2011)
9. Ehlers, W.: Foundation of multiphasic and porous material. In: Ehlers, W., Bluhm, J. (eds.) Foundation of Multiphasic and Porous Material. Springer (2002)
10. Reimann, P., Reiter, M., Schwarz, H., Karastoyanova, D., Leymann, F.: Simpl – a framework for accessing external data in simulation workflows. In: 14. GI-Fachtagung Datenbanksysteme für Business, Technologie und Web (2011)
11. Ehlers, W., Eipper, G.: Finite elastic deformations in liquid-saturated and empty porous solids. Transport in Porous Media (1999)
12. Heber, G., Gray, J.: Supporting finite element analysis with a relational database backend, part ii: Database design and access. CoRR (2007)
13. Zienkiewicz, O., Taylor, R., Zhu, J.: The Finite Element Method - Its Basis & Fundamentals, 6th edn. Elsevier Ltd., Butterworth-Heinemann (2005)
14. Shadid, J., Hutchinson, S., Hennigan, G., Moffat, H., Devine, K., Salinger, A.G.: Efficient parallel computation of unstructured finite element reacting flow solutions. Parallel Computing (9) (1997), Parallel computing methods in applied fluid mechanics

Performance Evaluation and Optimization of Nested High Resolution Weather Simulations

Preeti Malakar[1], Vaibhav Saxena[2], Thomas George[2], Rashmi Mittal[2],
Sameer Kumar[3], Abdul Ghani Naim[4], and Saiful Azmi bin Hj Husain[4]

[1] Indian Institute of Science
preeti@csa.iisc.ernet.in
[2] IBM Research - India
{vaibhavsaxena,thomasgeorge,rasmitta}@in.ibm.com
[3] IBM T.J. Watson Research Center
sameerk@us.ibm.com
[4] Universiti Brunei Darussalam, Brunei
{ghani.naim,saiful.husain}@ubd.edu.bn

Abstract. Weather models with high spatial and temporal resolutions
are required for accurate prediction of meso-micro scale weather phenomena. Using these models for operational purposes requires forecasts with
sufficient lead time, which in turn calls for large computational power.
There exists a lot of prior studies on the performance of weather models
on single domain simulations with a uniform horizontal resolution. However, there has not been much work on high resolution nested domains
that are essential for high-fidelity weather forecasts.

In this paper, we focus on improving and analyzing the performance
of nested domain simulations using WRF on IBM Blue Gene/P. We
demonstrate a significant reduction (up to 29%) in runtime via a combination of compiler optimizations, mapping of process topology to the
physical torus topology, overlapping communication with computation,
and parallel communications along torus dimensions. We also conduct a
detailed performance evaluation using four nested domain configurations
to assess the benefits of the different optimizations as well as the scalability of different WRF operations. Our analysis indicates that the choice
of nesting configuration is critical for good performance. To aid WRF
practitioners in making this choice, we describe a performance modeling approach that can predict the total simulation time in terms of the
domain and processor configurations with a very high accuracy ($< 8\%$)
using a regression-based model learned from empirical timing data.

1   Introduction

Operational weather forecasting is critical for planning operations in weather sensitive sectors such as energy, transportation, urban planning, and public safety.
Such weather forecasting is performed using fine resolution regional and global
atmospheric models that discretize the nonlinear partial differential equations
representing evolution of atmospheric flows in time, which entails a huge computational effort. It is also imperative that the forecasts are provided with sufficient
lead time (24-48 hrs) in order to allow actions that mitigate the socio-economic

C. Kaklamanis et al. (Eds.): Euro-Par 2012, LNCS 7484, pp. 805–817, 2012.
© Springer-Verlag Berlin Heidelberg 2012

impact. Hence, it is critical to have a highly efficient and scalable execution of the weather models on a high-performance computing platform.

Weather and Research Forecasting model (WRF) is a state-of-the-art regional to global-scale numerical weather prediction model that is used by weather agencies all over the world. WRF has been designed to perform well on massively parallel computers. It can be built in serial, parallel (MPI) and mixed-mode (OpenMP and MPI) forms and is available on various HPC machines. Motivated by earlier WRF performance studies on the IBM Blue Gene series [1,8], we explore optimizing WRF on the Blue Gene/P machine. Past studies on WRF are mainly focused on simple single domain benchmarks. These are not representative of real world short-term high-fidelity weather simulations[1,2] that often require nested domain configurations with one or more small high resolution domains (nests) embedded into a coarse resolution parent domain such as those in Figure 1. Fine resolution runs can effectively model weather at meso-micro scale because of higher granularity, but also require a proportionally smaller time step for numerical stability resulting in a quadratic increase in the net computational effort. Domain nesting is essential to achieve good prediction accuracy over small regions of interest (child domains) while avoiding expensive computation across the whole parent domain.

Nested domain simulations differ from single domain simulations in one key aspect. Modeling (i.e., the solve operation) needs to be performed at multiple (parent and child) spatial resolutions and the results need to be communicated and aligned at the points of overlap. The data for the finer resolution child domains are interpolated from the coarser domain by a process called *forcing* in WRF. In a two-way nest integration, the finer grid solution also overwrites the coarser grid solution for the coarse grid points that lie inside the finer grid by a process called *feedback* [11].

There are three key challenges faced by WRF users while performing nested domain simulations. First, nesting entails significant communication between the parent and child domains in the form of forcing and feedback operations, which results in an increased run time and poor scalability that in turn affect the forecast lead time. Second, there does not exist much work on scalability analysis of nested domain simulations that can provide guidance to WRF users on the potential benefits and trade-offs associated with extra computing resources. Lastly, there is risk of over-decomposition on a small-sized domain when the number of processors is large. Therefore, it is critical to choose the nesting configuration to ensure that the nest domain sizes are appropriate for the available processor configuration. Due to the relative opaqueness of the WRF code, most users typically employ a tedious and time-consuming trial and error approach that involves running the code multiple times in order to make this decision.

We make the following contributions:

(a) We significantly reduce the runtime of WRF nested domain simulations (up to 29%) via compiler optimizations on the IBM Blue Gene/P machine, efficient mapping of 2D process topology of WRF on to the 3D torus topology of BG/P to reduce the communication time in WRF and optimization of some

---

[1] http://www.emc.ncep.noaa.gov/mmb/mmbpll/nestpage/overlay_hiresw4km.jpg

[2] http://www.metoffice.gov.uk/research/modelling-systems/unified-model/weather-forecasting

critical portions of WRF source code. Unrolling the Z dimension loops and parallelizing the X and Y dimension communications leads to further optimization.

(b) We conduct a detailed performance evaluation of the original as well as the optimized WRF code by performing simulations with varying number of processors over four nested domain configurations with 2-level nesting and varying sibling domains at the innermost level. Our results indicate that forcing and feedback operations do not scale well. Further, performance comparison of similar nesting configurations indicates that optimal nesting choice varies with the number of processors.

(c) We propose a performance modeling approach for estimating the total integration time for any multi-level nested domain WRF simulation, which is based on learning regression models for each of the key WRF operations (solve, forcing, feedback) using empirical timing data. Practitioners can use such a model to determine the best (lowest runtime) nesting configuration among multiple choices for a given number of processors.

2   Related Work

WRF has been extensively studied in the HPC community since weather modeling is a key HPC application. Most of this work can be broadly grouped into three categories: (a) optimization for specific HPC architectures, (b) performance analysis, and (c) performance modeling.

Architecture-Specific Optimization. In the past, WRF code has been optimized for a number of HPC architectures such as BG/L [8], Cray XT4 [9]. For the BG/P architecture, Bhatele et al.[1], present a framework for automatic mapping of WRF onto the BG/P torus topology using the WRF communication graph. On a single domain configuration, they demonstrate that their mapping reduces the average number of hops per process by up to 60%, but increases the total communication time by 40% in certain cases. This approach is yet to be validated on nested domain configurations. Currently, the only existing work that deals with optimization for nested domain configurations is that of Porter et al. [9]. They studied compiler optimizations for improving WRF performance using the PathScale and PGI compilers on Cray XT4/5, which do not apply to BG/P. They also report improvement in WRF performance upon changing the default X-Y processor decomposition, which does not hold true in case of BG/P.

Performance Analysis. Wright et al. [12] examine the scalability of WRF (version 2.1.2) across different architectures using IPM to analyze the performance. They show that for most of the architectures, WRF exhibits a sublinear speedup of both computation and communication times with increasing number of cores and also identify bottlenecks such as MPI_Wait. There exists a number of other performance and profiling studies [10] focusing on other aspects of WRF. However, these studies only focus on a single domain configuration and do not consider forcing/feedback operations that are prominent in nested configurations.

Performance Modeling. Given the criticality of lead time, there is a strong interest in predicting the execution time of WRF runs. Kerbyson et al. [3] describe an analytical performance model with parameters such as the grid size and processor configuration. This model was developed via a careful manual inspection of the dynamic execution behaviour of the WRF application and was subsequently validated using performance measurements on real systems. Unlike [3], our model makes use of regression analysis to directly learn the coefficients of potential influencing factors using empirical timing data. Delgado et al. [2] also describe a regression-based approach for modeling WRF performance on systems with < 256 processors, but their primary focus is on capturing the system related factors such as clock speed, network bandwidth, which they do via a multiplicative effects model. Our modeling focuses on the interaction of domain and processor configurations assuming fixed architecture and simulation parameters allowing us to obtain much better accuracy even up to 1024 processors. Further, [2] and [3] only focus on single domain configurations whereas we learn separate models for solve/forcing/feedback operations, which can be used for predicting performance for even multi-level nested configurations.

3   Experimental Setup

Blue Gene/P Overview. The IBM Blue Gene/P (BG/P) is the second generation in the line of Blue Gene machines after Blue Gene/L. Each BG/P node has four 850 MHz embedded PowerPC 450 cores on a single ASIC and can achieve a peak floating point throughput of 13.6 GF/node. The software stack supports three modes: symmetric multi-processing mode (or SMP mode) with one process and up to four threads, dual mode with two processes, each with up to two threads and quad mode (also known as virtual node mode, or VN mode) with four processes. Systems software provides optimized MPI libraries [4] and an OpenMP runtime via the IBM XL compiler to take advantage of the 4-way SMP node. MPI point-to-point messages are sent on the 3D torus network, while global collective communication operations such as barrier, broadcast and allreduce on MPI_COMM_WORLD are executed on the collective network [5].

Nested Domain Configurations. For our experiments, we simulated a heavy rainfall event that occurred on 20 January 2009, between 00z to 04z UTC in the Borneo island. In order to capture this event, we started a 48-hour forecast run from 19 Jan 00z UTC to 21 Jan 2009 00z UTC and generated a restart file at 20 Jan 00z UTC. This choice of restart file at the end of a 24-hour simulation, ensured that the model is well spun off. Figure 1 shows the nested domain configurations used for our study. Figure 1(a) shows the 3-domain nested configuration with the innermost nest focused on the country of Brunei. Figure 1(b) shows the case which has 4 sibling nests of same size at the innermost level. The innermost nests were chosen such that some of the highly populated regions in the Borneo island are well represented. In order to study the effects of over-decomposition of innermost nests, we also created two sibling domain configurations that are formed by merging a subset of the domains in 1(b). More specifically, Figure 1(c) has 50% more grid points after combining, whereas, Figure 1(d) has 10% more grid points. The same spatial resolution is used for all sibling domains.

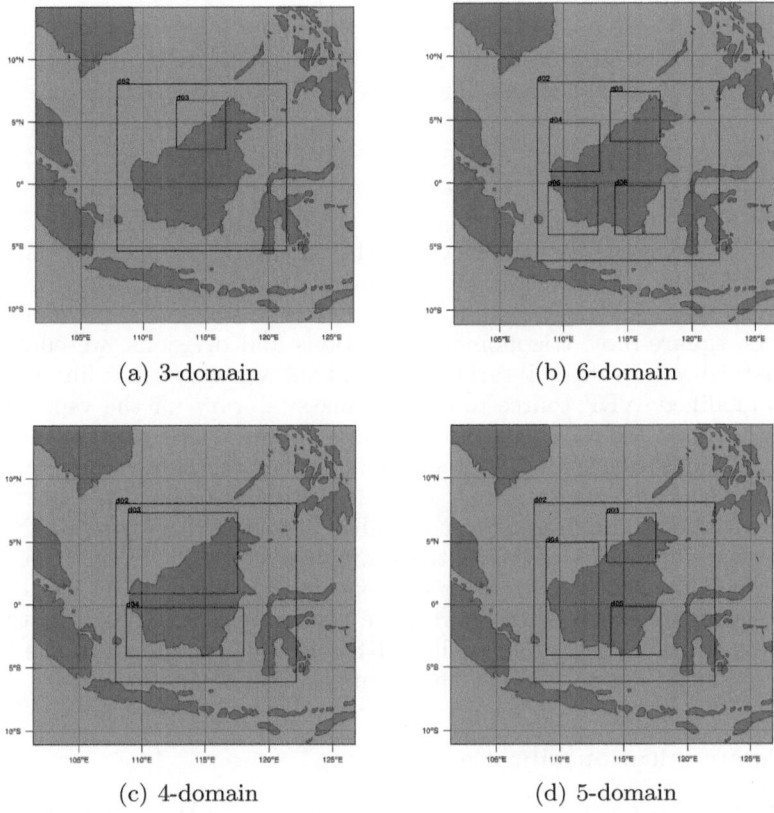

(a) 3-domain   (b) 6-domain

(c) 4-domain   (d) 5-domain

Fig. 1. Nested domain configurations used for the experiments

WRF Runtime Setup. WRF-ARW version 3.2.1, compiled in hybrid mode (dm+sm), was used for all the experiments. In all the simulations, Kain-Fritsch convection parameterization, Thompson microphysics scheme, RRTM long wave radiation, Yonsei University (YSU) boundary layer scheme, and Noah land surface model were used. File I/O was restricted to just the beginning and end of the run, however, for high resolution operational simulations that require forecast output very frequently, it is possible that I/O could become a bottleneck for scaling to large number of processors. This is especially true in production runs that typically involve two or more levels of nesting and a forecast output every ten minutes. We also explored the use of I/O quilting feature in WRF, however, we excluded quilting from the current study since parallel I/O gave the best performance. For this study we primarily use the total integration time since the I/O time is still a small fraction of the total simulation time when parallel I/O is used with moderate number of processors (up to 1024).

4   Optimizations

We explored several optimizations for the WRF application on BG/P that include compiler options, code modifications and tuning the MPI libraries.

4.1   Compiler Options

We modified the default WRF configure from UCAR to include parallel NetCDF for parallel I/O on BG/P. This serves as the *base* WRF configure to make *base* WRF run for our experiments. We enhanced the *base* WRF configure with a few new options. As the WRF moisture physics routines have several calls to exponents, square-root, trigonometric functions and divisions, we enabled the mathematical acceleration libraries mass and the vector variant library massv. We also modified WRF source to call the massv library for the vspow call. In addition, we also used the -qhot=vector compiler option that converts loops with exponents, square-root, division and trigonometric functions to massv library calls. The mass and massv library calls are optimized via SIMD instructions that have higher throughput. Most of WRF is compiled with the -O2 option, with select performance-critical routines compiled with -O3. To further optimize compiler performance, we added the -qmaxmem=128000 to enable the compiler to aggressively optimize WRF source. We also enabled OpenMP via the IBM XL compiler option -qsmp=omp. The WRF source can tile the X or the Y loops to enable the tile computation to be executed on different threads.

4.2   Communication Libraries

Communication overhead is a significant fraction of the WRF timestep. We explored mapfiles to efficiently map the 2D decomposition on to the 3D torus. Our mapfiles map the processors to diagonals on the 3D torus planes when the 3D torus dimensions cannot be folded to the WRF 2D stencil communication dimensions. In addition, we explored increasing the MPI eager limit and enabling the FIFO mode RZVANY in the DCMF library [4] that drives MPI communication on BG/P. This mode improves performance of messages to diagonal neighboring nodes by allocating more network resources to those messages. In addition, to minimize synchronization between nodes, we explored the Async Rectangle Broadcast that implements broadcast without forcing synchronization between the nodes. By default, the collective network is used that forces all nodes to synchronize before the broadcast data is transmitted on the network. We also replaced a call to MPI_Allgather in the forcing and feedback computation with an MPI_Alltoall call as MPI_Alltoall is efficient on BG/P.

4.3   Source Code Optimizations

Loop Unrolling. The XL compiler on BG/P typically only unrolls the innermost loops. WRF is a strong scaling application where the application domain is decomposed along the X and Y dimensions to MPI ranks. By default, the WRF compute loops execute in an XZY order that results in short X loops on

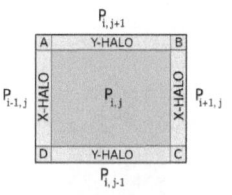

Fig. 2. Halo regions in X-Y dimensions

Table 1. Steps in WRF halo exchange

Pack data into buffer to send to the Y-neighbours.
Receive message from and send message to the Y-neighbours.
Wait for completion of communication with the Y-neighbours.
Unpack data received from Y-neighbours.
Pack data into buffer to send to the X-neighbours.
Receive message from and send message to the X-neighbours.
Wait for completion of communication with the X-neighbours.
Unpack data received from X-neighbours.

a large number of nodes. Hence, we explored unrolling the Z dimension loops in the pack unpack functions that serialize application buffers into MPI messages functions and WRF dynamics computation.

Parallel X and Y communication. The WRF solver sweeps over a 2D-XY-spatial grid performing nearest neighbour computations called *stencils*[7]. Due to the overlapping spatial decomposition, each process $P_{i,j}$ communicates its boundary regions, called *halo* regions, with its two X-neighbours $P_{i-1,j}$ and $P_{i+1,j}$ and two Y-neighbours $P_{i,j-1}$ and $P_{i,j+1}$ as shown in Figure 2. Each integration time-step in WRF involves a large number of halo exchanges, which are fairly expensive and comprise of the steps shown in Table 1. Using IBM's HPCT profiling tools, we found that the packing/unpacking of the halo regions is a computation-intensive step. Further, most of the time is spent in MPI_Wait since every process waits for the completion of the communications without doing any computation. To improve the performance, we modified the halo exchange sequence of steps, specifically delaying the MPI_Wait calls and performing the X and Y communications in parallel, which is favoured by the BG/P torus topology. In the new sequence, the Y-communications are overlapped with the packing of data for the X-neighbours and the X-communications are overlapped with the unpacking of data from the Y-neighbours. This affects only the corners of the halo regions (e.g. regions A,B,C,D in Figure 2) and does not make a substantial difference since these only comprise a tiny part of the subdomain. Note parallelizing X and Y communication does not affect bitwise reproducability in the application as we still maintain a deterministic order.

5 Performance Analysis

In this section, we present the results of our performance evaluation on multiple nested domain configurations. Specifically, we discuss the various BG/P execution modes, the benefits of various WRF optimizations, the scalability of the original and optimized WRF code, and also the importance of choosing a nesting configuration to match the processor configuration. We used 128 to 1024 BG/P nodes for performance evaluation to keep the simulation runtimes to reasonable limits as well as to prevent over-decomposition of the domains beyond 1024 nodes.

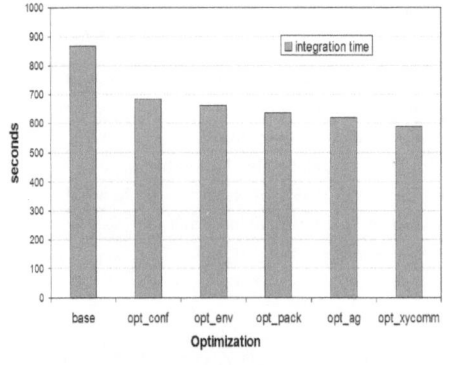
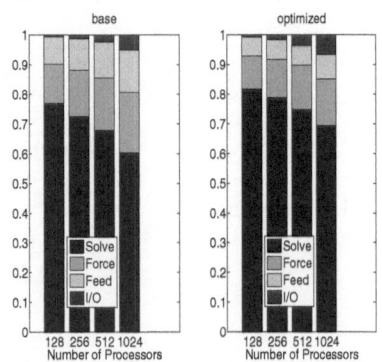

(a) Performance variation with different optimizations on 1 rack

(b) Variation of fractional times of solve, forcing, feedback, and I/O operations vs. number of processors

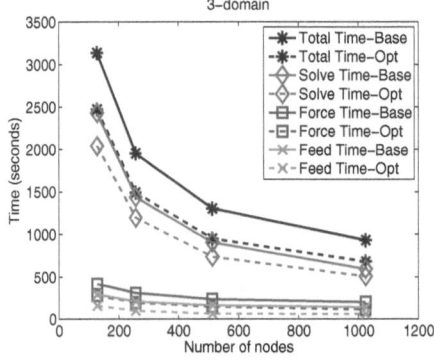
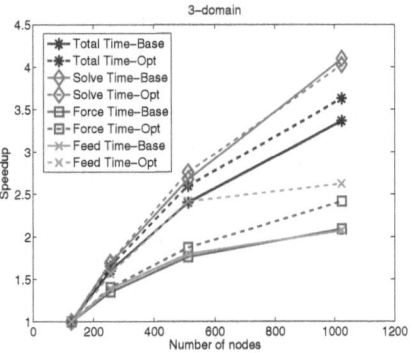

(c) Computation time of WRF operations

(d) Speedup of WRF operations

Fig. 3. WRF performance on nested domain configurations

BG/P Execution Modes. We compared the base WRF performance with different BG/P execution modes (SMP, DUAL and VN). The performance was best with SMP mode with 4 OpenMP threads compared to other modes. On 1024 nodes (1 rack) of BG/P, integration time for DUAL and VN modes increased by 4% and 48% over SMP mode respectively. The total simulation runtime with DUAL and VN modes increased by 12% and 65% over SMP mode respectively. The increase in total runtime was partly due to higher I/O overheads in DUAL and VN modes because of the increase in the number of MPI ranks. The I/O times increased 1.5x and 2.8x in DUAL and VN modes respectively.

WRF Optimizations. Figure 3(a) shows the incremental performance benefits of the various optimizations described in Section 4 on a single rack of BG/P. Each column bar in the figure also incorporates the optimizations indicated by the previous column bars. The performance of the base WRF configuration is presented in column bar *base*. The *opt_conf* column bar indicates the performance after incorporating compiler options mentioned in Section 4.1, that results in a 21% improvement over the *base* configuration. The column bar *opt_env* shows the benefit of environment variables (Section 4.2) that enable an optimized processor mapping, set the eager to rendezvous cutoff and the DCMF FIFO mode. We see a total improvement of 3% over *opt_conf* (23.7% over *base*). Loop unrolling of the pack and unpack routines (*opt_pack*) described in Section 4.3 results in an improvement of 3.6% over *opt_env* (26.5% over *base*). Replacing *MPI_Allgather* with *MPI_Alltoall* (*opt_ag*) provided 2.7% further improvement over *opt_pack* (28.4% over *base*). Finally, optimization for the parallel X and Y halo communication (*opt_xycomm*) results in a performance improvement of 5% over *opt_ag*. This corresponds to an overall improvement of 32% to the integration time and an improvement of 28.9% in the total simulation runtime over the *base* run.

WRF Scaling. Figure 3(b) shows the variation in fractional time for solve, forcing and feedback operations across number of processors for base and optimized codes. Observe the I/O and forcing components increase with the number of nodes suggesting that they are performance bottlenecks. Figures 3(c) and 3(d) show the scaling of solve, forcing and feedback operations in terms of actual timings and speedup respectively. The total time is the sum of the timings for these 3 operations. Though sub-linear, the observed speedups of solve operation in both the original and optimized WRF code (3.5x - 5x going from 128 to 1024 processors) exhibit the same general behavior as the theoretical model presented by Kerbyson et al. [3]. Relative to solve, the forcing and feedback operations exhibit much poorer scaling, attaining speedups of 2x - 2.7x for the same increase in the number of processors. From these figures, we also observe that with our optimizations, the solve time improves and the forcing and feedback components have lower overheads and better scaling.

Influence of Nesting Configuration. We also compare the performance of the three nested domain configurations in Figures 1(b), 1(c), and 1(d), which are equivalent in terms of practical utility and differ only in the sibling domains at the innermost level. Figure 4(a) shows the speedup (relative to 128 processors) as the number of processors increase. We observe that increasing the number of domains results in poorer scalability. One reason for this behavior is that more domains entail additional sequential forcing and feedback operations, which do not scale well. In addition, the innermost nests with small domain sizes get over-decomposed for higher number of processors resulting in lower computation to communication cost ratio. The superior scalability of the 4-domain and 5-domain cases suggest that as we increase the number of processors, the gap in the total integration time due to the additional number of innermost grid points in comparison to that of the 6-domain case should progressively decrease. Figure 4(b) shows the total integration time for the three nested domain configurations with varying number of processors. Even though the 6-domain case has the least number of grid points, for 128 processors, the 5-domain case provides the best performance (8984s in comparison to 10701s and 9108s for 4-domain and 6-domain cases). As the number of processors increases to 1024, the 4-domain configuration exhibits superior scaling and gives the best performance (2411s in comparison to 2459s and 2794s for 5-domain and 6-domain cases). Therefore, it is sometimes beneficial to consider a smaller number of consolidated domains instead of a large number of small focused domains.

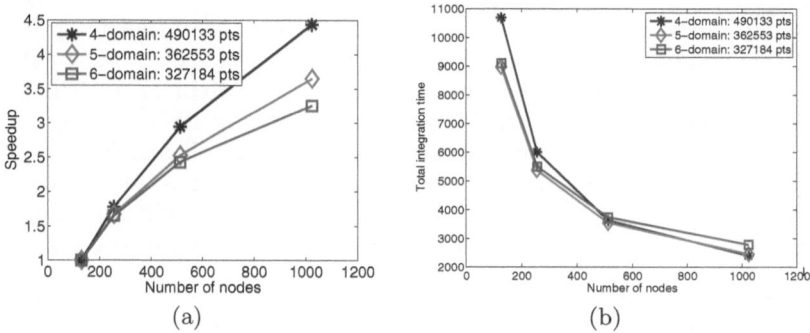

Fig. 4. Performance over the three sibling configurations across varying number of processors

6   Empirical Modeling of Computation Time

Observations from Section 5 highlight the importance of carefully choosing the nesting configuration that can extract the best performance for a given processor configuration. To aid in this effort, we use the empirical timing data from our evaluation to learn a statistical model of the computation time for the three key operations in WRF (solve, forcing and feedback). We then describe how these individual models can be combined to estimate the total modeling time for complex nesting scenarios that involve multiple levels of nesting.

Regression Model. For each operation and domain/processor configuration, the runtime for each iteration is computed as the maximum time across the different nodes since they are operating in parallel. Since the different iterations are executed sequentially and the number of iterations vary due to nesting or simulation configuration, the mean runtime across the different iterations is a good target. We choose the median value across the different iterations as the target variable to be modeled instead of the mean to disregard outlier cases, but the results are comparable in either case. Each domain/processor configuration is represented as a vector of features (column headings in Table 2) based on sizes of the domain(s) involved and the processor grid. Let the domain grid size be denoted by $n_x \times n_y$ and processor grid be denoted by $p_x \times p_y$. In case of forcing and feedback, we also consider the parent grid size denoted by $n_x^p \times n_y^p$. Using the median iteration time as the target response and the feature vector as independent variables, we learn a least squares linear regression model [6]. Table 2 shows the model coefficients. For instance, the computation time for solve operation is given by $T_{solve} = 7.69\mathrm{e}{-4}\frac{n_x n_y}{p_x p_y} + 3.37\mathrm{e}{-3}\frac{n_x}{p_x} + 2.35\mathrm{e}{-3}\frac{n_y}{p_y} + 4.047\mathrm{e}{-2}$. A similar model was computed for the optimized WRF code as well and for each operation, the coefficients of all the features were found to be positive, but lower than that of the base WRF timing model. Note that the model coefficients

Table 2. Coefficients of the model for base WRF version

Op	$\frac{n_x n_y}{p_x p_y}$	$\frac{n_x^p n_y^p}{p_x p_y}$	$\frac{n_x}{p_x}$	$\frac{n_y}{p_y}$	constant
Solve	7.693e-4	NA	3.3676e-3	2.3472e-3	4.04716e-2
Forcing	6.2493e-5	2.0758e-4	2.78171e-3	1.7226e-3	1.4445e-1
Feedback	9.8716e-5	5.2550e-5	2.4233e-3	1.2726e-3	8.5323e-2

Table 3. Actual (s), predicted (s) and relative error (%) of integration time for the base WRF model

Processors	128			256			512			1024		
Domain	Act	Pred	Err(%)	Act	Pred	Err(%)	Act	Pred	Err(%)	Act	Pred	Err(%)
3-domain	2976	2877	3.3	1809	1753	3.1	1200	1143	4.7	867	815	5.9
6-domain	9109	9064	0.5	5508	5524	0.29	3745	3603	3.8	2794	2574	7.9
4-domain	10701	10225	4.5	6011	5909	1.7	3632	3455	4.9	2412	2276	5.6
5-domain	8984	8989	0.06	5374	5266	2.01	3553	3367	5.2	2460	2283	7.2

are specific to choice of BG/P and the other fixed parameters of the simulation (e.g., mp_physics = 8), but the methodology can be adapted as needed.

Computation Time for Multilevel Nested Domains. Let $\mathcal{D} = \{D_0, \cdots, D_n\}$ be a set of nested domains with D_0 being the root domain and the domain hierarchy specified by the $p(\cdot)$ function. Let $T_{solve}(D)$ denote the computation time for a solve iteration on domain D and $T_{force}(D, D_p)$ and $T_{feedback}(D, D_p)$ denote the computation time for forcing and feedback operations between domain D and its parent domain D_p. Let $T_{solve}^{SH}(D)$ denote the solve time for the entire sub-hierarchy under the domain D. Given the constraints of the current WRF code, the forcing, feedback, and solve operations associated with the child domains have to be done sequentially. For two-way nesting, one can obtain the following recursive equation,

$$T_{solve}^{SH}(D) = T_{solve}(D) + \sum_{D_c | p(D_c) = D} (T_{force}(D_c, D) + rT_{solve}(D_c) + T_{feedback}(D_c, D))$$

where r is the parent-child time-step ratio (=3). The total integration time is then given by $T_{solve}^{SH}(D_0)$ multiplied by the number of timesteps.

Using the above recursion and the prediction models for $T_{solve}(\cdot)$, $T_{force}(\cdot, \cdot)$ and $T_{feedback}(\cdot, \cdot)$, one can obtain a reasonable estimate of the computation time for a specified nested domain configuration and processor configuration without having to actually run the code. Table 3 shows the actual integration timings as well as the timings predicted by the model and the absolute relative error for different domain and feasible processor configurations for the base WRF code. We observe that the prediction error is fairly low ($0.06 - 7.9\%$). Note that the 5-domain case is in fact a new configuration which was not used for training the statistical model, but the predicted timings are still fairly accurate. This timing prediction model can be very useful for practitioners since they can choose the domains and nesting configurations based on the estimated run times. We plan to do a more exhaustive evaluation over multiple nested domain cases in future.

7   Conclusions and Future Work

We performed a detailed study of nested domain weather simulations that are critical for operational weather forecasting. We described several optimizations to the base WRF code that reduces the total runtime by 29%. We also conducted a performance evaluation using four test configurations and demonstrated that high resolution nested weather simulation presents a number of challenging opportunities in terms of scaling to large number of processors especially in the case of multiple small-sized sibling domains. This is partly due to the design of the WRF code wherein, multiple nests at the same level are handled by all the processors in a sequential manner resulting in over-decomposition. The current design of the WRF code makes it both critical and challenging for practitioners to choose a good nesting configuration. Accounting for the constraints of the current WRF code, we also presented a regression-based model for predicting integration time for multi-level nested weather simulations that can be used by WRF users to determine the domain configurations best suited for a given processor grid. Future directions include improving the performance model by incorporating additional features based on the network topology, modifying WRF code to allow subsets of processors working in parallel over sibling domains and designing algorithms that can effectively balance the load under such circumstances. We also plan to explore MPI non contiguous datatypes to optimize the pack/unpack operations in WRF. In addition, weak scaling studies could potentially assist in further understanding of the effects of the proposed optimizations. Optimizing file I/O performance in WRF is another potential future work.

Acknowledgements. We would like to thank Yogish Sabharwal, Bob Walkup, Dong Chen, Sathish S. Vadhiyar, Vijay Natarajan and Lloyd A. Treinish for their help, technical support and valuable suggestions. The work presented in this paper was funded in part by the US Government contract No. B554331.

References

1. Bhatele, A., Gupta, G.R., Kale, L.V., Chung, I.H.: Automated Mapping of Regular Communication Graphs on Mesh Interconnects. In: HiPC (2010)
2. Delgado, J., et al.: Performance Prediction of Weather Forecasting Software on Multicore Systems. In: IPDPS, Workshops and PhD Forum (2010)
3. Kerbyson, D.J., Barker, K.J., Davis, K.: Analysis of the Weather Research and Forecasting (WRF) Model on Large-Scale Systems. In: PARCO, pp. 89–98 (2007)
4. Kumar, S., et al.: The Deep Computing Messaging Framework: Generalized Scalable Message passing on the Blue Gene/P Supercomputer. In: ICS 2008 (2008)
5. Kumar, S., et al.: Architecture of the Component Collective Messaging Interface. IJHPCA 24(1), 16–33 (2010)
6. Kutner, M.H., Nachtsheim, C.J., Neter, J.: Applied Linear Regression Models, Fourth International edn. McGraw-Hill (September 2004)
7. Michalakes, J.: RSL: A Parallel Runtime System Library For Regional Atmospheric Models With Nesting. Tech. Rep. ANL/MCS-TM-197, Mathematics and Computer Science Division, Argonne National Laboratory, Argonne (1997)
8. Michalakes, J., et al.: WRF Nature Run. In: SC, p. 59 (2007)

9. Porter, A.R., et al.: WRF code Optimisation for Mesoscale Process Studies (WOMPS) dCSE Project Report (June 2010)
10. Shainer, G., et al.: Weather Research and Forecast (WRF) Model Performance and Profiling Analysis on Advanced Multi-core HPC Clusters. In: 10th LCI ICHPCC (2009)
11. Skamarock, W.C., et al.: A Description of the Advanced Research WRF version 3. NCAR Technical Note TN-475 (2008)
12. Wright, N.J., Pfeiffer, W., Snavely, A.: Characterizing Parallel Scaling of Scientific Applications using IPM. In: 10th LCI International Conference on High-Performance Clustered Computing (2009)

Optimized Hybrid Parallel Lattice Boltzmann Fluid Flow Simulations on Complex Geometries

Jonas Fietz[2], Mathias J. Krause[2], Christian Schulz[1],
Peter Sanders[1], and Vincent Heuveline[2]

[1] Karlsruhe Institute of Technology (KIT),
Institute for Theoretical Informatics, Algorithmics II
[2] Karlsruhe Institute of Technology (KIT),
Engineering Mathematics and Computing Lab (EMCL)

Abstract. Computational fluid dynamics (CFD) have become more and more important in the last decades, accelerating research in many different areas for a variety of applications. In this paper, we present an optimized hybrid parallelization strategy capable of solving large-scale fluid flow problems on complex computational domains. The approach relies on the combination of lattice Boltzmann methods (LBM) for the fluid flow simulation, octree data structures for a sparse block-wise representation and decomposition of the geometry as well as graph partitioning methods optimizing load balance and communication costs. The approach is realized in the framework of the open source library OpenLB and evaluated for the simulation of respiration in a subpart of a human lung. The efficiency gains are discussed by comparing the results of the full optimized approach with those of more simpler ones realized prior.

Keywords: Computational Fluid Dynamics, Numerical Simulation, Lattice Boltzmann Method, Parallelization, Graph Partitioning, High Performance Computing, Human Lungs, Domain Decomposition.

1 Introduction

The importance of *computational fluid dynamics* (CFD) for medical applications have risen tremendously in the past few years. For example, the function of the human respiratory system has not yet been fully understood, and its complete description can be considered byzantine. Due to highly intricate multi-physics phenomena involving multi-scale features and ramified, complex geometries, it is considered one of the *Grand Challenges* in scientific computing today. One day, numerical simulation of fluid flows is hoped to enable surgeons to analyze possible implications prior to or even during surgery. Widely automated preprocessing as well as efficient numerical methods are both necessary conditions for enabling real-time simulations.

In the last decades, *lattice Boltzmann methods* (LBM) have evolved into a mature tool in CFD and related topics in the landscape of both commercial and academic software. The simplicity of the core algorithms as well as the locality properties resulting from the underlying kinetic approach lead to methods

C. Kaklamanis et al. (Eds.): Euro-Par 2012, LNCS 7484, pp. 818–829, 2012.

which are very attractive in the context of parallel computing and high performance computing [7,8,13]. In this context, it is of great importance to take advantage of nowadays available hardware architectures like Graphic Processing Units (GPUs), multi-core processors and especially hybrid high performance technologies that blur the line of separation between architectures with shared and distributed memory. A concept to use LBM dedicated for hybrid platforms has been described before in [3]. It relies on spatial domain decomposition where each domain represents a basic block entity which is solved on a symmetric multi-processing (SMP) system. The regularity of the data structure of each block allows a highly optimized implementation dedicated to the particular SMP hardware. Load balancing is achieved by assigning the same number of equally-sized blocks to each of the available SMP nodes. This concept has been extended and applied for fluid flows simulation on complex geometries [5].

The goal of this work is to optimize the hybrid parallelization approach for LBM simulations on complex geometries. The basic idea is to drop the equally-sized block constraint thereby enabling a sparse representation of the computational domain. Therefore, two domain decomposition strategies are proposed as well as the application of graph based load balancing techniques to the load distribution problem for LBM. The first domain decomposition strategy is a heuristic, which we further improve by a shrinking step. The second strategy is a geometry aware decomposition using octrees. This results in a sparse domain decomposition with larger computational domains. Both of these strategies require sophisticated load balancing. We propose a graph partitioning based approach optimizing the load and minimizing communication costs. While graph based load balancing has been done before by [1], we propose to apply this not on a fluid cell level but at block level. Finally, we evaluate the presented measures on a subset of the human lung, showing performance improvements for all of them.

2 Lattice Boltzmann Fluid Flow Simulations

The here considered subclass of *lattice Boltzmann methods* (LBM) enable to simulate the dynamics of *incompressible Newtonian fluids* which is usually described macroscopically by an initial value problem governed by a *Navier-Stokes equation*. Instead of directly computing the quantities of interests, which are the fluid velocity $u = u(t, r)$ and fluid pressure $p = p(t, r)$ where $r \in \Omega \subseteq \mathbb{R}^d$ and $t \in I = [t_0, t_1) \subseteq \mathbb{R}_{\geq 0}$, a lattice Boltzmann (LB) numerical model simulates the dynamics of particle distribution functions $f = f(t, r, v)$ in a phase space $\Omega \times \mathbb{R}^d$ with position $r \in \Omega$ and particle velocity $v \in \mathbb{R}^d$. The continuous transient phase space is replaced by a discrete space with a spacing of $\delta r = h$ for the positions, a set of $q \in \mathbb{N}$ vectors $v_i \in \mathcal{O}(h^{-1})$ for the velocities and a spacing of $\delta t = h^2$ for time. The resulting discrete phase space is called the lattice and is labeled with the term $DdQq$. To reflect the discretization of the velocity space, the continuous distribution function f is replaced by a set of q distribution functions f_i $(q = 0, 1, ..., q - 1)$, representing an average value of f in the vicinity of the velocity $\boldsymbol{v}_i$. Detailed derivations of various LBM can be found in the literature, e.g. in [11]. The iterative process in an LB algorithm can be written in two steps as follows, the *collision step* (1) and the *streaming step* (2):

$$\widetilde{f}_i(t, \boldsymbol{r}) = f_i(t, \boldsymbol{r}) - \frac{1}{3\nu + 1/2} \left(f_i(t, \boldsymbol{r}) - M_{f_i}^{eq}(t, \boldsymbol{r}) \right), \qquad (1)$$

$$f_i(t + h^2, \boldsymbol{r} + h^2 \boldsymbol{v_i}) = \widetilde{f}_i(t, \boldsymbol{r}) \qquad\qquad\qquad\qquad (2)$$

for $i = 0, .., q-1$. $M_{f_i}^{eq}(t, \boldsymbol{r}) := \frac{w_i}{w} \rho_{f_i} \left(1 + 3h^2 \, \boldsymbol{v_i} \cdot \boldsymbol{u_{f_i}} - \frac{3}{2} h^2 \boldsymbol{u}_{f_i}^2 + \frac{9}{2} h^4 \left(\boldsymbol{v_i} \cdot \boldsymbol{u_{f_i}} \right)^2 \right)$ is a discretized *Maxwell distribution* with moments ρ and $\boldsymbol{u}$ which are given according to $\rho := \sum_{i=0}^{q-1} f_i$ and $\rho\boldsymbol{u} := \sum_{i=0}^{q-1} \boldsymbol{v_i} f_i$. The variable $\boldsymbol{u}$ is the discrete fluid velocity and ρ the discrete mass density. The kinematic fluid viscosity is ν which is assumed to be given, and the terms w_i/w, $\boldsymbol{v_i}h$ $(i = 0, 1, ..., q - 1)$ are model dependent constants. The discrete fluid velocity $\boldsymbol{u}$ and the discrete mass density ρ can be related to the solution of a macroscopic initial value problem governed by an incompressible Navier-Stokes equation as shown by Junk and Klar [4].

3   Domain Decomposition for Hybrid Parallelization

The most time demanding steps in LB simulations are usually the collision (1) and the streaming (2) operations. Since the collision step is purely local and the streaming step only requires data of the neighboring nodes, parallelization has mostly been done by domain partitioning [7,8,13]. To take advantage of hybrid architectures, a multi-block approach is used [3]: the computational domain is partitioned into sub-grids with possibly different levels of resolution, and the interface between those sub-grids is handled appropriately. This leads to imple-

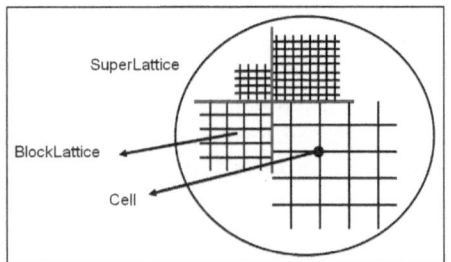

Fig. 1. Data structures used in OpenLB: BlockLattices consist of Cells and make up a SuperLattice enabling higher level software constructs

mentation that are both elegant and efficient since the execution on a set of regular blocks is much faster compared to an unstructured grid representation of the whole geometry. For complex domains a multi-block approach also yields sparse memory consumption. Furthermore, it encourages a particularly efficient form of data parallelism, in which an array is cut into regular pieces. This is a good mapping to hybrid architectures.

In *OpenLB*, the basic data-structure is a BlockLattice representing a regular array of Cells. In each Cell, the q variables for the storage of the discrete velocity distribution functions f_i, $(i = 0, 1, ..., q - 1)$ are contiguous in memory. Required memory is allocated only once since no temporary memory is needed in the applied algorithm. This data structure is encapsulated by a higher level, object-oriented layer. The purpose of this layer is to handle groups of `BlockLattice`s, and to build higher level software constructs in a relatively transparent way. Those constructs are called `SuperLattice`s and include multi-block, grid refined lattices as well as parallel lattices.

3.1 Heuristic Domain Decomposition with Shrinking Step

In this section, we describe our heuristic domain partitioning strategy. We further improve this by shrinking each of the partitions, so that it achieves a closer fit to the underlying geometry.

The hybrid parallelization strategy proposes to partition the data of a considered discrete position space Ω_h, which is a uniform mesh with spacing $h > 0$, according to their geometrical origin into $n \in \mathbb{N}$ disjoint, preferably cube-shaped sub-lattices Ω_h^k ($k = 0, 1, ..., n - 1$) of almost similar sizes. This becomes feasible by extending Ω_h to a cuboid-shaped lattice $\widetilde{\Omega}_h$ through the introduction of ghost cells. Then, $\widetilde{\Omega}_h$ is split into $m \in \mathbb{N}$ disjoint, cuboid-shaped extended sub-lattices $\widetilde{\Omega}_h^l$ ($l = 0, 1, ..., m - 1$) of approximately similar size and as cube-shaped as possible. Afterwards, all extended sub-lattices $\widetilde{\Omega}_h^l$ which consist solely of ghost cells are neglected. The number of the remaining extended sub-lattices $\widetilde{\Omega}_h^l$ ($l_0, l_1, ..., l_{n-1}$) defines n. Finally, for each $k \in \{0, 1, ..., n - 1\}$ one gets the Ω_h^k as a subspace of $\widetilde{\Omega}_h^{l_k}$ by neglecting the existing ghost cells.

For the number $p \in \mathbb{N}$ of available processing units (PUs) of a considered hybrid high performance computer, an even load balance will be assured for complex geometries in particular if the domain Ω_h is partitioned into a sufficiently large number $n \in \mathbb{N}$ of sub-lattices. Then, several of the sub-lattices Ω_h^k ($k = 0, 1, ..., n - 1$) can be assigned to each of the available PUs. To find a good value for n, we introduce a factor k for the amount of sub-lattices with the relation $n = p \times k$. This factor can be adjusted for a specific problem by evaluating run-times for a few hundred time steps to achieve better performance.

After removing all empty cuboids, we then optimize the fit of the cuboids to the underlying geometry. To find out if a cuboid can be shrunk, we start running through each layer in all 6 directions beginning at the respective faces of the cube. For each layer, we check if it is completely empty and stopping the iteration in this direction when a full cell is found. In the next step, all empty layers are removed. This shrinking is executed for all cuboids. Note that the same shrinking step can also be applied to the above mentioned octree domain decomposition and works in exactly the same way. An example decomposition can be seen in Fig. 2a.

3.2 Sparse Octree Domain Decomposition with Shrinking

A key part of load balancing is the decomposition of the complete domain in sub-domains. Here, one has to optimize for multiple, sometimes opposing properties of the sub-domains. As more cuboids mean more communication, cuboids should be as large as possible. The surface of each cuboid should be minimal, as this

(a) Heuristic Decomposition without additional steps

(b) Octree Decomposition with additional shrinking step

(c) Stream lines and a cut plane of velocity distribution.

determines the amount of communication for this cuboid. This usually implies a shape as close to a ball as possible. As ball-shaped objects are not space filling, and due to the current implementation in our library supporting only rectangular shapes, the optimal shape is a cube.

The streaming step is executed without respect for the underlying geometry information. Therefore, even non-fluid cells use some processing power. Because of this, a tight fit of the domain decomposition with respect to the specific geometry is desirable. This is where octrees come into play to adjust the size of cubes depending on the geometry.

The general concept starts with embedding the problem domain in a cube. As we want the boundaries to be exactly on the boundaries between the different cells, we use a size of $2^l \times \delta r$ for some $l \in \mathbb{N}$ as the side length of that cube. As described above, domain decomposition should always result in cuboids that are neither too small nor too large. E.g. using the surrounding cube by itself would not be very useful for load balancing, while using single cells would create a massive overhead. So the implementation allows for limiting the smallest and the biggest cube sizes.

Having defined the root cube now, one recursively divides the cube into smaller cubes as long as the geometry in this part is interesting. In our case this means that it contains empty cells at the same time as boundary or fluid cells. If this is not the case, for example if we are completely on the inside or outside of the geometry, we keep the cube at this size as long as it is smaller or equal to the maximum size. Additionally, we limit the size to the low end, not splitting further when the cubes would become smaller than the minimum size. This minimum size can be defined as the side length of the minimum cube, a number c. The shrinking procedure can be applied to the octree domain decomposition as well (combination is abbreviated as *ODS*). An example decomposition can be seen in Fig. 2b.

4   Load Balancing

As we explained in the introduction of Section 3, the most commonly used approach to load balancing LBM is based on an even decomposition of the computational domain. This section describes our alternative approach using techniques from graph partitioning.

4.1   Graph Partitioning Using KaFFPa

Our parallelization strategy employs the graph partitioning framework KaFFPa [9]. We shortly describe the graph partitioning problem. and introduce notations used. Consider an undirected graph $G = (V, E, c, \omega)$ with edge weights $\omega : E \to \mathbb{R}_{>0}$, node weights $c : V \to \mathbb{R}_{\geq 0}$, $n = |V|$, and $m = |E|$. Given a number $k \in \mathbb{N}$ (in our case the number of processors) the graph partitioning problem demands to partition V into *blocks* of nodes $V_1, \ldots, V_k$ such that $V_1 \cup \cdots \cup V_k = V$ and $V_i \cap V_j = \emptyset$ for $i \neq j$. A *balancing constraint* demands that all blocks have roughly equal size, i.e. the maximum load of a processing element is bounded. The *objective* is to minimize the total *cut*, i.e. the sum of the weight of the edges that run between blocks. We have tested and shown that the edge cut models the communication very well, because the edge weights correlate linearly with the amount of communication between two cuboids [2]. For more details on graph partitioning with KaFFPa we refer the reader to [9].

4.2   Graph-Based Parallelization Strategy for LBM

As described in the prior sections, the LBM algorithm is divided into two parts. The first part is the local collision and streaming for each cuboid. Afterwards, the information is exchanged between different cuboids, i.e. transmitting the information of the border cells for neighboring cuboids assigned to different processors. Logically, the perfect load balancer would always achieve minimal communication while achieving perfect load balancing, i.e. each processor would need exactly the same time for the compute step of all its cuboids combined. It is obvious that this can only be the case for the most trivial situations and geometries, because as soon as there are empty cells, computing times will differ.

   To map this problem to graph partitioning, we associate the work amount or needed CPU time for each cuboid with the weight of a node for this cuboid, and associate the needed communication between two cuboids with the edge weight of the edge between their respective nodes in the graph. Applying the graph partitioner to this graph will yield subsets of nodes, such that the subsets have approximately the same sum of node weights and therefore computing time. The edge-cut – the inter-processor communication – will be minimized. As the problem of graph partitioning is NP-complete, this will not necessarily be the minimal communication for this specific problem and domain decomposition, but it will be good enough in general.

4.3   Determining Node and Edge Weights

The mapping of the problem to the graph has become clear now. But one still needs to find the exact numbers for the amount of work to be done for each cuboid and the amount of communication between two cuboids.

   We begin with the latter. The edge weight is either the byte count or the number of cells to be communicated. This information is often already present, as every implementation of LBM has to find the border cells that need to be communicated, anyway. The case is not as simple for non-symmetrical communication between different cuboids. One can either choose the maximum or the sum of both parts as the edge weight. Since the data transfer between two cuboids is serial in our implementation – i.e. we first transfer in one direction, then in the opposite – we pick the sum.

Calculating the work to be done for each cuboid is not as easy, as the amount of work for empty cells, boundary cells and normal fluid cells differs. As the number of boundary cells is usually quite small, and as they are treated as an extra step, this special case is ignored; they are assumed to be normal fluid cells. Empty cell in our case are either cells in a solid area or that this cell is outside of the fluid filled body being simulated. While the collision step is not executed for the empty cells, the streaming still is. As it is very possible for a cuboid to consist largely of empty cells, it is important to know how much processing time the empty cells use when compared to the normal fluid cells. For this we introduce a factor χ. We measured χ for several different architectures. Unfortunately, it is not a specific constant valid even for the limited types we tested. Instead, it varies from 1.8 to 4.5 [2] for the differing machines used in our preliminary work. These dispartities are most likely due to the different memory and cache hierarchies and resulting diverse memory access speeds, as the streaming part of the LB algorithm is memory bound. To calculate the work to be done for each cuboid, using the symbols for the work ω, for the number of fluid cells n_f, and for the number of empty, non-fluid cells n_e, we get the formula $\omega = n_f + \chi n_e$.

In the end, the graph load balancer is now able to balance the work load to a set number of processing nodes or cores and to find a solution for a certain load imbalance with accordingly small communication overhead.

5 Experiments

The aim of this section is to illustrate the effectiveness of the presented options considering a practical problem with an underlying complex geometry, namely the expiration in a human lung. The geometry we use is a subset of the bronchi of the lungs, with bifurcation of the bronchi to the third level (see Fig. 2a). The air for the simulation is assumed to be at normal conditions ($1013hPa$, $20°C$), i.e. $\rho = 1.225 kg/m^3$ and its kinematic viscosity is $\nu = 1.4 \times 10^{-5} m^2/s$. The outflow region is set at the trachea with a pressure boundary condition with constant pressure of $1013hPa$. The inflow regions are the bronchioles. There, a velocity boundary condition is set as a Poiseuille distribution with a maximum speed of $1m/s$. The characteristical length is set to $2cm$, which is the diameter of the trachea. With a characteristical speed of $1m/s$, we get a Reynolds number of around 1400. To solve the problems numerically, we use a D3Q19 LB model with the pressure and velocity boundary conditions as proposed by Skordos [10]. No-slip conditions for the walls are realized as a bounce-back boundary. For the LB simulation, we set the Mach number to 0.05 and δr to 3.91×10^{-4}. We obtain the dimensions of $402 \times 54 \times 343$ cells, with about 1.06 million filled cells, i.e. a fill grade of approximately 14.5%.

Table 1. Comparison of balancers with 512 processors. The best value of k is emphasized.

k	DBLB	GBLB
1	0.117	0.067
2	0.223	0.044
4	0.198	**0.303**
8	**0.226**	0.297
16	0.199	0.260
32	0.134	0.137
64	0.022	0.068

All benchmarks were run on a cluster at the Karlsruhe Institute of Technology. It consists of 200 Intel Xeon X5355 nodes. Each of these nodes contains two quad-core Intel Xeon processors with a clock speed of 2.667 GHz and 2x4 MB of level-2 cache each with 16 GB of RAM. The nodes are connected to each other via an Infiniband 4X DDR interconnect. The Infiniband interconnect has a latency from node to node below 2 microseconds and a point to point bandwidth of 1300 MB/s. Programs on the IC1 were compiled with the GCC 4.5.3 with an optimization level of O3 and using the MPI library OpenMPI 1.5.4.

To compare the performance of LB, in most cases one uses the measurement unit of *million fluid-lattice-site updates per second* MLUP/s, e.g. [12]. This idea can be extended to the unit MLUP/ps for *million fluid-lattice-site updates per process and second* [6]. The latter unit is used in all examples. One calculates the amount of fluid cells N_c for each of the examples. With the run-time t_p for p processor cores, the number i of iterations, the result is given as $P_{LB} := 10^{-6} \frac{iN_c}{t_p p}$

5.1   Decomposition vs. Graph Based Load Balancing

The first comparison is between the *Decomposition Based Load Balancer* (DBLB) and the new *Graph Based Load Balancer* (GBLB). Small values of k are not very efficient, because they allow some processors to run empty. Therefore, only higher values of k are shown in Fig. 2.

This benchmark shows a steep initial decline of computation speed when scaling from one to eight computing cores. This is due to the memory bound characteristic of the LBM algorithm and the limited amount of faster caches on the target architecture and its shared memory buses.

While one can see that the results are not that far apart, starting in the range of approximately 128 processors the GBLB becomes more efficient by a margin. To show how much more efficient the load balancer performs for higher numbers of CPUs, see Tab. 1. For 512 cores, the GBLB solution only takes about two-thirds of the execution time of the DBLB one.

5.2   Effects of Using the Shrinking Step

The *shrinking step* as a step to optimize the size of the cuboids showed itself to be the most effective strategy of all, despite its seemingly simple nature. Performance for the test benchmark increased by over 100% in certain cases (see Tab. 2). All test cases with shrinking were run with the graph based load

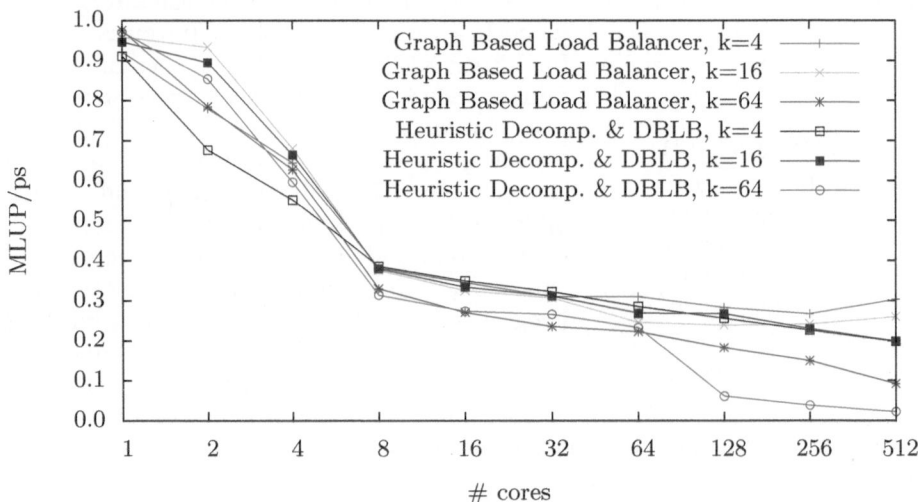

Fig. 2. Comparison of the DBLB and the GBLB for certain k. For larger numbers of cores, one can see the performance improvement of using the GBLB without any change to the decomposition algorithm.

balancer. These tests included core counts between 1 and 256 and the factor $k \in \{2^0, \ldots, 2^6\}$. An excerpt of the performance for the best decomposition based load balancer and the best graph based load balancer test runs with an additional shrinking step are shown in Fig. 3. One can see that the solution with the shrinking step and the GBLB outperforms the DBLB for all values of k, respectively. This speed-up can be attributed to multiple effects. Because empty cells are excluded from the streaming step, less streaming is required. One can get an impression of the possible reduction in the amount of empty cells from Tab. 3. As one

Table 2. Speeds in MLUP/ps for the *best* variant for each processor with DBLB compared to the *shrinking step* and GBLB.

# Cores	DBLB	Shrinking	Speed-up
1	1.023	1.562	52.7%
2	0.895	1.466	63.8%
4	0.696	1.247	79.2%
8	0.385	0.786	104.2%
16	0.349	0.708	102.9%
32	0.322	0.659	104,7%
64	0.326	0.569	74.5%
128	0.303	0.587	93.7%
256	0.247	0.473	91.5%

can deduce from these numbers, the speed-up can not solely be explained by the smaller amount of empty cells. The graph load balancer certainly has its part as was shown in the prior tests. But most likely several secondary effects are at work as well. The ratio of memory accesses to CPU work shifts towards the calculation side, as empty cells are removed, because the empty cells require no computation and are mainly memory intensive. Hence, the memory hierarchy

Table 3. Comparison of amount of cells that are computed before and after executing the *shrinking* step on a heuristically decompositioned geometry

# Cuboids Before Remove	# Cuboids	# Cells Before Shrinking	# Cells After Shrinking
64	33	3 776 144	1 628 975
128	58	3 303 225	1 577 444
512	185	2 659 763	1 410 635

is put under less strain, so the caches work more effectively. Another effect is that the communication between cuboids assigned to different nodes is reduced as when the cuboids are shrunk, their surface area shrinks, too. Therefore, the amount of communication needed for these cuboid is reduced as well.

5.3 Octree Domain Decomposition

For smaller number of cores the octree decomposition combined with the graph based load balancer turns out not to improve performance significantly over the original approach. This is due to the amount of cuboids generated by this approach which creates inefficiencies for small numbers of cores and small minimum cuboids. It is only when combined with the shrinking step that performance improves significantly, although not all across the board. This is because the sizing of the minimum cuboid is too coarse, as it is restricted to powers of two. In certain cases, this might lead to too many or to not enough cuboids for efficient load balancing, exactly the situation where the factor k for the heuristic decomposition

Table 4. Comparing DBLB to GBLB with heuristic decomposition (HD) and to GBLB with octree decomposition and shrinking step (ODS) for a randomly chosen example subset. Performance varies depending on the number of cuboids, but GBLB solutions always achieve a speed-up.

# Cores	k	HD & DBLB	HD& GBLB	c	GBLB & ODS
32	4	**0.322**	0.310	16	0.243
32	8	0.299	**0.319**	32	**0.421**
32	16	0.312	0.307	64	0.357
256	4	0.226	**0.267**	8	0.147
256	8	**0.247**	0.261	16	**0.319**
256	16	0.230	0.241	32	0.152
512	4	0.198	**0.303**	8	0.058
512	8	**0.226**	0.297	16	**0.264**
512	16	0.199	0.260	32	0.077

shows its strengths. Another detrimental effect can also be due to the specific structure of the tested geometry. Because the diameter of the bronchi is small, the middle coordinates have to align perfectly to get bigger cuboids with the octree decomposition. Yet in certain situations, the GBLB & ODS is the fastest solution (see Tab. 4).

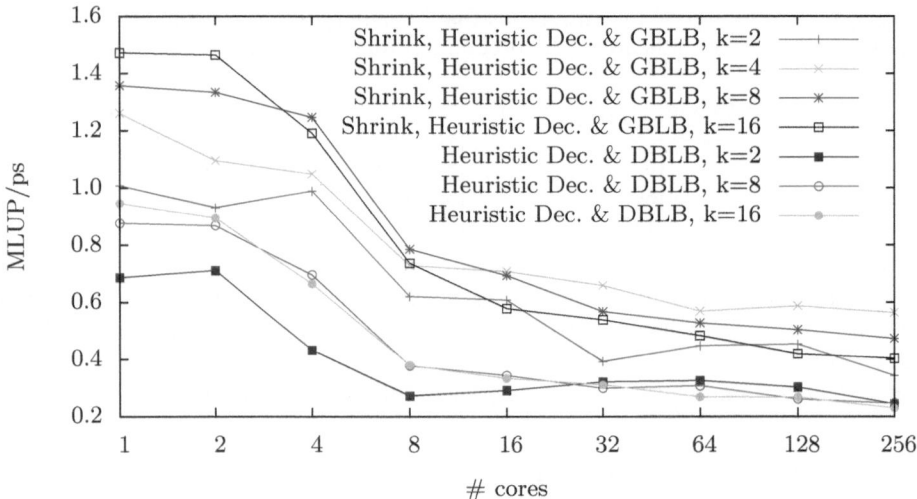

Fig. 3. Comparing the DBLB to the heuristic domain decomposition with the additional shrinking step and GBLB. One can see that performance approximately doubles with the new shrinking and GBLB solution.

6   Conclusions

We have given a successful example for a general technique that will become more and more important in the simulation of unstructured systems: Use partitioning of weighted graphs to do high level load balancing of computational grids where each node represents a regular grid that can be handled efficiently by modern hardware.

Specifically, we examined potential optimizations for Lattice Boltzmann Methods on the example of the OpenLB implementation. We identified two areas with potential for major improvement. First, the current load balancer, and second the simple heuristic sparse domain decomposition. The decomposition based load balancer *only* equalizes the computational complexity and limits potential optimizations for sparse domain decomposition. Therefore, we designed and implemented two alternatives which additionally allow us to improve the sparse domain decomposition.

Of the multitude of different improvement strategies, we propose and evaluate these: *shrinking of cuboids* and *Octree Domain Decomposition*. The *graph load balancer* performs at least as well as the original load balancer for nearly all cases, while outperforming it on most non-trivial geometries. The *decomposition based load balancer* (DBLB) does not achieve the efficiency of the graph based load balancer, but still permits to utilize some of the gains due to the domain decomposition improvements. As for the domain decomposition strategies, the octree allows scaling of the cuboids to the complexity of the geometry at each point. Octree decomposition creates better fitting domain decompositions, but measurements show that it sometimes results in higher overhead. Nevertheless, the results hint at a better performance with more processors. The *shrinking* strategy improves performance for the real world example from 75% up to 105%. Further improvements are expected to be made by combining other measures and fine-tuning settings. The achieved speed-up translates directly into time, money and energy savings for research and industrial applications. It moves boundaries for the problem size and geometry size even further, providing opportunities for ever more complex simulations.

References

1. Bisson, M., Bernaschi, M., Melchionna, S., Succi, S., Kaxiras, E.: Multiscale hemo-dynamics using GPU clusters. Communications in Computational Physics (2011)
2. Fietz, J.: Performance Optimization of Parallel Lattice Boltzmann Fluid Flow Simulations on Complex Geometries. Diplomarbeit, Karlsruhe Institute of Technology (KIT), Department of Mathematics (December 2011)
3. Heuveline, V., Krause, M.J., Latt, J.: Towards a Hybrid Parallelization of Lattice Boltzmann Methods. Computers & Mathematics with Applications 58, 1071–1080 (2009)
4. Junk, M., Klar, A.: Discretizations for the Incompressible Navier-Stokes Equations Based on the Lattice Boltzmann Method. SIAM J. Sci. Comput. 22(1), 1–19 (2000)
5. Krause, M.J., Gengenbach, T., Heuveline, V.: Hybrid Parallel Simulations of Fluid Flows in Complex Geometries: Application to the Human Lungs. In: Guarracino, M.R., Vivien, F., Träff, J.L., Cannataro, M., Danelutto, M., Hast, A., Perla, F., Knüpfer, A., Di Martino, B., Alexander, M. (eds.) Euro-Par-Workshop 2010. LNCS, vol. 6586, pp. 209–216. Springer, Heidelberg (2011)
6. Krause, M.J.: Fluid Flow Simulation and Optimisation with Lattice Boltzmann Methods on High Performance Computers: Application to the Human Respiratory System. Karlsruhe Institute of Technology, KIT (2010)
7. Massaioli, F., Amati, G.: Achieving high performance in a LBM code using OpenMP. Unknown
8. Pohl, T., Deserno, F., Thurey, N., Rude, U., Lammers, P., Wellein, G., Zeiser, T.: Performance Evaluation of Parallel Large-Scale Lattice Boltzmann Applications on Three Supercomputing Architectures. In: Proceedings of the ACM/IEEE SC 2004 Conference Supercomputing 2004, p. 21 (2004)
9. Sanders, P., Schulz, C.: Engineering Multilevel Graph Partitioning Algorithms. In: 19th European Symposium on Algorithms (2011)
10. Skordos, P.: Initial and boundary conditions for the Lattice Boltzmann Method. Phys. Rev. E 48(6), 4823–4842 (1993)
11. Sukop, M.C., Thorne, D.T.: Lattice Boltzmann modeling. Springer (2006)
12. Wellein, G., Zeiser, T., Hager, G., Donath, S.: On the single processor performance of simple lattice Boltzmann kernels. Comput. Fluids 35(8-9), 910–919 (2006)
13. Zeiser, T., Götz, J., Stürmer, M.: On performance and accuracy of lattice Boltzmann approaches for single phase flow in porous media: A toy became an accepted tool - how to maintain its features despite more and mor complex (physical) models and changing trends in high performance computing!?. In: Shokina, N., Resch, M., Shokin, Y. (eds.) Proceedings of 3rd Russian-German Workshop on High Performance Computing, Novosibirsk. Springer (July 2007)

Topology-Aware Mappings
for Large-Scale Eigenvalue Problems

Hasan Metin Aktulga[1], Chao Yang[1], Esmond G. Ng[1],
Pieter Maris[2], and James P. Vary[2]

[1] Lawrence Berkeley National Laboratory, Berkeley CA 94720, USA
[2] Iowa State University, Ames IA 50011, USA

Abstract. Obtaining highly accurate predictions for properties of light
atomic nuclei using the Configuration Interaction (CI) approach requires
computing the lowest eigenvalues and associated eigenvectors of a large
many-body nuclear Hamiltonian matrix, $\hat{H}$. Since $\hat{H}$ is a large sparse
matrix, a parallel iterative eigensolver designed for multi-core clusters
is used. Due to the extremely large size of $\hat{H}$, thousands of compute
nodes are required. Communication overhead may hinder the scalability
of the eigensolver at such scales. In this paper, we discuss how to reduce
such overhead. In particular, we quantitatively show that topology-aware
mapping of computational tasks to physical processors on large-scale
multi-core clusters may have a significant impact on efficiency. For typical large-scale eigenvalue calculations, we obtain up to a factor of 2.5
improvement in overall performance by using a topology-aware mapping.

1 Introduction

High fidelity scientific simulations are nowadays carried out on multi-core machines that consist of thousands or tens of thousands of nodes. Hopper, a Cray
XE6 machine at the National Energy Research Scientific Center (NERSC), has
6,384 nodes and 24 cores per node. Kraken, a Cray XT5 platform at the National
Institute for Computational Sciences (NICS), has 9,408 nodes and 12 cores per
node. These nodes and cores are connected by a sophisticated communication
network that has a limited bandwidth. The bandwidth per flop ratio (BPF) of
these machines has exhibited a declining trend. For the successive Cray models
XT4, XT5 and XE6, the BPF ratios are 1.36, 0.23 and 0.10 bytes per flop, respectively [2,3]. On exascale platforms, the BPF ratio is anticipated to be much
lower. This trend puts enormous pressure on scientific software developers to
devise new algorithms and implementations that have minimal communication
overhead. One way to achieve this goal is to develop algorithms that have a
lower communication volume and fewer number of messages. Another strategy,
which we will focus on in this paper, is to change the way processes are mapped
to physical processors so that the load on the interconnection network, which is
characterized by a number of metrics such as the average link dilation, traffic
and congestion, can be reduced. With this strategy, it is possible to reduce the
communication overhead in large-scale parallel computations [1].

C. Kaklamanis et al. (Eds.): Euro-Par 2012, LNCS 7484, pp. 830–842, 2012.

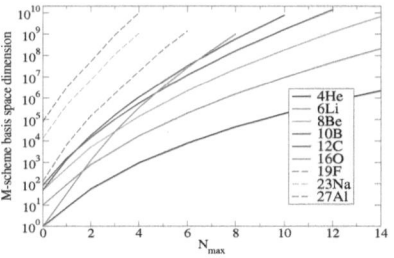 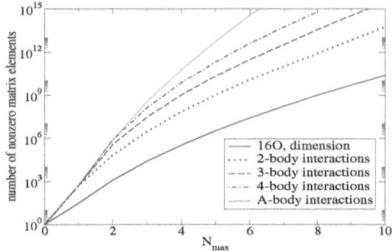

Fig. 1. The dimension and the number of non-zero matrix elements of the nuclear Hamiltonian with respect to N_{max} and the number of particles A

In this paper, we show that changes in task-to-processor mapping can have a drastic effect on the performance of large sparse eigenvalue calculations for predicting nuclear structures. We explain the observed effects quantitatively by reporting various load metrics associated with different mappings. Our results confirm the need to seek a topology-aware mapping to reduce communication overhead for large-scale parallel computations.

2 Eigensolver for the CI Approach

The key problem to be solved in nuclear structure calculations is the nuclear many-body Schrödinger's equation $H\psi = E\psi$, where ψ is a many-body wave-function and H is a nuclear many-body Hamiltonian. One way to solve the problem is to expand ψ in terms of Slater determinants of single-particle basis functions that satisfy a number of contraints. A particular choice of single-particle basis suitable for nuclear structure calculation is the harmonic oscillator basis. The representation of H under this basis expansion, which is often referred to as the configuration interaction (CI) approach, is a sparse symmetric matrix $\hat{H}$. The dimension of $\hat{H}$, which we denote by n, is defined by the number of Slater determinants used in the expansion, which is in turn determined by the number of nucleons A and a constraint on the harmonic oscillator quanta, which is often denoted by N_{max}. Higher N_{max} values yield more accurate results for the same nucleus, but at the expense of an exponential growth in problem size, see Fig. 1.

Due to the large dimension and the sparsity of $\hat{H}$, an iterative algorithm such as the Lanczos algorithm is preferred to solve the eigenvalue problem described above. The basic steps of the Lanczos algorithm are outlined in Alg. 1. The computational cost of the algorithm is generally dominated by the first two steps within the *while* loop: (i) multiplication of the sparse matrix $\hat{H}$ with the most recent Lanczos vector v, (ii) orthogonalization of the new vector w with respect to previous Lanczos vectors stored in V (a renormalization step may also be desirable). In this section, we discuss how these two tasks are decomposed into subtasks and how the subtasks are mapped to processing units for achieving a load balanced parallel implementation of Alg. 1.

Algorithm 1. The basic steps of the Lanczos Algorithm

> **Input:** $\hat{H}$, v_0;
> **Output:** ψ and E such that $\|\hat{H}\psi - \psi E\|_F$ is small.
> $v \leftarrow v_0/\|v_0\|$;
> $V \leftarrow (v)$;
> **while** not converged **do**
> $\quad w \leftarrow \hat{H}v$;
> $\quad w \leftarrow w - VV^T w$;
> $\quad v \leftarrow w/\|w\|$;
> $\quad V \leftarrow (V, v)$;
> $\quad T \leftarrow V^T \hat{H} V$;
> $\quad$ Diagonalize T to obtain (U, E);
> $\quad$ Check convergence;
> **end while**
> $\psi = VU$;

2.1 Sparse Matrix Vector Multiplication (SpMV)

On a distributed memory machine, the SpMV multiplication $w \leftarrow \hat{H}v$ can be carried out in parallel by partitioning the rows and columns of $\hat{H}$ and distributing the nonzero elements of $\hat{H}$ among different processing units. This is the strategy taken by the state-of-the-art nuclear CI calculation software package MFDn (Many-body Fermion Dynamics for nuclear physics) [6,7]. An even distribution of non-zero matrix elements over all processors is crucial for optimizing the use of available memory and achieving good load-balance. This is accomplished in MFDn by an appropriate matrix reordering through row/column permutation.

Since $\hat{H}$ is symmetric, we store only its lower triangular part to reduce memory usage. Consequently, it is natural to organize processing units into an $n_d \times n_d$ triangular grid over the $\hat{H}$ matrix, as shown in Fig. 2. Each processing unit, which stores the (i, j)th portion of the sparse matrix $\hat{H}$, can be labeled by its row and column positions on the grid. The total number of processing units n_p in this triangular grid is $n_d(n_d + 1)/2$, where n_d is also the number of processing units along the diagonal.

A simple way to perform the SpMV multiplication in parallel is to partition the vector v by rows into $\{v_i\}$, where $i = 1, 2, ..., n_d$. in a way that is conformal with the column partitioning of $\hat{H}$ as shown in Fig. 2. Row and column communication groups, labeled by the communicator C_{row} and C_{col} respectively, are set up to allow v_i to be broadcast among processing units that lie on the ith row or column of the triangular grid. If we denote the submatrix of $\hat{H}$ assigned to the (i, j)th processing unit by $\hat{H}_{i,j}$, each processing unit performs two SpMVs of the form $w_i = \hat{H}_{i,j}v_j$ and $w_j = \hat{H}_{i,j}^T v_i$. Two reductions are required (one along the row communication groups and one along the column communication groups) to merge local products w_i and w_j to form the global result vector w.

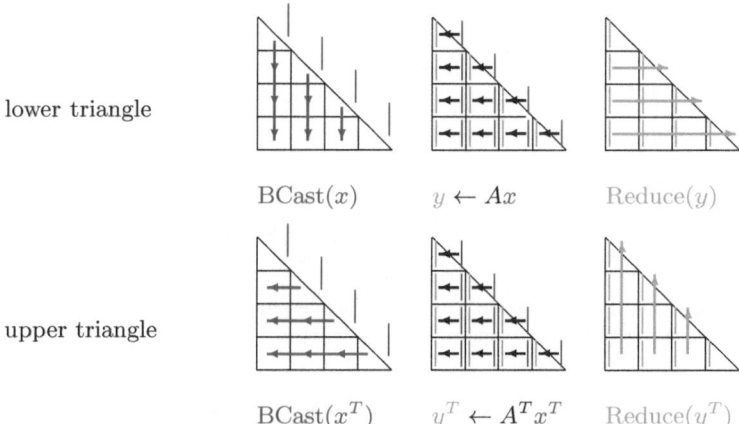

lower triangle

BCast(x) $y \leftarrow Ax$ Reduce(y)

upper triangle

BCast(x^T) $y^T \leftarrow A^T x^T$ Reduce(y^T)

Fig. 2. A visual illustration of the communication pattern for the distributed SpMV procedure, as implemented in MFDn

If n is the dimension of $\hat{H}$, the length of each distributed Lanczos vector is roughly n/n_d. The communication volume associated with the broadcast and reduction operations required for SpMV multiplication is $O(nn_d)$ along the C_{col} and C_{row} groups. The dependence of the communication volume on the number of diagonal processors n_d suggests that a multi-threaded (hybrid MPI/OpenMP parallel) implementation running on the same number of cores as a pure MPI implementation would have less communication overhead. To be precise, a multi-threaded implementation with t threads would generate $\sqrt{t}$ times less communication volume compared to the pure MPI version. Because in the hybrid MPI/OpenMP implementation with t-way thread parallelism, n_d decreases by a factor of $\sqrt{t}$, while n does not change.

2.2   Basis Orthogonalization

In MFDn, the orthogonalization of a new basis vector w against a previously constructed orthonormal basis contained in the matrix V is parallelized on a $n_d \times (n_d+1)/2$ 2D grid reconfigured from the $n_d \times n_d$ triangular grid as shown in Fig. 3. To maximize the amount of parallelism while minimizing communication volume, we distribute V by using a hierarchical 1D distribution scheme. A basis vector v is first divided into n_d subvectors v_j, $j = 1, 2, ..., n_d$, each associated with the jth row group of the 2D grid. Each subvector v_j is then further partitioned into $(n_d + 1)/2$ parts and distributed among the processing units in the same row group of the 2D grid.

After the SpMV multiplication $w = \hat{H}v$ is completed, a scattering operation is required to distribute w_j conforming to the distribution of the v_j subvectors among the jth row group of the 2D grid for $j = 1, 2, ...n_d$. Once w has been distributed among all processing units, an all-reduce operation is required to complete the orthogonalization $w - VV^Tw$. Communication volume associated with this all-reduce operation is typically small, when V does not contain too

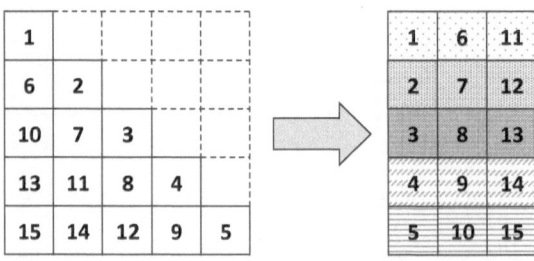

Fig. 3. Reconfiguring a $n_d \times n_d$ lower triangular processing grid (left) into a $n_d \times (n_d + 1)/2$ rectangular grid (right) for basis orthogonalization

many basis vectors. Finally, since the next SpMV in our iterative eigensolver requires the new basis vector v_i to be available on all processing units within the same column (or row) communication group, a gathering operation is required. The scattering and gathering operations involve a communication volume of $\mathcal{O}(n)$ only. Therefore, when the basis vectors are partitioned hierarchically in 1D, the total communication volume of the basis orthogonalization part is $\mathcal{O}(n)$, which is considerably smaller than that of the SpMV part.

3   Estimating the Communication Overhead

While the communication volume often provides a good estimate of the efficiency of a parallel program, the actual performance of the program will depend on other factors. The mapping between computational tasks and physical processing units has a strong influence, too. Therefore we need additional metrics that can capture the effect of process topology on the actual performance of the parallel program.

3.1   Network Load Model

We use a simplified version of the framework suggested by Hoefler and Snir [1]. A communication graph $\mathcal{G} = (\mathcal{V}_\mathcal{G}, \mathcal{E}_\mathcal{G})$ is a directed graph, where $\mathcal{V}_\mathcal{G}$ is the set of processes and an edge $e = \{u, v\} \in \mathcal{E}_\mathcal{G}$ denotes a message sent from process $u \in \mathcal{V}_\mathcal{G}$ to $v \in \mathcal{V}_\mathcal{G}$. We define two communication graphs, $\mathcal{G}_{col}$ and $\mathcal{G}_{row}$, which are associated with the column and row communication groups of the triangular grid, respectively.

Similarly, the physical interconnection network is represented as a directed, weighted graph $\mathcal{H} = (\mathcal{V}_\mathcal{H}, \mathcal{E}_\mathcal{H}, c_\mathcal{H})$, where $\mathcal{V}_\mathcal{H}$ is the set of compute nodes, $\mathcal{E}_\mathcal{H}$ is the set of links between these nodes, and $c_\mathcal{H}(e)$ corresponds to the bandwidth of a link $e = \{u, v\}$. We assume that messages are routed between nodes using the shortest path. We denote such a path by $p(u, v)$, the set of links connecting nodes u and v. Since there is usually more than one such shortest path, $\mathcal{P}(u, v)$ is used to refer to the set of all shortest paths between u and v. We assume that each path in $\mathcal{P}(u, v)$ is used with equal probability for sending a message.

We also assume static routing, *i.e.*, messages are not redirected when congestion is detected on certain parts of the network.

Under the model described above, we define $\Gamma : \mathcal{V}_\mathcal{G} \rightarrow \mathcal{V}_\mathcal{H}$ as a function that maps the vertices of a communication graph to the physical nodes in the network graph. Three quality measures can be defined for a mapping Γ: (average) dilation $\mathcal{D}(\Gamma)$, (average) traffic $\mathcal{T}(\Gamma)$ and (maximum) congestion $\mathcal{X}(\Gamma)$. These definitions are similar to those given by Hoefler and Snir [1], but slightly simpler due to the assumptions stated above. $\mathcal{D}(\Gamma)$ is the average number of links traveled by a message:

$$\mathcal{D}(\Gamma) = \frac{\sum_{\{u,v\}\in\mathcal{E}_\mathcal{G}} |p(\Gamma(u),\Gamma(v))|}{|\mathcal{E}_\mathcal{G}|} \tag{1}$$

Dilation is a measure of the total communication work that needs to be performed by the interconnection network. As dilation increases, the load on the interconnection network also increases.

We define traffic on a link $\mathcal{T}_\Gamma(e)$ as the number of messages that passes through the link e. Network traffic is the average traffic over all the links in an interconnect:

$$\mathcal{T}_\Gamma(e) = \sum_{\{u,v\}\in\mathcal{E}_\mathcal{G}} \frac{|\mathcal{S} = \{p : p \in \mathcal{P}(\Gamma(u),\Gamma(v)) \wedge e \in \mathcal{P}\}|}{|\mathcal{P}(\Gamma(u),\Gamma(v))|} \tag{2}$$

$$\mathcal{T}(\Gamma) = \frac{\sum_{e\in\mathcal{E}_\mathcal{H}} \mathcal{T}_\Gamma(e)}{|\mathcal{E}_\mathcal{H}|} \tag{3}$$

Finally, congestion on a link is defined as $\mathcal{X}(e) = \mathcal{T}_\Gamma(e)/c_\mathcal{H}(e)$ and network congestion is defined in terms of the maximum congestion on any of the network links, $\mathcal{X}(\Gamma) = \max_{e\in\mathcal{E}_\mathcal{H}} \mathcal{X}(e)$. Since a communication graph $\mathcal{G}$ does not contain any time related information, $\mathcal{X}(\Gamma)$ may not be a good approximation to the actual network congestion, in general. However, both $\mathcal{G}_{\mathrm{col}}$ and $\mathcal{G}_{\mathrm{row}}$ capture the communication happening in a small time window which begins and ends with a single collective call. Therefore, $\mathcal{X}(\Gamma)$ is a good approximation to the network congestions during the column and row communications of the eigensolver described in Sect. 2.

3.2 Practical Considerations

The performance results presented in Sect. 5 were obtained using the Hopper super-computer, which is a Cray XE6 machine at NERSC. Each compute node on Hopper contains 2 twelve-core AMD "MagnyCours" processors (24 cores per node) with 32 GBs of memory. Hopper uses the Cray "Gemini" interconnect for internode communication. The interconnection network has a 3D torus topology with dimensions 17×8×24. Two nodes share a single Gemini Network Interface Card (NIC), which has a total of 10 network connections, two each in +x, −x, +z, −z, and one +y and one −y links [3]. Consequently, the capacity of a link (c_H) in +y or −y direction is half the capacity of a link in other directions.

In order to compute the distance between two processing units, the physical coordinates of these units must be known. This machine-specific information can be obtained through `xtprocadmin` and `xtdb2proc` utilities available in the Cray Linux Environment (CLE). Since Hopper's network has a 3D torus topology, there is a physical link between a node and its 6 neighbors located at `+x`, `-x`, `+y`, `-y`, `+z` and `-z` directions. So the number of links that a message needs to hop through is simply the Manhattan distance [1] (in 3D) between the physical coordinates of its start and destination nodes. On Hopper, a message is routed from the start node to the destination node through the unique shortest path using the links in the `x` dimension first, then the links in the `y` dimension, and finally those in the `z` dimension.

In Cray's MPICH2-based MPI library on Hopper, the collective operations of `MPI_Bcast` and `MPI_Reduce` are implemented using a binomial tree algorithm [5]. $\mathcal{E}_{\mathcal{G}_{\mathrm{col}}}$ and $\mathcal{E}_{\mathcal{G}_{\mathrm{row}}}$ are constructed by identifying the the binomial trees associated with the column and row communication groups of the triangular grid.

On Hopper, multiple processes would be mapped to the same physical node. As a result, Γ is a many-to-one function in this case. We assume that the intra-node communication bandwidth is infinite. Therefore, we do not include edges that correspond to intra-node communications in our model.

4   Heuristic for Task-to-Processor Mapping

Given a communication graph $\mathcal{G}$ and a physical interconnection network $\mathcal{H}$, the problem of finding a mapping $\Gamma : \mathcal{V}_{\mathcal{G}} \to \mathcal{V}_{\mathcal{H}}$ that minimizes the effective load on the network measured in terms of $\mathcal{D}(\Gamma)$, $\mathcal{T}(\Gamma)$ and $\mathcal{X}(\Gamma)$ is an NP-hard problem [1]. If we label the vertices in $\mathcal{V}_{\mathcal{G}}$ by 1, 2,..., n_p, constructing Γ is equivalent to assigning these numbers to processors placed on a triangular grid. For example, we may assign $1, 2, ..., n_{\mathrm{d}}$ to the diagonal processors first, and continue the assignment for each of the subdiagonals until all processors on the grid are labeled. This gives what we call the diagonal-major (DM) ordering of the processors. Alternatively, we may go through the triangular processor grid column by column. This gives the column-major (CM) ordering. Row and column groups are created by grouping the numbered processing units based on their column and row positions. A clear drawback of this type grouping scheme is that the number of processing units in each row/column communication group is different, and the difference can be quite large. For example, while the largest C_{col} group contains n_{d} processing units, the smallest C_{col} group contains a single processing unit. As a result, there is a significant amount of imbalance in terms of communication volume among different communication groups.

To create a better mapping and grouping strategy, we extend the triangular grid to a square grid (note that $\hat{H}$ is symmetric), but require each processing unit to take either the (i, j)th grid point or the (j, i)th grid point, but not both. We modify the DM and CM orderings by limiting the number of processing units in each row or column of the grid to $(n_{\mathrm{d}} + 1)/2$. If assigning a task number to the (i, j)th grid point violates this rule, we map the task to the (j, i)th grid point in the modified DM scheme which we refer to as balanced diagonal major

---

[1] See http://en.wikipedia.org/wiki/Manhattan_distance for a definition.

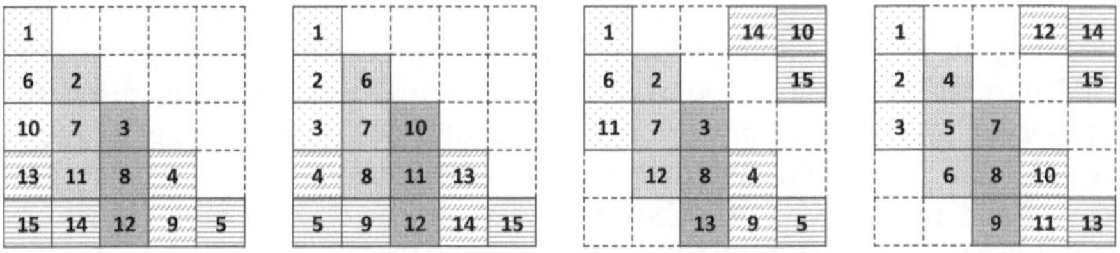

Fig. 4. Process orderings from left to right: DM, CM, BDM and BCM. Tasks mapped to the same column (row) of the grid belong to the same column (row) communication groups. Tasks with the same fill patterns belong to the same groups created for basis orthogonalization

(BDM) ordering. In the modified CM scheme, which we refer to as balanced column major (BCM) ordering, we start from the diagonal grid point in each column and assign a task to $(i - n_d, j)$th grid point when $i > n_d$. Figure 4 gives a schematic illustration of how DM, CM, BDM and BCM look for a 5×5 grid.

5   Performance Evaluation

In this section, we compare the mapping schemes described above on a few nuclear CI test problems involving ^{10}B. Different combinations of truncation ($N_{\max}$) and total magnetic angular momentum (M_j) parameters are used. Table 1 gives the size and sparsity characteristics for these problems and the number of cores used to solve these problems. The size of the problem (indicated by $nnz(\hat{H})$) increases roughly by a factor of 4 each time we change the ($N_{\max}, M_j$) parameters. Since we are mainly interested in the weak scaling of MFDn, we increase the number of cores used by approximately a factor of 4, too.

Table 1. Matrix dimensions n and number of non-zero matrix elements nnz of the Hamiltonian $\hat{H}$ associated with nuclear structure calculations of ^{10}B using different parameter pairs ($N_{\max}, M_j$)

Test Name	($N_{\max}, M_j$)	$n(\hat{H})$	$nnz(\hat{H})$	n_p	$n(\hat{H})/n_d$	$nnz(\hat{H})/n_p$
test$_{276}$	(7,0)	4.66×10^7	2.81×10^{10}	276	2.0×10^6	1.1×10^8
test$_{1128}$	(8,1)	1.60×10^8	1.24×10^{11}	1128	3.4×10^6	1.1×10^8
test$_{4560}$	(9,2)	4.82×10^8	4.62×10^{11}	4560	5.1×10^6	1.0×10^8
test$_{18336}$	(10,3)	1.30×10^9	1.51×10^{12}	18336	6.8×10^6	0.9×10^8

5.1   Performance Results with Pure MPI Implementation

Our experiments were performed on Hopper. Figure 5 shows the observed communication overhead associated with DM, CM, BDM and BCM ordering schemes for all four test problems. In each case, the wallclock time spent for communication in DM (t_{DM}) is taken as the baseline (shown at 100%), and those associated with other ordering schemes are shown as percentages of t_{DM}. Each bar in Fig. 5 shows three values: t_{col}, t_{row}, t_{orth}. They correspond to the wallclock times elapsed during communications within the column and row groups of the SpMV part and communication done for basis orthogonalization, respectively. We observe that the communication overhead of BDM is surprisingly higher than that associated with DM. We dropped the BDM scheme from larger test cases (test$_{4560}$ and test$_{18336}$.) Both CM and BCM produce significant reductions in communication time compared to DM. The reduction ranges from about a factor of 2 for the smallest test-case to a factor of 5 for larger ones. Although keeping the same number of processing units in each communication group seems to be desirable, the improvement of BCM over CM is small in larger test cases.

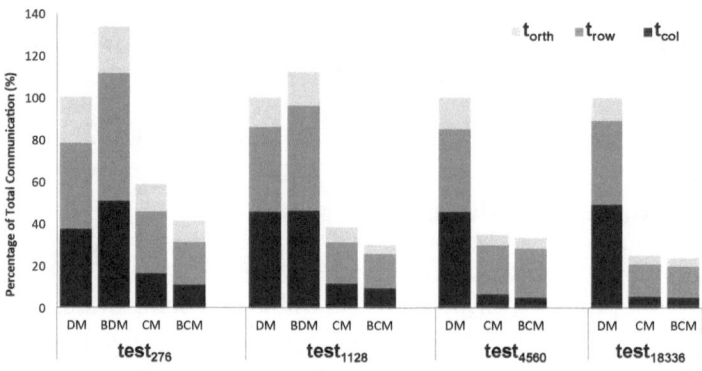

Fig. 5. Communication overhead associated with DM, CM, BDM and BCM ordering schemes with the pure MPI implementation

As discussed in Sect. 2, communication volume required in basis orthogonal-izaton is relatively small compared to that required in SpMV. Therefore it is not surprising to see that t_{col} and t_{row} dominate the communication overhead in Fig. 5. In the largest test case involving 18,336 cores, running 99 Lanczos iterations takes 1,260 seconds when the DM ordering is used. On average 80% (or 1,010 seconds) of the total runtime is spent in communication. As can be seen, in the DM column of test$_{18336}$ in Fig. 5, column group communications during SpMV is responsible for about 50% of the total communication time, and row group communications of SpMV is responsible for about 40%. The communication required in orthogonalization accounts for only 10%. Mapping tasks to processors according to the BCM ordering reduces the communication overhead by a factor of 5. While the overhead in all communication groups decreases sharply, the largest gain is seen in the column group, in which communication time drops from roughly 500 seconds for DM to only 50 seconds in BCM.

Since communication takes a significant portion of the total running time, the reductions in communication overhead achieved by BCM ordering translate to considerable speed-ups as summarized in Tab. 2. The last row in this table shows that the overall impact is as high as a factor of 2.57 speed-up in the total running time.

Table 2. Single-threaded performance improvement using different orderings

Ordering	Stats	test$_{276}$	test$_{1128}$	test$_{4560}$	test$_{18336}$
DM	t_{total} (sec)	211	410	567	1260
	$t_{\text{comm}}/t_{\text{total}}$	24%	47%	56%	80%
BDM	speed-up	0.93	0.95	−	−
	$t_{\text{comm}}/t_{\text{total}}$	30%	50%	−	−
CM	speed-up	1.12	1.41	1.57	2.52
	$t_{\text{comm}}/t_{\text{total}}$	16%	26%	31%	50%
BCM	speed-up	1.18	1.50	1.59	2.57
	$t_{\text{comm}}/t_{\text{total}}$	12%	21%	30%	49%

Table 3. Communication analysis for test$_{276}$ and test$_{1128}$

Ordering	Stats	test$_{276}$		test$_{1128}$	
		C_{col}	C_{row}	C_{col}	C_{row}
DM	t_{comm} (sec)	19	21	89	78
	$\{\mathcal{D}, \mathcal{T}, \mathcal{X}\}$	{0.9, 20, 56}	{0.8, 21, 62}	{5.5, 24, 148}	{5.6, 26, 136}
BDM	t_{comm} (sec)	26	31	90	97
	$\{\mathcal{D}, \mathcal{T}, \mathcal{X}\}$	{1.1, 34, 86}	{1.1, 34, 86}	{7.5, 49, 163}	{7.4, 48, 167}
CM	t_{comm} (sec)	8.5	15	22	39
	$\{\mathcal{D}, \mathcal{T}, \mathcal{X}\}$	{0.1, 3, 8}	{0.7, 20, 56}	{0.3, 5, 20}	{5.5, 23, 167}
BCM	t_{comm} (sec)	5.5	10.5	18	32
	$\{\mathcal{D}, \mathcal{T}, \mathcal{X}\}$	{0.0, 0, 0}	{0.8, 14, 44}	{0.0, 0, 0}	{5.2, 20, 122}

Table 4. Communication analysis for test$_{4560}$ and test$_{18336}$

Ordering	Stats	test$_{4560}$		test$_{18336}$	
		C_{col}	C_{row}	C_{col}	C_{row}
DM	t_{comm} (sec)	145	125	500	400
	$\{\mathcal{D}, \mathcal{T}, \mathcal{X}\}$	{2.1, 97, 538}	{2.1, 98, 538}	{3.7, 121, 1489}	{3.6, 120, 1557}
CM	t_{comm} (sec)	20	75	55	155
	$\{\mathcal{D}, \mathcal{T}, \mathcal{X}\}$	{0.1, 11, 25}	{2.1, 88, 290}	{0.1, 15, 48}	{3.6, 120, 632}
BCM	t_{comm} (sec)	15	75	50	150
	$\{\mathcal{D}, \mathcal{T}, \mathcal{X}\}$	{0.0, 4, 5}	{2.4, 81, 330}	{0.0, 6, 20}	{3.6, 111, 626}

Tables 3 and 4 summarize the wallclock time used for communication within column and row groups for four different orderings schemes and two test problems each. All timing figures are accompanied by a triplet of numbers $\{\mathcal{D}, \mathcal{T}, \mathcal{X}\}$ that correspond to the dilation, average traffic and congestion metrics defined in Sect. 3.1. Note that message sizes vary significantly between test cases (see Tab. 1). Therefore we take the message size in test$_{276}$ as our base unit and scale all metrics reported accordingly by dividing this message size to ensure a fair comparison across all test cases. A lower $\{\mathcal{D}, \mathcal{T}, \mathcal{X}\}$ value indicates lower load on the network, hence lower communication overhead. For CM and BCM orderings, the network load due to communication within column groups is considerably less than that of row groups. This observation explains the lower communication overhead seen in column groups. In fact, this also indicates that our simple heuristic of mapping the processes in the same column communication group into "nearby" nodes actually works well.

In Fig. 5, it is not intuitive at first to see that BDM performs worse than DM. Even though the BDM ordering balances communication volume over all groups, Table 3 shows that the dilation associated with BDM is higher than that of the DM. Consequently, the increased network load in BDM leads to a poor overall performance. This observation suggests that topology-awareness is more important than simply keeping communication volume balanced among different groups when we construct a task-to-processor map.

As discussed above, computational load per processor is roughly the same across all test cases. However, as seen in Tab. 3 and 4, communication overhead increases sharply as the test problem becomes larger. This sharp increase is partially caused by a fast increase in communication volume (which is of magnitude $O(nn_d)$) which exceeds the linear increase in the network bandwidth with respect to n_p. However, network dilation and, perhaps more importantly, network congestion increases are also important factors to consider.

We can gauge the severity of the network congestion in the row communication group by comparing the $\{\mathcal{D}, \mathcal{T}, \mathcal{X}\}$ triplets associated with the DM ordering with that associated with the BCM orderings in test$_{18336}$. Even though dilation and average traffic seems to be roughly the same for both orderings, t_{comm} of DM ordering is significantly higher than that of BCM (400 vs. 150 seconds). So is its network congestion (1557 vs. 626).

5.2   Performance Results with the MPI/OpenMP Implementation

Table 5 compares the communication overhead of the pure MPI and hybrid MPI/OpenMP implementations of the Lanczos algorithm for the large test cases. Both implementations use the BCM ordering. Despite the reduction in communication volume, the communication time used in the column groups is much higher in the multi-threaded implementation for the test$_{4560}$ problem. This is due to the increased dilation between communicating pairs in the column groups of the multi-threaded implementation. However, this difference is likely to vanish with

Table 5. Comparison of the pure MPI and hybrid MPI/OpenMP implementations for large testcases using approximately the same number of cores

| Threading | Stats | test$_{4560}$ | | test$_{18336}$ | |
		C_{col}	C_{row}	C_{col}	C_{row}
single	t_{comm} (sec)	15	75	50	150
	$\{\mathcal{D}, \mathcal{T}, \mathcal{X}\}$	{0.0, 4, 5}	{2.4, 81, 330}	{0.0, 6, 20}	{3.6, 111, 626}
multi	t_{comm} (sec)	40	60	58	110
	$\{\mathcal{D}, \mathcal{T}, \mathcal{X}\}$	{0.3, 24, 37}	{3.4, 44, 159}	{0.4, 29, 100}	{4.3, 89, 365}

increasing problem sizes, as indicated by the test$_{18336}$ results. Multi-threaded implementation performs clearly better along the unoptimized C_{row} communicator, where the reduced number of messages and communication volume results in less traffic and congestion on the network.

6   Conclusions and Future Work

We developed topology-aware task-to-processor mappings to reduce the communication overhead in a parallel implementation of the Lanczos algorithm used to solve the nuclear many-body Schrödinger's equation. The effectiveness of a mapping can be assessed by examining the average network dilation ($\mathcal{D}$), average traffic ($\mathcal{T}$) and the maximum network congestion ($\mathcal{X}$) associated with the mapping. Each mapping corresponds to a particular ordering of the distributed tasks. We compared several mapping strategies and showed that the balanced column major (BCM) ordering of tasks gives the best performance through a number of computational experiments. This observation is consistent with our network load model which is defined in terms ($\mathcal{D}, \mathcal{T}, \mathcal{X}$). However, even in the case of BCM ordering, our optimization of the task-to-processor mapping is not performed globally among all processors. Therefore, we believe further improvement to the BCM ordering scheme is possible by applying topology mapping techniques described in [1,2] to all three communication groups created in our implementation of the Lanczos algorithm. We will focus on this approach in the future. Another factor that may affect the choice of an optimal mapping is the combination of thread-level parallelism with message passing based parallelism. This is an important issue for multi-core/many-core platforms.

Acknowledgments. This work was supported in part through the Scientific Discovery through Advanced Computing (SciDAC) program funded by the U.S. DOE Office of Advanced Scientific Computing Research and Office of Nuclear Physics, by U.S. DOE Grants DE-FC02-09ER41582 (SciDAC/UNEDF) and DE-FG02-87ER40371, and by the US NSF grant 0904782. Computational resources were provided by NERSC, which is supported by the U.S. DOE Office of Science.

References

1. Hoefler, T., Snir, M.: Generic Topology Mapping Strategies for Large-scale Parallel Architectures. In: Proceedings of the 2011 ACM International Conference on Supercomputing (ICS), Tucson, AZ (June 2011)
2. Bhatele, A., Gupta, G., Kale, L.V., Chung, I.-H.: Automated Mapping of Regular Communication Graphs on Mesh Interconnects. In: Proceedings of International Conference on High Performance Computing, HiPC (2010)
3. NERSC, Hopper, NERSC's Cray XE6 System (January 2012), Web. (February 15, 2012), http://www.nersc.gov/users/computational-systems/hopper/.
4. Demmel, J.: Applied Numerical Linear Algebra, 1st edn. SIAM (1997)

5. MPICH2, MPICH2: High-performance and Widely Portable MPI,
   http://www.mcs.anl.gov/research/projects/mpich2
6. Sternberg, P., Ng, E.G., Yang, C., Maris, P., Vary, J.P., Sosonkina, M., Le, H.V.: Accelerating Configuration Interaction Calculations for Nuclear Structure. In: The Proceedings of the 2008 ACM/IEEE Conference on Supercomputing (SC 2008) (2008)
7. Maris, P., Sosonkina, M., Vary, J.P., Ng, E.G., Yang, C.: Scaling of ab-initio nuclear physics calculations on multicore computer architectures. Procedia CS 1, 97–106 (2010)

Fast and Effective Lossy Compression Algorithms for Scientific Datasets

Jeremy Iverson[1], Chandrika Kamath[2], and George Karypis[1]

[1] University of Minnesota, Minneapolis MN 55455, USA
[2] Lawrence Livermore National Laboratory, Livermore CA 94550, USA

Abstract. This paper focuses on developing effective and efficient algorithms for compressing scientific simulation data computed on structured and unstructured grids. A paradigm for lossy compression of this data is proposed in which the data computed on the grid is modeled as a graph, which gets decomposed into sets of vertices which satisfy a user defined error constraint ϵ. Each set of vertices is replaced by a constant value with reconstruction error bounded by ϵ. A comprehensive set of experiments is conducted by comparing these algorithms and other state-of-the-art scientific data compression methods. Over our benchmark suite, our methods obtained compression of 1% of the original size with average PSNR of 43.00 and 3% of the original size with average PSNR of 63.30. In addition, our schemes outperform other state-of-the-art lossy compression approaches and require on the average 25% of the space required by them for similar or better PSNR levels.

1 Introduction

The process of scientific discovery often requires scientists to run simulations, analyze the output, draw conclusions, then re-run the simulations to confirm or expand hypothesis. One of the most significant bottlenecks for current and future extreme-scale systems is I/O. In order to facilitate the scientific process described above, it is necessary for scientists to have efficient means to output and store data for offline analysis. To facilitate this, data compression is turned to, to create reduced representations of the resulting data for output, in such a way that the original result data can be reconstructed off-line for further analysis.

Straightforward approaches for scientific data compression exist in lossless techniques designed specifically for floating-point data. However, due to the high variability of the representation of floating-point numbers at the hardware level, the compression factors realized by these schemes are often very modest [4,10]. Since most post-run analysis is robust in the presence of some degree of error, it is possible to employ lossy compression techniques rather than lossless, which are capable of achieving much higher compression rates at the cost of a small amount of reconstruction error. As a result, a number of approaches have been investigated for lossy compression of scientific simulation datasets including classical [7] and diffusion wavelets [3], spectral methods [5], and methods based on the techniques used for transmission of HDTV signals [2]. However, these approaches

C. Kaklamanis et al. (Eds.): Euro-Par 2012, LNCS 7484, pp. 843–856, 2012.

are either applicable only to simulations performed on structured grids or have high computational requirements for *in situ* data compression applications.

In this paper we investigate the effectiveness of a class of lossy compression approaches that replace the actual values associated with sets of grid-nodes with a constant value whose difference from the actual value is bounded by a user-supplied error tolerance parameter. We develop approaches for obtaining these sets by considering only the nodes and their values and approaches that constrain these sets to connected subgraphs in order to further reduce the amount of information that needs to be stored. To ensure that these methods are applicable for *in situ* compression applications, our work focuses on methods that have near-linear complexity and are equally applicable to structured and unstructured grids. We experimentally evaluate the performance of our approaches and compare it against that of other state-of-the-art data compression methods for scientific simulation datasets. Over our benchmark suite, our methods obtained compression of 1% of the original size with average PSNR of 43.00 and 3% of the original size with average PSNR of 63.30. Our experiments show that our methods achieve compressed representations, which on average, require 50%–75% less space than competing schemes at similar or lower reconstruction errors.

2   Definitions and Notations

The methods developed in this paper are designed for scientific simulations in which the underlying physical domain is modeled by a grid. Here we assume that the grid topology is fixed and thus can be compressed and stored separately from the data which is computed on it. Each node of a grid has one or more values associated with it that correspond to the quantities being computed in the course of the simulation. The grid can be either structured or unstructured. A *structured grid* is a collection of elements which have an implicit geometric structure. That structure is a basic rectangular matrix structure, such that in $\mathbb{R}^3$, the nodes can be indexed by a triplet (x, y, z). Thus, the grid topology can be described simply by the number of nodes in each of the three dimensions. An *unstructured grid* has no implicit structure. Since there is no implicit structure, the topology is described by identifying the elements which each node belongs to. In this work, we model these grids via a *graph* $G = (V, E, L)$. The set of vertices V, models the nodes of the grid for which values are computed. The set of edges E, models the connectivity of adjacent nodes. Two nodes are adjacent if they belong to the same element in the grid. The set of vertex-labels L, models the values computed at each node of the grid such that l_i stores the value computed for node v_i. In this work we assume there is only one value being computed for each node of the grid.

An ϵ-*bounded set-based decomposition* of G is a partitioning of its set of vertices into non-overlapping sets $\{V_1, \ldots, V_k\}$ such that for each V_i, $\forall v_q, v_r \in V_i$, $|l_q - l_r| \leq \epsilon$ (i.e., each set contains vertices whose values differ at most by ϵ). When the induced subgraph $R_i = (V_i, E_i)$ of G is connected, the set V_i will also be referred to as a *region* of G. When all sets in an ϵ-bounded set-based decomposition form regions, then the decomposition will be referred to as an *ϵ-bounded region-based decomposition* of G. Given a set of vertices V_i, the average value of its vertices will be referred to as its *mean value* and will be denoted by $\mu(V_i)$. Given a region V_i, its *boundary vertices* are its subset of vertices $B_i \subseteq V_i$ that are adjacent to at least one other vertex not in V_i, and its *interior vertices* are the subset of vertices $I_i \subseteq V_i$ that are adjacent only to vertices in V_i. Note that $I_i \cup B_i = V_i$.

3   Related Work

Most of the work on lossy compression of scientific datasets has focused on compressing the simulation output for visualization purposes. The most popular techniques in this area are based on wavelet theory [7] that produces a compression-friendly sparse representation of the original data. To further sparsify this representation, coefficients with small magnitude are dropped with little impact on the reconstruction error [8,9]. Due to the nature of the wavelet transform, classical wavelet methods apply only to structured grids. An alternative to wavelet compression is Adaptive Coarsening (AC) [11]. AC is an extension of the adaptive sub-sampling technique first introduced for transmitting HDTV signals [2], which is based on down-sampling a mesh in areas which can be reconstructed within some error tolerance and storing at full resolution the others. In [12], the authors use AC to compress data on structured grids and compare the results to wavelet methods. Even though AC can potentially be extended for unstructured grids [11], current implementations are limited to structured grids.

Another approach is spectral compression that extends the discrete cosine transform used in JPEG, from 2D regular grids to the space of any dimensional unstructured grids [5]. This method uses the Laplacian matrix of the grid to compute topology aware basis functions. The basis functions serve the same purpose as those in the wavelet methods and define a space where the data can be projected to, in order to obtain a sparse representation. Since the Laplacian matrix can be defined for the nodes of any grid, this method is not limited to structured grids. However, deriving the basis functions from the Laplacian matrix of large graphs is computationally prohibitive. For this reason, practical approaches first use a graph partitioning algorithm to decompose the underlying graph into small parts, and each partition is then compressed independently using spectral compression [5]. Finally, another approach, introduced in [3], is diffusion wavelets. The motivation for diffusion wavelets is the same as that of spectral compression, and is used to generate basis functions for a graph. However, instead of using the eigenvectors of the Laplacian matrix to derive these basis functions, diffusion wavelets generate them by taking powers of a diffusion operator. The advantage of diffusion wavelet is that its basis functions capture characteristics of the graph at multiple resolutions, while spectral basis functions only capture global characteristics.

4   Methods

In this work we investigated the effectiveness of a lossy compression paradigm for grid-based scientific simulation datasets that replaces the values associated with a set of nodes with a constant value whose difference from the actual values is bounded. Specifically, given a graph $G = (V, E, L)$ modeling the underling grid, this paradigm computes an ϵ-bounded set-based decomposition $\{V_1, \ldots, V_k\}$ of G and replaces the values associated with all the nodes of each set V_i, with its mean value $\mu(V_i)$. This paradigm bounds the point-wise error to be no more than ϵ, whose actual value is explicitly controlled by the users based on their subsequent analysis requirements. Since the values associated with the nodes tend to exhibit local smoothness [1], these value substitutions increase the degree of redundancy, which can potentially lead to better compression.

Following this paradigm, we developed two classes of approaches for obtaining the ϵ-bounded set-based decomposition of G. The first class focuses entirely on the vertices of the grid and their values, where the second class also takes into account the connectivity of these vertices in the graph. In addition, we developed different approaches for encoding the information that needs to be stored on the disk in order to maximize the overall compression. The description of these algorithms is provided in the subsequent sections.

In developing these approaches, our research focused on algorithms whose underlying computational complexity is low because we are interested in being able to perform the compression *in-situ* with the execution of the scientific simulation on future exascale-class parallel systems. As a result of this design choice, the algorithms that we present tend to find sub-optimal solutions but do so in time that in most cases is bounded by $O(|V| \log |V| + |E|)$.

4.1   Set-Based Decomposition

This class of methods derives the ϵ-bounded set-based decomposition $\{V_1, \ldots, V_k\}$ of the vertices by focusing entirely on their values. Towards this end, we developed two different approaches. The first is designed to find the decomposition that has the smallest cardinality (i.e., minimize k), whereas the second is designed to find a decomposition that contains large-size sets.

The first approach, referred to as *SBD1*, operates as follows. The vertices of G are sorted in non-decreasing order based on their values. Let $\langle v_{i_1}, \ldots, v_{i_n} \rangle$ be the sequence of the vertices according to this ordering, where n is the number of vertices in G. The vertices are then scanned sequentially from v_{i_1} up to vertex v_{i_j} such that $l_{i_j} - l_{i_1} \leq \epsilon$ and $l_{i_{j+1}} - l_{i_1} > \epsilon$. The vertices in the set $\{v_{i_1}, \ldots, v_{i_j}\}$ satisfy the constraint of an ϵ-bounded set and are used to form a set of the set-based decomposition. These vertices are then removed from the sorted sequence and the above procedure is repeated on the remaining part of the sequence until it becomes empty. It can be easily shown that the above greedy algorithm will produce a set-based decomposition that has the smallest number of sets for a given ϵ.

The second approach, referred to as *SBD2*, utilizes the same sorted sequence of vertices $\langle v_{i_1}, \ldots, v_{i_n} \rangle$ but it uses a different greedy strategy for constructing the ϵ-bounded sets. Specifically, it identifies the pair of vertices v_{i_q} and v_{i_r} such that $l_{i_r} - l_{i_q} \leq \epsilon$ and $r - q$ is maximized. The vertices in the set $\{v_{i_q}, \ldots, v_{i_r}\}$ satisfy the constraint of an ϵ-bounded set and are used to form a set of the set-based decomposition. The original sequence is then partitioned into two parts: $\langle v_{i_1}, \ldots v_{i_{q-1}} \rangle$ and $\langle v_{i_{r+1}}, \ldots, v_{i_n} \rangle$, and the above procedure is repeated recursively on each of these subsequences. Note that the greedy decision in this approach is that of finding a set that has the most vertices (by maximizing $r - q$). It can be shown that SDB2 will lead to a decomposition whose maximum cardinality set will be at least as large as the maximum cardinality set of SBD1 and that the cardinality of the decomposition can be greater than that of SDB1's decomposition.

Decomposition Encoding. We developed two approaches for encoding the vertex values derived from the ϵ-bounded set-based decomposition. In both of these approaches, the encoded information is then further compressed using standard lossless compression methods such as GZIP, BZIP2, and LZMA.

The first approach uses scalar quantization and utilizes a pair of arrays Q and M. Array Q is of size k (the cardinality of the decomposition) and $Q[i]$ stores the mean value $\mu(V_i)$ of V_i. Array M is of size n (the number of vertices) and $M[j]$ stores the number of the set that vertex v_j belongs to. During reconstruction, the value of v_j is given by $Q[M[j]]$. Since for reasonable values of ϵ, $k \ll n$, the number of distinct values in M will be small, leading to a high degree of redundancy that can be exploited by the subsequent lossless compression step. We will refer to this approach as *scalar quantization encoding* and denote it by *SQE*.

The second approach encodes the information by sequentially storing the vertices that belong to each set of the decomposition. Specifically, it uses three arrays Q, S, and P, of sizes k, k, and n, respectively. Array Q is identical to the Q array of SQE and array S stores the number of vertices in each set (i.e., $S[i] = |V_i|$). Array P is used to store the vertices of each set in consecutive positions, starting with those of set V_1, followed by V_2, and so on. The vertices of each set are stored by first sorting them in increasing order based on their number and then representing them using a differential encoding scheme. The smallest numbered vertex of each set is stored as is and the number of each successive vertex is stored as the difference from the preceding vertex number. Since each vertex-set will likely have a large number of vertices, the differential encoding of the sorted vertex lists will tend to consist of many small values, and thus increase the amount of redundancy that can be exploited by the subsequent lossless compression step. We will refer to this approach as *differential encoding* and denote it by *DE*.

Vertex Ordering. To achieve good compression using the above encoding schemes, vertices which are close in the vertex ordering should have similar values. Towards this end, we investigate three vertex orderings which are as follows. The first is the original ordering of the nodes, that is often derived by the grid generator and tends to have a spatial coherence. The second ordering is a breadth first traversal of the graph starting from a randomly selected vertex. The third ordering is a priority first traversal, in which priority is given to those vertices which are adjacent to the most vertices which have been previously visited. Arranging the vertices according to their visit order is intended to put together in the ordering vertices that are close in the graph topology. Due to the local smoothness of values, vertices that appear close in the ordering will share similar values.

4.2   Region-Based Decomposition

This class of methods derives an ϵ-bounded set-based decomposition $\{V_1, \ldots, V_k\}$ by requiring that each set V_i also forms a region (i.e., its induced subgraph of G is connected). The motivation behind this region-based decomposition is to reduce the amount of data that needs to be stored by only writing information about V_i's boundary vertices and a select few of its interior vertices. During reconstruction, by taking advantage of V_i's connectivity, its non-saved interior vertices can be identified by a depth- or breadth-first traversal of G starting at the saved interior vertices and terminating at its boundary vertices. The set of vertices visited in the course of this traversal will be exactly those in V_i. From this discussion, we see that the amount of compression that can be achieved by this class of methods is directly impacted by the number of boundary vertices that must be stored. Thus, the region identification approaches must try to reduce the number of boundary vertices. Towards this end, we developed three different heuristic approaches whose description follows.

The first approach, referred to as *RBD1* , is designed to compute a decomposition that minimizes the number of regions. The motivation behind this approach is that by increasing the average size of each region (due to a reduction in the decomposition's cardinality), the number of interior vertices will also increase. RBD1 initially sorts the vertices in a way identical to SBD1, leading to the sorted sequence $s = \langle v_{i_1}, \ldots, v_{i_n} \rangle$. Then, it selects the first vertex in the sequence (v_{i_1}), assigns it to the first region V_1, and removes it from s. It then proceeds to select from s a vertex v_{i_j} that is adjacent to at least one vertex in V_1 and $l_{v_{i_j}} - l_{v_1} \le \epsilon$, inserts it into V_1, and removes it from s. This step is repeated until no such vertex can be selected or s becomes empty. The above algorithm ensures that V_1 is an ϵ-bounded set and that the subgraph of G induced by V_1 is connected. Thus, V_1 is a region and is included in the region-based decomposition. The above procedure is then repeated on the vertices remaining in s, each time identifying an additional region that is included in the decomposition. Note that unlike the algorithm for SBD1, the above algorithm does not guarantee that it will identify the ϵ-bounded region-based decomposition that has the minimum number of regions.

The second approach, referred to as *RBD2* , is designed to compute a decomposition that contains large regions, as the regions that contain a large number of vertices will also tend to contain many interior vertices. One way of developing such an algorithm is to use the greedy approach similar to that employed by SBD2 to repeatedly find the largest region from the unassigned vertices and include it in the decomposition. However, due to the region's connectivity requirement, this is computationally prohibitive. For this reason, we developed an algorithm that consists of two steps. The first step is to obtain an ϵ-bounded set-based decomposition $\{V_1, \ldots, V_k\}$ using SBD1. The second step is to compute an ϵ-bounded region-based decomposition of each set V_i. The union of these regions over $V_1, \ldots, V_k$ is then used as the region-based decomposition computed by RBD2. This two-step approach is motivated by the following observation. One of the reasons that prevents RBD1 from identifying large regions is that it starts growing each successive region from the lowest-valued unassigned vertex and does not stop until all of the unassigned vertices adjacent to that region have values that will violate the ϵ bound. This will tend to fragment subsequent regions as the are constrained by the initial vertices that have low values. RBD2, by forcing RBD1's region identification algorithm to stay within each set V_i, prevents this from happening and as our experiments will later show, lead to a decomposition that has smaller number of boundary vertices and better compression.

Finally, the third approach, referred to as *RBD3* , is designed to directly compute a decomposition whose regions have a large number of interior vertices. It consists of three distinct phases. The first phase identifies a set of *core* regions that contain at least one interior vertex, the second phase expands these regions by including additional vertices to them, and the third phase creates non-core regions. Let V' be the subset of vertices of V such that $\forall v \in V', v \cup \mathrm{adj}(v)$ is an ϵ-bounded set, where $\mathrm{adj}(v)$ is the set of vertices adjacent to v. A core region, V_i, is created as follows. An unassigned vertex $v \in V'$ whose adjacent vertices are also unassigned is randomly selected and $v \cup \mathrm{adj}(v)$ is inserted into V_i. Then the algorithm proceeds to identify an unassigned vertex $u \in V'$ such that: (i) it is connected to at least one vertex in V_i, (ii) all the vertices in $\mathrm{adj}(u) \setminus V_i$ are also unassigned, and (iii) $V_i \cup \{u\} \cup \mathrm{adj}(u)$ is an ϵ-bounded set. If such a vertex u exists, then u and $\mathrm{adj}(u) \setminus V_i$ are inserted into V_i. If no such vertex exists, then V_i's expansion stops. The above procedure is repeated until no more core regions can be created. Note that by including u and its $\mathrm{adj}(u) \setminus V_i$ vertices into V_i, we ensure that u becomes an interior vertex of V_i. During the second phase of the algorithm, the vertices that have not been assigned to any region are considered. If a vertex v can be included to an existing region while the resulting region remains an ϵ-bounded set, then it is assigned to that. Finally, the third phase is used to create additional regions containing the remaining unassigned vertices (if they exist), which is done using RBD1.

Decomposition Encoding. As discussed earlier, the region-based decomposition allows us to reduce the storage requirements by storing only the boundary vertices along with the interior vertices that are used as the *seeds* of the (depth- or breadth-first) traversals. For each region V_i, the set of seed-vertices I_i^s is determined as follows. An interior vertex is randomly selected, added to I_i^s, and a traversal from that vertex is performed terminating at V_i's boundary vertices. If any of V_i's interior vertices has not been visited, then the above procedure is repeated on the unvisited vertices, each time adding an additional source vertex into I_i^s. In most cases, one seed vertex will be sufficient to traverse all the interior vertices, but when regions are contained within other regions, multiple seed vertices may be required. Also, in the cases in which V_i consists of only boundary vertices, I_i^s will be empty.

An additional storage optimization is possible, as there is no need to store the boundary vertices for all the regions. In particular, consider a region V_i and let $\{V_{i_1}, \ldots, V_{i_m}\}$ be the set of its adjacent regions in the graph. We can then identify V_i by performing a traversal from the vertices in I_i^s that terminates at the boundary vertices of V_i's adjacent regions. All the vertices visited during that traversal (excluding the boundary vertices) along with I_i^s will be exactly the vertices of V_i. Thus, we can choose not to store V_i's boundary vertices as long as we store the boundary vertices for all of its adjacent regions. In our algorithm, we choose the regions whose boundary information will not be stored in a greedy fashion based on the size of their boundaries. Specifically, we construct the region-to-region adjacency graph (i.e., two regions are connected if they contain vertices that are adjacent to each other), assign a weight to the vertex corresponding to V_i that is equal to $|B_i|$ (i.e., the size of its boundary), and then identify the regions whose boundary information will not be stored by finding a maximal weight independent set of vertices in this graph using a greedy algorithm.

Given the above, we can now precisely describe how the region-based decomposition is stored. Let $\{V_1, \ldots, V_k\}$ be the ϵ-bounded region-based decomposition, $B_1, \ldots, B_k$ be the sets of boundary vertices that need to be stored (if no boundary information is stored for a region due to the earlier optimization, then the corresponding boundary set is empty), and $I_1^s, \ldots, I_k^s$ be the sets of internal seed-vertices that have been identified. Our method stores five arrays, Q, N_I, N_B, I_I, and I_B. The first three arrays are of length k, I_I is of length equal to the total number of seed vertices ($\sum_i |I_i^s|$), and I_B is of length equal to the total number of boundary vertices ($\sum_i |B_i|$). Array Q stores the mean values of each region, whereas arrays N_I and N_B store the number of seed and boundary vertices of each region, respectively. Array I_I stores the indices of the regions in consecutive order starting from I_1^s, whereas array I_B is used to store the boundary vertices of each region in consecutive positions starting from B_1. These indices are stored using the same differential encoding approach described in Sect. 4.1 and like that approach, the results of this encoding are further compressed using a standard lossless compression method.

5   Experimental Design and Results

Datasets. We evaluated our algorithms using seven real world datasets obtained from researchers at UMN and our colleagues at NASA and LLNL. These datasets correspond to fluid turbulence and combustion simulations and contain both structured and unstructured grids. Their characteristics are shown in Table 1.

Table 1. Information about the various datasets

Dataset	$|V|$	$|E|$	$\mu(V)$	Grid Type	Dataset	$|V|$	$|E|$	$\mu(V)$	Grid Type
d1	486051	4335611	0.9958	unstruct.	d5	31590144	94562224	0.0176	unstruct.
d2	589824	1744896	0.5430	struct.	d6	41472000	123926400	0.2107	struct.
d3	1936470	15399496	0.9874	unstruct.	d7	100663296	300744704	4.5644	struct.
d4	16777216	50102272	163.70	struct.					

Evaluation Methodology and Metrics. We measured the performance of the various approaches along two dimensions. The first is the error introduced by the lossy compression and the second is the degree of compression that was achieved. The error was measured using three different metrics: (i) the root mean squared error (RMSE), (ii) the maximum point-wise error (MPE), and (iii) the peak signal-to-noise ratio (PSNR). The RMSE is defined as

$$RMSE = \sqrt{\frac{1}{|V|} \sum_{i=1}^{|V|} |l_j - \hat{l}_j|^2},$$ (1)

where l_j is the original value of vertex v_j and $\hat{l}_j$, is its reconstructed value. The MPE is defined as

$$MPE = \max(|l_1 - \hat{l}_1|, ..., |l_n - \hat{l}_n|),$$ (2)

which is the ℓ_∞-norm of the point-wise error vector. The MPE measure is presented in tandem with RMSE to identify those algorithms which achieve low RMSE, but sustain high point-wise errors. Finally, the PSNR is defined as

$$PSNR = 20 \cdot \log_{10} \left(\frac{\max(x_1, ..., x_n)}{RMSE} \right),$$ (3)

which is a normalized error measure; thus, facilitating comparisons of error between datasets with values that differ greatly in magnitude. The compression effectiveness was measured by computing the compression ratio (CR) of each method, which is defined as follows:

$$CR = \frac{\text{compressed size}}{\text{uncompressed size}}.$$ (4)

The wavelet and spectral methods were implemented in Matlab®. The spectral method uses METIS [6] as a pre-processing step to partition the graph before compressing. The adaptive coarsening implementation was acquired from the authors of [12] and modified to provide the statistics necessary for these experiments. All algorithms described in Sect. 4 were implemented in C++. Finally, for the lossless compression of the decomposition encodings, we used LZMA compression (7-zip's implementation) as it resulted in better compression than either GZIP or BZIP2. In addition, the same LZMA-based compression was applied to the output of the spectral and wavelet-based compressions. Note that AC does not need that because it achieves its compression by coarsening the graph and reducing the data output.

6   Results

Our experimental evaluation is done in two parts. First, we select a fixed set of values for RMSE and compare the various algorithmic choices for the set- and region-based decomposition approaches in terms of their compression ability. Second, we compare the compression performance of the best combinations of these schemes against that achieved by other approaches for two different levels of lossy compression errors.

6.1   Set-Based Decomposition

Figure 1 shows the compression performance achieved by SBD1 and SBD2 for the different datasets across the different decomposition encoding schemes described in Section 4.1. These results show that SBD1 tends to perform better than SBD2 and on average, it requires 5% less storage for each specific combination of decomposition encoding and vertex ordering scheme. This can be attributed to the fact that the cardinality of its decomposition is often considerably lower than SBD2's, which tends to outweigh the benefits achieved by the few larger sets identified by SBD2.

Comparing the performance of the decomposition encoding schemes (SQE and DE), we see that SQE performs considerably better across both decomposition methods and ordering schemes. On the average, SQE requires only 75% of the storage of DE. These results suggest that when compared to scalar quantization, the differential encoding of the vertices in each set is not as effective in introducing redundancy in the encoding, which in turn reduces the compression that can be obtained by the lossless LZMA compression. Finally, comparing the performance of the three vertex ordering schemes, we found that the original ordering leads to greater compression than either of the breadth first traversal or the priority first traversal. As discussed in Section 4.1, this ordering utilizes

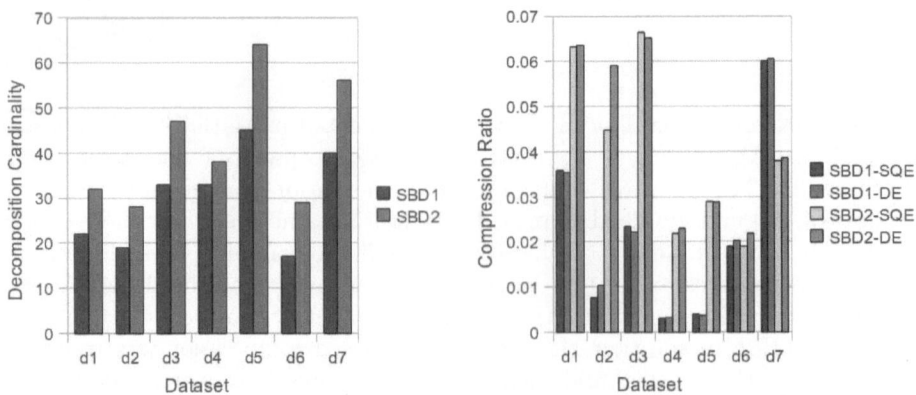

Fig. 1. Statistics for set-based decomposition

information from the underlying grid geometry, and as such it has a higher degree of regularity, leading to better compression. With respect to the other two methods, we found that the priority first traversal tends to perform better than breadth first.

6.2   Region-Based Decomposition

Figure 2 shows various statistics of the decompositions computed by RBD1, RBD2, and RBD3 for the different datasets and their compression performance for the three vertex ordering schemes. In terms of the number of regions into which G is decomposed, we see that RBD1 results in the least number of regions, whereas RBD3 identifies a considerably greater number of regions (often 2–7 times more regions than RBD1). We also see that RBD2 only identifies slightly more regions than RBD1 (about 18% more on average). In terms of the number of boundary vertices that need to be stored by each decomposition, we see an inversion of the previous results. RBD2 and RBD3 produce the smallest boundary sets, typically being within about 5% of each other, whereas RBD1 produces boundary sets which are considerably larger, in some cases, more than twice the size of those required by RBD2 and RBD3. These results suggest that the region identification heuristics employed by RBD2 and RBD3 are quite effective in minimizing the total number of boundary vertices, even though they find more regions.

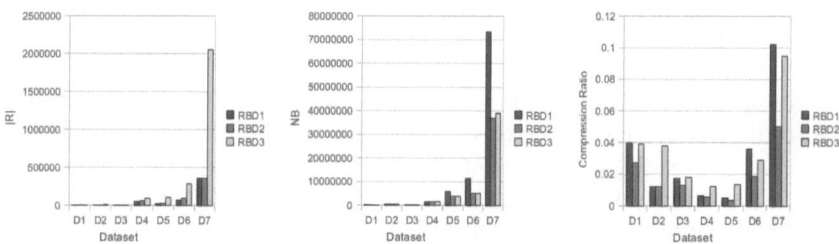

Fig. 2. Statistics for region-based decomposition, $|R|$ refers to number of regions identified, and NB refers to number of boundary vertices after storage optimization described in Sect. 4.1.

In terms of compression performance, we see that across all datasets RBD2 results in the lowest compression ratio. On the average, RBD2 requires only 70% of the storage of RBD1 and 56% of RBD3. Contrasting this with the number of boundary vertices identified by each approach, we see that there is a direct correlation, based on the size of the boundary vertex set, between RBD1 and RBD2 in terms of which approach results in lower compression ratio and by how much. RBD3 does not share in this correlation, due to its significantly higher number of regions.

Table 2. Comparison of scientific data compression algorithms for two different rmse

info		high error tolerance				low error tolerance			
Dataset	Algorithm	RMSE	PSNR	MPE	CR	RMSE	PSNR	MPE	CR
	SBD1	6.30E-03	4.64E+01	1.89E-02	2.39E-02	6.66E-04	6.60E+01	1.86E-03	**7.80E-02**
d1	RBD2	6.28E-03	4.65E+01	1.89E-02	**2.52E-02**	6.38E-04	6.63E+01	2.12E-03	1.28E-01
	Spctrl	6.37E-03	4.63E+01	1.11E-01	4.00E-02	3.90E-03	5.06E+01	7.14E-02	1.05E-01
	SBD1	2.92E-02	3.60E+01	7.33E-02	**2.51E-03**	2.50E-03	5.74E+01	7.71E-03	**1.27E-02**
	RBD2	2.88E-02	3.61E+01	8.02E-02	5.02E-03	1.91E-03	5.97E+01	7.71E-03	6.57E-02
d2	Wvlt	3.10E-02	3.55E+01	2.34E-01	2.00E-02	2.59E-03	5.70E+01	2.38E-02	1.15E-01
	Spctrl	3.17E-02	3.53E+01	7.34E-01	4.50E-02	7.04E-03	4.84E+01	3.56E-01	1.30E-01
	AC	3.31E-02	3.49E+01	1.50E-01	1.86E-02	6.80E-03	4.87E+01	7.19E-01	5.17E-02
	SBD1	5.22E-03	4.88E+01	1.91E-02	**1.27E-02**	4.79E-04	6.96E+01	2.05E-03	**3.56E-02**
d3	RBD2	5.18E-03	4.89E+01	1.93E-02	1.33E-02	4.54E-04	7.00E+01	2.07E-03	4.33E-02
	Spctrl	5.27E-03	4.87E+01	2.14E-01	4.50E-02	3.31E-03	5.28E+01	1.35E-01	1.00E-01
	SBD1	2.36E+01	4.70E+01	1.63E+02	**2.63E-03**	2.43E+00	6.68E+01	1.34E+01	**1.02E-02**
	RBD2	2.05E+01	4.83E+01	1.65E+02	6.30E-03	2.00E+00	6.85E+01	1.36E+01	3.51E-02
d4	Wvlt	2.47E+01	4.66E+01	6.86E+02	7.50E-03	2.64E+00	6.61E+01	4.87E+01	2.50E-02
	Spctrl	2.57E+01	4.63E+01	1.78E+03	3.50E-02	3.92E+00	6.26E+01	3.59E+02	1.95E-01
	AC	2.30E+01	4.73E+01	3.01E+02	2.15E-02	–	–	–	–
	SBD1	4.97E-04	4.59E+01	1.78E-03	**3.22E-03**	5.43E-05	6.51E+01	1.28E-04	**1.42E-02**
d5	RBD2	4.88E-04	4.61E+01	1.76E-03	4.47E-03	5.32E-05	6.53E+01	1.69E-04	4.96E-02
	Spctrl	5.84E-04	4.45E+01	4.56E-02	5.00E-03	5.87E-05	6.45E+01	8.74E-03	6.50E-02
	SBD1	1.21E-02	3.82E+01	5.70E-02	9.30E-03	1.05E-03	5.94E+01	4.87E-03	**2.96E-02**
	RBD2	1.20E-02	3.82E+01	5.71E-02	1.28E-02	8.75E-04	6.10E+01	4.87E-03	1.74E-01
d6	Wvlt	9.48E-03	4.03E+01	1.56E-01	**5.00E-03**	1.05E-03	5.94E+01	1.17E-02	5.50E-02
	Spctrl	1.60E-02	3.57E+01	6.37E-01	**5.00E-03**	1.05E-03	5.94E+01	4.32E-02	6.50E-02
	AC	1.82E-02	3.46E+01	1.50E-01	1.11E-02	–	–	–	–
	SBD1	2.72E-01	4.27E+01	5.50E-01	2.82E-02	2.74E-02	6.26E+01	6.37E-02	**8.70E-02**
	RBD2	2.70E-01	4.28E+01	7.41E-01	3.43E-02	2.17E-02	6.47E+01	7.99E-02	5.16E-01
d7	Wvlt	2.76E-01	4.26E+01	2.75E+00	**1.00E-02**	3.05E-02	6.17E+01	2.00E-01	1.60E-01
	AC	2.76E-01	4.26E+01	1.00E+00	1.82E-02	–	–	–	–

bold indicates the lowest CR for a given dataset and error tolerance

6.3   Comparison with Other Methods

In our last set of experiments, we compare the performance of the best-performing combinations of the set- and region-based decomposition approaches (SBD1 with SQE encoding and original vertex ordering, and RBD2 with original vertex ordering) against wavelet compression (Wvlt), spectral compression (Spctrl), and adaptive coarsening (AC). Among these techniques, the wavelet compression and adaptive coarsening can only be applied to structured grids and are only presented for the d2, d4, d6, and d7 datasets. Also, due to its high computational requirements, we were not able to obtain results for the spectral compression for the largest problem (d7). In addition to these schemes, we also experimented with diffusion wavelets [3]. However, we obtained poor compression and we omitted those results.

Table 2 shows the results of these experiments for two different compression levels, labeled "high error tolerance" and "low error tolerance". These compression levels result in RMSEs and MPEs that differ by approximately an order of magnitude, and were obtained by experimenting with the parameters of the various schemes so that to match their RMSEs for each of the datasets. However, for AC we were unable to achieve the desired RMSEs at all error tolerance levels. In the case that we could not achieve a desired RMSE, the results were omitted.

The results show that on average, our algorithms compress the simulation datasets to 2–5% of their original size. Compared with just lossless compression only, which results in storage costs of 40–80% of the original size, this is a big improvement. The results also show that for all but two experiments, SBD1 performs the best and that on average it required only 36% of the storage of the next best algorithm. For unstructured grids it requires on average 25% of the storage of Spctrl whereas for structured grids it requires on average 48% and 38% of the space of Wvlt and AC, respectively. Moreover, we see that as the amount of allowable error is lowered, the performance gap between SBD1 and the other methods grows. In addition, for unstructured grids, RBD2 performs the second best overall and requiring 61% of the space required by the Spctrl on average. We also see that due to the ϵ constraint placed on the our methods, they consistently result in MPE values which are much lower than those of the competing algorithms. These results suggest that in the context of grid-based simulation, SBD1 and RBD2 are consistently good choices for compression, providing low point-wise and global reconstruction error, high compression ratio, and low computational complexity.

7   Conclusion

In this paper, we introduced a paradigm for lossy compression of grid-based simulation data that achieves compression by modeling the grid data via a graph and identifying vertex-sets which can be approximated by a constant value within a user provided error constraint. Our comprehensive set of experiments showed that for structured and unstructured grids, these algorithms achieve compression which results in storage requirements that on average, are up to 75% lower than that other methods. Moreover, the near linear complexity of these algorithms makes them ideally suited for performing *in situ* compression in future exascale-class parallel systems.

References

1. Baldwin, C., Abdulla, G., Critchlow, T.: Multi-resolution Modeling of Large Scale Scientific Simulation Data. In: Proceedings of the Twelfth International Conference on Information and Knowledge Management - CIKM 2003, p. 40 (2003)
2. Belfor, R.A.F., Hesp, M.P.A., Lagendijk, R.L., Biemond, J.: Spatially Adaptive Subsampling of Image Sequences. IEEE Transactions on Image Processing 3(5), 492–500 (1994)
3. Coifman, R., Maggioni, M.: Diffusion wavelets. Applied and Computational Harmonic Analysis 21(1), 53–94 (2006)
4. Engelson, V., Fritzson, D., Fritzson, P.: Lossless Compression of High-volume Numerical Data from Simulations. In: Data Compression Conference, pp. 574–586 (2000)
5. Karni, Z., Gotsman, C.: Spectral Compression of Mesh Geometry. In: Proceedings of the 27th Annual Conference on Computer Graphics and Interactive Techniques - SIGGRAPH 2000, pp. 279–286 (2000)

 6. Karypis, G.: METIS˜5.0: Unstructured graph partitioning and sparse matrix ordering system. Tech. rep., Department of Computer Science, University of Minnesota (2011)
 7. Mallat, S.: A Theory for Multiresolution Signal Decomposition: The Wavelet Representation. IEEE Transactions on Pattern Analysis and Machine Intelligence 11(7), 674–693 (1989)
 8. Muraki, S.: Approximation and Rendering of Volume Data Using Wavelet Transforms. In: Proceedings Visualization 1992, pp. 21–28 (1992)
 9. Muraki, S.: Volume Data and Wavelet Transforms. IEEE Computer Graphics and Applications 13(4), 50–56 (1993)
10. Ratanaworabhan, P., Ke, J., Burtscher, M.: Fast Lossless Compression of Scientific Floating-Point Data. In: Data Compression Conference (DCC 2006), pp. 133–142 (2006)
11. Shafaat, T.M., Baden, S.B.: A Method of Adaptive Coarsening for Compressing Scientific Datasets. In: Kågström, B., Elmroth, E., Dongarra, J., Waśniewski, J. (eds.) PARA 2006. LNCS, vol. 4699, pp. 774–780. Springer, Heidelberg (2007)
12. Unat, D., Hromadka III, T., Baden, S.B.: An Adaptive Sub-sampling Method for In-memory Compression of Scientific Data. In: 2009 Data Compression Conference, pp. 262–271 (March 2009)

Topic 16: GPU and Accelerators Computing

Alex Ramirez, Dimitrios S. Nikolopoulos, David Kaeli, and Satoshi Matsuoka

Topic Committee

Accelerator-based computing systems invest significant fractions of hardware real estate to execute critical computation with vastly higher efficiency than general-purpose CPUs. Amdahl's Law of the Multi-core Era suggests that such an heterogeneous approach to parallel computing is bound to deliver better scalability and power-efficiency than homogeneous system scaling. While General Purpose Graphics Processing Units (GPGPUs) have catalyzed research in this area, new ideas emerge to help us model, deconstruct and analyze the performance of accelerators, develop new standards for programming accelerators at a high level of abstraction, and port end-to-end applications on accelerator-based systems. Topic 16 provides a forum to discuss advances in all aspects of GPU- and accelerator-based computing.

This year, eight papers have been accepted for publication in the GPU and accelerator computing track. Besides the important theme of scalable parallelization of applications on accelerator-based systems, the papers in the track explore emerging themes, including new standards for programming accelerator-based systems, new paradigms for scheduling and synchronization on accelerators, performance models, and novel uses of accelerators in scientific applications.

Two papers in the track *"OpenACC - First Experiences with Real-World Applications"*, by Wienke, Springer, Terboven and an Mey, and *"accull: An OpenACC Implementation with CUDA and OpenCL Support"* by Reyes, Rodríguez, Furnero and de Sande explore early commercial and academic implementations of the new OpenACC standard for parallel programming on systems with accelerators. *"Understanding the Performance of Concurrent Data Structures on Graphics Processors"* by Cederman, Chatterjee and Tsigas investigates the implementation of concurrent data structures on GPUs. Toss and Gautier introduce a new execution paradigm for GPUs, using lazy parallelization and work stealing, in *"A New Programming Paradigm for GPGPU"*. Anzt, Luszczek, Dongarra, and Heuveline propose a new application of accelerators for scientific computing in *"GPU-Accelerated Asynchronous Error Correction for Mixed Precision Iterative Refinement"*. Jia, Zhang, Guoping, Xu, Yan and Li present an adaptation of the Roofline model for GPUs in *"GPURoofline: A Model for Guiding Performance Optimizations on GPUs"*. Finally, *"Building a Collision for 75-Round Reduced SHA-1 Using GPU Clusters"* by Adinetz and Grechnikov and *"GPU-vote: A Framework for Accelerating Voting Algorithms on GPU"* by van den Braak, Nugteren, Mesman and Corporaal present methods for the efficient parallelization of applications on GPUs.

C. Kaklamanis et al. (Eds.): Euro-Par 2012, LNCS 7484, pp. 857–858, 2012.

We wish to thank all authors who submitted a paper to this topic, all external reviews for delivering quality reviews on time and the Euro-Par Organizing Committee for their constructive comments during the entire reviewing and paper selection process.

OpenACC — First Experiences
with Real-World Applications

Sandra Wienke, Paul Springer, Christian Terboven, and Dieter an Mey

JARA, RWTH Aachen University, Germany
Center for Computing and Communication
{wienke,springer,terboven,anmey}@rz.rwth-aachen.de

Abstract. Today's trend to use accelerators like GPGPUs in heterogeneous computer systems has entailed several low-level APIs for accelerator programming. However, programming these APIs is often tedious and therefore unproductive. To tackle this problem, recent approaches employ directive-based high-level programming for accelerators. In this work, we present our first experiences with OpenACC, an API consisting of compiler directives to offload loops and regions of C/C++ and Fortran code to accelerators. We compare the performance of OpenACC to PGI Accelerator and OpenCL for two real-world applications and evaluate programmability and productivity. We find that OpenACC offers a promising ratio of development effort to performance and that a directive-based approach to program accelerators is more efficient than low-level APIs, even if suboptimal performance is achieved.

1 Introduction

Due to a promising performance per watt ratio and an attractive price, HPC architectures prevailingly tend towards heterogeneous computer systems comprising general-purpose cores with attached accelerator devices. However, programming accelerators such as general-purpose graphic processing units (GPGPUs) with low-level APIs is difficult, may complicate the software design and usually couples the code to a device of a particular vendor. This leads to an unproductive development process with error-prone programming tasks and highly hardware-specific implementations, which is not acceptable for large development projects with a long projected code lifetime.

Recent approaches promise to make the compiler responsible for many of the low-level programming tasks by offering a directive-based high-level API. As finding parts of an algorithm that can efficiently be executed on an accelerator device is still up to the programmer, these approaches do not simplify programming accelerators in general. However, they improve the development productivity and simplify code maintenance. Unifying the syntax of various directive-based approaches for accelerators with the intention to make it available across multiple vendors, a group of members of the OpenMP Language Committee published OpenACC in November 2011. OpenACC enables the offloading of loops and regions of C/C++ and Fortran code to accelerators and is initiated by the companies CAPS, CRAY, NVIDIA and PGI. In this work, we present our first experiences with OpenACC applied to two real-world simulation codes from the fields of engineering and medicine that can both benefit from GPU acceleration.

C. Kaklamanis et al. (Eds.): Euro-Par 2012, LNCS 7484, pp. 859–870, 2012.

Our work comprehends a performance analysis of these codes, as well as an evaluation of programmability and productivity.

The paper is structured as follows: Section 2 covers related work and Sect. 3 gives an overview of OpenACC. In Sect. 4, the two real-world applications and the undertaking for their porting to OpenACC are explained. The measured performance values with OpenACC are evaluated in Sect. 5 and compared to the ones gained with OpenCL and PGI Accelerator. In Sect. 6, we discuss programmability and productivity aspects. Finally, we assess OpenACC's effort-performance ratio regarding our real-world applications in Sect. 7.

2 Related Work

The desire for general-purpose computations on GPUs caused the advance of new programming paradigms. Nowadays, the dominant GPU programming models are CUDA [13] and OpenCL [10]. Both empower the programmer to exploit performance from the accelerator by porting code to GPU kernel functions at a low level. CUDA is coupled to NVIDIA GPUs, while OpenCL as a standard is portable across vendors and targets different hardware architectures. As using low-level APIs often results in an unproductive development process due to repeatedly written code portions and error-prone programming [15], several directive-based approaches for accelerator computing have been proposed in which compilers undertake the implementation guided by hints from the user. OpenACC [7] is an industry standard for this directive-based accelerator programming that may contribute to the specification of OpenMP for accelerators [2].

The Portland Group provides the PGI Accelerator programming model [14] for C and Fortran which enables compiler-aided and directive-based work offloading to NVIDIA GPUs and specifies a broad range of features. It served as the foundation for the OpenACC specification. Furthermore, CAPS has previously established its hybrid multicore parallel programming (HMPP) environment [6] providing directives to declare codelets, which are functions suitable for hardware acceleration, and targeting a variety of accelerators. Intel also relies on directives to offload code to its Many Integrated Core (MIC) accelerator [11]. hiCUDA [9] defines a high-level abstraction of CUDA with kernel directives, data transfer clauses and function calls. While supporting CUDA-specific concepts, hiCUDA leaves more responsibilities to the programmer than OpenACC. OpenMPC [12] approaches the generation of CUDA code by translating OpenMP regions. Additional directives control CUDA-related parameters and the compiler may find an appropriate tuning configuration for the program. It is also restricted to NVIDIA GPUs due to the CUDA affiliation. The Barcelona Supercomputing Center has developed the StarSs programming model [1] which provides extensions to the OpenMP language to exploit several architectures, for instance GPUs. It focuses on OpenMP tasks and their distribution to different targets during runtime by forcing the programmer to specify all data input and output dependencies of a task.

Since the productive development is one main issue of directive-based models, we examine the programmability and productivity with OpenACC. Only few studies consider productivity aspects. While [3] provides a general overview of coding effort, [8] looks mainly at the effort from a computing center's point of view. In our previous work

[15], we concluded that the PGI Accelerator model has a good effort-performance ratio and [2] also approaches productivity with respect to a ratio estimation of performance to time effort. In this work, we do not only elaborate on productivity in general, but also specify the number of modified code lines as an indication of the development effort.

3   OpenACC Overview

The directive-based OpenACC API for C/C++ and Fortran delegates the responsibility for low-level GPU programming tasks to the compiler, while providing portability across operating systems, host CPUs and accelerator devices. Concerning accelerator types, up to now, the existing OpenACC implementations only support NVIDIA GPUs. In this section, we provide a brief overview of OpenACC, point out the most important use cases with respect to GPUs and map the terminology to the OpenCL nomenclature.

The OpenACC API assumes a host-directed execution model in which the main program runs on the host and compute-intensive regions are offloaded to an attached accelerator. The memory model is based on separate host and device memories which do not synchronize automatically. GPU devices implement a weak memory model that prevents coherence between operations on different compute units and enables coherence within the same compute unit only by using explicit synchronization.

The execution and data management is guided by the programmer using OpenACC directives. Some basic constructs are illustrated in Listing 1.1.

Listing 1.1. Brief use case of basic OpenACC constructs

```
// Initialization :  |x[i]| < 1,  i = 0 ,..., size −1
#pragma acc data copy(x[0: size ]) // Data movement to/from device
{    while( error > eps ) {
        error = 0.0;
#pragma acc parallel present(x[0: size ]) // Kernel execution
#pragma acc loop gang vector reduction (+: error ) // Loop schedule
        for(int i=0; i<size ; ++i) {
            x[i] *= x[i];
            error += fabs(x[i]);
}    } }
```

The most important directives are `parallel` and `kernels` which describe regions of code to be accelerated asynchronously or synchronously. Here, we focus on the `parallel` directive due to some restrictions on the `kernels` construct in the recent implementation. A `parallel` region maps to an OpenCL kernel function which can be enqueued for execution on the device in an n-dimensional range of work-items. To improve the performance, it is possible to specify the so-called number of `gangs`, the number of `workers` or the `vector_length`. The terms `gang` and `vector` correspond to work-groups and work-items (usually) within a work-group, respectively, in OpenCL. A `worker` defines a certain union of work-items, i.e. a *warp* on CUDA architectures. Within `parallel` regions, a `loop` directive instructs the worksharing of a loop among the accelerator's workers. The programmer can insert additional clauses in `parallel`, `kernels` or `loop` directives to optimize or correct the implicit data management chosen by the compiler. Furthermore, the data movement can be decoupled from these compute regions by using an enclosing explicit `data` region. Corresponding data clauses can specify the kind and direction of data movement, e.g. `copyin` or `copyout`. It is also possible to `create` arrays only on the device, define device `private` data, tell the compiler that data is already `present` on the device or specify different kinds of `reductions`. Since the user has to manually manage the data coherence between host and device, the executive `update` directive can be applied to synchronize the separate memories of host and device. The GPU's memory hierarchy also often supports a low-latency local memory for which the compiler may optimize the program. To guide the compiler, the developer can apply the `cache` directive which specifies (sub-)arrays to be stored in this fast memory.

Besides directives, the OpenACC API provides runtime library calls and environments variables too. For instance, library calls can gather information about the device, initialize it or allocate data on the device.

4   Applications

For our investigations on performance, productivity and programmability of OpenACC, we chose two real-world applications from the fields of engineering and medicine. In the following, we point out their contributions to their domains, describe implementation relevant features and explain how we carried out implementations with OpenACC.

4.1   Simulation of Bevel Gear Cutting

The engineering application *KegelSpan* [4] written in Fortran and developed by the Laboratory for Machine Tools and Production Engineering (WZL) at RWTH Aachen University is a 3D simulation software of the bevel gear cutting process and a leading-edge application in the automotive industry. It aims at minimizing the number of tool changes in the production process of bevel gears and contributing to a cost-efficient manufacturing by enabling a detailed tool load and wear analysis. The module under investigation computes the intersection of tool and gear. It contains one loop nest where outer and inner loop each iterates tens of thousands times (small dataset). The inner loop of the nest contains dependencies due to a minimum computation which is needed for the key value of chip thickness. The intersection module has to be executed repeatedly to optimize the manufacturing parameters.

We approached the *basic* implementation of the intersection module with OpenACC analogously to the one with PGI Accelerator in our previous work [15]: We distributed the work of the outer loop amongst all work-items and executed the inner loop serially. For that, we mainly applied a `parallel` region with numerous data clauses and a `loop` directive with `gang vector` schedule. Using the `vector_length` attribute, we optimized the number of work-items within a work-group. A second variant (*restruct*) comprises a restructured data format for an optimized, coalesced data access and serves for comparing performances rather than development efforts of the different programming models. In comparison to the basic version, just the adaption of OpenACC data clauses was necessary. Furthermore, we examined a more complex type of parallelism by distributing the outer loop to the work-groups and the inner loop to the work-items within a work-group. This approach required the addition of a minimum reduction, denoted by `reduction(min:varlist)`, to the inner loop due to the chip thickness computation. However, the position of the minimum, i.e. the array index, is needed as well. Since manual extraction of this index was tedious and delivered low performance, we omitted a detailed analysis. In OpenCL, we further leveraged the work-group's local memory (*locMem*) by storing intermediate data and input data that is needed multiple times in the fast software cache of the GPU (compare [15]). A similar approach was possible neither with PGI Accelerator nor with OpenACC, but the results gained with OpenCL show the potential of the caching technique in Sect. 5.1.

4.2 Neuromagnetic Inverse Problem

The second application comes from the field of medicine or more precisely magnetoencephalography (MEG). In MEG, the magnetic field induced by the current density inside the human brain is measured outside the head. To reconstruct the focal activity in the brain, the neuromagnetic inverse problem can be solved by means of a minimum p-norm solution. Since this unconstrained nonlinear optimization problem is challenging in terms of computational efficiency and accuracy effecting the convergence behavior [5], first- and second-order derivatives are computed by automatic differentiation (AD) in the software package of Bücker, Beucker and Rupp [5]. The software package is implemented primarily in MATLAB. To enable AD combined with parallel computing, the objective function of the optimization problem, as well as its first- and second-order derivatives are written in C. For our investigations, we concentrate on these three kernels which include the computations of matrix-vector products using a matrix of dimensions 128×512000. Additionally, the matrix can be divided into a big dense and a small sparse part. Each kernel contains a single loop or loop nest with summation reductions.

The MATLAB program calls the kernels about thousand times during the optimization process (simple configuration). For simplification, we established a C framework that mimics the original call hierarchy: Implemented with an explicit `data` region, the matrix is copied to the accelerator once per optimization run and then each kernel is called thousand times. The operands needed for the computations are copied into or `updated` on the device and the results are transferred to the host in between and at the end of the explicit data region. Porting the three kernels with OpenACC to the GPU, we first implemented a *basic* approach again. That means that only the outer loops run in parallel and the inner loops are executed serially. Each kernel consists of two or more `parallel` regions to distinguish at least between matrix-vector multiplications, vector-vector operations and summation reductions. Additionally, the implementations of first- and second-order-derivative evaluations require resolving race conditions. Therefore, intermediate values are stored in auxiliary arrays and the access to these arrays is globally synchronized by creating an additional parallel region. A loop interchange is applied as well. The corresponding basic PGI Accelerator implementation looks similarly. In a second variant (*l2par*), we added a level of parallelism to the OpenACC kernels which distributes the outer loops to the `gangs` and runs the inner loops in `vector` mode. This approach needed the usage of reduction clauses on outer and inner loops. A similar approach was not possible with PGI Accelerator due to a limitation in its implementation. The third approach aims at leveraging the cache of the GPU and improving the data access pattern by chunking the matrix-vector multiplications into blocks of size `vector_length`. Each block is mapped to a work-group, whereas the work-items within a work-group execute the multiplications row-wise. With PGI Accelerator, the *blocked* version was implemented analogously except that only one level of parallelism could be applied to the loop nests that do not access the matrix. The OpenCL version also employs a blocked matrix-vector multiplication. Furthermore, it is highly optimized with respect to the usage of device local memory and host pinned memory, asynchronous data transfer and kernel execution, the specification of constant values as preprocessed macros or loop unrolling. Loop unrolling is also applied to all PGI and OpenACC versions. The OpenACC compiler automatically unrolls loops within accelerated code regions or can be guided by the `#pragma unroll(size)`. Although the PGI Accelerator API includes an `unroll` clause, here, it is ignored by the compiler. Therefore, we had to unroll most of the loops manually up to a level of 32 to achieve comparable performance. `cache` and `async` optimizations could not yet successfully applied with both directive-based models.

5   Performance Evaluation

In this section, we present performance results of the OpenACC implementations of both applications and compare them to implementations made with OpenCL[1] and PGI Accelerator. Since tool support across all three programming models is limited, our evaluation also contains assumptions of result explanations. This analysis is based on NVIDIA's Visual Profiler. All reported runtimes include the accelerator's setup time, data transfers between host and device, kernel execution times and the overhead introduced by the need of manual management (OpenCL) or adapting the data structure. Each runtime is the minimum value of five program runs.

For all measurements, we used an NVIDIA Tesla C2050 GPU with ECC enabled and CUDA toolkit 4.0. The host system[2] on which the OpenACC results were gathered consists of one AMD Magny-Cours 12-core processor and runs SUSE Enterprise Server 11. There, we use the Cray 8.1.0 compiler[3] with the optimization flag *-O3*. For the results gained with OpenCL and PGI Accelerator, we worked on an Intel Westmere 4-core host processor and Scientific Linux 6.1. The OpenCL and PGI Accelerator implementations are compiled with the Intel 12.1.2 compiler and the PGI 12.3 compiler (*-ta=nvidia,4.0,cc20,fastmath(,nofma)*), respectively.

---

[1] OpenCL and CUDA performance results are comparable in almost all our application versions.

[2] Since this machine is an experimental system from Cray, performance should be better on the Cray XK6 product.

[3] At time of developing, Cray provided the first OpenACC release. The used Cray compiler comprises an early implementation of OpenACC which does not contain all features of OpenACC yet. Furthermore, the performance is likely to increase with future releases. First partly implementations by PGI and CAPS were released in March and May 2012, respectively, and will be subject to further investigations.

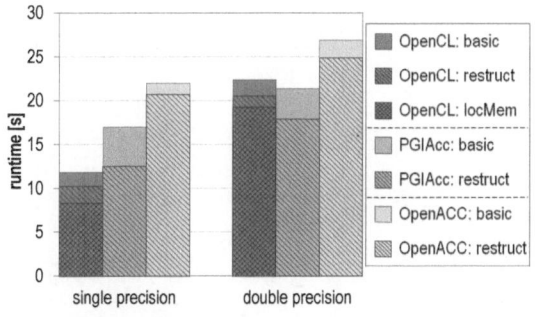 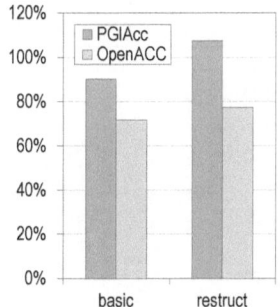

Fig. 1. Runtimes of OpenCL, PGI Accelerator and OpenACC (engineering application)

Fig. 2. Percentages of OpenCL performance in double precision

5.1 Simulation of Bevel Gear Cutting

The results for the engineering application can be found in Fig. 1. Each color represents one programming model, whereas the patterns correspond to different versions as mentioned in Sect. 4. For single and double precision, the absolute difference between PGI Accelerator and OpenACC runtimes of the *basic* and the *restructured* version remain approximately the same. A significant fraction of this difference is introduced by a great amount of L2 read misses in the OpenACC programs which extends the one of PGI Accelerator by a factor of 1.5. These L2 read misses must be resolved by long-latency global memory accesses that hurt the performance. The PGI compiler feedback shows that constant memory is used, whereas the OpenACC compiler does not elaborate on this. We assume that the constant cache is ignored by the current OpenACC compiler which would lead to the high number of L2 read misses. Additionally, the best-effort PGI Accelerator versions using double precision omit the generation of FMA/MAD instructions. However, neither the OpenCL nor the OpenACC compiler provide an easy way to disable FMA operations without losing other optimizations to verify this result. Figure 2 illustrates the double precision results in percentages. Here, the runtime of the best-effort OpenCL implementation serves as reference. With the *restruct* version, PGI Accelerator outperforms OpenCL, and OpenACC achieves a considerable fraction of approximately 80 % of OpenCL.

Furthermore, the OpenCL bars in Fig. 1 show that the usage of local memory was beneficial for this application. The PGI Accelerator API provides a cache clause for the loop schedule, but the compiler takes it as a hint instead of a rule. In this case, the compiler apparently cannot apply the hint successfully. In OpenACC, the cache directive is not yet fully implemented in the Cray compiler. Given Cray's ongoing implementation work, we hope to examine OpenACC's cache capabilities in the near future.

5.2 Neuromagnetic Inverse Problem

For the neuromagnetic inverse problem, we measured the runtime of the objective function evaluation, the first- and second-order-derivative computations and the whole program. The latter contains an equal number of calls to the three functional units as

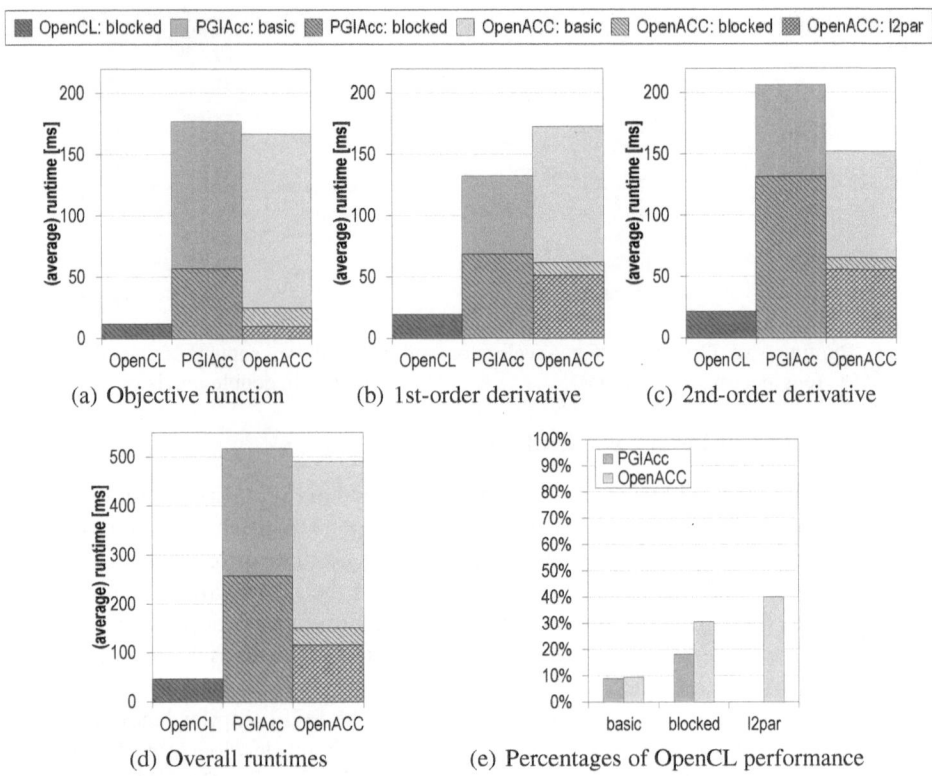

Fig. 3. Comparison of OpenCL, PGI Accelerator and OpenACC in double precision (medical application)

approximation of the actual optimization process. All runtimes represent an average over 1000 iterations including data transfers. Computations are done in double precision only due to the high impact of numerical inaccuracy.

Figure 3 illustrates the results of the various implemented versions as described in Sect. 4. The colors refer to the programming model and the bar pattern to the implementation variants. In all cases, the *basic* versions have the longest runtimes and are outperformed by the corresponding *blocked* versions. Although the blocking algorithm in OpenCL, PGI Accelerator and OpenACC is almost the same, the OpenCL versions contain much less read requests to the L2 cache and the device global memory. This is mainly due to the coalesced prefetching of matrix and vectors to the on-chip local memory and the consequent, cache-optimized computations that we implemented with OpenCL. Contrary, both directive-based models lack either of an implementation or a successful recognition of the caching clause. The runtime difference is intensified by an additional global synchronization in the directive-based models. With OpenACC, a performance improvement is surprisingly gained by using the algorithmic-simpler version *l2par* with two levels of parallelism. For the objective function evaluation (see Fig. 3(a)), the OpenACC performance even catches up with the OpenCL performance,

Table 1. Number of modified source code lines in host/kernel code with respect to the serial versions of the engineering application ($\sim$ 150 kernel code lines) and medical application ($\sim$ 100 kernel code lines)

	engineering			medical		
	basic	restruct	locMem	basic	blocked	l2par
OpenCL	106/35	183/39	183/58	-	330/300	-
PGI Acc	0/3	84/14	-	9/121	12/109	-
OpenACC	0/3	84/14	-	9/31	12/85	9/37

both containing an equal number of global memory accesses. In contrast, the runtimes of evaluating first- and second-order derivatives (Figs. 3(b) and 3(c)) are about 2.5 times higher than the ones from OpenCL. Here, OpenACC's number of global memory reads and writes exceeds the ones of OpenCL, possibly due to less local-memory usage. In general, the OpenCL implementations may also perform better due to asynchronous data transfer and kernel execution, extensive loop unrolling, usage of pinned memory or specification of constant values as preprocessed macros. Figure 3(e) presents the overall performance in percentages of the OpenCL version. The best-effort PGI Accelerator and OpenACC versions achieve about 20 % and 40 %, respectively.

Comparing OpenACC to PGI Accelerator, in most cases, the OpenACC versions run faster. We can see by profiling that OpenACC calls CUDA's internal 32-bit alignment routine. An optimal alignment may be the main reason for less memory accesses and the better performance. Furthermore, loop unrolling was essential for improving performance. While the OpenACC compiler automatically adapts the unroll level nicely with `#pragma unroll(size)`, the PGI compiler ignores any `unroll` clause so that a manual and assumingly suboptimal loop unrolling must be applied. PGI's only performance gain (see Fig. 3(b)) results from a better cache access pattern evoked by the reuse of the *pow* function.

6   Programmability and Productivity

Good programmability is the foundation of a productive development process. In this section, we examine OpenACC's ease-of-use in comparison with PGI's Accelerator model and OpenCL. We discuss OpenACC's capabilities, restrictions and our suggestions for improvement to decrease development effort.

Both applications show that the current OpenACC implementation allows successful porting to GPUs in a productive way. Table 1 lists the number of modified source code lines for each kernel and programming model. These values indicate how much time a programmer spent to port the application to GPUs and how maintainable the ported code is for further development.

The main features of OpenACC like data movement and loop acceleration can intuitively be applied, just as with PGI Accelerator. An inexperienced user can rely on an automatic loop scheduling, whereas more experienced users can tune it by using different levels of parallelism or by specifying the sizes of parallel chunks. Here, OpenACC is even capable of explicitly managing *warps*, while OpenCL and PGI Accelerator do not provide this degree of freedom. On the other hand, OpenACC is restricted to a single gang or vector dimension, whereas OpenCL and PGI Accelerator offer multiple dimensions to intuitively map e.g. a matrix to a two dimensional working space.

Synchronization. We find the automatic synchronization between work-items at any level of parallelism that is entirely within a work-group convenient. However, to synchronize between loops that are located within the same parallel region and whose work is distributed with different schedules, the user has to manually synchronize the data by splitting the parallel region. An additional executable directive that acts like a global barrier might be useful to circumvent manual splitting of parallel regions. An use-case of this directive would be the synchronization of GPU-local data after its initialization. [9] provides the `singular` directive for the purpose of compact initialization.

Reduction. In OpenCL, a manual implementation of the reduction operation often leads to poor performance. In contrast, the PGI compiler automatically recognizes a reduction (sometimes adding auxiliary variables is needed) and it creates well-optimized machine code for it. In OpenACC, the user has to specify the `reduction`, but the freedom to choose any typical kind of reduction (e.g. `min` reduction) and any scalar variable to reduce without adding intermediate variables. Here, the addition of user-defined reductions to the OpenACC standard would further improve the development. In our case study, the collection of the minimum along with its (array) index would be an appropriate application.

Function Calls. A major restriction of both directive-based models is that non-inlined function calls in accelerated code are not supported at the moment. Function calls in our investigated applications could be resolved by explicitly inling them. However, this is not bearable for bigger software packages. PGI Accelerator also does not support this feature currently, but in general, it seems possible to integrate it in high-level APIs [9].

Atomics and Critical Region. OpenMP-like atomics and critical regions for OpenACC would help to avoid race conditions. For instance, detecting a critical region, the compiler could undertake the needed addition of an auxiliary array, the split of parallel regions and the interchange of loops. The compiler could apply the most efficient implementation and prevent programming errors.

Asynchronous Data Transfers. Asynchronous data copies would be beneficial for OpenACC, not only regarding `update` directives, but also explicit `data` regions.

Multiple GPUs. To use multiple GPGPUs on a single host, the programmer must explicitly specify the particular device for work-offloading and manually manage data synchronization between the devices and the host using a low-level API. This distribution of workload and data could also be left to the compiler and runtime in future.

At the time of writing, several features of OpenACC were not yet fully implemented in the Cray compiler, but look promising. The `kernels` construct (combined with guided `loop` execution) will hopefully improve the productivity of accelerating code regions containing multiple loops. Aiming at increasing performance, the `async` clause of `parallel`, `kernels` and `update` constructs will be applied to the investigated medical application. Employing multiple command queues, it will enable the start of separate parallel regions simultaneously to independently compute matrix-vector multiplications and vector-vector operations and the overlap of kernel execution and data movement. In OpenCL, the usage of the GPU's software cache improved the performance of both software packages. The application of OpenACC's `cache` clause may also lead to further acceleration in the future. Moreover, we will examine the combination of OpenACC and MATLAB. With OpenCL, the communication to MATLAB was employed by exchanging the OpenCL context. However, at the moment, it is unclear whether a realization with OpenACC is possible.

7   Conclusion

In the context of two real-world applications, we examined the performance, programmability and productivity of OpenACC in comparison to OpenCL and PGI Accelerator.

With OpenACC, we find that the performance of the moderately complex kernel of the simulation software for bevel gear cutting is about 80 % of the best-effort OpenCL performance regarding double precision computations, although the implementation of OpenACC that we used is still incomplete. This result matches the expectations of the performance of a directive-based programming model. In contrast, the OpenACC performance of the more complex medical program is only approximately 40 % of the best-effort OpenCL implementation. Although this value is rather distressing at first sight, we still believe that OpenACC is a promising approach: We assume that the loss of performance is mainly due to the current lack of the ability to leverage the local memory of the GPU intensified by our manually implemented global synchronization to prevent race conditions. These trade-offs may be eliminated by the ongoing implementation work on OpenACC or by introducing additional directives. Moreover, it must be taken into account that the highly optimized OpenCL code is quite verbose, hard to read and requires 630 modified code lines, whereas the best OpenACC version only needed 46 modified lines of code. Basing on these numbers, the OpenACC's ratio of development effort to performance is encouraging.

Thus, in terms of programmability and productivity, the OpenACC API is generally convincing. It can be intuitively applied if the programmer has certain knowledge of the accelerator's hardware architecture. If adopting features like device function calls, user-defined reductions or critical regions, the programming efficiency may be further improved while simultaneously reducing sources of errors.

In our view, the move to directive-based accelerator programming is essential for the further growth and acceptance of accelerator devices. To this end, OpenACC is an important step as it standardizes a directive-based API for accelerators for the first time. It is intended to integrate the feedback and the lessons learned from OpenACC into the OpenMP specification. The inclusion of corresponding functionality into the OpenMP standard will be technically demanding though and may slow down further development of the standard. However, from the user's point of view, OpenMP for accelerators is certainly promising.

References

1. Ayguadé, E., Badia, R., Bellens, P., Cabrera, D., Duran, A., Ferrer, R., Gonzàlez, M., Igual, F., Jiménez-González, D., Labarta, J., Martinell, L., Martorell, X., Mayo, R., Pérez, J., Planas, J., Quintana-Ortí, E.: Extending OpenMP to Survive the Heterogeneous Multi-Core Era. International Journal of Parallel Programming 38, 440–459 (2010), doi:10.1007/s10766-010-0135-4
2. Beyer, J.C., Stotzer, E.J., Hart, A., de Supinski, B.R.: OpenMP for Accelerators. In: Chapman, B.M., Gropp, W.D., Kumaran, K., Müller, M.S. (eds.) IWOMP 2011. LNCS, vol. 6665, pp. 108–121. Springer, Heidelberg (2011)
3. Bordawekar, R., Bondhugula, U., Rao, R.: Can CPUs Match GPUs on Performance with Productivity?: Experiences with Optimizing a FLOP-intensive Application on CPUs and GPU. Technical report, IBM Res. Division (2010)
4. Brecher, C., Gorgels, C., Hardjosuwito, A.: Simulation based Tool Wear Analysis in Bevel Gear Cutting. In: International Conference on Gears, Düsseldorf. VDI-Berichte, vol. 2108.2, pp. 1381–1384. VDI Verlag (2010)
5. Bücker, M., Beucker, R., Rupp, A.: Parallel Minimum p-Norm Solution of the Neuromagnetic Inverse Problem for Realistic Signals Using Exact Hessian-Vector Products. SIAM Journal on Scientific Computing 30(6), 2905–2921 (2008)
6. Dolbeau, R., Bihan, S., Bodin, F.: HMPP: A Hybrid Multi-core Parallel Programming Evironment. In: First Workshop on General Purpose Processing on Graphics Processing Units (2007)
7. CAPS Enterprise, Cray Inc., NVIDIA, and the Portland Group. The OpenACC Application Programming Interface, v1.0 (November 2011)
8. Hacker, H., Trinitis, C., Weidendorfer, J., Brehm, M.: Considering GPGPU for HPC Centers: Is It Worth the Effort? In: Keller, R., Kramer, D., Weiss, J.-P. (eds.) Facing the Multicore-Challenge. LNCS, vol. 6310, pp. 118–130. Springer, Heidelberg (2010)
9. Han, T.D., Abdelrahman, T.S.: hiCUDA: High-Level GPGPU Programming. IEEE Transactions on Parallel and Distributed Systems 22(1), 78–90 (2011)
10. Khronos OpenCL Working Group. The OpenCL Specification, v1.1.44 (2011)
11. Koesterke, L., Boisseau, J., Cazes, J., Milfeld, K., Stanzione, D.: Early Experiences with the Intel Many Integrated Cores Accelerated Computing Technology. In: Proceedings of the 2011 TeraGrid Conference: Extreme Digital Discovery, TG 2011, pp. 21:1–21:8. ACM, New York (2011)
12. Lee, S., Eigenmann, R.: OpenMPC: Extended OpenMP Programming and Tuning for GPUs. In: 2010 International Conference for High Performance Computing, Networking, Storage and Analysis (SC), pp. 1–11 (November 2010)
13. NVIDIA. CUDA C Programming Guide, v4.0 (2011)
14. The Portland Group. PGI Accelerator Programming Model for Fortran & C, v1.3 (2010)
15. Wienke, S., Plotnikov, D., an Mey, D., Bischof, C., Hardjosuwito, A., Gorgels, C., Brecher, C.: Simulation of bevel gear cutting with GPGPUs-performance and productivity. Computer Science - Research and Development 26, 165–174 (2011), doi:10.1007/s00450-011-0158-0

accULL: An OpenACC Implementation with CUDA and OpenCL Support*

Ruymán Reyes, Iván López-Rodríguez,
Juan J. Fumero, and Francisco de Sande

Dept. de E.I.O. y Computación
Universidad de La Laguna, 38271–La Laguna, Spain
{rreyes,ilopezro,jfumeroa,fsande}@ull.es

Abstract. The irruption in the HPC scene of hardware accelerators, like GPUs, has made available unprecedented performance to developers. However, even expert developers may not be ready to exploit the new complex processor hierarchies. We need to find a way to leverage the programming effort in these devices at programming language level, otherwise, developers will spend most of their time focusing on device-specific code instead of implementing algorithmic enhancements. The recent advent of the OpenACC standard for heterogeneous computing represents an effort in this direction. This initiative, combined with future releases of the OpenMP standard, will converge into a fully heterogeneous framework that will cope the programming requirements of future computer architectures. In this work we present accULL, a novel implementation of the OpenACC standard, based on the combination of a source to source compiler and a runtime library. To our knowledge, our approach is the first providing support for both OpenCL and CUDA platforms under this new standard.

1 Introduction

The widespread use of graphics accelerators for general purpose computing has leveraged the entry cost of high performance computer systems. A modest commodity computer in combination with a graphic card constitutes a powerful tool which empowers users to solve problems with a significant size so far unavailable without the aid of large scale computers.

Despite the improvements achieved in the hardware field, there is still a lack of parallel problem solving environments that can help scientists to use easily and efficiently these hybrid architectures. From our point of view, at this moment, efforts have to be directed towards the development of high level abstractions of these heterogeneous environments. This will allow more users to take advantage of these architectures without the need of detailed hardware knowledge.

* This work has been partially supported by the EU (FEDER), the Spanish MEC (Plan Nacional de I+D+I, contracts TIN2008-06570-C04-03 and TIN2011-24598), HPC-EUROPA2 (project number 228398) and the Canary Islands Government, ACIISI (contract PI2008/285).

C. Kaklamanis et al. (Eds.): Euro-Par 2012, LNCS 7484, pp. 871–882, 2012.

CUDA is the most mature and extended approach to GPU programming, although currently only supports Nvidia devices. Despite of being partially simple to build a code using this tool, achieving good performance rate usually requires a noticeable coding and optimisation effort.

The OpenCL standard represents an effort to create a common programming interface for heterogeneous devices, which many manufacturers have joined. However, its programming model is not simple.

The presentation during Seattle SC2011 of the new OpenACC standard for heterogeneous computing [4] clearly represents a major effort in the aforementioned direction of leveraging the development effort. Following the OpenMP approach, in the OpenACC API, the programmer annotates its sequential code with compiler directives, indicating those regions of code susceptible to be executed in the the GPU. The simplicity of the model, its ease of adoption by non expert users and the support received from the leading companies in this field make us believe that it is a long-term standard.

Prior to OpenACC, the PGI Accelerator model [6] proposed a high-level programming model for accelerators, such as GPUs, similar in design and scope to the widely-used OpenMP directive approach. Also, the CAPS HMPP [1] toolkit is a set is a set of compiler directives, tools and software runtime that supports parallel programming in C and Fortran. Both PGI and CAPS are founders of the OpenACC standard, and have recently announced versions of their tools compliant to the standard.

As a continuation of our recent years work [5], here we present a first release of our implementation of the OpenACC standard. We offer support for the most common used constructs, and we are able to run in both CUDA and OpenCL platforms. To the best of our knowledge, ours is the first open-available implementation of the standard that supports OpenCL. In addition, we present results using both CPU and GPU OpenCL platforms. User can select the desired platform using the appropriate environment variable, conforming to the standard.

The contributions of this work are manifold: (a) It represents one of the first non-commercial implementation of the OpenACC standard. (b) This is the first implementation, as far as we know, with support for both OpenCL and CUDA platforms. (c) We present a runtime suitable to be decoupled from our compiler and used together with a different compiler infrastructure. (d) We validate our approach using codes from widely available benchmarks and using both GPU and CPU devices.

The rest of the paper is organized as follows. Section 2 discusses the implementation of our compilation framework. In Section 3 we expose the key ideas behind our approach. We guide our explanations through the use of a code example. Also in Section 3 we present computational results for the guiding example and three additional well-known algorithms. Finally, Section 4 includes the conclusions we have been able to achieve so far and ideas about future work regarding this framework.

2   The Implementation

Our approach is a two-layer based implementation composed by a source to source compiler and a runtime library, in a similar fashion to other compiler infrastructures. However, instead of generating a final binary file, the result of our compilation stage is a a project tree hierarchy with compilation instructions, suitable to be modified by advanced end-users. Default compilation instructions enable average users to generate an executable without additional effort. The aim of this approach is to maintain a low development effort in the programmer side, while keeping the opportunity window for further optimizations performed by high-skilled developers.

The compiler is based on our YaCF research compiler framework, while the runtime (Frangollo) has been designed from scratch. We have named accULL to the combination of our compiler driver and the runtime.

YaCF translates the annotated C+OpenACC source code into a C code with calls to the Frangollo API. The YaCF compiler framework [5] has been designed to create source to source translations. It is intended to be a fast-prototyping tool which allows compiler developers to write portable source to source transformations in just a a few lines of Python code. The framework is available as an open source tool [1]. On top of the YaCF infrastructure, we have built a set of Python modules, capable of extracting the kernel code from the annotated source and replace it with the appropriate runtime calls. Both OpenCL and CUDA kernels are generated from the extracted block statements

User annotations are validated against data dependency analysis. A warning is emitted if variables are missing. Also, we can check whether a variable is read-only or not, to allocate the appropriate type of memory. Source to source translation injects a set of Frangollo calls within the serial code. Whenever these calls are issued, control is deferred to Frangollo runtime, who will execute the code of the proper API call or whatever other code it might require (for example, to handle previous asynchronous operations). Frangollo deals with two major issues of any OpenACC implementation: memory management and kernel execution.

It is important to take into account that nowadays compute accelerator devices do not share the host processor address space. Therefore, it is critical to transparently handle the existence of several instances of an user variable on different devices. To address this situation, our runtime uses a base pointer address detection mechanism to match each host variable to its device counterpart. Using this mechanism, we are able to track accesses to variables across interprocedural calls. The only exception to this behaviour is the implementation of the acc host construct, which, by specification definition, requires a device specific pointer, and the deviceptr clauses, which are not currently implemented.

Memory transfers are handled on demand by Frangollo. No assumption can be done with respect to the time ordering of these transfers, apart from their completion before kernel execution. It is possible to use Frangollo without our

---

[1] http://code.google.com/p/yacf/

Table 1. Compliance with the OpenACC 1.0 standard (constructs)

Construct	Status	Description
`kernels`	Implemented	Kernels for OpenCL and CUDA are generated for each loop inside the scope
`loop`	Implemented	Indicates a potential accelerator kernel. Some restrictions apply (e.g., no external definitions)
`kernels loop`	Implemented	A kernel will be extracted. Dependency analysis is used to check and allocate RO variables if possible.
`parallel`	Not implemented	-
`update`	Implemented	Mixing host and device clauses in the same construct does not work, they must be separated
`copy, copyin, copyout, ...`	Implemented	Runtime dynamically handles memory transfers
`pcopy, pcopyin, pcopyout ,...`	Implemented	Runtime dynamically handles memory transfers when required
`async`	Not implemented	-
`deviceptr` clause	Not implemented	-
`host`	Partially implemented	Our framework generates the right code, but we still have to solve portability issues between OpenCL and CUDA
`name`	Not in standard	Optional clause to name a particular acc region or loop and refer it from an external optimization file at compile time.

compiler framework, and the software architecture based on components and interfaces would facilitate porting the runtime to other kind of devices or creating new bindings for different languages.

Frangollo is divided into separate pluggable components. A common component serves as an abstract interface to all kind of components. Generic operations over devices, like memory transfers or kernel execution, are mapped on top of an abstract interface. Operations at this level refer to three main objects: *Context*, *Devices* and *Variables*. Components instantiate the basic operations to perform the actual work. Interfaces access the abstract layer without requiring to know which component is enabled or not. Frangollo's API provide high level entry points to the runtime, independent from the destination platform. The compiler can emit these generic runtime calls, like `registerVar`, `createContext` or `launchKernel`, and Frangollo can handle the rest of the work, i.e, choosing the appropriate platform, load the kernel file, estimate the best grid configuration, copyin/out the result and even perform reductions over the selected variables.

YaCF supports most of the syntactic constructs in the OpenACC 1.0 specification, but some of them are silently ignored. In addition, although some operations inside Frangollo runtime are handled asynchronously, support for the `async` OpenACC clause has not been implemented yet. Table 1 describes some of the constructs implemented in `accULL`.

Being an initial release, our approach allows translating to CUDA/OpenCL a comprehensive set of codes properly annotated, as we show in Section 3. Nevertheless, at this point, we do not aim to create a full commercial implementation of the standard, but a research tool to demonstrate its potential.

3   Evaluation

To evaluate our `accULL` OpenACC implementation, we have used codes from different benchmarks and tested them on the next four different platforms:

- M1: Desktop computer with an Intel(R) Core(TM) i7 930 processor (2.80 GHz), with 1MB of L2 cache, 8MB of L3 cache, shared by the four cores. The machine has 4 GB RAM and two GPU devices are attached:
  - M1a: Tesla C1060 with 240 multiprocessors, 3 GB memory
    * Bandwidth from host to device: 2.40 GB/s
    * Bandwidth from device to host: 2.29 GB/s
  - M1b: Tesla C2050 with 448 multiprocessors, 4 GB memory
    * Bandwidth from host to device: 2.35 GB/s
    * Bandwidth from device to host: 2.20 GB/s
- M2: One cluster node consisting on two quad core Intel Xeon E5410 (2.25GHz) processors, 24 GB memory and an attached Fermi C2050 card with 448 multiprocessors and 4 GB memory.
- M3: Laptop computer with one Intel(R) Core(TM) i3 CPU M 350, using Hyperthreading to enable four virtual processors, 3GB RAM, and an integrated Nvidia OPTIMUS graphic card.
- M4: A second cluster node. M4 is a shared memory system, with 4 Intel(R) Xeon(R) E7 4850 CPU, with 2.50MB L2 cache and 24MB L3 cache (for all its 10 cores). 6GB of memory are available per core.

With platforms M1a / M1b we mimic the usual scenario of an OpenACC developer: A slightly experienced user interested in improving the performance a scientific code can purchase a new GPU card and plug in it into her desktop computer. It is a relatively cheap platform as opposed to a multinode cluster and could achieve a combined peak theoretical performance 478.36 GFLOPs in double performance (77.76 GFLOPs from Tesla C1060 + 345.6 GFLOPs from Tesla C2060 + 55 GFLOPs from main processor). This kind of user might have some insight in programming and even in GPU computing, but she is not an expert. Starting with his own serial code and using an OpenACC compliant compiler, this user will take advantage of the GPUs without investing excessive time in low level programming.

M2 is a node of a common multinode cluster. Nowadays clusters are composed by multicore processors and GPU devices, thus it is possible to take advantage of OpenACC in these platforms. Moreover, our implementation integrates seamlessly with MPI programs, and can be used to take advantage of the attached GPU devices without additional effort.

M3 represents a usual nowadays medium-end laptop computer. It uses reduced versions of desktop GPUs that support GPGPU computing. Laptop computers are not relevant in terms of HPC, however, `accULL` is suitable for other environments wherever GPU computing could be beneficial.

M4 is a shared memory system that showcases an alternative case use of OpenCL. Nowadays shared memory machines feature several CPUs with several cores on each. These cores also contain vector processing units that require particular compiler support (or a deep understanding of these technologies) to unleash their potential. There are implementations of OpenCL, like the Intel OpenCL SDK or the AMD APP SDK, targeting these shared memory machines. Writing algorithms in OpenCL is not an effortless task, but it allows a better mapping of hardware resources and improve thread scheduling. Using CPU-targeted OpenCL platforms along with OpenACC represents an interesting alternative to traditional OpenMP programming that we will explore in different examples. Our runtime detects whether the platform is a GPU or a CPU and uses the appropriate variable register and copy method, without requiring to know this parameter at compile time.

3.1   Molecular Dynamic simulation

Given positions, masses and velocities of np particles, the pseudo code shown in Listing 1.1 computes the energy of the system and the forces on each particle. The code is a C implementation of a simple Molecular Dynamics (MD) simulation. It employs an iterative numerical procedure to obtain an approximate solution whose accuracy is determined by the time step of the simulation. Particles are represented by three three-dimensional double precision matrices: Position, Velocity and Force (parameters). Rows of each matrix represent a particle, whereas columns represent a dimension. For example, the coordinate $\{3, 1\}$ contains the parameter value for the particle number three in dimension one.

```
1   int main (...) {
2       ...
3       // Initial energy calculation
4       compute(position, velocity, mass, force, &potential, &kinetic);
5       ...
6       // (S) Simulation
7       for (i = 0; i < NSTEPS; i++) {
8           compute(position, velocity, mass, force, &potential, &kinetic);
9           printf(..., potential, kinetic);
10          update (position, velocity, mass, force, &potential, &kinetic);
11      }
12      ...
13  }
14  void compute (...) {
15      // (C) Compute forces
16      for (...) {
17      }
18  }
19  void update (...) {
20      // (U) Update velocity/position
21      for (...)
22        for (...) {
23          ...
24      }
25  }
```

Listing 1.1. Sketch of MD simulation in OpenACC

Table 2. Time per phase and speedup for each incremental optimization over the naive implementation, as measured in M3 using the Intel OpenCL SDK over the CPU. In this situation, using the data clause does not represent and important performance benefit, due to the fact that (1) Frangollo OpenCL implementation uses the native pointer whenever possible and (2) Intel OpenCL features lower initialization time than GPU approaches.

Version	Time transfer in	Time transfer out	Kernel Time	Total time	% Speedup
Naive Approach	$< 0.02s$	$0.0127834s$	$5.69122s$	$5.791388s$	-
Using a data clause	$< 0.02s$	$0.0121639s$	$5.63023s$	$5.729317s$	1%
Splitting C loops	$< 0.02s$	$0.0120155s$	$3.87633s$	4.046456	30.1%

After an initial forces computation, on each simulation step, the algorithm performs two basic operations: *compute* (C) and *update* (U). C operation consists of several nested loops computing the forces for each position. An external loop iterates over all particles computing its forces in the current simulation step. This requires computing the distance among all other particles, hence accessing the **position** matrix, and computes the total potential and kinetic energy of the system, which requires access to the **velocity** matrix. In terms of the data access pattern, the code is highly un-coalesced, requiring several non-contiguous loads to compute each particle. In addition, it features several costly double precision operations (*sqrt*, *sin* and *cos*) which traditionally perform badly on GPU devices. The U operation is simply a for loop that runs over the particles, updating their positions, velocities and accelerations. C is more compute–intensive than U.

A naive porting of this code using OpenACC directives would consist into adding the **kernels loop** construct to the top of the outermost loops in both routines (before C and U in Listing 1.1), and writing the appropriate copy clauses to indicate variable directionality related to the loop.

In this case, our compiler would extract the kernel from the loops and inject the appropriate runtime calls. Each time these functions are executed, memory transfers between host and GPU will take place. Transfer time between host and GPU could represent a significant percentage of the total time. Developers should take into account that the outstanding performance achieved by accelerator devices can be easily hidden by an excessive memory transfer time. Usage of profiling tools is highly recommended to detect bottlenecks.

OpenACC features a **data** directive that enables to create a data region where the information of the indicated variables is transferred into the GPU, and back to the host at the end of the data region. accULL creates a context at this point, and the directionality information provided through the copy clauses is used to register the variables in the runtime. In this case, we precede the S loop with the aforementioned **data** construct, indicating that the parameters Force, Position, Velocity and Acceleration can be transferred into the device at this point. From now on, all references inside a kernel to these variables will not require a memory transfer from the host, as they are all already stored on the device. When entering kernels inside C and U functions, the runtime will not create a new context, but it creates a new scope level within the existent context. With this method, we

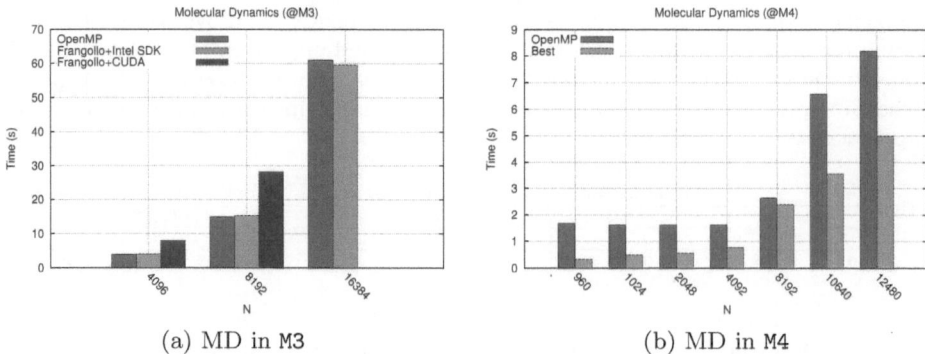

(a) MD in M3   (b) MD in M4

Fig. 1. (a) Performance comparison of the (best) OpenACC implementation vs. OpenMP in M3. OpenCL uses Intel OpenCL SDK over the CPU. CUDA version uses the CUDA 4.0 driver using the laptop's GPU. The largest problem size could not be executed in GPU due to the lack of graphics memory. It is worth to note the flexibility of the runtime, and how it is able to match OpenMP performance despite of the runtime overhead. (b) Performance comparison of the (best) OpenACC implementation with OpenMP in M4. OpenCL uses Intel OpenCL SDK over the CPU. OpenMP implementation was provided by GCC version 4.4.5.

ensure that variables are registered once and directionality from higher scopes is preserved. However, new variables might be added to these nested constructs. Existence and directionality of these variables inside the device is restricted to the scope of the current scope. In the MD code example, we require the variables `pot` and `kin` to be transferred in/out between iterations in order to show the appropriate information to the user. However, as both `pot` and `kin` are registered within an inner scope, whenever these inner scopes are exhausted, variables are transferred back from the device to the host.

`accULL` enables users to perform incremental parallelization over GPU devices with minor effort. Traditional GPU performance tools can be used with the resulting codes. For example, in the MD code, the Nvidia profiler shows that more than 80% of the time is devoted to the C kernel. As stated before, this kernel is highly compute-intensive, as it features un-coalesced memory accesses and costly/non parallel floating point operations. One possible solution is to split this kernel into several smaller ones, to increase coalescence. This could be considered counterintuitive in traditional CPU programming, where processor features large caches, but in GPUs it is a good idea. In order to rewrite this kernel into smaller ones, a CUDA developer would require a considerable effort, as she would be forced to write additional kernel calls, memory transfers, etc.. In OpenACC, the programmer only needs to split the sequential code and put the appropriate directives on the new loops. The compiler will extract the required kernels.

Table 3. Time per phase and speedup for each incremental optimization over the naive implementation, as measured in M2 using the Nvidia CUDA platform over the GPU. The cost of the CUDA calls, context initialization and memory transfers were noticeable in this case, thus using the data clause improved performance.

Version	Time transfer in	Time transfer out	Kernel time	Total time	% Speedup
Naive Approach	0.02524s	0.016229s	1.03017s	3.747910s	-
Using a data clause	0.01133s	0.016193s	1.02849s	1.433504s	61%
Splitting C loops	> 0.01s	0.016176s	0.23832s	0.434439s	88.4%

accULL provide the means to execute our codes on several platforms, not only restricted to GPU devices, with a single source code. Performance figures for the low-end system M3 are shown in Figure 1a. On the other hand, Figure 1b showcases the benefit of using an OpenCL implementation using an high-end shared memory multiprocessor (M4).

Tables 2 and 3 show detailed timing information for transfers, kernel and total time obtained using Frangollo's internal tracing module. In both Tables, the problem size was 4096 particles and 20 iteration steps were performed. Results were validated against the sequential implementation.

Users can turn this tracing feature when building the runtime and produce these statistics through an internal Frangollo call. In addition to this simple profiler module, we have experimental support for the Extrae tracing library, that enables us to perform detailed performance analysis with the Paraver [3] tool.

3.2 Mandelbrot Computation Set

Listing 1.2 shows an implementation using a Monte-Carlo method to compute the area of the Mandelbrot set using OpenACC.

When creating the parameters for the kernel launch, YaCF indicates to the runtime that the **numoutside** parameter requires a reduction operation and expands the scalar variable to a vector. This vector stores a private copy of the variable in each thread. Later, both CUDA and OpenCL components of the runtime, using a separated and optimized kernel, perform the reduction operation. The reduction operation is not performed during kernel execution, but later on when the variable is transferred back to the device, or if the variable were required by another kernel.

```
1   #pragma acc kernels loop reduction(+:numoutside) private(i,j)
         copyin(npoints, c[npoints]) copy(numoutside)
2   for(i = 0; i < npoints; i++) {
3       z.creal = c[i].creal; z.cimag = c[i].cimag;
4       for (j = 0; j < MAXITER; j++) {
5           ....
6           if ( z is outside set ) {
7               numoutside++;
8               break;
9           }
10      } /* for j */
11  } /* for i */
```

Listing 1.2. The Mandelbrot set computation in OpenACC

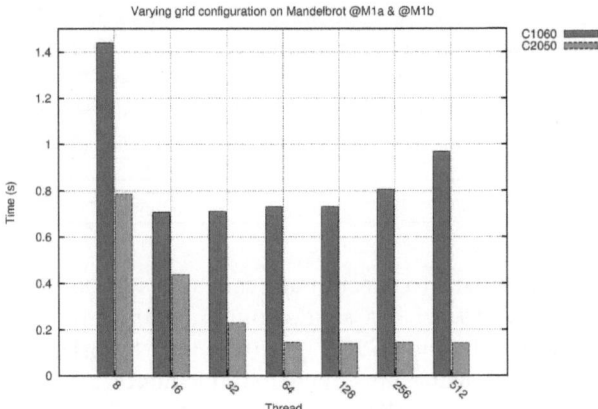

Fig. 2. Execution time of the implementation greatly varies in terms of the number of threads, using $N = 32768$ points. Besides, the optimal number of threads varies from Tesla C1060 to Tesla C2050.

Another issue that has a large impact on the performance of the CUDA code is the number of threads per block, particularly in the presence of irregular computations. Figure 2 shows the variability of the execution time while changing the number of threads. This variability reflects the significance of a proper launch grid configuration. Our compiler extracts compute intensity information from the kernel (i.e, relation between floating point operations and memory accesses) and passes the information to the runtime through the `launchkernel` API call, together with additional information about the loop, like boundaries, number of iterations, etc. This information is used to guess a first estimation to the number of threads per block. In case the user wants to influence this choice, an environment variable which varies the relation between floating point and memory operations is available.

For the CUDA component, a second estimator, which attempts to maximize the occupancy rate of the multiprocessors, is used. Information extracted from the PTX feeds this estimator. Using this two-tier system, we can guess a suitable number of threads for the target platform without user intervention. Current implementation features an environment variable which enables user to force a particular threads per block number and disables the thread estimation.

3.3 Rodinia Benchmarks

In order to complete our computational experience, in this Section we present performance comparisons for two benchmarks taken from the Rodinia suite [2]. Rodinia comprise compute-heavy applications meant to be run in the massively parallel environment of a GPU, and cover a wide range of applications. From this suite we have selected SRAD and NW for our experiments, and we present results for M1 and M4 platforms in Figures 3 and 4.

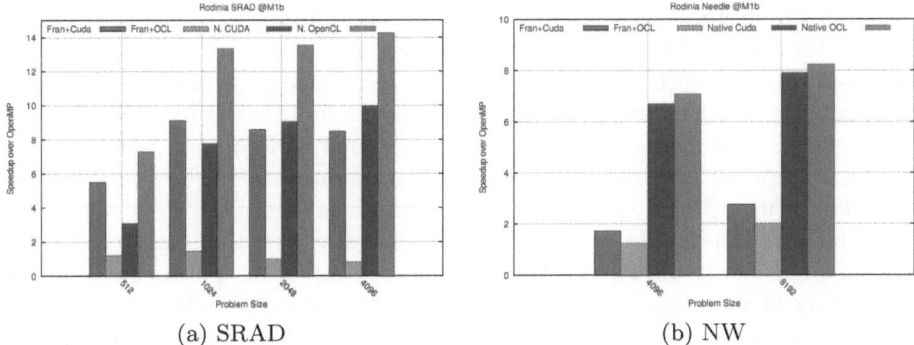

Fig. 3. Performance comparison of accULL using M1b versus native implementation, showing the speedup against OpenMP. Although native implementation clearly outperforms both OpenMP and accULL implementations, the coding effort of OpenACC is lower than the required to write both CUDA and OpenCL implementations.

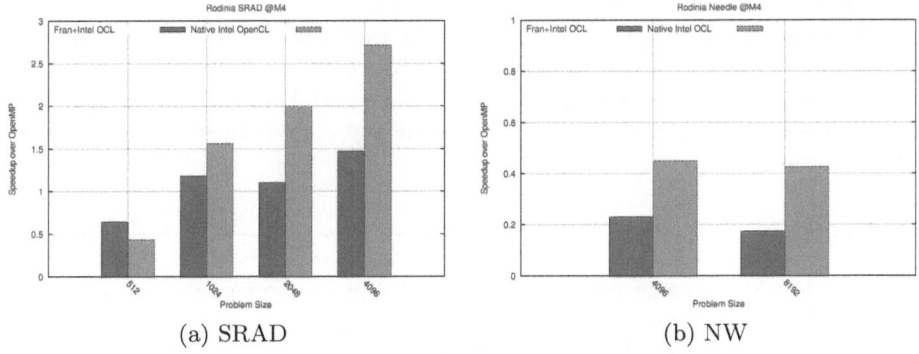

Fig. 4. Performance comparison using M4 of accULL versus native implementation, showing the speedup against OpenMP gcc implementation. It seems that using Intel OpenCL we are capable of extracting more performance from the shared memory machine. Native implementations of OpenCL also outperforms OpenMP.

4   Conclusions and Future Work

As we demonstrate in Section 3, the current status of the accULL implementation meets the requirements of a non-expert developer, and will improve the time to solution by decreasing the overall development effort. There are several implementations of the OpenACC standard. However, they are commercial solutions and, to our knowledge, do not currently feature support for OpenCL platforms. At the time of writing we could not access a commercial OpenACC implementation to compare the results of our CUDA support. We have made our best effort to compare our codes with a native CUDA or OpenCL implementation whenever possible.

Our compiler implementation of the OpenACC standard can be used as a fast-prototyping tool to explore optimizations and alternative runtime environments. Our runtime library can be fully detached from the compiler environment and used together with a commercial or production-ready compiler, like LLVM or Open64, to implement the OpenACC standard in a short time. Memory allocation, kernel scheduling, data splitting, overlapping of computation and communications or parallel reduction implementation are some of the issues that can be tackled within Frangollo independently from the compiler.

We believe that `accULL` is a good choice for non-expert users to exploit GPUs in HPC. The results we have shown in this work represent a clear improvement in the way of increase programmability of heterogeneous architectures. These preliminary results make us believe that our approach is worth to be explored more deeply.

Work in progress within the framework of the `accULL` project includes integration with a commercial compiler, taking advantage of pre-existing autovectorization support, and improvement of the support for memory allocation. We have work in progress to implement two dimensional arrays as cudaMatrix or OCLImages to improve non-contiguous memory access. Also we are exploring several possibilities of integration with MPI. The OpenCL component is capable of sharing pointers with MPI buffers, and we would like to use the new features the latest CUDA release, like GPU Direct, in the same direction. We believe that there are still plenty of opportunities to improve performance from this point, as this work settles the foundations of a dynamic and detachable compiler+runtime infrastructure.

References

1. Bihan, F.B.S.: Heterogeneous multicore parallel programming for graphics processing units. Sci. Program. 17, 325–336 (2009)
2. Che, S., Sheaffer, J.W., Boyer, M., Szafaryn, L.G., Wang, L., Skadron, K.: A characterization of the rodinia benchmark suite with comparison to contemporary cmp workloads. In: Proceedings of the IEEE International Symposium on Workload Characterization, IISWC 2010, pp. 1–11. IEEE Computer Society, Washington, DC (2010)
3. Giménez, J., Labarta, J., Pegenaute, F.X., Wen, H.-F., Klepacki, D., Chung, I.-H., Cong, G., Voigtländer, F., Mohr, B.: Guided Performance Analysis Combining Profile and Trace Tools. In: Guarracino, M.R., Vivien, F., Träff, J.L., Cannataro, M., Danelutto, M., Hast, A., Perla, F., Knüpfer, A., Di Martino, B., Alexander, M. (eds.) Euro-Par-Workshop 2010. LNCS, vol. 6586, pp. 513–521. Springer, Heidelberg (2011)
4. OpenACC directives for accelerators (2011), http://www.openacc-standard.org
5. Reyes, R., de Sande, F.: Optimization strategies in different CUDA architectures using. Microprocessors and Microsystems - Embedded Hardware Design 36(2), 78–87 (2012)
6. Wolfe, M.: Implementing the PGI accelerator model. In: Proceedings of the 3rd Workshop on General-Purpose Computation on Graphics Processing Units, GPGPU 2010, pp. 43–50. ACM, New York (2010)

Understanding the Performance of Concurrent Data Structures on Graphics Processors

Daniel Cederman, Bapi Chatterjee, and Philippas Tsigas*

Chalmers University of Technology, Sweden
{cederman,bapic,tsigas}@chalmers.se

Abstract. In this paper we revisit the design of concurrent data structures – specifically queues – and examine their performance portability with regard to the move from conventional CPUs to graphics processors. We have looked at both lock-based and lock-free algorithms and have, for comparison, implemented and optimized the same algorithms on both graphics processors and multi-core CPUs. Particular interest has been paid to study the difference between the old Tesla and the new Fermi and Kepler architectures in this context. We provide a comprehensive evaluation and analysis of our implementations on all examined platforms. Our results indicate that the queues are in general performance portable, but that platform specific optimizations are possible to increase performance. The Fermi and Kepler GPUs, with optimized atomic operations, are observed to provide excellent scalability for both lock-based and lock-free queues.

1 Introduction

While multi-core CPUs have been available for years, the use of GPUs as efficient programmable processing units is more recent. The advent of CUDA [1] and OpenCL [2] made general purpose programming on graphics processors more accessible to the non-graphics programmers. But still the problem of efficient algorithmic design and implementation of generic concurrent data structures for GPUs remains as challenging as ever.

Much research has been done in the area of concurrent data structures. There are efficient concurrent implementations of a variety of common data structures, such as stacks [3], queues [4–9] and skip-lists [10]. For a good overview of several concurrent data structures we refer to the chapter by Cederman et al. [11].

But while the aforementioned algorithms have all been implemented and evaluated on many different multi-core architectures, very little work has been done to evaluate them on graphics processors. Data structures targeting graphics applications have been implemented on GPUs, such as the kd-tree [12] and octree [13]. A C++ and Cg based template library [14] has been provided for random

* This work was partially supported by the EU as part of FP7 Project PEPPHER (www.peppher.eu) under grant 248481 and the Swedish Foundation for Strategic Research as part of the project RIT-10-0033 "Software Abstractions for Heterogeneous Multi-core Computer".

C. Kaklamanis et al. (Eds.): Euro-Par 2012, LNCS 7484, pp. 883–894, 2012.

access data structures for GPUs. Load balancing schemes on GPUs [15] using different data structures have been designed. A set of blocking synchronization primitives for GPUs [16] has been presented that could aid in the development or porting of data structures.

With the introduction of atomic primitives on graphics processors, we hypothesize that many of the existing concurrent data structures for multi-core CPUs could be transferred to graphics processors, perhaps without much change in the design. To evaluate how performance portable the designs of already existing common data structure algorithms are, we have, for this paper, implemented a set of concurrent FIFO queues with different synchronization mechanisms on both graphics processors and on multi-core CPUs. We have performed experiments comparing and analyzing the performance and cache behavior of the algorithms. We have specifically looked at how the performance changes by the move from NVIDIA's Tesla architecture to the newer Fermi [17] and Kepler (GK104) [18] architectures.

The paper is organized as follows. In section 2, we introduce the concurrent data structures and describe the distinguishing features of the algorithms considered. Section 3 presents a brief description of the CUDA programming model and different GPU architectures. In section 4, we present the experimental setup. A detailed performance analysis is presented in section 5. Section 6 concludes the paper.

2   Concurrent Data Structures

Depending on the synchronization mechanism, we broadly classify concurrent data structures into two categories, namely *blocking* and *non-blocking*. In blocking synchronization, no progress guarantees are made. For non-blocking synchronization, there are a number of different types of progress guarantees that can be assured. The two most important ones are known as *wait-free* and *lock-free*. Wait-free synchronization ensures that all the non-faulty processes eventually succeed in finite number of processing steps. Lock-free synchronization guarantees that at least one of the non-faulty processes out of the contending set will succeed in a finite number of processing steps. In practice, wait-free synchronization is usually more expensive and is mostly used in real-time settings with high demands on predictability, while lock-free synchronization targets high-performance computing.

Lock-free algorithms for multiple threads require the use of atomic primitives, such as Compare-And-Swap (CAS). CAS can conditionally set the value of a memory word, in one atomic step, if at the time, it holds a value specified as a parameter to the operation. It is a powerful synchronization primitive, but is unfortunately also expensive compared to normal read and write operations.

In this paper we have looked at different types of queues to evaluate their performance portability when moved from the CPU domain to the GPU domain. The queue data structures that we have chosen to implement are representative of several different design choices, such as being array-based or linked-list-based, cache-aware or not, lock-free or blocking. We have divided them up into two main categories, Single-Producer Single-Consumer (SPSC) and Multiple-Producer Multiple-Consumer (MPMC).

2.1 SPSC Queues

In '83, **Lamport** presented a lock-free array-based concurrent queue for the SPSC case [19]. For this case, synchronization can be achieved using only atomic read and write operations on shared head and tail pointers. No CAS operations are necessary. Having shared pointers cause a lot of cache thrashing however, as both the producer and consumer need to access the same variables in every operation.

The **FastForward** algorithm lowered the amount of cache thrashing by keeping the head and tail variables private to the consumer and producer, respectively [4]. The synchronization was instead performed using a special empty element that was inserted into the queue when an element was dequeued. The producer would then only insert elements when the next slot in the array contained such an element. Cache thrashing does however still occur when the producer catches up with the consumer. To lower this problem it was suggested to use a small delay to keep the threads apart. The settings used for the delay function are however so application dependant that we decided not to use it in our experiments.

The **BatchQueue** algorithm divides the array into two batches [5]. When the producer is writing to one batch, the consumer can read from the other. This removes much of the cache thrashing and also lowers the frequency at which the producer and consumer need to synchronize. The major disadvantage of this design is that a batch must be full before it can be read, leading to large latencies if elements are not enqueued fast enough. A suggested solution to this problem was to at regular intervals insert null elements into the queue. We deemed this as a poor solution and it is not used in the experiments. To take better advantage of the graphics hardware, we have also implemented a version of the BatchQueue where we copy the entire batch to the local shared memory, before reading individual elements from it. We call this version **Buffered BatchQueue**.

The **MCRingBuffer** algorithm is similar to the BatchQueue, but instead of having just two batches, it can handle an arbitrary number of batches. This can be used to find a balance between the latency caused by waiting for the other threads and the latency caused by synchronization. As for the BatchQueue we provide a version that copies the batches to the local shared memory. We call this version **Buffered MCRingBuffer**.

2.2 MPMC Queues

For the MPMC case we used the lock-free queue by Michael and Scott, henceforth called the **MS-Queue** [7]. It is based on a linked-list and adds items to the queue by using CAS to swap in a pointer at the tail node. The tail pointer is then moved to point to the new element, with the use of a CAS operation. This second step can be performed by the thread invoking the operation, or by another thread that needs to help the original thread to finish before it can continue. This helping behavior is an important part of what makes the queue lock-free, as a thread never has to wait for another thread to finish.

We also used the lock-free queue by Tsigas and Zhang, henceforth called the **TZ-Queue**, which is an array-based queue [8]. Elements are here inserted into the array using CAS. The head and tail pointers are also moved using CAS, but it is done lazily, after every x:th element instead of after every element. In the experiments we got the best performance doing it every second operation.

To compare lock-free synchronization with blocking, we used the **lock-based** queue by Michael and Scott, which stores elements in a linked-list [7]. We used both the standard version, with separate locks for the enqueue and dequeue operation, and a simpler version with a common lock for both operations. For locks we used a basic spinlock, which spins on a variable protecting a critical section, until it can acquire it using CAS. As CAS operations are expensive, we also implemented a lock that does not use CAS, the bakery-lock by Lamport [20].

3   GPU Architectures

Graphics processors are massively parallel shared memory architectures excellently suitable for data parallel programs. A GPU has a number of stream multiprocessors (SMs), each having many cores. The SMs have registers and a local fast shared memory available for access to threads and thread blocks (group of threads) respectively, executing on them. The global memory, the main graphics memory, is shared by all the thread blocks and the access is relatively slow compared to that of the local shared memory.

In this work we have used CUDA for all GPU implementations. CUDA is a mature programming environment for programming on GPUs. In CUDA threads are grouped into blocks where all threads in a specific block execute on the same SM. Threads in a block are in turn grouped into so called warps of 32 consecutive threads. These warps are then scheduled by the hardware scheduler. Exactly how the scheduler schedules warps is unspecified. This is problematic when using locks, as there is a potential for deadlocks if the scheduler is unfair. For lock-free algorithms this is not an issue, as they are guaranteed to make progress regardless of the scheduler.

The different generations of CUDA programmable GPUs are categorized in compute capabilities (CC) and are identified more popularly by their architecture's codename. CC 1.x are Tesla, 2.x are Fermi and 3.x are Kepler. The architectural features depend on the compute capability of the GPU. In particular the availability of atomic functions has been varying with the compute capabilities. In CC 1.0 there were no atomic operations available, from CC 1.1 onwards there are atomic operations available on the global memory and from CC 1.2 also for the shared memory. An important addition to the GPUs in the Fermi and Kepler architectures is the availability of a unified L2 cache and a configurable L1 cache. The performance of the atomic operations significantly increased in Fermi, with the atomic unit working on the L2 cache, instead of on the global memory [16]. The bandwidth of L2 cache increased in Kepler so that it is now 73% faster than that in Fermi [18]. The speed of atomic operations has also been significantly increased in Kepler as compared to Fermi.

4   Experimental Setup

The experiments were performed on four different types of graphics processors, with different memory clock rates, multiprocessor counts and compute capabilities. To explore the difference in performance between CPUs and GPUs, the same experiments were also performed on a conventional multi-core system, a 12-core Intel system (24 cores with HyperThreading). See Table 1 for an overview of the platforms used.

Table 1. Platforms used in experiments. Counting multiprocessors as cores in GPU.

Name	Clock speed	Memory clock rate	Cores	Cache	Architecture(CC)
GeForce 8800 GT	1.6 GHz	1.0 GHz	14	0	1.1 (Tesla)
GeForce GTX 280	1.3 GHz	1.1 GHz	30	0	1.3 (Tesla)
Tesla C2050	1.2 GHz	1.5 GHz	14	786 kB	2.0 (Fermi)
GeForce GTX 680	1.1 GHz	3.0 GHz	8	512 kB	3.0 (Kepler)
Intel E5645 (2x)	2.4 GHz	0.7 GHz	24	12 MB	

In the experiments we only consider communication between thread blocks, not between individual threads in a thread block.

For the SPSC experiments, a thread from one thread block was assigned the role of the producer and another thread from a second block the role of the consumer. The performance was measured by counting the number of successful enqueue/dequeue operations per ms that could be achieved when communicating a set of integers from the producer to the consumer. Enqueue operations on full queues or dequeue operations on empty queues were not counted. Local variables, variables that are only accessed by either the consumer or the producer, are placed in the shared memory to remove unnecessary communication with the global memory. For buffered queues, 32 threads were used for memory transfer between global and shared memory to take advantage of the hardware's coalescing of memory accesses. All array-based queues had a maximum length of 4096 elements. The MCRingBuffer used a batch size of 128 whereas the BatchQueue by design has batches of size as of half the queue size, in this case 2048. For the CPU experiments care was taken to place the consumer and producer on different sockets, to emphasize the penalty taken by using an inefficient memory access strategy.

For the MPMC experiments a varying number of thread blocks were used, from 2 up to 60. Each thread block performed 25% enqueue operations and 75% dequeue operations randomly, using a uniform distribution. Two scenarios were used, one with high contention, where operations were performed one after another, and one with low contention, in which a small amount of work was performed between the operations. The performance was measured in the number of successful operations per ms in total.

5   Performance Analysis

5.1   SPSC Queues

Figure 1(a) depicts the result from the experiments on the CPU system. It is clear from the figure that even the small difference in access pattern between the Lamport and the FastForward algorithms has a significant impact on the performance. The number of operations per ms differ by a factor of four between the two algorithms. The cache access profile in Figure 1(b) shows that the number of cache misses goes down dramatically when the head and tail variables are no longer shared between processes. It goes down even further when the producer and the consumer are forced to work on different memory locations. The figure also shows that the number of stalled cycles per instructions matches the cache misses relatively well. The reason for the performance difference between the BatchQueue and the MCRingBuffer, which both have a similar number of cache misses, lies in the difference between the size of the batches. This causes more frequent reads and writes of shared variables compared to the BatchQueue. It was observed that increasing the batch size lowers the synchronization overhead and the number of stalled cycles and improves the performance of the MCRingBuffer and brings it close to that of the BatchQueue.

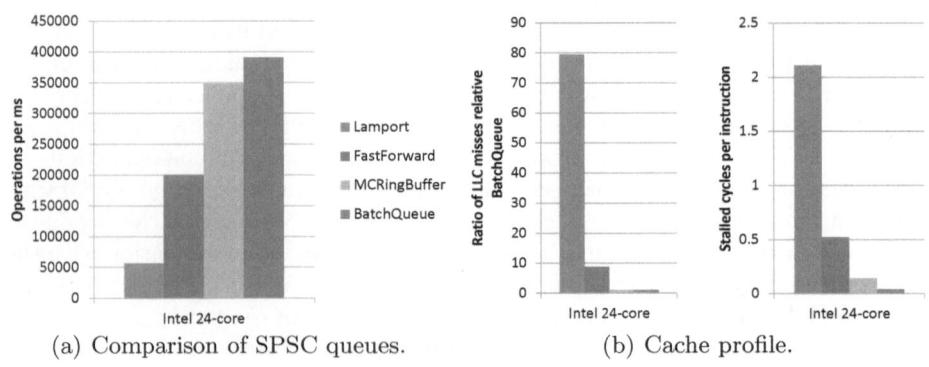

(a) Comparison of SPSC queues.   (b) Cache profile.

Fig. 1. Comparison of SPSC queues on the CPU based system

Figure 2 shows the results for the same experiment performed on the graphics processors. On the Tesla processors there are no cache memories available, which removes the problem of cache thrashing and causes the Lamport and FastForward algorithms to give similar results. In contrast to the CPU implementations, here the MCRingBuffer is actually faster than the BatchQueue. This is due to the fact that the BatchQueue enqueuer is faster than the dequeuer and has to

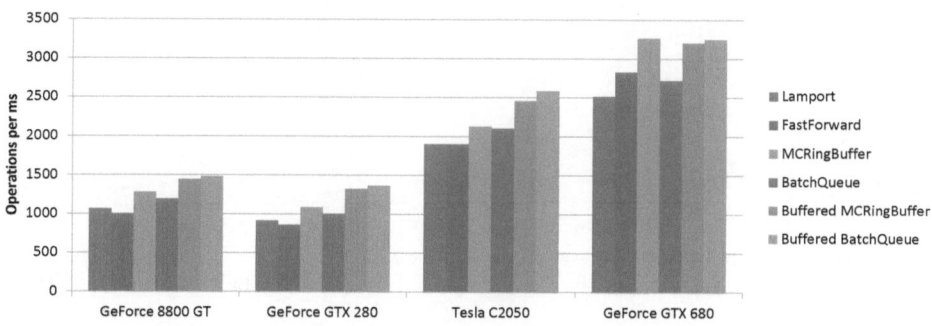

Fig. 2. Comparison of SPSC queues on four different GPUs

wait for a longer time for the larger batches to be processed. The smaller batch size in MCRingBuffer thus has an advantage here. The two buffered versions lower the overhead, as for most operations the data will be available locally from the shared memory. It is only at the end of a batch that the shared variables and the elements stored in the queue need to be accessed. This access is done using coalesced reads and writes, which speeds up the operation. When the queues are buffered, the BatchQueue becomes faster than the MCRingBuffer. Thus the overhead of the more frequent batch copies became more dominant. The performance on the Fermi and Kepler graphics processor is significantly better compared to the other processors, benefiting from the faster memory clock rate and the cache memory. The speed of the L2 cache is however not enough to make the unbuffered queues comparable with the buffered ones on the Fermi processor. On the Kepler processor, on the other hand, with its faster cache and higher memory clock rate, the unbuffered MCRingbuffer performs similarly to the buffered queues. The SPSC queues that we have examined thus need to be rewritten to achieve maximum performance on most GPUs. This might however change with the proliferation of the Kepler architecture.

5.2   MPMC Queues

All MPMC queue algorithms, except the ones that used the bakery-lock, make use of the CAS primitive. To visualize the behavior of the CAS primitive we measured the number of CAS operations that could be performed per thread block per ms for a varying number of thread blocks. The result is available in Figure 3. We see in Figure 3(a) that when the contention increases for the Tesla processors the number of CAS operations per ms drops quickly. However, it is observed that the CAS operations scale well on the Fermi, for up to 40 thread blocks, at high speed. The increased performance of the atomic primitives was one of the major improvements done when creating the Fermi architecture. The atomic operations are now handled at the L2 cache level and no longer need to access the global memory [16]. The Kepler processor has twice the memory clock rate of the Fermi processor and we can see that the CAS operations scales

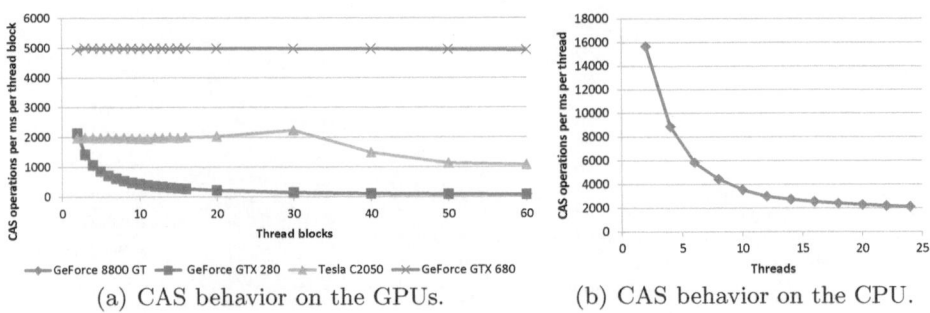

(a) CAS behavior on the GPUs.   (b) CAS behavior on the CPU.

Fig. 3. Visualization of the CAS behavior on the GPUs and the CPU

perfect despite increased contention. Figure 3(b) shows that on the conventional system the performance is quite high when few threads perform CAS operations, but the performance drops rapidly as the contention increases.

Figure 4 shows the performance of the MPMC queues on the CPU-based system. Looking first at the topmost graphs, which shows the result using just lock-based queues, we see that for a low number of threads the dual spinlock based queue clearly outperforms the bakery lock based queues. The bakery lock does not use any expensive CAS operation, but the overhead of the algorithm is still too high, until the number of threads goes above the number of cores and starts to use hyperthreading. The difference between dual and single spinlock is insignificant, however between the dual and the single bakery lock there is a noticeable difference.

The lower two graphs show the comparison results for the two lock-free queues together with the best lock-based one, the dual spinlock queue. The lock-free queues clearly outperform the lock-based one for all number of threads and for both contention levels. The array-based TZ-queue exhibits better results for the lower range of threads, but it is quickly overtaken by the linked-list based MS-queue. When hyperthreading kicks in, the performance does not drop any more for any of the queues.

The measurements taken for the lock-based queues on the Fermi and one of the Tesla graphics processors are shown in Figure 5. Just as in the CPU experiments the dual spinlock queue excels among the lock-based queues. There is however a much clearer performance difference between the dual and single spinlock queues in all graphs, although not for the low contention cases when using few thread blocks. The peak in the result in Figure 5(a) is due to the overhead of the benchmark and the non-atomic parts of the queue. When contention is lowered, as in Figure 5(b), the peak moves to the right. After the peak the cost of the atomic operations become dominant, and the performance drops. For the Fermi-processor, in Figure 5(c), the performance for the spinlock based queues is significantly higher, while at the same time scaling much better. As we could see in Figure 3(a), this is due to the much improved atomic operations of the Fermi-architecture.

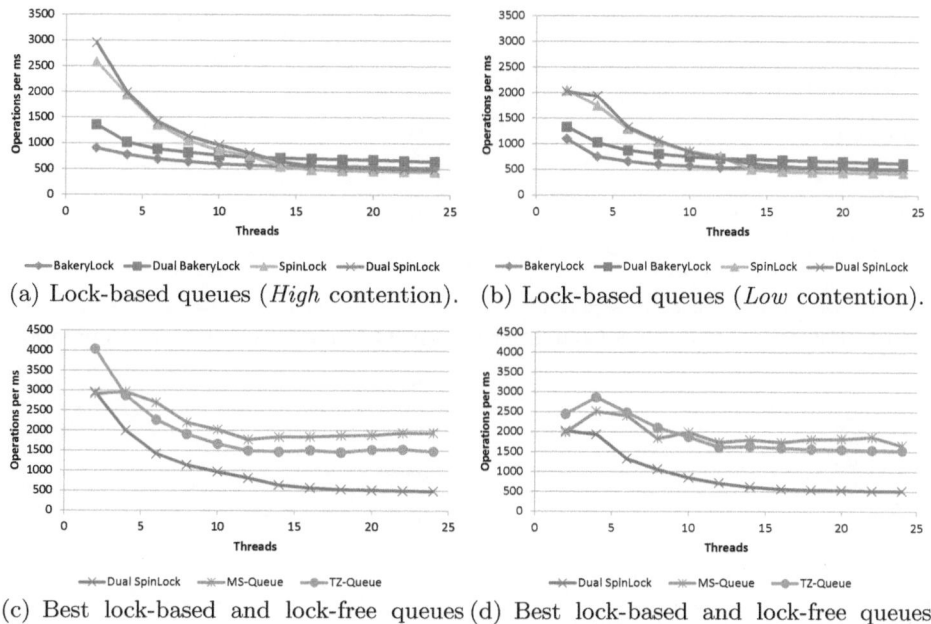

(a) Lock-based queues (*High* contention).   (b) Lock-based queues (*Low* contention).

(c) Best lock-based and lock-free queues (*High* contention).   (d) Best lock-based and lock-free queues (*Low* contention).

Fig. 4. Comparison of MPMC queues on the Intel 24-core system under *high* and *low* contention scenarios

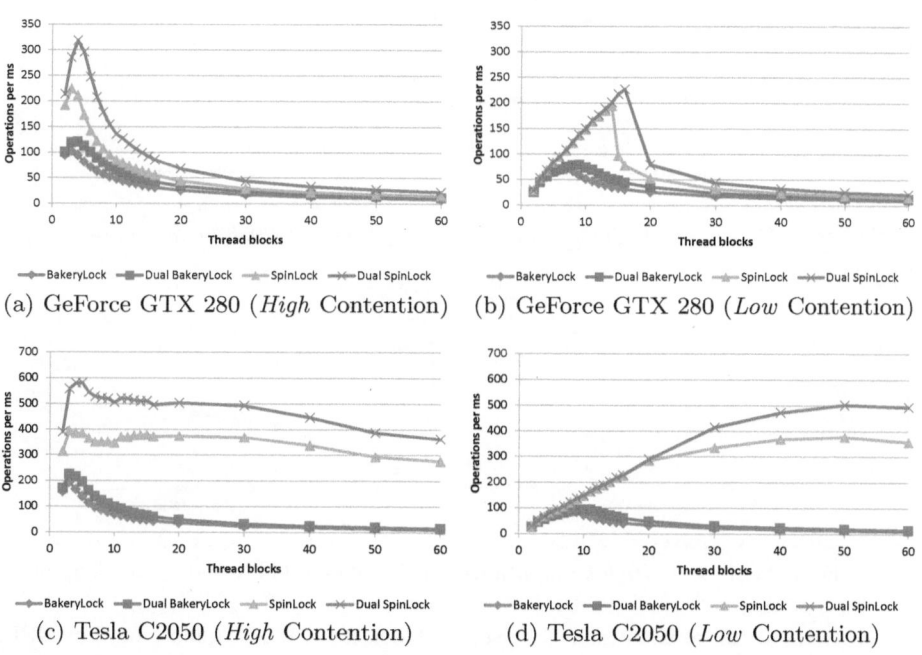

(a) GeForce GTX 280 (*High* Contention)   (b) GeForce GTX 280 (*Low* Contention)

(c) Tesla C2050 (*High* Contention)   (d) Tesla C2050 (*Low* Contention)

Fig. 5. Comparison of lock-based MPMC queues on two GPUs under *high* and *low* contention scenarios

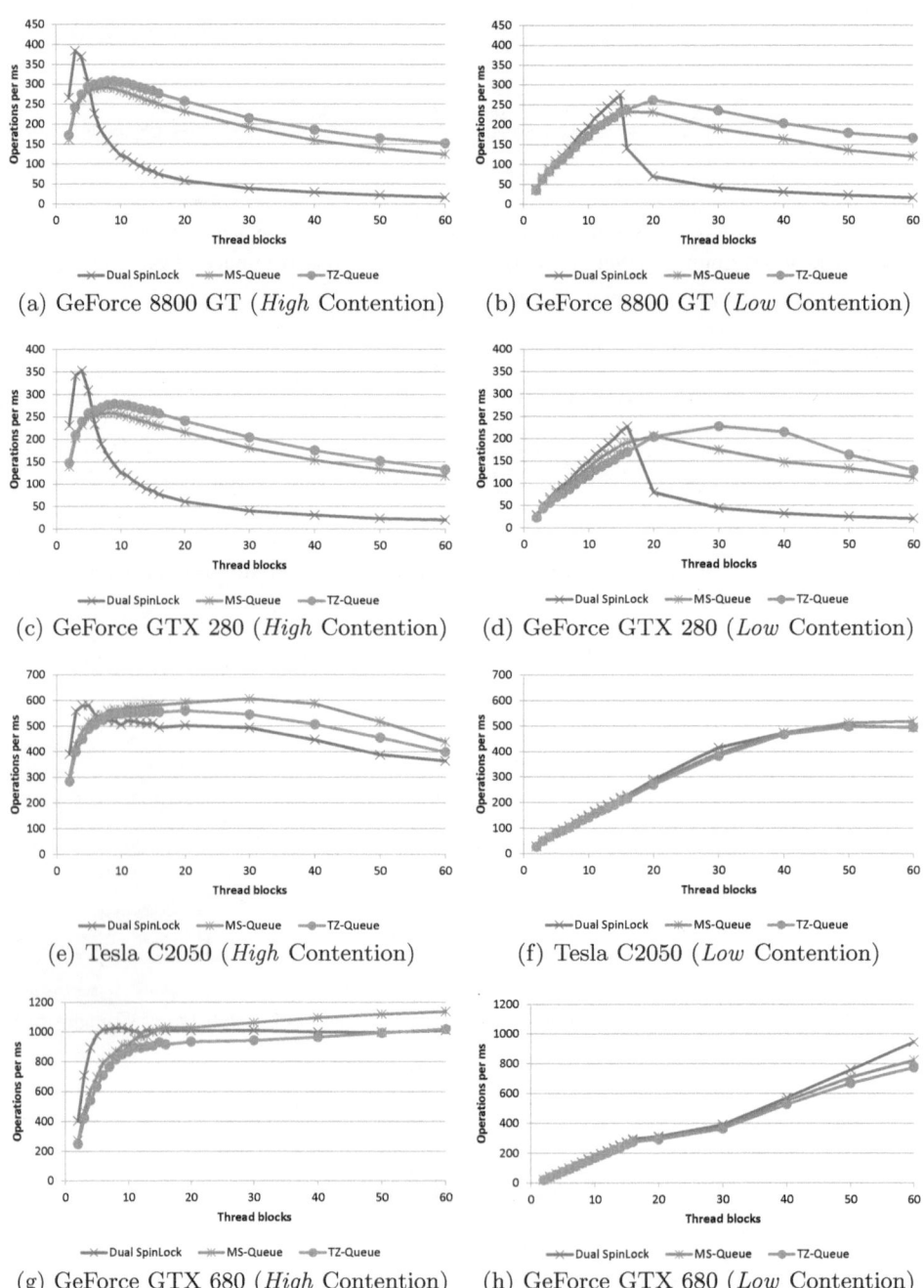

(a) GeForce 8800 GT (*High* Contention)   (b) GeForce 8800 GT (*Low* Contention)

(c) GeForce GTX 280 (*High* Contention)   (d) GeForce GTX 280 (*Low* Contention)

(e) Tesla C2050 (*High* Contention)   (f) Tesla C2050 (*Low* Contention)

(g) GeForce GTX 680 (*High* Contention)   (h) GeForce GTX 680 (*Low* Contention)

Fig. 6. Comparison of the best lock-based and lock-free MPMC queues on four GPUs under *high* and *low* contention scenarios

Comparing the dual spinlock queue with the lock-free queues, in Figure 6 we see that the lock-free queues scale much better than the lock-based one and provides the best performance when the thread block count is high. The spinlock queue does however achieve a better result on all graphics processors for a low number of thread blocks. As the contention is lowered, it remains useful for a higher number of threads. The array-based TZ-queue outperforms the linked-list based MS-queue on both the Tesla processors, but falls short on the Fermi and Kepler processors, Figure 6(e). When contention is lowered on the Fermi-processor, Figure 6(f), there is no longer any difference between the lock-based and the lock-free queues.

6 Conclusion and Future Work

In this paper we have examined the performance portability of common SPSC and MPMC queues. From our experiments on the SPSC queues we found that the best performing queues on the CPU were also the ones that performed well on the GPUs. It was however clear that the cache on the Fermi-architecture was not enough to remove the benefit of redesigning the algorithms to take advantage of the local shared memory. For the MPMC queue experiments we saw similar results in scalability for the GPU-versions on the Tesla processors as we did for the CPU-version. On the Fermi processor the result was surprising however. The scalability was close to perfect and for low contention there was no difference between the lock-based and the lock-free queues. The Fermi architecture has significantly improved the performance of atomic operations and this is an indication that new algorithmic designs should be considered to more properly take advantage of this new behavior. The Kepler architecture has continued in this direction and now provides atomic operations with performance competitive to that of conventional CPUs.

We will continue this work by studying the behavior of other concurrent data structures with higher potential to scale than queues, such as dictionaries and trees. Most queue data structures suffer from the fact that only two operations can succeed concurrently in the best case, whereas for a dictionary there are no such limitations.

References

1. NVIDIA: NVIDIA CUDA C Programming Guide. 4.0 edn. (2011)
2. The Khronos Group Inc.: OpenCl Reference Pages. 1.2 edn. (2011)
3. Treiber, R.: System programming: Coping with parallelism. Technical Report RJ5118, IBM Almaden Research Center (1986)
4. Giacomoni, J., Moseley, T., Vachharajani, M.: FastForward for efficient pipeline parallelism: a cache-optimized concurrent lock-free queue. In: Proceedings of the 13th ACM SIGPLAN Symposium on Principles and Practice of Parallel Programming, pp. 43–52. ACM (2008)

5. Preud'homme, T., Sopena, J., Thomas, G., Folliot, B.: BatchQueue: Fast and Memory-Thrifty Core to Core Communication. In: 22nd International Symposium on Computer Architecture and High Performance Computing (SBAC-PAD), pp. 215–222 (2010)
6. Lee, P.P.C., Bu, T., Chandranmenon, G.: A lock-free, cache-efficient shared ring buffer for multi-core architectures. In: Proceedings of the 5th ACM/IEEE Symposium on Architectures for Networking and Communications Systems, ANCS 2009, pp. 78–79. ACM, New York (2009)
7. Michael, M., Scott, M.: Simple, fast, and practical non-blocking and blocking concurrent queue algorithms. In: Proceedings of the 15th Annual ACM Symposium on Principles of Distributed Computing, pp. 267–275. ACM (1996)
8. Tsigas, P., Zhang, Y.: A simple, fast and scalable non-blocking concurrent fifo queue for shared memory multiprocessor systems. In: Proceedings of the 13th Annual ACM Symposium on Parallel Algorithms and Architectures, pp. 134–143. ACM (2001)
9. Gidenstam, A., Sundell, H., Tsigas, P.: Cache-Aware Lock-Free Queues for Multiple Producers/Consumers and Weak Memory Consistency. In: Proceedings of the 14th International Conference on Principles of Distributed Systems, pp. 302–317 (2010)
10. Sundell, H., Tsigas, P.: Fast and Lock-Free Concurrent Priority Queues for Multi-Thread Systems. In: Proceedings of the 17th IEEE/ACM International Parallel and Distributed Processing Symposium (IPDPS), pp. 84–94. IEEE Press (2003)
11. Cederman, D., Gidenstam, A., Ha, P., Sundell, H., Papatriantafilou, M., Tsigas, P.: Lock-free Concurrent Data Structures. In: Pllana, S., et al. (eds.) Programming Multi-core and Many-core Computing Systems. John Wiley & Sons (2012)
12. Zhou, K., Hou, Q., Wang, R., Guo, B.: Real-time kd-tree construction on graphics hardware. ACM Transactions on Graphics 27(5), 1–11 (2008)
13. Zhou, K., Gong, M., Huang, X., Guo, B.: Data-Parallel Octrees for Surface Reconstruction. IEEE Transactions on Visualization and Computer Graphics 17(5), 669–681 (2011)
14. Lefohn, A.E., Sengupta, S., Kniss, J., Strzodka, R., Owens, J.D.: Glift: Generic, efficient, random-access GPU data structures. ACM Transactions on Graphics 25(1), 60–99 (2006)
15. Cederman, D., Tsigas, P.: On dynamic load balancing on graphics processors. In: Proceedings of the 23rd Symposium on Graphics Hardware, GH 2008, pp. 57–64. Eurographics Association (2008)
16. Stuart, J., Owens, J.: Efficient synchronization primitives for gpus. Arxiv preprint arXiv:1110.4623 (2011)
17. NVIDIA: Whitepaper NVIDIA Next Generation CUDATM Compute Architecture: FermiTM. 1.1 edn. (2009)
18. NVIDIA: Whitepaper NVIDIA GeForce GTX 680. 1.0 edn. (2012)
19. Lamport, L.: Specifying concurrent program modules. ACM Transactions on Programming Languages and Systems 5, 190–222 (1983)
20. Lamport, L.: A new solution of Dijkstra's concurrent programming problem. Communications of the ACM 17, 453–455 (1974)

A New Programming Paradigm for GPGPU

Julio Toss[1,*] and Thierry Gautier[2]

[1] Institute of Informatics, UFRGS, Porto Alegre - RS, Brasil
`jtoss@inf.ufrgs.br`
[2] INRIA, MOAIS, LIG, Grenoble, France
`thierry.gautier@inrialpes.fr`

Abstract. Graphics Processing units (GPU) have become a valuable support for High Performance Computing (HPC) applications. However, despite the many improvements of General Purpose GPUs, the current programming paradigms available, such as NVIDIA's CUDA, are still low-level and require strong programming effort, especially for irregular applications where dynamic load balancing is a key point to reach high performances.

This paper introduces a new hybrid programming scheme for general purpose graphics processors using two levels of parallelism. In the upper level, a program creates, in a lazy fashion, tasks to be scheduled on the different *Streaming Multiprocessors* (MP), as defined in the NVIDIA's architecture. We have embedded inside GPU a well-known work stealing algorithm to dynamically balance the workload. At lower level, tasks exploit each *Streaming Processor* (SP) following a data-parallel approach. Preliminary comparisons on data-parallel iteration over vectors show that this approach is competitive on regular workload over the standard CUDA library Thrust, based on a static scheduling. Nevertheless, our approach outperforms Thrust-based scheduling on irregular workloads.

Keywords: Work Stealing, GPU, Task Parallelism.

1 Introduction

Nowadays, Graphical Processing Units have acquired great importance on the scenario of the High Performance Computing (HPC). Several HPC applications use this kind of hardware support to achieve better performances on parallel algorithms. The hardware is widely available and continues to evolve very fast, adding new capabilities and increasing its programmability. Programming models like OpenCL and Nvidia's CUDA allow developers to program and exploit parallelism on GPUs at the expense of a strong programming effort. Nevertheless, due to their important performances, GPUs have motivated the industry and the scientific community to port increasingly more applications to the GPU platform. At the same time, the generalization of the hardware reveals new challenges to be solved. Classical problems from the multicore-CPU architectures like

---

* Partially supported by FAPERGS and CNPq grants, through the project "Green-Grid".

C. Kaklamanis et al. (Eds.): Euro-Par 2012, LNCS 7484, pp. 895–907, 2012.

load balancing, synchronization and the need of abstract programming models, are now also present on GPUs.

Despite the enhancement of the programming capability provided by existing GPU programming solutions, for instance CUDA, the programming model they propose can only be directly exploited by sufficiently regular applications. Typically for those working over vectors or matrices. Nevertheless, there are several other kinds of applications where the parallelism is expressed recursively by creating tasks. For such applications it is necessary to provide a suitable runtime system to exploit the different cores present inside the GPUs.

The contribution of this paper is to propose and validate by experimentation a new paradigm for GPU programming based on data parallel tasks and work stealing. We show how task parallelism can be supported on graphics processors and how to deal with problems like load balancing and synchronization. Our preliminary results show that our approach is highly competitive with state of the art programming software for either data parallel programming or for pure task parallel programming.

The outline of this paper is the following: in Section 2 we briefly discuss the related works about work stealing algorithms and scheduling on GPUs. Section 3 presents the design and implementation of our approach to support work stealing with CUDA on GPUs. Then, on Section 4, we evaluate our model with several load balance patterns analyzing performance and overheads. On Section 5 we conclude and point future directions to improve the model.

2   Related Works

Task parallel applications are well suited to deal with irregular parallel algorithms. Scheduling of tasks among the computing resources must be effective to balance the workload. If scheduling is not performed well some processing units may be overloaded with work while others stay idle. Additionally, the scheduler implementation must be efficient to minimize overheads that come from the lock contention to access the work queue and to avoid stopping the computation for workload re-balancing.

2.1   Work Stealing

A well-known scheduling technique with several implementations on multi-core processors is work stealing. Here, each processing unit has its own queue of tasks to process. Whenever one gets idle it will, itself, look for tasks from other queues to steal. This technique is particularly interesting for applications that create tasks at execution time in an unpredictable way. Work stealing is, therefore, known as a dynamic load balancing method.

Blumofe et al. [4] give the first provably efficient work stealing scheduler. It proves that a parallel execution, on uniform P processors machine, using their work-stealing scheduler, has an expected run time of $\frac{T_1}{P} + O(T_\infty)$, where T_1 is the serial execution time of the multi-threaded computation and T_∞ is the minimum execution time with an infinite number of processors.

A work-stealing algorithm relies on a work queue data structure owned by each thread of the system. Threads can call three functions: Pop, Push and Steal on a work queue. Push and Pop functions are called only by the owner thread of a work queue to enqueue / dequeue tasks. Steal function is called by an idle thread on a victim work queue to get tasks to execute. The Cilk-5 runtime system [9] implements a *lock-based* work-stealing scheduler. It employs the Dijkstra's THE protocol for mutual exclusion [8] which greatly reduces the lock overhead [9] by only using systematic locks on steal operations to serialize thieves on the same victim. Arora et al. [2] present a completely *lock-free* work-stealing algorithm which uses array-based dequeues and minimizes the need of costly Compare-And-Swap (CAS) operations. Hendler et al. [11] overcame the potential overflow problem on ABP's algorithm [2] with a *dynamic memory* work-stealing algorithm. Chase and Lev [6] came with a simpler solution to this same problem by implementing unbounded dequeues as dynamic-cyclic-arrays. Hendler and Shavit [12] generalize the ABP algorithm to allow the processing units to steal, instead of one, up to half of the items in a given queue at a time. Their algorithm provides better load balancing than ABP while preserving the lock-free and CAS minimization properties.

2.2   Scheduling on GPUs

Recently, with the advent of the use of graphical processors for general purpose computing, those classical CPU load balancing methods started to be studied on GPUs. Chen et al. [7] use molecular dynamics simulation to evaluate a centralized method of dynamic load balancing on single and multi-GPUs systems. Their results showed that, for unbalanced workloads, task-based models can utilize the GPU hardware more efficiently than the standard CUDA scheduler.

Cederman and Tsigas [5] use the task of creating an octree partitioning to compare four different methods for dynamic load balancing. A centralized blocking task queue; a centralized non-blocking task queue; a static list and a distributed task queue using the ABP [2] work stealing protocol. Their results clearly show that centralized blocking methods are not suitable for GPUs as they perform poorly and cannot scale with the increase of processing units. The non-blocking task queue do perform better but, as a centralized approach, scales very poorly. The best performances and scalability were achieved with the work stealing method with distributed work queues.

In Angel et al. [1] the shortest-path problem is used as an irregular application to evaluate a framework for dynamic work scheduling based on Blumofe and Leiserson's work stealing algorithm[4]. They exploit the performance and synchronization characteristics of the GPU memory hierarchy by using a combination of shared and global memory queues. The overhead found was by a factor of 3 for queues on shared memory and 15 for queues on global memory.

In [16], Tzeng et al. propose a dynamic load balancing method based on task-donation, which shows similar performances to previous work-stealing approaches but uses less memory.

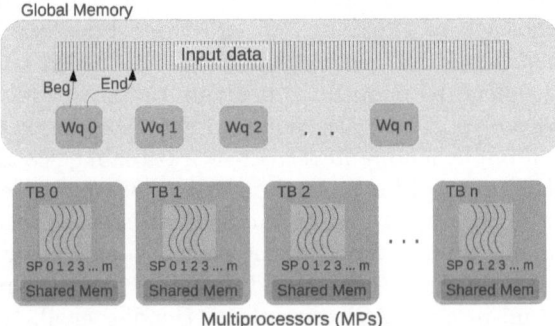

Fig. 1. Scheme of the work stealing scheduler on the GPU: one work queue on global memory for each Tread Block running on the Multiprocessors

Previous work shows that scheduling inside GPUs is necessary to improve performance in task-parallel applications. On the CPUs, several parallel programming tools like CILK+ (Frigo et al. [9]), Intel TBB (Pheatt [13]), KAAPI (Gautier et al. [10]), use work stealing as standard scheduling technique in parallel **for** loops. Additionally, recent architectures like the Intel Many Integrated Core (Intel MIC) use CILK+ as standard programming model reinforcing the trend of work stealing schedulers on massively parallel architectures. On the other hand, GPU programming tools like the Thrust Template library for CUDA (Bell [3]), provides generic templates (e.g. array Transform) to enhance programmer's productivity. However, CUDA does not have a dynamic scheduler, and for some types of workloads it cannot extract the best performance of the GPU. In this context, our model extends the use of work stealing in GPUs to a broader range of parallel applications. We implemented an hybrid programming model combining tasks and data parallelism. Our benchmarks showed comparable performance to State of Art on typical task-parallel problems (Octree Partitioning, Sec. 4.5) and we outperform Thrust on the array transform problem (Sec. 4.4).

3   Mixing Task Parallelism and Data Parallelism on CUDA

As showed before, parallel tasks with work stealing is being used on multi-core architectures and on GPU as standard paradigm for irregular divide and conquer parallel applications, which are the target applications for work stealing.

Here we present an unified programming paradigm for general parallel applications on GPU. This paradigm deals well with irregular and regular workloads on data parallel application as well as task parallel application. The paper focuses on scheduling data parallel GPU application using a novel approach.

CUDA is limited by the absence of a runtime scheduler to support dynamic load balancing. We show next how we can implement an efficient and generic support to load balancing with CUDA based on work stealing. Section 3.3 presents how to do an efficient scheduling of data parallel application. Section 4 provides experimental evaluation with different types of workloads.

3.1 Design of Our Approach

The abstraction provided by the CUDA model allows us to exploit two levels of parallelism. At first level, a program can be divided in coarse sub-problems that can be solved independently in parallel (*Thread Blocks* in CUDA), and then into finer threads that cooperate for the same task.

Independent Thread Blocks (*TB*) are mapped to multiprocessors[1] on the GPU. In practice, CUDA kernels use much more thread blocks than the number of multiprocessors available. The CUDA runtime has a very basic scheduler that assigns thread blocks to MPs. In our approach (Fig. 1), we consider a fixed number of thread blocks, which is independent from the size of the input data. Each thread block stays persistently on a multiprocessor and manages a task queue on the GPU global memory using the work stealing algorithm. When a TB does not have any more tasks to execute, it steals some from another's queue.

3.2 Work Queue Implementation

The work queue implemented consists of a work queue data structure named `workqueue_t`. This structure consists of two integers, `beg` to the beginning and `end` to the end of the interval `[beg,end)`. Additionally, each work queue is associated to a `mutex` variable used to control its access in conflicting cases.

Each Thread Block owns a work queue. This structure is accessed by the following three functions.

Push : `int push( workqueue_t* kwq, int* beg)`
The push function is called by the thread to add a new task to its own work queue. The value `beg` must be less than `kwq->beg`. This operation is always non-blocking and extends the work queue.

Pop : `int pop( workqueue_t* kwq, int* i, int* j, int size)`
The pop function is called by the thread to pop range `[*i, *j)` from its own work queue. The size of the returned range is at most `size`. The function returns a non zero value in case of success, *i.e.* iff the returned range is non empty. The pop increments the field `beg` of the work queue.

Steal : `int steal( workqueue_t* kwq, int* i, int* j, int size)`
The steal function is called by the thread to steal range `[*i, *j)` from another work queue than its own. It decrements the field `end` of the work queue. The size of the returned range is at most `size`. The function returns a non zero value in case of success, *i.e.* iff the returned range is non empty.

The work queue implementation relies on a Dijkstra's protocol and is similar to the T.H.E protocol as described in Frigo et al. [9]. The main difference is

---

[1] CUDA definition.

```
1   int *my_wq = workqueue_getown(blockIdx);
2   while (1) {
3     if ( IamTheMasterThread(threadIdx)) {
4       if( !workqueue_pop(my_wq, locbeg, locend, popSize) ) {
5         workqueue_t *victim_wq = workqueue_getrandom();
6         int stealSize = workqueue_size(victim_wq)/2;
7         if ( stealSize >= popSize ) {
8           workqueue_lock(victim_wq.mutex);
9           if (workqueue_steal(victim_wq, stealBeg, stealEnd, stealSize))
10            workqueue_push( my_wq, stealBeg, stealEnd );
11          workqueue_unlock(victim_wq.mutex);
12          continue;
13        }
14      }
15    }
16    __syncthreads();
17    TASKCall(locbeg, locend);
18    __syncthreads();
19    if ( terminate ) break;
20  }
```

Fig. 2. Cuda kernel of the work stealing scheduler loop

that pop or steal functions increment or decrement **beg** and **end** not only by 1 but by **size**. Steal function calls are serialized on the **mutex** lock: the runtime guarantees that the concurrency on a work queue structure is at most 2.

Our work queue data structure allows to steal range of indexes. For task parallelism, the runtime stores tasks into an array container: a task is identified by its index and the work queue can trivially be used to implement work stealing scheduler.

3.3   Data Parallel Application Scheduling

Moreover, our work queue can also be used to lazily create task. Let us consider the **foreach** parallel algorithm where the same functor is applied on each entries of an array. In that case, a task is then only identified by the sub range where it acts on. Stealing a task is equivalent to steal a sub range of the initial interval given by the initial **foreach** call. Our work queue implementation lets, at runtime, the scheduler to steal tasks by simply calling the **steal** function. To apply our method on more complex problems, one should define a linearization of the computations in an interval homomorphic to $[0, N)$, which is the case for almost all STL's algorithms on random access iterators Traore et al. [14].

To mix both task and data parallelisms our runtime implements the concept of the *malleable task* Turek et al. [15]. In our implementation, a *malleable task* exports a function that is called to extract work on work stealing scheduler decision. Data parallel task, such as the **foreach**, is a malleable task that represents its work using our work queue. The exported function calls the Split operation on the work queue. Therefore, after a successful steal operation, an idle TB receives a sub range of the initial range to perform.

3.4   CUDA Work Stealing Algorithm

In work stealing, each multiprocessor has its own work queue structure in global memory. Multiprocessors pop from their own queues, if there are no more tasks they randomly choose another work queue to steal from. Our work stealing scheduler is embedded within a CUDA kernel. The main scheduler loop is sketched in Fig. 2. The loop is executed by all TB on the GPU. While the work queue of a TB has enough local work (line 4), the master thread of the block pops a sub range and all threads ,synchronized at line 16, perform the data parallel task (at line 17). If there is no local work, the master thread selects at random a victim work queue (line 5), tries to steal half of its contents (line 9) and, if the steal successes, populates its own work queue with the theft range [stealBeg, stealEnd) (line 10).

4   Evaluation

The experiments were realized on a NVIDIA GTX 280 GPU running at 1.3 GHz with 1GB of global memory. This model of GPU contains 30 Multiprocessors (MP), each one with eight scalar processors (SP) giving a total of 240 cores. All the applications were tested using version 4.0 of the CUDA driver and runtime . Additionally, some experiments were also tested on a Tesla C2050 GPU (Fermi architecture) running at 1.15 GHz with 3GB of global memory.

Every measure presented in the following benchmarks is an average of 10 executions of the kernel. This number showed to be sufficient to get reliable results, with negligible standard deviation (which were omitted on our plots). The time is measured using GPU timers without counting overheads of the kernel launch nor PCI data transfers between host (CPU) and device (GPU).

4.1   Elementary Overhead

Work stealing adds an intrinsic overhead to the program due to the pop operation that is always done before starting the actual computation of the task (line 6 of Fig. 2). This benchmark used a kernel configuration of 1 block of 512 threads and varied the pop size from 2^9 to 2^{20}. Figure 3 reports the total execution time with several numbers of pop operations. Please note that the pop operation handles only work queue indexes, therefore the cost of a pop is independent of its size. By fitting the obtained curves to an affine function, the pop overhead found was 5186.72 cycles (3.52 μs) on the GTX280 and 2187.32 cycles (1.92 μs) on the Fermi GPU.

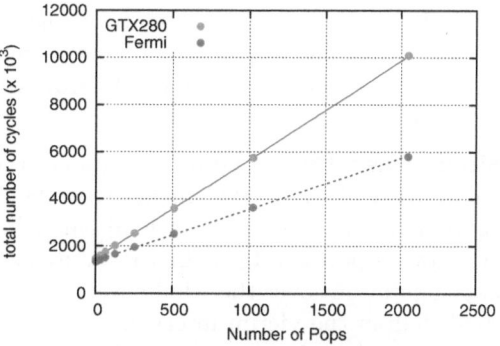

Fig. 3. Pop cost estimation on a GTX280 and a C2050(Fermi) GPU

4.2   Benchmark Application

Our target application consists in a simple transformation over an array of Floats. The program applies a function $x \longrightarrow f(x)$ to each element i of the input array and stores the result in the output array. In this benchmark we consider a task an instance of the **transform** function on a range of the initial interval.

The reference parallel implementation was taken from NVIDIA's Thrust library (Bell [3]). Next sections report experiments with regular then irregular workloads.

4.3   Load Balancing on Regular Workloads

This section compares three load balancing methods when used to manage regular workloads. Our benchmark transform application generates a regular workload when it applies a constant function to every position of the input array.

Load balancing is about managing tasks on processors. More precisely, in the experiments presented here, **the data parallel task updates at most 512 positions of the array**. This task size was chosen in conjunction with the block size, which also contains 512 threads. This way each task, except for a few ones at the end of the sub-ranges, is fully parallel and makes all the threads of the block to work. This number of threads per block showed the best performance for a transform on a sufficiently large array. Additionally, the same block size of 512 threads and 60 thread blocks is used by the static transform kernel in Thrust library, whose strategy is to optimize occupancy.

Our work stealing method is compared to two classical scheduling method:

- The *Static Scheduling* is the default load balancing method that CUDA uses when it schedules blocks on multiprocessors. Blocks are evenly distributed among the multiprocessor of the device until they reach the limit of active blocks. When an active block completes its job, the next blocks are scheduled.
- The *List Scheduling* uses a centralized work queue that every processing unit have to access to get new tasks to process. Tasks are assigned in a FIFO manner where idle processors get the task at the beginning of the list. Note that there is a lock on the work queue that serializes every access to it.

Single Task Pop. Figure 4a presents the total execution time of Transform on an array of 5120000 elements (*i.e.* 10000 tasks of 512 array positions). Each curve represents one different load balancing method. In this experiment LS and WS always pop one task (or 512 array elements) at a time which totals 10000 pops operations over the whole execution (one pop by task). WS steal operation steals half of the victim interval.

This graph makes evident the drawback of centralized load balancing methods. Even with very regular workload, the List Scheduling method quickly reaches a limit where it stops scaling. Actually, with more than 10 MPs the performance always gets worse. We attribute this behavior to a lock contention problem. As

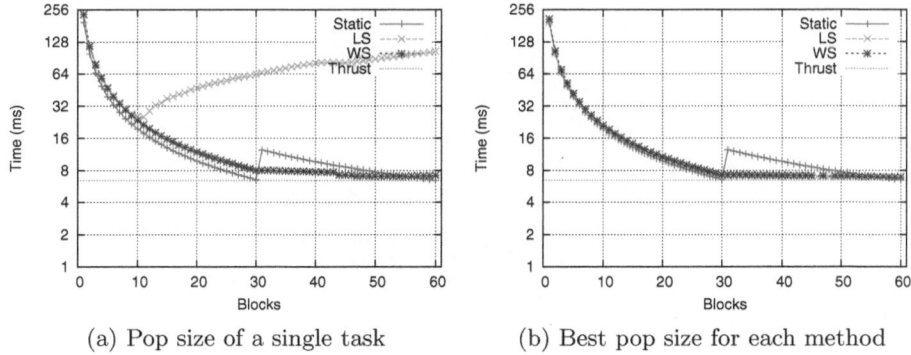

(a) Pop size of a single task   (b) Best pop size for each method

Fig. 4. Comparison of the Static, the List Scheduling (LS) and the Work Stealing (WS) methods

the time of execution of a single task is very short, blocks are often trying to acquire the lock which increases lock contention.

The static method shows the best absolute performance at 60 blocks, the maximum number of blocks that the GPU scheduler runs concurrently on the hardware (tests with more blocks didn't enhance the execution time). However, note that the static method does not scale in a regular way as does work stealing. This can be seen on the performance drop at 31 blocks. This drop is due to the fact that with 31 blocks, one multiprocessor has two active blocks and twice more tasks which causes the load imbalance. The work stealing (WS) does not suffer of this problem because an idle TB on a multiprocessor can steal task from overloaded multiprocessors.

The overhead of accessing the work queue can be reduced by popping more tasks at a time. Figure 4b shows the best performances found for each method when tuning the number of tasks retrieved by pop (the Pop Size). For work stealing, the optimal pop has a size of 3 tasks (i.e 1536 elements of the array) and the best time, 6.90ms, is achieved with 60 blocks of 512 threads. List scheduling achieves 6.92ms of peak performance when the pop size is equal to 7 tasks (i.e 7168 array elements) with 30 blocks of 512 threads. The best static time is 6.49ms at 60 block of 512 threads.

Pop Size Variation. Figure 5 shows how WS and LS behave with the variation of two parameters: pop size and number of blocks. The y axis represents the size of each pop in number of array elements (number of tasks x 512). The values plotted correspond to the difference of execution time between LS (List Scheduling) and WS (Work Stealing). Lighter tonalities means smaller differences.

We can identify two regions A and B. In region A, LS performs better than WS but the biggest difference is only 5.63 ms (37,75% speedup over WS). In region B, WS outperforms LS achieving a gain of 98.53 ms (93,03% of speedup over LS). Therefore, even if LS is simpler to design than WS, it suffers from its

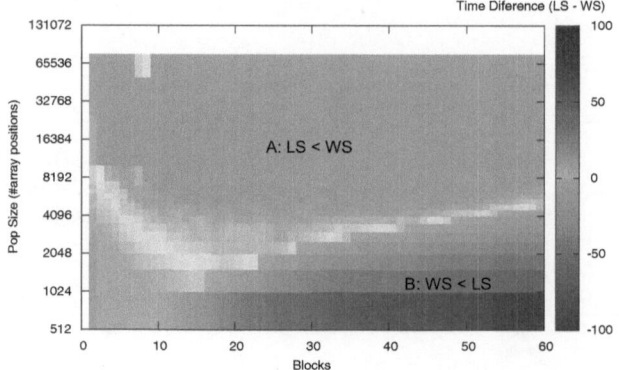

Fig. 5. Time difference between LS and WS

grain size selection where small value may degrade strongly the performance due to contention and where big values limit the parallelism (with LS, once popped a sub-range cannot be stolen anymore).

4.4   Load Balancing on Irregular Workloads

This section evaluates the load balancing methods with two different patterns of workloads. Like on Chen et al. [7], patterns of irregularity were created by nullifying the work of some tasks of the input array. These patterns are:

1. Pattern 1 = 0 1 0 1 0 1 0 1 : one task each other is nullified (50% of workload reduction).
2. Pattern 2 = 0 0 0 1 0 0 0 1 : one on each three task is nullified (75% of workload reduction).

Figure 6 shows the best results of each method of load balancing for the two patterns of irregularity. These tests represent the best configuration of pop size found for each method. LS uses a pop size of 10240 (20 tasks) with pattern 1 and 25600 (50 tasks) for pattern 2. WS uses pop sizes of 4096 (8 tasks) for both patterns. These results clearly show the instability of the static method for irregular workloads and how good dynamic scheduling approaches deal with it.

4.5   Octree Partitioning

The transform benchmark shows how to use our work stealing model to schedule array-based applications. However, this same model can be used for classical task-based problems. For instance, the octree partitioning problem create tasks to recursively separate particles in a 3D space into octants. We used the octree implementation provided by Cederman and Tsigas [5] and adapted it to use our scheduler. We then compared their load balancing method to ours.

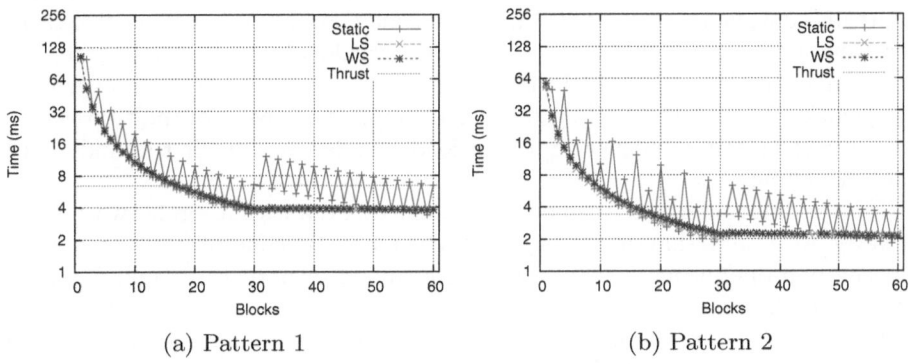

(a) Pattern 1   (b) Pattern 2

Fig. 6. Transform problem on irregular workloads

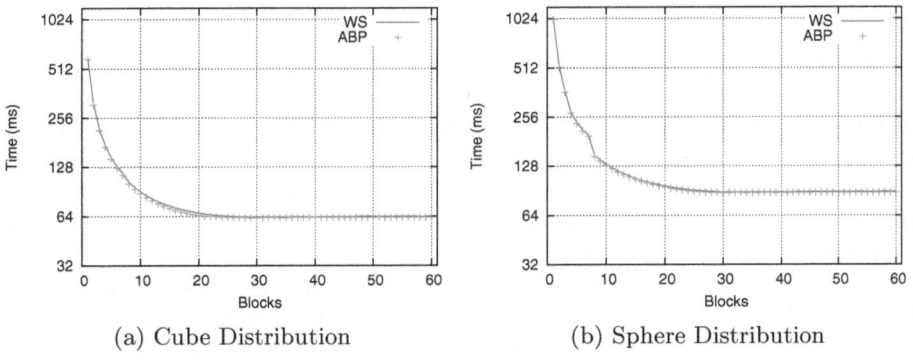

(a) Cube Distribution   (b) Sphere Distribution

Fig. 7. Load Balancing methods on Octree partitioning problem

The following benchmarks consists in creating an octree partitioning of a 3D space containing 500000 particles. The algorithm recursively subdivides the space until the threshold of 20 particles per subspace is reached. Every time a sub-space needs to be split, a new task is created.

Figure 7 shows a comparison between our algorithm (labeled WS) and Cederman et Tsigas (labeled ABP) with two different particles distribution. One where the particles are all randomly picked from a cubic space and other where they are randomly picked from the surface of a sphere. As shown in Fig. 7, WS and ABP presented similar performance. The best times found were $63.26ms$ with 31 blocks for WS and 62.94 ms for ABP with 35 blocks.

4.6   Discussion

The static scheduling showed to perform quite well for the regular array transform. This is mainly due to the cyclic algorithm used to assign array elements (tasks) to threads. The cyclic distribution makes a good division of the work because it gives to all the blocks the same amount of tasks and at the same time it spreads contiguous parts of the array that may contain more expensive tasks. However, this model is vulnerable to specific workload models (see section 4.4). Additionally, it should be noted that multiprocessors on a single GPU execute at same clock. On a multi-GPU system for example, the speed of multiprocessors may variate creating another source of load imbalance difficult to handle with static scheduling.

List scheduling can achieve good performances, even with irregular work loads. However we found that it is very dependent on the computation time spent between pop operations. Thus an accurate tuning of pop size is mandatory to get good performance.

Work stealing was the method that showed the best adaptability over all of the presented benchmarks. Even if it didn't have the best absolute performance, the difference from the other methods was very small. This method is also less sensible to lock contention than List Scheduling in which pop size has to be carefully tuned to overcome contention.

5   Conclusion and Future Work

In this work we considered a new programming model for general purpose GPUs based on work stealing. This model allows the programmer to express the parallelism of a GPGPU application in a hybrid manner taking benefit, at the same time, from an efficient task scheduling algorithm and from the highly SIMD computation power of graphics processors.

We presented empirical results that attest the effectiveness of our model and, to the extent of our knowledge this is the first work to evaluate a regular problem with dynamic load balancing on GPU. Our results confirm that work stealing is a generic scheduling method and performs well in both regular and irregular problems. We compared our scheduler on regular array transform micro-benchmark with respect to the static implementation of the Thrust well-known GPU library and found little degradation with uniform load, and better performances on unbalanced load.

Ongoing work is to explore in more details how this model behaves on the new Fermi GPU architecture and what optimizations can take favor of it. Preliminary tests (section 3), suggest that new hardware capabilities notably, the presence of a full cache memory hierarchy, could be better exploited by our work queue implementation. At long-term, we envision the integration of this model in the KAAPI library (Gautier et al. [10]) which lacks the ability of scheduling inside GPUs.

References

[1] Angel, M., Michael, M.M., Bivens, J.A.: Dynamic Work Scheduling for GPU Systems. Memory, 57–64 (2010)
[2] Arora, N.S., Blumofe, R.D., Plaxton, C.G.: Thread Scheduling for Multiprogrammed Multiprocessors. Theory of Computing Systems 34(2), 115–144 (2001), http://www.springerlink.com/openurl.asp?genre=rticle &id=doi:10.1007/s002240011004

[3] Bell, N.: Thrust: A Productivity-Oriented Library for CUDA. sbel.wisc.edu. 359–373 (2012),
http://sbel.wisc.edu/Courses/ME964/Literature/thrustGPUgems2011.pdf

[4] Blumofe, D.R., Leiserson, E.C.: Scheduling multithreaded computations by work stealing. Journal of the ACM 46(5), 720–748 (1999),
http://portal.acm.org/citation.cfm?doid=324133.324234

[5] Cederman, D., Tsigas, P.: On dynamic load balancing on graphics processors. In: Proceedings of the 23rd ACM SIGGRAPH/EUROGRAPHICS Symposium on Graphics Hardware, pp. 57–64. Eurographics Association (2008)

[6] Chase, D., Lev, Y.: Dynamic circular work-stealing deque. In: Proceedings of the 17th Annual ACM Symposium on Parallelism in Algorithms and Architectures - SPAA 2005 (c), vol. 21 (2005),
http://portal.acm.org/citation.cfm?doid=1073970.1073974

[7] Chen, L., Villa, O., Krishnamoorthy, S., Gao, G.R.: Dynamic load balancing on single- and multi-GPU systems. In: 2010 IEEE International Symposium on Parallel & Distributed Processing (IPDPS), pp. 1–12 (2010),
http://ieeexplore.ieee.org/lpdocs/epic03/wrapper.htm?arnumber=5470413

[8] Dijkstra, E.W.: Solution of a problem in concurrent programming control. Commun. ACM 8, 569 (1965), http://doi.acm.org/10.1145/365559.365617

[9] Frigo, M., Leiserson, C.E., Randall, K.H.: The implementation of the Cilk-5 multithreaded language. In: Proceedings of the ACM SIGPLAN 1998 Conference on Programming Language Design and Implementation - PLDI 1998, pp. 212–223 (1998), http://portal.acm.org/citation.cfm?doid=277650.277725

[10] Gautier, T., Besseron, X., Pigeon, L.: Kaapi: A thread scheduling runtime system for data flow computations on cluster of multi-processors. In: Proceedings of the 2007 International Workshop on Parallel Symbolic Computation, pp. 15–23. ACM (2007), http://portal.acm.org/citation.cfm?id=1278182

[11] Hendler, D., Lev, Y., Moir, M., Shavit, N.: A dynamic-sized nonblocking work stealing deque. Distributed Computing 18(3), 189–207 (2005),
http://www.springerlink.com/index/10.1007/s00446-005-0144-5

[12] Hendler, D., Shavit, N.: Non-blocking steal-half work queues. In: Proceedings of the Twenty-First Annual Symposium on Principles of Distributed Computing - PODC 2002, p. 280 (2002),
http://portal.acm.org/citation.cfm?doid=571825.571876

[13] Pheatt, C.: Intel threading building blocks. J. Comput. Sci. Coll. 23, 298–298 (2008),
http://portal.acm.org/citation.cfm?id=1352079.1352134

[14] Traoré, D., Roch, J.-L., Maillard, N., Gautier, T., Bernard, J.: Deque-Free Work-Optimal Parallel STL Algorithms. In: Luque, E., Margalef, T., Benítez, D. (eds.) Euro-Par 2008. LNCS, vol. 5168, pp. 887–897. Springer, Heidelberg (2008)

[15] Turek, J., Wolf, J.L., Yu, P.S.: Approximate algorithms scheduling parallelizable tasks. In: Proceedings of the Fourth Annual ACM Symposium on Parallel Algorithms and Architectures, SPAA 1992, pp. 323–332. ACM, New York (1992)

[16] Tzeng, S., Patney, A., Owens, J.D.: Task management for irregular-parallel workloads on the GPU. In: Proceedings of the Conference on High Performance Graphics, pp. 29–37. Eurographics Association (2010),
http://portal.acm.org/citation.cfm?id=1921485

GPU-Accelerated Asynchronous Error Correction for Mixed Precision Iterative Refinement

Hartwig Anzt[1], Piotr Luszczek[2], Jack Dongarra[2,3,4], and Vincent Heuveline[1]

[1] Karlsruhe Institute of Technology, Karlsruhe, Germany
[2] University of Tennessee, Knoxville, USA
[3] Oak Ridge National Laboratory, Oak Ridge, USA
[4] University of Manchester, Manchester, UK
{hartwig.anzt,vincent.heuveline}@kit.edu,
{luszczek,dongarra}@eecs.utk.edu

Abstract. In hardware-aware high performance computing, block-asynchronous iteration and mixed precision iterative refinement are two techniques that may be used to leverage the computing power of SIMD accelerators like GPUs in the iterative solution of linear equation systems. Although they use a very different approach for this purpose, they share the basic idea of compensating the convergence properties of an inferior numerical algorithm by a more efficient usage of the provided computing power. In this paper, we analyze the potential of combining both techniques. Therefore, we derive a mixed precision iterative refinement algorithm using a block-asynchronous iteration as an error correction solver, and compare its performance with a pure implementation of a block-asynchronous iteration and an iterative refinement method using double precision for the error correction solver. For matrices from the University of Florida Matrix collection, we report the convergence behaviour and provide the total solver runtime using different GPU architectures.

Keywords: mixed precision iterative refinement, block-asynchronous iteration, GPU, linear system, relaxation.

1   Introduction

Classical relaxation methods such as Gauss-Seidel and Jacobi usually require data transfer between each iteration which constitutes a synchronization point. This implies a severe restriction for parallel implementations. Block-asynchronous iteration removes this synchronization barrier, updating components using the latest available values. It allows a large freedom in the update order and the number of updates per component, while every component update uses the latest available values for the other components. In the end, the obtained algorithm is neither deterministic nor does it imply convergence for all systems that can be solved by the classical Jacobi approach, in fact it requires the linear equation

C. Kaklamanis et al. (Eds.): Euro-Par 2012, LNCS 7484, pp. 908–919, 2012.

system to fulfill additional conditions. While, due to the poor convergence rate, they may seem to be very unattractive from the mathematical point of view, the block-asynchronous iteration is, in contrast to most other iterative methods, usually able to exploit the high computational power of modern hardware platforms that are often accelerated by GPUs.

Another well-known technique that may be used to leverage the potential of accelerators is mixed precision iterative refinement. The basic idea is to use a lower precision format for the error correction solver inside an iterative refinement method at full precision. Without impacting the accuracy of the final solution approximation, the acceleration of the solving process is possible since the computations in the less complex floating point format can be conducted faster on the respective device. While the time for computations in the usually implemented single and double precision formats differs by a factor of two for most devices, additional acceleration may be possible since using single precision reduces the pressure on the memory bandwidth, that is often crucial in scientific computing on hybrid hardware.

An open question is how a combination of these two techniques impacts the convergence and properties and the performance. On the one hand the methods are similar: they both compensate their low complexity by leveraging the high computational power of GPUs. But on the other hand, they are contradictory since the iterative refinement artificially introduces synchronization points that we try to avoid in asynchronous iteration methods. For the latter ones, the most suitable applications are linear systems with condition numbers which require high iteration counts of the error correction solver. The mixed precision approach may suffer from these, since the error correction is impacted by them.

The paper is organized as follows. First, we provide some mathematical background by outlining the algorithms for iterative refinement and the mixed precision variant, and block-asynchronous iteration. We then introduce the hardware platforms used for the experiments and give details about the linear equation systems we target. Additionally, we outline the GPU implementation we use for the tests. In the numerical experiment section, we compare the convergence behaviour of iterative refinement using a double-precision and a single-precision error correction solver. We report the total solver runtimes and compare it with a plain implementation of the block-asynchronous iteration for various GPUs. In the last section we conclude and provide ideas for further optimization.

2   Mathematical Background

Block-Asynchronous Iteration. A possible motivation for asynchronous iteration is modern hardware, which provides a large number of cores that achieve excellent performance when running in parallel, but suffer when synchronizing or exchanging data. Therefore, algorithms that lack any synchronization would achieve outstanding performance on these devices, while most of the numerical algorithms are poorly parallel and require regular data exchange. For computing the next iteration in relaxation methods, one usually requires the latest values of all components. For some algorithms, e.g., Gauss-Seidel [16], even the already computed values of the current iteration step are used. This requires a strict order of the component updates, limiting the parallelization potential to a stage, where a component cannot be updated several times before all the other components are updated.

If this order is not adhered to, i.e., the individual components are updated independently and without consideration of the current state of the other components, the resulting algorithm is called a chaotic or asynchronous iteration method. In the past, the convergence behaviour and performance of these methods were analyzed in several papers [13,12,4,10]. Due to the superior convergence properties of synchronized iteration methods, they came out of the main focus of high performance computing, while research was put on investigating the convergence properties [20,5]. Today, due to the complexity of heterogeneous hardware platforms and the large number of computing units in parallel devices like GPUs, these schemes may become interesting again for applications like multigrid methods, where highly parallel smoothers are required on the distinct grid levels [9]. While traditional relaxation methods like the sequential Gauss-Seidel obtain their efficiency from their fast convergence, an asynchronous iteration scheme may compensate for its inferior convergence behavior by superior scalability [3]. We proposed [2] a block-asynchronous iteration, that, in addition to the global iterations, iterates on the subdomains determined by the iteration components that are handled by the same stream in the GPU implementation. The motivation for this is due to the design of graphics processing units and the CUDA programming language. As the subdomains are relatively small and the data needed largely fits into the multiprocessor's cache, these additional iterations on the subdomains come for almost free. During these local iterations, the iteration values used from outside the block are kept constant, equal to their values at the beginning of the global iteration. After the local iterations, the updated values are communicated. This approach is inspired by the well known hybrid relaxation schemes [9,8]. The obtained algorithm, visualized in Figure 1, can be written as a component-wise update of the solution approximation to form $x_k^{(m+1)}$:

$$
x_k^{(m)} + \frac{b_k}{a_{kk}} - \underbrace{\sum_{j=1}^{T_S} \frac{a_{kj}x_j^{(m-\nu(m+1,j))}}{a_{kk}}}_{\text{global part}} - \underbrace{\sum_{j=T_S+1}^{T_E} \frac{a_{kj}x_j^{(m)}}{a_{kk}}}_{\text{local part}} - \underbrace{\sum_{j=T_E+1}^{n} \frac{a_{kj}x_j^{(m-\nu(m+1,j))}}{a_{kk}}}_{\text{global part}},
$$

$$(1)$$

where T_S and T_E denote the starting and the ending indexes of the matrix/vector part in the thread block. Furthermore, for the local components, the antecedent values are always used, while for the global part, the values from the beginning of the iteration are used. The shift function $\nu(m+1,j)$ denotes the iteration shift for the component j – this can be positive or negative, depending on whether the respective other thread block has already conducted more or less iterations. Note that this gives a block Gauss-Seidel flavor to the updates. It should also be mentioned that the shift function may not be the same in different thread

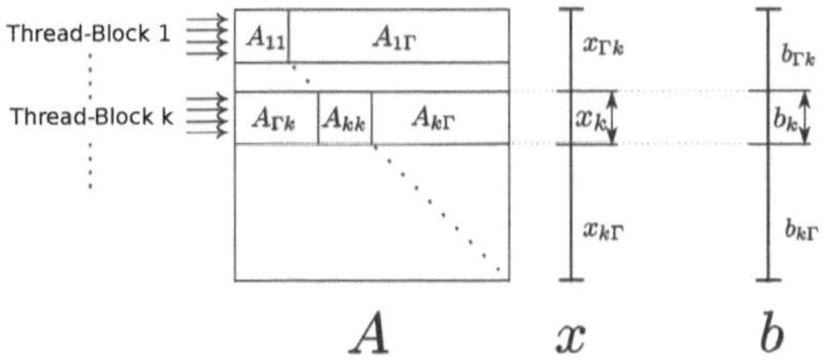

$$x_k + = D_{kk}^{-1} \left(b_k - A_{\Gamma k} x_{\Gamma k} - A_{kk} x_k - A_{k\Gamma} x_{k\Gamma} \right)$$

Fig. 1. Visualizing the asynchronous iteration in block description used for the GPU implementation. A_{kk} denotes the k-th diagonal block, $A_{\Gamma k}$ and $A_{k\Gamma}$ the block left, respectively right, of A_{kk}. Consistent notation is used for the block decomposition of the vectors.

blocks. While the GPU hardware encourages this approach, the idea is similar to a two-staged asynchronous iteration [7].

Mixed Precision Iterative Refinement.

Although iterative refinement methods have been known for long time, they have enjoyed a revival with the rise of computer systems in the middle of the last century. The core idea is to use the residual of a computed solution as the right-hand side to solve a correction equation. The algorithm then updates the solution approximation in every iteration by adding an error correction term computed by an error correction solver. Note that this error correction solver can be chosen independent: direct solvers as well as another iterative method are possible options. This implies the possibility of cascading iterative refinement methods. Denoting the solution update with $c^{(i)} := A^{-1} r^{(i)}$, the algorithm reads:

1: initial guess as starting vector: $x^{(0)}$
2: compute initial residual: $r^{(0)} = b - Ax^{(0)}$
3: **while** $(\| Ax^{(i)} - b \|_2 > \varepsilon \| r^{(0)} \|_2)$ **do**
4:     $r^{(i)} = b - Ax^{(i)}$
5:     solve: $Ac^{(i)} = r^{(i)}$
6:     update solution: $x^{(i+1)} = x^{(i)} + c^{(i)}$
7: **end while**

Algorithm 1. Error Correction Method

The underlying idea of mixed precision error correction methods is to use different precision formats within the algorithm of the error correction method, updating the solution approximation in high precision, but computing the error correction term in lower precision which has been suggested before [15,14,6,11].

Hence, one regards the inner correction solver as a black box, computing a solution update in lower precision. The term high precision refers to the precision

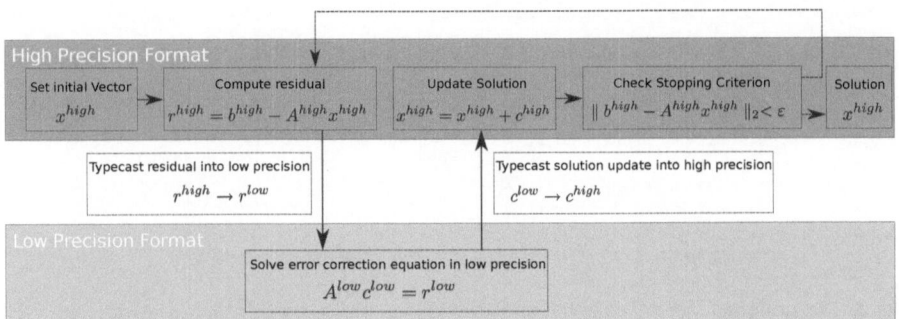

Fig. 2. Visualizing the mixed precision approach to an iterative refinement method

format that is necessary to display the accuracy of the final solution, and we can obtain the following algorithm where x^{high} denotes the high precision value and x^{low} denotes the value in low precision for the variable x. The conversion between the formats will be left abstract throughout this paper. Because the conversion of the matrix A is especially expensive, it should be stored in both precision formats, high and low precision. This leads to the drawback of a higher memory need.

Using the displayed algorithm (Figure 2), we obtain a mixed precision solver. If the final accuracy does not exceed the smallest number ε_{low} that can be represented in the lower precision, it may generate the same approximation quality as if the solver was performed in the high precision format. It should be mentioned, that the solution update of the error correction solver is usually not optimal for the outer system, since the representation of the problem in the lower precision format contains rounding errors, and it therefore solves a perturbed problem. When comparing the algorithm of an error correction solver to a plain solver, it is obvious, that the error correction method has more computations to execute. Each outer loop consists of the computation of the residual error term, a typecast, a vector update, the scaling process, the inner solver for the correction term, the reconversion of the data and the solution update. The computation of the residual error itself consists of a matrix-vector multiplication, a vector addition and a scalar product. The mixed precision refinement approach to a certain solver is superior to the plain solver in high precision, if the additional computations and typecasts are overcompensated by the cheaper inner correction solver using a lower precision format [1,11].

3   Experiment Setup

Linear Equation Systems. In our experiments, we search for the approximate solutions of linear systems of equations, where the respective matrices are taken from the University of Florida Matrix Collection (UFMC; see http://www.cise.ufl.edu/research/sparse/matrices/).

Table 1. Dimension and characteristics of the SPD test matrices and the corresponding iteration matrices where $\#n$ denotes the dimension and $\#nnz$ the number of nonzeros, respectively

Matrix name	$\#n$	$\#nnz$	cond(A)	cond($D^{-1}A$)	$\rho(M)$
CHEM97ZTZ	2,541	7,361	1.3e+03	7.2e+03	0.7889
FV1	9,604	85,264	9.3e+04	12.76	0.8541
FV3	9,801	87,025	3.6e+07	4.4e+03	0.9993
TREFETHEN_2000	2,000	41,906	5.1e+04	6.1579	0.8601

Due to the convergence properties of the iterative methods we analyze, the experiment matrices have to be chosen properly, fulfilling the necessary and sufficient convergence condition [12].

The matrix properties and sparsity plots are in Table 1. and Figure 3.

The first matrix, CHEM97ZTZ, comes from statistics[1]. Matrices FV1 and FV3 are finite element discretizations of the Laplace equation on a 2D mesh. Therefore, they share a common sparsity structure, but differ in dimension and condition number. The matrix TREFETHEN_2000 [21] is a 2000×2000 matrix where all entries are zero except for the ones at the positions (i,j) where $|i-j| = 2, 4, 8, 16 \ldots$. Furthermore, the main diagonal is filled with the primes $2, 3, 5, 7, 11 \ldots 17389$. Hence, this matrix has many off-diagonal entries distributed over the diagonals that are by a power of 2 distant to the main diagonal.

Implementation Issues. The GPU implementations of the block-asynchronous iteration is based on CUDA [19], while the respective libraries used are from CUDA 2.3 for the C1060 and the GTX280, and CUDA 4.0.17 [18] for the C2070 and GTX580 implementation. The kernels updating the respective components, launched through different streams, use thread blocks of size 512. The thread block size, the number of streams, along with other parameters, were determined through empirically based tuning. For the iterative refinement implementation

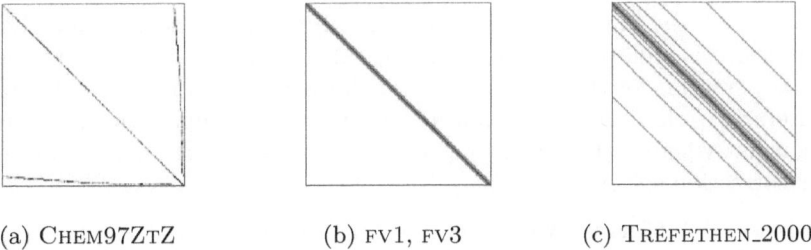

(a) CHEM97ZTZ   (b) FV1, FV3   (c) TREFETHEN_2000

Fig. 3. Sparsity plots of test matrices

---

[1] For more details see
http://www.cise.ufl.edu/research/sparse/mat/Bates/README.txt

Table 2. Key system characteristics of the four GPUs used. Computation rate and memory bandwidth are theoretical peak values [17].

Name	GTX280a	GTX580	Tesla C1060	Tesla C2070
Chip	GT200	GF110	T10	T20
Transistors	$1.4 \cdot 10^9$	$3 \cdot 10^9$	$1.4 \cdot 10^9$	$3 \cdot 10^9$
Core frequency	1.3 GHz	1.5 GHz	1.15 GHz	1.3 GHz
Thread Processors	240	512	240	448
GFLOPS (single)	933	1580	933	1030
GFLOPS (double)	78	790	78	515
Shared Memory/L1	16 KB	64 KB	16 KB	64 KB
L2 Cache	-	768 KB	-	768 KB
Memory	1 GB GDDR3	1.5 GB GDDR5	4 GB GDDR3	6 GB GDDR5
Memory Frequency	1.1GHz	2.0 GHz	0.8 GHz	1.5 GHz
Memory Bandwidth	141.0 GB/s	192.4 GB/s	102.0 GB/s	144.0 GB/s
ECC Memory	no	yes	no	yes
Power Consumption	236 W	244 W	200 W	190
IEEE double/single	yes/partial	yes/yes	yes/partial	yes/yes

we use a first outer iteration to analyze the residual improvement and then adapt the number of inner iterations such that the residual improvement equals the accuracy of floating point precision in every outer update. Hence, while the first error correction loop may provide different improvement for the distinct test cases, the further loops all decrease the residual by 6 to 8 digits.

In case of the mixed precision implementations, the error correction solver is implemented using single precision. Hence, due to the low precision representation of the linear equation system, additional rounding errors may be expected, slowing down the convergence of the iterative refinement. To analyze this issue, we compare in a first experiment the convergence behaviour of the iterative refinement method using a double- and a single- precision error correction solver, respectively. Using different precision formats, the vectors and the linear system have to be converted from double to single precision. This typecast, handled by the GPU, triggers some overhead and may be crucial for problems where only very few iterations of the error correction solver are executed.

To analyze the impact of the overhead of iterative refinement and the use of different precision formats we provide the solver runtimes for the different linear equation systems for the plain block-asynchronous iteration in double precision, the iterative refinement in double precision and the mixed precision iterative refinement, whereas the latter ones use the block-asynchronous iteration as an error correction solver.

Hardware Platforms. We target four GPU architectures located at the Engineering Mathematics and Computing Lab (EMCL)[2] at the Karlsruhe Institute of Technology, Germany, to analyze the potential of mixed precision block-asynchronous iteration. They are taken from the Fermi and the Tesla line of

---

[2] Supported by NVIDIA as Cuda Research Center.

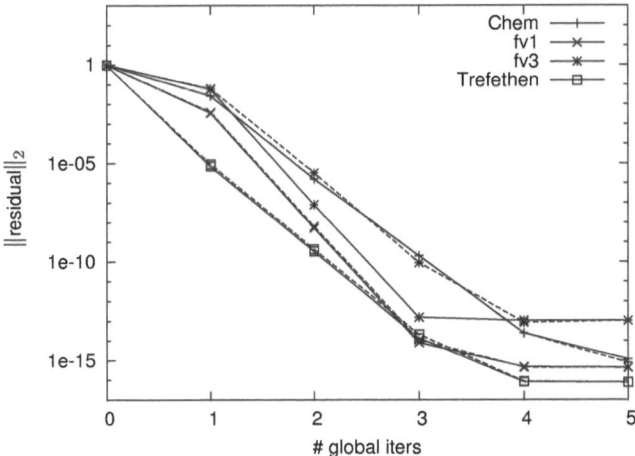

Fig. 4. Iterative refinement convergence, solid lines are double-precision error correction, dashed lines are single-precision error correction

Nvidia. The C2070 and the C1060 are the server versions of the line, the GTX580 and the GTX280 are the consumer version, respectively. While the chip and onboard memory specifications are given in table 2, the host system may have minor influence on the performance, since all computations are exclusively handled by the graphics. Note that the price for the larger (ECC protected) memory in the server versions is a lower memory bandwidth.

4   Numerical Experiments

In the first experiment, we analyze how using lower precision for the block-asynchronous iteration error correction solver impacts the iterative refinement convergence rate. Therefore, we report the relative residual depending on the iteration number for the different linear equation systems introduced in Section 3. Note that due to the implementation, the first outer loop is used to determine the residual improvement, while the further iterations improve the approximation iterate by 6 to 8 digits, depending on the rounding error.

The results reported in Figure 4 show that for the test matrices CHEM97ZTZ, FV1, and TREFETHEN_2000, using single precision for the error correction solver has a nearly negligible impact on the convergence of the iterative refinement. Only for the FV3 test case, does the convergence rate suffer. This was expected since the high condition number triggers representation errors in the low precision format that make the approximation updates less beneficial.

But while the convergence behaviour is interesting from the theoretical point of view, the next experiment is dedicated to analyzing how handling the error correction equation in single precision impacts the performance. The motivation is that using single instead of double as working precision, triggers at least a

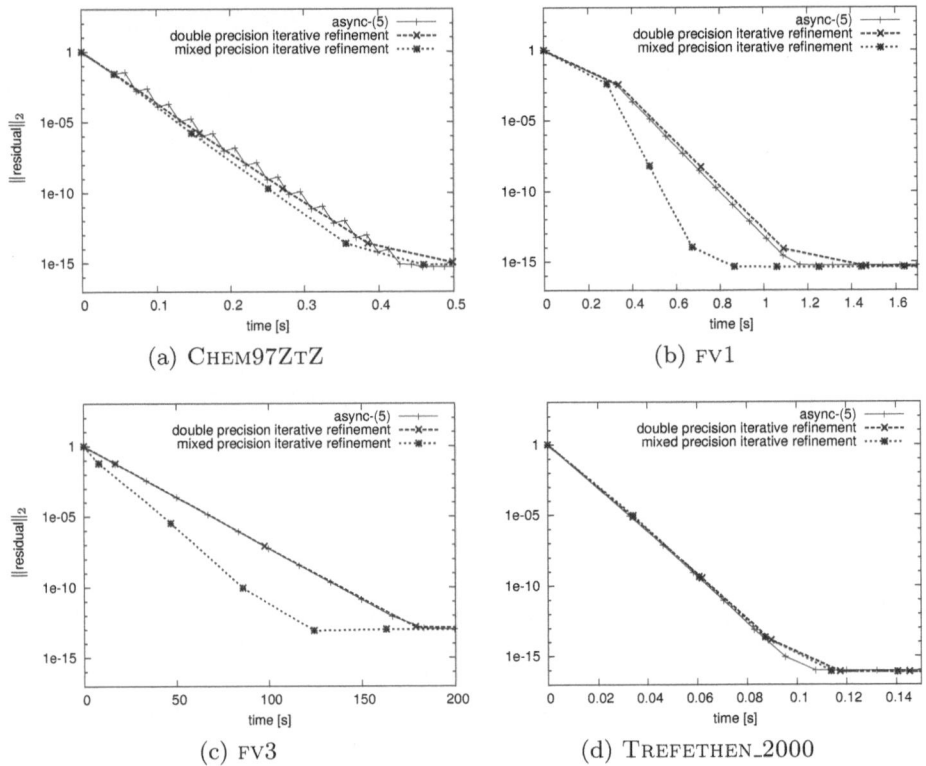

Fig. 5. Iterative refinement performance, time-dependent relative residual

speedup of two, and may potentially overcompensate for the overhead associated with the typecast between the formats.

While the convergence, with respect to iteration number, is independent of the hardware used, the performance depends on the architecture. We use the C2070 for this experiment, as this 'Fermi' generation is the state of the art from in scientific computing the Nvidia GPU manufacturer. In addition to the convergence performance of the iterative refinement, using a double or single precision error correction solver, we report the results for the plain block-asynchronous iteration in double precision. We observe in Figure 5, that for all test cases, the overhead is negligible when embedding the block-asynchronous iteration in double precision into the iterative refinement framework. For the small test cases CHEM97ZTZ and TREFETHEN_2000, switching to the mixed precision iterative refinement approach gives no improvement. For the larger matrices e.g. FV1, the improvement by using low precision for the error correction solver is relevant. The mixed precision implementation converges almost twice as fast. Even for the test case FV3, where we observed a slower convergence rate for the mixed precision approach in Figure 4, we benefit in terms of performance.

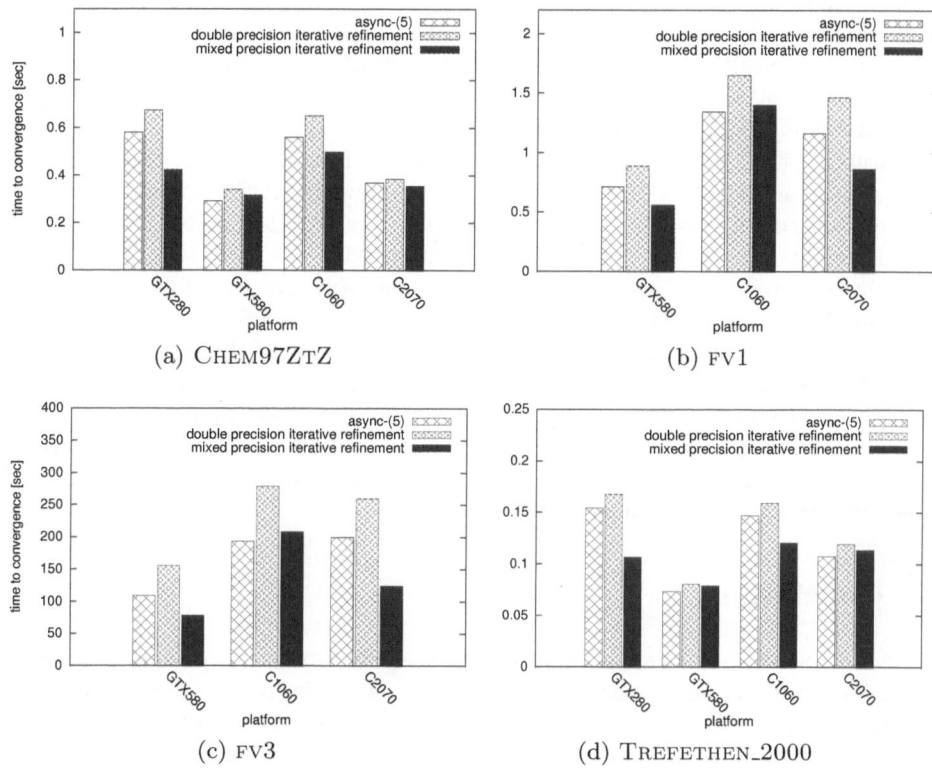

(a) CHEM97ZTZ   (b) FV1

(c) FV3   (d) TREFETHEN_2000

Fig. 6. Total solver runtime

Targeting different hardware architectures, we report in Figure 6 the respective time-to solution. For the test cases FV1 and FV3, despite the performance difference between single and double precision of around 10 on the Tesla line, the mixed precision iterative refinement performs inferior to the plain double implementation of async-(5). The reason is, that for these systems, the memory bandwidth is the limiting factor and the overhead, due to the iterative refinement framework, can not be compensated for by the single precision performance. For the small matrices, things are different. Since the size of CHEM97ZTZ and TREFETHEN_2000 allows for the caching of the iteration vector as well as the right-hand side, the C1060 and the GTX280 are able to leverage the single precision performance more efficiently. Still, the bandwidth remains the limiting factor, since the complete matrix cannot be loaded into cache, and the higher memory bandwidth of the consumer version explains the better performance for the mixed precision approach. Using double precision, the server version is superior, probably due to the more sophisticated memory structure. Unfortunately, the very limited main memory on the GTX280 does not allow for the handling of the large systems.

Note that the total solver runtime for TREFETHEN_2000 is on the GTX280 even smaller than on the server version of the Fermi line. An explanation may be that the overall runtime also includes the initialization process, which has to be taken into account for this system, and is longer for systems using CUDA in version 4.0 and equipped with more memory.

Targeting the Fermi generation, we observe that, especially for the large systems, we benefit from the mixed precision framework. Although we may only expect a factor of two concerning the floating point performance, the sophisticated memory hierarchy enables even higher speedups.

This speedup stems from the fact that, not only are we able to keep the iteration vector and the right-hand side local due to the larger L1 cache, but also because the L2 cache allows for the efficient data access of the iteration matrix. Note that for the test case FV3, the iterative refinement in double precision fulfills the critical stopping criterion after 4 iterations, while we could observe in Figure 4 that it is already very close after 3 iterations. Hence, the double precision iterative refinement runtime would benefit from choosing a smaller number of inner iterations for the last global iteration.

5   Conclusions

We were able to show that embedding block-asynchronous iteration into a mixed precision iterative refinement method not only retains its convergence properties, but may even be beneficial with respect to the runtime performance. Depending on the GPU architecture, we were able to achieve a performance increase of up to a factor of two for linear equation systems taken from the University of Florida Matrix Collection. The trade-off between the synchronization points introduced by iterative refinement and the desired asynchronism is not necessarily crucial, and for problems fulfilling the constraints, given by an upper and lower bound for the condition number of the linear system, the performance increase may be considerable. Concerning the hardware, the potential lies within systems that have large differences in the double–single precision performance, and a sophisticated memory hierarchy enabling them to transfer this performance factor to speedups of the asynchronous iteration solver.

While our analysis focused on the typically used single- and double precision formats, especially when targeting artificially created extended formats, the mixed precision iterative refinement approach is inevitable.

Aside from this, further research should focus on determining a priori, whether embedding the block-asynchronous iteration into the mixed precision iterative refinement framework is beneficial for a given problem. This may depend not only on the problem characteristics, i.e. the condition number, but also on the hardware platform used, potentially accelerated by multiple, even different GPUs.

References

1. Anzt, H., Heuveline, V., Rocker, B.: An Error Correction Solver for Linear Systems: Evaluation of Mixed Precision Implementations. In: Palma, J.M.L.M., Daydé, M., Marques, O., Lopes, J.C. (eds.) VECPAR 2010. LNCS, vol. 6449, pp. 58–70. Springer, Heidelberg (2011)
2. Anzt, H., Tomov, S., Dongarra, J., Heuveline, V.: A Block-Asynchronous Relaxation Method for Graphics Processing Units. Technical report, Innovative Computing Laboratory, University of Tennessee, UT-CS-11-687 (2011)
3. Anzt, H., Tomov, S., Gates, M., Dongarra, J., Heuveline, V.: Block-asynchronous Multigrid Smoothers for GPU-accelerated Systems. Technical report, Innovative Computing Laboratory, University of Tennessee, UT-CS-11-689 (2011)
4. Aydin, U., Dubois, M.: Generalized asynchronous iterations, pp. 272–278 (1986)
5. Aydin, U., Dubois, M.: Sufficient conditions for the convergence of asynchronous iterations. Parallel Computing 10(1), 83–92 (1989)
6. Baboulin, M., Buttari, A., Dongarra, J.J., Langou, J., Langou, J., Luszczek, P., Kurzak, J., Tomov, S.: Accelerating scientific computations with mixed precision algorithms. Computer Physics Communications 180(12), 2526–2533 (2009)
7. Bai, Z.-Z., Migallón, V., Penadés, J., Szyld, D.B.: Block and asynchronous two-stage methods for mildly nonlinear systems. Num. Math. 82, 1–20 (1999)
8. Baker, A.H., Falgout, R.D., Gamblin, T., Kolev, T.V., Martin, S., Meier Yang, U.: Scaling algebraic multigrid solvers: On the road to exascale. In: Proceedings of Competence in High Performance Computing CiHPC (2010)
9. Baker, A.H., Falgout, R.D., Kolev, T.V., Meier Yang, U.: Multigrid smoothers for ultra-parallel computing, LLNL-JRNL-435315 (2011)
10. Bertsekas, D.P., Eckstein, J.: Distributed asynchronous relaxation methods for linear network flow problems. In: Proceedings of IFAC 1987 (1986)
11. Buttari, A., Dongarra, J.J., Langou, J., Langou, J., Luszczek, P., Kurzak, J.: Mixed precision iterative refinement techniques for the solution of dense linear systems. Int. J. of High Perf. Comp. & Appl. 21(4), 457–486 (2007)
12. Chazan, D., Miranker, W.: Chaotic Relaxation. Linear Algebra and Its Applications 2(7), 199–222 (1969)
13. Frommer, A., Szyld, D.B.: On asynchronous iterations. Journal of Computational and Applied Mathematics 123, 201–216 (2000)
14. Göddeke, D., Strzodka, R.: Performance and accuracy of hardware–oriented native–, emulated– and mixed–precision solvers in FEM simulations (part 2: Double precision GPUs). Technical report, TU Dortmund (July 2008)
15. Göddeke, D., Strzodka, R., Turek, S.: Performance and accuracy of hardware-oriented native–, emulated– and mixed–precision solvers in FEM simulations. Int. J. of Parallel, Emergent and Distributed Systems 22(4), 221–256 (2007)
16. Kelley, C.T.: Iterative Methods for Linear and Nonlinear Equations. SIAM (1995)
17. NVIDIA Corporation. Whitepaper: NVIDIA's Next Generation CUDA Compute Architecture: Fermi
18. NVIDIA Corporation. CUDA Toolkit 4.0 Readiness For CUDA Applications, 4.0 edition (March 2011)
19. NVIDIA Corporation. NVIDIA CUDA Compute Unified Device Architecture C Programming Guide, 4.2 edition (April 2012)
20. Szyld, D.B.: The mystery of asynchronous iterations convergence when the spectral radius is one. Technical Report 98-102, Department of Mathematics, Temple University, Philadelphia, Pa. (October 1998)
21. Trefethen, N.: Hundred-dollar, hundred-digit challenge problems. SIAM News 35(1) (January 2, 2002), Problem no. 7.

GPURoofline: A Model for Guiding Performance Optimizations on GPUs

Haipeng Jia[1,2], Yunquan Zhang[1,3], Guoping Long[1], Jianliang Xu[2],
Shengen Yan[1,3,4], and Yan Li[1,3,4]

[1] Lab. of Parallel Software and Computational Science, Institute of Software,
Chinese Academy of Sciences
[2] College of Information Science and Engineering, The Ocean University of China
[3] State Key Laboratory of Computing Science, The Chinese Academy of Sciences
[4] Graduate University of Chinese Academy of Sciences
jiahaipeng95@gmail.com, zyq@mail.rdcps.ac.cn, guoping@iscas.ac.cn

Abstract. Performance optimization on GPUs requires deep technical knowledge of the underlying hardware. Modern GPU architectures
are becoming more and more diversified, which further exacerbates the
already difficult problem. This paper presents GPURoofline, an empirical model for guiding optimizations on GPUs. The goal is to help
non-expert programmers with limited knowledge of GPU architectures
implement high performance GPU kernels. The model addresses this
problem by exploring potential performance bottlenecks and evaluating
whether specific optimization techniques bring any performance improvement. To demonstrate the usage of the model, we optimize four representative kernels with different computation densities, namely matrix
transpose, Laplace transform, integral and face-dection, on both NVIDIA
and AMD GPUs. Experimental results show that under the guidance
of GPURoofline, performance of those kernels achieves 3.74~14.8 times
speedup compared to their naïve implementations on both NVIDIA and
AMD GPU platforms.

Keywords: GPURoofline, Threshold Carving, Tradeoff Carving,
Little's Law.

1 Introduction

More and more application developers have been adopting GPUs as standard
computing accelerators because of their increasing computing power and programmability. However, we won't get the required performance without careful optimizations because the performance problem has shifted from hardware
designers to compiler writers and application developers. Unfortunately, performance optimizations of GPU programs are difficult, because this process requires
deep technical knowledge of the underlying hardware architecture. Modern GPU
architectures are becoming more and more diversified, which further exacerbates
the already difficult problem of performance optimization. For programmers, it

C. Kaklamanis et al. (Eds.): Euro-Par 2012, LNCS 7484, pp. 920–932, 2012.

will be helpful to have a structured and insightful model that guides performance optimizations on GPUs. To make the model even more useful, it needs to be understandable by most programmers.

Our research addresses this problem by proposing GPURoofline, a model guiding performance optimizations on GPUs. The goal of the model is to facilitate the best match between algorithmic features and underlying hardware characteristics. On both NVIDIA and AMD GPUs, the model can help identify performance bottlenecks, and evaluate whether a particular optimization technique can achieve performance improvements. Instead of trying to predict performance, we choose a simpler approach called "bound and bottleneck analysis". The approach provides valuable insights into primary factors affecting the performance. In particular, critical performance bottlenecks are highlighted and quantified [12]. The proposed model provides three functionalities. Firstly it provides valuable insights on primary factors that affect the performance. Secondly it identifies performance bottlenecks and allows programmers and architectures to predict the benefits of potential optimizations and architecture improvements. Thirdly it can be incorporated into a tool to provide performance information to an auto-tuning compiler by narrowing the search space.

We also demonstrate the usage of our model through optimizing four representative programs with different compute intensity: Matrix Transpose, Laplace Transform, Integral and FaceDection. All evaluations are performed on both NVIDIA and AMD GPUs. Experimental results demonstrate that under the guidance of GPURoofline, performance of those kernels achieves 3.74~14.8 times speedup compared to their naïve implementations on both platforms.

In summary, we make the following contributions in this paper. Firstly, We build the first Roofline model for GPU, called GPURoofline, to guild GPU program optimization. Secondly, We demonstrate how the model can help programmers do GPU performance optimizations. Thirdly, to the best of our knowledge, this is the first performance model that takes global memory channel conflicts and load balancing into consideration.

The rest of the paper is organized as follows. We begin by discussing related works in section 2. Section 3 presents how to build our GPU model. Section 4 discusses experiment results and analysis. Section 5 concludes this paper.

2   Related Work

Enormous works have been invested on building GPU performance analysis and prediction models. Architecture-aware performance analysis methods were proposed in[3][7]. Ryoo et al. [5]used Pareto-optimal curves to narraw the optimization space of GPU programs and introduce efficiency and utilization as single number metrics. N. K. Govindaraju[10]presented a memory model to analyze and improve the performance of nested loops on GPUs. S. Hong[6]presents a simple performance analytical model to capture a rough estimate of the cost of memory operations by considering the number of running threads and memory bandwidth. Baghsorkhi[2]introduced an abstract interpretation of a GPU kernel to identify performance bottlenecks and used work flow graph to predict execution time. Kothapalli[4]presented a performance prediction model to analyze pseudo code for a GPU kernel to obtain a performance estimate. However, because of the complexity of the underlying hardware architecture, it is difficult to predict performance accurately.

Certainly, these performance models are powerful tools for optimizing. However, for a given kernel, they do not provide any insight into how to identify performance bottleneck and evaluate the benefits of potential optimization methods. Compared to them, our work can guide programmers to write high performance program directly, rather than write a naïve version first and then tune it again and again. There are also similar works to us: Yao Zhang[1]provided a quantitative way to analyzes GPU program performance, however, they didn't provide an easy-to-understand model; Samuel Williams[11]provided an insightful visual performance model, however, their works only for multi-core CPUs.

3   GPURoofline

Using bound and bottleneck analysis [8], the attainable performance on a given GPU architecture is restricted by two factors: peak performance and peak bandwidth. Performance depends on how well kernel features map to architectural characteristics. There is a single variable, Compute Intensity, which is defined as operations per byte of off-chip memory traffic. So the proposed GPURoofline model should integrate these three factors together. In this paper, although our work focuses on the NVIDIA Tesla C2050 and AMD Radeon HD5850 GPU, we believe our performance modeling methodology is also applicable to any other GPU architectures.

For simplicity, in this paper, we use peak performance refers to the peak performance of single-precision floating-point, peak bandwidth refers to the peak bandwidth of off-chip memory, NVIDIA GPU refers to the NVIDIA Tesla C2050 GPU and AMD GPU refers to the AMD Radeon HD5850 GPU.

3.1   Naïve GPURoofline

Fig.1a outlines a naïve GpuRoofline model for AMD GPU with peak performance of 2.09TFlopps/sec and peak bandwidth of 128GB/sec. Fig.1b outlines a naïve GpuRoofline model for NVIDIA GPU with peak performance of 1.03TFlopps/sec and peak bandwidth 144GB/sec. The graph is log-log scale and sets an upper bound on the performance of GPU kernels. The max attainable performance equals to min {peak performance, peak bandwidth * Compute Intensity}.

As shown in Fig.1, the vertical purple dashed line represents the Compute Intensity of hardware, calculated by peak performance dividing peak bandwidth. Two vertical red dashed lines represent two kernels with different Compute Intensity: the left one which Compute Intensity smaller than hardware Compute Intensity called memory-bound kernel; and the right one which Compute Intensity larger than hardware Compute Intensity called instruction-bound kernel. As will be explained later, the hardware Compute Intensity suggests the level of difficulty to achieve peak performance.

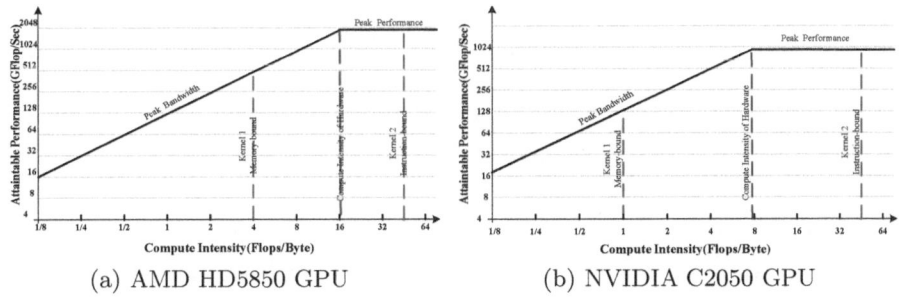

(a) AMD HD5850 GPU   (b) NVIDIA C2050 GPU

Fig. 1. Naïve GPURoofline for GPUs

As we see, we must build a unique GPURoofline for each of the different GPU architecture. Fortunately, given a GPURoofline, we can use it repeatedly on different kernels.

3.2   Threshold Optimizations

We introduce Little's Law to guide our designs on communication. We also define the three components included in Little's Law: memory access latency, concurrency and the utilization of the peak bandwidth. The utilization of the peak bandwidth will drop if Little's is not satisfied.

Optimization Space. According to Little's Law, we defined optimization spaces as follows:

Eliminating Channel Conflict (ECC), just as local memory, global memory is divided into 8 partitions of 256-byte width on both AMD and NVIDIA GPU. Channel conflict occurs when concurrent global memory access requests queue up at some partitions while other partitions go unused. Rearrange data structure to ensure adjacent work-items access adjacent memory address is a common optimization technique.

Reducing Memory Transactions (RMT), coalescing global memory access requests into as few memory transactions as possible. Alignment, vector and coalesced access are the main methods to achieve this.

Using Software Prefetching (USP), the highest performance usually requires keeping many memory operations in flight, which is easier to do via prefetching than by waiting until the data is actually requested by the program.

Using FastPath (UFP), this is for AMD GPU specially. Examine the code to ensure you are using FastPath not CompletePath, can improve performance significantly.

Threshold Carving. In this section, we will perform a sensitivity analysis to examine the impact of optimization methods on performance .We design a highly optimized implementation of copy micro-benchmark which the utilization

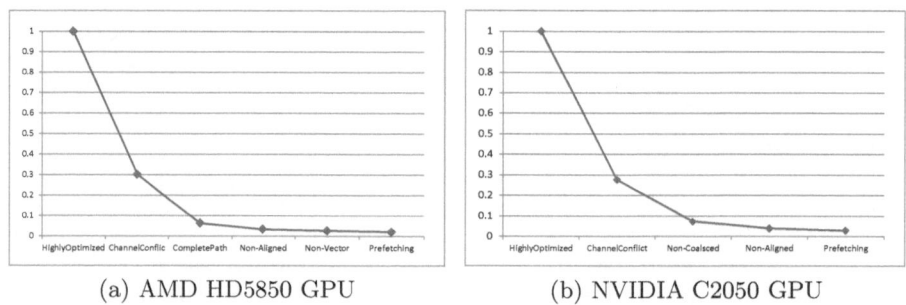

(a) AMD HD5850 GPU   (b) NVIDIA C2050 GPU

Fig. 2. Performance changes along with the optimizations removed one by one

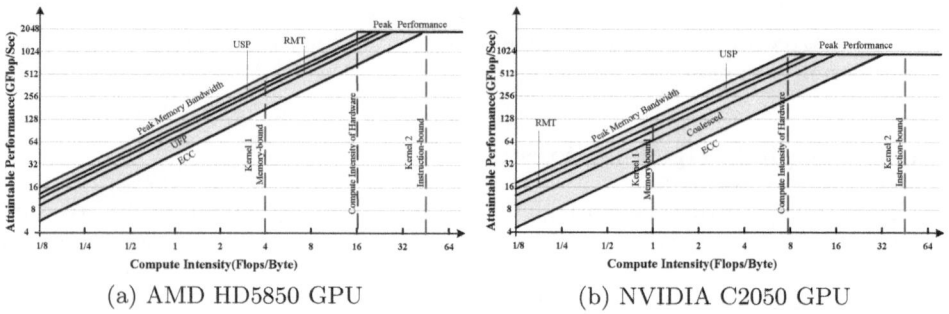

(a) AMD HD5850 GPU   (b) NVIDIA C2050 GPU

Fig. 3. GPURoofline model with threshold carvings

of peak bandwidth can achieve 90% on both NVIDIA and AMD GPU. And then remove those optimization methods one by one in a particular order, Fig.2 shows performance changes.

From Fig. 2, we can see that for both NVIDIA and AMD GPU, the most important optimization method is eliminating channel conflict which was ignored in previous work. However, the second important optimization method is different: using FastPath for AMD GPU and coalesced access for NVIDIA GPU, respectively. Changing access patterns to allow data alignment is also important for both NVIDIA and AMD GPU. We add those optimization methods to our GPURoofline model:

As shown in Fig. 3, similar to performance changes, as we remove these optimization methods, new bandwidth curves will be formed below the peak bandwidth curve. We call these interior GPURoofline-like structures Memory Carvings. These Carvings not only provide some reasonable bounds on performance but also provide some suggestions for the optimizations. You cannot break through a ceiling without performing the associated optimization method first, so these memory carvings are called threshold carvings. We rank the Carvings from bottom to top as the order we remove the optimization methods.

3.3   Tradeoff Optimizations

We also introduce Little's Law to guide our designs on computation and define the three components included in Little's Law: concurrency, latency and throughput of effective instruction. Performance will drop if Little's is not satisfied.

Optimization Space. According to Little's Law, we defined optimization spaces as follows:

Reducing Dynamic Instructions (RDIS), increase the efficiency of instruction stream. There are four methods for this: minimizing divergent threads within a warp or a wavefront; eliminating common subexpression; loop-invariant code motion and loop unrolling. However, these optimizations must be balanced against the increased usage of hardware resources.

Instruction Selection Optimizations (INS), throughputs of GPU instructions are very different. Selecting instructions with lower latency as much as possible is a very desirable method for instruction-bound kernels.

Increasing Thread-level Parallelism (TLP), GPUs hide latency based on a large number of threads. Exploiting TLB, providing enough threads for each compute unit is a basic optimize method for GPUs.

Increasing Instruction-level Parallelism (ILP), ensure the availability of independent instructions within a thread. This is usually achieved by loop unrolling, reordering the code and using vector instructions.

Work-redistribution (WRD), redistribute workloads across threads when there are workload imbalance. We can achieve it through four techniques: persistent thread, global queue, local queue and task stealing.

Tradeoff Carving. Using a similar analysis discussed in section3.2, we obtain some new GpuRoofline-like Carvings below GpuRoofline called Compute Carving. However, because of the discontinuous of optimization spaces, it is not clear that one should maximize or minimize an optimization method. So the Compute Carving is called tradeoff carving which just provides the insights into the performance improvement but not accurately, this is very different from the Memory Carving. The desired of accurate Compute Carving is the future work.

As shown in Fig. 4, when the Compute Intensity of a kernel greater than 0.81 for AMD GPU or 1.8 for NVIDIA GPU(calculated by hardware Compute Intensity divides process elements per stream core, then divides instruction cycles) we should consider the optimization of computation. We can also conclude from Fig 4 that, computation optimization for AMD GPU is more difficult than NVIDIA GPU, That is because AMD GPU is vector architecture and we cannot translate all the scalar instructions into vector instructions with appropriate length.

As Show in Fig.4, for both NVIDIA and AMD GPU, the most effective method is to exploiting TLP to hide latency. Exploiting ILP is the most obvious difference in the process of optimizing which is more effective for AMD GPU than NVIDIA GPU, because of AMD GPU's vector architecture. Additionally, RDIC is also an

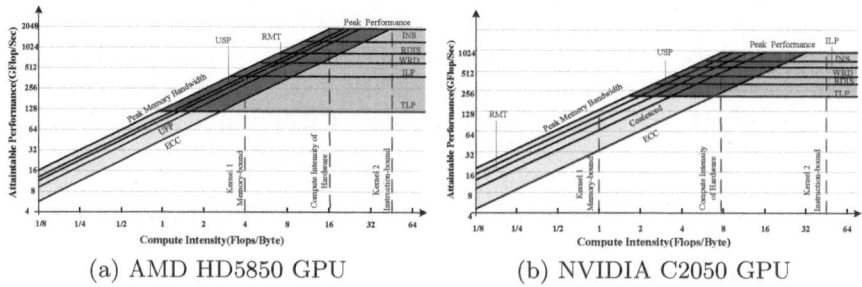

(a) AMD HD5850 GPU   (b) NVIDIA C2050 GPU

Fig. 4. GPURoofline model with tradeoff carvings

important optimization method. However, we must balance against the increased usage of hardware resources. Using Work-redistribution to enable load balance among threads can improve performance significantly on both NVIDA and AMD GPU for the irregular-parallel algorithm.

3.4   Data Locality

The main purpose of data locality is to increase kernel Compute Intensity. By increasing data reuse and decreasing the traffic of off-chip memory, this approach can improve performance significantly, especially for memory-bound kernel. Like memory access and computation constrained performance through performance carving, Compute Intensity also constrain performance like a wall, is called Compute Intensity Wall. We cannot achieve higher performance without improving kernel Compute Intensity especially for memory-bound kernels. So when you use GPURoofline model to guide your optimization and the performance is not achieve your expectation, the first optimization method you should think is increasing kernel Compute Intensity through data locality.

3.5   Interaction with Program Optimization

According to the GpuRoofline model, we can optimize kernels easily according to four rules:

Firstly, the Compute Intensity of a kernel determines the optimization region, and thus which optimization method to try. As shown in Fig.4, if the kernel dashed line falls into the green area, programmers should work only on the memory optimizations. If the dashed line falls into the blue area, programmers should work only on the computation optimizations. If the dashed line falls into the brown area, programmers should try both types of optimizations.

Secondly, optimization carvings suggest the corresponding methods that programmer should perform. And the gap between them represents the potential (Memory Carving) or relative potential (Computation Carving) benefits of related optimization method.

Thirdly, the order of the optimization carving suggests the optimization order.

Finally, the ridge point marks the minimum Compute Intensity required to achieve peak performance.

4   Evaluation

In this section we demonstrate the usage of GPURoofline model through four kernels with different Compute Intensity. Table 1 shows the configuration of GPUs in our experiments in detail. Fig.5 shows optimization regions of these four kernels in GpuRoofline model. Note that, when calculating kernel Compute Intensity, we consider all the calculations, including address calculations.

Table 1. Configuration of the GPUs in our experiments

GPU	Clock Rate	PE	CU	Peak performance	Memory	Peak BW	Regisgers/CU	LDS/CU
AMD HD5850	0.725GHZ	288	18	2090GFlops	1.0GB	128GB/s	16K	32K
NVIDIA C2050	1.15GHZ	448	14	1030GFlops	3.0GB	144GB/s	16K	48K

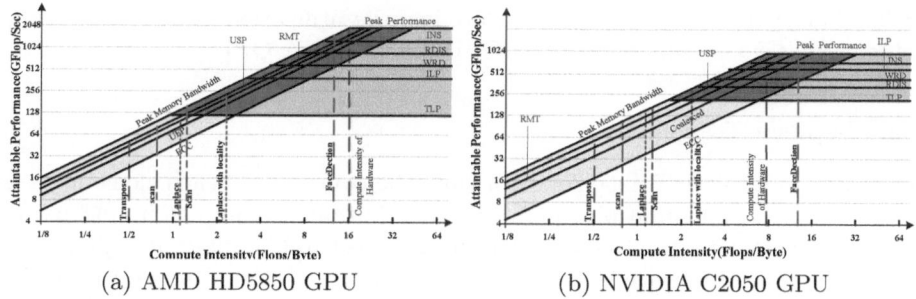

(a) AMD HD5850 GPU   (b) NVIDIA C2050 GPU

Fig. 5. Optimization regions of these four kernels in GpuRoofline model

4.1   Matrix Transpose

In this section, we optimize matrix transpose under the guidance of GpuRoofline model. The transpose operation of each element performs two address calculations, and each address calculation performs 2 floating-point operations, so the compute intensity of matrix transpose is 2*2/8=0.5. According to optimization chain, our optimization work should only focus on the off-chip memory bandwidth optimizations.

As our GPURoofline model suggests, for both NVIDIA and AMD GPU, the first method to consider is eliminating channel conflict, and we achieve it by using a technique called Diagonal Block Reordering method. We also use vector memory access pattern to exploit ILP and data alignment to reduce memory transactions. In addition, we use local memory to make its global memory access pattern coalesced. Fig.6 shows the performance results when satisfies desired optimization methods one by one. The performance of this kernel is on a 2560 * 2560 matrix of float and uses memory bandwidth as the performance metric.

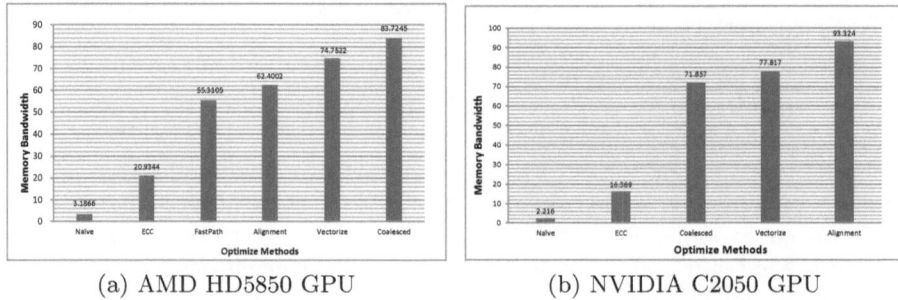

(a) AMD HD5850 GPU   (b) NVIDIA C2050 GPU

Fig. 6. Performance changes when satisfies optimization methods one by one

As shown in Fig. 6, eliminating channel conflict and using FastPath are the first two optimization methods for AMD GPU. However, for NVIDIA GPU, the first two optimize methods are eliminating channel conflict and coalesced memory access pattern. We can also see that, optimization on NVIDIA GPU is a little easier than AMD GPU. Using GPURoofline, the utilization of peak bandwidth achieves 65.4% and 64.8% on AMD GPU and NVIDIA GPU respectively.

4.2 Laplace Transform

According to Laplace Transform algorithm, the transform of each element needs to perform 9 add and multiply operations. In addition, it requires 9 iterations and 10 address calculations. Each calculation contains two floating-point operations. So the Compute Intensity is 47/36=1.3. However, with this Compute Intensity, we can't obtain a satisfied performance. So we consider to use data locality. If the work-group size is 16*16, calculating these 256 elements need to transfer17*17 = 289 elements from off-chip memory to local memory. Furthermore, we put the Laplacian matrix into the constant memory, further reduces the dependence of the off-chip memory bandwidth. After data locality, the compute intensity of this kernel reaches to 3.2. Just as shown in Fig 5.

According to GPURoofline model, optimization works should focus on both memory and computation optimization. Fig.7 shows performance results when satisfies desired optimization methods one by one. The performance of this kernel is on a 1024 * 1024 matrix of float and uses execution time as the performance metric.

As shown in Fig. 7, data locality is a common optimization method for memory-bound kernels. Using vector instruction to exploit ILP can improve performance significantly for both NVIDIA and AMD GPU, however, when the vector length exceed a value, 8 for AMD GPU and 4 for NVIDIA GPU respectively, the performance decreases. This is because vector instructions need more register files, limits the number of threads that can be executed simultaneously. We can also see that, exploiting ILP is more efficient for AMD GPU than NVIDIA GPU. Reduce dynamic instructions through eliminating divergent and

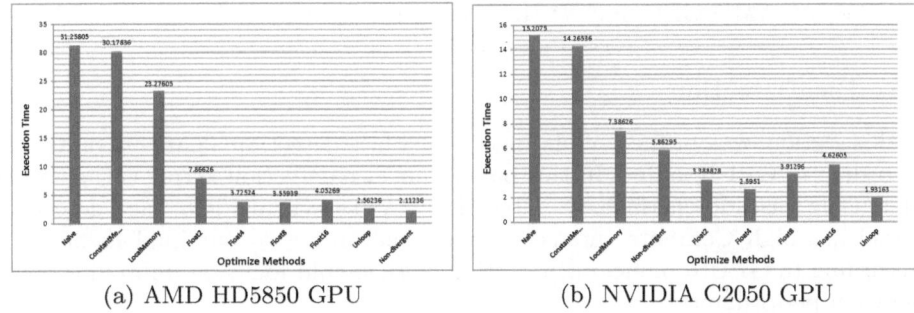

(a) AMD HD5850 GPU   (b) NVIDIA C2050 GPU

Fig. 7. Performance changes when satisfies optimization methods one by one

loop unrolling, are also efficient methods. Using GpuRoofline, the performance improved by 14.1 and 7.8 times on AMD GPU and NVIDIA GPU respectively.

4.3   Integral

According to our implementation of integral algorithm, the Compute Intensity of this kernel is 4.2 after using locality as we have to execute so many memory address calculation and iterations.According to GPURoofline model, optimization works should focus on both memory and computation optimization.

In order to improve the efficiency of instructions, we use a more work-efficient parallel scan algorithm that performs $O(n)$ operations instead of a naïve version that performs $O(nlog2n)$ operations. We also optimize this kernel a step further under the guidance of GPURoofline. Fig.8 shows the performance results when satisfies desired optimization methods one by one. The performance of this kernel is on a 1024 * 1024 matrix of float and uses execution time as performance metric.

As shown in Fig.8, data locality optimization is the most important for memory-bound kernels. Work-effective scan algorithm can improve performance by improving the utilization of thread. As we discussed previously, because of AMD GPU's vector architecture, exploiting ILP can improve performance more

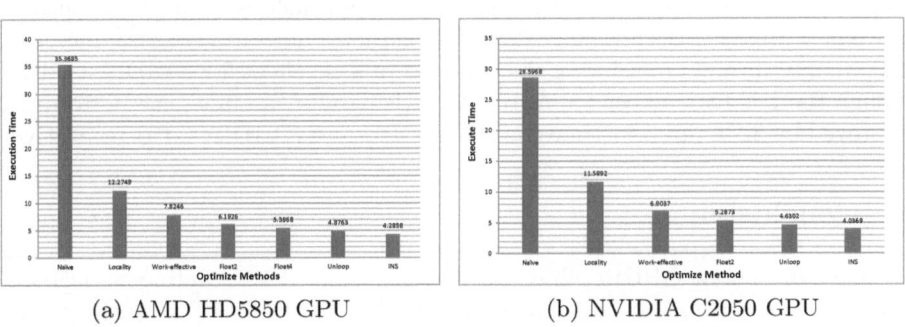

(a) AMD HD5850 GPU   (b) NVIDIA C2050 GPU

Fig. 8. Performance changes when satisfies optimization methods one by one

than NVIDIA GPU. Using GpuRoofline, the performance improved by 14.8 and 8.1 times on AMD GPU and NVIDIA GPU respectively.

4.4   FaceDetection

In this section, we optimize Viola-jones based face detection algorithm on GPUs according to GPURoofline. In this paper, our face detection kernel is the kernel that using cascade classifier to detect face. As shown in Fig.5, face detection kernel has high Compute Intensity, to 14.2 according to our implementation. According to GPURoofline, optimization work should focus on improving computation performance.

Different from algorithms discussed above, the face detection kernel is an irregular-parallel algorithm. There are serious load imbalance among threads. So we should use work-redistribution to address this problem. We use speedup to the naïve implementation as performance metric. Fig.9 shows the performance results when satisfies desired optimization methods one by one.

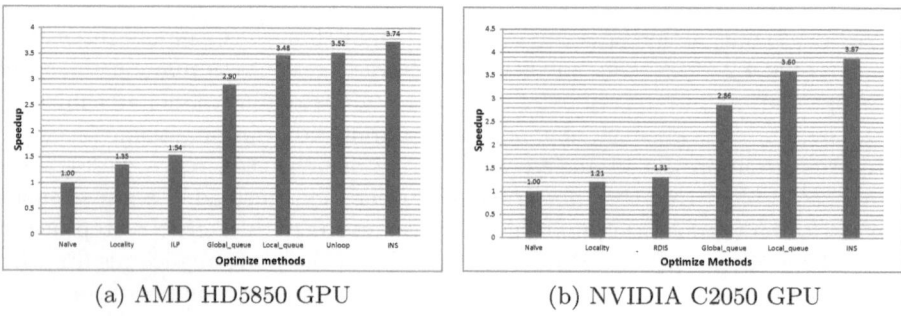

(a) AMD HD5850 GPU   (b) NVIDIA C2050 GPU

Fig. 9. Performance changes when satisfies optimization methods one by one

As shown in Fig.9, Work-redistribution is the most effective optimize method for this kernel. In additional, using data locality to increase Compute Intensity and selecting instructions with higher throughput such as mad24 can also improve performance. Because face detection kernel is hard to vectorize, increasing ILP, mainly through reordering the code, there is no effect for NVIDIA GPU, although there is little effect for AMD GPU.Using GpuRoofline, the performance improved by 3.74 and 3.87 times on AMD GPU and NVIDIA GPU respectively.

5   Conclusion

We have presented GPURoofline, an empirical model for guiding performance optimizations on both NVIDIA and AMD GPU platforms. The goal is to help non-expert programmers with limited knowledge of GPU architectures implement high performance GPU kernels. Programmers can identify performance bottleneck and select appropriate optimization methods. Furthermore, we have observed that for best performance, optimization strategies are closely related to hardware architectures. Although the model is not designed to achieve perfect accuracy, it captures primary performance characteristics of GPUs.

We also demonstrated the usage of the model through four kernels with different compute densities. Experimental results show that under the guidance of the GPURoofline, performance of those kernels achieves 3.74~14.8 times speedup compared to their naïve implementations on both NVIDIA and AMD GPU platforms.

Acknowledgements. We would like to thank reviewers for their helpful comments to our work. This work is supported by the National High-tech R&D Program of China (No. 2012AA 010902, No. 2012AA010903), the National Natural Science Foundation of China (No. 61133005, No.61100066) and ISCAS-AMD Fusion Software Center. Dr. Guoping Long is supported by National Natural Science Foundation of China (Grant No. 61100072).

References

1. Zhang, Y., Owens, J.D.: A quantitative performance analysis model for GPU architectures. In: High Performance Computer Architecture, pp. 382–393 (February 2011)
2. Baghsorkhi, S., Delahaye, M., Patel, S.J., Gropp, W.D., Hwu, W.-M.W.: An Adaptive Performance Modeling Tool for GPU Architectures. In: Principles and Practice of Parallel Programming, pp. 105–114 (January 2010)
3. Daga, M., Scogland, T.R.W., Feng, W-C.: Architecture-Aware Optimization on a 1600-core Graphics Processor. Technical Report TR-11-08, Computer Science, Virginia Tech.
4. Kothapalli, K., Mukherjee, R., Rehman, M.S., Patidar, S., Narayanan, P.J., Srinathan, K.: A performance prediction model for the CUDA GPGPU platform. In: International Conference on High Performance Computing, pp. 463–472 (2009)
5. Ryoo, S., Rodrigues, C.I., Stone, S.S., Baghsorkhi, S.S., Ueng, S., Stratton, J.A.: Program Optimization Space Pruning for a Multithreaded GPU. In: International Symposium on Code Generation and Optimization, pp. 195–204 (April 2008)
6. Hong, S., Kim, H.: An analytical model for a gpu architecture with memory-level and thread-level parallelism awareness. In: International Conference on Computer Architecture, pp. 152–163 (2009)
7. Jang, B., Do, S., Pien, H.: Architecture-Aware Optimization Targeting Multithreaded Stream Computing. In: Second Workshop on General-Purpose on Graphics Processing Units (2009)
8. Meng, J., Morozov, V.A., Kumaran, K., Vishwanath, V., Uram, T.D.: GROPHECY: GPU Performance Projection from CPU Code Skeletons. In: Conference on High Performance Computing (2011)
9. Bauer, M., Cook, H., Khailany, B.: CudaDMA: optimizing GPU memory bandwidth via warp specialization. In: Conference on High Performance Computing(Supercomputing) (2011)
10. Govindaraju, N.K., Larsen, S., Gray, J., Manocha, D.: A Memory Model for Scientific Algorithms on Graphics Processors. In: ACM/IEEE Conference on Supercomputing (November 2006)

11. Williams, S., Waterman, A., Patterson, D.: Roofline: An Insightful Visual Performance Model for Multicore Architectures. Communications of the ACM, 65–76 (2009)
12. Lazowska, E.D., Zahorjan, J., Scott Graham, G., Sevcik, K.C.: Quantitative System Performance: Computer System Analysis using Queueing Network Models. Prentice-Hall. Inc., Upper Saddle River (1984)
13. Fatahalian, K., Sugerman, J., Hanrahan, P.: Understanding the Efficiency of GPU Algorithms for Matrix-matrix Multiplication. In: Conference on Graphics Hardware, pp. 133–137 (August 2004)
14. Taylor, R., Li, X.: A Micro-benchmark Suite for AMD GPUs. In: International Conference on Parallel Processing Workshops, pp. 387–396 (2010)
15. Liu, W., Muller-Wittig, W., Schmidt, B.: Performance Predictions for General-Purpose Computation on GPUs. In: International Conference on Parallel Processing, pp. 50–57 (September 2007)
16. Viola, P., Jones, M.: Robust Real-time object Detection. In: Second International Workshop on Statistical and Computation, pp (July 2011)

Building a Collision for 75-Round Reduced SHA-1 Using GPU Clusters

Andrew V. Adinetz[1,2] and Evgeny A. Grechnikov[3]

[1] Lomonosov Moscow State University, Research Computing Center
adinetz@gmail.com
[2] Joint Institute for Nuclear Research
[3] Lomonosov Moscow State University, Faculty of Mechanics and Mathematics
grechnik@mccme.ru

Abstract. SHA-1 is one of the most widely used cryptographic hash functions. An important property of all cryptographic hash functions is collision resistance, that is, infeasibility of finding two different input messages such that they have the same hash values. Our work improves on differential attacks on SHA-1 and its reduced variants. In this work we describe porting collision search using method of characteristics to a GPU cluster. Method of characteristics employs backtracking search, which leads to low GPU performance due to branch divergence if implemented naively. Using a number of optimizations, we reduce branch divergence and achieve GPU usage efficiency of 50%, which gives 39× acceleration over a single CPU core. With the help of our application running on a 512-GPU cluster, we were able to find a collision for a version of SHA-1 reduced to 75 rounds, which is currently (February 2012) the world's best result in terms of number of rounds for SHA-1.

1 Introduction

A *cryptographic hash function* is a function which maps *messages* (bit-strings of arbitrary length) into *hash values*, or *hashes* (bit strings of fixed length). Such functions are widely used in modern cryptography and information security. A hash serves as a fingerprint for a message. An important property for practical applications of cryptographic hash functions is *computational infeasibility of finding a message with a given hash value*. A *collision* is a pair of different messages which give the same hash value. Due to limited size of hash value, collisions exist for any hash function; however, they are hard to find. If a collision has been built, then the cryptographic hash function is considered to be *compromised*, and is no longer suitable for practical applications. Collision search is therefore an important part of cryptoanalysis of hash functions.

Hash function called SHA-1 (Secure Hash Algorithm 1) maps messages of any length (maximum of $2^{64} - 1$ specified by the standard) into 160-bit hashes. It was published by NIST (National Institute of Standards and Technology) in 1995 and is now widely used in different government and industrial security standards, such as electronic digital signature, user authentication, key exchange

C. Kaklamanis et al. (Eds.): Euro-Par 2012, LNCS 7484, pp. 933–944, 2012.
© Springer-Verlag Berlin Heidelberg 2012

and generation of pseudo-random sequences. SHA-1 is available in almost all commercial security systems.

Attempts to compromise SHA-1 have been performed for a number of years. They advanced far enough, though as of February 2012, no full SHA-1 collision has been built. Currently, NIST is holding the competition for a new cryptographic hash function to replace SHA-1. The new function is expected to be announced in 2012.

As a rule, cryptoanalytic problems are easily parallelized and scale well to any available computational resources. It seems therefore logical to solve them using GPUs. And though GPUs are quite widely used to solve problems such as password cracking [1], so far we haven't found any working application of GPUs to collision search.

The contribution of this paper can be summarized as follows:

– We have ported collision search using method of characteristics for SHA-1 to GPUs, and after performing optimizations we proposed, obtained 39× acceleration compared to a single CPU core
– With our application running on a GPU cluster, we have found a collision for reduced 75-round SHA-1, which is, as of February 2012, world's best result in terms of number of rounds for SHA-1.

This paper is organized as follows. We describe SHA-1 hash function and differential attacks in section 2. Section 3 describes the characteristic search algorithm. GPU implementation of message search are described in section 4. We describe computational experiments in section 5 and conclude in section 6.

2   SHA-1 and Differential Attacks

Notational conventions used in this paper are presented in Table 1. SHA-1 hash function [2] works as follows. First, the message is padded with bits, including message length, and split into 512-bit message blocks $M_1, \ldots, M_k$. The *compression function* $g(M, H)$ is then applied sequentially to compute $H_i = H_{i-1} + g(M_i, H_{i-1})$. H_0 is the *initial value* provided by the standard,

Table 1. Notational Conventions Used in This Paper

Notation	Description
X	32-bit unsigned integer related to 1st message
X^*	32-unsigned integer related to 2nd message
X^2	a pair of 32-bit unsigned integers (X, X^*)
$X \oplus Y$	exclusive OR (XOR)
$X + Y$	2^{32} wrap-around addition
$[X]_i$	i-th bit of X ($i = 0$ — least significant bit)
$X \lll i$	left rotation by i bits
$X \ggg i$	right rotation by i bits

and H_k is the hash value of the message. For building a collision, it is sufficient
to provide two messages $(M_1, \ldots, M_k)$ and $(M_1^*, \ldots, M_k^*)$ of equal length so that
$H_k = H_k^*$.

The compression function consists of 80 rounds and maps a 160-bit *input vec-*
tor H and 512-bit message block M into the new 160-bit value. Input vectors consist of 5 32-bit unsigned integers $H = (A_0, B_0, C_0, D_0, E_0)$, $M = (M_0, \ldots, M_{15})$,
$g(M, H) = (A_{80}, B_{80}, C_{80}, D_{80}, E_{80})$. Computing the compression function consists of the *message expansion* and the *state update transformation*. 16-uint mes-
sagge M_i is expanded to 80 variables W_i as described by (1)

$$
\begin{array}{ll}
W_i = M_i & 0 \leqslant i < 16 \\
W_i = (W_{i-3} \oplus W_{i-8} \oplus W_{i-14} \oplus W_{i-16}) \lll 1 & i \geqslant 16
\end{array}
\tag{1}
$$

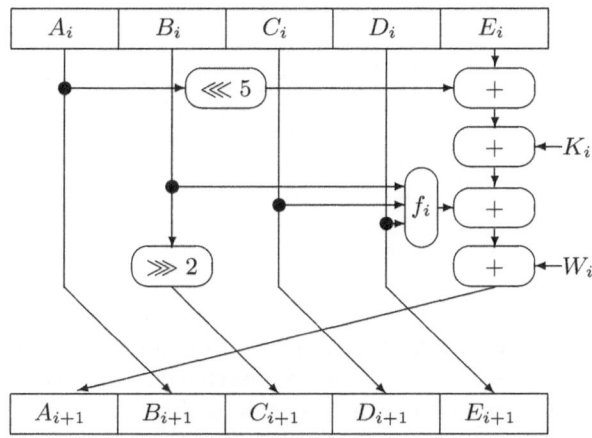

Fig. 1. One Round of SHA-1's Compression Function

One round of the state update transformation is described in Fig. 1. Constants
K_i and functions f_i are defined by (2)

$$
\begin{array}{llll}
K_i = \texttt{0x5A827999}, & f_i(b, c, d) = (b \wedge c) \vee (\overline{b} \wedge d), & 0 < i \leq 20 \\
K_i = \texttt{0x6ED9EBA1}, & f_i(b, c, d) = b \oplus c \oplus d, & 20 < i \leq 40 \\
K_i = \texttt{0x8F1BBCDC}, & f_i(b, c, d) = (b \wedge c) \vee (b \wedge d) \vee (c \wedge d), & 40 < i \leq 60 \\
K_i = \texttt{0xCA62C1D6}, & f_i(b, c, d) = b \oplus c \oplus d, & 60 < i \leq 80
\end{array}
\tag{2}
$$

It's obvious that $B_i = A_{i-1}$, $C_i = A_{i-2} \ggg 2$, $D_i = A_{i-3} \ggg 2$, $E_i = A_{i-4} \ggg$
2, so having only A_i is enough. This is the notation used for the rest of the
paper. $A_{-4}, \ldots, A_0$ give initial values while $A_{76}, \ldots, A_{80}$ can be used to compute
the hash value. As building a collision for full 80 rounds requires very large
computational resources which are not currently available, in our case we reduce
the compression function to 75 rounds.

Differential attacks have been developed for some time. The main stages of their development (including attacks on other hash functions) are described in [4] (MD4), [7] (35-step SHA-0), [8] (full SHA-0), [5] (MD5), [6] (58-step SHA-1), [9] (64-step SHA-1), [10] (70-step SHA-1). We improve on the method described in works on 64 and 70-step SHA-1.

The key idea of differential attacks is to restrict the search to message pairs with a fixed difference modulo 2 $\delta M_i = M_i \oplus M_i^*$, hence the name. It turns out to be convenient to fix some bits also in M_i, A_i, and δA_i. Precisely, a *characteristic* is a set of $(80 + 85) \cdot 32$ elementary conditions on bit pairs $([W_i]_j, [W_i^*]_j)$ and $([A_i]_j, [A_i^*]_j)$, each allowing only certain combinations of bit pair values. There are $2^{2^2} = 16$ possible bit-pair conditions, the six actually used for collision search are described in Table 2.

Table 2. Bit Conditions Used in Characteristics

∇_i	(0,0)	(1,0)	(0,1)	(1,1)
-	✓	—	—	✓
x	—	✓	✓	—
0	✓	—	—	—
u	—	✓	—	—
n	—	—	✓	—
1	—	—	—	✓

Let ∇X be the set of pairs (X, X^*) satisfying all 32 bit-pair conditions for a variable. We want to perform exhaustive search over a given characteristic to find a collision. For each i, we search through values of M_i^2 allowed by chararacteristic, compute A_{i+1}^2 and check it against characteristic for state. If a suitable value is found, the search proceeds to round $i + 1$; if not, it backtracks to $i - 1$. After finding M_{16}^2, the message is fully defined and further steps perform only checking. If the *input freedom* for states A_{i+1}^2 is less than for messages M_i^2, we search through the values of state instead, as there is one-to-one correspondence between message and state once values for previous rounds A_i^2 are fixed. The search continues either until a collision is found, or the search space is exhausted.

We will now estimate complexity of the search, assuming that it is successful. A set $(W_0^2, \ldots, W_i^2)$ is *consistent* if it can be extended to a full set of expanded messages satisfying the characteristic. *Input freedom for the message $\tilde{F}_W(i)$ at step i* is the number of consistent sets $(W_0, \ldots, W_i)$ which extend the consistent set $(W_0, \ldots, W_{i-1})$. It is obvious that $\tilde{F}_W(i) = 1$ when $i \geqslant 16$. When conditions for $W_{16}, \ldots, W_{79}$ are trivial, for $i < 16$ we have $\tilde{F}_W(i) = |\nabla W_i|$. In general case, conditions for $W_{16}, \ldots, W_{79}$ impose linear equations on bits of $[M_i]_j$. When there are m independent equations, $\tilde{F}_W(i) = \frac{|\nabla W_i|}{2^m}$. Input freedom for the state is $\tilde{F}_A(i) = |\nabla A_{i+1}^2|$. When $\tilde{F}_A(i) \geqslant \tilde{F}_W(i)$, we search through M_i^2 and compute A_{i+1}^2. Otherwise, we search through A_{i+1}^2 and compute M_i^2. In the first case we assume $F_W(i) = \tilde{F}_W(i)$, and in the second $F_W(i) = \frac{\tilde{F}_A(i)}{2^m}$. Thus defined, $F_W(i)$

is the number of children nodes of the search tree at step i when implicit linear equations are taken into account.

For SHA-1, A_{i+1} is computed at each step based on $A_{i-j}, 0 \leqslant j \leqslant 4$, and W_i. For our estimation, we assume that $A_{i-j}, 0 \leqslant j \leqslant 4$, and W_i are simply independent random variables (irrespective to hash function) which satisfy the characteristic.

The *uncontrolled probability* $P_u(i)$ *at step* i is the probability that the result of step i satisfies the characteristic if all state and extended message values at previous steps satisfy the characteristic. That is, for $\tilde{F}_A(i) \geqslant \tilde{F}_W(i)$ and $\tilde{F}_A(i) < \tilde{F}_W(i)$ by it is defined by (3) and (4), respectively.

$$P_u(i) := Pr(A_{i+1}^2 \in \nabla A_{i+1} | A_{i-j}^2 \in \nabla A_{i-j}, 0 \leqslant j \leqslant 4, W_i^2 \in \nabla W_i) \qquad (3)$$

$$P_u(i) := Pr(W_i^2 \in \nabla W_i | A_{i-j}^2 \in \nabla A_{i-j}, 0 \leqslant j \leqslant 4, A_{i+1}^2 \in \nabla A_{i+1}) \qquad (4)$$

The *controlled probability* $P_c(i)$ *at step* i is the probability that at least one pair W_i^2 satisfying the characteristic exists, such that the result of step i satisfies the characteristic on the condition that state values at all previous steps satisfy the characteristic. Formally (independent of whether A or W is enumerated) it is defiend by (5)

$$P_c(i) := Pr(\exists W_i^2 \in \nabla W_i : A_{i+1}^2 \in \nabla A_{i+1} | A_{i-j}^2 \in \nabla A_{i-j}, 0 \leqslant j \leqslant 4). \qquad (5)$$

We now estimate the complexity of a successful search. At step i the number of nodes $N_S(i)$ that must be traversed is, on average:

- $N_S(80) = 1$ (we need just a single collision),
- $N_S(i) = \max\left\{ \frac{N_S(i+1)}{F_W(i)P_u(i)}, \frac{1}{P_c(i)} \right\}$ (on the one hand, a search tree node has on average $F_W(i)$ children, among which the fraction of $P_u(i)$ give the next level node; on the other hand, with probability $P_c(i)$ the node won't give any next level nodes).

We call the value defined by (6)

$$\sum_{i=0}^{80} N_S(i), \qquad (6)$$

which depends on the characteristic only, the *work factor* of the characteristic. The less the work factor is, the better the characteristic is.

3   Finding a Characteristic

Finding a characteristic consists of three stages. At the first stage, a *linear* characteristic is searched for, which consists only of –x conditions; it fixes differences, but not bits. To do that, we construct a *linearization* of the hash functions by replacing non-linear operations with their linear "approximations". The goal of this stage is to minimize the number of x in the characteristic, which lead to differences between the function and linearization. A search for a linear characteristic with small x conditions is expressed as searching for small-weight vector in some linear code, which is a known problem from the coding theory.

We construct a 2-block collision. The characteristic for each block is different, but is constructed based on the same linear characteristic. Resulting hash is given by $H_2 = H_1 + g(M_2, H_1) = H_0 + g(M_1, H_0) + g(M_2, H_1)$, $H_2^* = H_0 + g(M_1^*, H_0) + g(M_2^*, H_1^*)$. Linear characteristic gives $g(M_1, H_0) \oplus g(M_1^*, H_0)$ and $g(M_2, H_1) \oplus g(M_2^*, H_1^*)$; as it is the same for both blocks, we can make differences of first and second block values differ only in sign by fixing bits that differ. This leads to $H_2 = H_2^*$, that is, a collision.

The second stage begins with discarding conditions for A_i at first 12 steps and replacing them with conditions for $A_{-4}^2, \ldots, A_0^2$. The initial condition is H_0 for the 1st block, and the result of the first block for the 2nd block. Therefore, we can construct 2nd block characteristic only after finding the 1st block of the collision. We also replace xx condition pairs for successive bits with -x, if the difference can be satisfied due to the carry. This is not always true because rotations are involved. At the second stage we need to find some "path" (a consistent set of conditions) from initial conditions to the linear characteristic. To do this, we choose random positions in A_i^2 which have no conditions, add - condition and find which additional conditions are satisfied based on ones already enforced. Also, when x-type conditons appear in A_i^2 it is useful to fix values of differing bits. If we find a contradiction, we backtrack to the last fixing x and choose an alternative fixing. The second stage finishes when all conditions have the form -xun01.

The third stage iteratively improves the work factor of the characteristic. To do this, we search through possible tightenings of conditions, *propagate* the new conditions, i.e. look at the additional conditions which follow from the new set of conditions, and compute the work factor for the new characteristic. At the end of the search the characteristic with the smallest work factor is chosen.

We'll note several important aspects of characteristic search here:

- F_W, P_u, and P_c are computed sequentially, from least to most significant bits by searching through elementary conditions and possible carries.
- The propagation of conditions is calculated in two passes. First, possible carries are evaluated from least to most significant bits, and then new conditions are evaluated taking carries into account. This is fast, but sometimes doesn't find all possible conditions (due to an interference between consecutive steps). To propagate further, we loop through bit positions, fix possible bit values and check if the fast procedure finds any contradictions. In the second stage we check only those bits who are close to some bit that was changed. In the third stage we check all bits.
- *Coherency*, i.e. similarity of control flow and memory accesses in neighbouring threads, in important for efficient GPU execution. Coherency can be improved by concentrating strong conditions in the middle of the initial

rounds of the characteristic. This is achieved by choosing these positions for - conditions with less probability at the second stage. This is the first GPU-related optimization, and it improves GPU search efficiency by 80%.

4   Message Search Implementation on GPU

Searching for a message which satisfies the characteristic is the most computationally intensive part of collision search. There is a number of points which do not depend on hardware:

- Characteristics always consist of conditions of type -xun01. Therefore, condition set for each 32-bit variable can be expressed as a pair of equations $X \oplus X' = a$, $X \wedge b = c$, where a, b, c are 32-bit values which depend only on characteristic.
- The following procedure is an efficient way to enumerate the set $\{X : X \wedge b = c\}$. The first element is $X := c$, every next element is given by the equation $X := (((X \vee b) + 1) \wedge \overline{b}) + c$, the enumeration is over when this equation gives $X = c$ due to overflow.
- As A_i at two last steps are not used in computing f_i, it is sufficient to check on those steps that $X - X' = a$. Moreover, for the 1st collision block conditions at two last steps can simply be ignored, as any difference due to them could be compensated by the 2nd block without increasing the number of conditions.
- Linear equations on M_i, appearing due to conditions on W_k for $k \geq 16$, can either express a bit $[M_i]_j$ through bits of previous message words, or give an equation involving values of 2 or more bits of M_i. In the first case, a, b, and c depend also on previous messages, and must be recomputed for each round. In the second case, these equations can be removed by adding some artificial conditions (e.g. imposing an additional restriction $[M_i]_j = 0$), without significantly changing the work factor.

The search is naturally divided into *generation* phase, which searches through message pairs, and *check* phase, which checks the rest of the characteristic for the pair of messages. Generation phase is a back-tracking search, and check is simply a function which is called at the last round of generation. Generation can in turn be divided into *host* part and *device* or *GPU* part. On the host, the search tree is expanded to a certain *host depth* to generate enough *search stacks* to make use of GPU parallelism. Host depth is specified individually based on the characteristic and available computational resources. Too little depth leads to insufficient parallelism, while with too large depth, search stacks won't fit into GPU memory. To utilize a single GPU efficiently, about 10^5 search stacks are needed.

During GPU part, the search is performed in parallel on a large number of GPUs. Stacks for which the search is finished are removed, and no new stacks are generated. The main GPU kernel implements back-tracking search and message check. In this kernel, each GPU thread processes only a single search stack for a fixed number of search iterations. The main kernel also collects statistics on the number of traversed nodes, check rounds and maximum depth reached by the search. Between kernel calls, the depth is checked and the defunct stacks are removed from the array of search stacks.

The computation is distributed among cluster nodes using MPI. Each MPI process uses only a single GPU. Search stacks are distributed between nodes in block-cyclic way. During host part, each MPI process generates all search stacks and discards those belonging to other processes. There is a global barrier at the end of host part, but after that, all MPI processes run independently and asynchronously. This means that if search is finished on some stacks and some processes will have less search stacks than the others, there will be no load imbalance; just some of the stacks will be searched through quicker. The only communication involved is sending statistics to statistics collection thread of master process. The master process also spawns one more thread, which prints out aggregated statistics at fixed time intervals.

The application is implemented using Nemerle, an extensible .NET language, and NUDA (Nemerle Unified Device Architecture), a system of Nemerle extensions [11] for programming GPUs. NUDA was chosen due to its free availability and support for high-level GPU programming, including automatic host-GPU data synchronization and generation of kernels. Internally, OpenCL is used to interact with GPUs and as a target for GPU code generation. mono is used to run .NET applications on Linux, and MPI .NET provides .NET bindings for MPI.

We first implemented GPU back-tracking search as a single loop, with branches inside the body handling specific conditions. There were 2 such conditions: round switch and message check. A round switch can arise when a successful message word is found, when all words are exhausted or when the kernel starts. In any case, large number of additional precomputing is required. Message check was implemented as a separate function, with loop on rounds fully unrolled using `inline` annotation available in NUDA.

The performane of our initial implementation, however, was unsatisfactory. While the application scaled well due to little communication, computational efficiency was only 15% on some characteristics. We define computational efficiency as the ratio of really executed integer operations to peak GPU performance in terms of integer operations. Low efficiency was due to low coherency between threads in a single warp, so we concentrated on improving coherency. The first optimization, described in section 3, was modifying search algorithm, which gave 1.8× improvement of efficiency.

The second optimization was sorting search stacks after each GPU pass. We quickly found out that stable sort was better than unstable (quicksort) in maintaining coherency. We also tried different keys, including round number, search value, round change direction (delta), number of search steps to nearest round change (njd) and their combinations. Additionally, we modified the search loop to allow exiting the kernel only on round change; we call this *snapping*. Results of our experiments for characteristic for 2nd block for 72-round collision (72-2) are presented in Fig. 2. We have finally chosen stable sort by search value and snapping, which gave 1.87× efficiency improvement. This was implemented using GPU radix sorting [12], and experiments have shown that sorting takes less than 1% of total computing time.

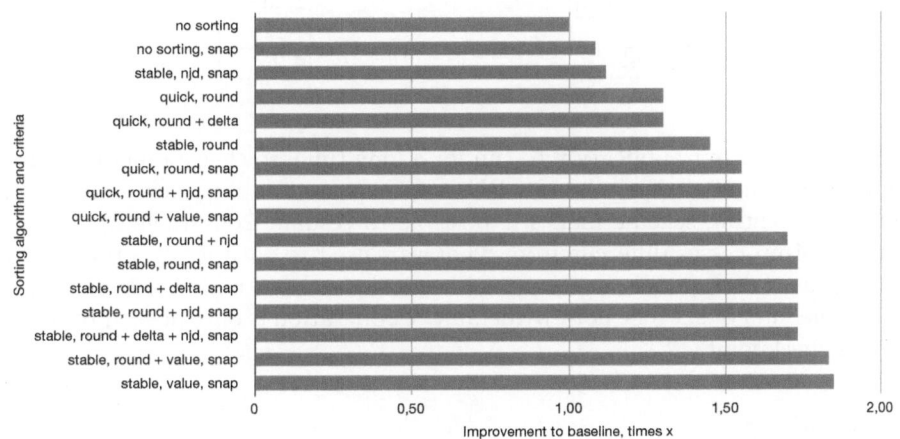

Fig. 2. Comparison of different sorting and snapping approaches

The third optimization was replacing one-loop implementation of backtracking search with a nest of 3 loops. Innermost loop iterates over search values of a single round until either all are exhausted, or a successful message word is found. The second loop works only for 16th round, and iterates over messages that must be checked. Our experiments have shown that more than 75% is spent is message check for some characteristics, so doing that in a separate loop improves performance. The outermost loop switches between rounds, and also checks thread termination condition. As search can be exited from outermost loop only, this ensures automatic snapping. On 75-1 characteristic, triple loop gave 1.25× improvement compared to stable sort and snapping only. Together, triple loop and stable sort give more than 2-fold performance improvement.

Other optimizations included using constant and shared on-chip GPU memory. Quite unexpectedly, using on-chip shared memory gave only 2.5% performance improvement (1.025×). This indicates that previous optimizations did a good job of improving memory access coherency, so that most memory accesses hit the GPU cache available on Fermi GPUs. Using constant memory gave additional improvement of about 9%. Effects of individual optimizations and all of them combined are presented in Table 3. The final version uses all of the optimizations described above, and has efficiency of 63% on 75-1 in the short run. For the long run, efficiency is lower, but still remains above 50%, which we consider sufficient for our purposes.

Table 3. Effects of Different Optimizations on Performance of GPU Backtracking Search

Optimization	Effects
Characteristic search modification	1.8×
Stable sort by value + snap	1.87×
Triple loop	1.25×
Other	1.12×
Total	**4.2×**

We expected our application to run for a long time, so checkpoints were used. And as the number of available GPUs was expected to fluctuate significantly, our checkpointing scheme makes it possible to resume from a checkpoint with number of processes different from what was used to save it. As processes are independent, each process just writes its search stacks to a file independently at fixed time intervals, every hour by default. Cooperation is only needed to resume from a checkpoint.

5   Results

Final computation of 1st and 2nd collision blocks were performed at GPU partition of "Lomonosov" supercomputer installed in Research Computing Center, Moscow State University (RCC MSU). Each GPU node has 2 NVidia Fermi X2070 GPUs with 6 GB RAM, of which only 5.25 GB is available because of ECC. As the GPU partition was still in beta stage, not all GPU nodes were available, and the number of available nodes fluctuated. Characteristics used to search for messages are presented in [3].

The search factor for the 1st block was 2^{58}. 264 GPUs were used, and the computation took 11000 seconds. $2^{54.06}$ nodes were actually traversed. Here, "node" is either a search step or message check round; the latter requires 2.5× more computations than the former. About 40% of nodes were check rounds.

The search factor for the 2nd block was $2^{63.01}$ nodes. The computation started with 320 GPUs and finished with 512 GPUs, with 455 GPUs being used on average. It took 1904252 seconds, or 22 days and 45 minutes. $2^{61.92}$ nodes were actually traversed, about 58.8% of them were check rounds. We achieved 52% efficiency (of GPU peak performance). The resources required were really enormous: were all 1554 GPUs available, the computation would still take about a week.

The actual number of nodes traversed were smaller than estimates. It was 16 times smaller for the 1st block and 2 times smaller for the 2nd block. Were it not the case, the entire collision search would have taken 1.5 month.

We also compared the GPU code with our previous CPU implementation [13]. A single Intel Nehalem CPU core traverses about $2^{27.85}$ search nodes per second. A single check round requires 2.5× more operations than a search node; that is, our 22-day run is equivalent to $2^{62.83}$ search nodes, or $2^{33.14}$ search nodes traversed per second per GPU. Based on those numbers, a single GPU is $39\times$ faster than a single CPU core, or $9.75\times$ faster than a 4-core CPU, i.e. a single CPU socket. The number varies slightly from one characteristic to the other, but the order remains the same. This means that our computation would have taken the same time if done on 17745 CPU cores, which is not much larger than the number of cores used for the previous computation. Obtaining this many "Lomonosov"'s cores for 3 weeks would be problematic. The GPU partition, however, wasn't oversubscribed, so we could easily use all the GPUs available.

The 75-round reduced SHA-1 collision we built is presented in Table 4.

Table 4. 75-Round Reduced SHA-1 Collision

i	Message 1, Block 1				Message 1, Block 2			
1–4	F01EE8EE	BDDFF313	B2F59EE4	BB37F2BB	F072633F	0D32226A	DFF74459	98507743
5–8	2F472A36	1C052F6A	96403EF0	F144298B	EEFE63DD	FE10D5C5	AFE33902	EF74984E
9–12	DAF5519C	7A90DD71	2BF3718E	A7E3DE6D	350272F7	DB382ABC	155B0414	B800179D
13–16	EFFA975E	9B00AA95	6056E3EE	2BA4483A	18ECD4BC	15497213	1505284C	60C4F869
i	Message 2, Block 1				Message 2, Block 2			
1–4	001EE884	3DDFF353	22F59E94	0B37F2E8	00726355	8D32222A	4FF74429	28507710
5–8	1F472A3E	1C052F29	46403E82	4144299B	DEFE63D5	FE10D586	7FE33970	5F74985E
9–12	2AF551FE	BA90DD33	2BF371BE	47E3DE2F	C5027295	1B382AFE	155B0424	580017DF
13–16	CFFA973E	7B00AAD4	4056E3BE	EBA4487B	38ECD4DC	F5497252	3505281C	A0C4F828
i	Colliding Hash Values							
1–5	3DF7F21E 130079F3 C2E6EFFF FD9C4141 9AA8723A							

6   Conclusion

We have proposed a GPU implementation of SHA-1 collision search using the method of characteristics. Based on our previous work and with GPU optimizations proposed, we were able to achieve 50% computational efficiency and $39\times$ acceleration compared to a single CPU core. Using our implementation on a cluster of GPUs, we have found a collision for 75-round reduced SHA-1, which is world's best result in terms of number of rounds for SHA-1 as of February 2012.

However, as search complexity increases $8\times$ with each additional round, searching for collisions with larger number of rounds would require modifications to the method of characteristics. While we are working in this direction, it is too early to talk about results.

Acknowledgements. We are thankful to Research Computing Center of Lomonosov Moscow State University for providing us with access to "Lomonosov" supercomputer. We are also thankful to "Lomonosov" support team and personally Anton Korzh for promptly resolving issues which appeared our using of the cluster. This work was supported by T-Platforms, Russian Fund for Basic Research (RFBR) grant 11-07-93960-SAR-a and CUDA Center of Excellence at Moscow State University.

References

1. Teat, C., Peltsverger, S.: The security of cryptographic hashes. In: Proceedings of the 49th Annual Southeast Regional Conference, pp. 103–108. ACM (2011)
2. National Institute of Standards and Technology (NIST). FIPS-180-2: Secure Hash Standard (August 2002), http://www.itl.nist.gov/fipspubs/
3. Grechnikov, E.A., Adinetz, A.V.: Collision for 75-step SHA-1: Intensive Parallelization with GPU // Cryptology ePrint Archive: Report 2011/641, http://eprint.iacr.org/2011/641
4. Dobbertin, H.: Cryptanalysis of MD4. In: Gollmann, D. (ed.) FSE 1996. LNCS, vol. 1039, pp. 53–69. Springer, Heidelberg (1996)
5. Wang, X., Yu, H.: How to Break MD5 and Other Hash Functions. In: Cramer, R. (ed.) EUROCRYPT 2005. LNCS, vol. 3494, pp. 19–35. Springer, Heidelberg (2005)
6. Wang, X., Yin, Y.L., Yu, H.: Finding Collisions in the Full SHA-1. In: Shoup, V. (ed.) CRYPTO 2005. LNCS, vol. 3621, pp. 17–36. Springer, Heidelberg (2005)
7. Chabaud, F., Joux, A.: Differential Collisions in SHA-0. In: Krawczyk, H. (ed.) CRYPTO 1998. LNCS, vol. 1462, pp. 56–71. Springer, Heidelberg (1998)
8. Biham, E., Chen, R., Joux, A., Carribault, P., Lemuet, C., Jalby, W.: Collisions of SHA-0 and Reduced SHA-1. In: Cramer, R. (ed.) EUROCRYPT 2005. LNCS, vol. 3494, pp. 36–57. Springer, Heidelberg (2005)
9. De Cannière, C., Rechberger, C.: Finding SHA-1 Characteristics: General Results and Applications. In: Lai, X., Chen, K. (eds.) ASIACRYPT 2006. LNCS, vol. 4284, pp. 1–20. Springer, Heidelberg (2006)
10. De Cannière, C., Mendel, F., Rechberger, C.: Collisions for 70-Step SHA-1: On the Full Cost of Collision Search. In: Adams, C., Miri, A., Wiener, M. (eds.) SAC 2007. LNCS, vol. 4876, pp. 56–73. Springer, Heidelberg (2007)
11. Adinetz, A.V.: NUDA Programmer's Guide, http://nuda.sf.net
12. Satish, N., Kim, C., Chhugani, J., Nguyen, A.D., Lee, V.W., Kim, D., Dubey, P.: Fast sort on CPUs and GPUs: a case for bandwidth oblivious SIMD sort. In: Proceedings of the 2010 International Conference on Management of Data (SIGMOD 2010), pp. 351–362. ACM, New York (2010)
13. Grechnikov, E.A.: Collisions for 72-step and 73-step SHA-1: Improvements in the Method of Characteristics. Cryptology ePrint Archive: Report 2010/413, http://eprint.iacr.org/2010/413

GPU-Vote: A Framework
for Accelerating Voting Algorithms on GPU

Gert-Jan van den Braak, Cedric Nugteren, Bart Mesman, and Henk Corporaal

Eindhoven University of Technology, The Netherlands
{g.j.w.v.d.braak,c.nugteren,b.mesman,h.corporaal}@tue.nl

Abstract. Voting algorithms, such as histogram and Hough transforms, are frequently used algorithms in various domains, such as statistics and image processing. Algorithms in these domains may be accelerated using GPUs. Implementing voting algorithms efficiently on a GPU however is far from trivial due to irregularities and unpredictable memory accesses. Existing GPU implementations therefore target only specific voting algorithms while we propose in this work a methodology which targets voting algorithms in general.

This methodology is used in GPU-VOTE, a framework to accelerate current and future voting algorithms on a GPU without significant programming effort. We classify voting algorithms into four categories. We describe a transformation to merge categories which enables GPU-VOTE to have a single implementation for all voting algorithms. Despite the generality of GPU-VOTE, being able to handle various voting algorithms, its performance is not compromised. Compared to recently published GPU implementations of the Hough transform and the histogram algorithms, GPU-VOTE yields a 11% and 38% lower execution time respectively. Additionally, we give an accurate and intuitive performance prediction model for the generalized GPU voting algorithm. Our model can predict the execution time of GPU-VOTE within an average absolute error of 5%.

1 Introduction

Accelerating applications with GPUs (Graphical Processing Units) has become increasingly popular from 2006 on, when GPUs became programmable with the introduction of "CUDA" by NVIDIA and "Close to Metal" by AMD. Just on NVIDIA's website over 1000 applications are listed which use a GPU for acceleration. These applications originate from various domains, such as image and signal processing, finance, statistics and electronic design automation.

GPUs are used in consumer desktop computers and notebooks as well as in embedded systems and industrial machines such as professional printers. Together with the (power) efficiency and the off-the-shelf availability, GPUs are interesting for large companies as well as for small and medium-sized enterprises. This motivates programmers to spend time and effort on making libraries, tools and generic (skeleton) implementations.

A number of algorithms in the image processing domain, such as color space conversion and low-level (pixel) filtering operations, are fairly straight forward

C. Kaklamanis et al. (Eds.): Euro-Par 2012, LNCS 7484, pp. 945–956, 2012.

to implement on a GPU due to their inherent parallelism. Voting algorithms on the other hand are far from trivial to implement efficiently on a GPU due to irregularities and unpredictable memory accesses.

Existing GPU implementations only target a single specific voting algorithm, such as histogram [1,2,3] and 2-D Hough transform [4,5,6,7]. In this work we propose a generic methodology which targets voting algorithms in general. We also introduce a framework called GPU-VOTE which can be used to accelerate a large range of voting algorithms. With this framework the time consuming and cumbersome implementation and optimization of voting algorithms is a thing of the past. Measurements show that GPU-VOTE is not just more generic than previous dedicated implementations, but also gives a performance improvement. To predict the execution time of GPU-VOTE, we also give a model based on the parameters of the voting algorithm, such as input size and number of bins.

This paper is organized as follows. First related work and background information about the GPU architecture is presented in Section 2. In Section 3, voting algorithms are categorized, the generic methodology for voting algorithms and the implementation in GPU-VOTE is discussed. Section 4 evaluates the proposed methodology and describes performance results. A model to predict execution time is given in Section 5. Finally, conclusions and future work are presented in Section 6.

2   Background and Related Work

In this section histogram and Hough transform, two common voting algorithms, are described in more detail. Also related work on GPU implementations of these algorithms is discussed in Section 2.3, as well as details on the NVIDIA GPU architecture in Section 2.4.

2.1   Histogram

In the histogram algorithm a set of bins is filled according to the frequency of occurrence in the input data. For example, a 1-D histogram of an 8-bit gray-scale image usually has 256 bins (due to the 256 possible shades of gray in the image).

A histogram can be used to enhance the contrast of an image by applying histogram equalization. In [8] a 2-D 30×30 histogram of normalized red and green is used for locating a road in an image. In [9] larger 2D-histograms of 256×256 bins are used for image registration purposes. Histogram is also an important statistical tool for displaying and summarizing data [10].

2.2   Hough Transform

The Hough transform is a popular technique to locate shapes in images, such as lines and circles, but also other arbitrary shapes. It is used in many computer vision and image processing applications, such as robot navigation [11], industrial inspection and object recognition [12].

The Hough transform for lines [13] is a 2-D voting algorithm for which each feature (edge) point in an image votes for all possible lines passing through that point. All votes are stored in the so called Hough space, which size is determined by the input image and the accuracy for the parameterization of the lines.

The Cartesian coordinate system is not well suited for the Hough transform, therefore a polar representation of a line Eq. 1 is used in which a line is parameterized with ρ and θ [14]. Parameter θ represents the angle of a line normal to the line in the image and parameter ρ represents the shortest distance between the origin and the line in the image. The angle θ ranges from $0°$ to $180°$ and the radius ρ ranges from $-W$ to $\sqrt{W^2 + H^2}$, where W and H are the width and height of the image respectively.

$$\rho = x\cos(\theta) + y\sin(\theta) \tag{1}$$

By selecting a step size for the angle parameter θ, the number of sets of output bins (or independent vote spaces) is chosen; by selecting a value for N in Eq. 2, the number of bins in each independent vote space is set. With an input image size of 1920×1080 pixels, and a resolution for ρ and θ of one pixel and one degree, the total vote space consists of $(\sqrt{1920^2 + 1080^2} + 1920) \times 180 = 742{,}140$ bins.

$$\rho' = \frac{x\cos(\theta) + y\sin(\theta) + W}{\sqrt{W^2 + H^2} + W} N \tag{2}$$

2.3   Related Work

Two implementations for histogramming are described by Podlozhnyuk in [1], one for 64-bin histograms and one for 256-bin histograms. Shams and Kennedy present two other histogramming methods in [2], which support a range of bin sizes. Nugteren et al. introduce two new histogramming implementations in [3] which are faster than the work of Podlozhnyuk and Shams and Kennedy. Their fastest implementation has a fixed processing time for a given input size.

The 2-D Hough transform has been implemented on a GPU in OpenGL and in CUDA. Two OpenGL implementations can be found in [4] and [5]. With the availability of CUDA nowadays, using OpenGL to program GPUs for general purpose computations is deprecated. A CUDA implementation of the Hough transform can be found in CuviLib [6], a proprietary computer vision library. Van den Braak et al. [7] introduced two new CUDA implementations of the Hough transform, one which focuses on minimizing processing time, while the other has an input data independent processing time.

2.4   GPU Architecture

In NVIDIA's latest GPU architecture named Fermi [15], 32 CUDA cores are grouped into a processing cluster called a Streaming Multiprocessor (SM). Each SM has an on-chip memory of 48 kB. The GPU used for timing measurements in this paper is an NVIDIA GTX 470 which has 14 SMs and is connected to 1280 MB of off-chip memory.

The code executed on a GPU is called a kernel. Kernels run on the GPU as thousands or even millions of threads. Each thread executes the same kernel, but not necessarily the same instruction at the same time. Threads are organized into thread blocks. All threads in a thread block are executed on the same processing cluster (SM) and can communicate via its shared memory. Threads within a thread block are arranged in warps of (at most) 32 threads, with each thread in a warp executing the same instruction at the same clock cycle [16].

3   GPU Voting Methodology

To enable a generic methodology for voting algorithms on GPU, we classify voting algorithms. Two properties are distinguished which leads to a classification of four categories:

- The first property characterizes if either the element **value** or the **location** in the input data is used to determine in which bin the vote will be placed.
- The second property describes if a voting algorithm increases a bin in the vote space by one (**unity vote**) or by a calculated number (**number vote**).

An algorithm which uses the input element location can be converted to an algorithm which uses the input element value by applying a transformation on the input data. By building an array of input element locations that are used by the voting algorithm (e.g. less than 10% in the Hough transform [7]), a location based voting algorithm can be transformed into a value based voting algorithm.

An overview of this two stage generic GPU voting methodology can be found in Fig. 1. First the building of the array of input element locations is described in Section 3.1, and in Section 3.2 the final stage where votes are placed in the vote space is described. Improvements to this final stage are described in Section 3.3.

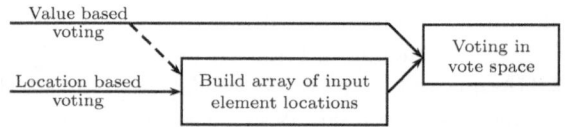

Fig. 1. Overview of the two stage generic GPU voting methodology

3.1   Building an Array of Locations

Building the arrays of input element locations and values is done similarly to [7], which is in turn inspired by the work in [3]. To build these arrays in a parallel way on the GPU, small arrays are created at warp-level granularity. The technique to make an array per warp is summarized in pseudo-code in Lst. 1. Note that all threads in a warp execute the same instruction at the same time in parallel, but some threads may be disabled due to branching conditions. More detailed information on the creation of the array can be found in [7].

```
 1  input_value = input[x,y]
 2  if(element_test(x,y,input_value)) {
 3     do {
 4        index++
 5        SMEM_index = index
 6        SMEM_array_L[index] = (x,y)
 7     } while(SMEM_array_L[index] != (x,y))
 8     SMEM_array_V[index] = input_value
 9  }
10  index = SMEM_index
```

Listing 1. Building an array of input element locations (SMEM_array_L) and an array with the corresponding values (SMEM_array_V) in shared (on-chip) memory (SMEM).

3.2   Voting in Vote Space

After the arrays of element location and element value have been constructed for the location based voting algorithms, the votes can be placed in the vote space[1]. The voting process is executed on the on-chip memory of an SM using atomic memory operations. In case the vote space can be divided in independent vote spaces (e.g. Hough transform), each independent vote space is calculated by a thread block on an SM. The number of bins in a vote space is limited by the amount of on-chip memory in an SM. For the Fermi GPU architecture, the amount of on-chip memory per SM is 48 kB, resulting in maximum 12,288 32-bit bins (N_{max}). In case more than N_{max} bins are required, this second stage of the voting process is executed multiple times in order to calculate all bins.

When the number of independent vote spaces is less than the number of SMs on the GPU (e.g. histogram algorithm, one independent vote space), the input data is split such that each SM will process an equal part. In a final step, the results of all parts are added together to make the final vote space.

3.3   Bin Stretching

As described in Section 2.2, the Hough transform can be implemented with a parameterizable number of bins per independent vote space. The effect of varying this number on the execution time of the Hough transform implementation from [7] is shown in Fig. 2(a) with blue squares. The graph shows that the Hough transform can be calculated faster when more bins are used, which leads to the unintuitive conclusion that more accurate results can be calculated quicker than less precise results.

This effect is caused by contention in the on-chip memory of the GPU. According to Eq. 2, pixels that are close together in the input image, will also end-up in bins close together in the vote space. If a small number of bins is used, more votes are placed in the same bin (through atomic memory operations), causing contention in the memory.

---

[1] For value based voting algorithms, the input itself is directly used as location and value array. Only when not all elements in the input are used in the voting process, an array can be build first, indicated by the dashed arrow in Fig. 1.

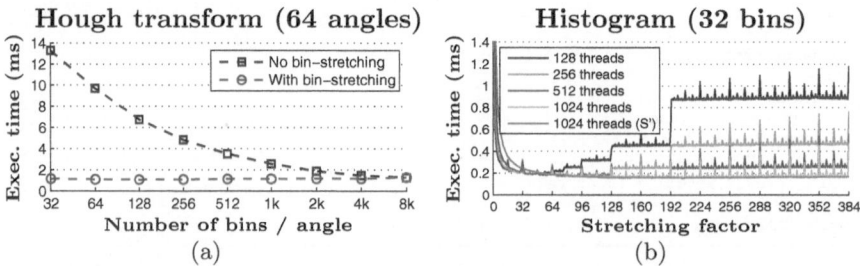

Fig. 2. a: Execution time measurements of the Hough transform algorithm with 64 angles for θ. Blue squares: no bin stretching, red circles with bin stretching. **b:** Execution time measurements of a 32-bin histogram for four different numbers of GPU threads per thread block. The last line shows the results of the updated stretching factor S'.

To solve this memory contention, and thus to improve performance, we introduce a technique called bin stretching. We explain the bin stretching technique and show the results based on two examples.

Bin Stretching Technique. To reduce the memory contention, the number of bins (N in Eq. 2) can be increased to the maximum number of bins allowed by the hardware (N_{max}). Since the maximum number of bins and the number of bins in a voting algorithm are both known, a stretching factor (S) can be calculated as shown in Eq. 3. To make sure that consecutive threads use as many different memory locations as possible, the new bin-index is calculated as shown in Eq. 4. After all votes are placed in the $S \times N$ bins, the results need to be compacted back into the desired N bins by summing each group of S consecutive bins together.

$$S = \max\left(\left\lfloor \frac{N_{max}}{N} \right\rfloor, 1\right) \tag{3}$$

$$bin' = bin \cdot S + (\textit{thread-id} \bmod S) \tag{4}$$

The Effect of Bin Stretching by Two Examples. Applying the bin stretching method on the Hough transform example from Fig. 2(a), the performance improvement between the original approach (blue squares) and the bin stretching approach (red circles) is clear. With bin stretching applied, the maximum amount of on-chip memory is used in the voting process for any number of bins per angle, which reduces memory contention to a minimum.

The effect of bin stretching on a 32-bin histogram algorithm is shown in Fig. 2(b). When using 1024 threads (which results in the overall lowest execution time compared to smaller numbers of threads per thread block), a stretching factor of 383 results in a speed-up of over 20× compared to an implementation without bin stretching.

In Fig. 2(b) the input is split into a number of parts to fill up all SMs as much as possible, as explained in Section 3.2. Each SM can process a few parts at once, given enough resources. The resource limitation can be either the number of resident threads per SM (maximum of 1536), or the number of resident thread blocks per SM (maximum of 8) or the on-chip shared memory per SM (48 kB) [16]. For example, when 128 threads are used (top blue line in Fig. 2(b)), and the stretching factor is 192, two parts can be processed by two thread blocks on one SM. But when the stretching factor is increased to 193, there is only enough on-chip memory for one part per SM, resulting in a significantly increased execution time.

One thing to notice in Fig. 2(b) are the peaks in execution time for certain stretching factors. These peaks occur at stretching factors which are a multiple of 32. This is due to the number of banks in the on-chip memory of an SM, which is also 32 [16]. Careful inspection of Fig. 2(b) shows that odd stretching factors give a lower execution time compared to an even stretching factor one value larger. But for stretching factors smaller than 64, the improvement in execution time due to the reduction in memory conflicts (caused by the stretching) is larger than the decline in execution time due to the even stretching factor.

Taking this into account, the optimal stretching factor is calculated with Eq. 5, and the corresponding bin-index with Eq. 6. This is in accordance with [16], where an odd step size is suggested for strided on-chip memory accesses.

$$S' = \begin{cases} S & \text{if } S < 64 \\ S - (1 - S \bmod 2) & \text{if } S \geq 64 \end{cases} \tag{5}$$

$$bin'' = bin \cdot S' + (thread\text{-}id \bmod S \bmod S') \tag{6}$$

The effects of this updated stretching factor on the execution time is shown in the last (purple) line in Fig. 2(b). By using only odd stretching factors with S', the peaks in execution time as shown before are removed.

```
1 #include "gpu_vote.h"
2 class Histogram : public CVoteFunction<uchar, uint> {
3   public:
4     __device__ uint vote_index(uint index, uchar value, uint nbins) {
5       return value * (nbins / 256.0f);
6     }
7 };
```

Listing 2. Implementation of a histogram algorithm using GPU-vote.

3.4   Implementation with GPU-Vote

The generic GPU voting methodology is implemented in a framework named GPU-VOTE. To support a large range of voting algorithms, the variable types for the input and output and four functions have to be defined for each voting algorithm. One function is required for the first stage in the methodology (array building), the other three are used in the second stage (voting in the vote space).

One example implementation of a gray-scale histogram algorithm is shown in Lst. 2 On line 1 the GPU-VOTE framework is loaded. The variable types of the input and output (uchar and uint) are specified on line 2. Only the standard vote index function, which determines in which location a vote is placed, is overloaded on lines 4-6. The other functions' default implementation is sufficient.

4 Evaluation

To evaluate the proposed methodology, the execution time of GPU-VOTE is compared with state of the art implementations. In Fig. 3 the performance of various approaches to implement the 2-D Hough transform on a GPU is shown. The performance of [3] is indicated with blue circles. This approach is optimized for histogramming and can only handle up to 512 bins. The voting method from [7] focuses on the Hough transform and is indicated with green squares. The best performance for this algorithm is achieved with a large number of bins.

Hough transform (64 angles)

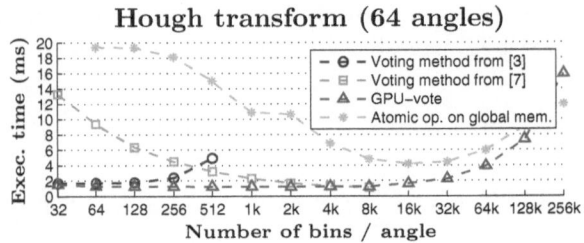

Fig. 3. Execution time of four different voting methods for a range of number of bins

The proposed generic voting methodology is shown as red triangles in Fig. 3. It outperforms the previous two methods and supports a range of number of bins with equal performance. Only in case the required number of bins is larger than the maximum number of bins the implementation can support (N_{max}), the execution time increases. This increase is caused by the second stage of the voting methodology being executed multiple times (see also Section 3.2). To prevent the necessity of executing the second stage multiple times the on-chip memory size of the GPU has to be increased for future hardware, or taking a hierarchical approach for the voting algorithm could be investigated.

For comparison, a GPU voting method where all votes are placed directly in off-chip memory is also included in Fig. 3 (cyan stars). Although this method has a large performance penalty for a number of bins less than 8k, its execution time gets closer to the proposed generic voting methodology when the number of bins increases. For number of bins larger than 128k this approach even outperforms the proposed generic voting methodology.

The performance of the proposed generic voting algorithm is evaluated by implementing four algorithms with GPU-VOTE on an NVIDIA GTX 470. The input for all algorithms in this section are 1920 by 1080 pixels gray scale images.

For histogram and Hough transform, the results of GPU-VOTE are compared with the best known GPU implementation from [3] and [7] respectively. Calculating a 256-bin histogram with GPU-VOTE is 38% faster on average for 110 randomly selected test images compared to the the implementation in [3]. Calculating a 2-D Hough transform with a total of 64 angles and 4200 bins per angle takes 11% less time on average for the same 110 images compared to the implementation in [7]. The main improvement of GPU-VOTE over [7] is the bin-stretching technique introduced in Section 3.3.

Other algorithms, which may not be identified as voting algorithms directly, also fit our proposed framework. We show performance results to indicate the wide applicability of GPU-VOTE. For example, calculating the sum of all elements in a matrix can be seen as a voting algorithm with only a single bin. Compared to the heavily optimized implementation in the reduction example in the NVIDIA CUDA software development kit, the results of GPU-VOTE are 12% and 14% slower when summing 8-bit integer inputs and 32-bit float inputs respectively.

In image sub-sampling, the location of the input pixel determines in which bin the pixel value has to be added, which makes it a location-based number-voting algorithm. GPU-VOTE has a 2.6× higher execution time compared to a manual optimized GPU implementation. Since the structure of this algorithm is very regular, using a generalized voting method is clearly not the best approach in terms of execution time for implementing this algorithm. However the manual optimized implementation requires detailed knowledge about the GPU architecture. Using GPU-VOTE does not require this knowledge, and the implementation consist of only four lines of straight forward C-code, in contrast to the 20 lines of highly optimized CUDA kernel code for the manual optimized implementation.

5   Performance Prediction

To predict the performance of a voting algorithm, we introduce a model which is based on properties of the image (e.g. size), the voting algorithm (e.g. number of bins) and GPU parameters (e.g number of SMs). The prediction can be used to select a suitable GPU or to choose algorithm parameters for example. Both stages of the generalized GPU voting algorithm are modeled separately. We first explain how these models are derived. Following, we evaluate the quality of the performance prediction.

5.1   Prediction Model

The first stage of the algorithm builds an array of input element locations. A part of the input elements (which pass the element test) are stored into two arrays, one of element locations and one of element values. The execution time of this stage of the algorithm is a combination of the total number of input elements which need to be evaluated (E) and the relative number of elements which have to be put in an array (ηE) as shown in Eq. 7.

$$T_1 = a \times E + b \times \eta E + c \qquad (7)$$

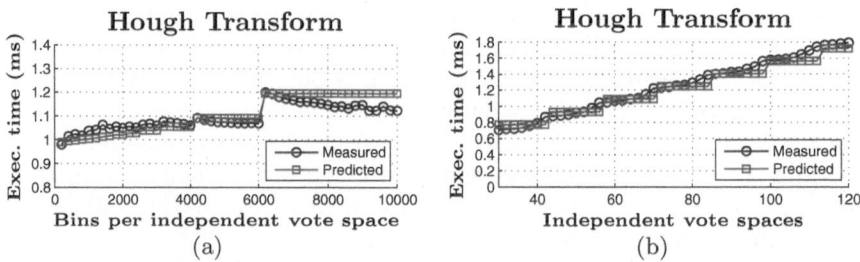

Fig. 4. Hough transform execution times for **a:** a range of bins per angle (N in Eq. 2 and Eq. 3) and **b:** a range of number of angles (G)

The second stage of the algorithm places votes in the vote space. The execution time of this stage of the algorithm depends on the relative number of elements used in this second stage (ηE) and the number of votes that need to be placed in the vote space ($G \times \eta E$), where G is the number of independent vote spaces. Since every independent vote space is calculated by one thread block, which is mapped to a single SM, G has to be rounded up to the nearest multiple of the available number of SMs (SM). In case the number of bins (N) in an independent vote space is larger than the maximum the hardware can supply (N_{max}), this second stage is executed multiple times and the product $G \times \lceil N/N_{max} \rceil$ has to be rounded up to the nearest multiple of SM.

When the product of the number of independent vote spaces and $\lceil N/N_{max} \rceil$ is less than the number of SMs, the input is split in parts (P in Eq. 8) to reduce the execution time of this second stage. The impact of G, N, N_{max} and SM on the execution time can be taken into account with the factor F in Eq. 9.

$$P = \max \left(\left\lfloor \frac{SM}{G \times \lceil N/N_{max} \rceil} \right\rfloor, 1 \right) \tag{8}$$

$$F = \left\lceil \frac{G \times \lceil N/N_{max} \rceil}{SM} \right\rceil \frac{1}{P} \tag{9}$$

The execution time of the second stage also depends on the number of bins (N) through the stretching factor (S'), as explained in Section 3.3. The following equation can be composed to estimate the execution time of the second stage:

$$T_2 = d \times \frac{\eta E}{S'} \times F + e \times \eta E \times F + f \tag{10}$$

The parameters $a - f$ in the model are estimated by performing experiments in which training images and algorithm parameters in Eq. 7 and Eq. 10 are varied and a least-squares minimization on the predicted execution time is applied.

5.2   Evaluation of the Prediction

The model of the prediction of the execution time is evaluated with two experiments. In each experiment the Hough transform is calculated on a 1920 × 1080

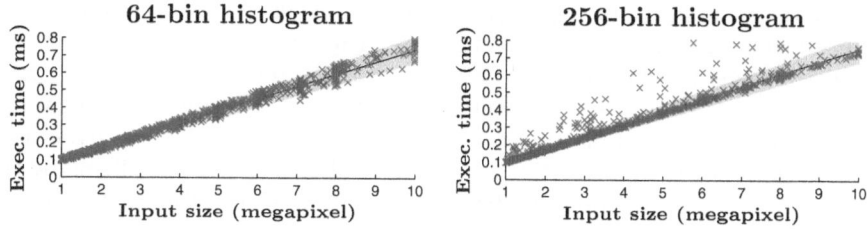

Fig. 5. Measured and predicted execution times for the 64-bin and 256-bin histogram algorithm executed on 1295 images

input image (distinct from the training images), on which first edge-detection and thresholding are applied as described in [7]. Two aspects of the Hough transform are varied: the number of bins (N) and the number of independent vote spaces (G). As shown in Fig. 4, the model clearly follows the steps in execution time caused by the bin stretching factor S (Fig. 4(a)) and the steps caused by the rounding of $G \times \lceil N/N_{max} \rceil$ up to the nearest multiple of SM (Fig. 4(b)).

The model is also evaluated on two other experiments, a 64-bin and a 256-bin histogram on 1295 gray scale images, varying in size from 1 to 10 megapixel. The average absolute error of the predicted execution time compared to the real execution time is 5%. For 95% of the images the execution time is predicted within an error range of -10% to 10%, as shown with the gray marked area in Fig. 5. For the 256-bin histogram, the execution time for some images is underestimated significantly. These images contain large surfaces with a single color, causing memory collisions. These can be resolved by applying bin-stretching in the 64-bin histogram, but not in the 256-bin histogram, since the stretching factor is limited by the available on-chip memory and the number of bins used.

6   Conclusions

In this work we have introduced a generic methodology for implementing voting algorithms on a GPU. A classification of voting algorithms is presented, which arranges voting algorithms into four groups. We also gave a transformation on the input data of a voting algorithm to be able to merge categories. This enables the development of a methodology to solve voting problems in all categories with a single unified solution. The proposed bin stretching technique forms the base of the performance improvements of the generic methodology.

The generic methodology is implemented in GPU-VOTE, a framework which can be used to accelerate a range of voting algorithms on a GPU. With GPU-VOTE two examples, histogram and Hough transform have been implemented on a GPU. As shown in the evaluation section, GPU-VOTE is not just generic but even yields a lower execution time compared to previously published GPU implementations which targeted only a single voting algorithm. The performance of histogram and Hough transform is improved by 38% and 11% respectively.

To show the wide range of applicability, algorithms such as sum reduction and image sub-sampling have been implemented with GPU-VOTE. Although the performance of these implementations cannot match the reference implementations, the programming effort to implement such algorithms is reduced significantly.

To estimate the execution time of a voting algorithm a model is is given based on parameters of the input, voting algorithm and GPU used. Results show that in 95% of the cases the model predicts the execution time within a 10% range.

As part of future work the proposed methodology can be implemented in OpenCL to enable its use on other architectures, such as AMD GPUs and multi-core CPUs, making the methodology even more useful. We also plan to to improve performance by making GPU-VOTE multi-GPU enabled and to develop a hierarchical approach for voting algorithms to support a larger number of bins.

References

1. Podlozhnyuk, V.: Histogram Calculation in CUDA (2007)
2. Shams, R., Kennedy, R.A.: Efficient Histogram Algorithms for NVIDIA CUDA Compatible Devices. In: International Conference on Signal Processing and Communications Systems (2007)
3. Nugteren, C., Van den Braak, G.J., Corporaal, H., Mesman, B.: High Performance Predictable Histogramming on GPUs: Exploring and Evaluating Algorithm Trade-offs. GPGPU 4 (2011)
4. Fung, J., Mann, S.: OpenVIDIA: Parallel GPU Computer Vision. In: 13th ACM International Conference on Multimedia (2005)
5. Ujaldón, M., Ruiz, A., Guil, N.: On the computation of the Circle Hough Transform by a GPU rasterizer. Pattern Recognition Letters (2008)
6. TunaCode (Ltd): CUDA Vision and Imaging Library, http://www.cuvilib.com/
7. Van den Braak, G.J., Nugteren, C., Mesman, B., Corporaal, H.: Fast Hough Transform on GPUs: Exploration of Algorithm Trade-Offs. In: Blanc-Talon, J., Kleihorst, R., Philips, W., Popescu, D., Scheunders, P. (eds.) ACIVS 2011. LNCS, vol. 6915, pp. 611–622. Springer, Heidelberg (2011)
8. Tan, C., Hong, T., Chang, T., Shneier, M.: Color Model-Based Real-Time Learning for Road Following. In: Intelligent Transportation Systems Conference (2006)
9. Maes, F., Collignon, A., Vandermeulen, D., Marchal, G., Suetens, P.: Multimodality Image Registration by Maximization of Mutual Information. IEEE Transactions on Medical Imaging (1997)
10. Scott, D.W.: On Optimal and Data-Based Histograms. Biometrika (1979)
11. Forsberg, J., Larsson, U., Wernersson, A.: Mobile Robot Navigation using the Range-Weighted Hough Transform. IEEE Robotics Automation Magazine (1995)
12. Wang, Y., Shi, M., Wu, T.: A Method of Fast and Robust for Traffic Sign Recognition. In: Fifth International Conference on Image and Graphics (2009)
13. Hough, P.: Method and Means for Recognising Complex Patterns. US Patent No. 3,069,654 (1962)
14. Duda, R.O., Hart, P.E.: Use of the Hough Transformation to Detect Lines and Curves in Pictures. Commun. ACM 15 (1972)
15. NVIDIA Corporation: NVIDIA's Next Generation CUDA Compute Architecture: Fermi (2009)
16. NVIDIA Corporation: NVIDIA CUDA C Programming Guide - Version 4.0 (2011)

Author Index